HEAT AND MASS TRANSFER

■ ■ ■ ■ ■ ■

A PRACTICAL APPROACH

Celebrating World Year of Physics 2005 to commemorate the pioneering contributions of Albert Einstein in 1905 with his own inspiring words.

Most of the fundamental ideas of science are essentially simple, and may, as a rule, be expressed in a language comprehensible to everyone.

Everything should be made as simple as possible, but not simpler.

Imagination is more important than knowledge.

Everything that is really great and inspiring is created by the individual who can labor in freedom.

Education is that which remains when one has forgotten everything learned in school.

Since the mathematicians have invaded the theory of relativity, I do not understand it myself anymore.

God does not care about our mathematical difficulties. He integrates empirically.

The most incomprehensible fact about the universe is that it is comprehensible.

The search for truth is more precious than its possession.

The truth of a theory is in your mind, not in your eyes.

Science without religion is lame; religion without science is blind.

Great spirits have always encountered violent opposition from mediocre minds.

If at first the idea is not absurd, then there is no hope for it.

I have never belonged wholeheartedly to a country, a state, nor to a circle of friends, nor even to my own family. When I was still a rather precocious young man, I already realized most vividly the futility of the hopes and aspirations that most men pursue throughout their lives. Well-being and happiness never appeared to me as an absolute aim. I am even inclined to compare such moral aims to the ambitions of a pig.

For more quotes with credits, visit http://www-groups.dcs.st-and.ac.uk/~history/Quotations/Einstein.html

DIMENSION	METRIC	METRIC/ENGLISH
Specific volume	$1 \ m^3/kg = 1000 \ L/kg = 1000 \ cm^3/g$	$1 \ m^3/kg = 16.02 \ ft^3/lbm$ $1 \ ft^3/lbm = 0.062428 \ m^3/kg$
Temperature	$T(K) = T(°C) + 273.15$ $\Delta T(K) = \Delta T(°C)$	$T(R) = T(°F) + 459.67 = 1.8T(K)$ $T(°F) = 1.8 \ T(°C) + 32$ $\Delta T(°F) = \Delta T(R) = 1.8 \ \Delta T(K)$
Thermal conductivity	$1 \ W/m \cdot °C = 1 \ W/m \cdot K$	$1 \ W/m \cdot °C = 0.57782 \ Btu/h \cdot ft \cdot °F$
Thermal resistance	$1°C/W = 1 \ K/W$	$1 \ K/W = 0.52750°F/h \cdot Btu$
Velocity	$1 \ m/s = 3.60 \ km/h$	$1 \ m/s = 3.2808 \ ft/s = 2.237 \ mi/h$ $1 \ mi/h = 1.46667 \ ft/s$ $1 \ mi/h = 1.609 \ km/h$
Viscosity, dynamic	$1 \ kg/m \cdot s = 1 \ N \cdot s/m^2 = 1 \ Pa \cdot s = 10 \ poise$	$1 \ kg/m \cdot s = 2419.1 \ lbm/ft \cdot h$ $= 0.020886 \ lbf \cdot s/ft^2$ $= 0.67197 \ lbm/ft \cdot s$
Viscosity, kinematic	$1 \ m^2/s = 10^4 \ cm^2/s$ $1 \ stoke = 1 \ cm^2/s = 10^{-4} \ m^2/s$	$1 \ m^2/s = 10.764 \ ft^2/s = 3.875 \times 10^4 \ ft^2/h$ $1 \ m^2/s = 10.764 \ ft^2/s$
Volume	$1 \ m^3 = 1000 \ L = 10^6 \ cm^3 \ (cc)$	$1 \ m^3 = 6.1024 \times 10^4 \ in^3 = 35.315 \ ft^3$ $= 264.17 \ gal \ (U.S.)$ $1 \ U.S. \ gallon = 231 \ in^3 = 3.7854 \ L$ $1 \ fl \ ounce = 29.5735 \ cm^3 = 0.0295735 \ L$ $1 \ U.S. \ gallon = 128 \ fl \ ounces$
Volume flow rate	$1 \ m^3/s = 60,000 \ L/min = 10^6 \ cm^3/s$	$1 \ m^3/s = 15,850 \ gal/min \ (gpm) = 35.315 \ ft^3/s$ $= 2118.9 \ ft^3/min \ (cfm)$

‡Mechanical horsepower. The electrical horsepower is taken to be exactly 746 W.

Some Physical Constants

Universal gas constant	$R_u = 8.31447 \ kJ/kmol \cdot K$ $= 8.31447 \ kPa \cdot m^3/kmol \cdot K$ $= 0.0831447 \ bar \cdot m^3/kmol \cdot K$ $= 82.05 \ L \cdot atm/kmol \cdot K$ $= 1.9858 \ Btu/lbmol \cdot R$ $= 1545.35 \ ft \cdot lbf/lbmol \cdot R$ $= 10.73 \ psia \cdot ft^3/lbmol \cdot R$
Standard acceleration of gravity	$g = 9.80665 \ m/s^2$ $= 32.174 \ ft/s^2$
Standard atmospheric pressure	$1 \ atm = 101.325 \ kPa$ $= 1.01325 \ bar$ $= 14.696 \ psia$ $= 760 \ mm \ Hg \ (0°C)$ $= 29.9213 \ in \ Hg \ (32°F)$ $= 10.3323 \ m \ H_2O \ (4°C)$
Stefan–Boltzmann constant	$\sigma = 5.670 \times 10^{-8} \ W/m^2 \cdot K^4$ $= 0.1714 \times 10^{-8} \ Btu/h \cdot ft^2 \cdot R^4$
Boltzmann's constant	$k = 1.380650 \times 10^{-23} \ J/K$
Speed of light in vacuum	$c = 2.9979 \times 10^8 \ m/s$ $= 9.836 \times 10^8 \ ft/s$
Speed of sound in dry air at 0°C and 1 atm	$c = 331.36 \ m/s$ $= 1089 \ ft/s$
Heat of fusion of water at 1 atm	$h_{if} = 333.7 \ kJ/kg$ $= 143.5 \ Btu/lbm$
Enthalpy of vaporization of water at 1 atm	$h_{fg} = 2256.5 \ kJ/kg$ $= 970.12 \ Btu/lbm$

HEAT AND MASS TRANSFER

■ ■ ■ ■ ■ ■ ■

A PRACTICAL APPROACH

MCGRAW-HILL SERIES IN MECHANICAL ENGINEERING

HEAT AND MASS TRANSFER

A PRACTICAL APPROACH

THIRD EDITION

YUNUS A. ÇENGEL
University of Nevada, Reno

Boston   Burr Ridge, IL   Dubuque, IA   Madison, WI   New York   San Francisco   St. Louis
Bangkok   Bogotá   Caracas   Kuala Lumpur   Lisbon   London   Madrid   Mexico City
Milan   Montreal   New Delhi   Santiago   Seoul   Singapore   Sydney   Taipei   Toronto

Higher Education

HEAT AND MASS TRANSFER: A PRACTICAL APPROACH, THIRD EDITION

This book is printed on acid-free paper.

4 5 6 7 8 9 0 DOW/DOW 0 9 8 7

ISBN 978–0–07–312930–3
MHID 0–07–312930–5

Publisher: *Suzanne Jeans*
Senior Sponsoring Editor: *Bill Stenquist*
Developmental Editor: *Amanda J. Green*
Executive Marketing Manager: *Michael Weitz*
Senior Project Manager: *Sheila M. Frank*
Senior Production Supervisor: *Sherry L. Kane*
Associate Media Producer: *Christina Nelson*
Senior Designer: *David W. Hash*
Cover Image: Altocumulus clouds at sunset, © *Royalty-Free/Corbis*
Senior Photo Research Coordinator: *Lori Hancock*
Compositor: *RPK Editorial Services, Inc.*
Typeface: *10.5/12 Times Roman*
Printer: *R. R. Donnelley Willard, OH*

Library of Congress Cataloging-in-Publication Data

Çengel, Yunus A.
   Heat and mass transfer : a practical approach / Yunus A. Çengel. — 3rd ed.
      p.   cm. — (McGraw-Hill series in mechanical engineering)
   Includes index.
   ISBN 978–0–07–312930–3 — ISBN 0–07–312930–5 (hard copy : alk. paper)
   1. Heat—Transmission. 2. Mass transfer. I. Title. II. Series.
   TJ260.C38   2007
   621.402'2—dc22   2005031039
                                                          CIP

www.mhhe.com

Yunus A. Çengel is Professor Emeritus of Mechanical Engineering at the University of Nevada, Reno. He received his B.S. in mechanical engineering from Istanbul Technical University and his M.S. and Ph.D. in mechanical engineering from North Carolina State University. He conducted research in radiation heat transfer, heat transfer enhancement, renewable energy, desalination, exergy analysis, and energy conservation. He served as the director of the Industrial Assessment Center (IAC) at the University of Nevada, Reno, from 1996 to 2000. He has led teams of engineering students to numerous manufacturing facilities in Northern Nevada and California to do industrial assessments, and has prepared energy conservation, waste minimization, and productivity enhancement reports for them.

Dr. Çengel is the coauthor of the widely adopted textbooks *Thermodynamics: An Engineering Approach,* 5th edition (©2006), *Fundamentals of Thermal-Fluid Sciences,* 2nd edition (©2005), and *Fluid Mechanics: Fundamentals and Applications* (©2006), all published by McGraw-Hill. He is the author of the textbook *Introduction to Thermodynamics and Heat Transfer* (©1997), also published by McGraw-Hill. Some of his textbooks have been translated into Chinese, Japanese, Korean, Thai, Spanish, Portuguese, Turkish, Italian, and Greek.

Dr. Çengel is the recipient of several outstanding teacher awards, and he has received the ASEE Meriam/Wiley Distinguished Author Award in 1992 and again in 2000 for excellence in authorship. Dr. Çengel is a registered professional engineer in the state of Nevada, and is a member of the American Society of Mechanical Engineers (ASME) and the American Society for Engineering Education (ASEE).

Brief Contents

CONTENTS

APPENDIX 2
PROPERTY TABLES AND CHARTS (ENGLISH UNITS)   869

BACKGROUND

Heat and mass transfer is a basic science that deals with the rate of transfer of thermal energy. It has a broad application area ranging from biological systems to common household appliances, residential and commercial buildings, industrial processes, electronic devices, and food processing. Students are assumed to have an adequate background in calculus and physics. The completion of first courses in thermodynamics, fluid mechanics, and differential equations prior to taking heat transfer is desirable. However, relevant concepts from these topics are introduced and reviewed as needed.

OBJECTIVES

This book is intended for undergraduate engineering students in their sophomore or junior year, and as a reference book by practicing engineers. The objectives of this text are

- To cover the *basic principles* of heat transfer.
- To present a wealth of real-world *engineering examples* to give students a feel for how heat transfer is applied in engineering practice.
- To develop an *intuitive understanding* of heat transfer by emphasizing the physics and physical arguments.

It is our hope that this book, through its careful explanations of concepts and its use of numerous practical examples and figures, helps the students develop the necessary skills to bridge the gap between knowledge and the confidence for proper application of that knowledge.

In engineering practice, an understanding of the mechanisms of heat transfer is becoming increasingly important since heat transfer plays a crucial role in the design of vehicles, power plants, refrigerators, electronic devices, buildings, and bridges, among other things. Even a chef needs to have an intuitive understanding of the heat transfer mechanism in order to cook the food "right" by adjusting the rate of heat transfer. We may not be aware of it, but we already use the principles of heat transfer when seeking thermal comfort. We insulate our bodies by putting on heavy coats in winter, and we minimize heat gain by radiation by staying in shady places in summer. We speed up the cooling of hot food by blowing on it and keep warm in cold weather by cuddling up and thus minimizing the exposed surface area. That is, we already use heat transfer whether we realize it or not.

GENERAL APPROACH

This text is the outcome of an attempt to have a textbook for a practically oriented heat transfer course for engineering students. The text covers the standard topics of heat transfer with an emphasis on physics and real-world applications. This approach is more in line with students' intuition, and makes learning the subject matter enjoyable.

The philosophy that contributed to the overwhelming popularity of the prior editions of this book has remained unchanged in this edition. Namely, our goal has been to offer an engineering textbook that

- Communicates directly to the minds of tomorrow's engineers in a *simple yet precise* manner.
- Leads students toward a clear understanding and firm grasp of the *basic principles* of heat transfer.
- Encourages *creative thinking* and development of a *deeper understanding* and *intuitive feel* for heat transfer.
- Is *read* by students with *interest* and *enthusiasm* rather than being used as an aid to solve problems.

Special effort has been made to appeal to students' natural curiosity and to help them explore the various facets of the exciting subject area of heat transfer. The enthusiastic response we received from the users of prior editions—from small colleges to large universities all over the world—indicates that our objectives have largely been achieved. It is our philosophy that the best way to learn is by practice. Therefore, special effort is made throughout the book to reinforce material that was presented earlier.

Yesterday's engineer spent a major portion of his or her time substituting values into the formulas and obtaining numerical results. However, now formula manipulations and number crunching are being left mainly to the computers. Tomorrow's engineer will have to have a clear understanding and a firm grasp of the *basic principles* so that he or she can understand even the most complex problems, formulate them, and interpret the results. A conscious effort is made to emphasize these basic principles while also providing students with a perspective at how computational tools are used in engineering practice.

NEW IN THIS EDITION

All the popular features of the previous edition are retained while new ones are added. With the exception of the coverage of the theoretical foundations of transient heat conduction and moving the chapter "Cooling of Electronic Equipment" to the Online Learning Center, the main body of the text remains largely unchanged. The most significant changes in this edition are highlighted below.

A NEW TITLE

The title of the book is changed to *Heat and Mass Transfer: A Practical Approach* to attract attention to the coverage of mass transfer. All topics related to mass transfer, including mass convection and vapor migration through building materials, are introduced in one comprehensive chapter (Chapter 14).

EXPANDED COVERAGE OF TRANSIENT CONDUCTION

The coverage of Chapter 4, Transient Heat Conduction, is now expanded to include (1) the derivation of the dimensionless Biot and Fourier numbers by nondimensionalizing the heat conduction equation and the boundary and initial conditions, (2) the derivation of the analytical solutions of a one-dimensional transient conduction equation using the method of separation of variables, (3) the derivation of the solution of a transient conduction equation in the semi-infinite medium using a similarity variable, and (4) the solutions of transient heat conduction in semi-infinite mediums for different boundary conditions such as specified heat flux and energy pulse at the surface.

FUNDAMENTALS OF ENGINEERING (FE) EXAM PROBLEMS

To prepare students for the Fundamentals of Engineering Exam (that is becoming more important for the outcome-based ABET 2000 criteria) and to facilitate multiple-choice tests, about 250 *multiple-choice problems* are included in the end-of-chapter problem sets. They are placed under the title "Fundamentals of Engineering (FE) Exam Problems" for easy recognition. These problems are intended to check the understanding of fundamentals and to help readers avoid common pitfalls.

MICROSCALE HEAT TRANSFER

Recent inventions in micro and nano-scale systems and the development of micro and nano-scale devices continues to pose new challenges, and the understanding of the fluid flow and heat transfer at such scales is becoming more and more important. In Chapter 6, microscale heat transfer is presented as a Topic of Special Interest.

THREE ONLINE APPLICATION CHAPTERS

The application chapter "Cooling of Electronic Equipment" (Chapter 15) is now moved to the Online Learning Center together with two new chapters "Heating and Cooling of Buildings" (Chapter 16) and "Refrigeration and Freezing of Foods" (Chapter 17).

CONTENT CHANGES AND REORGANIZATION

With the exception of the changes already mentioned, minor changes are made in the main body of the text. Nearly 400 new problems are added, and many of the existing problems are revised. The noteworthy changes in various chapters are summarized here for those who are familiar with the previous edition.

- The title of Chapter 1 is changed to "Introduction and Basic Concepts." Some artwork is replaced by photos, and several review problems on the first law of thermodynamics are deleted.
- Chapter 4 "Transient Heat Conduction" is revised greatly, as explained previously, by including the theoretical background and the mathematical details of the analytical solutions.
- Chapter 6 now has the Topic of Special Interest "Microscale Heat Transfer" contributed by Dr. Subrata Roy of Kettering University.
- Chapter 8 now has the Topic of Special Interest "Transitional Flow in Tubes" contributed by Dr. Afshin Ghajar of Oklahoma State University.
- Chapter 13 "Heat Exchangers" is moved up as Chapter 11 to succeed "Boiling and Condensation" and to precede "Radiation."
- In the appendices, the values of some physical constants are updated, and Appendix 3 "Introduction to EES" is moved to the enclosed CD and the Online Learning Center.

SUPPLEMENTS

The following supplements are available to the adopters of the book.

ENGINEERING EQUATION SOLVER (EES) CD-ROM

(Limited Academic Version packaged free with every new copy of the text)
Developed by Sanford Klein and William Beckman from the University of Wisconsin–Madison, this software combines equation-solving capability and engineering property data. EES can do optimization, parametric analysis, and linear and nonlinear regression, and provides publication-quality plotting capabilities. Thermodynamic and transport properties for air, water, and many other fluids are built in, and EES allows the user to enter property data or functional relationships. Some problems are solved using EES, and complete solutions together with parametric studies are included on the enclosed CD-ROM. To obtain the full version of EES, contact your McGraw-Hill representative or visit *www.mhhe.com/ees*.

ONLINE LEARNING CENTER (www.mhhe.com/cengel)

Web support is provided for the text on our Online Learning Center. Visit this web site for general text information, errata, and author information. The site also includes resources for students including a list of helpful web links. The instructor side of the site includes the solutions manual, the text's images in PowerPoint form, and more!

COSMOS CD-ROM

(Available to instructors only)
The instructor CD provides electronic solutions delivered via our database management tool. McGraw-Hill's COSMOS (Complete Online Solutions Manual Organization System) allows instructors to streamline the creation of assignments, quizzes, and tests by using problems and solutions from the textbook—as well as their own custom material. Contact your McGraw-Hill representative to obtain a copy.

ACKNOWLEDGMENTS

I would like to acknowledge with appreciation the numerous and valuable comments, suggestions, constructive criticisms, and praise from the following evaluators and reviewers:

Suresh Advani,
University of Delaware

Mark Barker,
Louisiana Tech University

John R. Biddle,
California State Polytechnic University, Pomona

Sanjeev Chandra,
University of Toronto

Shaochen Chen,
University of Texas, Austin

Fan-Bill Cheung,
Pennsylvania State University

Vic A. Cundy,
Montana State University

Radu Danescu,
North Dakota State University

Prashanta Dutta,
Washington State University

Richard A. Gardner,
Washington University

Afshin J. Ghajar,
Oklahoma State University

S. M. Ghiaasiaan,
Georgia Institute of Technology

Alain Kassab,
University of Central Florida

Roy W. Knight,
Auburn University

Milivoje Kostic,
Northern Illinois University

Wayne Krause,
*South Dakota School of Mines and
Technology*

Feng C. Lai,
University of Oklahoma

Charles Y. Lee,
University of North Carolina, Charlotte

Alistair Macpherson,
Lehigh University

Saeed Manafzadeh,
University of Illinois

A.K. Mehrotra,
University of Calgary

Abhijit Mukherjee,
Rochester Institute of Technology

Yoav Peles,
Rensselaer Polytechnic Institute

Ahmad Pourmovahed,
Kettering University

Paul Ricketts,
New Mexico State University

Subrata Roy,
Kettering University

Brian Sangeorzan,
Oakland University

Michael Thompson,
McMaster University

Their suggestions have greatly helped to improve the quality of this text.

Special thanks are due to Afshin J. Ghajar of Oklahoma State University and Subrata Roy of Kettering University for contributing new sections and problems, and to the following for contributing problems for this edition:

Edward Anderson, *Texas Tech University*
Radu Danescu, *General Electric (GE) Energy*
Ibrahim Dincer, *University of Ontario Institute of Technology, Canada*
Mehmet Kanoglu, *University of Gaziantep, Turkey*
Wayne Krause, *South Dakota School of Mines*
Anil Mehrotra, *University of Calgary, Canada*

I also would like to thank my students and instructors from all over the globe, who provided plenty of feedback from students' and users' perspectives. Finally, I would like to express my appreciation to my wife and children for their continued patience, understanding, and support throughout the preparation of this text.

Yunus A. Çengel

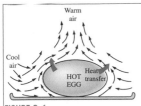

FIGURE 9–1
The cooling of a boiled egg in a cooler environment by natural convection.

The temperature of the air adjacent to the egg is higher and thus its density is lower, since at constant pressure the density of a gas is inversely proportional to its temperature. Thus, we have a situation in which some low-density or "light" gas is surrounded by a high-density or "heavy" gas, and the natural laws dictate that the *light gas rise*. This is no different than the oil in a vinegar-and-oil salad dressing rising to the top (since $\rho_{oil} < \rho_{vinegar}$). This phenomenon is characterized incorrectly by the phrase "heat rises," which is understood to mean *heated air rises*. The space vacated by the warmer air in the vicinity of the egg is replaced by the cooler air nearby, and the presence of cooler air in the vicinity of the egg speeds up the cooling process. The rise of warmer air and the flow of cooler air into its place continues until the egg is cooled to the temperature of the surrounding air.

EMPHASIS ON PHYSICS

The author believes that the emphasis in undergraduate education should remain on *developing a sense of underlying physical mechanisms* and a *mastery of solving practical problems* that an engineer is likely to face in the real world.

EFFECTIVE USE OF ASSOCIATION

An observant mind should have no difficulty understanding engineering sciences. After all, the principles of engineering sciences are based on our *everyday experiences* and *experimental observations*. The process of cooking, for example, serves as an excellent vehicle to demonstrate the basic principles of heat transfer.

EXAMPLE 4–3   **Boiling Eggs**

An ordinary egg can be approximated as a 5-cm-diameter sphere (Fig. 4–21). The egg is initially at a uniform temperature of 5°C and is dropped into boiling water at 95°C. Taking the convection heat transfer coefficient to be $h = 1200$ W/m$^2 \cdot$ °C, determine how long it will take for the center of the egg to reach 70°C.

SOLUTION   An egg is cooked in boiling water. The cooking time of the egg is to be determined.

Assumptions   **1** The egg is spherical in shape with a radius of $r_0 = 2.5$ cm. **2** Heat conduction in the egg is one-dimensional because of thermal symmetry about the midpoint. **3** The thermal properties of the egg and the heat transfer coefficient are constant. **4** The Fourier number is $\tau > 0.2$ so that the one-term approximate solutions are applicable.

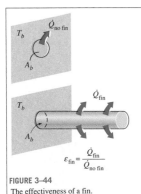

FIGURE 3–44
The effectiveness of a fin.

Fin Effectiveness

Fins are used to *enhance* heat transfer, and the use of fins on a surface cannot be recommended unless the enhancement in heat transfer justifies the added cost and complexity associated with the fins. In fact, there is no assurance that adding fins on a surface will *enhance* heat transfer. The performance of the fins is judged on the basis of the enhancement in heat transfer relative to the no-fin case. The performance of fins is expressed in terms of the *fin effectiveness* ε_{fin} defined as Fig. 3–44.

SELF-INSTRUCTING

The material in the text is introduced at a level that an average student can follow comfortably. It speaks *to* students, not *over* students. In fact, it is *self-instructive*. The order of coverage is from *simple* to *general*.

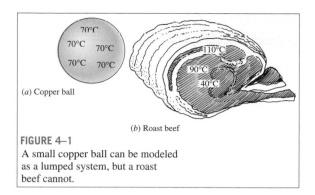

FIGURE 4–1
A small copper ball can be modeled as a lumped system, but a roast beef cannot.

EXTENSIVE USE OF ARTWORK

Art is an important learning tool that helps students "get the picture." The third edition of *Heat and Mass Transfer: A Practical Approach* contains more figures and illustrations than any other book in this category.

LEARNING OBJECTIVES AND SUMMARIES

Each chapter begins with an *Overview* of the material to be covered and chapter-specific *Learning Objectives*. A *Summary* is included at the end of each chapter, providing a quick review of basic concepts and important relations, and pointing out the relevance of the material.

CHAPTER 1

INTRODUCTION AND BASIC CONCEPTS

The science of thermodynamics deals with the *amount* of heat transfer as a system undergoes a process from one equilibrium state to another, and makes no reference to *how long* the process will take. But in engineering, we are often interested in the *rate* of heat transfer, which is the topic of the science of *heat transfer*.

We start this chapter with a review of the fundamental concepts of thermodynamics that form the framework for heat transfer. We first present the relation of heat to other forms of energy and review the energy balance. We then present the three basic mechanisms of heat transfer, which are conduction, convection, and radiation, and discuss thermal conductivity. *Conduction* is the transfer of energy from the more energetic particles of a substance to the adjacent, less energetic ones as a result of interactions between the particles. *Convection* is the mode of heat transfer between a solid surface and the adjacent liquid or gas that is in motion, and it involves the combined effects of conduction and fluid motion. *Radiation* is the energy emitted by matter in the form of electromagnetic waves (or photons) as a result of the changes in the electronic configurations of the atoms or molecules. We close this chapter with a discussion of simultaneous heat transfer.

OBJECTIVES

When you finish studying this chapter, you should be able to:

■ Understand how thermodynamics and heat transfer are related to each other,

■ Distinguish thermal energy from other forms of energy, and heat transfer from other forms of energy transfer,

■ Perform general energy balances as well as surface energy balances,

■ Understand the basic mechanisms of heat transfer, which are conduction, convection, and radiation, and Fourier's law of heat conduction, Newton's law of cooling, and the Stefan–Boltzmann law of radiation,

■ Identify the mechanisms of heat transfer that occur simultaneously in practice,

■ Develop an awareness of the cost associated with heat losses, and

■ Solve various heat transfer problems encountered in practice.

EXAMPLE 1–9   Radiation Effect on Thermal Comfort

It is a common experience to feel "chilly" in winter and "warm" in summer in our homes even when the thermostat setting is kept the same. This is due to the so called "radiation effect" resulting from radiation heat exchange between our bodies and the surrounding surfaces of the walls and the ceiling.

Consider a person standing in a room maintained at 22°C at all times. The inner surfaces of the walls, floors, and the ceiling of the house are observed to be at an average temperature of 10°C in winter and 25°C in summer. Determine the rate of radiation heat transfer between this person and the surrounding surfaces if the exposed surface area and the average outer surface temperature of the person are 1.4 m² and 30°C, respectively (Fig. 1–38).

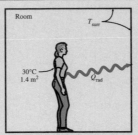

FIGURE 1–38
Schematic for Example 1–9.

SOLUTION The rates of radiation heat transfer between a person and the surrounding surfaces at specified temperatures are to be determined in summer and winter.

Assumptions 1 Steady operating conditions exist. 2 Heat transfer by convection is not considered. 3 The person is completely surrounded by the interior surfaces of the room. 4 The surrounding surfaces are at a uniform temperature.

Properties The emissivity of a person is $\varepsilon = 0.95$ (Table 1–6).

Analysis The net rates of radiation heat transfer from the body to the surrounding walls, ceiling, and floor in winter and summer are

$$\dot{Q}_{\text{rad, winter}} = \varepsilon \sigma A_s \left(T_s^4 - T_{\text{surr, winter}}^4 \right)$$

NUMEROUS WORKED-OUT EXAMPLES WITH A SYSTEMATIC SOLUTIONS PROCEDURE

Each chapter contains several worked-out *examples* that clarify the material and illustrate the use of the basic principles. An *intuitive* and *systematic* approach is used in the solution of the example problems, while maintaining an informal conversational style. The problem is first stated, and the objectives are identified. The assumptions are then stated, together with their justifications. The properties needed to solve the problem are listed separately, if appropriate. This approach is also used consistently in the solutions presented in the instructor's solutions manual.

A WEALTH OF REAL-WORLD END-OF-CHAPTER PROBLEMS

The end-of-chapter problems are grouped under specific topics to make problem selection easier for both instructors and students. Within each group of problems are:

- **Concept Questions**, indicated by "**C**," to check the students' level of understanding of basic concepts.
- **Review Problems** are more comprehensive in nature and are not directly tied to any specific section of a chapter—in some cases they require review of material learned in previous chapters.

1–94C We often turn the fan on in summer to help us cool. Explain how a fan makes us feel cooler in the summer. Also explain why some people use ceiling fans also in winter.

- **_Fundamentals of Engineering Exam_** problems are clearly marked and intended to check the understanding of fundamentals, to help students avoid common pitfalls, and to prepare students for the FE Exam that is becoming more important for the outcome based ABET 2000 criteria.

 These problems are solved using EES, and complete solutions together with parametric studies are included on the enclosed CD-ROM.

 These problems are comprehensive in nature and are intended to be solved with a computer, preferably using the EES software that accompanies this text.

- **_Design and Essay_** are intended to encourage students to make engineering judgments, to conduct independent exploration of topics of interest, and to communicate their findings in a professional manner.

Several economics- and safety-related problems are incorporated throughout to enhance cost and safety awareness among engineering students. Answers to selected problems are listed immediately following the problem for convenience to students.

1–152 A 30-cm-long, 0.5-cm-diameter electric resistance wire is used to determine the convection heat transfer coefficient in air at 25°C experimentally. The surface temperature of the wire is measured to be 230°C when the electric power consumption is 180 W. If the radiation heat loss from the wire is calculated to be 60 W, the convection heat transfer coefficient is

(a) 186 W/m² · °C   (b) 158 W/m² · °C
(c) 124 W/m² · °C   (d) 248 W/m² · °C
(e) 390 W/m² · °C

3–33 Reconsider Prob. 3–31. Using EES (or other) software, investigate the effect of thermal conductivity on the required insulation thickness. Plot the thickness of insulation as a function of the thermal conductivity of the insulation in the range of 0.02 W/m · °C to 0.08 W/m · °C, and discuss the results.

3–77 Consider a cold aluminum canned drink that is initially at a uniform temperature of 4°C. The can is 12.5 cm high and has a diameter of 6 cm. If the combined convection/radiation heat transfer coefficient between the can and the surrounding air at 25°C is 10 W/m² · °C, determine how long it will take for the average temperature of the drink to rise to 15°C.

In an effort to slow down the warming of the cold drink, a person puts the can in a perfectly fitting 1-cm-thick cylindrical rubber insulator ($k = 0.13$ W/m · °C). Now how long will it take for the average temperature of the drink to rise to 15°C? Assume the top of the can is not covered.

3–27 Consider a person standing in a room at 20°C with an exposed surface area of 1.7 m². The deep body temperature of the human body is 37°C, and the thermal conductivity of the human tissue near the skin is about 0.3 W/m · °C. The body is losing heat at a rate of 150 W by natural convection and radiation to the surroundings. Taking the body temperature 0.5 cm beneath the skin to be 37°C, determine the skin temperature of the person. _Answer:_ 35.5°C

3–29E A wall is constructed of two layers of 0.7-in-thick sheetrock ($k = 0.10$ Btu/h · ft · °F), which is a plasterboard made of two layers of heavy paper separated by a layer of gypsum, placed 7 in apart. The space between the sheetrocks is filled with fiberglass insulation ($k = 0.020$ Btu/h · ft · °F). Determine (a) the thermal resistance of the wall and (b) its R-value of insulation in English units.

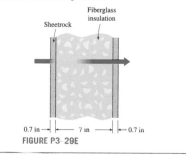

FIGURE P3–29E

A CHOICE OF SI ALONE OR SI/ENGLISH UNITS

In recognition of the fact that English units are still widely used in some industries, both SI and English units are used in this text, with an emphasis on SI. The material in this text can be covered using combined SI/English units or SI units alone, depending on the preference of the instructor. The property tables and charts in the appendices are presented in both units, except the ones that involve dimensionless quantities. Problems, tables, and charts in English units are designated by "**E**" after the number for easy recognition, and they can be ignored by SI users.

TOPIC OF SPECIAL INTEREST*

Heat Transfer through Windows

Windows are *glazed apertures* in the building envelope that typically consist of single or multiple glazing (glass or plastic), framing, and shading. In a building envelope, windows offer the least resistance to heat transfer. In a typical house, about one-third of the total heat loss in winter occurs through the windows. Also, most air infiltration occurs at the edges of the windows. The solar heat gain through the windows is responsible for much of the cooling load in summer. The net effect of a window on the heat balance of a building depends on the characteristics and orientation of the window as well as the solar and weather data. Workmanship is very important in the construction and installation of windows to provide effective sealing around the edges while allowing them to be opened and closed easily.

Despite being so undesirable from an energy conservation point of view, windows are an essential part of any building envelope since they enhance the appearance of the building, allow daylight and solar heat to come in, and allow people to view and observe outside without leaving their home. For low-rise buildings, windows also provide easy exit areas during emergencies such as fire. Important considerations in the selection of windows are *thermal comfort* and *energy conservation*. A window should have a good light transmittance while providing effective resistance to heat transfer. The lighting requirements of a building can be minimized by maximizing the use of natural daylight. Heat loss in winter through the windows can be minimized by using airtight double- or triple-pane windows with spectrally selective films or coatings, and letting in as much solar radiation as possible. Heat gain and thus cooling load in summer can be minimized by using effective internal or external shading on the windows.

TOPICS OF SPECIAL INTEREST

Most chapters contain a real world application, end-of-chapter optional section called "Topic of Special Interest" where interesting applications of heat transfer are discussed such as *Thermal Comfort* in Chapter 1, *A Brief Review of Differential Equations* in Chapter 2, *Heat Transfer through the Walls and Roofs* in Chapter 3, and *Heat Transfer through Windows* in Chapter 9.

CONVERSION FACTORS

Frequently used conversion factors and physical constants are listed on the inner cover pages of the text for easy reference.

Conversion Factors

DIMENSION	METRIC	METRIC/ENGLISH
Acceleration	$1 \text{ m/s}^2 = 100 \text{ cm/s}^2$	$1 \text{ m/s}^2 = 3.2808 \text{ ft/s}^2$ $1 \text{ ft/s}^2 = 0.3048^* \text{ m/s}^2$
Area	$1 \text{ m}^2 = 10^4 \text{ cm}^2 = 10^6 \text{ mm}^2$ $= 10^{-6} \text{ km}^2$	$1 \text{ m}^2 = 1550 \text{ in}^2 = 10.764 \text{ ft}^2$ $1 \text{ ft}^2 = 144 \text{ in}^2 = 0.09290304^* \text{ m}^2$
Density	$1 \text{ g/cm}^3 = 1 \text{ kg/L} = 1000 \text{ kg/m}^3$	$1 \text{ g/cm}^3 = 62.428 \text{ lbm/ft}^3 = 0.036127 \text{ lbm/in}^3$ $1 \text{ lbm/in}^3 = 1728 \text{ lbm/ft}^3$ $1 \text{ kg/m}^3 = 0.062428 \text{ lbm/ft}^3$
Energy, heat, work, internal energy, enthalpy	$1 \text{ kJ} = 1000 \text{ J} = 1000 \text{ Nm} = 1 \text{ kPa} \cdot \text{m}^3$ $1 \text{ kJ/kg} = 1000 \text{ m}^2/\text{s}^2$ $1 \text{ kWh} = 3600 \text{ kJ}$ $1 \text{ cal}^\dagger = 4.184 \text{ J}$ $1 \text{ IT cal}^\dagger = 4.1868 \text{ J}$ $1 \text{ Cal}^\dagger = 4.1868 \text{ kJ}$	$1 \text{ kJ} = 0.94782 \text{ Btu}$ $1 \text{ Btu} = 1.055056 \text{ kJ}$ $= 5.40395 \text{ psia} \cdot \text{ft}^3 = 778.169 \text{ lbf} \cdot \text{ft}$ $1 \text{ Btu/lbm} = 25{,}037 \text{ ft}^2/\text{s}^2 = 2.326^* \text{ kJ/kg}$ $1 \text{ kJ/kg} = 0.430 \text{ Btu/lbm}$ $1 \text{ kWh} = 3412.14 \text{ Btu}$ $1 \text{ therm} = 10^5 \text{ Btu} = 1.055 \times 10^5 \text{ kJ}$ (natural gas)

INTRODUCTION AND BASIC CONCEPTS

T he science of thermodynamics deals with the *amount* of heat transfer as a system undergoes a process from one equilibrium state to another, and makes no reference to *how long* the process will take. But in engineering, we are often interested in the *rate* of heat transfer, which is the topic of the science of *heat transfer.*

We start this chapter with a review of the fundamental concepts of thermodynamics that form the framework for heat transfer. We first present the relation of heat to other forms of energy and review the energy balance. We then present the three basic mechanisms of heat transfer, which are conduction, convection, and radiation, and discuss thermal conductivity. *Conduction* is the transfer of energy from the more energetic particles of a substance to the adjacent, less energetic ones as a result of interactions between the particles. *Convection* is the mode of heat transfer between a solid surface and the adjacent liquid or gas that is in motion, and it involves the combined effects of conduction and fluid motion. *Radiation* is the energy emitted by matter in the form of electromagnetic waves (or photons) as a result of the changes in the electronic configurations of the atoms or molecules. We close this chapter with a discussion of simultaneous heat transfer.

OBJECTIVES

When you finish studying this chapter, you should be able to:

- Understand how thermodynamics and heat transfer are related to each other,
- Distinguish thermal energy from other forms of energy, and heat transfer from other forms of energy transfer,
- Perform general energy balances as well as surface energy balances,
- Understand the basic mechanisms of heat transfer, which are conduction, convection, and radiation, and Fourier's law of heat conduction, Newton's law of cooling, and the Stefan–Boltzmann law of radiation,
- Identify the mechanisms of heat transfer that occur simultaneously in practice,
- Develop an awareness of the cost associated with heat losses, and
- Solve various heat transfer problems encountered in practice.

CONTENTS

1–1 · THERMODYNAMICS AND HEAT TRANSFER

We all know from experience that a cold canned drink left in a room warms up and a warm canned drink left in a refrigerator cools down. This is accomplished by the transfer of *energy* from the warm medium to the cold one. The energy transfer is always from the higher temperature medium to the lower temperature one, and the energy transfer stops when the two mediums reach the same temperature.

You will recall from thermodynamics that energy exists in various forms. In this text we are primarily interested in **heat**, which is *the form of energy that can be transferred from one system to another as a result of temperature difference*. The science that deals with the determination of the *rates* of such energy transfers is **heat transfer**.

You may be wondering why we need to undertake a detailed study on heat transfer. After all, we can determine the amount of heat transfer for any system undergoing any process using a thermodynamic analysis alone. The reason is that thermodynamics is concerned with the *amount* of heat transfer as a system undergoes a process from one equilibrium state to another, and it gives no indication about *how long* the process will take. A thermodynamic analysis simply tells us how much heat must be transferred to realize a specified change of state to satisfy the conservation of energy principle.

In practice we are more concerned about the rate of heat transfer (heat transfer per unit time) than we are with the amount of it. For example, we can determine the amount of heat transferred from a thermos bottle as the hot coffee inside cools from 90°C to 80°C by a thermodynamic analysis alone. But a typical user or designer of a thermos bottle is primarily interested in *how long* it will be before the hot coffee inside cools to 80°C, and a thermodynamic analysis cannot answer this question. Determining the rates of heat transfer to or from a system and thus the times of heating or cooling, as well as the variation of the temperature, is the subject of *heat transfer* (Fig. 1–1).

Thermodynamics deals with equilibrium states and changes from one equilibrium state to another. Heat transfer, on the other hand, deals with systems that lack thermal equilibrium, and thus it is a *nonequilibrium* phenomenon. Therefore, the study of heat transfer cannot be based on the principles of thermodynamics alone. However, the laws of thermodynamics lay the framework for the science of heat transfer. The *first law* requires that the rate of energy transfer into a system be equal to the rate of increase of the energy of that system. The *second law* requires that heat be transferred in the direction of decreasing temperature (Fig. 1–2). This is like a car parked on an inclined road must go downhill in the direction of decreasing elevation when its brakes are released. It is also analogous to the electric current flowing in the direction of decreasing voltage or the fluid flowing in the direction of decreasing total pressure.

The basic requirement for heat transfer is the presence of a *temperature difference*. There can be no net heat transfer between two bodies that are at the same temperature. The temperature difference is the *driving force* for heat transfer, just as the *voltage difference* is the driving force for electric current flow and *pressure difference* is the driving force for fluid flow. The rate of heat transfer in a certain direction depends on the magnitude of the *temperature gradient* (the temperature difference per unit length or the rate of change of temperature) in that direction. The larger the temperature gradient, the higher the rate of heat transfer.

FIGURE 1–1
We are normally interested in how long it takes for the hot coffee in a thermos bottle to cool to a certain temperature, which cannot be determined from a thermodynamic analysis alone.

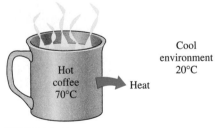

FIGURE 1–2
Heat flows in the direction of decreasing temperature.

Application Areas of Heat Transfer

Heat transfer is commonly encountered in engineering systems and other aspects of life, and one does not need to go very far to see some application areas of heat transfer. In fact, one does not need to go anywhere. The human body is constantly rejecting heat to its surroundings, and human comfort is closely tied to the rate of this heat rejection. We try to control this heat transfer rate by adjusting our clothing to the environmental conditions.

Many ordinary household appliances are designed, in whole or in part, by using the principles of heat transfer. Some examples include the electric or gas range, the heating and air-conditioning system, the refrigerator and freezer, the water heater, the iron, and even the computer, the TV, and the DVD player. Of course, energy-efficient homes are designed on the basis of minimizing heat loss in winter and heat gain in summer. Heat transfer plays a major role in the design of many other devices, such as car radiators, solar collectors, various components of power plants, and even spacecraft (Fig. 1–3). The optimal insulation thickness in the walls and roofs of the houses, on hot water or steam pipes, or on water heaters is again determined on the basis of a heat transfer analysis with economic consideration.

Historical Background

Heat has always been perceived to be something that produces in us a sensation of warmth, and one would think that the nature of heat is one of the first

The human body

Air conditioning systems

Airplanes

Car radiators

Power plants

Refrigeration systems

FIGURE 1–3
Some application areas of heat transfer.

A/C unit, fridge, radiator: © The McGraw-Hill Companies, Inc./Jill Braaten, photographer; Plane: © Vol. 14/PhotoDisc; Humans: © Vol. 121/PhotoDisc; Power plant: © Corbis Royalty Free

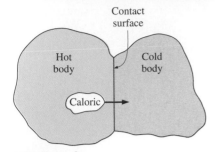

FIGURE 1–4

In the early nineteenth century, heat was thought to be an invisible fluid called the *caloric* that flowed from warmer bodies to the cooler ones.

things understood by mankind. But it was only in the middle of the nineteenth century that we had a true physical understanding of the nature of heat, thanks to the development at that time of the **kinetic theory**, which treats molecules as tiny balls that are in motion and thus possess kinetic energy. Heat is then defined as the energy associated with the random motion of atoms and molecules. Although it was suggested in the eighteenth and early nineteenth centuries that heat is the manifestation of motion at the molecular level (called the *live force*), the prevailing view of heat until the middle of the nineteenth century was based on the **caloric theory** proposed by the French chemist Antoine Lavoisier (1743–1794) in 1789. The caloric theory asserts that heat is a fluid-like substance called the **caloric** that is a massless, colorless, odorless, and tasteless substance that can be poured from one body into another (Fig. 1–4). When caloric was added to a body, its temperature increased; and when caloric was removed from a body, its temperature decreased. When a body could not contain any more caloric, much the same way as when a glass of water could not dissolve any more salt or sugar, the body was said to be saturated with caloric. This interpretation gave rise to the terms *saturated liquid* and *saturated vapor* that are still in use today.

The caloric theory came under attack soon after its introduction. It maintained that heat is a substance that could not be created or destroyed. Yet it was known that heat can be generated indefinitely by rubbing one's hands together or rubbing two pieces of wood together. In 1798, the American Benjamin Thompson (Count Rumford) (1753–1814) showed in his papers that heat can be generated continuously through friction. The validity of the caloric theory was also challenged by several others. But it was the careful experiments of the Englishman James P. Joule (1818–1889) published in 1843 that finally convinced the skeptics that heat was not a substance after all, and thus put the caloric theory to rest. Although the caloric theory was totally abandoned in the middle of the nineteenth century, it contributed greatly to the development of thermodynamics and heat transfer.

1–2 · ENGINEERING HEAT TRANSFER

Heat transfer equipment such as heat exchangers, boilers, condensers, radiators, heaters, furnaces, refrigerators, and solar collectors are designed primarily on the basis of heat transfer analysis. The heat transfer problems encountered in practice can be considered in two groups: (1) *rating* and (2) *sizing* problems. The rating problems deal with the determination of the heat transfer rate for an existing system at a specified temperature difference. The sizing problems deal with the determination of the size of a system in order to transfer heat at a specified rate for a specified temperature difference.

An engineering device or process can be studied either *experimentally* (testing and taking measurements) or *analytically* (by analysis or calculations). The experimental approach has the advantage that we deal with the actual physical system, and the desired quantity is determined by measurement, within the limits of experimental error. However, this approach is expensive, timeconsuming, and often impractical. Besides, the system we are analyzing may not even exist. For example, the entire heating and plumbing systems of a building must usually be sized *before* the building is actually built on the basis of the specifications given. The analytical approach (including the numerical approach) has the advantage that it is fast and inexpensive, but the results obtained are subject to the accuracy of the assumptions, approximations, and idealizations made in the analysis. In engineering studies, often a good compromise is reached by reducing the choices to just a few by analysis, and then verifying the findings experimentally.

Modeling in Engineering

The descriptions of most scientific problems involve equations that relate the changes in some key variables to each other. Usually the smaller the increment chosen in the changing variables, the more general and accurate the description. In the limiting case of infinitesimal or differential changes in variables, we obtain differential equations that provide precise mathematical formulations for the physical principles and laws by representing the rates of change as derivatives. Therefore, differential equations are used to investigate a wide variety of problems in sciences and engineering (Fig. 1–5). However, many problems encountered in practice can be solved without resorting to differential equations and the complications associated with them.

The study of physical phenomena involves two important steps. In the first step, all the variables that affect the phenomena are identified, reasonable assumptions and approximations are made, and the interdependence of these variables is studied. The relevant physical laws and principles are invoked, and the problem is formulated mathematically. The equation itself is very instructive as it shows the degree of dependence of some variables on others, and the relative importance of various terms. In the second step, the problem is solved using an appropriate approach, and the results are interpreted.

Many processes that seem to occur in nature randomly and without any order are, in fact, being governed by some visible or not-so-visible physical laws. Whether we notice them or not, these laws are there, governing consistently and predictably what seem to be ordinary events. Most of these laws are well defined and well understood by scientists. This makes it possible to predict the course of an event before it actually occurs, or to study various aspects of an event mathematically without actually running expensive and timeconsuming experiments. This is where the power of analysis lies. Very accurate results to meaningful practical problems can be obtained with relatively little effort by using a suitable and realistic mathematical model. The preparation of such models requires an adequate knowledge of the natural phenomena involved and the relevant laws, as well as a sound judgment. An unrealistic model will obviously give inaccurate and thus unacceptable results.

An analyst working on an engineering problem often finds himself or herself in a position to make a choice between a very accurate but complex model, and a simple but not-so-accurate model. The right choice depends on the situation at hand. The right choice is usually the simplest model that yields adequate results. For example, the process of baking potatoes or roasting a round chunk of beef in an oven can be studied analytically in a simple way by modeling the potato or the roast as a spherical solid ball that has the properties of water (Fig. 1–6). The model is quite simple, but the results obtained are sufficiently accurate for most practical purposes. As another example, when we analyze the heat losses from a building in order to select the right size for a heater, we determine the heat losses under anticipated worst conditions and select a furnace that will provide sufficient energy to make up for those losses.

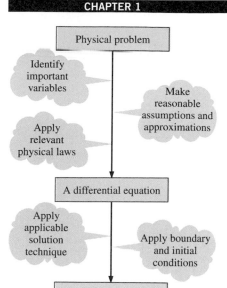

FIGURE 1–5
Mathematical modeling of physical problems.

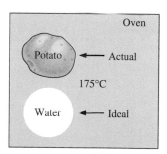

FIGURE 1–6
Modeling is a powerful engineering tool that provides great insight and simplicity at the expense of some accuracy.

Often we tend to choose a larger furnace in anticipation of some future expansion, or just to provide a factor of safety. A very simple analysis is adequate in this case.

When selecting heat transfer equipment, it is important to consider the actual operating conditions. For example, when purchasing a heat exchanger that will handle hard water, we must consider that some calcium deposits will form on the heat transfer surfaces over time, causing fouling and thus a gradual decline in performance. The heat exchanger must be selected on the basis of operation under these adverse conditions instead of under new conditions.

Preparing very accurate but complex models is usually not so difficult. But such models are not much use to an analyst if they are very difficult and time-consuming to solve. At the minimum, the model should reflect the essential features of the physical problem it represents. There are many significant real-world problems that can be analyzed with a simple model. But it should always be kept in mind that the results obtained from an analysis are as accurate as the assumptions made in simplifying the problem. Therefore, the solution obtained should not be applied to situations for which the original assumptions do not hold.

A solution that is not quite consistent with the observed nature of the problem indicates that the mathematical model used is too crude. In that case, a more realistic model should be prepared by eliminating one or more of the questionable assumptions. This will result in a more complex problem that, of course, is more difficult to solve. Thus any solution to a problem should be interpreted within the context of its formulation.

1–3 · HEAT AND OTHER FORMS OF ENERGY

Energy can exist in numerous forms such as thermal, mechanical, kinetic, potential, electrical, magnetic, chemical, and nuclear, and their sum constitutes the **total energy** E (or e on a unit mass basis) of a system. The forms of energy related to the molecular structure of a system and the degree of the molecular activity are referred to as the *microscopic energy*. The sum of all microscopic forms of energy is called the **internal energy** of a system, and is denoted by U (or u on a unit mass basis).

The international unit of energy is *joule* (J) or *kilojoule* (1 kJ = 1000 J). In the English system, the unit of energy is the *British thermal unit* (Btu), which is defined as the energy needed to raise the temperature of 1 lbm of water at 60°F by 1°F. The magnitudes of kJ and Btu are almost identical (1 Btu = 1.055056 kJ). Another well-known unit of energy is the *calorie* (1 cal = 4.1868 J), which is defined as the energy needed to raise the temperature of 1 gram of water at 14.5°C by 1°C.

Internal energy may be viewed as the sum of the kinetic and potential energies of the molecules. The portion of the internal energy of a system associated with the kinetic energy of the molecules is called **sensible energy** or **sensible heat**. The average velocity and the degree of activity of the molecules are proportional to the temperature. Thus, at higher temperatures the molecules possess higher kinetic energy, and as a result, the system has a higher internal energy.

The internal energy is also associated with the intermolecular forces between the molecules of a system. These are the forces that bind the molecules to each other, and, as one would expect, they are strongest in solids and weakest in gases. If sufficient energy is added to the molecules of a solid or liquid, they will overcome these molecular forces and simply break away, turning the system to a gas. This is a *phase change* process and because of this added energy, a system in the gas phase is at a higher internal energy level than it is in the solid or the liquid phase. The internal energy associated with the phase of a system is called **latent energy** or **latent heat**.

The changes mentioned above can occur without a change in the chemical composition of a system. Most heat transfer problems fall into this category, and one does not need to pay any attention to the forces binding the atoms in a molecule together. The internal energy associated with the atomic bonds in a molecule is called **chemical** (or **bond**) **energy**, whereas the internal energy associated with the bonds within the nucleus of the atom itself is called **nuclear energy**. The chemical and nuclear energies are absorbed or released during chemical or nuclear reactions, respectively.

In the analysis of systems that involve fluid flow, we frequently encounter the combination of properties u and Pv. For the sake of simplicity and convenience, this combination is defined as **enthalpy** h. That is, $h = u + Pv$ where the term Pv represents the *flow energy* of the fluid (also called the *flow work*), which is the energy needed to push a fluid and to maintain flow. In the energy analysis of flowing fluids, it is convenient to treat the flow energy as part of the energy of the fluid and to represent the microscopic energy of a fluid stream by enthalpy h (Fig. 1–7).

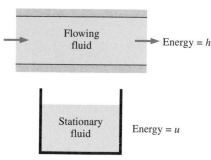

FIGURE 1–7
The *internal energy u* represents the microscopic energy of a nonflowing fluid, whereas *enthalpy h* represents the microscopic energy of a flowing fluid.

Specific Heats of Gases, Liquids, and Solids

You may recall that an **ideal gas** is defined as a gas that obeys the relation

$$Pv = RT \quad \text{or} \quad P = \rho RT \quad (1\text{--}1)$$

where P is the absolute pressure, v is the specific volume, T is the thermodynamic (or absolute) temperature, ρ is the density, and R is the gas constant. It has been experimentally observed that the ideal gas relation given above closely approximates the P-v-T behavior of real gases at low densities. At low pressures and high temperatures, the density of a gas decreases and the gas behaves like an ideal gas. In the range of practical interest, many familiar gases such as air, nitrogen, oxygen, hydrogen, helium, argon, neon, and krypton and even heavier gases such as carbon dioxide can be treated as ideal gases with negligible error (often less than one percent). Dense gases such as water vapor in steam power plants and refrigerant vapor in refrigerators, however, should not always be treated as ideal gases since they usually exist at a state near saturation.

You may also recall that **specific heat** is defined as *the energy required to raise the temperature of a unit mass of a substance by one degree* (Fig. 1–8). In general, this energy depends on how the process is executed. We are usually interested in two kinds of specific heats: specific heat at constant volume c_v and specific heat at constant pressure c_p. The **specific heat at constant volume** c_v can be viewed as the energy required to raise the temperature of a unit mass of a substance by one degree as the volume is held constant. The energy required to do the same as the pressure is held constant is the **specific heat at constant pressure** c_p. The specific heat at constant pressure c_p is

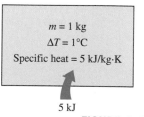

FIGURE 1–8
Specific heat is the energy required to raise the temperature of a unit mass of a substance by one degree in a specified way.

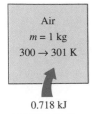
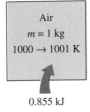

0.718 kJ   0.855 kJ

FIGURE 1–9

The specific heat of a substance changes with temperature.

greater than c_v because at constant pressure the system is allowed to expand and the energy for this expansion work must also be supplied to the system. For ideal gases, these two specific heats are related to each other by $c_p = c_v + R$.

A common unit for specific heats is kJ/kg · °C or kJ/kg · K. Notice that these two units are *identical* since $\Delta T(°C) = \Delta T(K)$, and 1°C change in temperature is equivalent to a change of 1 K. Also,

$$1 \text{ kJ/kg} \cdot °C \equiv 1 \text{ J/g} \cdot °C \equiv 1 \text{ kJ/kg} \cdot K \equiv 1 \text{ J/g} \cdot K$$

The specific heats of a substance, in general, depend on two independent properties such as temperature and pressure. For an *ideal gas,* however, they depend on *temperature* only (Fig. 1–9). At low pressures all real gases approach ideal gas behavior, and therefore their specific heats depend on temperature only.

The differential changes in the internal energy u and enthalpy h of an ideal gas can be expressed in terms of the specific heats as

$$du = c_v \, dT \quad \text{and} \quad dh = c_p \, dT \tag{1–2}$$

The finite changes in the internal energy and enthalpy of an ideal gas during a process can be expressed approximately by using specific heat values at the average temperature as

$$\Delta u = c_{v, \text{avg}} \Delta T \quad \text{and} \quad \Delta h = c_{p, \text{avg}} \Delta T \quad \text{(J/g)} \tag{1–3}$$

or

$$\Delta U = m c_{v, \text{avg}} \Delta T \quad \text{and} \quad \Delta H = m c_{p, \text{avg}} \Delta T \quad \text{(J)} \tag{1–4}$$

IRON
25°C
$c = c_v = c_p$
$= 0.45 \text{ kJ/kg·K}$

FIGURE 1–10

The c_v and c_p values of incompressible substances are identical and are denoted by c.

where m is the mass of the system.

A substance whose specific volume (or density) does not change with temperature or pressure is called an **incompressible substance**. The specific volumes of solids and liquids essentially remain constant during a process, and thus they can be approximated as incompressible substances without sacrificing much in accuracy.

The constant-volume and constant-pressure specific heats are identical for incompressible substances (Fig. 1–10). Therefore, for solids and liquids the subscripts on c_v and c_p can be dropped and both specific heats can be represented by a single symbol, c. That is, $c_p \cong c_v \cong c$. This result could also be deduced from the physical definitions of constant-volume and constant-pressure specific heats. Specific heats of several common gases, liquids, and solids are given in the Appendix.

The specific heats of incompressible substances depend on temperature only. Therefore, the change in the internal energy of solids and liquids can be expressed as

$$\Delta U = m c_{\text{avg}} \Delta T \quad \text{(J)} \tag{1–5}$$

where c_{avg} is the average specific heat evaluated at the average temperature. Note that the internal energy change of the systems that remain in a single phase (liquid, solid, or gas) during the process can be determined very easily using average specific heats.

Energy Transfer

Energy can be transferred to or from a given mass by two mechanisms: *heat transfer Q* and *work W*. An energy interaction is heat transfer if its driving force is a temperature difference. Otherwise, it is work. A rising piston, a rotating shaft, and an electrical wire crossing the system boundaries are all associated with work interactions. Work done *per unit time* is called **power**, and is denoted by $\dot{W}$. The unit of power is W or hp (1 hp = 746 W). Car engines and hydraulic, steam, and gas turbines produce work; compressors, pumps, and mixers consume work. Notice that the energy of a system decreases as it does work, and increases as work is done on it.

In daily life, we frequently refer to the sensible and latent forms of internal energy as **heat**, and we talk about the heat content of bodies (Fig. 1–11). In thermodynamics, however, those forms of energy are usually referred to as **thermal energy** to prevent any confusion with *heat transfer.*

The term *heat* and the associated phrases such as *heat flow, heat addition, heat rejection, heat absorption, heat gain, heat loss, heat storage, heat generation, electrical heating, latent heat, body heat,* and *heat source* are in common use today, and the attempt to replace *heat* in these phrases by *thermal energy* had only limited success. These phrases are deeply rooted in our vocabulary and they are used by both the ordinary people and scientists without causing any misunderstanding. For example, the phrase *body heat* is understood to mean the *thermal energy content* of a body. Likewise, *heat flow* is understood to mean the *transfer of thermal energy,* not the flow of a fluid-like substance called *heat,* although the latter incorrect interpretation, based on the caloric theory, is the origin of this phrase. Also, the transfer of heat into a system is frequently referred to as *heat addition* and the transfer of heat out of a system as *heat rejection.*

Keeping in line with current practice, we will refer to the thermal energy as *heat* and the transfer of thermal energy as *heat transfer.* The amount of heat transferred during the process is denoted by Q. The amount of heat transferred per unit time is called **heat transfer rate**, and is denoted by $\dot{Q}$. The overdot stands for the time derivative, or "per unit time." The heat transfer rate $\dot{Q}$ has the unit J/s, which is equivalent to W.

When the *rate* of heat transfer $\dot{Q}$ is available, then the total amount of heat transfer Q during a time interval Δt can be determined from

$$Q = \int_0^{\Delta t} \dot{Q}\, dt \qquad \text{(J)} \qquad (1\text{–}6)$$

provided that the variation of $\dot{Q}$ with time is known. For the special case of $\dot{Q}$ = constant, the equation above reduces to

$$Q = \dot{Q}\Delta t \qquad \text{(J)} \qquad (1\text{–}7)$$

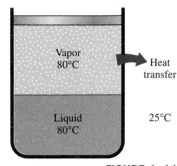

FIGURE 1–11

The sensible and latent forms of internal energy can be transferred as a result of a temperature difference, and they are referred to as *heat* or *thermal energy.*

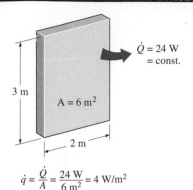

$$q = \frac{\dot{Q}}{A} = \frac{24\ \text{W}}{6\ \text{m}^2} = 4\ \text{W/m}^2$$

FIGURE 1–12
Heat flux is heat transfer *per unit time* and *per unit area,* and is equal to $\dot{q} = \dot{Q}/A$ when $\dot{Q}$ is uniform over the area A.

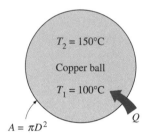

FIGURE 1–13
Schematic for Example 1–1.

The rate of heat transfer per unit area normal to the direction of heat transfer is called **heat flux**, and the average heat flux is expressed as (Fig. 1–12)

$$\dot{q} = \frac{\dot{Q}}{A} \qquad (\text{W/m}^2) \qquad\qquad (1\text{–}8)$$

where A is the heat transfer area. The unit of heat flux in English units is Btu/h · ft². Note that heat flux may vary with time as well as position on a surface.

EXAMPLE 1–1   Heating of a Copper Ball

A 10-cm-diameter copper ball is to be heated from 100°C to an average temperature of 150°C in 30 minutes (Fig. 1–13). Taking the average density and specific heat of copper in this temperature range to be $\rho = 8950$ kg/m³ and $c_p = 0.395$ kJ/kg · °C, respectively, determine (*a*) the total amount of heat transfer to the copper ball, (*b*) the average rate of heat transfer to the ball, and (*c*) the average heat flux.

SOLUTION The copper ball is to be heated from 100°C to 150°C. The total heat transfer, the average rate of heat transfer, and the average heat flux are to be determined.
Assumptions Constant properties can be used for copper at the average temperature.
Properties The average density and specific heat of copper are given to be $\rho = 8950$ kg/m³ and $c_p = 0.395$ kJ/kg · °C.
Analysis (*a*) The amount of heat transferred to the copper ball is simply the change in its internal energy, and is determined from

Energy transfer to the system = Energy increase of the system

$$Q = \Delta U = mc_{\text{avg}}\,(T_2 - T_1)$$

where

$$m = \rho V = \frac{\pi}{6}\rho D^3 = \frac{\pi}{6}(8950\ \text{kg/m}^3)(0.1\ \text{m})^3 = 4.686\ \text{kg}$$

Substituting,

$$Q = (4.686\ \text{kg})(0.395\ \text{kJ/kg} \cdot {}^\circ\text{C})(150 - 100){}^\circ\text{C} = \textbf{92.6 kJ}$$

Therefore, 92.6 kJ of heat needs to be transferred to the copper ball to heat it from 100°C to 150°C.

(*b*) The rate of heat transfer normally changes during a process with time. However, we can determine the *average* rate of heat transfer by dividing the total amount of heat transfer by the time interval. Therefore,

$$\dot{Q}_{\text{avg}} = \frac{Q}{\Delta t} = \frac{92.6\ \text{kJ}}{1800\ \text{s}} = 0.0514\ \text{kJ/s} = \textbf{51.4 W}$$

(c) Heat flux is defined as the heat transfer per unit time per unit area, or the rate of heat transfer per unit area. Therefore, the average heat flux in this case is

$$\dot{q}_{avg} = \frac{\dot{Q}_{avg}}{A} = \frac{\dot{Q}_{avg}}{\pi D^2} = \frac{51.4 \text{ W}}{\pi (0.1 \text{ m})^2} = \textbf{1636 W/m}^2$$

Discussion Note that heat flux may vary with location on a surface. The value calculated above is the *average* heat flux over the entire surface of the ball.

1–4 · THE FIRST LAW OF THERMODYNAMICS

The **first law of thermodynamics**, also known as the **conservation of energy principle**, states that *energy can neither be created nor destroyed during a process; it can only change forms.* Therefore, every bit of energy must be accounted for during a process. The conservation of energy principle (or the energy balance) for *any system* undergoing *any process* may be expressed as follows: *The net change (increase or decrease) in the total energy of the system during a process is equal to the difference between the total energy entering and the total energy leaving the system during that process.* That is,

$$\begin{pmatrix} \text{Total energy} \\ \text{entering the} \\ \text{system} \end{pmatrix} - \begin{pmatrix} \text{Total energy} \\ \text{leaving the} \\ \text{system} \end{pmatrix} = \begin{pmatrix} \text{Change in the} \\ \text{total energy of} \\ \text{the system} \end{pmatrix} \qquad \textbf{(1–9)}$$

Noting that energy can be transferred to or from a system by *heat, work,* and *mass flow,* and that the total energy of a simple compressible system consists of internal, kinetic, and potential energies, the **energy balance** for any system undergoing any process can be expressed as

$$\underbrace{E_{in} - E_{out}}_{\substack{\text{Net energy transfer} \\ \text{by heat, work, and mass}}} = \underbrace{\Delta E_{system}}_{\substack{\text{Change in internal, kinetic,} \\ \text{potential, etc., energies}}} \qquad \text{(J)} \qquad \textbf{(1–10)}$$

or, in the **rate form**, as

$$\underbrace{\dot{E}_{in} - \dot{E}_{out}}_{\substack{\text{Rate of net energy transfer} \\ \text{by heat, work, and mass}}} = \underbrace{dE_{system}/dt}_{\substack{\text{Rate of change in internal} \\ \text{kinetic, potential, etc., energies}}} \qquad \text{(W)} \qquad \textbf{(1–11)}$$

Energy is a property, and the value of a property does not change unless the state of the system changes. Therefore, the energy change of a system is zero ($\Delta E_{system} = 0$) if the state of the system does not change during the process, that is, the process is steady. The energy balance in this case reduces to (Fig. 1–14)

Steady, rate form:
$$\underbrace{\dot{E}_{in}}_{\substack{\text{Rate of net energy transfer in} \\ \text{by heat, work, and mass}}} = \underbrace{\dot{E}_{out}}_{\substack{\text{Rate of net energy transfer out} \\ \text{by heat, work, and mass}}} \qquad \textbf{(1–12)}$$

In the absence of significant electric, magnetic, motion, gravity, and surface tension effects (i.e., for stationary simple compressible systems), the change

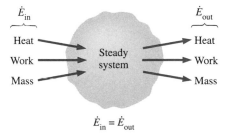

$\dot{E}_{in} = \dot{E}_{out}$

FIGURE 1–14
In steady operation, the rate of energy transfer to a system is equal to the rate of energy transfer from the system.

in the *total energy* of a system during a process is simply the change in its *internal energy*. That is, $\Delta E_{\text{system}} = \Delta U_{\text{system}}$.

In heat transfer analysis, we are usually interested only in the forms of energy that can be transferred as a result of a temperature difference, that is, heat or thermal energy. In such cases it is convenient to write a **heat balance** and to treat the conversion of nuclear, chemical, mechanical, and electrical energies into thermal energy as *heat generation*. The *energy balance* in that case can be expressed as

$$\underbrace{Q_{\text{in}} - Q_{\text{out}}}_{\substack{\text{Net heat} \\ \text{transfer}}} + \underbrace{E_{\text{gen}}}_{\substack{\text{Heat} \\ \text{generation}}} = \underbrace{\Delta E_{\text{thermal, system}}}_{\substack{\text{Change in thermal} \\ \text{energy of the system}}} \quad \text{(J)} \qquad \textbf{(1–13)}$$

Energy Balance for Closed Systems *(Fixed Mass)*

A closed system consists of a *fixed mass*. The total energy E for most systems encountered in practice consists of the internal energy U. This is especially the case for stationary systems since they don't involve any changes in their velocity or elevation during a process. The energy balance relation in that case reduces to

Stationary closed system: $\qquad E_{\text{in}} - E_{\text{out}} = \Delta U = mc_v \Delta T \qquad \text{(J)} \qquad \textbf{(1–14)}$

where we expressed the internal energy change in terms of mass m, the specific heat at constant volume c_v, and the temperature change ΔT of the system. When the system involves heat transfer only and no work interactions across its boundary, the energy balance relation further reduces to (Fig. 1–15)

Stationary closed system, no work: $\qquad Q = mc_v \Delta T \qquad \text{(J)} \qquad \textbf{(1–15)}$

where Q is the net amount of heat transfer to or from the system. This is the form of the energy balance relation we will use most often when dealing with a fixed mass.

Energy Balance for Steady-Flow Systems

A large number of engineering devices such as water heaters and car radiators involve mass flow in and out of a system, and are modeled as *control volumes*. Most control volumes are analyzed under steady operating conditions. The term *steady* means *no change with time* at a specified location. The opposite of steady is *unsteady* or *transient*. Also, the term *uniform* implies *no change with position* throughout a surface or region at a specified time. These meanings are consistent with their everyday usage (steady girlfriend, uniform distribution, etc.). The total energy content of a control volume during a *steady-flow process* remains constant (E_{CV} = constant). That is, the change in the total energy of the control volume during such a process is zero ($\Delta E_{\text{CV}} = 0$). Thus the amount of energy entering a control volume in all forms (heat, work, mass transfer) for a steady-flow process must be equal to the amount of energy leaving it.

The amount of mass flowing through a cross section of a flow device per unit time is called the **mass flow rate**, and is denoted by $\dot{m}$. A fluid may flow in and out of a control volume through pipes or ducts. The mass flow rate of a fluid flowing in a pipe or duct is proportional to the cross-sectional area A_c of

Specific heat = c_v
Mass = m
Initial temp = T_1
Final temp = T_2

$Q = mc_v(T_1 - T_2)$

FIGURE 1–15

In the absence of any work interactions, the change in the energy content of a closed system is equal to the net heat transfer.

the pipe or duct, the density ρ, and the velocity V of the fluid. The mass flow rate through a differential area dA_c can be expressed as $\delta \dot{m} = \rho V_n \, dA_c$ where V_n is the velocity component normal to dA_c. The mass flow rate through the entire cross-sectional area is obtained by integration over A_c.

The flow of a fluid through a pipe or duct can often be approximated to be *one-dimensional*. That is, the properties can be assumed to vary in one direction only (the direction of flow). As a result, all properties are assumed to be uniform at any cross section normal to the flow direction, and the properties are assumed to have *bulk average values* over the entire cross section. Under the one-dimensional flow approximation, the mass flow rate of a fluid flowing in a pipe or duct can be expressed as (Fig. 1–16)

$$\dot{m} = \rho V A_c \qquad (\text{kg/s}) \qquad (1\text{--}16)$$

where ρ is the fluid density, V is the average fluid velocity in the flow direction, and A_c is the cross-sectional area of the pipe or duct.

The volume of a fluid flowing through a pipe or duct per unit time is called the **volume flow rate** $\dot{V}$, and is expressed as

$$\dot{V} = V A_c = \frac{\dot{m}}{\rho} \qquad (\text{m}^3/\text{s}) \qquad (1\text{--}17)$$

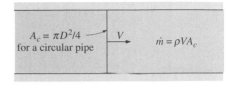

FIGURE 1–16

The mass flow rate of a fluid at a cross section is equal to the product of the fluid density, average fluid velocity, and the cross-sectional area.

Note that the mass flow rate of a fluid through a pipe or duct remains constant during steady flow. This is not the case for the volume flow rate, however, unless the density of the fluid remains constant.

For a steady-flow system with one inlet and one exit, the rate of mass flow into the control volume must be equal to the rate of mass flow out of it. That is, $\dot{m}_{\text{in}} = \dot{m}_{\text{out}} = \dot{m}$. When the changes in kinetic and potential energies are negligible, which is usually the case, and there is no work interaction, the energy balance for such a steady-flow system reduces to (Fig. 1–17)

$$\dot{Q} = \dot{m} \Delta h = \dot{m} c_p \Delta T \qquad (\text{kJ/s}) \qquad (1\text{--}18)$$

where $\dot{Q}$ is the rate of net heat transfer into or out of the control volume. This is the form of the energy balance relation that we will use most often for steady-flow systems.

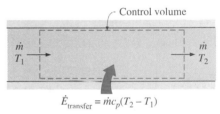

FIGURE 1–17

Under steady conditions, the net rate of energy transfer to a fluid in a control volume is equal to the rate of increase in the energy of the fluid stream flowing through the control volume.

Surface Energy Balance

As mentioned in the chapter opener, heat is transferred by the mechanisms of conduction, convection, and radiation, and heat often changes vehicles as it is transferred from one medium to another. For example, the heat conducted to the outer surface of the wall of a house in winter is convected away by the cold outdoor air while being radiated to the cold surroundings. In such cases, it may be necessary to keep track of the energy interactions at the surface, and this is done by applying the conservation of energy principle to the surface.

A surface contains no volume or mass, and thus no energy. Thereore, a surface can be viewed as a fictitious system whose energy content remains constant during a process (just like a steady-state or steady-flow system). Then the energy balance for a surface can be expressed as

Surface energy balance: $\qquad\qquad \dot{E}_{\text{in}} = \dot{E}_{\text{out}} \qquad\qquad (1\text{--}19)$

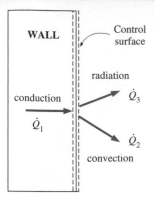

FIGURE 1–18

Energy interactions at the outer wall surface of a house.

This relation is valid for both steady and transient conditions, and the surface energy balance does not involve heat generation since a surface does not have a volume. The energy balance for the outer surface of the wall in Fig. 1–18, for example, can be expressed as

$$\dot{Q}_1 = \dot{Q}_2 + \dot{Q}_3 \qquad (1\text{–}20)$$

where $\dot{Q}_1$ is conduction through the wall to the surface, $\dot{Q}_2$ is convection from the surface to the outdoor air, and $\dot{Q}_3$ is net radiation from the surface to the surroundings.

When the directions of interactions are not known, all energy interactions can be assumed to be towards the surface, and the surface energy balance can be expressed as $\sum \dot{E}_{in} = 0$. Note that the interactions in opposite direction will end up having negative values, and balance this equation.

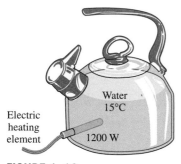

FIGURE 1–19

Schematic for Example 1–2.

EXAMPLE 1–2   **Heating of Water in an Electric Teapot**

1.2 kg of liquid water initially at 15°C is to be heated to 95°C in a teapot equipped with a 1200-W electric heating element inside (Fig. 1–19). The teapot is 0.5 kg and has an average specific heat of 0.7 kJ/kg · K. Taking the specific heat of water to be 4.18 kJ/kg · K and disregarding any heat loss from the teapot, determine how long it will take for the water to be heated.

SOLUTION   Liquid water is to be heated in an electric teapot. The heating time is to be determined.

Assumptions   **1** Heat loss from the teapot is negligible. **2** Constant properties can be used for both the teapot and the water.

Properties   The average specific heats are given to be 0.7 kJ/kg · K for the teapot and 4.18 kJ/kg · K for water.

Analysis   We take the teapot and the water in it as the system, which is a closed system (fixed mass). The energy balance in this case can be expressed as

$$E_{in} - E_{out} = \Delta E_{system}$$
$$E_{in} = \Delta U_{system} = \Delta U_{water} + \Delta U_{teapot}$$

Then the amount of energy needed to raise the temperature of water and the teapot from 15°C to 95°C is

$$
\begin{aligned}
E_{in} &= (mc_p\Delta T)_{water} + (mc_p\Delta T)_{teapot} \\
&= (1.2 \text{ kg})(4.18 \text{ kJ/kg} \cdot °\text{C})(95 - 15)°\text{C} + (0.5 \text{ kg})(0.7 \text{ kJ/kg} \cdot °\text{C}) \\
&\quad (95 - 15)°\text{C} \\
&= 429.3 \text{ kJ}
\end{aligned}
$$

The 1200-W electric heating unit will supply energy at a rate of 1.2 kW or 1.2 kJ per second. Therefore, the time needed for this heater to supply 429.3 kJ of heat is determined from

$$\Delta t = \frac{\text{Total energy transferred}}{\text{Rate of energy transfer}} = \frac{E_{in}}{\dot{E}_{transfer}} = \frac{429.3 \text{ kJ}}{1.2 \text{ kJ/s}} = 358 \text{ s} = \mathbf{6.0 \text{ min}}$$

Discussion In reality, it will take more than 6 minutes to accomplish this heating process since some heat loss is inevitable during heating. Also, the specific heat units kJ/kg · °C and kJ/kg · K are equivalent, and can be interchanged.

EXAMPLE 1–3   Heat Loss from Heating Ducts in a Basement

A 5-m-long section of an air heating system of a house passes through an unheated space in the basement (Fig. 1–20). The cross section of the rectangular duct of the heating system is 20 cm × 25 cm. Hot air enters the duct at 100 kPa and 60°C at an average velocity of 5 m/s. The temperature of the air in the duct drops to 54°C as a result of heat loss to the cool space in the basement. Determine the rate of heat loss from the air in the duct to the basement under steady conditions. Also, determine the cost of this heat loss per hour if the house is heated by a natural gas furnace that has an efficiency of 80 percent, and the cost of the natural gas in that area is $1.60/therm (1 therm = 100,000 Btu = 105,500 kJ).

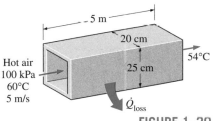

FIGURE 1–20
Schematic for Example 1–3.

SOLUTION The temperature of the air in the heating duct of a house drops as a result of heat loss to the cool space in the basement. The rate of heat loss from the hot air and its cost are to be determined.
Assumptions **1** Steady operating conditions exist. **2** Air can be treated as an ideal gas with constant properties at room temperature.
Properties The constant pressure specific heat of air at the average temperature of $(54 + 60)/2 = 57°C$ is 1.007 kJ/kg · K (Table A–15).
Analysis We take the basement section of the heating system as our system, which is a steady-flow system. The rate of heat loss from the air in the duct can be determined from

$$\dot{Q} = \dot{m} c_p \Delta T$$

where $\dot{m}$ is the mass flow rate and ΔT is the temperature drop. The density of air at the inlet conditions is

$$\rho = \frac{P}{RT} = \frac{100 \text{ kPa}}{(0.287 \text{ kPa} \cdot \text{m}^3/\text{kg} \cdot \text{K})(60 + 273)\text{K}} = 1.046 \text{ kg/m}^3$$

The cross-sectional area of the duct is

$$A_c = (0.20 \text{ m})(0.25 \text{ m}) = 0.05 \text{ m}^2$$

Then the mass flow rate of air through the duct and the rate of heat loss become

$$\dot{m} = \rho V A_c = (1.046 \text{ kg/m}^3)(5 \text{ m/s})(0.05 \text{ m}^2) = 0.2615 \text{ kg/s}$$

and

$$\dot{Q}_{\text{loss}} = \dot{m} c_p (T_{\text{in}} - T_{\text{out}})$$
$$= (0.2615 \text{ kg/s})(1.007 \text{ kJ/kg} \cdot °C)(60 - 54)°C$$
$$= 1.58 \text{ kJ/s}$$

or 5688 kJ/h. The cost of this heat loss to the home owner is

$$\text{Cost of heat loss} = \frac{(\text{Rate of heat loss})(\text{Unit cost of energy input})}{\text{Furnace efficiency}}$$

$$= \frac{(5688 \text{ kJ/h})(\$1.60/\text{therm})}{0.80}\left(\frac{1 \text{ therm}}{105{,}500 \text{ kJ}}\right)$$

$$= \$0.108/\text{h}$$

Discussion The heat loss from the heating ducts in the basement is costing the home owner 10.8 cents per hour. Assuming the heater operates 2000 hours during a heating season, the annual cost of this heat loss adds up to $216. Most of this money can be saved by insulating the heating ducts in the unheated areas.

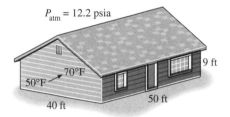

$P_{\text{atm}} = 12.2$ psia

9 ft

70°F

50°F

40 ft

50 ft

FIGURE 1–21

Schematic for Example 1–4.

EXAMPLE 1–4   Electric Heating of a House at High Elevation

Consider a house that has a floor space of 2000 ft² and an average height of 9 ft at 5000 ft elevation where the standard atmospheric pressure is 12.2 psia (Fig. 1–21). Initially the house is at a uniform temperature of 50°F. Now the electric heater is turned on, and the heater runs until the air temperature in the house rises to an average value of 70°F. Determine the amount of energy transferred to the air assuming (a) the house is air-tight and thus no air escapes during the heating process and (b) some air escapes through the cracks as the heated air in the house expands at constant pressure. Also determine the cost of this heat for each case if the cost of electricity in that area is $0.075/kWh.

SOLUTION   The air in the house is heated by an electric heater. The amount and cost of the energy transferred to the air are to be determined for constant-volume and constant-pressure cases.

Assumptions   1 Air can be treated as an ideal gas with constant properties. **2** Heat loss from the house during heating is negligible. **3** The volume occupied by the furniture and other things is negligible.

Properties   The specific heats of air at the average temperature of $(50 + 70)/2 = 60$°F are $c_p = 0.240$ Btu/lbm · R and $c_v = c_p - R = 0.171$ Btu/lbm · R (Tables A–1E and A–15E).

Analysis   The volume and the mass of the air in the house are

$$V = (\text{Floor area})(\text{Height}) = (2000 \text{ ft}^2)(9 \text{ ft}) = 18{,}000 \text{ ft}^3$$

$$m = \frac{PV}{RT} = \frac{(12.2 \text{ psia})(18{,}000 \text{ ft}^3)}{(0.3704 \text{ psia} \cdot \text{ft}^3/\text{lbm} \cdot \text{R})(50 + 460)\text{R}} = 1162 \text{ lbm}$$

(a) The amount of energy transferred to air at constant volume is simply the change in its internal energy, and is determined from

$$E_{\text{in}} - E_{\text{out}} = \Delta E_{\text{system}}$$

$$E_{\text{in, constant volume}} = \Delta U_{\text{air}} = mc_v\Delta T$$

$$= (1162 \text{ lbm})(0.171 \text{ Btu/lbm} \cdot °\text{F})(70 - 50)°\text{F}$$

$$= \textbf{3974 Btu}$$

At a unit cost of $0.075/kWh, the total cost of this energy is

$$\text{Cost of energy} = (\text{Amount of energy})(\text{Unit cost of energy})$$
$$= (3974 \text{ Btu})(\$0.075/\text{kWh})\left(\frac{1 \text{ kWh}}{3412 \text{ Btu}}\right)$$
$$= \mathbf{\$0.087}$$

(b) The amount of energy transferred to air at constant pressure is the change in its enthalpy, and is determined from

$$E_{\text{in, constant pressure}} = \Delta H_{\text{air}} = mc_p\Delta T$$
$$= (1162 \text{ lbm})(0.240 \text{ Btu/lbm} \cdot °\text{F})(70 - 50)°\text{F}$$
$$= \mathbf{5578 \text{ Btu}}$$

At a unit cost of $0.075/kWh, the total cost of this energy is

$$\text{Cost of energy} = (\text{Amount of energy})(\text{Unit cost of energy})$$
$$= (5578 \text{ Btu})(\$0.075/\text{kWh})\left(\frac{1 \text{ kWh}}{3412 \text{ Btu}}\right)$$
$$= \mathbf{\$0.123}$$

Discussion It costs about 9 cents in the first case and 12 cents in the second case to raise the temperature of the air in this house from 50°F to 70°F. The second answer is more realistic since every house has cracks, especially around the doors and windows, and the pressure in the house remains essentially constant during a heating process. Therefore, the second approach is used in practice. This conservative approach somewhat overpredicts the amount of energy used, however, since some of the air escapes through the cracks before it is heated to 70°F.

1–5 · HEAT TRANSFER MECHANISMS

In Section 1–1, we defined *heat* as the form of energy that can be transferred from one system to another as a result of temperature difference. A thermodynamic analysis is concerned with the *amount* of heat transfer as a system undergoes a process from one equilibrium state to another. The science that deals with the determination of the *rates* of such energy transfers is the *heat transfer*. The transfer of energy as heat is always from the higher-temperature medium to the lower-temperature one, and heat transfer stops when the two mediums reach the same temperature.

Heat can be transferred in three different modes: *conduction, convection,* and *radiation.* All modes of heat transfer require the existence of a temperature difference, and all modes are from the high-temperature medium to a lower-temperature one. Below we give a brief description of each mode. A detailed study of these modes is given in later chapters of this text.

1–6 · CONDUCTION

Conduction is the transfer of energy from the more energetic particles of a substance to the adjacent less energetic ones as a result of interactions between the particles. Conduction can take place in solids, liquids, or gases.

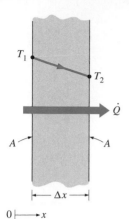

FIGURE 1–22

Heat conduction through a large plane wall of thickness Δx and area A.

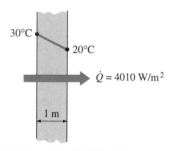

(a) Copper ($k = 401$ W/m·°C)

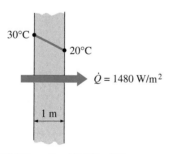

(b) Silicon ($k = 148$ W/m·°C)

FIGURE 1–23

The rate of heat conduction through a solid is directly proportional to its thermal conductivity.

In gases and liquids, conduction is due to the *collisions* and *diffusion* of the molecules during their random motion. In solids, it is due to the combination of *vibrations* of the molecules in a lattice and the energy transport by *free electrons*. A cold canned drink in a warm room, for example, eventually warms up to the room temperature as a result of heat transfer from the room to the drink through the aluminum can by conduction.

The *rate* of heat conduction through a medium depends on the *geometry* of the medium, its *thickness,* and the *material* of the medium, as well as the *temperature difference* across the medium. We know that wrapping a hot water tank with glass wool (an insulating material) reduces the rate of heat loss from the tank. The thicker the insulation, the smaller the heat loss. We also know that a hot water tank loses heat at a higher rate when the temperature of the room housing the tank is lowered. Further, the larger the tank, the larger the surface area and thus the rate of heat loss.

Consider steady heat conduction through a large plane wall of thickness $\Delta x = L$ and area A, as shown in Fig. 1–22. The temperature difference across the wall is $\Delta T = T_2 - T_1$. Experiments have shown that the rate of heat transfer $\dot{Q}$ through the wall is *doubled* when the temperature difference ΔT across the wall or the area A normal to the direction of heat transfer is doubled, but is *halved* when the wall thickness L is doubled. Thus we conclude that *the rate of heat conduction through a plane layer is proportional to the temperature difference across the layer and the heat transfer area, but is inversely proportional to the thickness of the layer.* That is,

$$\text{Rate of heat conduction} \propto \frac{(\text{Area})(\text{Temperature difference})}{\text{Thickness}}$$

or,

$$\dot{Q}_{\text{cond}} = kA\frac{T_1 - T_2}{\Delta x} = -kA\frac{\Delta T}{\Delta x} \qquad \text{(W)} \qquad \text{(1–21)}$$

where the constant of proportionality k is the **thermal conductivity** of the material, which is a *measure of the ability of a material to conduct heat* (Fig. 1–23). In the limiting case of $\Delta x \to 0$, the equation above reduces to the differential form

$$\dot{Q}_{\text{cond}} = -kA\frac{dT}{dx} \qquad \text{(W)} \qquad \text{(1–22)}$$

which is called **Fourier's law of heat conduction** after J. Fourier, who expressed it first in his heat transfer text in 1822. Here dT/dx is the **temperature gradient**, which is the slope of the temperature curve on a T-x diagram (the rate of change of T with x), at location x. The relation above indicates that the rate of heat conduction in a given direction is proportional to the temperature gradient in that direction. Heat is conducted in the direction of decreasing temperature, and the temperature gradient becomes negative when temperature decreases with increasing x. The *negative sign* in Eq. 1–22 ensures that heat transfer in the positive x direction is a positive quantity.

The heat transfer area A is always *normal* to the direction of heat transfer. For heat loss through a 5-m-long, 3-m-high, and 25-cm-thick wall, for example, the heat transfer area is $A = 15$ m². Note that the thickness of the wall has no effect on A (Fig. 1–24).

EXAMPLE 1–5   The Cost of Heat Loss through a Roof

The roof of an electrically heated home is 6 m long, 8 m wide, and 0.25 m thick, and is made of a flat layer of concrete whose thermal conductivity is $k = 0.8$ W/m · °C (Fig. 1–25). The temperatures of the inner and the outer surfaces of the roof one night are measured to be 15°C and 4°C, respectively, for a period of 10 hours. Determine (a) the rate of heat loss through the roof that night and (b) the cost of that heat loss to the home owner if the cost of electricity is $0.08/kWh.

SOLUTION   The inner and outer surfaces of the flat concrete roof of an electrically heated home are maintained at specified temperatures during a night. The heat loss through the roof and its cost that night are to be determined.

Assumptions   **1** Steady operating conditions exist during the entire night since the surface temperatures of the roof remain constant at the specified values. **2** Constant properties can be used for the roof.

Properties   The thermal conductivity of the roof is given to be $k = 0.8$ W/m · °C.

Analysis   (a) Noting that heat transfer through the roof is by conduction and the area of the roof is $A = 6$ m $\times$ 8 m = 48 m², the steady rate of heat transfer through the roof is

$$\dot{Q} = kA \frac{T_1 - T_2}{L} = (0.8 \text{ W/m} \cdot \text{°C})(48 \text{ m}^2)\frac{(15 - 4)\text{°C}}{0.25 \text{ m}} = \mathbf{1690 \text{ W}} = \mathbf{1.69 \text{ kW}}$$

(b) The amount of heat lost through the roof during a 10-hour period and its cost is

$$Q = \dot{Q} \, \Delta t = (1.69 \text{ kW})(10 \text{ h}) = 16.9 \text{ kWh}$$

$$\text{Cost} = (\text{Amount of energy})(\text{Unit cost of energy})$$

$$= (16.9 \text{ kWh})(\$0.08/\text{kWh}) = \mathbf{\$1.35}$$

Discussion   The cost to the home owner of the heat loss through the roof that night was $1.35. The total heating bill of the house will be much larger since the heat losses through the walls are not considered in these calculations.

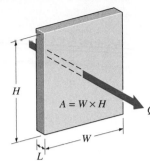

FIGURE 1–24
In heat conduction analysis, A represents the area *normal* to the direction of heat transfer.

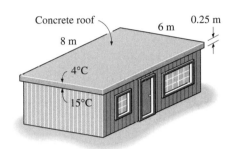

FIGURE 1–25
Schematic for Example 1–5.

Thermal Conductivity

We have seen that different materials store heat differently, and we have defined the property specific heat c_p as a measure of a material's ability to store thermal energy. For example, $c_p = 4.18$ kJ/kg · °C for water and $c_p = 0.45$ kJ/kg · °C for iron at room temperature, which indicates that water can store almost 10 times the energy that iron can per unit mass. Likewise, the thermal conductivity k is a measure of a material's ability to conduct heat. For example, $k = 0.607$ W/m · °C for water and $k = 80.2$ W/m · °C for iron at room temperature, which indicates that iron conducts heat more than 100 times faster than water can. Thus we say that water is a poor heat conductor relative to iron, although water is an excellent medium to store thermal energy.

Equation 1–21 for the rate of conduction heat transfer under steady conditions can also be viewed as the defining equation for thermal conductivity. Thus the **thermal conductivity** of a material can be defined as *the rate of*

TABLE 1–1

The thermal conductivities of some materials at room temperature

Material	k, W/m · °C*
Diamond	2300
Silver	429
Copper	401
Gold	317
Aluminum	237
Iron	80.2
Mercury (l)	8.54
Glass	0.78
Brick	0.72
Water (l)	0.607
Human skin	0.37
Wood (oak)	0.17
Helium (g)	0.152
Soft rubber	0.13
Glass fiber	0.043
Air (g)	0.026
Urethane, rigid foam	0.026

*Multiply by 0.5778 to convert to Btu/h · ft · °F.

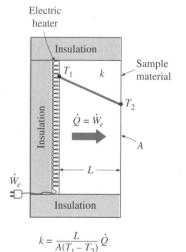

$$k = \frac{L}{A(T_1 - T_2)} \dot{Q}$$

FIGURE 1–26

A simple experimental setup to determine the thermal conductivity of a material.

heat transfer through a unit thickness of the material per unit area per unit temperature difference. The thermal conductivity of a material is a measure of the ability of the material to conduct heat. A high value for thermal conductivity indicates that the material is a good heat conductor, and a low value indicates that the material is a poor heat conductor or *insulator.* The thermal conductivities of some common materials at room temperature are given in Table 1–1. The thermal conductivity of pure copper at room temperature is $k = 401$ W/m · °C, which indicates that a 1-m-thick copper wall will conduct heat at a rate of 401 W per m² area per °C temperature difference across the wall. Note that materials such as copper and silver that are good electric conductors are also good heat conductors, and have high values of thermal conductivity. Materials such as rubber, wood, and Styrofoam are poor conductors of heat and have low conductivity values.

A layer of material of known thickness and area can be heated from one side by an electric resistance heater of known output. If the outer surfaces of the heater are well insulated, all the heat generated by the resistance heater will be transferred through the material whose conductivity is to be determined. Then measuring the two surface temperatures of the material when steady heat transfer is reached and substituting them into Eq. 1–21 together with other known quantities give the thermal conductivity (Fig. 1–26).

The thermal conductivities of materials vary over a wide range, as shown in Fig. 1–27. The thermal conductivities of gases such as air vary by a factor of 10^4 from those of pure metals such as copper. Note that pure crystals and metals have the highest thermal conductivities, and gases and insulating materials the lowest.

Temperature is a measure of the kinetic energies of the particles such as the molecules or atoms of a substance. In a liquid or gas, the kinetic energy of the molecules is due to their random translational motion as well as their vibrational and rotational motions. When two molecules possessing different kinetic energies collide, part of the kinetic energy of the more energetic (higher-temperature) molecule is transferred to the less energetic (lower-temperature) molecule, much the same as when two elastic balls of the same mass at different velocities collide, part of the kinetic energy of the faster ball is transferred to the slower one. The higher the temperature, the faster the molecules move and the higher the number of such collisions, and the better the heat transfer.

The *kinetic theory* of gases predicts and the experiments confirm that the thermal conductivity of gases is proportional to the *square root of the thermodynamic temperature T,* and inversely proportional to the *square root of the molar mass M.* Therefore, the thermal conductivity of a gas increases with increasing temperature and decreasing molar mass. So it is not surprising that the thermal conductivity of helium ($M = 4$) is much higher than those of air ($M = 29$) and argon ($M = 40$).

The thermal conductivities of *gases* at 1 atm pressure are listed in Table A–16. However, they can also be used at pressures other than 1 atm, since the thermal conductivity of gases is *independent of pressure* in a wide range of pressures encountered in practice.

The mechanism of heat conduction in a *liquid* is complicated by the fact that the molecules are more closely spaced, and they exert a stronger intermolecular force field. The thermal conductivities of liquids usually lie between those

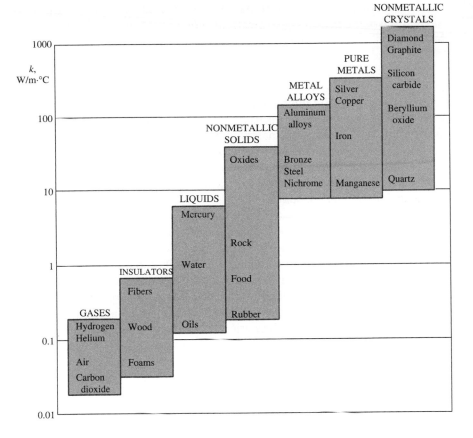

FIGURE 1–27
The range of thermal conductivity of
various materials at room temperature.

of solids and gases. The thermal conductivity of a substance is normally highest in the solid phase and lowest in the gas phase. Unlike gases, the thermal conductivities of most liquids decrease with increasing temperature, with water being a notable exception. Like gases, the conductivity of liquids decreases with increasing molar mass. Liquid metals such as mercury and sodium have high thermal conductivities and are very suitable for use in applications where a high heat transfer rate to a liquid is desired, as in nuclear power plants.

In *solids*, heat conduction is due to two effects: the *lattice vibrational waves* induced by the vibrational motions of the molecules positioned at relatively fixed positions in a periodic manner called a lattice, and the energy transported via the *free flow of electrons* in the solid (Fig. 1–28). The thermal conductivity of a solid is obtained by adding the lattice and electronic components. The relatively high thermal conductivities of pure metals are primarily due to the electronic component. The lattice component of thermal conductivity strongly depends on the way the molecules are arranged. For example, diamond, which is a highly ordered crystalline solid, has the highest known thermal conductivity at room temperature.

Unlike metals, which are good electrical and heat conductors, *crystalline solids* such as diamond and semiconductors such as silicon are good heat conductors but poor electrical conductors. As a result, such materials find widespread use in the electronics industry. Despite their higher price, diamond heat sinks are used in the cooling of sensitive electronic components because of the

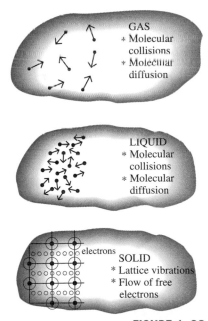

FIGURE 1–28
The mechanisms of heat conduction in
different phases of a substance.

TABLE 1–2

The thermal conductivity of an alloy is usually much lower than the thermal conductivity of either metal of which it is composed

Pure metal or alloy	k, W/m · °C at 300 K
Copper	401
Nickel	91
Constantan (55% Cu, 45% Ni)	23
Copper	401
Aluminum	237
Commercial bronze (90% Cu, 10% Al)	52

TABLE 1–3

Thermal conductivities of materials vary with temperature

	k, W/m · °C	
T, K	Copper	Aluminum
100	482	302
200	413	237
300	401	237
400	393	240
600	379	231
800	366	218

excellent thermal conductivity of diamond. Silicon oils and gaskets are commonly used in the packaging of electronic components because they provide both good thermal contact and good electrical insulation.

Pure metals have high thermal conductivities, and one would think that *metal alloys* should also have high conductivities. One would expect an alloy made of two metals of thermal conductivities k_1 and k_2 to have a conductivity k between k_1 and k_2. But this turns out not to be the case. The thermal conductivity of an alloy of two metals is usually much lower than that of either metal, as shown in Table 1–2. Even small amounts in a pure metal of "foreign" molecules that are good conductors themselves seriously disrupt the transfer of heat in that metal. For example, the thermal conductivity of steel containing just 1 percent of chrome is 62 W/m · °C, while the thermal conductivities of iron and chromium are 83 and 95 W/m · °C, respectively.

The thermal conductivities of materials vary with temperature (Table 1–3). The variation of thermal conductivity over certain temperature ranges is negligible for some materials, but significant for others, as shown in Fig. 1–29. The thermal conductivities of certain solids exhibit dramatic increases at temperatures near absolute zero, when these solids become *superconductors*. For example, the conductivity of copper reaches a maximum value of about 20,000 W/m · °C at 20 K, which is about 50 times the conductivity at room temperature. The thermal conductivities and other thermal properties of various materials are given in Tables A–3 to A–16.

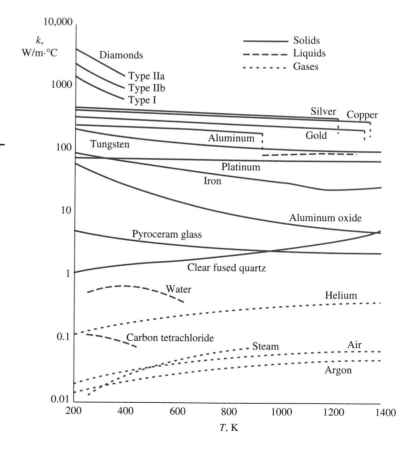

FIGURE 1–29

The variation of the thermal conductivity of various solids, liquids, and gases with temperature.

The temperature dependence of thermal conductivity causes considerable complexity in conduction analysis. Therefore, it is common practice to evaluate the thermal conductivity k at the *average temperature* and treat it as a *constant* in calculations.

In heat transfer analysis, a material is normally assumed to be *isotropic*; that is, to have uniform properties in all directions. This assumption is realistic for most materials, except those that exhibit different structural characteristics in different directions, such as laminated composite materials and wood. The thermal conductivity of wood across the grain, for example, is different than that parallel to the grain.

Thermal Diffusivity

The product ρc_p, which is frequently encountered in heat transfer analysis, is called the **heat capacity** of a material. Both the specific heat c_p and the heat capacity ρc_p represent the heat storage capability of a material. But c_p expresses it *per unit mass* whereas ρc_p expresses it *per unit volume,* as can be noticed from their units J/kg · °C and J/m^3 · °C, respectively.

Another material property that appears in the transient heat conduction analysis is the **thermal diffusivity**, which represents how fast heat diffuses through a material and is defined as

$$\alpha = \frac{\text{Heat conducted}}{\text{Heat stored}} = \frac{k}{\rho c_p} \qquad (\text{m}^2/\text{s}) \qquad (1\text{--}23)$$

Note that the thermal conductivity k represents how well a material conducts heat, and the heat capacity ρc_p represents how much energy a material stores per unit volume. Therefore, the thermal diffusivity of a material can be viewed as the ratio of the *heat conducted* through the material to the *heat stored* per unit volume. A material that has a high thermal conductivity or a low heat capacity will obviously have a large thermal diffusivity. The larger the thermal diffusivity, the faster the propagation of heat into the medium. A small value of thermal diffusivity means that heat is mostly absorbed by the material and a small amount of heat is conducted further.

The thermal diffusivities of some common materials at 20°C are given in Table 1–4. Note that the thermal diffusivity ranges from $\alpha = 0.14 \times 10^{-6}$ m^2/s for water to 149×10^{-6} m^2/s for silver, which is a difference of more than a thousand times. Also note that the thermal diffusivities of beef and water are the same. This is not surprising, since meat as well as fresh vegetables and fruits are mostly water, and thus they possess the thermal properties of water.

EXAMPLE 1–6   Measuring the Thermal Conductivity of a Material

A common way of measuring the thermal conductivity of a material is to sandwich an electric thermofoil heater between two identical samples of the material, as shown in Fig. 1–30. The thickness of the resistance heater, including its cover, which is made of thin silicon rubber, is usually less than 0.5 mm. A circulating fluid such as tap water keeps the exposed ends of the samples at constant temperature. The lateral surfaces of the samples are well insulated to ensure that heat transfer through the samples is one-dimensional. Two thermocouples are embedded into each sample some distance L apart, and a

---

TABLE 1–4

The thermal diffusivities of some materials at room temperature

Material	α, m^2/s*
Silver	149×10^{-6}
Gold	127×10^{-6}
Copper	113×10^{-6}
Aluminum	97.5×10^{-6}
Iron	22.8×10^{-6}
Mercury (l)	4.7×10^{-6}
Marble	1.2×10^{-6}
Ice	1.2×10^{-6}
Concrete	0.75×10^{-6}
Brick	0.52×10^{-6}
Heavy soil (dry)	0.52×10^{-6}
Glass	0.34×10^{-6}
Glass wool	0.23×10^{-6}
Water (l)	0.14×10^{-6}
Beef	0.14×10^{-6}
Wood (oak)	0.13×10^{-6}

*Multiply by 10.76 to convert to ft^2/s.

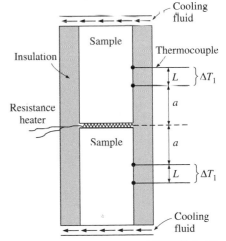

FIGURE 1–30

Apparatus to measure the thermal conductivity of a material using two identical samples and a thin resistance heater (Example 1–6).

differential thermometer reads the temperature drop ΔT across this distance along each sample. When steady operating conditions are reached, the total rate of heat transfer through both samples becomes equal to the electric power drawn by the heater.

In a certain experiment, cylindrical samples of diameter 5 cm and length 10 cm are used. The two thermocouples in each sample are placed 3 cm apart. After initial transients, the electric heater is observed to draw 0.4 A at 110 V, and both differential thermometers read a temperature difference of 15°C. Determine the thermal conductivity of the sample.

SOLUTION   The thermal conductivity of a material is to be determined by ensuring one-dimensional heat conduction, and by measuring temperatures when steady operating conditions are reached.

Assumptions   **1** Steady operating conditions exist since the temperature readings do not change with time. **2** Heat losses through the lateral surfaces of the apparatus are negligible since those surfaces are well insulated, and thus the entire heat generated by the heater is conducted through the samples. **3** The apparatus possesses thermal symmetry.

Analysis   The electrical power consumed by the resistance heater and converted to heat is

$$\dot{W}_e = \mathbf{V}I = (110 \text{ V})(0.4 \text{ A}) = 44 \text{ W}$$

The rate of heat flow through each sample is

$$\dot{Q} = \tfrac{1}{2}\dot{W}_e = \tfrac{1}{2} \times (44 \text{ W}) = 22 \text{ W}$$

since only half of the heat generated flows through each sample because of symmetry. Reading the same temperature difference across the same distance in each sample also confirms that the apparatus possesses thermal symmetry. The heat transfer area is the area normal to the direction of heat transfer, which is the cross-sectional area of the cylinder in this case:

$$A = \tfrac{1}{4}\pi D^2 = \tfrac{1}{4}\pi(0.05 \text{ m})^2 = 0.001963 \text{ m}^2$$

Noting that the temperature drops by 15°C within 3 cm in the direction of heat flow, the thermal conductivity of the sample is determined to be

$$\dot{Q} = kA\frac{\Delta T}{L} \quad \rightarrow \quad k = \frac{\dot{Q}\,L}{A\,\Delta T} = \frac{(22 \text{ W})(0.03 \text{ m})}{(0.001963 \text{ m}^2)(15°C)} = \mathbf{22.4\ W/m \cdot °C}$$

Discussion   Perhaps you are wondering if we really need to use two samples in the apparatus, since the measurements on the second sample do not give any additional information. It seems like we can replace the second sample by insulation. Indeed, we do not need the second sample; however, it enables us to verify the temperature measurements on the first sample and provides thermal symmetry, which reduces experimental error.

EXAMPLE 1–7   **Conversion between SI and English Units**

An engineer who is working on the heat transfer analysis of a brick building in English units needs the thermal conductivity of brick. But the only value he can find from his handbooks is 0.72 W/m · °C, which is in SI units. To make matters worse, the engineer does not have a direct conversion factor between the two unit systems for thermal conductivity. Can you help him out?

SOLUTION The situation this engineer is facing is not unique, and most engineers often find themselves in a similar position. A person must be very careful during unit conversion not to fall into some common pitfalls and to avoid some costly mistakes. Although unit conversion is a simple process, it requires utmost care and careful reasoning.

The conversion factors for W and m are straightforward and are given in conversion tables to be

$$1 \text{ W} = 3.41214 \text{ Btu/h}$$
$$1 \text{ m} = 3.2808 \text{ ft}$$

But the conversion of °C into °F is not so simple, and it can be a source of error if one is not careful. Perhaps the first thought that comes to mind is to replace °C by (°F − 32)/1.8 since $T(°C) = [T(°F) − 32]/1.8$. But this will be wrong since the °C in the unit W/m · °C represents *per °C change in temperature.* Noting that 1°C change in temperature corresponds to 1.8°F, the proper conversion factor to be used is

$$1°C = 1.8°F$$

Substituting, we get

$$1 \text{ W/m} \cdot °C = \frac{3.41214 \text{ Btu/h}}{(3.2808 \text{ ft})(1.8°F)} = 0.5778 \text{ Btu/h} \cdot \text{ft} \cdot °F$$

which is the desired conversion factor. Therefore, the thermal conductivity of the brick in English units is

$$k_{\text{brick}} = 0.72 \text{ W/m} \cdot °C$$
$$= 0.72 \times (0.5778 \text{ Btu/h} \cdot \text{ft} \cdot °F)$$
$$= \textbf{0.42 Btu/h} \cdot \textbf{ft} \cdot °\textbf{F}$$

Discussion Note that the thermal conductivity value of a material in English units is about half that in SI units (Fig. 1–31). Also note that we rounded the result to two significant digits (the same number in the original value) since expressing the result in more significant digits (such as 0.4160 instead of 0.42) would falsely imply a more accurate value than the original one.

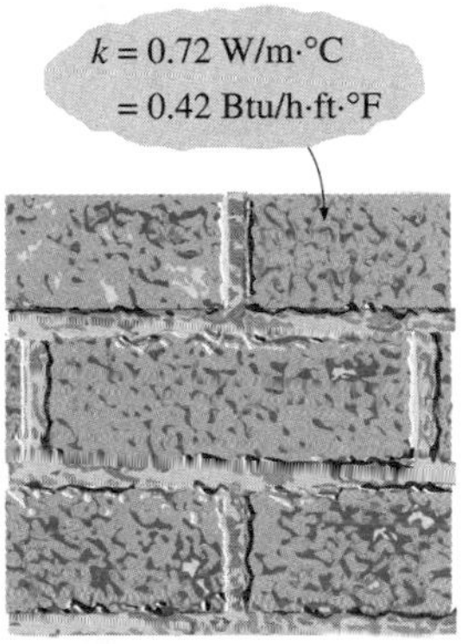

$k = 0.72 \text{ W/m·°C}$
$= 0.42 \text{ Btu/h·ft·°F}$

FIGURE 1–31
The thermal conductivity value in English units is obtained by multiplying the value in SI units by 0.5778.

1–7 · CONVECTION

Convection is the mode of energy transfer between a solid surface and the adjacent liquid or gas that is in motion, and it involves the combined effects of *conduction* and *fluid motion.* The faster the fluid motion, the greater the convection heat transfer. In the absence of any bulk fluid motion, heat transfer between a solid surface and the adjacent fluid is by pure conduction. The presence of bulk motion of the fluid enhances the heat transfer between the solid surface and the fluid, but it also complicates the determination of heat transfer rates.

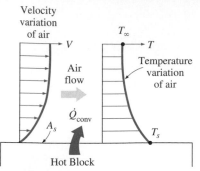

FIGURE 1–32

Heat transfer from a hot surface to air by convection.

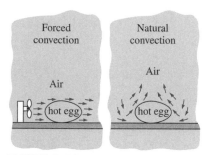

FIGURE 1–33

The cooling of a boiled egg by forced and natural convection.

TABLE 1–5

Typical values of convection heat transfer coefficient

Type of convection	h, W/m² · °C*
Free convection of gases	2–25
Free convection of liquids	10–1000
Forced convection of gases	25–250
Forced convection of liquids	50–20,000
Boiling and condensation	2500–100,000

*Multiply by 0.176 to convert to Btu/h · ft² · °F.

Consider the cooling of a hot block by blowing cool air over its top surface (Fig. 1–32). Heat is first transferred to the air layer adjacent to the block by conduction. This heat is then carried away from the surface by convection, that is, by the combined effects of conduction within the air that is due to random motion of air molecules and the bulk or macroscopic motion of the air that removes the heated air near the surface and replaces it by the cooler air.

Convection is called **forced convection** if the fluid is forced to flow over the surface by external means such as a fan, pump, or the wind. In contrast, convection is called **natural** (or **free**) **convection** if the fluid motion is caused by buoyancy forces that are induced by density differences due to the variation of temperature in the fluid (Fig. 1–33). For example, in the absence of a fan, heat transfer from the surface of the hot block in Fig. 1–32 is by natural convection since any motion in the air in this case is due to the rise of the warmer (and thus lighter) air near the surface and the fall of the cooler (and thus heavier) air to fill its place. Heat transfer between the block and the surrounding air is by conduction if the temperature difference between the air and the block is not large enough to overcome the resistance of air to movement and thus to initiate natural convection currents.

Heat transfer processes that involve *change of phase* of a fluid are also considered to be convection because of the fluid motion induced during the process, such as the rise of the vapor bubbles during boiling or the fall of the liquid droplets during condensation.

Despite the complexity of convection, the rate of *convection heat transfer* is observed to be proportional to the temperature difference, and is conveniently expressed by **Newton's law of cooling** as

$$\dot{Q}_{\text{conv}} = hA_s (T_s - T_\infty) \qquad \text{(W)} \qquad \text{(1–24)}$$

where h is the *convection heat transfer coefficient* in W/m² · °C or Btu/h · ft² · °F, A_s is the surface area through which convection heat transfer takes place, T_s is the surface temperature, and T_∞ is the temperature of the fluid sufficiently far from the surface. Note that at the surface, the fluid temperature equals the surface temperature of the solid.

The convection heat transfer coefficient h is not a property of the fluid. It is an experimentally determined parameter whose value depends on all the variables influencing convection such as the surface geometry, the nature of fluid motion, the properties of the fluid, and the bulk fluid velocity. Typical values of h are given in Table 1–5.

Some people do not consider convection to be a fundamental mechanism of heat transfer since it is essentially heat conduction in the presence of fluid motion. But we still need to give this combined phenomenon a name, unless we are willing to keep referring to it as "conduction with fluid motion." Thus, it is practical to recognize convection as a separate heat transfer mechanism despite the valid arguments to the contrary.

EXAMPLE 1–8 **Measuring Convection Heat Transfer Coefficient**

A 2-m-long, 0.3-cm-diameter electrical wire extends across a room at 15°C, as shown in Fig. 1–34. Heat is generated in the wire as a result of resistance heating, and the surface temperature of the wire is measured to be 152°C in steady operation. Also, the voltage drop and electric current through the wire are measured to be 60 V and 1.5 A, respectively. Disregarding any heat transfer by radiation, determine the convection heat transfer coefficient for heat transfer between the outer surface of the wire and the air in the room.

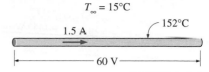

FIGURE 1–34
Schematic for Example 1–8.

SOLUTION The convection heat transfer coefficient for heat transfer from an electrically heated wire to air is to be determined by measuring temperatures when steady operating conditions are reached and the electric power consumed.

Assumptions **1** Steady operating conditions exist since the temperature readings do not change with time. **2** Radiation heat transfer is negligible.

Analysis When steady operating conditions are reached, the rate of heat loss from the wire equals the rate of heat generation in the wire as a result of resistance heating. That is,

$$\dot{Q} = \dot{E}_{generated} = \mathbf{V}I = (60 \text{ V})(1.5 \text{ A}) = 90 \text{ W}$$

The surface area of the wire is

$$A_s = \pi DL = \pi(0.003 \text{ m})(2 \text{ m}) = 0.01885 \text{ m}^2$$

Newton's law of cooling for convection heat transfer is expressed as

$$\dot{Q}_{conv} = hA_s (T_s - T_\infty)$$

Disregarding any heat transfer by radiation and thus assuming all the heat loss from the wire to occur by convection, the convection heat transfer coefficient is determined to be

$$h = \frac{\dot{Q}_{conv}}{A_s(T_s - T_\infty)} = \frac{90 \text{ W}}{(0.01885 \text{ m}^2)(152 - 15)°\text{C}} = \mathbf{34.9 \text{ W/m}^2 \cdot °C}$$

Discussion Note that the simple setup described above can be used to determine the average heat transfer coefficients from a variety of surfaces in air. Also, heat transfer by radiation can be eliminated by keeping the surrounding surfaces at the temperature of the wire.

1–8 ▪ RADIATION

Radiation is the energy emitted by matter in the form of *electromagnetic waves* (or *photons*) as a result of the changes in the electronic configurations of the atoms or molecules. Unlike conduction and convection, the transfer of heat by radiation does not require the presence of an *intervening medium*. In fact, heat transfer by radiation is fastest (at the speed of light) and it suffers no attenuation in a vacuum. This is how the energy of the sun reaches the earth.

In heat transfer studies we are interested in *thermal radiation,* which is the form of radiation emitted by bodies because of their temperature. It differs from other forms of electromagnetic radiation such as x-rays, gamma rays, microwaves, radio waves, and television waves that are not related to temperature. All bodies at a temperature above absolute zero emit thermal radiation.

Radiation is a *volumetric phenomenon,* and all solids, liquids, and gases emit, absorb, or transmit radiation to varying degrees. However, radiation is

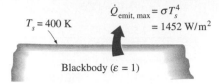

$T_s = 400 \text{ K}$

$\dot{Q}_{emit, \, max} = \sigma T_s^4$
$= 1452 \text{ W/m}^2$

Blackbody ($\varepsilon = 1$)

FIGURE 1–35
Blackbody radiation represents the *maximum amount of radiation that can be emitted from a surface at a specified temperature.*

TABLE 1–6

Emissivities of some materials at 300 K

Material	Emissivity
Aluminum foil	0.07
Anodized aluminum	0.82
Polished copper	0.03
Polished gold	0.03
Polished silver	0.02
Polished stainless steel	0.17
Black paint	0.98
White paint	0.90
White paper	0.92–0.97
Asphalt pavement	0.85–0.93
Red brick	0.93–0.96
Human skin	0.95
Wood	0.82–0.92
Soil	0.93–0.96
Water	0.96
Vegetation	0.92–0.96

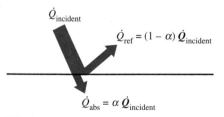

$\dot{Q}_{incident}$

$\dot{Q}_{ref} = (1 - \alpha) \, \dot{Q}_{incident}$

$\dot{Q}_{abs} = \alpha \, \dot{Q}_{incident}$

FIGURE 1–36
The absorption of radiation incident on an opaque surface of absorptivity α.

---

usually considered to be a *surface phenomenon* for solids that are opaque to thermal radiation such as metals, wood, and rocks since the radiation emitted by the interior regions of such material can never reach the surface, and the radiation incident on such bodies is usually absorbed within a few microns from the surface.

The maximum rate of radiation that can be emitted from a surface at a thermodynamic temperature T_s (in K or R) is given by the **Stefan–Boltzmann law** as

$$\dot{Q}_{emit, \, max} = \sigma A_s T_s^4 \qquad \text{(W)} \qquad \text{(1–25)}$$

where $\sigma = 5.670 \times 10^{-8} \text{ W/m}^2 \cdot \text{K}^4$ or $0.1714 \times 10^{-8} \text{ Btu/h} \cdot \text{ft}^2 \cdot \text{R}^4$ is the *Stefan–Boltzmann constant.* The idealized surface that emits radiation at this maximum rate is called a **blackbody**, and the radiation emitted by a blackbody is called **blackbody radiation** (Fig. 1–35). The radiation emitted by all real surfaces is less than the radiation emitted by a blackbody at the same temperature, and is expressed as

$$\dot{Q}_{emit} = \varepsilon \sigma A_s T_s^4 \qquad \text{(W)} \qquad \text{(1–26)}$$

where ε is the **emissivity** of the surface. The property emissivity, whose value is in the range $0 \le \varepsilon \le 1$, is a measure of how closely a surface approximates a blackbody for which $\varepsilon = 1$. The emissivities of some surfaces are given in Table 1–6.

Another important radiation property of a surface is its **absorptivity** α, which is the fraction of the radiation energy incident on a surface that is absorbed by the surface. Like emissivity, its value is in the range $0 \le \alpha \le 1$. A blackbody absorbs the entire radiation incident on it. That is, a blackbody is a perfect absorber ($\alpha = 1$) as it is a perfect emitter.

In general, both ε and α of a surface depend on the temperature and the wavelength of the radiation. **Kirchhoff's law** of radiation states that the emissivity and the absorptivity of a surface at a given temperature and wavelength are equal. In many practical applications, the surface temperature and the temperature of the source of incident radiation are of the same order of magnitude, and the average absorptivity of a surface is taken to be equal to its average emissivity. The rate at which a surface absorbs radiation is determined from (Fig. 1–36)

$$\dot{Q}_{absorbed} = \alpha \dot{Q}_{incident} \qquad \text{(W)} \qquad \text{(1–27)}$$

where $\dot{Q}_{incident}$ is the rate at which radiation is incident on the surface and α is the absorptivity of the surface. For opaque (nontransparent) surfaces, the portion of incident radiation not absorbed by the surface is reflected back.

The difference between the rates of radiation emitted by the surface and the radiation absorbed is the *net* radiation heat transfer. If the rate of radiation absorption is greater than the rate of radiation emission, the surface is said to be *gaining* energy by radiation. Otherwise, the surface is said to be *losing* energy by radiation. In general, the determination of the net rate of heat transfer by radiation between two surfaces is a complicated matter since it depends on the properties of the surfaces, their orientation relative to each other, and the interaction of the medium between the surfaces with radiation.

When a surface of emissivity ε and surface area A_s at a *thermodynamic temperature* T_s is *completely enclosed* by a much larger (or black) surface at thermodynamic temperature T_{surr} separated by a gas (such as air) that does not intervene with radiation, the net rate of radiation heat transfer between these two surfaces is given by (Fig. 1–37)

$$\dot{Q}_{rad} = \varepsilon \sigma A_s (T_s^4 - T_{surr}^4) \qquad (\text{W}) \qquad (1\text{–}28)$$

In this special case, the emissivity and the surface area of the surrounding surface do not have any effect on the net radiation heat transfer.

Radiation heat transfer to or from a surface surrounded by a gas such as air occurs *parallel* to conduction (or convection, if there is bulk gas motion) between the surface and the gas. Thus the total heat transfer is determined by *adding* the contributions of both heat transfer mechanisms. For simplicity and convenience, this is often done by defining a **combined heat transfer coefficient** $h_{combined}$ that includes the effects of both convection and radiation. Then the *total* heat transfer rate to or from a surface by convection and radiation is expressed as

$$\dot{Q}_{total} = h_{combined} A_s (T_s - T_\infty) \qquad (\text{W}) \qquad (1\text{–}29)$$

Note that the combined heat transfer coefficient is essentially a convection heat transfer coefficient modified to include the effects of radiation.

Radiation is usually significant relative to conduction or natural convection, but negligible relative to forced convection. Thus radiation in forced convection applications is usually disregarded, especially when the surfaces involved have low emissivities and low to moderate temperatures.

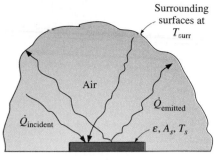

$$\dot{Q}_{rad} = \varepsilon \sigma A_s (T_s^4 - T_{surr}^4)$$

FIGURE 1–37
Radiation heat transfer between a surface and the surfaces surrounding it.

EXAMPLE 1–9   Radiation Effect on Thermal Comfort

It is a common experience to feel "chilly" in winter and "warm" in summer in our homes even when the thermostat setting is kept the same. This is due to the so called "radiation effect" resulting from radiation heat exchange between our bodies and the surrounding surfaces of the walls and the ceiling.

Consider a person standing in a room maintained at 22°C at all times. The inner surfaces of the walls, floors, and the ceiling of the house are observed to be at an average temperature of 10°C in winter and 25°C in summer. Determine the rate of radiation heat transfer between this person and the surrounding surfaces if the exposed surface area and the average outer surface temperature of the person are 1.4 m² and 30°C, respectively (Fig. 1–38).

SOLUTION The rates of radiation heat transfer between a person and the surrounding surfaces at specified temperatures are to be determined in summer and winter.
Assumptions **1** Steady operating conditions exist. **2** Heat transfer by convection is not considered. **3** The person is completely surrounded by the interior surfaces of the room. **4** The surrounding surfaces are at a uniform temperature.
Properties The emissivity of a person is $\varepsilon = 0.95$ (Table 1–6).

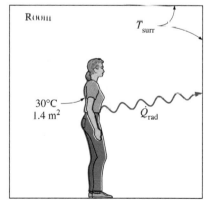

FIGURE 1–38
Schematic for Example 1–9.

Analysis The net rates of radiation heat transfer from the body to the surrounding walls, ceiling, and floor in winter and summer are

$$\dot{Q}_{\text{rad, winter}} = \varepsilon \sigma A_s \left(T_s^4 - T_{\text{surr, winter}}^4 \right)$$
$$= (0.95)(5.67 \times 10^{-8} \text{ W/m}^2 \cdot \text{K}^4)(1.4 \text{ m}^2)$$
$$\times \left[(30 + 273)^4 - (10 + 273)^4 \right] \text{K}^4$$
$$= \textbf{152 W}$$

and

$$\dot{Q}_{\text{rad, summer}} = \varepsilon \sigma A_s \left(T_s^4 - T_{\text{surr, summer}}^4 \right)$$
$$= (0.95)(5.67 \times 10^{-8} \text{ W/m}^2 \cdot \text{K}^4)(1.4 \text{ m}^2)$$
$$\times \left[(30 + 273)^4 - (25 + 273)^4 \right] \text{K}^4$$
$$= \textbf{40.9 W}$$

Discussion Note that we must use *thermodynamic (i.e., absolute) temperatures* in radiation calculations. Also note that the rate of heat loss from the person by radiation is almost four times as large in winter than it is in summer, which explains the "chill" we feel in winter even if the thermostat setting is kept the same.

1–9 ▪ SIMULTANEOUS HEAT TRANSFER MECHANISMS

We mentioned that there are three mechanisms of heat transfer, but not all three can exist simultaneously in a medium. For example, heat transfer is only by conduction in *opaque solids,* but by conduction and radiation in *semitransparent solids.* Thus, a solid may involve conduction and radiation but not convection. However, a solid may involve heat transfer by convection and/or radiation on its surfaces exposed to a fluid or other surfaces. For example, the outer surfaces of a cold piece of rock will warm up in a warmer environment as a result of heat gain by convection (from the air) and radiation (from the sun or the warmer surrounding surfaces). But the inner parts of the rock will warm up as this heat is transferred to the inner region of the rock by conduction.

Heat transfer is by conduction and possibly by radiation in a *still fluid* (no bulk fluid motion) and by convection and radiation in a *flowing fluid.* In the absence of radiation, heat transfer through a fluid is either by conduction or convection, depending on the presence of any bulk fluid motion. Convection can be viewed as combined conduction and fluid motion, and conduction in a fluid can be viewed as a special case of convection in the absence of any fluid motion (Fig. 1–39).

Thus, when we deal with heat transfer through a *fluid,* we have either *conduction* or *convection,* but not both. Also, gases are practically transparent to radiation, except that some gases are known to absorb radiation strongly at certain wavelengths. Ozone, for example, strongly absorbs ultraviolet radiation. But in most cases, a gas between two solid surfaces does not interfere with radiation and acts effectively as a vacuum. Liquids, on the other hand, are usually strong absorbers of radiation.

Finally, heat transfer through a *vacuum* is by radiation only since conduction or convection requires the presence of a material medium.

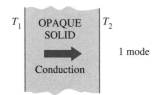

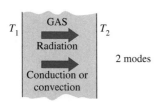

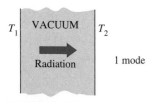

FIGURE 1–39
Although there are three mechanisms of heat transfer, a medium may involve only two of them simultaneously.

EXAMPLE 1–10   Heat Loss from a Person

Consider a person standing in a breezy room at 20°C. Determine the total rate of heat transfer from this person if the exposed surface area and the average outer surface temperature of the person are 1.6 m² and 29°C, respectively, and the convection heat transfer coefficient is 6 W/m² · K (Fig. 1–40).

SOLUTION   The total rate of heat transfer from a person by both convection and radiation to the surrounding air and surfaces at specified temperatures is to be determined.

Assumptions   **1** Steady operating conditions exist. **2** The person is completely surrounded by the interior surfaces of the room. **3** The surrounding surfaces are at the same temperature as the air in the room. **4** Heat conduction to the floor through the feet is negligible.

Properties   The emissivity of a person is $\varepsilon = 0.95$ (Table 1–6).

Analysis   The heat transfer between the person and the air in the room is by convection (instead of conduction) since it is conceivable that the air in the vicinity of the skin or clothing warms up and rises as a result of heat transfer from the body, initiating natural convection currents. It appears that the experimentally determined value for the rate of convection heat transfer in this case is 6 W per unit surface area (m²) per unit temperature difference (in K or °C) between the person and the air away from the person. Thus, the rate of convection heat transfer from the person to the air in the room is

$$\dot{Q}_{conv} = hA_s\,(T_s - T_\infty)$$
$$= (6\ \text{W/m}^2 \cdot \text{°C})(1.6\ \text{m}^2)(29 - 20)\text{°C}$$
$$= 86.4\ \text{W}$$

The person also loses heat by radiation to the surrounding wall surfaces. We take the temperature of the surfaces of the walls, ceiling, and floor to be equal to the air temperature in this case for simplicity, but we recognize that this does not need to be the case. These surfaces may be at a higher or lower temperature than the average temperature of the room air, depending on the outdoor conditions and the structure of the walls. Considering that air does not intervene with radiation and the person is completely enclosed by the surrounding surfaces, the net rate of radiation heat transfer from the person to the surrounding walls, ceiling, and floor is

$$\dot{Q}_{rad} = \varepsilon\sigma A_s\,(T_s^4 - T_{surr}^4)$$
$$= (0.95)(5.67 \times 10^{-8}\ \text{W/m}^2 \cdot \text{K}^4)(1.6\ \text{m}^2)$$
$$\times\,[(29 + 273)^4 - (20 + 273)^4]\ \text{K}^4$$
$$= 81.7\ \text{W}$$

Note that we must use *thermodynamic* temperatures in radiation calculations. Also note that we used the emissivity value for the skin and clothing at room temperature since the emissivity is not expected to change significantly at a slightly higher temperature.

Then the rate of total heat transfer from the body is determined by adding these two quantities:

$$\dot{Q}_{total} = \dot{Q}_{conv} + \dot{Q}_{rad} = (86.4 + 81.7)\ \text{W} \cong \textbf{168 W}$$

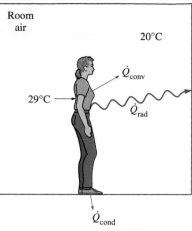

FIGURE 1–40
Heat transfer from the person described in Example 1–10.

Discussion The heat transfer would be much higher if the person were not dressed since the exposed surface temperature would be higher. Thus, an important function of the clothes is to serve as a barrier against heat transfer.

In these calculations, heat transfer through the feet to the floor by conduction, which is usually very small, is neglected. Heat transfer from the skin by perspiration, which is the dominant mode of heat transfer in hot environments, is not considered here.

Also, the units $W/m^2 \cdot °C$ and $W/m^2 \cdot K$ for heat transfer coefficient are equivalent, and can be interchanged.

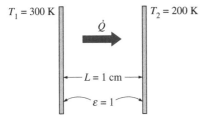

FIGURE 1–41
Schematic for Example 1–11.

EXAMPLE 1–11 Heat Transfer between Two Isothermal Plates

Consider steady heat transfer between two large parallel plates at constant temperatures of $T_1 = 300$ K and $T_2 = 200$ K that are $L = 1$ cm apart, as shown in Fig. 1–41. Assuming the surfaces to be black (emissivity $\varepsilon = 1$), determine the rate of heat transfer between the plates per unit surface area assuming the gap between the plates is (a) filled with atmospheric air, (b) evacuated, (c) filled with urethane insulation, and (d) filled with superinsulation that has an apparent thermal conductivity of 0.00002 W/m · K.

SOLUTION The total rate of heat transfer between two large parallel plates at specified temperatures is to be determined for four different cases.

Assumptions **1** Steady operating conditions exist. **2** There are no natural convection currents in the air between the plates. **3** The surfaces are black and thus $\varepsilon = 1$.

Properties The thermal conductivity at the average temperature of 250 K is $k = 0.0219$ W/m · K for air (Table A–15), 0.026 W/m · K for urethane insulation (Table A–6), and 0.00002 W/m · K for the superinsulation.

Analysis (a) The rates of conduction and radiation heat transfer between the plates through the air layer are

$$\dot{Q}_{cond} = kA \frac{T_1 - T_2}{L} = (0.0219 \text{ W/m} \cdot \text{K})(1 \text{ m}^2)\frac{(300 - 200)\text{K}}{0.01 \text{ m}} = 219 \text{ W}$$

and

$$\dot{Q}_{rad} = \varepsilon\sigma A(T_1^4 - T_2^4)$$
$$= (1)(5.67 \times 10^{-8} \text{ W/m}^2 \cdot \text{K}^4)(1 \text{ m}^2)[(300 \text{ K})^4 - (200 \text{ K})^4] = 369 \text{ W}$$

Therefore,

$$\dot{Q}_{total} = \dot{Q}_{cond} + \dot{Q}_{rad} = 219 + 369 = \textbf{588 W}$$

The heat transfer rate in reality will be higher because of the natural convection currents that are likely to occur in the air space between the plates.

(b) When the air space between the plates is evacuated, there will be no conduction or convection, and the only heat transfer between the plates will be by radiation. Therefore,

$$\dot{Q}_{total} = \dot{Q}_{rad} = \textbf{369 W}$$

(c) An opaque solid material placed between two plates blocks direct radiation heat transfer between the plates. Also, the thermal conductivity of an insulating material accounts for the radiation heat transfer that may be occurring

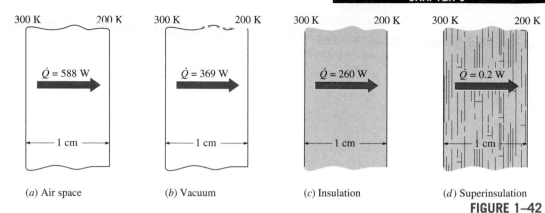

FIGURE 1–42
Different ways of reducing heat transfer between two isothermal plates, and their effectiveness.

through the voids in the insulating material. The rate of heat transfer through the urethane insulation is

$$\dot{Q}_{\text{total}} = \dot{Q}_{\text{cond}} = kA\,\frac{T_1 - T_2}{L} = (0.026\ \text{W/m} \cdot \text{K})(1\ \text{m}^2)\,\frac{(300-200)\text{K}}{0.01\ \text{m}} = \mathbf{260\ W}$$

Note that heat transfer through the urethane material is less than the heat transfer through the air determined in (a), although the thermal conductivity of the insulation is higher than that of air. This is because the insulation blocks the radiation whereas air transmits it.

(d) The layers of the superinsulation prevent any direct radiation heat transfer between the plates. However, radiation heat transfer between the sheets of superinsulation does occur, and the apparent thermal conductivity of the superinsulation accounts for this effect. Therefore,

$$\dot{Q}_{\text{total}} = kA\,\frac{T_1 - T_2}{L} = (0.00002\ \text{W/m} \cdot \text{K})(1\ \text{m}^2)\,\frac{(300-200)\text{K}}{0.01\ \text{m}} = \mathbf{0.2\ W}$$

which is $\frac{1}{1845}$ of the heat transfer through the vacuum. The results of this example are summarized in Fig. 1–42 to put them into perspective.

Discussion This example demonstrates the effectiveness of superinsulations and explains why they are the insulation of choice in critical applications despite their high cost.

EXAMPLE 1–12   **Heat Transfer in Conventional and Microwave Ovens**

The fast and efficient cooking of microwave ovens made them one of the essential appliances in modern kitchens (Fig. 1–43). Discuss the heat transfer mechanisms associated with the cooking of a chicken in microwave and conventional ovens, and explain why cooking in a microwave oven is more efficient.

SOLUTION   Food is cooked in a microwave oven by absorbing the electromagnetic radiation energy generated by the microwave tube, called the magnetron.

FIGURE 1–43
A chicken being cooked in a microwave oven (Example 1–12).

The radiation emitted by the magnetron is not thermal radiation, since its emission is not due to the temperature of the magnetron; rather, it is due to the conversion of electrical energy into electromagnetic radiation at a specified wavelength. The wavelength of the microwave radiation is such that it is *reflected* by metal surfaces; *transmitted* by the cookware made of glass, ceramic, or plastic; and *absorbed* and converted to internal energy by food (especially the water, sugar, and fat) molecules.

In a microwave oven, the *radiation* that strikes the chicken is absorbed by the skin of the chicken and the outer parts. As a result, the temperature of the chicken at and near the skin rises. Heat is then *conducted* toward the inner parts of the chicken from its outer parts. Of course, some of the heat absorbed by the outer surface of the chicken is lost to the air in the oven by *convection*.

In a conventional oven, the air in the oven is first heated to the desired temperature by the electric or gas heating element. This preheating may take several minutes. The heat is then transferred from the air to the skin of the chicken by *natural convection* in older ovens or by *forced convection* in the newer convection ovens that utilize a fan. The air motion in convection ovens increases the convection heat transfer coefficient and thus decreases the cooking time. Heat is then *conducted* toward the inner parts of the chicken from its outer parts as in microwave ovens.

Microwave ovens replace the slow convection heat transfer process in conventional ovens by the instantaneous radiation heat transfer. As a result, microwave ovens transfer energy to the food at full capacity the moment they are turned on, and thus they cook faster while consuming less energy.

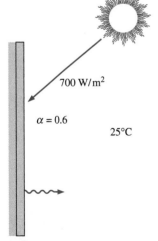

FIGURE 1–44
Schematic for Example 1–13.

EXAMPLE 1–13   Heating of a Plate by Solar Energy

A thin metal plate is insulated on the back and exposed to solar radiation at the front surface (Fig. 1–44). The exposed surface of the plate has an absorptivity of 0.6 for solar radiation. If solar radiation is incident on the plate at a rate of 700 W/m^2 and the surrounding air temperature is 25°C, determine the surface temperature of the plate when the heat loss by convection and radiation equals the solar energy absorbed by the plate. Assume the combined convection and radiation heat transfer coefficient to be 50 W/m$^2 \cdot$ °C.

SOLUTION   The back side of the thin metal plate is insulated and the front side is exposed to solar radiation. The surface temperature of the plate is to be determined when it stabilizes.
Assumptions   **1** Steady operating conditions exist. **2** Heat transfer through the insulated side of the plate is negligible. **3** The heat transfer coefficient remains constant.
Properties   The solar absorptivity of the plate is given to be $\alpha = 0.6$.
Analysis   The absorptivity of the plate is 0.6, and thus 60 percent of the solar radiation incident on the plate is absorbed continuously. As a result, the temperature of the plate rises, and the temperature difference between the plate and the surroundings increases. This increasing temperature difference causes the rate of heat loss from the plate to the surroundings to increase. At some point, the rate of heat loss from the plate equals the rate of solar

energy absorbed, and the temperature of the plate no longer changes. The temperature of the plate when steady operation is established is determined from

$$\dot{E}_{gained} = \dot{E}_{lost} \qquad \text{or} \qquad \alpha A_s \, \dot{q}_{incident, \, solar} = h_{combined} A_s (T_s - T_\infty)$$

Solving for T_s and substituting, the plate surface temperature is determined to be

$$T_s = T_\infty + \alpha \frac{\dot{q}_{\,incident, \, solar}}{h_{combined}} = 25°C + \frac{0.6 \times (700 \text{ W/m}^2)}{50 \text{ W/m}^2 \cdot °C} = \textbf{33.4°C}$$

Discussion Note that the heat losses prevent the plate temperature from rising above 33.4°C. Also, the combined heat transfer coefficient accounts for the effects of both convection and radiation, and thus it is very convenient to use in heat transfer calculations when its value is known with reasonable accuracy.

1–10 ▪ PROBLEM-SOLVING TECHNIQUE

The first step in learning any science is to grasp the fundamentals and to gain a sound knowledge of it. The next step is to master the fundamentals by testing this knowledge. This is done by solving significant real-world problems. Solving such problems, especially complicated ones, require a systematic approach. By using a step-by-step approach, an engineer can reduce the solution of a complicated problem into the solution of a series of simple problems (Fig. 1–45). When you are solving a problem, we recommend that you use the following steps zealously as applicable. This will help you avoid some of the common pitfalls associated with problem solving.

Step 1: Problem Statement
In your own words, briefly state the problem, the key information given, and the quantities to be found. This is to make sure that you understand the problem and the objectives before you attempt to solve the problem.

Step 2: Schematic
Draw a realistic sketch of the physical system involved, and list the relevant information on the figure. The sketch does not have to be something elaborate, but it should resemble the actual system and show the key features. Indicate any energy and mass interactions with the surroundings. Listing the given information on the sketch helps one to see the entire problem at once.

Step 3: Assumptions and Approximations
State any appropriate assumptions and approximations made to simplify the problem to make it possible to obtain a solution. Justify the questionable assumptions. Assume reasonable values for missing quantities that are necessary. For example, in the absence of specific data for atmospheric pressure, it can be taken to be 1 atm. However, it should be noted in the analysis that the

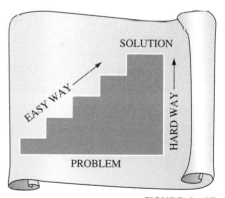

FIGURE 1–45
A step-by-step approach can greatly simplify problem solving.

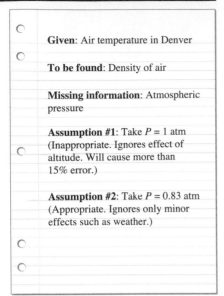

Given: Air temperature in Denver

To be found: Density of air

Missing information: Atmospheric pressure

Assumption #1: Take $P = 1$ atm (Inappropriate. Ignores effect of altitude. Will cause more than 15% error.)

Assumption #2: Take $P = 0.83$ atm (Appropriate. Ignores only minor effects such as weather.)

FIGURE 1–46
The assumptions made while solving an engineering problem must be reasonable and justifiable.

| Energy use: | $80/yr |
| Energy saved by insulation: | $200/yr |

IMPOSSIBLE!

FIGURE 1–47
The results obtained from an engineering analysis must be checked for reasonableness.

atmospheric pressure decreases with increasing elevation. For example, it drops to 0.83 atm in Denver (elevation 1610 m) (Fig. 1–46).

Step 4: Physical Laws

Apply all the relevant basic physical laws and principles (such as the conservation of energy), and reduce them to their simplest form by utilizing the assumptions made. However, the region to which a physical law is applied must be clearly identified first.

Step 5: Properties

Determine the unknown properties necessary to solve the problem from property relations or tables. List the properties separately, and indicate their source, if applicable.

Step 6: Calculations

Substitute the known quantities into the simplified relations and perform the calculations to determine the unknowns. Pay particular attention to the units and unit cancellations, and remember that a dimensional quantity without a unit is meaningless. Also, don't give a false implication of high precision by copying all the digits from the calculator—round the results to an appropriate number of significant digits (see p. 39).

Step 7: Reasoning, Verification, and Discussion

Check to make sure that the results obtained are reasonable and intuitive, and verify the validity of the questionable assumptions. Repeat the calculations that resulted in unreasonable values. For example, insulating a water heater that uses $80 worth of natural gas a year cannot result in savings of $200 a year (Fig. 1–47).

Also, point out the significance of the results, and discuss their implications. State the conclusions that can be drawn from the results, and any recommendations that can be made from them. Emphasize the limitations under which the results are applicable, and caution against any possible misunderstandings and using the results in situations where the underlying assumptions do not apply. For example, if you determined that wrapping a water heater with a $20 insulation jacket will reduce the energy cost by $30 a year, indicate that the insulation will pay for itself from the energy it saves in less than a year. However, also indicate that the analysis does not consider labor costs, and that this will be the case if you install the insulation yourself.

Keep in mind that the solutions you present to your instructors, and any engineering analysis presented to others, is a form of communication. Therefore neatness, organization, completeness, and visual appearance are of utmost importance for maximum effectiveness. Besides, neatness also serves as a great checking tool since it is very easy to spot errors and inconsistencies in neat work. Carelessness and skipping steps to save time often end up costing more time and unnecessary anxiety.

The approach described here is used in the solved example problems without explicitly stating each step, as well as in the Solutions Manual of this text.

For some problems, some of the steps may not be applicable or necessary. However, we cannot overemphasize the importance of a logical and orderly approach to problem solving. Most difficulties encountered while solving a problem are not due to a lack of knowledge; rather, they are due to a lack of organization. You are strongly encouraged to follow these steps in problem solving until you develop your own approach that works best for you.

Engineering Software Packages

You may be wondering why we are about to undertake an in-depth study of the fundamentals of another engineering science. After all, almost all such problems we are likely to encounter in practice can be solved using one of several sophisticated software packages readily available in the market today. These software packages not only give the desired numerical results, but also supply the outputs in colorful graphical form for impressive presentations. It is unthinkable to practice engineering today without using some of these packages. This tremendous computing power available to us at the touch of a button is both a blessing and a curse. It certainly enables engineers to solve problems easily and quickly, but it also opens the door for abuses and misinformation. In the hands of poorly educated people, these software packages are as dangerous as sophisticated powerful weapons in the hands of poorly trained soldiers.

Thinking that a person who can use the engineering software packages without proper training on fundamentals can practice engineering is like thinking that a person who can use a wrench can work as a car mechanic. If it were true that the engineering students do not need all these fundamental courses they are taking because practically everything can be done by computers quickly and easily, then it would also be true that the employers would no longer need high salaried engineers since any person who knows how to use a word-processing program can also learn how to use those software packages. However, the statistics show that the need for engineers is on the rise, not on the decline, despite the availability of these powerful packages.

We should always remember that all the computing power and the engineering software packages available today are just tools, and tools have meaning only in the hands of masters. Having the best word-processing program does not make a person a good writer, but it certainly makes the job of a good writer much easier and makes the writer more productive (Fig. 1–48). Hand calculators did not eliminate the need to teach our children how to add or subtract, and the sophisticated medical software packages did not take the place of medical school training. Neither will engineering software packages replace the traditional engineering education. They will simply cause a shift in emphasis in the courses from mathematics to physics. That is, more time will be spent in the classroom discussing the physical aspects of the problems in greater detail, and less time on the mechanics of solution procedures.

All these marvelous and powerful tools available today put an extra burden on today's engineers. They must still have a thorough understanding of the fundamentals, develop a "feel" of the physical phenomena, be able to put the data into proper perspective, and make sound engineering judgments, just like their predecessors. However, they must do it much better, and much

FIGURE 1–48
An excellent word-processing program does not make a person a good writer; it simply makes a good writer a better and more efficient writer.
© Vol. 80/PhotoDisc

faster, using more realistic models because of the powerful tools available today. The engineers in the past had to rely on hand calculations, slide rules, and later hand calculators and computers. Today they rely on software packages. The easy access to such power and the possibility of a simple misunderstanding or misinterpretation causing great damage make it more important today than ever to have solid training in the fundamentals of engineering. In this text we make an extra effort to put the emphasis on developing an intuitive and physical understanding of natural phenomena instead of on the mathematical details of solution procedures.

Engineering Equation Solver (EES)

EES is a program that solves systems of linear or nonlinear algebraic or differential equations numerically. It has a large library of built-in thermophysical property functions as well as mathematical functions, and allows the user to supply additional property data. Unlike some software packages, EES does not solve engineering problems; it only solves the equations supplied by the user. Therefore, the user must understand the problem and formulate it by applying any relevant physical laws and relations. EES saves the user considerable time and effort by simply solving the resulting mathematical equations. This makes it possible to attempt significant engineering problems not suitable for hand calculations, and to conduct parametric studies quickly and conveniently. EES is a very powerful yet intuitive program that is very easy to use, as shown in Example 1–14. The use and capabilities of EES are explained in Appendix 3 on the Online Learning Center.

EXAMPLE 1–14   **Solving a System of Equations with EES**

The difference of two numbers is 4, and the sum of their squares is equal to their sum plus 20. Determine these two numbers.

SOLUTION   Relations are given for the difference and the sum of the squares of two numbers. They are to be determined.

Analysis   We start the EES program by double-clicking on its icon, open a new file, and type the following on the blank screen that appears:

$$x - y = 4$$
$$x^2 + y^2 = x + y + 20$$

which is an exact mathematical expression of the problem statement with *x* and *y* denoting the unknown numbers. The solution to this system of two nonlinear equations with two unknowns is obtained by a single click on the "calculator" symbol on the taskbar. It gives

$$x = 5 \quad \text{and} \quad y = 1$$

Discussion   Note that all we did is formulate the problem as we would on paper; EES took care of all the mathematical details of solution. Also note that equations can be linear or nonlinear, and they can be entered in any order with unknowns on either side. Friendly equation solvers such as EES allow the user to concentrate on the physics of the problem without worrying about the mathematical complexities associated with the solution of the resulting system of equations.

A Remark on Significant Digits

In engineering calculations, the information given is not known to more than a certain number of significant digits, usually three digits. Consequently, the results obtained cannot possibly be accurate to more significant digits. Reporting results in more significant digits implies greater accuracy than exists, and it should be avoided.

For example, consider a 3.75-L container filled with gasoline whose density is 0.845 kg/L, and try to determine its mass. Probably the first thought that comes to your mind is to multiply the volume and density to obtain 3.16875 kg for the mass, which falsely implies that the mass determined is accurate to six significant digits. In reality, however, the mass cannot be more accurate than three significant digits since both the volume and the density are accurate to three significant digits only. Therefore, the result should be rounded to three significant digits, and the mass should be reported to be 3.17 kg instead of what appears in the screen of the calculator. The result 3.16875 kg would be correct only if the volume and density were given to be 3.75000 L and 0.845000 kg/L, respectively. The value 3.75 L implies that we are fairly confident that the volume is accurate within ±0.01 L, and it cannot be 3.74 or 3.76 L. However, the volume can be 3.746, 3.750, 3.753, etc., since they all round to 3.75 L (Fig. 1–49). It is more appropriate to retain all the digits during intermediate calculations, and to do the rounding in the final step since this is what a computer will normally do.

When solving problems, we will assume the given information to be accurate to at least three significant digits. Therefore, if the length of a pipe is given to be 40 m, we will assume it to be 40.0 m in order to justify using three significant digits in the final results. You should also keep in mind that all experimentally determined values are subject to measurement errors, and such

FIGURE 1–49
A result with more significant digits than that of given data falsely implies more accuracy.

errors are reflected in the results obtained. For example, if the density of a substance has an uncertainty of 2 percent, then the mass determined using this density value will also have an uncertainty of 2 percent.

You should also be aware that we sometimes knowingly introduce small errors in order to avoid the trouble of searching for more accurate data. For example, when dealing with liquid water, we just use the value of 1000 kg/m³ for density, which is the density value of pure water at 0°C. Using this value at 75°C will result in an error of 2.5 percent since the density at this temperature is 975 kg/m³. The minerals and impurities in the water introduce additional error. This being the case, you should have no reservation in rounding the final results to a reasonable number of significant digits. Besides, having a few percent uncertainty in the results of engineering analysis is usually the norm, not the exception.

TOPIC OF SPECIAL INTEREST*

FIGURE 1–50

Most animals come into this world with built-in insulation, but human beings come with a delicate skin.

Thermal Comfort

Unlike animals such as a fox or a bear that are born with built-in furs, human beings come into this world with little protection against the harsh environmental conditions (Fig. 1–50). Therefore, we can claim that the search for thermal comfort dates back to the beginning of human history. It is believed that early human beings lived in caves that provided shelter as well as protection from extreme thermal conditions. Probably the first form of heating system used was *open fire,* followed by fire in dwellings through the use of a *chimney* to vent out the combustion gases. The concept of *central heating* dates back to the times of the Romans, who heated homes by utilizing double-floor construction techniques and passing the fire's fumes through the opening between the two floor layers. The Romans were also the first to use *transparent windows* made of mica or glass to keep the wind and rain out while letting the light in. Wood and coal were the primary energy sources for heating, and oil and candles were used for lighting. The ruins of south-facing houses indicate that the value of *solar heating* was recognized early in the history.

The term **air-conditioning** is usually used in a restricted sense to imply cooling, but in its broad sense it means *to condition* the air to the desired level by heating, cooling, humidifying, dehumidifying, cleaning, and deodorizing. The purpose of the air-conditioning system of a building is to provide *complete thermal comfort* for its occupants. Therefore, we need to understand the thermal aspects of the *human body* in order to design an effective air-conditioning system.

The building blocks of living organisms are *cells,* which resemble miniature factories performing various functions necessary for the survival of organisms. The human body contains about 100 trillion cells with an average

*This section can be skipped without a loss in continuity.

diameter of 0.01 mm. In a typical cell, thousands of chemical reactions occur every second during which some molecules are broken down and energy is released and some new molecules are formed. The high level of chemical activity in the cells that maintain the human body temperature at a temperature of 37.0°C (98.6°F) while performing the necessary bodily functions is called the **metabolism**. In simple terms, metabolism refers to the burning of foods such as carbohydrates, fat, and protein. The metabolizable energy content of foods is usually expressed by nutritionists in terms of the capitalized Calorie. One Calorie is equivalent to 1 Cal = 1 kcal = 4.1868 kJ.

The rate of metabolism at the resting state is called the *basal metabolic rate,* which is the rate of metabolism required to keep a body performing the necessary bodily functions such as breathing and blood circulation at zero external activity level. The metabolic rate can also be interpreted as the energy consumption rate for a body. For an *average man* (30 years old, 70 kg, 1.73 m high, 1.8 m² surface area), the basal metabolic rate is 84 W. That is, the body converts chemical energy of the food (or of the body fat if the person had not eaten) into heat at a rate of 84 J/s, which is then dissipated to the surroundings. The metabolic rate increases with the *level of activity,* and it may exceed 10 times the basal metabolic rate when someone is doing strenuous exercise. That is, two people doing heavy exercising in a room may be supplying more energy to the room than a 1-kW resistance heater (Fig. 1–51). An average man generates heat at a rate of 108 W while reading, writing, typing, or listening to a lecture in a classroom in a seated position. The maximum metabolic rate of an average man is 1250 W at age 20 and 730 at age 70. The corresponding rates for women are about 30 percent lower. Maximum metabolic rates of trained athletes can exceed 2000 W.

Metabolic rates during various activities are given in Table 1–7 per unit body surface area. The **surface area** of a nude body was given by D. DuBois in 1916 as

$$A_s = 0.202 m^{0.425}\, h^{0.725} \qquad (m^2) \qquad (1\text{–}30)$$

where *m* is the mass of the body in kg and *h* is the height in m. *Clothing* increases the exposed surface area of a person by up to about 50 percent. The metabolic rates given in the table are sufficiently accurate for most purposes, but there is considerable uncertainty at high activity levels. More accurate values can be determined by measuring the rate of respiratory *oxygen consumption,* which ranges from about 0.25 L/min for an average resting man to more than 2 L/min during extremely heavy work. The entire energy released during metabolism can be assumed to be released as *heat* (in sensible or latent forms) since the external mechanical work done by the muscles is very small. Besides, the work done during most activities such as walking or riding an exercise bicycle is eventually converted to heat through friction.

The comfort of the human body depends primarily on three environmental factors: the temperature, relative humidity, and air motion. The temperature of the environment is the single most important index of comfort.

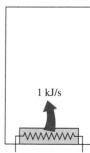

1.2 kJ/s

1 kJ/s

FIGURE 1–51
Two fast-dancing people supply more heat to a room than a 1-kW resistance heater.

TABLE 1–7

Metabolic rates during various activities (from ASHRAE *Handbook of Fundamentals,* Chap. 8, Table 4)

Activity	Metabolic rate* W/m²
Resting:	
Sleeping	40
Reclining	45
Seated, quiet	60
Standing, relaxed	70
Walking (on the level):	
2 mph (0.89 m/s)	115
3 mph (1.34 m/s)	150
4 mph (1.79 m/s)	220
Office Activities:	
Reading, seated	55
Writing	60
Typing	65
Filing, seated	70
Filing, standing	80
Walking about	100
Lifting/packing	120
Driving/Flying:	
Car	60–115
Aircraft, routine	70
Heavy vehicle	185
Miscellaneous Occupational Activities:	
Cooking	95–115
Cleaning house	115–140
Machine work:	
Light	115–140
Heavy	235
Handling 50-kg bags	235
Pick and shovel work	235–280
Miscellaneous Leisure Activities:	
Dancing, social	140–255
Calisthenics/exercise	175–235
Tennis, singles	210–270
Basketball	290–440
Wrestling, competitive	410–505

*Multiply by 1.8 m² to obtain metabolic rates for an average man. Multiply by 0.3171 to convert to Btu/h · ft².

Extensive research is done on human subjects to determine the "**thermal comfort zone**" and to identify the conditions under which the body feels comfortable in an environment. It has been observed that most normally clothed people resting or doing light work feel comfortable in the *operative temperature* (roughly, the average temperature of air and surrounding surfaces) range of 23°C to 27°C or 73°F to 80°F (Fig. 1–52). For unclothed people, this range is 29°C to 31°C. Relative humidity also has a considerable effect on comfort since it is a measure of air's ability to absorb moisture and thus it affects the amount of heat a body can dissipate by evaporation. High relative humidity slows down heat rejection by evaporation, especially at high temperatures, and low relative humidity speeds it up. The desirable level of *relative humidity* is the broad range of 30 to 70 percent, with 50 percent being the most desirable level. Most people at these conditions feel neither hot nor cold, and the body does not need to activate any of the defense mechanisms to maintain the normal body temperature (Fig. 1–53).

Another factor that has a major effect on thermal comfort is **excessive air motion** or **draft**, which causes undesired local cooling of the human body. Draft is identified by many as a most annoying factor in work places, automobiles, and airplanes. Experiencing discomfort by draft is most common among people wearing indoor clothing and doing light sedentary work, and least common among people with high activity levels. The air velocity should be kept below 9 m/min (30 ft/min) in winter and 15 m/min (50 ft/min) in summer to minimize discomfort by draft, especially when the air is cool. A low level of air motion is desirable as it removes the warm, moist air that builds around the body and replaces it with fresh air. Therefore, air motion should be strong enough to remove heat and moisture from the vicinity of the body, but gentle enough to be unnoticed. High speed air motion causes discomfort outdoors as well. For example, an environment at 10°C (50°F) with 48 km/h winds feels as cold as an environment at −7°C (20°F) with 3 km/h winds because of the chilling effect of the air motion (the wind-chill factor).

A comfort system should provide *uniform conditions* throughout the living space to avoid discomfort caused by nonuniformities such as *drafts, asymmetric thermal radiation, hot or cold floors,* and *vertical temperature stratification.* **Asymmetric thermal radiation** is caused by the *cold surfaces* of large windows, uninsulated walls, or cold products and the *warm surfaces* of gas or electric radiant heating panels on the walls or ceiling, solar-heated masonry walls or ceilings, and warm machinery. Asymmetric radiation causes discomfort by exposing different sides of the body to surfaces at different temperatures and thus to different heat loss or gain by radiation. A person whose left side is exposed to a cold window, for example, will feel like heat is being drained from that side of his or her body (Fig. 1–54). For thermal comfort, the radiant temperature asymmetry should not exceed 5°C in the vertical direction and 10°C in the horizontal direction. The unpleasant effect of radiation asymmetry can be minimized by properly sizing and installing heating panels, using double-pane windows, and providing generous insulation at the walls and the roof.

Direct contact with **cold or hot floor surfaces** also causes localized discomfort in the feet. The temperature of the floor depends on the way it is *constructed* (being directly on the ground or on top of a heated room, being made of wood or concrete, the use of insulation, etc.) as well as the *floor covering used* such as pads, carpets, rugs, and linoleum. A floor temperature of 23 to 25°C is found to be comfortable to most people. The floor asymmetry loses its significance for people with footwear. An effective and economical way of raising the floor temperature is to use radiant heating panels instead of turning the thermostat up. Another nonuniform condition that causes discomfort is **temperature stratification** in a room that exposes the head and the feet to different temperatures. For thermal comfort, the temperature difference between the head and foot levels should not exceed 3°C. This effect can be minimized by using destratification fans.

It should be noted that no thermal environment will please everyone. No matter what we do, some people will express some discomfort. The thermal comfort zone is based on a 90 percent acceptance rate. That is, an environment is deemed comfortable if only 10 percent of the people are dissatisfied with it. Metabolism decreases somewhat with *age,* but it has no effect on the comfort zone. Research indicates that there is no appreciable difference between the environments preferred by old and young people. Experiments also show that *men* and *women* prefer almost the same environment. The metabolism rate of women is somewhat lower, but this is compensated by their slightly lower skin temperature and evaporative loss. Also, there is no significant variation in the comfort zone from one part of the world to another and from winter to summer. Therefore, the same thermal comfort conditions can be used *throughout the world* in any season. Also, people cannot *acclimatize* themselves to prefer different comfort conditions.

In a **cold environment**, the rate of heat loss from the body may exceed the rate of metabolic heat generation. Average specific heat of the human body is 3.49 kJ/kg · °C, and thus each 1°C drop in body temperature corresponds to a deficit of 244 kJ in body heat content for an average 70-kg man. A drop of 0.5°C in mean body temperature causes noticeable but acceptable discomfort. A drop of 2.6°C causes extreme discomfort. A sleeping person wakes up when his or her mean body temperature drops by 1.3°C (which normally shows up as a 0.5°C drop in the deep body and 3°C in the skin area). The drop of deep body temperature below 35°C may damage the body temperature regulation mechanism, while a drop below 28°C may be fatal. Sedentary people reported to feel *comfortable* at a *mean skin temperature* of 33.3°C, *uncomfortably cold* at 31°C, *shivering cold* at 30°C, and *extremely cold* at 29°C. People doing heavy work reported to feel comfortable at much lower temperatures, which shows that the activity level affects human performance and comfort. The extremities of the body such as hands and feet are most easily affected by cold weather, and their temperature is a better indication of comfort and performance. A hand-skin temperature of 20°C is perceived to be uncomfortably cold, 15°C to be extremely cold, and 5°C to be painfully cold. Useful work can be performed by hands without difficulty as long as the skin temperature of

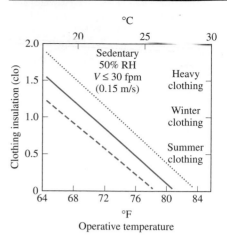

.... Upper acceptability limit
———— Optimum
– – – Lower acceptability limit

FIGURE 1–52
The effect of clothing on the environment temperature that feels comfortable (1 clo = 0.155 m² · °C/W = 0.880 ft² · °F · h/Btu).
(from ASHRAE Standard 55-1981)

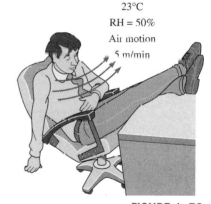

FIGURE 1–53
A thermally comfortable environment.

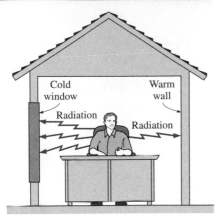

FIGURE 1–54
Cold surfaces cause excessive heat loss from the body by radiation, and thus discomfort on that side of the body.

FIGURE 1–55
The rate of metabolic heat generation may go up by six times the resting level during total body shivering in cold weather.

fingers remains above 16°C (ASHRAE *Handbook of Fundamentals,* Chapter 8).

The first line of defense of the body against excessive heat loss in a cold environment is *to reduce the skin temperature* and thus the rate of heat loss from the skin by constricting the veins and decreasing the blood flow to the skin. This measure decreases the temperature of the tissues subjacent to the skin, but maintains the inner body temperature. The next preventive measure is increasing the rate of *metabolic heat generation* in the body by *shivering,* unless the person does it voluntarily by increasing his or her level of activity or puts on additional clothing. Shivering begins slowly in small muscle groups and may double the rate of metabolic heat production of the body at its initial stages. In the extreme case of total body shivering, the rate of heat production may reach six times the resting levels (Fig. 1–55). If this measure also proves inadequate, the deep body temperature starts *falling.* Body parts furthest away from the core such as the hands and feet are at greatest danger for tissue damage.

In **hot environments**, the rate of heat loss from the body may drop below the metabolic heat generation rate. This time the body activates the opposite mechanisms. First the body increases the *blood flow* and thus heat transport to the skin, causing the temperature of the skin and the subjacent tissues to rise and approach the deep body temperature. Under extreme heat conditions, the *heart rate* may reach 180 beats per minute in order to maintain adequate blood supply to the brain and the skin. At higher heart rates, the *volumetric efficiency* of the heart drops because of the very short time between the beats to fill the heart with blood, and the blood supply to the skin and more importantly to the brain drops. This causes the person to faint as a result of *heat exhaustion.* Dehydration makes the problem worse. A similar thing happens when a person working very hard for a long time stops suddenly. The blood that has flooded the skin has difficulty returning to the heart in this case since the relaxed muscles no longer force the blood back to the heart, and thus there is less blood available for pumping to the brain.

The next line of defense is releasing water from sweat glands and resorting to *evaporative cooling,* unless the person removes some clothing and reduces the activity level (Fig. 1–56). The body can maintain its core temperature at 37°C in this evaporative cooling mode indefinitely, even in environments at higher temperatures (as high as 200°C during military endurance tests), if the person drinks plenty of liquids to replenish his or her water reserves and the ambient air is sufficiently dry to allow the sweat to evaporate instead of rolling down the skin. If this measure proves inadequate, the body will have to start absorbing the metabolic heat and the deep body temperature will rise. A person can tolerate a temperature rise of 1.4°C without major discomfort but may *collapse* when the temperature rise reaches 2.8°C. People feel sluggish and their efficiency drops considerably when the core body temperature rises above 39°C. A core temperature above 41°C may damage hypothalamic proteins, resulting in cessation of sweating, increased heat production by shivering, and a *heat stroke* with irreversible and life-threatening damage. Death can occur above 43°C.

A surface temperature of 46°C causes pain on the skin. Therefore, direct contact with a metal block at this temperature or above is painful. However, a person can stay in a room at 100°C for up to 30 min without any damage or pain on the skin because of the convective resistance at the skin surface and evaporative cooling. We can even put our hands into an oven at 200°C for a short time without getting burned.

Another factor that affects thermal comfort, health, and productivity is **ventilation**. Fresh outdoor air can be provided to a building *naturally* by doing nothing, or *forcefully* by a mechanical ventilation system. In the first case, which is the norm in residential buildings, the necessary ventilation is provided by *infiltration through cracks and leaks* in the living space and by the opening of the windows and doors. The additional ventilation needed in the bathrooms and kitchens is provided by *air vents with dampers* or *exhaust fans.* With this kind of uncontrolled ventilation, however, the fresh air supply will be either too high, wasting energy, or too low, causing poor indoor air quality. But the current practice is not likely to change for residential buildings since there is not a public outcry for energy waste or air quality, and thus it is difficult to justify the cost and complexity of mechanical ventilation systems.

Mechanical ventilation systems are part of any heating and air conditioning system in *commercial buildings,* providing the necessary amount of fresh outdoor air and distributing it uniformly throughout the building. This is not surprising since many rooms in large commercial buildings have no windows and thus rely on mechanical ventilation. Even the rooms with windows are in the same situation since the windows are tightly sealed and cannot be opened in most buildings. It is not a good idea to oversize the ventilation system just to be on the "safe side" since exhausting the heated or cooled indoor air wastes energy. On the other hand, reducing the ventilation rates below the required minimum to conserve energy should also be avoided so that the indoor air quality can be maintained at the required levels. The minimum fresh air ventilation requirements are listed in Table 1–8. The values are based on controlling the CO_2 and other contaminants with an adequate margin of safety, which requires each person be supplied with at least 7.5 L/s (15 ft^3/min) of fresh air.

Another function of the mechanical ventilation system is to **clean** the air by filtering it as it enters the building. Various types of filters are available for this purpose, depending on the cleanliness requirements and the allowable pressure drop.

FIGURE 1–56

In hot environments, a body can dissipate a large amount of metabolic heat by sweating since the sweat absorbs the body heat and evaporates.

TABLE 1–8

Minimum fresh air requirements in buildings (from ASHRAE Standard 62-1989)

Application	Requirement (per person)	
	L/s	ft³/min
Classrooms, libraries, supermarkets	8	15
Dining rooms, conference rooms, offices	10	20
Hospital rooms	13	25
Hotel rooms	15 (per room)	30 (per room)
Smoking lounges	30	60
Retail stores	1.0–1.5 (per m²)	0.2–0.3 (per ft²)
Residential buildings	0.35 air change per hour, but not less than 7.5 L/s (or 15 ft³/min) per person	

SUMMARY

In this chapter, the basics of heat transfer are introduced and discussed. The science of *thermodynamics* deals with the amount of heat transfer as a system undergoes a process from one equilibrium state to another, whereas the science of *heat transfer* deals with the rate of heat transfer, which is the main quantity of interest in the design and evaluation of heat transfer equipment. The sum of all forms of energy of a system is called *total energy,* and it includes the internal, kinetic, and potential energies. The *internal energy* represents the molecular energy of a system, and it consists of sensible, latent, chemical, and nuclear forms. The sensible and latent forms of internal energy can be transferred from one medium to another as a result of a temperature difference, and are referred to as *heat* or *thermal energy*. Thus, *heat transfer* is the exchange of the sensible and latent forms of internal energy between two mediums as a result of a temperature difference. The amount of heat transferred per unit time is called *heat transfer rate* and is denoted by $\dot{Q}$. The rate of heat transfer per unit area is called *heat flux, $\dot{q}$.*

A system of fixed mass is called a *closed system* and a system that involves mass transfer across its boundaries is called an *open system* or *control volume*. The *first law of thermodynamics* or the *energy balance* for any system undergoing any process can be expressed as

$$E_{in} - E_{out} = \Delta E_{system}$$

When a stationary closed system involves heat transfer only and no work interactions across its boundary, the energy balance relation reduces to

$$Q = mc_v\Delta T$$

where Q is the amount of net heat transfer to or from the system. When heat is transferred at a constant rate of $\dot{Q}$, the amount of heat transfer during a time interval Δt can be determined from $Q = \dot{Q}\Delta t$.

Under steady conditions and in the absence of any work interactions, the conservation of energy relation for a control volume with one inlet and one exit with negligible changes in kinetic and potential energies can be expressed as

$$\dot{Q} = \dot{m}c_p\Delta T$$

where $\dot{m} = \rho V A_c$ is the mass flow rate and $\dot{Q}$ is the rate of net heat transfer into or out of the control volume.

Heat can be transferred in three different modes: conduction, convection, and radiation. *Conduction* is the transfer of heat from the more energetic particles of a substance to the adjacent less energetic ones as a result of interactions between the particles, and is expressed by *Fourier's law of heat conduction* as

$$\dot{Q}_{cond} = -kA\frac{dT}{dx}$$

where k is the *thermal conductivity* of the material, A is the *area* normal to the direction of heat transfer, and dT/dx is the *temperature gradient*. The magnitude of the rate of heat conduction across a plane layer of thickness L is given by

$$\dot{Q}_{cond} = kA\frac{\Delta T}{L}$$

where ΔT is the temperature difference across the layer.

Convection is the mode of heat transfer between a solid surface and the adjacent liquid or gas that is in motion, and involves the combined effects of conduction and fluid motion. The rate of convection heat transfer is expressed by *Newton's law of cooling* as

$$\dot{Q}_{convection} = hA_s\,(T_s - T_\infty)$$

where h is the *convection heat transfer coefficient* in $W/m^2 \cdot K$ or $Btu/h \cdot ft^2 \cdot R$, A_s is the *surface area* through which convection heat transfer takes place, T_s is the *surface temperature,* and T_∞ is the *temperature of the fluid* sufficiently far from the surface.

Radiation is the energy emitted by matter in the form of electromagnetic waves (or photons) as a result of the changes in the electronic configurations of the atoms or molecules. The maximum rate of radiation that can be emitted from a surface at a thermodynamic temperature T_s is given by the *Stefan–Boltzmann law* as $\dot{Q}_{emit,\,max} = \sigma A_s T_s^4$, where $\sigma = 5.67 \times 10^{-8}$ $W/m^2 \cdot K^4$ or 0.1714×10^{-8} $Btu/h \cdot ft^2 \cdot R^4$ is the *Stefan–Boltzmann constant.*

When a surface of emissivity ε and surface area A_s at a temperature T_s is completely enclosed by a much larger (or black) surface at a temperature T_{surr} separated by a gas (such as air) that does not intervene with radiation, the net rate of radiation heat transfer between these two surfaces is given by

$$\dot{Q}_{rad} = \varepsilon\sigma A_s\,(T_s^4 - T_{surr}^4)$$

In this case, the emissivity and the surface area of the surrounding surface do not have any effect on the net radiation heat transfer.

The rate at which a surface absorbs radiation is determined from $\dot{Q}_{absorbed} = \alpha\dot{Q}_{incident}$ where $\dot{Q}_{incident}$ is the rate at which radiation is incident on the surface and α is the absorptivity of the surface.

REFERENCES AND SUGGESTED READING

1. American Society of Heating, Refrigeration, and Air-Conditioning Engineers, *Handbook of Fundamentals.* Atlanta: ASHRAE, 1993.

2. Y. A. Çengel and R. H. Turner. *Fundamentals of Thermal-Fluid Sciences.* 2nd ed. New York: McGraw-Hill, 2005.

3. Y. A. Çengel and M. A. Boles. *Thermodynamics—An Engineering Approach.* 5th ed. New York: McGraw-Hill, 2006.

4. Robert J. Ribando. *Heat Transfer Tools.* New York: McGraw-Hill, 2002.

PROBLEMS*

Thermodynamics and Heat Transfer

1–1C How does the science of heat transfer differ from the science of thermodynamics?

1–2C What is the driving force for (*a*) heat transfer, (*b*) electric current flow, and (*c*) fluid flow?

1–3C What is the caloric theory? When and why was it abandoned?

1–4C How do rating problems in heat transfer differ from the sizing problems?

1–5C What is the difference between the analytical and experimental approach to heat transfer? Discuss the advantages and disadvantages of each approach.

1–6C What is the importance of modeling in engineering? How are the mathematical models for engineering processes prepared?

1–7C When modeling an engineering process, how is the right choice made between a simple but crude and a complex but accurate model? Is the complex model necessarily a better choice since it is more accurate?

Heat and Other Forms of Energy

1–8C What is heat flux? How is it related to the heat transfer rate?

1–9C What are the mechanisms of energy transfer to a closed system? How is heat transfer distinguished from the other forms of energy transfer?

1–10C How are heat, internal energy, and thermal energy related to each other?

1–11C An ideal gas is heated from 50°C to 80°C (*a*) at constant volume and (*b*) at constant pressure. For which case do you think the energy required will be greater? Why?

1–12 A cylindrical resistor element on a circuit board dissipates 0.8 W of power. The resistor is 2 cm long, and has a diameter of 0.4 cm. Assuming heat to be transferred uniformly from all surfaces, determine (*a*) the amount of heat this resistor dissipates during a 24-hour period, (*b*) the heat flux, and (*c*) the fraction of heat dissipated from the top and bottom surfaces.

1–13E A logic chip used in a computer dissipates 3 W of power in an environment at 120°F, and has a heat transfer surface area of 0.08 in². Assuming the heat transfer from the surface to be uniform, determine (*a*) the amount of heat this chip dissipates during an eight-hour work day, in kWh, and (*b*) the heat flux on the surface of the chip, in W/in².

1–14 Consider a 150-W incandescent lamp. The filament of the lamp is 5-cm long and has a diameter of 0.5 mm. The

*Problems designated by a "C" are concept questions, and students are encouraged to answer them all. Problems designated by an "E" are in English units, and the SI users can ignore them. Problems with the icon ⊛ are solved using EES, and complete solutions together with parametric studies are included on the enclosed CD. Problems with the icon ▣ are the comprehensive in nature and are intended to be solved with a computer, preferably using the EES software that accompanies this text.

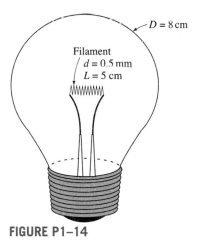

$D = 8$ cm

Filament
$d = 0.5$ mm
$L = 5$ cm

FIGURE P1–14

diameter of the glass bulb of the lamp is 8 cm. Determine the heat flux, in W/m², (a) on the surface of the filament and (b) on the surface of the glass bulb, and (c) calculate how much it will cost per year to keep that lamp on for eight hours a day every day if the unit cost of electricity is $0.08/kWh.

Answers: (a) 1.91×10^6 W/m², (b) 7500 W/m², (c) $35.04/yr

1–15 A 1200-W iron is left on the ironing board with its base exposed to the air. About 85 percent of the heat generated in the iron is dissipated through its base whose surface area is 150 cm², and the remaining 15 percent through other surfaces. Assuming the heat transfer from the surface to be uniform, determine (a) the amount of heat the iron dissipates during a 2-hour period, in kWh, (b) the heat flux on the surface of the iron base, in W/m², and (c) the total cost of the electrical energy consumed during this 2-hour period. Take the unit cost of electricity to be $0.07/kWh.

1–16 A 15-cm × 20-cm circuit board houses on its surface 120 closely spaced logic chips, each dissipating 0.12 W. If the heat transfer from the back surface of the board is negligible, determine (a) the amount of heat this circuit board dissipates during a 10-hour period, in kWh, and (b) the heat flux on the surface of the circuit board, in W/m².

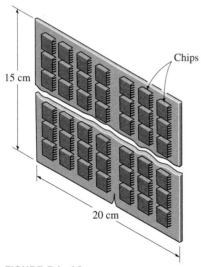

FIGURE P1–16

1–17 A 15-cm-diameter aluminum ball is to be heated from 80°C to an average temperature of 200°C. Taking the average density and specific heat of aluminum in this temperature range to be $\rho = 2700$ kg/m³ and $c_p = 0.90$ kJ/kg · °C, respectively, determine the amount of energy that needs to be transferred to the aluminum ball.   *Answer:* 515 kJ

1–18 The average specific heat of the human body is 3.6 kJ/kg · °C. If the body temperature of a 80-kg man rises from 37°C to 39°C during strenuous exercise, determine the

increase in the thermal energy content of the body as a result of this rise in body temperature.

1–19 Infiltration of cold air into a warm house during winter through the cracks around doors, windows, and other openings is a major source of energy loss since the cold air that enters needs to be heated to the room temperature. The infiltration is often expressed in terms of ACH (air changes per hour). An ACH of 2 indicates that the entire air in the house is replaced twice every hour by the cold air outside.

Consider an electrically heated house that has a floor space of 200 m² and an average height of 3 m at 1000 m elevation, where the standard atmospheric pressure is 89.6 kPa. The house is maintained at a temperature of 22°C, and the infiltration losses are estimated to amount to 0.7 ACH. Assuming the pressure and the temperature in the house remain constant, determine the amount of energy loss from the house due to infiltration for a day during which the average outdoor temperature is 5°C. Also, determine the cost of this energy loss for that day if the unit cost of electricity in that area is $0.082/kWh.

Answers: 53.8 kWh/day, $4.41/day

1–20 Consider a house with a floor space of 200 m² and an average height of 3 m at sea level, where the standard atmospheric pressure is 101.3 kPa. Initially the house is at a uniform temperature of 10°C. Now the electric heater is turned on, and the heater runs until the air temperature in the house rises to an average value of 22°C. Determine how much heat is absorbed by the air assuming some air escapes through the cracks as the heated air in the house expands at constant pressure. Also, determine the cost of this heat if the unit cost of electricity in that area is $0.075/kWh.

1–21E Consider a 60-gallon water heater that is initially filled with water at 45°F. Determine how much energy needs to be transferred to the water to raise its temperature to 120°F. Take the density and specific heat of water to be 62 lbm/ft³ and 1.0 Btu/lbm · °F, respectively.

Energy Balance

1–22C On a hot summer day, a student turns his fan on when he leaves his room in the morning. When he returns in the evening, will his room be warmer or cooler than the neighboring rooms? Why? Assume all the doors and windows are kept closed.

1–23C Consider two identical rooms, one with a refrigerator in it and the other without one. If all the doors and windows are closed, will the room that contains the refrigerator be cooler or warmer than the other room? Why?

1–24 Two 800-kg cars moving at a velocity of 90 km/h have a head-on collision on a road. Both cars come to a complete rest after the crash. Assuming all the kinetic energy of cars is converted to thermal energy, determine the average temperature rise of the remains of the cars immediately after the crash. Take the average specific heat of the cars to be 0.45 kJ/kg · °C.

1–25 A classroom that normally contains 40 people is to be air-conditioned using window air-conditioning units of 5-kW cooling capacity. A person at rest may be assumed to dissipate heat at a rate of 360 kJ/h. There are 10 lightbulbs in the room, each with a rating of 100 W. The rate of heat transfer to the classroom through the walls and the windows is estimated to be 15,000 kJ/h. If the room air is to be maintained at a constant temperature of 21°C, determine the number of window air-conditioning units required. *Answer: two units*

1–26 A 4-m × 5-m × 6-m room is to be heated by a baseboard resistance heater. It is desired that the resistance heater be able to raise the air temperature in the room from 7°C to 25°C within 15 minutes. Assuming no heat losses from the room and an atmospheric pressure of 100 kPa, determine the required power rating of the resistance heater. Assume constant specific heats at room temperature. *Answer: 3.01 kW*

1–27 A 4-m × 5-m × 7-m room is heated by the radiator of a steam heating system. The steam radiator transfers heat at a rate of 12,500 kJ/h and a 100-W fan is used to distribute the warm air in the room. The heat losses from the room are estimated to be at a rate of about 5000 kJ/h. If the initial temperature of the room air is 10°C, determine how long it will take for the air temperature to rise to 20°C. Assume constant specific heats at room temperature.

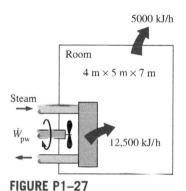

FIGURE P1–27

1–28 A student living in a 4-m × 6-m × 6-m dormitory room turns his 150-W fan on before she leaves her room on a summer day hoping that the room will be cooler when she

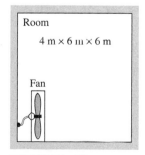

FIGURE P1–28

comes back in the evening. Assuming all the doors and windows are tightly closed and disregarding any heat transfer through the walls and the windows, determine the temperature in the room when she comes back 10 hours later. Use specific heat values at room temperature and assume the room to be at 100 kPa and 15°C in the morning when she leaves. *Answer: 58.1°C*

1–29 A room is heated by a baseboard resistance heater. When the heat losses from the room on a winter day amount to 7000 kJ/h, it is observed that the air temperature in the room remains constant even though the heater operates continuously. Determine the power rating of the heater, in kW.

1–30 A 5-m × 6-m × 8-m room is to be heated by an electrical resistance heater placed in a short duct in the room. Initially, the room is at 15°C, and the local atmospheric pressure is 98 kPa. The room is losing heat steadily to the outside at a rate of 200 kJ/min. A 300-W fan circulates the air steadily through the duct and the electric heater at an average mass flow rate of 50 kg/min. The duct can be assumed to be adiabatic, and there is no air leaking in or out of the room. If it takes 18 minutes for the room air to reach an average temperature of 25°C, find (a) the power rating of the electric heater and (b) the temperature rise that the air experiences each time it passes through the heater.

1–31 A house has an electric heating system that consists of a 300-W fan and an electric resistance heating element placed in a duct. Air flows steadily through the duct at a rate of 0.6 kg/s and experiences a temperature rise of 5°C. The rate of heat loss from the air in the duct is estimated to be 250 W. Determine the power rating of the electric resistance heating element.

1–32 A hair dryer is basically a duct in which a few layers of electric resistors are placed. A small fan pulls the air in and forces it to flow over the resistors where it is heated. Air enters a 1200-W hair dryer at 100 kPa and 22°C, and leaves at 47°C. The cross-sectional area of the hair dryer at the exit is 60 cm². Neglecting the power consumed by the fan and the heat losses through the walls of the hair dryer, determine (a) the volume flow rate of air at the inlet and (b) the velocity of the air at the exit. *Answers: (a) 0.0404 m³/s, (b) 7.30 m/s*

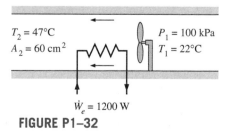

FIGURE P1–32

1–33 The ducts of an air heating system pass through an unheated area. As a result of heat losses, the temperature of the air in the duct drops by 3°C. If the mass flow rate of air is 90 kg/min, determine the rate of heat loss from the air to the cold environment.

1–34E Air enters the duct of an air-conditioning system at 15 psia and 50°F at a volume flow rate of 450 ft³/min. The diameter of the duct is 10 inches and heat is transferred to the air in the duct from the surroundings at a rate of 2 Btu/s. Determine (a) the velocity of the air at the duct inlet and (b) the temperature of the air at the exit.   *Answers:* (a) 825 ft/min, (b) 64°F

1–35 Water is heated in an insulated, constant diameter tube by a 7-kW electric resistance heater. If the water enters the heater steadily at 15°C and leaves at 70°C, determine the mass flow rate of water.

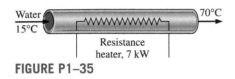

FIGURE P1–35

Heat Transfer Mechanisms

1–36C Consider two houses that are identical, except that the walls are built using bricks in one house, and wood in the other. If the walls of the brick house are twice as thick, which house do you think will be more energy efficient?

1–37C Define thermal conductivity and explain its significance in heat transfer.

1–38C What are the mechanisms of heat transfer? How are they distinguished from each other?

1–39C What is the physical mechanism of heat conduction in a solid, a liquid, and a gas?

1–40C Consider heat transfer through a windowless wall of a house on a winter day. Discuss the parameters that affect the rate of heat conduction through the wall.

1–41C Write down the expressions for the physical laws that govern each mode of heat transfer, and identify the variables involved in each relation.

1–42C How does heat conduction differ from convection?

1–43C Does any of the energy of the sun reach the earth by conduction or convection?

1–44C How does forced convection differ from natural convection?

1–45C Define emissivity and absorptivity. What is Kirchhoff's law of radiation?

1–46C What is a blackbody? How do real bodies differ from blackbodies?

1–47C Judging from its unit W/m · K, can we define thermal conductivity of a material as the rate of heat transfer through the material per unit thickness per unit temperature difference? Explain.

1–48C Consider heat loss through the two walls of a house on a winter night. The walls are identical, except that one of them has a tightly fit glass window. Through which wall will the house lose more heat? Explain.

1–49C Which is a better heat conductor, diamond or silver?

1–50C Consider two walls of a house that are identical except that one is made of 10-cm-thick wood, while the other is made of 25-cm-thick brick. Through which wall will the house lose more heat in winter?

1–51C How do the thermal conductivity of gases and liquids vary with temperature?

1–52C Why is the thermal conductivity of superinsulation orders of magnitude lower than the thermal conductivity of ordinary insulation?

1–53C Why do we characterize the heat conduction ability of insulators in terms of their apparent thermal conductivity instead of the ordinary thermal conductivity?

1–54C Consider an alloy of two metals whose thermal conductivities are k_1 and k_2. Will the thermal conductivity of the alloy be less than k_1, greater than k_2, or between k_1 and k_2?

1–55 The inner and outer surfaces of a 4-m × 7-m brick wall of thickness 30 cm and thermal conductivity 0.69 W/m · K are maintained at temperatures of 20°C and 5°C, respectively. Determine the rate of heat transfer through the wall, in W.

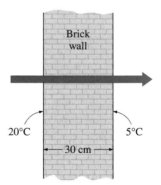

FIGURE P1–55

1–56 The inner and outer surfaces of a 0.5-cm thick 2-m × 2-m window glass in winter are 10°C and 3°C, respectively. If the thermal conductivity of the glass is 0.78 W/m · K, determine the amount of heat loss through the glass over a period of 5 h. What would your answer be if the glass were 1 cm thick? *Answers:* 78.6 MJ, 39.3 MJ

1–57 Reconsider Prob. 1–56. Using EES (or other) software, plot the amount of heat loss through the glass as a function of the window glass thickness in the range of 0.1 cm to 1.0 cm. Discuss the results.

1–58 An aluminum pan whose thermal conductivity is 237 W/m · °C has a flat bottom with diameter 15 cm and thickness 0.4 cm. Heat is transferred steadily to boiling water in the pan through its bottom at a rate of 800 W. If the inner surface of the bottom of the pan is at 105°C, determine the temperature of the outer surface of the bottom of the pan.

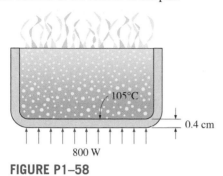

FIGURE P1–58

1–59E   The north wall of an electrically heated home is 20 ft long, 10 ft high, and 1 ft thick, and is made of brick whose thermal conductivity is $k = 0.42$ Btu/h · ft · °F. On a certain winter night, the temperatures of the inner and the outer surfaces of the wall are measured to be at about 62°F and 25°F, respectively, for a period of 8 h. Determine (*a*) the rate of heat loss through the wall that night and (*b*) the cost of that heat loss to the home owner if the cost of electricity is $0.07/kWh.

1–60   In a certain experiment, cylindrical samples of diameter 4 cm and length 7 cm are used (see Fig. 1–30). The two thermocouples in each sample are placed 3 cm apart. After initial transients, the electric heater is observed to draw 0.6 A at 110 V, and both differential thermometers read a temperature difference of 10°C. Determine the thermal conductivity of the sample.   *Answer:* 78.8 W/m · °C

1–61   One way of measuring the thermal conductivity of a material is to sandwich an electric thermofoil heater between two identical rectangular samples of the material and to heavily insulate the four outer edges, as shown in the figure. Thermocouples attached to the inner and outer surfaces of the samples record the temperatures.

During an experiment, two 0.5 cm thick samples 10 cm × 10 cm in size are used. When steady operation is reached, the heater is observed to draw 25 W of electric power, and the temperature of each sample is observed to drop from 82°C at the inner surface to 74°C at the outer surface.

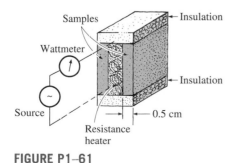

FIGURE P1–61

Determine the thermal conductivity of the material at the average temperature.

1–62   Repeat Prob. 1–61 for an electric power consumption of 20 W.

1–63   A heat flux meter attached to the inner surface of a 3-cm-thick refrigerator door indicates a heat flux of 25 W/m² through the door. Also, the temperatures of the inner and the outer surfaces of the door are measured to be 7°C and 15°C, respectively. Determine the average thermal conductivity of the refrigerator door.   *Answer:* 0.0938 W/m · °C

1–64   Consider a person standing in a room maintained at 20°C at all times. The inner surfaces of the walls, floors, and ceiling of the house are observed to be at an average temperature of 12°C in winter and 23°C in summer. Determine the rates of radiation heat transfer between this person and the surrounding surfaces in both summer and winter if the exposed surface area, emissivity, and the average outer surface temperature of the person are 1.6 m², 0.95, and 32°C, respectively.

1–65   Reconsider Prob. 1–64. Using EES (or other) software, plot the rate of radiation heat transfer in winter as a function of the temperature of the inner surface of the room in the range of 8°C to 18°C. Discuss the results.

1–66   For heat transfer purposes, a standing man can be modeled as a 30-cm-diameter, 170-cm-long vertical cylinder with both the top and bottom surfaces insulated and with the side surface at an average temperature of 34°C. For a convection heat transfer coefficient of 20 W/m² · °C, determine the rate of heat loss from this man by convection in an environment at 18°C.   *Answer:* 513 W

1–67   Hot air at 80°C is blown over a 2-m × 4-m flat surface at 30°C. If the average convection heat transfer coefficient is 55 W/m² · °C, determine the rate of heat transfer from the air to the plate, in kW.   *Answer:* 22 kW

1–68   Reconsider Prob. 1–67. Using EES (or other) software, plot the rate of heat transfer as a function of the heat transfer coefficient in the range of 20 W/m² · °C to 100 W/m² · °C. Discuss the results.

1–69   The heat generated in the circuitry on the surface of a silicon chip ($k = 130$ W/m · °C) is conducted to the ceramic substrate to which it is attached. The chip is 6 mm × 6 mm in

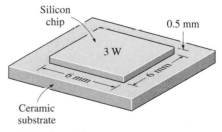

FIGURE P1–69

size and 0.5 mm thick and dissipates 3 W of power. Disregarding any heat transfer through the 0.5 mm high side surfaces, determine the temperature difference between the front and back surfaces of the chip in steady operation.

1–70 A 40-cm-long, 800-W electric resistance heating element with diameter 0.5 cm and surface temperature 120°C is immersed in 75 kg of water initially at 20°C. Determine how long it will take for this heater to raise the water temperature to 80°C. Also, determine the convection heat transfer coefficients at the beginning and at the end of the heating process.

1–71 A 5-cm-external-diameter, 10-m-long hot-water pipe at 80°C is losing heat to the surrounding air at 5°C by natural convection with a heat transfer coefficient of 25 W/m² · °C. Determine the rate of heat loss from the pipe by natural convection. *Answer:* 2945 W

1–72 A hollow spherical iron container with outer diameter 20 cm and thickness 0.4 cm is filled with iced water at 0°C. If the outer surface temperature is 5°C, determine the approximate rate of heat loss from the sphere, in kW, and the rate at which ice melts in the container. The heat of fusion of water is 333.7 kJ/kg.

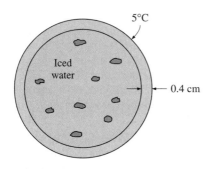

5°C

Iced
water

0.4 cm

FIGURE P1–72

1–73 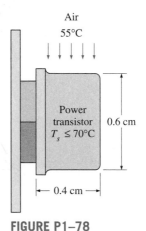 Reconsider Prob. 1–72. Using EES (or other) software, plot the rate at which ice melts as a function of the container thickness in the range of 0.2 cm to 2.0 cm. Discuss the results.

1–74E The inner and outer glasses of a 4-ft × 4-ft double-pane window are at 60°F and 48°F, respectively. If the 0.25-in. space between the two glasses is filled with still air, determine the rate of heat transfer through the window.
Answer: 131 Btu/h

1–75 Two surfaces of a 2-cm-thick plate are maintained at 0°C and 80°C, respectively. If it is determined that heat is transferred through the plate at a rate of 500 W/m², determine its thermal conductivity.

1–76 Four power transistors, each dissipating 15 W, are mounted on a thin vertical aluminum plate 22 cm × 22 cm in size. The heat generated by the transistors is to be dissipated by both surfaces of the plate to the surrounding air at 25°C, which is blown over the plate by a fan. The entire plate can be assumed to be nearly isothermal, and the exposed surface area of the transistor can be taken to be equal to its base area. If the average convection heat transfer coefficient is 25 W/m² · °C, determine the temperature of the aluminum plate. Disregard any radiation effects.

1–77 An ice chest whose outer dimensions are 30 cm × 40 cm × 40 cm is made of 3-cm-thick Styrofoam ($k = 0.033$ W/m · °C). Initially, the chest is filled with 28 kg of ice at 0°C, and the inner surface temperature of the ice chest can be taken to be 0°C at all times. The heat of fusion of ice at 0°C is 333.7 kJ/kg, and the surrounding ambient air is at 25°C. Disregarding any heat transfer from the 40-cm × 40-cm base of the ice chest, determine how long it will take for the ice in the chest to melt completely if the outer surfaces of the ice chest are at 8°C. *Answer:* 22.9 days

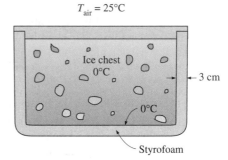

$T_{air} = 25°C$

Ice chest
0°C

3 cm

0°C

Styrofoam

FIGURE P1–77

1–78 A transistor with a height of 0.4 cm and a diameter of 0.6 cm is mounted on a circuit board. The transistor is cooled by air flowing over it with an average heat transfer coefficient of 30 W/m² · °C. If the air temperature is 55°C and the transistor case temperature is not to exceed 70°C, determine the amount of power this transistor can dissipate safely. Disregard any heat transfer from the transistor base.

Air
55°C

Power
transistor
$T_s \leq 70°C$

0.6 cm

0.4 cm

FIGURE P1–78

1–79 Reconsider Prob. 1–78. Using EES (or other) software, plot the amount of power the transistor can dissipate safely as a function of the maximum case temperature in the range of 60°C to 90°C. Discuss the results.

1–80E A 200-ft-long section of a steam pipe whose outer diameter is 4 in passes through an open space at 50°F. The average temperature of the outer surface of the pipe is measured to be 280°F, and the average heat transfer coefficient on that surface is determined to be 6 Btu/h · ft² · °F. Determine (a) the rate of heat loss from the steam pipe and (b) the annual cost of this energy loss if steam is generated in a natural gas furnace having an efficiency of 86 percent, and the price of natural gas is $1.10/therm (1 therm = 100,000 Btu).
Answers: (a) 289,000 Btu/h, (b) $32,380/yr

1–81 The boiling temperature of nitrogen at atmospheric pressure at sea level (1 atm) is −196°C. Therefore, nitrogen is commonly used in low temperature scientific studies since the temperature of liquid nitrogen in a tank open to the atmosphere remains constant at −196°C until the liquid nitrogen in the tank is depleted. Any heat transfer to the tank results in the evaporation of some liquid nitrogen, which has a heat of vaporization of 198 kJ/kg and a density of 810 kg/m³ at 1 atm.

Consider a 4-m-diameter spherical tank initially filled with liquid nitrogen at 1 atm and −196°C. The tank is exposed to 20°C ambient air with a heat transfer coefficient of 25 W/m² · °C. The temperature of the thin-shelled spherical tank is observed to be almost the same as the temperature of the nitrogen inside. Disregarding any radiation heat exchange, determine the rate of evaporation of the liquid nitrogen in the tank as a result of the heat transfer from the ambient air.

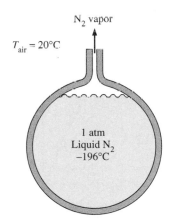

$T_{air} = 20°C$

N₂ vapor

1 atm
Liquid N₂
−196°C

FIGURE P1–81

1–82 Repeat Prob. 1–81 for liquid oxygen, which has a boiling temperature of −183°C, a heat of vaporization of 213 kJ/kg, and a density of 1140 kg/m³ at 1 atm pressure.

1–83 Reconsider Prob. 1–81. Using EES (or other) software, plot the rate of evaporation of liquid

nitrogen as a function of the ambient air temperature in the range of 0°C to 35°C. Discuss the results.

1–84 Consider a person whose exposed surface area is 1.7 m², emissivity is 0.5, and surface temperature is 32°C. Determine the rate of heat loss from that person by radiation in a large room having walls at a temperature of (a) 300 K and (b) 280 K. *Answers: (a)* 26.7 W, (b) 121 W

1–85 A 0.3-cm-thick, 12-cm-high, and 18-cm-long circuit board houses 80 closely spaced logic chips on one side, each dissipating 0.06 W. The board is impregnated with copper fillings and has an effective thermal conductivity of 16 W/m · °C. All the heat generated in the chips is conducted across the circuit board and is dissipated from the back side of the board to the ambient air. Determine the temperature difference between the two sides of the circuit board. *Answer:* 0.042°C

1–86 Consider a sealed 20-cm-high electronic box whose base dimensions are 40 cm × 40 cm placed in a vacuum chamber. The emissivity of the outer surface of the box is 0.95. If the electronic components in the box dissipate a total of 100 W of power and the outer surface temperature of the box is not to exceed 55°C, determine the temperature at which the surrounding surfaces must be kept if this box is to be cooled by radiation alone. Assume the heat transfer from the bottom surface of the box to the stand to be negligible.

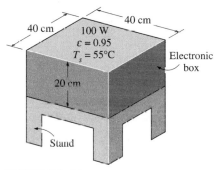

40 cm

40 cm

100 W
$\varepsilon = 0.95$
$T_s = 55°C$

Electronic box

20 cm

Stand

FIGURE P1–86

1–87E Using the conversion factors between W and Btu/h, m and ft, and K and R, express the Stefan–Boltzmann constant $\sigma = 5.67 \times 10^{-8}$ W/m² · K⁴ in the English unit Btu/h · ft² · R⁴.

1–88E An engineer who is working on the heat transfer analysis of a house in English units needs the convection heat transfer coefficient on the outer surface of the house. But the only value he can find from his handbooks is 14 W/m² · °C, which is in SI units. The engineer does not have a direct conversion factor between the two unit systems for the convection heat transfer coefficient. Using the conversion factors between W and Btu/h, m and ft, and °C and °F, express the given convection heat transfer coefficient in Btu/h · ft² · °F.
Answer: 2.47 Btu/h · ft² · °F

1–89 A 2.5-cm-diameter and 8-cm-long cylindrical sample of a material is used to determine its thermal conductivity experimentally. In the thermal conductivity apparatus, the sample is placed in a well-insulated cylindrical cavity to ensure one-dimensional heat transfer in the axial direction, and a heat flux generated by a resistance heater whose electricity consumption is measured is applied on one of its faces (say, the left face). A total of 9 thermocouples are imbedded into the sample, 1 cm apart, to measure the temperatures along the sample and on its faces. When the power consumption was fixed at 83.45 W, it is observed that the thermocouple readings are stabilized at the following values:

Distance from left face, cm	Temperature, °C
0	89.38
1	83.25
2	78.28
3	74.10
4	68.25
5	63.73
6	49.65
7	44.40
8	40.00

Plot the variation of temperature along the sample, and calculate the thermal conductivity of the sample material. Based on these temperature readings, do you think steady operating conditions are established? Are there any temperature readings that do not appear right and should be discarded? Also, discuss when and how the temperature profile in a plane wall will deviate from a straight line.

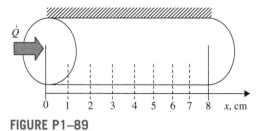

FIGURE P1–89

1–90 Water at 0°C releases 333.7 kJ/kg of heat as it freezes to ice ($\rho = 920$ kg/m³) at 0°C. An aircraft flying under icing conditions maintains a heat transfer coefficient of 150 W/m² · °C between the air and wing surfaces. What temperature must the wings be maintained at to prevent ice from forming on them during icing conditions at a rate of 1 mm/min or less?

Simultaneous Heat Transfer Mechanisms

1–91C Can all three modes of heat transfer occur simultaneously (in parallel) in a medium?

1–92C Can a medium involve (a) conduction and convection, (b) conduction and radiation, or (c) convection and radiation simultaneously? Give examples for the "yes" answers.

1–93C The deep human body temperature of a healthy person remains constant at 37°C while the temperature and the humidity of the environment change with time. Discuss the heat transfer mechanisms between the human body and the environment both in summer and winter, and explain how a person can keep cooler in summer and warmer in winter.

1–94C We often turn the fan on in summer to help us cool. Explain how a fan makes us feel cooler in the summer. Also explain why some people use ceiling fans also in winter.

1–95 Consider a person standing in a room at 23°C. Determine the total rate of heat transfer from this person if the exposed surface area and the skin temperature of the person are 1.7 m² and 32°C, respectively, and the convection heat transfer coefficient is 5 W/m² · °C. Take the emissivity of the skin and the clothes to be 0.9, and assume the temperature of the inner surfaces of the room to be the same as the air temperature.
  Answer: 161 W

1–96 Consider steady heat transfer between two large parallel plates at constant temperatures of $T_1 = 290$ K and $T_2 = 150$ K that are $L = 2$ cm apart. Assuming the surfaces to be black (emissivity $\varepsilon = 1$), determine the rate of heat transfer between the plates per unit surface area assuming the gap between the plates is (a) filled with atmospheric air, (b) evacuated, (c) filled with fiberglass insulation, and (d) filled with superinsulation having an apparent thermal conductivity of 0.00015 W/m · °C.

1–97 The inner and outer surfaces of a 25-cm-thick wall in summer are at 27°C and 44°C, respectively. The outer surface of the wall exchanges heat by radiation with surrounding surfaces at 40°C, and convection with ambient air also at 40°C with a convection heat transfer coefficient of 8 W/m² · °C. Solar radiation is incident on the surface at a rate of 150 W/m². If both the emissivity and the solar absorptivity of the outer surface are 0.8, determine the effective thermal conductivity of the wall.

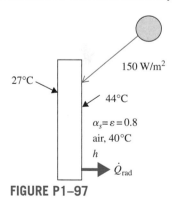

FIGURE P1–97

1–98 A 1.4-m-long, 0.2-cm-diameter electrical wire extends across a room that is maintained at 20°C. Heat is generated in the wire as a result of resistance heating, and the surface temperature of the wire is measured to be 240°C in steady operation. Also, the voltage drop and electric current through the wire are measured to be 110 V and 3 A, respectively.

Disregarding any heat transfer by radiation, determine the convection heat transfer coefficient for heat transfer between the outer surface of the wire and the air in the room.
Answer: 170.5 W/m² · °C

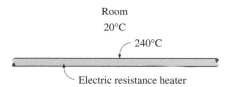

Room
20°C
240°C
Electric resistance heater

FIGURE P1–98

1–99 Reconsider Prob. 1–98. Using EES (or other) software, plot the convection heat transfer coefficient as a function of the wire surface temperature in the range of 100°C to 300°C. Discuss the results.

1–100E A 2-in-diameter spherical ball whose surface is maintained at a temperature of 170°F is suspended in the middle of a room at 70°F. If the convection heat transfer coefficient is 15 Btu/h · ft² · °F and the emissivity of the surface is 0.8, determine the total rate of heat transfer from the ball.

1–101 A 1000-W iron is left on the iron board with its base exposed to the air at 20°C. The convection heat transfer coefficient between the base surface and the surrounding air is 35 W/m² · °C. If the base has an emissivity of 0.6 and a surface area of 0.02 m², determine the temperature of the base of the iron. *Answer:* 674°C

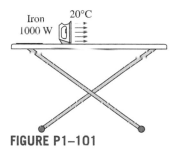

Iron
1000 W
20°C

FIGURE P1–101

1–102 The outer surface of a spacecraft in space has an emissivity of 0.8 and a solar absorptivity of 0.3. If solar radiation is incident on the spacecraft at a rate of 950 W/m², determine the surface temperature of the spacecraft when the radiation emitted equals the solar energy absorbed.

1–103 A 3-m-internal-diameter spherical tank made of 1-cm-thick stainless steel is used to store iced water at 0°C. The tank is located outdoors at 25°C. Assuming the entire steel tank to be at 0°C and thus the thermal resistance of the tank to be negligible, determine (*a*) the rate of heat transfer to the iced water in the tank and (*b*) the amount of ice at 0°C that melts during a 24-hour period. The heat of fusion of water at atmospheric pressure is h_{if} = 333.7 kJ/kg. The emissivity of the outer surface of the tank is 0.75, and the convection heat transfer coefficient on the outer surface can be taken to be 30 W/m² · °C. Assume the

average surrounding surface temperature for radiation exchange to be 15°C. *Answers: (a)* 23.1 kW, *(b)* 5980 kg

1–104 The roof of a house consists of a 15-cm-thick concrete slab (k = 2 W/m · °C) that is 15 m wide and 20 m long. The emissivity of the outer surface of the roof is 0.9, and the convection heat transfer coefficient on that surface is estimated to be 15 W/m² · °C. The inner surface of the roof is maintained at 15°C. On a clear winter night, the ambient air is reported to be at 10°C while the night sky temperature for radiation heat transfer is 255 K. Considering both radiation and convection heat transfer, determine the outer surface temperature and the rate of heat transfer through the roof.

If the house is heated by a furnace burning natural gas with an efficiency of 85 percent, and the unit cost of natural gas is $0.60/therm (1 therm = 105,500 kJ of energy content), determine the money lost through the roof that night during a 14-hour period.

1–105E Consider a flat-plate solar collector placed horizontally on the flat roof of a house. The collector is 5 ft wide and 15 ft long, and the average temperature of the exposed surface of the collector is 100°F. The emissivity of the exposed surface of the collector is 0.9. Determine the rate of heat loss from the collector by convection and radiation during a calm day when the ambient air temperature is 70°F and the effective sky temperature for radiation exchange is 50°F. Take the convection heat transfer coefficient on the exposed surface to be 2.5 Btu/h · ft² · °F.

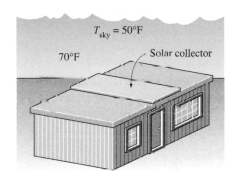

T_{sky} = 50°F
70°F
Solar collector

FIGURE P1–105E

Problem Solving Technique and EES

1–106C What is the value of the engineering software packages in (*a*) engineering education and (*b*) engineering practice?

1–107 Determine a positive real root of the following equation using EES:

$$2x^3 - 10x^{0.5} - 3x = -3$$

1–108 Solve the following system of two equations with two unknowns using EES:

$$x^3 - y^2 = 7.75$$
$$3xy + y = 3.5$$

1–109 Solve the following system of three equations with three unknowns using EES:

$$2x - y + z = 5$$
$$3x^2 + 2y = z + 2$$
$$xy + 2z = 8$$

1–110 Solve the following system of three equations with three unknowns using EES:

$$x^2 y - z = 1$$
$$x - 3y^{0.5} + xz = -2$$
$$x + y - z = 2$$

Special Topic: Thermal Comfort

1–111C What is metabolism? What is the range of metabolic rate for an average man? Why are we interested in metabolic rate of the occupants of a building when we deal with heating and air conditioning?

1–112C Why is the metabolic rate of women, in general, lower than that of men? What is the effect of clothing on the environmental temperature that feels comfortable?

1–113C What is asymmetric thermal radiation? How does it cause thermal discomfort in the occupants of a room?

1–114C How do (*a*) draft and (*b*) cold floor surfaces cause discomfort for a room's occupants?

1–115C What is stratification? Is it likely to occur at places with low or high ceilings? How does it cause thermal discomfort for a room's occupants? How can stratification be prevented?

1–116C Why is it necessary to ventilate buildings? What is the effect of ventilation on energy consumption for heating in winter and for cooling in summer? Is it a good idea to keep the bathroom fans on all the time? Explain.

Review Problems

1–117 It is well known that wind makes the cold air feel much colder as a result of the *windchill* effect that is due to the increase in the convection heat transfer coefficient with increasing air velocity. The windchill effect is usually expressed in terms of the *windchill factor,* which is the difference between the actual air temperature and the equivalent calm-air temperature. For example, a windchill factor of 20°C for an actual air temperature of 5°C means that the windy air at 5°C feels as cold as the still air at −15°C. In other words, a person will lose as much heat to air at 5°C with a windchill factor of 20°C as he or she would in calm air at −15°C.

For heat transfer purposes, a standing man can be modeled as a 30-cm-diameter, 170-cm-long vertical cylinder with both the top and bottom surfaces insulated and with the side surface at an average temperature of 34°C. For a convection heat transfer coefficient of 15 W/m² · °C, determine the rate of heat loss from this man by convection in still air at 20°C. What would

your answer be if the convection heat transfer coefficient is increased to 50 W/m² · °C as a result of winds? What is the windchill factor in this case? *Answers:* 336 W, 1120 W, 32.7°C

1–118 A thin metal plate is insulated on the back and exposed to solar radiation on the front surface. The exposed surface of the plate has an absorptivity of 0.7 for solar radiation. If solar radiation is incident on the plate at a rate of 550 W/m² and the surrounding air temperature is 10°C, determine the surface temperature of the plate when the heat loss by convection equals the solar energy absorbed by the plate. Take the convection heat transfer coefficient to be 25 W/m² · °C, and disregard any heat loss by radiation.

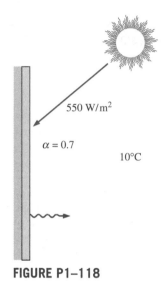

FIGURE P1–118

1–119 A 4-m × 5-m × 6-m room is to be heated by one ton (1000 kg) of liquid water contained in a tank placed in the room. The room is losing heat to the outside at an average rate of 10,000 kJ/h. The room is initially at 20°C and 100 kPa, and is maintained at an average temperature of 20°C at all times. If the hot water is to meet the heating requirements of this room for a 24-h period, determine the minimum temperature of the water when it is first brought into the room. Assume constant specific heats for both air and water at room temperature. *Answer:* 77.4°C

1–120 Consider a 3-m × 3-m × 3-m cubical furnace whose top and side surfaces closely approximate black surfaces at a temperature of 1200 K. The base surface has an emissivity of $\varepsilon = 0.7$, and is maintained at 800 K. Determine the net rate of radiation heat transfer to the base surface from the top and side surfaces. *Answer:* 594 kW

1–121 Consider a refrigerator whose dimensions are 1.8 m × 1.2 m × 0.8 m and whose walls are 3 cm thick. The refrigerator consumes 600 W of power when operating and has a COP of 1.5. It is observed that the motor of the refrigerator remains on for 5 min and then is off for 15 min periodically. If the average temperatures at the inner and outer surfaces of the refrigerator are 6°C and 17°C, respectively, determine the average thermal conductivity of the refrigerator walls. Also, determine the annual cost of operating this refrigerator if the unit cost of electricity is $0.08/kWh.

FIGURE P1–121

1–122 Engine valves (c_p = 440 J/kg · °C and ρ = 7840 kg/m³) are to be heated from 40°C to 800°C in 5 min in the heat treatment section of a valve manufacturing facility. The valves have a cylindrical stem with a diameter of 8 mm and a length of 10 cm. The valve head and the stem may be assumed to be of equal surface area, with a total mass of 0.0788 kg. For a single valve, determine (*a*) the amount of heat transfer, (*b*) the average rate of heat transfer, (*c*) the average heat flux, and (*d*) the number of valves that can be heat treated per day if the heating section can hold 25 valves and it is used 10 h per day.

1–123 Consider a flat-plate solar collector placed on the roof of a house. The temperatures at the inner and outer surfaces of the glass cover are measured to be 28°C and 25°C, respectively. The glass cover has a surface area of 2.5 m², a thickness of 0.6 cm, and a thermal conductivity of 0.7 W/m · °C. Heat is lost from the outer surface of the cover by convection and radiation with a convection heat transfer coefficient of 10 W/m² · °C and an ambient temperature of 15°C. Determine the fraction of heat lost from the glass cover by radiation.

1–124 The rate of heat loss through a unit surface area of a window per unit temperature difference between the indoors and the outdoors is called the *U*-factor. The value of the *U*-factor ranges from about 1.25 W/m² · °C (or 0.22 Btu/h · ft² · °F) for low-*e* coated, argon-filled, quadruple-pane windows to 6.25 W/m² · °C (or 1.1 Btu/h · ft² · °F) for a single-pane window with aluminum frames. Determine the range for the rate of heat loss through a 1.2-m × 1.8-m window of a house that is maintained at 20°C when the outdoor air temperature is −8°C.

1–125 ![EES] Reconsider Prob. 1–124. Using EES (or other) software, plot the rate of heat loss through the window as a function of the *U*-factor. Discuss the results.

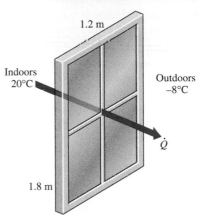

FIGURE P1–124

1–126 Consider a house in Atlanta, Georgia, that is maintained at 22°C and has a total of 20 m² of window area. The windows are double-door type with wood frames and metal spacers and have a *U*-factor of 2.5 W/m² · °C (see Prob. 1–124 for the definition of *U*-factor). The winter average temperature of Atlanta is 11.3°C. Determine the average rate of heat loss through the windows in winter.

1–127 A 50-cm-long, 2-mm-diameter electric resistance wire submerged in water is used to determine the boiling heat transfer coefficient in water at 1 atm experimentally. The wire temperature is measured to be 130°C when a wattmeter indicates the electric power consumed to be 4.1 kW. Using Newton's law of cooling, determine the boiling heat transfer coefficient.

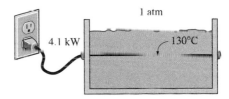

FIGURE P1–127

1–128 An electric heater with the total surface area of 0.25 m² and emissivity 0.75 is in a room where the air has a temperature of 20°C and the walls are at 10°C. When the heater consumes 500 W of electric power, its surface has a steady temperature of 120°C. Determine the temperature of the heater surface when it consumes 700 W. Solve the problem (*a*) assuming negligible

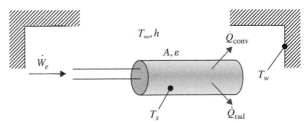

FIGURE P1–128

radiation and (b) taking radiation into consideration. Based on your results, comment on the assumption made in part (a).

1–129 An ice skating rink is located in a building where the air is at $T_{air} = 20°C$ and the walls are at $T_w = 25°C$. The convection heat transfer coefficient between the ice and the surrounding air is $h = 10$ W/m² · K. The emissivity of ice is $\varepsilon = 0.95$. The latent heat of fusion of ice is $h_{if} = 333.7$ kJ/kg and its density is 920 kg/m³. (a) Calculate the refrigeration load of the system necessary to maintain the ice at $T_s = 0°C$ for an ice rink of 12 m by 40 m. (b) How long would it take to melt $\delta = 3$ mm of ice from the surface of the rink if no cooling is supplied and the surface is considered insulated on the back side?

Fundamentals of Engineering (FE) Exam Problems

1–130 Which equation below is used to determine the heat flux for conduction?

(a) $-kA\dfrac{dT}{dx}$ (b) $-k$grad T (c) $h(T_2-T_1)$ (d) $\varepsilon\sigma T^4$

(e) None of them

1–131 Which equation below is used to determine the heat flux for convection?

(a) $-kA\dfrac{dT}{dx}$ (b) $-k$grad T (c) $h(T_2-T_1)$ (d) $\varepsilon\sigma T^4$

(e) None of them

1–132 Which equation below is used to determine the heat flux emitted by thermal radiation from a surface?

(a) $-kA\dfrac{dT}{dx}$ (b) $-k$grad T (c) $h(T_2-T_1)$ (d) $\varepsilon\sigma T^4$

(e) None of them

1–133 A 1-kW electric resistance heater in a room is turned on and kept on for 50 minutes. The amount of energy transferred to the room by the heater is

(a) 1 kJ   (b) 50 kJ   (c) 3000 kJ
(d) 3600 kJ (e) 6000 kJ

1–134 A hot 16-cm × 16-cm × 16-cm cubical iron block is cooled at an average rate of 80 W. The heat flux is

(a) 195 W/m²   (b) 521 W/m²   (c) 3125 W/m²
(d) 7100 W/m² (e) 19,500 W/m²

1–135 A 2-kW electric resistance heater submerged in 30-kg water is turned on and kept on for 10 min. During the process, 500 kJ of heat is lost from the water. The temperature rise of water is

(a) 5.6°C   (b) 9.6°C   (c) 13.6°C
(d) 23.3°C (e) 42.5°C

1–136 Eggs with a mass of 0.15 kg per egg and a specific heat of 3.32 kJ/kg · °C are cooled from 32°C to 10°C at a rate of 300 eggs per minute. The rate of heat removal from the eggs is

(a) 11 kW   (b) 80 kW   (c) 25 kW
(d) 657 kW   (e) 55 kW

1–137 Steel balls at 140°C with a specific heat of 0.50 kJ/kg · °C are quenched in an oil bath to an average temperature of 85°C at a rate of 35 balls per minute. If the average mass of steel balls is 1.2 kg, the rate of heat transfer from the balls to the oil is

(a) 33 kJ/s   (b) 1980 kJ/s   (c) 49 kJ/s
(d) 30 kJ/s   (e) 19 kJ/s

1–138 A cold bottled drink ($m = 2.5$ kg, $c_p = 4200$ J/kg · °C) at 5°C is left on a table in a room. The average temperature of the drink is observed to rise to 15°C in 30 minutes. The average rate of heat transfer to the drink is

(a) 23 W   (b) 29 W   (c) 58 W
(d) 88 W   (e) 122 W

1–139 Water enters a pipe at 20°C at a rate of 0.25 kg/s and is heated to 60°C. The rate of heat transfer to the water is

(a) 10 kW   (b) 20.9 kW   (c) 41.8 kW
(d) 62.7 kW   (e) 167.2 kW

1–140 Air enters a 12-m-long, 7-cm-diameter pipe at 50°C at a rate of 0.06 kg/s. The air is cooled at an average rate of

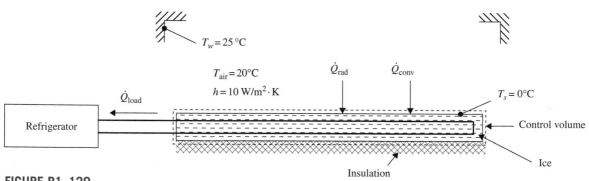

FIGURE P1–129

400 W per m² surface area of the pipe. The air temperature at the exit of the pipe is

(a) 4.3°C   (b) 17.5°C   (c) 32.5°C
(d) 43.4°C   (e) 45.8°C

1–141 Heat is lost steadily through a 0.5-cm thick 2 m × 3 m window glass whose thermal conductivity is 0.7 W/m · °C. The inner and outer surface temperatures of the glass are measured to be 12°C to 9°C. The rate of heat loss by conduction through the glass is

(a) 420 W   (b) 5040 W   (c) 17,600 W
(d) 1256 W   (e) 2520 W

1–142 The East wall of an electrically heated house is 6 m long, 3 m high, and 0.35 m thick, and it has an effective thermal conductivity of 0.7 W/m · °C. If the inner and outer surface temperatures of wall are 15°C and 6°C, the rate of heat loss through the wall is

(a) 324 W   (b) 40 W   (c) 756 W
(d) 648 W   (e) 1390 W

1–143 Steady heat conduction occurs through a 0.3-m-thick 9 m × 3 m composite wall at a rate of 1.2 kW. If the inner and outer surface temperatures of the wall are 15°C and 7°C, the effective thermal conductivity of the wall is

(a) 0.61 W/m · °C   (b) 0.83 W/m · °C   (c) 1.7 W/m · °C
(d) 2.2 W/m · °C   (e) 5.1 W/m · °C

1–144 Heat is lost through a brick wall ($k = 0.72$ W/m · °C), which is 4 m long, 3 m wide, and 25 cm thick at a rate of 500 W. If the inner surface of the wall is at 22°C, the temperature at the midplane of the wall is

(a) 0°C   (b) 7.5°C   (c) 11.0°C
(d) 14.8°C   (e) 22°C

1–145 Consider two different materials, A and B. The ratio of thermal conductivities is $k_A/k_B = 13$, the ratio of the densities is $\rho_A/\rho_B = 0.045$, and the ratio of specific heats is $c_{p,A}/c_{p,B} = 16.9$. The ratio of the thermal diffusivities α_A/α_B is

(a) 4882   (b) 17.1   (c) 0.06   (d) 0.1   (e) 0.03

1–146 A 10-cm-high and 20-cm-wide circuit board houses on its surface 100 closely spaced chips, each generating heat at a rate of 0.08 W and transferring it by convection and radiation to the surrounding medium at 40°C. Heat transfer from the back surface of the board is negligible. If the combined convection and radiation heat transfer coefficient on the surface of the board is 22 W/m² · °C, the average surface temperature of the chips is

(a) 72.4°C   (b) 66.5°C   (c) 40.4°C
(d) 58.2°C   (e) 49.1°C

1–147 A 40-cm-long, 0.4-cm-diameter electric resistance wire submerged in water is used to determine the convection heat transfer coefficient in water during boiling at 1 atm pressure. The surface temperature of the wire is measured to be 114°C when a wattmeter indicates the electric power consumption to be 7.6 kW. The heat transfer coefficient is

(a) 108 kW/m² · °C   (b) 13.3 kW/m² · °C
(c) 68.1 kW/m² · °C   (d) 0.76 kW/m² · °C
(e) 256 kW/m² · °C

1–148 A 10-cm × 12-cm × 14-cm rectangular prism object made of hardwood ($\rho = 721$ kg/m³, $c_p = 1.26$ kJ/kg · °C) is cooled from 100°C to the room temperature of 20°C in 54 minutes. The approximate heat transfer coefficient during this process is

(a) 0.47 W/m² · °C   (b) 5.5 W/m² · °C
(c) 8 W/m² · °C   (d) 11 W/m² · °C
(e) 17,830 W/m² · °C

1–149 A 30-cm-diameter black ball at 120°C is suspended in air, and is losing heat to the surrounding air at 25°C by convection with a heat transfer coefficient of 12 W/m² · °C, and by radiation to the surrounding surfaces at 15°C. The total rate of heat transfer from the black ball is

(a) 322 W   (b) 595 W   (c) 234 W
(d) 472 W   (e) 2100 W

1–150 A 3-m² black surface at 140°C is losing heat to the surrounding air at 35°C by convection with a heat transfer coefficient of 16 W/m² · °C, and by radiation to the surrounding surfaces at 15°C. The total rate of heat loss from the surface is

(a) 5105 W   (b) 2940 W   (c) 3779 W
(d) 8819 W   (e) 5040 W

1–151 A person's head can be approximated as a 25-cm diameter sphere at 35°C with an emissivity of 0.95. Heat is lost from the head to the surrounding air at 25°C by convection with a heat transfer coefficient of 11 W/m² · °C, and by radiation to the surrounding surfaces at 10°C. Disregarding the neck, determine the total rate of heat loss from the head.

(a) 22 W   (b) 27 W   (c) 49 W
(d) 172 W   (e) 249 W

1–152 A 30-cm-long, 0.5-cm-diameter electric resistance wire is used to determine the convection heat transfer coefficient in air at 25°C experimentally. The surface temperature of the wire is measured to be 230°C when the electric power consumption is 180 W. If the radiation heat loss from the wire is calculated to be 60 W, the convection heat transfer coefficient is

(a) 186 W/m² · °C   (b) 158 W/m² · °C
(c) 124 W/m² · °C   (d) 248 W/m² · °C
(e) 390 W/m² · °C

1–153 A room is heated by a 1.2 kW electric resistance heater whose wires have a diameter of 4 mm and a total length of 3.4 m. The air in the room is at 23°C and the interior surfaces of the room are at 17°C. The convection heat transfer coefficient on the surface of the wires is 8 W/m² · °C. If the rates of heat transfer from the wires to the room by convection and by radiation are equal, the surface temperature of the wire is

(a) 3534°C   (b) 1778°C   (c) 1772°C
(d) 98°C   (e) 25°C

1–154 A person standing in a room loses heat to the air in the room by convection and to the surrounding surfaces by radiation. Both the air in the room and the surrounding surfaces are at 20°C. The exposed surfaces of the person is 1.5 m² and has an average temperature of 32°C, and an emissivity of 0.90. If the rates of heat transfer from the person by convection and by radiation are equal, the combined heat transfer coefficient is

(a) 0.008 W/m² · °C (b) 3.0 W/m² · °C
(c) 5.5 W/m² · °C (d) 8.3 W/m² · °C
(e) 10.9 W/m² · °C

1–155 While driving down a highway early in the evening, the air flow over an automobile establishes an overall heat transfer coefficient of 25 W/m² · K. The passenger cabin of this automobile exposes 8 m² of surface to the moving ambient air. On a day when the ambient temperature is 33°C, how much cooling must the air conditioning system supply to maintain a temperature of 20°C in the passenger cabin?

(a) 0.65 MW (b) 1.4 MW (c) 2.6 MW
(d) 3.5 MW (e) 0.94 MW

1–156 On a still clear night, the sky appears to be a blackbody with an equivalent temperature of 250 K. What is the air temperature when a strawberry field cools to 0°C and freezes if the heat transfer coefficient between the plants and air is 6 W/m² · °C because of a light breeze and the plants have an emissivity of 0.9?

(a) 14°C (b) 7°C (c) 3°C (d) 0°C (e) −3°C

1–157 Over 90 percent of the energy dissipated by an incandescent light bulb is in the form of heat, not light. What is the temperature of a vacuum-enclosed tungsten filament with an exposed surface area of 2.03 cm² in a 100 W incandescent light bulb? The emissivity of tungsten at the anticipated high temperatures is about 0.35. Note that the light bulb consumes 100 W of electrical energy, and dissipates all of it by radiation.

(a) 1870 K (b) 2230 K (c) 2640 K
(d) 3120 K (e) 2980 K

1–158 Commercial surface-coating processes often use infrared lamps to speed the curing of the coating. A 2-mm-thick teflon ($k = 0.45$ W/m · K) coating is applied to a 4-m × 4-m surface using this process. Once the coating reaches steady-state, the temperature of its two surfaces are 50°C and 45°C. What is the minimum rate at which power must be supplied to the infrared lamps steadily?

(a) 18 kW (b) 20 kW (c) 22 kW (d) 24 kW
(e) 26 kW

Design and Essay Problems

1–159 Write an essay on how microwave ovens work, and explain how they cook much faster than conventional ovens. Discuss whether conventional electric or microwave ovens consume more electricity for the same task.

1–160 Using information from the utility bills for the coldest month last year, estimate the average rate of heat loss from your house for that month. In your analysis, consider the contribution of the internal heat sources such as people, lights, and appliances. Identify the primary sources of heat loss from your house and propose ways of improving the energy efficiency of your house.

1–161 Conduct this experiment to determine the combined heat transfer coefficient between an incandescent lightbulb and the surrounding air and surfaces using a 60-W lightbulb. You will need a thermometer, which can be purchased in a hardware store, and a metal glue. You will also need a piece of string and a ruler to calculate the surface area of the lightbulb. First, measure the air temperature in the room, and then glue the tip of the thermocouple wire of the thermometer to the glass of the lightbulb. Turn the light on and wait until the temperature reading stabilizes. The temperature reading will give the surface temperature of the lightbulb. Assuming 10 percent of the rated power of the bulb is converted to light and is transmitted by the glass, calculate the heat transfer coefficient from Newton's law of cooling.

HEAT CONDUCTION EQUATION

Heat transfer has *direction* as well as *magnitude.* The rate of heat conduction in a specified direction is proportional to the *temperature gradient,* which is the rate of change in temperature with distance in that direction. Heat conduction in a medium, in general, is three-dimensional and time dependent, and the temperature in a medium varies with position as well as time, that is, $T = T(x, y, z, t)$. Heat conduction in a medium is said to be *steady* when the temperature does not vary with time, and *unsteady* or *transient* when it does. Heat conduction in a medium is said to be *one-dimensional* when conduction is significant in one dimension only and negligible in the other two primary dimensions, *two-dimensional* when conduction in the third dimension is negligible, and *three-dimensional* when conduction in all dimensions is significant.

We start this chapter with a description of steady, unsteady, and multidimensional heat conduction. Then we derive the differential equation that governs heat conduction in a large plane wall, a long cylinder, and a sphere, and generalize the results to three-dimensional cases in rectangular, cylindrical, and spherical coordinates. Following a discussion of the boundary conditions, we present the formulation of heat conduction problems and their solutions. Finally, we consider heat conduction problems with variable thermal conductivity.

This chapter deals with the theoretical and mathematical aspects of heat conduction, and it can be covered selectively, if desired, without causing a significant loss in continuity. The more practical aspects of heat conduction are covered in the following two chapters.

OBJECTIVES

When you finish studying this chapter, you should be able to:

- Understand multidimensionality and time dependence of heat transfer, and the conditions under which a heat transfer problem can be approximated as being one-dimensional,

- Obtain the differential equation of heat conduction in various coordinate systems, and simplify it for steady one-dimensional case,

- Identify the thermal conditions on surfaces, and express them mathematically as boundary and initial conditions,

- Solve one-dimensional heat conduction problems and obtain the temperature distributions within a medium and the heat flux,

- Analyze one-dimensional heat conduction in solids that involve heat generation, and

- Evaluate heat conduction in solids with temperature-dependent thermal conductivity.

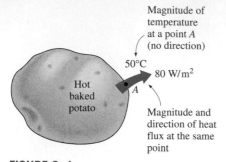

Magnitude of temperature at a point A (no direction)

50°C

80 W/m²

A

Magnitude and direction of heat flux at the same point

Hot baked potato

FIGURE 2–1

Heat transfer has direction as well as magnitude, and thus it is a *vector* quantity.

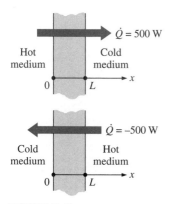

$\dot{Q} = 500$ W

Hot medium

Cold medium

0 L x

$\dot{Q} = -500$ W

Cold medium

Hot medium

0 L x

FIGURE 2–2

Indicating direction for heat transfer (positive in the positive direction; negative in the negative direction).

2–1 · INTRODUCTION

In Chapter 1 heat conduction was defined as the transfer of thermal energy from the more energetic particles of a medium to the adjacent less energetic ones. It was stated that conduction can take place in liquids and gases as well as solids provided that there is no bulk motion involved.

Although heat transfer and temperature are closely related, they are of a different nature. Unlike temperature, heat transfer has direction as well as magnitude, and thus it is a *vector* quantity (Fig. 2–1). Therefore, we must specify both direction and magnitude in order to describe heat transfer completely at a point. For example, saying that the temperature on the inner surface of a wall is 18°C describes the temperature at that location fully. But saying that the heat flux on that surface is 50 W/m² immediately prompts the question "in what direction?" We can answer this question by saying that heat conduction is toward the inside (indicating heat gain) or toward the outside (indicating heat loss).

To avoid such questions, we can work with a coordinate system and indicate direction with plus or minus signs. The generally accepted convention is that heat transfer in the positive direction of a coordinate axis is positive and in the opposite direction it is negative. Therefore, a positive quantity indicates heat transfer in the positive direction and a negative quantity indicates heat transfer in the negative direction (Fig. 2–2).

The driving force for any form of heat transfer is the *temperature difference,* and the larger the temperature difference, the larger the rate of heat transfer. Some heat transfer problems in engineering require the determination of the *temperature distribution* (the variation of temperature) throughout the medium in order to calculate some quantities of interest such as the local heat transfer rate, thermal expansion, and thermal stress at some critical locations at specified times. The specification of the temperature at a point in a medium first requires the specification of the location of that point. This can be done by choosing a suitable coordinate system such as the *rectangular, cylindrical,* or *spherical* coordinates, depending on the geometry involved, and a convenient reference point (the origin).

The *location* of a point is specified as (x, y, z) in rectangular coordinates, as (r, ϕ, z) in cylindrical coordinates, and as (r, ϕ, θ) in spherical coordinates, where the distances x, y, z, and r and the angles ϕ and θ are as shown in Fig. 2–3. Then the temperature at a point (x, y, z) at time t in rectangular coordinates is expressed as $T(x, y, z, t)$. The best coordinate system for a given geometry is the one that describes the surfaces of the geometry best. For example, a parallelepiped is best described in rectangular coordinates since each surface can be described by a constant value of the x-, y-, or z-coordinates. A cylinder is best suited for cylindrical coordinates since its lateral surface can be described by a constant value of the radius. Similarly, the entire outer surface of a spherical body can best be described by a constant value of the radius in spherical coordinates. For an arbitrarily shaped body, we normally use rectangular coordinates since it is easier to deal with distances than with angles.

The notation just described is also used to identify the variables involved in a heat transfer problem. For example, the notation $T(x, y, z, t)$ implies that the temperature varies with the space variables x, y, and z as well as time. The

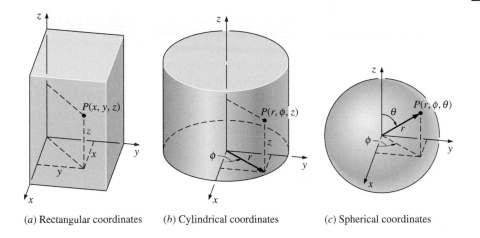

FIGURE 2–3
The various distances and angles involved when describing the location of a point in different coordinate systems.

(*a*) Rectangular coordinates (*b*) Cylindrical coordinates (*c*) Spherical coordinates

notation $T(x)$, on the other hand, indicates that the temperature varies in the x-direction only and there is no variation with the other two space coordinates or time.

Steady versus Transient Heat Transfer

Heat transfer problems are often classified as being **steady** (also called *steady-state*) or **transient** (also called *unsteady*). The term *steady* implies *no change with time* at any point within the medium, while *transient* implies *variation with time* or *time dependence*. Therefore, the temperature or heat flux remains unchanged with time during steady heat transfer through a medium at any location, although both quantities may vary from one location to another (Fig. 2–4). For example, heat transfer through the walls of a house is steady when the conditions inside the house and the outdoors remain constant for several hours. But even in this case, the temperatures on the inner and outer surfaces of the wall will be different unless the temperatures inside and outside the house are the same. The cooling of an apple in a refrigerator, on the other hand, is a transient heat transfer process since the temperature at any fixed point within the apple will change with time during cooling. During transient heat transfer, the temperature normally varies with time as well as position. In the special case of variation with time but not with position, the temperature of the medium changes *uniformly* with time. Such heat transfer systems are called **lumped systems**. A small metal object such as a thermocouple junction or a thin copper wire, for example, can be analyzed as a lumped system during a heating or cooling process.

Most heat transfer problems encountered in practice are *transient* in nature, but they are usually analyzed under some presumed *steady* conditions since steady processes are easier to analyze, and they provide the answers to our questions. For example, heat transfer through the walls and ceiling of a typical house is never steady since the outdoor conditions such as the temperature, the speed and direction of the wind, the location of the sun, and so on, change constantly. The conditions in a typical house are not so steady either. Therefore, it is almost impossible to perform a heat transfer analysis of a house accurately. But then, do we really need an in-depth heat transfer analysis? If the

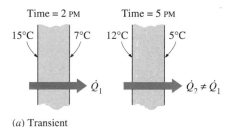

(*a*) Transient

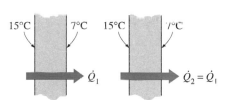

(*b*) Steady

FIGURE 2–4
Transient and steady heat conduction in a plane wall.

purpose of a heat transfer analysis of a house is to determine the proper size of a heater, which is usually the case, we need to know the *maximum* rate of heat loss from the house, which is determined by considering the heat loss from the house under *worst* conditions for an extended period of time, that is, during *steady* operation under worst conditions. Therefore, we can get the answer to our question by doing a heat transfer analysis under steady conditions. If the heater is large enough to keep the house warm under most demanding conditions, it is large enough for all conditions. The approach described above is a common practice in engineering.

Multidimensional Heat Transfer

Heat transfer problems are also classified as being *one-dimensional, two-dimensional,* or *three-dimensional,* depending on the relative magnitudes of heat transfer rates in different directions and the level of accuracy desired. In the most general case, heat transfer through a medium is **three-dimensional**. That is, the temperature varies along all three primary directions within the medium during the heat transfer process. The temperature distribution throughout the medium at a specified time as well as the heat transfer rate at any location in this general case can be described by a set of three coordinates such as the x, y, and z in the rectangular (or Cartesian) coordinate system; the r, ϕ, and z in the cylindrical coordinate system; and the r, ϕ, and θ in the spherical (or polar) coordinate system. The temperature distribution in this case is expressed as $T(x, y, z, t)$, $T(r, \phi, z, t)$, and $T(r, \phi, \theta, t)$ in the respective coordinate systems.

The temperature in a medium, in some cases, varies mainly in two primary directions, and the variation of temperature in the third direction (and thus heat transfer in that direction) is negligible. A heat transfer problem in that case is said to be **two-dimensional**. For example, the steady temperature distribution in a long bar of rectangular cross section can be expressed as $T(x, y)$ if the temperature variation in the z-direction (along the bar) is negligible and there is no change with time (Fig. 2–5).

A heat transfer problem is said to be **one-dimensional** if the temperature in the medium varies in one direction only and thus heat is transferred in one direction, and the variation of temperature and thus heat transfer in other directions are negligible or zero. For example, heat transfer through the glass of a window can be considered to be one-dimensional since heat transfer through the glass occurs predominantly in one direction (the direction normal to the surface of the glass) and heat transfer in other directions (from one side edge to the other and from the top edge to the bottom) is negligible (Fig. 2–6). Likewise, heat transfer through a hot water pipe can be considered to be one-dimensional since heat transfer through the pipe occurs predominantly in the radial direction from the hot water to the ambient, and heat transfer along the pipe and along the circumference of a cross section (z- and ϕ-directions) is typically negligible. Heat transfer to an egg dropped into boiling water is also nearly one-dimensional because of symmetry. Heat is transferred to the egg in this case in the radial direction, that is, along straight lines passing through the midpoint of the egg.

We mentioned in Chapter 1 that the rate of heat conduction through a medium in a specified direction (say, in the x-direction) is proportional to the temperature difference across the medium and the area normal to the direction

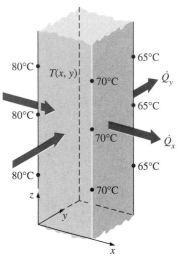

FIGURE 2–5

Two-dimensional heat transfer in a long rectangular bar.

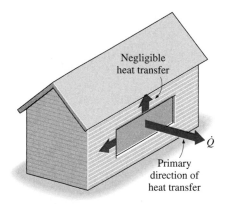

FIGURE 2–6

Heat transfer through the window of a house can be taken to be one-dimensional.

of heat transfer, but is inversely proportional to the distance in that direction. This was expressed in the differential form by **Fourier's law of heat conduction** for one-dimensional heat conduction as

$$\dot{Q}_{cond} = -kA\frac{dT}{dx} \qquad (W) \qquad (2\text{–}1)$$

where k is the *thermal conductivity* of the material, which is a measure of the ability of a material to conduct heat, and dT/dx is the *temperature gradient,* which is the slope of the temperature curve on a *T-x* diagram (Fig. 2–7). The thermal conductivity of a material, in general, varies with temperature. But sufficiently accurate results can be obtained by using a constant value for thermal conductivity at the *average* temperature.

Heat is conducted in the direction of decreasing temperature, and thus the temperature gradient is negative when heat is conducted in the positive *x*-direction. The *negative sign* in Eq. 2–1 ensures that heat transfer in the positive *x*-direction is a positive quantity.

To obtain a general relation for Fourier's law of heat conduction, consider a medium in which the temperature distribution is three-dimensional. Fig. 2–8 shows an isothermal surface in that medium. The heat flux vector at a point *P* on this surface must be perpendicular to the surface, and it must point in the direction of decreasing temperature. If *n* is the normal of the isothermal surface at point *P*, the rate of heat conduction at that point can be expressed by Fourier's law as

$$\dot{Q}_n = -kA\frac{\partial T}{\partial n} \qquad (W) \qquad (2\text{–}2)$$

In rectangular coordinates, the heat conduction vector can be expressed in terms of its components as

$$\vec{Q}_n = \dot{Q}_x \vec{i} + \dot{Q}_y \vec{j} + \dot{Q}_z \vec{k} \qquad (2\text{–}3)$$

where $\vec{i}$, $\vec{j}$, and $\vec{k}$ are the unit vectors, and $\dot{Q}_x$, $\dot{Q}_y$, and $\dot{Q}_z$ are the magnitudes of the heat transfer rates in the *x*-, *y*-, and *z*-directions, which again can be determined from Fourier's law as

$$\dot{Q}_x = -kA_x\frac{\partial T}{\partial x}, \qquad \dot{Q}_y = -kA_y\frac{\partial T}{\partial y}, \qquad \text{and} \qquad \dot{Q}_z = -kA_z\frac{\partial T}{\partial z} \qquad (2\text{–}4)$$

Here A_x, A_y and A_z are heat conduction areas normal to the *x*-, *y*-, and *z*-directions, respectively (Fig. 2–8).

Most engineering materials are *isotropic* in nature, and thus they have the same properties in all directions. For such materials we do not need to be concerned about the variation of properties with direction. But in *anisotropic* materials such as the fibrous or composite materials, the properties may change with direction. For example, some of the properties of wood along the grain are different than those in the direction normal to the grain. In such cases the thermal conductivity may need to be expressed as a tensor quantity to account for the variation with direction. The treatment of such advanced topics is beyond the scope of this text, and we will assume the thermal conductivity of a material to be independent of direction.

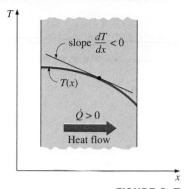

FIGURE 2–7

The temperature gradient dT/dx is simply the slope of the temperature curve on a *T-x* diagram.

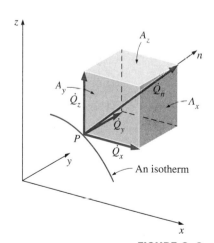

FIGURE 2–8

The heat transfer vector is always normal to an isothermal surface and can be resolved into its components like any other vector.

FIGURE 2–9
Heat is generated in the heating coils
of an electric range as a result of the
conversion of electrical energy to heat.

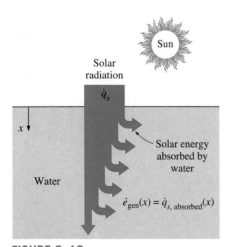

FIGURE 2–10
The absorption of solar radiation by
water can be treated as heat
generation.

Heat Generation

A medium through which heat is conducted may involve the conversion of
mechanical, electrical, nuclear, or chemical energy into heat (or thermal energy). In heat conduction analysis, such conversion processes are characterized as **heat** (or **thermal energy**) **generation**.

For example, the temperature of a resistance wire rises rapidly when electric current passes through it as a result of the electrical energy being converted to heat at a rate of I^2R, where I is the current and R is the electrical
resistance of the wire (Fig. 2–9). The safe and effective removal of this heat
away from the sites of heat generation (the electronic circuits) is the subject
of *electronics cooling,* which is one of the modern application areas of heat
transfer.

Likewise, a large amount of heat is generated in the fuel elements of nuclear
reactors as a result of nuclear fission that serves as the *heat source* for the nuclear power plants. The natural disintegration of radioactive elements in nuclear waste or other radioactive material also results in the generation of heat
throughout the body. The heat generated in the sun as a result of the fusion of
hydrogen into helium makes the sun a large nuclear reactor that supplies heat
to the earth.

Another source of heat generation in a medium is exothermic chemical reactions that may occur throughout the medium. The chemical reaction in this
case serves as a *heat source* for the medium. In the case of endothermic reactions, however, heat is absorbed instead of being released during reaction, and
thus the chemical reaction serves as a *heat sink*. The heat generation term becomes a negative quantity in this case.

Often it is also convenient to model the absorption of radiation such as solar energy or gamma rays as heat generation when these rays penetrate deep
into the body while being absorbed gradually. For example, the absorption of
solar energy in large bodies of water can be treated as heat generation
throughout the water at a rate equal to the rate of absorption, which varies with
depth (Fig. 2–10). But the absorption of solar energy by an opaque body
occurs within a few microns of the surface, and the solar energy that penetrates into the medium in this case can be treated as specified heat flux on the
surface.

Note that heat generation is a *volumetric phenomenon*. That is, it occurs
throughout the body of a medium. Therefore, the rate of heat generation in a
medium is usually specified *per unit volume* and is denoted by $\dot{e}_{gen}$, whose unit
is W/m^3 or Btu/h · ft^3.

The rate of heat generation in a medium may vary with time as well as position within the medium. When the variation of heat generation with position
is known, the *total* rate of heat generation in a medium of volume V can be determined from

$$\dot{E}_{gen} = \int_V \dot{e}_{gen} dV \qquad \text{(W)} \qquad \text{(2–5)}$$

In the special case of *uniform* heat generation, as in the case of electric resistance heating throughout a homogeneous material, the relation in Eq. 2–5
reduces to $\dot{E}_{gen} = \dot{e}_{gen}V$, where $\dot{e}_{gen}$ is the constant rate of heat generation per
unit volume.

EXAMPLE 2–1   Heat Gain by a Refrigerator

In order to size the compressor of a new refrigerator, it is desired to determine the rate of heat transfer from the kitchen air into the refrigerated space through the walls, door, and the top and bottom section of the refrigerator (Fig. 2–11). In your analysis, would you treat this as a transient or steady-state heat transfer problem? Also, would you consider the heat transfer to be one-dimensional or multidimensional? Explain.

SOLUTION   Heat transfer from the kitchen air to a refrigerator is considered. It is to be determined whether this heat transfer is steady or transient, and whether it is one- or multidimensional.

Analysis   The heat transfer process from the kitchen air to the refrigerated space is transient in nature since the thermal conditions in the kitchen and the refrigerator, in general, change with time. However, we would analyze this problem as a steady heat transfer problem under the worst anticipated conditions such as the lowest thermostat setting for the refrigerated space, and the anticipated highest temperature in the kitchen (the so-called design conditions). If the compressor is large enough to keep the refrigerated space at the desired temperature setting under the presumed worst conditions, then it is large enough to do so under all conditions by cycling on and off.

Heat transfer into the refrigerated space is three-dimensional in nature since heat will be entering through all six sides of the refrigerator. However, heat transfer through any wall or floor takes place in the direction normal to the surface, and thus it can be analyzed as being one-dimensional. Therefore, this problem can be simplified greatly by considering the heat transfer to be one-dimensional at each of the four sides as well as the top and bottom sections, and then by adding the calculated values of heat transfer at each surface.

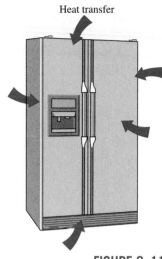

Heat transfer

FIGURE 2–11
Schematic for Example 2–1.

EXAMPLE 2–2   Heat Generation in a Hair Dryer

The resistance wire of a 1200-W hair dryer is 80 cm long and has a diameter of $D = 0.3$ cm (Fig. 2–12). Determine the rate of heat generation in the wire per unit volume, in W/cm³, and the heat flux on the outer surface of the wire as a result of this heat generation.

SOLUTION   The power consumed by the resistance wire of a hair dryer is given. The heat generation and the heat flux are to be determined.

Assumptions   Heat is generated uniformly in the resistance wire.

Analysis   A 1200-W hair dryer converts electrical energy into heat in the wire at a rate of 1200 W. Therefore, the rate of heat generation in a resistance wire is equal to the power consumption of a resistance heater. Then the rate of heat generation in the wire per unit volume is determined by dividing the total rate of heat generation by the volume of the wire,

$$\dot{e}_{gen} = \frac{\dot{E}_{gen}}{V_{wire}} = \frac{\dot{E}_{gen}}{(\pi D^2/4)L} = \frac{1200 \text{ W}}{[\pi(0.3 \text{ cm})^2/4](80 \text{ cm})} = \textbf{212 W/cm}^3$$

Similarly, heat flux on the outer surface of the wire as a result of this heat generation is determined by dividing the total rate of heat generation by the surface area of the wire,

Hair dryer
1200 W

FIGURE 2–12
Schematic for Example 2–2.

$$\dot{Q}_s = \frac{\dot{E}_{gen}}{A_{wire}} = \frac{\dot{E}_{gen}}{\pi DL} = \frac{1200 \text{ W}}{\pi (0.3 \text{ cm})(80 \text{ cm})} = \textbf{15.9 W/cm}^2$$

Discussion Note that heat generation is expressed per unit volume in W/cm^3 or Btu/h $\cdot$ ft^3, whereas heat flux is expressed per unit surface area in W/cm^2 or Btu/h $\cdot$ ft^2.

2–2 ▪ ONE-DIMENSIONAL HEAT CONDUCTION EQUATION

Consider heat conduction through a large plane wall such as the wall of a house, the glass of a single pane window, the metal plate at the bottom of a pressing iron, a cast-iron steam pipe, a cylindrical nuclear fuel element, an electrical resistance wire, the wall of a spherical container, or a spherical metal ball that is being quenched or tempered. Heat conduction in these and many other geometries can be approximated as being *one-dimensional* since heat conduction through these geometries is dominant in one direction and negligible in other directions. Next we develop the one-dimensional heat conduction equation in rectangular, cylindrical, and spherical coordinates.

Heat Conduction Equation in a Large Plane Wall

Consider a thin element of thickness Δx in a large plane wall, as shown in Fig. 2–13. Assume the density of the wall is ρ, the specific heat is c, and the area of the wall normal to the direction of heat transfer is A. An *energy balance* on this thin element during a small time interval Δt can be expressed as

$$\begin{pmatrix} \text{Rate of heat} \\ \text{conduction} \\ \text{at } x \end{pmatrix} - \begin{pmatrix} \text{Rate of heat} \\ \text{conduction} \\ \text{at } x + \Delta x \end{pmatrix} + \begin{pmatrix} \text{Rate of heat} \\ \text{generation} \\ \text{inside the} \\ \text{element} \end{pmatrix} = \begin{pmatrix} \text{Rate of change} \\ \text{of the energy} \\ \text{content of the} \\ \text{element} \end{pmatrix}$$

or

$$\dot{Q}_x - \dot{Q}_{x + \Delta x} + \dot{E}_{gen, \, element} = \frac{\Delta E_{element}}{\Delta t} \qquad \textbf{(2–6)}$$

But the change in the energy content of the element and the rate of heat generation within the element can be expressed as

$$\Delta E_{element} = E_{t + \Delta t} - E_t = mc(T_{t + \Delta t} - T_t) = \rho c A \Delta x (T_{t + \Delta t} - T_t) \qquad \textbf{(2–7)}$$

$$\dot{E}_{gen, \, element} = \dot{e}_{gen} V_{element} = \dot{e}_{gen} A \Delta x \qquad \textbf{(2–8)}$$

Substituting into Eq. 2–6, we get

$$\dot{Q}_x - \dot{Q}_{x + \Delta x} + \dot{e}_{gen} A \Delta x = \rho c A \Delta x \frac{T_{t + \Delta t} - T_t}{\Delta t} \qquad \textbf{(2–9)}$$

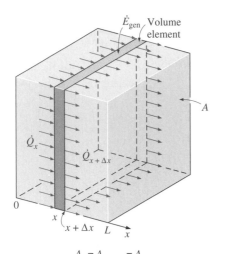

FIGURE 2–13
One-dimensional heat conduction through a volume element in a large plane wall.

$$A_x = A_{x + \Delta x} = A$$

Dividing by $A\Delta x$ gives

$$-\frac{1}{A}\frac{\dot{Q}_{x+\Delta x} - \dot{Q}_x}{\Delta x} + \dot{e}_{\text{gen}} = \rho c\frac{T_{t+\Delta t} - T_t}{\Delta t} \tag{2-10}$$

Taking the limit as $\Delta x \to 0$ and $\Delta t \to 0$ yields

$$\frac{1}{A}\frac{\partial}{\partial x}\left(kA\frac{\partial T}{\partial x}\right) + \dot{e}_{\text{gen}} = \rho c\frac{\partial T}{\partial t} \tag{2-11}$$

since, from the definition of the derivative and Fourier's law of heat conduction,

$$\lim_{\Delta x \to 0}\frac{\dot{Q}_{x+\Delta x} - \dot{Q}_x}{\Delta x} = \frac{\partial \dot{Q}}{\partial x} = \frac{\partial}{\partial x}\left(-kA\frac{\partial T}{\partial x}\right) \tag{2-12}$$

Noting that the area A is constant for a plane wall, the one-dimensional transient heat conduction equation in a plane wall becomes

Variable conductivity:
$$\frac{\partial}{\partial x}\left(k\frac{\partial T}{\partial x}\right) + \dot{e}_{\text{gen}} = \rho c\frac{\partial T}{\partial t} \tag{2-13}$$

The thermal conductivity k of a material, in general, depends on the temperature T (and therefore x), and thus it cannot be taken out of the derivative. However, the *thermal conductivity* in most practical applications can be assumed to remain *constant* at some average value. The equation above in that case reduces to

Constant conductivity:
$$\frac{\partial^2 T}{\partial x^2} + \frac{\dot{e}_{\text{gen}}}{k} = \frac{1}{\alpha}\frac{\partial T}{\partial t} \tag{2-14}$$

where the property $\alpha = k/\rho c$ is the **thermal diffusivity** of the material and represents how fast heat propagates through a material. It reduces to the following forms under specified conditions (Fig. 2–14):

(1) *Steady-state:*
(∂/∂t = 0)
$$\frac{d^2 T}{dx^2} + \frac{\dot{e}_{\text{gen}}}{k} = 0 \tag{2-15}$$

(2) *Transient, no heat generation:*
($\dot{e}_{\text{gen}} = 0$)
$$\frac{\partial^2 T}{\partial x^2} = \frac{1}{\alpha}\frac{\partial T}{\partial t} \tag{2-16}$$

(3) *Steady-state, no heat generation:*
(∂/∂t = 0 and $\dot{e}_{\text{gen}} = 0$)
$$\frac{d^2 T}{dx^2} = 0 \tag{2-17}$$

Note that we replaced the partial derivatives by ordinary derivatives in the one-dimensional steady heat conduction case since the partial and ordinary derivatives of a function are identical when the function depends on a single variable only [$T = T(x)$ in this case].

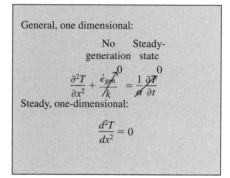

General, one dimensional:

$$\frac{\partial^2 T}{\partial x^2} + \underset{\overset{\uparrow}{0}}{\frac{\dot{e}_{\text{gen}}}{k}} = \underset{\overset{\uparrow}{0}}{\frac{1}{\alpha}\frac{\partial T}{\partial t}}$$

No generation   Steady-state

Steady, one-dimensional:

$$\frac{d^2 T}{dx^2} = 0$$

FIGURE 2–14
The simplification of the one-dimensional heat conduction equation in a plane wall for the case of constant conductivity for steady conduction with no heat generation.

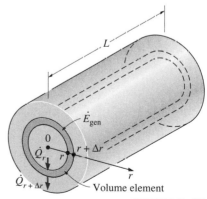

FIGURE 2–15
One-dimensional heat conduction through a volume element in a long cylinder.

Heat Conduction Equation in a Long Cylinder

Now consider a thin cylindrical shell element of thickness Δr in a long cylinder, as shown in Fig. 2–15. Assume the density of the cylinder is ρ, the specific heat is c, and the length is L. The area of the cylinder normal to the direction of heat transfer at any location is $A = 2\pi r L$ where r is the value of the radius at that location. Note that the heat transfer area A depends on r in this case, and thus it varies with location. An *energy balance* on this thin cylindrical shell element during a small time interval Δt can be expressed as

$$
\begin{pmatrix} \text{Rate of heat} \\ \text{conduction} \\ \text{at } r \end{pmatrix} - \begin{pmatrix} \text{Rate of heat} \\ \text{conduction} \\ \text{at } r + \Delta r \end{pmatrix} + \begin{pmatrix} \text{Rate of heat} \\ \text{generation} \\ \text{inside the} \\ \text{element} \end{pmatrix} = \begin{pmatrix} \text{Rate of change} \\ \text{of the energy} \\ \text{content of the} \\ \text{element} \end{pmatrix}
$$

or

$$
\dot{Q}_r - \dot{Q}_{r + \Delta r} + \dot{E}_{\text{gen, element}} = \frac{\Delta E_{\text{element}}}{\Delta t} \tag{2–18}
$$

The change in the energy content of the element and the rate of heat generation within the element can be expressed as

$$
\Delta E_{\text{element}} = E_{t + \Delta t} - E_t = mc(T_{t + \Delta t} - T_t) = \rho c A \Delta r (T_{t + \Delta t} - T_t) \tag{2–19}
$$

$$
\dot{E}_{\text{gen, element}} = \dot{e}_{\text{gen}} V_{\text{element}} = \dot{e}_{\text{gen}} A \Delta r \tag{2–20}
$$

Substituting into Eq. 2–18, we get

$$
\dot{Q}_r - \dot{Q}_{r + \Delta r} + \dot{e}_{\text{gen}} A \Delta r = \rho c A \Delta r \frac{T_{t + \Delta t} - T_t}{\Delta t} \tag{2–21}
$$

where $A = 2\pi r L$. You may be tempted to express the area at the *middle* of the element using the *average* radius as $A = 2\pi(r + \Delta r/2)L$. But there is nothing we can gain from this complication since later in the analysis we will take the limit as $\Delta r \to 0$ and thus the term $\Delta r/2$ will drop out. Now dividing the equation above by $A\Delta r$ gives

$$
-\frac{1}{A} \frac{\dot{Q}_{r + \Delta r} - \dot{Q}_r}{\Delta r} + \dot{e}_{\text{gen}} = \rho c \frac{T_{t + \Delta t} - T_t}{\Delta t} \tag{2–22}
$$

Taking the limit as $\Delta r \to 0$ and $\Delta t \to 0$ yields

$$
\frac{1}{A} \frac{\partial}{\partial r}\left(kA \frac{\partial T}{\partial r} \right) + \dot{e}_{\text{gen}} = \rho c \frac{\partial T}{\partial t} \tag{2–23}
$$

since, from the definition of the derivative and Fourier's law of heat conduction,

$$
\lim_{\Delta r \to 0} \frac{\dot{Q}_{r + \Delta r} - \dot{Q}_r}{\Delta r} = \frac{\partial \dot{Q}}{\partial r} = \frac{\partial}{\partial r}\left(-kA \frac{\partial T}{\partial r} \right) \tag{2–24}
$$

Noting that the heat transfer area in this case is $A = 2\pi r L$, the one-dimensional transient heat conduction equation in a cylinder becomes

Variable conductivity:
$$
\frac{1}{r} \frac{\partial}{\partial r}\left(rk \frac{\partial T}{\partial r} \right) + \dot{e}_{\text{gen}} = \rho c \frac{\partial T}{\partial t} \tag{2–25}
$$

For the case of constant thermal conductivity, the previous equation reduces to

Constant conductivity:
$$\frac{1}{r}\frac{\partial}{\partial r}\left(r\frac{\partial T}{\partial r}\right) + \frac{\dot{e}_{gen}}{k} = \frac{1}{\alpha}\frac{\partial T}{\partial t} \qquad (2\text{–}26)$$

where again the property $\alpha = k/\rho c$ is the thermal diffusivity of the material. Eq. 2–26 reduces to the following forms under specified conditions (Fig. 2–16):

(1) *Steady-state:*
($\partial/\partial t = 0$)
$$\frac{1}{r}\frac{d}{dr}\left(r\frac{dT}{dr}\right) + \frac{\dot{e}_{gen}}{k} = 0 \qquad (2\text{–}27)$$

(2) *Transient, no heat generation:*
($\dot{e}_{gen} = 0$)
$$\frac{1}{r}\frac{\partial}{\partial r}\left(r\frac{\partial T}{\partial r}\right) = \frac{1}{\alpha}\frac{\partial T}{\partial t} \qquad (2\text{–}28)$$

(3) *Steady-state, no heat generation:*
($\partial/\partial t = 0$ and $e_{gen} = 0$)
$$\frac{d}{dr}\left(r\frac{dT}{dr}\right) = 0 \qquad (2\text{–}29)$$

Note that we again replaced the partial derivatives by ordinary derivatives in the one-dimensional steady heat conduction case since the partial and ordinary derivatives of a function are identical when the function depends on a single variable only [$T = T(r)$ in this case].

Heat Conduction Equation in a Sphere

Now consider a sphere with density ρ, specific heat c, and outer radius R. The area of the sphere normal to the direction of heat transfer at any location is $A = 4\pi r^2$, where r is the value of the radius at that location. Note that the heat transfer area A depends on r in this case also, and thus it varies with location. By considering a thin spherical shell element of thickness Δr and repeating the approach described above for the cylinder by using $A = 4\pi r^2$ instead of $A = 2\pi rL$, the one-dimensional transient heat conduction equation for a sphere is determined to be (Fig. 2–17)

Variable conductivity:
$$\frac{1}{r^2}\frac{\partial}{\partial r}\left(r^2 k\frac{\partial T}{\partial r}\right) + \dot{e}_{gen} = \rho c\frac{\partial T}{\partial t} \qquad (2\text{–}30)$$

which, in the case of constant thermal conductivity, reduces to

Constant conductivity:
$$\frac{1}{r^2}\frac{\partial}{\partial r}\left(r^2\frac{\partial T}{\partial r}\right) + \frac{\dot{e}_{gen}}{k} = \frac{1}{\alpha}\frac{\partial T}{\partial t} \qquad (2\text{–}31)$$

where again the property $\alpha = k/\rho c$ is the thermal diffusivity of the material. It reduces to the following forms under specified conditions:

(1) *Steady-state:*
($\partial/\partial t = 0$)
$$\frac{1}{r^2}\frac{d}{dr}\left(r^2\frac{dT}{dr}\right) + \frac{\dot{e}_{gen}}{k} = 0 \qquad (2\text{–}32)$$

(2) *Transient,
no heat generation:*
($\dot{e}_{gen} = 0$)
$$\frac{1}{r^2}\frac{\partial}{\partial r}\left(r^2\frac{\partial T}{\partial r}\right) = \frac{1}{\alpha}\frac{\partial T}{\partial t} \qquad (2\text{–}33)$$

(3) *Steady-state,
no heat generation:*
($\partial/\partial t = 0$ and $\dot{e}_{gen} = 0$)
$$\frac{d}{dr}\left(r^2\frac{dT}{dr}\right) = 0 \qquad \text{or} \qquad r\frac{d^2T}{dr^2} + 2\frac{dT}{dr} = 0 \qquad (2\text{–}34)$$

where again we replaced the partial derivatives by ordinary derivatives in the one-dimensional steady heat conduction case.

(a) The form that is ready to integrate
$$\frac{d}{dr}\left(r\frac{dT}{dr}\right) = 0$$

(b) The equivalent alternative form
$$r\frac{d^2T}{dr^2} + \frac{dT}{dr} = 0$$

FIGURE 2–16
Two equivalent forms of the differential equation for the one-dimensional steady heat conduction in a cylinder with no heat generation.

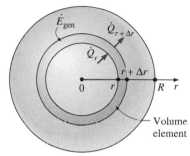

FIGURE 2–17
One-dimensional heat conduction through a volume element in a sphere.

Combined One-Dimensional Heat Conduction Equation

An examination of the one-dimensional transient heat conduction equations for the plane wall, cylinder, and sphere reveals that all three equations can be expressed in a compact form as

$$\frac{1}{r^n}\frac{\partial}{\partial r}\left(r^n k \frac{\partial T}{\partial r}\right) + \dot{e}_{\text{gen}} = \rho c \frac{\partial T}{\partial t} \tag{2–35}$$

where $n = 0$ for a plane wall, $n = 1$ for a cylinder, and $n = 2$ for a sphere. In the case of a plane wall, it is customary to replace the variable r by x. This equation can be simplified for steady-state or no heat generation cases as described before.

800 W

FIGURE 2–18
Schematic for Example 2–3.

EXAMPLE 2–3   Heat Conduction through the Bottom of a Pan

Consider a steel pan placed on top of an electric range to cook spaghetti (Fig. 2–18). The bottom section of the pan is 0.4 cm thick and has a diameter of 18 cm. The electric heating unit on the range top consumes 800 W of power during cooking, and 80 percent of the heat generated in the heating element is transferred uniformly to the pan. Assuming constant thermal conductivity, obtain the differential equation that describes the variation of the temperature in the bottom section of the pan during steady operation.

SOLUTION   A steel pan placed on top of an electric range is considered. The differential equation for the variation of temperature in the bottom of the pan is to be obtained.

Analysis   The bottom section of the pan has a large surface area relative to its thickness and can be approximated as a large plane wall. Heat flux is applied to the bottom surface of the pan uniformly, and the conditions on the inner surface are also uniform. Therefore, we expect the heat transfer through the bottom section of the pan to be from the bottom surface toward the top, and heat transfer in this case can reasonably be approximated as being one-dimensional. Taking the direction normal to the bottom surface of the pan to be the x-axis, we will have $T = T(x)$ during steady operation since the temperature in this case will depend on x only.

The thermal conductivity is given to be constant, and there is no heat generation in the medium (within the bottom section of the pan). Therefore, the differential equation governing the variation of temperature in the bottom section of the pan in this case is simply Eq. 2–17,

$$\frac{d^2T}{dx^2} = 0$$

which is the steady one-dimensional heat conduction equation in rectangular coordinates under the conditions of constant thermal conductivity and no heat generation.

Discussion   Note that the conditions at the surface of the medium have no effect on the differential equation.

EXAMPLE 2–4   Heat Conduction in a Resistance Heater

A 2-kW resistance heater wire with thermal conductivity $k = 15$ W/m · K, diameter $D = 0.4$ cm, and length $L = 50$ cm is used to boil water by immersing it in water (Fig. 2–19). Assuming the variation of the thermal conductivity of the wire with temperature to be negligible, obtain the differential equation that describes the variation of the temperature in the wire during steady operation.

SOLUTION   The resistance wire of a water heater is considered. The differential equation for the variation of temperature in the wire is to be obtained.
Analysis   The resistance wire can be considered to be a very long cylinder since its length is more than 100 times its diameter. Also, heat is generated uniformly in the wire and the conditions on the outer surface of the wire are uniform. Therefore, it is reasonable to expect the temperature in the wire to vary in the radial r direction only and thus the heat transfer to be one-dimensional. Then we have $T = T(r)$ during steady operation since the temperature in this case depends on r only.

The rate of heat generation in the wire per unit volume can be determined from

$$\dot{e}_{gen} = \frac{\dot{E}_{gen}}{V_{wire}} = \frac{\dot{E}_{gen}}{(\pi D^2/4)L} = \frac{2000 \text{ W}}{[\pi(0.004 \text{ m})^2/4](0.5 \text{ m})} = 0.318 \times 10^9 \text{ W/m}^3$$

Noting that the thermal conductivity is given to be constant, the differential equation that governs the variation of temperature in the wire is simply Eq. 2–27,

$$\frac{1}{r}\frac{d}{dr}\left(r\frac{dT}{dr}\right) + \frac{\dot{e}_{gen}}{k} = 0$$

which is the steady one-dimensional heat conduction equation in cylindrical coordinates for the case of constant thermal conductivity.
Discussion   Note again that the conditions at the surface of the wire have no effect on the differential equation.

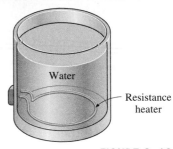

FIGURE 2–19
Schematic for Example 2–4.

EXAMPLE 2–5   Cooling of a Hot Metal Ball in Air

A spherical metal ball of radius R is heated in an oven to a temperature of 600°F throughout and is then taken out of the oven and allowed to cool in ambient air at $T_\infty = 75$°F by convection and radiation (Fig. 2–20). The thermal conductivity of the ball material is known to vary linearly with temperature. Assuming the ball is cooled uniformly from the entire outer surface, obtain the differential equation that describes the variation of the temperature in the ball during cooling.

SOLUTION   A hot metal ball is allowed to cool in ambient air. The differential equation for the variation of temperature within the ball is to be obtained.
Analysis   The ball is initially at a uniform temperature and is cooled uniformly from the entire outer surface. Also, the temperature at any point in the ball changes with time during cooling. Therefore, this is a one-dimensional transient heat conduction problem since the temperature within the ball changes with the radial distance r and the time t. That is, $T = T(r, t)$.

The thermal conductivity is given to be variable, and there is no heat generation in the ball. Therefore, the differential equation that governs the variation

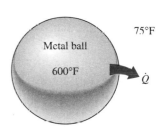

FIGURE 2–20
Schematic for Example 2–5.

of temperature in the ball in this case is obtained from Eq. 2–30 by setting the heat generation term equal to zero. We obtain

$$\frac{1}{r^2}\frac{\partial}{\partial r}\left(r^2 k \frac{\partial T}{\partial r}\right) = \rho c \frac{\partial T}{\partial t}$$

which is the one-dimensional transient heat conduction equation in spherical coordinates under the conditions of variable thermal conductivity and no heat generation.

Discussion Note again that the conditions at the outer surface of the ball have no effect on the differential equation.

2–3 ▪ GENERAL HEAT CONDUCTION EQUATION

In the last section we considered one-dimensional heat conduction and assumed heat conduction in other directions to be negligible. Most heat transfer problems encountered in practice can be approximated as being one-dimensional, and we mostly deal with such problems in this text. However, this is not always the case, and sometimes we need to consider heat transfer in other directions as well. In such cases heat conduction is said to be *multidimensional,* and in this section we develop the governing differential equation in such systems in rectangular, cylindrical, and spherical coordinate systems.

Rectangular Coordinates

Consider a small rectangular element of length Δx, width Δy, and height Δz, as shown in Fig. 2–21. Assume the density of the body is ρ and the specific heat is c. An *energy balance* on this element during a small time interval Δt can be expressed as

$$\begin{pmatrix}\text{Rate of heat}\\\text{conduction at}\\x, y, \text{ and } z\end{pmatrix} - \begin{pmatrix}\text{Rate of heat}\\\text{conduction}\\\text{at } x + \Delta x,\\y + \Delta y, \text{and } z + \Delta z\end{pmatrix} + \begin{pmatrix}\text{Rate of heat}\\\text{generation}\\\text{inside the}\\\text{element}\end{pmatrix} = \begin{pmatrix}\text{Rate of change}\\\text{of the energy}\\\text{content of}\\\text{the element}\end{pmatrix}$$

or

$$\dot{Q}_x + \dot{Q}_y + \dot{Q}_z - \dot{Q}_{x + \Delta x} - \dot{Q}_{y + \Delta y} - \dot{Q}_{z + \Delta z} + \dot{E}_{\text{gen, element}} = \frac{\Delta E_{\text{element}}}{\Delta t} \qquad \textbf{(2–36)}$$

Noting that the volume of the element is $V_{\text{element}} = \Delta x \Delta y \Delta z$, the change in the energy content of the element and the rate of heat generation within the element can be expressed as

$$\Delta E_{\text{element}} = E_{t + \Delta t} - E_t = mc(T_{t + \Delta t} - T_t) = \rho c \Delta x \Delta y \Delta z (T_{t + \Delta t} - T_t)$$
$$\dot{E}_{\text{gen, element}} = \dot{e}_{\text{gen}} V_{\text{element}} = \dot{e}_{\text{gen}} \Delta x \Delta y \Delta z$$

Substituting into Eq. 2–36, we get

$$\dot{Q}_x + \dot{Q}_y + \dot{Q}_z - \dot{Q}_{x + \Delta x} - \dot{Q}_{y + \Delta y} - \dot{Q}_{z + \Delta z} + \dot{e}_{\text{gen}} \Delta x \Delta y \Delta z = \rho c \Delta x \Delta y \Delta z \frac{T_{t+\Delta t} - T_t}{\Delta t}$$

Dividing by $\Delta x \Delta y \Delta z$ gives

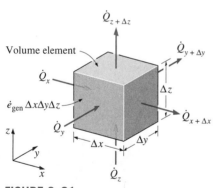

FIGURE 2–21
Three-dimensional heat conduction through a rectangular volume element.

$$-\frac{1}{\Delta y \Delta z}\frac{\dot{Q}_{x+\Delta x}-\dot{Q}_x}{\Delta x}-\frac{1}{\Delta x \Delta z}\frac{\dot{Q}_{y+\Delta y}-\dot{Q}_y}{\Delta y}-\frac{1}{\Delta x \Delta y}\frac{\dot{Q}_{z+\Delta z}-\dot{Q}_z}{\Delta z}+\dot{e}_{gen}=$$

$$\rho c \frac{T_{t+\Delta t}-T_t}{\Delta t} \tag{2-37}$$

Noting that the heat transfer areas of the element for heat conduction in the x, y, and z directions are $A_x = \Delta y \Delta z$, $A_y = \Delta x \Delta z$, and $A_z = \Delta x \Delta y$, respectively, and taking the limit as Δx, Δy, Δz and $\Delta t \to 0$ yields

$$\frac{\partial}{\partial x}\left(k\frac{\partial T}{\partial x}\right)+\frac{\partial}{\partial y}\left(k\frac{\partial T}{\partial y}\right)+\frac{\partial}{\partial z}\left(k\frac{\partial T}{\partial z}\right)+\dot{e}_{gen}=\rho c\frac{\partial T}{\partial t} \tag{2-38}$$

since, from the definition of the derivative and Fourier's law of heat conduction,

$$\lim_{\Delta x \to 0}\frac{1}{\Delta y \Delta z}\frac{\dot{Q}_{x+\Delta x}-\dot{Q}_x}{\Delta x}=\frac{1}{\Delta y \Delta z}\frac{\partial Q_x}{\partial x}=\frac{1}{\Delta y \Delta z}\frac{\partial}{\partial x}\left(-k\Delta y \Delta z\frac{\partial T}{\partial x}\right)=-\frac{\partial}{\partial x}\left(k\frac{\partial T}{\partial x}\right)$$

$$\lim_{\Delta y \to 0}\frac{1}{\Delta x \Delta z}\frac{\dot{Q}_{y+\Delta y}-\dot{Q}_y}{\Delta y}=\frac{1}{\Delta x \Delta z}\frac{\partial Q_y}{\partial y}=\frac{1}{\Delta x \Delta z}\frac{\partial}{\partial y}\left(-k\Delta x \Delta z\frac{\partial T}{\partial y}\right)=-\frac{\partial}{\partial y}\left(k\frac{\partial T}{\partial y}\right)$$

$$\lim_{\Delta z \to 0}\frac{1}{\Delta x \Delta y}\frac{\dot{Q}_{z+\Delta z}-\dot{Q}_z}{\Delta z}=\frac{1}{\Delta x \Delta y}\frac{\partial Q_z}{\partial z}=\frac{1}{\Delta x \Delta y}\frac{\partial}{\partial z}\left(-k\Delta x \Delta y\frac{\partial T}{\partial z}\right)=-\frac{\partial}{\partial z}\left(k\frac{\partial T}{\partial z}\right)$$

Eq. 2–38 is the general heat conduction equation in rectangular coordinates. In the case of constant thermal conductivity, it reduces to

$$\frac{\partial^2 T}{\partial x^2}+\frac{\partial^2 T}{\partial y^2}+\frac{\partial^2 T}{\partial z^2}+\frac{\dot{e}_{gen}}{k}=\frac{1}{\alpha}\frac{\partial T}{\partial t} \tag{2-39}$$

where the property $\alpha = k/\rho c$ is again the *thermal diffusivity* of the material. Eq. 2–39 is known as the **Fourier-Biot equation**, and it reduces to these forms under specified conditions:

(1) *Steady-state:* (called the **Poisson equation**)

$$\frac{\partial^2 T}{\partial x^2}+\frac{\partial^2 T}{\partial y^2}+\frac{\partial^2 T}{\partial z^2}+\frac{\dot{e}_{gen}}{k}=0 \tag{2-40}$$

(2) *Transient, no heat generation:* (called the **diffusion equation**)

$$\frac{\partial^2 T}{\partial x^2}+\frac{\partial^2 T}{\partial y^2}+\frac{\partial^2 T}{\partial z^2}=\frac{1}{\alpha}\frac{\partial T}{\partial t} \tag{2-41}$$

(3) *Steady-state, no heat generation:* (called the **Laplace equation**)

$$\frac{\partial^2 T}{\partial x^2}+\frac{\partial^2 T}{\partial y^2}+\frac{\partial^2 T}{\partial z^2}=0 \tag{2-42}$$

Note that in the special case of one-dimensional heat transfer in the x-direction, the derivatives with respect to y and z drop out and the equations above reduce to the ones developed in the previous section for a plane wall (Fig. 2–22).

Cylindrical Coordinates

The general heat conduction equation in cylindrical coordinates can be obtained from an energy balance on a volume element in cylindrical coordinates, shown in Fig. 2–23, by following the steps just outlined. It can also be obtained directly from Eq. 2–38 by coordinate transformation using the

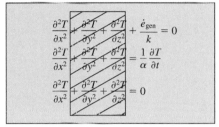

FIGURE 2–22
The three-dimensional heat conduction equations reduce to the one-dimensional ones when the temperature varies in one dimension only.

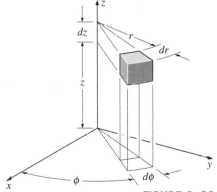

FIGURE 2–23
A differential volume element in cylindrical coordinates.

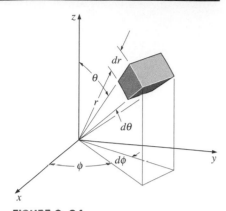

FIGURE 2–24
A differential volume element in spherical coordinates.

following relations between the coordinates of a point in rectangular and cylindrical coordinate systems:

$$x = r \cos \phi, \qquad y = r \sin \phi, \qquad \text{and} \qquad z = z$$

After lengthy manipulations, we obtain

$$\frac{1}{r} \frac{\partial}{\partial r}\left(kr \frac{\partial T}{\partial r}\right) + \frac{1}{r^2} \frac{\partial T}{\partial \phi}\left(k \frac{\partial T}{\partial \phi}\right) + \frac{\partial}{\partial z}\left(k \frac{\partial T}{\partial z}\right) + \dot{e}_{gen} = \rho c \frac{\partial T}{\partial t} \qquad \textbf{(2–43)}$$

Spherical Coordinates

The general heat conduction equations in spherical coordinates can be obtained from an energy balance on a volume element in spherical coordinates, shown in Fig. 2–24, by following the steps outlined above. It can also be obtained directly from Eq. 2–38 by coordinate transformation using the following relations between the coordinates of a point in rectangular and spherical coordinate systems:

$$x = r \cos \phi \sin \theta, \qquad y = r \sin \phi \sin \theta, \qquad \text{and} \qquad z = \cos \theta$$

Again after lengthy manipulations, we obtain

$$\frac{1}{r^2} \frac{\partial}{\partial r}\left(kr^2 \frac{\partial T}{\partial r}\right) + \frac{1}{r^2 \sin^2 \theta} \frac{\partial}{\partial \phi}\left(k \frac{\partial T}{\partial \phi}\right) + \frac{1}{r^2 \sin \theta} \frac{\partial}{\partial \theta}\left(k \sin \theta \frac{\partial T}{\partial \theta}\right) + \dot{e}_{gen} = \rho c \frac{\partial T}{\partial t}$$

$$\textbf{(2–44)}$$

Obtaining analytical solutions to these differential equations requires a knowledge of the solution techniques of partial differential equations, which is beyond the scope of this introductory text. Here we limit our consideration to one-dimensional steady-state cases, since they result in ordinary differential equations.

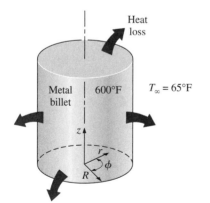

FIGURE 2–25
Schematic for Example 2–6.

EXAMPLE 2–6   Heat Conduction in a Short Cylinder

A short cylindrical metal billet of radius R and height h is heated in an oven to a temperature of 600°F throughout and is then taken out of the oven and allowed to cool in ambient air at $T_\infty = 65°F$ by convection and radiation. Assuming the billet is cooled uniformly from all outer surfaces and the variation of the thermal conductivity of the material with temperature is negligible, obtain the differential equation that describes the variation of the temperature in the billet during this cooling process.

SOLUTION   A short cylindrical billet is cooled in ambient air. The differential equation for the variation of temperature is to be obtained.
Analysis   The billet shown in Fig. 2–25 is initially at a uniform temperature and is cooled uniformly from the top and bottom surfaces in the z-direction as well as the lateral surface in the radial r-direction. Also, the temperature at any point in the ball changes with time during cooling. Therefore, this is a two-dimensional transient heat conduction problem since the temperature within the billet changes with the radial and axial distances r and z and with time t. That is, $T = T(r, z, t)$.

The thermal conductivity is given to be constant, and there is no heat generation in the billet. Therefore, the differential equation that governs the variation of temperature in the billet in this case is obtained from Eq. 2–43 by setting the heat generation term and the derivatives with respect to ϕ equal to zero. We obtain

$$\frac{1}{r}\frac{\partial}{\partial r}\left(kr\frac{\partial T}{\partial r}\right) + \frac{\partial}{\partial z}\left(k\frac{\partial T}{\partial z}\right) = \rho c\frac{\partial T}{\partial t}$$

In the case of constant thermal conductivity, it reduces to

$$\frac{1}{r}\frac{\partial}{\partial r}\left(r\frac{\partial T}{\partial r}\right) + \frac{\partial^2 T}{\partial z^2} = \frac{1}{\alpha}\frac{\partial T}{\partial t}$$

which is the desired equation.

Discussion Note that the boundary and initial conditions have no effect on the differential equation.

2–4 · BOUNDARY AND INITIAL CONDITIONS

The heat conduction equations above were developed using an energy balance on a differential element inside the medium, and they remain the same regardless of the *thermal conditions* on the *surfaces* of the medium. That is, the differential equations do not incorporate any information related to the conditions on the surfaces such as the surface temperature or a specified heat flux. Yet we know that the heat flux and the temperature distribution in a medium depend on the conditions at the surfaces, and the description of a heat transfer problem in a medium is not complete without a full description of the thermal conditions at the bounding surfaces of the medium. The *mathematical expressions* of the thermal conditions at the boundaries are called the **boundary conditions**.

From a mathematical point of view, solving a differential equation is essentially a process of *removing derivatives*, or an *integration* process, and thus the solution of a differential equation typically involves arbitrary constants (Fig. 2–26). It follows that to obtain a unique solution to a problem, we need to specify more than just the governing differential equation. We need to specify some conditions (such as the value of the function or its derivatives at some value of the independent variable) so that forcing the solution to satisfy these conditions at specified points will result in unique values for the arbitrary constants and thus a *unique solution*. But since the differential equation has no place for the additional information or conditions, we need to supply them separately in the form of boundary or initial conditions.

Consider the variation of temperature along the wall of a brick house in winter. The temperature at any point in the wall depends on, among other things, the conditions at the two surfaces of the wall such as the air temperature of the house, the velocity and direction of the winds, and the solar energy incident on the outer surface. That is, the temperature distribution in a medium depends on the conditions at the boundaries of the medium as well as the heat transfer mechanism inside the medium. To describe a heat transfer problem completely, *two boundary conditions* must be given for *each direction* of the coordinate system along which heat transfer is significant (Fig. 2–27). Therefore, we need to

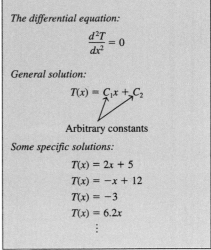

FIGURE 2–26

The general solution of a typical differential equation involves arbitrary constants, and thus an infinite number of solutions.

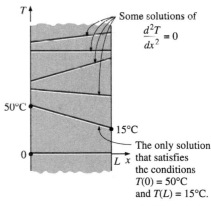

FIGURE 2–27

To describe a heat transfer problem completely, two boundary conditions must be given for each direction along which heat transfer is significant.

specify *two boundary conditions* for one-dimensional problems, *four boundary conditions* for two-dimensional problems, and *six boundary conditions* for three-dimensional problems. In the case of the wall of a house, for example, we need to specify the conditions at two locations (the inner and the outer surfaces) of the wall since heat transfer in this case is one-dimensional. But in the case of a parallelepiped, we need to specify six boundary conditions (one at each face) when heat transfer in all three dimensions is significant.

The physical argument presented above is consistent with the mathematical nature of the problem since the heat conduction equation is second order (i.e., involves second derivatives with respect to the space variables) in all directions along which heat conduction is significant, and the general solution of a second-order linear differential equation involves two arbitrary constants for each direction. That is, the number of boundary conditions that needs to be specified in a direction is equal to the order of the differential equation in that direction.

Reconsider the brick wall already discussed. The temperature at any point on the wall at a specified time also depends on the condition of the wall at the beginning of the heat conduction process. Such a condition, which is usually specified at time $t = 0$, is called the **initial condition**, which is a mathematical expression for the temperature distribution of the medium initially. Note that we need only one initial condition for a heat conduction problem regardless of the dimension since the conduction equation is first order in time (it involves the first derivative of temperature with respect to time).

In rectangular coordinates, the initial condition can be specified in the general form as

$$T(x, y, z, 0) = f(x, y, z) \tag{2–45}$$

where the function $f(x, y, z)$ represents the temperature distribution throughout the medium at time $t = 0$. When the medium is initially at a uniform temperature of T_i, the initial condition in Eq. 2–45 can be expressed as $T(x, y, z, 0) = T_i$. Note that under *steady* conditions, the heat conduction equation does not involve any time derivatives, and thus we do not need to specify an initial condition.

The heat conduction equation is first order in time, and thus the initial condition cannot involve any derivatives (it is limited to a specified temperature). However, the heat conduction equation is second order in space coordinates, and thus a boundary condition may involve first derivatives at the boundaries as well as specified values of temperature. Boundary conditions most commonly encountered in practice are the *specified temperature, specified heat flux, convection,* and *radiation* boundary conditions.

1   Specified Temperature Boundary Condition

The *temperature* of an exposed surface can usually be measured directly and easily. Therefore, one of the easiest ways to specify the thermal conditions on a surface is to specify the temperature. For one-dimensional heat transfer through a plane wall of thickness L, for example, the specified temperature boundary conditions can be expressed as (Fig. 2–28)

$$T(0, t) = T_1$$
$$T(L, t) = T_2 \tag{2–46}$$

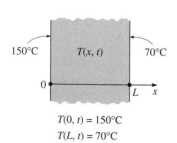

$T(0, t) = 150°C$

$T(L, t) = 70°C$

FIGURE 2–28
Specified temperature boundary conditions on both surfaces of a plane wall.

where T_1 and T_2 are the specified temperatures at surfaces at $x = 0$ and $x = L$, respectively. The specified temperatures can be constant, which is the case for steady heat conduction, or may vary with time.

2 Specified Heat Flux Boundary Condition

When there is sufficient information about energy interactions at a surface, it may be possible to determine the rate of heat transfer and thus the *heat flux* $\dot{q}$ (heat transfer rate per unit surface area, W/m²) on that surface, and this information can be used as one of the boundary conditions. The heat flux in the positive x-direction anywhere in the medium, including the boundaries, can be expressed by *Fourier's law* of heat conduction as

$$\dot{q} = -k \frac{\partial T}{\partial x} = \left(\begin{array}{c} \text{Heat flux in the} \\ \text{positive } x\text{-direction} \end{array} \right) \quad \text{(W/m}^2\text{)} \qquad \textbf{(2–47)}$$

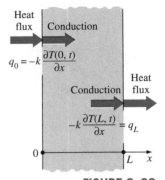

$$q_0 = -k \frac{\partial T(0, t)}{\partial x}$$

$$-k \frac{\partial T(L, t)}{\partial x} = q_L$$

FIGURE 2–29
Specified heat flux boundary conditions on both surfaces of a plane wall.

Then the boundary condition at a boundary is obtained by setting the specified heat flux equal to $-k(\partial T/\partial x)$ at that boundary. The sign of the specified heat flux is determined by inspection: *positive* if the heat flux is in the positive direction of the coordinate axis, and *negative* if it is in the opposite direction. Note that it is extremely important to have the *correct sign* for the specified heat flux since the wrong sign will invert the direction of heat transfer and cause the heat gain to be interpreted as heat loss (Fig. 2–29).

For a plate of thickness L subjected to heat flux of 50 W/m² into the medium from both sides, for example, the specified heat flux boundary conditions can be expressed as

$$-k \frac{\partial T(0, t)}{\partial x} = 50 \quad \text{and} \quad -k \frac{\partial T(L, t)}{\partial x} = -50 \qquad \textbf{(2–48)}$$

Note that the heat flux at the surface at $x = L$ is in the *negative* x-direction, and thus it is -50 W/m².

Special Case: Insulated Boundary

Some surfaces are commonly insulated in practice in order to minimize heat loss (or heat gain) through them. Insulation reduces heat transfer but does not totally eliminate it unless its thickness is infinity. However, heat transfer through a properly insulated surface can be taken to be zero since adequate insulation reduces heat transfer through a surface to negligible levels. Therefore, a well-insulated surface can be modeled as a surface with a specified heat flux of zero. Then the boundary condition on a perfectly insulated surface (at $x = 0$, for example) can be expressed as (Fig. 2–30)

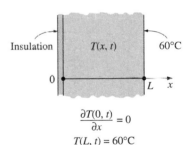

$$\frac{\partial T(0, t)}{\partial x} = 0$$

$$T(L, t) = 60°C$$

FIGURE 2–30
A plane wall with insulation and specified temperature boundary conditions.

$$k \frac{\partial T(0, t)}{\partial x} = 0 \quad \text{or} \quad \frac{\partial T(0, t)}{\partial x} = 0 \qquad \textbf{(2–49)}$$

That is, *on an insulated surface, the first derivative of temperature with respect to the space variable (the temperature gradient) in the direction normal to the insulated surface is zero.* This also means that the temperature function must be perpendicular to an insulated surface since the slope of temperature at the surface must be zero.

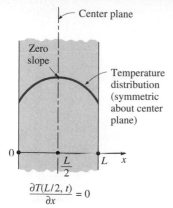

$$\frac{\partial T(L/2, t)}{\partial x} = 0$$

FIGURE 2–31
Thermal symmetry boundary condition at the center plane of a plane wall.

Another Special Case: Thermal Symmetry

Some heat transfer problems possess *thermal symmetry* as a result of the symmetry in imposed thermal conditions. For example, the two surfaces of a large hot plate of thickness L suspended vertically in air is subjected to the same thermal conditions, and thus the temperature distribution in one half of the plate is the same as that in the other half. That is, the heat transfer problem in this plate possesses thermal symmetry about the center plane at $x = L/2$. Also, the direction of heat flow at any point in the plate is toward the surface closer to the point, and there is no heat flow across the center plane. Therefore, the center plane can be viewed as an insulated surface, and the thermal condition at this plane of symmetry can be expressed as (Fig. 2–31)

$$\frac{\partial T(L/2, t)}{\partial x} = 0 \qquad (2\text{–}50)$$

which resembles the *insulation* or *zero heat flux* boundary condition. This result can also be deduced from a plot of temperature distribution with a maximum, and thus zero slope, at the center plane.

In the case of cylindrical (or spherical) bodies having thermal symmetry about the center line (or midpoint), the thermal symmetry boundary condition requires that the first derivative of temperature with respect to r (the radial variable) be zero at the centerline (or the midpoint).

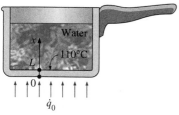

FIGURE 2–32
Schematic for Example 2–7.

EXAMPLE 2–7   Heat Flux Boundary Condition

Consider an aluminum pan used to cook beef stew on top of an electric range. The bottom section of the pan is $L = 0.3$ cm thick and has a diameter of $D = 20$ cm. The electric heating unit on the range top consumes 800 W of power during cooking, and 90 percent of the heat generated in the heating element is transferred to the pan. During steady operation, the temperature of the inner surface of the pan is measured to be 110°C. Express the boundary conditions for the bottom section of the pan during this cooking process.

SOLUTION An aluminum pan on an electric range top is considered. The boundary conditions for the bottom of the pan are to be obtained.
Analysis The heat transfer through the bottom section of the pan is from the bottom surface toward the top and can reasonably be approximated as being one-dimensional. We take the direction normal to the bottom surfaces of the pan as the x axis with the origin at the outer surface, as shown in Fig. 2–32. Then the inner and outer surfaces of the bottom section of the pan can be represented by $x = 0$ and $x = L$, respectively. During steady operation, the temperature will depend on x only and thus $T = T(x)$.

The boundary condition on the outer surface of the bottom of the pan at $x = 0$ can be approximated as being specified heat flux since it is stated that 90 percent of the 800 W (i.e., 720 W) is transferred to the pan at that surface. Therefore,

$$-k\frac{dT(0)}{dx} = \dot{q}_0$$

where

$$\dot{q}_0 = \frac{\text{Heat transfer rate}}{\text{Bottom surface area}} = \frac{0.720 \text{ kW}}{\pi(0.1 \text{ m})^2} = 22.9 \text{ kW/m}^2$$

The temperature at the inner surface of the bottom of the pan is specified to be 110°C. Then the boundary condition on this surface can be expressed as

$$T(L) = 110°C$$

where $L = 0.003$ m.

Discussion Note that the determination of the boundary conditions may require some reasoning and approximations.

3 Convection Boundary Condition

Convection is probably the most common boundary condition encountered in practice since most heat transfer surfaces are exposed to an environment at a specified temperature. The convection boundary condition is based on a *surface energy balance* expressed as

$$\begin{pmatrix} \text{Heat conduction} \\ \text{at the surface in a} \\ \text{selected direction} \end{pmatrix} = \begin{pmatrix} \text{Heat convection} \\ \text{at the surface in} \\ \text{the same direction} \end{pmatrix}$$

For one-dimensional heat transfer in the x-direction in a plate of thickness L, the convection boundary conditions on both surfaces can be expressed as

$$-k\frac{\partial T(0, t)}{\partial x} = h_1[T_{\infty 1} - T(0, t)] \qquad \textbf{(2-51a)}$$

and

$$-k\frac{\partial T(L, t)}{\partial x} = h_2[T(L, t) - T_{\infty 2}] \qquad \textbf{(2-51b)}$$

where h_1 and h_2 are the convection heat transfer coefficients and $T_{\infty 1}$ and $T_{\infty 2}$ are the temperatures of the surrounding mediums on the two sides of the plate, as shown in Fig. 2–33.

In writing Eqs. 2–51 for convection boundary conditions, we have selected the direction of heat transfer to be the positive x-direction at both surfaces. But those expressions are equally applicable when heat transfer is in the opposite direction at one or both surfaces since reversing the direction of heat transfer at a surface simply reverses the signs of *both* conduction and convection terms at that surface. This is equivalent to multiplying an equation by -1, which has no effect on the equality (Fig. 2–34). Being able to select either direction as the direction of heat transfer is certainly a relief since often we do not know the surface temperature and thus the direction of heat transfer at a surface in advance. This argument is also valid for other boundary conditions such as the radiation and combined boundary conditions discussed shortly.

Note that a surface has zero thickness and thus no mass, and it cannot store any energy. Therefore, the entire net heat entering the surface from one side must leave the surface from the other side. The convection boundary condition simply states that heat continues to flow from a body to the surrounding medium at the same rate, and it just changes vehicles at the surface from conduction to convection (or vice versa in the other direction). This is analogous to people traveling on buses on land and transferring to the ships at the shore. If the passengers are not allowed to wander around at the shore, then the rate at which

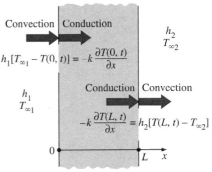

FIGURE 2–33

Convection boundary conditions on the two surfaces of a plane wall.

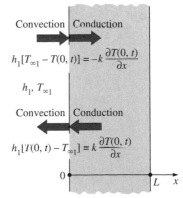

FIGURE 2–34

The assumed direction of heat transfer at a boundary has no effect on the boundary condition expression.

the people are unloaded at the shore from the buses must equal the rate at which they board the ships. We may call this the conservation of "people" principle.

Also note that the surface temperatures $T(0, t)$ and $T(L, t)$ are not known (if they were known, we would simply use them as the specified temperature boundary condition and not bother with convection). But a surface temperature can be determined once the solution $T(x, t)$ is obtained by substituting the value of x at that surface into the solution.

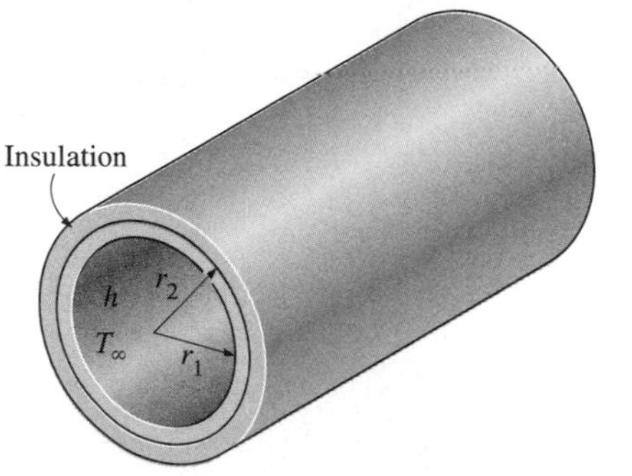

FIGURE 2–35
Schematic for Example 2–8.

EXAMPLE 2–8   **Convection and Insulation Boundary Conditions**

Steam flows through a pipe shown in Fig. 2–35 at an average temperature of $T_\infty = 200°C$. The inner and outer radii of the pipe are $r_1 = 8$ cm and $r_2 = 8.5$ cm, respectively, and the outer surface of the pipe is heavily insulated. If the convection heat transfer coefficient on the inner surface of the pipe is $h = 65$ W/m^2 · K, express the boundary conditions on the inner and outer surfaces of the pipe during transient periods.

SOLUTION   The flow of steam through an insulated pipe is considered. The boundary conditions on the inner and outer surfaces of the pipe are to be obtained.

Analysis   During initial transient periods, heat transfer through the pipe material predominantly is in the radial direction, and thus can be approximated as being one-dimensional. Then the temperature within the pipe material changes with the radial distance r and the time t. That is, $T = T(r, t)$.

It is stated that heat transfer between the steam and the pipe at the inner surface is by convection. Then taking the direction of heat transfer to be the positive r direction, the boundary condition on that surface can be expressed as

$$-k\frac{\partial T(r_1, t)}{\partial r} = h[T_\infty - T(r_1)]$$

The pipe is said to be well insulated on the outside, and thus heat loss through the outer surface of the pipe can be assumed to be negligible. Then the boundary condition at the outer surface can be expressed as

$$\frac{\partial T(r_2, t)}{\partial r} = 0$$

Discussion   Note that the temperature gradient must be zero on the outer surface of the pipe at all times.

---

4   Radiation Boundary Condition

In some cases, such as those encountered in space and cryogenic applications, a heat transfer surface is surrounded by an evacuated space and thus there is no convection heat transfer between a surface and the surrounding medium. In such cases, *radiation* becomes the only mechanism of heat transfer between the surface under consideration and the surroundings. Using an energy balance, the radiation boundary condition on a surface can be expressed as

$$\begin{pmatrix} \text{Heat conduction} \\ \text{at the surface in a} \\ \text{selected direction} \end{pmatrix} = \begin{pmatrix} \text{Radiation exchange} \\ \text{at the surface in} \\ \text{the same direction} \end{pmatrix}$$

For one-dimensional heat transfer in the x-direction in a plate of thickness L, the radiation boundary conditions on both surfaces can be expressed as (Fig. 2–36)

$$-k\frac{\partial T(0, t)}{\partial x} = \varepsilon_1\sigma[T_{surr, 1}^4 - T(0, t)^4] \tag{2–52a}$$

and

$$-k\frac{\partial T(L, t)}{\partial x} = \varepsilon_2\sigma[T(L, t)^4 - T_{surr, 2}^4] \tag{2–52b}$$

where ε_1 and ε_2 are the emissivities of the boundary surfaces, $\sigma = 5.67 \times 10^{-8}\ \text{W/m}^2 \cdot \text{K}^4$ is the Stefan–Boltzmann constant, and $T_{surr, 1}$ and $T_{surr, 2}$ are the average temperatures of the surfaces surrounding the two sides of the plate, respectively. Note that the temperatures in radiation calculations must be expressed in K or R (not in °C or °F).

The radiation boundary condition involves the fourth power of temperature, and thus it is a *nonlinear* condition. As a result, the application of this boundary condition results in powers of the unknown coefficients, which makes it difficult to determine them. Therefore, it is tempting to ignore radiation exchange at a surface during a heat transfer analysis in order to avoid the complications associated with nonlinearity. This is especially the case when heat transfer at the surface is dominated by convection, and the role of radiation is minor.

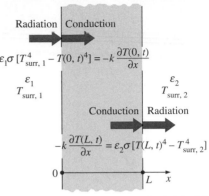

FIGURE 2–36
Radiation boundary conditions on both surfaces of a plane wall.

5   Interface Boundary Conditions

Some bodies are made up of layers of different materials, and the solution of a heat transfer problem in such a medium requires the solution of the heat transfer problem in each layer. This, in turn, requires the specification of the boundary conditions at each *interface*.

The boundary conditions at an interface are based on the requirements that (1) two bodies in contact must have the *same temperature* at the area of contact and (2) an interface (which is a surface) cannot store any energy, and thus the *heat flux* on the two sides of an interface *must be the same*. The boundary conditions at the interface of two bodies A and B in perfect contact at $x = x_0$ can be expressed as (Fig. 2–37)

$$T_A(x_0, t) = T_B(x_0, t) \tag{2–53}$$

and

$$-k_A\frac{\partial T_A(x_0, t)}{\partial x} = -k_B\frac{\partial T_B(x_0, t)}{\partial x} \tag{2–54}$$

where k_A and k_B are the thermal conductivities of the layers A and B, respectively. The case of imperfect contact results in thermal contact resistance, which is considered in the next chapter.

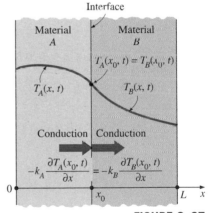

FIGURE 2–37
Boundary conditions at the interface of two bodies in perfect contact.

6 Generalized Boundary Conditions

So far we have considered surfaces subjected to *single mode* heat transfer, such as the specified heat flux, convection, or radiation for simplicity. In general, however, a surface may involve convection, radiation, *and* specified heat flux simultaneously. The boundary condition in such cases is again obtained from a surface energy balance, expressed as

$$\begin{pmatrix} \text{Heat transfer} \\ \text{to the surface} \\ \text{in all modes} \end{pmatrix} = \begin{pmatrix} \text{Heat transfer} \\ \text{from the surface} \\ \text{in all modes} \end{pmatrix} \qquad \textbf{(2–55)}$$

This is illustrated in Examples 2–9 and 2–10.

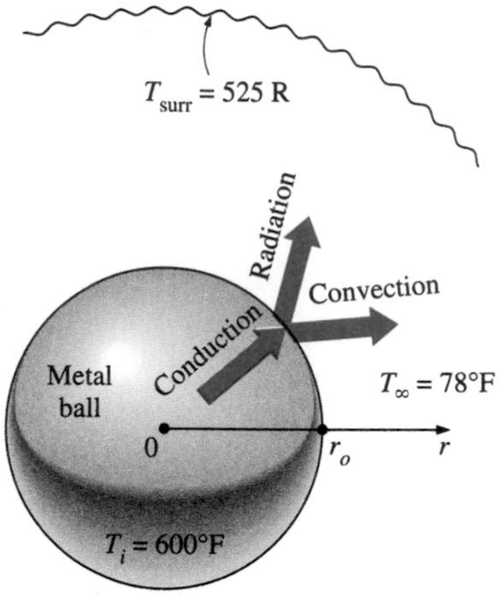

$T_{\text{surr}} = 525$ R

Radiation

Convection

Conduction

Metal ball

$T_\infty = 78°F$

0   r_o r

$T_i = 600°F$

FIGURE 2–38
Schematic for Example 2–9.

EXAMPLE 2–9   Combined Convection and Radiation Condition

A spherical metal ball of radius r_o is heated in an oven to a temperature of 600°F throughout and is then taken out of the oven and allowed to cool in ambient air at $T_\infty = 78°F$, as shown in Fig. 2–38. The thermal conductivity of the ball material is $k = 8.3$ Btu/h · ft · R, and the average convection heat transfer coefficient on the outer surface of the ball is evaluated to be $h = 4.5$ Btu/h · ft² · R. The emissivity of the outer surface of the ball is $\varepsilon = 0.6$, and the average temperature of the surrounding surfaces is $T_{\text{surr}} = 525$ R. Assuming the ball is cooled uniformly from the entire outer surface, express the initial and boundary conditions for the cooling process of the ball.

SOLUTION   The cooling of a hot spherical metal ball is considered. The initial and boundary conditions are to be obtained.

Analysis   The ball is initially at a uniform temperature and is cooled uniformly from the entire outer surface. Therefore, this is a one-dimensional transient heat transfer problem since the temperature within the ball changes with the radial distance r and the time t. That is, $T = T(r, t)$. Taking the moment the ball is removed from the oven to be $t = 0$, the initial condition can be expressed as

$$T(r, 0) = T_i = 600°F$$

The problem possesses symmetry about the midpoint $(r = 0)$ since the isotherms in this case are concentric spheres, and thus no heat is crossing the midpoint of the ball. Then the boundary condition at the midpoint can be expressed as

$$\frac{\partial T(0, t)}{\partial r} = 0$$

The heat conducted to the outer surface of the ball is lost to the environment by convection and radiation. Then taking the direction of heat transfer to be the positive r direction, the boundary condition on the outer surface can be expressed as

$$-k\frac{\partial T(r_o, t)}{\partial r} = h[T(r_o) - T_\infty] + \varepsilon\sigma[T(r_o)^4 - T_{\text{surr}}^4]$$

Discussion   All the quantities in the above relations are known except the temperatures and their derivatives at $r = 0$ and r_o. Also, the radiation part of the boundary condition is often ignored for simplicity by modifying the convection heat transfer coefficient to account for the contribution of radiation. The convection coefficient h in that case becomes the combined heat transfer coefficient.

EXAMPLE 2-10 **Combined Convection, Radiation, and Heat Flux**

Consider the south wall of a house that is $L = 0.2$ m thick. The outer surface of the wall is exposed to solar radiation and has an absorptivity of $\alpha = 0.5$ for solar energy. The interior of the house is maintained at $T_{\infty 1} = 20°C$, while the ambient air temperature outside remains at $T_{\infty 2} = 5°C$. The sky, the ground, and the surfaces of the surrounding structures at this location can be modeled as a surface at an effective temperature of $T_{sky} = 255$ K for radiation exchange on the outer surface. The radiation exchange between the inner surface of the wall and the surfaces of the walls, floor, and ceiling it faces is negligible. The convection heat transfer coefficients on the inner and the outer surfaces of the wall are $h_1 = 6$ W/m$^2 \cdot$ °C and $h_2 = 25$ W/m$^2 \cdot$ °C, respectively. The thermal conductivity of the wall material is $k = 0.7$ W/m $\cdot$ °C, and the emissivity of the outer surface is $\varepsilon_2 = 0.9$. Assuming the heat transfer through the wall to be steady and one-dimensional, express the boundary conditions on the inner and the outer surfaces of the wall.

SOLUTION The wall of a house subjected to solar radiation is considered. The boundary conditions on the inner and outer surfaces of the wall are to be obtained.

Analysis We take the direction normal to the wall surfaces as the x-axis with the origin at the inner surface of the wall, as shown in Fig. 2–39. The heat transfer through the wall is given to be steady and one-dimensional, and thus the temperature depends on x only and not on time. That is, $T = T(x)$.

The boundary condition on the inner surface of the wall at $x = 0$ is a typical convection condition since it does not involve any radiation or specified heat flux. Taking the direction of heat transfer to be the positive x-direction, the boundary condition on the inner surface can be expressed as

$$-k\frac{dT(0)}{dx} = h_1[T_{\infty 1} - T(0)]$$

The boundary condition on the outer surface at $x = 0$ is quite general as it involves conduction, convection, radiation, and specified heat flux. Again taking the direction of heat transfer to be the positive x-direction, the boundary condition on the outer surface can be expressed as

$$-k\frac{dT(L)}{dx} = h_2[T(L) - T_{\infty 2}] + \varepsilon_2\sigma[T(L)^4 - T_{sky}^4] - \alpha\dot{q}_{solar}$$

where $\dot{q}_{solar}$ is the incident solar heat flux.

Discussion Assuming the opposite direction for heat transfer would give the same result multiplied by -1, which is equivalent to the relation here. All the quantities in these relations are known except the temperatures and their derivatives at the two boundaries.

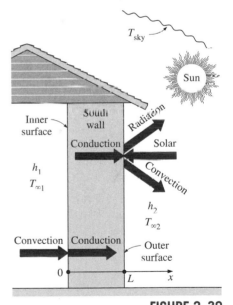

FIGURE 2–39
Schematic for Example 2–10.

Note that a heat transfer problem may involve different kinds of boundary conditions on different surfaces. For example, a plate may be subject to *heat flux* on one surface while losing or gaining heat by *convection* from the other surface. Also, the two boundary conditions in a direction may be specified *at the same boundary*, while no condition is imposed on the other boundary. For example, specifying the temperature and heat flux at $x = 0$ of a plate of thickness L will result in a unique solution for the one-dimensional steady temperature distribution in the plate, including the value of temperature at the surface $x = L$. Although not necessary, there is nothing wrong with specifying more than two boundary conditions in a specified direction, provided that there is no contradiction. The extra conditions in this case can be used to verify the results.

2–5 ▪ SOLUTION OF STEADY ONE-DIMENSIONAL HEAT CONDUCTION PROBLEMS

So far we have derived the differential equations for heat conduction in various coordinate systems and discussed the possible boundary conditions. A heat conduction problem can be formulated by specifying the applicable differential equation and a set of proper boundary conditions.

In this section we will solve a wide range of heat conduction problems in rectangular, cylindrical, and spherical geometries. We will limit our attention to problems that result in *ordinary differential equations* such as the *steady one-dimensional* heat conduction problems. We will also assume *constant thermal conductivity,* but will consider variable conductivity later in this chapter. If you feel rusty on differential equations or haven't taken differential equations yet, no need to panic. *Simple integration* is all you need to solve the steady one-dimensional heat conduction problems.

The solution procedure for solving heat conduction problems can be summarized as (1) *formulate* the problem by obtaining the applicable differential equation in its simplest form and specifying the boundary conditions, (2) obtain the *general solution* of the differential equation, and (3) apply the *boundary conditions* and determine the arbitrary constants in the general solution (Fig. 2–40). This is demonstrated below with examples.

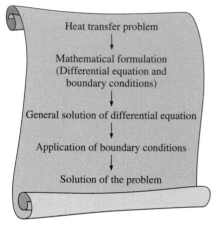

FIGURE 2–40
Basic steps involved in the solution of heat transfer problems.

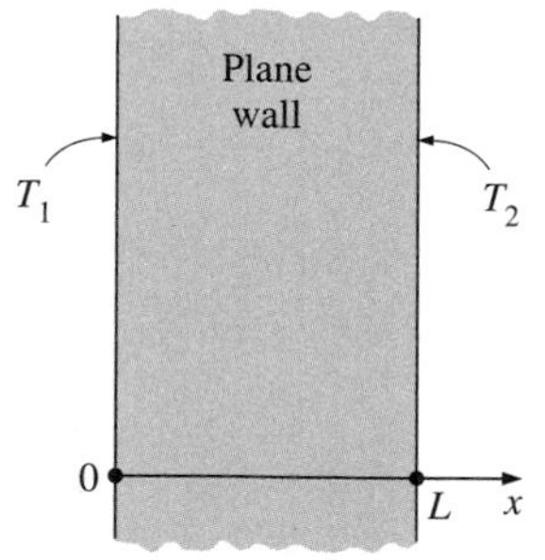

FIGURE 2–41
Schematic for Example 2–11.

EXAMPLE 2–11   **Heat Conduction in a Plane Wall**

Consider a large plane wall of thickness $L = 0.2$ m, thermal conductivity $k = 1.2$ W/m · °C, and surface area $A = 15$ m^2. The two sides of the wall are maintained at constant temperatures of $T_1 = 120$°C and $T_2 = 50$°C, respectively, as shown in Fig. 2–41. Determine (*a*) the variation of temperature within the wall and the value of temperature at $x = 0.1$ m and (*b*) the rate of heat conduction through the wall under steady conditions.

SOLUTION   A plane wall with specified surface temperatures is given. The variation of temperature and the rate of heat transfer are to be determined. **Assumptions 1** Heat conduction is steady. **2** Heat conduction is one-dimensional since the wall is large relative to its thickness and the thermal conditions on both sides are uniform. **3** Thermal conductivity is constant. **4** There is no heat generation.

Properties The thermal conductivity is given to be $k = 1.2$ W/m · °C.

Analysis (*a*) Taking the direction normal to the surface of the wall to be the x-direction, the differential equation for this problem can be expressed as

$$\frac{d^2T}{dx^2} = 0$$

with boundary conditions

$$T(0) = T_1 = 120°C$$
$$T(L) = T_2 = 50°C$$

The differential equation is linear and second order, and a quick inspection of it reveals that it has a single term involving derivatives and no terms involving the unknown function T as a factor. Thus, it can be solved by direct integration. Noting that an integration reduces the order of a derivative by one, the general solution of the differential equation above can be obtained by two simple successive integrations, each of which introduces an integration constant.

Integrating the differential equation once with respect to x yields

$$\frac{dT}{dx} = C_1$$

where C_1 is an arbitrary constant. Notice that the order of the derivative went down by one as a result of integration. As a check, if we take the derivative of this equation, we will obtain the original differential equation. This equation is not the solution yet since it involves a derivative.

Integrating one more time, we obtain

$$T(x) = C_1 x + C_2$$

which is the general solution of the differential equation (Fig. 2–42). The general solution in this case resembles the general formula of a straight line whose slope is C_1 and whose value at $x = 0$ is C_2. This is not surprising since the second derivative represents the change in the slope of a function, and a zero second derivative indicates that the slope of the function remains constant. Therefore, *any straight line* is a solution of this differential equation.

The general solution contains two unknown constants C_1 and C_2, and thus we need two equations to determine them uniquely and obtain the specific solution. These equations are obtained by forcing the general solution to satisfy the specified boundary conditions. The application of each condition yields one equation, and thus we need to specify two conditions to determine the constants C_1 and C_2.

When applying a boundary condition to an equation, *all occurrences of the dependent and independent variables and any derivatives are replaced by the specified values.* Thus the only unknowns in the resulting equations are the arbitrary constants.

The first boundary condition can be interpreted as *in the general solution, replace all the x's by zero and T(x) by T_1*. That is (Fig. 2–43),

$$T(0) = C_1 \times 0 + C_2 \quad \rightarrow \quad C_2 = T_1$$

FIGURE 2–42
Obtaining the general solution of a simple second order differential equation by integration.

FIGURE 2–43
When applying a boundary condition to the general solution at a specified point, all occurrences of the dependent and independent variables should be replaced by their specified values at that point.

The second boundary condition can be interpreted as *in the general solution, replace all the x's by L and T(x) by T_2. That is,*

$$T(L) = C_1 L + C_2 \quad \rightarrow \quad T_2 = C_1 L + T_1 \quad \rightarrow \quad C_1 = \frac{T_2 - T_1}{L}$$

Substituting the C_1 and C_2 expressions into the general solution, we obtain

$$T(x) = \frac{T_2 - T_1}{L} x + T_1 \tag{2–56}$$

which is the desired solution since it satisfies not only the differential equation but also the two specified boundary conditions. That is, differentiating Eq. 2–56 with respect to x twice will give d^2T/dx^2, which is the given differential equation, and substituting $x = 0$ and $x = L$ into Eq. 2–56 gives $T(0) = T_1$ and $T(L) = T_2$, respectively, which are the specified conditions at the boundaries.

Substituting the given information, the value of the temperature at $x = 0.1$ m is determined to be

$$T(0.1 \text{ m}) = \frac{(50 - 120)°C}{0.2 \text{ m}} (0.1 \text{ m}) + 120°C = \textbf{85°C}$$

(*b*) The rate of heat conduction anywhere in the wall is determined from Fourier's law to be

$$\dot{Q}_{wall} = -kA \frac{dT}{dx} = -kAC_1 = -kA \frac{T_2 - T_1}{L} = kA \frac{T_1 - T_2}{L} \tag{2–57}$$

The numerical value of the rate of heat conduction through the wall is determined by substituting the given values to be

$$\dot{Q} = kA \frac{T_1 - T_2}{L} = (1.2 \text{ W/m} \cdot °C)(15 \text{ m}^2) \frac{(120 - 50)°C}{0.2 \text{ m}} = \textbf{6300 W}$$

Discussion Note that under steady conditions, the rate of heat conduction through a plane wall is constant.

EXAMPLE 2–12   **A Wall with Various Sets of Boundary Conditions**

Consider steady one-dimensional heat conduction in a large plane wall of thickness L and constant thermal conductivity k with no heat generation. Obtain expressions for the variation of temperature within the wall for the following pairs of boundary conditions (Fig. 2–44):

(*a*) $\quad -k \dfrac{dT(0)}{dx} = \dot{q}_0 = 40 \text{ W/cm}^2 \quad$ and $\quad T(0) = T_0 = 15°C$

(*b*) $\quad -k \dfrac{dT(0)}{dx} = \dot{q}_0 = 40 \text{ W/cm}^2 \quad$ and $\quad -k \dfrac{dT(L)}{dx} = \dot{q}_L = -25 \text{ W/cm}^2$

(*c*) $\quad -k \dfrac{dT(0)}{dx} = \dot{q}_0 = 40 \text{ W/cm}^2 \quad$ and $\quad -k \dfrac{dT(L)}{dx} = \dot{q}_0 = 40 \text{ W/cm}^2$

SOLUTION Steady one-dimensional heat conduction in a large plane wall is considered. The variation of temperature is to be determined for different sets of boundary conditions.

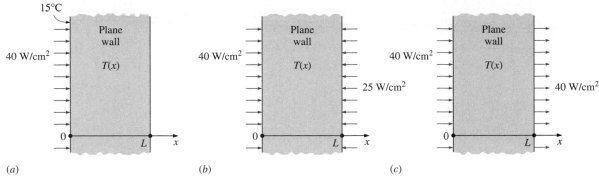

FIGURE 2–44
Schematic for Example 2–12.

Analysis This is a steady one-dimensional heat conduction problem with constant thermal conductivity and no heat generation in the medium, and the heat conduction equation in this case can be expressed as (Eq. 2–17)

$$\frac{d^2T}{dx^2} = 0$$

whose general solution was determined in the previous example by direct integration to be

$$T(x) = C_1 x + C_2$$

where C_1 and C_2 are two arbitrary integration constants. The specific solutions corresponding to each specified pair of boundary conditions are determined as follows.

(*a*) In this case, both boundary conditions are specified at the same boundary at $x = 0$, and no boundary condition is specified at the other boundary at $x = L$. Noting that

$$\frac{dT}{dx} = C_1$$

the application of the boundary conditions gives

$$-k\frac{dT(0)}{dx} = \dot{q}_0 \quad \rightarrow \quad -kC_1 = \dot{q}_0 \quad \rightarrow \quad C_1 = -\frac{\dot{q}_0}{k}$$

and

$$T(0) = T_0 \quad \rightarrow \quad T_0 = C_1 \times 0 + C_2 \quad \rightarrow \quad C_2 = T_0$$

Substituting, the specific solution in this case is determined to be

$$T(x) = -\frac{\dot{q}_0}{k}x + T_0$$

Therefore, the two boundary conditions can be specified at the same boundary, and it is not necessary to specify them at different locations. In fact, the fundamental theorem of linear ordinary differential equations guarantees that a unique solution exists when both conditions are specified at the same location.

Differential equation:
$$T''(x) = 0$$

General solution:
$$T(x) = C_1 x + C_2$$

(a) Unique solution:
$$\left.\begin{array}{r} -kT'(0) = \dot{q}_0 \\ T(0) = T_0 \end{array}\right\} \quad T(x) = -\frac{\dot{q}_0}{k} x + T_0$$

(b) No solution:
$$\left.\begin{array}{r} -kT'(0) = \dot{q}_0 \\ -kT'(L) = \dot{q}_L \end{array}\right\} \quad T(x) = \text{None}$$

(c) Multiple solutions:
$$\left.\begin{array}{r} -kT'(0) = \dot{q}_0 \\ -kT'(L) = \dot{q}_0 \end{array}\right\} \quad T(x) = -\frac{\dot{q}_0}{k} x + \underset{\underset{\text{Arbitrary}}{\uparrow}}{C_2}$$

FIGURE 2–45
A boundary-value problem may have a unique solution, infinitely many solutions, or no solutions at all.

But no such guarantee exists when the two conditions are specified at different boundaries, as you will see below.

(*b*) In this case different heat fluxes are specified at the two boundaries. The application of the boundary conditions gives

$$-k\frac{dT(0)}{dx} = \dot{q}_0 \quad \rightarrow \quad -kC_1 = \dot{q}_0 \quad \rightarrow \quad C_1 = -\frac{\dot{q}_0}{k}$$

and

$$-k\frac{dT(L)}{dx} = \dot{q}_L \quad \rightarrow \quad -kC_1 = \dot{q}_L \quad \rightarrow \quad C_1 = -\frac{\dot{q}_L}{k}$$

Since $\dot{q}_0 \neq \dot{q}_L$ and the constant C_1 cannot be equal to two different things at the same time, there is no solution in this case. This is not surprising since this case corresponds to supplying heat to the plane wall from both sides and expecting the temperature of the wall to remain steady (not to change with time). This is impossible.

(*c*) In this case, the same values for heat flux are specified at the two boundaries. The application of the boundary conditions gives

$$-k\frac{dT(0)}{dx} = \dot{q}_0 \quad \rightarrow \quad -kC_1 = \dot{q}_0 \quad \rightarrow \quad C_1 = -\frac{\dot{q}_0}{k}$$

and

$$-k\frac{dT(L)}{dx} = \dot{q}_0 \quad \rightarrow \quad -kC_1 = \dot{q}_0 \quad \rightarrow \quad C_1 = -\frac{\dot{q}_0}{k}$$

Thus, both conditions result in the same value for the constant C_1, but no value for C_2. Substituting, the specific solution in this case is determined to be

$$T(x) = -\frac{\dot{q}_0}{k} x + C_2$$

which is not a unique solution since C_2 is arbitrary.

Discussion The last solution represents a family of straight lines whose slope is $-\dot{q}_0/k$. Physically, this problem corresponds to requiring the rate of heat supplied to the wall at $x = 0$ be equal to the rate of heat removal from the other side of the wall at $x = L$. But this is a consequence of the heat conduction through the wall being steady, and thus the second boundary condition does not provide any new information. So it is not surprising that the solution of this problem is not unique. The three cases discussed above are summarized in Fig. 2–45.

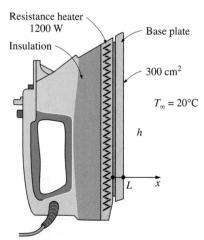

FIGURE 2–46
Schematic for Example 2–13.

EXAMPLE 2–13   **Heat Conduction in the Base Plate of an Iron**

Consider the base plate of a 1200-W household iron that has a thickness of $L = 0.5$ cm, base area of $A = 300$ cm^2, and thermal conductivity of $k = 15$ W/m · °C. The inner surface of the base plate is subjected to uniform heat flux generated by the resistance heaters inside, and the outer surface loses heat to the surroundings at $T_\infty = 20°$C by convection, as shown in Fig. 2–46.

Taking the convection heat transfer coefficient to be $h = 80$ W/m² · °C and disregarding heat loss by radiation, obtain an expression for the variation of temperature in the base plate, and evaluate the temperatures at the inner and the outer surfaces.

SOLUTION The base plate of an iron is considered. The variation of temperature in the plate and the surface temperatures are to be determined.

Assumptions **1** Heat transfer is steady since there is no change with time. **2** Heat transfer is one-dimensional since the surface area of the base plate is large relative to its thickness, and the thermal conditions on both sides are uniform. **3** Thermal conductivity is constant. **4** There is no heat generation in the medium. **5** Heat transfer by radiation is negligible. **6** The upper part of the iron is well insulated so that the entire heat generated in the resistance wires is transferred to the base plate through its inner surface.

Properties The thermal conductivity is given to be $k = 15$ W/m · °C.

Analysis The inner surface of the base plate is subjected to uniform heat flux at a rate of

$$\dot{q}_0 = \frac{\dot{Q}_0}{A_{\text{base}}} = \frac{1200 \text{ W}}{0.03 \text{ m}^2} = 40{,}000 \text{ W/m}^2$$

The outer side of the plate is subjected to the convection condition. Taking the direction normal to the surface of the wall as the x-direction with its origin on the inner surface, the differential equation for this problem can be expressed as (Fig. 2–47)

$$\frac{d^2 T}{dx^2} = 0$$

with the boundary conditions

$$-k\frac{dT(0)}{dx} = \dot{q}_0 = 40{,}000 \text{ W/m}^2$$

$$-k\frac{dT(L)}{dx} = h[T(L) - T_\infty]$$

The general solution of the differential equation is again obtained by two successive integrations to be

$$\frac{dT}{dx} = C_1$$

and

$$T(x) = C_1 x + C_2 \qquad \text{(a)}$$

where C_1 and C_2 are arbitrary constants. Applying the first boundary condition,

$$-k\frac{dT(0)}{dx} = \dot{q}_0 \quad \rightarrow \quad -kC_1 = \dot{q}_0 \quad \rightarrow \quad C_1 = -\frac{\dot{q}_0}{k}$$

Noting that $dT/dx = C_1$ and $T(L) = C_1 L + C_2$, the application of the second boundary condition gives

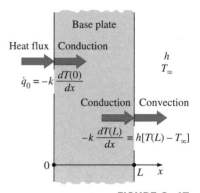

FIGURE 2–47
The boundary conditions on the base plate of the iron discussed in Example 2–13.

$$-k\frac{dT(L)}{dx} = h[T(L) - T_\infty] \quad \rightarrow \quad -kC_1 = h[(C_1L + C_2) - T_\infty]$$

Substituting $C_1 = -\dot{q}_0/k$ and solving for C_2, we obtain

$$C_2 = T_\infty + \frac{\dot{q}_0}{h} + \frac{\dot{q}_0}{k}L$$

Now substituting C_1 and C_2 into the general solution (a) gives

$$T(x) = T_\infty + \dot{q}_0\left(\frac{L - x}{k} + \frac{1}{h}\right) \qquad (b)$$

which is the solution for the variation of the temperature in the plate. The temperatures at the inner and outer surfaces of the plate are determined by substituting $x = 0$ and $x = L$, respectively, into the relation (b):

$$T(0) = T_\infty + \dot{q}_0\left(\frac{L}{k} + \frac{1}{h}\right)$$

$$= 20°C + (40,000 \text{ W/m}^2)\left(\frac{0.005 \text{ m}}{15 \text{ W/m} \cdot °C} + \frac{1}{80 \text{ W/m}^2 \cdot °C}\right) = \textbf{533°C}$$

and

$$T(L) = T_\infty + \dot{q}_0\left(0 + \frac{1}{h}\right) = 20°C + \frac{40,000 \text{ W/m}^2}{80 \text{ W/m}^2 \cdot °C} = \textbf{520°C}$$

Discussion Note that the temperature of the inner surface of the base plate is 13°C higher than the temperature of the outer surface when steady operating conditions are reached. Also note that this heat transfer analysis enables us to calculate the temperatures of surfaces that we cannot even reach. This example demonstrates how the heat flux and convection boundary conditions are applied to heat transfer problems.

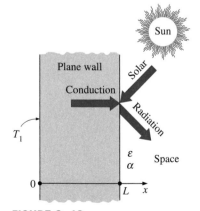

FIGURE 2–48
Schematic for Example 2–14.

EXAMPLE 2–14   Heat Conduction in a Solar Heated Wall

Consider a large plane wall of thickness $L = 0.06$ m and thermal conductivity $k = 1.2$ W/m · °C in space. The wall is covered with white porcelain tiles that have an emissivity of $\varepsilon = 0.85$ and a solar absorptivity of $\alpha = 0.26$, as shown in Fig. 2–48. The inner surface of the wall is maintained at $T_1 = 300$ K at all times, while the outer surface is exposed to solar radiation that is incident at a rate of $\dot{q}_{solar} = 800$ W/m². The outer surface is also losing heat by radiation to deep space at 0 K. Determine the temperature of the outer surface of the wall and the rate of heat transfer through the wall when steady operating conditions are reached. What would your response be if no solar radiation was incident on the surface?

SOLUTION A plane wall in space is subjected to specified temperature on one side and solar radiation on the other side. The outer surface temperature and the rate of heat transfer are to be determined.

Assumptions **1** Heat transfer is steady since there is no change with time. **2** Heat transfer is one-dimensional since the wall is large relative to its thickness, and the thermal conditions on both sides are uniform. **3** Thermal conductivity is constant. **4** There is no heat generation.

Properties The thermal conductivity is given to be $k = 1.2$ W/m · °C.

Analysis Taking the direction normal to the surface of the wall as the x-direction with its origin on the inner surface, the differential equation for this problem can be expressed as

$$\frac{d^2T}{dx^2} = 0$$

with boundary conditions

$$T(0) = T_1 = 300 \text{ K}$$

$$-k\frac{dT(L)}{dx} = \varepsilon\sigma[T(L)^4 - T_{\text{space}}^4] - \alpha\dot{q}_{\text{solar}}$$

where $T_{\text{space}} = 0$. The general solution of the differential equation is again obtained by two successive integrations to be

$$T(x) = C_1 x + C_2 \qquad\qquad\text{(a)}$$

where C_1 and C_2 are arbitrary constants. Applying the first boundary condition yields

$$T(0) = C_1 \times 0 + C_2 \quad\rightarrow\quad C_2 = T_1$$

Noting that $dT/dx = C_1$ and $T(L) = C_1 L + C_2 = C_1 L + T_1$, the application of the second boundary conditions gives

$$-k\frac{dT(L)}{dx} - \varepsilon\sigma T(L)^4 - \alpha\dot{q}_{\text{solar}} \quad\rightarrow\quad -kC_1 = \varepsilon\sigma(C_1 L + T_1)^4 - \alpha\dot{q}_{\text{solar}}$$

Although C_1 is the only unknown in this equation, we cannot get an explicit expression for it because the equation is nonlinear, and thus we cannot get a closed-form expression for the temperature distribution. This should explain why we do our best to avoid nonlinearities in the analysis, such as those associated with radiation.

Let us back up a little and denote the outer surface temperature by $T(L) = T_L$ instead of $T(L) = C_1 L + T_1$. The application of the second boundary condition in this case gives

$$-k\frac{dT(L)}{dx} = \varepsilon\sigma T(L)^4 - \alpha\dot{q}_{\text{solar}} \quad\rightarrow\quad -kC_1 = \varepsilon\sigma T_L^4 - \alpha\dot{q}_{\text{solar}}$$

Solving for C_1 gives

$$C_1 = \frac{\alpha\dot{q}_{\text{solar}} - \varepsilon\sigma T_L^4}{k} \qquad\qquad\text{(b)}$$

Now substituting C_1 and C_2 into the general solution (a), we obtain

$$T(x) = \frac{\alpha\dot{q}_{\text{solar}} - \varepsilon\sigma T_L^4}{k} x + T_1 \qquad\qquad\text{(c)}$$

(1) *Rearrange the equation to be solved:*

$$T_L = 310.4 - 0.240975\left(\frac{T_L}{100}\right)^4$$

The equation is in the proper form since the left side consists of T_L only.

(2) *Guess the value of T_L, say 300 K, and substitute into the right side of the equation. It gives*

$$T_L - 290.2 \text{ K}$$

(3) *Now substitute this value of T_L into the right side of the equation and get*

$$T_L = 293.1 \text{ K}$$

(4) *Repeat step (3) until convergence to desired accuracy is achieved. The subsequent iterations give*

$$T_L = 292.6 \text{ K}$$
$$T_L = 292.7 \text{ K}$$
$$T_L = 292.7 \text{ K}$$

Therefore, the solution is $T_L = 292.7$ K. The result is independent of the initial guess.

FIGURE 2–49
A simple method of solving a nonlinear equation is to arrange the equation such that the unknown is alone on the left side while everything else is on the right side, and to iterate after an initial guess until convergence.

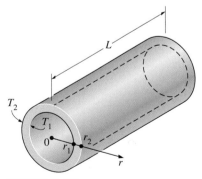

FIGURE 2–50
Schematic for Example 2–15.

which is the solution for the variation of the temperature in the wall in terms of the unknown outer surface temperature T_L. At $x = L$ it becomes

$$T_L = \frac{\alpha \dot{q}_{\text{solar}} - \varepsilon \sigma T_L^4}{k} L + T_1 \qquad (d)$$

which is an implicit relation for the outer surface temperature T_L. Substituting the given values, we get

$$T_L = \frac{0.26 \times (800 \text{ W/m}^2) - 0.85 \times (5.67 \times 10^{\pm 8} \text{ W/m}^2 \cdot \text{K}^4)\, T_L^4}{1.2 \text{ W/m} \cdot \text{K}} (0.06 \text{ m}) + 300 \text{ K}$$

which simplifies to

$$T_L = 310.4 - 0.240975\left(\frac{T_L}{100}\right)^4$$

This equation can be solved by one of the several nonlinear equation solvers available (or by the old fashioned trial-and-error method) to give (Fig. 2–49)

$$T_L = \mathbf{292.7 \text{ K}}$$

Knowing the outer surface temperature and knowing that it must remain constant under steady conditions, the temperature distribution in the wall can be determined by substituting the T_L value above into Eq. (*c*):

$$T(x) = \frac{0.26 \times (800 \text{ W/m}^2) - 0.85 \times (5.67 \times 10^{-8} \text{ W/m}^2 \cdot \text{K}^4)(292.7 \text{ K})^4}{1.2 \text{ W/m} \cdot \text{K}} x + 300 \text{ K}$$

which simplifies to

$$T(x) = (-121.5 \text{ K/m})x + 300 \text{ K}$$

Note that the outer surface temperature turned out to be lower than the inner surface temperature. Therefore, the heat transfer through the wall is toward the outside despite the absorption of solar radiation by the outer surface. Knowing both the inner and outer surface temperatures of the wall, the steady rate of heat conduction through the wall can be determined from

$$\dot{q} = k \frac{T_1 - T_L}{L} = (1.2 \text{ W/m} \cdot \text{K})\frac{(300 - 292.7) \text{ K}}{0.06 \text{ m}} = 146 \text{ W/m}^2$$

Discussion In the case of no incident solar radiation, the outer surface temperature, determined from Eq. (*d*) by setting $\dot{q}_{\text{solar}} = 0$, is $T_L = \mathbf{284.3 \text{ K}}$. It is interesting to note that the solar energy incident on the surface causes the surface temperature to increase by about 8 K only when the inner surface temperature of the wall is maintained at 300 K.

EXAMPLE 2–15   **Heat Loss through a Steam Pipe**

Consider a steam pipe of length $L = 20$ m, inner radius $r_1 = 6$ cm, outer radius $r_2 = 8$ cm, and thermal conductivity $k = 20$ W/m · °C, as shown in Fig. 2–50. The inner and outer surfaces of the pipe are maintained at average temperatures of $T_1 = 150$°C and $T_2 = 60$°C, respectively. Obtain a general relation for the temperature distribution inside the pipe under steady conditions, and determine the rate of heat loss from the steam through the pipe.

SOLUTION A steam pipe is subjected to specified temperatures on its surfaces. The variation of temperature and the rate of heat transfer are to be determined.

Assumptions **1** Heat transfer is steady since there is no change with time. **2** Heat transfer is one-dimensional since there is thermal symmetry about the centerline and no variation in the axial direction, and thus $T = T(r)$. **3** Thermal conductivity is constant. **4** There is no heat generation.

Properties The thermal conductivity is given to be $k = 20$ W/m · °C.

Analysis The mathematical formulation of this problem can be expressed as

$$\frac{d}{dr}\left(r\frac{dT}{dr}\right) = 0$$

with boundary conditions

$$T(r_1) = T_1 = 150°C$$
$$T(r_2) = T_2 = 60°C$$

Integrating the differential equation once with respect to r gives

$$r\frac{dT}{dr} = C_1$$

where C_1 is an arbitrary constant. We now divide both sides of this equation by r to bring it to a readily integrable form,

$$\frac{dT}{dr} = \frac{C_1}{r}$$

Again integrating with respect to r gives (Fig. 2–51)

$$T(r) = C_1 \ln r + C_2 \qquad (a)$$

We now apply both boundary conditions by replacing all occurrences of r and $T(r)$ in Eq. (a) with the specified values at the boundaries. We get

$$T(r_1) = T_1 \quad \rightarrow \quad C_1 \ln r_1 + C_2 = T_1$$
$$T(r_2) = T_2 \quad \rightarrow \quad C_1 \ln r_2 + C_2 = T_2$$

which are two equations in two unknowns, C_1 and C_2. Solving them simultaneously gives

$$C_1 = \frac{T_2 - T_1}{\ln(r_2/r_1)} \quad \text{and} \quad C_2 = T_1 - \frac{T_2 - T_1}{\ln(r_2/r_1)} \ln r_1$$

Substituting them into Eq. (a) and rearranging, the variation of temperature within the pipe is determined to be

$$T(r) = \frac{\ln(r/r_1)}{\ln(r_2/r_1)}(T_2 - T_1) + T_1 \qquad (2\text{–}58)$$

The rate of heat loss from the steam is simply the total rate of heat conduction through the pipe, and is determined from Fourier's law to be

Differential equation:

$$\frac{d}{dr}\left(r\frac{dT}{dr}\right) = 0$$

Integrate:

$$r\frac{dT}{dr} = C_1$$

Divide by r ($r \neq 0$):

$$\frac{dT}{dr} = \frac{C_1}{r}$$

Integrate again:

$$T(r) = C_1 \ln r + C_2$$

which is the general solution.

FIGURE 2–51
Basic steps involved in the solution of the steady one-dimensional heat conduction equation in cylindrical coordinates.

$$\dot{Q}_{cylinder} = -kA\frac{dT}{dr} = -k(2\pi rL)\frac{C_1}{r} = -2\pi kLC_1 = 2\pi kL\frac{T_1 - T_2}{\ln(r_2/r_1)} \quad (2\text{–}59)$$

The numerical value of the rate of heat conduction through the pipe is determined by substituting the given values

$$\dot{Q} = 2\pi(20 \text{ W/m} \cdot {}^\circ\text{C})(20 \text{ m})\frac{(150 - 60){}^\circ\text{C}}{\ln(0.08/0.06)} = \textbf{786 kW}$$

Discussion   Note that the total rate of heat transfer through a pipe is constant, but the heat flux $\dot{q} = \dot{Q}/(2\pi rL)$ is not since it decreases in the direction of heat transfer with increasing radius.

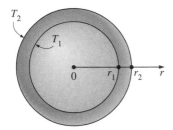

FIGURE 2–52
Schematic for Example 2–16.

EXAMPLE 2–16   **Heat Conduction through a Spherical Shell**

Consider a spherical container of inner radius $r_1 = 8$ cm, outer radius $r_2 = 10$ cm, and thermal conductivity $k = 45$ W/m · °C, as shown in Fig. 2–52. The inner and outer surfaces of the container are maintained at constant temperatures of $T_1 = 200{}^\circ$C and $T_2 = 80{}^\circ$C, respectively, as a result of some chemical reactions occurring inside. Obtain a general relation for the temperature distribution inside the shell under steady conditions, and determine the rate of heat loss from the container.

SOLUTION   A spherical container is subjected to specified temperatures on its surfaces. The variation of temperature and the rate of heat transfer are to be determined.
Assumptions   **1** Heat transfer is steady since there is no change with time. **2** Heat transfer is one-dimensional since there is thermal symmetry about the midpoint, and thus $T = T(r)$. **3** Thermal conductivity is constant. **4** There is no heat generation.
Properties   The thermal conductivity is given to be $k = 45$ W/m · °C.
Analysis   The mathematical formulation of this problem can be expressed as

$$\frac{d}{dr}\left(r^2\frac{dT}{dr}\right) = 0$$

with boundary conditions

$$T(r_1) = T_1 = 200{}^\circ\text{C}$$
$$T(r_2) = T_2 = 80{}^\circ\text{C}$$

Integrating the differential equation once with respect to r yields

$$r^2\frac{dT}{dr} = C_1$$

where C_1 is an arbitrary constant. We now divide both sides of this equation by r^2 to bring it to a readily integrable form,

$$\frac{dT}{dr} = \frac{C_1}{r^2}$$

Again integrating with respect to r gives

$$T(r) = -\frac{C_1}{r} + C_2 \qquad (a)$$

We now apply both boundary conditions by replacing all occurrences of r and $T(r)$ in the relation above by the specified values at the boundaries. We get

$$T(r_1) = T_1 \quad \rightarrow \quad -\frac{C_1}{r_1} + C_2 = T_1$$

$$T(r_2) = T_2 \quad \rightarrow \quad -\frac{C_1}{r_2} + C_2 = T_2$$

which are two equations in two unknowns, C_1 and C_2. Solving them simultaneously gives

$$C_1 = -\frac{r_1 r_2}{r_2 - r_1}(T_1 - T_2) \qquad \text{and} \qquad C_2 = \frac{r_2 T_2 - r_1 T_1}{r_2 - r_1}$$

Substituting into Eq. (a), the variation of temperature within the spherical shell is determined to be

$$T(r) = \frac{r_1 r_2}{r(r_2 - r_1)}(T_1 - T_2) + \frac{r_2 T_2 - r_1 T_1}{r_2 - r_1} \qquad \textbf{(2-60)}$$

The rate of heat loss from the container is simply the total rate of heat conduction through the container wall and is determined from Fourier's law

$$\dot{Q}_{\text{sphere}} = -kA\frac{dT}{dr} = -k(4\pi r^2)\frac{C_1}{r^2} = -4\pi kC_1 = 4\pi kr_1 r_2\frac{T_1 - T_2}{r_2 - r_1} \qquad \textbf{(2-61)}$$

The numerical value of the rate of heat conduction through the wall is determined by substituting the given values to be

$$\dot{Q} = 4\pi(45 \text{ W/m} \cdot {}^\circ\text{C})(0.08 \text{ m})(0.10 \text{ m})\frac{(200 - 80){}^\circ\text{C}}{(0.10 - 0.08) \text{ m}} = 27.1 \text{ kW}$$

Discussion Note that the total rate of heat transfer through a spherical shell is constant, but the heat flux $\dot{q} = \dot{Q}/4\pi r^2$ is not since it decreases in the direction of heat transfer with increasing radius as shown in Fig. 2–53.

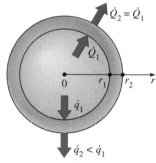

$$\dot{q}_1 = \frac{\dot{Q}_1}{A_1} = \frac{27.1 \text{ kW}}{4\pi(0.08 \text{ m})^2} = 337 \text{ kW/m}^2$$

$$\dot{q}_2 = \frac{\dot{Q}_2}{A_2} = \frac{27.1 \text{ kW}}{4\pi(0.10 \text{ m})^2} = 216 \text{ kW/m}^2$$

FIGURE 2–53
During steady one-dimensional heat conduction in a spherical (or cylindrical) container, the total rate of heat transfer remains constant, but the heat flux decreases with increasing radius.

2–6 ■ HEAT GENERATION IN A SOLID

Many practical heat transfer applications involve the conversion of some form of energy into *thermal* energy in the medium. Such mediums are said to involve internal *heat generation,* which manifests itself as a rise in temperature throughout the medium. Some examples of heat generation are *resistance heating* in wires, exothermic *chemical reactions* in a solid, and *nuclear reactions* in nuclear fuel rods where electrical, chemical, and nuclear energies are converted to heat, respectively (Fig. 2–54). The absorption of radiation throughout the volume of a semitransparent medium such as water can also be considered as heat generation within the medium, as explained earlier.

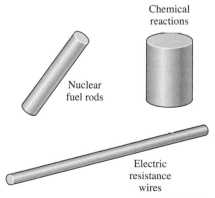

FIGURE 2–54
Heat generation in solids is commonly encountered in practice.

Heat generation is usually expressed *per unit volume* of the medium, and is denoted by $\dot{e}_{gen}$, whose unit is W/m³. For example, heat generation in an electrical wire of outer radius r_o and length L can be expressed as

$$\dot{e}_{gen} = \frac{\dot{E}_{gen, \, electric}}{V_{wire}} = \frac{I^2 R_e}{\pi r_o^2 L} \qquad (\text{W/m}^3) \qquad \text{(2–62)}$$

where I is the electric current and R_e is the electrical resistance of the wire.

The temperature of a medium *rises* during heat generation as a result of the absorption of the generated heat by the medium during transient start-up period. As the temperature of the medium increases, so does the heat transfer from the medium to its surroundings. This continues until steady operating conditions are reached and the rate of heat generation equals the rate of heat transfer to the surroundings. Once steady operation has been established, the temperature of the medium at any point no longer changes.

The *maximum temperature* T_{max} in a solid that involves uniform heat generation occurs at a location *farthest away* from the outer surface when the outer surface of the solid is maintained at a constant temperature T_s. For example, the maximum temperature occurs at the *midplane* in a plane wall, at the *centerline* in a long cylinder, and at the *midpoint* in a sphere. The temperature distribution within the solid in these cases is *symmetrical* about the center of symmetry.

The quantities of major interest in a medium with heat generation are the surface temperature T_s and the maximum temperature T_{max} that occurs in the medium in *steady* operation. Below we develop expressions for these two quantities for common geometries for the case of *uniform* heat generation ($\dot{e}_{gen}$ = constant) within the medium.

Consider a solid medium of surface area A_s, volume V, and constant thermal conductivity k, where heat is generated at a constant rate of $\dot{e}_{gen}$ per unit volume. Heat is transferred from the solid to the surrounding medium at T_∞, with a constant heat transfer coefficient of h. All the surfaces of the solid are maintained at a common temperature T_s. Under *steady* conditions, the energy balance for this solid can be expressed as (Fig. 2–55)

$$\begin{pmatrix} \text{Rate of} \\ heat \ transfer \\ \text{from the solid} \end{pmatrix} = \begin{pmatrix} \text{Rate of} \\ energy \ generation \\ \text{within the solid} \end{pmatrix} \qquad \text{(2–63)}$$

or

$$\dot{Q} = \dot{e}_{gen} V \qquad (\text{W}) \qquad \text{(2–64)}$$

Disregarding radiation (or incorporating it in the heat transfer coefficient h), the heat transfer rate can also be expressed from Newton's law of cooling as

$$\dot{Q} = hA_s \, (T_s - T_\infty) \qquad (\text{W}) \qquad \text{(2–65)}$$

Combining Eqs. 2–64 and 2–65 and solving for the surface temperature T_s gives

$$T_s = T_\infty + \frac{\dot{e}_{gen} V}{hA_s} \qquad \text{(2–66)}$$

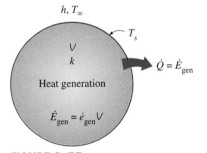

h, T_∞

T_s

V

k

Heat generation

$\dot{E}_{gen} = \dot{e}_{gen} V$

$\dot{Q} = \dot{E}_{gen}$

FIGURE 2–55

At steady conditions, the entire heat generated in a solid must leave the solid through its outer surface.

For a large *plane wall* of thickness $2L$ ($A_s = 2A_{wall}$ and $V = 2LA_{wall}$), a long solid *cylinder* of radius r_o ($A_s = 2\pi r_o L$ and $V = \pi r_o^2 L$), and a solid *sphere* of radius r_o ($A_s = 4\pi r_o^2$ and $V = \frac{4}{3}\pi r_o^3$), Eq. 2–66 reduces to

$$T_{s,\,\text{plane wall}} = T_\infty + \frac{\dot{e}_{gen}L}{h} \qquad (2\text{–}67)$$

$$T_{s,\,\text{cylinder}} = T_\infty + \frac{\dot{e}_{gen}r_o}{2h} \qquad (2\text{–}68)$$

$$T_{s,\,\text{sphere}} = T_\infty + \frac{\dot{e}_{gen}r_o}{3h} \qquad (2\text{–}69)$$

Note that the rise in surface temperature T_s is due to heat generation in the solid.

Reconsider heat transfer from a long solid cylinder with heat generation. We mentioned above that, under *steady* conditions, the entire heat generated within the medium is conducted through the outer surface of the cylinder. Now consider an imaginary inner cylinder of radius r within the cylinder (Fig. 2–56). Again the *heat generated* within this inner cylinder must be equal to the *heat conducted* through its outer surface. That is, from Fourier's law of heat conduction,

$$-kA_r \frac{dT}{dr} = \dot{e}_{gen}V_r \qquad (2\text{–}70)$$

where $A_r = 2\pi r L$ and $V_r = \pi r^2 L$ at any location r. Substituting these expressions into Eq. 2–70 and separating the variables, we get

$$-k(2\pi r L)\frac{dT}{dr} = \dot{e}_{gen}(\pi r^2 L) \quad \rightarrow \quad dT = -\frac{\dot{e}_{gen}}{2k}r\,dr$$

Integrating from $r = 0$ where $T(0) = T_0$ to $r = r_o$ where $T(r_o) = T_s$ yields

$$\Delta T_{max,\,\text{cylinder}} = T_0 - T_s = \frac{\dot{e}_{gen}r_o^2}{4k} \qquad (2\text{–}71)$$

where T_0 is the centerline temperature of the cylinder, which is the *maximum temperature*, and ΔT_{max} is the difference between the centerline and the surface temperatures of the cylinder, which is the *maximum temperature rise* in the cylinder above the surface temperature. Once ΔT_{max} is available, the centerline temperature can easily be determined from (Fig. 2–57)

$$T_{center} = T_0 = T_s + \Delta T_{max} \qquad (2\text{–}72)$$

The approach outlined above can also be used to determine the *maximum temperature rise* in a plane wall of thickness $2L$ and a solid sphere of radius r_o, with these results:

$$\Delta T_{max,\,\text{plane wall}} = \frac{\dot{e}_{gen}L^2}{2k} \qquad (2\text{–}73)$$

$$\Delta T_{max,\,\text{sphere}} = \frac{\dot{e}_{gen}r_o^2}{6k} \qquad (2\text{–}74)$$

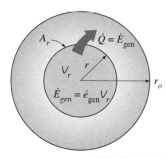

FIGURE 2–56

Heat conducted through a cylindrical shell of radius r is equal to the heat generated within a shell.

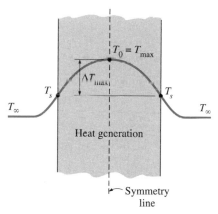

FIGURE 2–57

The maximum temperature in a symmetrical solid with uniform heat generation occurs at its center.

Again the maximum temperature at the center can be determined from Eq. 2–72 by adding the maximum temperature rise to the surface temperature of the solid.

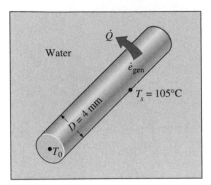

FIGURE 2–58
Schematic for Example 2–17.

EXAMPLE 2–17   Centerline Temperature of a Resistance Heater

A 2-kW resistance heater wire whose thermal conductivity is $k = 15$ W/m · K has a diameter of $D = 4$ mm and a length of $L = 0.5$ m, and is used to boil water (Fig. 2–58). If the outer surface temperature of the resistance wire is $T_s = 105°C$, determine the temperature at the center of the wire.

SOLUTION   The center temperature of a resistance heater submerged in water is to be determined.
Assumptions **1** Heat transfer is steady since there is no change with time. **2** Heat transfer is one-dimensional since there is thermal symmetry about the centerline and no change in the axial direction. **3** Thermal conductivity is constant. **4** Heat generation in the heater is uniform.
Properties   The thermal conductivity is given to be $k = 15$ W/m · K.
Analysis   The 2-kW resistance heater converts electric energy into heat at a rate of 2 kW. The heat generation per unit volume of the wire is

$$\dot{e}_{gen} = \frac{\dot{E}_{gen}}{V_{wire}} = \frac{\dot{E}_{gen}}{\pi r_o^2 L} = \frac{2000 \text{ W}}{\pi (0.002 \text{ m})^2 (0.5 \text{ m})} = 0.318 \times 10^9 \text{ W/m}^3$$

Then the center temperature of the wire is determined from Eq. 2–71 to be

$$T_0 = T_s + \frac{\dot{e}_{gen} r_o^2}{4k} = 105°C + \frac{(0.318 \times 10^9 \text{ W/m}^3)(0.002 \text{ m})^2}{4 \times (15 \text{ W/m} \cdot °C)} = \mathbf{126°C}$$

Discussion   Note that the temperature difference between the center and the surface of the wire is 21°C. Also, the thermal conductivity units W/m · °C and W/m · K are equivalent.

We have developed these relations using the intuitive *energy balance* approach. However, we could have obtained the same relations by setting up the appropriate *differential equations* and solving them, as illustrated in Examples 2–18 and 2–19.

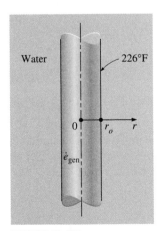

FIGURE 2–59
Schematic for Example 2–18.

EXAMPLE 2–18   Variation of Temperature in a Resistance Heater

A long homogeneous resistance wire of radius $r_o = 0.2$ in and thermal conductivity $k = 7.8$ Btu/h · ft · °F is being used to boil water at atmospheric pressure by the passage of electric current, as shown in Fig. 2–59. Heat is generated in the wire uniformly as a result of resistance heating at a rate of $\dot{e}_{gen} = 2400$ Btu/h · in³. If the outer surface temperature of the wire is measured to be $T_s = 226°F$, obtain a relation for the temperature distribution, and determine the temperature at the centerline of the wire when steady operating conditions are reached.

SOLUTION This heat transfer problem is similar to the problem in Example 2–17, except that we need to obtain a relation for the variation of temperature within the wire with r. Differential equations are well suited for this purpose.

Assumptions 1 Heat transfer is steady since there is no change with time. 2 Heat transfer is one-dimensional since there is no thermal symmetry about the centerline and no change in the axial direction. 3 Thermal conductivity is constant. 4 Heat generation in the wire is uniform.

Properties The thermal conductivity is given to be $k = 7.8$ Btu/h · ft · °F.

Analysis The differential equation which governs the variation of temperature in the wire is simply Eq. 2–27,

$$\frac{1}{r}\frac{d}{dr}\left(r\frac{dT}{dr}\right) + \frac{\dot{e}_{gen}}{k} = 0$$

This is a second-order linear ordinary differential equation, and thus its general solution contains two arbitrary constants. The determination of these constants requires the specification of two boundary conditions, which can be taken to be

$$T(r_o) = T_s = 226°F$$

and

$$\frac{dT(0)}{dr} = 0$$

The first boundary condition simply states that the temperature of the outer surface of the wire is 226°F. The second boundary condition is the symmetry condition at the centerline, and states that the maximum temperature in the wire occurs at the centerline, and thus the slope of the temperature at $r = 0$ must be zero (Fig. 2–60) This completes the mathematical formulation of the problem.

Although not immediately obvious, the differential equation is in a form that can be solved by direct integration. Multiplying both sides of the equation by r and rearranging, we obtain

$$\frac{d}{dr}\left(r\frac{dT}{dr}\right) = -\frac{\dot{e}_{gen}}{k}r$$

Integrating with respect to r gives

$$r\frac{dT}{dr} = -\frac{\dot{e}_{gen}}{k}\frac{r^2}{2} + C_1 \qquad (a)$$

since the heat generation is constant, and the integral of a derivative of a function is the function itself. That is, integration removes a derivative. It is convenient at this point to apply the second boundary condition, since it is related to the first derivative of the temperature, by replacing all occurrences of r and dT/dr in Eq. (a) by zero. It yields

$$0 \times \frac{dT(0)}{dr} = -\frac{\dot{e}_{gen}}{2k} \times 0 + C_1 \quad \rightarrow \quad C_1 = 0$$

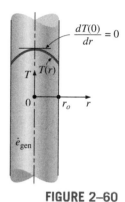

FIGURE 2–60
The thermal symmetry condition at the centerline of a wire in which heat is generated uniformly.

Thus C_1 cancels from the solution. We now divide Eq. (a) by r to bring it to a readily integrable form,

$$\frac{dT}{dr} = -\frac{\dot{e}_{\text{gen}}}{2k} r$$

Again integrating with respect to r gives

$$T(r) = -\frac{\dot{e}_{\text{gen}}}{4k} r^2 + C_2 \qquad (b)$$

We now apply the first boundary condition by replacing all occurrences of r by r_0 and all occurrences of T by T_s. We get

$$T_s = -\frac{\dot{e}_{\text{gen}}}{4k} r_o^2 + C_2 \quad \rightarrow \quad C_2 = T_s + \frac{\dot{e}_{\text{gen}}}{4k} r_o^2$$

Substituting this C_2 relation into Eq. (b) and rearranging give

$$T(r) = T_s + \frac{\dot{e}_{\text{gen}}}{4k} (r_o^2 - r^2) \qquad (c)$$

which is the desired solution for the temperature distribution in the wire as a function of r. The temperature at the centerline ($r = 0$) is obtained by replacing r in Eq. (c) by zero and substituting the known quantities,

$$T(0) = T_s + \frac{\dot{e}_{\text{gen}}}{4k} r_o^2 = 226°\text{F} + \frac{2400 \text{ Btu/h} \cdot \text{in}^3}{4 \times (7.8 \text{ Btu/h} \cdot \text{ft} \cdot °\text{F})} \left(\frac{12 \text{ in}}{1 \text{ ft}}\right)(0.2 \text{ in})^2 = \textbf{263°F}$$

Discussion The temperature of the centerline is 37°F above the temperature of the outer surface of the wire. Note that the expression above for the centerline temperature is identical to Eq. 2–71, which was obtained using an energy balance on a control volume.

EXAMPLE 2–19   **Heat Conduction in a Two-Layer Medium**

Consider a long resistance wire of radius $r_1 = 0.2$ cm and thermal conductivity $k_{\text{wire}} = 15$ W/m · °C in which heat is generated uniformly as a result of resistance heating at a constant rate of $\dot{e}_{\text{gen}} = 50$ W/cm³ (Fig. 2–61). The wire is embedded in a 0.5-cm-thick layer of ceramic whose thermal conductivity is $k_{\text{ceramic}} = 1.2$ W/m · °C. If the outer surface temperature of the ceramic layer is measured to be $T_s = 45$°C, determine the temperatures at the center of the resistance wire and the interface of the wire and the ceramic layer under steady conditions.

SOLUTION   The surface and interface temperatures of a resistance wire covered with a ceramic layer are to be determined.
Assumptions **1** Heat transfer is steady since there is no change with time. **2** Heat transfer is one-dimensional since this two-layer heat transfer problem possesses symmetry about the centerline and involves no change in the axial direction, and thus $T = T(r)$. **3** Thermal conductivities are constant. **4** Heat generation in the wire is uniform.
Properties It is given that $k_{\text{wire}} = 15$ W/m · °C and $k_{\text{ceramic}} = 1.2$ W/m · ° C.

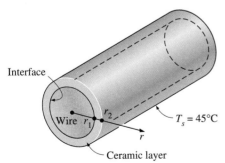

FIGURE 2–61
Schematic for Example 2–19.

Analysis Letting T_I denote the unknown interface temperature, the heat transfer problem in the wire can be formulated as

$$\frac{1}{r}\frac{d}{dr}\left(r\frac{dT_{\text{wire}}}{dr}\right) + \frac{\dot{e}_{\text{gen}}}{k} = 0$$

with

$$T_{\text{wire}}(r_1) = T_I$$
$$\frac{dT_{\text{wire}}(0)}{dr} = 0$$

This problem was solved in Example 2–18, and its solution was determined to be

$$T_{\text{wire}}(r) = T_I + \frac{\dot{e}_{\text{gen}}}{4k_{\text{wire}}}(r_1^2 - r^2) \qquad (a)$$

Noting that the ceramic layer does not involve any heat generation and its outer surface temperature is specified, the heat conduction problem in that layer can be expressed as

$$\frac{d}{dr}\left(r\frac{dT_{\text{ceramic}}}{dr}\right) = 0$$

with

$$T_{\text{ceramic}}(r_1) = T_I$$
$$T_{\text{ceramic}}(r_2) = T_s = 45°C$$

This problem was solved in Example 2–15, and its solution was determined to be

$$T_{\text{ceramic}}(r) = \frac{\ln(r/r_1)}{\ln(r_2/r_1)}(T_s - T_I) + T_I \qquad (b)$$

We have already utilized the first interface condition by setting the wire and ceramic layer temperatures equal to T_I at the interface $r = r_1$. The interface temperature T_I is determined from the second interface condition that the heat flux in the wire and the ceramic layer at $r = r_1$ must be the same:

$$-k_{\text{wire}}\frac{dT_{\text{wire}}(r_1)}{dr} = -k_{\text{ceramic}}\frac{dT_{\text{ceramic}}(r_1)}{dr} \rightarrow \frac{\dot{e}_{\text{gen}}r_1}{2} = -k_{\text{ceramic}}\frac{T_s - T_I}{\ln(r_2/r_1)}\left(\frac{1}{r_1}\right)$$

Solving for T_I and substituting the given values, the interface temperature is determined to be

$$T_I = \frac{\dot{e}_{\text{gen}}r_1^2}{2k_{\text{ceramic}}}\ln\frac{r_2}{r_1} + T_s$$
$$= \frac{(50 \times 10^6 \text{ W/m}^3)(0.002 \text{ m})^2}{2(1.2 \text{ W/m} \cdot °C)}\ln\frac{0.007 \text{ m}}{0.002 \text{ m}} + 45° \text{ C} = \mathbf{149.4°C}$$

Knowing the interface temperature, the temperature at the centerline ($r = 0$) is obtained by substituting the known quantities into Eq. (*a*),

$$T_{\text{wire}}(0) = T_I + \frac{\dot{e}_{\text{gen}}r_1^2}{4k_{\text{wire}}} = 149.4°C + \frac{(50 \times 10^6 \text{ W/m}^3)(0.002 \text{ m})^2}{4 \times (15 \text{ W/m} \cdot °C)} = \mathbf{152.7°C}$$

Thus the temperature of the centerline is slightly above the interface temperature.

Discussion This example demonstrates how steady one-dimensional heat conduction problems in composite media can be solved. We could also solve this problem by determining the heat flux at the interface by dividing the total heat generated in the wire by the surface area of the wire, and then using this value as the specifed heat flux boundary condition for both the wire and the ceramic layer. This way the two problems are decoupled and can be solved separately.

2–7 ▪ VARIABLE THERMAL CONDUCTIVITY, $k(T)$

You will recall from Chapter 1 that the thermal conductivity of a material, in general, varies with temperature (Fig. 2–62). However, this variation is mild for many materials in the range of practical interest and can be disregarded. In such cases, we can use an average value for the thermal conductivity and treat it as a constant, as we have been doing so far. This is also common practice for other temperature-dependent properties such as the density and specific heat.

When the variation of thermal conductivity with temperature in a specified temperature interval is large, however, it may be necessary to account for this variation to minimize the error. Accounting for the variation of the thermal conductivity with temperature, in general, complicates the analysis. But in the case of simple one-dimensional cases, we can obtain heat transfer relations in a straightforward manner.

When the variation of thermal conductivity with temperature $k(T)$ is known, the average value of the thermal conductivity in the temperature range between T_1 and T_2 can be determined from

$$k_{avg} = \frac{\int_{T_1}^{T_2} k(T)dT}{T_2 - T_1} \tag{2–75}$$

This relation is based on the requirement that the rate of heat transfer through a medium with constant average thermal conductivity k_{avg} equals the rate of heat transfer through the same medium with variable conductivity $k(T)$. Note that in the case of constant thermal conductivity $k(T) = k$, Eq. 2–75 reduces to $k_{avg} = k$, as expected.

Then the rate of steady heat transfer through a plane wall, cylindrical layer, or spherical layer for the case of variable thermal conductivity can be determined by replacing the constant thermal conductivity k in Eqs. 2–57, 2–59, and 2–61 by the k_{avg} expression (or value) from Eq. 2–75:

$$\dot{Q}_{plane\ wall} = k_{avg} A \frac{T_1 - T_2}{L} = \frac{A}{L} \int_{T_2}^{T_1} k(T)dT \tag{2–76}$$

$$\dot{Q}_{cylinder} = 2\pi k_{avg} L \frac{T_1 - T_2}{\ln(r_2/r_1)} = \frac{2\pi L}{\ln(r_2/r_1)} \int_{T_2}^{T_1} k(T)dT \tag{2–77}$$

$$\dot{Q}_{sphere} = 4\pi k_{avg} r_1 r_2 \frac{T_1 - T_2}{r_2 - r_1} = \frac{4\pi r_1 r_2}{r_2 - r_1} \int_{T_2}^{T_1} k(T)dT \tag{2–78}$$

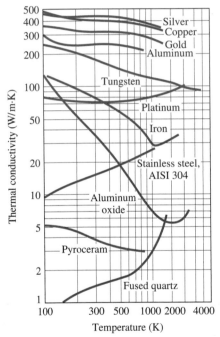

FIGURE 2–62
Variation of the thermal conductivity of some solids with temperature.

The variation in thermal conductivity of a material with temperature in the temperature range of interest can often be approximated as a linear function and expressed as

$$k(T) = k_0(1 + \beta T) \tag{2-79}$$

where β is called the **temperature coefficient of thermal conductivity**. The *average* value of thermal conductivity in the temperature range T_1 to T_2 in this case can be determined from

$$k_{\text{avg}} = \frac{\int_{T_1}^{T_2} k_0(1 + \beta T)dT}{T_2 - T_1} = k_0\left(1 + \beta \frac{T_2 + T_1}{2}\right) = k(T_{\text{avg}}) \tag{2-80}$$

Note that the *average thermal conductivity* in this case is equal to the thermal conductivity value at the *average temperature*.

We have mentioned earlier that in a plane wall the temperature varies linearly during steady one-dimensional heat conduction when the thermal conductivity is constant. But this is no longer the case when the thermal conductivity changes with temperature, even linearly, as shown in Fig. 2–63.

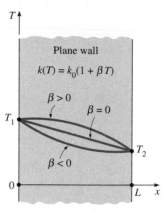

FIGURE 2–63
The variation of temperature in a plane wall during steady one-dimensional heat conduction for the cases of constant and variable thermal conductivity.

EXAMPLE 2–20   Variation of Temperature in a Wall with $k(T)$

Consider a plane wall of thickness L whose thermal conductivity varies linearly in a specified temperature range as $k(T) = k_0(1 + \beta T)$ where k_0 and β are constants. The wall surface at $x = 0$ is maintained at a constant temperature of T_1 while the surface at $x = L$ is maintained at T_2, as shown in Fig. 2–64. Assuming steady one-dimensional heat transfer, obtain a relation for (a) the heat transfer rate through the wall and (b) the temperature distribution $T(x)$ in the wall.

SOLUTION A plate with variable conductivity is subjected to specified temperatures on both sides. The variation of temperature and the rate of heat transfer are to be determined.

Assumptions **1** Heat transfer is given to be steady and one-dimensional. **2** Thermal conductivity varies linearly. **3** There is no heat generation.

Properties The thermal conductivity is given to be $k(T) = k_0(1 + \beta T)$.

Analysis (a) The rate of heat transfer through the wall can be determined from

$$\dot{Q} = k_{\text{avg}} A \frac{T_1 - T_2}{L}$$

where A is the heat conduction area of the wall and

$$k_{\text{avg}} = k(T_{\text{avg}}) = k_0\left(1 + \beta \frac{T_2 + T_1}{2}\right)$$

is the average thermal conductivity (Eq. 2–80).

(b) To determine the temperature distribution in the wall, we begin with Fourier's law of heat conduction, expressed as

$$\dot{Q} = -k(T) A \frac{dT}{dx}$$

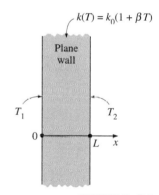

FIGURE 2–64
Schematic for Example 2–20.

where the rate of conduction heat transfer $\dot{Q}$ and the area A are constant. Separating variables and integrating from $x = 0$ where $T(0) = T_1$ to any x where $T(x) = T$, we get

$$\int_0^x \dot{Q}\,dx = -A \int_{T_1}^{T} k(T)\,dT$$

Substituting $k(T) = k_0(1 + \beta T)$ and performing the integrations we obtain

$$\dot{Q}x = -Ak_0[(T - T_1) + \beta(T^2 - T_1^2)/2]$$

Substituting the $\dot{Q}$ expression from part (a) and rearranging give

$$T^2 + \frac{2}{\beta}T + \frac{2k_{\text{avg}}}{\beta k_0}\frac{x}{L}(T_1 - T_2) - T_1^2 - \frac{2}{\beta}T_1 = 0$$

which is a *quadratic* equation in the unknown temperature T. Using the quadratic formula, the temperature distribution $T(x)$ in the wall is determined to be

$$T(x) = -\frac{1}{\beta} \pm \sqrt{\frac{1}{\beta^2} - \frac{2k_{\text{avg}}}{\beta k_0}\frac{x}{L}(T_1 - T_2) + T_1^2 + \frac{2}{\beta}T_1}$$

Discussion The proper sign of the square root term (+ or −) is determined from the requirement that the temperature at any point within the medium must remain between T_1 and T_2. This result explains why the temperature distribution in a plane wall is no longer a straight line when the thermal conductivity varies with temperature.

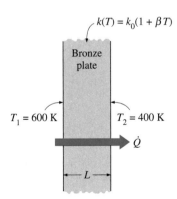

FIGURE 2–65
Schematic for Example 2–21.

EXAMPLE 2–21   **Heat Conduction through a Wall with $k(T)$**

Consider a 2-m-high and 0.7-m-wide bronze plate whose thickness is 0.1 m. One side of the plate is maintained at a constant temperature of 600 K while the other side is maintained at 400 K, as shown in Fig. 2–65. The thermal conductivity of the bronze plate can be assumed to vary linearly in that temperature range as $k(T) = k_0(1 + \beta T)$ where $k_0 = 38$ W/m · K and $\beta = 9.21 \times 10^{-4}$ K^{-1}. Disregarding the edge effects and assuming steady one-dimensional heat transfer, determine the rate of heat conduction through the plate.

SOLUTION A plate with variable conductivity is subjected to specified temperatures on both sides. The rate of heat transfer is to be determined.
Assumptions **1** Heat transfer is given to be steady and one-dimensional. **2** Thermal conductivity varies linearly. **3** There is no heat generation.
Properties The thermal conductivity is given to be $k(T) = k_0(1 + \beta T)$.
Analysis The average thermal conductivity of the medium in this case is simply the value at the average temperature and is determined from

$$k_{\text{avg}} = k(T_{\text{avg}}) = k_0\left(1 + \beta\frac{T_2 + T_1}{2}\right)$$

$$= (38 \text{ W/m} \cdot \text{K})\left[1 + (9.21 \times 10^{-4} \text{ K}^{-1})\frac{(600 + 400)\text{ K}}{2}\right]$$

$$= 55.5 \text{ W/m} \cdot \text{K}$$

Then the rate of heat conduction through the plate can be determined from Eq. 2–76 to be

$$\dot{Q} = k_{avg} A \frac{T_1 - T_2}{L}$$

$$= (55.5 \text{ W/m} \cdot \text{K})(2 \text{ m} \times 0.7 \text{ m}) \frac{(600 - 400)\text{K}}{0.1 \text{ m}} = \mathbf{155 \text{ kW}}$$

Discussion We would have obtained the same result by substituting the given $k(T)$ relation into the second part of Eq. 2–76 and performing the indicated integration.

TOPIC OF SPECIAL INTEREST*

A Brief Review of Differential Equations

As we mentioned in Chapter 1, the description of most scientific problems involves relations that involve changes in some key variables with respect to each other. Usually the smaller the increment chosen in the changing variables, the more general and accurate the description. In the limiting case of infinitesimal or differential changes in variables, we obtain *differential equations,* which provide precise mathematical formulations for the physical principles and laws by representing the rates of change as *derivatives.* Therefore, differential equations are used to investigate a wide variety of problems in sciences and engineering, including heat transfer.

Differential equations arise when relevant *physical laws* and *principles* are applied to a problem by considering infinitesimal changes in the variables of interest. Therefore, obtaining the governing differential equation for a specific problem requires an adequate knowledge of the nature of the problem, the variables involved, appropriate simplifying assumptions, and the applicable physical laws and principles involved, as well as a careful analysis.

An equation, in general, may involve one or more variables. As the name implies, a **variable** is a quantity that may assume various values during a study. A quantity whose value is fixed during a study is called a **constant**. Constants are usually denoted by the earlier letters of the alphabet such as a, b, c, and d, whereas variables are usually denoted by the later ones such as t, x, y, and z. A variable whose value can be changed arbitrarily is called an **independent variable** (or argument). A variable whose value depends on the value of other variables and thus cannot be varied independently is called a **dependent variable** (or a function).

A dependent variable y that depends on a variable x is usually denoted as $y(x)$ for clarity. However, this notation becomes very inconvenient and cumbersome when y is repeated several times in an expression. In such cases it is desirable to denote $y(x)$ simply as y when it is clear that y is a function of x. This shortcut in notation improves the appearance and the

*This section can be skipped if desired without a loss in continuity.

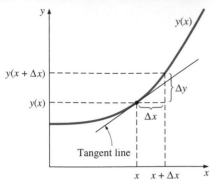

FIGURE 2–66
The derivative of a function at a point represents the slope of the tangent line of the function at that point.

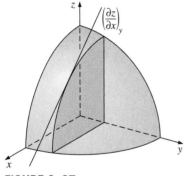

FIGURE 2–67
Graphical representation of partial derivative $\partial z / \partial x$.

readability of the equations. The value of y at a fixed number a is denoted by $y(a)$.

The **derivative** of a function $y(x)$ at a point is equivalent to the *slope* of the tangent line to the graph of the function at that point and is defined as (Fig. 2–66)

$$y'(x) = \frac{dy(x)}{dx} = \lim_{\Delta x \to 0} \frac{\Delta y}{\Delta x} = \lim_{\Delta x \to 0} \frac{y(x + \Delta x) - y(x)}{\Delta x} \quad \textbf{(2–81)}$$

Here Δx represents a (small) change in the independent variable x and is called an *increment* of x. The corresponding change in the function y is called an increment of y and is denoted by Δy. Therefore, the derivative of a function can be viewed as the ratio of the increment Δy of the function to the increment Δx of the independent variable for very small Δx. Note that Δy and thus $y'(x)$ are zero if the function y does not change with x.

Most problems encountered in practice involve quantities that change with time t, and their first derivatives with respect to time represent the rate of change of those quantities with time. For example, if $N(t)$ denotes the population of a bacteria colony at time t, then the first derivative $N' = dN/dt$ represents the rate of change of the population, which is the amount the population increases or decreases per unit time.

The derivative of the first derivative of a function y is called the *second derivative* of y, and is denoted by y'' or d^2y/dx^2. In general, the derivative of the $(n - 1)$st derivative of y is called the nth derivative of y and is denoted by $y^{(n)}$ or d^ny/dx^n. Here, n is a positive integer and is called the *order* of the derivative. The order n should not be confused with the *degree* of a derivative. For example, y''' is the third-order derivative of y, but $(y')^3$ is the third degree of the first derivative of y. Note that the first derivative of a function represents the *slope* or the *rate of change* of the function with the independent variable, and the second derivative represents the *rate of change of the slope* of the function with the independent variable.

When a function y depends on two or more independent variables such as x and t, it is sometimes of interest to examine the dependence of the function on one of the variables only. This is done by taking the derivative of the function with respect to that variable while holding the other variables constant. Such derivatives are called **partial derivatives**. The first partial derivatives of the function $y(x, t)$ with respect to x and t are defined as (Fig. 2–67)

$$\frac{\partial y}{\partial x} = \lim_{\Delta x \to 0} \frac{y(x + \Delta x, t) - y(x, t)}{\Delta x} \quad \textbf{(2–82)}$$

$$\frac{\partial y}{\partial t} = \lim_{\Delta t \to 0} \frac{y(x, t + \Delta t) - y(x, t)}{\Delta t} \quad \textbf{(2–83)}$$

Note that when finding $\partial y/\partial x$ we treat t as a constant and differentiate y with respect to x. Likewise, when finding $\partial y/\partial t$ we treat x as a constant and differentiate y with respect to t.

Integration can be viewed as the inverse process of differentiation. Integration is commonly used in solving differential equations since solving a differential equation is essentially a process of removing the derivatives from the equation. Differentiation is the process of finding $y'(x)$ when a function $y(x)$ is given, whereas integration is the process of finding the function $y(x)$ when its derivative $y'(x)$ is given. The integral of this derivative is expressed as

$$\int y'(x)dx = \int dy = y(x) + C \tag{2-84}$$

since $y'(x)dx = dy$ and the integral of the differential of a function is the function itself (plus a constant, of course). In Eq. 2–84, x is the integration variable and C is an arbitrary constant called the **integration constant**.

The derivative of $y(x) + C$ is $y'(x)$ no matter what the value of the constant C is. Therefore, two functions that differ by a constant have the same derivative, and we always add a constant C during integration to recover this constant that is lost during differentiation. The integral in Eq. 2–84 is called an **indefinite integral** since the value of the arbitrary constant C is indefinite. The described procedure can be extended to higher-order derivatives (Fig. 2–68). For example,

$$\int y''(x)dx = y'(x) + C \tag{2-85}$$

This can be proved by defining a new variable $u(x) - y'(x)$, differentiating it to obtain $u'(x) = y''(x)$, and then applying Eq. 2–84. Therefore, the order of a derivative decreases by one each time it is integrated.

Classification of Differential Equations

A differential equation that involves only ordinary derivatives is called an **ordinary differential equation**, and a differential equation that involves partial derivatives is called a **partial differential equation**. Then it follows that problems that involve a single independent variable result in ordinary differential equations, and problems that involve two or more independent variables result in partial differential equations. A differential equation may involve several derivatives of various orders of an unknown function. The order of the highest derivative in a differential equation is the order of the equation. For example, the order of $y''' + (y'')^4 = 7x^5$ is 3 since it contains no fourth or higher order derivatives.

You will remember from algebra that the equation $3x - 5 = 0$ is much easier to solve than the equation $x^4 + 3x - 5 = 0$ because the first equation is linear whereas the second one is nonlinear. This is also true for differential equations. Therefore, before we start solving a differential equation, we usually check for linearity. A differential equation is said to be **linear** if the dependent variable and all of its derivatives are of the first degree and their coefficients depend on the independent variable only. In other words, a differential equation is linear if it can be written in a form that does not involve (1) any powers of the dependent variable or its derivatives such as y^3 or $(y')^2$, (2) any products of the dependent variable or its derivatives such as yy' or $y'y'''$, and (3) any other nonlinear functions of the dependent variable such as $\sin y$ or e^y. If any of these conditions apply, it is **nonlinear** (Fig. 2–69).

$$\int dy = y + C$$

$$\int y'\, dx = y + C$$

$$\int y''\, dx = y' + C$$

$$\int y'''\, dx = y'' + C$$

$$\int y^{(n)}\, dx = y^{(n-1)} + C$$

FIGURE 2–68
Some indefinite integrals that involve derivatives.

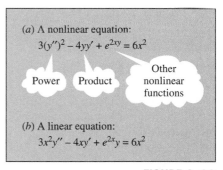

(a) A nonlinear equation:
$$3(y'')^2 - 4yy' + e^{2x}y = 6x^2$$

Power   Product   Other nonlinear functions

(b) A linear equation:
$$3x^2y'' - 4xy' + e^{2x}y = 6x^2$$

FIGURE 2–69
A differential equation that is (a) nonlinear and (b) linear. When checking for linearity, we examine the dependent variable only.

A linear differential equation, however, may contain (1) powers or nonlinear functions of the independent variable, such as x^2 and cos x and (2) products of the dependent variable (or its derivatives) and functions of the independent variable, such as x^3y', x^2y, and $e^{-2x}y''$. A linear differential equation of order n can be expressed in the most general form as

$$y^{(n)} + f_1(x)y^{(n-1)} + \cdots + f_{n-1}(x)y' + f_n(x)y = R(x) \qquad \textbf{(2–86)}$$

A differential equation that cannot be put into this form is nonlinear. A linear differential equation in y is said to be **homogeneous** as well if $R(x) = 0$. Otherwise, it is nonhomogeneous. That is, each term in a linear homogeneous equation contains the dependent variable or one of its derivatives after the equation is cleared of any common factors. The term $R(x)$ is called the *nonhomogeneous term.*

Differential equations are also classified by the nature of the coefficients of the dependent variable and its derivatives. A differential equation is said to have **constant coefficients** if the coefficients of all the terms that involve the dependent variable or its derivatives are constants. If, after clearing any common factors, any of the terms with the dependent variable or its derivatives involve the independent variable as a coefficient, that equation is said to have **variable coefficients** (Fig. 2–70). Differential equations with constant coefficients are usually much easier to solve than those with variable coefficients.

(*a*) *With constant coefficients:*

$$y'' + 6y' - 2y = xe^{-2x}$$

Constant

(*b*) *With variable coefficients:*

$$y'' - 6x^2y' - \frac{2}{x-1}y = xe^{-2x}$$

Variable

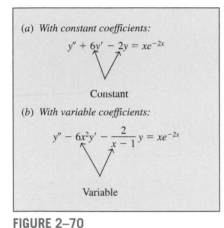

FIGURE 2–70
A differential equation with (*a*) constant coefficients and (*b*) variable coefficients.

Solutions of Differential Equations

Solving a differential equation can be as easy as performing one or more integrations; but such simple differential equations are usually the exception rather than the rule. There is no single general solution method applicable to all differential equations. There are different solution techniques, each being applicable to different classes of differential equations. Sometimes solving a differential equation requires the use of two or more techniques as well as ingenuity and mastery of solution methods. Some differential equations can be solved only by using some very clever tricks. Some cannot be solved analytically at all.

In algebra, we usually seek discrete values that satisfy an algebraic equation such as $x^2 - 7x - 10 = 0$. When dealing with differential equations, however, we seek functions that satisfy the equation in a specified interval. For example, the algebraic equation $x^2 - 7x - 10 = 0$ is satisfied by two numbers only: 2 and 5. But the differential equation $y' - 7y = 0$ is satisfied by the function e^{7x} for any value of x (Fig. 2–71).

(*a*) *An algebraic equation:*

$$y^2 - 7y - 10 = 0$$

Solution: $y = 2$ and $y = 5$

(*b*) *A differential equation:*

$$y' - 7y = 0$$

Solution: $y = e^{7x}$

FIGURE 2–71
Unlike those of algebraic equations, the solutions of differential equations are typically functions instead of discrete values.

Consider the algebraic equation $x^3 - 6x^2 + 11x - 6 = 0$. Obviously, $x = 1$ satisfies this equation, and thus it is a solution. However, it is not the only solution of this equation. We can easily show by direct substitution that $x = 2$ and $x = 3$ also satisfy this equation, and thus they are solutions as well. But there are no other solutions to this equation. Therefore, we say that the set 1, 2, and 3 forms the complete solution to this algebraic equation.

The same line of reasoning also applies to differential equations. Typically, differential equations have multiple solutions that contain at least one arbitrary constant. Any function that satisfies the differential equation on an interval is called a *solution* of that differential equation in that interval.

A solution that involves one or more arbitrary constants represents a family of functions that satisfy the differential equation and is called **a general solution** of that equation. Not surprisingly, a differential equation may have more than one general solution. A general solution is usually referred to as **the general solution** or the **complete solution** if every solution of the equation can be obtained from it as a special case. A solution that can be obtained from a general solution by assigning particular values to the arbitrary constants is called a **specific solution**.

You will recall from algebra that a number is a solution of an algebraic equation if it satisfies the equation. For example, 2 is a solution of the equation $x^3 - 8 = 0$ because the substitution of 2 for x yields identically zero. Likewise, a function is a solution of a differential equation if that function satisfies the differential equation. In other words, a solution function yields identity when substituted into the differential equation. For example, it can be shown by direct substitution that the function $3e^{-2x}$ is a solution of $y'' - 4y = 0$ (Fig. 2–72).

Function: $f = 3e^{-2x}$

Differential equation: $y'' - 4y = 0$

Derivatives of f:
$$f' = -6e^{-2x}$$
$$f'' = 12e^{-2x}$$

Substituting into $y'' - 4y = 0$:
$$f'' - 4f \stackrel{?}{=} 0$$
$$12e^{-2x} - 4 \times 3e^{-2x} \stackrel{?}{=} 0$$
$$0 = 0$$

Therefore, the function $3e^{-2x}$ is a solution of the differential equation $y'' - 4y = 0$.

FIGURE 2–72
Verifying that a given function is a solution of a differential equation.

SUMMARY

In this chapter we have studied the heat conduction equation and its solutions. Heat conduction in a medium is said to be *steady* when the temperature does not vary with time and *unsteady* or *transient* when it does. Heat conduction in a medium is said to be *one-dimensional* when conduction is significant in one dimension only and negligible in the other two dimensions. It is said to be *two-dimensional* when conduction in the third dimension is negligible and *three-dimensional* when conduction in all dimensions is significant. In heat transfer analysis, the conversion of electrical, chemical, or nuclear energy into heat (or thermal) energy is characterized as *heat generation*.

The heat conduction equation can be derived by performing an energy balance on a differential volume element. The one-dimensional heat conduction equation in rectangular, cylindrical, and spherical coordinate systems for the case of constant thermal conductivities are expressed as

$$\frac{\partial^2 T}{\partial x^2} + \frac{\dot{e}_{gen}}{k} = \frac{1}{\alpha}\frac{\partial T}{\partial t}$$

$$\frac{1}{r}\frac{\partial}{\partial r}\left(r\frac{\partial T}{\partial r}\right) + \frac{\dot{e}_{gen}}{k} = \frac{1}{\alpha}\frac{\partial T}{\partial t}$$

$$\frac{1}{r^2}\frac{\partial}{\partial r}\left(r^2\frac{\partial T}{\partial r}\right) + \frac{\dot{e}_{gen}}{k} = \frac{1}{\alpha}\frac{\partial T}{\partial t}$$

where the property $\alpha = k/\rho c$ is the *thermal diffusivity* of the material.

The solution of a heat conduction problem depends on the conditions at the surfaces, and the mathematical expressions for the thermal conditions at the boundaries are called the *boundary conditions*. The solution of transient heat conduction problems also depends on the condition of the medium at the beginning of the heat conduction process. Such a condition, which is usually specified at time $t = 0$, is called the *initial condition*, which is a mathematical expression for the temperature distribution of the medium initially. Complete mathematical description of a heat conduction problem requires the specification of two boundary conditions for each dimension along which heat conduction is significant, and an initial condition when the problem is transient. The most common boundary conditions are the *specified temperature, specified heat flux, convection,* and *radiation* boundary conditions. A boundary surface, in general, may involve specified heat flux, convection, and radiation at the same time.

For steady one-dimensional heat transfer through a plate of thickness L, the various types of boundary conditions at the surfaces at $x = 0$ and $x = L$ can be expressed as

Specified temperature:

$$T(0) = T_1 \qquad \text{and} \qquad T(L) = T_2$$

where T_1 and T_2 are the specified temperatures at surfaces at $x = 0$ and $x = L$.

Specified heat flux:

$$-k\frac{dT(0)}{dx} = \dot{q}_0 \qquad \text{and} \qquad -k\frac{dT(L)}{dx} = \dot{q}_L$$

where $\dot{q}_0$ and $\dot{q}_L$ are the specified heat fluxes at surfaces at $x = 0$ and $x = L$.

Insulation or thermal symmetry:

$$\frac{dT(0)}{dx} = 0 \quad \text{and} \quad \frac{dT(L)}{dx} = 0$$

Convection:

$$-k\frac{dT(0)}{dx} = h_1[T_{\infty 1} - T(0)] \quad \text{and} \quad -k\frac{dT(L)}{dx} = h_2[T(L) - T_{\infty 2}]$$

where h_1 and h_2 are the convection heat transfer coefficients and $T_{\infty 1}$ and $T_{\infty 2}$ are the temperatures of the surrounding mediums on the two sides of the plate.

Radiation:

$$-k\frac{dT(0)}{dx} = \varepsilon_1\sigma[T_{\text{surr, 1}}^4 - T(0)^4] \quad \text{and}$$

$$-k\frac{dT(L)}{dx} = \varepsilon_2\sigma[T(L)^4 - T_{\text{surr, 2}}^4]$$

where ε_1 and ε_2 are the emissivities of the boundary surfaces, $\sigma = 5.67 \times 10^{-8}$ W/m$^2 \cdot$ K^4 is the Stefan–Boltzmann constant, and $T_{\text{surr, 1}}$ and $T_{\text{surr, 2}}$ are the average temperatures of the surfaces surrounding the two sides of the plate. In radiation calculations, the temperatures must be in K or R.

Interface of two bodies A and B in perfect contact at $x = x_0$:

$$T_A(x_0) = T_B(x_0) \quad \text{and} \quad -k_A\frac{dT_A(x_0)}{dx} = -k_B\frac{dT_B(x_0)}{dx}$$

where k_A and k_B are the thermal conductivities of the layers A and B.

Heat generation is usually expressed *per unit volume* of the medium and is denoted by $\dot{e}_{\text{gen}}$, whose unit is W/m^3. Under steady conditions, the surface temperature T_s of a plane wall of thickness $2L$, a cylinder of outer radius r_o, and a sphere of radius r_o in which heat is generated at a constant rate of $\dot{e}_{\text{gen}}$ per unit volume in a surrounding medium at T_∞ can be expressed as

$$T_{s, \text{plane wall}} = T_\infty + \frac{\dot{e}_{\text{gen}}L}{h}$$

$$T_{s, \text{cylinder}} = T_\infty + \frac{\dot{e}_{\text{gen}}r_o}{2h}$$

$$T_{s, \text{sphere}} = T_\infty + \frac{\dot{e}_{\text{gen}}r_o}{3h}$$

where h is the convection heat transfer coefficient. The maximum temperature rise between the surface and the midsection of a medium is given by

$$\Delta T_{\text{max, plane wall}} = \frac{\dot{e}_{\text{gen}}L^2}{2k}$$

$$\Delta T_{\text{max, cylinder}} = \frac{\dot{e}_{\text{gen}}r_o^2}{4k}$$

$$\Delta T_{\text{max, sphere}} = \frac{\dot{e}_{\text{gen}}r_o^2}{6k}$$

When the variation of thermal conductivity with temperature $k(T)$ is known, the average value of the thermal conductivity in the temperature range between T_1 and T_2 can be determined from

$$k_{\text{avg}} = \frac{\int_{T_1}^{T_2} k(T)dT}{T_2 - T_1}$$

Then the rate of steady heat transfer through a plane wall, cylindrical layer, or spherical layer can be expressed as

$$\dot{Q}_{\text{plane wall}} = k_{\text{avg}}A\frac{T_1 - T_2}{L} = \frac{A}{L}\int_{T_2}^{T_1} k(T)dT$$

$$\dot{Q}_{\text{cylinder}} = 2\pi k_{\text{avg}}L\frac{T_1 - T_2}{\ln(r_2/r_1)} = \frac{2\pi L}{\ln(r_2/r_1)}\int_{T_2}^{T_1} k(T)dT$$

$$\dot{Q}_{\text{sphere}} = 4\pi k_{\text{avg}}r_1r_2\frac{T_1 - T_2}{r_2 - r_1} = \frac{4\pi r_1 r_2}{r_2 - r_1}\int_{T_2}^{T_1} k(T)dT$$

The variation of thermal conductivity of a material with temperature can often be approximated as a linear function and expressed as

$$k(T) = k_0(1 + \beta T)$$

where β is called the *temperature coefficient of thermal conductivity.*

REFERENCES AND SUGGESTED READING

1. W. E. Boyce and R. C. Diprima. *Elementary Differential Equations and Boundary Value Problems.* 4th ed. New York: John Wiley & Sons, 1986.

2. S. S. Kutateladze. *Fundamentals of Heat Transfer.* New York: Academic Press, 1963.

PROBLEMS*

Introduction

2–1C Is heat transfer a scalar or vector quantity? Explain. Answer the same question for temperature.

2–2C How does transient heat transfer differ from steady heat transfer? How does one-dimensional heat transfer differ from two-dimensional heat transfer?

2–3C Consider a cold canned drink left on a dinner table. Would you model the heat transfer to the drink as one-, two-, or three-dimensional? Would the heat transfer be steady or transient? Also, which coordinate system would you use to analyze this heat transfer problem, and where would you place the origin? Explain.

2–4C Consider a round potato being baked in an oven. Would you model the heat transfer to the potato as one-, two-, or three-dimensional? Would the heat transfer be steady or transient? Also, which coordinate system would you use to solve this problem, and where would you place the origin? Explain.

2–5C Consider an egg being cooked in boiling water in a pan. Would you model the heat transfer to the egg as one-, two-, or three-dimensional? Would the heat transfer be steady or transient? Also, which coordinate system would you use to solve this problem, and where would you place the origin? Explain.

2–6C Consider a hot dog being cooked in boiling water in a pan. Would you model the heat transfer to the hot dog as one-, two-, or three-dimensional? Would the heat transfer be steady or transient? Also, which coordinate system would you use to solve this problem, and where would you place the origin? Explain.

FIGURE P2–6C

---

*Problems designated by a "C" are concept questions, and students are encouraged to answer them all. Problems designated by an "E" are in English units, and the SI users can ignore them. Problems with the icon ❀ are solved using EES, and the complete solutions together with parametric studies are included on the enclosed CD. Problems with the icon ▨ are comprehensive in nature, and are intended to be solved with a computer, preferably using the EES software that accompanies this text.

2–7C Consider the cooking process of a roast beef in an oven. Would you consider this to be a steady or transient heat transfer problem? Also, would you consider this to be one-, two-, or three-dimensional? Explain.

2–8C Consider heat loss from a 200-L cylindrical hot water tank in a house to the surrounding medium. Would you consider this to be a steady or transient heat transfer problem? Also, would you consider this heat transfer problem to be one-, two-, or three-dimensional? Explain.

2–9C Does a heat flux vector at a point P on an isothermal surface of a medium have to be perpendicular to the surface at that point? Explain.

2–10C From a heat transfer point of view, what is the difference between isotropic and unisotropic materials?

2–11C What is heat generation in a solid? Give examples.

2–12C Heat generation is also referred to as energy generation or thermal energy generation. What do you think of these phrases?

2–13C In order to determine the size of the heating element of a new oven, it is desired to determine the rate of heat loss through the walls, door, and the top and bottom section of the oven. In your analysis, would you consider this to be a steady or transient heat transfer problem? Also, would you consider the heat transfer to be one-dimensional or multidimensional? Explain.

2–14E The resistance wire of a 1000-W iron is 15 in long and has a diameter of $D - 0.08$ in. Determine the rate of heat generation in the wire per unit volume, in Btu/h · ft³, and the heat flux on the outer surface of the wire, in Btu/h · ft², as a result of this heat generation.

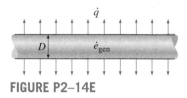

FIGURE P2–14E

2–15E ⬚ Reconsider Prob. 2–14E. Using EES (or other) software, evaluate and plot the surface heat flux as a function of wire diameter as the diameter varies from 0.02 to 0.20 in. Discuss the results.

2–16 Heat flux meters use a very sensitive device known as a thermopile to measure the temperature difference across a thin, heat conducting film made of kapton ($k = 0.345$ W/m · K). If the thermopile can detect temperature differences of 0.1°C or more and the film thickness is 2 mm, what is the minimum heat flux this meter can detect? *Answer:* 17.3 W/m²

2–17 In a nuclear reactor, heat is generated uniformly in the 5-cm-diameter cylindrical uranium rods at a rate of 7×10^7 W/m³. If the length of the rods is 1 m, determine the rate of heat generation in each rod. *Answer:* 137 kW

2–18 In a solar pond, the absorption of solar energy can be modeled as heat generation and can be approximated by $\dot{e}_{gen} = \dot{e}_0 e^{-bx}$, where $\dot{e}_0$ is the rate of heat absorption at the top surface per unit volume and b is a constant. Obtain a relation for the total rate of heat generation in a water layer of surface area A and thickness L at the top of the pond.

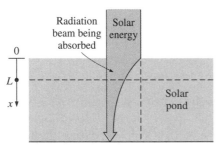

FIGURE P2–18

2–19 Consider a large 3-cm-thick stainless steel plate in which heat is generated uniformly at a rate of 5×10^6 W/m³. Assuming the plate is losing heat from both sides, determine the heat flux on the surface of the plate during steady operation. *Answer:* 75 kW/m²

Heat Conduction Equation

2–20C Write down the one-dimensional transient heat conduction equation for a plane wall with constant thermal conductivity and heat generation in its simplest form, and indicate what each variable represents.

2–21C Write down the one-dimensional transient heat conduction equation for a long cylinder with constant thermal conductivity and heat generation, and indicate what each variable represents.

2–22 Starting with an energy balance on a rectangular volume element, derive the one-dimensional transient heat conduction equation for a plane wall with constant thermal conductivity and no heat generation.

2–23 Starting with an energy balance on a cylindrical shell volume element, derive the steady one-dimensional heat conduction equation for a long cylinder with constant thermal conductivity in which heat is generated at a rate of $\dot{e}_{gen}$.

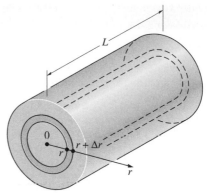

FIGURE P2–23

2–24 Starting with an energy balance on a spherical shell volume element, derive the one-dimensional transient heat conduction equation for a sphere with constant thermal conductivity and no heat generation.

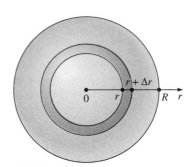

FIGURE P2–24

2–25 Consider a medium in which the heat conduction equation is given in its simplest form as

$$\frac{\partial^2 T}{\partial x^2} = \frac{1}{\alpha} \frac{\partial T}{\partial t}$$

(a) Is heat transfer steady or transient?
(b) Is heat transfer one-, two-, or three-dimensional?
(c) Is there heat generation in the medium?
(d) Is the thermal conductivity of the medium constant or variable?

2–26 Consider a medium in which the heat conduction equation is given in its simplest form as

$$\frac{1}{r} \frac{d}{dr}\left(rk\frac{dT}{dr}\right) + \dot{e}_{gen} = 0$$

(a) Is heat transfer steady or transient?
(b) Is heat transfer one-, two-, or three-dimensional?
(c) Is there heat generation in the medium?
(d) Is the thermal conductivity of the medium constant or variable?

2–27 Consider a medium in which the heat conduction equation is given in its simplest form as

$$\frac{1}{r^2}\frac{\partial}{\partial r}\left(r^2\frac{\partial T}{\partial r}\right) = \frac{1}{\alpha}\frac{\partial T}{\partial t}$$

(a) Is heat transfer steady or transient?
(b) Is heat transfer one-, two-, or three-dimensional?
(c) Is there heat generation in the medium?
(d) Is the thermal conductivity of the medium constant or variable?

2–28 Consider a medium in which the heat conduction equation is given in its simplest form as

$$r\frac{d^2T}{dr^2} + \frac{dT}{dr} = 0$$

(a) Is heat transfer steady or transient?
(b) Is heat transfer one-, two-, or three-dimensional?
(c) Is there heat generation in the medium?
(d) Is the thermal conductivity of the medium constant or variable?

2–29 Starting with an energy balance on a volume element, derive the two-dimensional transient heat conduction equation in rectangular coordinates for $T(x, y, t)$ for the case of constant thermal conductivity and no heat generation.

2–30 Starting with an energy balance on a ring-shaped volume element, derive the two-dimensional steady heat conduction equation in cylindrical coordinates for $T(r, z)$ for the case of constant thermal conductivity and no heat generation.

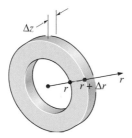

FIGURE P2–30

2–31 Starting with an energy balance on a disk volume element, derive the one-dimensional transient heat conduction equation for $T(z, t)$ in a cylinder of diameter D with an insulated side surface for the case of constant thermal conductivity with heat generation.

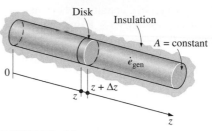

FIGURE P2–31

2–32 Consider a medium in which the heat conduction equation is given in its simplest form as

$$\frac{\partial^2 T}{\partial x^2} + \frac{\partial^2 T}{\partial y^2} = \frac{1}{\alpha}\frac{\partial T}{\partial t}$$

(a) Is heat transfer steady or transient?
(b) Is heat transfer one-, two-, or three-dimensional?
(c) Is there heat generation in the medium?
(d) Is the thermal conductivity of the medium constant or variable?

2–33 Consider a medium in which the heat conduction equation is given in its simplest form as

$$\frac{1}{r}\frac{\partial}{\partial r}\left(kr\frac{\partial T}{\partial r}\right) + \frac{\partial}{\partial z}\left(k\frac{\partial T}{\partial z}\right) + \dot{e}_{gen} = 0$$

(a) Is heat transfer steady or transient?
(b) Is heat transfer one-, two-, or three-dimensional?
(c) Is there heat generation in the medium?
(d) Is the thermal conductivity of the medium constant or variable?

2–34 Consider a medium in which the heat conduction equation is given in its simplest form as

$$\frac{1}{r^2}\frac{\partial}{\partial r}\left(r^2\frac{\partial T}{\partial r}\right) + \frac{1}{r^2\sin^2\theta}\frac{\partial^2 T}{\partial\phi^2} = \frac{1}{\alpha}\frac{\partial T}{\partial t}$$

(a) Is heat transfer steady or transient?
(b) Is heat transfer one-, two-, or three-dimensional?
(c) Is there heat generation in the medium?
(d) Is the thermal conductivity of the medium constant or variable?

Boundary and Initial Conditions; Formulation of Heat Conduction Problems

2–35C What is a boundary condition? How many boundary conditions do we need to specify for a two-dimensional heat conduction problem?

2–36C What is an initial condition? How many initial conditions do we need to specify for a two-dimensional heat conduction problem?

2–37C What is a thermal symmetry boundary condition? How is it expressed mathematically?

2–38C How is the boundary condition on an insulated surface expressed mathematically?

2–39C It is claimed that the temperature profile in a medium must be perpendicular to an insulated surface. Is this a valid claim? Explain.

2–40C Why do we try to avoid the radiation boundary conditions in heat transfer analysis?

2–41 Consider a spherical container of inner radius r_1, outer radius r_2, and thermal conductivity k. Express the boundary condition on the inner surface of the container for steady one-dimensional conduction for the following cases: (*a*) specified temperature of 50°C, (*b*) specified heat flux of 30 W/m² toward the center, (*c*) convection to a medium at T_∞ with a heat transfer coefficient of h.

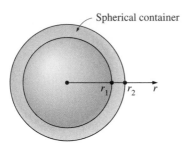

FIGURE P2–41

2–42 Heat is generated in a long wire of radius r_o at a constant rate of $\dot{e}_{gen}$ per unit volume. The wire is covered with a plastic insulation layer. Express the heat flux boundary condition at the interface in terms of the heat generated.

2–43 Consider a long pipe of inner radius r_1, outer radius r_2, and thermal conductivity k. The outer surface of the pipe is subjected to convection to a medium at T_∞ with a heat transfer coefficient of h, but the direction of heat transfer is not known. Express the convection boundary condition on the outer surface of the pipe.

2–44 Consider a spherical shell of inner radius r_1, outer radius r_2, thermal conductivity k, and emissivity ε. The outer surface of the shell is subjected to radiation to surrounding surfaces at T_{surr}, but the direction of heat transfer is not known. Express the radiation boundary condition on the outer surface of the shell.

2–45 A container consists of two spherical layers, *A* and *B*, that are in perfect contact. If the radius of the interface is r_o, express the boundary conditions at the interface.

2–46 Consider a steel pan used to boil water on top of an electric range. The bottom section of the pan is $L = 0.5$ cm thick and has a diameter of $D = 20$ cm. The electric heating unit on the range top consumes 1250 W of power during cooking, and 85 percent of the heat generated in the heating element is transferred uniformly to the pan. Heat transfer from the top surface of the bottom section to the water is by convection with a heat transfer coefficient of h. Assuming constant thermal conductivity and one-dimensional heat transfer, express the mathematical formulation (the differential equation and the boundary conditions) of this heat conduction problem during steady operation. Do not solve.

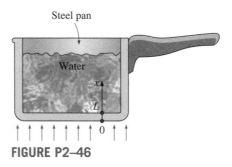

FIGURE P2–46

2–47E A 2-kW resistance heater wire whose thermal conductivity is $k = 10.4$ Btu/h · ft · R has a radius of $r_o = 0.06$ in. and a length of $L = 15$ in, and is used for space heating. Assuming constant thermal conductivity and one-dimensional heat transfer, express the mathematical formulation (the differential equation and the boundary conditions) of this heat conduction problem during steady operation. Do not solve.

2–48 Consider an aluminum pan used to cook stew on top of an electric range. The bottom section of the pan is $L = 0.25$ cm thick and has a diameter of $D = 18$ cm. The electric heating unit on the range top consumes 900 W of power during cooking, and 90 percent of the heat generated in the heating element is transferred to the pan. During steady operation, the temperature of the inner surface of the pan is measured to be 108°C.

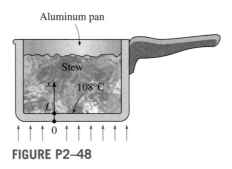

FIGURE P2–48

Assuming temperature-dependent thermal conductivity and one-dimensional heat transfer, express the mathematical formulation (the differential equation and the boundary conditions) of this heat conduction problem during steady operation. Do not solve.

2–49 Water flows through a pipe at an average temperature of $T_\infty = 70°C$. The inner and outer radii of the pipe are $r_1 = 6$ cm and $r_2 = 6.5$ cm, respectively. The outer surface of the pipe is wrapped with a thin electric heater that consumes 300 W per m length of the pipe. The exposed surface of the heater is heavily insulated so that the entire heat generated in the heater is transferred to the pipe. Heat is transferred from the inner surface of the pipe to the water by convection with a heat transfer coefficient of $h = 85$ W/m² · K. Assuming constant thermal conductivity and one-dimensional heat transfer, express the mathematical formulation (the differential equation and the boundary conditions) of the heat conduction in the pipe during steady operation. Do not solve.

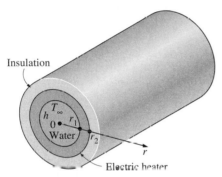

FIGURE P2–49

2–50 A spherical metal ball of radius r_o is heated in an oven to a temperature of T_i throughout and is then taken out of the oven and dropped into a large body of water at T_∞ where it is cooled by convection with an average convection heat transfer coefficient of h. Assuming constant thermal conductivity and transient one-dimensional heat transfer, express the mathematical formulation (the differential equation and the boundary and initial conditions) of this heat conduction problem. Do not solve.

2–51 A spherical metal ball of radius r_o is heated in an oven to a temperature of T_i throughout and is then taken out of the oven and allowed to cool in ambient air at T_∞ by convection and radiation. The emissivity of the outer surface of the cylinder is ε, and the temperature of the surrounding surfaces is T_{surr}. The average convection heat transfer coefficient is estimated to be h. Assuming variable thermal conductivity and transient one-dimensional heat transfer, express the mathematical formulation (the differential equation and the boundary

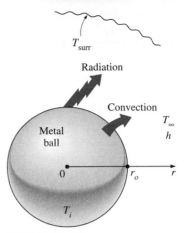

FIGURE P2–51

and initial conditions) of this heat conduction problem. Do not solve.

2–52 Consider the East wall of a house of thickness L. The outer surface of the wall exchanges heat by both convection and radiation. The interior of the house is maintained at $T_{\infty 1}$, while the ambient air temperature outside remains at $T_{\infty 2}$. The sky, the ground, and the surfaces of the surrounding structures at this location can be modeled as a surface at an effective temperature of T_{sky} for radiation exchange on the outer surface. The radiation exchange between the inner surface of the wall and the surfaces of the walls, floor, and ceiling it faces is negligible. The convection heat transfer coefficients on the inner and outer surfaces of the wall are h_1 and h_2, respectively. The thermal conductivity of the wall material is k and the emissivity of the outer surface is ε_2. Assuming the heat transfer through the wall to be steady and one-dimensional, express the mathematical formulation (the differential equation and the boundary and initial conditions) of this heat conduction problem. Do not solve.

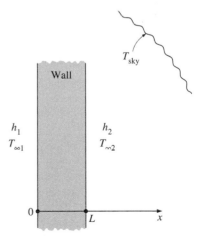

FIGURE P2–52

Solution of Steady One-Dimensional Heat Conduction Problems

2–53C Consider one-dimensional heat conduction through a large plane wall with no heat generation that is perfectly insulated on one side and is subjected to convection and radiation on the other side. It is claimed that under steady conditions, the temperature in a plane wall must be uniform (the same everywhere). Do you agree with this claim? Why?

2–54C It is stated that the temperature in a plane wall with constant thermal conductivity and no heat generation varies linearly during steady one-dimensional heat conduction. Will this still be the case when the wall loses heat by radiation from its surfaces?

2–55C Consider a solid cylindrical rod whose ends are maintained at constant but different temperatures while the side surface is perfectly insulated. There is no heat generation. It is claimed that the temperature along the axis of the rod varies linearly during steady heat conduction. Do you agree with this claim? Why?

2–56C Consider a solid cylindrical rod whose side surface is maintained at a constant temperature while the end surfaces are perfectly insulated. The thermal conductivity of the rod material is constant and there is no heat generation. It is claimed that the temperature in the radial direction within the rod will not vary during steady heat conduction. Do you agree with this claim? Why?

2–57 Consider a large plane wall of thickness $L = 0.4$ m, thermal conductivity $k = 2.3$ W/m · °C, and surface area $A = 30$ m². The left side of the wall is maintained at a constant temperature of $T_1 = 90$°C while the right side loses heat by convection to the surrounding air at $T_\infty = 25$°C with a heat transfer coefficient of $h = 24$ W/m² · °C. Assuming constant thermal conductivity and no heat generation in the wall, (a) express the differential equation and the boundary conditions for steady one-dimensional heat conduction through the wall, (b) obtain a relation for the variation of temperature in the wall by solving the differential equation, and (c) evaluate the rate of heat transfer through the wall.   *Answer: (c) 9045 W*

2–58 Consider a solid cylindrical rod of length 0.15 m and diameter 0.05 m. The top and bottom surfaces of the rod are maintained at constant temperatures of 20°C and 95°C, respectively, while the side surface is perfectly insulated. Determine the rate of heat transfer through the rod if it is made of (a) copper, $k = 380$ W/m · °C, (b) steel, $k = 18$ W/m · °C, and (c) granite, $k = 1.2$ W/m · °C.

2–59 Reconsider Prob. 2–58. Using EES (or other) software, plot the rate of heat transfer as a function of the thermal conductivity of the rod in the range of 1 W/m · °C to 400 W/m · °C. Discuss the results.

2–60 Consider the base plate of an 800-W household iron with a thickness of $L = 0.6$ cm, base area of $A = 160$ cm², and

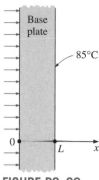

FIGURE P2–60

thermal conductivity of $k = 20$ W/m · °C. The inner surface of the base plate is subjected to uniform heat flux generated by the resistance heaters inside. When steady operating conditions are reached, the outer surface temperature of the plate is measured to be 85°C. Disregarding any heat loss through the upper part of the iron, (a) express the differential equation and the boundary conditions for steady one-dimensional heat conduction through the plate, (b) obtain a relation for the variation of temperature in the base plate by solving the differential equation, and (c) evaluate the inner surface temperature.
   Answer: (c) 100°C

2–61 Repeat Prob. 2–60 for a 1200-W iron.

2–62 Reconsider Prob. 2–60. Using the relation obtained for the variation of temperature in the base plate, plot the temperature as a function of the distance x in the range of $x = 0$ to $x = L$, and discuss the results. Use the EES (or other) software.

2–63 Consider a chilled-water pipe of length L, inner radius r_1, outer radius r_2, and thermal conductivity k. Water flows in the pipe at a temperature T_f and the heat transfer coefficient at the inner surface is h. If the pipe is well-insulated on the outer surface, (a) express the differential equation and the boundary conditions for steady one-dimensional heat conduction through the pipe and (b) obtain a relation for the variation of temperature in the pipe by solving the differential equation.

2–64E Consider a steam pipe of length $L = 30$ ft, inner radius $r_1 = 2$ in, outer radius $r_2 = 2.4$ in, and thermal conductivity $k = 7.2$ Btu/h · ft · °F. Steam is flowing through the pipe at an average temperature of 250°F, and the average convection heat transfer coefficient on the inner surface is given to be $h = 12.5$ Btu/h · ft² · °F . If the average temperature on the outer surfaces of the pipe is $T_2 = 160$°F, (a) express the differential equation and the boundary conditions for steady one-dimensional heat conduction through the pipe, (b) obtain a relation for the variation of temperature in the pipe by solving the differential equation, and (c) evaluate the rate of heat loss from the steam through the pipe.
   Answer: (c) 33,600 Btu/h

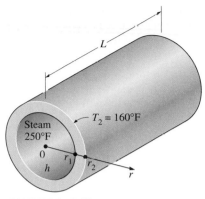

FIGURE P2–64E

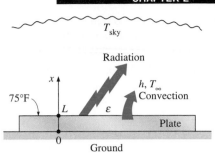

FIGURE P2–68E

2–65 A spherical container of inner radius $r_1 = 2$ m, outer radius $r_2 = 2.1$ m, and thermal conductivity $k = 30$ W/m · °C is filled with iced water at 0°C. The container is gaining heat by convection from the surrounding air at $T_\infty = 25$°C with a heat transfer coefficient of $h = 18$ W/m² · °C. Assuming the inner surface temperature of the container to be 0°C, (a) express the differential equation and the boundary conditions for steady one-dimensional heat conduction through the container, (b) obtain a relation for the variation of temperature in the container by solving the differential equation, and (c) evaluate the rate of heat gain to the iced water.

2–66 Consider a large plane wall of thickness $L = 0.3$ m, thermal conductivity $k = 2.5$ W/m · °C, and surface area $A = 12$ m². The left side of the wall at $x = 0$ is subjected to a net heat flux of $\dot{q}_0 = 700$ W/m² while the temperature at that surface is measured to be $T_1 = 80$°C. Assuming constant thermal conductivity and no heat generation in the wall, (a) express the differential equation and the boundary conditions for steady one-dimensional heat conduction through the wall, (b) obtain a relation for the variation of temperature in the wall by solving the differential equation, and (c) evaluate the temperature of the right surface of the wall at $x = L$. *Answer:* (c) −4°C

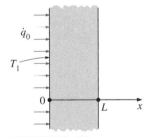

FIGURE P2–66

2–67 Repeat Prob. 2–66 for a heat flux of 1050 W/m² and a surface temperature of 90°C at the left surface at $x = 0$.

2–68E A large steel plate having a thickness of $L = 4$ in, thermal conductivity of $k = 7.2$ Btu/h · ft · °F, and an emissivity of $\varepsilon = 0.7$ is lying on the ground. The exposed surface of

the plate at $x = L$ is known to exchange heat by convection with the ambient air at $T_\infty = 90$°F with an average heat transfer coefficient of $h = 12$ Btu/h · ft² · °F as well as by radiation with the open sky with an equivalent sky temperature of $T_{sky} = 480$ R. Also, the temperature of the upper surface of the plate is measured to be 75°F. Assuming steady one-dimensional heat transfer, (a) express the differential equation and the boundary conditions for heat conduction through the plate, (b) obtain a relation for the variation of temperature in the plate by solving the differential equation, and (c) determine the value of the lower surface temperature of the plate at $x = 0$.

2–69E Repeat Prob. 2–68E by disregarding radiation heat transfer.

2–70 When a long section of a compressed air line passes through the outdoors, it is observed that the moisture in the compressed air freezes in cold weather, disrupting and even completely blocking the air flow in the pipe. To avoid this problem, the outer surface of the pipe is wrapped with electric strip heaters and then insulated.

Consider a compressed air pipe of length $L = 6$ m, inner radius $r_1 = 3.7$ cm, outer radius $r_2 = 4.0$ cm, and thermal conductivity $k = 14$ W/m · °C equipped with a 300-W strip heater. Air is flowing through the pipe at an average temperature of -10°C, and the average convection heat transfer coefficient on the inner surface is $h = 30$ W/m² · °C. Assuming 15 percent of the heat generated in the strip heater is lost through the insulation, (a) express the differential equation and the boundary conditions for steady one-dimensional heat conduction through the pipe, (b) obtain a relation for the variation of temperature in the

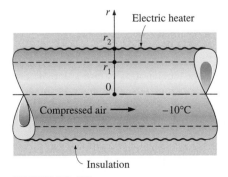

FIGURE P2–70

pipe material by solving the differential equation, and (c) evaluate the inner and outer surface temperatures of the pipe.
Answers: (c) −3.91°C, −3.87°C

2–71 [EES] Reconsider Prob. 2–70. Using the relation obtained for the variation of temperature in the pipe material, plot the temperature as a function of the radius r in the range of $r = r_1$ to $r = r_2$, and discuss the results. Use the EES (or other) software.

2–72 In a food processing facility, a spherical container of inner radius $r_1 = 40$ cm, outer radius $r_2 = 41$ cm, and thermal conductivity $k = 1.5$ W/m · °C is used to store hot water and to keep it at 100°C at all times. To accomplish this, the outer surface of the container is wrapped with a 500-W electric strip heater and then insulated. The temperature of the inner surface of the container is observed to be nearly 100°C at all times. Assuming 10 percent of the heat generated in the heater is lost through the insulation, (a) express the differential equation and the boundary conditions for steady one-dimensional heat conduction through the container, (b) obtain a relation for the variation of temperature in the container material by solving the differential equation, and (c) evaluate the outer surface temperature of the container. Also determine how much water at 100°C this tank can supply steadily if the cold water enters at 20°C.

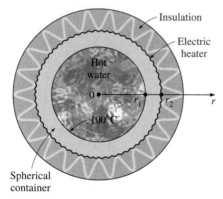

Insulation
Electric heater
Hot water
0 r_1 r_2 r
100°C
Spherical container

FIGURE P2–72

2–73 [EES] Reconsider Prob. 2–72. Using the relation obtained for the variation of temperature in the container material, plot the temperature as a function of the radius r in the range of $r = r_1$ to $r = r_2$, and discuss the results. Use the EES (or other) software.

Heat Generation in a Solid

2–74C Does heat generation in a solid violate the first law of thermodynamics, which states that energy cannot be created or destroyed? Explain.

2–75C What is heat generation? Give some examples.

2–76C An iron is left unattended and its base temperature rises as a result of resistance heating inside. When will the rate of heat generation inside the iron be equal to the rate of heat loss from the iron?

2–77C Consider the uniform heating of a plate in an environment at a constant temperature. Is it possible for part of the heat generated in the left half of the plate to leave the plate through the right surface? Explain.

2–78C Consider uniform heat generation in a cylinder and a sphere of equal radius made of the same material in the same environment. Which geometry will have a higher temperature at its center? Why?

2–79 A 2-kW resistance heater wire with thermal conductivity of $k = 20$ W/m · °C, a diameter of $D = 4$ mm, and a length of $L = 0.9$ m is used to boil water. If the outer surface temperature of the resistance wire is $T_s = 110$°C, determine the temperature at the center of the wire.

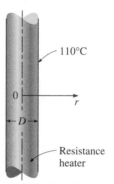

110°C
0 r
D
Resistance heater

FIGURE P2–79

2–80 Consider a long solid cylinder of radius $r_o = 4$ cm and thermal conductivity $k = 25$ W/m · °C. Heat is generated in the cylinder uniformly at a rate of $\dot{e}_{gen} = 35$ W/cm³. The side surface of the cylinder is maintained at a constant temperature of $T_s = 80$°C. The variation of temperature in the cylinder is given by

$$T(r) = \frac{\dot{e}_{gen}r_o^2}{k}\left[1 - \left(\frac{r}{r_o}\right)^2\right] + T_s$$

Based on this relation, determine (a) if the heat conduction is steady or transient, (b) if it is one-, two-, or three-dimensional, and (c) the value of heat flux on the side surface of the cylinder at $r = r_o$.

2–81 [EES] Reconsider Prob. 2–80. Using the relation obtained for the variation of temperature in the cylinder, plot the temperature as a function of the radius r in the range of $r = 0$ to $r = r_o$, and discuss the results. Use the EES (or other) software.

2–82 Consider a large plate of thickness L and thermal conductivity k in which heat is generated uniformly at a rate of $\dot{e}_{gen}$. One side of the plate is insulated while the other side is exposed to an environment at T_∞ with a heat transfer coefficient of h. (a) Express the differential equation and the boundary conditions for steady one-dimensional heat conduction through the plate, (b) determine the variation of temperature in the plate, and (c) obtain relations for the temperatures on both surfaces and the maximum temperature rise in the plate in terms of given parameters.

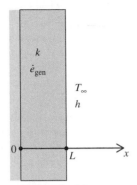

FIGURE P2–82

2–83E A long homogeneous resistance wire of radius $r_o = 0.25$ in and thermal conductivity $k = 8.6$ Btu/h · ft · °F is being used to boil water at atmospheric pressure by the passage of electric current. Heat is generated in the wire uniformly as a result of resistance heating at a rate of 1800 Btu/h · in³. The heat generated is transferred to water at 212°F by convection with an average heat transfer coefficient of $h = 820$ Btu/h · ft² · °F. Assuming steady one-dimensional heat transfer, (a) express the differential equation and the boundary conditions for heat conduction through the wire, (b) obtain a relation for the variation of temperature in the wire by solving the differential equation, and (c) determine the temperature at the centerline of the wire. *Answer:* (c) 290.8°F

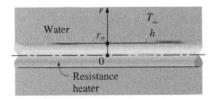

FIGURE P2–83E

2–84E Reconsider Prob. 2–83E. Using the relation obtained for the variation of temperature in the wire, plot the temperature at the centerline of the wire as a function of the heat generation $\dot{e}_{gen}$ in the range of 400 Btu/h · in³ to 2400 Btu/h · in³, and discuss the results. Use the EES (or other) software.

2–85 In a nuclear reactor, 1-cm-diameter cylindrical uranium rods cooled by water from outside serve as the fuel. Heat is

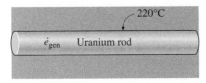

FIGURE P2–85

generated uniformly in the rods ($k = 29.5$ W/m · °C) at a rate of 4×10^7 W/m³. If the outer surface temperature of rods is 220°C, determine the temperature at their center.

2–86 Consider a large 3-cm-thick stainless steel plate ($k = 15.1$ W/m · °C) in which heat is generated uniformly at a rate of 5×10^5 W/m³. Both sides of the plate are exposed to an environment at 30°C with a heat transfer coefficient of 60 W/m² · °C. Explain where in the plate the highest and the lowest temperatures will occur, and determine their values.

2–87 Consider a large 5-cm-thick brass plate ($k = 111$ W/m · °C) in which heat is generated uniformly at a rate of 2×10^5 W/m³. One side of the plate is insulated while the other side is exposed to an environment at 25°C with a heat transfer coefficient of 44 W/m² · °C. Explain where in the plate the highest and the lowest temperatures will occur, and determine their values.

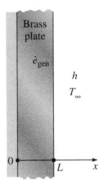

FIGURE P2–87

2–88 Reconsider Prob. 2–87. Using EES (or other) software, investigate the effect of the heat transfer coefficient on the highest and lowest temperatures in the plate. Let the heat transfer coefficient vary from 20 W/m² · °C to 100 W/m² · °C. Plot the highest and lowest temperatures as a function of the heat transfer coefficient, and discuss the results.

2–89 A 6-m-long 2-kW electrical resistance wire is made of 0.2-cm-diameter stainless steel ($k = 15.1$ W/m · °C). The resistance wire operates in an environment at 20°C with a heat transfer coefficient of 175 W/m² · °C at the outer surface. Determine the surface temperature of the wire (a) by using the applicable relation and (b) by setting up the proper differential equation and solving it. *Answers:* (a) 323°C, (b) 323°C

2–90E Heat is generated uniformly at a rate of 3 kW per ft length in a 0.08-in-diameter electric resistance wire made of nickel steel ($k = 5.8$ Btu/h · ft · °F). Determine the temperature difference between the centerline and the surface of the wire.

2–91E Repeat Prob. 2–90E for a manganese wire ($k = 4.5$ Btu/h · ft · °F).

2–92 Consider a homogeneous spherical piece of radioactive material of radius $r_o = 0.04$ m that is generating heat at a constant rate of $\dot{e}_{gen} = 4 \times 10^7$ W/m³. The heat generated is dissipated to the environment steadily. The outer surface of the sphere is maintained at a uniform temperature of 80°C and the thermal conductivity of the sphere is $k = 15$ W/m · °C. Assuming steady one-dimensional heat transfer, (a) express the differential equation and the boundary conditions for heat conduction through the sphere, (b) obtain a relation for the variation of temperature in the sphere by solving the differential equation, and (c) determine the temperature at the center of the sphere.

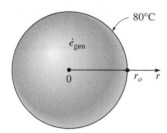

FIGURE P2–92

2–93 [EES] Reconsider Prob. 2–92. Using the relation obtained for the variation of temperature in the sphere, plot the temperature as a function of the radius r in the range of $r = 0$ to $r = r_o$. Also, plot the center temperature of the sphere as a function of the thermal conductivity in the range of 10 W/m · °C to 400 W/m · °C. Discuss the results. Use the EES (or other) software.

2–94 A long homogeneous resistance wire of radius $r_o = 5$ mm is being used to heat the air in a room by the passage of electric current. Heat is generated in the wire uniformly at a rate of 5×10^7 W/m³ as a result of resistance heating. If the temperature of the outer surface of the wire remains at 180°C, determine the temperature at $r = 3.5$ mm after steady operation conditions are reached. Take the thermal conductivity of the wire to be $k = 8$ W/m · °C. *Answer:* 200°C

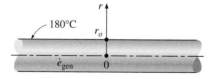

FIGURE P2–94

2–95 Consider a large plane wall of thickness $L = 0.05$ m. The wall surface at $x = 0$ is insulated, while the surface at $x = L$ is maintained at a temperature of 30°C. The thermal conductivity of the wall is $k = 30$ W/m · °C, and heat is generated in the wall at a rate of $\dot{e}_{gen} = \dot{e}_0 e^{-0.5x/L}$ W/m³ where $\dot{e}_0 = 8 \times 10^6$ W/m³. Assuming steady one-dimensional heat transfer, (a) express the differential equation and the boundary conditions for heat conduction through the wall, (b) obtain a relation for the variation of temperature in the wall by solving the differential

equation, and (c) determine the temperature of the insulated surface of the wall. *Answer:* (c) 314°C

2–96 [EES] Reconsider Prob. 2–95. Using the relation given for the heat generation in the wall, plot the heat generation as a function of the distance x in the range of $x = 0$ to $x = L$, and discuss the results. Use the EES (or other) software.

Variable Thermal Conductivity, $k(T)$

2–97C Consider steady one-dimensional heat conduction in a plane wall, long cylinder, and sphere with constant thermal conductivity and no heat generation. Will the temperature in any of these mediums vary linearly? Explain.

2–98C Is the thermal conductivity of a medium, in general, constant or does it vary with temperature?

2–99C Consider steady one-dimensional heat conduction in a plane wall in which the thermal conductivity varies linearly. The error involved in heat transfer calculations by assuming constant thermal conductivity at the average temperature is (a) none, (b) small, or (c) significant.

2–100C The temperature of a plane wall during steady one-dimensional heat conduction varies linearly when the thermal conductivity is constant. Is this still the case when the thermal conductivity varies linearly with temperature?

2–101C When the thermal conductivity of a medium varies linearly with temperature, is the average thermal conductivity always equivalent to the conductivity value at the average temperature?

2–102 Consider a plane wall of thickness L whose thermal conductivity varies in a specified temperature range as $k(T) = k_0(1 + \beta T^2)$ where k_0 and β are two specified constants. The wall surface at $x = 0$ is maintained at a constant temperature of T_1, while the surface at $x = L$ is maintained at T_2. Assuming steady one-dimensional heat transfer, obtain a relation for the heat transfer rate through the wall.

2–103 Consider a cylindrical shell of length L, inner radius r_1, and outer radius r_2 whose thermal conductivity varies linearly in a specified temperature range as $k(T) = k_0(1 + \beta T)$

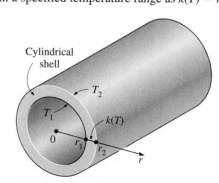

FIGURE P2–103

where k_0 and β are two specified constants. The inner surface of the shell is maintained at a constant temperature of T_1, while the outer surface is maintained at T_2. Assuming steady one-dimensional heat transfer, obtain a relation for (a) the heat transfer rate through the wall and (b) the temperature distribution $T(r)$ in the shell.

2–104 Consider a spherical shell of inner radius r_1 and outer radius r_2 whose thermal conductivity varies linearly in a specified temperature range as $k(T) = k_0(1 + \beta T)$ where k_0 and β are two specified constants. The inner surface of the shell is maintained at a constant temperature of T_1 while the outer surface is maintained at T_2. Assuming steady one-dimensional heat transfer, obtain a relation for (a) the heat transfer rate through the shell and (b) the temperature distribution $T(r)$ in the shell.

2–105 Consider a 1.5-m-high and 0.6-m-wide plate whose thickness is 0.15 m. One side of the plate is maintained at a constant temperature of 500 K while the other side is maintained at 350 K. The thermal conductivity of the plate can be assumed to vary linearly in that temperature range as $k(T) = k_0(1 + \beta T)$ where $k_0 = 25$ W/m · K and $\beta = 8.7 \times 10^{-4}$ K^{-1}. Disregarding the edge effects and assuming steady one-dimensional heat transfer, determine the rate of heat conduction through the plate. *Answer:* 30.8 kW

2–106 ⬛EES⬛ Reconsider Prob. 2–105. Using EES (or other) software, plot the rate of heat conduction through the plate as a function of the temperature of the hot side of the plate in the range of 400 K to 700 K. Discuss the results.

Special Topic: Review of Differential Equations

2–107C Why do we often utilize simplifying assumptions when we derive differential equations?

2–108C What is a variable? How do you distinguish a dependent variable from an independent one in a problem?

2–109C Can a differential equation involve more than one independent variable? Can it involve more than one dependent variable? Give examples.

2–110C What is the geometrical interpretation of a derivative? What is the difference between partial derivatives and ordinary derivatives?

2–111C What is the difference between the degree and the order of a derivative?

2–112C Consider a function $f(x, y)$ and its partial derivative $\partial f/\partial x$. Under what conditions will this partial derivative be equal to the ordinary derivative df/dx?

2–113C Consider a function $f(x)$ and its derivative df/dx. Does this derivative have to be a function of x?

2–114C How is integration related to derivation?

2–115C What is the difference between an algebraic equation and a differential equation?

2–116C What is the difference between an ordinary differential equation and a partial differential equation?

2–117C How is the order of a differential equation determined?

2–118C How do you distinguish a linear differential equation from a nonlinear one?

2–119C How do you recognize a linear homogeneous differential equation? Give an example and explain why it is linear and homogeneous.

2–120C How do differential equations with constant coefficients differ from those with variable coefficients? Give an example for each type.

2–121C What kind of differential equations can be solved by direct integration?

2–122C Consider a third order linear and homogeneous differential equation. How many arbitrary constants will its general solution involve?

Review Problems

2–123 Consider a small hot metal object of mass m and specific heat c that is initially at a temperature of T_i. Now the object is allowed to cool in an environment at T_∞ by convection with a heat transfer coefficient of h. The temperature of the metal object is observed to vary uniformly with time during cooling. Writing an energy balance on the entire metal object, derive the differential equation that describes the variation of temperature of the ball with time, $T(t)$. Assume constant thermal conductivity and no heat generation in the object. Do not solve.

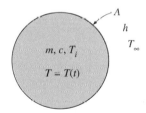

FIGURE P2–123

2–124 Consider a long rectangular bar of length a in the x-direction and width b in the y-direction that is initially at a uniform temperature of T_i. The surfaces of the bar at $x = 0$ and $y = 0$ are insulated, while heat is lost from the other two surfaces by convection to the surrounding medium at temperature T_∞ with a heat transfer coefficient of h. Assuming constant thermal conductivity and transient two-dimensional heat transfer with no heat generation, express the mathematical formulation (the differential equation and the boundary and initial conditions) of this heat conduction problem. Do not solve.

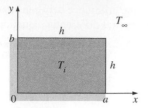

FIGURE P2–124

2–125 Consider a short cylinder of radius r_o and height H in which heat is generated at a constant rate of $\dot{e}_{gen}$. Heat is lost from the cylindrical surface at $r = r_o$ by convection to the surrounding medium at temperature T_∞ with a heat transfer coefficient of h. The bottom surface of the cylinder at $z = 0$ is insulated, while the top surface at $z = H$ is subjected to uniform heat flux $\dot{q}_H$. Assuming constant thermal conductivity and steady two-dimensional heat transfer, express the mathematical formulation (the differential equation and the boundary conditions) of this heat conduction problem. Do not solve.

2–126E Consider a large plane wall of thickness $L = 0.8$ ft and thermal conductivity $k = 1.2$ Btu/h $\cdot$ ft $\cdot$ °F. The wall is covered with a material that has an emissivity of $\varepsilon = 0.80$ and a solar absorptivity of $\alpha = 0.60$. The inner surface of the wall is maintained at $T_1 = 520$ R at all times, while the outer surface is exposed to solar radiation that is incident at a rate of $\dot{q}_{solar} = 300$ Btu/h $\cdot$ ft². The outer surface is also losing heat by radiation to deep space at 0 K. Determine the temperature of the outer surface of the wall and the rate of heat transfer through the wall when steady operating conditions are reached.

Answers: 554 R, 50.9 Btu/h $\cdot$ ft²

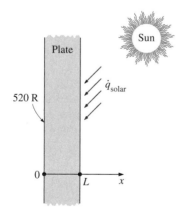

FIGURE P2–126E

2–127E Repeat Prob. 2–126E for the case of no solar radiation incident on the surface.

2–128 Consider a steam pipe of length L, inner radius r_1, outer radius r_2, and constant thermal conductivity k. Steam flows inside the pipe at an average temperature of T_i with a convection heat transfer coefficient of h_i. The outer surface of the pipe is exposed to convection to the surrounding air at a temperature of T_0 with a heat transfer coefficient of h_o. Assuming steady one-dimensional heat conduction through the pipe, (a) express the differential equation and the boundary conditions for heat conduction through the pipe material, (b) obtain a relation for the variation of temperature in the pipe material by solving the differential equation, and (c) obtain a relation for the temperature of the outer surface of the pipe.

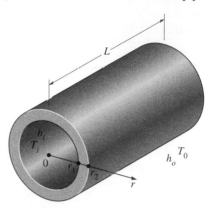

FIGURE P2–128

2–129 The boiling temperature of nitrogen at atmospheric pressure at sea level (1 atm pressure) is -196°C. Therefore, nitrogen is commonly used in low temperature scientific studies since the temperature of liquid nitrogen in a tank open to the atmosphere remains constant at -196°C until the liquid nitrogen in the tank is depleted. Any heat transfer to the tank results in the evaporation of some liquid nitrogen, which has a heat of vaporization of 198 kJ/kg and a density of 810 kg/m³ at 1 atm.

Consider a thick-walled spherical tank of inner radius $r_1 = 2$ m, outer radius $r_2 = 2.1$ m, and constant thermal conductivity $k = 18$ W/m $\cdot$ °C. The tank is initially filled with liquid nitrogen at 1 atm and -196°C, and is exposed to ambient air at $T_\infty = 20$°C with a heat transfer coefficient of $h = 25$ W/m² $\cdot$ °C. The inner surface temperature of the spherical tank is observed to be almost the same as the temperature of the nitrogen inside. Assuming steady one-dimensional heat transfer, (a) express the differential equation and the boundary conditions for heat conduction through the tank, (b) obtain a relation for the variation of temperature in the tank material by solving the differential equation, and (c) determine the rate of evaporation of the liquid nitrogen in the tank as a result of the heat transfer from the ambient air. *Answer:* (c) 1.32 kg/s

2–130 Repeat Prob. 2–129 for liquid oxygen, which has a boiling temperature of -183°C, a heat of vaporization of 213 kJ/kg, and a density of 1140 kg/m³ at 1 atm.

2–131 Consider a large plane wall of thickness $L = 0.4$ m and thermal conductivity $k = 8.4$ W/m $\cdot$ °C. There is no access to the inner side of the wall at $x = 0$ and thus the thermal conditions on that surface are not known. However, the outer surface of the wall at $x = L$, whose emissivity is $\varepsilon - 0.7$, is known to exchange heat by convection with ambient air at $T_\infty = 25°C$ with an average heat transfer coefficient of $h = 14$ W/m$^2 \cdot °C$ as well as by radiation with the surrounding surfaces at an average temperature of $T_{surr} = 290$ K. Further, the temperature of the outer surface is measured to be $T_2 = 45°C$. Assuming steady one-dimensional heat transfer, (a) express the differential equation and the boundary conditions for heat conduction through the plate, (b) obtain a relation for the temperature of the outer surface of the plate by solving the differential equation, and (c) evaluate the inner surface temperature of the wall at $x = 0$.   *Answer:* (c) 64.3°C

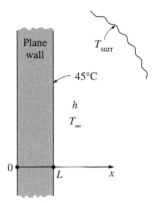

FIGURE P2–131

2–132   A 1000-W iron is left on the iron board with its base exposed to ambient air at 26°C. The base plate of the iron has a thickness of $L = 0.5$ cm, base area of $A = 150$ cm^2, and thermal conductivity of $k = 18$ W/m $\cdot °C$. The inner surface of the base plate is subjected to uniform heat flux generated by the resistance heaters inside. The outer surface of the base plate whose emissivity is $\varepsilon = 0.7$, loses heat by convection to ambient air with an average heat transfer coefficient of $h = 30$ W/m$^2 \cdot °C$ as well as by radiation to the surrounding

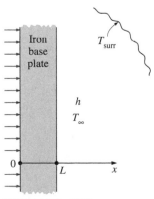

FIGURE P2–132

surfaces at an average temperature of $T_{surr} = 295$ K. Disregarding any heat loss through the upper part of the iron, (a) express the differential equation and the boundary conditions for steady one-dimensional heat conduction through the plate, (b) obtain a relation for the temperature of the outer surface of the plate by solving the differential equation, and (c) evaluate the outer surface temperature.

2–133   Repeat Prob. 2–132 for a 1500-W iron.

2–134E   The roof of a house consists of a 0.8-ft-thick concrete slab ($k = 1.1$ Btu/h $\cdot$ ft $\cdot$ °F) that is 25 ft wide and 35 ft long. The emissivity of the outer surface of the roof is 0.8, and the convection heat transfer coefficient on that surface is estimated to be 3.2 Btu/h $\cdot$ ft$^2 \cdot$ °F. On a clear winter night, the ambient air is reported to be at 50°F, while the night sky temperature for radiation heat transfer is 310 R. If the inner surface temperature of the roof is $T_1 = 62°F$, determine the outer surface temperature of the roof and the rate of heat loss through the roof when steady operating conditions are reached.

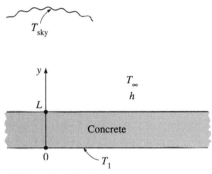

FIGURE P2–134E

2–135   Consider a long resistance wire of radius $r_1 = 0.3$ cm and thermal conductivity $k_{wire} = 18$ W/m $\cdot °C$ in which heat is generated uniformly at a constant rate of $\dot{e}_{gen} = 1.5$ W/cm^3 as a result of resistance heating. The wire is embedded in a 0.4-cm-thick layer of plastic whose thermal conductivity is $k_{plastic} = 1.8$ W/m $\cdot °C$. The outer surface of the plastic cover loses heat by convection to the ambient air at $T_\infty = 25°C$ with an average combined heat transfer coefficient of $h = 14$ W/m$^2 \cdot °C$.

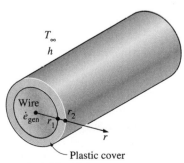

FIGURE P2–135

Assuming one-dimensional heat transfer, determine the temperatures at the center of the resistance wire and the wire-plastic layer interface under steady conditions.

Answers: 97.1°C, 97.3°C

2–136 Consider a cylindrical shell of length L, inner radius r_1, and outer radius r_2 whose thermal conductivity varies in a specified temperature range as $k(T) = k_0(1 + \beta T^2)$ where k_0 and β are two specified constants. The inner surface of the shell is maintained at a constant temperature of T_1 while the outer surface is maintained at T_2. Assuming steady one-dimensional heat transfer, obtain a relation for the heat transfer rate through the shell.

2–137 In a nuclear reactor, heat is generated in 1-cm-diameter cylindrical uranium fuel rods at a rate of 4×10^7 W/m³. Determine the temperature difference between the center and the surface of the fuel rod. *Answer:* 9.0°C

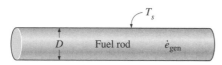

FIGURE P2–137

2–138 Consider a 20-cm-thick large concrete plane wall ($k = 0.77$ W/m · °C) subjected to convection on both sides with $T_{\infty 1} = 27$°C and $h_1 = 5$ W/m² · °C on the inside, and $T_{\infty 2} = 8$°C and $h_2 = 12$ W/m² · °C on the outside. Assuming constant thermal conductivity with no heat generation and negligible radiation, (a) express the differential equations and the boundary conditions for steady one-dimensional heat conduction through the wall, (b) obtain a relation for the variation of temperature in the wall by solving the differential equation, and (c) evaluate the temperatures at the inner and outer surfaces of the wall.

2–139 Consider a water pipe of length $L = 17$ m, inner radius $r_1 = 15$ cm, outer radius $r_2 = 20$ cm, and thermal conductivity $k = 14$ W/m · °C. Heat is generated in the pipe material uniformly by a 25-kW electric resistance heater. The inner and outer surfaces of the pipe are at $T_1 = 60$°C and $T_2 = 80$°C, respectively. Obtain a general relation for temperature distribution inside the pipe under steady conditions and determine the temperature at the center plane of the pipe.

2–140 A plane wall of thickness $L = 4$ cm has a thermal conductivity of $k = 20$ W/m · K. A chemical reaction takes place inside the wall resulting in a uniform heat generation at a rate of $\dot{e}_{gen} = 10^5$ W/m³. Sandwiched between the wall and an insulating layer is a film heater of negligible thickness that generates a heat flux $\dot{q}_s = 16$ kW/m². The opposite side of the wall is in contact with water at temperature $T_\infty = 40$°C. A thermocouple mounted on the surface of the wall in contact with the water reads $T_s = 90$°C.

   (a) Determine the convection coefficient between the wall and water.

(b) Show that the steady-state temperature distribution has the form $T(x) = ax^2 + bx + c$, and determine the values and units of a, b, and c. The origin of x is shown in the figure.

(c) Determine the location and value of the maximum temperature in the wall. Could this location be found without knowing a, b, and c, but knowing that $T(x)$ is a quadratic function? Explain.

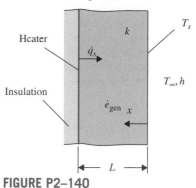

FIGURE P2–140

2–141 A plane wall of thickness $2L = 40$ mm and constant thermal conductivity $k = 5$ W/m · K experiences uniform heat generation at a rate $\dot{e}_{gen}$. Under steady conditions, the temperature distribution in the wall is of the form $T(x) = a - bx^2$, where $a = 80$°C and $b = 2 \times 10^4$ °C/m², and x is in meters. The origin of the x coordinate is at the midplane of the wall.

   (a) Determine the surface temperatures and sketch the temperature distribution in the wall.

   (b) What is the volumetric rate of heat generation, $\dot{e}_{gen}$?

   (c) Determine the surface heat fluxes $\dot{q}_s(-L)$ and $\dot{q}_s(L)$.

   (d) What is the relationship between these fluxes, the heat generation rate and the geometry of the wall?

2–142 Steady one-dimensional heat conduction takes place in a long slab of width W (in the direction of heat flow, x) and thickness Z. The slab's thermal conductivity varies with temperature as $k = k^*/(T^* + T)$, where T is the temperature (in K), and k^* (in W/m) and T^* (in K) are two constants. The temperatures at $x = 0$ and $x = W$ are T_0 and T_W, respectively. Show that the heat flux in steady operation is given by

$$\dot{q} = \frac{k^*}{W} \ln \left(\frac{T^* + T_0}{T^* + T_W} \right)$$

Also, calculate the heat flux for $T^* = 1000$ K, $T_0 = 600$ K, $T_W = 400$ K, $k^* = 7 \times 10^4$ W/m, and $W = 20$ cm.

2–143 Heat is generated uniformly at a rate of 2.6×10^6 W/m³ in a spherical ball ($k = 45$ W/m · °C) of diameter 24 cm. The ball is exposed to iced-water at 0°C with a heat transfer coefficient of 1200 W/m² · °C. Determine the temperatures at the center and the surface of the ball.

Fundamentals of Engineering (FE) Exam Problems

2–144 The heat conduction equation in a medium is given in its simplest form as

$$\frac{1}{r}\frac{d}{dr}\left(rk\frac{dT}{dr}\right) + \dot{e}_{gen} = 0$$

Select the wrong statement below.

(a) The medium is of cylindrical shape.
(b) The thermal conductivity of the medium is constant.
(c) Heat transfer through the medium is steady.
(d) There is heat generation within the medium.
(e) Heat conduction through the medium is one-dimensional.

2–145 Consider a medium in which the heat conduction equation is given in its simplest forms as

$$\frac{1}{r^2}\frac{\partial}{\partial r}\left(r^2\frac{\partial T}{\partial r}\right) = \frac{1}{\alpha}\frac{\partial T}{\partial t}$$

(a) Is heat transfer steady or transient?
(b) Is heat transfer one-, two-, or three-dimensional?
(c) Is there heat generation in the medium?
(d) Is the thermal conductivity of the medium constant or variable?
(e) Is the medium a plane wall, a cylinder, or a sphere?
(f) Is this differential equation for heat conduction linear or nonlinear?

2–146 An apple of radius R is losing heat steadily and uniformly from its outer surface to the ambient air at temperature T_∞ with a convection coefficient of h, and to the surrounding surfaces at temperature T_{surr} (all temperatures are absolute temperatures). Also, heat is generated within the apple uniformly at a rate of $\dot{e}_{gen}$ per unit volume. If T_s denotes the outer surface temperature, the boundary condition at the outer surface of the apple can be expressed as

(a) $-k\dfrac{dT}{dr}\bigg|_{r=R} = h(T_s - T_\infty) + \varepsilon\sigma(T_s^4 - T_{surr}^4)$

(b) $-k\dfrac{dT}{dr}\bigg|_{r=R} = h(T_s - T_\infty) + \varepsilon\sigma(T_s^4 - T_{surr}^4) + \dot{e}_{gen}$

(c) $k\dfrac{dT}{dr}\bigg|_{r=R} = h(T_s - T_\infty) + \varepsilon\sigma(T_s^4 - T_{surr}^4)$

(d) $k\dfrac{dT}{dr}\bigg|_{r=R} = h(T_s - T_\infty) + \varepsilon\sigma(T_s^4 - T_{surr}^4) + \dfrac{4\pi R^3/3}{4\pi R^2}\dot{e}_{gen}$

(e) None of them

2–147 A furnace of spherical shape is losing heat steadily and uniformly from its outer surface of radius R to the ambient air at temperature T_∞ with a convection coefficient of h, and to the surrounding surfaces at temperature T_{surr} (all temperatures are absolute temperatures). If T_0 denotes the outer surface temperature, the boundary condition at the outer surface of the furnace can be expressed as

(a) $-k\dfrac{dT}{dr}\bigg|_{r=R} = h(T_0 - T_\infty) + \varepsilon\sigma(T_0^4 - T_{surr}^4)$

(b) $-k\dfrac{dT}{dr}\bigg|_{r=R} = h(T_0 - T_\infty) - \varepsilon\sigma(T_0^4 - T_{surr}^4)$

(c) $k\dfrac{dT}{dr}\bigg|_{r=R} = h(T_0 - T_\infty) + \varepsilon\sigma(T_0^4 - T_{surr}^4)$

(d) $k\dfrac{dT}{dr}\bigg|_{r=R} = h(T_0 - T_\infty) - \varepsilon\sigma(T_0^4 - T_{surr}^4)$

(e) $k(4\pi R^2)\dfrac{dT}{dr}\bigg|_{r=R} = h(T_0 - T_\infty) + \varepsilon\sigma(T_0^4 - T_{surr}^4)$

2–148 A plane wall of thickness L is subjected to convection at both surfaces with ambient temperature $T_{\infty 1}$ and heat transfer coefficient h_1 at inner surface, and corresponding $T_{\infty 2}$ and h_2 values at the outer surface. Taking the positive direction of x to be from the inner surface to the outer surface, the correct expression for the convection boundary condition is

(a) $k\dfrac{dT(0)}{dx} = h_1[T(0) - T_{\infty 1})]$

(b) $k\dfrac{dT(L)}{dx} = h_2[T(L) - T_{\infty 2})]$

(c) $-k\dfrac{dT(0)}{dx} = h_1[T_{\infty 1} - T_{\infty 2})]$

(d) $-k\dfrac{dT(L)}{dx} = h_2[T_{\infty 1} - T_{\infty 2})]$

(e) None of them

2–149 Consider steady one-dimensional heat conduction through a plane wall, a cylindrical shell, and a spherical shell of uniform thickness with constant thermophysical properties and no thermal energy generation. The geometry in which the variation of temperature in the direction of heat transfer will be linear is

(a) plane wall (b) cylindrical shell (c) spherical shell
(d) all of them (e) none of them

2–150 Consider a large plane wall of thickness L, thermal conductivity k, and surface area A. The left surface of the wall is exposed to the ambient air at T_∞ with a heat transfer coefficient of h while the right surface is insulated. The variation of temperature in the wall for steady one-dimensional heat conduction with no heat generation is

(a) $T(x) = \dfrac{h(L-x)}{k}T_\infty$

(b) $T(x) = \dfrac{k}{h(x+0.5L)}T_\infty$

(c) $T(x) = \left(1 - \dfrac{xh}{k}\right)T_\infty$

(d) $T(x) = (L - x)T_\infty$
(e) $T(x) = T_\infty$

2–151 The variation of temperature in a plane wall is determined to be $T(x) = 65x + 25$ where x is in m and T is in °C. If the temperature at one surface is 38°C, the thickness of the wall is
(a) 2 m   (b) 0.4 m   (c) 0.2 m   (d) 0.1 m   (e) 0.05 m

2–152 The variation of temperature in a plane wall is determined to be $T(x) = 110 - 48x$ where x is in m and T is in °C. If the thickness of the wall is 0.75 m, the temperature difference between the inner and outer surfaces of the wall is
(a) 110°C   (b) 74°C   (c) 55°C   (d) 36°C   (e) 18°C

2–153 The temperatures at the inner and outer surfaces of a 15-cm-thick plane wall are measured to be 40°C and 28°C, respectively. The expression for steady, one-dimensional variation of temperature in the wall is
(a) $T(x) = 28x + 40$ (b) $T(x) = -40x + 28$
(c) $T(x) = 40x + 28$ (d) $T(x) = -80x + 40$
(e) $T(x) = 40x - 80$

2–154 Heat is generated in a long 0.3-cm-diameter cylindrical electric heater at a rate of 150 W/cm³. The heat flux at the surface of the heater in steady operation is
(a) 42.7 W/cm²   (b) 159 W/cm²   (c) 150 W/cm²
(d) 10.6 W/cm²   (e) 11.3 W/cm²

2–155 Heat is generated in a 8-cm-diameter spherical radioactive material whose thermal conductivity is 25 W/m · °C uniformly at a rate of 15 W/cm³. If the surface temperature of the material is measured to be 120°C, the center temperature of the material during steady operation is
(a) 160°C   (b) 280°C   (c) 212°C
(d) 360°C   (e) 600°C

2–156 Heat is generated in a 3-cm-diameter spherical radioactive material uniformly at a rate of 15 W/cm³. Heat is dissipated to the surrounding medium at 25°C with a heat transfer coefficient of 120 W/m² · °C. The surface temperature of the material in steady operation is
(a) 56°C   (b) 84°C   (c) 494°C   (d) 650°C   (e) 108°C

2–157 Heat is generated uniformly in a 4-cm-diameter, 16-cm-long solid bar ($k = 2.4$ W/m · °C). The temperatures at the center and at the surface of the bar are measured to be 210°C and 45°C, respectively. The rate of heat generation within the bar is
(a) 240 W   (b) 796 W   (c) 1013 W
(d) 79,620 W   (e) 3.96 × 10⁶ W

2–158 A solar heat flux $\dot{q}_s$ is incident on a sidewalk whose thermal conductivity is k, solar absorptivity is α_s, and convective heat transfer coefficient is h. Taking the positive x direction to be towards the sky and disregarding radiation exchange with the surroundings surfaces, the correct boundary condition for this sidewalk surface is
(a) $-k\dfrac{dT}{dx} = \alpha_s \dot{q}_s$ (b) $-k\dfrac{dT}{dx} = h(T - T_\infty)$
(c) $-k\dfrac{dT}{dx} = h(T - T_\infty) - \alpha_s \dot{q}_s$ (d) $h(T - T_\infty) = \alpha_s \dot{q}_s$
(e) None of them

2–159 Hot water flows through a PVC ($k = 0.092$ W/m · K) pipe whose inner diameter is 2 cm and outer diameter is 2.5 cm. The temperature of the interior surface of this pipe is 35°C and the temperature of the exterior surface is 20°C. The rate of heat transfer per unit of pipe length is
(a) 22.8 W/m   (b) 38.9 W/m   (c) 48.7 W/m
(d) 63.6 W/m   (e) 72.6 W/m

2–160 The thermal conductivity of a solid depends upon the solid's temperature as $k = aT + b$ where a and b are constants. The temperature in a planar layer of this solid as it conducts heat is given by
(a) $aT + b = x + C_2$ (b) $aT + b = C_1x^2 + C_2$
(c) $aT^2 + bT = C_1x + C_2$ (d) $aT^2 + bT = C_1x^2 + C_2$
(e) None of them

2–161 Harvested grains, like wheat, undergo a volumetric exothermic reaction while they are being stored. This heat generation causes these grains to spoil or even start fires if not controlled properly. Wheat ($k = 0.5$ W/m · K) is stored on the ground (effectively an adiabatic surface) in 5-m-thick layers. Air at 20°C contacts the upper surface of this layer of wheat with $h = 3$ W/m² · K. The temperature distribution inside this layer is given by

$$\frac{T - T_s}{T_0 - T_s} = 1 - \left(\frac{x}{L}\right)^2$$

where T_s is the upper surface temperature, T_0 is the lower surface temperature, x is measured upwards from the ground, and L is the thickness of the layer. When the temperature of the upper surface is 24°C, what is the temperature of the wheat next to the ground?
(a) 39°C   (b) 51°C   (c) 72°C   (d) 84°C   (e) 91°C

2–162 The conduction equation boundary condition for an adiabatic surface with direction n being normal to the surface is
(a) $T = 0$ (b) $dT/dn = 0$ (c) $d^2T/dn^2 = 0$
(d) $d^3T/dn^3 = 0$ (e) $-kdT/dn = 1$

2–163 Which one of the followings is the correct expression for one-dimensional, steady-state, constant thermal conductivity heat conduction equation for a cylinder with heat generation?

(a) $\dfrac{1}{r}\dfrac{\partial}{\partial r}\left(rk\dfrac{\partial T}{\partial r}\right) + \dot{e}_{gen} = \rho c\dfrac{\partial T}{\partial t}$

(b) $\dfrac{1}{r}\dfrac{\partial}{\partial r}\left(r\dfrac{\partial T}{\partial r}\right) + \dfrac{\dot{e}_{gen}}{k} = \dfrac{1}{\alpha}\dfrac{\partial T}{\partial t}$

(c) $\dfrac{1}{r}\dfrac{\partial}{\partial r}\left(r\dfrac{\partial T}{\partial r}\right) = \dfrac{1}{\alpha}\dfrac{\partial T}{\partial t}$

(d) $\dfrac{1}{r}\dfrac{d}{dr}\left(r\dfrac{dT}{dr}\right) + \dfrac{\dot{e}_{gen}}{k} = 0$

(e) $\dfrac{d}{dr}\left(r\dfrac{dT}{dr}\right) = 0$

Design and Essay Problems

2–164 Write an essay on heat generation in nuclear fuel rods. Obtain information on the ranges of heat generation, the variation of heat generation with position in the rods, and the absorption of emitted radiation by the cooling medium.

2–165 Write an interactive computer program to calculate the heat transfer rate and the value of temperature anywhere in the medium for steady one-dimensional heat conduction in a long cylindrical shell for any combination of specified temperature, specified heat flux, and convection boundary conditions. Run the program for five different sets of specified boundary conditions.

2–166 Write an interactive computer program to calculate the heat transfer rate and the value of temperature anywhere in the medium for steady one-dimensional heat conduction in a spherical shell for any combination of specified temperature, specified heat flux, and convection boundary conditions. Run the program for five different sets of specified boundary conditions.

2–167 Write an interactive computer program to calculate the heat transfer rate and the value of temperature anywhere in the medium for steady one-dimensional heat conduction in a plane wall whose thermal conductivity varies linearly as $k(T) = k_0(1 + \beta T)$ where the constants k_0 and β are specified by the user for specified temperature boundary conditions.

STEADY HEAT CONDUCTION

I n heat transfer analysis, we are often interested in the rate of heat transfer through a medium under steady conditions and surface temperatures. Such problems can be solved easily without involving any differential equations by the introduction of the *thermal resistance concept* in an analogous manner to electrical circuit problems. In this case, the thermal resistance corresponds to electrical resistance, temperature difference corresponds to voltage, and the heat transfer rate corresponds to electric current.

We start this chapter with *one-dimensional steady heat conduction* in a plane wall, a cylinder, and a sphere, and develop relations for *thermal resistances* in these geometries. We also develop thermal resistance relations for convection and radiation conditions at the boundaries. We apply this concept to heat conduction problems in *multilayer* plane walls, cylinders, and spheres and generalize it to systems that involve heat transfer in two or three dimensions. We also discuss the *thermal contact resistance* and the *overall heat transfer coefficient* and develop relations for the critical radius of insulation for a cylinder and a sphere. Finally, we discuss steady heat transfer from *finned surfaces* and some complex geometries commonly encountered in practice through the use of *conduction shape factors*.

OBJECTIVES

When you finish studying this chapter, you should be able to:

■ Understand the concept of thermal resistance and its limitations, and develop thermal resistance networks for practical heat conduction problems,

■ Solve steady conduction problems that involve multilayer rectangular, cylindrical, or spherical geometries,

■ Develop an intuitive understanding of thermal contact resistance, and circumstances under which it may be significant,

■ Identify applications in which insulation may actually increase heat transfer,

■ Analyze finned surfaces, and assess how efficiently and effectively fins enhance heat transfer, and

■ Solve multidimensional practical heat conduction problems using conduction shape factors.

CONTENTS

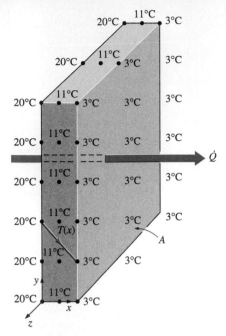

FIGURE 3–1
Heat transfer through a wall is one-dimensional when the temperature of the wall varies in one direction only.

3–1 · STEADY HEAT CONDUCTION IN PLANE WALLS

Consider steady heat conduction through the walls of a house during a winter day. We know that heat is continuously lost to the outdoors through the wall. We intuitively feel that heat transfer through the wall is in the *normal direction* to the wall surface, and no significant heat transfer takes place in the wall in other directions (Fig. 3–1).

Recall that heat transfer in a certain direction is driven by the *temperature gradient* in that direction. There is no heat transfer in a direction in which there is no change in temperature. Temperature measurements at several locations on the inner or outer wall surface will confirm that a wall surface is nearly *isothermal*. That is, the temperatures at the top and bottom of a wall surface as well as at the right and left ends are almost the same. Therefore, there is no heat transfer through the wall from the top to the bottom, or from left to right, but there is considerable temperature difference between the inner and the outer surfaces of the wall, and thus significant heat transfer in the direction from the inner surface to the outer one.

The small thickness of the wall causes the temperature gradient in that direction to be large. Further, if the air temperatures in and outside the house remain constant, then heat transfer through the wall of a house can be modeled as *steady* and *one-dimensional*. The temperature of the wall in this case depends on one direction only (say the *x*-direction) and can be expressed as $T(x)$.

Noting that heat transfer is the only energy interaction involved in this case and there is no heat generation, the *energy balance* for the wall can be expressed as

$$
\begin{pmatrix} \text{Rate of} \\ \text{heat transfer} \\ \text{into the wall} \end{pmatrix} - \begin{pmatrix} \text{Rate of} \\ \text{heat transfer} \\ \text{out of the wall} \end{pmatrix} = \begin{pmatrix} \text{Rate of change} \\ \text{of the energy} \\ \text{of the wall} \end{pmatrix}
$$

or

$$
\dot{Q}_{\text{in}} - \dot{Q}_{\text{out}} = \frac{dE_{\text{wall}}}{dt} \tag{3–1}
$$

But $dE_{\text{wall}}/dt = 0$ for *steady* operation, since there is no change in the temperature of the wall with time at any point. Therefore, the rate of heat transfer into the wall must be equal to the rate of heat transfer out of it. In other words, *the rate of heat transfer through the wall must be constant*, $\dot{Q}_{\text{cond, wall}}$ = constant.

Consider a plane wall of thickness L and average thermal conductivity k. The two surfaces of the wall are maintained at constant temperatures of T_1 and T_2. For one-dimensional steady heat conduction through the wall, we have $T(x)$. Then Fourier's law of heat conduction for the wall can be expressed as

$$
\dot{Q}_{\text{cond, wall}} = -kA \frac{dT}{dx} \qquad \text{(W)} \tag{3–2}
$$

where the rate of conduction heat transfer $\dot{Q}_{\text{cond, wall}}$ and the wall area A are constant. Thus dT/dx = constant, which means that *the temperature through the wall varies linearly with x.* That is, the temperature distribution in the wall under steady conditions is a *straight line* (Fig. 3–2).

Separating the variables in the preceding equation and integrating from $x = 0$, where $T(0) = T_1$, to $x = L$, where $T(L) = T_2$, we get

$$\int_{x=0}^{L} \dot{Q}_{\text{cond, wall}} \, dx = -\int_{T=T_1}^{T_2} kA \, dT$$

Performing the integrations and rearranging gives

$$\dot{Q}_{\text{cond, wall}} = kA \frac{T_1 - T_2}{L} \qquad \text{(W)} \qquad \text{(3–3)}$$

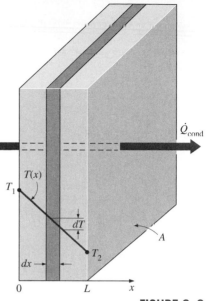

which is identical to Eq. 1–21. Again, *the rate of heat conduction through a plane wall is proportional to the average thermal conductivity, the wall area, and the temperature difference, but is inversely proportional to the wall thickness.* Also, once the rate of heat conduction is available, the temperature $T(x)$ at any location x can be determined by replacing T_2 in Eq. 3–3 by T, and L by x.

Thermal Resistance Concept

Equation 3–3 for heat conduction through a plane wall can be rearranged as

$$\dot{Q}_{\text{cond, wall}} = \frac{T_1 - T_2}{R_{\text{wall}}} \qquad \text{(W)} \qquad \text{(3–4)}$$

where

$$R_{\text{wall}} = \frac{L}{kA} \qquad \text{(°C/W)} \qquad \text{(3–5)}$$

is the *thermal resistance* of the wall against heat conduction or simply the **conduction resistance** of the wall. Note that the thermal resistance of a medium depends on the *geometry* and the *thermal properties* of the medium.

This equation for heat transfer is analogous to the relation for *electric current flow I*, expressed as

$$I = \frac{V_1 - V_2}{R_e} \qquad \text{(3–6)}$$

where $R_e = L/\sigma_e A$ is the *electric resistance* and $V_1 - V_2$ is the *voltage difference* across the resistance (σ_e is the electrical conductivity). Thus, the *rate of heat transfer* through a layer corresponds to the *electric current,* the *thermal resistance* corresponds to *electrical resistance,* and the *temperature difference* corresponds to *voltage difference* across the layer (Fig. 3–3).

Consider convection heat transfer from a solid surface of area A_s and temperature T_s to a fluid whose temperature sufficiently far from the surface is T_∞, with a convection heat transfer coefficient h. Newton's law of cooling for convection heat transfer rate $\dot{Q}_{\text{conv}} = hA_s(T_s - T_\infty)$ can be rearranged as

$$\dot{Q}_{\text{conv}} = \frac{T_s - T_\infty}{R_{\text{conv}}} \qquad \text{(W)} \qquad \text{(3–7)}$$

FIGURE 3–2
Under steady conditions, the temperature distribution in a plane wall is a straight line.

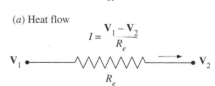

(a) Heat flow

(b) Electric current flow

FIGURE 3–3
Analogy between thermal and electrical resistance concepts.

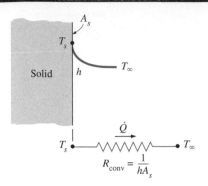

FIGURE 3–4
Schematic for convection
resistance at a surface.

where

$$R_{\text{conv}} = \frac{1}{hA_s} \qquad (°C/W) \qquad \text{(3–8)}$$

is the *thermal resistance* of the surface against heat convection, or simply the **convection resistance** of the surface (Fig. 3–4). Note that when the convection heat transfer coefficient is very large ($h \to \infty$), the convection resistance becomes *zero* and $T_s \approx T_\infty$. That is, the surface offers *no resistance to convection,* and thus it does not slow down the heat transfer process. This situation is approached in practice at surfaces where boiling and condensation occur. Also note that the surface does not have to be a plane surface. Equation 3–8 for convection resistance is valid for surfaces of any shape, provided that the assumption of $h = $ constant and uniform is reasonable.

When the wall is surrounded by a gas, the *radiation effects,* which we have ignored so far, can be significant and may need to be considered. The rate of radiation heat transfer between a surface of emissivity ε and area A_s at temperature T_s and the surrounding surfaces at some average temperature T_{surr} can be expressed as

$$\dot{Q}_{\text{rad}} = \varepsilon \sigma A_s (T_s^4 - T_{\text{surr}}^4) = h_{\text{rad}} A_s (T_s - T_{\text{surr}}) = \frac{T_s - T_{\text{surr}}}{R_{\text{rad}}} \qquad (W) \qquad \text{(3–9)}$$

where

$$R_{\text{rad}} = \frac{1}{h_{\text{rad}} A_s} \qquad (K/W) \qquad \text{(3–10)}$$

is the *thermal resistance* of a surface against radiation, or the **radiation resistance**, and

$$h_{\text{rad}} = \frac{\dot{Q}_{\text{rad}}}{A_s (T_s - T_{\text{surr}})} = \varepsilon \sigma (T_s^2 + T_{\text{surr}}^2)(T_s + T_{\text{surr}}) \qquad (W/m^2 \cdot K) \qquad \text{(3–11)}$$

is the **radiation heat transfer coefficient**. Note that both T_s and T_{surr} *must* be in K in the evaluation of h_{rad}. The definition of the radiation heat transfer coefficient enables us to express radiation conveniently in an analogous manner to convection in terms of a temperature difference. But h_{rad} depends strongly on temperature while h_{conv} usually does not.

A surface exposed to the surrounding air involves convection and radiation simultaneously, and the total heat transfer at the surface is determined by adding (or subtracting, if in the opposite direction) the radiation and convection components. The convection and radiation resistances are parallel to each other, as shown in Fig. 3–5, and may cause some complication in the thermal resistance network. When $T_{\text{surr}} \approx T_\infty$, the radiation effect can properly be accounted for by replacing h in the convection resistance relation by

$$h_{\text{combined}} = h_{\text{conv}} + h_{\text{rad}} \qquad (W/m^2 \cdot K) \qquad \text{(3–12)}$$

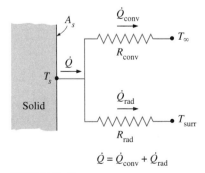

FIGURE 3–5
Schematic for convection and
radiation resistances at a surface.

where h_{combined} is the **combined heat transfer coefficient**. This way all complications associated with radiation are avoided.

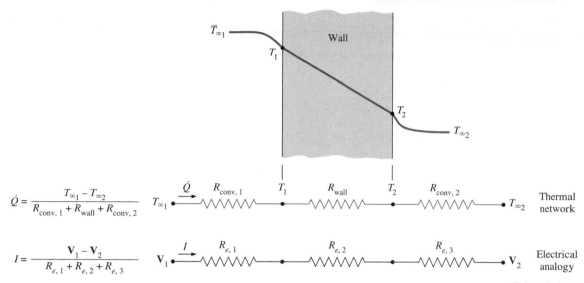

FIGURE 3–6
The thermal resistance network for heat transfer through a plane wall subjected to convection on both sides, and the electrical analogy.

Thermal Resistance Network

Now consider steady one-dimensional heat transfer through a plane wall of thickness L, area A, and thermal conductivity k that is exposed to convection on both sides to fluids at temperatures $T_{\infty 1}$ and $T_{\infty 2}$ with heat transfer coefficients h_1 and h_2, respectively, as shown in Fig. 3–6. Assuming $T_{\infty 2} < T_{\infty 1}$, the variation of temperature will be as shown in the figure. Note that the temperature varies linearly in the wall, and asymptotically approaches $T_{\infty 1}$ and $T_{\infty 2}$ in the fluids as we move away from the wall.

Under steady conditions we have

$$\begin{pmatrix} \text{Rate of} \\ \textit{heat convection} \\ \text{into the wall} \end{pmatrix} = \begin{pmatrix} \text{Rate of} \\ \textit{heat conduction} \\ \text{through the wall} \end{pmatrix} = \begin{pmatrix} \text{Rate of} \\ \textit{heat convection} \\ \text{from the wall} \end{pmatrix}$$

or

$$\dot{Q} = h_1 A(T_{\infty 1} - T_1) = kA \frac{T_1 - T_2}{L} = h_2 A(T_2 - T_{\infty 2}) \qquad \textbf{(3–13)}$$

which can be rearranged as

$$\dot{Q} = \frac{T_{\infty 1} - T_1}{1/h_1 A} = \frac{T_1 - T_2}{L/kA} = \frac{T_2 - T_{\infty 2}}{1/h_2 A}$$

$$= \frac{T_{\infty 1} - T_1}{R_{\text{conv}, 1}} = \frac{T_1 - T_2}{R_{\text{wall}}} = \frac{T_2 - T_{\infty 2}}{R_{\text{conv}, 2}} \qquad \textbf{(3–14)}$$

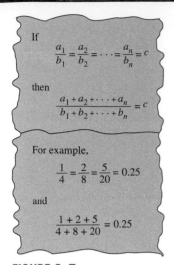

If

$$\frac{a_1}{b_1} = \frac{a_2}{b_2} = \cdots = \frac{a_n}{b_n} = c$$

then

$$\frac{a_1 + a_2 + \cdots + a_n}{b_1 + b_2 + \cdots + b_n} = c$$

For example,

$$\frac{1}{4} = \frac{2}{8} = \frac{5}{20} = 0.25$$

and

$$\frac{1 + 2 + 5}{4 + 8 + 20} = 0.25$$

FIGURE 3–7
A useful mathematical identity.

Adding the numerators and denominators yields (Fig. 3–7)

$$\dot{Q} = \frac{T_{\infty 1} - T_{\infty 2}}{R_{\text{total}}} \qquad \text{(W)} \qquad (3\text{–}15)$$

where

$$R_{\text{total}} = R_{\text{conv, 1}} + R_{\text{wall}} + R_{\text{conv, 2}} = \frac{1}{h_1 A} + \frac{L}{kA} + \frac{1}{h_2 A} \qquad \text{(°C/W)} \qquad (3\text{–}16)$$

Note that the heat transfer area A is constant for a plane wall, and the rate of heat transfer through a wall separating two mediums is equal to the temperature difference divided by the total thermal resistance between the mediums. Also note that the thermal resistances are in *series,* and the equivalent thermal resistance is determined by simply *adding* the individual resistances, just like the electrical resistances connected in series. Thus, the electrical analogy still applies. We summarize this as *the rate of steady heat transfer between two surfaces is equal to the temperature difference divided by the total thermal resistance between those two surfaces.*

Another observation that can be made from Eq. 3–15 is that the ratio of the temperature drop to the thermal resistance across any layer is constant, and thus the temperature drop across any layer is proportional to the thermal resistance of the layer. The larger the resistance, the larger the temperature drop. In fact, the equation $\dot{Q} = \Delta T/R$ can be rearranged as

$$\Delta T = \dot{Q} R \qquad \text{(°C)} \qquad (3\text{–}17)$$

which indicates that the *temperature drop* across any layer is equal to the *rate of heat transfer* times the *thermal resistance* across that layer (Fig. 3–8). You may recall that this is also true for voltage drop across an electrical resistance when the electric current is constant.

It is sometimes convenient to express heat transfer through a medium in an analogous manner to Newton's law of cooling as

$$\dot{Q} = UA\,\Delta T \qquad \text{(W)} \qquad (3\text{–}18)$$

where U is the **overall heat transfer coefficient**. A comparison of Eqs. 3–15 and 3–18 reveals that

$$UA = \frac{1}{R_{\text{total}}} \qquad \text{(°C/K)} \qquad (3\text{–}19)$$

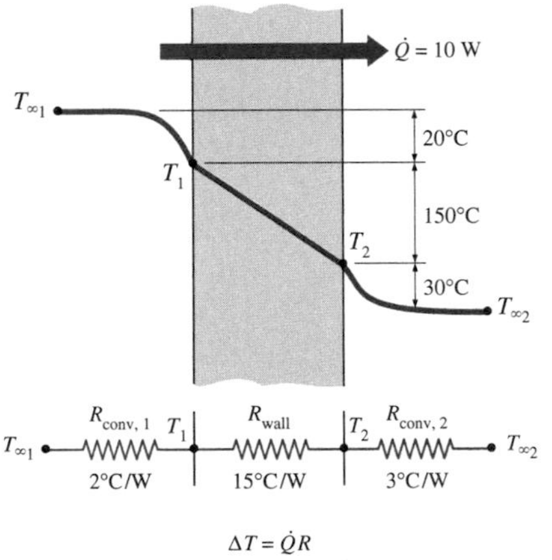

Therefore, for a unit area, the overall heat transfer coefficient is equal to the inverse of the total thermal resistance.

Note that we do not need to know the surface temperatures of the wall in order to evaluate the rate of steady heat transfer through it. All we need to know is the convection heat transfer coefficients and the fluid temperatures on both sides of the wall. The *surface temperature* of the wall can be determined as

$\dot{Q} = 10$ W

$T_{\infty 1}$

$20°C$

T_1

$150°C$

T_2

$30°C$

$T_{\infty 2}$

$T_{\infty 1}$ — $R_{\text{conv, 1}}$ T_1 — R_{wall} T_2 — $R_{\text{conv, 2}}$ — $T_{\infty 2}$

$2°C/W$ $15°C/W$ $3°C/W$

$\Delta T = \dot{Q} R$

FIGURE 3–8
The temperature drop across a layer is proportional to its thermal resistance.

described above using the thermal resistance concept, but by taking the surface at which the temperature is to be determined as one of the terminal surfaces. For example, once $\dot{Q}$ is evaluated, the surface temperature T_1 can be determined from

$$\dot{Q} = \frac{T_{\infty 1} - T_1}{R_{\text{conv, 1}}} = \frac{T_{\infty 1} - T_1}{1/h_1 A} \qquad (3\text{--}20)$$

Multilayer Plane Walls

In practice we often encounter plane walls that consist of several layers of different materials. The thermal resistance concept can still be used to determine the rate of steady heat transfer through such *composite* walls. As you may have already guessed, this is done by simply noting that the conduction resistance of each wall is L/kA connected in series, and using the electrical analogy. That is, by dividing the *temperature difference* between two surfaces at known temperatures by the *total thermal resistance* between them.

Consider a plane wall that consists of two layers (such as a brick wall with a layer of insulation). The rate of steady heat transfer through this two-layer composite wall can be expressed as (Fig. 3–9)

$$\dot{Q} = \frac{T_{\infty 1} - T_{\infty 2}}{R_{\text{total}}} \qquad (3\text{--}21)$$

where R_{total} is the *total thermal resistance*, expressed as

$$R_{\text{total}} = R_{\text{conv, 1}} + R_{\text{wall, 1}} + R_{\text{wall, 2}} + R_{\text{conv, 2}}$$
$$= \frac{1}{h_1 A} + \frac{L_1}{k_1 A} + \frac{L_2}{k_2 A} + \frac{1}{h_2 A} \qquad (3\text{--}22)$$

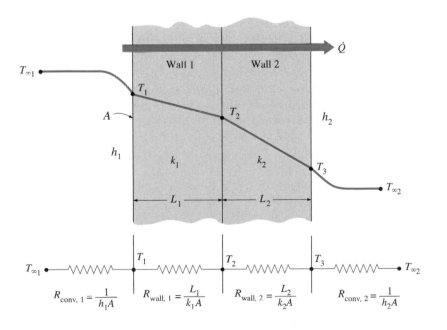

FIGURE 3–9

The thermal resistance network for heat transfer through a two-layer plane wall subjected to convection on both sides.

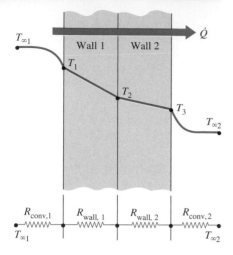

To find T_1: $\dot{Q} = \dfrac{T_{\infty 1} - T_1}{R_{\text{conv},1}}$

To find T_2: $\dot{Q} = \dfrac{T_{\infty 1} - T_2}{R_{\text{conv},1} + R_{\text{wall},1}}$

To find T_3: $\dot{Q} = \dfrac{T_3 - T_{\infty 2}}{R_{\text{conv},2}}$

FIGURE 3–10

The evaluation of the surface and interface temperatures when $T_{\infty 1}$ and $T_{\infty 2}$ are given and $\dot{Q}$ is calculated.

The subscripts 1 and 2 in the R_{wall} relations above indicate the first and the second layers, respectively. We could also obtain this result by following the approach already used for the single-layer case by noting that the rate of steady heat transfer $\dot{Q}$ through a multilayer medium is constant, and thus it must be the same through each layer. Note from the thermal resistance network that the resistances are *in series*, and thus the *total thermal resistance* is simply the *arithmetic sum* of the individual thermal resistances in the path of heat transfer.

This result for the *two-layer* case is analogous to the *single-layer* case, except that an *additional resistance* is added for the *additional layer*. This result can be extended to plane walls that consist of *three* or *more layers* by adding an *additional resistance* for each *additional layer*.

Once $\dot{Q}$ is *known*, an unknown surface temperature T_j at any surface or interface j can be determined from

$$\dot{Q} = \frac{T_i - T_j}{R_{\text{total}, i-j}} \tag{3–23}$$

where T_i is a *known* temperature at location i and $R_{\text{total}, i-j}$ is the total thermal resistance between locations i and j. For example, when the fluid temperatures $T_{\infty 1}$ and $T_{\infty 2}$ for the two-layer case shown in Fig. 3–9 are available and $\dot{Q}$ is calculated from Eq. 3–21, the interface temperature T_2 between the two walls can be determined from (Fig. 3–10)

$$\dot{Q} = \frac{T_{\infty 1} - T_2}{R_{\text{conv}, 1} + R_{\text{wall}, 1}} = \frac{T_{\infty 1} - T_2}{\dfrac{1}{h_1 A} + \dfrac{L_1}{k_1 A}} \tag{3–24}$$

The temperature drop across a layer is easily determined from Eq. 3–17 by multiplying $\dot{Q}$ by the thermal resistance of that layer.

The thermal resistance concept is widely used in practice because it is intuitively easy to understand and it has proven to be a powerful tool in the solution of a wide range of heat transfer problems. But its use is limited to systems through which the rate of heat transfer $\dot{Q}$ remains *constant*; that is, to systems involving *steady* heat transfer with *no heat generation* (such as resistance heating or chemical reactions) within the medium.

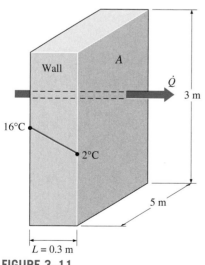

FIGURE 3–11
Schematic for Example 3–1.

EXAMPLE 3–1   Heat Loss through a Wall

Consider a 3-m-high, 5-m-wide, and 0.17-m-thick wall whose thermal conductivity is $k = 0.9$ W/m · K (Fig. 3–11). On a certain day, the temperatures of the inner and the outer surfaces of the wall are measured to be 16°C and 2°C, respectively. Determine the rate of heat loss through the wall on that day.

SOLUTION The two surfaces of a wall are maintained at specified temperatures. The rate of heat loss through the wall is to be determined.

Assumptions **1** Heat transfer through the wall is steady since the surface temperatures remain constant at the specified values. **2** Heat transfer is one-dimensional since any significant temperature gradients exist in the direction from the indoors to the outdoors. **3** Thermal conductivity is constant.

Properties The thermal conductivity is given to be $k = 0.9$ W/m · K.

Analysis Noting that heat transfer through the wall is by conduction and the area of the wall is $A = 3$ m $\times$ 5 m $= 15$ m^2, the steady rate of heat transfer through the wall can be determined from Eq. 3–3 to be

$$\dot{Q} = kA\frac{T_1 - T_2}{L} = (0.9\ \text{W/m} \cdot {}^\circ\text{C})(15\ \text{m}^2)\frac{(16 - 2){}^\circ\text{C}}{0.3\ \text{m}} = \mathbf{630\ W}$$

We could also determine the steady rate of heat transfer through the wall by making use of the thermal resistance concept from

$$\dot{Q} = \frac{\Delta T_{\text{wall}}}{R_{\text{wall}}}$$

where

$$R_{\text{wall}} = \frac{L}{kA} = \frac{0.3\ \text{m}}{(0.9\ \text{W/m} \cdot {}^\circ\text{C})(15\ \text{m}^2)} = 0.02222{}^\circ\text{C/W}$$

Substituting, we get

$$\dot{Q} = \frac{(16 - 2){}^\circ\text{C}}{0.02222{}^\circ\text{C/W}} = 630\ \text{W}$$

Discussion This is the same result obtained earlier. Note that heat conduction through a plane wall with specified surface temperatures can be determined directly and easily without utilizing the thermal resistance concept. However, the thermal resistance concept serves as a valuable tool in more complex heat transfer problems, as you will see in the following examples. Also, the units W/m · °C and W/m · K for thermal conductivity are equivalent, and thus interchangeable. This is also the case for °C and K for temperature differences.

EXAMPLE 3–2   **Heat Loss through a Single-Pane Window**

Consider a 0.8-m-high and 1.5-m-wide glass window with a thickness of 8 mm and a thermal conductivity of $k = 0.78$ W/m · K. Determine the steady rate of heat transfer through this glass window and the temperature of its inner surface for a day during which the room is maintained at 20°C while the temperature of the outdoors is −10°C. Take the heat transfer coefficients on the inner and outer surfaces of the window to be $h_1 = 10$ W/m^2 · °C and $h_2 = 40$ W/m^2 · °C, which includes the effects of radiation.

SOLUTION Heat loss through a window glass is considered. The rate of heat transfer through the window and the inner surface temperature are to be determined.

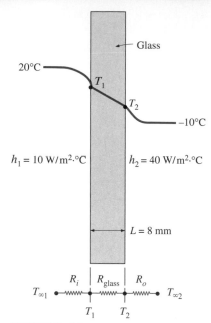

20°C

T_1

T_2

−10°C

Glass

$h_1 = 10$ W/m²·°C

$h_2 = 40$ W/m²·°C

$L = 8$ mm

$T_{\infty 1}$ —www—•—www—•—www—• $T_{\infty 2}$

R_i R_{glass} R_o

T_1 T_2

FIGURE 3–12
Schematic for Example 3–2.

Assumptions **1** Heat transfer through the window is steady since the surface temperatures remain constant at the specified values. **2** Heat transfer through the wall is one-dimensional since any significant temperature gradients exist in the direction from the indoors to the outdoors. **3** Thermal conductivity is constant.

Properties The thermal conductivity is given to be $k = 0.78$ W/m · K.

Analysis This problem involves conduction through the glass window and convection at its surfaces, and can best be handled by making use of the thermal resistance concept and drawing the thermal resistance network, as shown in Fig. 3–12. Noting that the area of the window is $A = 0.8$ m × 1.5 m = 1.2 m², the individual resistances are evaluated from their definitions to be

$$R_i = R_{conv, 1} = \frac{1}{h_1 A} = \frac{1}{(10 \text{ W/m}^2 \cdot \text{°C})(1.2 \text{ m}^2)} = 0.08333\text{°C/W}$$

$$R_{glass} = \frac{L}{kA} = \frac{0.008 \text{ m}}{(0.78 \text{ W/m} \cdot \text{°C})(1.2 \text{ m}^2)} = 0.00855\text{°C/W}$$

$$R_o = R_{conv, 2} = \frac{1}{h_2 A} = \frac{1}{(40 \text{ W/m}^2 \cdot \text{°C})(1.2 \text{ m}^2)} = 0.02083\text{°C/W}$$

Noting that all three resistances are in series, the total resistance is

$$R_{total} = R_{conv, 1} + R_{glass} + R_{conv, 2} = 0.08333 + 0.00855 + 0.02083$$
$$= 0.1127\text{°C/W}$$

Then the steady rate of heat transfer through the window becomes

$$\dot{Q} = \frac{T_{\infty 1} - T_{\infty 2}}{R_{total}} = \frac{[20 - (-10)]\text{°C}}{0.1127\text{°C/W}} = \textbf{266 W}$$

Knowing the rate of heat transfer, the inner surface temperature of the window glass can be determined from

$$\dot{Q} = \frac{T_{\infty 1} - T_1}{R_{conv, 1}} \quad \longrightarrow \quad T_1 = T_{\infty 1} - \dot{Q}R_{conv, 1}$$
$$= 20\text{°C} - (266 \text{ W})(0.08333\text{°C/W})$$
$$= \textbf{−2.2°C}$$

Discussion Note that the inner surface temperature of the window glass is −2.2°C even though the temperature of the air in the room is maintained at 20°C. Such low surface temperatures are highly undesirable since they cause the formation of fog or even frost on the inner surfaces of the glass when the humidity in the room is high.

EXAMPLE 3–3   Heat Loss through Double-Pane Windows

Consider a 0.8-m-high and 1.5-m-wide double-pane window consisting of two 4-mm-thick layers of glass ($k = 0.78$ W/m · K) separated by a 10-mm-wide stagnant air space ($k = 0.026$ W/m · K). Determine the steady rate of heat transfer through this double-pane window and the temperature of its inner surface for a day during which the room is maintained at 20°C while the temperature of the outdoors is −10°C. Take the convection heat transfer coefficients on the inner and outer surfaces of the window to be $h_1 = 10$ W/m² · °C and $h_2 = 40$ W/m² · °C, which includes the effects of radiation.

SOLUTION A double-pane window is considered. The rate of heat transfer through the window and the inner surface temperature are to be determined.
Analysis This example problem is identical to the previous one except that the single 8-mm-thick window glass is replaced by two 4-mm-thick glasses that enclose a 10-mm-wide stagnant air space. Therefore, the thermal resistance network of this problem involves two additional conduction resistances corresponding to the two additional layers, as shown in Fig. 3–13. Noting that the area of the window is again $A = 0.8$ m × 1.5 m = 1.2 m², the individual resistances are evaluated from their definitions to be

$$R_i = R_{conv, 1} = \frac{1}{h_1 A} = \frac{1}{(10 \text{ W/m}^2 \cdot °C)(1.2 \text{ m}^2)} = 0.08333°C/W$$

$$R_1 = R_3 = R_{glass} = \frac{L_1}{k_1 A} = \frac{0.004 \text{ m}}{(0.78 \text{ W/m} \cdot °C)(1.2 \text{ m}^2)} = 0.00427°C/W$$

$$R_2 = R_{air} = \frac{L_2}{k_2 A} = \frac{0.01 \text{ m}}{(0.026 \text{ W/m} \cdot °C)(1.2 \text{ m}^2)} = 0.3205°C/W$$

$$R_o = R_{conv, 2} = \frac{1}{h_2 A} = \frac{1}{(40 \text{ W/m}^2 \cdot °C)(1.2 \text{ m}^2)} = 0.02083°C/W$$

Noting that all three resistances are in series, the total resistance is

$$R_{total} = R_{conv, 1} + R_{glass, 1} + R_{air} + R_{glass, 2} + R_{conv, 2}$$
$$= 0.08333 + 0.00427 + 0.3205 + 0.00427 + 0.02083$$
$$= 0.4332°C/W$$

Then the steady rate of heat transfer through the window becomes

$$\dot{Q} = \frac{T_{\infty 1} - T_{\infty 2}}{R_{total}} = \frac{[20 - (-10)]°C}{0.4332°C/W} = \textbf{69.2 W}$$

which is about one-fourth of the result obtained in the previous example. This explains the popularity of the double- and even triple-pane windows in cold climates. The drastic reduction in the heat transfer rate in this case is due to the large thermal resistance of the air layer between the glasses.
    The inner surface temperature of the window in this case will be

$$T_1 = T_{\infty 1} - \dot{Q} R_{conv, 1} = 20°C - (69.2 \text{ W})(0.08333°C/W) = \textbf{14.2°C}$$

which is considerably higher than the −2.2°C obtained in the previous example. Therefore, a double-pane window will rarely get fogged. A double-pane window will also reduce the heat gain in summer, and thus reduce the air-conditioning costs.

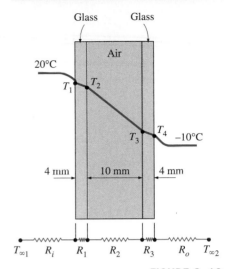

FIGURE 3–13
Schematic for Example 3–3.

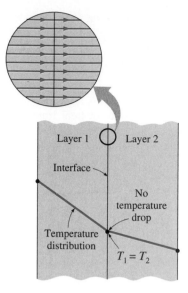
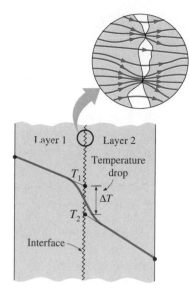

(a) Ideal (perfect) thermal contact

(b) Actual (imperfect) thermal contact

FIGURE 3–14

Temperature distribution and heat flow lines along two solid plates pressed against each other for the case of perfect and imperfect contact.

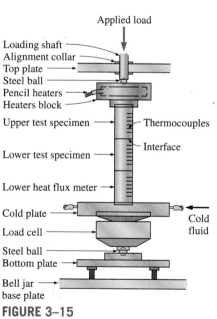

FIGURE 3–15

A typical experimental setup for the determination of thermal contact resistance (from Song et al.).

3–2 · THERMAL CONTACT RESISTANCE

In the analysis of heat conduction through multilayer solids, we assumed "perfect contact" at the interface of two layers, and thus no temperature drop at the interface. This would be the case when the surfaces are perfectly smooth and they produce a perfect contact at each point. In reality, however, even flat surfaces that appear smooth to the eye turn out to be rather rough when examined under a microscope, as shown in Fig. 3–14, with numerous peaks and valleys. That is, a surface is *microscopically rough* no matter how smooth it appears to be.

When two such surfaces are pressed against each other, the peaks form good material contact but the valleys form voids filled with air. As a result, an interface contains numerous *air gaps* of varying sizes that act as *insulation* because of the low thermal conductivity of air. Thus, an interface offers some resistance to heat transfer, and this resistance per unit interface area is called the **thermal contact resistance**, R_c. The value of R_c is determined experimentally using a setup like the one shown in Fig. 3–15, and as expected, there is considerable scatter of data because of the difficulty in characterizing the surfaces.

Consider heat transfer through two metal rods of cross-sectional area A that are pressed against each other. Heat transfer through the interface of these two rods is the sum of the heat transfers through the *solid contact spots* and the *gaps* in the noncontact areas and can be expressed as

$$\dot{Q} = \dot{Q}_{\text{contact}} + \dot{Q}_{\text{gap}} \tag{3–25}$$

It can also be expressed in an analogous manner to Newton's law of cooling as

$$\dot{Q} = h_c A \, \Delta T_{\text{interface}} \tag{3–26}$$

where A is the apparent interface area (which is the same as the cross-sectional area of the rods) and $\Delta T_{\text{interface}}$ is the effective temperature difference at the interface. The quantity h_c, which corresponds to the convection heat transfer coefficient, is called the **thermal contact conductance** and is expressed as

$$h_c = \frac{\dot{Q}/A}{\Delta T_{\text{interface}}} \qquad (\text{W/m}^2 \cdot {}^\circ\text{C}) \qquad \textbf{(3–27)}$$

It is related to thermal contact resistance by

$$R_c = \frac{1}{h_c} = \frac{\Delta T_{\text{interface}}}{\dot{Q}/A} \qquad (\text{m}^2 \cdot {}^\circ\text{C/W}) \qquad \textbf{(3–28)}$$

That is, thermal contact resistance is the inverse of thermal contact conductance. Usually, thermal contact conductance is reported in the literature, but the concept of thermal contact resistance serves as a better vehicle for explaining the effect of interface on heat transfer. Note that R_c represents thermal contact resistance *per unit area*. The thermal resistance for the entire interface is obtained by dividing R_c by the apparent interface area A.

The thermal contact resistance can be determined from Eq. 3–28 by measuring the temperature drop at the interface and dividing it by the heat flux under steady conditions. The value of thermal contact resistance depends on the *surface roughness* and the *material properties* as well as the *temperature* and *pressure* at the interface and the *type of fluid* trapped at the interface. The situation becomes more complex when plates are fastened by bolts, screws, or rivets since the interface pressure in this case is nonuniform. The thermal contact resistance in that case also depends on the plate thickness, the bolt radius, and the size of the contact zone. Thermal contact resistance is observed to *decrease* with *decreasing surface roughness* and *increasing interface pressure,* as expected. Most experimentally determined values of the thermal contact resistance fall between 0.000005 and 0.0005 m² · °C/W (the corresponding range of thermal contact conductance is 2000 to 200,000 W/m² · °C).

When we analyze heat transfer in a medium consisting of two or more layers, the first thing we need to know is whether the thermal contact resistance is *significant* or not. We can answer this question by comparing the magnitudes of the thermal resistances of the layers with typical values of thermal contact resistance. For example, the thermal resistance of a 1-cm-thick layer of an insulating material per unit surface area is

$$R_{c,\,\text{insulation}} = \frac{L}{k} = \frac{0.01\ \text{m}}{0.04\ \text{W/m} \cdot {}^\circ\text{C}} = 0.25\ \text{m}^2 \cdot {}^\circ\text{C/W}$$

whereas for a 1-cm-thick layer of copper, it is

$$R_{c,\,\text{copper}} = \frac{L}{k} = \frac{0.01\ \text{m}}{386\ \text{W/m} \cdot {}^\circ\text{C}} = 0.000026\ \text{m}^2 \cdot {}^\circ\text{C/W}$$

Comparing the values above with typical values of thermal contact resistance, we conclude that thermal contact resistance is significant and can even dominate the heat transfer for good heat conductors such as metals, but can be

TABLE 3–1

Thermal contact conductance for aluminum plates with different fluids at the interface for a surface roughness of 10 μm and interface pressure of 1 atm (from Fried).

Fluid at the interface	Contact conductance, h_c, W/m² · K
Air	3640
Helium	9520
Hydrogen	13,900
Silicone oil	19,000
Glycerin	37,700

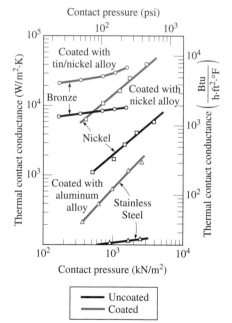

FIGURE 3–16
Effect of metallic coatings on thermal contact conductance (from Peterson).

disregarded for poor heat conductors such as insulations. This is not surprising since insulating materials consist mostly of air space just like the interface itself.

The thermal contact resistance can be minimized by applying a thermally conducting liquid called a *thermal grease* such as silicon oil on the surfaces before they are pressed against each other. This is commonly done when attaching electronic components such as power transistors to heat sinks. The thermal contact resistance can also be reduced by replacing the air at the interface by a *better conducting gas* such as helium or hydrogen, as shown in Table 3–1.

Another way to minimize the contact resistance is to insert a *soft metallic foil* such as tin, silver, copper, nickel, or aluminum between the two surfaces. Experimental studies show that the thermal contact resistance can be reduced by a factor of up to 7 by a metallic foil at the interface. For maximum effectiveness, the foils must be very thin. The effect of metallic coatings on thermal contact conductance is shown in Fig. 3–16 for various metal surfaces.

There is considerable uncertainty in the contact conductance data reported in the literature, and care should be exercised when using them. In Table 3–2 some experimental results are given for the contact conductance between similar and dissimilar metal surfaces for use in preliminary design calculations. Note that the *thermal contact conductance* is *highest* (and thus the contact resistance is lowest) for *soft metals* with *smooth surfaces* at *high pressure*.

EXAMPLE 3–4   **Equivalent Thickness for Contact Resistance**

The thermal contact conductance at the interface of two 1-cm-thick aluminum plates is measured to be 11,000 W/m² · K. Determine the thickness of the aluminum plate whose thermal resistance is equal to the thermal resistance of the interface between the plates (Fig. 3–17).

SOLUTION   The thickness of the aluminum plate whose thermal resistance is equal to the thermal contact resistance is to be determined.
Properties   The thermal conductivity of aluminum at room temperature is $k = 237$ W/m · K (Table A–3).
Analysis   Noting that thermal contact resistance is the inverse of thermal contact conductance, the thermal contact resistance is

$$R_c = \frac{1}{h_c} = \frac{1}{11,000 \text{ W/m}^2 \cdot \text{K}} = 0.909 \times 10^{-4} \text{ m}^2 \cdot \text{K/W}$$

For a unit surface area, the thermal resistance of a flat plate is defined as

$$R = \frac{L}{k}$$

where L is the thickness of the plate and k is the thermal conductivity. Setting $R = R_c$, the equivalent thickness is determined from the relation above to be

$$L = kR_c = (237 \text{ W/m} \cdot \text{K})(0.909 \times 10^{-4} \text{ m}^2 \cdot \text{K/W}) = 0.0215 \text{ m} = \textbf{2.15 cm}$$

TABLE 3-2

Thermal contact conductance of some metal surfaces in air (from various sources)

Material	Surface condition	Roughness, μm	Temperature, °C	Pressure, MPa	h_c,* W/m² · °C
Identical Metal Pairs					
416 Stainless steel	Ground	2.54	90–200	0.17–2.5	3800
304 Stainless steel	Ground	1.14	20	4–7	1900
Aluminum	Ground	2.54	150	1.2–2.5	11,400
Copper	Ground	1.27	20	1.2–20	143,000
Copper	Milled	3.81	20	1–5	55,500
Copper (vacuum)	Milled	0.25	30	0.17–7	11,400
Dissimilar Metal Pairs					
Stainless steel–				10	2900
Aluminum		20–30	20	20	3600
Stainless steel–				10	16,400
Aluminum		1.0–2.0	20	20	20,800
Steel Ct-30–				10	50,000
Aluminum	Ground	1.4–2.0	20	15–35	59,000
Steel Ct-30–				10	4800
Aluminum	Milled	4.5–7.2	20	30	8300
				5	42,000
Aluminum-Copper	Ground	1.17–1.4	20	15	56,000
				10	12,000
Aluminum-Copper	Milled	4.4–4.5	20	20–35	22,000

*Divide the given values by 5.678 to convert to Btu/h · ft² · °F.

Discussion Note that the interface between the two plates offers as much resistance to heat transfer as a 2.15-cm-thick aluminum plate. It is interesting that the thermal contact resistance in this case is greater than the sum of the thermal resistances of both plates.

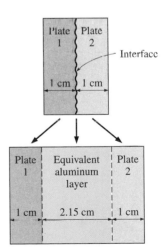

FIGURE 3–17
Schematic for Example 3–4.

EXAMPLE 3–5   Contact Resistance of Transistors

Four identical power transistors with aluminum casing are attached on one side of a 1-cm-thick 20-cm × 20-cm square copper plate ($k = 386$ W/m · °C) by screws that exert an average pressure of 6 MPa (Fig. 3–18). The base area of each transistor is 8 cm², and each transistor is placed at the center of a 10-cm × 10-cm quarter section of the plate. The interface roughness is estimated to be about 1.5 μm. All transistors are covered by a thick Plexiglas layer, which is a poor conductor of heat, and thus all the heat generated at the junction of the transistor must be dissipated to the ambient at 20°C through the back surface of the copper plate. The combined convection/radiation heat transfer coefficient at the back surface can be taken to be 25 W/m² · °C. If the case temperature of the

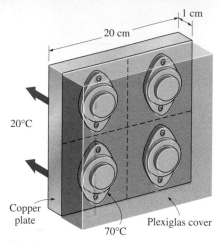

1 cm

20 cm

20°C

Copper
plate

70°C

Plexiglas cover

FIGURE 3–18
Schematic for Example 3–5.

transistor is not to exceed 70°C, determine the maximum power each transistor can dissipate safely, and the temperature jump at the case-plate interface.

SOLUTION Four identical power transistors are attached on a copper plate. For a maximum case temperature of 70°C, the maximum power dissipation and the temperature jump at the interface are to be determined.

Assumptions **1** Steady operating conditions exist. **2** Heat transfer can be approximated as being one-dimensional, although it is recognized that heat conduction in some parts of the plate will be two-dimensional since the plate area is much larger than the base area of the transistor. But the large thermal conductivity of copper will minimize this effect. **3** All the heat generated at the junction is dissipated through the back surface of the plate since the transistors are covered by a thick Plexiglas layer. **4** Thermal conductivities are constant.

Properties The thermal conductivity of copper is given to be $k = 386$ W/m · °C. The contact conductance is obtained from Table 3–2 to be $h_c = 42,000$ W/m² · °C, which corresponds to copper-aluminum interface for the case of 1.17–1.4 μm roughness and 5 MPa pressure, which is sufficiently close to what we have.

Analysis The contact area between the case and the plate is given to be 8 cm², and the plate area for each transistor is 100 cm². The thermal resistance network of this problem consists of three resistances in series (interface, plate, and convection), which are determined to be

$$R_{\text{interface}} = \frac{1}{h_c A_c} = \frac{1}{(42,000 \text{ W/m}^2 \cdot \text{°C})(8 \times 10^{-4} \text{ m}^2)} = 0.030\text{°C/W}$$

$$R_{\text{plate}} = \frac{L}{kA} = \frac{0.01 \text{ m}}{(386 \text{ W/m} \cdot \text{°C})(0.01 \text{ m}^2)} = 0.0026\text{°C/W}$$

$$R_{\text{conv}} = \frac{1}{h_o A} = \frac{1}{(25 \text{ W/m}^2 \cdot \text{°C})(0.01 \text{ m}^2)} = 4.0\text{°C/W}$$

The total thermal resistance is then

$$R_{\text{total}} = R_{\text{interface}} + R_{\text{plate}} + R_{\text{ambient}} = 0.030 + 0.0026 + 4.0 = 4.0326\text{°C/W}$$

Note that the thermal resistance of a copper plate is very small and can be ignored altogether. Then the rate of heat transfer is determined to be

$$\dot{Q} = \frac{\Delta T}{R_{\text{total}}} = \frac{(70 - 20)\text{°C}}{4.0326\text{°C/W}} = \textbf{12.4 W}$$

Therefore, the power transistor should not be operated at power levels greater than 12.4 W if the case temperature is not to exceed 70°C.

The temperature jump at the interface is determined from

$$\Delta T_{\text{interface}} = \dot{Q} R_{\text{interface}} = (12.4 \text{ W})(0.030\text{°C/W}) = \textbf{0.37°C}$$

which is not very large. Therefore, even if we eliminate the thermal contact resistance at the interface completely, we lower the operating temperature of the transistor in this case by less than 0.4°C.

3–3 ▪ GENERALIZED THERMAL RESISTANCE NETWORKS

The *thermal resistance* concept or the *electrical analogy* can also be used to solve steady heat transfer problems that involve parallel layers or combined series-parallel arrangements. Although such problems are often two- or even three-dimensional, approximate solutions can be obtained by assuming one-dimensional heat transfer and using the thermal resistance network.

Consider the composite wall shown in Fig. 3–19, which consists of two parallel layers. The thermal resistance network, which consists of two parallel resistances, can be represented as shown in the figure. Noting that the total heat transfer is the sum of the heat transfers through each layer, we have

$$\dot{Q} = \dot{Q}_1 + \dot{Q}_2 = \frac{T_1 - T_2}{R_1} + \frac{T_1 - T_2}{R_2} = (T_1 - T_2)\left(\frac{1}{R_1} + \frac{1}{R_2}\right) \qquad \text{(3–29)}$$

Utilizing electrical analogy, we get

$$\dot{Q} = \frac{T_1 - T_2}{R_{\text{total}}} \qquad \text{(3–30)}$$

where

$$\frac{1}{R_{\text{total}}} = \frac{1}{R_1} + \frac{1}{R_2} \longrightarrow R_{\text{total}} = \frac{R_1 R_2}{R_1 + R_2} \qquad \text{(3–31)}$$

since the resistances are in parallel.

Now consider the combined series-parallel arrangement shown in Fig. 3–20. The total rate of heat transfer through this composite system can again be expressed as

$$\dot{Q} = \frac{T_1 - T_\infty}{R_{\text{total}}} \qquad \text{(3–32)}$$

where

$$R_{\text{total}} = R_{12} + R_3 + R_{\text{conv}} = \frac{R_1 R_2}{R_1 + R_2} + R_3 + R_{\text{conv}} \qquad \text{(3–33)}$$

and

$$R_1 = \frac{L_1}{k_1 A_1} \qquad R_2 = \frac{L_2}{k_2 A_2} \qquad R_3 = \frac{L_3}{k_3 A_3} \qquad R_{\text{conv}} = \frac{1}{h A_3} \qquad \text{(3–34)}$$

Once the individual thermal resistances are evaluated, the total resistance and the total rate of heat transfer can easily be determined from the relations above.

The result obtained is somewhat approximate, since the surfaces of the third layer are probably not isothermal, and heat transfer between the first two layers is likely to occur.

Two assumptions commonly used in solving complex multidimensional heat transfer problems by treating them as one-dimensional (say, in the

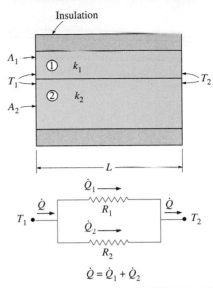

FIGURE 3–19
Thermal resistance network for two parallel layers.

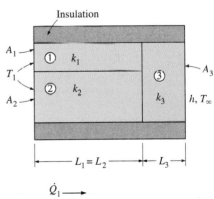

FIGURE 3–20
Thermal resistance network for combined series-parallel arrangement.

x-direction) using the thermal resistance network are (1) any plane wall normal to the *x*-axis is *isothermal* (i.e., to assume the temperature to vary in the *x*-direction only) and (2) any plane parallel to the *x*-axis is *adiabatic* (i.e., to assume heat transfer to occur in the *x*-direction only). These two assumptions result in different resistance networks, and thus different (but usually close) values for the total thermal resistance and thus heat transfer. The actual result lies between these two values. In geometries in which heat transfer occurs predominantly in one direction, either approach gives satisfactory results.

Foam   **Plaster**

h_2
$T_{\infty 2}$

1.5 cm

Brick

22 cm

h_1
$T_{\infty 1}$

1.5 cm

x

3 · 2 · 16 cm · 2

R_3

$T_{\infty 1}$ — R_i — R_1 — R_2 — R_4 — R_6 — R_o — $T_{\infty 2}$

R_5

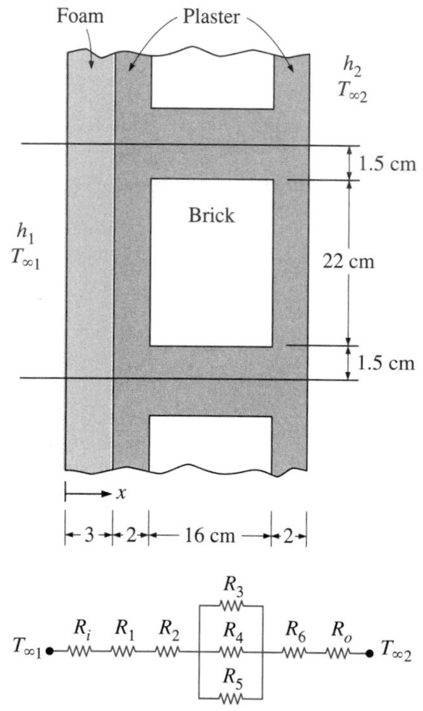

FIGURE 3–21
Schematic for Example 3–6.

EXAMPLE 3–6   **Heat Loss through a Composite Wall**

A 3-m-high and 5-m-wide wall consists of long 16-cm × 22-cm cross section horizontal bricks ($k = 0.72$ W/m · °C) separated by 3-cm-thick plaster layers ($k = 0.22$ W/m · °C). There are also 2-cm-thick plaster layers on each side of the brick and a 3-cm-thick rigid foam ($k = 0.026$ W/m · °C) on the inner side of the wall, as shown in Fig. 3–21. The indoor and the outdoor temperatures are 20°C and −10°C, respectively, and the convection heat transfer coefficients on the inner and the outer sides are $h_1 = 10$ W/m² · °C and $h_2 = 25$ W/m² · °C, respectively. Assuming one-dimensional heat transfer and disregarding radiation, determine the rate of heat transfer through the wall.

SOLUTION   The composition of a composite wall is given. The rate of heat transfer through the wall is to be determined.

Assumptions   **1** Heat transfer is steady since there is no indication of change with time. **2** Heat transfer can be approximated as being one-dimensional since it is predominantly in the *x*-direction. **3** Thermal conductivities are constant. **4** Heat transfer by radiation is negligible.

Properties   The thermal conductivities are given to be $k = 0.72$ W/m · °C for bricks, $k = 0.22$ W/m · °C for plaster layers, and $k = 0.026$ W/m · °C for the rigid foam.

Analysis   There is a pattern in the construction of this wall that repeats itself every 25-cm distance in the vertical direction. There is no variation in the horizontal direction. Therefore, we consider a 1-m-deep and 0.25-m-high portion of the wall, since it is representative of the entire wall.

Assuming any cross section of the wall normal to the *x*-direction to be *isothermal*, the thermal resistance network for the representative section of the wall becomes as shown in Fig. 3–21. The individual resistances are evaluated as:

$$R_i = R_{\text{conv, 1}} = \frac{1}{h_1 A} = \frac{1}{(10 \text{ W/m}^2 \cdot °C)(0.25 \times 1 \text{ m}^2)} = 0.40 °C/W$$

$$R_1 = R_{\text{foam}} = \frac{L}{kA} = \frac{0.03 \text{ m}}{(0.026 \text{ W/m} \cdot °C)(0.25 \times 1 \text{ m}^2)} = 4.62 °C/W$$

$$R_2 = R_6 = R_{\text{plaster, side}} = \frac{L}{kA} = \frac{0.02 \text{ m}}{(0.22 \text{ W/m} \cdot °C)(0.25 \times 1 \text{ m}^2)}$$
$$= 0.36 °C/W$$

$$R_3 = R_5 = R_{\text{plaster, center}} = \frac{L}{kA} = \frac{0.16 \text{ m}}{(0.22 \text{ W/m} \cdot °C)(0.015 \times 1 \text{ m}^2)}$$
$$= 48.48 °C/W$$

$$R_4 = R_{brick} = \frac{L}{kA} = \frac{0.16 \text{ m}}{(0.72 \text{ W/m} \cdot {}^\circ\text{C})(0.22 \times 1 \text{ m}^2)} = 1.01{}^\circ\text{C/W}$$

$$R_o = R_{conv, 2} = \frac{1}{h_2 A} = \frac{1}{(25 \text{ W/m}^2 \cdot {}^\circ\text{C})(0.25 \times 1 \text{ m}^2)} = 0.16{}^\circ\text{C/W}$$

The three resistances R_3, R_4, and R_5 in the middle are parallel, and their equivalent resistance is determined from

$$\frac{1}{R_{mid}} = \frac{1}{R_3} + \frac{1}{R_4} + \frac{1}{R_5} = \frac{1}{48.48} + \frac{1}{1.01} + \frac{1}{48.48} = 1.03 \text{ W/}{}^\circ\text{C}$$

which gives

$$R_{mid} = 0.97{}^\circ\text{C/W}$$

Now all the resistances are in series, and the total resistance is

$$\begin{aligned} R_{total} &= R_i + R_1 + R_2 + R_{mid} + R_6 + R_o \\ &= 0.40 + 4.62 + 0.36 + 0.97 + 0.36 + 0.16 \\ &= 6.87{}^\circ\text{C/W} \end{aligned}$$

Then the steady rate of heat transfer through the wall becomes

$$\dot{Q} = \frac{T_{\infty 1} - T_{\infty 2}}{R_{total}} = \frac{[20 - (-10)]{}^\circ\text{C}}{6.87{}^\circ\text{C/W}} = 4.37 \text{ W} \qquad \text{(per 0.25 m}^2 \text{ surface area)}$$

or 4.37/0.25 = 17.5 W per m² area. The total area of the wall is A = 3 m × 5 m = 15 m². Then the rate of heat transfer through the entire wall becomes

$$\dot{Q}_{total} = (17.5 \text{ W/m}^2)(15 \text{ m}^2) = \mathbf{263 \text{ W}}$$

Of course, this result is approximate, since we assumed the temperature within the wall to vary in one direction only and ignored any temperature change (and thus heat transfer) in the other two directions.

Discussion In the above solution, we assumed the temperature at any cross section of the wall normal to the x-direction to be *isothermal*. We could also solve this problem by going to the other extreme and assuming the surfaces parallel to the x-direction to be *adiabatic*. The thermal resistance network in this case will be as shown in Fig. 3–22. By following the approach outlined above, the total thermal resistance in this case is determined to be R_{total} = 6.97°C/W, which is very close to the value 6.85°C/W obtained before. Thus either approach gives roughly the same result in this case. This example demonstrates that either approach can be used in practice to obtain satisfactory results.

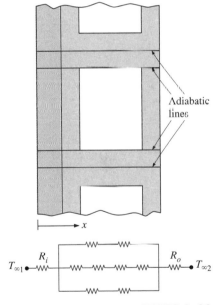

FIGURE 3–22
Alternative thermal resistance network for Example 3–6 for the case of surfaces parallel to the primary direction of heat transfer being adiabatic.

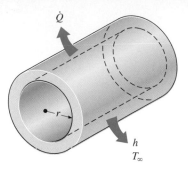

FIGURE 3–23

Heat is lost from a hot-water pipe to the air outside in the radial direction, and thus heat transfer from a long pipe is one-dimensional.

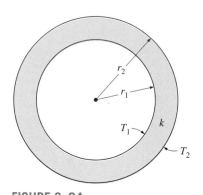

FIGURE 3–24

A long cylindrical pipe (or spherical shell) with specified inner and outer surface temperatures T_1 and T_2.

3–4 ▪ HEAT CONDUCTION IN CYLINDERS AND SPHERES

Consider steady heat conduction through a hot-water pipe. Heat is continuously lost to the outdoors through the wall of the pipe, and we intuitively feel that heat transfer through the pipe is in the normal direction to the pipe surface and no significant heat transfer takes place in the pipe in other directions (Fig. 3–23). The wall of the pipe, whose thickness is rather small, separates two fluids at different temperatures, and thus the temperature gradient in the radial direction is relatively large. Further, if the fluid temperatures inside and outside the pipe remain constant, then heat transfer through the pipe is *steady*. Thus heat transfer through the pipe can be modeled as *steady* and *one-dimensional*. The temperature of the pipe in this case depends on one direction only (the radial r-direction) and can be expressed as $T = T(r)$. The temperature is independent of the azimuthal angle or the axial distance. This situation is approximated in practice in long cylindrical pipes and spherical containers.

In *steady* operation, there is no change in the temperature of the pipe with time at any point. Therefore, the rate of heat transfer into the pipe must be equal to the rate of heat transfer out of it. In other words, heat transfer through the pipe must be constant, $\dot{Q}_{\text{cond, cyl}} = $ constant.

Consider a long cylindrical layer (such as a circular pipe) of inner radius r_1, outer radius r_2, length L, and average thermal conductivity k (Fig. 3–24). The two surfaces of the cylindrical layer are maintained at constant temperatures T_1 and T_2. There is no heat generation in the layer and the thermal conductivity is constant. For one-dimensional heat conduction through the cylindrical layer, we have $T(r)$. Then Fourier's law of heat conduction for heat transfer through the cylindrical layer can be expressed as

$$\dot{Q}_{\text{cond, cyl}} = -kA\frac{dT}{dr} \qquad \text{(W)} \qquad (3\text{–}35)$$

where $A = 2\pi rL$ is the heat transfer area at location r. Note that A depends on r, and thus it *varies* in the direction of heat transfer. Separating the variables in the above equation and integrating from $r = r_1$, where $T(r_1) = T_1$, to $r = r_2$, where $T(r_2) = T_2$, gives

$$\int_{r=r_1}^{r_2} \frac{\dot{Q}_{\text{cond, cyl}}}{A}\,dr = -\int_{T=T_1}^{T_2} k\,dT \qquad (3\text{–}36)$$

Substituting $A = 2\pi rL$ and performing the integrations give

$$\dot{Q}_{\text{cond, cyl}} = 2\pi Lk\frac{T_1 - T_2}{\ln(r_2/r_1)} \qquad \text{(W)} \qquad (3\text{–}37)$$

since $\dot{Q}_{\text{cond, cyl}} = $ constant. This equation can be rearranged as

$$\dot{Q}_{\text{cond, cyl}} = \frac{T_1 - T_2}{R_{\text{cyl}}} \qquad \text{(W)} \qquad (3\text{–}38)$$

where

$$R_{cyl} = \frac{\ln(r_2/r_1)}{2\pi Lk} = \frac{\ln(\text{Outer radius/Inner radius})}{2\pi \times \text{Length} \times \text{Thermal conductivity}} \qquad (3\text{--}39)$$

is the *thermal resistance* of the cylindrical layer against heat conduction, or simply the *conduction resistance* of the cylinder layer.

We can repeat the analysis for a *spherical layer* by taking $A = 4\pi r^2$ and performing the integrations in Eq. 3–36. The result can be expressed as

$$\dot{Q}_{cond,\,sph} = \frac{T_1 - T_2}{R_{sph}} \qquad (3\text{--}40)$$

where

$$R_{sph} = \frac{r_2 - r_1}{4\pi r_1 r_2 k} = \frac{\text{Outer radius} - \text{Inner radius}}{4\pi(\text{Outer radius})(\text{Inner radius})(\text{Thermal conductivity})} \qquad (3\text{--}41)$$

is the *thermal resistance* of the spherical layer against heat conduction, or simply the *conduction resistance* of the spherical layer.

Now consider steady one-dimensional heat transfer through a cylindrical or spherical layer that is exposed to convection on both sides to fluids at temperatures $T_{\infty1}$ and $T_{\infty2}$ with heat transfer coefficients h_1 and h_2, respectively, as shown in Fig. 3–25. The thermal resistance network in this case consists of one conduction and two convection resistances in series, just like the one for the plane wall, and the rate of heat transfer under steady conditions can be expressed as

$$\dot{Q} = \frac{T_{\infty1} - T_{\infty2}}{R_{total}} \qquad (3\text{--}42)$$

where

$$
\begin{aligned}
R_{total} &= R_{conv,\,1} + R_{cyl} + R_{conv,\,2} \\
&= \frac{1}{(2\pi r_1 L)h_1} + \frac{\ln(r_2/r_1)}{2\pi Lk} + \frac{1}{(2\pi r_2 L)h_2}
\end{aligned} \qquad (3\text{--}43)
$$

for a *cylindrical* layer, and

$$
\begin{aligned}
R_{total} &= R_{conv,\,1} + R_{sph} + R_{conv,\,2} \\
&= \frac{1}{(4\pi r_1^2)h_1} + \frac{r_2 - r_1}{4\pi r_1 r_2 k} + \frac{1}{(4\pi r_2^2)h_2}
\end{aligned} \qquad (3\text{--}44)
$$

for a *spherical* layer. Note that A in the convection resistance relation $R_{conv} = 1/hA$ is the *surface area at which convection occurs*. It is equal to $A = 2\pi rL$ for a cylindrical surface and $A = 4\pi r^2$ for a spherical surface of radius r. Also note that the thermal resistances are in series, and thus the total thermal resistance is determined by simply adding the individual resistances, just like the electrical resistances connected in series.

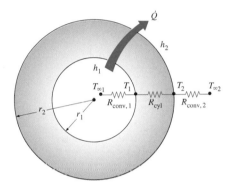

$$R_{total} = R_{conv,1} + R_{cyl} + R_{conv,2}$$

FIGURE 3–25
The thermal resistance network for a cylindrical (or spherical) shell subjected to convection from both the inner and the outer sides.

Multilayered Cylinders and Spheres

Steady heat transfer through multilayered cylindrical or spherical shells can be handled just like multilayered plane walls discussed earlier by simply adding an *additional resistance* in series for each *additional layer*. For example, the steady heat transfer rate through the three-layered composite cylinder of length L shown in Fig. 3–26 with convection on both sides can be expressed as

$$\dot{Q} = \frac{T_{\infty 1} - T_{\infty 2}}{R_{\text{total}}} \tag{3–45}$$

where R_{total} is the *total thermal resistance,* expressed as

$$R_{\text{total}} = R_{\text{conv, 1}} + R_{\text{cyl, 1}} + R_{\text{cyl, 2}} + R_{\text{cyl, 3}} + R_{\text{conv, 2}}$$
$$= \frac{1}{h_1 A_1} + \frac{\ln(r_2/r_1)}{2\pi L k_1} + \frac{\ln(r_3/r_2)}{2\pi L k_2} + \frac{\ln(r_4/r_3)}{2\pi L k_3} + \frac{1}{h_2 A_4} \tag{3–46}$$

where $A_1 = 2\pi r_1 L$ and $A_4 = 2\pi r_4 L$. Equation 3–46 can also be used for a three-layered spherical shell by replacing the thermal resistances of cylindrical layers by the corresponding spherical ones. Again, note from the thermal resistance network that the resistances are in series, and thus the total thermal resistance is simply the *arithmetic sum* of the individual thermal resistances in the path of heat flow.

Once $\dot{Q}$ is known, we can determine any intermediate temperature T_j by applying the relation $\dot{Q} = (T_i - T_j)/R_{\text{total}, i-j}$ across any layer or layers such that T_i is a *known* temperature at location i and $R_{\text{total}, i-j}$ is the total thermal resistance between locations i and j (Fig. 3–27). For example, once $\dot{Q}$ has been calculated, the interface temperature T_2 between the first and second cylindrical layers can be determined from

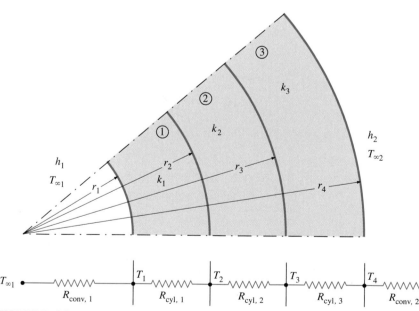

FIGURE 3–26

The thermal resistance network for heat transfer through a three-layered composite cylinder subjected to convection on both sides.

$$\dot{Q} = \frac{T_{\infty 1} - T_2}{R_{conv, 1} + R_{cyl, 1}} = \frac{T_{\infty 1} - T_2}{\dfrac{1}{h_1(2\pi r_1 L)} + \dfrac{\ln(r_2/r_1)}{2\pi L k_1}} \qquad (3\text{-}47)$$

We could also calculate T_2 from

$$\dot{Q} = \frac{T_2 - T_{\infty 2}}{R_2 + R_3 + R_{conv, 2}} = \frac{T_2 - T_{\infty 2}}{\dfrac{\ln(r_3/r_2)}{2\pi L k_2} + \dfrac{\ln(r_4/r_3)}{2\pi L k_3} + \dfrac{1}{h_o(2\pi r_4 L)}} \qquad (3\text{-}48)$$

Although both relations give the same result, we prefer the first one since it involves fewer terms and thus less work.

The thermal resistance concept can also be used for *other geometries*, provided that the proper conduction resistances and the proper surface areas in convection resistances are used.

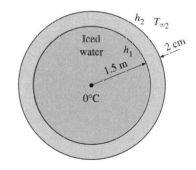

FIGURE 3–27

The ratio $\Delta T/R$ across any layer is equal to $\dot{Q}$, which remains constant in one-dimensional steady conduction.

$$\dot{Q} = \frac{T_{\infty 1} - T_1}{R_{conv,1}}$$
$$= \frac{T_{\infty 1} - T_2}{R_{conv,1} + R_1}$$
$$= \frac{T_1 - T_3}{R_1 + R_2}$$
$$= \frac{T_2 - T_3}{R_2}$$
$$= \frac{T_2 - T_{\infty 2}}{R_2 + R_{conv,2}}$$
$$= \cdots$$

EXAMPLE 3–7   Heat Transfer to a Spherical Container

A 3-m internal diameter spherical tank made of 2-cm-thick stainless steel ($k = 15$ W/m · °C) is used to store iced water at $T_{\infty 1} = 0$°C. The tank is located in a room whose temperature is $T_{\infty 2} = 22$°C. The walls of the room are also at 22°C. The outer surface of the tank is black and heat transfer between the outer surface of the tank and the surroundings is by natural convection and radiation. The convection heat transfer coefficients at the inner and the outer surfaces of the tank are $h_1 = 80$ W/m² · °C and $h_2 = 10$ W/m² · °C, respectively. Determine (a) the rate of heat transfer to the iced water in the tank and (b) the amount of ice at 0°C that melts during a 24-h period.

SOLUTION   A spherical container filled with iced water is subjected to convection and radiation heat transfer at its outer surface. The rate of heat transfer and the amount of ice that melts per day are to be determined.
Assumptions   1 Heat transfer is steady since the specified thermal conditions at the boundaries do not change with time. 2 Heat transfer is one-dimensional since there is thermal symmetry about the midpoint. 3 Thermal conductivity is constant.
Properties   The thermal conductivity of steel is given to be $k = 15$ W/m · °C. The heat of fusion of water at atmospheric pressure is $h_{if} = 333.7$ kJ/kg. The outer surface of the tank is black and thus its emissivity is $\varepsilon = 1$.
Analysis   (a) The thermal resistance network for this problem is given in Fig. 3–28. Noting that the inner diameter of the tank is $D_1 = 3$ m and the outer diameter is $D_2 = 3.04$ m, the inner and the outer surface areas of the tank are

$$A_1 = \pi D_1^2 = \pi(3 \text{ m})^2 = 28.3 \text{ m}^2$$
$$A_2 = \pi D_2^2 = \pi(3.04 \text{ m})^2 = 29.0 \text{ m}^2$$

Also, the radiation heat transfer coefficient is given by

$$h_{rad} = \varepsilon \sigma (T_2^2 + T_{\infty 2}^2)(T_2 + T_{\infty 2})$$

But we do not know the outer surface temperature T_2 of the tank, and thus we cannot calculate h_{rad}. Therefore, we need to assume a T_2 value now and check

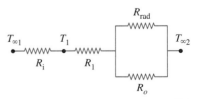

FIGURE 3–28

Schematic for Example 3–7.

the accuracy of this assumption later. We will repeat the calculations if necessary using a revised value for T_2.

We note that T_2 must be between 0°C and 22°C, but it must be closer to 0°C, since the heat transfer coefficient inside the tank is much larger. Taking $T_2 = 5°C = 278$ K, the radiation heat transfer coefficient is determined to be

$$h_{rad} = (1)(5.67 \times 10^{-8} \text{ W/m}^2 \cdot \text{K}^4)[(295 \text{ K})^2 + (278 \text{ K})^2][(295 + 278) \text{ K}]$$
$$= 5.34 \text{ W/m}^2 \cdot \text{K} = 5.34 \text{ W/m}^2 \cdot °\text{C}$$

Then the individual thermal resistances become

$$R_i = R_{conv, 1} = \frac{1}{h_1 A_1} = \frac{1}{(80 \text{ W/m}^2 \cdot °\text{C})(28.3 \text{ m}^2)} = 0.000442°\text{C/W}$$

$$R_1 = R_{sphere} = \frac{r_2 - r_1}{4\pi k r_1 r_2} = \frac{(1.52 - 1.50) \text{ m}}{4\pi (15 \text{ W/m} \cdot °\text{C})(1.52 \text{ m})(1.50 \text{ m})}$$
$$= 0.000047°\text{C/W}$$

$$R_o = R_{conv, 2} = \frac{1}{h_2 A_2} = \frac{1}{(10 \text{ W/m}^2 \cdot °\text{C})(29.0 \text{ m}^2)} = 0.00345°\text{C/W}$$

$$R_{rad} = \frac{1}{h_{rad} A_2} = \frac{1}{(5.34 \text{ W/m}^2 \cdot °\text{C})(29.0 \text{ m}^2)} = 0.00646°\text{C/W}$$

The two parallel resistances R_o and R_{rad} can be replaced by an equivalent resistance R_{equiv} determined from

$$\frac{1}{R_{equiv}} = \frac{1}{R_o} + \frac{1}{R_{rad}} = \frac{1}{0.00345} + \frac{1}{0.00646} = 444.7 \text{ W/°C}$$

which gives

$$R_{equiv} = 0.00225°\text{C/W}$$

Now all the resistances are in series, and the total resistance is

$$R_{total} = R_i + R_1 + R_{equiv} = 0.000442 + 0.000047 + 0.00225 = 0.00274°\text{C/W}$$

Then the steady rate of heat transfer to the iced water becomes

$$\dot{Q} = \frac{T_{\infty 2} - T_{\infty 1}}{R_{total}} = \frac{(22 - 0)°\text{C}}{0.00274°\text{C/W}} = \textbf{8029 W} \qquad (\text{or } \dot{Q} = 8.029 \text{ kJ/s})$$

To check the validity of our original assumption, we now determine the outer surface temperature from

$$\dot{Q} = \frac{T_{\infty 2} - T_2}{R_{equiv}} \longrightarrow T_2 = T_{\infty 2} - \dot{Q}R_{equiv}$$
$$= 22°\text{C} - (8029 \text{ W})(0.00225°\text{C/W}) = 4°\text{C}$$

which is sufficiently close to the 5°C assumed in the determination of the radiation heat transfer coefficient. Therefore, there is no need to repeat the calculations using 4°C for T_2.

(*b*) The total amount of heat transfer during a 24-h period is

$$Q = \dot{Q}\,\Delta t = (8.029 \text{ kJ/s})(24 \times 3600 \text{ s}) = 693,700 \text{ kJ}$$

Noting that it takes 333.7 kJ of energy to melt 1 kg of ice at 0°C, the amount of ice that will melt during a 24-h period is

$$m_{ice} = \frac{Q}{h_{if}} = \frac{693,700 \text{ kJ}}{333.7 \text{ kJ/kg}} = \textbf{2079 kg}$$

Therefore, about 2 metric tons of ice will melt in the tank every day.

Discussion An easier way to deal with combined convection and radiation at a surface when the surrounding medium and surfaces are at the same temperature is to add the radiation and convection heat transfer coefficients and to treat the result as the convection heat transfer coefficient. That is, to take $h = 10 + 5.34 = 15.34 \text{ W/m}^2 \cdot °C$ in this case. This way, we can ignore radiation since its contribution is accounted for in the convection heat transfer coefficient. The convection resistance of the outer surface in this case would be

$$R_{combined} = \frac{1}{h_{combined} A_2} = \frac{1}{(15.34 \text{ W/m}^2 \cdot °C)(29.0 \text{ m}^2)} = 0.00225°C/W$$

which is identical to the value obtained for equivalent resistance for the parallel convection and the radiation resistances.

EXAMPLE 3–8   Heat Loss through an Insulated Steam Pipe

Steam at $T_{\infty 1} = 320°C$ flows in a cast iron pipe ($k = 80 \text{ W/m} \cdot °C$) whose inner and outer diameters are $D_1 = 5$ cm and $D_2 = 5.5$ cm, respectively. The pipe is covered with 3-cm-thick glass wool insulation with $k = 0.05 \text{ W/m} \cdot °C$. Heat is lost to the surroundings at $T_{\infty 2} = 5°C$ by natural convection and radiation, with a combined heat transfer coefficient of $h_2 = 18 \text{ W/m}^2 \cdot °C$. Taking the heat transfer coefficient inside the pipe to be $h_1 = 60 \text{ W/m}^2 \cdot °C$, determine the rate of heat loss from the steam per unit length of the pipe. Also determine the temperature drops across the pipe shell and the insulation.

SOLUTION A steam pipe covered with glass wool insulation is subjected to convection on its surfaces. The rate of heat transfer per unit length and the temperature drops across the pipe and the insulation are to be determined.
Assumptions **1** Heat transfer is steady since there is no indication of any change with time. **2** Heat transfer is one-dimensional since there is thermal symmetry about the centerline and no variation in the axial direction. **3** Thermal conductivities are constant. **4** The thermal contact resistance at the interface is negligible.
Properties The thermal conductivities are given to be $k = 80 \text{ W/m} \cdot °C$ for cast iron and $k = 0.05 \text{ W/m} \cdot °C$ for glass wool insulation.

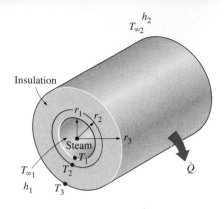

$T_{\infty 1}$ •—www—•—www—•—www—•—www—• $T_{\infty 2}$
$\quad\quad R_i \quad\quad R_1 \quad\quad R_2 \quad\quad R_o$
$\quad\quad\quad T_1 \quad\quad T_2 \quad\quad T_3$

FIGURE 3–29
Schematic for Example 3–8.

Analysis The thermal resistance network for this problem involves four resistances in series and is given in Fig. 3–29. Taking $L = 1$ m, the areas of the surfaces exposed to convection are determined to be

$$A_1 = 2\pi r_1 L = 2\pi(0.025\ m)(1\ m) = 0.157\ m^2$$
$$A_3 = 2\pi r_3 L = 2\pi(0.0575\ m)(1\ m) = 0.361\ m^2$$

Then the individual thermal resistances become

$$R_i = R_{conv,\ 1} = \frac{1}{h_1 A_1} = \frac{1}{(60\ W/m^2 \cdot °C)(0.157\ m^2)} = 0.106°C/W$$

$$R_1 = R_{pipe} = \frac{\ln(r_2/r_1)}{2\pi k_1 L} = \frac{\ln(2.75/2.5)}{2\pi(80\ W/m \cdot °C)(1\ m)} = 0.0002°C/W$$

$$R_2 = R_{insulation} = \frac{\ln(r_3/r_2)}{2\pi k_2 L} = \frac{\ln(5.75/2.75)}{2\pi(0.05\ W/m \cdot °C)(1\ m)} = 2.35°C/W$$

$$R_o = R_{conv,\ 2} = \frac{1}{h_2 A_3} = \frac{1}{(18\ W/m^2 \cdot °C)(0.361\ m^2)} = 0.154°C/W$$

Noting that all resistances are in series, the total resistance is determined to be

$$R_{total} = R_i + R_1 + R_2 + R_o = 0.106 + 0.0002 + 2.35 + 0.154 = 2.61°C/W$$

Then the steady rate of heat loss from the steam becomes

$$\dot{Q} = \frac{T_{\infty 1} - T_{\infty 2}}{R_{total}} = \frac{(320 - 5)°C}{2.61°C/W} = \mathbf{121\ W} \quad \text{(per m pipe length)}$$

The heat loss for a given pipe length can be determined by multiplying the above quantity by the pipe length L.

The temperature drops across the pipe and the insulation are determined from Eq. 3–17 to be

$$\Delta T_{pipe} = \dot{Q} R_{pipe} = (121\ W)(0.0002°C/W) = \mathbf{0.02°C}$$

$$\Delta T_{insulation} = \dot{Q} R_{insulation} = (121\ W)(2.35°C/W) = \mathbf{284°C}$$

That is, the temperatures between the inner and the outer surfaces of the pipe differ by 0.02°C, whereas the temperatures between the inner and the outer surfaces of the insulation differ by 284°C.

Discussion Note that the thermal resistance of the pipe is too small relative to the other resistances and can be neglected without causing any significant error. Also note that the temperature drop across the pipe is practically zero, and thus the pipe can be assumed to be isothermal. The resistance to heat flow in insulated pipes is primarily due to insulation.

3–5 · CRITICAL RADIUS OF INSULATION

We know that adding more insulation to a wall or to the attic always decreases heat transfer. The thicker the insulation, the lower the heat transfer rate. This is expected, since the heat transfer area A is constant, and adding insulation always increases the thermal resistance of the wall without increasing the convection resistance.

Adding insulation to a cylindrical pipe or a spherical shell, however, is a different matter. The additional insulation increases the conduction resistance of the insulation layer but decreases the convection resistance of the surface because of the increase in the outer surface area for convection. The heat transfer from the pipe may increase or decrease, depending on which effect dominates.

Consider a cylindrical pipe of outer radius r_1 whose outer surface temperature T_1 is maintained constant (Fig. 3–30). The pipe is now insulated with a material whose thermal conductivity is k and outer radius is r_2. Heat is lost from the pipe to the surrounding medium at temperature T_∞, with a convection heat transfer coefficient h. The rate of heat transfer from the insulated pipe to the surrounding air can be expressed as (Fig. 3–31)

$$\dot{Q} = \frac{T_1 - T_\infty}{R_{ins} + R_{conv}} = \frac{T_1 - T_\infty}{\frac{\ln(r_2/r_1)}{2\pi Lk} + \frac{1}{h(2\pi r_2 L)}} \tag{3-49}$$

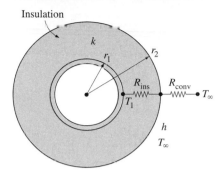

FIGURE 3–30
An insulated cylindrical pipe exposed to convection from the outer surface and the thermal resistance network associated with it.

The variation of $\dot{Q}$ with the outer radius of the insulation r_2 is plotted in Fig. 3–31. The value of r_2 at which $\dot{Q}$ reaches a maximum is determined from the requirement that $d\dot{Q}/dr_2 = 0$ (zero slope). Performing the differentiation and solving for r_2 yields the **critical radius of insulation** for a cylindrical body to be

$$r_{cr, cylinder} = \frac{k}{h} \quad \text{(m)} \tag{3-50}$$

Note that the critical radius of insulation depends on the thermal conductivity of the insulation k and the external convection heat transfer coefficient h. The rate of heat transfer from the cylinder increases with the addition of insulation for $r_2 < r_{cr}$, reaches a maximum when $r_2 = r_{cr}$, and starts to decrease for $r_2 > r_{cr}$. Thus, insulating the pipe may actually increase the rate of heat transfer from the pipe instead of decreasing it when $r_2 < r_{cr}$.

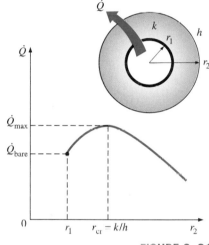

FIGURE 3–31

The important question to answer at this point is whether we need to be concerned about the critical radius of insulation when insulating hot-water pipes or even hot-water tanks. Should we always check and make sure that the outer radius of insulation sufficiently exceeds the critical radius before we install any insulation? Probably not, as explained here.

The value of the critical radius r_{cr} is the largest when k is large and h is small. Noting that the lowest value of h encountered in practice is about 5 W/m² · °C for the case of natural convection of gases, and that the thermal conductivity of common insulating materials is about 0.05 W/m² · °C, the largest value of the critical radius we are likely to encounter is

$$r_{cr, max} = \frac{k_{max, insulation}}{h_{min}} \approx \frac{0.05 \text{ W/m} \cdot °\text{C}}{5 \text{ W/m}^2 \cdot °\text{C}} = 0.01 \text{ m} = 1 \text{ cm}$$

This value would be even smaller when the radiation effects are considered. The critical radius would be much less in forced convection, often less than 1 mm, because of much larger h values associated with forced convection. Therefore, we can insulate hot-water or steam pipes freely without worrying about the possibility of increasing the heat transfer by insulating the pipes.

The radius of electric wires may be smaller than the critical radius. Therefore, the plastic electrical insulation may actually *enhance* the heat transfer from electric wires and thus keep their steady operating temperatures at lower and thus safer levels.

The discussions above can be repeated for a sphere, and it can be shown in a similar manner that the critical radius of insulation for a spherical shell is

$$r_{\text{cr, sphere}} = \frac{2k}{h} \tag{3-51}$$

where k is the thermal conductivity of the insulation and h is the convection heat transfer coefficient on the outer surface.

EXAMPLE 3–9　Heat Loss from an Insulated Electric Wire

A 3-mm-diameter and 5-m-long electric wire is tightly wrapped with a 2-mm-thick plastic cover whose thermal conductivity is $k = 0.15$ W/m · °C. Electrical measurements indicate that a current of 10 A passes through the wire and there is a voltage drop of 8 V along the wire. If the insulated wire is exposed to a medium at $T_\infty = 30$°C with a heat transfer coefficient of $h = 12$ W/m² · °C, determine the temperature at the interface of the wire and the plastic cover in steady operation. Also determine whether doubling the thickness of the plastic cover will increase or decrease this interface temperature.

SOLUTION　An electric wire is tightly wrapped with a plastic cover. The interface temperature and the effect of doubling the thickness of the plastic cover on the interface temperature are to be determined.

Assumptions　**1** Heat transfer is steady since there is no indication of any change with time. **2** Heat transfer is one-dimensional since there is thermal symmetry about the centerline and no variation in the axial direction. **3** Thermal conductivities are constant. **4** The thermal contact resistance at the interface is negligible. **5** Heat transfer coefficient incorporates the radiation effects, if any.

Properties　The thermal conductivity of plastic is given to be $k = 0.15$ W/m · °C.

Analysis　Heat is generated in the wire and its temperature rises as a result of resistance heating. We assume heat is generated uniformly throughout the wire and is transferred to the surrounding medium in the radial direction. In steady operation, the rate of heat transfer becomes equal to the heat generated within the wire, which is determined to be

$$\dot{Q} = \dot{W}_e = \mathbf{V}I = (8 \text{ V})(10 \text{ A}) = 80 \text{ W}$$

The thermal resistance network for this problem involves a conduction resistance for the plastic cover and a convection resistance for the outer surface in series, as shown in Fig. 3–32. The values of these two resistances are

$$A_2 = (2\pi r_2)L = 2\pi(0.0035 \text{ m})(5 \text{ m}) = 0.110 \text{ m}^2$$

$$R_{\text{conv}} = \frac{1}{hA_2} = \frac{1}{(12 \text{ W/m}^2 \cdot °\text{C})(0.110 \text{ m}^2)} = 0.76°\text{C/W}$$

$$R_{\text{plastic}} = \frac{\ln(r_2/r_1)}{2\pi kL} = \frac{\ln(3.5/1.5)}{2\pi(0.15 \text{ W/m} \cdot °\text{C})(5 \text{ m})} = 0.18°\text{C/W}$$

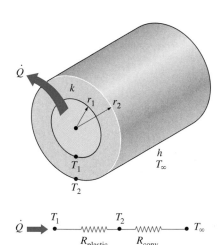

FIGURE 3–32
Schematic for Example 3–9.

and therefore

$$R_{\text{total}} = R_{\text{plastic}} + R_{\text{conv}} = 0.76 + 0.18 = 0.94°\text{C/W}$$

Then the interface temperature can be determined from

$$\dot{Q} = \frac{T_1 - T_\infty}{R_{\text{total}}} \longrightarrow T_1 = T_\infty + \dot{Q}R_{\text{total}}$$
$$= 30°\text{C} + (80\ \text{W})(0.94°\text{C/W}) = \mathbf{105°C}$$

Note that we did not involve the electrical wire directly in the thermal resistance network, since the wire involves heat generation.

To answer the second part of the question, we need to know the critical radius of insulation of the plastic cover. It is determined from Eq. 3–50 to be

$$r_{\text{cr}} = \frac{k}{h} = \frac{0.15\ \text{W/m} \cdot °\text{C}}{12\ \text{W/m}^2 \cdot °\text{C}} = 0.0125\ \text{m} = 12.5\ \text{mm}$$

which is larger than the radius of the plastic cover. Therefore, increasing the thickness of the plastic cover will *enhance* heat transfer until the outer radius of the cover reaches 12.5 mm. As a result, the rate of heat transfer $\dot{Q}$ will *increase* when the interface temperature T_1 is held constant, or T_1 will *decrease* when $\dot{Q}$ is held constant, which is the case here.

Discussion It can be shown by repeating the calculations above for a 4-mm-thick plastic cover that the interface temperature drops to 90.6°C when the thickness of the plastic cover is doubled. It can also be shown in a similar manner that the interface reaches a minimum temperature of 83°C when the outer radius of the plastic cover equals the critical radius.

3–6 ▪ HEAT TRANSFER FROM FINNED SURFACES

The rate of heat transfer from a surface at a temperature T_s to the surrounding medium at T_∞ is given by Newton's law of cooling as

$$\dot{Q}_{\text{conv}} = hA_s(T_s - T_\infty)$$

where A_s is the heat transfer surface area and h is the convection heat transfer coefficient. When the temperatures T_s and T_∞ are fixed by design considerations, as is often the case, there are *two ways* to increase the rate of heat transfer: to increase the *convection heat transfer coefficient* h or to increase the *surface area* A_s. Increasing h may require the installation of a pump or fan, or replacing the existing one with a larger one, but this approach may or may not be practical. Besides, it may not be adequate. The alternative is to increase the surface area by attaching to the surface *extended surfaces* called *fins* made of highly conductive materials such as aluminum. Finned surfaces are manufactured by extruding, welding, or wrapping a thin metal sheet on a surface. Fins enhance heat transfer from a surface by exposing a larger surface area to convection and radiation.

Finned surfaces are commonly used in practice to enhance heat transfer, and they often increase the rate of heat transfer from a surface severalfold.

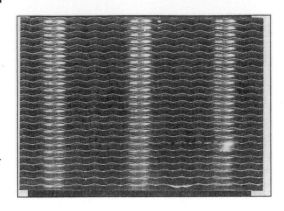

FIGURE 3–33
The thin plate fins of a car radiator greatly increase the rate of heat transfer to the air. (© Yunus Çengel, photo by James Kleiser.)

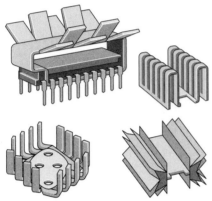

FIGURE 3–34
Some innovative fin designs.

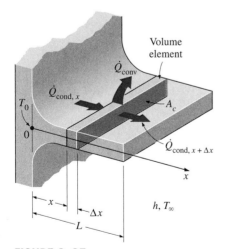

FIGURE 3–35
Volume element of a fin at location x having a length of Δx, cross-sectional area of A_c, and perimeter of p.

The car radiator shown in Fig. 3–33 is an example of a finned surface. The closely packed thin metal sheets attached to the hot-water tubes increase the surface area for convection and thus the rate of convection heat transfer from the tubes to the air many times. There are a variety of innovative fin designs available in the market, and they seem to be limited only by imagination (Fig. 3–34).

In the analysis of fins, we consider *steady* operation with *no heat generation* in the fin, and we assume the thermal conductivity k of the material to remain constant. We also assume the convection heat transfer coefficient h to be *constant* and *uniform* over the entire surface of the fin for convenience in the analysis. We recognize that the convection heat transfer coefficient h, in general, varies along the fin as well as its circumference, and its value at a point is a strong function of the *fluid motion* at that point. The value of h is usually much lower at the *fin base* than it is at the *fin tip* because the fluid is surrounded by solid surfaces near the base, which seriously disrupt its motion to the point of "suffocating" it, while the fluid near the fin tip has little contact with a solid surface and thus encounters little resistance to flow. Therefore, adding too many fins on a surface may actually decrease the overall heat transfer when the decrease in h offsets any gain resulting from the increase in the surface area.

Fin Equation

Consider a volume element of a fin at location x having a length of Δx, cross-sectional area of A_c, and a perimeter of p, as shown in Fig. 3–35. Under steady conditions, the energy balance on this volume element can be expressed as

$$\begin{pmatrix} \text{Rate of } heat \\ conduction \text{ into} \\ \text{the element at } x \end{pmatrix} = \begin{pmatrix} \text{Rate of } heat \\ conduction \text{ from the} \\ \text{element at } x + \Delta x \end{pmatrix} + \begin{pmatrix} \text{Rate of } heat \\ convection \text{ from} \\ \text{the element} \end{pmatrix}$$

or

$$\dot{Q}_{\text{cond}, x} = \dot{Q}_{\text{cond}, x + \Delta x} + \dot{Q}_{\text{conv}}$$

$$\dot{Q}_{cond,x} = \dot{Q}_{cond,x+\Delta x} + \dot{Q}_{conv}$$

$$\dot{Q}_{conv} = h(p\,\Delta x)(T-T_\infty) \qquad \dot{Q}_{cond} = -kA_c \frac{dT}{dx}$$

$$\frac{\dot{Q}_{cond,x+\Delta x} - \dot{Q}_{cond,x}}{\Delta x} + hp(T-T_\infty) = 0$$

taking the limit as $\Delta x \to 0$

$$\frac{d\dot{Q}_{cond}}{dx} + hp(T-T_\infty) = 0$$

$$\frac{d}{dx}\left(kA_c \frac{dT}{dx}\right) - hp(T-T_\infty) = 0$$

$$\theta = T - T_\infty \qquad \theta_b = T_b - T_\infty$$

$$kA_c \frac{d^2T}{dx^2} - hp(T-T_\infty) = 0$$

$$kA_c \frac{d^2\theta}{dx^2} - hp^2\theta = 0 \qquad \frac{d^2\theta}{dx^2} - \frac{hp}{kA_c}\theta = 0$$

$$\frac{d^2\theta}{dx^2} - m^2\theta = 0 \qquad m^2 = \frac{hp}{kA_c}$$

$$\boxed{\theta(x) = C_1 e^{mx} + C_2 e^{-mx}}$$

$$\theta(0) = \theta_b = T_b - T_\infty$$

$$T(x) = T_\infty + (T_b - T_\infty) e^{-x\sqrt{\frac{hp}{A_c k}}}$$

$$\dot{Q}_{long\ fin} = -kA_c \frac{dT}{dx}\bigg|_{x=0}$$

$$-kA \frac{dT}{dx} = -kA_c -\sqrt{\frac{hp}{kA_c}}\; e^{-0} = \sqrt{hp\, k A_c}\; (T_b - T_\infty)$$

$$(T_b - T_\infty)$$

$$\boxed{\dot{Q}_{long\ fin} = \sqrt{hp\, k A_c}\; (T_b - T_\infty)}$$

p: perimeter

where

$$\dot{Q}_{conv} = h(p\,\Delta x)(T - T_\infty)$$

Substituting and dividing by Δx, we obtain

$$\frac{\dot{Q}_{cond,\,x+\Delta x} - \dot{Q}_{cond,\,x}}{\Delta x} + hp(T - T_\infty) = 0 \tag{3–52}$$

Taking the limit as $\Delta x \to 0$ gives

$$\frac{d\dot{Q}_{cond}}{dx} + hp(T - T_\infty) = 0 \tag{3–53}$$

From Fourier's law of heat conduction we have

$$\dot{Q}_{cond} = -kA_c \frac{dT}{dx} \tag{3–54}$$

where A_c is the cross-sectional area of the fin at location x. Substitution of this relation into Eq. 3–53 gives the differential equation governing heat transfer in fins,

$$\frac{d}{dx}\left(kA_c \frac{dT}{dx}\right) - hp(T - T_\infty) = 0 \tag{3–55}$$

In general, the cross-sectional area A_c and the perimeter p of a fin vary with x, which makes this differential equation difficult to solve. In the special case of *constant cross section* and *constant thermal conductivity,* the differential equation 3–55 reduces to

$$\frac{d^2\theta}{dx^2} - m^2\theta = 0 \tag{3–56}$$

where

$$m^2 = \frac{hp}{kA_c} \tag{3–57}$$

and $\theta = T - T_\infty$ is the *temperature excess.* At the fin base we have $\theta_b = T_b - T_\infty$.

Equation 3–56 is a linear, homogeneous, second-order differential equation with constant coefficients. A fundamental theory of differential equations states that such an equation has two linearly independent solution functions, and its general solution is the linear combination of those two solution functions. A careful examination of the differential equation reveals that subtracting a constant multiple of the solution function θ from its second derivative yields zero. Thus we conclude that the function θ and its second derivative must be *constant multiples* of each other. The only functions whose derivatives are constant multiples of the functions themselves are the *exponential functions* (or a linear combination of exponential functions such as sine and cosine hyperbolic functions). Therefore, the solution functions of the differential equation above are the exponential functions e^{-mx} or e^{mx} or constant multiples of them. This can be verified by direct substitution. For example, the second derivative of e^{-mx} is $m^2 e^{-mx}$, and its substitution into Eq. 3–56

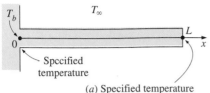

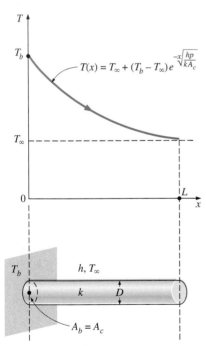

(a) Specified temperature
(b) Negligible heat loss
(c) Convection
(d) Convection and radiation

FIGURE 3–36

Boundary conditions at the fin base and the fin tip.

$T(x) = T_\infty + (T_b - T_\infty)e^{-x\sqrt{\frac{hp}{kA_c}}}$

$(p = \pi D, A_c = \pi D^2/4$ for a cylindrical fin)

FIGURE 3–37

A long circular fin of uniform cross section and the variation of temperature along it.

yields zero. Therefore, the general solution of the differential equation Eq. 3–56 is

$$\theta(x) = C_1 e^{mx} + C_2 e^{-mx} \qquad (3\text{–}58)$$

where C_1 and C_2 are arbitrary constants whose values are to be determined from the boundary conditions at the base and at the tip of the fin. Note that we need only two conditions to determine C_1 and C_2 uniquely.

The temperature of the plate to which the fins are attached is normally known in advance. Therefore, at the fin base we have a *specified temperature* boundary condition, expressed as

Boundary condition at fin base: $\qquad \theta(0) = \theta_b = T_b - T_\infty \qquad (3\text{–}59)$

At the fin tip we have several possibilities, including specified temperature, negligible heat loss (idealized as an adiabatic tip), convection, and combined convection and radiation (Fig. 3–36). Next, we consider each case separately.

1 Infinitely Long Fin ($T_{\text{fin tip}} = T_\infty$)

For a sufficiently long fin of *uniform* cross section (A_c = constant), the temperature of the fin at the fin tip approaches the environment temperature T_∞ and thus θ approaches zero. That is,

Boundary condition at fin tip: $\quad \theta(L) = T(L) - T_\infty = 0 \quad$ as $\quad L \to \infty$

This condition is satisfied by the function e^{-mx}, but not by the other prospective solution function e^{mx} since it tends to infinity as x gets larger. Therefore, the general solution in this case will consist of a constant multiple of e^{-mx}. The value of the constant multiple is determined from the requirement that at the fin base where $x = 0$ the value of θ is θ_b. Noting that $e^{-mx} = e^0 = 1$, the proper value of the constant is θ_b, and the solution function we are looking for is $\theta(x) = \theta_b e^{-mx}$. This function satisfies the differential equation as well as the requirements that the solution reduce to θ_b at the fin base and approach zero at the fin tip for large x. Noting that $\theta = T - T_\infty$ and $m = \sqrt{hp/kA_c}$, the variation of temperature along the fin in this case can be expressed as

Very long fin: $\qquad \dfrac{T(x) - T_\infty}{T_b - T_\infty} = e^{-mx} = e^{-x\sqrt{hp/kA_c}} \qquad (3\text{–}60)$

Note that the temperature along the fin in this case decreases *exponentially* from T_b to T_∞, as shown in Fig. 3–37. The steady rate of *heat transfer* from the entire fin can be determined from Fourier's law of heat conduction

Very long fin: $\qquad \dot{Q}_{\text{long fin}} = -kA_c \dfrac{dT}{dx}\bigg|_{x=0} = \sqrt{hpkA_c}\,(T_b - T_\infty) \qquad (3\text{–}61)$

where p is the perimeter, A_c is the cross-sectional area of the fin, and x is the distance from the fin base. Alternatively, the rate of heat transfer from the fin could also be determined by considering heat transfer from a differential volume element of the fin and integrating it over the entire surface of the fin:

$$\dot{Q}_{\text{fin}} = \int_{A_{\text{fin}}} h[T(x) - T_\infty]\, dA_{\text{fin}} = \int_{A_{\text{fin}}} h\theta(x)\, dA_{\text{fin}} \qquad (3\text{–}62)$$

The two approaches described are equivalent and give the same result since, under steady conditions, the heat transfer from the exposed surfaces of the fin is equal to the heat transfer to the fin at the base (Fig. 3–38).

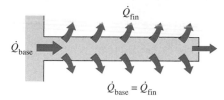

FIGURE 3–38
Under steady conditions, heat transfer from the exposed surfaces of the fin is equal to heat conduction to the fin at the base.

2 Negligible Heat Loss from the Fin Tip (Adiabatic fin tip, $Q_{\text{fin tip}} = 0$)

Fins are not likely to be so long that their temperature approaches the surrounding temperature at the tip. A more realistic situation is for heat transfer from the fin tip to be negligible since the heat transfer from the fin is proportional to its surface area, and the surface area of the fin tip is usually a negligible fraction of the total fin area. Then the fin tip can be assumed to be adiabatic, and the condition at the fin tip can be expressed as

Boundary condition at fin tip:
$$\frac{d\theta}{dx}\bigg|_{x=L} = 0 \qquad\qquad (3\text{–}63)$$

The condition at the fin base remains the same as expressed in Eq. 3–59. The application of these two conditions on the general solution (Eq. 3–58) yields, after some manipulations, this relation for the temperature distribution:

Adiabatic fin tip:
$$\frac{T(x) - T_\infty}{T_b - T_\infty} = \frac{\cosh m(L - x)}{\cosh mL} \qquad\qquad (3\text{–}64)$$

The rate of heat transfer from the fin can be determined again from Fourier's law of heat conduction:

Adiabatic fin tip:
$$\dot{Q}_{\text{adiabatic tip}} = -kA_c \frac{dT}{dx}\bigg|_{x=0}$$
$$= \sqrt{hpkA_c}\,(T_b - T_\infty)\tanh mL \qquad\qquad (3\text{–}65)$$

Note that the heat transfer relations for the very long fin and the fin with negligible heat loss at the tip differ by the factor tanh mL, which approaches 1 as L becomes very large.

3 Convection (or Combined Convection and Radiation) from Fin Tip

The fin tips, in practice, are exposed to the surroundings, and thus the proper boundary condition for the fin tip is convection that also includes the effects of radiation. The fin equation can still be solved in this case using the convection at the fin tip as the second boundary condition, but the analysis becomes more involved, and it results in rather lengthy expressions for the temperature distribution and the heat transfer. Yet, in general, the fin tip area is a small fraction of the total fin surface area, and thus the complexities involved can hardly justify the improvement in accuracy.

A practical way of accounting for the heat loss from the fin tip is to replace the *fin length L* in the relation for the *insulated tip* case by a **corrected length** defined as (Fig. 3–39)

Corrected fin length:
$$L_c = L + \frac{A_c}{p} \qquad\qquad (3\text{–}66)$$

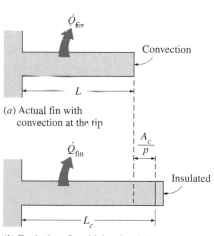

(a) Actual fin with convection at the tip

(b) Equivalent fin with insulated tip

FIGURE 3–39
Corrected fin length L_c is defined such that heat transfer from a fin of length L_c with insulated tip is equal to heat transfer from the actual fin of length L with convection at the fin tip.

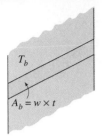

T_b

$A_b = w \times t$

(a) Surface without fins

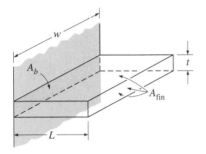

(b) Surface with a fin

$$A_{\text{fin}} = 2 \times w \times L + w \times t$$
$$\cong 2 \times w \times L$$

FIGURE 3–40

Fins enhance heat transfer from a surface by enhancing surface area.

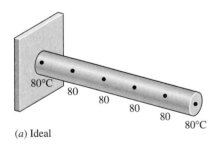

(a) Ideal

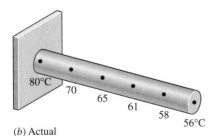

(b) Actual

FIGURE 3–41

Ideal and actual temperature distribution along a fin.

where A_c is the cross-sectional area and p is the perimeter of the fin at the tip. Multiplying the relation above by the perimeter gives $A_{\text{corrected}} = A_{\text{fin (lateral)}} + A_{\text{tip}}$, which indicates that the fin area determined using the corrected length is equivalent to the sum of the lateral fin area plus the fin tip area.

The corrected length approximation gives very good results when the variation of temperature near the fin tip is small (which is the case when $mL \geq 1$) and the heat transfer coefficient at the fin tip is about the same as that at the lateral surface of the fin. Therefore, *fins subjected to convection at their tips can be treated as fins with insulated tips by replacing the actual fin length by the corrected length in Eqs. 3–64 and 3–65.*

Using the proper relations for A_c and p, the corrected lengths for rectangular and cylindrical fins are easily determined to be

$$L_{c, \text{rectangular fin}} = L + \frac{t}{2} \quad \text{and} \quad L_{c, \text{cylindrical fin}} = L + \frac{D}{4}$$

where t is the thickness of the rectangular fins and D is the diameter of the cylindrical fins.

Fin Efficiency

Consider the surface of a *plane wall* at temperature T_b exposed to a medium at temperature T_∞. Heat is lost from the surface to the surrounding medium by convection with a heat transfer coefficient of h. Disregarding radiation or accounting for its contribution in the convection coefficient h, heat transfer from a surface area A_s is expressed as $\dot{Q} = hA_s(T_s - T_\infty)$.

Now let us consider a fin of constant cross-sectional area $A_c = A_b$ and length L that is attached to the surface with a perfect contact (Fig. 3–40). This time heat is transfered from the surface to the fin *by conduction* and from the fin to the surrounding medium *by convection* with the same heat transfer coefficient h. The temperature of the fin is T_b at the fin base and gradually decreases toward the fin tip. Convection from the fin surface causes the temperature at any cross section to drop somewhat from the midsection toward the outer surfaces. However, the cross-sectional area of the fins is usually very small, and thus the temperature at any cross section can be considered to be uniform. Also, the fin tip can be assumed for convenience and simplicity to be adiabatic by using the corrected length for the fin instead of the actual length.

In the limiting case of *zero thermal resistance* or *infinite thermal conductivity* ($k \rightarrow \infty$), the temperature of the fin is uniform at the base value of T_b. The heat transfer from the fin is *maximum* in this case and can be expressed as

$$\dot{Q}_{\text{fin, max}} = hA_{\text{fin}}(T_b - T_\infty) \tag{3–67}$$

In reality, however, the temperature of the fin drops along the fin, and thus the heat transfer from the fin is less because of the decreasing temperature difference $T(x) - T_\infty$ toward the fin tip, as shown in Fig. 3–41. To account for the effect of this decrease in temperature on heat transfer, we define a **fin efficiency** as

$$\eta_{\text{fin}} = \frac{\dot{Q}_{\text{fin}}}{\dot{Q}_{\text{fin, max}}} = \frac{\text{Actual heat transfer rate from the fin}}{\substack{\text{Ideal heat transfer rate from the fin} \\ \text{if the entire fin were at base temperature}}} \tag{3–68}$$

TABLE 3–3

Efficiency and surface areas of common fin configurations

Straight rectangular fins

$m = \sqrt{2h/kt}$
$L_c = L + t/2$
$A_{fin} = 2wL_c$

$$\eta_{fin} = \frac{\tanh mL_c}{mL_c}$$

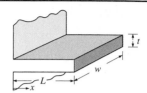

Straight triangular fins

$m = \sqrt{2h/kt}$
$A_{fin} = 2w\sqrt{L^2 + (t/2)^2}$

$$\eta_{fin} = \frac{1}{mL}\frac{I_1(2mL)}{I_0(2mL)}$$

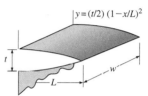

Straight parabolic fins

$m = \sqrt{2h/kt}$
$A_{fin} = wL[C_1 + (L/t)\ln(t/L + C_1)]$
$C_1 = \sqrt{1 + (t/L)^2}$

$$\eta_{fin} = \frac{2}{1 + \sqrt{(2mL)^2 + 1}}$$

Circular fins of rectangular profile

$m = \sqrt{2h/kt}$
$r_{2c} = r_2 + t/2$
$A_{fin} = 2\pi(r_{2c}^2 - r_1^2)$

$$\eta_{fin} = C_2\frac{K_1(mr_1)I_1(mr_{2c}) - I_1(mr_1)K_1(mr_{2c})}{I_0(mr_1)K_1(mr_{2c}) + K_0(mr_1)I_1(mr_{2c})}$$

$$C_2 = \frac{2r_1/m}{r_{2c}^2 - r_1^2}$$

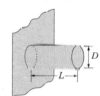

Pin fins of rectangular profile

$m = \sqrt{4h/kD}$
$L_c = L + D/4$
$A_{fin} = \pi DL_c$

$$\eta_{fin} = \frac{\tanh mL_c}{mL_c}$$

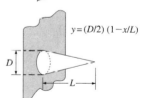

Pin fins of triangular profile

$m = \sqrt{4h/kD}$
$A_{fin} = \frac{\pi D}{2}\sqrt{L^2 + (D/2)^2}$

$$\eta_{fin} = \frac{2}{mL}\frac{I_2(2mL)}{I_1(2mL)}$$

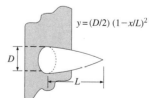

Pin fins of parabolic profile

$m = \sqrt{4h/kD}$
$A_{fin} = \frac{\pi L^3}{8D}[C_3C_4 - \frac{L}{2D}\ln(2DC_4/L + C_3)]$
$C_3 = 1 + 2(D/L)^2$
$C_4 = \sqrt{1 + (D/L)^2}$

$$\eta_{fin} = \frac{2}{1 + \sqrt{(2mL/3)^2 + 1}}$$

Pin fins of parabolic profile (blunt tip)

$m = \sqrt{4h/kD}$
$A_{fin} = \frac{\pi D^4}{96L^2}\{[16(L/D)^2 + 1]^{3/2} - 1\}$

$$\eta_{fin} = \frac{3}{2mL}\frac{I_1(4mL/3)}{I_0(4mL/3)}$$

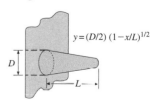

TABLE 3–4

Modified Bessel functions of the first and second kinds*

x	$e^{-x}I_0(x)$	$e^{-x}I_1(x)$	$e^{x}K_0(x)$	$e^{x}K_1(x)$
0.0	1.0000	0.0000	—	—
0.2	0.8269	0.0823	2.1408	5.8334
0.4	0.6974	0.1368	1.6627	3.2587
0.6	0.5993	0.1722	1.4167	2.3739
0.8	0.5241	0.1945	1.2582	1.9179
1.0	0.4658	0.2079	1.1445	1.6362
1.2	0.4198	0.2153	1.0575	1.4429
1.4	0.3831	0.2185	0.9881	1.3011
1.6	0.3533	0.2190	0.9309	1.1919
1.8	0.3289	0.2177	0.8828	1.1048
2.0	0.3085	0.2153	0.8416	1.0335
2.2	0.2913	0.2121	0.8057	0.9738
2.4	0.2766	0.2085	0.7740	0.9229
2.6	0.2639	0.2047	0.7459	0.8790
2.8	0.2528	0.2007	0.7206	0.8405
3.0	0.2430	0.1968	0.6978	0.8066
3.2	0.2343	0.1930	0.6770	0.7763
3.4	0.2264	0.1892	0.6580	0.7491
3.6	0.2193	0.1856	0.6405	0.7245
3.8	0.2129	0.1821	0.6243	0.7021
4.0	0.2070	0.1788	0.6093	0.6816
4.2	0.2016	0.1755	0.5953	0.6627
4.4	0.1966	0.1725	0.5823	0.6454
4.6	0.1919	0.1695	0.5701	0.6292
4.8	0.1876	0.1667	0.5586	0.6143
5.0	0.1835	0.1640	0.5478	0.6003
5.2	0.1797	0.1614	0.5376	0.5872
5.4	0.1762	0.1589	0.5280	0.5749
5.6	0.1728	0.1565	0.5188	0.5634
5.8	0.1697	0.1542	0.5101	0.5525
6.0	0.1667	0.1521	0.5019	0.5422
6.5	0.1598	0.1469	0.4828	0.5187
7.0	0.1537	0.1423	0.4658	0.4981
7.5	0.1483	0.1380	0.4505	0.4797
8.0	0.1434	0.1341	0.4366	0.4631
8.5	0.1390	0.1305	0.4239	0.4482
9.0	0.1350	0.1272	0.4123	0.4346
9.5	0.1313	0.1241	0.4016	0.4222
10.0	0.1278	0.1213	0.3916	0.4108

*Evaluated from EES using the mathematical functions Bessel_I(x) and Bessel_K(x)

or

$$\dot{Q}_{\text{fin}} = \eta_{\text{fin}} \dot{Q}_{\text{fin, max}} = \eta_{\text{fin}} hA_{\text{fin}} (T_b - T_\infty) \qquad (3\text{–}69)$$

where A_{fin} is the total surface area of the fin. This relation enables us to determine the heat transfer from a fin when its efficiency is known. For the cases of constant cross section of *very long fins* and *fins with adiabatic tips*, the fin efficiency can be expressed as

$$\eta_{\text{long fin}} = \frac{\dot{Q}_{\text{fin}}}{\dot{Q}_{\text{fin, max}}} = \frac{\sqrt{hpkA_c}\,(T_b - T_\infty)}{hA_{\text{fin}}\,(T_b - T_\infty)} = \frac{1}{L}\sqrt{\frac{kA_c}{hp}} = \frac{1}{mL} \qquad (3\text{–}70)$$

and

$$\eta_{\text{adiabatic tip}} = \frac{\dot{Q}_{\text{fin}}}{\dot{Q}_{\text{fin, max}}} = \frac{\sqrt{hpkA_c}\,(T_b - T_\infty)\tanh aL}{hA_{\text{fin}}\,(T_b - T_\infty)} = \frac{\tanh mL}{mL} \qquad (3\text{–}71)$$

since $A_{\text{fin}} = pL$ for fins with constant cross section. Equation 3–71 can also be used for fins subjected to convection provided that the fin length L is replaced by the corrected length L_c.

Fin efficiency relations are developed for fins of various profiles, listed in Table 3–3 on page 165. The mathematical functions I and K that appear in some of these relations are the *modified Bessel functions,* and their values are given in Table 3–4. Efficiencies are plotted in Fig. 3–42 for fins on a *plain surface* and in Fig. 3–43 for *circular fins* of constant thickness. For most fins of constant thickness encountered in practice, the fin thickness t is too small relative to the fin length L, and thus the fin tip area is negligible.

Note that fins with triangular and parabolic profiles contain less material and are more efficient than the ones with rectangular profiles, and thus are more suitable for applications requiring minimum weight such as space applications.

An important consideration in the design of finned surfaces is the selection of the proper *fin length L.* Normally the *longer* the fin, the *larger* the heat transfer area and thus the *higher* the rate of heat transfer from the fin. But also the larger the fin, the bigger the mass, the higher the price, and the larger the fluid friction. Therefore, increasing the length of the fin beyond a certain value cannot be justified unless the added benefits outweigh the added cost. Also, the fin efficiency decreases with increasing fin length because of the decrease in fin temperature with length. Fin lengths that cause the fin efficiency to drop below 60 percent usually cannot be justified economically and should be avoided. The efficiency of most fins used in practice is above 90 percent.

Fin Effectiveness

Fins are used to *enhance* heat transfer, and the use of fins on a surface cannot be recommended unless the enhancement in heat transfer justifies the added cost and complexity associated with the fins. In fact, there is no assurance that adding fins on a surface will *enhance* heat transfer. The performance of the fins is judged on the basis of the enhancement in heat transfer relative to the

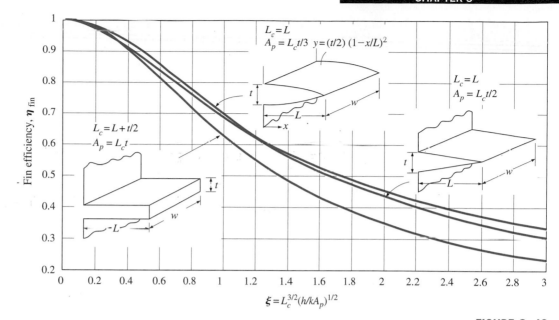

FIGURE 3–42
Efficiency of straight fins of rectangular, triangular, and parabolic profiles.

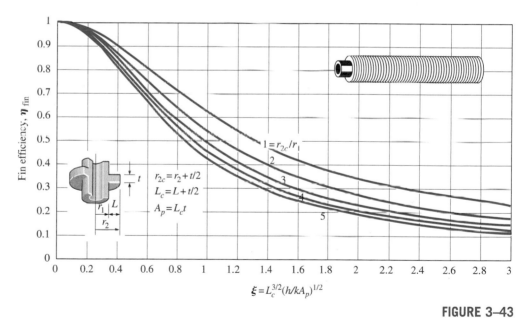

FIGURE 3–43
Efficiency of annular fins of constant thickness t.

no-fin case. The performance of fins is expressed in terms of the *fin effectiveness* ε_{fin} defined as (Fig. 3–44)

$$\varepsilon_{\text{fin}} = \frac{\dot{Q}_{\text{fin}}}{\dot{Q}_{\text{no fin}}} = \frac{\dot{Q}_{\text{fin}}}{hA_b\,(T_b - T_\infty)} = \frac{\text{Heat transfer rate from}}{\text{the fin of }base\ area\ A_b} \atop \frac{\text{Heat transfer rate from}}{\text{the surface of }area\ A_b} \qquad (3\text{–}72)$$

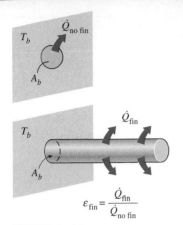

FIGURE 3–44
The effectiveness of a fin.

Here, A_b is the cross-sectional area of the fin at the base and $\dot{Q}_{\text{no fin}}$ represents the rate of heat transfer from this area if no fins are attached to the surface. An effectiveness of $\varepsilon_{\text{fin}} = 1$ indicates that the addition of fins to the surface does not affect heat transfer at all. That is, heat conducted to the fin through the base area A_b is equal to the heat transferred from the same area A_b to the surrounding medium. An effectiveness of $\varepsilon_{\text{fin}} < 1$ indicates that the fin actually acts as *insulation,* slowing down the heat transfer from the surface. This situation can occur when fins made of low thermal conductivity materials are used. An effectiveness of $\varepsilon_{\text{fin}} > 1$ indicates that fins are *enhancing* heat transfer from the surface, as they should. However, the use of fins cannot be justified unless ε_{fin} is sufficiently larger than 1. Finned surfaces are designed on the basis of *maximizing* effectiveness for a specified cost or *minimizing* cost for a desired effectiveness.

Note that both the fin efficiency and fin effectiveness are related to the performance of the fin, but they are different quantities. However, they are related to each other by

$$\varepsilon_{\text{fin}} = \frac{\dot{Q}_{\text{fin}}}{\dot{Q}_{\text{no fin}}} = \frac{\dot{Q}_{\text{fin}}}{hA_b\,(T_b - T_\infty)} = \frac{\eta_{\text{fin}}\,hA_{\text{fin}}\,(T_b - T_\infty)}{hA_b\,(T_b - T_\infty)} = \frac{A_{\text{fin}}}{A_b}\,\eta_{\text{fin}} \qquad (3\text{–}73)$$

Therefore, the fin effectiveness can be determined easily when the fin efficiency is known, or vice versa.

The rate of heat transfer from a sufficiently *long* fin of *uniform* cross section under steady conditions is given by Eq. 3–61. Substituting this relation into Eq. 3–72, the effectiveness of such a long fin is determined to be

$$\varepsilon_{\text{long fin}} = \frac{\dot{Q}_{\text{fin}}}{\dot{Q}_{\text{no fin}}} = \frac{\sqrt{hpkA_c}\,(T_b - T_\infty)}{hA_b\,(T_b - T_\infty)} = \sqrt{\frac{kp}{hA_c}} \qquad (3\text{–}74)$$

since $A_c = A_b$ in this case. We can draw several important conclusions from the fin effectiveness relation above for consideration in the design and selection of the fins:

- The *thermal conductivity k* of the fin material should be as high as possible. Thus it is no coincidence that fins are made from metals, with copper, aluminum, and iron being the most common ones. Perhaps the most widely used fins are made of aluminum because of its low cost and weight and its resistance to corrosion.

- The ratio of the *perimeter* to the *cross-sectional area* of the fin p/A_c should be as high as possible. This criterion is satisfied by *thin* plate fins and *slender* pin fins.

- The use of fins is *most effective* in applications involving a *low convection heat transfer coefficient.* Thus, the use of fins is more easily justified when the medium is a *gas* instead of a liquid and the heat transfer is by *natural convection* instead of by forced convection. Therefore, it is no coincidence that in liquid-to-gas heat exchangers such as the car radiator, fins are placed on the *gas* side.

When determining the rate of heat transfer from a finned surface, we must consider the *unfinned portion* of the surface as well as the *fins.* Therefore, the rate of heat transfer for a surface containing n fins can be expressed as

$$\dot{Q}_{\text{total, fin}} = \dot{Q}_{\text{unfin}} + \dot{Q}_{\text{fin}}$$
$$= hA_{\text{unfin}} (T_b - T_\infty) + \eta_{\text{fin}} hA_{\text{fin}} (T_b - T_\infty)$$
$$= h(A_{\text{unfin}} + \eta_{\text{fin}} A_{\text{fin}})(T_b - T_\infty) \tag{3-75}$$

We can also define an **overall effectiveness** for a finned surface as the ratio of the total heat transfer from the finned surface to the heat transfer from the same surface if there were no fins,

$$\varepsilon_{\text{fin, overall}} = \frac{\dot{Q}_{\text{total, fin}}}{\dot{Q}_{\text{total, no fin}}} = \frac{h(A_{\text{unfin}} + \eta_{\text{fin}} A_{\text{fin}})(T_b - T_\infty)}{hA_{\text{no fin}} (T_b - T_\infty)} \tag{3-76}$$

where $A_{\text{no fin}}$ is the area of the surface when there are no fins, A_{fin} is the total surface area of all the fins on the surface, and A_{unfin} is the area of the unfinned portion of the surface (Fig. 3–45). Note that the overall fin effectiveness depends on the fin density (number of fins per unit length) as well as the effectiveness of the individual fins. The overall effectiveness is a better measure of the performance of a finned surface than the effectiveness of the individual fins.

Proper Length of a Fin

An important step in the design of a fin is the determination of the appropriate length of the fin once the fin material and the fin cross section are specified. You may be tempted to think that the longer the fin, the larger the surface area and thus the higher the rate of heat transfer. Therefore, for maximum heat transfer, the fin should be infinitely long. However, the temperature drops along the fin exponentially and reaches the environment temperature at some length. The part of the fin beyond this length does not contribute to heat transfer since it is at the temperature of the environment, as shown in Fig. 3–46. Therefore, designing such an "extra long" fin is out of the question since it results in material waste, excessive weight, and increased size and thus increased cost with no benefit in return (in fact, such a long fin will hurt performance since it will suppress fluid motion and thus reduce the convection heat transfer coefficient). Fins that are so long that the temperature approaches the environment temperature cannot be recommended either since the little increase in heat transfer at the tip region cannot justify the disproportionate increase in the weight and cost.

To get a sense of the proper length of a fin, we compare heat transfer from a fin of finite length to heat transfer from an infinitely long fin under the same conditions. The ratio of these two heat transfers is

Heat transfer ratio:
$$\frac{\dot{Q}_{\text{fin}}}{\dot{Q}_{\text{long fin}}} = \frac{\sqrt{hpkA_c} (T_b - T_\infty) \tanh mL}{\sqrt{hpkA_c} (T_b - T_\infty)} = \tanh mL \tag{3-77}$$

Using a hand calculator, the values of $\tanh mL$ are evaluated for some values of mL and the results are given in Table 3–5. We observe from the table that heat transfer from a fin increases with mL almost linearly at first, but the curve reaches a plateau later and reaches a value for the infinitely long fin at about $mL = 5$. Therefore, a fin whose length is $L = \frac{1}{5}m$ can be considered to be an infinitely long fin. We also observe that reducing the fin length by half in that case (from $mL = 5$ to $mL = 2.5$) causes a drop of just 1 percent in heat trans-

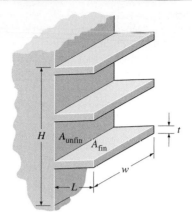

$A_{\text{no fin}} = w \times H$
$A_{\text{unfin}} = w \times H - 3 \times (t \times w)$
$A_{\text{fin}} = 2 \times L \times w + t \times w$
$\cong 2 \times L \times w$ (one fin)

FIGURE 3–45
Various surface areas associated with a rectangular surface with three fins.

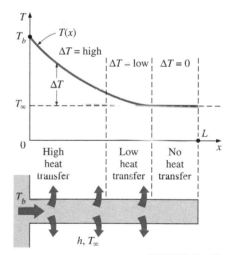

FIGURE 3–46
Because of the gradual temperature drop along the fin, the region near the fin tip makes little or no contribution to heat transfer.

TABLE 3–5

The variation of heat transfer from a fin relative to that from an infinitely long fin

mL	$\dfrac{\dot{Q}_{fin}}{\dot{Q}_{long\ fin}} = \tanh mL$
0.1	0.100
0.2	0.197
0.5	0.462
1.0	0.762
1.5	0.905
2.0	0.964
2.5	0.987
3.0	0.995
4.0	0.999
5.0	1.000

fer. We certainly would not hesitate sacrificing 1 percent in heat transfer performance in return for 50 percent reduction in the size and possibly the cost of the fin. In practice, a fin length that corresponds to about $mL = 1$ will transfer 76.2 percent of the heat that can be transferred by an infinitely long fin, and thus it should offer a good compromise between heat transfer performance and the fin size.

TABLE 3–6

Combined natural convection and radiation thermal resistance of various heat sinks used in the cooling of electronic devices between the heat sink and the surroundings. All fins are made of aluminum 6063T-5, are black anodized, and are 76 mm (3 in) long.

HS 5030

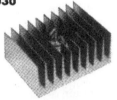

$R = 0.9°C/W$ (vertical)
$R = 1.2°C/W$ (horizontal)

Dimensions: 76 mm × 105 mm × 44 mm
Surface area: 677 cm^2

HS 6065

$R = 5°C/W$

Dimensions: 76 mm × 38 mm × 24 mm
Surface area: 387 cm^2

HS 6071

$R = 1.4°C/W$ (vertical)
$R = 1.8°C/W$ (horizontal)

Dimensions: 76 mm × 92 mm × 26 mm
Surface area: 968 cm^2

HS 6105

$R = 1.8°C/W$ (vertical)
$R = 2.1°C/W$ (horizontal)

Dimensions: 76 mm × 127 mm × 91 mm
Surface area: 677 cm^2

HS 6115

$R = 1.1°C/W$ (vertical)
$R = 1.3°C/W$ (horizontal)

Dimensions: 76 mm × 102 mm × 25 mm
Surface area: 929 cm^2

HS 7030

$R = 2.9°C/W$ (vertical)
$R = 3.1°C/W$ (horizontal)

Dimensions: 76 mm × 97 mm × 19 mm
Surface area: 290 cm^2

A common approximation used in the analysis of fins is to assume the fin temperature to vary in one direction only (along the fin length) and the temperature variation along other directions is negligible. Perhaps you are wondering if this one-dimensional approximation is a reasonable one. This is certainly the case for fins made of thin metal sheets such as the fins on a car radiator, but we wouldn't be so sure for fins made of thick materials. Studies have shown that the error involved in one-dimensional fin analysis is negligible (less than about 1 percent) when

$$\frac{h\delta}{k} < 0.2$$

where δ is the characteristic thickness of the fin, which is taken to be the plate thickness t for rectangular fins and the diameter D for cylindrical ones.

Specially designed finned surfaces called *heat sinks,* which are commonly used in the cooling of electronic equipment, involve one-of-a-kind complex geometries, as shown in Table 3–6. The heat transfer performance of heat sinks is usually expressed in terms of their *thermal resistances R* in °C/W, which is defined as

$$\dot{Q}_{\text{fin}} = \frac{T_b - T_\infty}{R} = hA_{\text{fin}}\,\eta_{\text{fin}}\,(T_b - T_\infty) \qquad \textbf{(3–78)}$$

A small value of thermal resistance indicates a small temperature drop across the heat sink, and thus a high fin efficiency.

EXAMPLE 3–10   Maximum Power Dissipation of a Transistor

Power transistors that are commonly used in electronic devices consume large amounts of electric power. The failure rate of electronic components increases almost exponentially with operating temperature. As a rule of thumb, the failure rate of electronic components is halved for each 10°C reduction in the junction operating temperature. Therefore, the operating temperature of electronic components is kept below a safe level to minimize the risk of failure.

The sensitive electronic circuitry of a power transistor at the junction is protected by its case, which is a rigid metal enclosure. Heat transfer characteristics of a power transistor are usually specified by the manufacturer in terms of the case-to-ambient thermal resistance, which accounts for both the natural convection and radiation heat transfers.

The case-to-ambient thermal resistance of a power transistor that has a maximum power rating of 10 W is given to be 20°C/W. If the case temperature of the transistor is not to exceed 85°C, determine the power at which this transistor can be operated safely in an environment at 25°C.

SOLUTION   The maximum power rating of a transistor whose case temperature is not to exceed 85°C is to be determined.

Assumptions   **1** Steady operating conditions exist. **2** The transistor case is isothermal at 85°C.

Properties   The case-to-ambient thermal resistance is given to be 20°C/W.

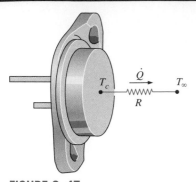

FIGURE 3–47
Schematic for Example 3–10.

Analysis The power transistor and the thermal resistance network associated with it are shown in Fig. 3–47. We notice from the thermal resistance network that there is a single resistance of 20°C/W between the case at $T_c = 85$°C and the ambient at $T_\infty = 25$°C, and thus the rate of heat transfer is

$$\dot{Q} = \left(\frac{\Delta T}{R}\right)_{\text{case-ambient}} = \frac{T_c - T_\infty}{R_{\text{case-ambient}}} = \frac{(85 - 25)°\text{C}}{20°\text{C/W}} = 3 \text{ W}$$

Therefore, this power transistor should not be operated at power levels above 3 W if its case temperature is not to exceed 85°C.

Discussion This transistor can be used at higher power levels by attaching it to a heat sink (which lowers the thermal resistance by increasing the heat transfer surface area, as discussed in the next example) or by using a fan (which lowers the thermal resistance by increasing the convection heat transfer coefficient).

EXAMPLE 3–11   Selecting a Heat Sink for a Transistor

A 60-W power transistor is to be cooled by attaching it to one of the commercially available heat sinks shown in Table 3–6. Select a heat sink that will allow the case temperature of the transistor not to exceed 90°C in the ambient air at 30°C.

SOLUTION A commercially available heat sink from Table 3–6 is to be selected to keep the case temperature of a transistor below 90°C.
Assumptions **1** Steady operating conditions exist. **2** The transistor case is isothermal at 90°C. **3** The contact resistance between the transistor and the heat sink is negligible.
Analysis The rate of heat transfer from a 60-W transistor at full power is $\dot{Q} = 60$ W. The thermal resistance between the transistor attached to the heat sink and the ambient air for the specified temperature difference is determined to be

$$\dot{Q} = \frac{\Delta T}{R} \longrightarrow R = \frac{\Delta T}{\dot{Q}} = \frac{(90 - 30)°\text{C}}{60 \text{ W}} = 1.0°\text{C/W}$$

Therefore, the thermal resistance of the heat sink should be below 1.0°C/W. An examination of Table 3–6 reveals that the HS 5030, whose thermal resistance is 0.9°C/W in the vertical position, is the only heat sink that will meet this requirement.

$r_2 = 3$ cm ⊢—⊣ $r_1 = 1.5$ cm

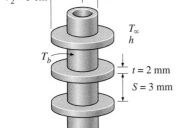

T_∞
h

T_b

$t = 2$ mm

$S = 3$ mm

FIGURE 3–48
Schematic for Example 3–12.

EXAMPLE 3–12   Effect of Fins on Heat Transfer from Steam Pipes

Steam in a heating system flows through tubes whose outer diameter is $D_1 = 3$ cm and whose walls are maintained at a temperature of 120°C. Circular aluminum alloy fins ($k = 180$ W/m · °C) of outer diameter $D_2 = 6$ cm and constant thickness $t = 2$ mm are attached to the tube, as shown in Fig. 3–48. The space between the fins is 3 mm, and thus there are 200 fins per meter length of the tube. Heat is transferred to the surrounding air at $T_\infty = 25$°C, with a combined heat transfer coefficient of $h = 60$ W/m$^2 \cdot$ °C. Determine the increase in heat transfer from the tube per meter of its length as a result of adding fins.

SOLUTION Circular aluminum alloy fins are to be attached to the tubes of a heating system. The increase in heat transfer from the tubes per unit length as a result of adding fins is to be determined.

Assumptions **1** Steady operating conditions exist. **2** The heat transfer coefficient is uniform over the entire fin surfaces. **3** Thermal conductivity is constant. **4** Heat transfer by radiation is negligible.

Properties The thermal conductivity of the fins is given to be $k = 180$ W/m $\cdot$ °C.

Analysis In the case of no fins, heat transfer from the tube per meter of its length is determined from Newton's law of cooling to be

$$A_{\text{no fin}} = \pi D_1 L = \pi (0.03 \text{ m})(1 \text{ m}) = 0.0942 \text{ m}^2$$
$$\dot{Q}_{\text{no fin}} = hA_{\text{no fin}}(T_b - T_\infty)$$
$$= (60 \text{ W/m}^2 \cdot \text{°C})(0.0942 \text{ m}^2)(120 - 25)\text{°C}$$
$$= 537 \text{ W}$$

The efficiency of the circular fins attached to a circular tube is plotted in Fig. 3–43. Noting that $L = \frac{1}{2}(D_2 - D_1) = \frac{1}{2}(0.06 - 0.03) = 0.015$ m in this case, we have

$$r_{2c} = r_2 + t/2 = 0.03 + 0.002/2 = 0.031 \text{ m}$$
$$L_c = L + t/2 = 0.015 + 0.002/2 = 0.016 \text{ m}$$
$$A_p = L_c t = (0.016 \text{ m})(0.002 \text{ m}) = 3.20 \times 10^{-5} \text{ m}^2$$

$$\frac{r_{2c}}{r_1} = \frac{0.031 \text{ m}}{0.015 \text{ m}} = 2.07$$

$$\left. L_c^{3/2} \sqrt{\frac{h}{kA_p}} = (0.016 \text{ m})^{3/2} \times \sqrt{\frac{60 \text{ W/m}^2 \cdot \text{°C}}{(180 \text{ W/m} \cdot \text{°C})(3.20 \times 10^{-5} \text{ m}^2)}} = 0.207 \right\} \eta_{\text{fin}} = 0.96$$

$$A_{\text{fin}} = 2\pi(r_{2c}^2 - r_1^2) = 2\pi[(0.031 \text{ m})^2 - (0.015 \text{ m})^2]$$
$$= 0.004624 \text{ m}^2$$

$$\dot{Q}_{\text{fin}} = \eta_{\text{fin}} \dot{Q}_{\text{fin, max}} = \eta_{\text{fin}} hA_{\text{fin}} (T_b - T_\infty)$$
$$= 0.96(60 \text{ W/m}^2 \cdot \text{°C})(0.004624 \text{ m}^2)(120 - 25)\text{°C}$$
$$= 25.3 \text{ W}$$

Heat transfer from the unfinned portion of the tube is

$$A_{\text{unfin}} = \pi D_1 S = \pi(0.03 \text{ m})(0.003 \text{ m}) = 0.000283 \text{ m}^2$$
$$\dot{Q}_{\text{unfin}} = hA_{\text{unfin}}(T_b - T_\infty)$$
$$= (60 \text{ W/m}^2 \cdot \text{°C})(0.000283 \text{ m}^2)(120 - 25)\text{°C}$$
$$= 1.6 \text{ W}$$

Noting that there are 200 fins and thus 200 interfin spacings per meter length of the tube, the total heat transfer from the finned tube becomes

$$\dot{Q}_{\text{total, fin}} = n(\dot{Q}_{\text{fin}} + \dot{Q}_{\text{unfin}}) = 200(25.3 + 1.6) \text{ W} = 5380 \text{ W}$$

Therefore, the increase in heat transfer from the tube per meter of its length as a result of the addition of fins is

$$\dot{Q}_{\text{increase}} = \dot{Q}_{\text{total, fin}} - \dot{Q}_{\text{no fin}} = 5380 - 537 = \mathbf{4843 \text{ W}} \qquad \text{(per m tube length)}$$

Discussion   The overall effectiveness of the finned tube is

$$\varepsilon_{\text{fin, overall}} = \frac{\dot{Q}_{\text{total, fin}}}{\dot{Q}_{\text{total, no fin}}} = \frac{5380 \text{ W}}{537 \text{ W}} = 10.0$$

That is, the rate of heat transfer from the steam tube increases by a factor of 10 as a result of adding fins. This explains the widespread use of finned surfaces.

3–7 ▪ HEAT TRANSFER IN COMMON CONFIGURATIONS

So far, we have considered heat transfer in *simple* geometries such as large plane walls, long cylinders, and spheres. This is because heat transfer in such geometries can be approximated as *one-dimensional,* and simple analytical solutions can be obtained easily. But many problems encountered in practice are two- or three-dimensional and involve rather complicated geometries for which no simple solutions are available.

An important class of heat transfer problems for which simple solutions are obtained encompasses those involving two surfaces maintained at *constant* temperatures T_1 and T_2. The steady rate of heat transfer between these two surfaces is expressed as

$$Q = Sk(T_1 - T_2) \tag{3–79}$$

where S is the **conduction shape factor**, which has the dimension of *length,* and k is the thermal conductivity of the medium between the surfaces. The conduction shape factor depends on the *geometry* of the system only.

Conduction shape factors have been determined for a number of configurations encountered in practice and are given in Table 3–7 for some common cases. More comprehensive tables are available in the literature. Once the value of the shape factor is known for a specific geometry, the total steady heat transfer rate can be determined from the equation above using the specified two constant temperatures of the two surfaces and the thermal conductivity of the medium between them. Note that conduction shape factors are applicable only when heat transfer between the two surfaces is by *conduction.* Therefore, they cannot be used when the medium between the surfaces is a liquid or gas, which involves natural or forced convection currents.

A comparison of Eqs. 3–4 and 3–79 reveals that the conduction shape factor S is related to the thermal resistance R by $R = 1/kS$ or $S = 1/kR$. Thus, these two quantities are the inverse of each other when the thermal conductivity of the medium is unity. The use of the conduction shape factors is illustrated with Examples 3–13 and 3–14.

TABLE 3–7

Conduction shape factors S for several configurations for use in $\dot{Q} = kS(T_1 - T_2)$ to determine the steady rate of heat transfer through a medium of thermal conductivity k between the surfaces at temperatures T_1 and T_2

(1) Isothermal cylinder of length L buried in a semi-infinite medium ($L \gg D$ and $z > 1.5D$)

$$S = \frac{2\pi L}{\ln(4z/D)}$$

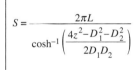

(2) Vertical isothermal cylinder of length L buried in a semi-infinite medium ($L \gg D$)

$$S = \frac{2\pi L}{\ln(4L/D)}$$

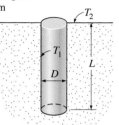

(3) Two parallel isothermal cylinders placed in an infinite medium ($L \gg D_1, D_2, z$)

$$S = \frac{2\pi L}{\cosh^{-1}\left(\frac{4z^2 - D_1^2 - D_2^2}{2D_1 D_2}\right)}$$

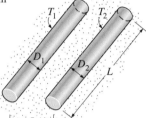

(4) A row of equally spaced parallel isothermal cylinders buried in a semi-infinite medium ($L \gg D$, z, and $w > 1.5D$)

$$S = \frac{2\pi L}{\ln\left(\frac{2w}{\pi D} \sinh \frac{2\pi z}{w}\right)}$$
(per cylinder)

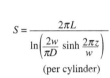
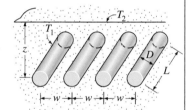

(5) Circular isothermal cylinder of length L in the midplane of an infinite wall ($z > 0.5D$)

$$S = \frac{2\pi L}{\ln(8z/\pi D)}$$

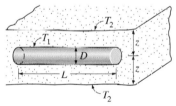

(6) Circular isothermal cylinder of length L at the center of a square solid bar of the same length

$$S = \frac{2\pi L}{\ln(1.08w/D)}$$

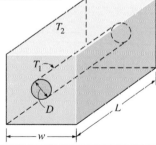

(7) Eccentric circular isothermal cylinder of length L in a cylinder of the same length ($L > D_2$)

$$S = \frac{2\pi L}{\cosh^{-1}\left(\frac{D_1^2 + D_2^2 - 4z^2}{2D_1 D_2}\right)}$$

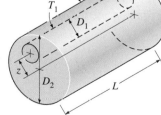

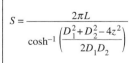

(8) Large plane wall

$$S = \frac{A}{L}$$

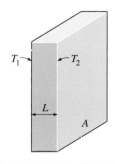

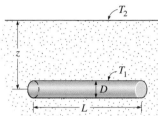

(continued)

TABLE 3–7 (*Continued*)

(9) A long cylindrical layer

$$S = \frac{2\pi L}{\ln (D_2/D_1)}$$

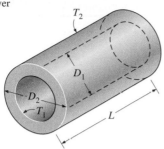

(10) A square flow passage

(*a*) For $a/b > 1.4$,

$$S = \frac{2\pi L}{0.93 \ln (0.948\, a/b)}$$

(*b*) For $a/b < 1.41$,

$$S = \frac{2\pi L}{0.785 \ln (a/b)}$$

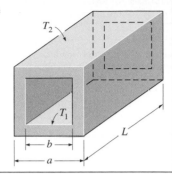

(11) A spherical layer

$$S = \frac{2\pi D_1 D_2}{D_2 - D_1}$$

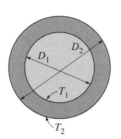

(12) Disk buried parallel to the surface in a semi-infinite medium ($z \gg D$)

$$S = 4D$$

($S = 2D$ when $z = 0$)

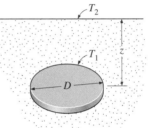

(13) The edge of two adjoining walls of equal thickness

$$S = 0.54w$$

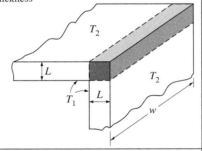

(14) Corner of three walls of equal thickness

$$S = 0.15L$$

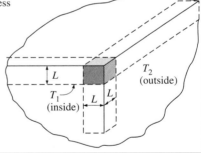

(15) Isothermal sphere buried in a semi-infinite medium

$$S = \frac{2\pi D}{1 - 0.25D/z}$$

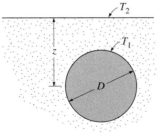

(16) Isothermal sphere buried in a semi-infinite medium at T_2 whose surface is insulated

$$S = \frac{2\pi D}{1 + 0.25D/z}$$

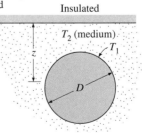

EXAMPLE 3–13   Heat Loss from Buried Steam Pipes

A 30-m-long, 10-cm-diameter hot-water pipe of a district heating system is buried in the soil 50 cm below the ground surface, as shown in Fig. 3–49. The outer surface temperature of the pipe is 80°C. Taking the surface temperature of the earth to be 10°C and the thermal conductivity of the soil at that location to be 0.9 W/m · °C, determine the rate of heat loss from the pipe.

SOLUTION   The hot-water pipe of a district heating system is buried in the soil. The rate of heat loss from the pipe is to be determined.
Assumptions   **1** Steady operating conditions exist. **2** Heat transfer is two-dimensional (no change in the axial direction). **3** Thermal conductivity of the soil is constant.
Properties   The thermal conductivity of the soil is given to be $k = 0.9$ W/m · °C.
Analysis   The shape factor for this configuration is given in Table 3–7 to be

$$S = \frac{2\pi L}{\ln(4z/D)}$$

since $z > 1.5D$, where z is the distance of the pipe from the ground surface, and D is the diameter of the pipe. Substituting,

$$S = \frac{2\pi \times (30 \text{ m})}{\ln(4 \times 0.5/0.1)} = 62.9 \text{ m}$$

Then the steady rate of heat transfer from the pipe becomes

$$\dot{Q} = Sk(T_1 - T_2) = (62.9 \text{ m})(0.9 \text{ W/m} \cdot \text{°C})(80 - 10)\text{°C} = \textbf{3963 W}$$

Discussion   Note that this heat is conducted from the pipe surface to the surface of the earth through the soil and then transferred to the atmosphere by convection and radiation.

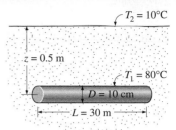

FIGURE 3–49
Schematic for Example 3–13.

EXAMPLE 3–14   Heat Transfer between Hot- and Cold-Water Pipes

A 5-m-long section of hot- and cold-water pipes run parallel to each other in a thick concrete layer, as shown in Fig. 3–50. The diameters of both pipes are 5 cm, and the distance between the centerline of the pipes is 30 cm. The surface temperatures of the hot and cold pipes are 70°C and 15°C, respectively. Taking the thermal conductivity of the concrete to be $k = 0.75$ W/m · °C, determine the rate of heat transfer between the pipes.

SOLUTION   Hot- and cold-water pipes run parallel to each other in a thick concrete layer. The rate of heat transfer between the pipes is to be determined.
Assumptions   **1** Steady operating conditions exist. **2** Heat transfer is two-dimensional (no change in the axial direction). **3** Thermal conductivity of the concrete is constant.
Properties   The thermal conductivity of concrete is given to be $k = 0.75$ W/m · °C.

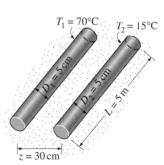

FIGURE 3–50
Schematic for Example 3–14.

Analysis The shape factor for this configuration is given in Table 3–7 to be

$$S = \frac{2\pi L}{\cosh^{-1}\left(\dfrac{4z^2 - D_1^2 - D_2^2}{2D_1 D_2}\right)}$$

where z is the distance between the centerlines of the pipes and L is their length. Substituting,

$$S = \frac{2\pi \times (5 \text{ m})}{\cosh^{-1}\left(\dfrac{4 \times 0.3^2 - 0.05^2 - 0.05^2}{2 \times 0.05 \times 0.05}\right)} = 6.34 \text{ m}$$

Then the steady rate of heat transfer between the pipes becomes

$$\dot{Q} = Sk(T_1 - T_2) = (6.34 \text{ m})(0.75 \text{ W/m} \cdot {}^\circ\text{C})(70 - 15^\circ)\text{C} = \textbf{262 W}$$

Discussion We can reduce this heat loss by placing the hot- and cold-water pipes further away from each other.

It is well known that insulation reduces heat transfer and saves energy and money. Decisions on the right amount of insulation are based on a heat transfer analysis, followed by an economic analysis to determine the "monetary value" of energy loss. This is illustrated with Example 3–15.

EXAMPLE 3–15 **Cost of Heat Loss through Walls in Winter**

Consider an electrically heated house whose walls are 9 ft high and have an R-value of insulation of 13 (i.e., a thickness-to-thermal conductivity ratio of $L/k = 13 \text{ h} \cdot \text{ft}^2 \cdot {}^\circ\text{F/Btu}$). Two of the walls of the house are 40 ft long and the others are 30 ft long. The house is maintained at 75°F at all times, while the temperature of the outdoors varies. Determine the amount of heat lost through the walls of the house on a certain day during which the average temperature of the outdoors is 45°F. Also, determine the cost of this heat loss to the home owner if the unit cost of electricity is $0.075/kWh. For combined convection and radiation heat transfer coefficients, use the ASHRAE (American Society of Heating, Refrigeration, and Air Conditioning Engineers) recommended values of $h_i = 1.46 \text{ Btu/h} \cdot \text{ft}^2 \cdot {}^\circ\text{F}$ for the inner surface of the walls and $h_o = 6.0 \text{ Btu/h} \cdot \text{ft}^2 \cdot {}^\circ\text{F}$ for the outer surface of the walls under 15 mph wind conditions in winter.

SOLUTION An electrically heated house with R-13 insulation is considered. The amount of heat lost through the walls and its cost are to be determined. **Assumptions** **1** The indoor and outdoor air temperatures have remained at the given values for the entire day so that heat transfer through the walls is steady. **2** Heat transfer through the walls is one-dimensional since any significant temperature gradients in this case exists in the direction from the indoors to the outdoors. **3** The radiation effects are accounted for in the heat transfer coefficients.

Analysis This problem involves conduction through the wall and convection at its surfaces and can best be handled by making use of the thermal resistance concept and drawing the thermal resistance network, as shown in Fig. 3–51. The heat transfer area of the walls is

$$A = \text{Circumference} \times \text{Height} = (2 \times 30 \text{ ft} + 2 \times 40 \text{ ft})(9 \text{ ft}) = 1260 \text{ ft}^2$$

Then the individual resistances are evaluated from their definitions to be

$$R_i = R_{\text{conv}, i} = \frac{1}{h_i A} = \frac{1}{(1.46 \text{ Btu/h} \cdot \text{ft}^2 \cdot {}^\circ\text{F})(1260 \text{ ft}^2)} = 0.00054 \text{ h} \cdot {}^\circ\text{F/Btu}$$

$$R_{\text{wall}} = \frac{L}{kA} = \frac{R\text{-value}}{A} = \frac{13 \text{ h} \cdot \text{ft}^2 \cdot {}^\circ\text{F/Btu}}{1260 \text{ ft}^2} = 0.01032 \text{ h} \cdot {}^\circ\text{F/Btu}$$

$$R_o = R_{\text{conv}, o} = \frac{1}{h_o A} = \frac{1}{(4.0 \text{ Btu/h} \cdot \text{ft}^2 \cdot {}^\circ\text{F})(1260 \text{ ft}^2)} = 0.00020 \text{ h} \cdot {}^\circ\text{F/Btu}$$

Noting that all three resistances are in series, the total resistance is

$$R_{\text{total}} = R_i + R_{\text{wall}} + R_o = 0.00054 + 0.01032 + 0.00020 = 0.01106 \text{ h} \cdot {}^\circ\text{F/Btu}$$

Then the steady rate of heat transfer through the walls of the house becomes

$$\dot{Q} = \frac{T_{\infty 1} - T_{\infty 2}}{R_{\text{total}}} = \frac{(75 - 45){}^\circ\text{F}}{0.01106 \text{ h} \cdot {}^\circ\text{F/Btu}} = 2712 \text{ Btu/h}$$

Finally, the total amount of heat lost through the walls during a 24-h period and its cost to the home owner are

$$Q = \dot{Q} \, \Delta t = (2712 \text{ Btu/h})(24\text{-h/day}) = \textbf{65,100 Btu/day} = \textbf{19.1 kWh/day}$$

since 1 kWh = 3412 Btu, and

Heating cost = (Energy lost)(Cost of energy) = (19.1 kWh/day)($0.075/kWh)
$$= \textbf{\$1.43/day}$$

Discussion The heat losses through the walls of the house that day cost the home owner $1.43 worth of electricity. Most of this loss can be saved by insulation.

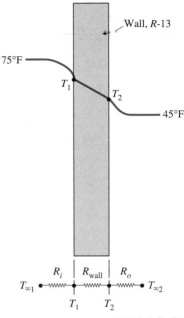

FIGURE 3–51
Schematic for Example 3–15.

Heat Transfer through Walls and Roofs

Under steady conditions, the rate of heat transfer through any section of a building wall or roof can be determined from

$$\dot{Q} = UA(T_i - T_o) = \frac{A(T_i - T_o)}{R} \tag{3–80}$$

where T_i and T_o are the indoor and outdoor air temperatures, A is the heat transfer area, U is the overall heat transfer coefficient (the U-factor), and

*This section can be skipped without a loss of continuity.

$R = 1/U$ is the overall unit thermal resistance (the R-value). Walls and roofs of buildings consist of various layers of materials, and the structure and operating conditions of the walls and the roofs may differ significantly from one building to another. Therefore, it is not practical to list the R-values (or U-factors) of different kinds of walls or roofs under different conditions. Instead, the overall R-value is determined from the thermal resistances of the individual components using the thermal resistance network. The overall thermal resistance of a structure can be determined most accurately in a lab by actually assembling the unit and testing it as a whole, but this approach is usually very time consuming and expensive. The analytical approach described here is fast and straightforward, and the results are usually in good agreement with the experimental values.

The unit thermal resistance of a plane layer of thickness L and thermal conductivity k can be determined from $R = L/k$. The thermal conductivity and other properties of common building materials are given in the appendix. The unit thermal resistances of various components used in building structures are listed in Table 3–8 for convenience.

Heat transfer through a wall or roof section is also affected by the convection and radiation heat transfer coefficients at the exposed surfaces. The effects of convection and radiation on the inner and outer surfaces of walls and roofs are usually combined into the *combined convection and radiation heat transfer coefficients* (also called *surface conductances*) h_i and h_o,

TABLE 3–8

Unit thermal resistance (the R-value) of common components used in buildings

Component	R-value m² · °C/W	R-value ft² · h · °F/Btu	Component	R-value m² · °C/W	R-value ft² · h · °F/Btu
Outside surface (winter)	0.030	0.17	Wood stud, nominal 2 in × 6 in		
Outside surface (summer)	0.044	0.25	(5.5 in or 140 mm wide)	0.98	5.56
Inside surface, still air	0.12	0.68	Clay tile, 100 mm (4 in)	0.18	1.01
Plane air space, vertical, ordinary surfaces ($\varepsilon_{eff} = 0.82$):			Acoustic tile	0.32	1.79
13 mm ($\frac{1}{2}$ in)	0.16	0.90	Asphalt shingle roofing	0.077	0.44
20 mm ($\frac{3}{4}$ in)	0.17	0.94	Building paper	0.011	0.06
40 mm (1.5 in)	0.16	0.90	Concrete block, 100 mm (4 in):		
90 mm (3.5 in)	0.16	0.91	Lightweight	0.27	1.51
Insulation, 25 mm (1 in):			Heavyweight	0.13	0.71
Glass fiber	0.70	4.00	Plaster or gypsum board,		
Mineral fiber batt	0.66	3.73	13 mm ($\frac{1}{2}$ in)	0.079	0.45
Urethane rigid foam	0.98	5.56	Wood fiberboard, 13 mm ($\frac{1}{2}$ in)	0.23	1.31
Stucco, 25 mm (1 in)	0.037	0.21	Plywood, 13 mm ($\frac{1}{2}$ in)	0.11	0.62
Face brick, 100 mm (4 in)	0.075	0.43	Concrete, 200 mm (8 in):		
Common brick, 100 mm (4 in)	0.12	0.79	Lightweight	1.17	6.67
Steel siding	0.00	0.00	Heavyweight	0.12	0.67
Slag, 13 mm ($\frac{1}{2}$ in)	0.067	0.38	Cement mortar, 13 mm ($\frac{1}{2}$ in)	0.018	0.10
Wood, 25 mm (1 in)	0.22	1.25	Wood bevel lapped siding,		
Wood stud, nominal 2 in × 4 in			13 mm × 200 mm		
(3.5 in or 90 mm wide)	0.63	3.58	($\frac{1}{2}$ in × 8 in)	0.14	0.81

respectively, whose values are given in Table 3–9 for ordinary surfaces ($\varepsilon = 0.9$) and reflective surfaces ($\varepsilon = 0.2$ or 0.05). Note that surfaces having a low emittance also have a low surface conductance due to the reduction in radiation heat transfer. The values in the table are based on a surface temperature of 21°C (72°F) and a surface–air temperature difference of 5.5°C (10°F). Also, the equivalent surface temperature of the environment is assumed to be equal to the ambient air temperature. Despite the convenience it offers, this assumption is not quite accurate because of the additional radiation heat loss from the surface to the clear sky. The effect of sky radiation can be accounted for approximately by taking the outside temperature to be the average of the outdoor air and sky temperatures.

The inner surface heat transfer coefficient h_i remains fairly constant throughout the year, but the value of h_o varies considerably because of its dependence on the orientation and wind speed, which can vary from less than 1 km/h in calm weather to over 40 km/h during storms. The commonly used values of h_i and h_o for peak load calculations are

$$h_i = 8.29 \text{ W/m}^2 \cdot °C = 1.46 \text{ Btu/h} \cdot \text{ft}^2 \cdot °F \qquad \text{(winter and summer)}$$

$$h_o = \begin{cases} 34.0 \text{ W/m}^2 \cdot °C = 6.0 \text{ Btu/h} \cdot \text{ft}^2 \cdot °F & \text{(winter)} \\ 22.7 \text{ W/m}^2 \cdot °C = 4.0 \text{ Btu/h} \cdot \text{ft}^2 \cdot °F & \text{(summer)} \end{cases}$$

which correspond to design wind conditions of 24 km/h (15 mph) for winter and 12 km/h (7.5 mph) for summer. The corresponding surface thermal resistances (R-values) are determined from $R_i = 1/h_i$ and $R_o = 1/h_o$. The surface conductance values under still air conditions can be used for interior surfaces as well as exterior surfaces in calm weather.

Building components often involve *trapped air spaces* between various layers. Thermal resistances of such air spaces depend on the thickness of the layer, the temperature difference across the layer, the mean air temperature, the emissivity of each surface, the orientation of the air layer, and the direction of heat transfer. The emissivities of surfaces commonly encountered in buildings are given in Table 3–10. The **effective emissivity** of a plane-parallel air space is given by

$$\frac{1}{\varepsilon_{\text{effective}}} = \frac{1}{\varepsilon_1} + \frac{1}{\varepsilon_2} - 1 \qquad (3\text{–}81)$$

where ε_1 and ε_2 are the emissivities of the surfaces of the air space. Table 3–10 also lists the effective emissivities of air spaces for the cases where (1) the emissivity of one surface of the air space is ε while the emissivity of the other surface is 0.9 (a building material) and (2) the emissivity of both surfaces is ε. Note that the effective emissivity of an air space between building materials is $0.82/0.03 = 27$ times that of an air space between surfaces covered with aluminum foil. For specified surface temperatures, radiation heat transfer through an air space is proportional to effective emissivity, and thus the rate of radiation heat transfer in the ordinary surface case is 27 times that of the reflective surface case.

Table 3–11 lists the thermal resistances of 20-mm-, 40-mm-, and 90-mm- (0.75-in, 1.5-in, and 3.5-in) thick air spaces under various conditions. The

TABLE 3–9

Combined convection and radiation heat transfer coefficients at window, wall, or roof surfaces (from ASHRAE *Handbook of Fundamentals*, Chap. 22, Table 1).

Position	Direction of Heat Flow	h, W/m² · °C* Surface Emittance, ε		
		0.90	0.20	0.05
Still air (both indoors and outdoors)				
Horiz.	Up ↑	9.26	5.17	4.32
Horiz.	Down ↓	6.13	2.10	1.25
45° slope	Up ↑	9.09	5.00	4.15
45° slope	Down ↓	7.50	3.41	2.56
Vertical	Horiz. →	8.29	4.20	3.35
Moving air (any position, any direction)				
Winter condition (winds at 15 mph or 24 km/h)		34.0	—	—
Summer condition (winds at 7.5 mph or 12 km/h)		22.7	—	—

*Multiply by 0.176 to convert to Btu/h · ft² · °F. Surface resistance can be obtained from $R = 1/h$.

TABLE 3–11

Unit thermal resistances (*R*-values) of well-sealed plane air spaces (from ASHRAE *Handbook of Fundamentals,* Chap. 22, Table 2)

(*a*) SI units (in m² · °C/W)

Position of Air Space	Direction of Heat Flow	Mean Temp., °C	Temp. Diff., °C	20-mm Air Space Effective Emissivity, ε_{eff}				40-mm Air Space Effective Emissivity, ε_{eff}				90-mm Air Space Effective Emissivity, ε_{eff}			
				0.03	0.05	0.5	0.82	0.03	0.05	0.5	0.82	0.03	0.05	0.5	0.82
Horizontal Up ↑		32.2	5.6	0.41	0.39	0.18	0.13	0.45	0.42	0.19	0.14	0.50	0.47	0.20	0.14
		10.0	16.7	0.30	0.29	0.17	0.14	0.33	0.32	0.18	0.14	0.27	0.35	0.19	0.15
		10.0	5.6	0.40	0.39	0.20	0.15	0.44	0.42	0.21	0.16	0.49	0.47	0.23	0.16
		−17.8	11.1	0.32	0.32	0.20	0.16	0.35	0.34	0.22	0.17	0.40	0.38	0.23	0.18
45° slope Up ↑		32.2	5.6	0.52	0.49	0.20	0.14	0.51	0.48	0.20	0.14	0.56	0.52	0.21	0.14
		10.0	16.7	0.35	0.34	0.19	0.14	0.38	0.36	0.20	0.15	0.40	0.38	0.20	0.15
		10.0	5.6	0.51	0.48	0.23	0.17	0.51	0.48	0.23	0.17	0.55	0.52	0.24	0.17
		−17.8	11.1	0.37	0.36	0.23	0.18	0.40	0.39	0.24	0.18	0.43	0.41	0.24	0.19
Vertical	Horizontal →	32.2	5.6	0.62	0.57	0.21	0.15	0.70	0.64	0.22	0.15	0.65	0.60	0.22	0.15
		10.0	16.7	0.51	0.49	0.23	0.17	0.45	0.43	0.22	0.16	0.47	0.45	0.22	0.16
		10.0	5.6	0.65	0.61	0.25	0.18	0.67	0.62	0.26	0.18	0.64	0.60	0.25	0.18
		−17.8	11.1	0.55	0.53	0.28	0.21	0.49	0.47	0.26	0.20	0.51	0.49	0.27	0.20
45° slope Down ↓		32.2	5.6	0.62	0.58	0.21	0.15	0.89	0.80	0.24	0.16	0.85	0.76	0.24	0.16
		10.0	16.7	0.60	0.57	0.24	0.17	0.63	0.59	0.25	0.18	0.62	0.58	0.25	0.18
		10.0	5.6	0.67	0.63	0.26	0.18	0.90	0.82	0.28	0.19	0.83	0.77	0.28	0.19
		−17.8	11.1	0.66	0.63	0.30	0.22	0.68	0.64	0.31	0.22	0.67	0.64	0.31	0.22
Horizontal Down ↓		32.2	5.6	0.62	0.58	0.21	0.15	1.07	0.94	0.25	0.17	1.77	1.44	0.28	0.18
		10.0	16.7	0.66	0.62	0.25	0.18	1.10	0.99	0.30	0.20	1.69	1.44	0.33	0.21
		10.0	5.6	0.68	0.63	0.26	0.18	1.16	1.04	0.30	0.20	1.96	1.63	0.34	0.22
		−17.8	11.1	0.74	0.70	0.32	0.23	1.24	1.13	0.39	0.26	1.92	1.68	0.43	0.29

(*b*) English units (in h · ft² · °F/Btu)

Position of Air Space	Direction of Heat Flow	Mean Temp., °F	Temp. Diff., °F	0.75-in Air Space Effective Emissivity, ε_{eff}				1.5-in Air Space Effective Emissivity, ε_{eff}				3.5-in Air Space Effective Emissivity, ε_{eff}			
				0.03	0.05	0.5	0.82	0.03	0.05	0.5	0.82	0.03	0.05	0.5	0.82
Horizontal Up ↑		90	10	2.34	2.22	1.04	0.75	2.55	2.41	1.08	0.77	2.84	2.66	1.13	0.80
		50	30	1.71	1.66	0.99	0.77	1.87	1.81	1.04	0.80	2.09	2.01	1.10	0.84
		50	10	2.30	2.21	1.16	0.87	2.50	2.40	1.21	0.89	2.80	2.66	1.28	0.93
		0	20	1.83	1.79	1.16	0.93	2.01	1.95	1.23	0.97	2.25	2.18	1.32	1.03
45° slope Up ↑		90	10	2.96	2.78	1.15	0.81	2.92	2.73	1.14	0.80	3.18	2.96	1.18	0.82
		50	30	1.99	1.92	1.08	0.82	2.14	2.06	1.12	0.84	2.26	2.17	1.15	0.86
		50	10	2.90	2.75	1.29	0.94	2.88	2.74	1.29	0.94	3.12	2.95	1.34	0.96
		0	20	2.13	2.07	1.28	1.00	2.30	2.23	1.34	1.04	2.42	2.35	1.38	1.06
Vertical	Horizontal →	90	10	3.50	3.24	1.22	0.84	3.99	3.66	1.27	0.87	3.69	3.40	1.24	0.85
		50	30	2.91	2.77	1.30	0.94	2.58	2.46	1.23	0.90	2.67	2.55	1.25	0.91
		50	10	3.70	3.46	1.43	1.01	3.79	3.55	1.45	1.02	3.63	3.40	1.42	1.01
		0	20	3.14	3.02	1.58	1.18	2.76	2.66	1.48	1.12	2.88	2.78	1.51	1.14
45° slope Down ↓		90	10	3.53	3.27	1.22	0.84	5.07	4.55	1.36	0.91	4.81	4.33	1.34	0.90
		50	30	3.43	3.23	1.39	0.99	3.58	3.36	1.42	1.00	3.51	3.30	1.40	1.00
		50	10	3.81	3.57	1.45	1.02	5.10	4.66	1.60	1.09	4.74	4.36	1.57	1.08
		0	20	3.75	3.57	1.72	1.26	3.85	3.66	1.74	1.27	3.81	3.63	1.74	1.27
Horizontal Down ↓		90	10	3.55	3.29	1.22	0.85	6.09	5.35	1.43	0.94	10.07	8.19	1.57	1.00
		50	30	3.77	3.52	1.44	1.02	6.27	5.63	1.70	1.14	9.60	8.17	1.88	1.22
		50	10	3.84	3.59	1.45	1.02	6.61	5.90	1.73	1.15	11.15	9.27	1.93	1.24
		0	20	4.18	3.96	1.81	1.30	7.03	6.43	2.19	1.49	10.90	9.52	2.47	1.62

thermal resistance values in the table are applicable to air spaces of uniform thickness bounded by plane, smooth, parallel surfaces with no air leakage. Thermal resistances for other temperatures, emissivities, and air spaces can be obtained by interpolation and moderate extrapolation. Note that the presence of a low-emissivity surface reduces radiation heat transfer across an air space and thus significantly increases the thermal resistance. The thermal effectiveness of a low-emissivity surface will decline, however, if the condition of the surface changes as a result of some effects such as condensation, surface oxidation, and dust accumulation.

The R-value of a wall or roof structure that involves layers of uniform thickness is determined easily by simply adding up the unit thermal resistances of the layers that are in series. But when a structure involves components such as wood studs and metal connectors, then the thermal resistance network involves parallel connections and possible two-dimensional effects. The overall R-value in this case can be determined by assuming (1) parallel heat flow paths through areas of different construction or (2) isothermal planes normal to the direction of heat transfer. The first approach usually overpredicts the overall thermal resistance, whereas the second approach usually underpredicts it. The parallel heat flow path approach is more suitable for wood frame walls and roofs, whereas the isothermal planes approach is more suitable for masonry or metal frame walls.

The thermal contact resistance between different components of building structures ranges between 0.01 and 0.1 m$^2 \cdot$ °C/W, which is negligible in most cases. However, it may be significant for metal building components such as steel framing members.

The construction of wood frame flat ceilings typically involve 2-in × 6-in joists on 400-mm (16-in) or 600-mm (24-in) centers. The fraction of framing is usually taken to be 0.10 for joists on 400-mm centers and 0.07 for joists on 600-mm centers.

Most buildings have a combination of a ceiling and a roof with an attic space in between, and the determination of the R-value of the roof–attic–ceiling combination depends on whether the attic is vented or not. For adequately ventilated attics, the attic air temperature is practically the same as the outdoor air temperature, and thus heat transfer through the roof is governed by the R-value of the ceiling only. However, heat is also transferred between the roof and the ceiling by radiation, and it needs to be considered (Fig. 3–52). The major function of the roof in this case is to serve as a radiation shield by blocking off solar radiation. Effectively ventilating the attic in summer should not lead one to believe that heat gain to the building through the attic is greatly reduced. This is because most of the heat transfer through the attic is by radiation.

Radiation heat transfer between the ceiling and the roof can be minimized by covering at least one side of the attic (the roof or the ceiling side) by a reflective material, called *radiant barrier,* such as aluminum foil or aluminum-coated paper. Tests on houses with R-19 attic floor insulation have shown that radiant barriers can reduce summer ceiling heat gains by 16 to 42 percent compared to an attic with the same insulation level and no

TABLE 3–10

Emissivities ε of various surfaces and the effective emissivity of air spaces (from ASHRAE *Handbook of Fundamentals,* Chap. 22, Table 3).

Surface	ε	$\varepsilon_1 = \varepsilon$ $\varepsilon_2 = 0.9$	$\varepsilon_1 = \varepsilon$ $\varepsilon_2 = \varepsilon$
Aluminum foil, bright	0.05*	0.05	0.03
Aluminum sheet	0.12	0.12	0.06
Aluminum-coated paper, polished	0.20	0.20	0.11
Steel, galvanized, bright	0.25	0.24	0.15
Aluminum paint	0.50	0.47	0.35
Building materials: Wood, paper, masonry, nonmetallic paints	0.90	0.82	0.82
Ordinary glass	0.84	0.77	0.72

*Surface emissivity of aluminum foil increases to 0.30 with barely visible condensation, and to 0.70 with clearly visible condensation.

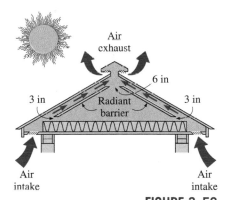

FIGURE 3–52
Ventilation paths for a naturally ventilated attic and the appropriate size of the flow areas around the radiant barrier for proper air circulation (from DOE/CE-0335P, U.S. Dept. of Energy).

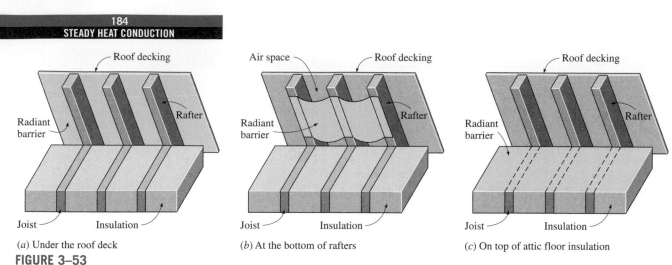

(a) Under the roof deck

(b) At the bottom of rafters

(c) On top of attic floor insulation

FIGURE 3–53
Three possible locations for an attic radiant barrier (from DOE/CE-0335P, U.S. Dept. of Energy).

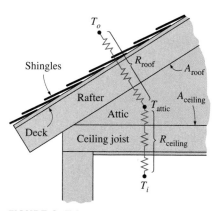

FIGURE 3–54
Thermal resistance network for a pitched roof–attic–ceiling combination for the case of an unvented attic.

radiant barrier. Considering that the ceiling heat gain represents about 15 to 25 percent of the total cooling load of a house, radiant barriers will reduce the air conditioning costs by 2 to 10 percent. Radiant barriers also reduce the heat loss in winter through the ceiling, but tests have shown that the percentage reduction in heat losses is less. As a result, the percentage reduction in heating costs will be less than the reduction in the air-conditioning costs. Also, the values given are for new and undusted radiant barrier installations, and percentages will be lower for aged or dusty radiant barriers.

Some possible locations for attic radiant barriers are given in Figure 3–53. In whole house tests on houses with R-19 attic floor insulation, radiant barriers have reduced the ceiling heat gain by an average of 35 percent when the radiant barrier is installed on the attic floor, and by 24 percent when it is attached to the bottom of roof rafters. Test cell tests also demonstrated that the best location for radiant barriers is the attic floor, provided that the attic is not used as a storage area and is kept clean.

For unvented attics, any heat transfer must occur through (1) the ceiling, (2) the attic space, and (3) the roof (Fig. 3–54). Therefore, the overall R-value of the roof–ceiling combination with an unvented attic depends on the combined effects of the R-value of the ceiling and the R-value of the roof as well as the thermal resistance of the attic space. The attic space can be treated as an air layer in the analysis. But a more practical way of accounting for its effect is to consider surface resistances on the roof and ceiling surfaces facing each other. In this case, the R-values of the ceiling and the roof are first determined separately (by using convection resistances for the still-air case for the attic surfaces). Then it can be shown that the overall R-value of the ceiling–roof combination per unit area of the ceiling can be expressed as

$$R = R_{\text{ceiling}} + R_{\text{roof}} \left(\frac{A_{\text{ceiling}}}{A_{\text{roof}}} \right) \qquad (3\text{–}82)$$

where A_{ceiling} and A_{roof} are the ceiling and roof areas, respectively. The area ratio is equal to 1 for flat roofs and is less than 1 for pitched roofs. For a 45° pitched roof, the area ratio is $A_{\text{ceiling}}/A_{\text{roof}} = 1/\sqrt{2} = 0.707$. Note that the pitched roof has a greater area for heat transfer than the flat ceiling, and the area ratio accounts for the reduction in the unit R-value of the roof when expressed per unit area of the ceiling. Also, the direction of heat flow is up in winter (heat loss through the roof) and down in summer (heat gain through the roof).

The R-value of a structure determined by analysis assumes that the materials used and the quality of workmanship meet the standards. Poor workmanship and substandard materials used during construction may result in R-values that deviate from predicted values. Therefore, some engineers use a safety factor in their designs based on experience in critical applications.

EXAMPLE 3–16   The *R*-Value of a Wood Frame Wall

Determine the overall unit thermal resistance (the R-value) and the overall heat transfer coefficient (the U-factor) of a wood frame wall that is built around 38-mm × 90-mm (2 × 4 nominal) wood studs with a center-to-center distance of 400 mm. The 90-mm-wide cavity between the studs is filled with glass fiber insulation. The inside is finished with 13-mm gypsum wallboard and the outside with 13-mm wood fiberboard and 13-mm × 200-mm wood bevel lapped siding. The insulated cavity constitutes 75 percent of the heat transmission area while the studs, plates, and sills constitute 21 percent. The headers constitute 4 percent of the area, and they can be treated as studs.

Also, determine the rate of heat loss through the walls of a house whose perimeter is 50 m and wall height is 2.5 m in Las Vegas, Nevada, whose winter design temperature is −2°C. Take the indoor design temperature to be 22°C and assume 20 percent of the wall area is occupied by glazing.

SOLUTION   The R-value and the U-factor of a wood frame wall as well as the rate of heat loss through such a wall in Las Vegas are to be determined.

Assumptions   **1** Steady operating conditions exist. **2** Heat transfer through the wall is one-dimensional. **3** Thermal properties of the wall and the heat transfer coefficients are constant.

Properties   The R-values of different materials are given in Table 3–8.

Analysis   The schematic of the wall as well as the different elements used in its construction are shown here. Heat transfer through the insulation and through the studs meets different resistances, and thus we need to analyze the thermal resistance for each path separately. Once the unit thermal resistances and the U-factors for the insulation and stud sections are available, the overall average thermal resistance for the entire wall can be determined from

$$R_{\text{overall}} = 1/U_{\text{overall}}$$

where

$$U_{\text{overall}} = (U \times f_{\text{area}})_{\text{insulation}} + (U \times f_{\text{area}})_{\text{stud}}$$

and the value of the area fraction f_{area} is 0.75 for the insulation section and 0.25 for the stud section since the headers that constitute a small part of the wall are to be treated as studs. Using the available R-values from Table 3–8 and calculating others, the total R-values for each section can be determined in a systematic manner in the table below.

Schematic

Construction	R-value, m² · °C/W	
	Between Studs	At Studs
1. Outside surface, 24 km/h wind	0.030	0.030
2. Wood bevel lapped siding	0.14	0.14
3. Wood fiberboard sheeting, 13 mm	0.23	0.23
4a. Glass fiber insulation, 90 mm	2.45	—
4b. Wood stud, 38 mm × 90 mm	—	0.63
5. Gypsum wallboard, 13 mm	0.079	0.079
6. Inside surface, still air	0.12	0.12
Total unit thermal resistance of each section, R (in m² · °C/W)	3.05	1.23
The U-factor of each section, $U = 1/R$, in W/m² · °C	0.328	0.813
Area fraction of each section, f_{area}	0.75	0.25

Overall U-factor: $U = \Sigma f_{area, i} U_i = 0.75 \times 0.328 + 0.25 \times 0.813$
$$= \textbf{0.449 W/m}^2 \cdot \textbf{°C}$$

Overall unit thermal resistance: $R = 1/U = \textbf{2.23 m}^2 \cdot \textbf{°C/W}$

We conclude that the overall unit thermal resistance of the wall is 2.23 m² · °C/W, and this value accounts for the effects of the studs and headers. It corresponds to an R-value of $2.23 \times 5.68 = 12.7$ (or nearly R-13) in English units. Note that if there were no wood studs and headers in the wall, the overall thermal resistance would be 3.05 m² · °C/W, which is 37 percent greater than 2.23 m² · °C/W. Therefore, the wood studs and headers in this case serve as thermal bridges in wood frame walls, and their effect must be considered in the thermal analysis of buildings.

The perimeter of the building is 50 m and the height of the walls is 2.5 m. Noting that glazing constitutes 20 percent of the walls, the total wall area is

$$A_{wall} = 0.80(\text{Perimeter})(\text{Height}) = 0.80(50 \text{ m})(2.5 \text{ m}) = 100 \text{ m}^2$$

Then the rate of heat loss through the walls under design conditions becomes

$$\dot{Q}_{wall} = (UA)_{wall} (T_i - T_o)$$
$$= (0.449 \text{ W/m}^2 \cdot \text{°C})(100 \text{ m}^2)[22 - (-2)\text{°C}]$$
$$= \textbf{1078 W}$$

Discussion Note that a 1-kW resistance heater in this house will make up almost all the heat lost through the walls, except through the doors and windows, when the outdoor air temperature drops to −2°C.

EXAMPLE 3–17   The R-Value of a Wall with Rigid Foam

The 13-mm-thick wood fiberboard sheathing of the wood stud wall discussed in the previous example is replaced by a 25-mm-thick rigid foam insulation. Determine the percent increase in the R-value of the wall as a result.

SOLUTION   The overall R-value of the existing wall was determined in Example 3–16 to be 2.23 m² · °C/W. Noting that the R-values of the fiberboard and the foam insulation are 0.23 m² · °C/W and 0.98 m² · °C/W, respectively, and the added and removed thermal resistances are in series, the overall R-value of the wall after modification becomes

$$R_{\text{new}} = R_{\text{old}} - R_{\text{removed}} + R_{\text{added}}$$
$$= 2.23 - 0.23 + 0.98$$
$$= 2.98 \text{ m}^2 \cdot {}^\circ\text{C/W}$$

This represents an increase of $(2.98 - 2.23)/2.23 = 0.34$ or **34 percent** in the R-value of the wall. This example demonstrated how to evaluate the new R-value of a structure when some structural members are added or removed.

EXAMPLE 3–18   The R-Value of a Masonry Wall

Determine the overall unit thermal resistance (the R-value) and the overall heat transfer coefficient (the U-factor) of a masonry cavity wall that is built around 6-in thick concrete blocks made of lightweight aggregate with 3 cores filled with perlite ($R = 4.2$ h · ft² · °F/Btu). The outside is finished with 4-in face brick with $\frac{1}{2}$-in cement mortar between the bricks and concrete blocks. The inside finish consists of $\frac{1}{2}$-in gypsum wallboard separated from the concrete block by $\frac{3}{4}$-in-thick (1-in × 3-in nominal) vertical furring ($R = 4.2$ h · ft² · °F/Btu) whose center-to-center distance is 16 in. Both sides of the $\frac{3}{4}$-in-thick air space between the concrete block and the gypsum board are coated with reflective aluminum foil ($\varepsilon = 0.05$) so that the effective emissivity of the air space is 0.03. For a mean temperature of 50°F and a temperature difference of 30°F, the R-value of the air space is 2.91 h · ft² · °F/Btu. The reflective air space constitutes 80 percent of the heat transmission area, while the vertical furring constitutes 20 percent.

SOLUTION   The R-value and the U-factor of a masonry cavity wall are to be determined.

Assumptions   **1** Steady operating conditions exist. **2** Heat transfer through the wall is one-dimensional. **3** Thermal properties of the wall and the heat transfer coefficients are constant.

Properties   The R-values of different materials are given in Table 3–8.

Analysis   The schematic of the wall as well as the different elements used in its construction are shown below. Following the approach described here and using the available R-values from Table 3–8, the overall R-value of the wall is determined in the following table.

Schematic

	Construction	R-value, h · ft² · °F/Btu Between Furring	At Furring
1.	Outside surface, 15 mph wind	0.17	0.17
2.	Face brick, 4 in	0.43	0.43
3.	Cement mortar, 0.5 in	0.10	0.10
4.	Concrete block, 6 in	4.20	4.20
5a.	Reflective air space, $\frac{3}{4}$ in	2.91	—
5b.	Nominal 1×3 vertical furring	—	0.94
6.	Gypsum wallboard, 0.5 in	0.45	0.45
7.	Inside surface, still air	0.68	0.68

	Between	At
Total unit thermal resistance of each section, R	8.94	6.97
The U-factor of each section, $U = 1/R$, in Btu/h · ft² · °F	0.112	0.143
Area fraction of each section, f_{area}	0.80	0.20

Overall U-factor: $U = \Sigma f_{area,\,i}\, U_i = 0.80 \times 0.112 + 0.20 \times 0.143$
$$= \mathbf{0.118\ Btu/h \cdot ft^2 \cdot °F}$$

Overall unit thermal resistance: $\qquad R = 1/U = \mathbf{8.46\ h \cdot ft^2 \cdot °F/Btu}$

Therefore, the overall unit thermal resistance of the wall is 8.46 h · ft² · °F/Btu and the overall U-factor is 0.118 Btu/h · ft² · °F. These values account for the effects of the vertical furring.

EXAMPLE 3–19   The R-Value of a Pitched Roof

Determine the overall unit thermal resistance (the R-value) and the overall heat transfer coefficient (the U-factor) of a 45° pitched roof built around nominal 2-in × 4-in wood studs with a center-to-center distance of 16 in. The 3.5-in-wide air space between the studs does not have any reflective surface and thus its effective emissivity is 0.84. For a mean temperature of 90°F and a temperature difference of 30°F, the R-value of the air space is 0.86 h · ft² · °F/Btu. The lower part of the roof is finished with $\frac{1}{2}$-in gypsum wallboard and the upper part with $\frac{5}{8}$-in plywood, building paper, and asphalt shingle roofing. The air space constitutes 75 percent of the heat transmission area, while the studs and headers constitute 25 percent.

SOLUTION The R-value and the U-factor of a 45° pitched roof are to be determined.

Assumptions **1** Steady operating conditions exist. **2** Heat transfer through the roof is one-dimensional. **3** Thermal properties of the roof and the heat transfer coefficients are constant.

Properties The *R* values of different materials are given in Table 3–8.

Analysis The schematic of the pitched roof as well as the different elements used in its construction are shown below. Following the approach described above and using the available *R*-values from Table 3–8, the overall *R*-value of the roof can be determined in the table here.

Schematic

	Construction	R-value, h · ft² · °F/Btu	
		Between Studs	At Studs
1.	Outside surface, 15 mph wind	0.17	0.17
2.	Asphalt shingle roofing	0.44	0.44
3.	Building paper	0.10	0.10
4.	Plywood deck, $\frac{5}{8}$ in	0.78	0.78
5a.	Nonreflective air space, 3.5 in	0.86	—
5b.	Wood stud, 2 in × 4 in	—	3.58
6.	Gypsum wallboard, 0.5 in	0.45	0.45
7.	Inside surface, 45° slope, still air	0.63	0.63
Total unit thermal resistance of each section, *R*		3.43	6.15
The *U*-factor of each section, *U* = 1/*R*, in Btu/h · ft² · °F		0.292	0.163
Area fraction of each section, f_{area}		0.75	0.25

Overall *U*-factor: $U = \Sigma f_{area,\,i} U_i = 0.75 \times 0.292 + 0.25 \times 0.163$
   $= 0.260$ Btu/h · ft² · °F

Overall unit thermal resistance: $R = 1/U = \textbf{3.85 h · ft² · °F/Btu}$

Therefore, the overall unit thermal resistance of this pitched roof is 3.85 h · ft² · °F/Btu and the overall *U*-factor is 0.260 Btu/h · ft² · °F. Note that the wood studs offer much larger thermal resistance to heat flow than the air space between the studs.

SUMMARY

One-dimensional heat transfer through a simple or composite body exposed to convection from both sides to mediums at temperatures $T_{\infty 1}$ and $T_{\infty 2}$ can be expressed as

$$\dot{Q} = \frac{T_{\infty 1} - T_{\infty 2}}{R_{total}}$$

where R_{total} is the total thermal resistance between the two mediums. For a plane wall exposed to convection on both sides, the total resistance is expressed as

$$R_{total} = R_{conv,\,1} + R_{wall} + R_{conv,\,2} = \frac{1}{h_1 A} + \frac{L}{kA} + \frac{1}{h_2 A}$$

This relation can be extended to plane walls that consist of two or more layers by adding an additional resistance for each additional layer. The elementary thermal resistance relations can be expressed as follows:

Conduction resistance (plane wall): $R_{wall} = \dfrac{L}{kA}$

Conduction resistance (cylinder): $R_{cyl} = \dfrac{\ln(r_2/r_1)}{2\pi Lk}$

Conduction resistance (sphere): $R_{sph} = \dfrac{r_2 - r_1}{4\pi r_1 r_2 k}$

Convection resistance: $R_{conv} = \dfrac{1}{hA}$

Interface resistance: $\quad R_{\text{interface}} = \dfrac{1}{h_c A} = \dfrac{R_c}{A}$

Radiation resistance: $\quad R_{\text{rad}} = \dfrac{1}{h_{\text{rad}} A}$

where h_c is the thermal contact conductance, R_c is the thermal contact resistance, and the radiation heat transfer coefficient is defined as

$$h_{\text{rad}} = \varepsilon\sigma(T_s^2 + T_{\text{surr}}^2)(T_s + T_{\text{surr}})$$

Once the rate of heat transfer is available, the *temperature drop* across any layer can be determined from

$$\Delta T = \dot{Q}R$$

The thermal resistance concept can also be used to solve steady heat transfer problems involving parallel layers or combined series-parallel arrangements.

Adding insulation to a cylindrical pipe or a spherical shell increases the rate of heat transfer if the outer radius of the insulation is less than the *critical radius of insulation,* defined as

$$r_{\text{cr, cylinder}} = \frac{k_{\text{ins}}}{h}$$

$$r_{\text{cr, sphere}} = \frac{2k_{\text{ins}}}{h}$$

The effectiveness of an insulation is often given in terms of its *R-value,* the thermal resistance of the material per unit surface area, expressed as

$$R\text{-value} = \frac{L}{k} \qquad \text{(flat insulation)}$$

where L is the thickness and k is the thermal conductivity of the material.

Finned surfaces are commonly used in practice to enhance heat transfer. Fins enhance heat transfer from a surface by exposing a larger surface area to convection. The temperature distribution along the fin for very long fins and for fins with negligible heat transfer at the fin tip are given by

Very long fin: $\quad \dfrac{T(x) - T_\infty}{T_b - T_\infty} = e^{-x\sqrt{hp/kA_c}}$

Adiabatic fin tip: $\quad \dfrac{T(x) - T_\infty}{T_b - T_\infty} = \dfrac{\cosh m(L - x)}{\cosh mL}$

where $m = \sqrt{hp/kA_c}$, p is the perimeter, and A_c is the cross-sectional area of the fin. The rates of heat transfer for both cases are given to be

Very long fin:
$$\dot{Q}_{\text{long fin}} = -kA_c \frac{dT}{dx}\bigg|_{x=0} = \sqrt{hpkA_c}\,(T_b - T_\infty)$$

Adiabatic fin tip:
$$\dot{Q}_{\text{adiabatic tip}} = -kA_c \frac{dT}{dx}\bigg|_{x=0} = \sqrt{hpkA_c}\,(T_b - T_\infty)\tanh mL$$

Fins exposed to convection at their tips can be treated as fins with adiabatic tips by using the corrected length $L_c = L + A_c/p$ instead of the actual fin length.

The temperature of a fin drops along the fin, and thus the heat transfer from the fin is less because of the decreasing temperature difference toward the fin tip. To account for the effect of this decrease in temperature on heat transfer, we define *fin efficiency* as

$$\eta_{\text{fin}} = \frac{\dot{Q}_{\text{fin}}}{\dot{Q}_{\text{fin, max}}} = \frac{\text{Actual heat transfer rate from the fin}}{\substack{\text{Ideal heat transfer rate from the fin if} \\ \text{the entire fin were at base temperature}}}$$

When the fin efficiency is available, the rate of heat transfer from a fin can be determined from

$$\dot{Q}_{\text{fin}} = \eta_{\text{fin}}\dot{Q}_{\text{fin, max}} = \eta_{\text{fin}}hA_{\text{fin}}\,(T_b - T_\infty)$$

The performance of the fins is judged on the basis of the enhancement in heat transfer relative to the no-fin case and is expressed in terms of the *fin effectiveness* ε_{fin}, defined as

$$\varepsilon_{\text{fin}} = \frac{\dot{Q}_{\text{fin}}}{\dot{Q}_{\text{no fin}}} = \frac{\dot{Q}_{\text{fin}}}{hA_b\,(T_b - T_\infty)} = \frac{\substack{\text{Heat transfer rate from} \\ \text{the fin of } base\ area\ A_b}}{\substack{\text{Heat transfer rate from} \\ \text{the surface of } area\ A_b}}$$

Here, A_b is the cross-sectional area of the fin at the base and $\dot{Q}_{\text{no fin}}$ represents the rate of heat transfer from this area if no fins are attached to the surface. The *overall effectiveness* for a finned surface is defined as the ratio of the total heat transfer from the finned surface to the heat transfer from the same surface if there were no fins,

$$\varepsilon_{\text{fin, overall}} = \frac{\dot{Q}_{\text{total, fin}}}{\dot{Q}_{\text{total, no fin}}} = \frac{h(A_{\text{unfin}} + \eta_{\text{fin}}A_{\text{fin}})(T_b - T_\infty)}{hA_{\text{no fin}}\,(T_b - T_\infty)}$$

Fin efficiency and fin effectiveness are related to each other by

$$\varepsilon_{\text{fin}} = \frac{A_{\text{fin}}}{A_b}\eta_{\text{fin}}$$

Certain multidimensional heat transfer problems involve two surfaces maintained at constant temperatures T_1 and T_2. The steady rate of heat transfer between these two surfaces is expressed as

$$\dot{Q} = Sk(T_1 - T_2)$$

where S is the *conduction shape factor* that has the dimension of *length* and k is the thermal conductivity of the medium between the surfaces.

REFERENCES AND SUGGESTED READINGS

1. American Society of Heating, Refrigeration, and Air Conditioning Engineers. *Handbook of Fundamentals*. Atlanta: ASHRAE, 1993.

2. R. V. Andrews. "Solving Conductive Heat Transfer Problems with Electrical-Analogue Shape Factors." *Chemical Engineering Progress* 5 (1955), p. 67.

3. R. Barron. *Cryogenic Systems*. New York: McGraw-Hill, 1967.

4. L. S. Fletcher. "Recent Developments in Contact Conductance Heat Transfer." *Journal of Heat Transfer* 110, no. 4B (1988), pp. 1059–79.

5. E. Fried. "Thermal Conduction Contribution to Heat Transfer at Contacts." *Thermal Conductivity,* vol. 2, ed. R. P. Tye. London: Academic Press, 1969.

6. K. A. Gardner. "Efficiency of Extended Surfaces." *Trans. ASME* 67 (1945), pp. 621–31. Reprinted by permission of ASME International.

7. D. Q. Kern and A. D. Kraus. *Extended Surface Heat Transfer.* New York: McGraw-Hill, 1972.

8. G. P. Peterson. "Thermal Contact Resistance in Waste Heat Recovery Systems." *Proceedings of the 18th ASME/ETCE Hydrocarbon Processing Symposium.* Dallas, TX, 1987, pp. 45–51. Reprinted by permission of ASME International.

9. S. Song, M. M. Yovanovich, and F. O. Goodman. "Thermal Gap Conductance of Conforming Surfaces in Contact." *Journal of Heat Transfer* 115 (1993), p. 533.

10. J. E. Sunderland and K. R. Johnson. "Shape Factors for Heat Conduction through Bodies with Isothermal or Convective Boundary Conditions." *Trans. ASME* 10 (1964), pp. 2317–41.

11. W. M. Edmunds. "Residential Insulation." *ASTM Standardization News* (Jan. 1989), pp. 36–39.

PROBLEMS*

Steady Heat Conduction in Plane Walls

3–1C Consider one-dimensional heat conduction through a cylindrical rod of diameter D and length L. What is the heat transfer area of the rod if (a) the lateral surfaces of the rod are insulated and (b) the top and bottom surfaces of the rod are insulated?

3–2C Consider heat conduction through a plane wall. Does the energy content of the wall change during steady heat conduction? How about during transient conduction? Explain.

3–3C Consider heat conduction through a wall of thickness L and area A. Under what conditions will the temperature distributions in the wall be a straight line?

3–4C What does the thermal resistance of a medium represent?

3–5C How is the combined heat transfer coefficient defined? What convenience does it offer in heat transfer calculations?

3–6C Can we define the convection resistance per unit surface area as the inverse of the convection heat transfer coefficient?

3–7C Why are the convection and the radiation resistances at a surface in parallel instead of being in series?

3–8C Consider a surface of area A at which the convection and radiation heat transfer coefficients are h_{conv} and h_{rad}, respectively. Explain how you would determine (a) the single equivalent heat transfer coefficient, and (b) the equivalent thermal resistance. Assume the medium and the surrounding surfaces are at the same temperature.

3–9C How does the thermal resistance network associated with a single-layer plane wall differ from the one associated with a five-layer composite wall?

3–10C Consider steady one-dimensional heat transfer through a multilayer medium. If the rate of heat transfer $\dot{Q}$ is known, explain how you would determine the temperature drop across each layer.

3–11C Consider steady one-dimensional heat transfer through a plane wall exposed to convection from both sides to environments at known temperatures $T_{\infty 1}$ and $T_{\infty 2}$ with known heat transfer coefficients h_1 and h_2. Once the rate of heat transfer $\dot{Q}$ has been evaluated, explain how you would determine the temperature of each surface.

---

*Problems designated by a "C" are concept questions, and students are encouraged to answer them all. Problems designated by an "E" are in English units, and the SI users can ignore them. Problems with the icon ⊛ are solved using EES, and complete solutions together with parametric studies are included on the enclosed CD. Problems with the icon ▧ are comprehensive in nature, and are intended to be solved with a computer, preferably using the EES software that accompanies this text.

3–12C Someone comments that a microwave oven can be viewed as a conventional oven with zero convection resistance at the surface of the food. Is this an accurate statement?

3–13C Consider a window glass consisting of two 4-mm-thick glass sheets pressed tightly against each other. Compare the heat transfer rate through this window with that of one consisting of a single 8-mm-thick glass sheet under identical conditions.

3–14C Consider steady heat transfer through the wall of a room in winter. The convection heat transfer coefficient at the outer surface of the wall is three times that of the inner surface as a result of the winds. On which surface of the wall do you think the temperature will be closer to the surrounding air temperature? Explain.

3–15C The bottom of a pan is made of a 4-mm-thick aluminum layer. In order to increase the rate of heat transfer through the bottom of the pan, someone proposes a design for the bottom that consists of a 3-mm-thick copper layer sandwiched between two 2-mm-thick aluminum layers. Will the new design conduct heat better? Explain. Assume perfect contact between the layers.

2 mm
3 mm
2 mm

Aluminum   Copper

FIGURE P3–15C

3–16C Consider two cold canned drinks, one wrapped in a blanket and the other placed on a table in the same room. Which drink will warm up faster?

3–17 Consider a 3-m-high, 6-m-wide, and 0.3-m-thick brick wall whose thermal conductivity is $k = 0.8$ W/m · °C . On a certain day, the temperatures of the inner and the outer surfaces of the wall are measured to be 14°C and 2°C, respectively. Determine the rate of heat loss through the wall on that day.

3–18 A 1.0 m × 1.5 m double-pane window consists of two 4-mm thick layers of glass ($k = 0.78$ W/m · K) that are the separated by a 5-mm air gap ($k_{air} = 0.025$ W/m · K). The heat flow through the air gap is assumed to be by condition. The inside and outside air temperatures are 20°C and −20°C, respectively, and the inside and outside heat transfer coefficients are 40 and 20 W/m² · K. Determine (a) the daily rate of heat loss

through the window in steady operation and (b) the temperature difference across the largest thermal resistence.

3–19 Consider a 1.2-m-high and 2-m-wide glass window whose thickness is 6 mm and thermal conductivity is $k = 0.78$ W/m · °C. Determine the steady rate of heat transfer through this glass window and the temperature of its inner surface for a day during which the room is maintained at 24°C while the temperature of the outdoors is −5°C. Take the convection heat transfer coefficients on the inner and outer surfaces of the window to be $h_1 = 10$ W/m² · °C and $h_2 = 25$ W/m² · °C, and disregard any heat transfer by radiation.

3–20 Consider a 1.2-m-high and 2-m-wide double-pane window consisting of two 3-mm-thick layers of glass ($k = 0.78$ W/m · °C) separated by a 12-mm-wide stagnant air space ($k = 0.026$ W/m · °C). Determine the steady rate of heat transfer through this double-pane window and the temperature of its inner surface for a day during which the room is maintained at 24°C while the temperature of the outdoors is −5°C. Take the convection heat transfer coefficients on the inner and outer surfaces of the window to be $h_1 = 10$ W/m² · °C and $h_2 = 25$ W/m² · °C, and disregard any heat transfer by radiation. *Answers:* 114 W, 19.2°C

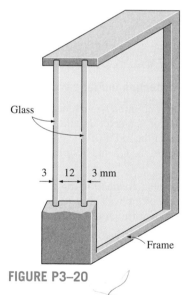

Glass

3 | 12 | 3 mm

Frame

FIGURE P3–20

3–21 Repeat Prob. 3–20, assuming the space between the two glass layers is evacuated.

3–22 Reconsider Prob. 3–20. Using EES (or other) software, plot the rate of heat transfer through the window as a function of the width of air space in the range of 2 mm to 20 mm, assuming pure conduction through the air. Discuss the results.

3–23E Consider an electrically heated brick house ($k = 0.40$ Btu/h · ft · °F) whose walls are 9 ft high and 1 ft thick. Two of the walls of the house are 50 ft long and the others are 35 ft long. The house is maintained at 70°F at all times while the temperature of the outdoors varies. On a certain day, the temperature of the inner surface of the walls is measured to be at 55°F while the average temperature of the outer surface is observed to remain at 45°F during the day for 10 h and at 35°F at night for 14 h. Determine the amount of heat lost from the house that day. Also determine the cost of that heat loss to the home owner for an electricity price of $0.09/kWh.

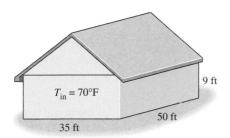

FIGURE P3–23E

3–24 A cylindrical resistor element on a circuit board dissipates 0.15 W of power in an environment at 40°C. The resistor is 1.2 cm long, and has a diameter of 0.3 cm. Assuming heat to be transferred uniformly from all surfaces, determine (*a*) the amount of heat this resistor dissipates during a 24-h period; (*b*) the heat flux on the surface of the resistor, in W/m²; and (*c*) the surface temperature of the resistor for a combined convection and radiation heat transfer coefficient of 9 W/m² · °C.

3–25 Consider a power transistor that dissipates 0.2 W of power in an environment at 30°C. The transistor is 0.4 cm long and has a diameter of 0.5 cm. Assuming heat to be transferred uniformly from all surfaces, determine (*a*) the amount of heat this transistor dissipates during a 24-h period, in kWh; (*b*) the heat flux on the surface of the transistor, in W/m²; and (*c*) the surface temperature of the transistor for a combined convection and radiation heat transfer coefficient of 18 W/m² · °C.

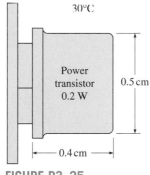

FIGURE P3–25

3–26 A 12-cm × 18-cm circuit board houses on its surface 100 closely spaced logic chips, each dissipating 0.06 W in an environment at 40°C. The heat transfer from the back surface of the board is negligible. If the heat transfer coefficient on the surface of the board is 10 W/m² · °C, determine (*a*) the heat flux on the surface of the circuit board, in W/m²; (*b*) the surface temperature of the chips; and (*c*) the thermal resistance between the surface of the circuit board and the cooling medium, in °C/W.

3–27 Consider a person standing in a room at 20°C with an exposed surface area of 1.7 m². The deep body temperature of the human body is 37°C, and the thermal conductivity of the human tissue near the skin is about 0.3 W/m · °C. The body is losing heat at a rate of 150 W by natural convection and radiation to the surroundings. Taking the body temperature 0.5 cm beneath the skin to be 37°C, determine the skin temperature of the person.   *Answer:* 35.5°C

3–28 Water is boiling in a 25-cm-diameter aluminum pan (*k* = 237 W/m · °C) at 95°C. Heat is transferred steadily to the boiling water in the pan through its 0.5-cm-thick flat bottom at a rate of 800 W. If the inner surface temperature of the bottom of the pan is 108°C, determine (*a*) the boiling heat transfer coefficient on the inner surface of the pan; and (*b*) the outer surface temperature of the bottom of the pan.

3–29E A wall is constructed of two layers of 0.7-in-thick sheetrock (*k* = 0.10 Btu/h · ft · °F), which is a plasterboard made of two layers of heavy paper separated by a layer of gypsum, placed 7 in apart. The space between the sheetrocks is filled with fiberglass insulation (*k* = 0.020 Btu/h · ft · °F). Determine (*a*) the thermal resistance of the wall and (*b*) its *R*-value of insulation in English units.

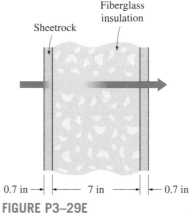

FIGURE P3–29E

3–30 The roof of a house consists of a 15-cm-thick concrete slab (*k* = 2 W/m · °C) that is 15 m wide and 20 m long. The convection heat transfer coefficients on the inner and outer surfaces of the roof are 5 and 12 W/m² · °C, respectively. On a clear winter night, the ambient air is reported to be at 10°C, while the night sky temperature is 100 K. The house and the interior surfaces of the wall are maintained at a constant temperature of 20°C. The emissivity of both surfaces of the concrete roof is 0.9. Considering both radiation and convection heat transfers, determine the rate of heat transfer through the roof, and the inner surface temperature of the roof.

If the house is heated by a furnace burning natural gas with an efficiency of 80 percent, and the price of natural gas is $1.20/therm (1 therm = 105,500 kJ of energy content), determine the money lost through the roof that night during a 14-h period.

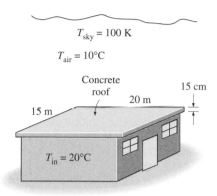

FIGURE P3–30

3–31 A 2-m × 1.5-m section of wall of an industrial furnace burning natural gas is not insulated, and the temperature at the outer surface of this section is measured to be 80°C. The temperature of the furnace room is 30°C, and the combined convection and radiation heat transfer coefficient at the surface of the outer furnace is 10 W/m² · °C. It is proposed to insulate this section of the furnace wall with glass wool insulation ($k = 0.038$ W/m · °C) in order to reduce the heat loss by 90 percent. Assuming the outer surface temperature of the metal section still remains at about 80°C, determine the thickness of the insulation that needs to be used.

The furnace operates continuously and has an efficiency of 78 percent. The price of the natural gas is $1.10/therm (1 therm = 105,500 kJ of energy content). If the installation of the insulation will cost $250 for materials and labor, determine how long it will take for the insulation to pay for itself from the energy it saves.

3–32 Repeat Prob. 3–31 for expanded perlite insulation assuming conductivity is $k = 0.052$ W/m · °C.

3–33 Reconsider Prob. 3–31. Using EES (or other) software, investigate the effect of thermal conductivity on the required insulation thickness. Plot the thickness of insulation as a function of the thermal conductivity of the insulation in the range of 0.02 W/m · °C to 0.08 W/m · °C, and discuss the results.

3–34E Consider a house whose walls are 12 ft high and 40 ft long. Two of the walls of the house have no windows, while each of the other two walls has four windows made of 0.25-in-thick glass ($k = 0.45$ Btu/h · ft · °F), 3 ft × 5 ft in size. The walls are certified to have an R-value of 19 (i.e., an L/k value of 19 h · ft² · °F/Btu). Disregarding any direct radiation gain or loss through the windows and taking the heat transfer coefficients at the inner and outer surfaces of the house to be 2 and 4 Btu/h · ft² · °F,

respectively, determine the ratio of the heat transfer through the walls with and without windows.

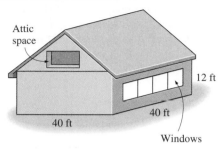

FIGURE P3–34E

3–35 Consider a house that has a 10-m × 20-m base and a 4-m-high wall. All four walls of the house have an R-value of 2.31 m² · °C/W. The two 10-m × 4-m walls have no windows. The third wall has five windows made of 0.5-cm-thick glass ($k = 0.78$ W/m · °C), 1.2 m × 1.8 m in size. The fourth wall has the same size and number of windows, but they are double-paned with a 1.5-cm-thick stagnant air space ($k = 0.026$ W/m · °C) enclosed between two 0.5-cm-thick glass layers. The thermostat in the house is set at 24°C and the average temperature outside at that location is 8°C during the seven-month-long heating season. Disregarding any direct radiation gain or loss through the windows and taking the heat transfer coefficients at the inner and outer surfaces of the house to be 7 and 18 W/m² · °C, respectively, determine the average rate of heat transfer through each wall.

If the house is electrically heated and the price of electricity is $0.08/kWh, determine the amount of money this household will save per heating season by converting the single-pane windows to double-pane windows.

3–36 The wall of a refrigerator is constructed of fiberglass insulation ($k = 0.035$ W/m · °C) sandwiched between two layers of 1-mm-thick sheet metal ($k = 15.1$ W/m · °C). The refrigerated space is maintained at 3°C, and the average heat transfer coefficients at the inner and outer surfaces of the wall

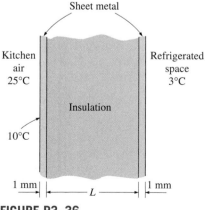

FIGURE P3–36

are 4 W/m² · °C and 9 W/m² · °C, respectively. The kitchen temperature averages 25°C. It is observed that condensation occurs on the outer surfaces of the refrigerator when the temperature of the outer surface drops to 20°C. Determine the minimum thickness of fiberglass insulation that needs to be used in the wall in order to avoid condensation on the outer surfaces.

3–37 Reconsider Prob. 3–36. Using EES (or other) software, investigate the effects of the thermal conductivities of the insulation material and the sheet metal on the thickness of the insulation. Let the thermal conductivity vary from 0.02 W/m · °C to 0.08 W/m · °C for insulation and 10 W/m · °C to 400 W/m · °C for sheet metal. Plot the thickness of the insulation as the functions of the thermal conductivities of the insulation and the sheet metal, and discuss the results.

3–38 Heat is to be conducted along a circuit board that has a copper layer on one side. The circuit board is 15 cm long and 15 cm wide, and the thicknesses of the copper and epoxy layers are 0.1 mm and 1.2 mm, respectively. Disregarding heat transfer from side surfaces, determine the percentages of heat conduction along the copper ($k = 386$ W/m · °C) and epoxy ($k = 0.26$ W/m · °C) layers. Also determine the effective thermal conductivity of the board.
Answers: 0.8 percent, 99.2 percent, and 29.9 W/m · °C

3–39E A 0.03-in-thick copper plate ($k = 223$ Btu/h · ft · °F) is sandwiched between two 0.15-in-thick epoxy boards ($k = 0.15$ Btu/h · ft · °F) that are 7 in × 9 in in size. Determine the effective thermal conductivity of the board along its 9-in-long side. What fraction of the heat conducted along that side is conducted through copper?

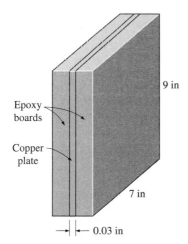

Epoxy boards
Copper plate

9 in

7 in

→ |← 0.03 in

FIGURE P3–39E

Thermal Contact Resistance

3–40C What is thermal contact resistance? How is it related to thermal contact conductance?

3–41C Will the thermal contact resistance be greater for smooth or rough plain surfaces?

3–42C A wall consists of two layers of insulation pressed against each other. Do we need to be concerned about the thermal contact resistance at the interface in a heat transfer analysis or can we just ignore it?

3–43C A plate consists of two thin metal layers pressed against each other. Do we need to be concerned about the thermal contact resistance at the interface in a heat transfer analysis or can we just ignore it?

3–44C Consider two surfaces pressed against each other. Now the air at the interface is evacuated. Will the thermal contact resistance at the interface increase or decrease as a result?

3–45C Explain how the thermal contact resistance can be minimized.

3–46 The thermal contact conductance at the interface of two 1-cm-thick copper plates is measured to be 18,000 W/m² · °C. Determine the thickness of the copper plate whose thermal resistance is equal to the thermal resistance of the interface between the plates.

3–47 Six identical power transistors with aluminum casing are attached on one side of a 1.2-cm-thick 20-cm × 30-cm copper plate ($k = 386$ W/m · °C) by screws that exert an average pressure of 10 MPa. The base area of each transistor is 9 cm², and each transistor is placed at the center of a 10-cm × 10-cm section of the plate. The interface roughness is estimated to be about 1.4 μm. All transistors are covered by a thick Plexiglas layer, which is a poor conductor of heat, and thus all the heat generated at the junction of the transistor must be dissipated to the ambient at 23°C through the back surface of the

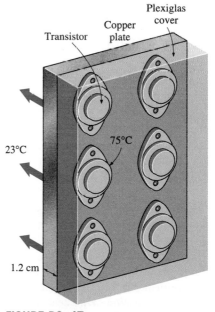

Plexiglas cover
Copper plate
Transistor

23°C

75°C

1.2 cm

FIGURE P3–47

copper plate. The combined convection/radiation heat transfer coefficient at the back surface can be taken to be 30 W/m² · °C. If the case temperature of the transistor is not to exceed 75°C, determine the maximum power each transistor can dissipate safely, and the temperature jump at the case-plate interface.

3–48 Two 5-cm-diameter, 15-cm-long aluminum bars ($k = 176$ W/m · °C) with ground surfaces are pressed against each other with a pressure of 20 atm. The bars are enclosed in an insulation sleeve and, thus, heat transfer from the lateral surfaces is negligible. If the top and bottom surfaces of the two-bar system are maintained at temperatures of 150°C and 20°C, respectively, determine (a) the rate of heat transfer along the cylinders under steady conditions and (b) the temperature drop at the interface.   *Answers: (a) 142.4 W, (b) 6.4°C*

3–49 A 1-mm-thick copper plate ($k = 386$ W/m · °C) is sandwiched between two 5-mm-thick epoxy boards ($k = 0.26$ W/m · °C) that are 15 cm × 20 cm in size. If the thermal contact conductance on both sides of the copper plate is estimated to be 6000 W/m · °C, determine the error involved in the total thermal resistance of the plate if the thermal contact conductances are ignored.

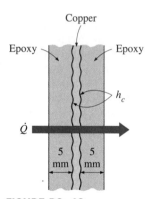

FIGURE P3–49

Generalized Thermal Resistance Networks

3–50C When plotting the thermal resistance network associated with a heat transfer problem, explain when two resistances are in series and when they are in parallel.

3–51C The thermal resistance networks can also be used approximately for multidimensional problems. For what kind of multidimensional problems will the thermal resistance approach give adequate results?

3–52C What are the two approaches used in the development of the thermal resistance network for two-dimensional problems?

3–53 A typical section of a building wall is shown in Fig. P3–53. This section extends in and out of the page and is repeated in the vertical direction. The wall support members are made of steel ($k = 50$ W/m · K). The support members are 8 cm (t_{23}) × 0.5 cm (L_B). The remainder of the inner wall space

is filled with insulation ($k = 0.03$ W/m · K) and measures 8 cm (t_{23}) × 60 cm (L_B). The inner wall is made of gypsum board ($k = 0.5$ W/m · K) that is 1 cm thick (t_{12}) and the outer wall is made of brick ($k = 1.0$ W/m · K) that is 10 cm thick (t_{34}). What is the average heat flux through this wall when $T_1 = 20$°C and $T_4 = 35$°C?

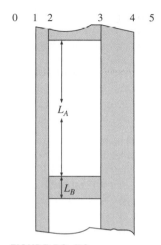

FIGURE P3–53

3–54 A 4-m-high and 6-m-wide wall consists of a long 18-cm × 30-cm cross section of horizontal bricks ($k = 0.72$ W/m · °C) separated by 3-cm-thick plaster layers ($k = 0.22$ W/m · °C). There are also 2-cm-thick plaster layers on each side of the wall, and a 2-cm-thick rigid foam ($k = 0.026$ W/m · °C) on the inner side of the wall. The indoor and the outdoor temperatures are 22°C and −4°C, and the convection heat transfer coefficients on the inner and the outer sides are $h_1 = 10$ W/m² · °C and $h_2 = 20$ W/m² · °C, respectively. Assuming one-dimensional heat

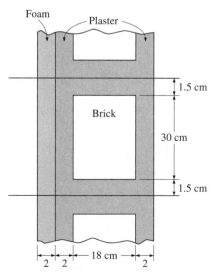

FIGURE P3–54

transfer and disregarding radiation, determine the rate of heat transfer through the wall.

3–55 [EES] Reconsider Prob. 3–54. Using EES (or other) software, plot the rate of heat transfer through the wall as a function of the thickness of the rigid foam in the range of 1 cm to 10 cm. Discuss the results.

3–56 A 10-cm-thick wall is to be constructed with 2.5-m-long wood studs ($k = 0.11$ W/m · °C) that have a cross section of 10 cm × 10 cm. At some point the builder ran out of those studs and started using pairs of 2.5-m-long wood studs that have a cross section of 5 cm × 10 cm nailed to eachother instead. The manganese steel nails ($k = 50$ W/m · °C) are 10 cm long and have a diameter of 0.4 cm. A total of 50 nails are used to connect the two studs, which are mounted to the wall such that the nails cross the wall. The temperature difference between the inner and outer surfaces of the wall is 8°C. Assuming the thermal contact resistance between the two layers to be negligible, determine the rate of heat transfer (a) through a solid stud and (b) through a stud pair of equal length and width nailed to each other. (c) Also determine the effective conductivity of the nailed stud pair.

3–57 A 12-m-long and 5-m-high wall is constructed of two layers of 1-cm-thick sheetrock ($k = 0.17$ W/m · °C) spaced 16 cm by wood studs ($k = 0.11$ W/m · °C) whose cross section is 12 cm × 5 cm. The studs are placed vertically 60 cm apart, and the space between them is filled with fiberglass insulation ($k = 0.034$ W/m · °C). The house is maintained at 20°C and the ambient temperature outside is −9°C. Taking the heat transfer coefficients at the inner and outer surfaces of the house to be 8.3 and 34 W/m² · °C, respectively, determine (a) the thermal resistance of the wall considering a representative section of it and (b) the rate of heat transfer through the wall.

3–58E A 10-in-thick, 30-ft-long, and 10-ft-high wall is to be constructed using 9-in-long solid bricks ($k =$

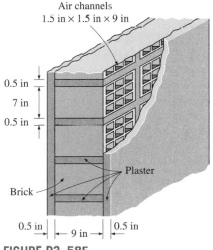

Air channels
1.5 in × 1.5 in × 9 in

0.5 in
7 in
0.5 in

Plaster

Brick

0.5 in ← 9 in → 0.5 in

FIGURE P3–58E

0.40 Btu/h · ft · °F) of cross section 7 in × 7 in, or identical size bricks with nine square air holes ($k = 0.015$ Btu/h · ft · °F) that are 9 in long and have a cross section of 1.5 in × 1.5 in. There is a 0.5-in-thick plaster layer ($k = 0.10$ Btu/h · ft · °F) between two adjacent bricks on all four sides and on both sides of the wall. The house is maintained at 80°F and the ambient temperature outside is 30°F. Taking the heat transfer coefficients at the inner and outer surfaces of the wall to be 1.5 and 4 Btu/h · ft² · °F, respectively, determine the rate of heat transfer through the wall constructed of (a) solid bricks and (b) bricks with air holes.

3–59 Consider a 5-m-high, 8-m-long, and 0.22-m-thick wall whose representative cross section is as given in the figure. The thermal conductivities of various materials used, in W/m · °C, are $k_A = k_F = 2$, $k_B = 8$, $k_C - 20$, $k_D = 15$, and $k_E = 35$. The left and right surfaces of the wall are maintained at uniform temperatures of 300°C and 100°C, respectively. Assuming heat transfer through the wall to be one-dimensional, determine (a) the rate of heat transfer through the wall; (b) the temperature at the point where the sections B, D, and E meet; and (c) the temperature drop across the section F. Disregard any contact resistances at the interfaces.

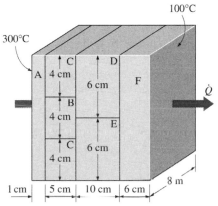

300°C

100°C

A | C 4 cm | D
  | B 4 cm | 6 cm | F | $\dot{Q}$
  | C 4 cm | E 6 cm |

1 cm | 5 cm | 10 cm | 6 cm | 8 m

FIGURE P3–59

3–60 Repeat Prob. 3–59 assuming that the thermal contact resistance at the interfaces D-F and E-F is 0.00012 m² · °C/W.

3–61 Clothing made of several thin layers of fabric with trapped air in between, often called ski clothing, is commonly used in cold climates because it is light, fashionable, and a very effective thermal insulator. So it is no surprise that such clothing has largely replaced thick and heavy old-fashioned coats.

Consider a jacket made of five layers of 0.1-mm-thick synthetic fabric ($k = 0.13$ W/m · °C) with 1.5-mm-thick air space ($k = 0.026$ W/m · °C) between the layers. Assuming the inner surface temperature of the jacket to be 28°C and the surface area to be 1.25 m², determine the rate of heat loss through the jacket when the temperature of the outdoors is 0°C and the heat transfer coefficient at the outer surface is 25 W/m² · °C.

What would your response be if the jacket is made of a single layer of 0.5-mm-thick synthetic fabric? What should be the thickness of a wool fabric ($k = 0.035$ W/m · °C) if the person is to achieve the same level of thermal comfort wearing a thick wool coat instead of a five-layer ski jacket?

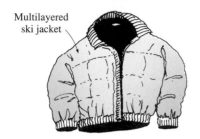

Multilayered ski jacket

FIGURE P3–61

3–62 Repeat Prob. 3–61 assuming the layers of the jacket are made of cotton fabric ($k = 0.06$ W/m · °C).

3–63 A 5-m-wide, 4-m-high, and 40-m-long kiln used to cure concrete pipes is made of 20-cm-thick concrete walls and ceiling ($k = 0.9$ W/m · °C). The kiln is maintained at 40°C by injecting hot steam into it. The two ends of the kiln, 4 m × 5 m in size, are made of a 3-mm-thick sheet metal covered with 2-cm-thick Styrofoam ($k = 0.033$ W/m · °C). The convection heat transfer coefficients on the inner and the outer surfaces of the kiln are 3000 W/m² · °C and 25 W/m² · °C, respectively. Disregarding any heat loss through the floor, determine the rate of heat loss from the kiln when the ambient air is at −4°C.

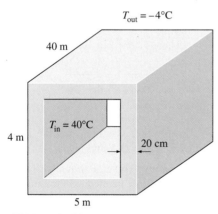

$T_{out} = -4°C$

40 m

$T_{in} = 40°C$

4 m

20 cm

5 m

FIGURE P3–63

3–64 Reconsider Prob. 3–63. Using EES (or other) software, investigate the effects of the thickness of the wall and the convection heat transfer coefficient on the outer surface of the rate of heat loss from the kiln. Let the thickness vary from 10 cm to 30 cm and the convection heat transfer coefficient from 5 W/m² · °C to 50 W/m² · °C. Plot the rate of heat transfer as functions of wall thickness and the convection heat transfer coefficient, and discuss the results.

3–65E Consider a 6-in × 8-in epoxy glass laminate ($k = 0.10$ Btu/h · ft · °F) whose thickness is 0.05 in. In order to reduce the thermal resistance across its thickness, cylindrical copper fillings ($k = 223$ Btu/h · ft · °F) of 0.02 in diameter are to be planted throughout the board, with a center-to-center distance of 0.06 in. Determine the new value of the thermal resistance of the epoxy board for heat conduction across its thickness as a result of this modification. *Answer:* 0.00064 h · °F/Btu

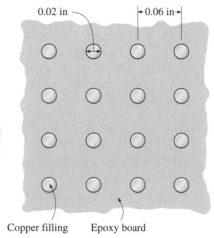

0.02 in ⊢0.06 in⊣

Copper filling   Epoxy board

FIGURE P3–65E

Heat Conduction in Cylinders and Spheres

3–66C What is an infinitely long cylinder? When is it proper to treat an actual cylinder as being infinitely long, and when is it not?

3–67C Consider a short cylinder whose top and bottom surfaces are insulated. The cylinder is initially at a uniform temperature T_i and is subjected to convection from its side surface to a medium at temperature T_∞, with a heat transfer coefficient of h. Is the heat transfer in this short cylinder one- or two-dimensional? Explain.

3–68C Can the thermal resistance concept be used for a solid cylinder or sphere in steady operation? Explain.

3–69 Chilled water enters a thin-shelled 5-cm-diameter, 150-m-long pipe at 7°C at a rate of 0.98 kg/s and leaves at 8°C. The pipe is exposed to ambient air at 30°C with a heat transfer coefficient of 9 W/m² · °C. If the pipe is to be insulated with glass wool insulation ($k = 0.05$ W/m · °C) in order to decrease the temperature rise of water to 0.25°C, determine the required thickness of the insulation.

3–70 Superheated steam at an average temperature 200°C is transported through a steel pipe ($k = 50$ W/m · K, $D_o = 8.0$ cm, $D_i = 6.0$ cm, and $L = 20.0$ m). The pipe is insulated with a 4-cm thick layer of gypsum plaster ($k = 0.5$ W/m · K). The insulated pipe is placed horizontally inside a warehouse where the average air temperature is 10°C. The steam and the air heat transfer coefficients are estimated to be 800 and 200 W/m² · K, respectively. Calculate (a) the daily rate of heat transfer from the superheated steam, and (b) the temperature on the outside surface of the gypsum plaster insulation.

3–71 An 8-m-internal-diameter spherical tank made of 1.5-cm-thick stainless steel ($k = 15$ W/m · °C) is used to store iced water at 0°C. The tank is located in a room whose temperature is 25°C. The walls of the room are also at 25°C. The outer surface of the tank is black (emissivity $\varepsilon = 1$), and heat transfer between the outer surface of the tank and the surroundings is by natural convection and radiation. The convection heat transfer coefficients at the inner and the outer surfaces of the tank are 80 W/m² · °C and 10 W/m² · °C, respectively. Determine (a) the rate of heat transfer to the iced water in the tank and (b) the amount of ice at 0°C that melts during a 24-h period. The heat of fusion of water at atmospheric pressure is $h_{if} = 333.7$ kJ/kg.

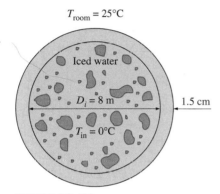

$T_{room} = 25°C$

Iced water

$D_i = 8$ m   1.5 cm

$T_{in} = 0°C$

FIGURE P3–71

3–72 Steam at 320°C flows in a stainless steel pipe ($k = 15$ W/m · °C) whose inner and outer diameters are 5 cm and 5.5 cm, respectively. The pipe is covered with 3-cm-thick glass wool insulation ($k = 0.038$ W/m · °C). Heat is lost to the surroundings at 5°C by natural convection and radiation, with a combined natural convection and radiation heat transfer coefficient of 15 W/m² · °C. Taking the heat transfer coefficient inside the pipe to be 80 W/m² · °C, determine the rate of heat loss from the steam per unit length of the pipe. Also determine the temperature drops across the pipe shell and the insulation.

3–73 Reconsider Prob. 3–72. Using EES (or other) software, investigate the effect of the thickness of the insulation on the rate of heat loss from the steam and the temperature drop across the insulation layer. Let the insulation thickness vary from 1 cm to 10 cm. Plot the rate of heat loss and the temperature drop as a function of insulation thickness, and discuss the results.

3–74 A 50-m-long section of a steam pipe whose outer diameter is 10 cm passes through an open space at 15°C. The average temperature of the outer surface of the pipe is measured to be 150°C. If the combined heat transfer coefficient on the outer surface of the pipe is 20 W/m² · °C, determine (a) the rate of heat loss from the steam pipe; (b) the

annual cost of this energy lost if steam is generated in a natural gas furnace that has an efficiency of 75 percent and the price of natural gas is $0.52/therm (1 therm = 105,500 kJ); and (c) the thickness of fiberglass insulation ($k = 0.035$ W/m · °C) needed in order to save 90 percent of the heat lost. Assume the pipe temperature to remain constant at 150°C.

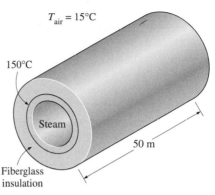

$T_{air} = 15°C$

150°C

Steam

50 m

Fiberglass insulation

FIGURE P3–74

3–75 Consider a 2-m-high electric hot-water heater that has a diameter of 40 cm and maintains the hot water at 55°C. The tank is located in a small room whose average temperature is 27°C, and the heat transfer coefficients on the inner and outer surfaces of the heater are 50 and 12 W/m² · °C, respectively. The tank is placed in another 46-cm-diameter sheet metal tank of negligible thickness, and the space between the two tanks is filled with foam insulation ($k = 0.03$ W/m · °C). The thermal resistances of the water tank and the outer thin sheet metal shell are very small and can be neglected. The price of electricity is $0.08/kWh, and the home owner pays $280 a year for water heating. Determine the fraction of the hot-water energy cost of this household that is due to the heat loss from the tank.

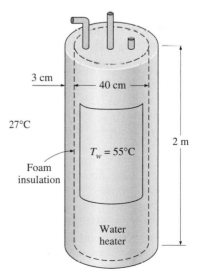

3 cm   40 cm

27°C

$T_w = 55°C$ 2 m

Foam insulation

Water heater

FIGURE P3–75

Hot-water tank insulation kits consisting of 3-cm-thick fiberglass insulation ($k = 0.035$ W/m · °C) large enough to wrap the entire tank are available in the market for about $30. If such an insulation is installed on this water tank by the home owner himself, how long will it take for this additional insulation to pay for itself? *Answers:* 17.5 percent, 1.5 years

3–76  Reconsider Prob. 3–75. Using EES (or other) software, plot the fraction of energy cost of hot water due to the heat loss from the tank as a function of the hot-water temperature in the range of 40°C to 90°C. Discuss the results.

3–77 Consider a cold aluminum canned drink that is initially at a uniform temperature of 4°C. The can is 12.5 cm high and has a diameter of 6 cm. If the combined convection/radiation heat transfer coefficient between the can and the surrounding air at 25°C is 10 W/m² · °C, determine how long it will take for the average temperature of the drink to rise to 15°C.

In an effort to slow down the warming of the cold drink, a person puts the can in a perfectly fitting 1-cm-thick cylindrical rubber insulator ($k = 0.13$ W/m · °C). Now how long will it take for the average temperature of the drink to rise to 15°C? Assume the top of the can is not covered.

4°C

12.5 cm

$T_{air} = 25$°C

6 cm

FIGURE P3–77

3–78 Repeat Prob. 3–77, assuming a thermal contact resistance of 0.00008 m² · °C/W between the can and the insulation.

3–79E Steam at 450°F is flowing through a steel pipe ($k = 8.7$ Btu/h · ft · °F) whose inner and outer diameters are 3.5 in and 4.0 in, respectively, in an environment at 55°F. The pipe is insulated with 2-in-thick fiberglass insulation ($k = 0.020$ Btu/h · ft · °F). If the heat transfer coefficients on the inside and the outside of the pipe are 30 and 5 Btu/h · ft² · °F, respectively, determine the rate of heat loss from the steam per foot length of the pipe. What is the error involved in neglecting the thermal resistance of the steel pipe in calculations?

3–80 Hot water at an average temperature of 70°C is flowing through a 15-m section of a cast iron pipe ($k = 52$ W/m · °C) whose inner and outer diameters are 4 cm and 4.6 cm, respectively. The outer surface of the pipe, whose emissivity is 0.7, is exposed to the cold air at 10°C in the basement, with a

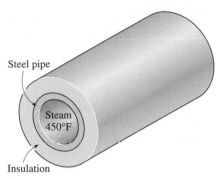

Steel pipe

Steam 450°F

Insulation

FIGURE P3–79E

heat transfer coefficient of 15 W/m² · °C. The heat transfer coefficient at the inner surface of the pipe is 120 W/m² · °C. Taking the walls of the basement to be at 10°C also, determine the rate of heat loss from the hot water. Also, determine the average velocity of the water in the pipe if the temperature of the water drops by 3°C as it passes through the basement.

3–81 Repeat Prob. 3–80 for a pipe made of copper ($k = 386$ W/m · °C) instead of cast iron.

3–82E Steam exiting the turbine of a steam power plant at 100°F is to be condensed in a large condenser by cooling water flowing through copper pipes ($k = 223$ Btu/h · ft · °F) of inner diameter 0.4 in and outer diameter 0.6 in at an average temperature of 70°F. The heat of vaporization of water at 100°F is 1037 Btu/lbm. The heat transfer coefficients are 1500 Btu/h · ft² · °F on the steam side and 35 Btu/h · ft² · °F on the water side. Determine the length of the tube required to condense steam at a rate of 120 lbm/h. *Answer:* 1148 ft

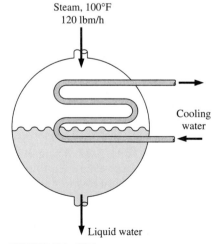

Steam, 100°F
120 lbm/h

Cooling water

Liquid water

FIGURE P3–82E

3–83E Repeat Prob. 3–82E, assuming that a 0.01-in-thick layer of mineral deposit ($k = 0.5$ Btu/h · ft · °F) has formed on the inner surface of the pipe.

3–84E  Reconsider Prob. 3–82E. Using EES (or other) software, investigate the effects of the thermal conductivity of the pipe material and the outer diameter of the pipe on the length of the tube required. Let the thermal conductivity vary from 10 Btu/h · ft · °F to 400 Btu/h · ft · °F and the outer diameter from 0.5 in to 1.0 in. Plot the length of the tube as functions of pipe conductivity and the outer pipe diameter, and discuss the results.

3–85 The boiling temperature of nitrogen at atmospheric pressure at sea level (1 atm pressure) is −196°C. Therefore, nitrogen is commonly used in low-temperature scientific studies since the temperature of liquid nitrogen in a tank open to the atmosphere will remain constant at −196°C until it is depleted. Any heat transfer to the tank will result in the evaporation of some liquid nitrogen, which has a heat of vaporization of 198 kJ/kg and a density of 810 kg/m³ at 1 atm.

Consider a 3-m-diameter spherical tank that is initially filled with liquid nitrogen at 1 atm and −196°C. The tank is exposed to ambient air at 15°C, with a combined convection and radiation heat transfer coefficient of 35 W/m² · °C. The temperature of the thin-shelled spherical tank is observed to be almost the same as the temperature of the nitrogen inside. Determine the rate of evaporation of the liquid nitrogen in the tank as a result of the heat transfer from the ambient air if the tank is (a) not insulated, (b) insulated with 5-cm-thick fiberglass insulation ($k = 0.035$ W/m · °C), and (c) insulated with 2-cm-thick superinsulation which has an effective thermal conductivity of 0.00005 W/m · °C.

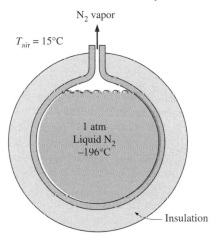

FIGURE P3–85

3–86 Repeat Prob. 3–85 for liquid oxygen, which has a boiling temperature of −183°C, a heat of vaporization of 213 kJ/kg, and a density of 1140 kg/m³ at 1 atm pressure.

Critical Radius of Insulation

3–87C What is the critical radius of insulation? How is it defined for a cylindrical layer?

3–88C A pipe is insulated such that the outer radius of the insulation is less than the critical radius. Now the insulation is taken off. Will the rate of heat transfer from the pipe increase or decrease for the same pipe surface temperature?

3–89C A pipe is insulated to reduce the heat loss from it. However, measurements indicate that the rate of heat loss has increased instead of decreasing. Can the measurements be right?

3–90C Consider a pipe at a constant temperature whose radius is greater than the critical radius of insulation. Someone claims that the rate of heat loss from the pipe has increased when some insulation is added to the pipe. Is this claim valid?

3–91C Consider an insulated pipe exposed to the atmosphere. Will the critical radius of insulation be greater on calm days or on windy days? Why?

3–92 A 2.2-mm-diameter and 10-m-long electric wire is tightly wrapped with a 1-mm-thick plastic cover whose thermal conductivity is $k = 0.15$ W/m · °C. Electrical measurements indicate that a current of 13 A passes through the wire and there is a voltage drop of 8 V along the wire. If the insulated wire is exposed to a medium at $T_\infty = 30°C$ with a heat transfer coefficient of $h = 24$ W/m² · °C, determine the temperature at the interface of the wire and the plastic cover in steady operation. Also determine if doubling the thickness of the plastic cover will increase or decrease this interface temperature.

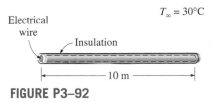

FIGURE P3–92

3–93E A 0.083-in-diameter electrical wire at 90°F is covered by 0.02-in-thick plastic insulation ($k = 0.075$ Btu/h · ft · °F). The wire is exposed to a medium at 50°F, with a combined convection and radiation heat transfer coefficient of 2.5 Btu/h · ft² · °F. Determine if the plastic insulation on the wire will increase or decrease heat transfer from the wire. *Answer:* It helps

3–94E Repeat Prob. 3–93E, assuming a thermal contact resistance of 0.001 h · ft² · °F/Btu at the interface of the wire and the insulation.

3–95 A 5-mm-diameter spherical ball at 50°C is covered by a 1-mm-thick plastic insulation ($k = 0.13$ W/m · °C). The ball is exposed to a medium at 15°C, with a combined convection and radiation heat transfer coefficient of 20 W/m² · °C. Determine if the plastic insulation on the ball will help or hurt heat transfer from the ball.

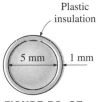

FIGURE P3–95

3–96 Reconsider Prob. 3–95. Using EES (or other) software, plot the rate of heat transfer from the ball as a function of the plastic insulation thickness in the range of 0.5 mm to 20 mm. Discuss the results.

Heat Transfer from Finned Surfaces

3–97C What is the reason for the widespread use of fins on surfaces?

3–98C What is the difference between the fin effectiveness and the fin efficiency?

3–99C The fins attached to a surface are determined to have an effectiveness of 0.9. Do you think the rate of heat transfer from the surface has increased or decreased as a result of the addition of these fins?

3–100C Explain how the fins enhance heat transfer from a surface. Also, explain how the addition of fins may actually decrease heat transfer from a surface.

3–101C How does the overall effectiveness of a finned surface differ from the effectiveness of a single fin?

3–102C Hot water is to be cooled as it flows through the tubes exposed to atmospheric air. Fins are to be attached in order to enhance heat transfer. Would you recommend attaching the fins inside or outside the tubes? Why?

3–103C Hot air is to be cooled as it is forced to flow through the tubes exposed to atmospheric air. Fins are to be added in order to enhance heat transfer. Would you recommend attaching the fins inside or outside the tubes? Why? When would you recommend attaching fins both inside and outside the tubes?

3–104C Consider two finned surfaces that are identical except that the fins on the first surface are formed by casting or extrusion, whereas they are attached to the second surface afterwards by welding or tight fitting. For which case do you think the fins will provide greater enhancement in heat transfer? Explain.

3–105C The heat transfer surface area of a fin is equal to the sum of all surfaces of the fin exposed to the surrounding medium, including the surface area of the fin tip. Under what conditions can we neglect heat transfer from the fin tip?

3–106C Does the (*a*) efficiency and (*b*) effectiveness of a fin increase or decrease as the fin length is increased?

3–107C Two pin fins are identical, except that the diameter of one of them is twice the diameter of the other. For which fin is the (*a*) fin effectiveness and (*b*) fin efficiency higher? Explain.

3–108C Two plate fins of constant rectangular cross section are identical, except that the thickness of one of them is twice the thickness of the other. For which fin is the (*a*) fin effectiveness and (*b*) fin efficiency higher? Explain.

3–109C Two finned surfaces are identical, except that the convection heat transfer coefficient of one of them is twice that of the other. For which finned surface is the (*a*) fin effectiveness and (*b*) fin efficiency higher? Explain.

3–110 Obtain a relation for the fin efficiency for a fin of constant cross-sectional area A_c, perimeter p, length L, and thermal conductivity k exposed to convection to a medium at T_∞ with a heat transfer coefficient h. Assume the fins are sufficiently long so that the temperature of the fin at the tip is nearly T_∞. Take the temperature of the fin at the base to be T_b and neglect heat transfer from the fin tips. Simplify the relation for (*a*) a circular fin of diameter D and (*b*) rectangular fins of thickness t.

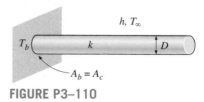

FIGURE P3–110

3–111 The case-to-ambient thermal resistance of a power transistor that has a maximum power rating of 15 W is given to be 25°C/W. If the case temperature of the transistor is not to exceed 80°C, determine the power at which this transistor can be operated safely in an environment at 40°C.

3–112 A 4-mm-diameter and 10-cm-long aluminum fin ($k = 237$ W/m · °C) is attached to a surface. If the heat transfer coefficient is 12 W/m^2 · °C, determine the percent error in the rate of heat transfer from the fin when the infinitely long fin assumption is used instead of the adiabatic fin tip assumption.

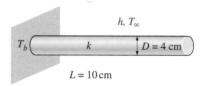

FIGURE P3–112

3–113 Consider a very long rectangular fin attached to a flat surface such that the temperature at the end of the fin is essentially that of the surrounding air, i.e. 20°C. Its width is 5.0 cm; thickness is 1.0 mm; thermal conductivity is 200 W/m · K; and base temperature is 40°C. The heat transfer coefficient is 20 W/m^2 · K. Estimate the fin temperature at a distance of 5.0 cm from the base and the rate of heat loss from the entire fin.

3–114 Circular cooling fins of diameter $D = 1$ mm and length $L = 25.4$ mm, made of copper ($k = 400$ W/m · K), are used to enhance heat transfer from a surface that is maintained at temperature $T_{s1} = 132$°C. Each rod has one end attached to this surface ($x = 0$), while the opposite end ($x = L$) is joined to a second surface, which is maintained at $T_{s2} = 0$°C. The air flowing between the surfaces and the rods is also at $T_\infty = 0$°C, and the convection coefficient is $h = 100$ W/m^2 · K. For fin with prescribed tip temperature, the temperature distribution and the rate of heat transfer are given by

$$\frac{\theta}{\theta_b} = \frac{\theta_L/\theta_b \sinh(mx) + \sinh[m(L-x)]}{\sinh(mL)} \quad \text{and}$$

$$\dot{Q} = \theta_b \sqrt{hpkA_c} \frac{\cosh(mL)}{\sinh(mL)}$$

(*a*) Express the function $\theta(x) = T(x) - T_\infty$ along a fin, and calculate the temperature at $x = L/2$.

(b) Determine the rate of heat transferred from the hot surface through each fin and the fin effectiveness. Is the use of fins justified? Why?

(c) What is the total rate of heat transfer from a 10-cm by 10-cm section of the wall, which has 625 uniformly distributed fins? Assume the same convection coefficient for the fin and for the unfinned wall surface.

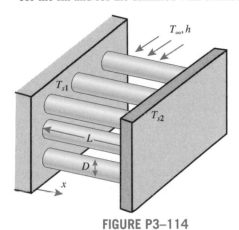

FIGURE P3–114

3–115 A 40-W power transistor is to be cooled by attaching it to one of the commercially available heat sinks shown in Table 3–6. Select a heat sink that will allow the case temperature of the transistor not to exceed 90°C in the ambient air at 20°C.

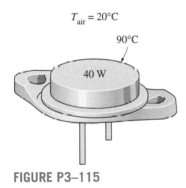

FIGURE P3–115

3–116 A 25-W power transistor is to be cooled by attaching it to one of the commercially available heat sinks shown in Table 3–6. Select a heat sink that will allow the case temperature of the transistor not to exceed 55°C in the ambient air at 18°C.

3–117 Steam in a heating system flows through tubes whose outer diameter is 5 cm and whose walls are maintained at a temperature of 180°C. Circular aluminum alloy 2024-T6 fins ($k = 186$ W/m · °C) of outer diameter 6 cm and constant thickness 1 mm are attached to the tube. The space between the fins is 3 mm, and thus there are 250 fins per meter length of the tube. Heat is transferred to the surrounding air at $T_\infty = 25°C$, with a heat transfer coefficient of 40 W/m² · °C. Determine the increase in heat transfer from the tube per meter of its length as a result of adding fins. *Answer: 2639 W*

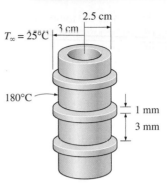

FIGURE P3–117

3–118E Consider a stainless steel spoon ($k = 8.7$ Btu/h · ft · °F) partially immersed in boiling water at 200°F in a kitchen at 75°F. The handle of the spoon has a cross section of 0.08 in × 0.5 in, and extends 7 in in the air from the free surface of the water. If the heat transfer coefficient at the exposed surfaces of the spoon handle is 3 Btu/h · ft² · °F, determine the temperature difference across the exposed surface of the spoon handle. State your assumptions. *Answer: 124.6°F*

FIGURE P3–118E

3–119E Repeat Prob. 3–118E for a silver spoon ($k = 247$ Btu/h · ft · °F).

3–120E Reconsider Prob. 3–118E. Using EES (or other) software, investigate the effects of the thermal conductivity of the spoon material and the length of its extension in the air on the temperature difference across the exposed surface of the spoon handle. Let the thermal conductivity vary from 5 Btu/h · ft · °F to 225 Btu/h · ft · °F and the length from 5 in to 12 in. Plot the temperature difference as the functions of thermal conductivity and length, and discuss the results.

3–121 A 0.3-cm-thick, 12-cm-high, and 18-cm-long circuit board houses 80 closely spaced logic chips on one side, each dissipating 0.04 W. The board is impregnated with copper fillings and has an effective thermal conductivity of 30 W/m · °C. All the heat generated in the chips is conducted across the circuit board and is dissipated from the back side of the board to a medium at 40°C, with a heat transfer coefficient of

40 W/m² · °C. (*a*) Determine the temperatures on the two sides of the circuit board. (*b*) Now a 0.2-cm-thick, 12-cm-high, and 18-cm-long aluminum plate ($k = 237$ W/m · °C) with 864 2-cm-long aluminum pin fins of diameter 0.25 cm is attached to the back side of the circuit board with a 0.02-cm-thick epoxy adhesive ($k = 1.8$ W/m · °C). Determine the new temperatures on the two sides of the circuit board.

3–122 Repeat Prob. 3–121 using a copper plate with copper fins ($k = 386$ W/m · °C) instead of aluminum ones.

3–123 A hot surface at 100°C is to be cooled by attaching 3-cm-long, 0.25-cm-diameter aluminum pin fins ($k = 237$ W/m · °C) to it, with a center-to-center distance of 0.6 cm. The temperature of the surrounding medium is 30°C, and the heat transfer coefficient on the surfaces is 35 W/m² · °C. Determine the rate of heat transfer from the surface for a 1-m × 1-m section of the plate. Also determine the overall effectiveness of the fins.

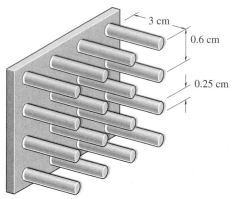

3 cm
0.6 cm
0.25 cm

FIGURE P3–123

3–124 Repeat Prob. 3–123 using copper fins ($k = 386$ W/m · °C) instead of aluminum ones.

3–125 Reconsider Prob. 3–123. Using EES (or other) software, investigate the effect of the center-to-center distance of the fins on the rate of heat transfer from the surface and the overall effectiveness of the fins. Let the center-to-center distance vary from 0.4 cm to 2.0 cm. Plot the rate of heat transfer and the overall effectiveness as a function of the center-to-center distance, and discuss the results.

3–126 Two 3-m-long and 0.4-cm-thick cast iron ($k = 52$ W/m · °C) steam pipes of outer diameter 10 cm are connected to each other through two 1-cm-thick flanges of outer diameter 20 cm. The steam flows inside the pipe at an average temperature of 200°C with a heat transfer coefficient of 180 W/m² · °C. The outer surface of the pipe is exposed to an ambient at 12°C, with a heat transfer coefficient of 25 W/m² · °C. (*a*) Disregarding the flanges, determine the average outer surface temperature of the pipe. (*b*) Using this temperature for the base of the flange and treating the flanges as the fins, determine the fin efficiency and the rate of heat trans-

fer from the flanges. (*c*) What length of pipe is the flange section equivalent to for heat transfer purposes?

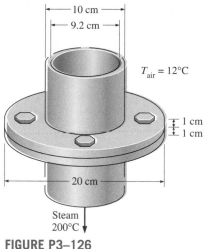

10 cm
9.2 cm
$T_{air} = 12°C$
1 cm
1 cm
20 cm
Steam
200°C

FIGURE P3–126

Heat Transfer in Common Configurations

3–127C What is a conduction shape factor? How is it related to the thermal resistance?

3–128C What is the value of conduction shape factors in engineering?

3–129 A 20-m-long and 8-cm-diameter hot-water pipe of a district heating system is buried in the soil 80 cm below the ground surface. The outer surface temperature of the pipe is 60°C. Taking the surface temperature of the earth to be 5°C and the thermal conductivity of the soil at that location to be 0.9 W/m · °C, determine the rate of heat loss from the pipe.

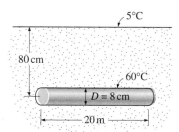

5°C
80 cm
60°C
$D = 8$ cm
20 m

FIGURE P3–129

3–130 Reconsider Prob. 3–129. Using EES (or other) software, plot the rate of heat loss from the pipe as a function of the burial depth in the range of 20 cm to 2.0 m. Discuss the results.

3–131 Hot- and cold-water pipes 8 m long run parallel to each other in a thick concrete layer. The diameters of both pipes are 5 cm, and the distance between the centerlines of the pipes is 40 cm. The surface temperatures of the hot and cold pipes are 60°C and 15°C, respectively. Taking the thermal conductivity of the concrete to be $k = 0.75$ W/m · °C, determine the rate of heat transfer between the pipes. *Answer: 306 W*

3–132 [EES] Reconsider Prob. 3–131. Using EES (or other) software, plot the rate of heat transfer between the pipes as a function of the distance between the centerlines of the pipes in the range of 10 cm to 1.0 m. Discuss the results.

3–133E A row of 3-ft-long and 1-in-diameter used uranium fuel rods that are still radioactive are buried in the ground parallel to each other with a center-to-center distance of 8 in at a depth of 15 ft from the ground surface at a location where the thermal conductivity of the soil is 0.6 Btu/h · ft · °F. If the surface temperature of the rods and the ground are 350°F and 60°F, respectively, determine the rate of heat transfer from the fuel rods to the atmosphere through the soil.

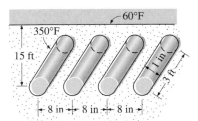

FIGURE P3–133E

3–134 Hot water at an average temperature of 53°C and an average velocity of 0.4 m/s is flowing through a 5-m section of a thin-walled hot-water pipe that has an outer diameter of 2.5 cm. The pipe passes through the center of a 14-cm-thick wall filled with fiberglass insulation ($k = 0.035$ W/m · °C). If the surfaces of the wall are at 18°C, determine (*a*) the rate of heat transfer from the pipe to the air in the rooms and (*b*) the temperature drop of the hot water as it flows through this 5-m-long section of the wall. *Answers: 19.6 W, 0.024°C*

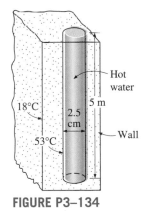

FIGURE P3–134

3–135 Hot water at an average temperature of 80°C and an average velocity of 1.5 m/s is flowing through a 25-m section of a pipe that has an outer diameter of 5 cm. The pipe extends 2 m in the ambient air above the ground, dips into the ground ($k = 1.5$ W/m · °C) vertically for 3 m, and continues horizontally at

this depth for 20 m more before it enters the next building. The first section of the pipe is exposed to the ambient air at 8°C, with a heat transfer coefficient of 22 W/m² · °C. If the surface of the ground is covered with snow at 0°C, determine (*a*) the total rate of heat loss from the hot water and (*b*) the temperature drop of the hot water as it flows through this 25-m-long section of the pipe.

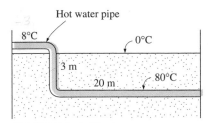

FIGURE P3–135

3–136 Consider a house with a flat roof whose outer dimensions are 12 m × 12 m. The outer walls of the house are 6 m high. The walls and the roof of the house are made of 20-cm-thick concrete ($k = 0.75$ W/m · °C). The temperatures of the inner and outer surfaces of the house are 15°C and 3°C, respectively. Accounting for the effects of the edges of adjoining surfaces, determine the rate of heat loss from the house through its walls and the roof. What is the error involved in ignoring the effects of the edges and corners and treating the roof as a 12 m × 12 m surface and the walls as 6 m × 12 m surfaces for simplicity?

3–137 Consider a 25-m-long thick-walled concrete duct ($k = 0.75$ W/m · °C) of square cross section. The outer dimensions of the duct are 20 cm × 20 cm, and the thickness of the duct wall is 2 cm. If the inner and outer surfaces of the duct are at 100°C and 30°C, respectively, determine the rate of heat transfer through the walls of the duct. *Answer: 47.1 kW*

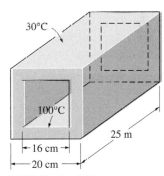

FIGURE P3–137

3–138 A 3-m-diameter spherical tank containing some radioactive material is buried in the ground ($k = 1.4$ W/m · °C). The distance between the top surface of the tank and the ground surface is 4 m. If the surface temperatures of the tank and the ground are 140°C and 15°C, respectively, determine the rate of heat transfer from the tank.

3–139 Reconsider Prob. 3–138. Using EES (or other) software, plot the rate of heat transfer from the tank as a function of the tank diameter in the range of 0.5 m to 5.0 m. Discuss the results.

3–140 Hot water at an average temperature of 85°C passes through a row of eight parallel pipes that are 4 m long and have an outer diameter of 3 cm, located vertically in the middle of a concrete wall ($k = 0.75$ W/m · °C) that is 4 m high, 8 m long, and 15 cm thick. If the surfaces of the concrete walls are exposed to a medium at 32°C, with a heat transfer coefficient of 12 W/m² · °C, determine the rate of heat loss from the hot water and the surface temperature of the wall.

Special Topics:
Heat Transfer through the Walls and Roofs

3–141C What is the R-value of a wall? How does it differ from the unit thermal resistance of the wall? How is it related to the U-factor?

3–142C What is effective emissivity for a plane-parallel air space? How is it determined? How is radiation heat transfer through the air space determined when the effective emissivity is known?

3–143C The unit thermal resistances (R-values) of both 40-mm and 90-mm vertical air spaces are given in Table 3–9 to be 0.22 m² · C/W, which implies that more than doubling the thickness of air space in a wall has no effect on heat transfer through the wall. Do you think this is a typing error? Explain.

3–144C What is a radiant barrier? What kind of materials are suitable for use as radiant barriers? Is it worthwhile to use radiant barriers in the attics of homes?

3–145C Consider a house whose attic space is ventilated effectively so that the air temperature in the attic is the same as the ambient air temperature at all times. Will the roof still have any effect on heat transfer through the ceiling? Explain.

3–146 Determine the summer R-value and the U-factor of a wood frame wall that is built around 38-mm × 140-mm wood studs with a center-to-center distance of 400 mm. The 140-mm-wide cavity between the studs is filled with mineral fiber batt insulation. The inside is finished with 13-mm gypsum wallboard and the outside with 13-mm wood fiberboard and 13-mm × 200-mm wood bevel lapped siding. The insulated cavity constitutes 80 percent of the heat transmission area, while the studs, headers, plates, and sills constitute 20 percent.
Answers: 3.213 m² · °C/W, 0.311 W/m² · °C

3–147 The 13-mm-thick wood fiberboard sheathing of the wood stud wall in Prob. 3–146 is replaced by a 25-mm-thick rigid foam insulation. Determine the percent increase in the R-value of the wall as a result.

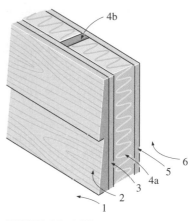

FIGURE P3–146

3–148E Determine the winter R-value and the U-factor of a masonry cavity wall that is built around 4-in-thick concrete blocks made of lightweight aggregate. The outside is finished with 4-in face brick with $\frac{1}{2}$-in cement mortar between the bricks and concrete blocks. The inside finish consists of $\frac{1}{2}$-in gypsum wallboard separated from the concrete block by $\frac{3}{4}$-in-thick (1-in by 3-in nominal) vertical furring whose center-to-center distance is 16 in. Neither side of the $\frac{3}{4}$-in-thick air space between the concrete block and the gypsum board is coated with any reflective film. When determining the R-value of the air space, the temperature difference across it can be taken to be 30°F with a mean air temperature of 50°F. The air space constitutes 80 percent of the heat transmission area, while the vertical furring and similar structures constitute 20 percent.

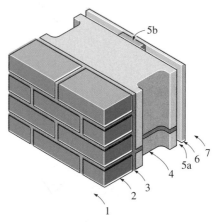

FIGURE P3–148E

3–149 Consider a flat ceiling that is built around 38-mm × 90-mm wood studs with a center-to-center distance of 400 mm. The lower part of the ceiling is finished with 13-mm gypsum wallboard, while the upper part consists of a wood subfloor ($R = 0.166$ m² · °C/W), a 13-mm plywood, a layer of felt ($R = 0.011$ m² · °C/W), and linoleum ($R = 0.009$ m² · °C/W). Both sides of the ceiling are exposed to still air. The air space constitutes 82 percent of the heat transmission area, while the studs and headers constitute 18 percent. Determine the winter R-value and the U-factor of the ceiling assuming the 90-mm-wide air space between the studs (a) does not have any reflective surface, (b) has a reflective surface with $\varepsilon = 0.05$ on one side, and (c) has reflective surfaces with $\varepsilon = 0.05$ on both sides. Assume a mean temperature of 10°C and a temperature difference of 5.6°C for the air space.

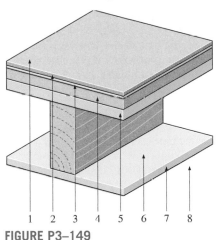

FIGURE P3–149

3–150 Determine the winter R-value and the U-factor of a masonry cavity wall that consists of 100-mm common bricks, a 90-mm air space, 100-mm concrete blocks made of light-weight aggregate, 20-mm air space, and 13-mm gypsum wall-board separated from the concrete block by 20-mm-thick (1-in × 3-in nominal) vertical furring whose center-to-center distance is 400 mm. Neither side of the two air spaces is coated with any reflective films. When determining the R-value of the air spaces, the temperature difference across them can be taken to be 16.7°C with a mean air temperature of 10°C. The air space constitutes 84 percent of the heat transmission area, while the vertical furring and similar structures constitute 16 percent.

Answers: 1.02 m² · °C/W, 0.978 W/m² · °C

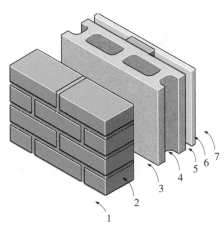

FIGURE P3–150

3–151 Repeat Prob. 3–150 assuming one side of both air spaces is coated with a reflective film of $\varepsilon = 0.05$.

3–152 Determine the winter R-value and the U-factor of a masonry wall that consists of the following layers: 100-mm face bricks, 100-mm common bricks, 25-mm urethane rigid foam insulation, and 13-mm gypsum wallboard.

Answers: 1.404 m² · °C/W, 0.712 W/m² · °C

3–153 The overall heat transfer coefficient (the U-value) of a wall under winter design conditions is $U = 1.40$ W/m² · °C. Determine the U-value of the wall under summer design conditions.

3–154 The overall heat transfer coefficient (the U-value) of a wall under winter design conditions is $U = 2.25$ W/m² · °C. Now a layer of 100-mm face brick is added to the outside, leaving a 20-mm air space between the wall and the bricks. Determine the new U-value of the wall. Also, determine the rate of heat transfer through a 3-m-high, 7-m-long section of the wall after modification when the indoor and outdoor temperatures are 22°C and –25°C, respectively.

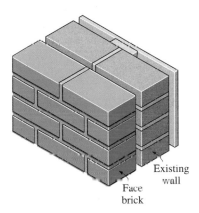

FIGURE P3–154

3–155 Determine the summer and winter R-values, in m² · °C/W, of a masonry wall that consists of 100-mm face bricks, 13-mm of cement mortar, 100-mm lightweight concrete block, 40-mm air space, and 20-mm plasterboard.

Answers: 0.809 and 0.795 m² · °C/W

3–156E The overall heat transfer coefficient of a wall is determined to be $U = 0.075$ Btu/h · ft² · F under the conditions of still air inside and winds of 7.5 mph outside. What will the U-factor be when the wind velocity outside is doubled?

Answer: 0.0755 Btu/h · ft² · °F

3–157 Two homes are identical, except that the walls of one house consist of 200-mm lightweight concrete blocks, 20-mm air space, and 20-mm plasterboard, while the walls of the other house involve the standard R-2.4 m² · °C/W frame wall construction. Which house do you think is more energy efficient?

3–158 Determine the R-value of a ceiling that consists of a layer of 19-mm acoustical tiles whose top surface is covered with a highly reflective aluminum foil for winter conditions. Assume still air below and above the tiles.

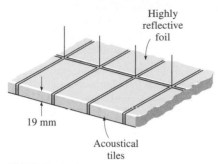

FIGURE P3–158

Review Problems

3–159E Steam is produced in the copper tubes ($k = 223$ Btu/h · ft · °F) of a heat exchanger at a temperature of 250°F by another fluid condensing on the outside surfaces of the tubes at 350°F. The inner and outer diameters of the tube are 1 in and 1.3 in, respectively. When the heat exchanger was new, the rate of heat transfer per foot length of the tube was 2×10^4 Btu/h. Determine the rate of heat transfer per foot length of the tube when a 0.01-in-thick layer of limestone ($k = 1.7$ Btu/h · ft · °F) has formed on the inner surface of the tube after extended use.

3–160E Repeat Prob. 3–159E, assuming that a 0.01-in-thick limestone layer has formed on both the inner and outer surfaces of the tube.

3–161 A 1.2-m-diameter and 6-m-long cylindrical propane tank is initially filled with liquid propane whose density is 581 kg/m³. The tank is exposed to the ambient air at 30°C, with a heat transfer coefficient of 25 W/m² · °C. Now a crack develops at the top of the tank and the pressure inside drops to 1 atm while the temperature drops to −42°C, which is the boiling temperature of propane at 1 atm. The heat of vaporization of propane at 1 atm is 425 kJ/kg. The propane is slowly vaporized as a result of the heat transfer from the ambient air into the tank, and the propane vapor escapes the tank at −42°C through the crack. Assuming the propane tank to be at about the same

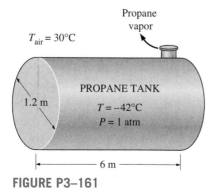

FIGURE P3–161

temperature as the propane inside at all times, determine how long it will take for the propane tank to empty if the tank is (a) not insulated and (b) insulated with 5-cm-thick glass wool insulation ($k = 0.038$ W/m · °C).

3–162 Hot water is flowing at an average velocity of 1.5 m/s through a cast iron pipe ($k = 52$ W/m · °C) whose inner and outer diameters are 3 cm and 3.5 cm, respectively. The pipe passes through a 15-m-long section of a basement whose temperature is 15°C. If the temperature of the water drops from 70°C to 67°C as it passes through the basement and the heat transfer coefficient on the inner surface of the pipe is 400 W/m² · °C, determine the combined convection and radiation heat transfer coefficient at the outer surface of the pipe. *Answer:* 272.5 W/m² · °C

3–163 Newly formed concrete pipes are usually cured first overnight by steam in a curing kiln maintained at a temperature of 45°C before the pipes are cured for several days outside. The heat and moisture to the kiln is provided by steam flowing in a pipe whose outer diameter is 12 cm. During a plant inspection, it was noticed that the pipe passes through a 10-m section that is completely exposed to the ambient air before it reaches the kiln. The temperature measurements indicate that the average temperature of the outer surface of the steam pipe is 82°C when the ambient temperature is 8°C. The combined convection and radiation heat transfer coefficient at the outer surface of the pipe is estimated to be 35 W/m² · °C. Determine the amount of heat lost from the steam during a 10-h curing process that night.

Steam is supplied by a gas-fired steam generator that has an efficiency of 85 percent, and the plant pays \$1.20/therm of natural gas (1 therm = 105,500 kJ). If the pipe is insulated and 90 percent of the heat loss is saved as a result, determine the amount of money this facility will save a year as a result of insulating the steam pipes. Assume that the concrete pipes are cured 110 nights a year. State your assumptions.

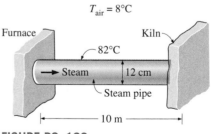

FIGURE P3–163

3–164 Consider an 18-cm × 18-cm multilayer circuit board dissipating 27 W of heat. The board consists of four layers of 0.2-mm-thick copper ($k = 386$ W/m · °C) and three layers of 1.5-mm-thick epoxy glass ($k = 0.26$ W/m · °C) sandwiched together, as shown in the figure. The circuit board is attached to a heat sink from both ends, and the temperature of the board at those ends is 35°C. Heat is considered to be uniformly generated in the epoxy layers of the board at a rate of 0.5 W per 1-cm ×

18-cm epoxy laminate strip (or 1.5 W per 1-cm × 18-cm strip of the board). Considering only a portion of the board because of symmetry, determine the magnitude and location of the maximum temperature that occurs in the board. Assume heat transfer from the top and bottom faces of the board to be negligible.

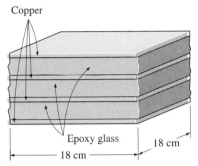

FIGURE P3–164

3–165 The plumbing system of a house involves a 0.5-m section of a plastic pipe ($k = 0.16$ W/m · °C) of inner diameter 2 cm and outer diameter 2.4 cm exposed to the ambient air. During a cold and windy night, the ambient air temperature remains at about −5°C for a period of 14 h. The combined convection and radiation heat transfer coefficient on the outer surface of the pipe is estimated to be 40 W/m² · °C, and the heat of fusion of water is 333.7 kJ/kg. Assuming the pipe to contain stationary water initially at 0°C, determine if the water in that section of the pipe will completely freeze that night.

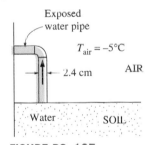

FIGURE P3–165

3–166 Repeat Prob. 3–165 for the case of a heat transfer coefficient of 10 W/m² · °C on the outer surface as a result of putting a fence around the pipe that blocks the wind.

3–167E The surface temperature of a 3-in-diameter baked potato is observed to drop from 300°F to 200°F in 5 min in an environment at 70°F. Determine the average heat transfer coefficient between the potato and its surroundings. Using this heat transfer coefficient and the same surface temperature, determine how long it will take for the potato to experience the same temperature drop if it is wrapped completely in a 0.12-in-thick towel ($k = 0.035$ Btu/h · ft · °F). You may use the properties of water for potato.

3–168E Repeat Prob. 3–167E assuming there is a 0.02-in-thick air space ($k = 0.015$ Btu/h · ft · °F) between the potato and the towel.

3–169 An ice chest whose outer dimensions are 30 cm × 40 cm × 50 cm is made of 3-cm-thick Styrofoam ($k = 0.033$ W/m · °C). Initially, the chest is filled with 50 kg of ice at 0°C, and the inner surface temperature of the ice chest can be taken to be 0°C at all times. The heat of fusion of ice at 0°C is 333.7 kJ/kg, and the heat transfer coefficient between the outer surface of the ice chest and surrounding air at 28°C is 18 W/m² · °C. Disregarding any heat transfer from the 40-cm × 50-cm base of the ice chest, determine how long it will take for the ice in the chest to melt completely.

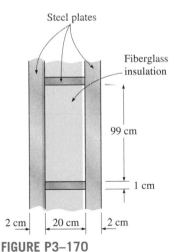

FIGURE P3–169

3–170 A 4-m-high and 6-m-long wall is constructed of two large 2-cm thick steel plates ($k = 15$ W/m · °C) separated by 1-cm-thick and 20-cm-wide steel bars placed 99 cm apart. The remaining space between the steel plates is filled with fiberglass insulation ($k = 0.035$ W/m · °C). If the temperature difference between the inner and the outer surfaces of the walls is 22°C, determine the rate of heat transfer through the wall. Can we ignore the steel bars between the plates in heat transfer analysis since they occupy only 1 percent of the heat transfer surface area?

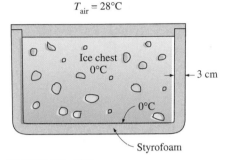

FIGURE P3–170

3–171 A 0.2-cm-thick, 10-cm-high, and 15-cm-long circuit board houses electronic components on one side that dissipate a total of 15 W of heat uniformly. The board is impregnated with conducting metal fillings and has an effective thermal conductivity of 12 W/m · °C. All the heat generated in the components is conducted across the circuit board and is dissipated from the back side of the board to a medium at 37°C, with a heat transfer coefficient of 45 W/m² · °C. (a) Determine the surface temperatures on the two sides of the circuit board. (b) Now a 0.1-cm-thick, 10-cm-high, and 15-cm-long aluminum plate ($k = 237$ W/m · °C) with 20 0.2-cm-thick, 2-cm-long, and 15-cm-wide aluminum fins of rectangular profile are attached to the back side of the circuit board with a 0.03-cm-thick epoxy adhesive ($k = 1.8$ W/m · °C). Determine the new temperatures on the two sides of the circuit board.

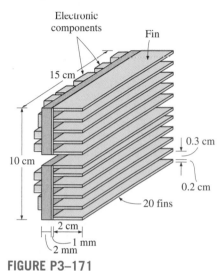

FIGURE P3–171

3–172 Repeat Prob. 3–171 using a copper plate with copper fins ($k = 386$ W/m · °C) instead of aluminum ones.

3–173 A row of 10 parallel pipes that are 5 m long and have an outer diameter of 6 cm are used to transport steam at 145°C through the concrete floor ($k = 0.75$ W/m · °C) of a 10-m × 5-m room that is maintained at 20°C. The combined convection and radiation heat transfer coefficient at the floor is 12 W/m² · °C. If the surface temperature of the concrete floor is not to exceed 35°C, determine how deep the steam pipes should be buried below the surface of the concrete floor.

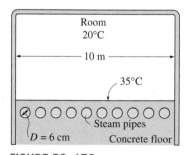

FIGURE P3–173

3–174 Consider two identical people each generating 60 W of metabolic heat steadily while doing sedentary work, and dissipating it by convection and perspiration. The first person is wearing clothes made of 1-mm-thick leather ($k = 0.159$ W/m · °C) that covers half of the body while the second one is wearing clothes made of 1-mm-thick synthetic fabric ($k = 0.13$ W/m · °C) that covers the body completely. The ambient air is at 30°C, the heat transfer coefficient at the outer surface is 15 W/m² · °C, and the inner surface temperature of the clothes can be taken to be 32°C. Treating the body of each person as a 25-cm-diameter, 1.7-m-long cylinder, determine the fractions of heat lost from each person by perspiration.

3–175 A 6-m-wide, 2.8-m-high wall is constructed of one layer of common brick ($k = 0.72$ W/m · °C) of thickness 20 cm, one inside layer of light-weight plaster ($k = 0.36$ W/m · °C) of thickness 1 cm, and one outside layer of cement based covering ($k = 1.40$ W/m · °C) of thickness 2 cm. The inner surface of the wall is maintained at 23°C while the outer surface is exposed to outdoors at 8°C with a combined convection and radiation heat transfer coefficient of 17 W/m² · °C. Determine the rate of heat transfer through the wall and temperature drops across the plaster, brick, covering, and surface-ambient air.

3–176 Reconsider Prob. 3–175. It is desired to insulate the wall in order to decrease the heat loss by 85 percent. For the same inner surface temperature, determine the thickness of insulation and the outer surface temperature if the wall is insulated with (a) polyurethane foam ($k = 0.025$ W/m · °C) and (b) glass fiber ($k = 0.036$ W/m · °C).

3–177 Cold conditioned air at 12°C is flowing inside a 1.5-cm-thick square aluminum ($k = 237$ W/m · °C) duct of inner cross section 22 cm × 22 cm at a mass flow rate of 0.8 kg/s. The duct is exposed to air at 33°C with a combined convection-radiation heat transfer coefficient of 13 W/m² · °C. The convection heat transfer coefficient at the inner surface is 75 W/m² · °C. If the air temperature in the duct should not increase by more than 1°C determine the maximum length of the duct.

3–178 When analyzing heat transfer through windows, it is important to consider the frame as well as the glass area. Consider a 2-m-wide, 1.5-m-high wood-framed window with 85 percent of the area covered by 3-mm-thick single-pane glass ($k = 0.7$ W/m · °C). The frame is 5 cm thick, and is made of pine wood ($k = 0.12$ W/m · °C). The heat transfer coefficient is 7 W/m² · °C inside and 13 W/m² · °C outside. The room is maintained at 24°C, and the outdoor temperature is 40°C. Determine the percent error involved in heat transfer when the window is assumed to consist of glass only.

3–179 Steam at 235°C is flowing inside a steel pipe ($k = 61$ W/m · °C) whose inner and outer diameters are 10 cm and 12 cm, respectively, in an environment at 20°C. The heat transfer coefficients inside and outside the pipe are 105 W/m² · °C and 14 W/m² · °C, respectively. Determine (a) the thickness of the insulation ($k = 0.038$ W/m · °C) needed to reduce the heat loss by 95 percent and (b) the thickness of the insulation needed to reduce the exposed surface temperature of insulated pipe to 40°C for safety reasons.

3–180 When the transportation of natural gas in a pipeline is not feasible for economic or other reasons, it is first liquefied at about $-160°C$, and then transported in specially insulated tanks placed in marine ships. Consider a 4-m-diameter spherical tank that is filled with liquefied natural gas (LNG) at $-160°C$. The tank is exposed to ambient air at 24°C with a heat transfer coefficient of 22 W/m^2 · °C. The tank is thin shelled and its temperature can be taken to be the same as the LNG temperature. The tank is insulated with 5-cm-thick super insulation that has an effective thermal conductivity of 0.00008 W/m · °C. Taking the density and the specific heat of LNG to be 425 kg/m^3 and 3.475 kJ/kg · °C, respectively, estimate how long it will take for the LNG temperature to rise to $-150°C$.

3–181 A 15-cm × 20-cm hot surface at 85°C is to be cooled by attaching 4-cm-long aluminum ($k = 237$ W/m · °C) fins of 2-mm × 2-mm square cross section. The temperature of surrounding medium is 25°C and the heat transfer coefficient on the surfaces can be taken to be 20 W/m^2 · °C. If it is desired to triple the rate of heat transfer from the bare hot surface, determine the number of fins that needs to be attached.

3–182 Reconsider Prob. 3–181. Using EES (or other) software, plot the number of fins as a function of the increase in the heat loss by fins relative to no fin case (i.e., overall effectiveness of the fins) in the range of 1.5 to 5. Discuss the results. Is it realistic to assume the heat transfer coefficient to remain constant?

3–183 A 1.4-m-diameter spherical steel tank filled with iced water at 0°C is buried underground at a location where the thermal conductivity of the soil is $k = 0.55$ W/m · °C. The distance between the tank center and the ground surface is 2.4 m. For ground surface temperature of 18°C, determine the rate of heat transfer to the iced water in the tank. What would your answer be if the soil temperature were 18°C and the ground surface were insulated?

3–184 A 0.6-m-diameter, 1.9-m-long cylindrical tank containing liquefied natural gas (LNG) at $-160°C$ is placed at the center of a 1.9-m-long 1.4-m × 1.4-m square solid bar made of an insulating material with $k = 0.0002$ W/m · °C. If the outer surface temperature of the bar is 12°C, determine the rate of heat transfer to the tank. Also, determine the LNG temperature after one month. Take the density and the specific heat of LNG to be 425 kg/m^3 and 3.475 kJ/kg · °C, respectively.

3–185 A typical section of a building wall is shown in Fig. P3–185. This section extends in and out of the page and is repeated in the vertical direction. The wall support members are made of steel ($k = 50$ W/m · K). The support members are 8 cm (t_{23}) × 0.5 cm (L_B). The remainder of the inner wall space is filled with insulation ($k = 0.03$ W/m · K) and measures 8 cm (t_{23}) × 60 cm (L_B). The inner wall is made of gypsum board

($k = 0.5$ W/m · K) that is 1 cm thick (t_{12}) and the outer wall is made of brick ($k = 1.0$ W/m · K) that is 10 cm thick (t_{34}). What is the temperature on the interior brick surface, 3, when $T_1 = 20$ °C and $T_4 = 35°C$?

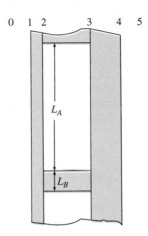

FIGURE P3–185

3–186 A total of 10 rectangular aluminum fins ($k = 203$ W/m · K) are placed on the outside flat surface of an electronic device. Each fin is 100 mm wide, 20 mm high and 4 mm thick. The fins are located parallel to each other at a center-to-center distance of 8 mm. The temperature at the outside surface of the electronic device is 60°C. The air is at 20°C, and the heat transfer coefficient is 100 W/m^2 · K. Determine (*a*) the rate of heat loss from the electronic device to the surrounding air and (*b*) the fin effectiveness.

3–187 One wall of a refrigerated warehouse is 10.0-m-high and 5.0-m-wide. The wall is made of three layers: 1.0-cm-thick aluminum ($k = 200$ W/m · K), 8.0-cm-thick fibreglass ($k = 0.038$ W/m · K), and 3.0-cm thick gypsum board ($k = 0.48$ W/m · K). The warehouse inside and outside temperatures are $-10°C$ and 20°C, respectively, and the average value of both inside and outside heat transfer coefficients is 40 W/m^2 · K.

(*a*) Calculate the rate of heat transfer across the warehouse wall in steady operation.

(*b*) Suppose that 400 metal bolts ($k = 43$ W/m · K), each 2.0 cm in diameter and 12.0 cm long, are used to fasten (i.e. hold together) the three wall layers. Calculate the rate of heat transfer for the "bolted" wall.

(*c*) What is the percent change in the rate of heat transfer across the wall due to metal bolts?

3–188 An agitated vessel is used for heating 500 kg/min of an aqueous solution at 15°C by saturated steam condensing in the jacket outside the vessel. The vessel can hold 6200 kg of the aqueous solution. It is fabricated from 15-mm-thick sheet of 1.0 percent carbon steel ($k = 43$ W/m · K), and it provides a heat transfer area of 12.0 m^2. The heat transfer coefficient due to agitation is 5.5 kW/m^2 · K, while the steam condensation at

115°C in the jacket gives a heat transfer coefficient of 10.0 kW/m$^2 \cdot$ K. All properties of the aqueous solution are comparable to those of pure water. Calculate the temperature of the outlet stream in steady operation.

3–189 A 10-cm long bar with a square cross-section, as shown in Fig. P3–189, consists of a 1-cm thick copper layer ($k = 400$ W/m $\cdot$ K) and a 1-cm thick epoxy composite layer ($k = 0.4$ W/m $\cdot$ K). Calculate the rate of heat transfer under a thermal driving force of 50°C, when the direction of steady one-dimensional heat transfer is (a) from front to back (i.e. along its length), (b) from left to right, and (c) from top to bottom.

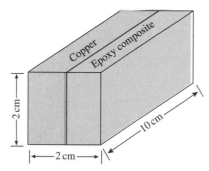

FIGURE P3–189

3–190 A spherical vessel, 3.0 m in diameter (and negligible wall thickness), is used for storing a fluid at a temperature of 0°C. The vessel is covered with a 5.0-cm-thick layer of an insulation ($k = 0.20$ W/m $\cdot$ K). The surrounding air is at 22°C. The inside and outside heat transfer coefficients are 40 and 10 W/m$^2 \cdot$ K, respectively. Calculate (a) all thermal resistances, in K/W, (b) the steady rate of heat transfer, and (c) the temperature difference across the insulation layer.

3–191 Air in a room is maintained at $T_\infty = 15$°C by a heated wall, which is 2 m wide, 3 m high, and 5 cm thick and is made of material with $k = 2$ W/m $\cdot$ K. The necessary heating power is $\dot{Q} = 5$ kW. The back of the wall is insulated. Two methods are considered to achieve heating: (a) a thin film heater at the back of the wall (surface heating), and (b) uniform volumetric heating within the wall at a rate of $\dot{e}_{gen}$ (W/m^3). The convection coefficient between the wall and the air is $h = 30$ W/m$^2 \cdot$ K.

(a) Plot qualitatively the variation of temperature T and heat flux $\dot{q}_s$ (W/m^2) across the wall in each case.

(b) Determine for each case the temperature at the surface of the wall, T_s.

(c) Determine for each case the temperature at the back of the wall, T_B.

Fundamentals of Engineering (FE) Exam Problems

3–192 A 2.5-m-high, 4-m-wide, and 20-cm-thick wall of a house has a thermal resistance of 0.0125°C/W. The thermal conductivity of the wall is

(a) 0.72 W/m $\cdot$ °C   (b) 1.1 W/m $\cdot$ °C   (c) 1.6 W/m $\cdot$ °C
(d) 16 W/m $\cdot$ °C   (e) 32 W/m $\cdot$ °C

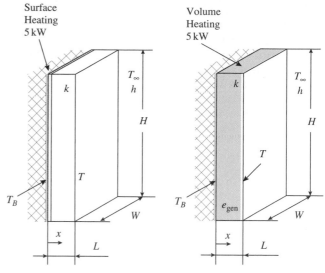

FIGURE P3–191

3–193 Consider two walls, A and B, with the same surface areas and the same temperature drops across their thicknesses. The ratio of thermal conductivities is $k_A/k_B = 4$ and the ratio of the wall thicknesses is $L_A/L_B = 2$. The ratio of heat transfer rates through the walls $\dot{Q}_A/\dot{Q}_B$ is

(a) 0.5   (b) 1   (c) 2   (d) 4   (e) 8

3–194 Heat is lost at a rate of 275 W per m^2 area of a 15-cm-thick wall with a thermal conductivity of $k = 1.1$ W/m $\cdot$ °C. The temperature drop across the wall is

(a) 37.5°C   (b) 27.5°C   (c) 16.0°C   (d) 8.0°C
(e) 4.0°C

3–195 Consider a wall that consists of two layers, A and B, with the following values: $k_A = 0.8$ W/m $\cdot$ °C, $L_A = 8$ cm, $k_B = 0.2$ W/m $\cdot$ °C, $L_B = 5$ cm. If the temperature drop across the wall is 18°C, the rate of heat transfer through the wall per unit area of the wall is

(a) 180 W/m^2 (b) 153 W/m^2 (c) 89.6 W/m^2
(d) 72.0 W/m^2 (e) 51.4 W/m^2

3–196 A plane furnace surface at 150°C covered with 1-cm-thick insulation is exposed to air at 30°C, and the combined heat transfer coefficient is 25 W/m$^2 \cdot$ °C. The thermal conductivity of insulation is 0.04 W/m $\cdot$ °C. The rate of heat loss from the surface per unit surface area is

(a) 35 W   (b) 414 W   (c) 300 W
(d) 480 W   (e) 128 W

3–197 A room at 20°C air temperature is loosing heat to the outdoor air at 0°C at a rate of 1000 W through a 2.5-m-high and 4-m-long wall. Now the wall is insulated with 2-cm thick insulation with a conductivity of 0.02 W/m $\cdot$ °C. Determine the rate of heat loss from the room through this wall after insulation. Assume the heat transfer coefficients on the inner and outer surface of the wall, the room air temperature, and the outdoor air temperature to remain unchanged. Also, disregard radiation.

(a) 20 W   (b) 561 W   (c) 388 W
(d) 167 W   (e) 200 W

3–198 Consider a 1.5-m-high and 2-m-wide triple pane window. The thickness of each glass layer ($k = 0.80$ W/m · °C) is 0.5 cm, and the thickness of each air space ($k = 0.025$ W/m · °C) is 1 cm. If the inner and outer surface temperatures of the window are 10°C and 0°C, respectively, the rate of heat loss through the window is

(a) 75 W   (b) 12 W   (c) 46 W
(d) 25 W   (e) 37 W

3–199 Consider a furnace wall made of sheet metal at an average temperature of 800°C exposed to air at 40°C. The combined heat transfer coefficient is 200 W/m² · °C inside the furnace, and 80 W/m² · °C outside. If the thermal resistance of the furnace wall is negligible, the rate of heat loss from the furnace per unit surface area is

(a) 48.0 kW/m²   (b) 213 kW/m²   (c) 91.2 kW/m²
(d) 151 kW/m²   (e) 43.4 kW/m²

3–200 Consider a jacket made of 5 layers of 0.1-mm-thick cotton fabric ($k = 0.060$ W/m · °C) with a total of 4 layers of 1-mm-thick air space ($k = 0.026$ W/m · °C) in between. Assuming the inner surface temperature of the jacket is 25°C and the surface area normal to the direction of heat transfer is 1.1 m², determine the rate of heat loss through the jacket when the temperature of the outdoors is 0°C and the heat transfer coefficient on the outer surface is 18 W/m² · °C.

(a) 6 W   (b) 115 W   (c) 126 W
(d) 287 W   (e) 170 W

3–201 Consider two metal plates pressed against each other. Other things being equal, which of the measures below will cause the thermal contact resistance to increase?

(a) Cleaning the surfaces to make them shinier.
(b) Pressing the plates against each other with a greater force.
(c) Filling the gab with a conducting fluid.
(d) Using softer metals.
(e) Coating the contact surfaces with a thin layer of soft metal such as tin.

3–202 A 10-m-long 5-cm-outer-radius cylindrical steam pipe is covered with 3-cm thick cylindrical insulation with a thermal conductivity of 0.05 W/m · °C. If the rate of heat loss from the pipe is 1000 W, the temperature drop across the insulation is

(a) 163°C   (b) 600°C   (c) 48°C
(d) 79°C   (e) 150°C

3–203 Steam at 200°C flows in a cast iron pipe ($k = 80$ W/m · °C) whose inner and outer diameters are $D_1 = 0.20$ m and $D_2 = 0.22$ m, respectively. The pipe is covered with 2-cm-thick glass wool insulation ($k = 0.05$ W/m · °C). The heat transfer coefficient at the inner surface is 75 W/m² · °C. If the

temperature at the interface of the iron pipe and the insulation is 194°C, the temperature at the outer surface of the insulation is

(a) 32 °C   (b) 45 °C   (c) 51 °C
(d) 75 °C   (e) 100 °C

3–204 A 6-m-diameter spherical tank is filled with liquid oxygen at -184°C. The tank is thin-shelled and its temperature can be taken to be the same as the oxygen temperature. The tank is insulated with 5-cm-thick super insulation that has an effective thermal conductivity of 0.00015 W/m · °C. The tank is exposed to ambient air at 15°C with a heat transfer coefficient of 14 W/m² · °C. The rate of heat transfer to the tank is

(a) 11 W   (b) 29 W   (c) 57 W
(d) 68 W   (e) 315,000 W

3–205 A 6-m-diameter spherical tank is filled with liquid oxygen ($\rho = 1141$ kg/m³, $c_p = 1.71$ kJ/kg · °C) at -184°C. It is observed that the temperature of oxygen increases to -183°C in a 144-hour period. The average rate of heat transfer to the tank is

(a) 249 W   (b) 426 W   (c) 570 W
(d) 1640 W   (e) 2207 W

3–206 A hot plane surface at 100°C is exposed to air at 25°C with a combined heat transfer coefficient of 20 W/m² · °C. The heat loss from the surface is to be reduced by half by covering it with sufficient insulation with a thermal conductivity of 0.10 W/m · °C. Assuming the heat transfer coefficient to remain constant, the required thickness of insulation is

(a) 0.1 cm   (b) 0.5 cm   (c) 1.0 cm
(d) 2.0 cm   (e) 5 cm

3–207 Consider a 4.5-m-long, 3.0-m-high, and 0.22-m-thick wall made of concrete ($k = 1.1$ W/m · °C). The design temperatures of the indoor and outdoor air are 24°C and 3°C, respectively, and the heat transfer coefficients on the inner and outer surfaces are 10 and 20 W/m² · °C. If a polyurethane foam insulation ($k = 0.03$ W/m · °C) is to be placed on the inner surface of the wall to increase the inner surface temperature of the wall to 22°C, the required thickness of the insulation is

(a) 3.3 cm   (b) 3.0 cm   (c) 2.7 cm
(d) 2.4 cm   (e) 2.1 cm

3–208 Steam at 200°C flows in a cast iron pipe ($k = 80$ W/m · °C) whose inner and outer diameters are $D_1 = 0.20$ m and $D_2 = 0.22$ m. The pipe is exposed to room air at 25°C. The heat transfer coefficients at the inner and outer surfaces of the pipe are 75 and 20 W/m² · °C, respectively. The pipe is to be covered with glass wool insulation ($k = 0.05$ W/m · °C) to decrease the heat loss from the stream by 90 percent. The required thickness of the insulation is

(a) 1.1 cm   (b) 3.4 cm   (c) 5.2 cm
(d) 7.9 cm   (e) 14.4 cm

3–209 A 50-cm-diameter spherical tank is filled with iced water at 0°C. The tank is thin-shelled and its temperature can be taken to be the same as the ice temperature. The tank is exposed to ambient air at 20°C with a heat transfer coefficient of 12 W/m² · °C. The tank is to be covered with glass wool insulation ($k = 0.05$ W/m · °C) to decrease the heat gain to the iced water by 90 percent. The required thickness of the insulation is

(a) 4.6 cm   (b) 6.7 cm   (c) 8.3 cm
(d) 25.0 cm   (e) 29.6 cm

3–210 Heat is generated steadily in a 3-cm-diameter spherical ball. The ball is exposed to ambient air at 26°C with a heat transfer coefficient of 7.5 W/m² · °C. The ball is to be covered with a material of thermal conductivity 0.15 W/m · °C. The thickness of the covering material that will maximize heat generation within the ball while maintaining ball surface temperature constant is

(a) 0.5 cm   (b) 1.0 cm   (c) 1.5 cm
(d) 2.0 cm   (e) 2.5 cm

3–211 A 1-cm-diameter, 30-cm-long fin made of aluminum ($k = 237$ W/m · °C) is attached to a surface at 80°C. The surface is exposed to ambient air at 22°C with a heat transfer coefficient of 11 W/m² · °C. If the fin can be assumed to be very long, the rate of heat transfer from the fin is

(a) 2.2 W   (b) 3.0 W   (c) 3.7 W
(d) 4.0 W   (e) 4.7 W

3–212 A 1-cm-diameter, 30-cm-long fin made of aluminum ($k = 237$ W/m · °C) is attached to a surface at 80°C. The surface is exposed to ambient air at 22°C with a heat transfer coefficient of 11 W/m² · °C. If the fin can be assumed to be very long, its efficiency is

(a) 0.60   (b) 0.67   (c) 0.72
(d) 0.77   (e) 0.88

3–213 A hot surface at 80°C in air at 20°C is to be cooled by attaching 10-cm-long and 1-cm-diameter cylindrical fins. The combined heat transfer coefficient is 30 W/m² · °C, and heat transfer from the fin tip is negligible. If the fin efficiency is 0.75, the rate of heat loss from 100 fins is

(a) 325 W   (b) 707 W   (c) 566 W
(d) 424 W   (e) 754 W

3–214 A cylindrical pin fin of diameter 1 cm and length 5 cm with negligible heat loss from the tip has an effectiveness of 15. If the fin base temperature is 280°C, the environment temperature is 20°C, and the heat transfer coefficient is 85 W/m² · °C, the rate of heat loss from this fin is

(a) 2 W   (b) 188 W   (c) 26 W
(d) 521 W   (e) 547 W

3–215 A cylindrical pin fin of diameter 0.6 cm and length of 3 cm with negligible heat loss from the tip has an efficiency of 0.7. The effectiveness of this fin is

(a) 0.3   (b) 0.7   (c) 2   (d) 8   (e) 14

3–216 A 3-cm-long, 2-mm × 2-mm rectangular cross-section aluminum fin ($k = 237$ W/m · °C) is attached to a surface. If the fin efficiency is 65 percent, the effectiveness of this single fin is

(a) 39   (b) 30   (c) 24   (d) 18   (e) 7

3–217 Aluminum square pin fins ($k = 237$ W/m · °C) of 3-cm-long, 2 mm × 2 mm cross-section with a total number of 150 are attached to an 8-cm-long, 6-cm-wide surface. If the fin efficiency is 65 percent, the overall fin effectiveness for the surface is

(a) 1.03   (b) 2.30   (c) 5.75   (d) 8.38   (e) 12.6

3–218 Two finned surfaces with long fins are identical, except that the convection heat transfer coefficient for the first finned surface is twice that of the second one. What statement below is accurate for the efficiency and effectiveness of the first finned surface relative to the second one?

(a) Higher efficiency and higher effectiveness
(b) Higher efficiency but lower effectiveness
(c) Lower efficiency but higher effectiveness
(d) Lower efficiency and lower effectiveness
(e) Equal efficiency and equal effectiveness

3–219 A 20-cm-diameter hot sphere at 120°C is buried in the ground with a thermal conductivity of 1.2 W/m · °C. The distance between the center of the sphere and the ground surface is 0.8 m and the ground surface temperature is 15°C. The rate of heat loss from the sphere is

(a) 169 W   (b) 20 W   (c) 217 W
(d) 312 W   (e) 1.8 W

3–220 A 25-cm-diameter, 2.4-m-long vertical cylinder containing ice at 0°C is buried right under the ground. The cylinder is thin-shelled and is made of a high thermal conductivity material. The surface temperature and the thermal conductivity of the ground are 18°C and 0.85 W/m · °C respectively. The rate of heat transfer to the cylinder is

(a) 37.2 W   (b) 63.2 W   (c) 158 W
(d) 480 W   (e) 1210 W

3–221 Hot water ($c_p = 4.179$ kJ/kg · K) flows through a 200-m-long PVC ($k = 0.092$ W/m · K) pipe whose inner diameter is 2 cm and outer diameter is 2.5 cm at a rate of 1 kg/s, entering at 40°C. If the entire interior surface of this pipe is maintained at 35°C and the entire exterior surface at 20°C, the outlet temperature of water is

(a) 39°C   (b) 38°C   (c) 37°C
(d) 36°C   (e) 35°C

3–222 Heat transfer rate through the wall of a circular tube with convection acting on the outer surface is given per unit of its length by

$$\dot{q} = \frac{2\pi L(T_i - T_o)}{\dfrac{\ln(r_o/r_i)}{k} + \dfrac{1}{r_o h}}$$

where i refers to the innertube surface and o the outer tube surface. Increasing r_o will reduce the heat transfer as long as

(a) $r_o < k/h$ (b) $r_o = k/h$
(c) $r_o > k/h$ (d) $r_o > 2k/h$
(e) Increasing r_o will always reduce the heat transfer.

3–223 The walls of a food storage facility are made of a 2-cm-thick layer of wood ($k = 0.1$ W/m · K) in contact with a 5-cm-thick layer of polyurethane foam ($k = 0.03$ W/m · K). If the temperature of the surface of the wood is $-10°C$ and the temperature of the surface of the polyurethane foam is $20°C$, the temperature of the surface where the two layers are in contact is

(a) $-7°C$ (b) $-2°C$ (c) $3°C$
(d) $8°C$ (e) $11°C$

3–224 A typical section of a building wall is shown in Fig. P3–224. This section extends in and out of the page and is repeated in the vertical direction. The correct thermal resistance circuit for this wall is

(a)
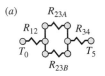

(b)
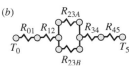

(c)
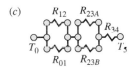

(d)
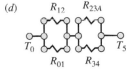

(e) None of them

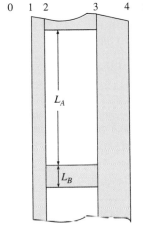

FIGURE P3–224

3–225 The 700 m² ceiling of a building has a thermal resistance of 0.2 m² · K/W. The rate at which heat is lost through this ceiling on a cold winter day when the ambient temperature is $-10°C$ and the interior is at $20°C$ is

(a) 56 MW   (b) 72 MW   (c) 87 MW
(d) 105 MW   (e) 118 MW

3–226 A 1-m-inner-diameter liquid-oxygen storage tank at a hospital keeps the liquid oxygen at 90 K. The tank consists of a 0.5-cm-thick aluminum ($k = 170$ W/m · K) shell whose exterior is covered with a 10-cm-thick layer of insulation ($k = 0.02$ W/m · K). The insulation is exposed to the ambient air at 20°C and the heat transfer coefficient on the exterior side of the insulation is 5 W/m² · K. The rate at which the liquid oxygen gains heat is

(a) 141 W   (b) 176 W   (c) 181 W
(d) 201 W   (e) 221 W

3–227 A 1-m-inner-diameter liquid-oxygen storage tank at a hospital keeps the liquid oxygen at 90 K. The tank consists of a 0.5-cm-thick aluminum ($k = 170$ W/m · K) shell whose exterior is covered with a 10-cm-thick layer of insulation ($k = 0.02$ W/m · K). The insulation is exposed to the ambient air at 20°C and the heat transfer coefficient on the exterior side of the insulation is 5 W/m² · K. The temperature of the exterior surface of the insulation is

(a) 13°C   (b) 9°C   (c) 2°C
(d) $-3°C$ (e) $-12°C$

3–228 The fin efficiency is defined as the ratio of the actual heat transfer from the fin to

(a) The heat transfer from the same fin with an adiabatic tip
(b) The heat transfer from an equivalent fin which is infinitely long
(c) The heat transfer from the same fin if the temperature along the entire length of the fin is the same as the base temperature
(d) The heat transfer through the base area of the same fin
(e) None of the above

3–229 Computer memory chips are mounted on a finned metallic mount to protect them from overheating. A 152 MB memory chip dissipates 5 W of heat to air at 25°C. If the temperature of this chip is to not exceed 50°C, the overall heat transfer coefficient–area product of the finned metal mount must be at least

(a) 0.2 W/°C   (b) 0.3 W/°C   (c) 0.4 W/°C
(d) 0.5 W/°C   (e) 0.6 W/°C

3–230 In the United States, building insulation is specified by the R-value (thermal resistance in h · ft² · °F/Btu units). A home owner decides to save on the cost of heating the home by adding additional insulation in the attic. If the total R-value is increased from 15 to 25, the home owner can expect the heat loss through the ceiling to be reduced by

(a) 25%   (b) 40%   (c) 50%   (d) 60%   (e) 75%

3–231 Coffee houses frequently serve coffee in a paper cup that has a corrugated paper jacket surrounding the cup like that shown here. This corrugated jacket:

(a) Serves to keep the coffee hot.

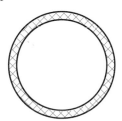

FIGURE P3–231

(b) Increases the coffee-to-surrounding thermal resistance.

(c) Lowers the temperature where the hand clasps the cup.

(d) All of the above.

(e) None of the above.

3–232 A triangular shaped fin on a motorcycle engine is 0.5-cm thick at its base and 3-cm long (normal distance between the base and the tip of the triangle), and is made of aluminum ($k = 150$ W/m · K). This fin is exposed to air with a convective heat transfer coefficient of 30 W/m² · K acting on its surfaces. The efficiency of the fin is 50 percent. If the fin base temperature is 130°C and the air temperature is 25°C, the heat transfer from this fin per unit width is

(a) 32 W/m   (b) 47 W/m   (c) 68 W/m

(d) 82 W/m   (e) 95 W/m

3–233 A plane brick wall ($k = 0.7$ W/m · K) is 10 cm thick. The thermal resistance of this wall per unit of wall area is

(a) 0.143 m² · K/W   (b) 0.250 m² · K/W

(c) 0.327 m² · K/W   (d) 0.448 m² · K/W

(e) 0.524 m² · K/W

3–234 The equivalent thermal resistance for the thermal circuit shown here is

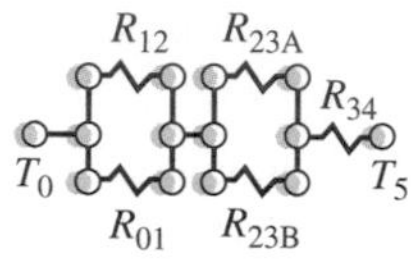

(a) $R_{12} R_{01} + R_{23A} R_{23B} + R_{34}$

(b) $R_{12} R_{01} + \left(\dfrac{R_{23A} R_{23B}}{R_{23A} + R_{23B}} \right) + R_{34}$

(c) $\left(\dfrac{R_{12} R_{01}}{R_{12} + R_{01}} \right) + \left(\dfrac{R_{23A} R_{23B}}{R_{23A} + R_{23B}} \right) + \dfrac{1}{R_{34}}$

(d) $\left(\dfrac{R_{12} R_{01}}{R_{12} + R_{01}} \right) + \left(\dfrac{R_{23A} R_{23B}}{R_{23A} + R_{23B}} \right) + R_{34}$

(e) None of them

FIGURE P3–234

Design and Essay Problems

3–235 The temperature in deep space is close to absolute zero, which presents thermal challenges for the astronauts who do space walks. Propose a design for the clothing of the astronauts that will be most suitable for the thermal environment in space. Defend the selections in your design.

3–236 In the design of electronic components, it is very desirable to attach the electronic circuitry to a substrate material that is a very good thermal conductor but also a very effective electrical insulator. If the high cost is not a major concern, what material would you propose for the substrate?

3–237 Using cylindrical samples of the same material, devise an experiment to determine the thermal contact resistance. Cylindrical samples are available at any length, and the thermal conductivity of the material is known.

3–238 Find out about the wall construction of the cabins of large commercial airplanes, the range of ambient conditions under which they operate, typical heat transfer coefficients on the inner and outer surfaces of the wall, and the heat generation rates inside. Determine the size of the heating and air-conditioning system that will be able to maintain the cabin at 20°C at all times for an airplane capable of carrying 400 people.

3–239 Repeat Prob. 3–238 for a submarine with a crew of 60 people.

3–240 A house with 200-m² floor space is to be heated with geothermal water flowing through pipes laid in the ground under the floor. The walls of the house are 4 m high, and there are 10 single-paned windows in the house that are 1.2 m wide and 1.8 m high. The house has R-19 (in h · ft² · °F/Btu) insulation in the walls and R-30 on the ceiling. The floor temperature is not to exceed 40°C. Hot geothermal water is available at 90°C, and the inner and outer diameter of the pipes to be used are 2.4 cm and 3.0 cm. Design such a heating system for this house in your area.

3–241 Using a timer (or watch) and a thermometer, conduct this experiment to determine the rate of heat gain of your refrigerator. First, make sure that the door of the refrigerator is not opened for at least a few hours to make sure that steady operating conditions are established. Start the timer when the refrigerator stops running and measure the time Δt_1 it stays off before it kicks in. Then measure the time Δt_2 it stays on. Noting that the heat removed during Δt_2 is equal to the heat gain of the refrigerator during $\Delta t_1 + \Delta t_2$ and using the power consumed by the refrigerator when it is running, determine the average rate of heat gain for your refrigerator, in watts. Take the COP (coefficient of performance) of your refrigerator to be 1.3 if it is not available.

Now, clean the condenser coils of the refrigerator and remove any obstacles on the way of airflow through the coils. By replacing these measurements, determine the improvement in the COP of the refrigerator.

TRANSIENT HEAT CONDUCTION

T he temperature of a body, in general, varies with time as well as position. In rectangular coordinates, this variation is expressed as $T(x, y, z, t)$, where (x, y, z) indicate variation in the x-, y-, and z-directions, and t indicates variation with time. In the preceding chapter, we considered heat conduction under *steady* conditions, for which the temperature of a body at any point does not change with time. This certainly simplified the analysis, especially when the temperature varied in one direction only, and we were able to obtain analytical solutions. In this chapter, we consider the variation of temperature with *time* as well as *position* in one- and multidimensional systems.

We start this chapter with the analysis of *lumped systems* in which the temperature of a body varies with time but remains uniform throughout at any time. Then we consider the variation of temperature with time as well as position for one-dimensional heat conduction problems such as those associated with a large plane wall, a long cylinder, a sphere, and a semi-infinite medium using *transient temperature charts* and analytical solutions. Finally, we consider transient heat conduction in multidimensional systems by utilizing the *product solution*.

OBJECTIVES

When you finish studying this chapter, you should be able to:

- Assess when the spatial variation of temperature is negligible, and temperature varies nearly uniformly with time, making the simplified lumped system analysis applicable,
- Obtain analytical solutions for transient one-dimensional conduction problems in rectangular, cylindrical, and spherical geometries using the method of separation of variables, and understand why a one-term solution is usually a reasonable approximation,
- Solve the transient conduction problem in large mediums using the similarity variable, and predict the variation of temperature with time and distance from the exposed surface, and
- Construct solutions for multi-dimensional transient conduction problems using the product solution approach.

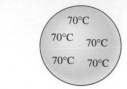

(a) Copper ball

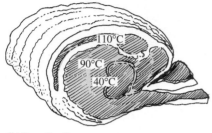

(b) Roast beef

FIGURE 4–1
A small copper ball can be modeled as a lumped system, but a roast beef cannot.

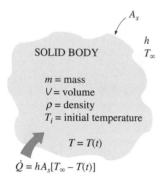

$$\dot{Q} = hA_s[T_\infty - T(t)]$$

FIGURE 4–2
The geometry and parameters involved in the lumped system analysis.

4–1 · LUMPED SYSTEM ANALYSIS

In heat transfer analysis, some bodies are observed to behave like a "lump" whose interior temperature remains essentially uniform at all times during a heat transfer process. The temperature of such bodies can be taken to be a function of time only, $T(t)$. Heat transfer analysis that utilizes this idealization is known as **lumped system analysis**, which provides great simplification in certain classes of heat transfer problems without much sacrifice from accuracy.

Consider a small hot copper ball coming out of an oven (Fig. 4–1). Measurements indicate that the temperature of the copper ball changes with time, but it does not change much with position at any given time. Thus the temperature of the ball remains nearly uniform at all times, and we can talk about the temperature of the ball with no reference to a specific location.

Now let us go to the other extreme and consider a large roast in an oven. If you have done any roasting, you must have noticed that the temperature distribution within the roast is not even close to being uniform. You can easily verify this by taking the roast out before it is completely done and cutting it in half. You will see that the outer parts of the roast are well done while the center part is barely warm. Thus, lumped system analysis is not applicable in this case. Before presenting a criterion about applicability of lumped system analysis, we develop the formulation associated with it.

Consider a body of arbitrary shape of mass m, volume V, surface area A_s, density ρ, and specific heat c_p initially at a uniform temperature T_i (Fig. 4–2). At time $t = 0$, the body is placed into a medium at temperature T_∞, and heat transfer takes place between the body and its environment, with a heat transfer coefficient h. For the sake of discussion, we assume that $T_\infty > T_i$, but the analysis is equally valid for the opposite case. We assume lumped system analysis to be applicable, so that the temperature remains uniform within the body at all times and changes with time only, $T = T(t)$.

During a differential time interval dt, the temperature of the body rises by a differential amount dT. An energy balance of the solid for the time interval dt can be expressed as

$$\left(\begin{array}{c}\text{Heat transfer into the body}\\\text{during } dt\end{array}\right) = \left(\begin{array}{c}\text{The increase in the}\\\text{energy of the body}\\\text{during } dt\end{array}\right)$$

or

$$hA_s(T_\infty - T)\,dt = mc_p\,dT \tag{4–1}$$

Noting that $m = \rho V$ and $dT = d(T - T_\infty)$ since $T_\infty = $ constant, Eq. 4–1 can be rearranged as

$$\frac{d(T - T_\infty)}{T - T_\infty} = -\frac{hA_s}{\rho V c_p}\,dt \tag{4–2}$$

Integrating from $t = 0$, at which $T = T_i$, to any time t, at which $T = T(t)$, gives

$$\ln\frac{T(t) - T_\infty}{T_i - T_\infty} = -\frac{hA_s}{\rho V c_p}\,t \tag{4–3}$$

Taking the exponential of both sides and rearranging, we obtain

$$\frac{T(t) - T_\infty}{T_i - T_\infty} = e^{-bt} \qquad (4\text{-}4)$$

where

$$b = \frac{hA_s}{\rho V c_p} \qquad (1/s) \qquad (4\text{-}5)$$

is a positive quantity whose dimension is $(\text{time})^{-1}$. The reciprocal of b has time unit (usually s), and is called the **time constant**. Equation 4–4 is plotted in Fig. 4–3 for different values of b. There are two observations that can be made from this figure and the relation above:

1. Equation 4–4 enables us to determine the temperature $T(t)$ of a body at time t, or alternatively, the time t required for the temperature to reach a specified value $T(t)$.
2. The temperature of a body approaches the ambient temperature T_∞ exponentially. The temperature of the body changes rapidly at the beginning, but rather slowly later on. A large value of b indicates that the body approaches the environment temperature in a short time. The larger the value of the exponent b, the higher the rate of decay in temperature. Note that b is proportional to the surface area, but inversely proportional to the mass and the specific heat of the body. This is not surprising since it takes longer to heat or cool a larger mass, especially when it has a large specific heat.

Once the temperature $T(t)$ at time t is available from Eq. 4–4, the *rate* of convection heat transfer between the body and its environment at that time can be determined from Newton's law of cooling as

$$\dot{Q}(t) = hA_s[T(t) - T_\infty] \qquad (\text{W}) \qquad (4\text{-}6)$$

The *total amount* of heat transfer between the body and the surrounding medium over the time interval $t = 0$ to t is simply the change in the energy content of the body:

$$Q = mc_p[T(t) - T_i] \qquad (\text{kJ}) \qquad (4\text{-}7)$$

The amount of heat transfer reaches its upper limit when the body reaches the surrounding temperature T_∞. Therefore, the *maximum* heat transfer between the body and its surroundings is (Fig. 4–4)

$$Q_{max} = mc_p(T_\infty - T_i) \qquad (\text{kJ}) \qquad (4\text{-}8)$$

We could also obtain this equation by substituting the $T(t)$ relation from Eq. 4–4 into the $\dot{Q}(t)$ relation in Eq. 4–6 and integrating it from $t = 0$ to $t \rightarrow \infty$.

Criteria for Lumped System Analysis

The lumped system analysis certainly provides great convenience in heat transfer analysis, and naturally we would like to know when it is appropriate

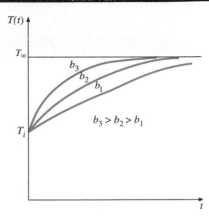

FIGURE 4–3
The temperature of a lumped system approaches the environment temperature as time gets larger.

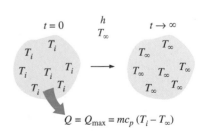

FIGURE 4–4
Heat transfer to or from a body reaches its maximum value when the body reaches the environment temperature.

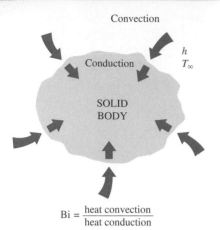

Convection

Conduction

h
T_∞

SOLID BODY

$Bi = \dfrac{\text{heat convection}}{\text{heat conduction}}$

FIGURE 4–5

The Biot number can be viewed as the ratio of the convection at the surface to conduction within the body.

to use it. The first step in establishing a criterion for the applicability of the lumped system analysis is to define a **characteristic length** as

$$L_c = \frac{V}{A_s}$$

and a **Biot number** Bi as

$$Bi = \frac{hL_c}{k} \tag{4–9}$$

It can also be expressed as (Fig. 4–5)

$$Bi = \frac{h}{k/L_c} \frac{\Delta T}{\Delta T} = \frac{\text{Convection at the surface of the body}}{\text{Conduction within the body}}$$

or

$$Bi = \frac{L_c/k}{1/h} = \frac{\text{Conduction resistance within the body}}{\text{Convection resistance at the surface of the body}}$$

When a solid body is being heated by the hotter fluid surrounding it (such as a potato being baked in an oven), heat is first *convected* to the body and subsequently *conducted* within the body. The Biot number is the *ratio* of the internal resistance of a body to *heat conduction* to its external resistance to *heat convection*. Therefore, a small Biot number represents small resistance to heat conduction, and thus small temperature gradients within the body.

Lumped system analysis assumes a *uniform* temperature distribution throughout the body, which is the case only when the thermal resistance of the body to heat conduction (the *conduction resistance*) is zero. Thus, lumped system analysis is *exact* when Bi = 0 and *approximate* when Bi > 0. Of course, the smaller the Bi number, the more accurate the lumped system analysis. Then the question we must answer is, How much accuracy are we willing to sacrifice for the convenience of the lumped system analysis?

Before answering this question, we should mention that a 15 percent uncertainty in the convection heat transfer coefficient h in most cases is considered "normal" and "expected." Assuming h to be *constant* and *uniform* is also an approximation of questionable validity, especially for irregular geometries. Therefore, in the absence of sufficient experimental data for the specific geometry under consideration, we cannot claim our results to be better than ±15 percent, even when Bi = 0. This being the case, introducing another source of uncertainty in the problem will not have much effect on the overall uncertainty, provided that it is minor. It is generally accepted that lumped system analysis is *applicable* if

$$Bi \le 0.1$$

When this criterion is satisfied, the temperatures within the body relative to the surroundings (i.e., $T - T_\infty$) remain within 5 percent of each other even for well-rounded geometries such as a spherical ball. Thus, when Bi < 0.1, the variation of temperature with location within the body is slight and can reasonably be approximated as being uniform.

The first step in the application of lumped system analysis is the calculation of the *Biot number,* and the assessment of the applicability of this approach. One may still wish to use lumped system analysis even when the criterion Bi < 0.1 is not satisfied, if high accuracy is not a major concern.

Note that the Biot number is the ratio of the *convection* at the surface to *conduction* within the body, and this number should be as small as possible for lumped system analysis to be applicable. Therefore, *small bodies* with *high thermal conductivity* are good candidates for lumped system analysis, especially when they are in a medium that is a poor conductor of heat (such as air or another gas) and motionless. Thus, the hot small copper ball placed in quiescent air, discussed earlier, is most likely to satisfy the criterion for lumped system analysis (Fig. 4–6).

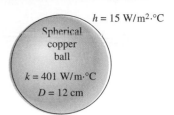

$$L_c = \frac{V}{A_s} = \frac{\frac{1}{6}\pi D^3}{\pi D^2} = \frac{1}{6}D = 0.02 \text{ m}$$

$$\text{Bi} = \frac{hL_c}{k} = \frac{15 \times 0.02}{401} = 0.00075 < 0.1$$

FIGURE 4–6
Small bodies with high thermal conductivities and low convection coefficients are most likely to satisfy the criterion for lumped system analysis.

Some Remarks on Heat Transfer in Lumped Systems

To understand the heat transfer mechanism during the heating or cooling of a solid by the fluid surrounding it, and the criterion for lumped system analysis, consider this analogy (Fig. 4–7). People from the mainland are to go *by boat* to an island whose entire shore is a harbor, and from the harbor to their destinations on the island *by bus.* The overcrowding of people at the harbor depends on the boat traffic to the island and the ground transportation system on the island. If there is an excellent ground transportation system with plenty of buses, there will be no overcrowding at the harbor, especially when the boat traffic is light. But when the opposite is true, there will be a huge overcrowding at the harbor, creating a large difference between the populations at the harbor and inland. The chance of overcrowding is much lower in a small island with plenty of fast buses.

In heat transfer, a poor ground transportation system corresponds to poor heat conduction in a body, and overcrowding at the harbor to the accumulation of thermal energy and the subsequent rise in temperature near the surface of the body relative to its inner parts. Lumped system analysis is obviously not applicable when there is overcrowding at the surface. Of course, we have disregarded radiation in this analogy and thus the air traffic to the island. Like passengers at the harbor, heat changes *vehicles* at the surface from *convection* to *conduction.* Noting that a surface has zero thickness and thus cannot store any energy, heat reaching the surface of a body by convection must continue its journey within the body by conduction.

Consider heat transfer from a hot body to its cooler surroundings. Heat is transferred from the body to the surrounding fluid as a result of a temperature difference. But this energy comes from the region near the surface, and thus the temperature of the body near the surface will drop. This creates a *temperature gradient* between the inner and outer regions of the body and initiates heat transfer by conduction from the interior of the body toward the outer surface.

When the convection heat transfer coefficient h and thus the rate of convection from the body are high, the temperature of the body near the surface drops quickly (Fig. 4–8). This creates a larger temperature difference between the inner and outer regions unless the body is able to transfer heat from the inner to the outer regions just as fast. Thus, the magnitude of the maximum temperature difference within the body depends strongly on the ability of a body to conduct heat toward its surface relative to the ability of the surrounding

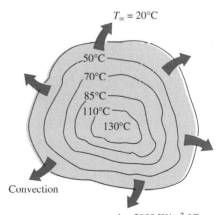

FIGURE 4–7
Analogy between heat transfer to a solid and passenger traffic to an island

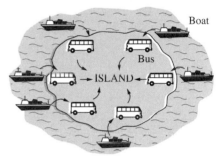

FIGURE 4–8
When the convection coefficient h is high and k is low, large temperature differences occur between the inner and outer regions of a large solid.

medium to convect heat away from the surface. The Biot number is a measure of the relative magnitudes of these two competing effects.

Recall that heat conduction in a specified direction n per unit surface area is expressed as $\dot{q} = -k \, \partial T / \partial n$, where $\partial T / \partial n$ is the temperature gradient and k is the thermal conductivity of the solid. Thus, the temperature distribution in the body will be *uniform* only when its thermal conductivity is *infinite,* and no such material is known to exist. Therefore, temperature gradients and thus temperature differences must exist within the body, no matter how small, in order for heat conduction to take place. Of course, the temperature gradient and the thermal conductivity are inversely proportional for a given heat flux. Therefore, the larger the thermal conductivity, the smaller the temperature gradient.

Thermocouple wire

Gas
T_∞, h → Junction
$D = 1$ mm
$T(t)$

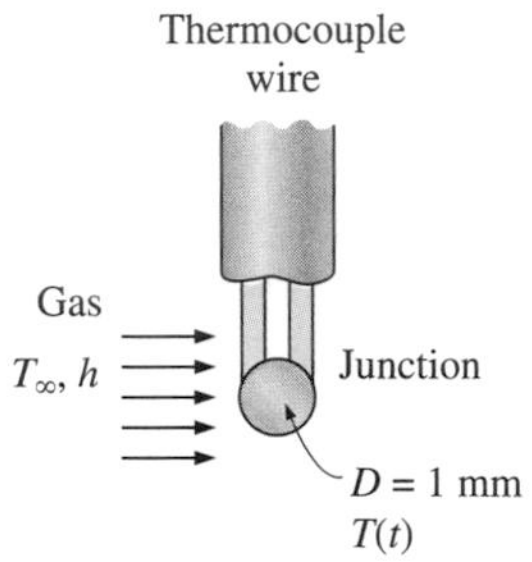

FIGURE 4–9
Schematic for Example 4–1.

EXAMPLE 4–1 Temperature Measurement by Thermocouples

The temperature of a gas stream is to be measured by a thermocouple whose junction can be approximated as a 1-mm-diameter sphere, as shown in Fig. 4–9. The properties of the junction are $k = 35$ W/m · °C, $\rho = 8500$ kg/m^3, and $c_p = 320$ J/kg · °C, and the convection heat transfer coefficient between the junction and the gas is $h = 210$ W/m^2 · °C. Determine how long it will take for the thermocouple to read 99 percent of the initial temperature difference.

SOLUTION The temperature of a gas stream is to be measured by a thermocouple. The time it takes to register 99 percent of the initial ΔT is to be determined.

Assumptions **1** The junction is spherical in shape with a diameter of $D = 0.001$ m. **2** The thermal properties of the junction and the heat transfer coefficient are constant. **3** Radiation effects are negligible.

Properties The properties of the junction are given in the problem statement.

Analysis The characteristic length of the junction is

$$L_c = \frac{V}{A_s} = \frac{\frac{1}{6}\pi D^3}{\pi D^2} = \frac{1}{6} D = \frac{1}{6}(0.001 \text{ m}) = 1.67 \times 10^{-4} \text{ m}$$

Then the Biot number becomes

$$\text{Bi} = \frac{hL_c}{k} = \frac{(210 \text{ W/m}^2 \cdot °\text{C})(1.67 \times 10^{-4} \text{ m})}{35 \text{ W/m} \cdot °\text{C}} = 0.001 < 0.1$$

Therefore, lumped system analysis is applicable, and the error involved in this approximation is negligible.

In order to read 99 percent of the initial temperature difference $T_i - T_\infty$ between the junction and the gas, we must have

$$\frac{T(t) - T_\infty}{T_i - T_\infty} = 0.01$$

For example, when $T_i = 0°$C and $T_\infty = 100°$C, a thermocouple is considered to have read 99 percent of this applied temperature difference when its reading indicates $T(t) = 99°$C.

The value of the exponent b is

$$b = \frac{hA_s}{\rho c_p V} = \frac{h}{\rho c_p L_c} = \frac{210 \text{ W/m}^2 \cdot \text{°C}}{(8500 \text{ kg/m}^3)(320 \text{ J/kg} \cdot \text{°C})(1.67 \times 10^{-4} \text{ m})} = 0.462 \text{ s}^{-1}$$

We now substitute these values into Eq. 4–4 and obtain

$$\frac{T(t) - T_\infty}{T_i - T_\infty} = e^{-bt} \longrightarrow 0.01 = e^{-(0.462 \text{ s}^{-1})t}$$

which yields

$$t = 10 \text{ s}$$

Therefore, we must wait at least 10 s for the temperature of the thermocouple junction to approach within 99 percent of the initial junction-gas temperature difference.

Discussion Note that conduction through the wires and radiation exchange with the surrounding surfaces affect the result, and should be considered in a more refined analysis.

EXAMPLE 4–2   Predicting the Time of Death

A person is found dead at 5 PM in a room whose temperature is 20°C. The temperature of the body is measured to be 25°C when found, and the heat transfer coefficient is estimated to be $h = 8$ W/m² · °C. Modeling the body as a 30-cm-diameter, 1.70-m-long cylinder, estimate the time of death of that person (Fig. 4–10).

SOLUTION A body is found while still warm. The time of death is to be estimated.

Assumptions 1 The body can be modeled as a 30-cm-diameter, 1.70-m-long cylinder. 2 The thermal properties of the body and the heat transfer coefficient are constant. 3 The radiation effects are negligible. 4 The person was healthy(!) when he or she died with a body temperature of 37°C.

Properties The average human body is 72 percent water by mass, and thus we can assume the body to have the properties of water at the average temperature of $(37 + 25)/2 = 31$°C; $k = 0.617$ W/m · °C, $\rho = 996$ kg/m³, and $c_p = 4178$ J/kg · °C (Table A–9).

Analysis The characteristic length of the body is

$$L_c = \frac{V}{A_s} = \frac{\pi r_o^2 L}{2\pi r_o L + 2\pi r_o^2} = \frac{\pi (0.15 \text{ m})^2 (1.7 \text{ m})}{2\pi (0.15 \text{ m})(1.7 \text{ m}) + 2\pi (0.15 \text{ m})^2} = 0.0689 \text{ m}$$

Then the Biot number becomes

$$\text{Bi} = \frac{hL_c}{k} = \frac{(8 \text{ W/m}^2 \cdot \text{°C})(0.0689 \text{ m})}{0.617 \text{ W/m} \cdot \text{°C}} = 0.89 > 0.1$$

FIGURE 4–10
Schematic for Example 4–2.

Therefore, lumped system analysis is *not* applicable. However, we can still use it to get a "rough" estimate of the time of death. The exponent b in this case is

$$b = \frac{hA_s}{\rho c_p V} = \frac{h}{\rho c_p L_c} = \frac{8 \text{ W/m}^2 \cdot {}^{\circ}\text{C}}{(996 \text{ kg/m}^3)(4178 \text{ J/kg} \cdot {}^{\circ}\text{C})(0.0689 \text{ m})}$$
$$= 2.79 \times 10^{-5} \text{ s}^{-1}$$

We now substitute these values into Eq. 4–4,

$$\frac{T(t) - T_{\infty}}{T_i - T_{\infty}} = e^{-bt} \quad \longrightarrow \quad \frac{25 - 20}{37 - 20} = e^{-(2.79 \times 10^{-5} \text{ s}^{-1})t}$$

which yields

$$t = 43,860 \text{ s} = \mathbf{12.2 \ h}$$

Therefore, as a rough estimate, the person died about 12 h before the body was found, and thus the time of death is 5 AM.

Discussion   This example demonstrates how to obtain "ball park" values using a simple analysis. A similar analysis is used in practice by incorporating constants to account for deviation from lumped system analysis.

4–2 ▪ TRANSIENT HEAT CONDUCTION IN LARGE PLANE WALLS, LONG CYLINDERS, AND SPHERES WITH SPATIAL EFFECTS

In Section 4–1, we considered bodies in which the variation of temperature within the body is negligible; that is, bodies that remain nearly *isothermal* during a process. Relatively small bodies of highly conductive materials approximate this behavior. In general, however, the temperature within a body changes from point to point as well as with time. In this section, we consider the variation of temperature with *time* and *position* in one-dimensional problems such as those associated with a large plane wall, a long cylinder, and a sphere.

Consider a plane wall of thickness $2L$, a long cylinder of radius r_o, and a sphere of radius r_o initially at a *uniform temperature T_i*, as shown in Fig. 4–11. At time $t = 0$, each geometry is placed in a large medium that is at a constant temperature T_{∞} and kept in that medium for $t > 0$. Heat transfer takes place between these bodies and their environments by convection with a *uniform* and *constant* heat transfer coefficient h. Note that all three cases possess geometric and thermal symmetry: the plane wall is symmetric about its *center plane* ($x = 0$), the cylinder is symmetric about its *centerline* ($r = 0$), and the sphere is symmetric about its *center point* ($r = 0$). We neglect *radiation* heat transfer between these bodies and their surrounding surfaces, or incorporate the radiation effect into the convection heat transfer coefficient h.

The variation of the temperature profile with *time* in the plane wall is illustrated in Fig. 4–12. When the wall is first exposed to the surrounding medium at $T_{\infty} < T_i$ at $t = 0$, the entire wall is at its initial temperature T_i. But the wall temperature at and near the surfaces starts to drop as a result of heat transfer from the wall to the surrounding medium. This creates a *temperature*

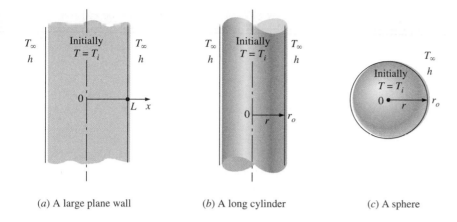

(a) A large plane wall   (b) A long cylinder   (c) A sphere

FIGURE 4–11
Schematic of the simple geometries in which heat transfer is one-dimensional.

gradient in the wall and initiates heat conduction from the inner parts of the wall toward its outer surfaces. Note that the temperature at the center of the wall remains at T_i until $t = t_2$, and that the temperature profile within the wall remains symmetric at all times about the center plane. The temperature profile gets flatter and flatter as time passes as a result of heat transfer, and eventually becomes uniform at $T = T_\infty$. That is, the wall reaches *thermal equilibrium* with its surroundings. At that point, heat transfer stops since there is no longer a temperature difference. Similar discussions can be given for the long cylinder or sphere.

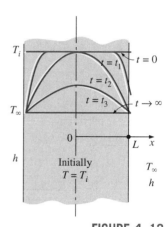

FIGURE 4–12
Transient temperature profiles in a plane wall exposed to convection from its surfaces for $T_i > T_\infty$.

Nondimensionalized One-Dimensional Transient Conduction Problem

The formulation of heat conduction problems for the determination of the one-dimensional transient temperature distribution in a plane wall, a cylinder, or a sphere results in a partial differential equation whose solution typically involves infinite series and transcendental equations, which are inconvenient to use. But the analytical solution provides valuable insight to the physical problem, and thus it is important to go through the steps involved. Below we demonstrate the solution procedure for the case of plane wall.

Consider a plane wall of thickness $2L$ initially at a uniform temperature of T_i, as shown in Fig. 4–11a. At time $t = 0$, the wall is immersed in a fluid at temperature T_∞ and is subjected to convection heat transfer from both sides with a convection coefficient of h. The height and the width of the wall are large relative to its thickness, and thus heat conduction in the wall can be approximated to be one-dimensional. Also, there is thermal symmetry about the midplane passing through $x = 0$, and thus the temperature distribution must be symmetrical about the midplane. Therefore, the value of temperature at any $-x$ value in $-L \leq x \leq 0$ at any time t must be equal to the value at $+x$ in $0 \leq x \leq L$ at the same time. This means we can formulate and solve the heat conduction problem in the positive half domain $0 \leq x \leq L$, and then apply the solution to the other half.

Under the conditions of constant thermophysical properties, no heat generation, thermal symmetry about the midplane, uniform initial temperature, and constant convection coefficient, the one-dimensional transient heat conduction problem in the half-domain $0 \le x \le L$ of the plain wall can be expressed as (see Chapter 2)

Differential equation:
$$\frac{\partial^2 T}{\partial x^2} = \frac{1}{\alpha} \frac{\partial T}{\partial t} \tag{4–10a}$$

Boundary conditions:
$$\frac{\partial T(0, t)}{\partial x} = 0 \quad \text{and} \quad -k \frac{\partial T(L, t)}{\partial x} = h[T(L, t) - T_\infty] \tag{4–10b}$$

Initial condition:
$$T(x, 0) = T_i \tag{4–10c}$$

where the property $\alpha = k/\rho c_p$ is the thermal diffusivity of the material.

We now attempt to nondimensionalize the problem by defining a dimensionless space variable $X = x/L$ and dimensionless temperature $\theta(x, t) = [T(x, t) - T_\infty]/[T_i - T_\infty]$. These are convenient choices since both X and θ vary between 0 and 1. However, there is no clear guidance for the proper form of the dimensionless time variable and the h/k ratio, so we will let the analysis indicate them. We note that

$$\frac{\partial \theta}{\partial X} = \frac{\partial \theta}{\partial(x/L)} = \frac{L}{T_i - T_\infty} \frac{\partial T}{\partial x}, \quad \frac{\partial^2 \theta}{\partial X^2} = \frac{L^2}{T_i - T_\infty} \frac{\partial T}{\partial x} \quad \text{and} \quad \frac{\partial \theta}{\partial t} = \frac{1}{T_i - T_\infty} \frac{\partial T}{\partial t}$$

Substituting into Eqs. 4–10a and 4–10b and rearranging give

$$\frac{\partial^2 \theta}{\partial X^2} = \frac{L^2}{\alpha} \frac{\partial \theta}{\partial t} \quad \text{and} \quad \frac{\partial \theta(1, t)}{\partial X} = \frac{hL}{k} \theta(1, t) \tag{4–11}$$

Therefore, the proper form of the dimensionless time is $\tau = \alpha t/L^2$, which is called the **Fourier number** Fo, and we recognize Bi $= k/hL$ as the Biot number defined in Section 4–1. Then the formulation of the one-dimensional transient heat conduction problem in a plane wall can be expressed in nondimensional form as

Dimensionless differential equation:
$$\frac{\partial^2 \theta}{\partial X^2} = \frac{\partial \theta}{\partial \tau} \tag{4–12a}$$

Dimensionless BC's:
$$\frac{\partial \theta(0, \tau)}{\partial X} = 0 \quad \text{and} \quad \frac{\partial \theta(1, \tau)}{\partial X} = -\text{Bi}\theta(1, \tau) \tag{4–12b}$$

Dimensionless initial condition:
$$\theta(X, 0) = 1 \tag{4–12c}$$

where

$$\theta(X, \tau) = \frac{T(x, t) - T_i}{T_\infty - T_i} \qquad \text{Dimensionless temperature}$$

$$X = \frac{x}{L} \qquad \text{Dimensionless distance from the center}$$

$$\text{Bi} = \frac{hL}{k} \qquad \text{Dimensionless heat transfer coefficient (Biot number)}$$

$$\tau = \frac{\alpha t}{L^2} = \text{Fo} \qquad \text{Dimensionless time (Fourier number)}$$

The heat conduction equation in cylindrical or spherical coordinates can be nondimensionalized in a similar way. Note that nondimensionalization

(a) Original heat conduction problem:

$$\frac{\partial^2 T}{\partial x^2} = \frac{1}{\alpha} \frac{\partial T}{\partial t}, \quad T(x, 0) = T_i$$

$$\frac{\partial T(0, t)}{\partial x} = 0, \quad -k \frac{\partial T(L, t)}{\partial x} = h[T(L, t) - T_\infty]$$

$$T = F(x, L, t, k, \alpha, h, T_i)$$

(b) Nondimensionalized problem:

$$\frac{\partial^2 \theta}{\partial X^2} = \frac{\partial \theta}{\partial \tau}, \quad \theta(X, 0) = 1$$

$$\frac{\partial \theta(0, \tau)}{\partial X} = 0, \quad \frac{\partial \theta(1, \tau)}{\partial X} = -\text{Bi}\theta(1, \tau)$$

$$\theta = f(X, \text{Bi}, \tau)$$

FIGURE 4–13
Nondimensionalization reduces the number of independent variables in one-dimensional transient conduction problems from 8 to 3, offering great convenience in the presentation of results.

reduces the number of independent variables and parameters from 8 to 3—from x, L, t, k, α, h, T_i, and T_∞ to X, Bi, and Fo (Fig. 4–13). That is,

$$\theta = f(X, \text{Bi}, \text{Fo}) \tag{4–13}$$

This makes it very practical to conduct parametric studies and to present results in graphical form. Recall that in the case of lumped system analysis, we had $\theta = f(\text{Bi}, \text{Fo})$ with no space variable.

Exact Solution of One-Dimensional Transient Conduction Problem*

The non-dimensionalized partial differential equation given in Eqs. 4–12 together with its boundary and initial conditions can be solved using several analytical and numerical techniques, including the Laplace or other transform methods, the method of separation of variables, the finite difference method, and the finite-element method. Here we use the method of **separation of variables** developed by J. Fourier in 1820s and is based on expanding an arbitrary function (including a constant) in terms of Fourier series. The method is applied by assuming the dependent variable to be a product of a number of functions, each being a function of a single independent variable. This reduces the partial differential equation to a system of ordinary differential equations, each being a function of a single independent variable. In the case of transient conduction in a plain wall, for example, the dependent variable is the solution function $\theta(X, \tau)$, which is expressed as $\theta(X, \tau) = F(X)G(\tau)$, and the application of the method results in two ordinary differential equation, one in X and the other in τ.

The method is applicable if (1) the geometry is simple and finite (such as a rectangular block, a cylinder, or a sphere) so that the boundary surfaces can be described by simple mathematical functions, and (2) the differential equation and the boundary and initial conditions in their most simplified form are linear (no terms that involve products of the dependent variable or its derivatives) and involve only one nonhomogeneous term (a term without the dependent variable or its derivatives). If the formulation involves a number of nonhomogeneous terms, the problem can be split up into an equal number of simpler problems each involving only one nonhomogeneous term, and then combining the solutions by superposition.

Now we demonstrate the use of the method of separation of variables by applying it to the one-dimensional transient heat conduction problem given in Eqs. 4–12. First, we express the dimensionless temperature function $\theta(X, \tau)$ as a product of a function of X only and a function of τ only as

$$\theta(X, \tau) = F(X)G(\tau) \tag{4–14}$$

Substituting Eq. 4–14 into Eq. 4–12a and dividing by the product FG gives

$$\frac{1}{F}\frac{d^2F}{dX^2} = \frac{1}{G}\frac{dG}{d\tau} \tag{4–15}$$

Observe that all the terms that depend on X are on the left-hand side of the equation and all the terms that depend on τ are on the right-hand side. That is,

---

*This section can be skipped if desired without a loss of continuity.

the terms that are function of different variables are *separated* (and thus the name *separation of variables*). The left-hand side of this equation is a function of X only and the right-hand side is a function of only τ. Considering that both X and τ can be varied independently, the equality in Eq. 4–15 can hold for any value of X and τ only if Eq. 4–15 is equal to a constant. Further, it must be a *negative* constant that we will indicate by $-\lambda^2$ since a positive constant will cause the function $G(\tau)$ to increase indefinitely with time (to be infinite), which is unphysical, and a value of zero for the constant means no time dependence, which is again inconsistent with the physical problem. Setting Eq. 4–15 equal to $-\lambda^2$ gives

$$\frac{d^2F}{dX^2} + \lambda^2 F = 0 \quad \text{and} \quad \frac{dG}{d\tau} + \lambda^2 G = 0 \tag{4–16}$$

whose general solutions are

$$F = C_1 \cos(\lambda X) + C_2 \sin(\lambda X) \quad \text{and} \quad G = C_3 e^{-\lambda^2 \tau} \tag{4–17}$$

and

$$\theta = FG = C_3 e^{-\lambda^2 \tau}[C_1 \cos(\lambda X) + C_2 \sin(\lambda X)] = e^{-\lambda^2 \tau}[A \cos(\lambda X) + B \sin(\lambda X)] \tag{4–18}$$

where $A = C_1 C_3$ and $B = C_2 C_3$ are arbitrary constants. Note that we need to determine only A and B to obtain the solution of the problem.

Applying the boundary conditions in Eq. 4–12b gives

$$\frac{\partial \theta(0, \tau)}{\partial X} = 0 \rightarrow -e^{-\lambda^2 \tau}(A\lambda \sin 0 + B\lambda \cos 0) = 0 \ \rightarrow \ B = 0 \ \rightarrow \ \theta = Ae^{-\lambda^2 \tau} \cos(\lambda X)$$

$$\frac{\partial \theta(1, \tau)}{\partial X} = -\text{Bi}\theta(1,\tau) \ \rightarrow -Ae^{-\lambda^2 \tau}\lambda \sin \lambda = -\text{Bi}Ae^{-\lambda^2 \tau} \cos \lambda \ \rightarrow \ \lambda \tan \lambda = \text{Bi}$$

But tangent is a periodic function with a period of π, and the equation $\lambda \tan \lambda = \text{Bi}$ has the root λ_1 between 0 and π, the root λ_2 between π and 2π, the root λ_n between $(n-1)\pi$ and $n\pi$, etc. To recognize that the transcendental equation $\lambda \tan \lambda = \text{Bi}$ has an infinite number of roots, it is expressed as

$$\lambda_n \tan \lambda_n = \text{Bi} \tag{4–19}$$

Eq. 4–19 is called the **characteristic equation** or **eigenfunction**, and its roots are called the **characteristic values** or **eigenvalues**. The characteristic equation is implicit in this case, and thus the characteristic values need to be determined numerically. Then it follows that there are an infinite number of solutions of the form $Ae^{-\lambda^2 \tau} \cos(\lambda X)$, and the solution of this linear heat conduction problem is a linear combination of them,

$$\theta = \sum_{n=1}^{\infty} A_n e^{-\lambda_n^2 \tau} \cos(\lambda_n X) \tag{4–20}$$

The constants A_n are determined from the initial condition, Eq. 4–12c,

$$\theta(X, 0) = 1 \ \rightarrow \ 1 = \sum_{n=1}^{\infty} A_n \cos(\lambda_n X) \tag{4–21}$$

This is a Fourier series expansion that expresses a constant in terms of an infinite series of cosine functions. Now we multiply both sides of Eq. 4–21 by $\cos(\lambda_m X)$, and integrate from $X = 0$ to $X = 1$. The right-hand side involves an infinite number of integrals of the form $\int_0^1 \cos(\lambda_m X) \cos(\lambda_n X) dx$. It can be shown that all of these integrals vanish except when $n = m$, and the coefficient A_n becomes

$$\int_0^1 \cos(\lambda_n X) dX = A_n \int_0^1 \cos^2(\lambda_n X) dx \quad \rightarrow \quad A_n = \frac{4 \sin \lambda_n}{2\lambda_n + \sin(2\lambda_n)} \qquad \textbf{(4–22)}$$

This completes the analysis for the solution of one-dimensional transient heat conduction problem in a plane wall. Solutions in other geometries such as a long cylinder and a sphere can be determined using the same approach. The results for all three geometries are summarized in Table 4–1. The solution for the plane wall is also applicable for a plane wall of thickness L whose left surface at $x = 0$ is insulated and the right surface at $x = L$ is subjected to convection since this is precisely the mathematical problem we solved.

The analytical solutions of transient conduction problems typically involve infinite series, and thus the evaluation of an infinite number of terms to determine the temperature at a specified location and time. This may look intimidating at first, but there is no need to worry. As demonstrated in Fig. 4–14, the terms in the summation decline rapidly as n and thus λ_n increases because of the exponential decay function $e^{-\lambda_n^2 \tau}$. This is especially the case when the dimensionless time τ is large. Therefore, the evaluation of the first few terms of the infinite series (in this case just the first term) is usually adequate to determine the dimensionless temperature θ.

$$\theta_n = A_n e^{-\lambda_n^2 \tau} \cos(\lambda_n X)$$

$$A_n = \frac{4 \sin \lambda_n}{2\lambda_n + \sin(2\lambda_n)}$$

$$\lambda_n \tan \lambda_n = \text{Bi}$$

For Bi = 5, X = 1, and t = 0.2:

n	λ_n	A_n	θ_n
1	1.3138	1.2402	0.22321
2	4.0336	−0.3442	0.00835
3	6.9096	0.1588	0.00001
4	9.8928	−0.876	0.00000

FIGURE 4–14

The term in the series solution of transient conduction problems decline rapidly as n and thus λ_n increases because of the exponential decay function with the exponent $-\lambda_n \tau$.

Approximate Analytical and Graphical Solutions

The analytical solution obtained above for one-dimensional transient heat conduction in a plane wall involves infinite series and implicit equations, which are difficult to evaluate. Therefore, there is clear motivation to simplify

TABLE 4–1

Summary of the solutions for one-dimensional transient conduction in a plane wall of thickness 2L, a cylinder of radius r_o and a sphere of radius r_o subjected to convention from all surfaces.*

Geometry	Solution	λ_n's are the roots of
Plane wall	$\theta = \sum_{n=1}^{\infty} \dfrac{4 \sin \lambda_n}{2\lambda_n + \sin(2\lambda_n)} e^{-\lambda_n^2 \tau} \cos(\lambda_n x/L)$	$\lambda_n \tan \lambda_n = \text{Bi}$
Cylinder	$\theta = \sum_{n=1}^{\infty} \dfrac{2}{\lambda_n} \dfrac{J_1(\lambda_n)}{J_0^2(\lambda_n) + J_1^2(\lambda_n)} e^{-\lambda_n^2 \tau} J_0(\lambda_n r/r_o)$	$\lambda_n \dfrac{J_1(\lambda_n)}{J_0(\lambda_n)} = \text{Bi}$
Sphere	$\theta = \sum_{n=1}^{\infty} \dfrac{4(\sin \lambda_n - \lambda_n \cos \lambda_n)}{2\lambda_n - \sin(2\lambda_n)} e^{-\lambda_n^2 \tau} \dfrac{\sin(\lambda_n x/L)}{\lambda_n x/L}$	$1 - \lambda_n \cot \lambda_n = \text{Bi}$

*Here $\theta = (T - T_i)/(T_\infty - T_i)$ is the dimensionless temperature, Bi = hL/k or hr_o/k is the Biot number, Fo = $\tau = \alpha t/L^2$ or $\alpha \tau/r_o^2$ is the Fourier number, and J_0 and J_1 are the Bessel functions of the first kind whose values are given in Table 4–3.

the analytical solutions and to present the solutions in *tabular* or *graphical* form using simple relations.

The dimensionless quantities defined above for a plane wall can also be used for a *cylinder* or *sphere* by replacing the space variable x by r and the half-thickness L by the outer radius r_o. Note that the characteristic length in the definition of the Biot number is taken to be the *half-thickness* L for the plane wall, and the *radius* r_o for the long cylinder and sphere instead of V/A used in lumped system analysis.

We mentioned earlier that the terms in the series solutions in Table 4–1 converge rapidly with increasing time, and for $\tau > 0.2$, keeping the first term and neglecting all the remaining terms in the series results in an error under 2 percent. We are usually interested in the solution for times with $\tau > 0.2$, and thus it is very convenient to express the solution using this **one-term approximation**, given as

Plane wall: $$\theta_{\text{wall}} = \frac{T(x, t) - T_\infty}{T_i - T_\infty} = A_1 e^{-\lambda_1^2 \tau} \cos (\lambda_1 x/L), \quad \tau > 0.2 \tag{4–23}$$

Cylinder: $$\theta_{\text{cyl}} = \frac{T(r, t) - T_\infty}{T_i - T_\infty} = A_1 e^{-\lambda_1^2 \tau} J_0(\lambda_1 r/r_o), \quad \tau > 0.2 \tag{4–24}$$

Sphere: $$\theta_{\text{sph}} = \frac{T(r, t) - T_\infty}{T_i - T_\infty} = A_1 e^{-\lambda_1^2 \tau} \frac{\sin(\lambda_1 r/r_o)}{\lambda_1 r/r_o}, \quad \tau > 0.2 \tag{4–25}$$

where the constants A_1 and λ_1 are functions of the Bi number only, and their values are listed in Table 4–2 against the Bi number for all three geometries. The function J_0 is the zeroth-order Bessel function of the first kind, whose value can be determined from Table 4–3. Noting that $\cos (0) = J_0(0) = 1$ and the limit of $(\sin x)/x$ is also 1, these relations simplify to the next ones at the center of a plane wall, cylinder, or sphere:

Center of plane wall ($x = 0$): $$\theta_{0,\,\text{wall}} = \frac{T_0 - T_\infty}{T_i - T_\infty} = A_1 e^{-\lambda_1^2 \tau} \tag{4–26}$$

Center of cylinder ($r = 0$): $$\theta_{0,\,\text{cyl}} = \frac{T_0 - T_\infty}{T_i - T_\infty} = A_1 e^{-\lambda_1^2 \tau} \tag{4–27}$$

Center of sphere ($r = 0$): $$\theta_{0,\,\text{sph}} = \frac{T_0 - T_\infty}{T_i - T_\infty} = A_1 e^{-\lambda_1^2 \tau} \tag{4–28}$$

Comparing the two sets of equations above, we notice that the dimensionless temperatures anywhere in a plane wall, cylinder, and sphere are related to the center temperature by

$$\frac{\theta_{\text{wall}}}{\theta_{0,\,\text{wall}}} = \cos\left(\frac{\lambda_1 x}{L}\right), \quad \frac{\theta_{\text{cyl}}}{\theta_{0,\,\text{cyl}}} = J_0\left(\frac{\lambda_1 r}{r_o}\right), \quad \text{and} \quad \frac{\theta_{\text{sph}}}{\theta_{0,\,\text{sph}}} = \frac{\sin(\lambda_1 r/r_o)}{\lambda_1 r/r_o} \tag{4-29}$$

which shows that time dependence of dimensionless temperature within a given geometry is the same throughout. That is, if the dimensionless center temperature θ_0 drops by 20 percent at a specified time, so does the dimensionless temperature θ_0 anywhere else in the medium at the same time.

Once the Bi number is known, these relations can be used to determine the temperature anywhere in the medium. The determination of the constants A_1

TABLE 4–2

Coefficients used in the one-term approximate solution of transient one-dimensional heat conduction in plane walls, cylinders, and spheres (Bi $= hL/k$ for a plane wall of thickness $2L$, and Bi $= hr_o/k$ for a cylinder or sphere of radius r_o)

	Plane Wall		Cylinder		Sphere	
Bi	λ_1	A_1	λ_1	A_1	λ_1	A_1
0.01	0.0998	1.0017	0.1412	1.0025	0.1730	1.0030
0.02	0.1410	1.0033	0.1995	1.0050	0.2445	1.0060
0.04	0.1987	1.0066	0.2814	1.0099	0.3450	1.0120
0.06	0.2425	1.0098	0.3438	1.0148	0.4217	1.0179
0.08	0.2791	1.0130	0.3960	1.0197	0.4860	1.0239
0.1	0.3111	1.0161	0.4417	1.0246	0.5423	1.0298
0.2	0.4328	1.0311	0.6170	1.0483	0.7593	1.0592
0.3	0.5218	1.0450	0.7465	1.0712	0.9208	1.0880
0.4	0.5932	1.0580	0.8516	1.0931	1.0528	1.1164
0.5	0.6533	1.0701	0.9408	1.1143	1.1656	1.1441
0.6	0.7051	1.0814	1.0184	1.1345	1.2644	1.1713
0.7	0.7506	1.0918	1.0873	1.1539	1.3525	1.1978
0.8	0.7910	1.1016	1.1490	1.1724	1.4320	1.2236
0.9	0.8274	1.1107	1.2048	1.1902	1.5044	1.2488
1.0	0.8603	1.1191	1.2558	1.2071	1.5708	1.2732
2.0	1.0769	1.1785	1.5995	1.3384	2.0288	1.4793
3.0	1.1925	1.2102	1.7887	1.4191	2.2889	1.6227
4.0	1.2646	1.2287	1.9081	1.4698	2.4556	1.7202
5.0	1.3138	1.2403	1.9898	1.5029	2.5704	1.7870
6.0	1.3496	1.2479	2.0490	1.5253	2.6537	1.8338
7.0	1.3766	1.2532	2.0937	1.5411	2.7165	1.8673
8.0	1.3978	1.2570	2.1286	1.5526	2.7654	1.8920
9.0	1.4149	1.2598	2.1566	1.5611	2.8044	1.9106
10.0	1.4289	1.2620	2.1795	1.5677	2.8363	1.9249
20.0	1.4961	1.2699	2.2880	1.5919	2.9857	1.9781
30.0	1.5202	1.2717	2.3261	1.5973	3.0372	1.9898
40.0	1.5325	1.2723	2.3455	1.5993	3.0632	1.9942
50.0	1.5400	1.2727	2.3572	1.6002	3.0788	1.9962
100.0	1.5552	1.2731	2.3809	1.6015	3.1102	1.9990
∞	1.5708	1.2732	2.4048	1.6021	3.1416	2.0000

TABLE 4–3

The zeroth- and first-order Bessel functions of the first kind

η	$J_0(\eta)$	$J_1(\eta)$
0.0	1.0000	0.0000
0.1	0.9975	0.0499
0.2	0.9900	0.0995
0.3	0.9776	0.1483
0.4	0.9604	0.1960
0.5	0.9385	0.2423
0.6	0.9120	0.2867
0.7	0.8812	0.3290
0.8	0.8463	0.3688
0.9	0.8075	0.4059
1.0	0.7652	0.4400
1.1	0.7196	0.4709
1.2	0.6711	0.4983
1.3	0.6201	0.5220
1.4	0.5669	0.5419
1.5	0.5118	0.5579
1.6	0.4554	0.5699
1.7	0.3980	0.5778
1.8	0.3400	0.5815
1.9	0.2818	0.5812
2.0	0.2239	0.5767
2.1	0.1666	0.5683
2.2	0.1104	0.5560
2.3	0.0555	0.5399
2.4	0.0025	0.5202
2.6	−0.0968	−0.4708
2.8	−0.1850	−0.4097
3.0	−0.2601	−0.3391
3.2	−0.3202	−0.2613

and λ_1 usually requires interpolation. For those who prefer reading charts to interpolating, these relations are plotted and the one-term approximation solutions are presented in graphical form, known as the *transient temperature charts*. Note that the charts are sometimes difficult to read, and they are subject to reading errors. Therefore, the relations above should be preferred to the charts.

The transient temperature charts in Figs. 4–15, 4–16, and 4–17 for a large plane wall, long cylinder, and sphere were presented by M. P. Heisler in 1947 and are called **Heisler charts**. They were supplemented in 1961 with transient heat transfer charts by H. Gröber. There are *three* charts associated with each geometry: the first chart is to determine the temperature T_0 at the *center* of the

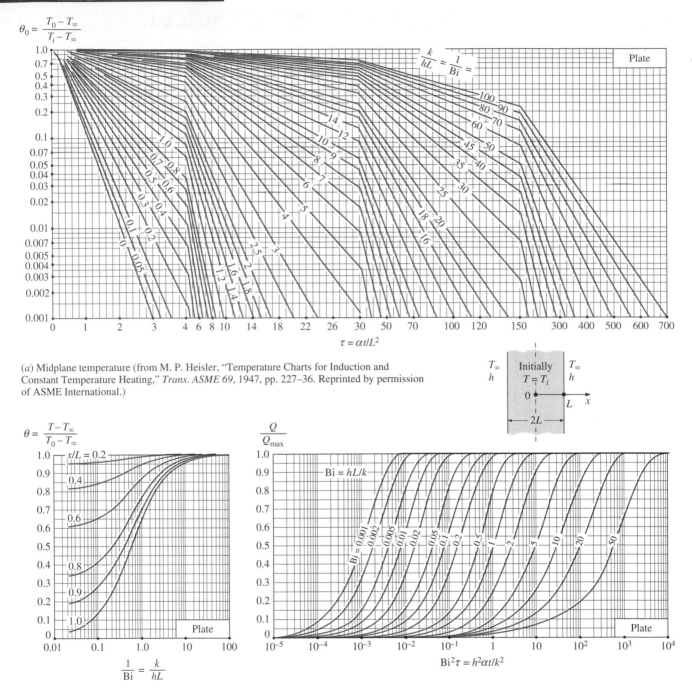

$$\theta_0 = \frac{T_0 - T_\infty}{T_i - T_\infty}$$

$$\tau = \alpha t / L^2$$

$$\frac{k}{hL} = \frac{1}{Bi} =$$

Plate

(a) Midplane temperature (from M. P. Heisler, "Temperature Charts for Induction and Constant Temperature Heating," *Trans. ASME 69*, 1947, pp. 227–36. Reprinted by permission of ASME International.)

$$\theta = \frac{T - T_\infty}{T_0 - T_\infty}$$

$$\frac{1}{Bi} = \frac{k}{hL}$$

Plate

$$\frac{Q}{Q_{max}}$$

Bi = hL/k

Bi² $\tau = h^2 \alpha t / k^2$

Plate

(b) Temperature distribution (from M. P. Heisler, "Temperature Charts for Induction and Constant Temperature Heating," *Trans. ASME 69*, 1947, pp. 227–36. Reprinted by permission of ASME International.)

(c) Heat transfer (from H. Gröber et al.)

FIGURE 4–15

Transient temperature and heat transfer charts for a plane wall of thickness $2L$ initially at a uniform temperature T_i subjected to convection from both sides to an environment at temperature T_∞ with a convection coefficient of h.

$$\theta_0 = \frac{T_0 - T_\infty}{T_i - T_\infty}$$

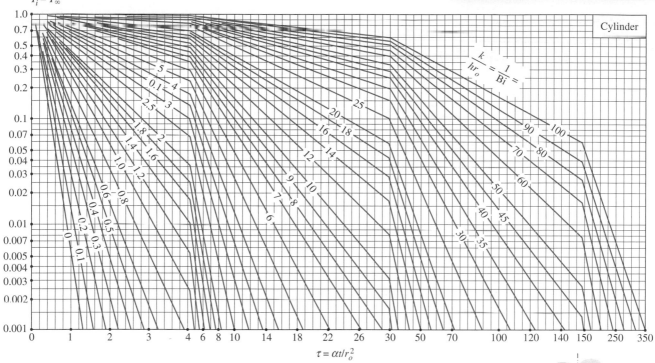

$$\tau = \alpha t / r_o^2$$

(a) Centerline temperature (from M. P. Heisler, "Temperature Charts for Induction and Constant Temperature Heating," *Trans. ASME 69*, 1947, pp. 227–36. Reprinted by permission of ASME International.)

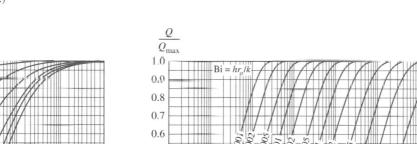

$$\theta = \frac{T - T_\infty}{T_0 - T_\infty}$$

$$\frac{1}{Bi} = \frac{k}{hr_o}$$

(b) Temperature distribution (from M. P. Heisler, "Temperature Charts for Induction and Constant Temperature Heating," *Trans. ASME 69*, 1947, pp. 227–36. Reprinted by permission of ASME International.)

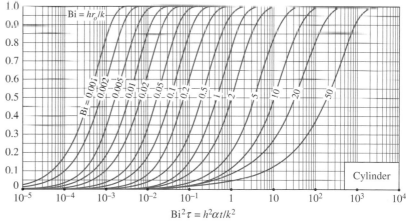

$$\frac{Q}{Q_{max}}$$

$$Bi^2\tau = h^2\alpha t / k^2$$

(c) Heat transfer (from H. Gröber et al.)

FIGURE 4–16

Transient temperature and heat transfer charts for a long cylinder of radius r_o initially at a uniform temperature T_i subjected to convection from all sides to an environment at temperature T_∞ with a convection coefficient of h.

$$\theta_0 = \frac{T_0 - T_\infty}{T_i - T_\infty}$$

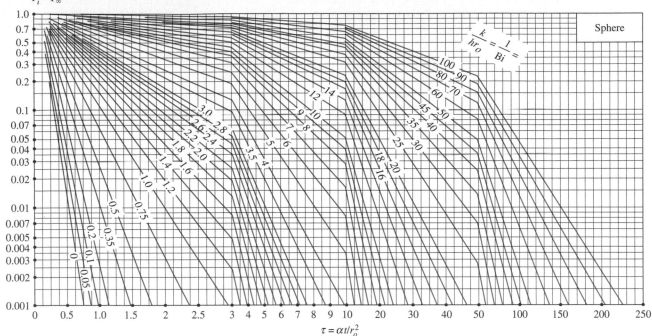

$$\tau = \alpha t / r_o^2$$

(a) Midpoint temperature (from M. P. Heisler, "Temperature Charts for Induction and Constant Temperature Heating," *Trans. ASME 69*, 1947, pp. 227–36. Reprinted by permission of ASME International.)

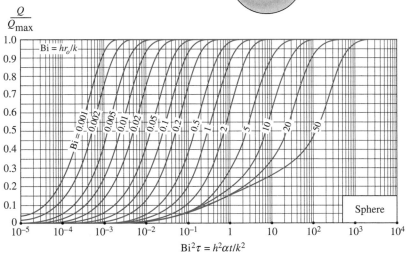

$$\theta = \frac{T - T_\infty}{T_0 - T_\infty}$$

$$\frac{Q}{Q_{max}}$$

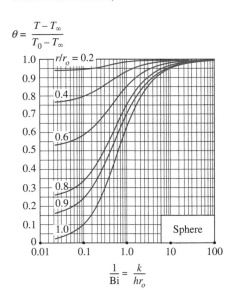

$$\frac{1}{Bi} = \frac{k}{hr_o}$$

$$Bi^2\tau = h^2\alpha t/k^2$$

(b) Temperature distribution (from M. P. Heisler, "Temperature Charts for Induction and Constant Temperature Heating," *Trans. ASME 69*, 1947,

(c) Heat transfer (from H. Gröber et al.)

FIGURE 4–17

Transient temperature and heat transfer charts for a sphere of radius r_o initially at a uniform temperature T_i subjected to convection from all sides to an environment at temperature T_∞ with a convection coefficient of h.

geometry at a given time t. The second chart is to determine the temperature at *other locations* at the same time in terms of T_0. The third chart is to determine the total amount of *heat transfer* up to the time t. These plots are valid for $\tau > 0.2$.

Note that the case $1/\text{Bi} = k/hL = 0$ corresponds to $h \rightarrow \infty$, which corresponds to the case of *specified surface temperature* T_∞. That is, the case in which the surfaces of the body are suddenly brought to the temperature T_∞ at $t = 0$ and kept at T_∞ at all times can be handled by setting h to infinity (Fig. 4–18).

The temperature of the body changes from the initial temperature T_i to the temperature of the surroundings T_∞ at the end of the transient heat conduction process. Thus, the *maximum* amount of heat that a body can gain (or lose if $T_i > T_\infty$) is simply the *change* in the *energy content* of the body. That is,

$$Q_{max} = mc_p(T_\infty - T_i) = \rho V c_p(T_\infty - T_i) \qquad \text{(kJ)} \qquad \text{(4–30)}$$

where m is the mass, V is the volume, ρ is the density, and c_p is the specific heat of the body. Thus, Q_{max} represents the amount of heat transfer for $t \rightarrow \infty$. The amount of heat transfer Q at a finite time t is obviously less than this maximum, and it can be expressed as the sum of the internal energy changes throughout the entire geometry as

$$Q = \int_V \rho c_p[T(x,t) - T_i]dV \qquad \text{(4–31)}$$

where $T(x, t)$ is the temperature distribution in the medium at time t. Assuming constant properties, the ratio of Q/Q_{max} becomes

$$\frac{Q}{Q_{max}} = \frac{\int_V \rho c_p[T(x,t) - T_i]dV}{\rho c_\pi(T_\infty - T_i)V} = \frac{1}{V}\int_V (1 - \theta)dV \qquad \text{(4–32)}$$

Using the appropriate nondimensional temperature relations based on the one-term approximation for the plane wall, cylinder, and sphere, and performing the indicated integrations, we obtain the following relations for the fraction of heat transfer in those geometries:

Plane wall: $$\left(\frac{Q}{Q_{max}}\right)_{wall} = 1 - \theta_{0,\,wall}\frac{\sin \lambda_1}{\lambda_1} \qquad \text{(4–33)}$$

Cylinder: $$\left(\frac{Q}{Q_{max}}\right)_{cyl} = 1 - 2\theta_{0,\,cyl}\frac{J_1(\lambda_1)}{\lambda_1} \qquad \text{(4–34)}$$

Sphere: $$\left(\frac{Q}{Q_{max}}\right)_{sph} = 1 - 3\theta_{0,\,sph}\frac{\sin \lambda_1 - \lambda_1 \cos \lambda_1}{\lambda_1^3} \qquad \text{(4–35)}$$

These Q/Q_{max} ratio relations based on the one-term approximation are also plotted in Figures 4–15c, 4–16c, and 4–17c, against the variables Bi and $h^2\alpha t/k^2$ for the large plane wall, long cylinder, and sphere, respectively. Note that once the *fraction* of heat transfer Q/Q_{max} has been determined from these charts or equations for the given t, the actual amount of heat transfer by that time can be evaluated by multiplying this fraction by Q_{max}. A *negative* sign for Q_{max} indicates that the body is *rejecting* heat (Fig. 4–19).

The use of the Heisler/Gröber charts and the one-term solutions already discussed is limited to the conditions specified at the beginning of this section:

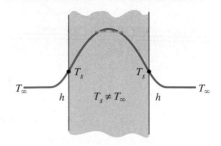

(a) Finite convection coefficient

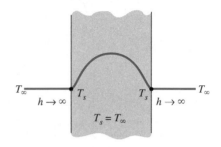

(b) Infinite convection coefficient

FIGURE 4–18
The specified surface temperature corresponds to the case of convection to an environment at T_∞ with a convection coefficient h that is *infinite*.

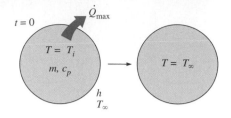

(a) Maximum heat transfer ($t \rightarrow \infty$)

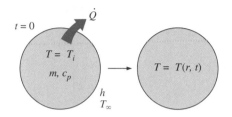

$$\text{Bi} = \cdots$$
$$\left.\frac{h^2\alpha t}{k^2} = \text{Bi}^2\tau = \cdots\right\} \quad \frac{Q}{Q_{\text{max}}} = \cdots$$
(Gröber chart)

(b) Actual heat transfer for time t

FIGURE 4–19
The fraction of total heat transfer Q/Q_{max} up to a specified time t is determined using the Gröber charts.

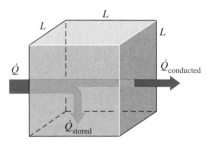

Fourier number: $\tau = \dfrac{\alpha t}{L^2} = \dfrac{\dot{Q}_{\text{conducted}}}{\dot{Q}_{\text{stored}}}$

FIGURE 4–20
Fourier number at time t can be viewed as the ratio of the rate of heat conducted to the rate of heat stored at that time.

the body is initially at a *uniform* temperature, the temperature of the medium surrounding the body and the convection heat transfer coefficient are *constant* and *uniform*, and there is no *heat generation* in the body.

We discussed the physical significance of the *Biot number* earlier and indicated that it is a measure of the relative magnitudes of the two heat transfer mechanisms: *convection* at the surface and *conduction* through the solid. A *small* value of Bi indicates that the inner resistance of the body to heat conduction is *small* relative to the resistance to convection between the surface and the fluid. As a result, the temperature distribution within the solid becomes fairly uniform, and lumped system analysis becomes applicable. Recall that when Bi < 0.1, the error in assuming the temperature within the body to be *uniform* is negligible.

To understand the physical significance of the *Fourier number* τ, we express it as (Fig. 4–20)

$$\tau = \frac{\alpha t}{L^2} = \frac{kL^2\,(1/L)}{\rho c_p L^3/t}\frac{\Delta T}{\Delta T} = \frac{\begin{array}{c}\text{The rate at which heat is } \textit{conducted}\\\text{across } L \text{ of a body of volume } L^3\end{array}}{\begin{array}{c}\text{The rate at which heat is } \textit{stored}\\\text{in a body of volume } L^3\end{array}} \quad \text{(4–36)}$$

Therefore, the Fourier number is a measure of *heat conducted* through a body relative to *heat stored*. Thus, a large value of the Fourier number indicates faster propagation of heat through a body.

Perhaps you are wondering about what constitutes an infinitely large plate or an infinitely long cylinder. After all, nothing in this world is infinite. A plate whose thickness is small relative to the other dimensions can be modeled as an infinitely large plate, except very near the outer edges. But the edge effects on large bodies are usually negligible, and thus a large plane wall such as the wall of a house can be modeled as an infinitely large wall for heat transfer purposes. Similarly, a long cylinder whose diameter is small relative to its length can be analyzed as an infinitely long cylinder. The use of the transient temperature charts and the one-term solutions is illustrated in Examples 4–3, 4–4, and 4–5.

EXAMPLE 4–3   Boiling Eggs

An ordinary egg can be approximated as a 5-cm-diameter sphere (Fig. 4–21). The egg is initially at a uniform temperature of 5°C and is dropped into boiling water at 95°C. Taking the convection heat transfer coefficient to be $h = 1200 \text{ W/m}^2 \cdot °\text{C}$, determine how long it will take for the center of the egg to reach 70°C.

SOLUTION An egg is cooked in boiling water. The cooking time of the egg is to be determined.
Assumptions **1** The egg is spherical in shape with a radius of $r_0 = 2.5$ cm. **2** Heat conduction in the egg is one-dimensional because of thermal symmetry about the midpoint. **3** The thermal properties of the egg and the heat transfer coefficient are constant. **4** The Fourier number is $\tau > 0.2$ so that the one-term approximate solutions are applicable.

Properties The water content of eggs is about 74 percent, and thus the thermal conductivity and diffusivity of eggs can be approximated by those of water at the average temperature of $(5 + 70)/2 = 37.5°C$; $k = 0.627$ W/m · °C and $\alpha = k/\rho c_p = 0.151 \times 10^{-6}$ m²/s (Table A–9).

Analysis The temperature within the egg varies with radial distance as well as time, and the temperature at a specified location at a given time can be determined from the Heisler charts or the one-term solutions. Here we use the latter to demonstrate their use. The Biot number for this problem is

$$Bi = \frac{hr_o}{k} = \frac{(1200 \text{ W/m}^2 \cdot °C)(0.025 \text{ m})}{0.627 \text{ W/m} \cdot °C} = 47.8$$

which is much greater than 0.1, and thus the lumped system analysis is not applicable. The coefficients λ_1 and A_1 for a sphere corresponding to this Bi are, from Table 4–2,

$$\lambda_1 = 3.0754, \qquad A_1 = 1.9958$$

Substituting these and other values into Eq. 4–28 and solving for τ gives

$$\frac{T_0 - T_\infty}{T_i - T_\infty} = A_1 e^{-\lambda_1^2 \tau} \longrightarrow \frac{70 - 95}{5 - 95} = 1.9958 e^{-(3.0754)^2 \tau} \longrightarrow \tau = 0.209$$

which is greater than 0.2, and thus the one-term solution is applicable with an error of less than 2 percent. Then the cooking time is determined from the definition of the Fourier number to be

$$t = \frac{\tau r_o^2}{\alpha} = \frac{(0.209)(0.025 \text{ m})^2}{0.151 \times 10^{-6} \text{ m}^2/\text{s}} = 865 \text{ s} \approx \textbf{14.4 min}$$

Therefore, it will take about 15 min for the center of the egg to be heated from 5°C to 70°C.

Discussion Note that the Biot number in lumped system analysis was defined differently as $Bi = hL_c/k = h(r_o/3)/k$. However, either definition can be used in determining the applicability of the lumped system analysis unless $Bi \approx 0.1$.

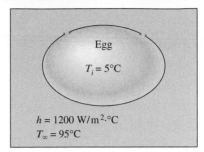

FIGURE 4–21
Schematic for Example 4–3.

EXAMPLE 4–4   Heating of Brass Plates in an Oven

In a production facility, large brass plates of 4-cm thickness that are initially at a uniform temperature of 20°C are heated by passing them through an oven that is maintained at 500°C (Fig. 4–22). The plates remain in the oven for a period of 7 min. Taking the combined convection and radiation heat transfer coefficient to be $h = 120$ W/m² · °C, determine the surface temperature of the plates when they come out of the oven.

SOLUTION Large brass plates are heated in an oven. The surface temperature of the plates leaving the oven is to be determined.

Assumptions 1 Heat conduction in the plate is one-dimensional since the plate is large relative to its thickness and there is thermal symmetry about the center plane. 2 The thermal properties of the plate and the heat transfer coefficient are constant. 3 The Fourier number is $\tau > 0.2$ so that the one-term approximate solutions are applicable.

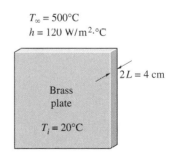

FIGURE 4–22
Schematic for Example 4–4.

Properties The properties of brass at room temperature are $k = 110$ W/m · °C, $\rho = 8530$ kg/m³, $c_p = 380$ J/kg · °C, and $\alpha = 33.9 \times 10^{-6}$ m²/s (Table A–3). More accurate results are obtained by using properties at average temperature.

Analysis The temperature at a specified location at a given time can be determined from the Heisler charts or one-term solutions. Here we use the charts to demonstrate their use. Noting that the half-thickness of the plate is $L = 0.02$ m, from Fig. 4–15 we have

$$\left. \begin{aligned} \frac{1}{\text{Bi}} &= \frac{k}{hL} = \frac{110 \text{ W/m} \cdot {}^{\circ}\text{C}}{(120 \text{ W/m}^2 \cdot {}^{\circ}\text{C})(0.02 \text{ m})} = 45.8 \\ \tau &= \frac{\alpha t}{L^2} = \frac{(33.9 \times 10^{-6} \text{ m}^2/\text{s})(7 \times 60 \text{ s})}{(0.02 \text{ m})^2} - 35.6 \end{aligned} \right\} \frac{T_0 - T_\infty}{T_i - T_\infty} = 0.46$$

Also,

$$\left. \begin{aligned} \frac{1}{\text{Bi}} &= \frac{k}{hL} = 45.8 \\ \frac{x}{L} &= \frac{L}{L} = 1 \end{aligned} \right\} \frac{T - T_\infty}{T_0 - T_\infty} = 0.99$$

Therefore,

$$\frac{T - T_\infty}{T_i - T_\infty} = \frac{T - T_\infty}{T_0 - T_\infty} \frac{T_0 - T_\infty}{T_i - T_\infty} = 0.46 \times 0.99 = 0.455$$

and

$$T = T_\infty + 0.455(T_i - T_\infty) = 500 + 0.455(20 - 500) = \mathbf{282°C}$$

Therefore, the surface temperature of the plates will be 282°C when they leave the oven.

Discussion We notice that the Biot number in this case is Bi = 1/45.8 = 0.022, which is much less than 0.1. Therefore, we expect the lumped system analysis to be applicable. This is also evident from $(T - T_\infty)/(T_0 - T_\infty) = 0.99$, which indicates that the temperatures at the center and the surface of the plate relative to the surrounding temperature are within 1 percent of each other. Noting that the error involved in reading the Heisler charts is typically a few percent, the lumped system analysis in this case may yield just as accurate results with less effort.

The heat transfer surface area of the plate is $2A$, where A is the face area of the plate (the plate transfers heat through both of its surfaces), and the volume of the plate is $V = (2L)A$, where L is the half-thickness of the plate. The exponent b used in the lumped system analysis is

$$b = \frac{hA_s}{\rho c_p V} = \frac{h(2A)}{\rho c_p (2LA)} = \frac{h}{\rho c_p L}$$

$$= \frac{120 \text{ W/m}^2 \cdot {}^{\circ}\text{C}}{(8530 \text{ kg/m}^3)(380 \text{ J/kg} \cdot {}^{\circ}\text{C})(0.02 \text{ m})} = 0.00185 \text{ s}^{-1}$$

Then the temperature of the plate at $t = 7$ min = 420 s is determined from

$$\frac{T(t) - T_\infty}{T_i - T_\infty} = e^{-bt} \longrightarrow \frac{T(t) - 500}{20 - 500} = e^{-(0.00185 \text{ s}^{-1})(420 \text{ s})}$$

It yields

$$T(t) = 279°C$$

which is practically identical to the result obtained above using the Heisler charts. Therefore, we can use lumped system analysis with confidence when the Biot number is sufficiently small.

EXAMPLE 4–5 Cooling of a Long Stainless Steel Cylindrical Shaft

A long 20-cm-diameter cylindrical shaft made of stainless steel 304 comes out of an oven at a uniform temperature of 600°C (Fig. 4–23). The shaft is then allowed to cool slowly in an environment chamber at 200°C with an average heat transfer coefficient of $h = 80$ W/m^2 · °C. Determine the temperature at the center of the shaft 45 min after the start of the cooling process. Also, determine the heat transfer per unit length of the shaft during this time period.

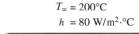
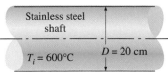

$T_\infty = 200°C$
$h = 80$ W/m^2·°C

Stainless steel shaft

$T_i = 600°C$ $D = 20$ cm

FIGURE 4–23
Schematic for Example 4–5.

SOLUTION A long cylindrical shaft is allowed to cool slowly. The center temperature and the heat transfer per unit length are to be determined.

Assumptions 1 Heat conduction in the shaft is one-dimensional since it is long and it has thermal symmetry about the centerline. 2 The thermal properties of the shaft and the heat transfer coefficient are constant. 3 The Fourier number is $\tau > 0.2$ so that the one-term approximate solutions are applicable.

Properties The properties of stainless steel 304 at room temperature are $k = 14.9$ W/m · °C, $\rho = 7900$ kg/m^3, $c_p = 477$ J/kg · °C, and $\alpha = 3.95 \times 10^{-6}$ m^2/s (Table A–3). More accurate results can be obtained by using properties at average temperature.

Analysis The temperature within the shaft may vary with the radial distance r as well as time, and the temperature at a specified location at a given time can be determined from the Heisler charts. Noting that the radius of the shaft is $r_o = 0.1$ m, from Fig. 4–16 we have

$$\left.\begin{array}{l} \dfrac{1}{\text{Bi}} = \dfrac{k}{hr_o} = \dfrac{14.9 \text{ W/m} \cdot °C}{(80 \text{ W/m}^2 \cdot °C)(0.1 \text{ m})} = 1.86 \\[3mm] \tau = \dfrac{\alpha t}{r_o^2} = \dfrac{(3.95 \times 10^{-6} \text{ m}^2/\text{s})(45 \times 60 \text{ s})}{(0.1 \text{ m})^2} = 1.07 \end{array}\right\} \quad \dfrac{T_0 - T_\infty}{T_i - T_\infty} = 0.40$$

and

$$T_0 = T_\infty + 0.4(T_i - T_\infty) = 200 + 0.4(600 - 200) = 360°C$$

Therefore, the center temperature of the shaft drops from 600°C to 360°C in 45 min.

To determine the actual heat transfer, we first need to calculate the maximum heat that can be transferred from the cylinder, which is the sensible energy of the cylinder relative to its environment. Taking $L = 1$ m,

$$m = \rho V = \rho \pi r_o^2 L = (7900 \text{ kg/m}^3)\pi(0.1 \text{ m})^2(1 \text{ m}) = 248.2 \text{ kg}$$
$$Q_{max} = mc_p(T_\infty - T_i) = (248.2 \text{ kg})(0.477 \text{ kJ/kg} \cdot °C)(600 - 200)°C$$
$$= 47,350 \text{ kJ}$$

The dimensionless heat transfer ratio is determined from Fig. 4–16c for a long cylinder to be

$$\left.\begin{array}{l} \text{Bi} = \dfrac{1}{1/\text{Bi}} = \dfrac{1}{1.86} = 0.537 \\[2mm] \dfrac{h^2 \alpha t}{k^2} = \text{Bi}^2 \tau = (0.537)^2 (1.07) = 0.309 \end{array}\right\} \quad \dfrac{Q}{Q_{max}} = 0.62$$

Therefore,

$$Q = 0.62 Q_{max} = 0.62 \times (47{,}350 \text{ kJ}) = \textbf{29,360 kJ}$$

which is the total heat transfer from the shaft during the first 45 min of the cooling.

Alternative solution We could also solve this problem using the one-term solution relation instead of the transient charts. First we find the Biot number

$$\text{Bi} = \frac{hr_o}{k} = \frac{(80 \text{ W/m}^2 \cdot {}^\circ\text{C})(0.1 \text{ m})}{14.9 \text{ W/m} \cdot {}^\circ\text{C}} = 0.537$$

The coefficients λ_1 and A_1 for a cylinder corresponding to this Bi are determined from Table 4–2 to be

$$\lambda_1 = 0.970, \qquad A_1 = 1.122$$

Substituting these values into Eq. 4–27 gives

$$\theta_0 = \frac{T_0 - T_\infty}{T_i - T_\infty} = A_1 e^{-\lambda_1^2 \tau} = 1.122 e^{-(0.970)^2(1.07)} = 0.41$$

and thus

$$T_0 = T_\infty + 0.41(T_i - T_\infty) = 200 + 0.41(600 - 200) = \textbf{364}^\circ\textbf{C}$$

The value of $J_1(\lambda_1)$ for $\lambda_1 = 0.970$ is determined from Table 4–3 to be 0.430. Then the fractional heat transfer is determined from Eq. 4–34 to be

$$\frac{Q}{Q_{max}} = 1 - 2\theta_0 \frac{J_1(\lambda_1)}{\lambda_1} = 1 - 2 \times 0.41 \frac{0.430}{0.970} = 0.636$$

and thus

$$Q = 0.636 Q_{max} = 0.636 \times (47{,}350 \text{ kJ}) = \textbf{30,120 kJ}$$

Discussion The slight difference between the two results is due to the reading error of the charts.

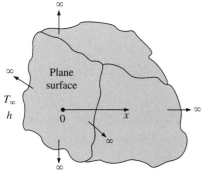

FIGURE 4–24
Schematic of a semi-infinite body.

4–3 ▪ TRANSIENT HEAT CONDUCTION IN SEMI-INFINITE SOLIDS

A semi-infinite solid is an idealized body that has a *single plane surface* and extends to infinity in all directions, as shown in Figure 4–24. This idealized body is used to indicate that the temperature change in the part of the body in which we are interested (the region close to the surface) is due to the thermal conditions on a single surface. The earth, for example, can be considered to be a semi-infinite medium in determining the variation of temperature near its surface. Also, a thick wall can be modeled as a semi-infinite medium if all we are interested in is the variation of temperature in the region near one of the surfaces, and the other surface is too far to have any impact on the region of interest during the time of observation. The temperature in the core region of the wall remains unchanged in this case.

For short periods of time, most bodies can be modeled as semi-infinite solids since heat does not have sufficient time to penetrate deep into the body, and the thickness of the body does not enter into the heat transfer analysis. A steel piece of any shape, for example, can be treated as a semi-infinite solid when it is quenched rapidly to harden its surface. A body whose surface is heated by a laser pulse can be treated the same way.

Consider a semi-infinite solid with constant thermophysical properties, no internal heat generation, uniform thermal conditions on its exposed surface, and initially a uniform temperature of T_i throughout. Heat transfer in this case occurs only in the direction normal to the surface (the x direction), and thus it is one-dimensional. Differential equations are independent of the boundary or initial conditions, and thus Eq. 4–10a for one-dimensional transient conduction in Cartesian coordinates applies. The depth of the solid is large ($x \to \infty$) compared to the depth that heat can penetrate, and these phenomena can be expressed mathematically as a boundary condition as $T(x \to \infty, t) = T_i$.

Heat conduction in a semi-infinite solid is governed by the thermal conditions imposed on the exposed surface, and thus the solution depends strongly on the boundary condition at $x = 0$. Below we present a detailed analytical solution for the case of constant temperature T_s on the surface, and give the results for other more complicated boundary conditions. When the surface temperature is changed to T_s at $t = 0$ and held constant at that value at all times, the formulation of the problem can be expressed as

Differential equation:
$$\frac{\partial^2 T}{\partial x^2} = \frac{1}{\alpha} \frac{\partial T}{\partial t} \qquad \text{(4–37a)}$$

Boundary conditions: $\quad T(0, t) = T_s \quad \text{and} \quad T(x \to \infty, t) = T_i \qquad \text{(4–37b)}$

Initial condition: $\qquad\qquad\qquad T(x, 0) = T_i \qquad \text{(4–37c)}$

The separation of variables technique does not work in this case since the medium is infinite. But another clever approach that converts the partial differential equation into an ordinary differential equation by combining the two independent variables x and t into a single variable η, called the **similarity variable**, works well. For transient conduction in a semi-infinite medium, it is defined as

Similarity variable: $\qquad\qquad \eta = \frac{x}{\sqrt{4\alpha t}} \qquad \text{(4–38)}$

Assuming $T = T(\eta)$ (to be verified) and using the chain rule, all derivatives in the heat conduction equation can be transformed into the new variable, as shown in Fig. 4–25. Noting that $\eta = 0$ at $x = 0$ and $\eta \to \infty$ as $x \to \infty$ (and also at $t = 0$) and substituting into Eqs. 4–37 give, after simplification,

$$\frac{\partial^2 T}{\partial x^2} = \frac{1}{\alpha} \frac{\partial T}{\partial t} \quad \text{and} \quad \eta = \frac{x}{\sqrt{4\alpha t}}$$

$$\frac{\partial T}{\partial t} = \frac{dT}{d\eta} \frac{\partial \eta}{\partial t} = -\frac{x}{2t\sqrt{4\alpha t}} \frac{dT}{d\eta}$$

$$\frac{\partial T}{\partial x} = \frac{dT}{d\eta} \frac{\partial \eta}{\partial x} = \frac{1}{\sqrt{4\alpha t}} \frac{dT}{d\eta}$$

$$\frac{\partial^2 T}{\partial x^2} = \frac{d}{d\eta}\left(\frac{\partial T}{\partial x}\right) \frac{\partial \eta}{\partial x} = \frac{1}{4\alpha t} \frac{d^2 T}{d\eta^2}$$

FIGURE 4–25

Transformation of variables in the derivatives of the heat conduction equation by the use of chain rule.

$$\frac{d^2T}{d\eta^2} = -2\eta \frac{dT}{d\eta} \tag{4–39a}$$

$$T(0) = T_s \quad \text{and} \quad T(\eta \to \infty) = T_i \tag{4–39b}$$

Note that the second boundary condition and the initial condition result in the same boundary condition. Both the transformed equation and the boundary conditions depend on η only and are independent of x and t. Therefore, transformation is successful, and η is indeed a similarity variable.

To solve the 2nd order ordinary differential equation in Eqs. 4–39, we define a new variable w as $w = dT/d\eta$. This reduces Eq. 4–39a into a first order differential equation than can be solved by separating variables,

$$\frac{dw}{d\eta} = -2\eta w \quad \to \quad \frac{dw}{w} = -2\eta d\eta \quad \to \quad \ln w = -\eta^2 + C_0 \quad \to \quad w = C_1 e^{-\eta^2}$$

where $C_1 = \ln C_0$. Back substituting $w = dT/d\eta$ and integrating again,

$$T = C_1 \int_0^\eta e^{-u^2} du + C_2 \tag{4–40}$$

where u is a dummy integration variable. The boundary condition at $\eta = 0$ gives $C_2 = T_s$, and the one for $\eta \to \infty$ gives

$$T_i = C_1 \int_0^\infty e^{-u^2}\, du + C_2 = C_1 \frac{\sqrt{\pi}}{2} + T_s \quad \to \quad C_1 = \frac{2(T_i - T_s)}{\sqrt{\pi}} \tag{4–41}$$

Substituting the C_1 and C_2 expressions into Eq. 4–40 and rearranging, the variation of temperature becomes

$$\frac{T - T_s}{T_i - T_s} = \frac{2}{\sqrt{\pi}} \int_0^\eta e^{-u^2} du = \mathrm{erf}(\eta) = 1 - \mathrm{erfc}(\eta) \tag{4–42}$$

where the mathematical functions

$$\mathrm{erf}(\eta) = \frac{2}{\sqrt{\pi}} \int_0^\eta e^{-u^2} du \quad \text{and} \quad \mathrm{erfc}(\eta) = 1 - \frac{2}{\sqrt{\pi}} \int_0^\eta \varepsilon^{-u^2} du \tag{4–43}$$

are called the **error function** and the **complementary error function**, respectively, of argument η (Fig. 4–26). Despite its simple appearance, the integral in the definition of the error function cannot be performed analytically. Therefore, the function $\mathrm{erfc}(\eta)$ is evaluated numerically for different values of η, and the results are listed in Table 4–4.

Knowing the temperature distribution, the heat flux at the surface can be determined from the Fourier's law to be

$$\dot{q}_s = -k \frac{\partial T}{\partial x}\bigg|_{x=0} = -k \frac{dT}{d\eta} \frac{\partial \eta}{\partial x}\bigg|_{\eta=0} = -kC_1 e^{-\eta^2} \frac{1}{\sqrt{4\alpha t}}\bigg|_{\eta=0} = \frac{k(T_s - T_i)}{\sqrt{\pi \alpha t}} \tag{4–44}$$

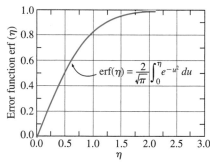

FIGURE 4–26
Error function is a standard mathematical function, just like the sinus and tangent functions, whose value varies between 0 and 1.

TABLE 4–4

The complementary error function

η	erfc (η)	η	erfc (η)	η	erfc (η)	η	erfc (η)	η	erfc (η)	η	erfc (η)
0.00	1.00000	0.38	0.5910	0.76	0.2825	1.14	0.1069	1.52	0.03159	1.90	0.00721
0.02	0.9774	0.40	0.5716	0.78	0.2700	1.16	0.10090	1.54	0.02941	1.92	0.00662
0.04	0.9549	0.42	0.5525	0.80	0.2579	1.18	0.09516	1.56	0.02737	1.94	0.00608
0.06	0.9324	0.44	0.5338	0.82	0.2462	1.20	0.08969	1.58	0.02545	1.96	0.00557
0.08	0.9099	0.46	0.5153	0.84	0.2349	1.22	0.08447	1.60	0.02365	1.98	0.00511
0.10	0.8875	0.48	0.4973	0.86	0.2239	1.24	0.07950	1.62	0.02196	2.00	0.00468
0.12	0.8652	0.50	0.4795	0.88	0.2133	1.26	0.07476	1.64	0.02038	2.10	0.00298
0.14	0.8431	0.52	0.4621	0.90	0.2031	1.28	0.07027	1.66	0.01890	2.20	0.00186
0.16	0.8210	0.54	0.4451	0.92	0.1932	1.30	0.06599	1.68	0.01751	2.30	0.00114
0.18	0.7991	0.56	0.4284	0.94	0.1837	1.32	0.06194	1.70	0.01612	2.40	0.00069
0.20	0.7773	0.58	0.4121	0.96	0.1746	1.34	0.05809	1.72	0.01500	2.50	0.00041
0.22	0.7557	0.60	0.3961	0.98	0.1658	1.36	0.05444	1.74	0.01387	2.60	0.00024
0.24	0.7343	0.62	0.3806	1.00	0.1573	1.38	0.05098	1.76	0.01281	2.70	0.00013
0.26	0.7131	0.64	0.3654	1.02	0.1492	1.40	0.04772	1.78	0.01183	2.80	0.00008
0.28	0.6921	0.66	0.3506	1.04	0.1413	1.42	0.04462	1.80	0.01091	2.90	0.00004
0.30	0.6714	0.68	0.3362	1.06	0.1339	1.44	0.04170	1.82	0.01006	3.00	0.00002
0.32	0.6509	0.70	0.3222	1.08	0.1267	1.46	0.03895	1.84	0.00926	3.20	0.00001
0.34	0.6306	0.72	0.3086	1.10	0.1198	1.48	0.03635	1.86	0.00853	3.40	0.00000
0.36	0.6107	0.74	0.2953	1.12	0.1132	1.50	0.03390	1.88	0.00784	3.60	0.00000

The solutions in Eqs. 4–42 and 4–44 correspond to the case when the temperature of the exposed surface of the medium is suddenly raised (or lowered) to T_s at $t = 0$ and is maintained at that value at all times. The specified surface temperature case is closely approximated in practice when condensation or boiling takes place on the surface. Using a similar approach or the Laplace transform technique, analytical solutions can be obtained for other boundary conditions on the surface, with the following results.

Case 1: Specified Surface Temperature, $T_s = $ constant (Fig. 4–27).

$$\frac{T(x, t) - T_i}{T_s - T_i} = \text{erfc}\left(\frac{x}{2\sqrt{\alpha t}}\right) \quad \text{and} \quad \dot{q}_s(t) = \frac{k(T_s - T_i)}{\sqrt{\pi \alpha t}} \tag{4–45}$$

Case 2: Specified Surface Heat Flux, $\dot{q}_s = $ constant.

$$T(x, t) - T_i = \frac{\dot{q}_s}{k}\left[\sqrt{\frac{4\alpha t}{\pi}}\exp\left(-\frac{x^2}{4\alpha t}\right) - x\,\text{erfc}\left(\frac{x}{2\sqrt{\alpha t}}\right)\right] \tag{4–46}$$

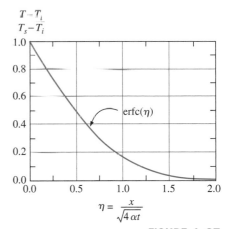

FIGURE 4–27
Dimensionless temperature distribution for transient conduction in a semi-infinite solid whose surface is maintained at a constant temperature T_s.

Case 3: Convection on the Surface, $\dot{q}_s(t) = h[T_\infty - T(0, t)]$.

$$\frac{T(x,t) - T_i}{T_\infty - T_i} = \text{erfc}\left(\frac{x}{2\sqrt{\alpha t}}\right) - \exp\left(\frac{hx}{k} + \frac{h^2\alpha t}{k^2}\right)\text{erfc}\left(\frac{x}{2\sqrt{\alpha t}} + \frac{h\sqrt{\alpha t}}{k}\right)$$

(4–47)

Case 4: Energy Pulse at Surface, e_s = constant.
Energy in the amount of e_s per unit surface area (in J/m^2) is supplied to the semi-infinite body instantaneously at time $t = 0$ (by a laser pulse, for example), and the entire energy is assumed to enter the body, with no heat loss from the surface.

$$T(x, t) - T_i = \frac{e_s}{k\sqrt{\pi t/\alpha}}\exp\left(-\frac{x^2}{4\alpha t}\right)$$

(4–48)

Note that Cases 1 and 3 are closely related. In Case 1, the surface $x = 0$ is brought to a temperature T_s at time $t = 0$, and kept at that value at all times. In Case 3, the surface is exposed to convection by a fluid at a constant temperature T_∞ with a heat transfer coefficient h.

The solutions for all four cases are plotted in Fig. 4–28 for a representative case using a large cast iron block initially at 0°C throughout. In Case 1, the surface temperature remains constant at the specified value of T_s, and temperature increases gradually within the medium as heat penetrates deeper into the solid. Note that during initial periods only a thin slice near the surface is affected by heat transfer. Also, the temperature gradient at the surface and thus the rate of heat transfer into the solid decreases with time. In Case 2, heat is continually supplied to the solid, and thus the temperature within the solid, including the surface, increases with time. This is also the case with convection (Case 3), except that the surrounding fluid temperature T_∞ is the highest temperature that the solid body can rise to. In Case 4, the surface is subjected to an instant burst of heat supply at time $t = 0$, such as heating by a laser pulse, and then the surface is covered with insulation. The result is an instant rise in surface temperature, followed by a temperature drop as heat is conducted deeper into the solid. Note that the temperature profile is always normal to the surface at all times.(Why?)

The variation of temperature with position and time in a semi-infinite solid subjected to convection heat transfer is plotted in Fig. 4–29 for the nondimensionalized temperature against the dimensionless similarity variable $\eta = x/\sqrt{4\alpha t}$ for various values of the parameter $h\sqrt{\alpha t}/k$. Although the graphical solution given in Fig. 4–29 is simply a plot of the exact analytical solution, it is subject to reading errors, and thus is of limited accuracy compared to the analytical solution. Also, the values on the vertical axis of Fig. 4–29 correspond to $x = 0$, and thus represent the surface temperature. The curve $h\sqrt{\alpha t}/k = \infty$ corresponds to $h \to \infty$, which corresponds to the case of *specified temperature* T_∞ at the surface at $x = 0$. That is, the case in which the surface of the semi-infinite body is suddenly brought to temperature T_∞ at $t = 0$ and kept at T_∞ at all times can be handled by setting h to infinity. For a *finite* heat transfer coefficient h, the surface temperature approaches the fluid temperature T_∞ as the time t approaches infinity.

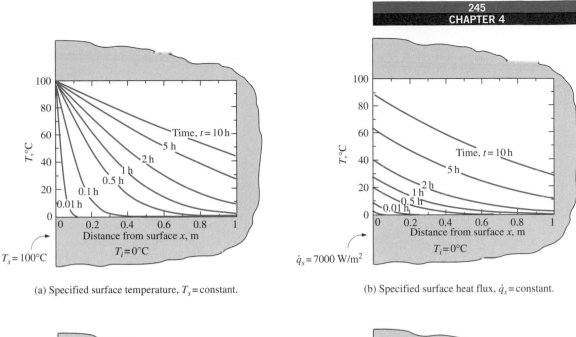

(a) Specified surface temperature, T_s = constant.

(b) Specified surface heat flux, $\dot{q}_s$ = constant.

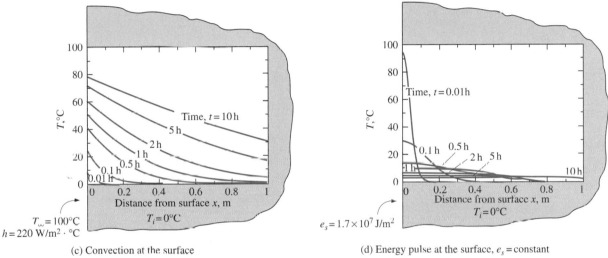

(c) Convection at the surface

(d) Energy pulse at the surface, e_s = constant

FIGURE 4–28

Variations of temperature with position and time in a large cast iron block ($\alpha = 2.31 \times 10^{-5}$ m²/s, $k = 80.2$ W/m °C) initially at 0 °C under different thermal conditions on the surface.

Contact of Two Semi-Infinite Solids

When two large bodies A and B, initially at uniform temperatures $T_{A,i}$ and $T_{B,i}$ are brought into contact, they instantly achieve temperature equality at the contact surface (temperature equality is achieved over the entire surface if the contact resistance is negligible). If the two bodies are of the same material with constant properties, thermal symmetry requires the contact surface temperature to be the arithmetic average, $T_s = (T_{A,i} + T_{B,i})/2$ and to remain constant at that value at all times.

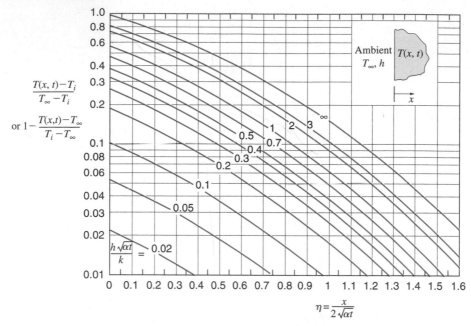

FIGURE 4–29
Variation of temperature with position and time in a semi-infinite solid initially at temperature T_i subjected to convection to an environment at T_∞ with a convection heat transfer coefficient of h (plotted using EES).

If the bodies are of different materials, they still achieve a temperature equality, but the surface temperature T_s in this case will be different than the arithmetic average. Noting that both bodies can be treated as semi-infinite solids with the same specified surface temperature, the energy balance on the contact surface gives, from Eq. 4–45,

$$\dot{q}_{s,A} = \dot{q}_{s,B} \rightarrow -\frac{k_A(T_s - T_{A,i})}{\sqrt{\pi \alpha_A t}} = \frac{k_B(T_s - T_{B,i})}{\sqrt{\pi \alpha_B t}} \rightarrow \frac{T_{A,i} - T_s}{T_s - T_{B,i}} = \sqrt{\frac{(k\rho c_p)_B}{(k\rho c_p)_A}}$$

Then T_s is determined to be (Fig. 4–30)

$$T_s = \frac{\sqrt{(k\rho c_p)_A}T_{A,i} + \sqrt{(k\rho c_p)_B}T_{B,i}}{\sqrt{(k\rho c_p)_A} + \sqrt{(k\rho c_p)_B}} \tag{4–49}$$

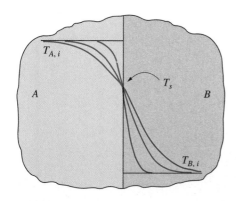

FIGURE 4–30
Contact of two semi-infinite solids of different initial temperatures.

Therefore, the interface temperature of two bodies brought into contact is dominated by the body with the larger $k\rho c_p$. This also explains why a metal at room temperature feels colder than wood at the same temperature. At room temperature, the $\sqrt{k\rho c_p}$ value is 24 kJ/m² · °C for aluminum, 0.38 kJ/m² · °C for wood, and 1.1 kJ/m² · °C for the human flesh. Using Eq. 4–49, if can be shown that when a person with a skin temperature of 35°C touches an aluminum block and then a wood block both at 15°C, the contact surface temperature will be 15.9°C in the case of aluminum and 30°C in the case of wood.

EXAMPLE 4–6 Minimum Burial Depth of Water Pipes to Avoid Freezing

In areas where the air temperature remains below 0°C for prolonged periods of time, the freezing of water in underground pipes is a major concern. Fortunately, the soil remains relatively warm during those periods, and it takes weeks for the subfreezing temperatures to reach the water mains in the ground. Thus, the soil effectively serves as an insulation to protect the water from subfreezing temperatures in winter.

The ground at a particular location is covered with snow pack at −10°C for a continuous period of three months, and the average soil properties at that location are $k = 0.4$ W/m · °C and $\alpha = 0.15 \times 10^{-6}$ m²/s (Fig. 4–31). Assuming an initial uniform temperature of 15°C for the ground, determine the minimum burial depth to prevent the water pipes from freezing.

SOLUTION The water pipes are buried in the ground to prevent freezing. The minimum burial depth at a particular location is to be determined.

Assumptions **1** The temperature in the soil is affected by the thermal conditions at one surface only, and thus the soil can be considered to be a semi-infinite medium. **2** The thermal properties of the soil are constant.

Properties The properties of the soil are as given in the problem statement.

Analysis The temperature of the soil surrounding the pipes will be 0°C after three months in the case of minimum burial depth. Therefore, from Fig. 4–29, we have

$$\left.\begin{array}{l} \dfrac{h\sqrt{\alpha t}}{k} = \infty \quad (\text{since } h \to \infty) \\[2mm] \dfrac{T(x, t) - T_i}{T_\infty - T_i} = \dfrac{0 - 15}{-10 - 15} = 0.6 \end{array}\right\} \eta = \dfrac{x}{2\sqrt{\alpha t}} = 0.36$$

We note that

$$t = (90 \text{ days})(24 \text{ h/day})(3600 \text{ s/h}) = 7.78 \times 10^6 \text{ s}$$

and thus

$$x = 2\eta\sqrt{\alpha t} = 2 \times 0.36\sqrt{(0.15 \times 10^{-6} \text{ m}^2/\text{s})(7.78 \times 10^6 \text{ s})} = \textbf{0.78 m}$$

Therefore, the water pipes must be buried to a depth of at least 78 cm to avoid freezing under the specified harsh winter conditions.

ALTERNATIVE SOLUTION The solution of this problem could also be determined from Eq. 4–45:

$$\frac{T(x, t) - T_i}{T_s - T_i} = \text{erfc}\left(\frac{x}{2\sqrt{\alpha t}}\right) \longrightarrow \frac{0 - 15}{-10 - 15} = \text{erfc}\left(\frac{x}{2\sqrt{\alpha t}}\right) = 0.60$$

The argument that corresponds to this value of the complementary error function is determined from Table 4–4 to be $\eta = 0.37$. Therefore,

$$x = 2\eta\sqrt{\alpha t} = 2 \times 0.37\sqrt{(0.15 \times 10^{-6} \text{ m}^2/\text{s})(7.78 \times 10^6 \text{ s})} = \textbf{0.80 m}$$

Again, the slight difference is due to the reading error of the chart.

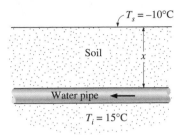

FIGURE 4–31
Schematic for Example 4–6.

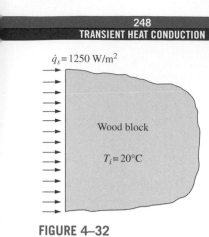

$\dot{q}_s = 1250 \text{ W/m}^2$

Wood block

$T_i = 20°C$

FIGURE 4–32
Schematic for Example 4–7.

EXAMPLE 4–7   Surface Temperature Rise of Heated Blocks

A thick black-painted wood block at 20°C is subjected to constant solar heat flux of 1250 W/m² (Fig. 4–32). Determine the exposed surface temperature of the block after 20 minutes. What would your answer be if the block were made of aluminum?

SOLUTION   A wood block is subjected to solar heat flux. The surface temperature of the block is to be determined, and to be compared to the value for an aluminum block.

Assumptions   **1** All incident solar radiation is absorbed by the block. **2** Heat loss from the block is disregarded (and thus the result obtained is the maximum temperature). **3** The block is sufficiently thick to be treated as a semi-infinite solid, and the properties of the block are constant.

Properties   Thermal conductivity and diffusivity values at room temperature are $k = 1.26$ W/m · K and $\alpha = 1.1 \times 10^{-5}$ m²/s for wood, and $k = 237$ W/m · K and $\alpha = 9.71 \times 10^{-5}$ m²/s for aluminum.

Analysis   This is a transient conduction problem in a semi-infinite medium subjected to constant surface heat flux, and the surface temperature can be expressed from Eq. 4–46 as

$$T_s = T(0, t) = T_i + \frac{\dot{q}_s}{k}\sqrt{\frac{4\alpha t}{\pi}}$$

Substituting the given values, the surface temperatures for both the wood and aluminum blocks are determined to be

$$T_{s,\text{wood}} = 20°C + \frac{1250\,\text{W/m}^2}{1.26\,\text{W/m}\cdot°C}\sqrt{\frac{4(1.1 \times 10^{-5}\,\text{m}^2/\text{s})(20 \times 60\,\text{s})}{\pi}} = \mathbf{149°C}$$

$$T_{s,\text{Al}} = 20°C + \frac{1250\,\text{W/m}^2}{237\,\text{W/m}\cdot°C}\sqrt{\frac{4(9.71 \times 10^{-5}\,\text{m}^2/\text{s})(20 \times 60\,\text{s})}{\pi}} = \mathbf{22.0°C}$$

Note that thermal energy supplied to the wood accumulates near the surface because of the low conductivity and diffusivity of wood, causing the surface temperature to rise to high values. Metals, on the other hand, conduct the heat they receive to inner parts of the block because of their high conductivity and diffusivity, resulting in minimal temperature rise at the surface. In reality, both temperatures will be lower because of heat losses.

Discussion   The temperature profiles for both the wood and aluminum blocks at $t = 20$ min are evaluated and plotted in Fig. 4–33 using EES. At a depth of $x = 0.41$ m, the temperature in both blocks is 20.6°C. At a depth of 0.5 m, the temperatures become 20.1°C for wood and 20.4°C for aluminum block, which confirms that heat penetrates faster and further in metals compared to nonmetals.

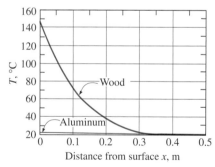

FIGURE 4–33
Variation of temperature within the wood and aluminum blocks at $t = 20$ min.

4–4 ▪ TRANSIENT HEAT CONDUCTION IN MULTIDIMENSIONAL SYSTEMS

The transient temperature charts and analytical solutions presented earlier can be used to determine the temperature distribution and heat transfer in *one-dimensional* heat conduction problems associated with a large plane wall, a long cylinder, a sphere, and a semi-infinite medium. Using a superposition approach called the **product solution**, these charts and solutions can also be used to construct solutions for the *two-dimensional* transient heat conduction problems encountered in geometries such as a short cylinder, a long rectangular bar, or a semi-infinite cylinder or plate, and even *three-dimensional* problems associated with geometries such as a rectangular prism or a semi-infinite rectangular bar, provided that *all* surfaces of the solid are subjected to convection to the *same* fluid at temperature T_∞, with the *same* heat transfer coefficient h, and the body involves no heat generation (Fig. 4–34). The solution in such multidimensional geometries can be expressed as the *product* of the solutions for the one-dimensional geometries whose intersection is the multi-dimensional geometry.

Consider a *short cylinder* of height a and radius r_o initially at a uniform temperature T_i. There is no heat generation in the cylinder. At time $t = 0$, the cylinder is subjected to convection from all surfaces to a medium at temperature T_∞ with a heat transfer coefficient h. The temperature within the cylinder will change with x as well as r and time t since heat transfer occurs from the top and bottom of the cylinder as well as its side surfaces. That is, $T = T(r, x, t)$ and thus this is a two-dimensional transient heat conduction problem. When the properties are assumed to be constant, it can be shown that the solution of this two-dimensional problem can be expressed as

$$\left(\frac{T(r, x, t) - T_\infty}{T_i - T_\infty}\right)_{\substack{\text{short} \\ \text{cylinder}}} = \left(\frac{T(x, t) - T_\infty}{T_i - T_\infty}\right)_{\substack{\text{plane} \\ \text{wall}}} \left(\frac{T(r, t) - T_\infty}{T_i - T_\infty}\right)_{\substack{\text{infinite} \\ \text{cylinder}}} \quad \textbf{(4–50)}$$

That is, the solution for the two-dimensional short cylinder of height a and radius r_o is equal to the *product* of the nondimensionalized solutions for the one-dimensional plane wall of thickness a and the long cylinder of radius r_o, which are the two geometries whose intersection is the short cylinder, as shown in Fig. 4–35. We generalize this as follows: *the solution for a multi-dimensional geometry is the product of the solutions of the one-dimensional geometries whose intersection is the multidimensional body.*

For convenience, the one-dimensional solutions are denoted by

$$\theta_{\text{wall}}(x, t) = \left(\frac{T(x, t) - T_\infty}{T_i - T_\infty}\right)_{\substack{\text{plane} \\ \text{wall}}}$$

$$\theta_{\text{cyl}}(r, t) = \left(\frac{T(r, t) - T_\infty}{T_i - T_\infty}\right)_{\substack{\text{infinite} \\ \text{cylinder}}}$$

$$\theta_{\text{semi-inf}}(x, t) = \left(\frac{T(x, t) - T_\infty}{T_i - T_\infty}\right)_{\substack{\text{semi-infinite} \\ \text{solid}}} \quad \textbf{(4–51)}$$

For example, the solution for a long solid bar whose cross section is an $a \times b$ rectangle is the intersection of the two infinite plane walls of thicknesses a and b, as shown in Fig. 4–36, and thus the transient temperature distribution for this rectangular bar can be expressed as

$$\left(\frac{T(x, y, t) - T_\infty}{T_i - T_\infty}\right)_{\substack{\text{rectangular} \\ \text{bar}}} = \theta_{\text{wall}}(x, t)\theta_{\text{wall}}(y, t) \quad \textbf{(4–52)}$$

The proper forms of the product solutions for some other geometries are given in Table 4–5. It is important to note that the x-coordinate is measured from the

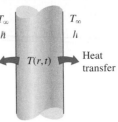

T_∞
h
T_∞
h
$T(r,t)$
Heat transfer

(a) Long cylinder

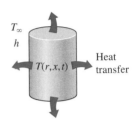

T_∞
h
$T(r,x,t)$
Heat transfer

(b) Short cylinder (two-dimensional)

FIGURE 4–34
The temperature in a short cylinder exposed to convection from all surfaces varies in both the radial and axial directions, and thus heat is transferred in both directions.

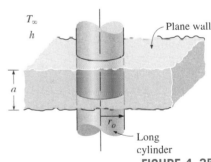

T_∞
h
Plane wall
a
r_o
Long cylinder

FIGURE 4–35
A short cylinder of radius r_o and height a is the *intersection* of a long cylinder of radius r_o and a plane wall of thickness a.

TABLE 4–5

Multidimensional solutions expressed as products of one-dimensional solutions for bodies that are initially at a uniform temperature T_i and exposed to convection from all surfaces to a medium at T_∞

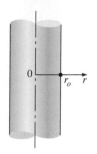

$\theta(r,t) = \theta_{cyl}(r,t)$
Infinite cylinder

$\theta(x,r,t) = \theta_{cyl}(r,t)\,\theta_{semi-inf}(x,t)$
Semi-infinite cylinder

$\theta(x,r,t) = \theta_{cyl}(r,t)\,\theta_{wall}(x,t)$
Short cylinder

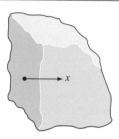

$\theta(x,t) = \theta_{semi-inf}(x,t)$
Semi-infinite medium

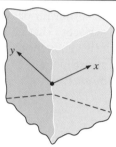

$\theta(x,y,t) = \theta_{semi-inf}(x,t)\,\theta_{semi-inf}(y,t)$
Quarter-infinite medium

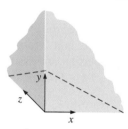

$\theta(x,y,z,t) =$
$\theta_{semi-inf}(x,t)\,\theta_{semi-inf}(y,t)\,\theta_{semi-inf}(z,t)$
Corner region of a large medium

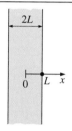

$\theta(x,t) = \theta_{wall}(x,t)$
Infinite plate (or plane wall)

$\theta(x,y,t) = \theta_{wall}(x,t)\,\theta_{semi-inf}(y,t)$
Semi-infinite plate

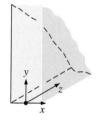

$\theta(x,y,z,t) =$
$\theta_{wall}(x,t)\,\theta_{semi-inf}(y,t)\,\theta_{semi-inf}(z,t)$
Quarter-infinite plate

$\theta(x,y,t) = \theta_{wall}(x,t)\,\theta_{wall}(y,t)$
Infinite rectangular bar

$\theta(x,y,z,t) =$
$\theta_{wall}(x,t)\,\theta_{wall}(y,t)\,\theta_{semi-inf}(z,t)$
Semi-infinite rectangular bar

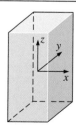

$\theta(x,y,z,t) =$
$\theta_{wall}(x,t)\,\theta_{wall}(y,t)\,\theta_{wall}(z,t)$
Rectangular parallelepiped

surface in a semi-infinite solid, and from the *midplane* in a plane wall. The radial distance r is always measured from the centerline.

Note that the solution of a *two-dimensional* problem involves the product of *two* one-dimensional solutions, whereas the solution of a *three-dimensional* problem involves the product of *three* one-dimensional solutions.

A modified form of the product solution can also be used to determine the total transient heat transfer to or from a multidimensional geometry by using the one-dimensional values, as shown by L. S. Langston in 1982. The transient heat transfer for a two-dimensional geometry formed by the intersection of two one-dimensional geometries 1 and 2 is

$$\left(\frac{Q}{Q_{max}}\right)_{total, 2D} = \left(\frac{Q}{Q_{max}}\right)_1 + \left(\frac{Q}{Q_{max}}\right)_2 \left[1 - \left(\frac{Q}{Q_{max}}\right)_1\right] \quad (4-53)$$

Transient heat transfer for a three-dimensional body formed by the intersection of three one-dimensional bodies 1, 2, and 3 is given by

$$\left(\frac{Q}{Q_{max}}\right)_{total, 3D} = \left(\frac{Q}{Q_{max}}\right)_1 + \left(\frac{Q}{Q_{max}}\right)_2 \left[1 - \left(\frac{Q}{Q_{max}}\right)_1\right]$$
$$+ \left(\frac{Q}{Q_{max}}\right)_3 \left[1 - \left(\frac{Q}{Q_{max}}\right)_1\right]\left[1 - \left(\frac{Q}{Q_{max}}\right)_2\right] \quad (4-54)$$

The use of the product solution in transient two- and three-dimensional heat conduction problems is illustrated in the following examples.

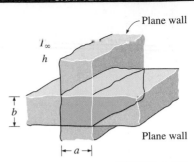

FIGURE 4–36
A long solid bar of rectangular profile $a \times b$ is the *intersection* of two plane walls of thicknesses a and b.

EXAMPLE 4–8   Cooling of a Short Brass Cylinder

A short brass cylinder of diameter $D = 10$ cm and height $H = 12$ cm is initially at a uniform temperature $T_i = 120°C$. The cylinder is now placed in atmospheric air at 25°C, where heat transfer takes place by convection, with a heat transfer coefficient of $h = 60$ W/m² · °C. Calculate the temperature at (a) the center of the cylinder and (b) the center of the top surface of the cylinder 15 min after the start of the cooling.

SOLUTION   A short cylinder is allowed to cool in atmospheric air. The temperatures at the centers of the cylinder and the top surface are to be determined.

Assumptions   **1** Heat conduction in the short cylinder is two-dimensional, and thus the temperature varies in both the axial x- and the radial r-directions. **2** The thermal properties of the cylinder and the heat transfer coefficient are constant. **3** The Fourier number is $\tau > 0.2$ so that the one-term approximate solutions are applicable.

Properties   The properties of brass at room temperature are $k = 110$ W/m · °C and $\alpha = 33.9 \times 10^{-6}$ m²/s (Table A–3). More accurate results can be obtained by using properties at average temperature.

Analysis   (a) This short cylinder can physically be formed by the intersection of a long cylinder of radius $r_o = 5$ cm and a plane wall of thickness $2L = 12$ cm,

$T_\infty = 25°C$
$h = 60$ W/m²·°C

$T_i = 120°C$

FIGURE 4–37
Schematic for Example 4–8.

as shown in Fig. 4–37. The dimensionless temperature at the center of the plane wall is determined from Fig. 4–15a to be

$$\left.\begin{array}{l} \tau = \dfrac{\alpha t}{L^2} = \dfrac{(3.39 \times 10^{-5} \text{ m}^2/\text{s})(900 \text{ s})}{(0.06 \text{ m})^2} = 8.48 \\[3mm] \dfrac{1}{\text{Bi}} = \dfrac{k}{hL} = \dfrac{110 \text{ W/m} \cdot {}^\circ\text{C}}{(60 \text{ W/m}^2 \cdot {}^\circ\text{C})(0.06 \text{ m})} = 30.6 \end{array}\right\} \quad \theta_\text{wall}(0, t) = \dfrac{T(0, t) - T_\infty}{T_i - T_\infty} = 0.8$$

Similarly, at the center of the cylinder, we have

$$\left.\begin{array}{l} \tau = \dfrac{\alpha t}{r_o^2} = \dfrac{(3.39 \times 10^{-5} \text{ m}^2/\text{s})(900 \text{ s})}{(0.05 \text{ m})^2} = 12.2 \\[3mm] \dfrac{1}{\text{Bi}} = \dfrac{k}{hr_o} = \dfrac{110 \text{ W/m} \cdot {}^\circ\text{C}}{(60 \text{ W/m}^2 \cdot {}^\circ\text{C})(0.05 \text{ m})} = 36.7 \end{array}\right\} \quad \theta_\text{cyl}(0, t) = \dfrac{T(0, t) - T_\infty}{T_i - T_\infty} = 0.5$$

Therefore,

$$\left(\dfrac{T(0, 0, t) - T_\infty}{T_i - T_\infty}\right)_{\substack{\text{short} \\ \text{cylinder}}} = \theta_\text{wall}(0, t) \times \theta_\text{cyl}(0, t) = 0.8 \times 0.5 = 0.4$$

and

$$T(0, 0, t) = T_\infty + 0.4(T_i - T_\infty) = 25 + 0.4(120 - 25) = \mathbf{63^\circ C}$$

This is the temperature at the center of the short cylinder, which is also the center of both the long cylinder and the plate.

(b) The center of the top surface of the cylinder is still at the center of the long cylinder ($r = 0$), but at the outer surface of the plane wall ($x = L$). Therefore, we first need to find the surface temperature of the wall. Noting that $x = L = 0.06$ m,

$$\left.\begin{array}{l} \dfrac{x}{L} = \dfrac{0.06 \text{ m}}{0.06 \text{ m}} = 1 \\[3mm] \dfrac{1}{\text{Bi}} = \dfrac{k}{hL} = \dfrac{110 \text{ W/m} \cdot {}^\circ\text{C}}{(60 \text{ W/m}^2 \cdot {}^\circ\text{C})(0.06 \text{ m})} = 30.6 \end{array}\right\} \quad \dfrac{T(L, t) - T_\infty}{T_0 - T_\infty} = 0.98$$

Then

$$\theta_\text{wall}(L, t) = \dfrac{T(L, t) - T_\infty}{T_i - T_\infty} = \left(\dfrac{T(L, t) - T_\infty}{T_0 - T_\infty}\right)\left(\dfrac{T_0 - T_\infty}{T_i - T_\infty}\right) = 0.98 \times 0.8 = 0.784$$

Therefore,

$$\left(\dfrac{T(L, 0, t) - T_\infty}{T_i - T_\infty}\right)_{\substack{\text{short} \\ \text{cylinder}}} = \theta_\text{wall}(L, t)\theta_\text{cyl}(0, t) = 0.784 \times 0.5 = 0.392$$

and

$$T(L, 0, t) = T_\infty + 0.392(T_i - T_\infty) = 25 + 0.392(120 - 25) = \mathbf{62.2^\circ C}$$

which is the temperature at the center of the top surface of the cylinder.

EXAMPLE 4-9   Heat Transfer from a Short Cylinder

Determine the total heat transfer from the short brass cylinder ($\rho = 8530$ kg/m^3, $c_p = 0.380$ kJ/kg · °C) discussed in Example 4–8.

SOLUTION   We first determine the maximum heat that can be transferred from the cylinder, which is the sensible energy content of the cylinder relative to its environment:

$$m = \rho V = \rho \pi r_o^2 H = (8530 \text{ kg/m}^3)\pi(0.05 \text{ m})^2(0.12 \text{ m}) = 8.04 \text{ kg}$$

$$Q_{max} = mc_p(T_i - T_\infty) = (8.04 \text{ kg})(0.380 \text{ kJ/kg} \cdot °\text{C})(120 - 25)°\text{C} = 290.2 \text{ kJ}$$

Then we determine the dimensionless heat transfer ratios for both geometries. For the plane wall, it is determined from Fig. 4–15c to be

$$\left.\begin{array}{l} \text{Bi} = \dfrac{1}{1/\text{Bi}} = \dfrac{1}{30.6} = 0.0327 \\[2em] \dfrac{h^2\alpha t}{k^2} = \text{Bi}^2\tau = (0.0327)^2(8.48) = 0.0091 \end{array}\right\} \left(\dfrac{Q}{Q_{max}}\right)_{\substack{\text{plane} \\ \text{wall}}} = 0.23$$

Similarly, for the cylinder, we have

$$\left.\begin{array}{l} \text{Bi} = \dfrac{1}{1/\text{Bi}} = \dfrac{1}{36.7} = 0.0272 \\[2em] \dfrac{h^2\alpha t}{k^2} = \text{Bi}^2\tau = (0.0272)^2(12.2) = 0.0090 \end{array}\right\} \left(\dfrac{Q}{Q_{max}}\right)_{\substack{\text{infinite} \\ \text{cylinder}}} = 0.47$$

Then the heat transfer ratio for the short cylinder is, from Eq. 4–53,

$$\left(\frac{Q}{Q_{max}}\right)_{\text{short cyl}} = \left(\frac{Q}{Q_{max}}\right)_1 + \left(\frac{Q}{Q_{max}}\right)_2\left[1 - \left(\frac{Q}{Q_{max}}\right)_1\right]$$
$$= 0.23 + 0.47(1 - 0.23) = 0.592$$

Therefore, the total heat transfer from the cylinder during the first 15 min of cooling is

$$Q = 0.592 Q_{max} = 0.592 \times (290.2 \text{ kJ}) = \mathbf{172 \text{ kJ}}$$

EXAMPLE 4-10   Cooling of a Long Cylinder by Water

A semi-infinite aluminum cylinder of diameter $D = 20$ cm is initially at a uniform temperature $T_i = 200°$C. The cylinder is now placed in water at 15°C where heat transfer takes place by convection, with a heat transfer coefficient of $h = 120$ W/m^2 · °C. Determine the temperature at the center of the cylinder 15 cm from the end surface 5 min after the start of the cooling.

SOLUTION   A semi-infinite aluminum cylinder is cooled by water. The temperature at the center of the cylinder 15 cm from the end surface is to be determined.

Assumptions **1** Heat conduction in the semi-infinite cylinder is two-dimensional, and thus the temperature varies in both the axial x- and the radial

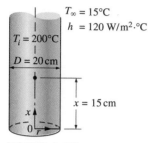

$T_\infty = 15°C$
$h = 120$ W/m²·°C
$T_i = 200°C$
$D = 20$ cm
$x = 15$ cm
x
0 r

FIGURE 4–38
Schematic for Example 4–10.

r-directions. **2** The thermal properties of the cylinder and the heat transfer coefficient are constant. **3** The Fourier number is $\tau > 0.2$ so that the one-term approximate solutions are applicable.

Properties The properties of aluminum at room temperature are $k = 237$ W/m·°C and $\alpha = 9.71 \times 10^{-6}$ m²/s (Table A–3). More accurate results can be obtained by using properties at average temperature.

Analysis This semi-infinite cylinder can physically be formed by the intersection of an infinite cylinder of radius $r_o = 10$ cm and a semi-infinite medium, as shown in Fig. 4–38.

We solve this problem using the one-term solution relation for the cylinder and the analytic solution for the semi-infinite medium. First we consider the infinitely long cylinder and evaluate the Biot number:

$$\text{Bi} = \frac{hr_o}{k} = \frac{(120 \text{ W/m}^2 \cdot °C)(0.1 \text{ m})}{237 \text{ W/m} \cdot °C} = 0.05$$

The coefficients λ_1 and A_1 for a cylinder corresponding to this Bi are determined from Table 4–2 to be $\lambda_1 = 0.3126$ and $A_1 = 1.0124$. The Fourier number in this case is

$$\tau = \frac{\alpha t}{r_o^2} = \frac{(9.71 \times 10^{-5} \text{ m}^2/\text{s})(5 \times 60 \text{ s})}{(0.1 \text{ m})^2} = 2.91 > 0.2$$

and thus the one-term approximation is applicable. Substituting these values into Eq. 4–27 gives

$$\theta_0 = \theta_{\text{cyl}}(0, t) = A_1 e^{-\lambda_1^2 \tau} = 1.0124 e^{-(0.3126)^2(2.91)} = 0.762$$

The solution for the semi-infinite solid can be determined from

$$1 - \theta_{\text{semi-inf}}(x, t) = \text{erfc}\left(\frac{x}{2\sqrt{\alpha t}}\right)$$
$$- \exp\left(\frac{hx}{k} + \frac{h^2 \alpha t}{k^2}\right)\left[\text{erfc}\left(\frac{x}{2\sqrt{\alpha t}} + \frac{h\sqrt{\alpha t}}{k}\right)\right]$$

First we determine the various quantities in parentheses:

$$\eta = \frac{x}{2\sqrt{\alpha t}} = \frac{0.15 \text{ m}}{2\sqrt{(9.71 \times 10^{-5} \text{ m}^2/\text{s})(5 \times 60 \text{ s})}} = 0.44$$

$$\frac{h\sqrt{\alpha t}}{k} = \frac{(120 \text{ W/m}^2 \cdot °C)\sqrt{(9.71 \times 10^{-5} \text{ m}^2/\text{s})(300 \text{ s})}}{237 \text{ W/m} \cdot °C} = 0.086$$

$$\frac{hx}{k} = \frac{(120 \text{ W/m}^2 \cdot °C)(0.15 \text{ m})}{237 \text{ W/m} \cdot °C} = 0.0759$$

$$\frac{h^2 \alpha t}{k^2} = \left(\frac{h\sqrt{\alpha t}}{k}\right)^2 = (0.086)^2 = 0.0074$$

Substituting and evaluating the complementary error functions from Table 4–4,

$$\theta_{\text{semi-inf}}(x, t) = 1 - \text{erfc}(0.44) + \exp(0.0759 + 0.0074)\text{ erfc}(0.44 + 0.086)$$
$$= 1 - 0.5338 + \exp(0.0833) \times 0.457$$
$$= 0.963$$

Now we apply the product solution to get

$$\left(\frac{T(x, 0, t) - T_\infty}{T_i - T_\infty}\right)_{\substack{\text{semi-infinite} \\ \text{cylinder}}} = \theta_{\text{semi-inf}}(x, t)\theta_{\text{cyl}}(0, t) = 0.963 \times 0.762 = 0.734$$

and

$$T(x, 0, t) = T_\infty + 0.734(T_i - T_\infty) = 15 + 0.734(200 - 15) = \mathbf{151°C}$$

which is the temperature at the center of the cylinder 15 cm from the exposed bottom surface.

EXAMPLE 4–11   Refrigerating Steaks while Avoiding Frostbite

In a meat processing plant, 1-in-thick steaks initially at 75°F are to be cooled in the racks of a large refrigerator that is maintained at 5°F (Fig. 4–39). The steaks are placed close to each other, so that heat transfer from the 1-in-thick edges is negligible. The entire steak is to be cooled below 45°F, but its temperature is not to drop below 35°F at any point during refrigeration to avoid "frostbite." The convection heat transfer coefficient and thus the rate of heat transfer from the steak can be controlled by varying the speed of a circulating fan inside. Determine the heat transfer coefficient h that will enable us to meet both temperature constraints while keeping the refrigeration time to a minimum. The steak can be treated as a homogeneous layer having the properties $\rho = 74.9$ lbm/ft^3, $c_p = 0.98$ Btu/lbm · °F, $k = 0.26$ Btu/h · ft · °F, and $\alpha = 0.0035$ ft^2/h.

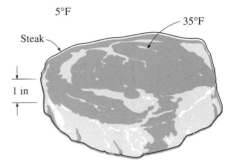

FIGURE 4–39
Schematic for Example 4–11.

SOLUTION Steaks are to be cooled in a refrigerator maintained at 5°F. The heat transfer coefficient that allows cooling the steaks below 45°F while avoiding frostbite is to be determined.

Assumptions **1** Heat conduction through the steaks is one-dimensional since the steaks form a large layer relative to their thickness and there is thermal symmetry about the center plane. **2** The thermal properties of the steaks and the heat transfer coefficient are constant. **3** The Fourier number is $\tau > 0.2$ so that the one-term approximate solutions are applicable.

Properties The properties of the steaks are as given in the problem statement.

Analysis The lowest temperature in the steak occurs at the surfaces and the highest temperature at the center at a given time, since the inner part is the last place to be cooled. In the limiting case, the surface temperature at $x = L = 0.5$ in from the center will be 35°F, while the midplane temperature is 45°F in an environment at 5°F. Then, from Fig. 4–15b, we obtain

$$\left. \begin{array}{l} \dfrac{x}{L} = \dfrac{0.5 \text{ in}}{0.5 \text{ in}} = 1 \\[2mm] \dfrac{T(L, t) - T_\infty}{T_0 - T_\infty} = \dfrac{35 - 5}{45 - 5} = 0.75 \end{array} \right\} \quad \dfrac{1}{\text{Bi}} = \dfrac{k}{hL} = 1.5$$

which gives

$$h = \frac{1}{1.5}\frac{k}{L} = \frac{0.26 \text{ Btu/h} \cdot \text{ft} \cdot \text{°F}}{1.5(0.5/12 \text{ ft})} = 4.16 \text{ Btu/h} \cdot \text{ft}^2 \cdot \text{°F}$$

Discussion The convection heat transfer coefficient should be kept below this value to satisfy the constraints on the temperature of the steak during refrigeration. We can also meet the constraints by using a lower heat transfer coefficient, but doing so would extend the refrigeration time unnecessarily.

The restrictions that are inherent in the use of Heisler charts and the one-term solutions (or any other analytical solutions) can be lifted by using the numerical methods discussed in Chapter 5.

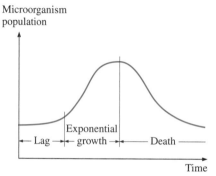

FIGURE 4–40
Typical growth curve of microorganisms.

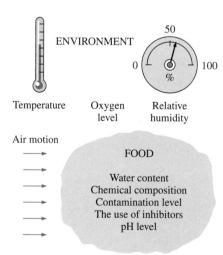

FIGURE 4–41
The factors that affect the rate of growth of microorganisms.

Refrigeration and Freezing of Foods

Control of Microorganisms in Foods

Microorganisms such as *bacteria, yeasts, molds,* and *viruses* are widely encountered in air, water, soil, living organisms, and unprocessed food items, and cause *off-flavors* and *odors, slime production, changes* in the texture and appearances, and the eventual *spoilage* of foods. Holding perishable foods at warm temperatures is the primary cause of spoilage, and the prevention of food spoilage and the premature degradation of quality due to microorganisms is the largest application area of refrigeration. The first step in controlling microorganisms is to understand what they are and the factors that affect their transmission, growth, and destruction.

Of the various kinds of microorganisms, *bacteria* are the prime cause for the spoilage of foods, especially moist foods. Dry and acidic foods create an undesirable environment for the growth of bacteria, but not for the growth of yeasts and molds. *Molds* are also encountered on moist surfaces, cheese, and spoiled foods. Specific *viruses* are encountered in certain animals and humans, and poor sanitation practices such as keeping processed foods in the same area as the uncooked ones and being careless about handwashing can cause the contamination of food products.

When *contamination* occurs, the microorganisms start to adapt to the new environmental conditions. This initial slow or no-growth period is called the **lag phase,** and the shelf life of a food item is directly proportional to the length of this phase (Fig. 4–40). The adaptation period is followed by an *exponential growth* period during which the population of microorganisms can double two or more times every hour under favorable conditions unless drastic sanitation measures are taken. The depletion of nutrients and the accumulation of toxins slow down the growth and start the *death* period.

The *rate of growth* of microorganisms in a food item depends on the characteristics of the food itself such as the chemical structure, pH level, presence of inhibitors and competing microorganisms, and water activity as well as the environmental conditions such as the temperature and relative humidity of the environment and the air motion (Fig. 4–41).

---

*This section can be skipped without a loss of continuity.

Microorganisms need *food* to grow and multiply, and their nutritional needs are readily provided by the carbohydrates, proteins, minerals, and vitamins in a food. Different types of microorganisms have different nutritional needs, and the types of nutrients in a food determine the types of microorganisms that may dwell on them. The preservatives added to the food may also inhibit the growth of certain microorganisms. Different kinds of microorganisms that exist compete for the same food supply, and thus the composition of microorganisms in a food at any time depends on the *initial make-up* of the microorganisms.

All living organisms need *water* to grow, and microorganisms cannot grow in foods that are not sufficiently moist. Microbiological growth in refrigerated foods such as fresh fruits, vegetables, and meats starts at the *exposed surfaces* where contamination is most likely to occur. Fresh meat in a package left in a room will spoil quickly, as you may have noticed. A meat carcass hung in a controlled environment, on the other hand, will age healthily as a result of *dehydration* on the outer surface, which inhibits microbiological growth there and protects the carcass.

Microorganism growth in a food item is governed by the combined effects of the *characteristics of the food* and the *environmental factors*. We cannot do much about the characteristics of the food, but we certainly can alter the environmental conditions to more desirable levels through *heating, cooling, ventilating, humidification, dehumidification,* and control of the *oxygen* levels. The growth rate of microorganisms in foods is a strong function of temperature, and temperature control is the single most effective mechanism for controlling the growth rate.

Microorganisms grow best at "warm" temperatures, usually between 20 and 60°C. The growth rate *declines* at high temperatures, and *death* occurs at still higher temperatures, usually above 70°C for most microorganisms. *Cooling* is an effective and practical way of reducing the growth rate of microorganisms and thus extending the *shelf life* of perishable foods. A temperature of 4°C or lower is considered to be a safe refrigeration temperature. Sometimes a small increase in refrigeration temperature may cause a large increase in the growth rate, and thus a considerable decrease in shelf life of the food (Fig. 4–42). The growth rate of some microorganisms, for example, doubles for each 3°C rise in temperature.

Another factor that affects microbiological growth and transmission is the *relative humidity* of the environment, which is a measure of the water content of the air. High humidity in *cold rooms* should be avoided since condensation that forms on the walls and ceiling creates the proper environment for *mold growth* and buildups. The drip of contaminated condensate onto food products in the room poses a potential health hazard.

Different microorganisms react differently to the presence of oxygen in the environment. Some microorganisms such as molds require oxygen for growth, while some others cannot grow in the presence of oxygen. Some grow best in low-oxygen environments, while others grow in environments regardless of the amount of oxygen. Therefore, the growth of certain microorganisms can be controlled by controlling the *amount of oxygen* in the environment. For example, vacuum packaging inhibits the growth of

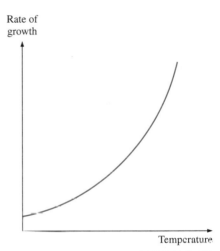

Rate of growth

Temperature

FIGURE 4–42

The rate of growth of microorganisms in a food product increases exponentially with increasing environmental temperature.

microorganisms that require oxygen. Also, the storage life of some fruits can be extended by reducing the oxygen level in the storage room.

Microorganisms in food products can be controlled by (1) *preventing* contamination by following strict sanitation practices, (2) *inhibiting* growth by altering the environmental conditions, and (3) *destroying* the organisms by heat treatment or chemicals. The best way to minimize contamination in food processing areas is to use fine air filters in ventilation systems to capture the *dust particles* that transport the bacteria in the air. Of course, the filters must remain dry since microorganisms can grow in wet filters. Also, the ventilation system must maintain a positive pressure in the food processing areas to prevent any airborne contaminants from entering inside by infiltration. The elimination of *condensation* on the walls and the ceiling of the facility and the diversion of *plumbing* condensation drip pans of refrigerators to the drain system are two other preventive measures against contamination. Drip systems must be cleaned regularly to prevent microbiological growth in them. Also, any *contact* between raw and cooked food products should be minimized, and cooked products must be stored in rooms with positive pressures. Frozen foods must be kept at −18°C or below, and utmost care should be exercised when food products are packaged after they are frozen to avoid contamination during packaging.

The growth of microorganisms is best controlled by keeping the *temperature* and *relative humidity* of the environment in the desirable range. Keeping the relative humidity below 60 percent, for example, prevents the growth of all microorganisms on the surfaces. Microorganisms can be destroyed by *heating* the food product to high temperatures (usually above 70°C), by treating them with *chemicals,* or by exposing them to *ultraviolet light* or solar radiation.

Distinction should be made between *survival* and *growth* of microorganisms. A particular microorganism that may not grow at some low temperature may be able to survive at that temperature for a very long time (Fig. 4–43). Therefore, freezing is not an effective way of killing microorganisms. In fact, some microorganism cultures are preserved by freezing them at very low temperatures. The *rate of freezing* is also an important consideration in the refrigeration of foods since some microorganisms adapt to low temperatures and grow at those temperatures when the cooling rate is very low.

Refrigeration and Freezing of Foods

The *storage life* of fresh perishable foods such as meats, fish, vegetables, and fruits can be extended by several days by storing them at temperatures just above freezing, usually between 1 and 4°C. The storage life of foods can be extended by several months by freezing and storing them at subfreezing temperatures, usually between −18 and −35°C, depending on the particular food (Fig. 4–44).

Refrigeration *slows down* the chemical and biological processes in foods, and the accompanying deterioration and loss of quality and nutrients. Sweet corn, for example, may lose half of its initial sugar content in one day at 21°C, but only 5 percent of it at 0°C. Fresh asparagus may lose 50 percent of its vitamin C content in one day at 20°C, but in 12 days

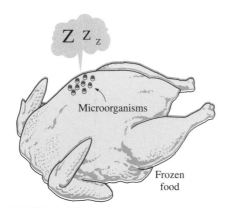

FIGURE 4–43
Freezing may stop the growth of microorganisms, but it may not necessarily kill them.

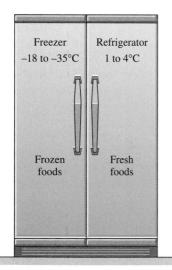

FIGURE 4–44
Recommended refrigeration and freezing temperatures for most perishable foods.

at 0°C. Refrigeration also extends the shelf life of products. The first appearance of unsightly yellowing of broccoli, for example, may be delayed by three or more days by refrigeration.

Early attempts to freeze food items resulted in poor-quality products because of the large ice crystals that formed. It was determined that the *rate of freezing* has a major effect on the size of ice crystals and the quality, texture, and nutritional and sensory properties of many foods. During *slow freezing,* ice crystals can grow to a large size, whereas during *fast freezing* a large number of ice crystals start forming at once and are much smaller in size. Large ice crystals are not desirable since they can *puncture* the walls of the cells, causing a degradation of texture and a loss of natural juices during thawing. A *crust* forms rapidly on the outer layer of the product and seals in the juices, aromatics, and flavoring agents. The product quality is also affected adversely by temperature fluctuations of the storage room.

The ordinary refrigeration of foods involves *cooling* only without any phase change. The *freezing* of foods, on the other hand, involves three stages: *cooling* to the freezing point (removing the sensible heat), *freezing* (removing the latent heat), and *further cooling* to the desired subfreezing temperature (removing the sensible heat of frozen food), as shown in Figure 4–45.

Beef Products

Meat carcasses in slaughterhouses should be cooled *as fast as possible* to a uniform temperature of about 1.7°C to reduce the growth rate of microorganisms that may be present on carcass surfaces, and thus minimize spoilage. The right level of *temperature, humidity,* and *air motion* should be selected to prevent excessive shrinkage, toughening, and discoloration.

The deep body temperature of an animal is about 39°C, but this temperature tends to rise a couple of degrees in the midsections after slaughter as a result of the *heat generated* during the biological reactions that occur in the cells. The temperature of the exposed surfaces, on the other hand, tends to drop as a result of heat losses. The thickest part of the carcass is the *round,* and the center of the round is the last place to cool during chilling. Therefore, the cooling of the carcass can best be monitored by inserting a thermometer deep into the central part of the round.

About 70 percent of the beef carcass is water, and the carcass is cooled mostly by *evaporative cooling* as a result of moisture migration toward the surface where evaporation occurs. But this shrinking translates into a loss of salable mass that can amount to 2 percent of the total mass during an overnight chilling. To prevent *excessive* loss of mass, carcasses are usually washed or sprayed with water prior to cooling. With adequate care, spray chilling can eliminate carcass cooling shrinkage almost entirely.

The average total mass of dressed beef, which is normally split into two sides, is about 300 kg, and the average specific heat of the carcass is about 3.14 kJ/kg · °C (Table 4–6). The *chilling room* must have a capacity equal to the daily kill of the slaughterhouse, which may be several hundred. A beef carcass is washed before it enters the chilling room and absorbs a large amount of water (about 3.6 kg) at its surface during the washing process. This does not represent a net mass gain, however, since it is lost by

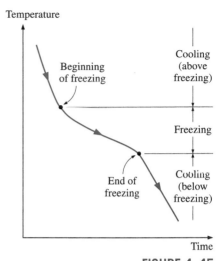

FIGURE 4–45
Typical freezing curve of a food item.

TABLE 4–6

Thermal properties of beef

Quantity	Typical value
Average density	1070 kg/m³
Specific heat:	
Above freezing	3.14 kJ/kg · °C
Below freezing	1.70 kJ/kg · °C
Freezing point	−2.7°C
Latent heat of fusion	249 kJ/kg
Thermal	0.41 W/m · °C
conductivity	(at 6°C)

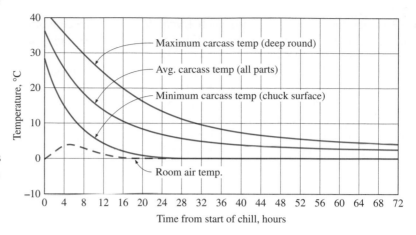

FIGURE 4–46

Typical cooling curve of a beef carcass in the chilling and holding rooms at an average temperature of 0°C (from ASHRAE, *Handbook: Refrigeration,* Chap. 11, Fig. 2).

dripping or evaporation in the chilling room during cooling. Ideally, the carcass does not lose or gain any net weight as it is cooled in the chilling room. However, it does lose about 0.5 percent of the total mass in the *holding room* as it continues to cool. The actual product loss is determined by first weighing the dry carcass before washing and then weighing it again after it is cooled.

The refrigerated air temperature in the chilling room of beef carcasses must be sufficiently high to avoid *freezing* and *discoloration* on the outer surfaces of the carcass. This means a long residence time for the massive beef carcasses in the chilling room to cool to the desired temperature. Beef carcasses are only partially cooled at the end of an overnight stay in the chilling room. The temperature of a beef carcass drops to 1.7 to 7°C at the surface and to about 15°C in mid parts of the round in 10 h. It takes another day or two in the *holding room* maintained at 1 to 2°C to complete *chilling* and *temperature equalization.* But hog carcasses are fully chilled during that period because of their smaller size. The *air circulation* in the holding room is kept at minimum levels to avoid excessive moisture loss and discoloration. The refrigeration load of the holding room is much smaller than that of the chilling room, and thus it requires a smaller refrigeration system.

Beef carcasses intended for distant markets are shipped the day after slaughter in refrigerated trucks, where the rest of the cooling is done. This practice makes it possible to deliver fresh meat long distances in a timely manner.

The variation in temperature of the beef carcass during cooling is given in Figure 4–46. Initially, the cooling process is dominated by *sensible* heat transfer. Note that the average temperature of the carcass is reduced by about 28°C (from 36 to 8°C) in 20 h. The cooling rate of the carcass could be increased by *lowering* the refrigerated air temperature and *increasing* the air velocity, but such measures also increase the risk of *surface freezing.*

Most meats are judged on their **tenderness**, and the preservation of tenderness is an important consideration in the refrigeration and freezing of meats. Meat consists primarily of bundles of tiny *muscle fibers* bundled together inside long strings of *connective tissues* that hold it together. The tenderness of a certain cut of beef depends on the location of the cut, the age, and the activity level of the animal. Cuts from the relatively inactive mid-backbone section of the animal such as short loins, sirloin, and prime ribs are more tender than the cuts from the active parts such as the legs and the neck (Fig. 4–47). The more active the animal, the more the connective tissue, and the tougher the meat. The meat of an older animal is more flavorful, however, and is preferred for stewing since the toughness of the meat does not pose a problem for moist-heat cooking such as boiling. The protein *collagen,* which is the main component of the connective tissue, softens and dissolves in hot and moist environments and gradually transforms into *gelatin,* and tenderizes the meat.

The old saying "one should either cook an animal immediately after slaughter or wait at least two days" has a lot of truth in it. The biomechanical reactions in the muscle continue after the slaughter until the energy supplied to the muscle to do work diminishes. The muscle then stiffens and goes into *rigor mortis.* This process begins several hours after the animal is slaughtered and continues for 12 to 36 h until an enzymatic action sets in and tenderizes the connective tissue, as shown in Figure 4–48. It takes about seven days to complete tenderization naturally in storage facilities maintained at 2°C. Electrical stimulation also causes the meat to be tender. To avoid toughness, fresh meat should not be frozen before rigor mortis has passed.

You have probably noticed that steaks are tender and rather tasty when they are hot but toughen as they cool. This is because the gelatin that formed during cooking thickens as it cools, and meat loses its tenderness. So it is no surprise that first-class restaurants serve their steak on hot thick plates that keep the steaks warm for a long time. Also, cooking *softens* the connective tissue but *toughens* the tender muscle fibers. Therefore, barbecuing on *low heat* for a long time results in a tough steak.

Variety meats intended for long-term storage must be frozen rapidly to reduce spoilage and preserve quality. Perhaps the first thought that comes to mind to freeze meat is to place the meat packages into the *freezer* and wait. But the freezing time is *too long* in this case, especially for large boxes. For example, the core temperature of a 4-cm-deep box containing 32 kg of variety meat can be as high as 16°C 24 h after it is placed into a −30°C freezer. The freezing time of large boxes can be shortened considerably by adding some *dry ice* into it.

A more effective method of freezing, called *quick chilling,* involves the use of lower air temperatures, −40 to −30°C, with higher velocities of 2.5 m/s to 5 m/s over the product (Fig. 4–49). The internal temperature should be lowered to −4°C for products to be transferred to a storage freezer and to −18°C for products to be shipped immediately. The *rate of freezing* depends on the *package material* and its insulating properties, the *thickness* of the largest box, the *type* of meat, and the *capacity* of the refrigeration system. Note that the air temperature will rise excessively during initial stages of freezing and increase the freezing time if the capacity of the system is inadequate. A smaller refrigeration system will be adequate if dry ice is to be used in packages. Shrinkage during freezing varies from about 0.5 to 1 percent.

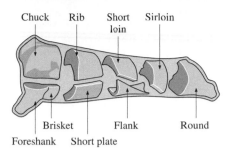

FIGURE 4–47
Various cuts of beef (from National Livestock and Meat Board).

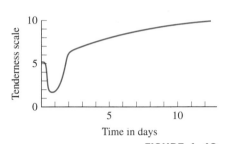

FIGURE 4–48
Variation of tenderness of meat stored at 2°C with time after slaughter.

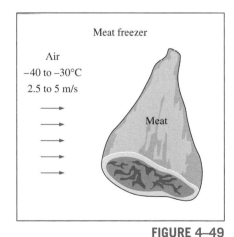

FIGURE 4–49
The freezing time of meat can be reduced considerably by using low temperature air at high velocity.

TABLE 4–7

Storage life of frozen meat products at different storage temperatures (from ASHRAE *Handbook: Refrigeration,* Chap. 10, Table 7)

Product	Storage Life, Months Temperature		
	−12°C	−18°C	−23°C
Beef	4–12	6–18	12–24
Lamb	3–8	6–16	12–18
Veal	3–4	4–14	8
Pork	2–6	4–12	8–15
Chopped beef	3–4	4–6	8
Cooked foods	2–3	2–4	

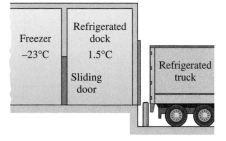

FIGURE 4–50

A refrigerated truck dock for loading frozen items to a refrigerated truck.

Although the average freezing point of lean meat can be taken to be −2°C with a latent heat of 249 kJ/kg, it should be remembered that freezing occurs over a *temperature range,* with most freezing occurring between −1 and −4°C. Therefore, cooling the meat through this temperature range and removing the latent heat takes the most time during freezing.

Meat can be kept at an internal temperature of −2 to −1°C for local use and storage for *under a week.* Meat must be frozen and stored at much lower temperatures for *long-term storage.* The lower the storage temperature, the longer the storage life of meat products, as shown in Table 4–7.

The *internal temperature* of carcasses entering the cooling sections varies from 38 to 41°C for hogs and from 37 to 39°C for lambs and calves. It takes about 15 h to cool the hogs and calves to the recommended temperature of 3 to 4°C. The cooling-room temperature is maintained at −1 to 0°C and the temperature difference between the refrigerant and the cooling air is kept at about 6°C. Air is circulated at a rate of about 7 to 12 air changes per hour. *Lamb carcasses* are cooled to an internal temperature of 1 to 2°C, which takes about 12 to 14 h, and are held at that temperature with 85 to 90 percent relative humidity until shipped or processed. The recommended rate of *air circulation* is 50 to 60 air changes per hour during the first 4 to 6 h, which is reduced to 10 to 12 changes per hour afterward.

Freezing does not seem to affect the *flavor* of meat much, but it affects the *quality* in several ways. The *rate* and *temperature* of freezing may influence color, tenderness, and drip. Rapid freezing increases tenderness and reduces the tissue damage and the amount of drip after thawing. Storage at low freezing temperatures causes significant changes in *animal fat.* Frozen pork experiences more undesirable changes during storage because of its fat structure, and thus its acceptable storage period is shorter than that of beef, veal, or lamb.

Meat storage facilities usually have a *refrigerated shipping dock* where the orders are assembled and shipped out. Such docks save valuable storage space from being used for shipping purposes and provide a more acceptable working environment for the employees. Packing plants that ship whole or half carcasses in bulk quantities may not need a shipping dock; a load-out door is often adequate for such cases.

A refrigerated *shipping dock,* as shown in Figure 4–50, reduces the *refrigeration load* of freezers or coolers and prevents *temperature fluctuations* in the storage area. It is often adequate to maintain the shipping docks at 4 to 7°C for the coolers and about 1.5°C for the freezers. The dew point of the dock air should be below the product temperature to avoid condensation on the surface of the products and loss of quality. The rate of *airflow* through the loading doors and other openings is proportional to the *square root* of the temperature difference, and thus reducing the temperature difference at the opening by half by keeping the shipping dock at the average temperature reduces the rate of airflow into the dock and thus into the freezer by $1 - \sqrt{0.5} \cong 0.3$, or 30 percent. Also, the air that flows into the freezer is already cooled to about 1.5°C by the refrigeration unit of the dock, which represents about 50 percent of the cooling load of the incoming air. Thus, the net effect of the refrigerated shipping dock is a reduction of the *infiltration load* of the freezer by about 65 percent since

$1 - 0.7 \times 0.5 = 0.65$. The net gain is equal to the difference between the reduction of the infiltration load of the freezer and the refrigeration load of the shipping dock. Note that the dock refrigerators operate at much higher temperatures ($1.5°C$ instead of about $-23°C$), and thus they consume much less power for the same amount of cooling.

Poultry Products

Poultry products can be preserved by *ice-chilling* to 1 to 2°C or *deep chilling* to about −2°C for short-term storage, or by *freezing* them to −18°C or below for long-term storage. Poultry processing plants are completely *automated,* and the small size of the birds makes continuous conveyor line operation feasible.

The birds are first electrically stunned before cutting to prevent struggling. Following a 90- to 120-s bleeding time, the birds are *scalded* by immersing them into a tank of warm water, usually at 51 to 55°C, for up to 120 s to loosen the feathers. Then the feathers are removed by feather-picking machines, and the eviscerated carcass is *washed* thoroughly before chilling. The internal temperature of the birds ranges from 24 to 35°C after washing, depending on the temperatures of the ambient air and the washing water as well as the extent of washing.

To control the microbial growth, the USDA regulations require that poultry be chilled to 4°C or below in less than 4 h for carcasses of less than 1.8 kg, in less than 6 h for carcasses of 1.8 to 3.6 kg, and in less than 8 h for carcasses more than 3.6 kg. Meeting these requirements today is not difficult since the slow *air chilling* is largely replaced by the rapid *immersion chilling* in tanks of slush ice. Immersion chilling has the added benefit that it not only prevents dehydration, but it causes a *net absorption of water* and thus increases the mass of salable product. Cool air chilling of unpacked poultry can cause a moisture loss of 1 to 2 percent, while water immersion chilling can cause a moisture absorption of 4 to 15 percent (Fig. 4–51). Water spray chilling can cause a moisture absorption of up to 4 percent. Most water absorbed is held between the flesh and the skin and the connective tissues in the skin. In immersion chilling, some soluble solids are lost from the carcass to the water, but the loss has no significant effect on flavor.

Many slush ice tank chillers today are replaced by *continuous* flow-type immersion slush ice chillers. Continuous slush ice-chillers can reduce the internal temperature of poultry from 32 to 4°C in about 30 minutes at a rate up to 10, 000 birds per hour. Ice requirements depend on the inlet and exit temperatures of the carcass and the water, but 0.25 kg of ice per kg of carcass is usually adequate. However, *bacterial contamination* such as salmonella remains a concern with this method, and it may be necessary to chloride the water to control contamination.

Tenderness is an important consideration for poultry products just as it is for red meat, and preserving tenderness is an important consideration in the cooling and freezing of poultry. Birds cooked or frozen before passing through rigor mortis remain very tough. Natural tenderization begins soon after slaughter and is completed within 24 h when birds are held at 4°C.

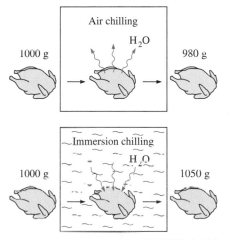

FIGURE 4–51

Air chilling causes dehydration and thus weight loss for poultry, whereas immersion chilling causes a weight gain as a result of water absorption.

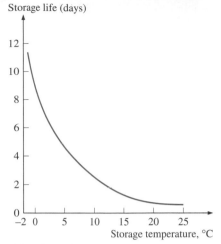

FIGURE 4–52

The storage life of fresh poultry decreases exponentially with increasing storage temperature.

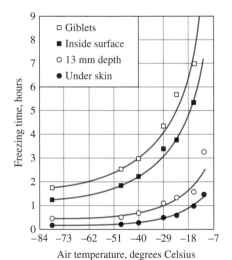

Note: Freezing time is the time required for temperature to fall from 0 to –4°C. The values are for 2.3 to 3.6 kg chickens with initial temperature of 0 to 2°C and with air velocity of 2.3 to 2.8 m/s.

FIGURE 4–53

The variation of freezing time of poultry with air temperature.

Tenderization is rapid during the first three hours and slows down thereafter. Immersion in hot water and cutting into the muscle adversely affect tenderization. Increasing the *scalding temperature* or the scalding time has been observed to increase toughness, and decreasing the scalding time has been observed to increase tenderness. The *beating action* of mechanical feather-picking machines causes considerable toughening. Therefore, it is recommended that any cutting be done after tenderization. *Cutting up* the bird into pieces before natural tenderization is completed reduces tenderness considerably. Therefore, it is recommended that any cutting be done after tenderization. *Rapid chilling* of poultry can also have a toughening effect. It is found that the tenderization process can be speeded up considerably by a patented *electrical stunning* process.

Poultry products are *highly perishable,* and thus they should be kept at the *lowest* possible temperature to maximize their shelf life. Studies have shown that the populations of certain bacteria double every 36 h at −2°C, 14 h at 0°C, 7 h at 5°C, and less than 1 h at 25°C (Fig. 4–52). Studies have also shown that the total bacterial counts on birds held at 2°C for 14 days are equivalent to those held at 10°C for 5 days or 24°C for 1 day. It has also been found that birds held at −1°C had 8 days of additional shelf life over those held at 4°C.

The growth of microorganisms on the *surfaces* of the poultry causes the development of an *off-odor* and *bacterial slime.* The higher the initial amount of bacterial contamination, the faster the sliming occurs. Therefore, good sanitation practices during processing such as cleaning the equipment frequently and washing the carcasses are as important as the storage temperature in extending shelf life.

Poultry must be frozen *rapidly* to ensure a light, pleasing appearance. Poultry that is frozen slowly appears dark and develops large ice crystals that damage the tissue. The ice crystals formed during rapid freezing are small. Delaying freezing of poultry causes the ice crystals to become larger. Rapid freezing can be accomplished by forced air at temperatures of −23 to −40°C and velocities of 1.5 to 5 m/s in *air-blast tunnel freezers.* Most poultry is frozen this way. Also, the packaged birds freeze much faster on open shelves than they do in boxes. If poultry packages must be frozen in boxes, then it is very desirable to leave the boxes open or to cut holes on the boxes in the direction of airflow during freezing. For best results, the blast tunnel should be fully loaded across its cross-section with even spacing between the products to assure uniform airflow around all sides of the packages. The freezing time of poultry as a function of refrigerated air temperature is given in Figure 4–53. Thermal properties of poultry are given in Table 4–8.

Other freezing methods for poultry include sandwiching between *cold plates, immersion* into a refrigerated liquid such as glycol or calcium chloride brine, and *cryogenic cooling* with liquid nitrogen. Poultry can be frozen in several hours by cold plates. Very high freezing rates can be obtained by *immersing* the packaged birds into a low-temperature brine. The freezing time of birds in −29°C brine can be as low as 20 min, depending on the size of the bird (Fig. 4–54). Also, immersion freezing produces a very appealing light appearance, and the high rates of heat transfer make

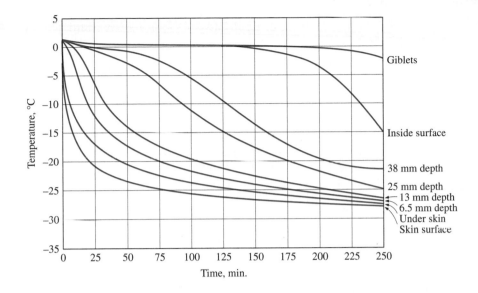

FIGURE 4–54

The variation of temperature of the breast of 6.8-kg turkeys initially at 1°C with depth during immersion cooling at −29°C (from van der Berg and Lentz, 1958).

continuous line operation feasible. It also has lower initial and maintenance costs than forced air, but *leaks* into the packages through some small holes or cracks remain a concern. The convection heat transfer coefficient is 17 W/m² · °C for air at −29°C and 2.5 m/s whereas it is 170 W/m² · °C for sodium chloride brine at −18°C and a velocity of 0.02 m/s. Sometimes liquid nitrogen is used to crust freeze the poultry products to −73°C. The freezing is then completed with air in a holding room at −23°C.

Properly packaged poultry products can be *stored* frozen for up to about a year at temperatures of −18°C or lower. The storage life drops considerably at higher (but still below-freezing) temperatures. Significant changes occur in flavor and juiciness when poultry is frozen for too long, and a stale rancid odor develops. Frozen poultry may become dehydrated and experience **freezer burn**, which may reduce the eye appeal of the product and cause toughening of the affected area. Dehydration and thus freezer burn can be controlled by *humidification, lowering* the storage temperature, and packaging the product in essentially *impermeable* film. The storage life can be extended by packing the poultry in an *oxygen-free* environment. The bacterial counts in precooked frozen products must be kept at safe levels since bacteria may not be destroyed completely during the reheating process at home.

Frozen poultry can be *thawed* in ambient air, water, refrigerator, or oven without any significant difference in taste. Big birds like turkey should be thawed safely by holding it in a refrigerator at 2 to 4°C for two to four days, depending on the size of the bird. They can also be thawed by immersing them into cool water in a large container for 4 to 6 h, or holding them in a paper bag. Care must be exercised to keep the bird's surface *cool* to minimize *microbiological growth* when thawing in air or water.

TABLE 4–8

Thermal properties of poultry

Quantity	Typical value
Average density:	
Muscle	1070 kg/m³
Skin	1030 kg/m³
Specific heat:	
Above freezing	2.94 kJ/kg · °C
Below freezing	1.55 kJ/kg · °C
Freezing point	−2.8°C
Latent heat of fusion	247 kJ/kg
Thermal conductivity: (in W/m · °C)	
Breast muscle	0.502 at 20°C
	1.384 at −20°C
	1.506 at −40°C
Dark muscle	1.557 at −40°C

EXAMPLE 4–11   Chilling of Beef Carcasses in a Meat Plant

The chilling room of a meat plant is 18 m × 20 m × 5.5 m in size and has a capacity of 450 beef carcasses. The power consumed by the fans and the lights of the chilling room are 26 and 3 kW, respectively, and the room gains heat through its envelope at a rate of 13 kW. The average mass of beef carcasses is 285 kg. The carcasses enter the chilling room at 36°C after they are washed to facilitate evaporative cooling and are cooled to 15°C in 10 h. The water is expected to evaporate at a rate of 0.080 kg/s. The air enters the evaporator section of the refrigeration system at 0.7°C and leaves at −2°C. The air side of the evaporator is heavily finned, and the overall heat transfer coefficient of the evaporator based on the air side is 20 W/m² · °C. Also, the average temperature difference between the air and the refrigerant in the evaporator is 5.5°C. Determine (a) the refrigeration load of the chilling room, (b) the volume flow rate of air, and (c) the heat transfer surface area of the evaporator on the air side, assuming all the vapor and the fog in the air freezes in the evaporator.

SOLUTION   The chilling room of a meat plant with a capacity of 450 beef carcasses is considered. The cooling load, the airflow rate, and the heat transfer area of the evaporator are to be determined.

Assumptions   **1** Water evaporates at a rate of 0.080 kg/s. **2** All the moisture in the air freezes in the evaporator.

Properties   The heat of fusion and the heat of vaporization of water at 0°C are 333.7 kJ/kg and 2501 kJ/kg (Table A–9). The density and specific heat of air at 0°C are 1.292 kg/m³ and 1.006 kJ/kg · °C (Table A–15). Also, the specific heat of beef carcass is determined from the relation in Table A–7b to be

$$c_p = 1.68 + 2.51 \times (\text{water content}) = 1.68 + 2.51 \times 0.58 = 3.14 \text{ kJ/kg} \cdot °C$$

Analysis   (a) A sketch of the chilling room is given in Figure 4–55. The amount of beef mass that needs to be cooled per unit time is

$$\dot{m}_{\text{beef}} = (\text{Total beef mass cooled})/(\text{Cooling time})$$
$$= (450 \text{ carcasses})(285 \text{ kg/carcass})/(10 \times 3600 \text{ s}) = 3.56 \text{ kg/s}$$

The product refrigeration load can be viewed as the energy that needs to be removed from the beef carcass as it is cooled from 36 to 15°C at a rate of 3.56 kg/s and is determined to be

$$\dot{Q}_{\text{beef}} = (\dot{m} c_p \Delta T)_{\text{beef}} = (3.56 \text{ kg/s})(3.14 \text{ kJ/kg} \cdot °C)(36 - 15)°C = 235 \text{ kW}$$

Then the total refrigeration load of the chilling room becomes

$$\dot{Q}_{\text{total, chillroom}} = \dot{Q}_{\text{beef}} + \dot{Q}_{\text{fan}} + \dot{Q}_{\text{lights}} + \dot{Q}_{\text{heat gain}} = 235 + 26 + 3 + 13$$
$$= 277 \text{ kW}$$

The amount of carcass cooling due to evaporative cooling of water is

$$\dot{Q}_{\text{beef, evaporative}} = (\dot{m} h_{fg})_{\text{water}} = (0.080 \text{ kg/s})(2501 \text{ kJ/kg}) = 200 \text{ kW}$$

which is 200/235 = 85 percent of the total product cooling load. The remaining 15 percent of the heat is transferred by convection and radiation.

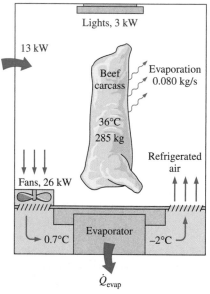

FIGURE 4–55
Schematic for Example 4–11.

Lights, 3 kW

13 kW

Beef carcass

Evaporation 0.080 kg/s

36°C
285 kg

Refrigerated air

Fans, 26 kW

0.7°C   Evaporator   −2°C

$\dot{Q}_{\text{evap}}$

(b) Heat Is transferred to air at the rate determined above, and the temperature of the air rises from $-2°C$ to $0.7°C$ as a result. Therefore, the mass flow rate of air is

$$\dot{m}_{air} = \frac{\dot{Q}_{air}}{(c_p \Delta T_{air})} = \frac{277 \text{ kW}}{(1.006 \text{ kJ/kg} \cdot °C)[0.7 - (-2)°C]} = 102.0 \text{ kg/s}$$

Then the volume flow rate of air becomes

$$\dot{V}_{air} = \frac{\dot{m}_{air}}{\rho_{air}} = \frac{102 \text{ kg/s}}{1.292 \text{ kg/m}^3} = 78.9 \text{ m}^3/\text{s}$$

(c) Normally the heat transfer load of the evaporator is the same as the refrigeration load. But in this case the water that enters the evaporator as a liquid is frozen as the temperature drops to $-2°C$, and the evaporator must also remove the latent heat of freezing, which is determined from

$$\dot{Q}_{freezing} = (\dot{m}h_{latent})_{water} = (0.080 \text{ kg/s})(333.7 \text{ kJ/kg}) = 27 \text{ kW}$$

Therefore, the total rate of heat removal at the evaporator is

$$\dot{Q}_{evaporator} = \dot{Q}_{total, \text{ chill room}} + \dot{Q}_{freezing} = 277 + 27 = 304 \text{ kW}$$

Then the heat transfer surface area of the evaporator on the air side is determined from $\dot{Q}_{evaporator} = (UA)_{\text{air side}} \Delta T$,

$$A = \frac{\dot{Q}_{evaporator}}{U\Delta T} = \frac{304,000 \text{ W}}{(20 \text{ W/m}^2 \cdot °C)(5.5°C)} = 2764 \text{ m}^2$$

Obviously, a finned surface must be used to provide such a large surface area on the air side.

SUMMARY

In this chapter, we considered the variation of temperature with time as well as position in one- or multidimensional systems. We first considered the *lumped systems* in which the temperature varies with time but remains uniform throughout the system at any time. The temperature of a lumped body of arbitrary shape of mass m, volume V, surface area A_s, density ρ, and specific heat c_p initially at a uniform temperature T_i that is exposed to convection at time $t = 0$ in a medium at temperature T_∞ with a heat transfer coefficient h is expressed as

$$\frac{T(t) - T_\infty}{T_i - T_\infty} = e^{-bt}$$

where

$$b = \frac{hA_s}{\rho c_p V} = \frac{h}{\rho c_p L_c}$$

is a positive quantity whose dimension is $(\text{time})^{-1}$. This relation can be used to determine the temperature $T(t)$ of a body at time t or, alternatively, the time t required for the temperature to reach a specified value $T(t)$. Once the temperature $T(t)$ at time t is available, the *rate* of convection heat transfer between the body and its environment at that time can be determined from Newton's law of cooling as

$$\dot{Q}(t) = hA_s[T(t) - T_\infty]$$

The *total amount* of heat transfer between the body and the surrounding medium over the time interval $t = 0$ to t is simply the change in the energy content of the body,

$$Q = mc_p[T(t) - T_i]$$

The *maximum* heat transfer between the body and its surroundings is

$$Q_{max} = mc_p(T_\infty - T_i)$$

The error involved in lumped system analysis is negligible when

$$\text{Bi} = \frac{hL_c}{k} < 0.1$$

where Bi is the *Biot number* and $L_c = V/A_s$ is the *characteristic length*.

When the lumped system analysis is not applicable, the variation of temperature with position as well as time can be determined using the *transient temperature charts* given in Figs. 4–15, 4–16, 4–17, and 4–29 for a large plane wall, a long cylinder, a sphere, and a semi-infinite medium, respectively. These charts are applicable for one-dimensional heat transfer in those geometries. Therefore, their use is limited to situations in which the body is initially at a uniform temperature, all surfaces are subjected to the same thermal conditions, and the body does not involve any heat generation. These charts can also be used to determine the total heat transfer from the body up to a specified time t.

Using the *one-term approximation*, the solutions of one-dimensional transient heat conduction problems are expressed analytically as

Plane wall: $\quad \theta_{\text{wall}} = \dfrac{T(x, t) - T_\infty}{T_i - T_\infty} = A_1 e^{-\lambda_1^2 \tau} \cos (\lambda_1 x/L)$

Cylinder: $\quad \theta_{\text{cyl}} = \dfrac{T(r, t) - T_\infty}{T_i - T_\infty} = A_1 e^{-\lambda_1^2 \tau} J_0(\lambda_1 r/r_o)$

Sphere: $\quad \theta_{\text{sph}} = \dfrac{T(r, t) - T_\infty}{T_i - T_\infty} = A_1 e^{-\lambda_1^2 \tau} \dfrac{\sin(\lambda_1 r/r_o)}{\lambda_1 r/r_o}$

where the constants A_1 and λ_1 are functions of the Bi number only, and their values are listed in Table 4–2 against the Bi number for all three geometries. The error involved in one-term solutions is less than 2 percent when $\tau > 0.2$.

Using the one-term solutions, the fractional heat transfers in different geometries are expressed as

Plane wall: $\quad \left(\dfrac{Q}{Q_{\max}}\right)_{\text{wall}} = 1 - \theta_{0,\,\text{wall}} \dfrac{\sin \lambda_1}{\lambda_1}$

Cylinder: $\quad \left(\dfrac{Q}{Q_{\max}}\right)_{\text{cyl}} = 1 - 2\theta_{0,\,\text{cyl}} \dfrac{J_1(\lambda_1)}{\lambda_1}$

Sphere: $\quad \left(\dfrac{Q}{Q_{\max}}\right)_{\text{sph}} = 1 - 3\theta_{0,\,\text{sph}} \dfrac{\sin \lambda_1 - \lambda_1 \cos \lambda_1}{\lambda_1^3}$

The solutions of transient heat conduction in a semi-infinite solid with constant properties under various boundary conditions at the surface are given as follows:

Specified Surface Temperature, T_s = constant:

$$\frac{T(x, t) - T_i}{T_s - T_i} = \text{erfc}\left(\frac{x}{2\sqrt{\alpha t}}\right) \quad \text{and} \quad \dot{q}_s(t) = \frac{k(T_s - T_i)}{\sqrt{\pi \alpha t}}$$

Specified Surface Heat Flux, $\dot{q}_s$ = constant:

$$T(x, t) - T_i = \frac{\dot{q}_s}{k}\left[\sqrt{\frac{4\alpha t}{\pi}} \exp\left(-\frac{x^2}{4\alpha t}\right) - x\,\text{erfc}\left(\frac{x}{2\sqrt{\alpha t}}\right)\right]$$

Convection on the Surface, $\dot{q}_s(t) = h[T_\infty - T(0, t)]$:

$$\frac{T(x, t) - T_i}{T_\infty - T_i} = \text{erfc}\left(\frac{x}{2\sqrt{\alpha t}}\right) - \exp\left(\frac{hx}{k} + \frac{h^2\alpha t}{k^2}\right)$$

$$\times \text{erfc}\left(\frac{x}{2\sqrt{\alpha t}} + \frac{h\sqrt{\alpha t}}{k}\right)$$

Energy Pulse at Surface, e_s = constant:

$$T(x, t) - T_i = \frac{e_s}{k\sqrt{\pi t/\alpha}} \exp\left(-\frac{x^2}{4\alpha t}\right)$$

where $\text{erfc}(\eta)$ is the *complementary error function* of argument η.

Using a superposition principle called the *product solution* these charts can also be used to construct solutions for the *two-dimensional* transient heat conduction problems encountered in geometries such as a short cylinder, a long rectangular bar, or a semi-infinite cylinder or plate, and even *three-dimensional* problems associated with geometries such as a rectangular prism or a semi-infinite rectangular bar, provided that all surfaces of the solid are subjected to convection to the same fluid at temperature T_∞, with the same convection heat transfer coefficient h, and the body involves no heat generation. The solution in such multidimensional geometries can be expressed as the product of the solutions for the one-dimensional geometries whose intersection is the multidimensional geometry.

The total heat transfer to or from a multidimensional geometry can also be determined by using the one-dimensional values. The transient heat transfer for a two-dimensional geometry formed by the intersection of two one-dimensional geometries 1 and 2 is

$$\left(\frac{Q}{Q_{\max}}\right)_{\text{total, 2D}} = \left(\frac{Q}{Q_{\max}}\right)_1 + \left(\frac{Q}{Q_{\max}}\right)_2 \left[1 - \left(\frac{Q}{Q_{\max}}\right)_1\right]$$

Transient heat transfer for a three-dimensional body formed by the intersection of three one-dimensional bodies 1, 2, and 3 is given by

$$\left(\frac{Q}{Q_{\max}}\right)_{\text{total, 3D}} = \left(\frac{Q}{Q_{\max}}\right)_1 + \left(\frac{Q}{Q_{\max}}\right)_2 \left[1 - \left(\frac{Q}{Q_{\max}}\right)_1\right]$$

$$+ \left(\frac{Q}{Q_{\max}}\right)_3 \left[1 - \left(\frac{Q}{Q_{\max}}\right)_1\right]\left[1 - \left(\frac{Q}{Q_{\max}}\right)_2\right]$$

REFERENCES AND SUGGESTED READINGS

1. ASHRAE. *Handbook of Fundamentals.* SI version. Atlanta, GA: American Society of Heating, Refrigerating, and Air-Conditioning Engineers, Inc., 1993.

2. ASHRAE. *Handbook of Fundamentals.* SI version. Atlanta, GA: American Society of Heating, Refrigerating, and Air-Conditioning Engineers, Inc., 1994.

3. H. S. Carslaw and J. C. Jaeger. *Conduction of Heat in Solids.* 2nd ed. London: Oxford University Press, 1959.

4. H. Gröber, S. Erk, and U. Grigull. *Fundamentals of Heat Transfer.* New York: McGraw-Hill, 1961.

5. M. P. Heisler. "Temperature Charts for Induction and Constant Temperature Heating." *ASME Transactions* 69 (1947), pp. 227–36.

6. H. Hillman. *Kitchen Science.* Mount Vernon, NY: Consumers Union, 1981.

7. S. Kakaç and Y. Yener, *Heat Conduction,* New York: Hemisphere Publishing Co., 1985.

8. L. S. Langston. "Heat Transfer from Multidimensional Objects Using One-Dimensional Solutions for Heat Loss." *International Journal of Heat and Mass Transfer* 25 (1982), pp. 149–50.

9. P. J. Schneider. *Conduction Heat Transfer.* Reading, MA: Addison-Wesley, 1955.

10. L. van der Berg and C. P. Lentz. "Factors Affecting Freezing Rate and Appearance of Eviscerated Poultry Frozen in Air." *Food Technology* 12 (1958).

PROBLEMS*

Lumped System Analysis

4–1C What is lumped system analysis? When is it applicable?

4–2C Consider heat transfer between two identical hot solid bodies and the air surrounding them. The first solid is being cooled by a fan while the second one is allowed to cool naturally. For which solid is the lumped system analysis more likely to be applicable? Why?

4–3C Consider heat transfer between two identical hot solid bodies and their environments. The first solid is dropped in a large container filled with water, while the second one is allowed to cool naturally in the air. For which solid is the lumped system analysis more likely to be applicable? Why?

4–4C Consider a hot baked potato on a plate. The temperature of the potato is observed to drop by 4°C during the first minute. Will the temperature drop during the second minute be less than, equal to, or more than 4°C? Why?

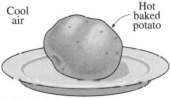

Cool air

Hot baked potato

FIGURE P4–4C

4–5C Consider a potato being baked in an oven that is maintained at a constant temperature. The temperature of the potato is observed to rise by 5°C during the first minute. Will the temperature rise during the second minute be less than, equal to, or more than 5°C? Why?

4–6C What is the physical significance of the Biot number? Is the Biot number more likely to be larger for highly conducting solids or poorly conducting ones?

4–7C Consider two identical 4-kg pieces of roast beef. The first piece is baked as a whole, while the second is baked after being cut into two equal pieces in the same oven. Will there be any difference between the cooking times of the whole and cut roasts? Why?

4–8C Consider a sphere and a cylinder of equal volume made of copper. Both the sphere and the cylinder are initially at the same temperature and are exposed to convection in the same environment. Which do you think will cool faster, the cylinder or the sphere? Why?

*Problems designated by a "C" are concept questions, and students are encouraged to answer them all. Problems designated by an "E" are in English units, and the SI users can ignore them. Problems with the icon 🌐 are solved using EES, and complete solutions together with parametric studies are included on the enclosed CD. Problems with the icon 📀 are comprehensive in nature, and are intended to be solved with a computer, preferably using the EES software that accompanies this text.

4–9C In what medium is the lumped system analysis more likely to be applicable: in water or in air? Why?

4–10C For which solid is the lumped system analysis more likely to be applicable: an actual apple or a golden apple of the same size? Why?

4–11C For which kind of bodies made of the same material is the lumped system analysis more likely to be applicable: slender ones or well-rounded ones of the same volume? Why?

4–12 Obtain relations for the characteristic lengths of a large plane wall of thickness $2L$, a very long cylinder of radius r_o, and a sphere of radius r_o.

4–13 Obtain a relation for the time required for a lumped system to reach the average temperature $\frac{1}{2}(T_i + T_\infty)$, where T_i is the initial temperature and T_∞ is the temperature of the environment.

4–14 The temperature of a gas stream is to be measured by a thermocouple whose junction can be approximated as a 1.2-mm-diameter sphere. The properties of the junction are $k = 35$ W/m · °C, $\rho = 8500$ kg/m³, and $c_p = 320$ J/kg · °C, and the heat transfer coefficient between the junction and the gas is $h = 90$ W/m² · °C. Determine how long it will take for the thermocouple to read 99 percent of the initial temperature difference. *Answer:* 27.8 s

4–15E In a manufacturing facility, 2-in-diameter brass balls ($k = 64.1$ Btu/h · ft · °F, $\rho = 532$ lbm/ft³, and $c_p = 0.092$ Btu/lbm · °F) initially at 250°F are quenched in a water bath at 120°F for a period of 2 min at a rate of 120 balls per minute. If the convection heat transfer coefficient is 42 Btu/h · ft² · °F, determine (*a*) the temperature of the balls after quenching and (*b*) the rate at which heat needs to be removed from the water in order to keep its temperature constant at 120°F.

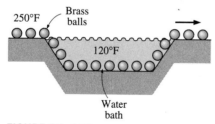

250°F Brass balls

120°F

Water bath

FIGURE P4–15E

4–16E Repeat Prob. 4–15E for aluminum balls.

4–17 To warm up some milk for a baby, a mother pours milk into a thin-walled glass whose diameter is 6 cm. The height of the milk in the glass is 7 cm. She then places the glass into a large pan filled with hot water at 60°C. The milk is stirred constantly, so that its temperature is uniform at all times. If the heat transfer coefficient between the water and the glass is 120 W/m² · °C, determine how long it will take for the milk to warm up from 3°C to 38°C. Take the properties of the milk

to be the same as those of water. Can the milk in this case be treated as a lumped system? Why? *Answer:* 5.8 min

4–18 Repeat Prob. 4–17 for the case of water also being stirred, so that the heat transfer coefficient is doubled to 240 W/m² · °C.

4–19 A long copper rod of diameter 2.0 cm is initially at a uniform temperature of 100°C. It is now exposed to an air stream at 20°C with a heat transfer coefficient of 200 W/m² · K. How long would it take for the copper road to cool to an average temperature of 25°C?

4–20 Consider a sphere of diameter 5 cm, a cube of side length 5 cm, and a rectangular prism of dimension 4 cm × 5 cm × 6 cm, all initially at 0°C and all made of silver ($k = 429$ W/m · °C, $\rho = 10,500$ kg/m³, $c_p = 0.235$ kJ/kg · °C). Now all three of these geometries are exposed to ambient air at 33°C on all of their surfaces with a heat transfer coefficient of 12 W/m² · °C. Determine how long it will take for the temperature of each geometry to rise to 25°C.

4–21E During a picnic on a hot summer day, all the cold drinks disappeared quickly, and the only available drinks were those at the ambient temperature of 90°F. In an effort to cool a 12-fluid-oz drink in a can, which is 5 in high and has a diameter of 2.5 in, a person grabs the can and starts shaking it in the iced water of the chest at 32°F. The temperature of the drink can be assumed to be uniform at all times, and the heat transfer coefficient between the iced water and the aluminum can is 30 Btu/h · ft² · °F. Using the properties of water for the drink, estimate how long it will take for the canned drink to cool to 40°F.

FIGURE P4–21E

4–22 Consider a 1000-W iron whose base plate is made of 0.5-cm-thick aluminum alloy 2024–T6 ($\rho = 2770$ kg/m³, $c_p = 875$ J/kg · °C, $\alpha = 7.3 \times 10^{-5}$ m²/s). The base plate has a surface area of 0.03 m². Initially, the iron is in thermal equilibrium with the ambient air at 22°C. Taking the heat transfer coefficient at the surface of the base plate to be 12 W/m² · °C and assuming 85 percent of the heat generated in the resistance wires is transferred to the plate, determine how long it will take for the plate temperature to reach 140°C. Is it realistic to assume the plate temperature to be uniform at all times?

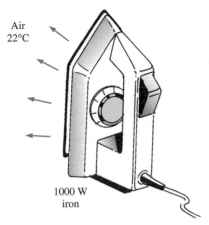

Air 22°C

1000 W iron

FIGURE P4–22

4–23 Reconsider Prob. 4–22. Using EES (or other) software, investigate the effects of the heat transfer coefficient and the final plate temperature on the time it will take for the plate to reach this temperature. Let the heat transfer coefficient vary from 5 W/m² · °C to 25 W/m² · °C and the temperature from 30°C to 200°C. Plot the time as functions of the heat transfer coefficient and the temperature, and discuss the results.

4–24 Stainless steel ball bearings ($\rho = 8085$ kg/m³, $k = 15.1$ W/m · °C, $c_p = 0.480$ kJ/kg · °C, and $\alpha = 3.91 \times 10^{-6}$ m²/s) having a diameter of 1.2 cm are to be quenched in water. The balls leave the oven at a uniform temperature of 900°C and are exposed to air at 30°C for a while before they are dropped into the water. If the temperature of the balls is not to fall below 850°C prior to quenching and the heat transfer coefficient in the air is 125 W/m² · °C, determine how long they can stand in the air before being dropped into the water. *Answer:* 3.7 s

4–25 Carbon steel balls ($\rho = 7833$ kg/m³, $k = 54$ W/m · °C, $c_p = 0.465$ kJ/kg · °C, and $\alpha = 1.474 \times 10^{-6}$ m²/s) 8 mm in diameter are annealed by heating them first to 900°C in a furnace and then allowing them to cool slowly to 100°C in ambient air at 35°C. If the average heat transfer coefficient is 75 W/m² · °C, determine how long the annealing process will take. If 2500 balls are to be annealed per hour, determine the total rate of heat transfer from the balls to the ambient air.

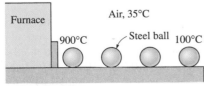

Furnace | **Air, 35°C**

900°C | Steel ball | 100°C

FIGURE P4–25

4–26 Reconsider Prob. 4–25. Using EES (or other) software, investigate the effect of the initial temperature of the balls on the annealing time and the total rate of heat transfer. Let the temperature vary from 500°C to 1000°C. Plot the time and the total rate of heat transfer as a function of the initial temperature, and discuss the results.

4–27 An electronic device dissipating 20 W has a mass of 20 g, a specific heat of 850 J/kg · °C, and a surface area of 4 cm². The device is lightly used, and it is on for 5 min and then off for several hours, during which it cools to the ambient temperature of 25°C. Taking the heat transfer coefficient to be 12 W/m² · °C, determine the temperature of the device at the end of the 5-min operating period. What would your answer be if the device were attached to an aluminum heat sink having a mass of 200 g and a surface area of 80 cm²? Assume the device and the heat sink to be nearly isothermal.

Transient Heat Conduction in Large Plane Walls, Long Cylinders, and Spheres with Spatial Effects

4–28C What is an infinitely long cylinder? When is it proper to treat an actual cylinder as being infinitely long, and when is it not? For example, is it proper to use this model when finding the temperatures near the bottom or top surfaces of a cylinder? Explain.

4–29C Can the transient temperature charts in Fig. 4–15 for a plane wall exposed to convection on both sides be used for a plane wall with one side exposed to convection while the other side is insulated? Explain.

4–30C Why are the transient temperature charts prepared using nondimensionalized quantities such as the Biot and Fourier numbers instead of the actual variables such as thermal conductivity and time?

4–31C What is the physical significance of the Fourier number? Will the Fourier number for a specified heat transfer problem double when the time is doubled?

4–32C How can we use the transient temperature charts when the surface temperature of the geometry is specified instead of the temperature of the surrounding medium and the convection heat transfer coefficient?

4–33C A body at an initial temperature of T_i is brought into a medium at a constant temperature of T_∞. How can you determine the maximum possible amount of heat transfer between the body and the surrounding medium?

4–34C The Biot number during a heat transfer process between a sphere and its surroundings is determined to be 0.02. Would you use lumped system analysis or the transient temperature charts when determining the midpoint temperature of the sphere? Why?

4–35 A student calculates that the total heat transfer from a spherical copper ball of diameter 18 cm initially at 200°C and its environment at a constant temperature of 25°C during the first 20 min of cooling is 3150 kJ. Is this result reasonable? Why?

4–36 An experiment is to be conducted to determine heat transfer coefficient on the surfaces of tomatoes that are placed in cold water at 7°C. The tomatoes ($k = 0.59$ W/m · °C, $\alpha = 0.141 \times 10^{-6}$ m^2/s, $\rho = 999$ kg/m^3, $c_p = 3.99$ kJ/kg · °C) with an initial uniform temperature of 30°C are spherical in shape with a diameter of 8 cm. After a period of 2 hours, the temperatures at the center and the surface of the tomatoes are measured to bc 10.0°C and 7.1°C, respectively. Using analytical one-term approximation method (not the Heisler charts), determine the heat transfer coefficient and the amount of heat transfer during this period if there are eight such tomatoes in water.

4–37 An ordinary egg can be approximated as a 5.5-cm-diameter sphere whose properties are roughly $k = 0.6$ W/m · °C and $\alpha = 0.14 \times 10^{-6}$ m^2/s. The egg is initially at a uniform temperature of 8°C and is dropped into boiling water at 97°C. Taking the convection heat transfer coefficient to be $h = 1400$ W/m^2 · °C, determine how long it will take for the center of the egg to reach 70°C.

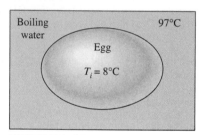

FIGURE P4–37

4–38 [EES] Reconsider Prob. 4–37. Using EES (or other) software, investigate the effect of the final center temperature of the egg on the time it will take for the center to reach this temperature. Let the temperature vary from 50°C to 95°C. Plot the time versus the temperature, and discuss the results.

4–39 In a production facility, 3-cm-thick large brass plates ($k = 110$ W/m · °C, $\rho = 8530$ kg/m^3, $c_p = 380$ J/kg · °C, and

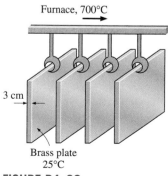

FIGURE P4–39

$\alpha = 33.9 \times 10^{-6}$ m^2/s) that are initially at a uniform temperature of 25°C are heated by passing them through an oven maintained at 700°C. The plates remain in the oven for a period of 10 min. Taking the convection heat transfer coefficient to be $h = 80$ W/m^2 · °C, determine the surface temperature of the plates when they come out of the oven.

4–40 [EES] Reconsider Prob. 4–39. Using EES (or other) software, investigate the effects of the temperature of the oven and the heating time on the final surface temperature of the plates. Let the oven temperature vary from 500°C to 900°C and the time from 2 min to 30 min. Plot the surface temperature as the functions of the oven temperature and the time, and discuss the results.

4–41 A long 35-cm-diameter cylindrical shaft made of stainless steel 304 ($k = 14.9$ W/m · °C, $\rho = 7900$ kg/m^3, $c_p = 477$ J/kg · °C, and $\alpha = 3.95 \times 10^{-6}$ m^2/s) comes out of an oven at a uniform temperature of 400°C. The shaft is then allowed to cool slowly in a chamber at 150°C with an average convection heat transfer coefficient of $h = 60$ W/m^2 · °C. Determine the temperature at the center of the shaft 20 min after the start of the cooling process. Also, determine the heat transfer per unit length of the shaft during this time period.

Answers: 390°C, 15,960 kJ/m

4–42 [EES] Reconsider Prob. 4–41. Using EES (or other) software, investigate the effect of the cooling time on the final center temperature of the shaft and the amount of heat transfer. Let the time vary from 5 min to 60 min. Plot the center temperature and the heat transfer as a function of the time, and discuss the results.

4–43E Long cylindrical AISI stainless steel rods ($k = 7.74$ Btu/h · ft · °F and $\alpha = 0.135$ ft^2/h) of 4-in-diameter are heat treated by drawing them at a velocity of 7 ft/min through a 21-ft-long oven maintained at 1700°F. The heat transfer coefficient in the oven is 20 Btu/h · ft^2 · °F. If the rods enter the oven at 70°F, determine their centerline temperature when they leave.

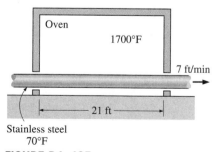

FIGURE P4–43E

4–44 In a meat processing plant, 2-cm-thick steaks ($k = 0.45$ W/m · °C and $\alpha = 0.91 \times 10^{-7}$ m^2/s) that are initially at 25°C are to be cooled by passing them through a refrigeration room at -11°C. The heat transfer coefficient on both sides of the steaks is 9 W/m² · °C. If both surfaces of the steaks are to be cooled to 2°C, determine how long the steaks should be kept in the refrigeration room.

4–45 A long cylindrical wood log ($k = 0.17$ W/m · °C and $\alpha = 1.28 \times 10^{-7}$ m²/s) is 10 cm in diameter and is initially at a uniform temperature of 15°C. It is exposed to hot gases at 550°C in a fireplace with a heat transfer coefficient of 13.6 W/m² · °C on the surface. If the ignition temperature of the wood is 420°C, determine how long it will be before the log ignites.

4–46 In *Betty Crocker's Cookbook,* it is stated that it takes 2 h 45 min to roast a 3.2-kg rib initially at 4.5°C "rare" in an oven maintained at 163°C. It is recommended that a meat thermometer be used to monitor the cooking, and the rib is considered rare done when the thermometer inserted into the center of the thickest part of the meat registers 60°C. The rib can be treated as a homogeneous spherical object with the properties $\rho = 1200$ kg/m³, $c_p = 4.1$ kJ/kg · °C, $k = 0.45$ W/m · °C, and $\alpha = 0.91 \times 10^{-7}$ m²/s. Determine (*a*) the heat transfer coefficient at the surface of the rib; (*b*) the temperature of the outer surface of the rib when it is done; and (*c*) the amount of heat transferred to the rib. (*d*) Using the values obtained, predict how long it will take to roast this rib to "medium" level, which occurs when the innermost temperature of the rib reaches 71°C. Compare your result to the listed value of 3 h 20 min.

If the roast rib is to be set on the counter for about 15 min before it is sliced, it is recommended that the rib be taken out of the oven when the thermometer registers about 4°C below the indicated value because the rib will continue cooking even after it is taken out of the oven. Do you agree with this recommendation?

Answers: (*a*) 156.9 W/m² · °C, (*b*) 159.5°C, (*c*) 1629 kJ, (*d*) 3.0 h

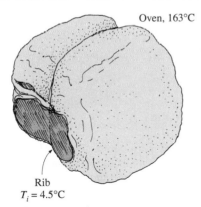

Oven, 163°C

Rib
$T_i = 4.5$°C

FIGURE P4–46

4–47 Repeat Prob. 4–46 for a roast rib that is to be "well-done" instead of "rare." A rib is considered to be well-done when its center temperature reaches 77°C, and the roasting in this case takes about 4 h 15 min.

4–48 For heat transfer purposes, an egg can be considered to be a 5.5-cm-diameter sphere having the properties of water. An egg that is initially at 8°C is dropped into the boiling water at 100°C. The heat transfer coefficient at the surface of the egg is estimated to be 800 W/m² · °C. If the egg is considered cooked when its center temperature reaches 60°C, determine how long the egg should be kept in the boiling water.

4–49 Repeat Prob. 4–48 for a location at 1610-m elevation such as Denver, Colorado, where the boiling temperature of water is 94.4°C.

4–50 The author and his then 6-year-old son have conducted the following experiment to determine the thermal conductivity of a hot dog. They first boiled water in a large pan and measured the temperature of the boiling water to be 94°C, which is not surprising, since they live at an elevation of about 1650 m in Reno, Nevada. They then took a hot dog that is 12.5 cm long and 2.2 cm in diameter and inserted a thermocouple into the midpoint of the hot dog and another thermocouple just under the skin. They waited until both thermocouples read 20°C, which is the ambient temperature. They then dropped the hot dog into boiling water and observed the changes in both temperatures. Exactly 2 min after the hot dog was dropped into the boiling water, they recorded the center and the surface temperatures to be 59°C and 88°C, respectively. The density of the hot dog can be taken to be 980 kg/m³, which is slightly less than the density of water, since the hot dog was observed to be floating in water while being almost completely immersed. The specific heat of a hot dog can be taken to be 3900 J/kg · °C, which is slightly less than that of water, since a hot dog is mostly water. Using transient temperature charts, determine (*a*) the thermal diffusivity of the hot dog; (*b*) the thermal conductivity of the hot dog; and (*c*) the convection heat transfer coefficient.

Answers: (*a*) 2.02×10^{-7} m²/s, (*b*) 0.771 W/m · °C, (*c*) 467 W/m² · °C.

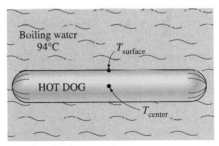

Boiling water
94°C

$T_{surface}$

HOT DOG

T_{center}

FIGURE P4–50

4–51 Using the data and the answers given in Prob. 4–50, determine the center and the surface temperatures of the hot dog 4 min after the start of the cooking. Also determine the amount of heat transferred to the hot dog.

4–52E In a chicken processing plant, whole chickens averaging 5 lbm each and initially at 65°F are to be cooled in the racks of a large refrigerator that is maintained at 5°F. The entire chicken is to be cooled below 45°F, but the temperature of the chicken is not to drop below 35°F at any point during refrigeration. The convection heat transfer coefficient and thus the rate of heat transfer from the chicken can be controlled by varying the speed of a circulating fan inside. Determine the heat transfer-coefficient that will enable us to meet both temperature constraints while keeping the refrigeration time to a minimum. The chicken can be treated as a homogeneous spherical object having the properties $\rho = 74.9$ lbm/ft³, $c_p = 0.98$ Btu/lbm · °F, $k = 0.26$ Btu/h · ft · °F, and $\alpha = 0.0035$ ft²/h.

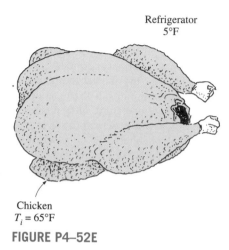

Refrigerator
5°F

Chicken
$T_i = 65°F$

FIGURE P4–52E

4–53 A person puts a few apples into the freezer at $-15°C$ to cool them quickly for guests who are about to arrive. Initially, the apples are at a uniform temperature of 20°C, and the heat transfer coefficient on the surfaces is 8 W/m² · °C. Treating the apples as 9-cm-diameter spheres and taking their properties to be $\rho = 840$ kg/m³, $c_p = 3.81$ kJ/kg · °C, $k = 0.418$ W/m · °C, and $\alpha = 1.3 \times 10^{-7}$ m²/s, determine the center and surface temperatures of the apples in 1 h. Also, determine the amount of heat transfer from each apple.

4–54 ⟨EES⟩ Reconsider Prob. 4–53. Using EES (or other) software, investigate the effect of the initial temperature of the apples on the final center and surface temperatures and the amount of heat transfer. Let the initial temperature vary from 2°C to 30°C. Plot the center temperature, the surface temperature, and the amount of heat transfer as a function of the initial temperature, and discuss the results.

4–55 Citrus fruits are very susceptible to cold weather, and extended exposure to subfreezing temperatures can destroy them. Consider an 8-cm-diameter orange that is initially at 15°C. A cold front moves in one night, and the ambient temperature suddenly drops to $-6°C$, with a heat transfer coefficient of 15 W/m² · °C. Using the properties of water for the orange and assuming the ambient conditions to remain constant for 4 h before the cold front moves out, determine if any part of the orange will freeze that night.

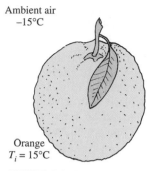

Ambient air
−15°C

Orange
$T_i = 15°C$

FIGURE P4–55

4–56 A 9-cm-diameter potato ($\rho = 1100$ kg/m³, $c_p = 3900$ J/kg · °C, $k = 0.6$ W/m · °C, and $\alpha = 1.4 \times 10^{-7}$ m²/s) that is initially at a uniform temperature of 25°C is baked in an oven at 170°C until a temperature sensor inserted to the center of the potato indicates a reading of 70°C. The potato is then taken out of the oven and wrapped in thick towels so that almost no heat is lost from the baked potato. Assuming the heat transfer coefficient in the oven to be 40 W/m² · °C, determine (a) how long the potato is baked in the oven and (b) the final equilibrium temperature of the potato after it is wrapped.

4–57 White potatoes ($k = 0.50$ W/m · °C and $\alpha = 0.13 \times 10^{-6}$ m²/s) that are initially at a uniform temperature of 25°C and have an average diameter of 6 cm are to be cooled by refrigerated air at 2°C flowing at a velocity of 4 m/s. The average heat transfer coefficient between the potatoes and the air is experimentally determined to be 19 W/m² · °C. Determine how long it will take for the center temperature of the potatoes to drop to 6°C. Also, determine if any part of the potatoes will experience chilling injury during this process.

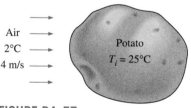

Air
2°C
4 m/s

Potato
$T_i = 25°C$

FIGURE P4–57

4–58E Oranges of 2.5-in-diameter ($k = 0.26$ Btu/h · ft · °F and $\alpha = 1.4 \times 10^{-6}$ ft²/s) initially at a uniform temperature of 78°F are to be cooled by refrigerated air at 25°F flowing at a velocity of 1 ft/s. The average heat transfer coefficient between the oranges and the air is experimentally determined to be 4.6 Btu/h · ft² · °F. Determine how long it will take for the center temperature of the oranges to drop to 40°F. Also, determine if any part of the oranges will freeze during this process.

4–59 A 65-kg beef carcass ($k = 0.47$ W/m · °C and $\alpha = 0.13 \times 10^{-6}$ m²/s) initially at a uniform temperature of 37°C is to be cooled by refrigerated air at $-10°C$ flowing at a velocity of 1.2 m/s. The average heat transfer coefficient

FIGURE P4–62

between the carcass and the air is 22 W/m² · °C. Treating the carcass as a cylinder of diameter 24 cm and height 1.4 m and disregarding heat transfer from the base and top surfaces, determine how long it will take for the center temperature of the carcass to drop to 4°C. Also, determine if any part of the carcass will freeze during this process.   *Answer:* 12.2 h

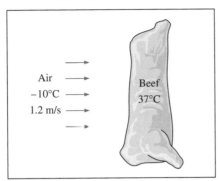

FIGURE P4–59

4–60 Layers of 23-cm-thick meat slabs ($k = 0.47$ W/m · °C and $\alpha = 0.13 \times 10^{-6}$ m²/s) initially at a uniform temperature of 7°C are to be frozen by refrigerated air at -30°C flowing at a velocity of 1.4 m/s. The average heat transfer coefficient between the meat and the air is 20 W/m² · °C. Assuming the size of the meat slabs to be large relative to their thickness, determine how long it will take for the center temperature of the slabs to drop to -18°C. Also, determine the surface temperature of the meat slab at that time.

4–61E Layers of 6-in-thick meat slabs ($k = 0.26$ Btu/h · ft · °F and $\alpha = 1.4 \times 10^{-6}$ ft²/s) initially at a uniform temperature of 50°F are cooled by refrigerated air at 23°F to a temperature of 36°F at their center in 12 h. Estimate the average heat transfer coefficient during this cooling process.
  Answer: 1.5 Btu/h · ft² · °F

4–62 Chickens with an average mass of 1.7 kg ($k = 0.45$ W/m · °C and $\alpha = 0.13 \times 10^{-6}$ m²/s) initially at a uniform temperature of 15°C are to be chilled in agitated brine at -7°C. The average heat transfer coefficient between the chicken and the brine is determined experimentally to be 440 W/m² · °C. Taking the average density of the chicken to be 0.95 g/cm³ and treating the chicken as a spherical lump, determine the center and the surface temperatures of the chicken in 2 h and 45 min. Also, determine if any part of the chicken will freeze during this process.

Transient Heat Conduction in Semi-Infinite Solids

4–63C What is a semi-infinite medium? Give examples of solid bodies that can be treated as semi-infinite mediums for heat transfer purposes.

4–64C Under what conditions can a plane wall be treated as a semi-infinite medium?

4–65C Consider a hot semi-infinite solid at an initial temperature of T_i that is exposed to convection to a cooler medium at a constant temperature of T_∞, with a heat transfer coefficient of h. Explain how you can determine the total amount of heat transfer from the solid up to a specified time t_o.

4–66 In areas where the air temperature remains below 0°C for prolonged periods of time, the freezing of water in underground pipes is a major concern. Fortunately, the soil remains relatively warm during those periods, and it takes weeks for the subfreezing temperatures to reach the water mains in the ground. Thus, the soil effectively serves as an insulation to protect the water from the freezing atmospheric temperatures in winter.

The ground at a particular location is covered with snow pack at -8°C for a continuous period of 60 days, and the average soil properties at that location are $k = 0.35$ W/m · °C and $\alpha = 0.15 \times 10^{-6}$ m²/s. Assuming an initial uniform temperature of 8°C for the ground, determine the minimum burial depth to prevent the water pipes from freezing.

4–67 The soil temperature in the upper layers of the earth varies with the variations in the atmospheric conditions. Before a cold front moves in, the earth at a location is initially at a uniform temperature of 10°C. Then the area is subjected to a temperature of -10°C and high winds that resulted in a convection heat transfer coefficient of 40 W/m² · °C on the earth's surface for a period of 10 h. Taking the properties of the soil at that location to be $k = 0.9$ W/m · °C and $\alpha = 1.6 \times 10^{-5}$ m²/s, determine the soil temperature at distances 0, 10, 20, and 50 cm from the earth's surface at the end of this 10-h period.

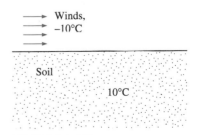

FIGURE P4–67

4–68 Reconsider Prob. 4–67. Using EES (or other) software, plot the soil temperature as a function of the distance from the earth's surface as the distance varies from 0 m to 1m, and discuss the results.

4–69 A thick aluminum block initially at 20°C is subjected to constant heat flux of 4000 W/m² by an electric resistance heater whose top surface is insulated. Determine how much the surface temperature of the block will rise after 30 minutes.

4–70 A bare-footed person whose feet are at 32°C steps on a large aluminum block at 20°C. Treating both the feet and the aluminum block as semi-infinite solids, determine the contact surface temperature. What would your answer be if the person stepped on a wood block instead? At room temperature, the $\sqrt{k\rho c_p}$ value is 24 kJ/m² · °C for aluminum, 0.38 kJ/m² · °C for wood, and 1.1 kJ/m² · °C for human flesh.

4–71E The walls of a furnace are made of 1.2-ft-thick concrete ($k = 0.64$ Btu/h · ft · °F and $\alpha = 0.023$ ft²/h). Initially, the furnace and the surrounding air are in thermal equilibrium at 70°F. The furnace is then fired, and the inner surfaces of the furnace are subjected to hot gases at 1800°F with a very large heat transfer coefficient. Determine how long it will take for the temperature of the outer surface of the furnace walls to rise to 70.1°F. *Answer:* 116 min

4–72 A thick wood slab ($k = 0.17$ W/m · °C and $\alpha = 1.28 \times 10^{-7}$ m²/s) that is initially at a uniform temperature of 25°C is exposed to hot gases at 550°C for a period of 5 min. The heat transfer coefficient between the gases and the wood slab is 35 W/m² · °C. If the ignition temperature of the wood is 450°C, determine if the wood will ignite.

4–73 A large cast iron container ($k = 52$ W/m · °C and $\alpha = 1.70 \times 10^{-5}$ m²/s) with 5-cm-thick walls is initially at a uniform temperature of 0°C and is filled with ice at 0°C. Now the outer surfaces of the container are exposed to hot water at 60°C with a very large heat transfer coefficient. Determine how long it will be before the ice inside the container starts melting. Also, taking the heat transfer coefficient on the inner surface of the container to be 250 W/m² · °C, determine the rate of heat transfer to the ice through a 1.2-m-wide and 2-m-high section of the wall when steady operating conditions are reached. Assume the ice starts melting when its inner surface temperature rises to 0.1°C.

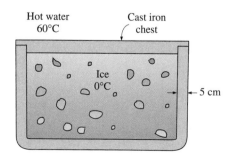

Hot water
60°C

Cast iron
chest

Ice
0°C

5 cm

FIGURE P4–73

Transient Heat Conduction in Multidimensional Systems

4–74C What is the product solution method? How is it used to determine the transient temperature distribution in a two-dimensional system?

4–75C How is the product solution used to determine the variation of temperature with time and position in three-dimensional systems?

4–76C A short cylinder initially at a uniform temperature T_i is subjected to convection from all of its surfaces to a medium at temperature T_∞. Explain how you can determine the temperature of the midpoint of the cylinder at a specified time t.

4–77C Consider a short cylinder whose top and bottom surfaces are insulated. The cylinder is initially at a uniform temperature T_i and is subjected to convection from its side surface to a medium at temperature T_∞ with a heat transfer coefficient of h. Is the heat transfer in this short cylinder one- or two-dimensional? Explain.

4–78 A short brass cylinder ($\rho = 8530$ kg/m³, $c_p = 0.389$ kJ/kg · °C, $k = 110$ W/m · °C, and $\alpha = 3.39 \times 10^{-5}$ m²/s) of diameter $D = 8$ cm and height $H = 15$ cm is initially at a uniform temperature of $T_i = 150°C$. The cylinder is now placed in atmospheric air at 20°C, where heat transfer takes place by convection with a heat transfer coefficient of $h = 40$ W/m² · °C. Calculate (*a*) the center temperature of the cylinder; (*b*) the center temperature of the top surface of the cylinder; and (*c*) the total heat transfer from the cylinder 15 min after the start of the cooling.

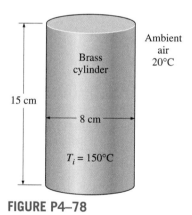

Ambient
air
20°C

Brass
cylinder

15 cm

8 cm

$T_i = 150°C$

FIGURE P4–78

4–79 Reconsider Prob. 4–78. Using EES (or other) software, investigate the effect of the cooling time on the center temperature of the cylinder, the center temperature of the top surface of the cylinder, and the total heat transfer. Let the time vary from 5 min to 60 min. Plot the center temperature of the cylinder, the center temperature of the top surface, and the total heat transfer as a function of the time, and discuss the results.

4–80 A semi-infinite aluminum cylinder ($k = 237$ W/m · °C, $\alpha = 9.71 \times 10^{-5}$ m²/s) of diameter $D = 15$ cm is initially at a uniform temperature of $T_i = 115$°C. The cylinder is now placed in water at 10°C, where heat transfer takes place by convection with a heat transfer coefficient of $h = 140$ W/m² · °C. Determine the temperature at the center of the cylinder 5 cm from the end surface 8 min after the start of cooling.

4–81E A hot dog can be considered to be a cylinder 5 in long and 0.8 in in diameter whose properties are $\rho = 61.2$ lbm/ft³, $c_p = 0.93$ Btu/lbm · °F, $k = 0.44$ Btu/h · ft · °F, and $\alpha = 0.0077$ ft²/h. A hot dog initially at 40°F is dropped into boiling water at 212°F. If the heat transfer coefficient at the surface of the hot dog is estimated to be 120 Btu/h · ft² · °F, determine the center temperature of the hot dog after 5, 10, and 15 min by treating the hot dog as (a) a finite cylinder and (b) an infinitely long cylinder.

4–82E Repeat Prob. 4–81E for a location at 5300-ft elevation such as Denver, Colorado, where the boiling temperature of water is 202°F.

4–83 A 5-cm-high rectangular ice block ($k = 2.22$ W/m · °C and $\alpha = 0.124 \times 10^{-7}$ m²/s) initially at −20°C is placed on a table on its square base 4 cm × 4 cm in size in a room at 18°C. The heat transfer coefficient on the exposed surfaces of the ice block is 12 W/m² · °C. Disregarding any heat transfer from the base to the table, determine how long it will be before the ice block starts melting. Where on the ice block will the first liquid droplets appear?

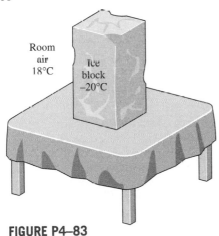

FIGURE P4–83

4–84 Reconsider Prob. 4–83. Using EES (or other) software, investigate the effect of the initial temperature of the ice block on the time period before the ice block starts melting. Let the initial temperature vary from −26°C to −4°C. Plot the time versus the initial temperature, and discuss the results.

4–85 A 2-cm-high cylindrical ice block ($k = 2.22$ W/m · °C and $\alpha = 0.124 \times 10^{-7}$ m²/s) is placed on a table on its base of diameter 2 cm in a room at 24°C. The heat transfer coefficient

on the exposed surfaces of the ice block is 13 W/m² · °C, and heat transfer from the base of the ice block to the table is negligible. If the ice block is not to start melting at any point for at least 3 h, determine what the initial temperature of the ice block should be.

4–86 Consider a cubic block whose sides are 5 cm long and a cylindrical block whose height and diameter are also 5 cm. Both blocks are initially at 20°C and are made of granite ($k = 2.5$ W/m · °C and $\alpha = 1.15 \times 10^{-6}$ m²/s). Now both blocks are exposed to hot gases at 500°C in a furnace on all of their surfaces with a heat transfer coefficient of 40 W/m² · °C. Determine the center temperature of each geometry after 10, 20, and 60 min.

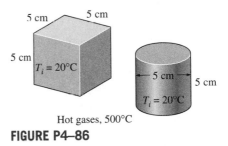

FIGURE P4–86

4–87 Repeat Prob. 4–86 with the heat transfer coefficient at the top and the bottom surfaces of each block being doubled to 80 W/m² · °C.

4–88 A 20-cm-long cylindrical aluminum block ($\rho = 2702$ kg/m³, $c_p = 0.896$ kJ/kg · °C, $k = 236$ W/m · °C, and $\alpha = 9.75 \times 10^{-5}$ m²/s), 15 cm in diameter, is initially at a uniform temperature of 20°C. The block is to be heated in a furnace at 1200°C until its center temperature rises to 300°C. If the heat transfer coefficient on all surfaces of the block is 80 W/m² · °C, determine how long the block should be kept in the furnace. Also, determine the amount of heat transfer from the aluminum block if it is allowed to cool in the room until its temperature drops to 20°C throughout.

4–89 Repeat Prob. 4–88 for the case where the aluminum block is inserted into the furnace on a low-conductivity material so that the heat transfer to or from the bottom surface of the block is negligible.

4–90 Reconsider Prob. 4–88. Using EES (or other) software, investigate the effect of the final center temperature of the block on the heating time and the amount of heat transfer. Let the final center temperature vary from 50°C to 1000°C. Plot the time and the heat transfer as a function of the final center temperature, and discuss the results.

Special Topic: Refrigeration and Freezing of Foods

4–91C What are the common kinds of microorganisms? What undesirable changes do microorganisms cause in foods?

4–92C How does refrigeration prevent or delay the spoilage of foods? Why does freezing extend the storage life of foods for months?

4–93C What are the environmental factors that affect the growth rate of microorganisms in foods?

4–94C What is the effect of cooking on the microorganisms in foods? Why is it important that the internal temperature of a roast in an oven be raised above 70°C?

4–95C How can the contamination of foods with micro-organisms be prevented or minimized? How can the growth of microorganisms in foods be retarded? How can the micro-organisms in foods be destroyed?

4–96C How does (a) the air motion and (b) the relative humidity of the environment affect the growth of microorganisms in foods?

4–97C The cooling of a beef carcass from 37°C to 5°C with refrigerated air at 0°C in a chilling room takes about 48 h. To reduce the cooling time, it is proposed to cool the carcass with refrigerated air at –10°C. How would you evaluate this proposal?

4–98C Consider the freezing of packaged meat in boxes with refrigerated air. How do (a) the temperature of air, (b) the velocity of air, (c) the capacity of the refrigeration system, and (d) the size of the meat boxes affect the freezing time?

4–99C How does the rate of freezing affect the tenderness, color, and the drip of meat during thawing?

4–100C It is claimed that beef can be stored for up to two years at –23°C but no more than one year at –12°C. Is this claim reasonable? Explain.

4–101C What is a refrigerated shipping dock? How does it reduce the refrigeration load of the cold storage rooms?

4–102C How does immersion chilling of poultry compare to forced-air chilling with respect to (a) cooling time, (b) moisture loss of poultry, and (c) microbial growth.

4–103C What is the proper storage temperature of frozen poultry? What are the primary methods of freezing for poultry?

4–104C What are the factors that affect the quality of frozen fish?

4–105 The chilling room of a meat plant is 15 m × 18 m × 5.5 m in size and has a capacity of 350 beef carcasses. The power consumed by the fans and the lights in the chilling room are 22 and 2 kW, respectively, and the room gains heat through its envelope at a rate of 14 kW. The average mass of beef carcasses is 220 kg. The carcasses enter the chilling room at 35°C, after they are washed to facilitate evaporative cooling, and are cooled to 16°C in 12 h. The air enters the chilling room at –2.2°C and leaves at 0.5°C. Determine (a) the refrigeration load of the chilling room and (b) the volume flow rate of air. The average specific heats of beef carcasses and air are 3.14 and 1.0 kJ/kg · °C, respectively, and the density of air can be taken to be 1.28 kg/m³.

4–106 Turkeys with a water content of 64 percent that are initially at 1°C and have a mass of about 7 kg are to be frozen by submerging them into brine at –29°C. Using Figure 4–54, determine how long it will take to reduce the temperature of the turkey breast at a depth of 3.8 cm to –18°C. If the temperature at a depth of 3.8 cm in the breast represents the average temperature of the turkey, determine the amount of heat transfer per turkey assuming (a) the entire water content of the turkey is frozen and (b) only 90 percent of the water content of the turkey is frozen at –18°C. Take the specific heats of turkey to be 2.98 and 1.65 kJ/kg · °C above and below the freezing point of –2.8°C, respectively, and the latent heat of fusion of turkey to be 214 kJ/kg.   *Answers: (a) 1753 kJ, (b) 1617 kJ*

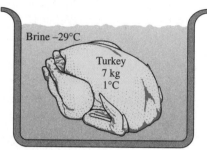

Brine –29°C

Turkey
7 kg
1°C

FIGURE P4–106

4–107 Chickens with an average mass of 2.2 kg and average specific heat of 3.54 kJ/kg · °C are to be cooled by chilled water that enters a continuous-flow-type immersion chiller at 0.5°C. Chickens are dropped into the chiller at a uniform temperature of 15°C at a rate of 500 chickens per hour and are cooled to an average temperature of 3°C before they are taken out. The chiller gains heat from the surroundings at a rate of 210 kJ/min. Determine (a) the rate of heat removal from the chicken, in kW, and (b) the mass flow rate of water, in kg/s, if the temperature rise of water is not to exceed 2°C.

4–108E Chickens with a water content of 74 percent, an initial temperature of 32°F, and a mass of about 7.5 lbm are to be frozen by refrigerated air at –40°F. Using Figure 4–53, determine how long it will take to reduce the inner surface temperature of chickens to 25°F. What would your answer be if the air temperature were –80°F?

4–109 In a meat processing plant, 10-cm-thick beef slabs (ρ = 1090 kg/m³, c_p = 3.54 kJ/kg · °C, k = 0.47 W/m · °C, and α = 0.13 × 10⁻⁶ m²/s) initially at 15°C are to be cooled in the racks of a large freezer that is maintained at –12°C. The meat slabs are placed close to each other so that heat transfer from the 10-cm-thick edges is negligible. The entire slab is to be cooled below 5°C, but the temperature of the steak is not to drop below –1°C anywhere during refrigeration to avoid "frost bite." The convection heat transfer coefficient and thus the rate of heat transfer from the steak can be controlled by varying the speed of a circulating fan inside. Determine the heat transfer coefficient h that will enable us to meet both temperature constraints while keeping the refrigeration time to a minimum.   *Answer:* 9.9 W/m² · °C.

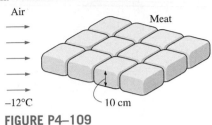

FIGURE P4–109

Review Problems

4–110   Consider two 2-cm-thick large steel plates ($k = 43$ W/m · °C and $\alpha = 1.17 \times 10^{-5}$ m²/s) that were put on top of each other while wet and left outside during a cold winter night at -15°C. The next day, a worker needs one of the plates, but the plates are stuck together because the freezing of the water between the two plates has bonded them together. In an effort to melt the ice between the plates and separate them, the worker takes a large hair dryer and blows hot air at 50°C all over the exposed surface of the plate on the top. The convection heat transfer coefficient at the top surface is estimated to be 40 W/m² · °C. Determine how long the worker must keep blowing hot air before the two plates separate.   *Answer:* 482 s

4–111   Consider a curing kiln whose walls are made of 30-cm-thick concrete whose properties are $k = 0.9$ W/m · °C and $\alpha = 0.23 \times 10^{-5}$ m²/s. Initially, the kiln and its walls are in equilibrium with the surroundings at 6°C. Then all the doors are closed and the kiln is heated by steam so that the temperature of the inner surface of the walls is raised to 42°C and is maintained at that level for 2.5 h. The curing kiln is then opened and exposed to the atmospheric air after the steam flow is turned off. If the outer surfaces of the walls of the kiln were insulated, would it save any energy that day during the period the kiln was used for curing for 2.5 h only, or would it make no difference? Base your answer on calculations.

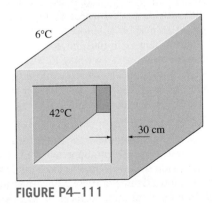

FIGURE P4–111

4–112   The water main in the cities must be placed at sufficient depth below the earth's surface to avoid freezing during extended periods of subfreezing temperatures. Determine the minimum depth at which the water main must be placed at a location where the soil is initially at 15°C and the earth's surface temperature under the worst conditions is expected to remain at -10°C for a period of 75 days. Take the properties of soil at that location to be $k = 0.7$ W/m · °C and $\alpha = 1.4 \times 10^{-5}$ m²/s.   *Answer:* 7.05 m

4–113   A hot dog can be considered to be a 12-cm-long cylinder whose diameter is 2 cm and whose properties are $\rho = 980$ kg/m³, $c_p = 3.9$ kJ/kg · °C, $k = 0.76$ W/m · °C, and $\alpha = 2 \times 10^{-7}$ m²/s. A hot dog initially at 5°C is dropped into boiling water at 100°C. The heat transfer coefficient at the surface of the hot dog is estimated to be 600 W/m² · °C. If the hot dog is considered cooked when its center temperature reaches 80°C, determine how long it will take to cook it in the boiling water.

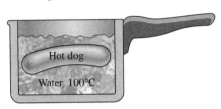

FIGURE P4–113

4–114   A long roll of 2-m-wide and 0.5-cm-thick 1-Mn manganese steel plate coming off a furnace at 820°C is to be quenched in an oil bath ($c_p = 2.0$ kJ/kg · °C) at 45°C. The metal sheet is moving at a steady velocity of 15 m/min, and the oil bath is 9 m long. Taking the convection heat transfer coefficient on both sides of the plate to be 860 W/m² · °C, determine the temperature of the sheet metal when it leaves the oil bath. Also, determine the required rate of heat removal from the oil to keep its temperature constant at 45°C.

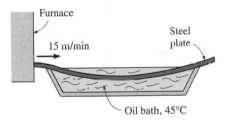

FIGURE P4–114

4–115E   In *Betty Crocker's Cookbook,* it is stated that it takes 5 h to roast a 14-lb stuffed turkey initially at 40°F in an oven maintained at 325°F. It is recommended that a meat thermometer be used to monitor the cooking, and the turkey is considered done when the thermometer inserted deep into the thickest part of the breast or thigh without touching the bone registers 185°F. The turkey can be treated as a homogeneous spherical object with the properties $\rho = 75$ lbm/ft³, $c_p = 0.98$ Btu/lbm · °F, $k = 0.26$ Btu/h · ft · °F, and $\alpha = 0.0035$ ft²/h. Assuming the tip of the thermometer is at one-third radial distance from the center of the turkey, determine (*a*) the average heat transfer coefficient at the surface of the turkey; (*b*) the temperature of the skin of the turkey when it is done; and (*c*) the total amount of heat transferred to the turkey in the oven. Will the reading of the thermometer be more or less than 185°F 5 min after the turkey is taken out of the oven?

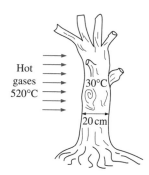

FIGURE P4–115E

4–116 During a fire, the trunks of some dry oak trees ($k = 0.17$ W/m · °C and $\alpha = 1.28 \times 10^{-7}$ m²/s) that are initially at a uniform temperature of 30°C are exposed to hot gases at 520°C for a period of 5 h, with a heat transfer coefficient of 65 W/m² · °C on the surface. The ignition temperature of the trees is 410°C. Treating the trunks of the trees as long cylindrical rods of diameter 20 cm, determine if these dry trees will ignite as the fire sweeps through them.

Hot gases 520°C 30°C 20 cm

FIGURE P4–116

4–117 We often cut a watermelon in half and put it into the freezer to cool it quickly. But usually we forget to check on it and end up having a watermelon with a frozen layer on the top. To avoid this potential problem a person wants to set the timer such that it will go off when the temperature of the exposed surface of the watermelon drops to 3°C.

Consider a 25-cm-diameter spherical watermelon that is cut into two equal parts and put into a freezer at −12°C. Initially, the entire watermelon is at a uniform temperature of 25°C, and the heat transfer coefficient on the surfaces is 22 W/m² · °C. Assuming the watermelon to have the properties of water, determine how long it will take for the center of the exposed cut surfaces of the watermelon to drop to 3°C.

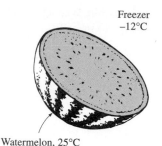

Freezer −12°C

Watermelon, 25°C

FIGURE P4–117

4–118 The thermal conductivity of a solid whose density and specific heat are known can be determined from the relation $k = \alpha / \rho c_p$ after evaluating the thermal diffusivity α.

Consider a 2-cm-diameter cylindrical rod made of a sample material whose density and specific heat are 3700 kg/m³ and 920 J/kg · °C, respectively. The sample is initially at a uniform temperature of 25°C. In order to measure the temperatures of the sample at its surface and its center, a thermocouple is inserted to the center of the sample along the centerline, and another thermocouple is welded into a small hole drilled on the surface. The sample is dropped into boiling water at 100°C. After 3 min, the surface and the center temperatures are recorded to be 93°C and 75°C, respectively. Determine the thermal diffusivity and the thermal conductivity of the material.

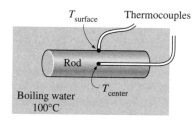

T_{surface} Thermocouples

Rod

T_{center}

Boiling water 100°C

FIGURE P4–118

4–119 In desert climates, rainfall is not a common occurrence since the rain droplets formed in the upper layer of the atmosphere often evaporate before they reach the ground. Consider a raindrop that is initially at a temperature of 5°C and has a diameter of 5 mm. Determine how long it will take for the diameter of the raindrop to reduce to 3 mm as it falls through ambient air at 18°C with a heat transfer coefficient of 400 W/m² · °C. The water temperature can be assumed to remain constant and uniform at 5°C at all times.

4–120E Consider a plate of thickness 1 in, a long cylinder of diameter 1 in, and a sphere of diameter 1 in, all initially at 400°F and all made of bronze ($k = 15.0$ Btu/h · ft · °F and $\alpha = 0.333$ ft²/h). Now all three of these geometries are exposed to cool air at 75°F on all of their surfaces, with a heat transfer coefficient of 7 Btu/h · ft² · °F. Determine the center temperature of each geometry after 5, 10, and 30 min. Explain why the center temperature of the sphere is always the lowest.

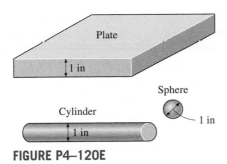

FIGURE P4–120E

4–121E Repeat Prob. 4–120E for cast iron geometries ($k = 29$ Btu/h · ft · °F and $\alpha = 0.61$ ft²/h).

4–122E [ES] Reconsider Prob. 4–120E. Using EES (or other) software, plot the center temperature of each geometry as a function of the cooling time as the time varies from 5 min to 60 min, and discuss the results.

4–123 Engine valves ($k = 48$ W/m · °C, $c_p = 440$ J/kg · °C, and $\rho = 7840$ kg/m³) are heated to 800°C in the heat treatment section of a valve manufacturing facility. The valves are then quenched in a large oil bath at an average temperature of 50°C. The heat transfer coefficient in the oil bath is 800 W/m² · °C. The valves have a cylindrical stem with a diameter of 8 mm and a length of 10 cm. The valve head and the stem may be assumed to be of equal surface area, and the volume of the valve head can be taken to be 80 percent of the volume of stem. Determine how long it will take for the valve temperature to drop to (a) 400°C, (b) 200°C, and (c) 51°C, and (d) the maximum heat transfer from a single valve.

4–124 A watermelon initially at 35°C is to be cooled by dropping it into a lake at 15°C. After 4 h and 40 min of cooling, the center temperature of watermelon is measured to be 20°C. Treating the watermelon as a 20-cm-diameter sphere and using the properties $k = 0.618$ W/m · °C, $\alpha = 0.15 \times 10^{-6}$ m²/s, $\rho = 995$ kg/m³, and $c_p = 4.18$ kJ/kg · °C, determine the average heat transfer coefficient and the surface temperature of watermelon at the end of the cooling period.

4–125 10-cm-thick large food slabs tightly wrapped by thin paper are to be cooled in a refrigeration room maintained at 0°C. The heat transfer coefficient on the box surfaces is 25 W/m² · °C and the boxes are to be kept in the refrigeration room for a period of 6 h. If the initial temperature of the boxes is 30°C determine the center temperature of the boxes if the boxes contain (a) margarine ($k = 0.233$ W/m · °C and $\alpha = 0.11 \times 10^{-6}$ m²/s), (b) white cake ($k = 0.082$ W/m · °C and $\alpha = 0.10 \times 10^{-6}$ m²/s), and (c) chocolate cake ($k = 0.106$ W/m · °C and $\alpha = 0.12 \times 10^{-6}$ m²/s).

4–126 A 30-cm-diameter, 4-m-high cylindrical column of a house made of concrete ($k = 0.79$ W/m · °C, $\alpha = 5.94 \times 10^{-7}$ m²/s, $\rho = 1600$ kg/m³, and $c_p = 0.84$ kJ/kg · °C) cooled to 14°C during a cold night is heated again during the day by being exposed to ambient air at an average temperature of 28°C with an average heat transfer coefficient of 14 W/m² · °C.

Determine (a) how long it will take for the column surface temperature to rise to 27°C, (b) the amount of heat transfer until the center temperature reaches to 28°C, and (c) the amount of heat transfer until the surface temperature reaches to 27°C.

4–127 Long aluminum wires of diameter 3 mm ($\rho = 2702$ kg/m³, $c_p = 0.896$ kJ/kg · °C, $k = 236$ W/m · °C, and $\alpha = 9.75 \times 10^{-5}$ m²/s) are extruded at a temperature of 350°C and exposed to atmospheric air at 30°C with a heat transfer coefficient of 35 W/m² · °C. (a) Determine how long it will take for the wire temperature to drop to 50°C. (b) If the wire is extruded at a velocity of 10 m/min, determine how far the wire travels after extrusion by the time its temperature drops to 50°C. What change in the cooling process would you propose to shorten this distance? (c) Assuming the aluminum wire leaves the extrusion room at 50°C, determine the rate of heat transfer from the wire to the extrusion room.

Answers: (a) 144 s, (b) 24 m, (c) 856 W

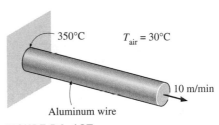

FIGURE P4–127

4–128 Repeat Prob. 4–127 for a copper wire ($\rho = 8950$ kg/m³, $c_p = 0.383$ kJ/kg · °C, $k = 386$ W/m · °C, and $\alpha = 1.13 \times 10^{-4}$ m²/s).

4–129 Consider a brick house ($k = 0.72$ W/m · °C and $\alpha = 0.45 \times 10^{-6}$ m²/s) whose walls are 10 m long, 3 m high, and 0.3 m thick. The heater of the house broke down one night, and the entire house, including its walls, was observed to be 5°C throughout in the morning. The outdoors warmed up as the day progressed, but no change was felt in the house, which was tightly sealed. Assuming the outer surface temperature of the house to remain constant at 15°C, determine how long it would take for the temperature of the inner surfaces of the walls to rise to 5.1°C.

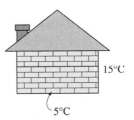

FIGURE P4–129

4–130 A 40-cm-thick brick wall ($k = 0.72$ W/m · °C, and $\alpha = 1.6 \times 10^{-6}$ m²/s) is heated to an average temperature of

18°C by the heating system and the solar radiation incident on it during the day. During the night, the outer surface of the wall is exposed to cold air at -3°C with an average heat transfer coefficient of 20 W/m$^2 \cdot$ °C, determine the wall temperatures at distances 15, 30, and 40 cm from the outer surface for a period of 2 h.

4–131 Consider the engine block of a car made of cast iron ($k = 52$ W/m $\cdot$ °C and $\alpha = 1.7 \times 10^{-5}$ m^2/s). The engine can be considered to be a rectangular block whose sides are 80 cm, 40 cm, and 40 cm. The engine is at a temperature of 150°C when it is turned off. The engine is then exposed to atmospheric air at 17°C with a heat transfer coefficient of 6 W/m$^2 \cdot$ °C. Determine (*a*) the center temperature of the top surface whose sides are 80 cm and 40 cm and (*b*) the corner temperature after 45 min of cooling.

4–132 A man is found dead in a room at 16°C. The surface temperature on his waist is measured to be 23°C and the heat transfer coefficient is estimated to be 9 W/m$^2 \cdot$ °C. Modeling the body as 28-cm diameter, 1.80-m-long cylinder, estimate how long it has been since he died. Take the properties of the body to be $k = 0.62$ W/m $\cdot$ °C and $\alpha = 0.15 \times 10^{-6}$ m^2/s, and assume the initial temperature of the body to be 36°C.

4–133 An exothermic process occurs uniformly throughout a 10-cm-diameter sphere ($k = 300$ W/m $\cdot$ K, $c_p = 400$ J/kg $\cdot$ K, $\rho = 7500$ kg/m^3), and it generates heat at a constant rate of 1.2 MW/m^3. The sphere initially is at a uniform temperature of 20°C, and the exothermic process is commenced at time $t = 0$. To keep the sphere temperature under control, it is submerged in a liquid bath maintained at 20°C. The heat transfer coefficient at the sphere surface is 250 W/m$^2 \cdot$ K.

Due to the high thermal conductivity of sphere, the conductive resistance within the sphere can be neglected in comparison to the convective resistance at its surface. Accordingly, this unsteady state heat transfer situation could be analyzed as a lumped system.

(*a*) Show that the variation of sphere temperature T with time t can be expressed as $dT/dt = 0.5 - 0.005T$.
(*b*) Predict the steady-state temperature of the sphere.
(*c*) Calculate the time needed for the sphere temperature to reach the average of its initial and final (steady) temperatures.

4–134 Large steel plates 1.0-cm in thickness are quenched from 600°C to 100°C by submerging them in an oil reservoir held at 30°C. The average heat transfer coefficient for both faces of steel plates is 400 W/m$^2 \cdot$ K. Average steel properties are $k = 45$ W/m $\cdot$ K, $\rho = 7800$ kg/m^3, and $c_p = 470$ J/kg $\cdot$ K. Calculate the quench time for steel plates.

4–135 Aluminium wires, 3 mm in diameter, are produced by extrusion. The wires leave the extruder at an average temperature of 350°C and at a linear rate of 10 m/min. Before leaving the extrusion room, the wires are cooled to an average temperature of 50°C by transferring heat to the surrounding air at 25°C with a

heat transfer coefficient of 50 W/m$^2 \cdot$ K. Calculate the necessary length of the wire cooling section in the extrusion room.

Fundamentals of Engineering (FE) Exam Problems

4–136 Copper balls ($\rho = 8933$ kg/m^3, $k = 401$ W/m $\cdot$ °C, $c_p = 385$ J/kg $\cdot$ °C, $\alpha = 1.166 \times 10^{-4}$ m^2/s) initially at 200°C are allowed to cool in air at 30°C for a period of 2 minutes. If the balls have a diameter of 2 cm and the heat transfer coefficient is 80 W/m$^2 \cdot$ °C, the center temperature of the balls at the end of cooling is

(*a*) 104°C   (*b*) 87°C   (*c*) 198°C
(*d*) 126°C   (*e*) 152°C

4–137 A 10-cm-inner diameter, 30-cm-long can filled with water initially at 25°C is put into a household refrigerator at 3°C. The heat transfer coefficient on the surface of the can is 14 W/m$^2 \cdot$ °C. Assuming that the temperature of the water remains uniform during the cooling process, the time it takes for the water temperature to drop to 5°C is

(*a*) 0.55 h   (*b*) 1.17 h   (*c*) 2.09 h
(*d*) 3.60 h   (*e*) 4.97 h

4–138 An 18-cm-long, 16-cm-wide, and 12-cm-high hot iron block ($\rho = 7870$ kg/m^3, $c_p = 447$ J/kg $\cdot$ °C) initially at 20°C is placed in an oven for heat treatment. The heat transfer coefficient on the surface of the block is 100 W/m$^2 \cdot$ °C. If it is required that the temperature of the block rises to 750°C in a 25-min period, the oven must be maintained at

(*a*) 750°C   (*b*) 830°C   (*c*) 875°C
(*d*) 910°C   (*e*) 1000°C

4–139 A small chicken ($k = 0.45$ W/m $\cdot$ °C, $\alpha = 0.15 \times 10^{-6}$ m^2/s) can be approximated as an 11.25-cm-diameter solid sphere. The chicken is initially at a uniform temperature of 8°C and is to be cooked in an oven maintained at 220°C with a heat transfer coefficient of 80 W/m$^2 \cdot$ °C. With this idealization, the temperature at the center of the chicken after a 90-min period is

(*a*) 25°C   (*b*) 61°C   (*c*) 89°C
(*d*) 122°C   (*e*) 168°C

4–140 In a production facility, large plates made of stainless steel ($k = 15$ W/m $\cdot$ °C, $\alpha = 3.91 \times 10^{-6}$ m^2/s) of 40 cm thickness are taken out of an oven at a uniform temperature of 750°C. The plates are placed in a water bath that is kept at a constant temperature of 20°C with a heat transfer coefficient of 600 W/m$^2 \cdot$ °C. The time it takes for the surface temperature of the plates to drop to 100°C is

(*a*) 0.28 h   (*b*) 0.99 h   (*c*) 2.05 h
(*d*) 3.55 h   (*e*) 5.33 h

4–141 A long 18-cm-diameter bar made of hardwood ($k = 0.159$ W/m $\cdot$ °C, $\alpha = 1.75 \times 10^{-7}$ m^2/s) is exposed to air at 30°C with a heat transfer coefficient of 8.83 W/m$^2 \cdot$ °C. If the center temperature of the bar is measured to be 15°C after a period of 3-hours, the initial temperature of the bar is

(*a*) 11.9°C   (*b*) 4.9°C   (*c*) 1.7°C
(*d*) 0°C   (*e*) -9.2°C

4–142 A potato may be approximated as a 5.7-cm-diameter solid sphere with the properties $\rho = 910$ kg/m^3, $c_p = 4.25$ kJ/kg · °C, $k = 0.68$ W/m · °C, and $\alpha = 1.76 \times 10^{-7}$ m^2/s. Twelve such potatoes initially at 25°C are to be cooked by placing them in an oven maintained at 250°C with a heat transfer coefficient of 95 W/m^2 · °C. The amount of heat transfer to the potatoes during a 30-minute period is

(a) 77 kJ   (b) 483 kJ   (c) 927 kJ
(d) 970 kJ   (e) 1012 kJ

4–143 A potato that may be approximated as a 5.7-cm solid sphere with the properties $\rho = 910$ kg/m^3, $c_p = 4.25$ kJ/kg · °C, $k = 0.68$ W/m · °C, and $\alpha = 1.76 \times 10^{-7}$ m^2/s. Twelve such potatoes initially at 25°C are to be cooked by placing them in an oven maintained at 250°C with a heat transfer coefficient of 95 W/m^2 · °C. The amount of heat transfer to the potatoes by the time the center temperature reaches 100°C is

(a) 56 kJ   (b) 666 kJ   (c) 838 kJ
(d) 940 kJ   (e) 1088 kJ

4–144 A large chunk of tissue at 35°C with a thermal diffusivity of 1×10^{-7} m^2/s is dropped into iced water The water is well-stirred so that the surface temperature of the tissue drops to 0 °C at time zero and remains at 0°C at all times. The temperature of the tissue after 4 minutes at a depth of 1 cm is

(a) 5°C   (b) 30°C   (c) 25°C
(d) 20°C   (e) 10°C

4–145 Consider a 7.6-cm-diameter cylindrical lamb meat chunk ($\rho = 1030$ kg/m^3, $c_p = 3.49$ kJ/kg · °C, $k = 0.456$ W/m · °C, $\alpha = 1.3 \times 10^{-7}$ m^2/s). Such a meat chunk intially at 2°C is dropped into boiling water at 95°C with a heat transfer coefficient of 1200 W/m^2 · °C. The time it takes for the center temperature of the meat chunk to rise to 75 °C is

(a) 136 min   (b) 21.2 min   (c) 13.6 min
(d) 11.0 min   (e) 8.5 min

4–146 Consider a 7.6-cm-long and 3-cm-diameter cylindrical lamb meat chunk ($\rho = 1030$ kg/m^3, $c_p = 3.49$ kJ/kg · °C, $k = 0.456$ W/m · °C, $\alpha = 1.3 \times 10^{-7}$ m^2/s). Fifteen such meat chunks initially at 2°C is dropped into boiling water at 95°C with a heat transfer coefficient of 1200 W/m^2 · °C. The amount of heat transfer during the first 8 minutes of cooking is

(a) 71 kJ   (b) 227 kJ   (c) 238 kJ
(d) 269 kJ   (e) 307 kJ

4–147 Carbon steel balls ($\rho = 7830$ kg/m^3, $k = 64$ W/m · °C, $c_p = 434$ J/kg · °C) initially at 150°C are quenched in an oil bath at 20°C for a period of 3 minutes. If the balls have a diameter of 5 cm and the convection heat transfer coefficient is 450 W/m^2 · °C. The center temperature of the balls after quenching will be (*Hint:* Check the Biot number).

(a) 27.4°C   (b) 143°C   (c) 12.7°C
(d) 48.2°C   (e) 76.9°C

4–148 A 6-cm-diameter 13-cm-high canned drink ($\rho = 977$ kg/m^3, $k = 0.607$ W/m · °C, $c_p = 4180$ J/kg · °C) initially at 25°C is to be cooled to 5°C by dropping it into iced water at 0°C. Total surface area and volume of the drink are $A_s = 301.6$ cm^2 and $V = 367.6$ cm^3. If the heat transfer coefficient is 120 W/m^2 · °C, determine how long it will take for the drink to cool to 5°C. Assume the can is agitated in water and thus the temperature of the drink changes uniformly with time.

(a) 1.5 min   (b) 8.7 min   (c) 11.1 min
(d) 26.6 min   (e) 6.7 min

4–149 Lumped system analysis of transient heat conduction situations is valid when the Biot number is

(a) very small   (b) approximately one
(c) very large   (d) any real number
(e) cannot say unless the Fourier number is also known.

4–150 Polyvinylchloride automotive body panels ($k = 0.092$ W/m · K, $c_p = 1.05$ kJ/kg · K, $\rho = 1714$ kg/m^3), 3-mm thick, emerge from an injection molder at 120°C. They need to be cooled to 40°C by exposing both sides of the panels to 20°C air before they can be handled. If the convective heat transfer coefficient is 30 W/m^2 · K and radiation is not considered, the time that the panels must be exposed to air before they can be handled is

(a) 1.6 min   (b) 2.4 min   (c) 2.8 min
(d) 3.5 min   (e) 4.2 min

4–151 A steel casting cools to 90 percent of the original temperature difference in 30 min in still air. The time it takes to cool this same casting to 90 percent of the original temperature difference in a moving air stream whose convective heat transfer coefficient is 5 times that of still air is

(a) 3 min   (b) 6 min   (c) 9 min
(d) 12 min   (e) 15 min

4–152 The Biot number can be thought of as the ratio of
(a) The conduction thermal resistance to the convective thermal resistance.
(b) The convective thermal resistance to the conduction thermal resistance.
(c) The thermal energy storage capacity to the conduction thermal resistance.
(d) The thermal energy storage capacity to the convection thermal resistance.
(e) None of the above.

4–153 When water, as in a pond or lake, is heated by warm air above it, it remains stable, does not move, and forms a warm layer of water on top of a cold layer. Consider a deep lake ($k = 0.6$ W/m · K, $c_p = 4.179$ kJ/kg · K) that is initially at a uniform temperature of 2°C and has its surface temperature suddenly increased to 20°C by a spring weather front. The temperature of the water 1 m below the surface 400 hours after this change is

(a) 2.1°C   (b) 4.2°C   (c) 6.3°C
(d) 8.4°C   (e) 10.2°C

4–154 The 40-cm-thick roof of a large room made of concrete ($k = 0.79$ W/m · °C, $\alpha = 5.88 \times 10^{-7}$ m²/s) is initially at a uniform temperature of 15°C. After a heavy snow storm, the outer surface of the roof remains covered with snow at -5°C. The roof temperature at 18.2 cm distance from the outer surface after a period of 2 hours is

(a) 14.0°C   (b) 12.5°C   (c) 7.8°C
(d) 0°C   (e) -5.0°C

Design and Essay Problems

4–155 Conduct the following experiment at home to determine the combined convection and radiation heat transfer coefficient at the surface of an apple exposed to the room air. You will need two thermometers and a clock.

First, weigh the apple and measure its diameter. You can measure its volume by placing it in a large measuring cup halfway filled with water, and measuring the change in volume when it is completely immersed in the water. Refrigerate the apple overnight so that it is at a uniform temperature in the morning and measure the air temperature in the kitchen. Then take the apple out and stick one of the thermometers to its middle and the other just under the skin. Record both temperatures every 5 min for an hour. Using these two temperatures, calculate the heat transfer coefficient for each interval and take their average. The result is the combined convection and radiation heat transfer coefficient for this heat transfer process. Using your experimental data, also calculate the thermal conductivity

and thermal diffusivity of the apple and compare them to the values given above.

4–156 Repeat Prob. 4–155 using a banana instead of an apple. The thermal properties of bananas are practically the same as those of apples.

4–157 Conduct the following experiment to determine the time constant for a can of soda and then predict the temperature of the soda at different times. Leave the soda in the refrigerator overnight. Measure the air temperature in the kitchen and the temperature of the soda while it is still in the refrigerator by taping the sensor of the thermometer to the outer surface of the can. Then take the soda out and measure its temperature again in 5 min. Using these values, calculate the exponent b. Using this b-value, predict the temperatures of the soda in 10, 15, 20, 30, and 60 min and compare the results with the actual temperature measurements. Do you think the lumped system analysis is valid in this case?

4–158 Citrus trees are very susceptible to cold weather, and extended exposure to subfreezing temperatures can destroy the crop. In order to protect the trees from occasional cold fronts with subfreezing temperatures, tree growers in Florida usually install water sprinklers on the trees. When the temperature drops below a certain level, the sprinklers spray water on the trees and their fruits to protect them against the damage the subfreezing temperatures can cause. Explain the basic mechanism behind this protection measure and write an essay on how the system works in practice.

NUMERICAL METHODS IN HEAT CONDUCTION

So far we have mostly considered relatively simple heat conduction problems involving *simple geometries* with simple boundary conditions because only such simple problems can be solved *analytically*. But many problems encountered in practice involve *complicated geometries* with complex boundary conditions or variable properties, and cannot be solved analytically. In such cases, sufficiently accurate approximate solutions can be obtained by computers using a *numerical method*.

Analytical solution methods such as those presented in Chapter 2 are based on solving the governing differential equation together with the boundary conditions. They result in solution functions for the temperature at *every point* in the medium. Numerical methods, on the other hand, are based on replacing the differential equation by a set of *n* algebraic equations for the unknown temperatures at *n* selected points in the medium, and the simultaneous solution of these equations results in the temperature values at those *discrete points*.

There are several ways of obtaining the numerical formulation of a heat conduction problem, such as the *finite difference* method, the *finite element* method, the *boundary element* method, and the *energy balance* (or control volume) method. Each method has its own advantages and disadvantages, and each is used in practice. In this chapter, we use primarily the *energy balance* approach since it is based on the familiar energy balances on control volumes instead of heavy mathematical formulations, and thus it gives a better physical feel for the problem. Besides, it results in the same set of algebraic equations as the finite difference method. In this chapter, the numerical formulation and solution of heat conduction problems are demonstrated for both steady and transient cases in various geometries.

OBJECTIVES

When you finish studying this chapter, you should be able to:

- Understand the limitations of analytical solutions of conduction problems, and the need for computation-intensive numerical methods,

- Express derivates as differences, and obtain finite difference formulations,

- Solve steady one- or two-dimensional conduction problems numerically using the finite difference method, and

- Solve transient one- or two-dimensional conduction problems using the finite difference method.

CONTENTS

5–1 · WHY NUMERICAL METHODS?

The ready availability of high-speed computers and easy-to-use powerful software packages has had a major impact on engineering education and practice in recent years. Engineers in the past had to rely on *analytical skills* to solve significant engineering problems, and thus they had to undergo a rigorous training in mathematics. Today's engineers, on the other hand, have access to a tremendous amount of *computation power* under their fingertips, and they mostly need to understand the physical nature of the problem and interpret the results. But they also need to understand how calculations are performed by the computers to develop an awareness of the processes involved and the limitations, while avoiding any possible pitfalls.

In Chapter 2, we solved various heat conduction problems in various geometries in a systematic but highly mathematical manner by (1) deriving the governing differential equation by performing an energy balance on a differential volume element, (2) expressing the boundary conditions in the proper mathematical form, and (3) solving the differential equation and applying the boundary conditions to determine the integration constants. This resulted in a solution function for the temperature distribution in the medium, and the solution obtained in this manner is called the analytical solution of the problem. For example, the mathematical formulation of one-dimensional steady heat conduction in a sphere of radius r_o whose outer surface is maintained at a uniform temperature of T_1 with uniform heat generation at a rate of $\dot{e}$ was expressed as (Fig. 5–1)

$$\frac{1}{r^2}\frac{d}{dr}\left(r^2\frac{dT}{dr}\right) + \frac{\dot{e}}{k} = 0$$

$$\frac{dT(0)}{dr} = 0 \quad \text{and} \quad T(r_o) = T_1 \tag{5-1}$$

whose (analytical) solution is

$$T(r) = T_1 + \frac{\dot{e}}{6k}(r_o^2 - r^2) \tag{5-2}$$

This is certainly a very desirable form of solution since the temperature at any point within the sphere can be determined simply by substituting the *r*-coordinate of the point into the analytical solution function above. The analytical solution of a problem is also referred to as the exact solution since it satisfies the differential equation and the boundary conditions. This can be verified by substituting the solution function into the differential equation and the boundary conditions. Further, the *rate of heat transfer* at any location within the sphere or its surface can be determined by taking the derivative of the solution function $T(r)$ and substituting it into Fourier's law as

$$\dot{Q}(r) = -kA\frac{dT}{dr} = -k(4\pi r^2)\left(-\frac{\dot{e}r}{3k}\right) = \frac{4\pi r^3 \dot{e}}{3} \tag{5-3}$$

The analysis above did not require any mathematical sophistication beyond the level of *simple integration,* and you are probably wondering why anyone

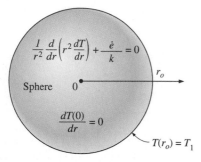

Solution:

$$T(r) = T_1 + \frac{\dot{e}}{6k}(r_o^2 - r^2)$$

$$\dot{Q}(r) = -kA\frac{dT}{dr} = \frac{4\pi r^3 \dot{e}}{3}$$

FIGURE 5–1
The analytical solution of a problem requires solving the governing differential equation and applying the boundary conditions.

would ask for something else. After all, the solutions obtained are exact and easy to use. Besides, they are instructive since they show clearly the functional dependence of temperature and heat transfer on the independent variable r. Well, there are several reasons for searching for alternative solution methods.

1 Limitations

Analytical solution methods are limited to *highly simplified problems* in *simple geometries* (Fig. 5–2). The geometry must be such that its entire surface can be described mathematically in a coordinate system by setting the variables equal to constants. That is, it must fit into a coordinate system *perfectly* with nothing sticking out or in. In the case of one-dimensional heat conduction in a solid sphere of radius r_o, for example, the entire outer surface can be described by $r = r_o$. Likewise, the surfaces of a finite solid cylinder of radius r_o and height H can be described by $r = r_o$ for the side surface and $z = 0$ and $z = H$ for the bottom and top surfaces, respectively. Even minor complications in geometry can make an analytical solution impossible. For example, a spherical object with an extrusion like a *handle* at some location is impossible to handle analytically since the boundary conditions in this case cannot be expressed in any familiar coordinate system.

Even in simple geometries, heat transfer problems cannot be solved analytically if the *thermal conditions* are not sufficiently simple. For example, the consideration of the variation of thermal conductivity with temperature, the variation of the heat transfer coefficient over the surface, or the radiation heat transfer on the surfaces can make it impossible to obtain an analytical solution. Therefore, analytical solutions are limited to problems that are simple or can be simplified with reasonable approximations.

2 Better Modeling

We mentioned earlier that analytical solutions are exact solutions since they do not involve any approximations. But this statement needs some clarification. Distinction should be made between an *actual real-world problem* and the *mathematical model* that is an idealized representation of it. The solutions we get are the solutions of mathematical models, and the degree of applicability of these solutions to the actual physical problems depends on the accuracy of the model. An "approximate" solution of a realistic model of a physical problem is usually more accurate than the "exact" solution of a crude mathematical model (Fig. 5–3).

When attempting to get an analytical solution to a physical problem, there is always the tendency to *oversimplify* the problem to make the mathematical model sufficiently simple to warrant an analytical solution. Therefore, it is common practice to ignore any effects that cause mathematical complications such as nonlinearities in the differential equation or the boundary conditions. So it comes as no surprise that *nonlinearities* such as temperature dependence of thermal conductivity and the radiation boundary conditions are seldom considered in analytical solutions. A mathematical model intended for a numerical solution is likely to represent the actual problem better. Therefore, the numerical solution of engineering problems has now become the norm rather than the exception even when analytical solutions are available.

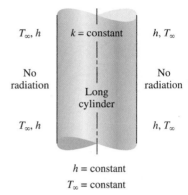

h = constant
T_∞ = constant

FIGURE 5–2
Analytical solution methods are limited to simplified problems in simple geometries.

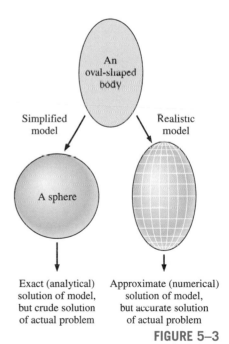

Exact (analytical) solution of model, but crude solution of actual problem

Approximate (numerical) solution of model, but accurate solution of actual problem

FIGURE 5–3
The approximate numerical solution of a real-world problem may be more accurate than the exact (analytical) solution of an oversimplified model of that problem.

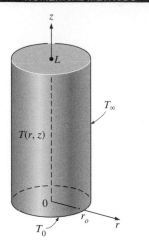

Analytical solution:

$$\frac{T(r, z) - T_\infty}{T_0 - T_\infty} = \sum_{n=1}^{\infty} \frac{J_0(\lambda_n r)}{\lambda_n J_1(\lambda_n r_o)} \frac{\sinh \lambda_n(L - z)}{\sinh (\lambda_n L)}$$

where λ_n's are roots of $J_0(\lambda_n r_o) = 0$

FIGURE 5–4
Some analytical solutions are very complex and difficult to use.

FIGURE 5–5
The ready availability of high-powered computers with sophisticated software packages has made numerical solution the norm rather than the exception.

3 Flexibility

Engineering problems often require extensive *parametric studies* to understand the influence of some variables on the solution in order to choose the right set of variables and to answer some "what-if" questions. This is an *iterative process* that is extremely tedious and time-consuming if done by hand. Computers and numerical methods are ideally suited for such calculations, and a wide range of related problems can be solved by minor modifications in the code or input variables. Today it is almost unthinkable to perform any significant optimization studies in engineering without the power and flexibility of computers and numerical methods.

4 Complications

Some problems can be solved analytically, but the solution procedure is so complex and the resulting solution expressions so complicated that it is not worth all that effort. With the exception of steady one-dimensional or transient lumped system problems, all heat conduction problems result in *partial* differential equations. Solving such equations usually requires mathematical sophistication beyond that acquired at the undergraduate level, such as orthogonality, eigenvalues, Fourier and Laplace transforms, Bessel and Legendre functions, and infinite series. In such cases, the evaluation of the solution, which often involves double or triple summations of infinite series at a specified point, is a challenge in itself (Fig. 5–4). Therefore, even when the solutions are available in some handbooks, they are intimidating enough to scare prospective users away.

5 Human Nature

As human beings, we like to sit back and make wishes, and we like our wishes to come true without much effort. The invention of TV remote controls made us feel like kings in our homes since the commands we give in our comfortable chairs by pressing buttons are immediately carried out by the obedient TV sets. After all, what good is cable TV without a remote control. We certainly would love to continue being the king in our little cubicle in the engineering office by solving problems at the press of a button on a computer (until they invent a remote control for the computers, of course). Well, this might have been a fantasy yesterday, but it is a reality today. Practically all engineering offices today are equipped with *high-powered computers* with *sophisticated software packages,* with impressive presentation-style colorful output in graphical and tabular form (Fig. 5–5). Besides, the results are as accurate as the analytical results for all practical purposes. The computers have certainly changed the way engineering is practiced.

The discussions above should not lead you to believe that analytical solutions are unnecessary and that they should be discarded from the engineering curriculum. On the contrary, insight to the *physical phenomena* and *engineering wisdom* is gained primarily through analysis. The "feel" that engineers develop during the analysis of simple but fundamental problems serves as an invaluable tool when interpreting a huge pile of results obtained from a computer when solving a complex problem. A simple analysis by hand for a limiting case can be used to check if the results are in the proper range. Also, nothing can take the place of getting "ball park" results on a piece of paper during preliminary discussions. The calculators made the basic arithmetic operations by hand a thing of the past, but they did not eliminate the need for instructing grade school children how to add or multiply.

In this chapter, you will learn how to *formulate* and *solve* heat transfer problems numerically using one or more approaches. In your professional life, you will probably solve the heat transfer problems you come across using a professional software package, and you are highly unlikely to write your own programs to solve such problems. (Besides, people will be highly skeptical of the results obtained using your own program instead of using a well-established commercial software package that has stood the test of time.) The insight you gain in this chapter by formulating and solving some heat transfer problems will help you better understand the available software packages and be an informed and responsible user.

5–2 ▪ FINITE DIFFERENCE FORMULATION OF DIFFERENTIAL EQUATIONS

The numerical methods for solving differential equations are based on replacing the *differential equations* by *algebraic equations*. In the case of the popular **finite difference** method, this is done by replacing the *derivatives* by *differences*. Below we demonstrate this with both first- and second-order derivatives. But first we give a motivational example.

Consider a man who deposits his money in the amount of $A_0 = \$100$ in a savings account at an annual interest rate of 18 percent, and let us try to determine the amount of money he will have after one year if interest is compounded continuously (or instantaneously). In the case of simple interest, the money will earn $18 interest, and the man will have $100 + 100 \times 0.18 = \$118.00$ in his account after one year. But in the case of compounding, the interest earned during a compounding period will also earn interest for the remaining part of the year, and the year-end balance will be greater than $118. For example, if the money is compounded twice a year, the balance will be $100 + 100 \times (0.18/2) = \$109$ after six months, and $109 + 109 \times (0.18/2) = \$118.81$ at the end of the year. We could also determine the balance A directly from

$$A = A_0(1 + i)^n = (\$100)(1 + 0.09)^2 = \$118.81 \tag{5-4}$$

where i is the interest rate for the compounding period and n is the number of periods. Using the same formula, the year-end balance is determined for monthly, daily, hourly, minutely, and even secondly compounding, and the results are given in Table 5–1.

Note that in the case of daily compounding, the year-end balance will be $119.72, which is $1.72 more than the simple interest case. (So it is no wonder that the credit card companies usually charge interest compounded daily when determining the balance.) Also note that compounding at smaller time intervals, even at the end of each second, does not change the result, and we suspect that instantaneous compounding using "differential" time intervals dt will give the same result. This suspicion is confirmed by obtaining the differential

TABLE 5–1

Year-end balance of a $100 account earning interest at an annual rate of 18 percent for various compounding periods

Compounding Period	Number of Periods, n	Year-End Balance
1 year	1	$118.00
6 months	2	118.81
1 month	12	119.56
1 week	52	119.68
1 day	365	119.72
1 hour	8760	119.72
1 minute	525,600	119.72
1 second	31,536,000	119.72
Instantaneous	∞	119.72

equation $dA/dt = iA$ for the balance A, whose solution is $A = A_0 \exp(it)$. Substitution yields

$$A = (\$100)\exp(0.18 \times 1) = \$119.72$$

which is identical to the result for daily compounding. Therefore, replacing a differential time interval dt by a finite time interval of $\Delta t = 1$ day gave the same result when rounded to the second decimal place for cents, which leads us into believing that *reasonably accurate results can be obtained by replacing differential quantities by sufficiently small differences.*

Next, we develop the finite difference formulation of heat conduction problems by replacing the derivatives in the differential equations by differences. In the following section we do it using the energy balance method, which does not require any knowledge of differential equations.

Derivatives are the building blocks of differential equations, and thus we first give a brief review of derivatives. Consider a function f that depends on x, as shown in Figure 5–6. The **first derivative** of $f(x)$ at a point is equivalent to the *slope* of a line tangent to the curve at that point and is defined as

$$\frac{df(x)}{dx} = \lim_{\Delta x \to 0} \frac{\Delta f}{\Delta x} = \lim_{\Delta x \to 0} \frac{f(x + \Delta x) - f(x)}{\Delta x} \tag{5–5}$$

which is the ratio of the increment Δf of the function to the increment Δx of the independent variable as $\Delta x \to 0$. If we don't take the indicated limit, we will have the following *approximate* relation for the derivative:

$$\frac{df(x)}{dx} \cong \frac{f(x + \Delta x) - f(x)}{\Delta x} \tag{5–6}$$

This approximate expression of the derivative in terms of differences is the **finite difference form** of the first derivative. The equation above can also be obtained by writing the *Taylor series expansion* of the function f about the point x,

$$f(x + \Delta x) = f(x) + \Delta x \frac{df(x)}{dx} + \frac{1}{2} \Delta x^2 \frac{d^2 f(x)}{dx^2} + \cdots \tag{5–7}$$

and neglecting all the terms in the expansion except the first two. The first term neglected is proportional to Δx^2, and thus the *error* involved in each step of this approximation is also proportional to Δx^2. However, the *commutative error* involved after M steps in the direction of length L is proportional to Δx since $M \Delta x^2 = (L/\Delta x)\Delta x^2 = L\Delta x$. Therefore, the smaller the Δx, the smaller the error, and thus the more accurate the approximation.

Now consider steady one-dimensional heat conduction in a plane wall of thickness L with heat generation. The wall is subdivided into M sections of equal thickness $\Delta x = L/M$ in the x-direction, separated by planes passing through $M + 1$ points $0, 1, 2, \ldots, m - 1, m, m + 1, \ldots, M$ called **nodes** or **nodal points**, as shown in Figure 5–7. The x-coordinate of any point m is simply $x_m = m\Delta x$, and the temperature at that point is simply $T(x_m) = T_m$.

The heat conduction equation involves the second derivatives of temperature with respect to the space variables, such as $d^2 T/dx^2$, and the finite difference formulation is based on replacing the second derivatives by appropriate

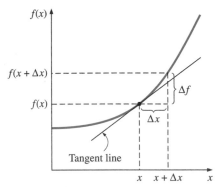

FIGURE 5–6

The derivative of a function at a point represents the slope of the function at that point.

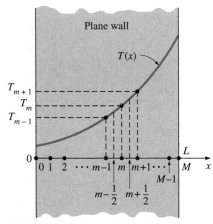

FIGURE 5–7

Schematic of the nodes and the nodal temperatures used in the development of the finite difference formulation of heat transfer in a plane wall.

differences. But we need to start the process with first derivatives. Using Eq. 5–6, the first derivative of temperature dT/dx at the midpoints $m - \frac{1}{2}$ and $m + \frac{1}{2}$ of the sections surrounding the node m can be expressed as

$$\frac{dT}{dx}\bigg|_{m-\frac{1}{2}} \cong \frac{T_m - T_{m-1}}{\Delta x} \quad \text{and} \quad \frac{dT}{dx}\bigg|_{m+\frac{1}{2}} \cong \frac{T_{m+1} - T_m}{\Delta x} \tag{5–8}$$

Noting that the second derivative is simply the derivative of the first derivative, the second derivative of temperature at node m can be expressed as

$$\frac{d^2T}{dx^2}\bigg|_m \cong \frac{\dfrac{dT}{dx}\bigg|_{m+\frac{1}{2}} - \dfrac{dT}{dx}\bigg|_{m-\frac{1}{2}}}{\Delta x} = \frac{\dfrac{T_{m+1} - T_m}{\Delta x} - \dfrac{T_m - T_{m-1}}{\Delta x}}{\Delta x} = \frac{T_{m-1} - 2T_m + T_{m+1}}{\Delta x^2}$$

$$\tag{5–9}$$

which is the *finite difference representation* of the *second derivative* at a general internal node m. Note that the second derivative of temperature at a node m is expressed in terms of the temperatures at node m and its two neighboring nodes. Then the differential equation

$$\frac{d^2T}{dx^2} + \frac{\dot{e}}{k} = 0 \tag{5–10}$$

which is the governing equation for *steady one-dimensional* heat transfer in a plane wall with heat generation and constant thermal conductivity, can be expressed in the *finite difference* form as (Fig. 5–8)

$$\frac{T_{m-1} - 2T_m + T_{m+1}}{\Delta x^2} + \frac{\dot{e}_m}{k} = 0, \qquad m = 1, 2, 3, \ldots, M - 1 \tag{5–11}$$

where $\dot{e}_m$ is the rate of heat generation per unit volume at node m. If the surface temperatures T_0 and T_M are specified, the application of this equation to each of the $M - 1$ interior nodes results in $M - 1$ equations for the determination of $M - 1$ unknown temperatures at the interior nodes. Solving these equations simultaneously gives the temperature values at the nodes. If the temperatures at the outer surfaces are not known, then we need to obtain two more equations in a similar manner using the specified boundary conditions. Then the unknown temperatures at $M + 1$ nodes are determined by solving the resulting system of $M + 1$ equations in $M + 1$ unknowns simultaneously.

Note that the *boundary conditions* have no effect on the finite difference formulation of interior nodes of the medium. This is not surprising since the control volume used in the development of the formulation does not involve any part of the boundary. You may recall that the boundary conditions had no effect on the differential equation of heat conduction in the medium either.

The finite difference formulation above can easily be extended to two- or three-dimensional heat transfer problems by replacing each second derivative by a difference equation in that direction. For example, the *finite difference formulation* for **steady two-dimensional heat conduction** in a region with

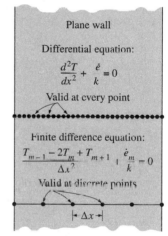

FIGURE 5–8

The differential equation is valid at every point of a medium, whereas the finite difference equation is valid at discrete points (the nodes) only.

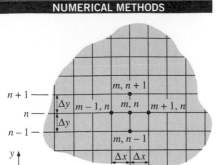

FIGURE 5–9
Finite difference mesh for two-dimensional conduction in rectangular coordinates.

heat generation and constant thermal conductivity can be expressed in rectangular coordinates as (Fig. 5–9)

$$\frac{T_{m+1,n} - 2T_{m,n} + T_{m-1,n}}{\Delta x^2} + \frac{T_{m,n+1} - 2T_{m,n} + T_{m,n-1}}{\Delta y^2} + \frac{\dot{e}_{m,n}}{k} = 0 \quad \textbf{(5–12)}$$

for $m = 1, 2, 3, \ldots, M-1$ and $n = 1, 2, 3, \ldots, N-1$ at any interior node (m, n). Note that a rectangular region that is divided into M equal subregions in the x-direction and N equal subregions in the y-direction has a total of $(M + 1)(N + 1)$ nodes, and Eq. 5–12 can be used to obtain the finite difference equations at $(M - 1)(N - 1)$ of these nodes (i.e., all nodes except those at the boundaries).

The finite difference formulation is given above to demonstrate how difference equations are obtained from differential equations. However, we use the *energy balance approach* in the following sections to obtain the numerical formulation because it is more *intuitive* and can handle *boundary conditions* more easily. Besides, the energy balance approach does not require having the differential equation before the analysis.

5–3 · ONE-DIMENSIONAL STEADY HEAT CONDUCTION

In this section we develop the finite difference formulation of heat conduction in a plane wall using the energy balance approach and discuss how to solve the resulting equations. The **energy balance method** is based on *subdividing* the medium into a sufficient number of volume elements and then applying an *energy balance* on each element. This is done by first *selecting* the nodal points (or nodes) at which the temperatures are to be determined and then *forming elements* (or control volumes) over the nodes by drawing lines through the midpoints between the nodes. This way, the interior nodes remain at the middle of the elements, and the properties *at the node* such as the temperature and the rate of heat generation represent the *average* properties of the element. Sometimes it is convenient to think of temperature as varying *linearly* between the nodes, especially when expressing heat conduction between the elements using Fourier's law.

To demonstrate the approach, again consider steady one-dimensional heat transfer in a plane wall of thickness L with heat generation $\dot{e}(x)$ and constant conductivity k. The wall is now subdivided into M equal regions of thickness $\Delta x = L/M$ in the x-direction, and the divisions between the regions are selected as the nodes. Therefore, we have $M + 1$ nodes labeled 0, 1, 2, . . . , $m -1, m, m + 1, \ldots, M,$ as shown in Figure 5–10. The x-coordinate of any node m is simply $x_m = m\Delta x$, and the temperature at that point is $T(x_m) = T_m$. Elements are formed by drawing vertical lines through the midpoints between the nodes. Note that all interior elements represented by interior nodes are full-size elements (they have a thickness of Δx), whereas the two elements at the boundaries are half-sized.

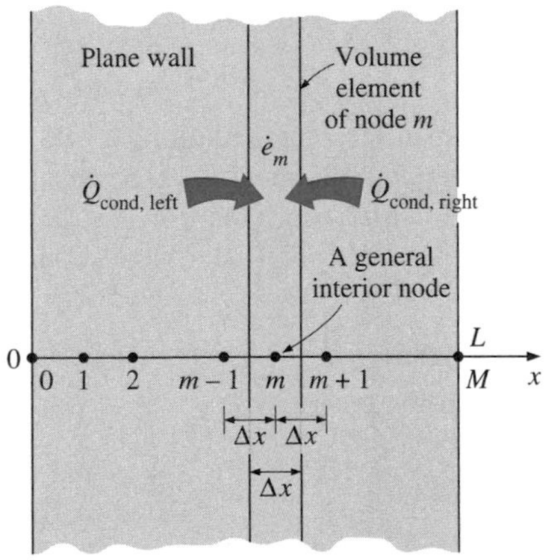

To obtain a general difference equation for the interior nodes, consider the element represented by node m and the two neighboring nodes $m - 1$ and $m + 1$. Assuming the heat conduction to be *into* the element on all surfaces, an energy balance on the element can be expressed as

FIGURE 5–10
The nodal points and volume elements for the finite difference formulation of one-dimensional conduction in a plane wall.

$$\begin{pmatrix} \text{Rate of heat} \\ \text{conduction} \\ \text{at the left} \\ \text{surface} \end{pmatrix} + \begin{pmatrix} \text{Rate of heat} \\ \text{conduction} \\ \text{at the right} \\ \text{surface} \end{pmatrix} + \begin{pmatrix} \text{Rate of heat} \\ \text{generation} \\ \text{inside the} \\ \text{element} \end{pmatrix} = \begin{pmatrix} \text{Rate of change} \\ \text{of the energy} \\ \text{content of} \\ \text{the element} \end{pmatrix}$$

or

$$\dot{Q}_{\text{cond, left}} + \dot{Q}_{\text{cond, right}} + \dot{E}_{\text{gen, element}} = \frac{\Delta E_{\text{element}}}{\Delta t} = 0 \qquad (5\text{-}13)$$

since the energy content of a medium (or any part of it) does not change under *steady* conditions and thus $\Delta E_{\text{element}} = 0$. The rate of *heat generation* within the element can be expressed as

$$\dot{E}_{\text{gen, element}} = \dot{e}_m V_{\text{element}} = \dot{e}_m A \Delta x \qquad (5\text{-}14)$$

where $\dot{e}_m$ is the rate of heat generation per unit volume in W/m^3 evaluated at node m and treated as a constant for the entire element, and A is heat transfer area, which is simply the inner (or outer) surface area of the wall.

Recall that when temperature varies *linearly*, the steady rate of heat conduction across a plane wall of thickness L can be expressed as

$$\dot{Q}_{\text{cond}} = kA \frac{\Delta T}{L} \qquad (5\text{-}15)$$

where ΔT is the temperature change across the wall and the direction of heat transfer is from the high temperature side to the low temperature. In the case of a plane wall with heat generation, the variation of temperature is not linear and thus the relation above is not applicable. However, the variation of temperature between the nodes can be *approximated* as being linear in the determination of heat conduction across a thin layer of thickness Δx between two nodes (Fig. 5–11). Obviously the smaller the distance Δx between two nodes, the more accurate is this approximation. (In fact, such approximations are the reason for classifying the numerical methods as approximate solution methods. In the limiting case of Δx approaching zero, the formulation becomes exact and we obtain a differential equation.) Noting that the direction of heat transfer on both surfaces of the element is assumed to be *toward* the node m, the rate of heat conduction at the left and right surfaces can be expressed as

$$\dot{Q}_{\text{cond, left}} = kA \frac{T_{m-1} - T_m}{\Delta x} \quad \text{and} \quad \dot{Q}_{\text{cond, right}} = kA \frac{T_{m+1} - T_m}{\Delta x} \qquad (5\text{-}16)$$

Substituting Eqs. 5–14 and 5–16 into Eq. 5–13 gives

$$kA \frac{T_{m-1} - T_m}{\Delta x} + kA \frac{T_{m+1} - T_m}{\Delta x} + \dot{e}_m A \Delta x = 0 \qquad (5\text{-}17)$$

which simplifies to

$$\frac{T_{m-1} - 2T_m + T_{m+1}}{\Delta x^2} + \frac{\dot{e}_m}{k} = 0, \qquad m = 1, 2, 3, \ldots, M - 1 \qquad (5\text{-}18)$$

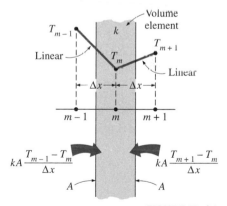

FIGURE 5–11
In finite difference formulation, the temperature is assumed to vary linearly between the nodes.

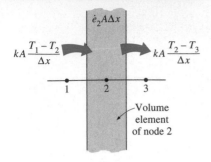

$$kA\frac{T_1 - T_2}{\Delta x} - kA\frac{T_2 - T_3}{\Delta x} + \dot{e}_2 A\Delta x = 0$$

or

$$T_1 - 2T_2 + T_3 + \dot{e}_2 A\Delta x^2 / k = 0$$

(*a*) Assuming heat transfer to be out of the volume element at the right surface.

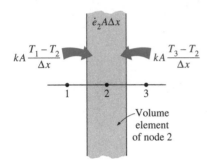

$$kA\frac{T_1 - T_2}{\Delta x} + kA\frac{T_3 - T_2}{\Delta x} + \dot{e}_2 A\Delta x = 0$$

or

$$T_1 - 2T_2 + T_3 + \dot{e}_2 A\Delta x^2 / k = 0$$

(*b*) Assuming heat transfer to be into the volume element at all surfaces.

FIGURE 5–12

The assumed direction of heat transfer at surfaces of a volume element has no effect on the finite difference formulation.

which is *identical* to the difference equation (Eq. 5–11) obtained earlier. Again, this equation is applicable to each of the $M - 1$ interior nodes, and its application gives $M - 1$ equations for the determination of temperatures at $M + 1$ nodes. The two additional equations needed to solve for the $M + 1$ unknown nodal temperatures are obtained by applying the energy balance on the two elements at the boundaries (unless, of course, the boundary temperatures are specified).

You are probably thinking that if heat is conducted into the element from both sides, as assumed in the formulation, the temperature of the medium will have to rise and thus heat conduction cannot be steady. Perhaps a more realistic approach would be to assume the heat conduction to be *into* the element on the left side and *out of* the element on the right side. If you repeat the formulation using this assumption, you will again obtain the same result since the heat conduction term on the right side in this case involves $T_m - T_{m+1}$ instead of $T_{m+1} - T_m$, which is subtracted instead of being added. Therefore, the assumed direction of heat conduction at the surfaces of the volume elements has no effect on the formulation, as shown in Figure 5–12. (Besides, the actual direction of heat transfer is usually not known.) However, it is convenient to assume heat conduction to be into the element at all surfaces and not worry about the sign of the conduction terms. Then all temperature differences in conduction relations are expressed as the temperature of the neighboring node minus the temperature of the node under consideration, and all conduction terms are added.

Boundary Conditions

Above we have developed a general relation for obtaining the finite difference equation for each interior node of a plane wall. This relation is not applicable to the nodes on the boundaries, however, since it requires the presence of nodes on both sides of the node under consideration, and a boundary node does not have a neighboring node on at least one side. Therefore, we need to obtain the finite difference equations of boundary nodes separately. This is best done by applying an *energy balance* on the volume elements of boundary nodes.

Boundary conditions most commonly encountered in practice are the *specified temperature, specified heat flux, convection,* and *radiation* boundary conditions, and here we develop the finite difference formulations for them for the case of steady one-dimensional heat conduction in a plane wall of thickness L as an example. The node number at the left surface at $x = 0$ is 0, and at the right surface at $x = L$ it is M. Note that the width of the volume element for either boundary node is $\Delta x/2$.

The **specified temperature** boundary condition is the simplest boundary condition to deal with. For one-dimensional heat transfer through a plane wall of thickness L, the *specified temperature boundary conditions* on both the left and right surfaces can be expressed as (Fig. 5–13)

$$T(0) = T_0 = \text{Specified value}$$
$$T(L) = T_M = \text{Specified value} \tag{5–19}$$

where T_0 and T_M are the specified temperatures at surfaces at $x = 0$ and $x = L$, respectively. Therefore, the specified temperature boundary conditions are incorporated by simply assigning the given surface temperatures to the boundary nodes. We do not need to write an energy balance in this case unless we decide to determine the rate of heat transfer into or out of the medium after the temperatures at the interior nodes are determined.

When other boundary conditions such as the *specified heat flux, convection, radiation,* or *combined convection and radiation* conditions are specified at a boundary, the finite difference equation for the node at that boundary is obtained by writing an *energy balance* on the volume element at that boundary. The energy balance is again expressed as

$$\sum_{\text{All sides}} \dot{Q} + \dot{E}_{\text{gen, element}} = 0 \tag{5–20}$$

for heat transfer under *steady* conditions. Again we assume all heat transfer to be *into* the volume element from all surfaces for convenience in formulation, except for specified heat flux since its direction is already specified. Specified heat flux is taken to be a *positive* quantity if into the medium and a *negative* quantity if out of the medium. Then the finite difference formulation at the node $m = 0$ (at the left boundary where $x = 0$) of a plane wall of thickness L during steady one-dimensional heat conduction can be expressed as (Fig. 5–14)

$$\dot{Q}_{\text{left surface}} + kA \frac{T_1 - T_0}{\Delta x} + \dot{e}_0(A\Delta x/2) = 0 \tag{5–21}$$

where $A\Delta x/2$ is the *volume* of the volume element (note that the boundary element has half thickness), $\dot{e}_0$ is the rate of heat generation per unit volume (in W/m³) at $x = 0$, and A is the heat transfer area, which is constant for a plane wall. Note that we have Δx in the denominator of the second term instead of $\Delta x/2$. This is because the ratio in that term involves the temperature difference between nodes 0 and 1, and thus we must use the distance between those two nodes, which is Δx.

The finite difference form of various boundary conditions can be obtained from Eq. 5–21 by replacing $\dot{Q}_{\text{left surface}}$ by a suitable expression. Next this is done for various boundary conditions at the left boundary.

1. Specified Heat Flux Boundary Condition

$$\dot{q}_0 A + kA \frac{T_1 - T_0}{\Delta x} + \dot{e}_0(A\Delta x/2) = 0 \tag{5–22}$$

Special case: **Insulated Boundary** ($\dot{q}_0 = 0$)

$$kA \frac{T_1 - T_0}{\Delta x} + \dot{e}_0(A\Delta x/2) = 0 \tag{5–23}$$

2. Convection Boundary Condition

$$hA(T_\infty - T_0) + kA \frac{T_1 - T_0}{\Delta x} + \dot{e}_0(A\Delta x/2) = 0 \tag{5–24}$$

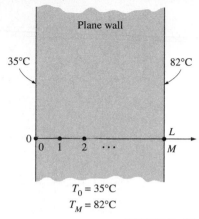

$T_0 = 35°C$
$T_M = 82°C$

FIGURE 5–13
Finite difference formulation of specified temperature boundary conditions on both surfaces of a plane wall.

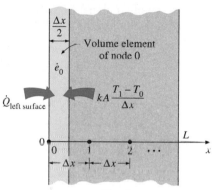

$$\dot{Q}_{\text{left surface}} + kA \frac{T_1 - T_0}{\Delta x} + \dot{e}_0 A \frac{\Delta x}{2} = 0$$

FIGURE 5–14
Schematic for the finite difference formulation of the left boundary node of a plane wall.

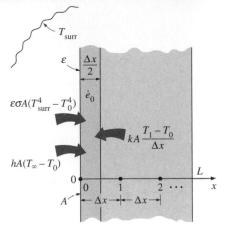

$$hA(T_\infty - T_0) + \varepsilon\sigma A(T_{surr}^4 - T_0^4)$$
$$+ kA\frac{T_1 - T_0}{\Delta x} + \dot{e}_0 A\frac{\Delta x}{2} = 0$$

FIGURE 5–15
Schematic for the finite difference formulation of combined convection and radiation on the left boundary of a plane wall.

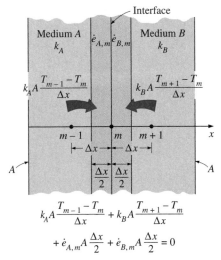

$$k_A A\frac{T_{m-1} - T_m}{\Delta x} + k_B A\frac{T_{m+1} - T_m}{\Delta x}$$
$$+ \dot{e}_{A,m} A\frac{\Delta x}{2} + \dot{e}_{B,m} A\frac{\Delta x}{2} = 0$$

FIGURE 5–16
Schematic for the finite difference formulation of the interface boundary condition for two mediums A and B that are in perfect thermal contact.

3. Radiation Boundary Condition

$$\varepsilon\sigma A(T_{surr}^4 - T_0^4) + kA\frac{T_1 - T_0}{\Delta x} + \dot{e}_0(A\Delta x/2) = 0 \qquad \text{(5–25)}$$

4. Combined Convection and Radiation Boundary Condition (Fig. 5–15)

$$hA(T_\infty - T_0) + \varepsilon\sigma A(T_{surr}^4 - T_0^4) + kA\frac{T_1 - T_0}{\Delta x} + \dot{e}_0(A\Delta x/2) = 0 \qquad \text{(5–26)}$$

or

$$h_{combined} A(T_\infty - T_0) + kA\frac{T_1 - T_0}{\Delta x} + \dot{e}_0(A\Delta x/2) = 0 \qquad \text{(5–27)}$$

5. Combined Convection, Radiation, and Heat Flux Boundary Condition

$$\dot{q}_0 A + hA(T_\infty - T_0) + \varepsilon\sigma A(T_{surr}^4 - T_0^4) + kA\frac{T_1 - T_0}{\Delta x} + \dot{e}_0(A\Delta x/2) = 0 \qquad \text{(5–28)}$$

6. Interface Boundary Condition Two different solid media A and B are assumed to be in perfect contact, and thus at the same temperature at the interface at node m (Fig. 5–16). Subscripts A and B indicate properties of media A and B, respectively.

$$k_A A\frac{T_{m-1} - T_m}{\Delta x} + k_B A\frac{T_{m+1} - T_m}{\Delta x} + \dot{e}_{A,m}(A\Delta x/2) + \dot{e}_{B,m}(A\Delta x/2) = 0 \qquad \text{(5–29)}$$

In these relations, $\dot{q}_0$ is the specified heat flux in W/m², h is the convection coefficient, $h_{combined}$ is the combined convection and radiation coefficient, T_∞ is the temperature of the surrounding medium, T_{surr} is the temperature of the surrounding surfaces, ε is the emissivity of the surface, and σ is the Stefan–Boltzman constant. The relations above can also be used for node M on the right boundary by replacing the subscript "0" by "M" and the subscript "1" by "$M - 1$".

Note that *thermodynamic temperatures* must be used in radiation heat transfer calculations, and all temperatures should be expressed in K or R when a boundary condition involves radiation to avoid mistakes. We usually try to avoid the *radiation boundary condition* even in numerical solutions since it causes the finite difference equations to be *nonlinear*, which are more difficult to solve.

Treating Insulated Boundary Nodes as Interior Nodes: The Mirror Image Concept

One way of obtaining the finite difference formulation of a node on an insulated boundary is to treat insulation as "zero" heat flux and to write an energy balance, as done in Eq. 5–23. Another and more practical way is to treat the node on an insulated boundary as an interior node. Conceptually this is done

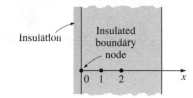

by replacing the insulation on the boundary by a *mirror* and considering the reflection of the medium as its extension (Fig. 5–17). This way the node next to the boundary node appears on both sides of the boundary node because of symmetry, converting it into an interior node. Then using the general formula (Eq. 5–18) for an interior node, which involves the sum of the temperatures of the adjoining nodes minus twice the node temperature, the finite difference formulation of a node $m = 0$ on an insulated boundary of a plane wall can be expressed as

$$\frac{T_{m+1} - 2T_m + T_{m-1}}{\Delta x^2} + \frac{\dot{e}_m}{k} = 0 \quad \rightarrow \quad \frac{T_1 - 2T_0 + T_1}{\Delta x^2} + \frac{\dot{e}_0}{k} = 0 \qquad \textbf{(5–30)}$$

which is equivalent to Eq. 5–23 obtained by the energy balance approach.

The mirror image approach can also be used for problems that possess thermal symmetry by replacing the plane of symmetry by a mirror. Alternately, we can replace the plane of symmetry by insulation and consider only half of the medium in the solution. The solution in the other half of the medium is simply the mirror image of the solution obtained.

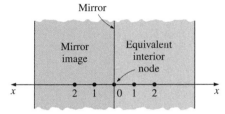

FIGURE 5–17
A node on an insulated boundary can be treated as an interior node by replacing the insulation by a mirror.

EXAMPLE 5–1   Steady Heat Conduction in a Large Uranium Plate

Consider a large uranium plate of thickness $L = 4$ cm and thermal conductivity $k = 28$ W/m · °C in which heat is generated uniformly at a constant rate of $\dot{e} = 5 \times 10^6$ W/m³. One side of the plate is maintained at 0°C by iced water while the other side is subjected to convection to an environment at $T_\infty = 30$°C with a heat transfer coefficient of $h = 45$ W/m² · °C, as shown in Figure 5–18. Considering a total of three equally spaced nodes in the medium, two at the boundaries and one at the middle, estimate the exposed surface temperature of the plate under steady conditions using the finite difference approach.

SOLUTION A uranium plate is subjected to specified temperature on one side and convection on the other. The unknown surface temperature of the plate is to be determined numerically using three equally spaced nodes.
Assumptions **1** Heat transfer through the wall is steady since there is no indication of any change with time. **2** Heat transfer is one-dimensional since the plate is large relative to its thickness. **3** Thermal conductivity is constant. **4** Radiation heat transfer is negligible.
Properties The thermal conductivity is given to be $k = 28$ W/m · °C.
Analysis The number of nodes is specified to be $M = 3$, and they are chosen to be at the two surfaces of the plate and the midpoint, as shown in the figure. Then the nodal spacing Δx becomes

$$\Delta x = \frac{L}{M - 1} = \frac{0.04 \text{ m}}{3 - 1} = 0.02 \text{ m}$$

We number the nodes 0, 1, and 2. The temperature at node 0 is given to be $T_0 = 0$°C, and the temperatures at nodes 1 and 2 are to be determined. This problem involves only two unknown nodal temperatures, and thus we need to have only two equations to determine them uniquely. These equations are obtained by applying the finite difference method to nodes 1 and 2.

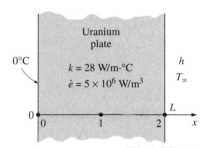

FIGURE 5–18
Schematic for Example 5–1.

Node 1 is an interior node, and the finite difference formulation at that node is obtained directly from Eq. 5–18 by setting $m = 1$:

$$\frac{T_0 - 2T_1 + T_2}{\Delta x^2} + \frac{\dot{e}_1}{k} = 0 \quad \rightarrow \quad \frac{0 - 2T_1 + T_2}{\Delta x^2} + \frac{\dot{e}_1}{k} = 0 \quad \rightarrow \quad 2T_1 - T_2 = \frac{\dot{e}_1 \Delta x^2}{k} \tag{1}$$

Node 2 is a boundary node subjected to convection, and the finite difference formulation at that node is obtained by writing an energy balance on the volume element of thickness $\Delta x/2$ at that boundary by assuming heat transfer to be into the medium at all sides:

$$hA(T_\infty - T_2) + kA \frac{T_1 - T_2}{\Delta x} + \dot{e}_2(A\Delta x/2) = 0$$

Canceling the heat transfer area A and rearranging give

$$T_1 - \left(1 + \frac{h\Delta x}{k}\right)T_2 = -\frac{h\Delta x}{k} T_\infty - \frac{\dot{e}_2 \Delta x^2}{2k} \tag{2}$$

Equations (1) and (2) form a system of two equations in two unknowns T_1 and T_2. Substituting the given quantities and simplifying gives

$$2T_1 - T_2 = 71.43 \qquad \text{(in °C)}$$
$$T_1 - 1.032T_2 = -36.68 \qquad \text{(in °C)}$$

This is a system of two algebraic equations in two unknowns and can be solved easily by the elimination method. Solving the first equation for T_1 and substituting into the second equation result in an equation in T_2 whose solution is

$$T_2 = 136.1°C$$

This is the temperature of the surface exposed to convection, which is the desired result. Substitution of this result into the first equation gives $T_1 = 103.8°C$, which is the temperature at the middle of the plate.

Discussion The purpose of this example is to demonstrate the use of the finite difference method with minimal calculations, and the accuracy of the result was not a major concern. But you might still be wondering how accurate the result obtained above is. After all, we used a mesh of only three nodes for the entire plate, which seems to be rather crude. This problem can be solved analytically as described in Chapter 2, and the analytical (exact) solution can be shown to be

$$T(x) = \frac{0.5\dot{e}hL^2/k + \dot{e}L + T_\infty h}{hL + k} x - \frac{\dot{e}x^2}{2k}$$

Substituting the given quantities, the temperature of the exposed surface of the plate at $x = L = 0.04$ m is determined to be 136.0°C, which is almost identical to the result obtained here with the approximate finite difference method (Fig. 5–19). Therefore, highly accurate results can be obtained with numerical methods by using a limited number of nodes.

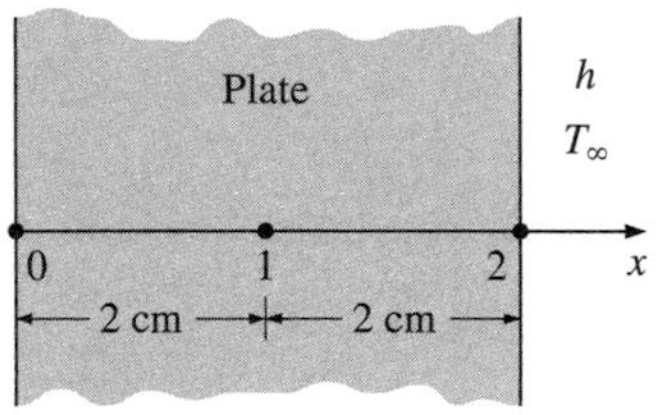

Finite difference solution:

$$T_2 = 136.1°C$$

Exact solution:

$$T_2 = 136.0°C$$

FIGURE 5–19
Despite being approximate in nature, highly accurate results can be obtained by numerical methods.

EXAMPLE 5–2 Heat Transfer from Triangular Fins

Consider an aluminum alloy fin (k = 180 W/m · °C) of triangular cross section with length L = 5 cm, base thickness b = 1 cm, and very large width w, as shown in Figure 5–20. The base of the fin is maintained at a temperature of T_0 = 200°C. The fin is losing heat to the surrounding medium at T_∞ = 25°C with a heat transfer coefficient of h = 15 W/m² · °C. Using the finite difference method with six equally spaced nodes along the fin in the x-direction, determine (a) the temperatures at the nodes, (b) the rate of heat transfer from the fin for w = 1 m, and (c) the fin efficiency.

SOLUTION A long triangular fin attached to a surface is considered. The nodal temperatures, the rate of heat transfer, and the fin efficiency are to be determined numerically using six equally spaced nodes.

Assumptions **1** Heat transfer is steady since there is no indication of any change with time. **2** The temperature along the fin varies in the x direction only. **3** Thermal conductivity is constant. **4** Radiation heat transfer is negligible.

Properties The thermal conductivity is given to be k = 180 W/m · °C.

Analysis (a) The number of nodes in the fin is specified to be M = 6, and their location is as shown in the figure. Then the nodal spacing Δx becomes

$$\Delta x = \frac{L}{M-1} = \frac{0.05 \text{ m}}{6-1} = 0.01 \text{ m}$$

The temperature at node 0 is given to be T_0 = 200°C, and the temperatures at the remaining five nodes are to be determined. Therefore, we need to have five equations to determine them uniquely. Nodes 1, 2, 3, and 4 are interior nodes, and the finite difference formulation for a general interior node m is obtained by applying an energy balance on the volume element of this node. Noting that heat transfer is steady and there is no heat generation in the fin and assuming heat transfer to be into the medium at all sides, the energy balance can be expressed as

$$\sum_{\text{All sides}} \dot{Q} = 0 \quad \rightarrow \quad kA_{\text{left}} \frac{T_{m-1} - T_m}{\Delta x} + kA_{\text{right}} \frac{T_{m+1} - T_m}{\Delta x} + hA_{\text{conv}}(T_\infty - T_m) = 0$$

Note that heat transfer areas are different for each node in this case, and using geometrical relations, they can be expressed as

$$A_{\text{left}} = (\text{Height} \times \text{Width})_{@m-\frac{1}{2}} = 2w[L - (m - 1/2)\Delta x]\tan\theta$$
$$A_{\text{right}} = (\text{Height} \times \text{Width})_{@m+\frac{1}{2}} = 2w[L - (m + 1/2)\Delta x]\tan\theta$$
$$A_{\text{conv}} = 2 \times \text{Length} \times \text{Width} = 2w(\Delta x/\cos\theta)$$

Substituting,

$$2kw[L - (m - \tfrac{1}{2})\Delta x]\tan\theta \frac{T_{m-1} - T_m}{\Delta x}$$
$$+ 2kw[L - (m + \tfrac{1}{2})\Delta x]\tan\theta \frac{T_{m+1} - T_m}{\Delta x} + h\frac{2w\Delta x}{\cos\theta}(T_\infty - T_m) = 0$$

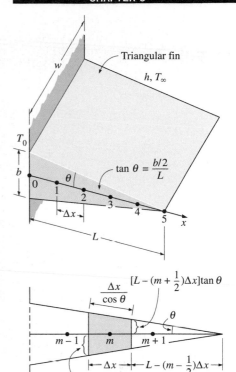

FIGURE 5–20
Schematic for Example 5–2 and the volume element of a general interior node of the fin.

Dividing each term by $2kwL \tan \theta/\Delta x$ gives

$$\left[1 - \left(m - \tfrac{1}{2}\right)\frac{\Delta x}{L}\right](T_{m-1} - T_m) + \left[1 - \left(m + \tfrac{1}{2}\right)\frac{\Delta x}{L}\right](T_{m+1} - T_m)$$
$$+ \frac{h(\Delta x)^2}{kL \sin \theta}(T_\infty - T_m) = 0$$

Note that

$$\tan \theta = \frac{b/2}{L} = \frac{0.5 \text{ cm}}{5 \text{ cm}} = 0.1 \quad \rightarrow \quad \theta = \tan^{-1} 0.1 = 5.71°$$

Also, $\sin 5.71° = 0.0995$. Then the substitution of known quantities gives

$$(5.5 - m)T_{m-1} - (10.008 - 2m)T_m + (4.5 - m)T_{m+1} = -0.209$$

Now substituting 1, 2, 3, and 4 for m results in these finite difference equations for the interior nodes:

$$
\begin{aligned}
m = 1: & \quad -8.008T_1 + 3.5T_2 = -900.209 & \textbf{(1)}\\
m = 2: & \quad 3.5T_1 - 6.008T_2 + 2.5T_3 = -0.209 & \textbf{(2)}\\
m = 3: & \quad 2.5T_2 - 4.008T_3 + 1.5T_4 = -0.209 & \textbf{(3)}\\
m = 4: & \quad 1.5T_3 - 2.008T_4 + 0.5T_5 = -0.209 & \textbf{(4)}
\end{aligned}
$$

The finite difference equation for the boundary node 5 is obtained by writing an energy balance on the volume element of length $\Delta x/2$ at that boundary, again by assuming heat transfer to be into the medium at all sides (Fig. 5–21):

$$kA_{\text{left}}\frac{T_4 - T_5}{\Delta x} + hA_{\text{conv}}(T_\infty - T_5) = 0$$

where

$$A_{\text{left}} = 2w\frac{\Delta x}{2}\tan \theta \quad \text{and} \quad A_{\text{conv}} = 2w\frac{\Delta x/2}{\cos \theta}$$

Canceling w in all terms and substituting the known quantities gives

$$T_4 - 1.008T_5 = -0.209 \tag{5}$$

Equations (1) through (5) form a linear system of five algebraic equations in five unknowns. Solving them simultaneously using an equation solver gives

$$
\begin{aligned}
T_1 &= \textbf{198.6°C}, & T_2 &= \textbf{197.1°C}, & T_3 &= \textbf{195.7°C},\\
T_4 &= \textbf{194.3°C}, & T_5 &= \textbf{192.9°C}
\end{aligned}
$$

which is the desired solution for the nodal temperatures.

(*b*) The total rate of heat transfer from the fin is simply the sum of the heat transfer from each volume element to the ambient, and for $w = 1$ m it is determined from

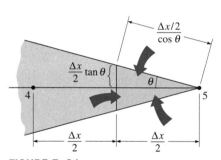

FIGURE 5–21
Schematic of the volume element of node 5 at the tip of a triangular fin.

$$\dot{Q}_{fin} = \sum_{m=0}^{5} \dot{Q}_{element, m} = \sum_{m=0}^{5} hA_{conv, m}(T_m - T_\infty)$$

Noting that the heat transfer surface area is $w\Delta x/\cos\theta$ for the boundary nodes 0 and 5, and twice as large for the interior nodes 1, 2, 3, and 4, we have

$$\dot{Q}_{fin} = h\frac{w\Delta x}{\cos\theta}[(T_0 - T_\infty) + 2(T_1 - T_\infty) + 2(T_2 - T_\infty) + 2(T_3 - T_\infty)$$

$$+ 2(T_4 - T_\infty) + (T_5 - T_\infty)]$$

$$= h\frac{w\Delta x}{\cos\theta}[T_0 + 2(T_1 + T_2 + T_3 + T_4) + T_5 - 10T_\infty]°C$$

$$= (15 \text{ W/m}^2 \cdot °C)\frac{(1 \text{ m})(0.01 \text{ m})}{\cos 5.71°}[200 + 2 \times 785.7$$

$$+ 192.9 - 10 \times 25] °C$$

$$= \textbf{258.4 W}$$

(c) If the entire fin were at the base temperature of $T_0 = 200°C$, the total rate of heat transfer from the fin for $w = 1$ m would be

$$\dot{Q}_{max} = hA_{fin, total}(T_0 - T_\infty) = h(2wL/\cos\theta)(T_0 - T_\infty)$$

$$= (15 \text{ W/m}^2 \cdot °C)[2(1 \text{ m})(0.05 \text{ m})/\cos 5.71°](200 - 25)°C$$

$$= 263.8 \text{ W}$$

Then the fin efficiency is determined from

$$\eta_{fin} = \frac{\dot{Q}_{fin}}{\dot{Q}_{max}} = \frac{258.4 \text{ W}}{263.8 \text{ W}} = \textbf{0.98}$$

which is less than 1, as expected. We could also determine the fin efficiency in this case from the proper fin efficiency curve in Chapter 3, which is based on the analytical solution. We would read 0.98 for the fin efficiency, which is identical to the value determined above numerically.

The finite difference formulation of steady heat conduction problems usually results in a system of N algebraic equations in N unknown nodal temperatures that need to be solved simultaneously. When N is small (such as 2 or 3), we can use the elementary *elimination method* to eliminate all unknowns except one and then solve for that unknown (see Example 5–1). The other unknowns are then determined by back substitution. When N is large, which is usually the case, the elimination method is not practical and we need to use a more systematic approach that can be adapted to computers.

There are numerous systematic approaches available in the literature, and they are broadly classified as **direct** and **iterative** methods. The direct methods are based on a fixed number of well-defined steps that result in the solution in a systematic manner. The iterative methods, on the other hand, are based on an initial guess for the solution that is refined by iteration until a specified convergence criterion is satisfied (Fig. 5–22). The direct methods usually require a large amount of computer memory and computation time,

Direct methods:
  Solve in a systematic manner following a series of well-defined steps.

Iterative methods:
  Start with an initial guess for the solution, and iterate until solution converges.

FIGURE 5–22
Two general categories of solution methods for solving systems of algebraic equations.

and they are more suitable for systems with a relatively small number of equations. The computer memory requirements for iterative methods are minimal, and thus they are usually preferred for large systems. The convergence of iterative methods to the desired solution, however, may pose a problem.

5–4 · TWO-DIMENSIONAL STEADY HEAT CONDUCTION

In Section 5–3 we considered one-dimensional heat conduction and assumed heat conduction in other directions to be negligible. Many heat transfer problems encountered in practice can be approximated as being one-dimensional, but this is not always the case. Sometimes we need to consider heat transfer in other directions as well when the variation of temperature in other directions is significant. In this section we consider the numerical formulation and solution of two-dimensional steady heat conduction in rectangular coordinates using the finite difference method. The approach presented below can be extended to three-dimensional cases.

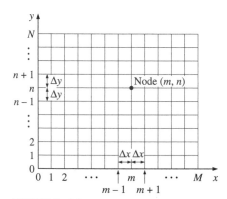

FIGURE 5–23

The nodal network for the finite difference formulation of two-dimensional conduction in rectangular coordinates.

Consider a *rectangular region* in which heat conduction is significant in the x- and y-directions. Now divide the x-y plane of the region into a rectangular mesh of nodal points spaced Δx and Δy apart in the x- and y-directions, respectively, as shown in Figure 5–23, and consider a unit depth of $\Delta z = 1$ in the z-direction. Our goal is to determine the temperatures at the nodes, and it is convenient to number the nodes and describe their position by the numbers instead of actual coordinates. A logical numbering scheme for two-dimensional problems is the *double subscript notation* (m, n) where $m = 0, 1, 2, \ldots, M$ is the node count in the x-direction and $n = 0, 1, 2, \ldots, N$ is the node count in the y-direction. The coordinates of the node (m, n) are simply $x = m\Delta x$ and $y = n\Delta y$, and the temperature at the node (m, n) is denoted by $T_{m, n}$.

Now consider a *volume element* of size $\Delta x \times \Delta y \times 1$ centered about a general interior node (m, n) in a region in which heat is generated at a rate of $\dot{e}$ and the thermal conductivity k is constant, as shown in Figure 5–24. Again assuming the direction of heat conduction to be *toward* the node under consideration at all surfaces, the energy balance on the volume element can be expressed as

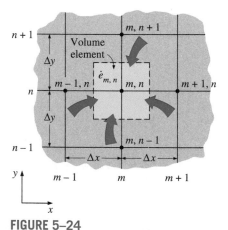

FIGURE 5–24

The volume element of a general interior node (m, n) for two-dimensional conduction in rectangular coordinates.

$$\begin{pmatrix} \text{Rate of heat conduction} \\ \text{at the left, top, right,} \\ \text{and bottom surfaces} \end{pmatrix} + \begin{pmatrix} \text{Rate of heat} \\ \text{generation inside} \\ \text{the element} \end{pmatrix} = \begin{pmatrix} \text{Rate of change of} \\ \text{the energy content} \\ \text{of the element} \end{pmatrix}$$

or

$$\dot{Q}_{\text{cond, left}} + \dot{Q}_{\text{cond, top}} + \dot{Q}_{\text{cond, right}} + \dot{Q}_{\text{cond, bottom}} + \dot{E}_{\text{gen, element}} = \frac{\Delta E_{\text{element}}}{\Delta t} = 0 \quad \textbf{(5–31)}$$

for the *steady* case. Again assuming the temperatures between the adjacent nodes to vary linearly and noting that the heat transfer area is $A_x = \Delta y \times 1 = \Delta y$ in the x-direction and $A_y = \Delta x \times 1 = \Delta x$ in the y-direction, the energy balance relation above becomes

$$k\Delta y \frac{T_{m-1,n} - T_{m,n}}{\Delta x} + k\Delta x \frac{T_{m,n+1} - T_{m,n}}{\Delta y} + k\Delta y \frac{T_{m+1,n} - T_{m,n}}{\Delta x}$$

$$+ k\Delta x \frac{T_{m,n-1} - T_{m,n}}{\Delta y} + \dot{e}_{m,n} \Delta x \Delta y = 0 \qquad \text{(5–32)}$$

Dividing each term by $\Delta x \times \Delta y$ and simplifying gives

$$\frac{T_{m-1,n} - 2T_{m,n} + T_{m+1,n}}{\Delta x^2} + \frac{T_{m,n-1} - 2T_{m,n} + T_{m,n+1}}{\Delta y^2} + \frac{\dot{e}_{m,n}}{k} = 0 \qquad \text{(5–33)}$$

for $m = 1, 2, 3, \ldots, M - 1$ and $n = 1, 2, 3, \ldots, N - 1$. This equation is identical to Eq. 5–12 obtained earlier by replacing the derivatives in the differential equation by differences for an interior node (m, n). Again a rectangular region M equally spaced nodes in the x-direction and N equally spaced nodes in the y-direction has a total of $(M + 1)(N + 1)$ nodes, and Eq. 5–33 can be used to obtain the finite difference equations at all interior nodes.

In finite difference analysis, usually a **square mesh** is used for simplicity (except when the magnitudes of temperature gradients in the x- and y-directions are very different), and thus Δx and Δy are taken to be the same. Then $\Delta x = \Delta y = l$, and the relation above simplifies to

$$T_{m-1,n} + T_{m+1,n} + T_{m,n+1} + T_{m,n-1} - 4T_{m,n} + \frac{\dot{e}_{m,n} l^2}{k} = 0 \qquad \text{(5–34)}$$

That is, the finite difference formulation of an interior node is obtained by *adding the temperatures of the four nearest neighbors of the node, subtracting four times the temperature of the node itself, and adding the heat generation term.* It can also be expressed in this form, which is easy to remember:

$$T_{\text{left}} + T_{\text{top}} + T_{\text{right}} + T_{\text{bottom}} - 4T_{\text{node}} + \frac{\dot{e}_{\text{node}} l^2}{k} - 0 \qquad \text{(5–35)}$$

When there is no heat generation in the medium, the finite difference equation for an interior node further simplifies to $T_{\text{node}} = (T_{\text{left}} + T_{\text{top}} + T_{\text{right}} + T_{\text{bottom}})/4$, *which has the interesting interpretation that the temperature of each interior node is the arithmetic average of the temperatures of the four neighboring nodes.* This statement is also true for the three-dimensional problems except that the interior nodes in that case will have six neighboring nodes instead of four.

Boundary Nodes

The development of finite difference formulation of *boundary* nodes in two- (or three-) dimensional problems is similar to the development in the one-dimensional case discussed earlier. Again, the region is partitioned between the nodes by forming *volume elements* around the nodes, and an *energy balance* is written for each boundary node. Various boundary conditions can be handled as discussed for a plane wall, except that the volume elements in the two-dimensional case involve heat transfer in the *y-direction* as well as the *x-direction*. Insulated surfaces can still be viewed as "mirrors," and the

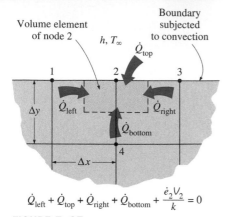

$$\dot{Q}_{left} + \dot{Q}_{top} + \dot{Q}_{right} + \dot{Q}_{bottom} + \frac{\dot{e}_2 V_2}{k} = 0$$

FIGURE 5–25
The finite difference formulation of
a boundary node is obtained by
writing an energy balance
on its volume element.

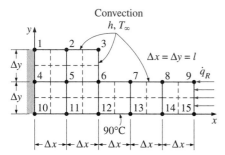

FIGURE 5–26
Schematic for Example 5–3 and
the nodal network (the boundaries
of volume elements of the nodes are
indicated by dashed lines).

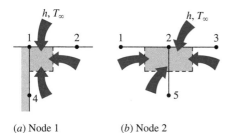

(a) Node 1   (b) Node 2

FIGURE 5–27
Schematics for energy balances on the
volume elements of nodes 1 and 2.

mirror image concept can be used to treat nodes on insulated boundaries as interior nodes.

For heat transfer under *steady* conditions, the basic equation to keep in mind when writing an *energy balance* on a volume element is (Fig. 5–25)

$$\sum_{\text{All sides}} \dot{Q} + \dot{e} V_{\text{element}} = 0 \qquad (5\text{–}36)$$

whether the problem is one-, two-, or three-dimensional. Again we assume, for convenience in formulation, all heat transfer to be *into* the volume element from all surfaces except for specified heat flux, whose direction is already specified. This is demonstrated in Example 5–3 for various boundary conditions.

EXAMPLE 5–3   Steady Two-Dimensional Heat Conduction in L-Bars

Consider steady heat transfer in an L-shaped solid body whose cross section is given in Figure 5–26. Heat transfer in the direction normal to the plane of the paper is negligible, and thus heat transfer in the body is two-dimensional. The thermal conductivity of the body is $k = 15$ W/m · °C, and heat is generated in the body at a rate of $\dot{e} = 2 \times 10^6$ W/m³. The left surface of the body is insulated, and the bottom surface is maintained at a uniform temperature of 90°C. The entire top surface is subjected to convection to ambient air at $T_\infty = 25$°C with a convection coefficient of $h = 80$ W/m² · °C, and the right surface is subjected to heat flux at a uniform rate of $\dot{q}_R = 5000$ W/m². The nodal network of the problem consists of 15 equally spaced nodes with $\Delta x = \Delta y = 1.2$ cm, as shown in the figure. Five of the nodes are at the bottom surface, and thus their temperatures are known. Obtain the finite difference equations at the remaining nine nodes and determine the nodal temperatures by solving them.

SOLUTION   Heat transfer in a long L-shaped solid bar with specified boundary conditions is considered. The nine unknown nodal temperatures are to be determined with the finite difference method.
Assumptions   **1** Heat transfer is steady and two-dimensional, as stated. **2** Thermal conductivity is constant. **3** Heat generation is uniform. **4** Radiation heat transfer is negligible.
Properties   The thermal conductivity is given to be $k = 15$ W/m · °C.
Analysis   We observe that all nodes are boundary nodes except node 5, which is an interior node. Therefore, we have to rely on energy balances to obtain the finite difference equations. But first we form the volume elements by partitioning the region among the nodes equitably by drawing dashed lines between the nodes. If we consider the volume element represented by an interior node to be *full size* (i.e., $\Delta x \times \Delta y \times 1$), then the element represented by a regular boundary node such as node 2 becomes *half size* (i.e., $\Delta x \times \Delta y/2 \times 1$), and a corner node such as node 1 is *quarter size* (i.e., $\Delta x/2 \times \Delta y/2 \times 1$). Keeping Eq. 5–36 in mind for the energy balance, the finite difference equations for each of the nine nodes are obtained as follows:

(a) Node 1. The volume element of this corner node is insulated on the left and subjected to convection at the top and to conduction at the right and bottom surfaces. An energy balance on this element gives (Fig. 5–27a)

$$0 + h\frac{\Delta x}{2}(T_\infty - T_1) + k\frac{\Delta y}{2}\frac{T_2 - T_1}{\Delta x} + k\frac{\Delta x}{2}\frac{T_4 - T_1}{\Delta y} + \dot{e}_1\frac{\Delta x}{2}\frac{\Delta y}{2} = 0$$

Taking $\Delta x = \Delta y = l$, it simplifies to

$$-\left(2 + \frac{hl}{k}\right)T_1 + T_2 + T_4 = -\frac{hl}{k}T_\infty - \frac{\dot{e}_1 l^2}{2k}$$

(b) Node 2. The volume element of this boundary node is subjected to convection at the top and to conduction at the right, bottom, and left surfaces. An energy balance on this element gives (Fig. 5–27b)

$$h\Delta x(T_\infty - T_2) + k\frac{\Delta y}{2}\frac{T_3 - T_2}{\Delta x} + k\Delta x\frac{T_5 - T_2}{\Delta y} + k\frac{\Delta y}{2}\frac{T_1 - T_2}{\Delta x} + \dot{e}_2\Delta x\frac{\Delta y}{2} = 0$$

Taking $\Delta x = \Delta y = l$, it simplifies to

$$T_1 - \left(4 + \frac{2hl}{k}\right)T_2 + T_3 + 2T_5 = -\frac{2hl}{k}T_\infty - \frac{\dot{e}_2 l^2}{k}$$

(c) Node 3. The volume element of this corner node is subjected to convection at the top and right surfaces and to conduction at the bottom and left surfaces. An energy balance on this element gives (Fig. 5–28a)

$$h\left(\frac{\Delta x}{2} + \frac{\Delta y}{2}\right)(T_\infty - T_3) + k\frac{\Delta x}{2}\frac{T_6 - T_3}{\Delta y} + k\frac{\Delta y}{2}\frac{T_2 - T_3}{\Delta x} + \dot{e}_3\frac{\Delta x}{2}\frac{\Delta y}{2} = 0$$

Taking $\Delta x = \Delta y = l$, it simplifies to

$$T_2 - \left(2 + \frac{2hl}{k}\right)T_3 + T_6 = -\frac{2hl}{k}T_\infty - \frac{\dot{e}_3 l^2}{2k}$$

(d) Node 4. This node is on the insulated boundary and can be treated as an interior node by replacing the insulation by a mirror. This puts a reflected image of node 5 to the left of node 4. Noting that $\Delta x = \Delta y = l$, the general interior node relation for the steady two-dimensional case (Eq. 5–35) gives (Fig. 5–28b)

$$T_5 + T_1 + T_5 + T_{10} - 4T_4 + \frac{\dot{e}_4 l^2}{k} = 0$$

or, noting that $T_{10} = 90°C$,

$$T_1 - 4T_4 + 2T_5 = -90 - \frac{\dot{e}_4 l^2}{k}$$

(e) Node 5. This is an interior node, and noting that $\Delta x = \Delta y = l$, the finite difference formulation of this node is obtained directly from Eq. 5–35 to be (Fig. 5–29a)

$$T_4 + T_2 + T_6 + T_{11} - 4T_5 + \frac{\dot{e}_5 l^2}{k} = 0$$

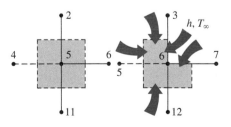

(a) Node 3   (b) Node 4

FIGURE 5–28

Schematics for energy balances on the volume elements of nodes 3 and 4.

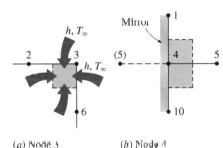

(a) Node 5   (b) Node 6

FIGURE 5–29

Schematics for energy balances on the volume elements of nodes 5 and 6.

or, noting that $T_{11} = 90°C$,

$$T_2 + T_4 - 4T_5 + T_6 = -90 - \frac{\dot{e}_5 l^2}{k}$$

(f) Node 6. The volume element of this inner corner node is subjected to convection at the L-shaped exposed surface and to conduction at other surfaces. An energy balance on this element gives (Fig. 5–29b)

$$h\left(\frac{\Delta x}{2} + \frac{\Delta y}{2}\right)(T_\infty - T_6) + k\frac{\Delta y}{2}\frac{T_7 - T_6}{\Delta x} + k\Delta x\frac{T_{12} - T_6}{\Delta y}$$
$$+ k\Delta y\frac{T_5 - T_6}{\Delta x} + k\frac{\Delta x}{2}\frac{T_3 - T_6}{\Delta y} + \dot{e}_6\frac{3\Delta x\Delta y}{4} = 0$$

Taking $\Delta x = \Delta y = l$ and noting that $T_{12} = 90°C$, it simplifies to

$$T_3 + 2T_5 - \left(6 + \frac{2hl}{k}\right)T_6 + T_7 = -180 - \frac{2hl}{k}T_\infty - \frac{3\dot{e}_6 l^2}{2k}$$

(g) Node 7. The volume element of this boundary node is subjected to convection at the top and to conduction at the right, bottom, and left surfaces. An energy balance on this element gives (Fig. 5–30a)

$$h\Delta x(T_\infty - T_7) + k\frac{\Delta y}{2}\frac{T_8 - T_7}{\Delta x} + k\Delta x\frac{T_{13} - T_7}{\Delta y}$$
$$+ k\frac{\Delta y}{2}\frac{T_6 - T_7}{\Delta x} + \dot{e}_7\Delta x\frac{\Delta y}{2} = 0$$

Taking $\Delta x = \Delta y = l$ and noting that $T_{13} = 90°C$, it simplifies to

$$T_6 - \left(4 + \frac{2hl}{k}\right)T_7 + T_8 = -180 - \frac{2hl}{k}T_\infty - \frac{\dot{e}_7 l^2}{k}$$

(h) Node 8. This node is identical to node 7, and the finite difference formulation of this node can be obtained from that of node 7 by shifting the node numbers by 1 (i.e., replacing subscript m by $m + 1$). It gives

$$T_7 - \left(4 + \frac{2hl}{k}\right)T_8 + T_9 = -180 - \frac{2hl}{k}T_\infty - \frac{\dot{e}_8 l^2}{k}$$

(i) Node 9. The volume element of this corner node is subjected to convection at the top surface, to heat flux at the right surface, and to conduction at the bottom and left surfaces. An energy balance on this element gives (Fig. 5–30b)

$$h\frac{\Delta x}{2}(T_\infty - T_9) + \dot{q}_R\frac{\Delta y}{2} + k\frac{\Delta x}{2}\frac{T_{15} - T_9}{\Delta y} + k\frac{\Delta y}{2}\frac{T_8 - T_9}{\Delta x} + \dot{e}_9\frac{\Delta x}{2}\frac{\Delta y}{2} = 0$$

Taking $\Delta x = \Delta y = l$ and noting that $T_{15} = 90°C$, it simplifies to

$$T_8 - \left(2 + \frac{hl}{k}\right)T_9 = -90 - \frac{\dot{q}_R l}{k} - \frac{hl}{k}T_\infty - \frac{\dot{e}_9 l^2}{2k}$$

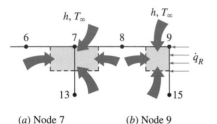

(a) Node 7   (b) Node 9

FIGURE 5–30
Schematics for energy balances on the volume elements of nodes 7 and 9.

This completes the development of finite difference formulation for this problem. Substituting the given quantities, the system of nine equations for the determination of nine unknown nodal temperatures becomes

$$-2.064T_1 + T_2 + T_4 = -11.2$$
$$T_1 - 4.128T_2 + T_3 + 2T_5 = -22.4$$
$$T_2 - 2.128T_3 + T_6 = -12.8$$
$$T_1 - 4T_4 + 2T_5 = -109.2$$
$$T_2 + T_4 - 4T_5 + T_6 = -109.2$$
$$T_3 + 2T_5 - 6.128T_6 + T_7 = -212.0$$
$$T_6 - 4.128T_7 + T_8 = -202.4$$
$$T_7 - 4.128T_8 + T_9 = -202.4$$
$$T_8 - 2.064T_9 = -105.2$$

which is a system of nine algebraic equations with nine unknowns. Using an equation solver, its solution is determined to be

$$T_1 = 112.1°C \quad T_2 = 110.8°C \quad T_3 = 106.6°C$$
$$T_4 = 109.4°C \quad T_5 = 108.1°C \quad T_6 = 103.2°C$$
$$T_7 = 97.3°C \quad T_8 = 96.3°C \quad T_9 = 97.6°C$$

Note that the temperature is the highest at node 1 and the lowest at node 8. This is consistent with our expectations since node 1 is the farthest away from the bottom surface, which is maintained at 90°C and has one side insulated, and node 8 has the largest exposed area relative to its volume while being close to the surface at 90°C.

Irregular Boundaries

In problems with simple geometries, we can fill the entire region using simple volume elements such as strips for a plane wall and rectangular elements for two-dimensional conduction in a rectangular region. We can also use cylindrical or spherical shell elements to cover the cylindrical and spherical bodies entirely. However, many geometries encountered in practice such as turbine blades or engine blocks do not have simple shapes, and it is difficult to fill such geometries having irregular boundaries with simple volume elements. A practical way of dealing with such geometries is to replace the irregular geometry by a series of simple volume elements, as shown in Figure 5–31. This simple approach is often satisfactory for practical purposes, especially when the nodes are closely spaced near the boundary. More sophisticated approaches are available for handling irregular boundaries, and they are commonly incorporated into the commercial software packages.

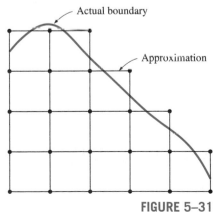

Actual boundary

Approximation

FIGURE 5–31
Approximating an irregular boundary with a rectangular mesh.

EXAMPLE 5–4   Heat Loss through Chimneys

Hot combustion gases of a furnace are flowing through a square chimney made of concrete ($k = 1.4$ W/m · °C). The flow section of the chimney is 20 cm × 20 cm, and the thickness of the wall is 20 cm. The average temperature of the hot gases in the chimney is $T_i = 300°C$, and the average convection heat transfer coefficient inside the chimney is $h_i = 70$ W/m$^2 \cdot$ °C. The chimney is losing heat from its outer surface to the ambient air at $T_o = 20°C$ by convection with a heat transfer coefficient of $h_o = 21$ W/m$^2 \cdot$ °C and to the sky by radiation. The emissivity of the outer surface of the wall is $\varepsilon = 0.9$, and the effective sky temperature is estimated to be 260 K. Using the finite difference method with $\Delta x = \Delta y = 10$ cm and taking full advantage of symmetry, determine the temperatures at the nodal points of a cross section and the rate of heat loss for a 1-m-long section of the chimney.

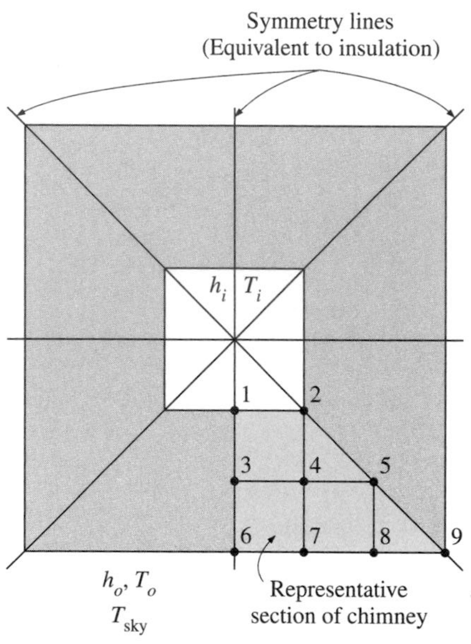

SOLUTION Heat transfer through a square chimney is considered. The nodal temperatures and the rate of heat loss per unit length are to be determined with the finite difference method.

Assumptions **1** Heat transfer is steady since there is no indication of change with time. **2** Heat transfer through the chimney is two-dimensional since the height of the chimney is large relative to its cross section, and thus heat conduction through the chimney in the axial direction is negligible. It is tempting to simplify the problem further by considering heat transfer in each wall to be one-dimensional, which would be the case if the walls were thin and thus the corner effects were negligible. This assumption cannot be justified in this case since the walls are very thick and the corner sections constitute a considerable portion of the chimney structure. **3** Thermal conductivity is constant.

Properties The properties of chimney are given to be $k = 1.4$ W/m $\cdot$ °C and $\varepsilon = 0.9$.

Analysis The cross section of the chimney is given in Figure 5–32. The most striking aspect of this problem is the apparent symmetry about the horizontal and vertical lines passing through the midpoint of the chimney as well as the diagonal axes, as indicated on the figure. Therefore, we need to consider only one-eighth of the geometry in the solution whose nodal network consists of nine equally spaced nodes.

No heat can cross a symmetry line, and thus symmetry lines can be treated as insulated surfaces and thus "mirrors" in the finite difference formulation. Then the nodes in the middle of the symmetry lines can be treated as interior nodes by using mirror images. Six of the nodes are boundary nodes, so we have to write energy balances to obtain their finite difference formulations. First we partition the region among the nodes equitably by drawing dashed lines between the nodes through the middle. Then the region around a node surrounded by the boundary or the dashed lines represents the volume element of the node. Considering a unit depth and using the energy balance approach for the boundary nodes (again assuming all heat transfer into the volume element for convenience) and the formula for the interior nodes, the finite difference equations for the nine nodes are determined as follows:

(*a*) Node 1. On the inner boundary, subjected to convection, Figure 5–33*a*

$$0 + h_i \frac{\Delta x}{2}(T_i - T_1) + k \frac{\Delta y}{2} \frac{T_2 - T_1}{\Delta x} + k \frac{\Delta x}{2} \frac{T_3 - T_1}{\Delta y} + 0 = 0$$

Taking $\Delta x = \Delta y = l$, it simplifies to

$$-\left(2 + \frac{h_i l}{k}\right) T_1 + T_2 + T_3 = -\frac{h_i l}{k} T_i$$

Symmetry lines
(Equivalent to insulation)

h_i, T_i

h_o, T_o
T_{sky}

Representative section of chimney

FIGURE 5–32
Schematic of the chimney discussed in Example 5–4 and the nodal network for a representative section.

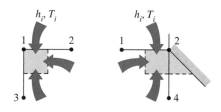

(*a*) Node 1   (*b*) Node 2

FIGURE 5–33
Schematics for energy balances on the volume elements of nodes 1 and 2.

(b) Node 2. On the inner boundary, subjected to convection, Figure 5–33b

$$k\frac{\Delta y}{2}\frac{T_1 - T_2}{\Delta x} + h_i\frac{\Delta x}{2}(T_i - T_2) + 0 + k\Delta x\frac{T_4 - T_2}{\Delta y} = 0$$

Taking $\Delta x = \Delta y = l$, it simplifies to

$$T_1 - \left(3 + \frac{h_i l}{k}\right)T_2 + 2T_4 = -\frac{h_i l}{k}T_i$$

(c) Nodes 3, 4, and 5. (Interior nodes, Fig. 5–34)

$$\text{Node 3: } T_4 + T_1 + T_4 + T_6 - 4T_3 = 0$$
$$\text{Node 4: } T_3 + T_2 + T_5 + T_7 - 4T_4 = 0$$
$$\text{Node 5: } T_4 + T_4 + T_8 + T_8 - 4T_5 = 0$$

(d) Node 6. (On the outer boundary, subjected to convection and radiation)

$$0 + k\frac{\Delta x}{2}\frac{T_3 - T_6}{\Delta y} + k\frac{\Delta y}{2}\frac{T_7 - T_6}{\Delta x}$$
$$+ h_o\frac{\Delta x}{2}(T_o - T_6) + \varepsilon\sigma\frac{\Delta x}{2}(T_{sky}^4 - T_6^4) = 0$$

Taking $\Delta x = \Delta y = l$, it simplifies to

$$T_2 + T_3 - \left(2 + \frac{h_o l}{k}\right)T_6 = -\frac{h_o l}{k}T_o - \frac{\varepsilon\sigma l}{k}(T_{sky}^4 - T_6^4)$$

(e) Node 7. (On the outer boundary, subjected to convection and radiation, Fig. 5–35)

$$k\frac{\Delta y}{2}\frac{T_6 - T_7}{\Delta x} + k\Delta x\frac{T_4 - T_7}{\Delta y} + k\frac{\Delta y}{2}\frac{T_8 - T_7}{\Delta x}$$
$$+ h_o\Delta x(T_o - T_7) + \varepsilon\sigma\Delta x(T_{sky}^4 - T_7^4) = 0$$

Taking $\Delta x = \Delta y = l$, it simplifies to

$$2T_4 + T_6 - \left(4 + \frac{2h_o l}{k}\right)T_7 + T_8 = -\frac{2h_o l}{k}T_o - \frac{2\varepsilon\sigma l}{k}(T_{sky}^4 - T_7^4)$$

(f) Node 8. Same as node 7, except shift the node numbers up by 1 (replace 4 by 5, 6 by 7, 7 by 8, and 8 by 9 in the last relation)

$$2T_5 + T_7 - \left(4 + \frac{2h_o l}{k}\right)T_8 + T_9 = -\frac{2h_o l}{k}T_o - \frac{2\varepsilon\sigma l}{k}(T_{sky}^4 - T_8^4)$$

(g) Node 9. (On the outer boundary, subjected to convection and radiation, Fig. 5–35)

$$k\frac{\Delta y}{2}\frac{T_8 - T_9}{\Delta x} + 0 + h_o\frac{\Delta x}{2}(T_o - T_9) + \varepsilon\sigma\frac{\Delta x}{2}(T_{sky}^4 - T_9^4) = 0$$

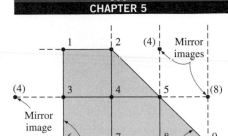

FIGURE 5–34
Converting the boundary nodes 3 and 5 on symmetry lines to interior nodes by using mirror images.

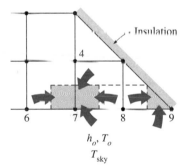

FIGURE 5–35
Schematics for energy balances on the volume elements of nodes 7 and 9.

Taking $\Delta x = \Delta y = l$, it simplifies to

$$T_8 - \left(1 + \frac{h_o l}{k}\right) T_9 = -\frac{h_o l}{k} T_o - \frac{\varepsilon \sigma l}{k} (T_{sky}^4 - T_9^4)$$

This problem involves radiation, which requires the use of absolute temperature, and thus all temperatures should be expressed in Kelvin. Alternately, we could use °C for all temperatures provided that the four temperatures in the radiation terms are expressed in the form $(T + 273)^4$. Substituting the given quantities, the system of nine equations for the determination of nine unknown nodal temperatures in a form suitable for use with an iteration method becomes

$$T_1 = (T_2 + T_3 + 2865)/7$$
$$T_2 = (T_1 + 2T_4 + 2865)/8$$
$$T_3 = (T_1 + 2T_4 + T_6)/4$$
$$T_4 = (T_2 + T_3 + T_5 + T_7)/4$$
$$T_5 = (2T_4 + 2T_8)/4$$
$$T_6 = (T_2 + T_3 + 456.2 - 0.3645 \times 10^{-9} T_6^4)/3.5$$
$$T_7 = (2T_4 + T_6 + T_8 + 912.4 - 0.729 \times 10^{-9} T_7^4)/7$$
$$T_8 = (2T_5 + T_7 + T_9 + 912.4 - 0.729 \times 10^{-9} T_8^4)/7$$
$$T_9 = (T_8 + 456.2 - 0.3645 \times 10^{-9} T_9^4)/2.5$$

which is a system of *nonlinear* equations. Using the Gauss-Seidel iteration method or an equation solver, its solution is determined to be

$T_1 = 545.7 \text{ K} = 272.6°C \quad T_2 = 529.2 \text{ K} = 256.1°C \quad T_3 = 425.2 \text{ K} = 152.1°C$
$T_4 = 411.2 \text{ K} = 138.0°C \quad T_5 = 362.1 \text{ K} = 89.0°C \quad T_6 = 332.9 \text{ K} = 59.7°C$
$T_7 = 328.1 \text{ K} = 54.9°C \quad T_8 = 313.1 \text{ K} = 39.9°C \quad T_9 = 296.5 \text{ K} = 23.4°C$

The variation of temperature in the chimney is shown in Figure 5–36.

Note that the temperatures are highest at the inner wall (but less than 300°C) and lowest at the outer wall (but more that 260 K), as expected.

The average temperature at the outer surface of the chimney weighed by the surface area is

$$T_{\text{wall, out}} = \frac{(0.5T_6 + T_7 + T_8 + 0.5T_9)}{(0.5 + 1 + 1 + 0.5)}$$

$$= \frac{0.5 \times 332.9 + 328.1 + 313.1 + 0.5 \times 296.5}{3} = 318.6 \text{ K}$$

Then the rate of heat loss through the 1-m-long section of the chimney can be determined approximately from

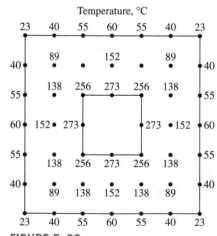

FIGURE 5–36

The variation of temperature in the chimney.

$$\dot{Q}_{\text{chimney}} = h_o A_o (T_{\text{wall, out}} - T_o) + \varepsilon \sigma A_v (T_{\text{wall, out}}^4 - T_{\text{sky}}^4)$$
$$= (21 \text{ W/m}^2 \cdot \text{K})[4 \times (0.6 \text{ m})(1 \text{ m})](318.6 - 293)\text{K}$$
$$+ 0.9(5.67 \times 10^{-8} \text{ W/m}^2 \cdot \text{K}^4)$$
$$[4 \times (0.6 \text{ m})(1 \text{ m})](318.6 \text{ K})^4 - (260 \text{ K})^4]$$
$$= 1291 + 702 = \mathbf{1993 \text{ W}}$$

We could also determine the heat transfer by finding the average temperature of the inner wall, which is (272.6 + 256.1)/2 = 264.4°C, and applying Newton's law of cooling at that surface:

$$\dot{Q}_{\text{chimney}} = h_i A_i (T_i - T_{\text{wall, in}})$$
$$= (70 \text{ W/m}^2 \cdot \text{K})[4 \times (0.2 \text{ m})(1 \text{ m})](300 - 264.4)°\text{C} = 1994 \text{ W}$$

The difference between the two results is due to the approximate nature of the numerical analysis.

Discussion We used a relatively crude numerical model to solve this problem to keep the complexities at a manageable level. The accuracy of the solution obtained can be improved by using a finer mesh and thus a greater number of nodes. Also, when radiation is involved, it is more accurate (but more laborious) to determine the heat losses for each node and add them up instead of using the average temperature.

5–5 · TRANSIENT HEAT CONDUCTION

So far in this chapter we have applied the finite difference method to *steady* heat transfer problems. In this section we extend the method to solve *transient* problems.

We applied the finite difference method to *steady* problems by *discretizing* the problem in the space variables and solving for temperatures at *discrete* points called the nodes. The solution obtained is valid for any time since under steady conditions the temperatures do not change with time. In transient problems, however, the temperatures change with time as well as position, and thus the finite difference solution of transient problems requires *discretization in time* in addition to discretization in space, as shown in Figure 5–37. This is done by selecting a suitable time step Δt and solving for the unknown nodal temperatures repeatedly for each Δt until the solution at the desired time is obtained. For example, consider a hot metal object that is taken out of the oven at an initial temperature of T_i at time $t = 0$ and is allowed to cool in ambient air. If a time step of $\Delta t = 5$ min is chosen, the determination of the temperature distribution in the metal piece after 3 h requires the determination of the temperatures $3 \times 60/5 = 36$ times, or in 36 time steps. Therefore, the computation time of this problem is 36 times that of a steady problem. Choosing a smaller Δt increases the accuracy of the solution, but it also increases the computation time.

In transient problems, the *superscript i* is used as the *index* or *counter* of time steps, with $i = 0$ corresponding to the specified initial condition. In the case of the hot metal piece discussed above, $i = 1$ corresponds to $t = 1 \times \Delta t = 5$ min, $i = 2$ corresponds to $t = 2 \times \Delta t = 10$ min, and a general

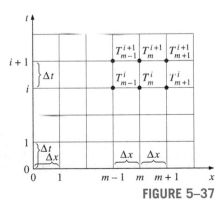

FIGURE 5–37

Finite difference formulation of time-dependent problems involves discrete points in time as well as space.

time step i corresponds to $t_i = i\Delta t$. The notation T_m^i is used to represent the temperature at the node m at time step i.

The formulation of transient heat conduction problems differs from that of steady ones in that the transient problems involve an *additional term* representing the *change in the energy content* of the medium with time. This additional term appears as a first derivative of temperature with respect to time in the differential equation, and as a change in the internal energy content during Δt in the energy balance formulation. The nodes and the volume elements in transient problems are selected as they are in the steady case, and, again assuming all heat transfer is *into* the element for convenience, the energy balance on a volume element during a time interval Δt can be expressed as

$$\left(\begin{array}{c} \text{Heat transferred into} \\ \text{the volume element} \\ \text{from all of its surfaces} \\ \text{during } \Delta t \end{array}\right) + \left(\begin{array}{c} \text{Heat generated} \\ \text{within the} \\ \text{volume element} \\ \text{during } \Delta t \end{array}\right) = \left(\begin{array}{c} \text{The change in the} \\ \text{energy content of} \\ \text{the volume element} \\ \text{during } \Delta t \end{array}\right)$$

or

$$\Delta t \times \sum_{\text{All sides}} \dot{Q} + \Delta t \times \dot{E}_{\text{gen, element}} = \Delta E_{\text{element}} \tag{5–37}$$

where the rate of heat transfer $\dot{Q}$ normally consists of conduction terms for interior nodes, but may involve convection, heat flux, and radiation for boundary nodes.

Noting that $\Delta E_{\text{element}} = mc_p\Delta T = \rho V_{\text{element}} c_p\Delta T$, where ρ is density and c_p is the specific heat of the element, dividing the earlier relation by Δt gives

$$\sum_{\text{All sides}} \dot{Q} + \dot{E}_{\text{gen, element}} = \frac{\Delta E_{\text{element}}}{\Delta t} = \rho V_{\text{element}} c_p \frac{\Delta T}{\Delta t} \tag{5–38}$$

or, for any node m in the medium and its volume element,

$$\sum_{\text{All sides}} \dot{Q} + \dot{E}_{\text{gen, element}} = \rho V_{\text{element}} c_p \frac{T_m^{i+1} - T_m^i}{\Delta t} \tag{5–39}$$

where T_m^i and T_m^{i+1} are the temperatures of node m at times $t_i = i\Delta t$ and $t_{i+1} = (i+1)\Delta t$, respectively, and $T_m^{i+1} - T_m^i$ represents the temperature change of the node during the time interval Δt between the time steps i and $i+1$ (Fig. 5–38).

Note that the ratio $(T_m^{i+1} - T_m^i)/\Delta t$ is simply the finite difference approximation of the partial derivative $\partial T/\partial t$ that appears in the differential equations of transient problems. Therefore, we would obtain the same result for the finite difference formulation if we followed a strict mathematical approach instead of the energy balance approach used above. Also note that the finite difference formulations of steady and transient problems differ by the single term on the right side of the equal sign, and the format of that term remains the same in all coordinate systems regardless of whether heat transfer is one-, two-, or three-dimensional. For the special case of $T_m^{i+1} = T_m^i$ (i.e., no change in temperature with time), the formulation reduces to that of steady case, as expected.

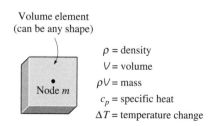

Volume element
(can be any shape)

ρ = density
V = volume
ρV = mass
c_p = specific heat
ΔT = temperature change

$\Delta U = \rho V c_p \Delta T = \rho V c_p (T_m^{i+1} - T_m^i)$

FIGURE 5–38

The change in the energy content of the volume element of a node during a time interval Δt.

The nodal temperatures in transient problems normally change during each time step, and you may be wondering whether to use temperatures at the *previous* time step i or the *new* time step $i + 1$ for the terms on the left side of Eq. 5–39. Well, both are reasonable approaches and both are used in practice. The finite difference approach is called the **explicit method** in the first case and the **implicit method** in the second case, and they are expressed in the general form as (Fig. 5–39)

Explicit method:
$$\sum_{\text{All sides}} \dot{Q}^i + \dot{E}^i_{\text{gen, element}} = \rho V_{\text{element}} c_p \frac{T_m^{i+1} - T_m^i}{\Delta t} \qquad (5\text{-}40)$$

Implicit method:
$$\sum_{\text{All sides}} \dot{Q}^{i+1} + \dot{E}^{i+1}_{\text{gen, element}} = \rho V_{\text{element}} c_p \frac{T_m^{i+1} - T_m^i}{\Delta t} \qquad (5\text{-}41)$$

It appears that the time derivative is expressed in *forward difference* form in the explicit case and *backward difference* form in the implicit case. Of course, it is also possible to mix the two fundamental formulations of Eqs. 5–40 and 5–41 and come up with more elaborate formulations, but such formulations offer little insight and are beyond the scope of this text. Note that both formulations are simply expressions between the nodal temperatures before and after a time interval and are based on determining the new temperatures T_m^{i+1} using the *previous* temperatures T_m^i. *The explicit and implicit formulations given here are quite general and can be used in any coordinate system regardless of the dimension of heat transfer.* The volume elements in multidimensional cases simply have more surfaces and thus involve more terms in the summation.

The explicit and implicit methods have their advantages and disadvantages, and one method is not necessarily better than the other one. Next you will see that the *explicit method* is easy to implement but imposes a limit on the allowable time step to avoid instabilities in the solution, and the *implicit method* requires the nodal temperatures to be solved simultaneously for each time step but imposes no limit on the magnitude of the time step. We limit the discussion to one- and two-dimensional cases to keep the complexities at a manageable level, but the analysis can readily be extended to three-dimensional cases and other coordinate systems.

Transient Heat Conduction in a Plane Wall

Consider transient one-dimensional heat conduction in a plane wall of thickness L with heat generation $\dot{e}(x, t)$ that may vary with time and position and constant conductivity k with a mesh size of $\Delta x = L/M$ and nodes 0, 1, 2, . . . , M in the x-direction, as shown in Figure 5–40. Noting that the volume element of a general interior node m involves heat conduction from two sides and the volume of the element is $V_{\text{element}} = A\Delta x$, the transient finite difference formulation for an interior node can be expressed on the basis of Eq. 5–39 as

$$kA \frac{T_{m-1} - T_m}{\Delta x} + kA \frac{T_{m+1} - T_m}{\Delta x} + \dot{e}_m A\Delta x = \rho A \Delta x c_p \frac{T_m^{i+1} - T_m^i}{\Delta t} \qquad (5\text{-}42)$$

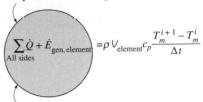

If expressed at $i + 1$: Implicit method

$$\sum_{\text{All sides}} \dot{Q} + \dot{E}_{\text{gen, element}} = \rho V_{\text{element}} c_p \frac{T_m^{i+1} - T_m^i}{\Delta t}$$

If expressed at i: Explicit method

FIGURE 5–39
The formulation of explicit and implicit methods differs at the time step (previous or new) at which the heat transfer and heat generation terms are expressed.

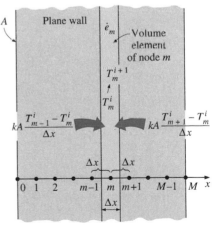

FIGURE 5–40
The nodal points and volume elements for the transient finite difference formulation of one-dimensional conduction in a plane wall.

Canceling the surface area A and multiplying by $\Delta x/k$, it simplifies to

$$T_{m-1} - 2T_m + T_{m+1} + \frac{\dot{e}_m \Delta x^2}{k} = \frac{\Delta x^2}{\alpha \Delta t}(T_m^{i+1} - T_m^i) \quad \text{(5-43)}$$

where $\alpha = k/\rho c_p$ is the *thermal diffusivity* of the wall material. We now define a dimensionless **mesh Fourier number** as

$$\tau = \frac{\alpha \Delta t}{\Delta x^2} \quad \text{(5-44)}$$

Then Eq. 5–43 reduces to

$$T_{m-1} - 2T_m + T_{m+1} + \frac{\dot{e}_m \Delta x^2}{k} = \frac{T_m^{i+1} - T_m^i}{\tau} \quad \text{(5-45)}$$

Note that the left side of this equation is simply the finite difference formulation of the problem for the steady case. This is not surprising since the formulation must reduce to the steady case for $T_m^{i+1} = T_m^i$. Also, we are still not committed to explicit or implicit formulation since we did not indicate the time step on the left side of the equation. We now obtain the *explicit* finite difference formulation by expressing the left side at time step i as

$$T_{m-1}^i - 2T_m^i + T_{m+1}^i + \frac{\dot{e}_m^i \Delta x^2}{k} = \frac{T_m^{i+1} - T_m^i}{\tau} \quad \text{(explicit)} \quad \text{(5-46)}$$

This equation can be solved *explicitly* for the new temperature T_m^{i+1} (and thus the name *explicit* method) to give

$$T_m^{i+1} = \tau(T_{m-1}^i + T_{m+1}^i) + (1 - 2\tau)T_m^i + \tau \frac{\dot{e}_m^i \Delta x^2}{k} \quad \text{(5-47)}$$

for all interior nodes $m = 1, 2, 3, \ldots, M - 1$ in a plane wall. Expressing the left side of Eq. 5–45 at time step $i + 1$ instead of i would give the *implicit* finite difference formulation as

$$T_{m-1}^{i+1} - 2T_m^{i+1} + T_{m+1}^{i+1} + \frac{\dot{e}_m^{i+1} \Delta x^2}{k} = \frac{T_m^{i+1} - T_m^i}{\tau} \quad \text{(implicit)} \quad \text{(5-48)}$$

which can be rearranged as

$$\tau T_{m-1}^{i+1} - (1 + 2\tau)T_m^{i+1} + \tau T_{m+1}^{i+1} + \tau \frac{\dot{e}_m^{i+1} \Delta x^2}{k} + T_m^i = 0 \quad \text{(5-49)}$$

The application of either the explicit or the implicit formulation to each of the $M - 1$ interior nodes gives $M - 1$ equations. The remaining two equations are obtained by applying the same method to the two boundary nodes unless, of course, the boundary temperatures are specified as constants (invariant with time). For example, the formulation of the convection boundary condition at the left boundary (node 0) for the explicit case can be expressed as (Fig. 5–41)

$$hA(T_\infty - T_0^i) + kA\frac{T_1^i - T_0^i}{\Delta x} + \dot{e}_0^i A\frac{\Delta x}{2} = \rho A\frac{\Delta x}{2}c_p\frac{T_0^{i+1} - T_0^i}{\Delta t} \quad \text{(5-50)}$$

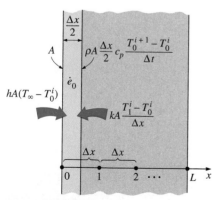

FIGURE 5–41
Schematic for the explicit finite difference formulation of the convection condition at the left boundary of a plane wall.

which simplifies to

$$T_0^{i+1} = \left(1 - 2\tau - 2\tau \frac{h\Delta x}{k}\right) T_0^i + 2\tau T_1^i + 2\tau \frac{h\Delta x}{k} T_\infty + \tau \frac{\dot{e}_0^i \Delta x^2}{k} \qquad \text{(5–51)}$$

Note that in the case of no heat generation and $\tau = 0.5$, the explicit finite difference formulation for a general interior node reduces to $T_m^{i+1} = (T_{m-1}^i + T_{m+1}^i)/2$, which has the interesting interpretation that *the temperature of an interior* node *at the new time step is simply the average of the temperatures of its neighboring* nodes *at the previous time step.*

Once the formulation (explicit or implicit) is complete and the initial condition is specified, the solution of a transient problem is obtained by *marching in time* using a step size of Δt as follows: select a suitable time step Δt and determine the nodal temperatures from the initial condition. Taking the initial temperatures as the *previous* solution T_m^i at $t = 0$, obtain the new solution T_m^{i+1} at all nodes at time $t = \Delta t$ using the transient finite difference relations. Now using the solution just obtained at $t = \Delta t$ as the *previous* solution T_m^i, obtain the new solution T_m^{i+1} at $t = 2\Delta t$ using the same relations. Repeat the process until the solution at the desired time is obtained.

Stability Criterion for Explicit Method: Limitation on Δt

The explicit method is easy to use, but it suffers from an undesirable feature that severely restricts its utility: the explicit method is not unconditionally stable, and the largest permissible value of the time step Δt is limited by the stability criterion. If the time step Δt is not sufficiently small, the solutions obtained by the explicit method may oscillate wildly and diverge from the actual solution. To avoid such divergent oscillations in nodal temperatures, the value of Δt must be maintained below a certain upper limit established by the **stability criterion**. It can be shown mathematically or by a physical argument based on the second law of thermodynamics that *the stability criterion is satisfied if the coefficients of all T_m^i in the T_m^{i+1} expressions (called the* **primary coefficients**) *are greater than or equal to zero for all* nodes m (Fig. 5–42). Of course, all the terms involving T_m^i for a particular node must be grouped together before this criterion is applied.

Different equations for different nodes may result in different restrictions on the size of the time step Δt, and the criterion that is most restrictive should be used in the solution of the problem. A practical approach is to identify the equation with the *smallest primary coefficient* since it is the most restrictive and to determine the allowable values of Δt by applying the stability criterion to that equation only. A Δt value obtained this way also satisfies the stability criterion for all other equations in the system.

For example, in the case of transient one-dimensional heat conduction in a plane wall with specified surface temperatures, the explicit finite difference equations for all the nodes (which are *interior nodes*) are obtained from Eq. 5–47. The coefficient of T_m^i in the T_m^{i+1} expression is $1 - 2\tau$, which is independent of the node number m, and thus the stability criterion for all nodes in this case is $1 - 2\tau \geq 0$ or

$$\tau = \frac{\alpha \Delta t}{\Delta x^2} \leq \frac{1}{2} \qquad \binom{\text{interior nodes, one-dimensional heat}}{\text{transfer in rectangular coordinates}} \qquad \text{(5–52)}$$

Explicit formulation:

$$T_0^{i+1} = a_0 T_0^i + \cdots$$

$$T_1^{i+1} = a_1 T_1^i + \cdots$$

$$\vdots$$

$$T_m^{i+1} = a_m T_m^i + \cdots$$

$$\vdots$$

$$T_M^{i+1} = a_M T_M^i + \cdots$$

Stability criterion:

$$a_m \geq 0, \quad m = 0, 1, 2, \ldots m, \ldots M$$

FIGURE 5–42

The stability criterion of the explicit method requires all primary coefficients to be positive or zero.

When the material of the medium and thus its thermal diffusivity α is known and the value of the mesh size Δx is specified, the largest allowable value of the time step Δt can be determined from this relation. For example, in the case of a brick wall ($\alpha = 0.45 \times 10^{-6}$ m²/s) with a mesh size of $\Delta x = 0.01$ m, the upper limit of the time step is

$$\Delta t \leq \frac{1}{2}\frac{\Delta x^2}{\alpha} = \frac{(0.01 \text{ m})^2}{2(0.45 \times 10^{-6} \text{ m}^2\text{/s})} = 111 \text{ s} = 1.85 \text{ min}$$

The boundary nodes involving convection and/or radiation are more restrictive than the interior nodes and thus require smaller time steps. Therefore, the most restrictive boundary node should be used in the determination of the maximum allowable time step Δt when a transient problem is solved with the explicit method.

To gain a better understanding of the stability criterion, consider the explicit finite difference formulation for an interior node of a plane wall (Eq. 5–47) for the case of no heat generation,

$$T_m^{i+1} = \tau(T_{m-1}^i + T_{m+1}^i) + (1 - 2\tau)T_m^i$$

Assume that at some time step i the temperatures T_{m-1}^i and T_{m+1}^i are equal but less than T_m^i (say, $T_{m-1}^i = T_{m+1}^i = 50°C$ and $T_m^i = 80°C$). At the next time step, we expect the temperature of node m to be between the two values (say, 70°C). However, if the value of τ exceeds 0.5 (say, $\tau = 1$), the temperature of node m at the next time step will be less than the temperature of the neighboring nodes (it will be 20°C), which is physically impossible and violates the second law of thermodynamics (Fig. 5–43). Requiring the new temperature of node m to remain above the temperature of the neighboring nodes is equivalent to requiring the value of τ to remain below 0.5.

The implicit method is *unconditionally stable,* and thus we can use any time step we please with that method (of course, the smaller the time step, the better the accuracy of the solution). The disadvantage of the implicit method is that it results in a set of equations that must be solved *simultaneously* for each time step. Both methods are used in practice.

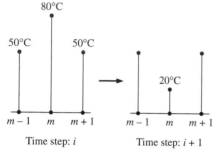

FIGURE 5–43
The violation of the stability criterion in the explicit method may result in the violation of the second law of thermodynamics and thus divergence of solution.

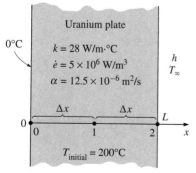

FIGURE 5–44
Schematic for Example 5–5.

EXAMPLE 5–5   Transient Heat Conduction in a Large Uranium Plate

Consider a large uranium plate of thickness $L = 4$ cm, thermal conductivity $k = 28$ W/m · °C, and thermal diffusivity $\alpha = 12.5 \times 10^{-6}$ m²/s that is initially at a uniform temperature of 200°C. Heat is generated uniformly in the plate at a constant rate of $\dot{e} = 5 \times 10^6$ W/m³. At time $t = 0$, one side of the plate is brought into contact with iced water and is maintained at 0°C at all times, while the other side is subjected to convection to an environment at $T_\infty = 30°C$ with a heat transfer coefficient of $h = 45$ W/m² · °C, as shown in Fig. 5–44. Considering a total of three equally spaced nodes in the medium, two at the boundaries and one at the middle, estimate the exposed surface temperature of the plate 2.5 min after the start of cooling using (a) the explicit method and (b) the implicit method.

SOLUTION We have solved this problem in Example 5–1 for the steady case, and here we repeat it for the transient case to demonstrate the application of the transient finite difference methods. Again we assume one-dimensional heat transfer in rectangular coordinates and constant thermal conductivity. The number of nodes is specified to be $M = 3$, and they are chosen to be at the two surfaces of the plate and at the middle, as shown in the figure. Then the nodal spacing Δx becomes

$$\Delta x = \frac{L}{M - 1} = \frac{0.04 \text{ m}}{3 - 1} = 0.02 \text{ m}$$

We number the nodes as 0, 1, and 2. The temperature at node 0 is given to be $T_0 = 0°C$ at all times, and the temperatures at nodes 1 and 2 are to be determined. This problem involves only two unknown nodal temperatures, and thus we need to have only two equations to determine them uniquely. These equations are obtained by applying the finite difference method to nodes 1 and 2.

(a) Node 1 is an interior node, and the *explicit* finite difference formulation at that node is obtained directly from Eq. 5–47 by setting $m = 1$:

$$T_1^{i+1} = \tau(T_0 + T_2^i) + (1 - 2\tau) T_1^i + \tau \frac{\dot{e}_1 \Delta x^2}{k} \qquad \text{(a)}$$

Node 2 is a boundary node subjected to convection, and the finite difference formulation at that node is obtained by writing an energy balance on the volume element of thickness $\Delta x/2$ at that boundary by assuming heat transfer to be into the medium at all sides (Fig. 5–45):

$$hA(T_\infty - T_2^i) + kA \frac{T_1^i - T_2^i}{\Delta x} + \dot{e}_2 A \frac{\Delta x}{2} = \rho A \frac{\Delta x}{2} c_p \frac{T_2^{i+1} - T_2^i}{\Delta x}$$

Dividing by $kA/2\Delta x$ and using the definitions of thermal diffusivity $\alpha = k/\rho c_p$ and the dimensionless mesh Fourier number $\tau = \alpha \Delta t/\Delta x^2$ gives

$$\frac{2h\Delta x}{k}(T_\infty - T_2^i) + 2(T_1^i - T_2^i) + \frac{\dot{e}_2 \Delta x^2}{k} = \frac{T_2^{i+1} - T_2^i}{\tau}$$

which can be solved for T_2^{i+1} to give

$$T_2^{i+1} = \left(1 - 2\tau - 2\tau \frac{h\Delta x}{k}\right) T_2^i + \tau\left(2T_1^i + 2\frac{h\Delta x}{k} T_\infty + \frac{\dot{e}_2 \Delta x^2}{k}\right) \qquad \text{(b)}$$

Note that we did not use the superscript i for quantities that do not change with time. Next we need to determine the upper limit of the time step Δt from the stability criterion, which requires the coefficient of T_1^i in Equation (a) and the coefficient of T_2^i in the second equation to be greater than or equal to zero. The coefficient of T_2^i is smaller in this case, and thus the stability criterion for this problem can be expressed as

$$1 - 2\tau - 2\tau \frac{h\Delta x}{k} \geq 0 \quad \rightarrow \quad \tau \leq \frac{1}{2(1 + h\Delta x/k)} \quad \rightarrow \quad \Delta t \leq \frac{\Delta x^2}{2\alpha(1 + h\Delta x/k)}$$

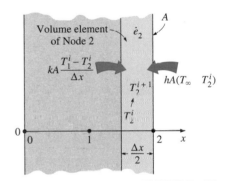

FIGURE 5–45
Schematic for the explicit finite difference formulation of the convection condition at the right boundary of a plane wall.

since $\tau = \alpha \Delta t/\Delta x^2$. Substituting the given quantities, the maximum allowable value of the time step is determined to be

$$\Delta t \leq \frac{(0.02 \text{ m})^2}{2(12.5 \times 10^{-6} \text{ m}^2/\text{s})[1 + (45 \text{ W/m}^2 \cdot °C)(0.02 \text{ m})/28 \text{ W/m} \cdot °C]} = 15.5 \text{ s}$$

Therefore, any time step less than 15.5 s can be used to solve this problem. For convenience, let us choose the time step to be $\Delta t = 15$ s. Then the mesh Fourier number becomes

$$\tau = \frac{\alpha \Delta t}{\Delta x^2} = \frac{(12.5 \times 10^{-6} \text{ m}^2/\text{s})(15 \text{ s})}{(0.02 \text{ m})^2} = 0.46875 \quad (\text{for } \Delta t = 15 \text{ s})$$

Substituting this value of τ and other quantities, the explicit finite difference Equations (a) and (b) reduce to

$$T_1^{i+1} = 0.0625T_1^i + 0.46875T_2^i + 33.482$$
$$T_2^{i+1} = 0.9375T_1^i + 0.032366T_2^i + 34.386$$

The initial temperature of the medium at $t = 0$ and $i = 0$ is given to be 200°C throughout, and thus $T_1^0 = T_2^0 = 200°C$. Then the nodal temperatures at T_1^1 and T_2^1 at $t = \Delta t = 15$ s are determined from these equations to be

$$T_1^1 = 0.0625T_1^0 + 0.46875T_2^0 + 33.482$$
$$= 0.0625 \times 200 + 0.46875 \times 200 + 33.482 = 139.7°C$$
$$T_2^1 = 0.9375T_1^0 + 0.032366T_2^0 + 34.386$$
$$= 0.9375 \times 200 + 0.032366 \times 200 + 34.386 = 228.4°C$$

Similarly, the nodal temperatures T_1^2 and T_2^2 at $t = 2\Delta t = 2 \times 15 = 30$ s are

$$T_1^2 = 0.0625T_1^1 + 0.46875T_2^1 + 33.482$$
$$= 0.0625 \times 139.7 + 0.46875 \times 228.4 + 33.482 = 149.3°C$$
$$T_2^2 = 0.9375T_1^1 + 0.032366T_2^1 + 34.386$$
$$= 0.9375 \times 139.7 + 0.032366 \times 228.4 + 34.386 = 172.8°C$$

Continuing in the same manner, the temperatures at nodes 1 and 2 are determined for $i = 1, 2, 3, 4, 5, \ldots, 40$ and are given in Table 5–2. Therefore, the temperature at the exposed boundary surface 2.5 min after the start of cooling is

$$T_L^{2.5 \text{ min}} = T_2^{10} = 139.0°C$$

(b) Node 1 is an interior node, and the *implicit* finite difference formulation at that node is obtained directly from Eq. 5–49 by setting $m = 1$:

$$\tau T_0 - (1 + 2\tau) T_1^{i+1} + \tau T_2^{i+1} + \tau \frac{\dot{e}_0 \Delta x^2}{k} + T_1^i = 0 \qquad (c)$$

Node 2 is a boundary node subjected to convection, and the implicit finite difference formulation at that node can be obtained from this formulation by expressing the left side of the equation at time step $i + 1$ instead of i as

TABLE 5–2

The variation of the nodal temperatures in Example 5–5 with time obtained by the *explicit* method

Time Step, i	Time, s	Node Temperature, °C T_1^i	Node Temperature, °C T_2^i
0	0	200.0	200.0
1	15	139.7	228.4
2	30	149.3	172.8
3	45	123.8	179.9
4	60	125.6	156.3
5	75	114.6	157.1
6	90	114.3	146.9
7	105	109.5	146.3
8	120	108.9	141.8
9	135	106.7	141.1
10	150	106.3	139.0
20	300	103.8	136.1
30	450	103.7	136.0
40	600	103.7	136.0

$$\frac{2h\Delta x}{k}(T_\infty - T_2^{i+1}) + 2(T_1^{i+1} - T_2^{i+1}) + \frac{\dot{e}_2 \Delta x^2}{k} = \frac{T_2^{i+1} - T_2^i}{\tau}$$

which can be rearranged as

$$2\tau T_1^{i+1} - \left(1 + 2\tau + 2\tau \frac{h\Delta x}{k}\right)T_2^{i+1} + 2\tau \frac{h\Delta x}{k}T_\infty + \tau \frac{\dot{e}_2 \Delta x^2}{k} + T_2^i = 0 \qquad (d)$$

Again we did not use the superscript i or $i + 1$ for quantities that do not change with time. The implicit method imposes no limit on the time step, and thus we can choose any value we want. However, we again choose $\Delta t = 15$ s, and thus $\tau = 0.46875$, to make a comparison with part (a) possible. Substituting this value of τ and other given quantities, the two implicit finite difference equations developed here reduce to

$$-1.9375T_1^{i+1} + 0.46875T_2^{i+1} + T_1^i + 33.482 = 0$$
$$0.9375T_1^{i+1} - 1.9676T_2^{i+1} + T_2^i + 34.386 = 0$$

Again $T_1^0 = T_2^0 = 200°C$ at $t = 0$ and $i = 0$ because of the initial condition, and for $i = 0$, these two equations reduce to

$$-1.9375T_1^1 + 0.46875T_2^1 + 200 + 33.482 = 0$$
$$0.9375T_1^1 - 1.9676T_2^1 + 200 + 34.386 = 0$$

The unknown nodal temperatures T_1^1 and T_2^1 at $t = \Delta t = 15$ s are determined by solving these two equations simultaneously to be

$$T_1^1 = 168.8°C \qquad \text{and} \qquad T_2^1 = 199.6°C$$

Similarly, for $i = 1$, these equations reduce to

$$-1.9375T_2^1 + 0.46875T_2^2 + 168.8 + 33.482 = 0$$
$$0.9375T_1^2 - 1.9676T_2^2 + 199.6 + 34.386 = 0$$

The unknown nodal temperatures T_1^2 and T_2^2 at $t = \Delta t = 2 \times 15 = 30$ s are determined by solving these two equations simultaneously to be

$$T_1^2 = 150.5°C \qquad \text{and} \qquad T_2^2 = 190.6°C$$

Continuing in this manner, the temperatures at nodes 1 and 2 are determined for $i = 2, 3, 4, 5, \ldots, 40$ and are listed in Table 5–3, and the temperature at the exposed boundary surface (node 2) 2.5 min after the start of cooling is obtained to be

$$T_L^{2.5 \text{ min}} = T_2^{10} = 143.9°C$$

which is close to the result obtained by the explicit method. Note that either method could be used to obtain satisfactory results to transient problems, except, perhaps, for the first few time steps. The implicit method is preferred when it is desirable to use large time steps, and the explicit method is preferred when one wishes to avoid the simultaneous solution of a system of algebraic equations.

TABLE 5–3

The variation of the nodal temperatures in Example 5–5 with time obtained by the *implicit* method

Time Step, i	Time, s	Node Temperature, °C T_1^i	T_2^i
0	0	200.0	200.0
1	15	168.8	199.6
2	30	150.5	190.6
3	45	138.6	180.4
4	60	130.3	171.2
5	75	124.1	163.6
6	90	119.5	157.6
7	105	115.9	152.8
8	120	113.2	149.0
9	135	111.0	146.1
10	150	109.4	143.9
20	300	104.2	136.7
30	450	103.8	136.1
40	600	103.8	136.1

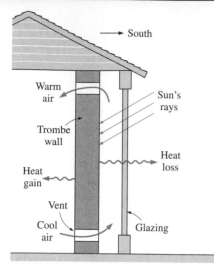

FIGURE 5–46
Schematic of a Trombe wall
(Example 5–6).

TABLE 5–4

The hourly variation of monthly average ambient temperature and solar heat flux incident on a vertical surface for January in Reno, Nevada

Time of Day	Ambient Temperature, °F	Solar Radiation, Btu/h · ft²
7 AM–10 AM	33	114
10 AM–1 PM	43	242
1 PM–4 PM	45	178
4 PM–7 PM	37	0
7 PM–10 PM	32	0
10 PM–1 AM	27	0
1 AM–4 AM	26	0
4 AM–7 AM	25	0

EXAMPLE 5–6   **Solar Energy Storage in Trombe Walls**

Dark painted thick masonry walls called Trombe walls are commonly used on south sides of passive solar homes to absorb solar energy, store it during the day, and release it to the house during the night (Fig. 5–46). The idea was proposed by E. L. Morse of Massachusetts in 1881 and is named after Professor Felix Trombe of France, who used it extensively in his designs in the 1970s. Usually a single or double layer of glazing is placed outside the wall and transmits most of the solar energy while blocking heat losses from the exposed surface of the wall to the outside. Also, air vents are commonly installed at the bottom and top of the Trombe walls so that the house air enters the parallel flow channel between the Trombe wall and the glazing, rises as it is heated, and enters the room through the top vent.

Consider a house in Reno, Nevada, whose south wall consists of a 1-ft-thick Trombe wall whose thermal conductivity is $k = 0.40$ Btu/h · ft · °F and whose thermal diffusivity is $\alpha = 4.78 \times 10^{-6}$ ft²/s. The variation of the ambient temperature T_{out} and the solar heat flux $\dot{q}_{solar}$ incident on a south-facing vertical surface throughout the day for a typical day in January is given in Table 5–4 in 3-h intervals. The Trombe wall has single glazing with an absorptivity-transmissivity product of $\kappa = 0.77$ (that is, 77 percent of the solar energy incident is absorbed by the exposed surface of the Trombe wall), and the average combined heat transfer coefficient for heat loss from the Trombe wall to the ambient is determined to be $h_{out} = 0.7$ Btu/h · ft² · °F. The interior of the house is maintained at $T_{in} = 70$°F at all times, and the heat transfer coefficient at the interior surface of the Trombe wall is $h_{in} = 1.8$ Btu/h · ft² · °F. Also, the vents on the Trombe wall are kept closed, and thus the only heat transfer between the air in the house and the Trombe wall is through the interior surface of the wall. Assuming the temperature of the Trombe wall to vary linearly between 70°F at the interior surface and 30°F at the exterior surface at 7 AM and using the explicit finite difference method with a uniform nodal spacing of $\Delta x = 0.2$ ft, determine the temperature distribution along the thickness of the Trombe wall after 12, 24, 36, and 48 h. Also, determine the net amount of heat transferred to the house from the Trombe wall during the first day and the second day. Assume the wall is 10 ft high and 25 ft long.

SOLUTION   The passive solar heating of a house through a Trombe wall is considered. The temperature distribution in the wall in 12-h intervals and the amount of heat transfer during the first and second days are to be determined.
Assumptions   **1** Heat transfer is one-dimensional since the exposed surface of the wall is large relative to its thickness. **2** Thermal conductivity is constant. **3** The heat transfer coefficients are constant.
Properties   The wall properties are given to be $k = 0.40$ Btu/h · ft · °F, $\alpha = 4.78 \times 10^{-6}$ ft²/s, and $\kappa = 0.77$.
Analysis   The nodal spacing is given to be $\Delta x = 0.2$ ft, and thus the total number of nodes along the Trombe wall is

$$M = \frac{L}{\Delta x} + 1 = \frac{1\,\text{ft}}{0.2\,\text{ft}} + 1 = 6$$

We number the nodes as 0, 1, 2, 3, 4, and 5, with node 0 on the interior surface of the Trombe wall and node 5 on the exterior surface, as shown in Figure 5–47. Nodes 1 through 4 are interior nodes, and the explicit finite difference formulations of these nodes are obtained directly from Eq. 5–47 to be

Node 1 ($m = 1$): $\quad T_1^{i+1} = \tau(T_0^i + T_2^i) + (1 - 2\tau)T_1^i \quad$ **(1)**

Node 2 ($m = 2$): $\quad T_2^{i+1} = \tau(T_1^i + T_3^i) + (1 - 2\tau)T_2^i \quad$ **(2)**

Node 3 ($m = 3$): $\quad T_3^{i+1} = \tau(T_2^i + T_4^i) + (1 - 2\tau)T_3^i \quad$ **(3)**

Node 4 ($m = 4$): $\quad T_4^{i+1} = \tau(T_3^i + T_5^i) + (1 - 2\tau)T_4^i \quad$ **(4)**

The interior surface is subjected to convection, and thus the explicit formulation of node 0 can be obtained directly from Eq. 5–51 to be

$$T_0^{i+1} = \left(1 - 2\tau - 2\tau\frac{h_{\text{in}}\Delta x}{k}\right)T_0^i + 2\tau T_1^i + 2\tau\frac{h_{\text{in}}\Delta x}{k}T_{\text{in}}$$

Substituting the quantities h_{in}, Δx, k, and T_{in}, which do not change with time, into this equation gives

$$T_0^{i+1} = (1 - 3.80\tau)\,T_0^i + \tau(2T_1^i + 126.0) \quad \textbf{(5)}$$

The exterior surface of the Trombe wall is subjected to convection as well as to heat flux. The explicit finite difference formulation at that boundary is obtained by writing an energy balance on the volume element represented by node 5,

$$h_{\text{out}}A(T_{\text{out}}^i - T_5^i) + \kappa A\dot{q}_{\text{solar}}^i + kA\frac{T_4^i - T_5^i}{\Delta x} = \rho A\frac{\Delta x}{2}c_p\frac{T_5^{i+1} - T_5^i}{\Delta t} \quad \textbf{(5-53)}$$

which simplifies to

$$T_5^{i+1} = \left(1 - 2\tau - 2\tau\frac{h_{\text{out}}\Delta x}{k}\right)T_5^i + 2\tau T_4^i + 2\tau\frac{h_{\text{out}}\Delta x}{k}T_{\text{out}}^i + 2\tau\frac{\kappa\dot{q}_{\text{solar}}^i\Delta x}{k} \quad \textbf{(5-54)}$$

where $\tau = \alpha\Delta t/\Delta x^2$ is the dimensionless mesh Fourier number. Note that we kept the superscript i for quantities that vary with time. Substituting the quantities h_{out}, Δx, k, and κ, which do not change with time, into this equation gives

$$T_5^{i+1} = (1 - 2.70\tau)\,T_5^i + \tau(2T_4^i + 0.70T_{\text{out}}^i + 0.770\dot{q}_{\text{solar}}^i) \quad \textbf{(6)}$$

where the unit of $\dot{q}_{\text{solar}}^i$ is Btu/h $\cdot$ ft^2.

Next we need to determine the upper limit of the time step Δt from the stability criterion since we are using the explicit method. This requires the identification of the smallest primary coefficient in the system. We know that the boundary nodes are more restrictive than the interior nodes, and thus we examine the formulations of the boundary nodes 0 and 5 only. The smallest and thus the most restrictive primary coefficient in this case is the coefficient of T_0^i in the formulation of node 0 since $1 - 3.8\tau < 1 - 2.7\tau$, and thus the stability criterion for this problem can be expressed as

$$1 - 3.80\tau \geq 0 \quad \rightarrow \quad \tau = \frac{\alpha\Delta x}{\Delta x^2} \leq \frac{1}{3.80}$$

Substituting the given quantities, the maximum allowable value of the time step is determined to be

$$\Delta t \leq \frac{\Delta x^2}{3.80\alpha} = \frac{(0.2\text{ ft})^2}{3.80 \times (4.78 \times 10^{-6}\text{ ft}^2/\text{s})} = 2202\text{ s}$$

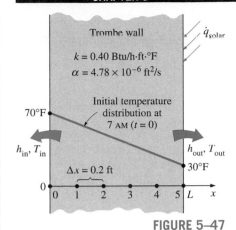

FIGURE 5–47

The nodal network for the Trombe wall discussed in Example 5–6.

Therefore, any time step less than 2202 s can be used to solve this problem. For convenience, let us choose the time step to be $\Delta t = 900$ s = 15 min. Then the mesh Fourier number becomes

$$\tau = \frac{\alpha \Delta t}{\Delta x^2} = \frac{(4.78 \times 10^{-6} \text{ ft}^2/\text{s})(900 \text{ s})}{(0.2 \text{ ft})^2} = 0.10755 \qquad (\text{for } \Delta t = 15 \text{ min})$$

Initially (at 7 AM or $t = 0$), the temperature of the wall is said to vary linearly between 70°F at node 0 and 30°F at node 5. Noting that there are five nodal spacings of equal length, the temperature change between two neighboring nodes is $(70 - 30)°\text{F}/5 = 8°\text{F}$. Therefore, the initial nodal temperatures are

$$T_0^0 = 70°\text{F}, \qquad T_1^0 = 62°\text{F}, \qquad T_2^0 = 54°\text{F},$$
$$T_3^0 = 46°\text{F}, \qquad T_4^0 = 38°\text{F}, \qquad T_5^0 = 30°\text{F}$$

Then the nodal temperatures at $t = \Delta t = 15$ min (at 7:15 AM) are determined from these equations to be

$$T_0^1 = (1 - 3.80\tau) T_0^0 + \tau(2T_1^0 + 126.0)$$
$$= (1 - 3.80 \times 0.10755) 70 + 0.10755(2 \times 62 + 126.0) = 68.3° \text{ F}$$
$$T_1^1 = \tau(T_0^0 + T_2^0) + (1 - 2\tau) T_1^0$$
$$= 0.10755(70 + 54) + (1 - 2 \times 0.10755)62 = 62°\text{F}$$
$$T_2^1 = \tau(T_1^0 + T_3^0) + (1 - 2\tau) T_2^0$$
$$= 0.10755(62 + 46) + (1 - 2 \times 0.10755)54 = 54°\text{F}$$
$$T_3^1 = \tau(T_2^0 + T_4^0) + (1 - 2\tau) T_3^0$$
$$= 0.10755(54 + 38) + (1 - 2 \times 0.10755)46 = 46°\text{F}$$
$$T_4^1 = \tau(T_3^0 + T_5^0) + (1 - 2\tau) T_4^0$$
$$= 0.10755(46 + 30) + (1 - 2 \times 0.10755)38 = 38°\text{F}$$
$$T_5^1 = (1 - 2.70\tau) T_5^0 + \tau(2T_4^0 + 0.70T_{\text{out}}^0 + 0.770\dot{q}_{\text{solar}}^0)$$
$$= (1 - 2.70 \times 0.10755)30 + 0.10755(2 \times 38 + 0.70 \times 33 + 0.770 \times 114)$$
$$= 41.4°\text{F}$$

Note that the inner surface temperature of the Trombe wall dropped by 1.7°F and the outer surface temperature rose by 11.4°F during the first time step while the temperatures at the interior nodes remained the same. This is typical of transient problems in mediums that involve no heat generation. The nodal temperatures at the following time steps are determined similarly with the help of a computer. Note that the data for ambient temperature and the incident solar radiation change every 3 hours, which corresponds to 12 time steps, and this must be reflected in the computer program. For example, the value of $\dot{q}_{\text{solar}}^i$ must be taken to be $\dot{q}_{\text{solar}}^i = 114$ for $i = 1$–12, $\dot{q}_{\text{solar}}^i = 242$ for $i = 13$–24, $\dot{q}_{\text{solar}}^i = 178$ for $i = 25$–36, and $\dot{q}_{\text{solar}}^i = 0$ for $i = 37$–96.

The results after 6, 12, 18, 24, 30, 36, 42, and 48 h are given in Table 5–5 and are plotted in Figure 5–48 for the first day. Note that the interior temperature of the Trombe wall drops in early morning hours, but then rises as the solar energy absorbed by the exterior surface diffuses through the wall. The exterior surface temperature of the Trombe wall rises from 30 to 142°F in just 6 h because of the solar energy absorbed, but then drops to 53°F by next morning as a result of heat loss at night. Therefore, it may be worthwhile to cover the outer surface at night to minimize the heat losses.

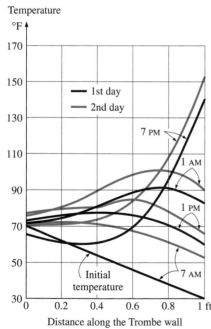

FIGURE 5–48

The variation of temperatures in the Trombe wall discussed in Example 5–6.

TABLE 5–5

The temperatures at the nodes of a Trombe wall at various times

Time	Time Step, i	Nodal Temperatures, °F					
		T_0	T_1	T_2	T_3	T_4	T_5
0 h (7 AM)	0	70.0	62.0	54.0	46.0	38.0	30.0
6 h (1 PM)	24	65.3	61.7	61.5	69.7	94.1	142.0
12 h (7 PM)	48	71.6	74.2	80.4	88.4	91.7	82.4
18 h (1 AM)	72	73.3	75.9	77.4	76.3	71.2	61.2
24 h (7 AM)	96	71.2	71.9	70.9	67.7	61.7	53.0
30 h (1 PM)	120	70.3	71.1	74.3	84.2	108.3	153.2
36 h (7 PM)	144	75.4	81.1	89.4	98.2	101.0	89.7
42 h (1 AM)	168	75.8	80.7	83.5	83.0	77.4	66.2
48 h (7 AM)	192	73.0	75.1	72.2	66.0	66.0	56.3

The rate of heat transfer from the Trombe wall to the interior of the house during each time step is determined from Newton's law using the average temperature at the inner surface of the wall (node 0) as

$$Q^i_{\text{Trombe wall}} = \dot{Q}^i_{\text{Trombe wall}} \, \Delta t = h_{\text{in}} A(T^i_0 - T_{\text{in}}) \, \Delta t = h_{\text{in}} A[(T^i_0 + T^{i-1}_0)/2 - T_{\text{in}}]\Delta t$$

Therefore, the amount of heat transfer during the first time step ($i = 1$) or during the first 15-min period is

$$\begin{aligned}
Q^1_{\text{Trombe wall}} &= h_{\text{in}} A[(T^1_0 + T^0_0)/2 - T_{\text{in}}] \, \Delta t \\
&= (1.8 \text{ Btu/h} \cdot \text{ft}^2 \cdot {}^\circ\text{F})(10 \times 25 \text{ ft}^2)[(68.3 + 70)/2 - 70{}^\circ\text{F}](0.25 \text{ h}) \\
&= -95.6 \text{ Btu}
\end{aligned}$$

The negative sign indicates that heat is transferred to the Trombe wall from the air in the house, which represents a heat loss. Then the total heat transfer during a specified time period is determined by adding the heat transfer amounts for each time step as

$$Q_{\text{Trombe wall}} = \sum_{i=1}^{I} \dot{Q}^i_{\text{Trombe wall}} - \sum_{i=1}^{I} h_{\text{in}} A[(T^i_0 + T^{i-1}_0)/2 - T_{\text{in}}] \, \Delta t \qquad \textbf{(5-55)}$$

where I is the total number of time intervals in the specified time period. In this case I = 48 for 12 h, 96 for 24 h, and so on. Following the approach described here using a computer, the amount of heat transfer between the Trombe wall and the interior of the house is determined to be

$Q_{\text{Trombe wall}} = -17,048$ Btu after 12 h	($-17,048$ Btu during the first 12 h)
$Q_{\text{Trombe wall}} = -2483$ Btu after 24 h	(14,565 Btu during the second 12 h)
$Q_{\text{Trombe wall}} = 5610$ Btu after 36 h	(8093 Btu during the third 12 h)
$Q_{\text{Trombe wall}} = 34,400$ Btu after 48 h	(28,790 Btu during the fourth 12 h)

Therefore, the house loses 2483 Btu through the Trombe wall the first day as a result of the low start-up temperature but delivers a total of 36,883 Btu of heat to the house the second day. It can be shown that the Trombe wall will deliver even more heat to the house during the third day since it will start the day at a higher average temperature.

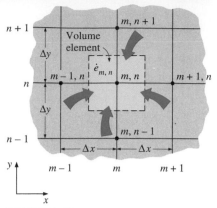

FIGURE 5–49

The volume element of a general interior node (m, n) for two-dimensional transient conduction in rectangular coordinates.

Two-Dimensional Transient Heat Conduction

Consider a rectangular region in which heat conduction is significant in the x- and y-directions, and consider a unit depth of $\Delta z = 1$ in the z-direction. Heat may be generated in the medium at a rate of $\dot{e}(x, y, t)$, which may vary with time and position, with the thermal conductivity k of the medium assumed to be constant. Now divide the x-y-plane of the region into a *rectangular mesh* of nodal points spaced Δx and Δy apart in the x- and y-directions, respectively, and consider a general interior node (m, n) whose coordinates are $x = m\Delta x$ and $y = n\Delta y$, as shown in Figure 5–49. Noting that the volume element centered about the general interior node (m, n) involves heat conduction from four sides (right, left, top, and bottom) and the volume of the element is $V_{\text{element}} = \Delta x \times \Delta y \times 1 = \Delta x \Delta y$, the transient finite difference formulation for a general interior node can be expressed on the basis of Eq. 5–39 as

$$k\Delta y \frac{T_{m-1, n} - T_{m, n}}{\Delta x} + k\Delta x \frac{T_{m, n+1} - T_{m, n}}{\Delta y} + k\Delta y \frac{T_{m+1, n} - T_{m, n}}{\Delta x}$$

$$+ k\Delta x \frac{T_{m, n-1} - T_{m, n}}{\Delta y} + \dot{e}_{m, n}\, \Delta x \Delta y = \rho \Delta x \Delta y\, c_p \frac{T_m^{i+1} - T_m^i}{\Delta t} \quad \textbf{(5–56)}$$

Taking a square mesh ($\Delta x = \Delta y = l$) and dividing each term by k gives after simplifying,

$$T_{m-1, n} + T_{m+1, n} + T_{m, n+1} + T_{m, n-1} - 4T_{m, n} + \frac{\dot{e}_{m, n}l^2}{k} = \frac{T_m^{i+1} - T_m^i}{\tau} \quad \textbf{(5–57)}$$

where again $\alpha = k/\rho c_p$ is the thermal diffusivity of the material and $\tau = \alpha\Delta t/l^2$ is the dimensionless mesh Fourier number. It can also be expressed in terms of the temperatures at the neighboring nodes in the following easy-to-remember form:

$$T_{\text{left}} + T_{\text{top}} + T_{\text{right}} + T_{\text{bottom}} - 4T_{\text{node}} + \frac{\dot{e}_{\text{node}}l^2}{k} = \frac{T_{\text{node}}^{i+1} - T_{\text{node}}^i}{\tau} \quad \textbf{(5–58)}$$

Again the left side of this equation is simply the finite difference formulation of the problem for the *steady case,* as expected. Also, we are still not committed to explicit or implicit formulation since we did not indicate the time step on the left side of the equation. We now obtain the *explicit* finite difference formulation by expressing the left side at time step i as

$$T_{\text{left}}^i + T_{\text{top}}^i + T_{\text{right}}^i + T_{\text{bottom}}^i - 4T_{\text{node}}^i + \frac{\dot{e}_{\text{node}}^i l^2}{k} = \frac{T_{\text{node}}^{i+1} - T_{\text{node}}^i}{\tau} \quad \textbf{(5–59)}$$

Expressing the left side at time step $i + 1$ instead of i would give the implicit formulation. This equation can be solved *explicitly* for the new temperature T_{node}^{i+1} to give

$$T_{\text{node}}^{i+1} = \tau(T_{\text{left}}^i + T_{\text{top}}^i + T_{\text{right}}^i + T_{\text{bottom}}^i) + (1 - 4\tau)\, T_{\text{node}}^i + \tau \frac{\dot{e}_{\text{node}}^i l^2}{k} \quad \textbf{(5–60)}$$

for all interior nodes (m, n) where $m = 1, 2, 3, \ldots, M - 1$ and $n = 1, 2, 3, \ldots, N - 1$ in the medium. In the case of no heat generation and $\tau = \frac{1}{4}$, the explicit finite difference formulation for a general interior node reduces to $T_{node}^{i+1} = (T_{left}^i + T_{top}^i + T_{right}^i + T_{bottom}^i)/4$, which has the interpretation that *the temperature of an interior node at the new time step is simply the average of the temperatures of its neighboring nodes at the previous time step* (Fig. 5–50).

The stability criterion that requires the coefficient of T_m^i in the T_m^{i+1} expression to be greater than or equal to zero for all nodes is equally valid for two- or three-dimensional cases and severely limits the size of the time step Δt that can be used with the explicit method. In the case of transient two-dimensional heat transfer in rectangular coordinates, the coefficient of T_m^i in the T_m^{i+1} expression is $1 - 4\tau$, and thus the stability criterion for all interior nodes in this case is $1 - 4\tau > 0$, or

$$\tau = \frac{\alpha \Delta t}{l^2} \leq \frac{1}{4} \qquad \text{(interior nodes, two-dimensional heat transfer in rectangular coordinates)} \qquad \textbf{(5-61)}$$

where $\Delta x = \Delta y = l$. When the material of the medium and thus its thermal diffusivity α are known and the value of the mesh size l is specified, the largest allowable value of the time step Δt can be determined from the relation above. Again the boundary nodes involving convection and/or radiation are more restrictive than the interior nodes and thus require smaller time steps. Therefore, the most restrictive boundary node should be used in the determination of the maximum allowable time step Δt when a transient problem is solved with the explicit method.

The application of Eq. 5–60 to each of the $(M - 1) \times (N - 1)$ interior nodes gives $(M - 1) \times (N - 1)$ equations. The remaining equations are obtained by applying the method to the boundary nodes unless, of course, the boundary temperatures are specified as being constant. The development of the transient finite difference formulation of boundary nodes in two- (or three-) dimensional problems is similar to the development in the one-dimensional case discussed earlier. Again the region is partitioned between the nodes by forming volume elements around the nodes, and an energy balance is written for each boundary node on the basis of Eq. 5–39. This is illustrated in Example 5–7.

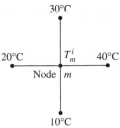

Time step i:

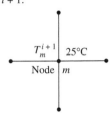

Time step $i + 1$:

FIGURE 5–50
In the case of no heat generation and $\tau = \frac{1}{4}$, the temperature of an interior node at the new time step is the average of the temperatures of its neighboring nodes at the previous time step.

EXAMPLE 5–7   **Transient Two-Dimensional Heat Conduction in L-Bars**

Consider two-dimensional transient heat transfer in an L-shaped solid body that is initially at a uniform temperature of 90°C and whose cross section is given in Fig. 5–51. The thermal conductivity and diffusivity of the body are $k = 15$ W/m · °C and $\alpha = 3.2 \times 10^{-6}$ m²/s, respectively, and heat is generated in the body at a rate of $\dot{e} = 2 \times 10^6$ W/m³. The left surface of the body is insulated, and the bottom surface is maintained at a uniform temperature of 90°C at all times. At time $t = 0$, the entire top surface is subjected to convection to ambient air at $T_\infty = 25$°C with a convection coefficient of $h = 80$ W/m² · °C, and the right surface is subjected to heat flux at a uniform rate of $\dot{q}_R = 5000$ W/m². The nodal network of the problem consists of 15 equally spaced nodes with $\Delta x = \Delta y = 1.2$ cm, as shown in the figure. Five of the nodes are at the bottom surface, and thus their temperatures are known. Using the explicit method, determine the temperature at the top corner (node 3) of the body after 1, 3, 5, 10, and 60 min.

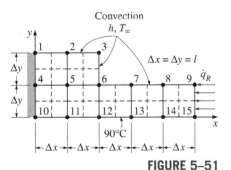

FIGURE 5–51
Schematic and nodal network for Example 5–7.

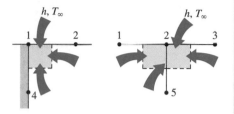

(a) Node 1   *(b)* Node 2

FIGURE 5–52

Schematics for energy balances on the volume elements of nodes 1 and 2.

SOLUTION This is a transient two-dimensional heat transfer problem in rectangular coordinates, and it was solved in Example 5–3 for the steady case. Therefore, the solution of this transient problem should approach the solution for the steady case when the time is sufficiently large. The thermal conductivity and heat generation rate are given to be constants. We observe that all nodes are boundary nodes except node 5, which is an interior node. Therefore, we have to rely on energy balances to obtain the finite difference equations. The region is partitioned among the nodes equitably as shown in the figure, and the explicit finite difference equations are determined on the basis of the energy balance for the transient case expressed as

$$\sum_{\text{All sides}} \dot{Q}^i + \dot{e} V_{\text{element}} = \rho V_{\text{element}} c_p \frac{T_m^{i+1} - T_m^i}{\Delta t}$$

The quantities h, T_∞, $\dot{e}$, and $\dot{q}_R$ do not change with time, and thus we do not need to use the superscript i for them. Also, the energy balance expressions are simplified using the definitions of thermal diffusivity $\alpha = k/\rho c_p$ and the dimensionless mesh Fourier number $\tau = \alpha \Delta t / l^2$, where $\Delta x = \Delta y = l$.

(a) Node 1. (Boundary node subjected to convection and insulation, Fig. 5–52*a*)

$$h \frac{\Delta x}{2} (T_\infty - T_1^i) + k \frac{\Delta y}{2} \frac{T_2^i - T_1^i}{\Delta x} + k \frac{\Delta x}{2} \frac{T_4^i - T_1^i}{\Delta y}$$
$$+ \dot{e}_1 \frac{\Delta x}{2} \frac{\Delta y}{2} = \rho \frac{\Delta x}{2} \frac{\Delta y}{2} c_p \frac{T_1^{i+1} - T_1^i}{\Delta t}$$

Dividing by $k/4$ and simplifying,

$$\frac{2hl}{k} (T_\infty - T_1^i) + 2(T_2^i - T_1^i) + 2(T_4^i - T_1^i) + \frac{\dot{e}_1 l^2}{k} = \frac{T_1^{i+1} - T_1^i}{\tau}$$

which can be solved for T_1^{i+1} to give

$$T_1^{i+1} = \left(1 - 4\tau - 2\tau \frac{hl}{k}\right) T_1^i + 2\tau \left(T_2^i + T_4^i + \frac{hl}{k} T_\infty + \frac{\dot{e}_1 l^2}{2k}\right)$$

(b) Node 2. (Boundary node subjected to convection, Fig. 5–52*b*)

$$h\Delta x(T_\infty - T_2^i) + k \frac{\Delta y}{2} \frac{T_3^i - T_2^i}{\Delta x} + k\Delta x \frac{T_5^i - T_2^i}{\Delta y}$$
$$+ k \frac{\Delta y}{2} \frac{T_1^i - T_2^i}{\Delta x} + \dot{e}_2 \Delta x \frac{\Delta y}{2} = \rho \Delta x \frac{\Delta y}{2} c_p \frac{T_2^{i+1} - T_2^i}{\Delta t}$$

Dividing by $k/2$, simplifying, and solving for T_2^{i+1} gives

$$T_2^{i+1} = \left(1 - 4\tau - 2\tau \frac{hl}{k}\right) T_2^i + \tau \left(T_1^i + T_3^i + 2T_5^i + \frac{2hl}{k} T_\infty + \frac{\dot{e}_2 l^2}{2k}\right)$$

(c) Node 3. (Boundary node subjected to convection on two sides, Fig. 5–53a)

$$h\left(\frac{\Delta x}{2} + \frac{\Delta y}{2}\right)(T_\infty - T_3^i) + k\frac{\Delta x}{2}\frac{T_6^i - T_3^i}{\Delta y}$$

$$+ k\frac{\Delta y}{2}\frac{T_2^i - T_3^i}{\Delta x} + \dot{e}_3\frac{\Delta x}{2}\frac{\Delta y}{2} = \rho\frac{\Delta x}{2}\frac{\Delta y}{2}c_p\frac{T_3^{i+1} - T_3^i}{\Delta t}$$

Dividing by $k/4$, simplifying, and solving for T_3^{i+1} gives

$$T_3^{i+1} = \left(1 - 4\tau - 4\tau\frac{hl}{k}\right)T_3^i + 2\tau\left(T_4^i + T_6^i + 2\frac{hl}{k}T_\infty + \frac{\dot{e}_3 l^2}{2k}\right)$$

(d) Node 4. (On the insulated boundary, and can be treated as an interior node, Fig. 5–53b). Noting that $T_{10} = 90°C$, Eq. 5–60 gives

$$T_4^{i+1} = (1 - 4\tau)\,T_4^i + \tau\left(T_1^i + 2T_5^i + 90 + \frac{\dot{e}_4 l^2}{k}\right)$$

(e) Node 5. (Interior node, Fig. 5–54a). Noting that $T_{11} = 90°C$, Eq. 5–60 gives

$$T_5^{i+1} = (1 - 4\tau)\,T_5^i + \tau\left(T_2^i + T_4^i + T_6^i + 90 + \frac{\dot{e}_5 l^2}{k}\right)$$

(f) Node 6. (Boundary node subjected to convection on two sides, Fig. 5–54b)

$$h\left(\frac{\Delta x}{2} + \frac{\Delta y}{2}\right)(T_\infty - T_6^i) + k\frac{\Delta y}{2}\frac{T_7^i - T_6^i}{\Delta x} + k\Delta x\frac{T_{12}^i - T_6^i}{\Delta y} + k\Delta y\frac{T_5^i - T_6^i}{\Delta x}$$

$$+ \frac{\Delta x}{2}\frac{T_3^i - T_6^i}{\Delta y} + \dot{e}_6\frac{3\Delta x\Delta y}{4} = \rho\frac{3\Delta x\Delta y}{4}c_p\frac{T_6^{i+1} - T_6^i}{\Delta t}$$

Dividing by $3k/4$, simplifying, and solving for T_6^{i+1} gives

$$T_6^{i+1} = \left(1 - 4\tau - 4\tau\frac{hl}{3k}\right)T_3^i$$

$$+ \frac{\tau}{3}\left[2T_3^i + 4T_5^i + 2T_7^i + 4\times 90 + 4\frac{hl}{k}T_\infty + 3\frac{\dot{e}_6 l^2}{k}\right]$$

(g) Node 7. (Boundary node subjected to convection, Fig. 5–55a)

$$h\Delta x(T_\infty - T_7^i) + k\frac{\Delta y}{2}\frac{T_8^i - T_7^i}{\Delta x} + k\Delta x\frac{T_{13}^i - T_7^i}{\Delta y}$$

$$+ k\frac{\Delta y}{2}\frac{T_6^i - T_7^i}{\Delta x} + \dot{e}_7\Delta x\frac{\Delta y}{2} = \rho\Delta x\frac{\Delta y}{2}c_p\frac{T_7^{i+1} - T_7^i}{\Delta t}$$

Dividing by $k/2$, simplifying, and solving for T_7^{i+1} gives

$$T_7^{i+1} = \left(1 - 4\tau - 2\tau\frac{hl}{k}\right)T_7^i + \tau\left[T_6^i + T_8^i + 2\times 90 + \frac{2hl}{k}T_\infty + \frac{\dot{e}_7 l^2}{k}\right]$$

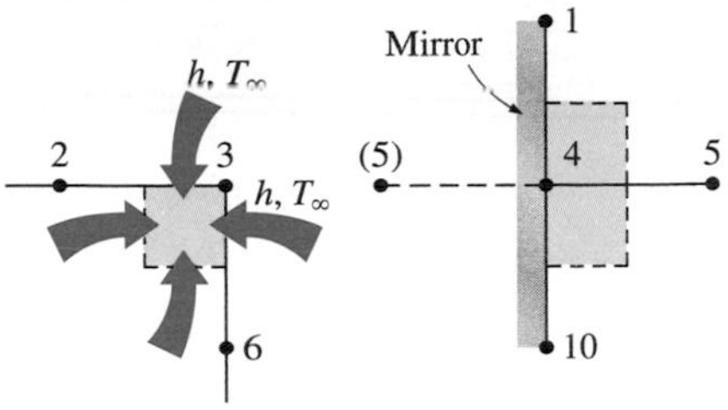

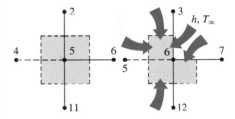

(a) Node 3   (b) Node 4

FIGURE 5–53

Schematics for energy balances on the volume elements of nodes 3 and 4.

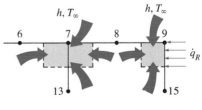

(a) Node 5   (b) Node 6

FIGURE 5–54

Schematics for energy balances on the volume elements of nodes 5 and 6.

(a) Node 7   (b) Node 9

FIGURE 5–55

Schematics for energy balances on the volume elements of nodes 7 and 9.

(*h*) Node 8. This node is identical to node 7, and the finite difference formulation of this node can be obtained from that of node 7 by shifting the node numbers by 1 (i.e., replacing subscript *m* by subscript *m* + 1). It gives

$$T_8^{i+1} = \left(1 - 4\tau - 2\tau \frac{hl}{k}\right) T_8^i + \tau\left[T_7^i + T_9^i + 2 \times 90 + \frac{2hl}{k} T_\infty + \frac{\dot{e}_8 l^2}{k}\right]$$

(*i*) Node 9. (Boundary node subjected to convection on two sides, Fig. 5–55*b*)

$$h\frac{\Delta x}{2}(T_\infty - T_9^i) + \dot{q}_R \frac{\Delta y}{2} + k\frac{\Delta x}{2}\frac{T_{15}^i - T_9^i}{\Delta y}$$

$$+ \frac{k\Delta y}{2}\frac{T_8^i - T_9^i}{\Delta x} + \dot{e}_9\frac{\Delta x}{2}\frac{\Delta y}{2} = \rho\frac{\Delta x}{2}\frac{\Delta y}{2}c_p\frac{T_9^{i+1} - T_9^i}{\Delta t}$$

Dividing by *k*/4, simplifying, and solving for T_9^{i+1} gives

$$T_9^{i+1} = \left(1 - 4\tau - 2\tau \frac{hl}{k}\right) T_9^i + 2\tau\left(T_8^i + 90 + \frac{\dot{q}_R l}{k} + \frac{hl}{k} T_\infty + \frac{\dot{e}_9 l^2}{2k}\right)$$

This completes the finite difference formulation of the problem. Next we need to determine the upper limit of the time step Δt from the stability criterion, which requires the coefficient of T_m^i in the T_m^{i+1} expression (the primary coefficient) to be greater than or equal to zero for all nodes. The smallest primary coefficient in the nine equations here is the coefficient of T_3^i in the expression, and thus the stability criterion for this problem can be expressed as

$$1 - 4\tau - 4\tau \frac{hl}{k} \geq 0 \quad \rightarrow \quad \tau \leq \frac{1}{4(1 + hl/k)} \quad \rightarrow \quad \Delta t \leq \frac{l^2}{4\alpha(1 + hl/k)}$$

since $\tau = \alpha\Delta t/l^2$. Substituting the given quantities, the maximum allowable value of the time step is determined to be

$$\Delta t \leq \frac{(0.012 \text{ m})^2}{4(3.2 \times 10^{\pm6} \text{ m}^2/\text{s})[1 + (80 \text{ W/m}^2 \cdot {}^\circ\text{C})(0.012 \text{ m})/(15 \text{ W/m} \cdot {}^\circ\text{C})]} = 10.6 \text{ s}$$

Therefore, any time step less than 10.6 s can be used to solve this problem. For convenience, let us choose the time step to be $\Delta t = 10$ s. Then the mesh Fourier number becomes

$$\tau = \frac{\alpha\Delta t}{l^2} = \frac{(3.2 \times 10^{-6} \text{ m}^2/\text{s})(10 \text{ s})}{(0.012 \text{ m})^2} = 0.222 \qquad (\text{for } \Delta t = 10 \text{ s})$$

Substituting this value of τ and other given quantities, the developed transient finite difference equations simplify to

$$T_1^{i+1} = 0.0836T_1^i + 0.444(T_2^i + T_4^i + 11.2)$$
$$T_2^{i+1} = 0.0836T_2^i + 0.222(T_1^i + T_3^i + 2T_5^i + 22.4)$$
$$T_3^{i+1} = 0.0552T_3^i + 0.444(T_2^i + T_6^i + 12.8)$$
$$T_4^{i+1} = 0.112T_4^i + 0.222(T_1^i + 2T_5^i + 109.2)$$
$$T_5^{i+1} = 0.112T_5^i + 0.222(T_2^i + T_4^i + T_6^i + 109.2)$$
$$T_6^{i+1} = 0.0931T_6^i + 0.074(2T_3^i + 4T_5^i + 2T_7^i + 424)$$

$$T_7^{i+1} = 0.0836T_7^j + 0.222(T_6^j + T_8^j + 202.4)$$
$$T_8^{i+1} = 0.0836T_8^j + 0.222(T_7^j + T_9^j + 202.4)$$
$$T_9^{i+1} = 0.0836T_9^j + 0.444(T_8^j + 105.2)$$

Using the specified initial condition as the solution at time $t = 0$ (for $i = 0$), sweeping through these nine equations gives the solution at intervals of 10 s. The solution at the upper corner node (node 3) is determined to be 100.2, 105.9, 106.5, 106.6, and 106.6°C at 1, 3, 5, 10, and 60 min, respectively. Note that the last three solutions are practically identical to the solution for the steady case obtained in Example 5–3. This indicates that steady conditions are reached in the medium after about 5 min.

Controlling the Numerical Error

A comparison of the numerical results with the exact results for temperature distribution in a cylinder would show that the results obtained by a numerical method are approximate, and they may or may not be sufficiently close to the exact (true) solution values. The difference between a numerical solution and the exact solution is the **error** involved in the numerical solution, and it is primarily due to two sources:

- The **discretization error** (also called the *truncation* or *formulation* error), which is caused by the approximations used in the formulation of the numerical method.

- The **round-off error**, which is caused by the computer's use of a limited number of significant digits and continuously rounding (or chopping) off the digits it cannot retain.

Below we discuss both types of errors.

Discretization Error

The discretization error involved in numerical methods is due to replacing the *derivatives* by *differences* in each step, or the actual temperature distribution between two adjacent nodes by a straight line segment.

Consider the variation of the solution of a transient heat transfer problem with time at a specified nodal point. Both the numerical and actual (exact) solutions coincide at the beginning of the first time step, as expected, but the numerical solution deviates from the exact solution as the time t increases. The difference between the two solutions at $t = \Delta t$ is due to the approximation at the first time step only and is called the *local discretization error*. One would expect the situation to get worse with each step since the second step uses the erroneous result of the first step as its starting point and adds a second local discretization error on top of it, as shown

*This section can be skipped without a loss in continuity.

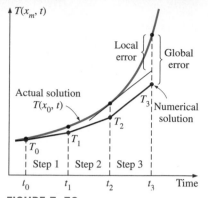

FIGURE 5–56

The local and global discretization errors of the finite difference method at the third time step at a specified nodal point.

in Fig. 5–56. The accumulation of the local discretization errors continues with the increasing number of time steps, and the total discretization error at any step is called the *global* or *accumulated discretization error*. Note that the local and global discretization errors are identical for the first time step. The global discretization error usually increases with the increasing number of steps, but the opposite may occur when the solution function changes direction frequently, giving rise to local discretization errors of opposite signs, which tend to cancel each other.

To have an idea about the magnitude of the local discretization error, consider the Taylor series expansion of the temperature at a specified nodal point m about time t_i,

$$T(x_m, t_i + \Delta t) = T(x_m, t_i) + \Delta t \frac{\partial T(x_m, t_i)}{\partial t} + \frac{1}{2} \Delta t^2 \frac{\partial^2 T(x_m, t_i)}{\partial t^2} + \cdots \quad \text{(5–62)}$$

The finite difference formulation of the time derivative at the same nodal point is expressed as

$$\frac{\partial T(x_m, t_i)}{\partial t} \cong \frac{T(x_m, t_i + \Delta t) - T(x_m, t_i)}{\Delta t} = \frac{T_m^{i+1} - T_m^i}{\Delta t} \quad \text{(5–63)}$$

or

$$T(x_m, t_i + \Delta t) \cong T(x_m, t_i) + \Delta t \frac{\partial T(x_m, t_i)}{\partial t} \quad \text{(5–64)}$$

which resembles the *Taylor series expansion* terminated after the first two terms. Therefore, the third and later terms in the Taylor series expansion represent the error involved in the finite difference approximation. For a sufficiently small time step, these terms decay rapidly as the order of derivative increases, and their contributions become smaller and smaller. The first term neglected in the Taylor series expansion is proportional to Δt^2, and thus the local discretization error of this approximation, which is the error involved in each step, is also proportional to Δt^2.

The local discretization error is the formulation error associated with a single step and gives an idea about the accuracy of the method used. However, the solution results obtained at every step except the first one involve the *accumulated error* up to that point, and the local error alone does not have much significance. What we really need to know is the global discretization error. At the worst case, the accumulated discretization error after I time steps during a time period t_0 is $i(\Delta t)^2 = (t_0/\Delta t)(\Delta t)^2 = t_0 \Delta t$, which is proportional to Δt. Thus, we conclude that the local discretization error is proportional to the square of the step size Δt^2 while the global discretization error is proportional to the step size Δt itself. Therefore, the smaller the mesh size (or the size of the time step in transient problems), the smaller the error, and thus the more accurate is the approximation. For example, halving the step size will reduce the global discretization error by half. It should be clear from the discussions above that the discretization error can be minimized by decreasing the step size in space or time as much as possible. The discretization error approaches zero as the difference quantities such as Δx and Δt approach the differential quantities such as dx and dt.

Round-Off Error

If we had a computer that could retain an infinite number of digits for all numbers, the difference between the exact solution and the approximate (numerical) solution at any point would entirely be due to discretization error. But we know that every computer (or calculator) represents numbers using a finite number of significant digits. The default value of the number of significant digits for many computers is 7, which is referred to as *single precision*. But the user may perform the calculations using 15 significant digits for the numbers, if he or she wishes, which is referred to as *double precision*. Of course, performing calculations in double precision will require more computer memory and a longer execution time.

In single precision mode with seven significant digits, a computer registers the number 44444.666666 as 44444.67 or 44444.66, depending on the method of rounding the computer uses. In the first case, the excess digits are said to be *rounded* to the closest integer, whereas in the second case they are said to be *chopped off*. Therefore, the numbers $a = 44444.12345$ and $b = 44444.12032$ are equivalent for a computer that performs calculations using seven significant digits. Such a computer would give $a - b = 0$ instead of the true value 0.00313.

The error due to retaining a limited number of digits during calculations is called the **round-off error**. This error is random in nature and there is no easy and systematic way of predicting it. It depends on the number of calculations, the method of rounding off, the type of computer, and even the sequence of calculations.

In algebra you learned that $a + b + c = a + c + b$, which seems quite reasonable. But this is not necessarily true for calculations performed with a computer, as demonstrated in Fig. 5–57. Note that changing the sequence of calculations results in an error of 30.8 percent in just two operations. Considering that any significant problem involves thousands or even millions of such operations performed in sequence, we realize that the accumulated round-off error has the potential to cause serious error without giving any warning signs. Experienced programmers are very much aware of this danger, and they structure their programs to prevent any buildup of the round-off error. For example, it is much safer to multiply a number by 10 than to add it 10 times. Also, it is much safer to start any addition process with the smallest numbers and continue with larger numbers. This rule is particularly important when evaluating series with a large number of terms with alternating signs.

The round-off error is proportional to the *number of computations* performed during the solution. In the finite difference method, the number of calculations increases as the mesh size or the time step size decreases. Halving the mesh or time step size, for example, doubles the number of calculations and thus the accumulated round-off error.

Controlling the Error in Numerical Methods

The total error in any result obtained by a numerical method is the sum of the *discretization error*, which decreases with decreasing step size, and the *round-off error*, which increases with decreasing step size, as shown in Fig. 5–58.

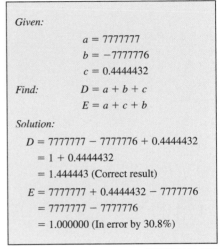

FIGURE 5–57
A simple arithmetic operation performed with a computer in single precision using seven significant digits, which results in 30.8 percent error when the order of operation is reversed.

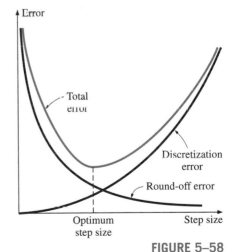

FIGURE 5–58
As the mesh or time step size decreases, the discretization error decreases but the round-off error increases.

Therefore, decreasing the step size too much in order to get more accurate results may actually backfire and give less accurate results because of a faster increase in the round-off error. We should be careful not to let round-off error get out of control by avoiding a large number of computations with very small numbers.

In practice, we do not know the exact solution of the problem, and thus we cannot determine the magnitude of the error involved in the numerical method. Knowing that the global discretization error is proportional to the step size is not much help either since there is no easy way of determining the value of the proportionality constant. Besides, the global discretization error alone is meaningless without a true estimate of the round-off error. Therefore, we recommend the following practical procedures to assess the accuracy of the results obtained by a numerical method.

- Start the calculations with a reasonable mesh size Δx (and time step size Δt for transient problems) based on experience. Then repeat the calculations using a mesh size of $\Delta x/2$. If the results obtained by halving the mesh size do not differ significantly from the results obtained with the full mesh size, we conclude that the discretization error is at an acceptable level. But if the difference is larger than we can accept, then we have to repeat the calculations using a mesh size $\Delta x/4$ or even a smaller one at regions of high temperature gradients. We continue in this manner until halving the mesh size does not cause any significant change in the results, which indicates that the discretization error is reduced to an acceptable level.

- Repeat the calculations using double precision holding the mesh size (and the size of the time step in transient problems) constant. If the changes are not significant, we conclude that the round-off error is not a problem. But if the changes are too large to accept, then we may try reducing the total number of calculations by increasing the mesh size or changing the order of computations. But if the increased mesh size gives unacceptable discretization errors, then we may have to find a reasonable compromise.

It should always be kept in mind that the results obtained by any numerical method may not reflect any trouble spots in certain problems that require special consideration such as hot spots or areas of high temperature gradients. The results that seem quite reasonable overall may be in considerable error at certain locations. This is another reason for always repeating the calculations at least twice with different mesh sizes before accepting them as the solution of the problem. Most commercial software packages have built-in routines that vary the mesh size as necessary to obtain highly accurate solutions. But it is a good engineering practice to be aware of any potential pitfalls of numerical methods and to examine the results obtained with a critical eye.

SUMMARY

Analytical solution methods are limited to highly simplified problems in simple geometries, and it is often necessary to use a numerical method to solve real world problems with complicated geometries or nonuniform thermal conditions. The numerical *finite difference method* is based on replacing derivatives by differences, and the finite difference formulation of a heat transfer problem is obtained by selecting a sufficient number of points in the region, called the *nodal points* or nodes, and writing *energy balances* on the volume elements centered about the nodes.

For *steady* heat transfer, the *energy balance* on a volume element can be expressed in general as

$$\sum_{\text{All sides}} \dot{Q} + \dot{e} V_{\text{element}} = 0$$

whether the problem is one-, two-, or three-dimensional. For convenience in formulation, we always assume all heat transfer to be *into* the volume element from all surfaces toward the node under consideration, except for specified heat flux whose direction is already specified. The finite difference formulations for a general interior node under *steady* conditions are expressed for some geometries as follows:

One-dimensional steady conduction in a plane wall:

$$\frac{T_{m-1} - 2T_m + T_{m+1}}{(\Delta x)^2} + \frac{\dot{e}_m}{k} = 0$$

Two-dimensional steady conduction in rectangular coordinates:

$$T_{\text{left}} + T_{\text{top}} + T_{\text{right}} + T_{\text{bottom}} - 4T_{\text{node}} + \frac{\dot{e}_{\text{node}}l^2}{k} = 0$$

where Δx is the nodal spacing for the plane wall and $\Delta x = \Delta y = l$ is the nodal spacing for the two-dimensional case. Insulated boundaries can be viewed as mirrors in formulation, and thus the nodes on insulated boundaries can be treated as interior nodes by using mirror images.

The finite difference formulation at node 0 at the left boundary of a plane wall for steady one-dimensional heat conduction can be expressed as

$$\dot{Q}_{\text{left surface}} + kA \frac{T_1 - T_0}{\Delta x} + \dot{e}_0 (A \Delta x / 2) = 0$$

where $A \Delta x / 2$ is the volume of the volume, $\dot{e}_0$ is the rate of heat generation per unit volume at $x = 0$, and A is the heat transfer area. The form of the first term depends on the boundary condition at $x = 0$ (convection, radiation, specified heat flux, etc.).

The finite difference formulation of heat conduction problems usually results in a system of N algebraic equations in N unknown nodal temperatures that need to be solved simultaneously.

The finite difference formulation of *transient* heat conduction problems is based on an energy balance that also accounts for the variation of the energy content of the volume element during a time interval Δt. The heat transfer and heat generation terms are expressed at the previous time step i in the *explicit method*, and at the new time step $i + 1$ in the *implicit method*. For a general node m, the finite difference formulations are expressed as

Explicit method:

$$\sum_{\text{All sides}} \dot{Q}^i + \dot{e}_m^i V_{\text{element}} = \rho V_{\text{element}} c_p \frac{T_m^{i+1} - T_m^i}{\Delta t}$$

Implicit method:

$$\sum_{\text{All sides}} \dot{Q}^{i+1} + \dot{e}_m^{i+1} V_{\text{element}} = \rho V_{\text{element}} c_p \frac{T_m^{i+1} - T_m^i}{\Delta t}$$

where T_m^i and T_m^{i+1} are the temperatures of node m at times $t_i = i\Delta t$ and $t_{i+1} = (i + 1)\Delta t$, respectively, and $T_m^{i+1} - T_m^i$ represents the temperature change of the node during the time interval Δt between the time steps i and $i + 1$. The explicit and implicit formulations given here are quite general and can be used in any coordinate system regardless of heat transfer being one-, two-, or three-dimensional.

The explicit formulation of a general interior node for one- and two-dimensional heat transfer in rectangular coordinates can be expressed as

One-dimensional case:

$$T_m^{i+1} = \tau(T_{m-1}^i + T_{m+1}^i) + (1 - 2\tau) T_m^i + \tau \frac{\dot{e}_m^i \Delta x^2}{k}$$

Two-dimensional case:

$$T_{\text{node}}^{i+1} = \tau(T_{\text{left}}^i + T_{\text{top}}^i + T_{\text{right}}^i + T_{\text{bottom}}^i)$$
$$+ (1 - 4\tau) T_{\text{node}}^i + \tau \frac{\dot{e}_{\text{node}}^i l^2}{k}$$

where $\tau = \alpha \Delta t / \Delta x^2$ is the dimensionless *mesh Fourier number* and $\alpha = k/\rho c_p$ is the *thermal diffusivity* of the medium.

The implicit method is inherently stable, and any value of Δt can be used with that method as the time step. The largest value of the time step Δt in the explicit method is limited by the *stability criterion*, expressed as: *the coefficients of all T_m^i in the*

T_m^{i+1} expressions (called the primary coefficients) must be greater than or equal to zero for all nodes m. The maximum value of Δt is determined by applying the stability criterion to the equation with the smallest primary coefficient since it is the most restrictive. For problems with specified temperatures or heat fluxes at all the boundaries, the stability criterion can be expressed as $\tau \leq \frac{1}{2}$ for one-dimensional problems and $\tau \leq \frac{1}{4}$ for the two-dimensional problems in rectangular coordinates.

REFERENCES AND SUGGESTED READING

1. D. A. Anderson, J. C. Tannehill, and R. H. Pletcher. *Computational Fluid Mechanics and Heat Transfer.* New York: Hemisphere, 1984.

2. C. A. Brebbia. *The Boundary Element Method for Engineers.* New York: Halsted Press, 1978.

3. G. E. Forsythe and W. R. Wasow. *Finite Difference Methods for Partial Differential Equations.* New York: John Wiley & Sons, 1960.

4. B. Gebhart. *Heat Conduction and Mass Diffusion.* New York: McGraw-Hill, 1993.

5. K. H. Huebner and E. A. Thornton. *The Finite Element Method for Engineers.* 2nd ed. New York: John Wiley & Sons, 1982.

6. Y. Jaluria and K. E. Torrance. *Computational Heat Transfer.* New York: Hemisphere, 1986.

7. W. J. Minkowycz, E. M. Sparrow, G. E. Schneider, and R. H. Pletcher. *Handbook of Numerical Heat Transfer.* New York: John Wiley & Sons, 1988.

8. G. E. Myers. *Analytical Methods in Conduction Heat Transfer.* New York: McGraw-Hill, 1971.

9. D. H. Norrie and G. DeVries. *An Introduction to Finite Element Analysis.* New York: Academic Press, 1978.

10. M. N. Özişik. *Finite Difference Methods in Heat Transfer.* Boca Raton, FL: CRC Press, 1994.

11. S. V. Patankhar. *Numerical Heat Transfer and Fluid Flow.* New York: Hemisphere, 1980.

12. T. M. Shih. *Numerical Heat Transfer.* New York: Hemisphere, 1984.

PROBLEMS*

Why Numerical Methods?

5–1C What are the limitations of the analytical solution methods?

5–2C How do numerical solution methods differ from analytical ones? What are the advantages and disadvantages of numerical and analytical methods?

5–3C What is the basis of the energy balance method? How does it differ from the formal finite difference method? For a specified nodal network, will these two methods result in the same or a different set of equations?

5–4C Consider a heat conduction problem that can be solved both analytically, by solving the governing differential equation and applying the boundary conditions, and numerically, by a software package available on your computer. Which approach would you use to solve this problem? Explain your reasoning.

5–5C Two engineers are to solve an actual heat transfer problem in a manufacturing facility. Engineer A makes the necessary simplifying assumptions and solves the problem analytically, while engineer B solves it numerically using a powerful software package. Engineer A claims he solved the problem exactly and thus his results are better, while engineer B claims that he used a more realistic model and thus his results are better. To resolve the dispute, you are asked to solve the problem experimentally in a lab. Which engineer do you think the experiments will prove right? Explain.

Finite Difference Formulation of Differential Equations

5–6C Define these terms used in the finite difference formulation: node, nodal network, volume element, nodal spacing, and difference equation.

5–7 Consider three consecutive nodes $n - 1$, n, and $n + 1$ in a plane wall. Using the finite difference form of the first derivative at the midpoints, show that the finite difference form of the second derivative can be expressed as

$$\frac{T_{n-1} - 2T_n + T_{n+1}}{\Delta x^2} = 0$$

---

*Problems designated by a "C" are concept questions, and students are encouraged to answer them all. Problems designated by an "E" are in English units, and the SI users can ignore them. Problems with the icon ⊛ are solved using EES, and complete solutions together with parametric studies are included on the enclosed CD. Problems with the icon ▨ are comprehensive in nature, and are intended to be solved with a computer, preferably using the EES software that accompanies this text.

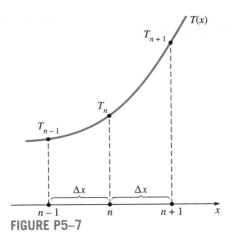

FIGURE P5–7

5–8 The finite difference formulation of steady two-dimensional heat conduction in a medium with heat generation and constant thermal conductivity is given by

$$\frac{T_{m-1, n} - 2T_{m, n} + T_{m+1, n}}{\Delta x^2} + \frac{T_{m, n-1} - 2T_{m, n} + T_{m, n+1}}{\Delta y^2}$$
$$+ \frac{\dot{e}_{m, n}}{k} = 0$$

in rectangular coordinates. Modify this relation for the three-dimensional case.

5–9 Consider steady one-dimensional heat conduction in a plane wall with variable heat generation and constant thermal conductivity. The nodal network of the medium consists of nodes 0, 1, 2, 3, and 4 with a uniform nodal spacing of Δx. Using the finite difference form of the first derivative (*not* the energy balance approach), obtain the finite difference formulation of the boundary nodes for the case of uniform heat flux $\dot{q}_0$ at the left boundary (node 0) and convection at the right boundary (node 4) with a convection coefficient of h and an ambient temperature of T_∞.

5–10 Consider steady one-dimensional heat conduction in a plane wall with variable heat generation and constant thermal conductivity. The nodal network of the medium consists of nodes 0, 1, 2, 3, 4, and 5 with a uniform nodal spacing of Δx.

Using the finite difference form of the first derivative (*not* the energy balance approach), obtain the finite difference formulation of the boundary nodes for the case of insulation at the left boundary (node 0) and radiation at the right boundary (node 5) with an emissivity of ε and surrounding temperature of T_{surr}.

One-Dimensional Steady Heat Conduction

5–11C Explain how the finite difference form of a heat conduction problem is obtained by the energy balance method.

5–12C In the energy balance formulation of the finite difference method, it is recommended that all heat transfer at the boundaries of the volume element be assumed to be into the volume element even for steady heat conduction. Is this a valid recommendation even though it seems to violate the conservation of energy principle?

5–13C How is an insulated boundary handled in the finite difference formulation of a problem? How does a symmetry line differ from an insulated boundary in the finite difference formulation?

5–14C How can a node on an insulated boundary be treated as an interior node in the finite difference formulation of a plane wall? Explain.

5–15C Consider a medium in which the finite difference formulation of a general interior node is given in its simplest form as

$$\frac{T_{m-1} - 2T_m + T_{m+1}}{\Delta x^2} + \frac{\dot{e}_m}{k} = 0$$

(*a*) Is heat transfer in this medium steady or transient?
(*b*) Is heat transfer one-, two-, or three-dimensional?
(*c*) Is there heat generation in the medium?
(*d*) Is the nodal spacing constant or variable?
(*e*) Is the thermal conductivity of the medium constant or variable?

5–16 Consider steady heat conduction in a plane wall whose left surface (node 0) is maintained at 30°C while the right surface (node 8) is subjected to a heat flux of 1200 W/m². Express the finite difference formulation of the boundary nodes 0 and 8 for the case of no heat generation. Also obtain the finite dif-

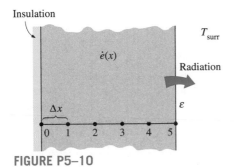

FIGURE P5–10

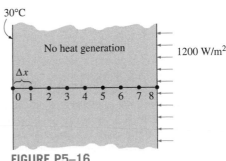

FIGURE P5–16

ference formulation for the rate of heat transfer at the left boundary.

5–17 Consider steady heat conduction in a plane wall with variable heat generation and constant thermal conductivity. The nodal network of the medium consists of nodes 0, 1, 2, 3, and 4 with a uniform nodal spacing of Δx. Using the energy balance approach, obtain the finite difference formulation of the boundary nodes for the case of uniform heat flux $\dot{q}_0$ at the left boundary (node 0) and convection at the right boundary (node 4) with a convection coefficient of h and an ambient temperature of T_∞.

5–18 Consider steady one-dimensional heat conduction in a plane wall with variable heat generation and constant thermal conductivity. The nodal network of the medium consists of nodes 0, 1, 2, 3, 4, and 5 with a uniform nodal spacing of Δx. Using the energy balance approach, obtain the finite difference formulation of the boundary nodes for the case of insulation at the left boundary (node 0) and radiation at the right boundary (node 5) with an emissivity of ε and surrounding temperature of T_{surr}.

5–19 Consider steady one-dimensional heat conduction in a plane wall with variable heat generation and constant thermal conductivity. The nodal network of the medium consists of nodes 0, 1, 2, 3, 4, and 5 with a uniform nodal spacing of Δx. The temperature at the right boundary (node 5) is specified. Using the energy balance approach, obtain the finite difference formulation of the boundary node 0 on the left boundary for the case of combined convection, radiation, and heat flux at the left boundary with an emissivity of ε, convection coefficient of h, ambient temperature of T_∞, surrounding temperature of T_{surr}, and uniform heat flux of $\dot{q}_0$. Also, obtain the finite difference formulation for the rate of heat transfer at the right boundary.

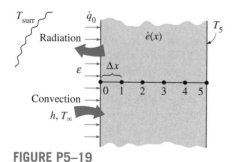

FIGURE P5–19

5–20 Consider steady one-dimensional heat conduction in a composite plane wall consisting of two layers A and B in perfect contact at the interface. The wall involves no heat generation. The nodal network of the medium consists of nodes 0, 1 (at the interface), and 2 with a uniform nodal spacing of Δx. Using the energy balance approach, obtain the finite difference formulation of this problem for the case of insulation at the left boundary (node 0) and radiation at the right boundary (node 2) with an emissivity of ε and surrounding temperature of T_{surr}.

5–21 Consider steady one-dimensional heat conduction in a plane wall with variable heat generation and variable thermal conductivity. The nodal network of the medium consists of nodes 0, 1, and 2 with a uniform nodal spacing of Δx. Using the energy balance approach, obtain the finite difference formulation of this problem for the case of specified heat flux $\dot{q}_0$ to the wall and convection at the left boundary (node 0) with a convection coefficient of h and ambient temperature of T_∞, and radiation at the right boundary (node 2) with an emissivity of ε and surrounding surface temperature of T_{surr}.

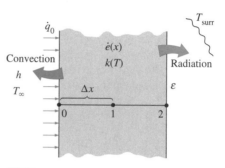

FIGURE P5–21

5–22 Consider steady one-dimensional heat conduction in a pin fin of constant diameter D with constant thermal conductivity. The fin is losing heat by convection to the ambient air at T_∞ with a heat transfer coefficient of h. The nodal network of the fin consists of nodes 0 (at the base), 1 (in the middle), and 2 (at the fin tip) with a uniform nodal spacing of Δx. Using the energy balance approach, obtain the finite difference formulation of this problem to determine T_1 and T_2 for the case of specified temperature at the fin base and negligible heat transfer at the fin tip. All temperatures are in °C.

5–23 Consider steady one-dimensional heat conduction in a pin fin of constant diameter D with constant thermal conductivity. The fin is losing heat by convection to the ambient air at T_∞ with a convection coefficient of h, and by radiation to the surrounding surfaces at an average temperature of T_{surr}.

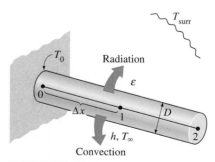

FIGURE P5–23

The nodal network of the fin consists of nodes 0 (at the base), 1 (in the middle), and 2 (at the fin tip) with a uniform nodal spacing of Δx. Using the energy balance approach, obtain the finite difference formulation of this problem to determine T_1 and T_2 for the case of specified temperature at the fin base and negligible heat transfer at the fin tip. All temperatures are in °C.

5–24   Consider a large uranium plate of thickness 5 cm and thermal conductivity $k = 28$ W/m · °C in which heat is generated uniformly at a constant rate of $\dot{e} = 6 \times 10^5$ W/m³. One side of the plate is insulated while the other side is subjected to convection to an environment at 30°C with a heat transfer coefficient of $h = 60$ W/m² · °C. Considering six equally spaced nodes with a nodal spacing of 1 cm, (a) obtain the finite difference formulation of this problem and (b) determine the nodal temperatures under steady conditions by solving those equations.

5–25   Consider an aluminum alloy fin ($k = 180$ W/m · °C) of triangular cross section whose length is $L = 5$ cm, base thickness is $b = 1$ cm, and width w in the direction normal to the plane of paper is very large. The base of the fin is maintained at a temperature of $T_0 = 180$°C. The fin is losing heat by convection to the ambient air at $T_\infty = 25$°C with a heat transfer coefficient of $h = 25$ W/m² · °C and by radiation to the surrounding surfaces at an average temperature of $T_{surr} = 290$ K. Using the finite difference method with six equally spaced nodes along the fin in the x-direction, determine (a) the temperatures at the nodes and (b) the rate of heat transfer from the fin for $w = 1$ m. Take the emissivity of the fin surface to be 0.9 and assume steady one-dimensional heat transfer in the fin.

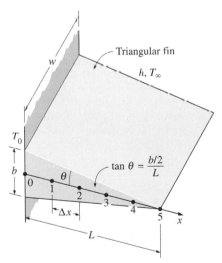

FIGURE P5–25

5–26   Reconsider Prob. 5–25. Using EES (or other) software, investigate the effect of the fin base temperature on the fin tip temperature and the rate of heat transfer from the fin. Let the temperature at the fin base vary from 100°C to 200°C. Plot the fin tip temperature and the rate of heat transfer as a function of the fin base temperature, and discuss the results.

5–27   Consider a large plane wall of thickness $L = 0.4$ m, thermal conductivity $k = 2.3$ W/m · °C, and surface area $A = 20$ m². The left side of the wall is maintained at a constant temperature of 95°C, while the right side loses heat by convection to the surrounding air at $T_\infty = 15$°C with a heat transfer coefficient of $h = 18$ W/m² · °C. Assuming steady one-dimensional heat transfer and taking the nodal spacing to be 10 cm, (a) obtain the finite difference formulation for all nodes, (b) determine the nodal temperatures by solving those equations, and (c) evaluate the rate of heat transfer through the wall.

5–28   Consider the base plate of a 800-W household iron having a thickness of $L = 0.6$ cm, base area of $A = 160$ cm², and thermal conductivity of $k = 20$ W/m · °C. The inner surface of the base plate is subjected to uniform heat flux generated by the resistance heaters inside. When steady operating conditions are reached, the outer surface temperature of the plate is measured to be 85°C. Disregarding any heat loss through the upper part of the iron and taking the nodal spacing to be 0.2 cm, (a) obtain the finite difference formulation for the nodes and (b) determine the inner surface temperature of the plate by solving those equations.   *Answer: (b) 100°C*

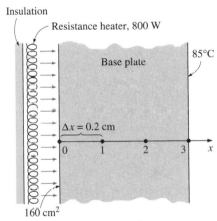

FIGURE P5–28

5–29   Consider a large plane wall of thickness $L = 0.3$ m, thermal conductivity $k = 2.5$ W/m · °C, and surface area $A = 24$ m². The left side of the wall is subjected to a heat flux of $\dot{q}_0 = 350$ W/m² while the temperature at that surface is measured to be $T_0 = 60$°C. Assuming steady one-dimensional heat transfer and taking the nodal spacing to be 6 cm, (a) obtain the finite difference formulation for the six nodes and (b) determine the temperature of the other surface of the wall by solving those equations.

5–30E   A large steel plate having a thickness of $L = 5$ in, thermal conductivity of $k = 7.2$ Btu/h · ft · °F, and an emissivity of $\varepsilon = 0.6$ is lying on the ground. The exposed surface of

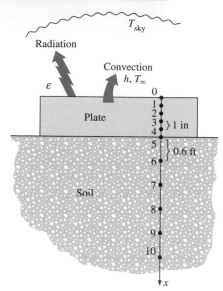

FIGURE P5–30E

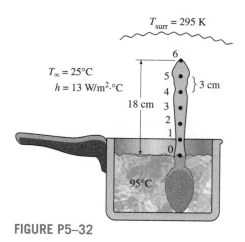

FIGURE P5–32

the plate exchanges heat by convection with the ambient air at $T_\infty = 80°F$ with an average heat transfer coefficient of $h = 3.5 \, \text{Btu/h} \cdot \text{ft}^2 \cdot °F$ as well as by radiation with the open sky at an equivalent sky temperature of $T_{\text{sky}} = 510 \, \text{R}$. The ground temperature below a certain depth (say, 3 ft) is not affected by the weather conditions outside and remains fairly constant at $50°F$ at that location. The thermal conductivity of the soil can be taken to be $k_{\text{soil}} = 0.49 \, \text{Btu/h} \cdot \text{ft} \cdot °F$, and the steel plate can be assumed to be in perfect contact with the ground. Assuming steady one-dimensional heat transfer and taking the nodal spacings to be 1 in in the plate and 0.6 ft in the ground, (a) obtain the finite difference formulation for all 11 nodes shown in Figure P5–30E and (b) determine the top and bottom surface temperatures of the plate by solving those equations.

5–31E Repeat Prob. 5–30E by disregarding radiation heat transfer from the upper surface.   *Answers: (b)* 78.7°F, 78.4°F

5–32 Consider a stainless steel spoon ($k = 15.1 \, \text{W/m} \cdot °C$, $\varepsilon = 0.6$) that is partially immersed in boiling water at 95°C in a kitchen at 25°C. The handle of the spoon has a cross section of about 0.2 cm $\times$ 1 cm and extends 18 cm in the air from the free surface of the water. The spoon loses heat by convection to the ambient air with an average heat transfer coefficient of $h = 13 \, \text{W/m}^2 \cdot °C$ as well as by radiation to the surrounding surfaces at an average temperature of $T_{\text{surr}} = 295 \, \text{K}$. Assuming steady one-dimensional heat transfer along the spoon and taking the nodal spacing to be 3 cm, (a) obtain the finite difference formulation for all nodes, (b) determine the temperature of the tip of the spoon by solving those equations, and (c) determine the rate of heat transfer from the exposed surfaces of the spoon.

5–33 Repeat Prob. 5–32 using a nodal spacing of 1.5 cm.

5–34 Reconsider Prob. 5–33. Using EES (or other) software, investigate the effects of the thermal conductivity and the emissivity of the spoon material on the temperature at the spoon tip and the rate of heat transfer from the exposed surfaces of the spoon. Let the thermal conductivity vary from 10 W/m · °C to 400 W/m · °C, and the emissivity from 0.1 to 1.0. Plot the spoon tip temperature and the heat transfer rate as functions of thermal conductivity and emissivity, and discuss the results.

5–35 One side of a 2-m-high and 3-m-wide vertical plate at 80°C is to be cooled by attaching aluminum fins ($k = 237 \, \text{W/m} \cdot °C$) of rectangular profile in an environment at 35°C. The fins are 2 cm long, 0.3 cm thick, and 0.4 cm apart. The heat transfer coefficient between the fins and the surrounding air for combined convection and radiation is estimated to be 30 W/m² · °C. Assuming steady one-dimensional heat transfer along the fin and taking the nodal spacing to be 0.5 cm, determine (a) the finite difference formulation of this problem, (b) the nodal temperatures along the fin by solving these equations, (c) the rate of heat transfer from a single fin,

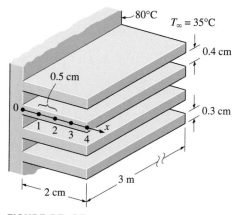

FIGURE P5–35

and (d) the rate of heat transfer from the entire finned surface of the plate.

5–36 A hot surface at 100°C is to be cooled by attaching 3-cm-long, 0.25-cm-diameter aluminum pin fins ($k = 237$ W/m · °C) with a center-to-center distance of 0.6 cm. The temperature of the surrounding medium is 30°C, and the combined heat transfer coefficient on the surfaces is 35 W/m² · °C. Assuming steady one-dimensional heat transfer along the fin and taking the nodal spacing to be 0.5 cm, determine (a) the finite difference formulation of this problem, (b) the nodal temperatures along the fin by solving these equations, (c) the rate of heat transfer from a single fin, and (d) the rate of heat transfer from a 1-m × 1-m section of the plate.

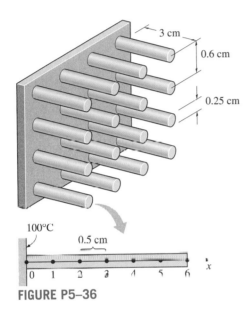

FIGURE P5–36

5–37 Repeat Prob. 5–36 using copper fins ($k - 386$ W/m · °C) instead of aluminum ones.

Answers: (b) 98.6°C, 97.5°C, 96.7°C, 96.0°C, 95.7°C, 95.5°C

5–38 Two 3-m-long and 0.4-cm-thick cast iron ($k = 52$ W/m · °C, $\varepsilon = 0.8$) steam pipes of outer diameter 10 cm are connected to each other through two 1-cm-thick flanges of outer diameter 20 cm, as shown in the figure. The steam flows inside the pipe at an average temperature of 200°C with a heat transfer coefficient of 180 W/m² · °C. The outer surface of the pipe is exposed to convection with ambient air at 8°C with a heat transfer coefficient of 25 W/m² · °C as well as radiation with the surrounding surfaces at an average temperature of $T_{surr} = 290$ K. Assuming steady one-dimensional heat conduction along the flanges and taking the nodal spacing to be 1 cm along the flange (a) obtain the finite difference formulation for all nodes, (b) determine the temperature at the tip of the flange by solving those equations, and (c) determine the rate of heat transfer from the exposed surfaces of the flange.

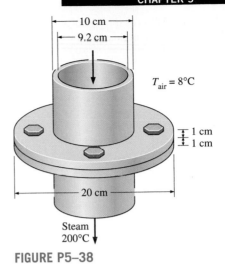

FIGURE P5–38

5–39 Reconsider Prob. 5–38. Using EES (or other) software, investigate the effects of the steam temperature and the outer heat transfer coefficient on the flange tip temperature and the rate of heat transfer from the exposed surfaces of the flange. Let the steam temperature vary from 150°C to 300°C and the heat transfer coefficient from 15 W/m² · °C to 60 W/m² · °C. Plot the flange tip temperature and the heat transfer rate as functions of steam temperature and heat transfer coefficient, and discuss the results.

5–40 Using EES (or other) software, solve these systems of algebraic equations.

(a) $3x_1 - x_2 + 3x_3 = 0$
$-x_1 + 2x_2 + x_3 = 3$
$2x_1 - x_2 - x_3 = 2$

(b) $4x_1 - 2x_2^2 + 0.5x_3 - -2$
$x_1^3 - x_2 + x_3 = 11.964$
$x_1 + x_2 + x_3 = 3$

Answers: (a) $x_1 = 2$, $x_2 = 3$, $x_3 = -1$, (b) $x_1 = 2.33$, $x_2 = 2.29$, $x_3 = -1.62$

5–41 Using EES (or other) software, solve these systems of algebraic equations.

(a) $3x_1 + 2x_2 - x_3 + x_4 = 6$
$x_1 + 2x_2 - x_4 = -3$
$-2x_1 + x_2 + 3x_3 + x_4 = 2$
$3x_2 + x_3 - 4x_4 = -6$

(b) $3x_1 + x_2^2 + 2x_3 = 8$
$-x_1^2 + 3x_2 + 2x_3 = -6.293$
$2x_1 - x_2^4 + 4x_3 = -12$

5–42 Using EES (or other) software, solve these systems of algebraic equations.

(a)
$$4x_1 - x_2 + 2x_3 + x_4 = -6$$
$$x_1 + 3x_2 - x_3 + 4x_4 = -1$$
$$-x_1 + 2x_2 + 5x_4 = 5$$
$$2x_2 - 4x_3 - 3x_4 = -5$$

(b)
$$2x_1 + x_2^4 - 2x_3 + x_4 = 1$$
$$x_1^2 + 4x_2 + 2x_3^2 - 2x_4 = -3$$
$$-x_1 + x_2^4 + 5x_3 = 10$$
$$3x_1 - x_3^2 + 8x_4 = 15$$

Two-Dimensional Steady Heat Conduction

5–43C Consider a medium in which the finite difference formulation of a general interior node is given in its simplest form as

$$T_{\text{left}} + T_{\text{top}} + T_{\text{right}} + T_{\text{bottom}} - 4T_{\text{node}} + \frac{\dot{e}_{\text{node}} l^2}{k} = 0$$

(a) Is heat transfer in this medium steady or transient?
(b) Is heat transfer one-, two-, or three-dimensional?
(c) Is there heat generation in the medium?
(d) Is the nodal spacing constant or variable?
(e) Is the thermal conductivity of the medium constant or variable?

5–44C Consider a medium in which the finite difference formulation of a general interior node is given in its simplest form as

$$T_{\text{node}} = (T_{\text{left}} + T_{\text{top}} + T_{\text{right}} + T_{\text{bottom}})/4$$

(a) Is heat transfer in this medium steady or transient?
(b) Is heat transfer one-, two-, or three-dimensional?
(c) Is there heat generation in the medium?
(d) Is the nodal spacing constant or variable?
(e) Is the thermal conductivity of the medium constant or variable?

5–45C What is an irregular boundary? What is a practical way of handling irregular boundary surfaces with the finite difference method?

5–46 The wall of a heat exchanger separates hot water at $T_A = 90°C$ from cold water at $T_B = 10°C$. To extend the heat transfer area, two-dimensional ridges are machined on the cold side of the wall, as shown in Fig. P5–46. This geometry causes non-uniform thermal stresses, which may become critical for crack initiation along the lines between two ridges. To predict thermal stresses, the temperature field inside the wall must be determined. Convection coefficients are high enough so that the surface temperature is equal to that of the water on each side of the wall.

(a) Identify the smallest section of the wall that can be analyzed in order to find the temperature field in the whole wall.
(b) For the domain found in part (a), construct a two-dimensional grid with $\Delta x = \Delta y = 5$ mm and write the matrix equation $AT = C$ (elements of matrices A and C must be numbers). Do not solve for T.
(c) A thermocouple mounted at point M reads 46.9°C. Determine the other unknown temperatures in the grid defined in part (b).

5–47 A long tube has a square cross section as shown in Fig. P5–47, with insulated sides, and the top and bottom surfaces maintained at T_A, and inner surface maintained at T_B. The thermal conductivity of the tube is k and heat generation occurs within the material at a rate of $\dot{e}$.

(a) Write the matrix equation $AT = C$ used to determine the steady temperature field T, for the discretization grid shown on the figure. Simplify the equation for $T_A = 20°C$, $T_B = 100°C$, $k = 10$ W/m · K, $L = 4$ cm, and $\dot{e} = 5 \times 10^5$ W/m³.
(b) The solution to the equation in part (a) is included in the table below. Determine the rate of heat loss by the tube through its outer surface per unit length.

Grid point	$T(°C)$
1	20
2	20
3	20
4	71.4
5	92.9
6	100
7	105.7
8	100

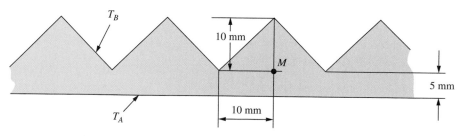

FIGURE P5–46

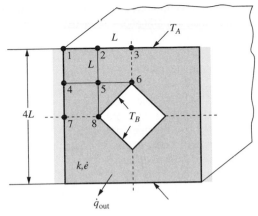

FIGURE P5–47

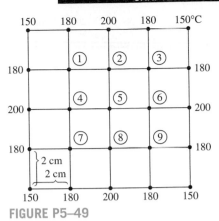

FIGURE P5–49

5–48 Consider steady two-dimensional heat transfer in a long solid body whose cross section is given in the figure. The temperatures at the selected nodes and the thermal conditions at the boundaries are as shown. The thermal conductivity of the body is $k = 45$ W/m · °C, and heat is generated in the body uniformly at a rate of $\dot{e} = 4 \times 10^6$ W/m³. Using the finite difference method with a mesh size of $\Delta x = \Delta y = 5.0$ cm, determine (a) the temperatures at nodes 1, 2, and 3 and (b) the rate of heat loss from the bottom surface through a 1-m-long section of the body.

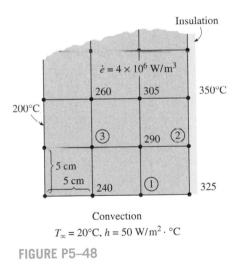

Convection
$T_\infty = 20°C, h = 50$ W/m² · °C

FIGURE P5–48

5–49 Consider steady two-dimensional heat transfer in a long solid body whose cross section is given in the figure. The measured temperatures at selected points of the outer surfaces are as shown. The thermal conductivity of the body is $k = 45$ W/m · °C, and there is no heat generation. Using the finite difference method with a mesh size of $\Delta x = \Delta y = 2.0$ cm, determine the temperatures at the indicated points in the medium. *Hint:* Take advantage of symmetry.

5–50 Consider steady two-dimensional heat transfer in a long solid bar of (a) square and (b) rectangular cross sections as shown in the figure. The measured temperatures at selected points of the outer surfaces are as shown. The thermal conductivity of the body is $k = 20$ W/m · °C, and there is no heat generation. Using the finite difference method with a mesh size of $\Delta x = \Delta y = 1.0$ cm, determine the temperatures at the indicated points in the medium.

Answers: (a) $T_1 = 185°C, T_2 = T_3 = T_4 = 190°C$

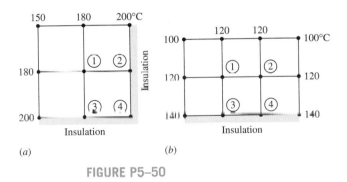

FIGURE P5–50

5–51 Starting with an energy balance on a volume element, obtain the steady two-dimensional finite difference equation for a general interior node in rectangular coordinates for $T(x, y)$ for the case of variable thermal conductivity and uniform heat generation.

5–52 Consider steady two-dimensional heat transfer in a long solid body whose cross section is given in Fig. P5–52 (on the next page). The temperatures at the selected nodes and the thermal conditions on the boundaries are as shown. The thermal conductivity of the body is $k = 180$ W/m · °C, and heat is generated in the body uniformly at a rate of $\dot{e} = 10^7$ W/m³. Using the finite difference method with a mesh size of $\Delta x = \Delta y = 10$ cm, determine (a) the temperatures at nodes 1, 2, 3, and 4 and (b) the rate of heat loss from the top surface through a 1-m-long section of the body.

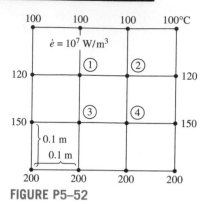

FIGURE P5–52

5–53 [EES] Reconsider Prob. 5–52. Using EES (or other) software, investigate the effects of the thermal conductivity and the heat generation rate on the temperatures at nodes 1 and 3, and the rate of heat loss from the top surface. Let the thermal conductivity vary from 10 W/m · °C to 400 W/m · °C and the heat generation rate from 10^5 W/m³ to 10^8 W/m³. Plot the temperatures at nodes 1 and 3, and the rate of heat loss as functions of thermal conductivity and heat generation rate, and discuss the results.

5–54 Consider steady two-dimensional heat transfer in two long solid bars whose cross sections are given in the figure. The measured temperatures at selected points on the outer surfaces are as shown. The thermal conductivity of the body is $k = 20$ W/m · °C, and there is no heat generation. Using the finite difference method with a mesh size of $\Delta x = \Delta y = 1.0$ cm,

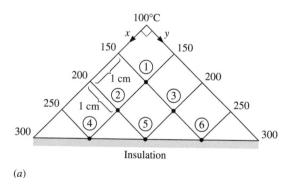

(a)

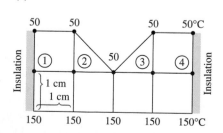

(b)

FIGURE P5–54

determine the temperatures at the indicated points in the medium. *Hint:* Take advantage of symmetry.

Answers: (b) $T_1 = T_4 = 93$°C, $T_2 = T_3 = 86$°C

5–55 Consider steady two-dimensional heat transfer in an L-shaped solid body whose cross section is given in the figure. The thermal conductivity of the body is $k = 45$ W/m · °C, and heat is generated in the body at a rate of $\dot{e} = 5 \times 10^6$ W/m³. The right surface of the body is insulated, and the bottom surface is maintained at a uniform temperature of 120°C. The entire top surface is subjected to convection with ambient air at $T_\infty = 30$°C with a heat transfer coefficient of $h = 55$ W/m² · °C, and the left surface is subjected to heat flux at a uniform rate of $\dot{q}_L = 8000$ W/m². The nodal network of the problem consists of 13 equally spaced nodes with $\Delta x = \Delta y = 1.5$ cm. Five of the nodes are at the bottom surface and thus their temperatures are known. (a) Obtain the finite difference equations at the remaining eight nodes and (b) determine the nodal temperatures by solving those equations.

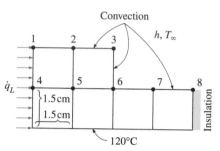

FIGURE P5–55

5–56E Consider steady two-dimensional heat transfer in a long solid bar of square cross section in which heat is generated uniformly at a rate of $\dot{e} = 0.19 \times 10^5$ Btu/h · ft³. The cross section of the bar is 0.5 ft × 0.5 ft in size, and its thermal conductivity is $k = 16$ Btu/h · ft · °F. All four sides of the bar are subjected to convection with the ambient air at $T_\infty = 70$°F with a heat transfer coefficient of $h = 7.9$ Btu/h · ft² · °F. Using the finite difference method with a mesh size of $\Delta x = \Delta y = 0.25$ ft, determine (a) the temperatures at the nine nodes and (b) the rate of heat loss from the bar through a 1-ft-long section.

Answer: (b) 4750 Btu/h

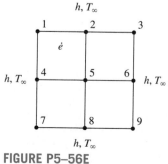

FIGURE P5–56E

5–57 Hot combustion gases of a furnace are flowing through a concrete chimney ($k = 1.4$ W/m · °C) of rectangular cross section. The flow section of the chimney is 20 cm × 40 cm, and the thickness of the wall is 10 cm. The average temperature of the hot gases in the chimney is $T_i = 280$°C, and the average convection heat transfer coefficient inside the chimney is $h_i = 75$ W/m² · °C. The chimney is losing heat from its outer surface to the ambient air at $T_o = 15$°C by convection with a heat transfer coefficient of $h_o = 18$ W/m² · °C and to the sky by radiation. The emissivity of the outer surface of the wall is $\varepsilon = 0.9$, and the effective sky temperature is estimated to be 250 K. Using the finite difference method with $\Delta x = \Delta y = 10$ cm and taking full advantage of symmetry, (a) obtain the finite difference formulation of this problem for steady two-dimensional heat transfer, (b) determine the temperatures at the nodal points of a cross section, and (c) evaluate the rate of heat loss for a 1-m-long section of the chimney.

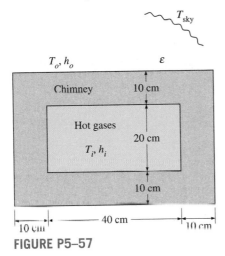

FIGURE P5–57

5–58 Repeat Prob. 5–57 by disregarding radiation heat transfer from the outer surfaces of the chimney.

5–59 Reconsider Prob. 5–57. Using EES (or other) software, investigate the effects of hot-gas temperature and the outer surface emissivity on the temperatures at the outer corner of the wall and the middle of the inner surface of the right wall, and the rate of heat loss. Let the temperature of the hot gases vary from 200°C to 400°C and the emissivity from 0.1 to 1.0. Plot the temperatures and the rate of heat loss as functions of the temperature of the hot gases and the emissivity, and discuss the results.

5–60 Consider a long concrete dam ($k = 0.6$ W/m · °C, $\alpha_s = 0.7$) of triangular cross section whose exposed surface is subjected to solar heat flux of $\dot{q}_s = 800$ W/m² and to convection and radiation to the environment at 25°C with a combined heat transfer coefficient of 30 W/m² · °C. The 2-m-high vertical section of the dam is subjected to convection by water at 15°C with a heat transfer coefficient of 150 W/m² · °C,

and heat transfer through the 2-m-long base is considered to be negligible. Using the finite difference method with a mesh size of $\Delta x = \Delta y = 1$ m and assuming steady two-dimensional heat transfer, determine the temperature of the top, middle, and bottom of the exposed surface of the dam. *Answers:* 21.3°C, 43.2°C, 43.6°C

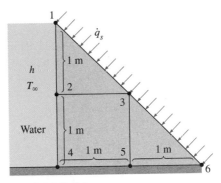

FIGURE P5–60

5–61E Consider steady two-dimensional heat transfer in a V-grooved solid body whose cross section is given in the figure. The top surfaces of the groove are maintained at 32°F while the bottom surface is maintained at 212°F. The side surfaces of the groove are insulated. Using the finite difference method with a mesh size of $\Delta x = \Delta y = 1$ ft and taking advantage of symmetry, determine the temperatures at the middle of the insulated surfaces.

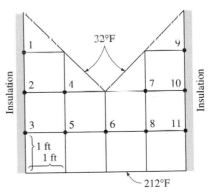

FIGURE P5–61E

5–62E Reconsider Prob. 5–61E. Using EES (or other) software, investigate the effects of the temperatures at the top and bottom surfaces on the temperature in the middle of the insulated surface. Let the temperatures at the top and bottom surfaces vary from 32°F to 212°F. Plot the temperature in the middle of the insulated surface as functions of the temperatures at the top and bottom surfaces, and discuss the results.

5–63 Consider a long solid bar whose thermal conductivity is $k = 5$ W/m · °C and whose cross section is given in the figure. The top surface of the bar is maintained at 50°C while the bottom surface is maintained at 120°C. The left surface is insulated and the remaining three surfaces are subjected to convection with ambient air at $T_\infty = 25$°C with a heat transfer coefficient of $h = 40$ W/m² · °C. Using the finite difference method with a mesh size of $\Delta x = \Delta y = 10$ cm, (a) obtain the finite difference formulation of this problem for steady two-dimensional heat transfer and (b) determine the unknown nodal temperatures by solving those equations.

Answers: (b) 78.8°C, 72.7°C, 64.6°C

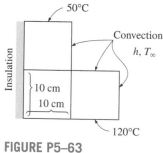

FIGURE P5–63

5–64 Consider a 5-m-long constantan block ($k = 23$ W/m · °C) 30 cm high and 50 cm wide. The block is completely submerged in iced water at 0°C that is well stirred, and the heat transfer coefficient is so high that the temperatures on both sides of the block can be taken to be 0°C. The bottom surface of the bar is covered with a low-conductivity material so that heat transfer through the bottom surface is negligible. The top surface of the block is heated uniformly by a 6-kW resistance heater. Using the finite difference method with a mesh size of $\Delta x = \Delta y = 10$ cm and taking advantage of symmetry, (a) obtain the finite difference formulation of this problem for steady two-dimensional heat transfer, (b) determine the unknown nodal temperatures by solving those equations, and (c) determine the rate of heat transfer from the block to the iced water.

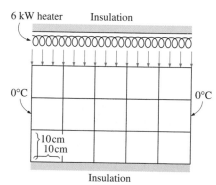

FIGURE P5–64

Transient Heat Conduction

5–65C How does the finite difference formulation of a transient heat conduction problem differ from that of a steady heat conduction problem? What does the term $\rho A \Delta x c_p (T_m^{i+1} - T_m^i)/\Delta t$ represent in the transient finite difference formulation?

5–66C What are the two basic methods of solution of transient problems based on finite differencing? How do heat transfer terms in the energy balance formulation differ in the two methods?

5–67C The explicit finite difference formulation of a general interior node for transient heat conduction in a plane wall is given by

$$T_{m-1}^i - 2T_m^i + T_{m+1}^i + \frac{\dot{e}_m^i \Delta x^2}{k} = \frac{T_m^{i+1} - T_m^i}{\tau}$$

Obtain the finite difference formulation for the steady case by simplifying the relation above.

5–68C The explicit finite difference formulation of a general interior node for transient two-dimensional heat conduction is given by

$$T_{node}^{i+1} = \tau(T_{left}^i + T_{top}^i + T_{right}^i + T_{bottom}^i)$$
$$+ (1 - 4\tau)T_{node}^i + \tau \frac{\dot{e}_{node}^i l^2}{k}$$

Obtain the finite difference formulation for the steady case by simplifying the relation above.

5–69C Is there any limitation on the size of the time step Δt in the solution of transient heat conduction problems using (a) the explicit method and (b) the implicit method?

5–70C Express the general stability criterion for the explicit method of solution of transient heat conduction problems.

5–71C Consider transient one-dimensional heat conduction in a plane wall that is to be solved by the explicit method. If both sides of the wall are at specified temperatures, express the stability criterion for this problem in its simplest form.

5–72C Consider transient one-dimensional heat conduction in a plane wall that is to be solved by the explicit method. If both sides of the wall are subjected to specified heat flux, express the stability criterion for this problem in its simplest form.

5–73C Consider transient two-dimensional heat conduction in a rectangular region that is to be solved by the explicit method. If all boundaries of the region are either insulated or at specified temperatures, express the stability criterion for this problem in its simplest form.

5–74C The implicit method is unconditionally stable and thus any value of time step Δt can be used in the solution of transient heat conduction problems. To minimize the computation time, someone suggests using a very large value of Δt since there is no danger of instability. Do you agree with this suggestion? Explain.

5–75 Consider transient heat conduction in a plane wall whose left surface (node 0) is maintained at 50°C while the right surface (node 6) is subjected to a solar heat flux of 600 W/m². The wall is initially at a uniform temperature of 50°C. Express the explicit finite difference formulation of the boundary nodes 0 and 6 for the case of no heat generation. Also, obtain the finite difference formulation for the total amount of heat transfer at the left boundary during the first three time steps.

5–76 Consider transient heat conduction in a plane wall with variable heat generation and constant thermal conductivity. The nodal network of the medium consists of nodes 0, 1, 2, 3, and 4 with a uniform nodal spacing of Δx. The wall is initially at a specified temperature. Using the energy balance approach, obtain the explicit finite difference formulation of the boundary nodes for the case of uniform heat flux $\dot{q}_0$ at the left boundary (node 0) and convection at the right boundary (node 4) with a convection coefficient of h and an ambient temperature of T_∞. Do not simplify.

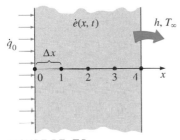

FIGURE P5–76

5–77 Repeat Prob. 5–76 for the case of implicit formulation.

5–78 Consider transient heat conduction in a plane wall with variable heat generation and constant thermal conductivity. The nodal network of the medium consists of nodes 0, 1, 2, 3, 4, and 5 with a uniform nodal spacing of Δx. The wall is initially at a specified temperature. Using the energy balance approach, obtain the explicit finite difference formulation of the boundary nodes for the case of insulation at the left boundary (node 0) and radiation at the right boundary (node 5) with an emissivity of ε and surrounding temperature of T_{surr}.

5–79 Consider transient heat conduction in a plane wall with variable heat generation and constant thermal conductivity. The nodal network of the medium consists of nodes 0, 1, 2, 3, and 4 with a uniform nodal spacing of Δx. The wall is initially at a specified temperature. The temperature at the right boundary (node 4) is specified. Using the energy balance approach, obtain the explicit finite difference formulation of the boundary

node 0 for the case of combined convection, radiation, and heat flux at the left boundary with an emissivity of ε, convection coefficient of h, ambient temperature of T_∞, surrounding temperature of T_{surr}, and uniform heat flux of $\dot{q}_0$ toward the wall. Also, obtain the finite difference formulation for the total amount of heat transfer at the right boundary for the first 20 time steps.

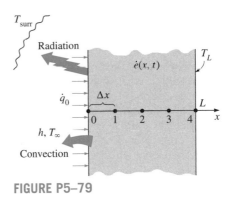

FIGURE P5–79

5–80 Starting with an energy balance on a volume element, obtain the two-dimensional transient explicit finite difference equation for a general interior node in rectangular coordinates for $T(x, y, t)$ for the case of constant thermal conductivity and no heat generation.

5–81 Starting with an energy balance on a volume element, obtain the two-dimensional transient implicit finite difference equation for a general interior node in rectangular coordinates for $T(x, y, t)$ for the case of constant thermal conductivity and no heat generation.

5–82 Starting with an energy balance on a disk volume element, derive the one-dimensional transient explicit finite difference equation for a general interior node for $T(z, t)$ in a cylinder whose side surface is insulated for the case of constant thermal conductivity with uniform heat generation.

5–83 Consider one-dimensional transient heat conduction in a composite plane wall that consists of two layers A and B with perfect contact at the interface. The wall involves no heat

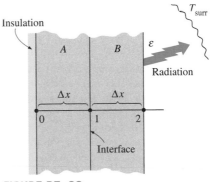

FIGURE P5–83

generation and initially is at a specified temperature. The nodal network of the medium consists of nodes 0, 1 (at the interface), and 2 with a uniform nodal spacing of Δx. Using the energy balance approach, obtain the explicit finite difference formulation of this problem for the case of insulation at the left boundary (node 0) and radiation at the right boundary (node 2) with an emissivity of ε and surrounding temperature of T_{surr}.

5–84 Consider transient one-dimensional heat conduction in a pin fin of constant diameter D with constant thermal conductivity. The fin is losing heat by convection to the ambient air at T_∞ with a heat transfer coefficient of h and by radiation to the surrounding surfaces at an average temperature of T_{surr}. The nodal network of the fin consists of nodes 0 (at the base), 1 (in the middle), and 2 (at the fin tip) with a uniform nodal spacing of Δx. Using the energy balance approach, obtain the explicit finite difference formulation of this problem for the case of a specified temperature at the fin base and negligible heat transfer at the fin tip.

5–85 Repeat Prob. 5–84 for the case of implicit formulation.

5–86 Consider a large uranium plate of thickness $L = 8$ cm, thermal conductivity $k = 28$ W/m · °C, and thermal diffusivity $\alpha = 12.5 \times 10^{-6}$ m²/s that is initially at a uniform temperature of 100°C. Heat is generated uniformly in the plate at a constant rate of $\dot{e} = 10^6$ W/m³. At time $t = 0$, the left side of the plate is insulated while the other side is subjected to convection with an environment at $T_\infty = 20$°C with a heat transfer coefficient of $h = 35$ W/m² · °C. Using the explicit finite difference approach with a uniform nodal spacing of $\Delta x = 2$ cm, determine (a) the temperature distribution in the plate after 5 min and (b) how long it will take for steady conditions to be reached in the plate.

5–87 Reconsider Prob. 5–86. Using EES (or other) software, investigate the effect of the cooling time on the temperatures of the left and right sides of the plate. Let the time vary from 5 min to 60 min. Plot the temperatures at the left and right surfaces as a function of time, and discuss the results.

5–88 Consider a house whose south wall consists of a 30-cm-thick Trombe wall whose thermal conductivity is $k = 0.70$ W/m · °C and whose thermal diffusivity is $\alpha = 0.44 \times 10^{-6}$ m²/s. The variations of the ambient temperature T_{out} and the solar heat flux $\dot{q}_{solar}$ incident on a south-facing vertical surface throughout the day for a typical day in February are given in the table in 3-h intervals. The Trombe wall has single glazing with an absorptivity-transmissivity product of $\kappa = 0.76$ (that is, 76 percent of the solar energy incident is absorbed by the exposed surface of the Trombe wall), and the average combined heat transfer coefficient for heat loss from the Trombe wall to the ambient is determined to be $h_{out} = 3.4$ W/m² · °C. The interior of the house is maintained at $T_{in} = 20$°C at all times, and the heat transfer coefficient at the interior surface of the Trombe wall is $h_{in} = 9.1$ W/m² · °C. Also, the vents on the Trombe wall are kept closed, and thus the only heat transfer

between the air in the house and the Trombe wall is through the interior surface of the wall. Assuming the temperature of the Trombe wall to vary linearly between 20°C at the interior surface and 0°C at the exterior surface at 7 AM and using the explicit finite difference method with a uniform nodal spacing of $\Delta x = 5$ cm, determine the temperature distribution along the thickness of the Trombe wall after 6, 12, 18, 24, 30, 36, 42, and 48 hours and plot the results. Also, determine the net amount of heat transferred to the house from the Trombe wall during the first day if the wall is 2.8 m high and 7 m long.

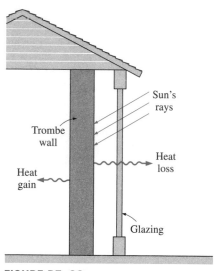

FIGURE P5–88

TABLE P5–88

The hourly variations of the monthly average ambient temperature and solar heat flux incident on a vertical surface

Time of Day	Ambient Temperature, °C	Solar Insolation, W/m²
7 AM–10 AM	0	375
10 AM–1 PM	4	750
1 PM–4 PM	6	580
4 PM–7 PM	1	95
7 PM–10 PM	−2	0
10 PM–1 AM	−3	0
1 AM–4 AM	−4	0
4 AM–7 AM	4	0

5–89 Consider two-dimensional transient heat transfer in an L-shaped solid bar that is initially at a uniform temperature of 140°C and whose cross section is given in the figure. The thermal conductivity and diffusivity of the body are $k = 15$ W/m · °C and $\alpha = 3.2 \times 10^{-6}$ m²/s, respectively, and heat is generated in the body at a rate of $\dot{e} = 2 \times 10^7$ W/m³. The right surface of the body is insulated, and the bottom surface is maintained at a uniform temperature of 140°C at all times. At time $t = 0$, the entire top surface is subjected to convection with ambient air at $T_\infty = 25°C$ with a heat transfer coefficient of $h = 80$ W/m² · °C, and the left surface is subjected to uniform heat flux at a rate of $\dot{q}_L = 8000$ W/m². The nodal network of the problem consists of 13 equally spaced nodes with $\Delta x = \Delta y = 1.5$ cm. Using the explicit method, determine the temperature at the top corner (node 3) of the body after 2, 5, and 30 min.

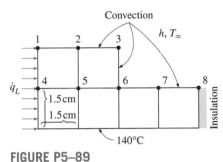

FIGURE P5–89

5–90 Reconsider Prob. 5–89. Using EES (or other) software, plot the temperature at the top corner as a function of heating time as it varies from 2 min to 30 min, and discuss the results.

5–91 Consider a long solid bar ($k = 28$ W/m · °C and $\alpha = 12 \times 10^{-6}$ m²/s) of square cross section that is initially at a uniform temperature of 20°C. The cross section of the bar is 20 cm × 20 cm in size, and heat is generated in it uniformly at a rate of $\dot{e} = 8 \times 10^5$ W/m³. All four sides of the bar are subjected to convection to the ambient air at $T_\infty = 30°C$ with a heat transfer coefficient of $h = 45$ W/m² · °C. Using the explicit finite difference method with a mesh size of $\Delta x = \Delta y = 10$ cm, determine the centerline temperature of the bar (a) after 20 min and (b) after steady conditions are established.

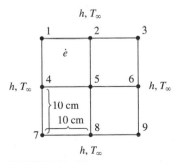

FIGURE P5–91

5–92E Consider a house whose windows are made of 0.375-in-thick glass ($k = 0.48$ Btu/h · ft · °F and $\alpha = 4.2 \times$

10^{-6} ft²/s). Initially, the entire house, including the walls and the windows, is at the outdoor temperature of $T_o = 35°F$. It is observed that the windows are fogged because the indoor temperature is below the dew-point temperature of 54°F. Now the heater is turned on and the air temperature in the house is raised to $T_i = 72°F$ at a rate of 2°F rise per minute. The heat transfer coefficients at the inner and outer surfaces of the wall can be taken to be $h_i = 1.2$ and $h_o = 2.6$ Btu/h · ft² · °F, respectively, and the outdoor temperature can be assumed to remain constant. Using the explicit finite difference method with a mesh size of $\Delta x = 0.125$ in, determine how long it will take for the fog on the windows to clear up (i.e., for the inner surface temperature of the window glass to reach 54°F).

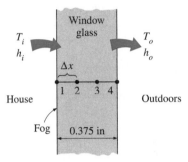

FIGURE P5–92E

5–93 A common annoyance in cars in winter months is the formation of fog on the glass surfaces that blocks the view. A practical way of solving this problem is to blow hot air or to attach electric resistance heaters to the inner surfaces. Consider the rear window of a car that consists of a 0.4-cm-thick glass ($k = 0.84$ W/m · °C and $\alpha = 0.39 \times 10^{-6}$ m²/s). Strip heater wires of negligible thickness are attached to the inner surface

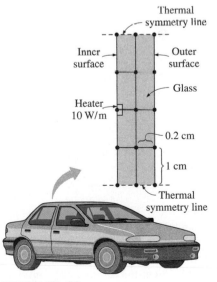

FIGURE P5–93

of the glass, 4 cm apart. Each wire generates heat at a rate of 10 W/m length. Initially the entire car, including its windows, is at the outdoor temperature of $T_o = -3°C$. The heat transfer coefficients at the inner and outer surfaces of the glass can be taken to be $h_i = 6$ and $h_o = 20$ W/m$^2 \cdot °C$, respectively. Using the explicit finite difference method with a mesh size of $\Delta x = 0.2$ cm along the thickness and $\Delta y = 1$ cm in the direction normal to the heater wires, determine the temperature distribution throughout the glass 15 min after the strip heaters are turned on. Also, determine the temperature distribution when steady conditions are reached.

5–94 Repeat Prob. 5–93 using the implicit method with a time step of 1 min.

5–95 The roof of a house consists of a 15-cm-thick concrete slab ($k = 1.4$ W/m $\cdot °C$ and $\alpha = 0.69 \times 10^{-6}$ m^2/s) that is 18 m wide and 32 m long. One evening at 6 PM, the slab is observed to be at a uniform temperature of 18°C. The average ambient air and the night sky temperatures for the entire night are predicted to be 6°C and 260 K, respectively. The convection heat transfer coefficients at the inner and outer surfaces of the roof can be taken to be $h_i = 5$ and $h_o = 12$ W/m$^2 \cdot °C$, respectively. The house and the interior surfaces of the walls and the floor are maintained at a constant temperature of 20°C during the night, and the emissivity of both surfaces of the concrete roof is 0.9. Considering both radiation and convection heat transfers and using the explicit finite difference method with a time step of $\Delta t = 5$ min and a mesh size of $\Delta x = 3$ cm, determine the temperatures of the inner and outer surfaces of the roof at 6 AM. Also, determine the average rate of heat transfer through the roof during that night.

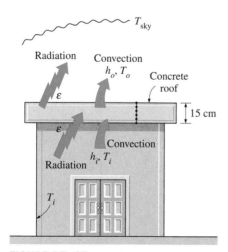

FIGURE P5–95

5–96 Consider a refrigerator whose outer dimensions are 1.80 m × 0.8 m × 0.7 m. The walls of the refrigerator are constructed of 3-cm-thick urethane insulation ($k = 0.026$ W/m $\cdot °$ C and $\alpha = 0.36 \times 10^{-6}$ m^2/s) sandwiched

between two layers of sheet metal with negligible thickness. The refrigerated space is maintained at 3°C and the average heat transfer coefficients at the inner and outer surfaces of the wall are 6 W/m$^2 \cdot °C$ and 9 W/m$^2 \cdot °C$, respectively. Heat transfer through the bottom surface of the refrigerator is negligible. The kitchen temperature remains constant at about 25°C. Initially, the refrigerator contains 15 kg of food items at an average specific heat of 3.6 kJ/kg $\cdot °C$. Now a malfunction occurs and the refrigerator stops running for 6 h as a result. Assuming the temperature of the contents of the refrigerator, including the air inside, rises uniformly during this period, predict the temperature inside the refrigerator after 6 h when the repair-man arrives. Use the explicit finite difference method with a time step of $\Delta t = 1$ min and a mesh size of $\Delta x = 1$ cm and disregard corner effects (i.e., assume one-dimensional heat transfer in the walls).

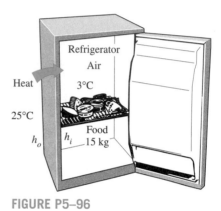

FIGURE P5–96

5–97 Reconsider Prob. 5–96. Using EES (or other) software, plot the temperature inside the refrigerator as a function of heating time as time varies from 1 h to 10 h, and discuss the results.

Special Topic: Controlling the Numerical Error

5–98C Why do the results obtained using a numerical method differ from the exact results obtained analytically? What are the causes of this difference?

5–99C What is the cause of the discretization error? How does the global discretization error differ from the local discretization error?

5–100C Can the global (accumulated) discretization error be less than the local error during a step? Explain.

5–101C How is the finite difference formulation for the first derivative related to the Taylor series expansion of the solution function?

5–102C Explain why the local discretization error of the finite difference method is proportional to the square of the step size. Also explain why the global discretization error is proportional to the step size itself.

5–103C What causes the round-off error? What kind of calculations are most susceptible to round-off error?

5–104C What happens to the discretization and the round-off errors as the step size is decreased?

5–105C Suggest some practical ways of reducing the round-off error.

5–106C What is a practical way of checking if the round-off error has been significant in calculations?

5–107C What is a practical way of checking if the discretization error has been significant in calculations?

Review Problems

5–108 Starting with an energy balance on the volume element, obtain the steady three-dimensional finite difference equation for a general interior node in rectangular coordinates for $T(x, y, z)$ for the case of constant thermal conductivity and uniform heat generation.

5–109 Starting with an energy balance on the volume element, obtain the three-dimensional transient explicit finite difference equation for a general interior node in rectangular coordinates for $T(x, y, z, t)$ for the case of constant thermal conductivity and no heat generation.

5–110 Consider steady one-dimensional heat conduction in a plane wall with variable heat generation and constant thermal conductivity. The nodal network of the medium consists of nodes 0, 1, 2, and 3 with a uniform nodal spacing of Δx. The temperature at the left boundary (node 0) is specified. Using the energy balance approach, obtain the finite difference formulation of boundary node 3 at the right boundary for the case of combined convection and radiation with an emissivity of ε, convection coefficient of h, ambient temperature of T_∞, and surrounding temperature of T_{surr}. Also, obtain the finite difference formulation for the rate of heat transfer at the left boundary.

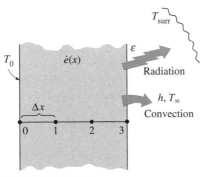

FIGURE P5–110

5–111 Consider one-dimensional transient heat conduction in a plane wall with variable heat generation and variable

thermal conductivity. The nodal network of the medium consists of nodes 0, 1, and 2 with a uniform nodal spacing of Δx. Using the energy balance approach, obtain the explicit finite difference formulation of this problem for the case of specified heat flux $\dot{q}_0$ and convection at the left boundary (node 0) with a convection coefficient of h and ambient temperature of T_∞, and radiation at the right boundary (node 2) with an emissivity of ε and surrounding temperature of T_{surr}.

5–112 Repeat Prob. 5–111 for the case of implicit formulation.

5–113 Consider steady one-dimensional heat conduction in a pin fin of constant diameter D with constant thermal conductivity. The fin is losing heat by convection with the ambient air at T_∞ (in °C) with a convection coefficient of h, and by radiation to the surrounding surfaces at an average temperature of T_{surr} (in K). The nodal network of the fin consists of nodes 0 (at the base), 1 (in the middle), and 2 (at the fin tip) with a uniform nodal spacing of Δx. Using the energy balance approach, obtain the finite difference formulation of this problem for the case of a specified temperature at the fin base and convection and radiation heat transfer at the fin tip.

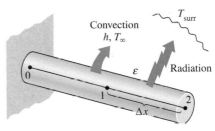

FIGURE P5–113

5–114 Starting with an energy balance on the volume element, obtain the two-dimensional transient explicit finite difference equation for a general interior node in rectangular coordinates for $T(x, y, t)$ for the case of constant thermal conductivity and uniform heat generation.

5–115 Starting with an energy balance on a disk volume element, derive the one-dimensional transient implicit finite difference equation for a general interior node for $T(z, t)$ in a cylinder whose side surface is subjected to convection with a convection coefficient of h and an ambient temperature of T_∞ for the case of constant thermal conductivity with uniform heat generation.

5–116E The roof of a house consists of a 5-in-thick concrete slab ($k = 0.81$ Btu/h · ft · °F and $\alpha = 7.4 \times 10^{-6}$ ft²/s) that is 30 ft wide and 50 ft long. One evening at 6 PM, the slab is observed to be at a uniform temperature of 70°F. The ambient air temperature is predicted to be at about 50°F from 6 PM to 10 PM, 42°F from 10 PM to 2 AM, and 38°F from 2 AM to 6 AM, while the night sky temperature is expected to be about 445 R for the entire night. The convection heat transfer coefficients at the inner and outer surfaces of the roof can be taken to be $h_i = 0.9$ and $h_o = 2.1$ Btu/h · ft^2 · °F, respectively. The house and the interior surfaces of the walls and the floor are maintained at a constant temperature of 70°F during the night, and the emissivity of both surfaces of the concrete roof is 0.9.

Considering both radiation and convection heat transfers and using the explicit finite difference method with a mesh size of $\Delta x = 1$ in and a time step of $\Delta t = 5$ min, determine the temperatures of the inner and outer surfaces of the roof at 6 AM. Also, determine the average rate of heat transfer through the roof during that night.

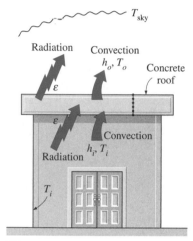

FIGURE P5–116E

5–117 A two-dimensional bar has the geometry shown in Fig. P5–117 with specified temperature T_A on the upper surface and T_B on the lower surfaces, and insulation on the sides. The thermal conductivity of the upper part of the bar is k_A while that of the lower part is k_B. For a grid defined by $\Delta x = \Delta y = l$, write the simplest form of the matrix equation, $AT = C$, used to find the steady-state temperature field in the cross section of the bar. Identify on the figure the grid nodes where you write the energy balance.

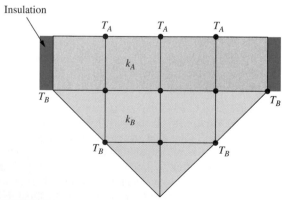

FIGURE P5–117

5–118 A long steel bar has the cross section shown in Fig. P5–118. The bar is removed from a heat treatment oven at $T_i = 700°C$ and placed on the bottom of a tank filled with water at 10°C. To intensify the heat transfer, the water is vigorously circulated, which creates a virtually constant temperature $T_s = 10°C$ on all sides of the bar, except for the bottom side, which is adiabatic. The properties of the bar are $c_p = 430$ J/kg · K, $k = 40$ W/m · K, and $\rho = 8000$ kg/m^3.

(a) Write the finite difference equations for the unknown temperatures in the grid using the explicit method. Group all constant quantities in one term. Identify dimensionless parameters such as Bi and Fo if applicable.

(b) Determine the range of time steps for which the explicit scheme is numerically stable.

(c) For $\Delta t = 10$ s, determine the temperature field at $t = 10$ s and $t = 20$ s. Fill in the table below.

Node	$T(10\text{ s})$	$T(20\text{ s})$
1	——	——
2	——	——
3	——	——
4	——	——
5	——	——
6	——	——
7	——	——

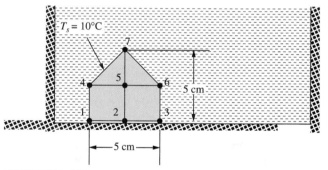

FIGURE P5–118

5–119 Solar radiation incident on a large body of clean water ($k = 0.61$ W/m · °C and $\alpha = 0.15 \times 10^{-6}$ m^2/s) such as a lake, a river, or a pond is mostly absorbed by water, and the amount of absorption varies with depth. For solar radiation incident at a 45° angle on a 1-m-deep large pond whose bottom surface is black (zero reflectivity), for example, 2.8 percent of the solar energy is reflected back to the atmosphere, 37.9 percent is absorbed by the bottom surface, and the remaining 59.3 percent is absorbed by the water body. If the pond is considered to be four layers of equal thickness (0.25 m in this case), it can be shown that 47.3 percent of the incident solar energy is absorbed by the top layer, 6.1 percent by the upper mid layer, 3.6 percent by the lower mid layer, and 2.4 percent by the bottom layer [for more information see Çengel and Özişik, *Solar*

Energy, 33, no. 6 (1984), pp. 581–591]. The radiation absorbed by the water can be treated conveniently as heat generation in the heat transfer analysis of the pond.

Consider a large 1-m-deep pond that is initially at a uniform temperature of 15°C throughout. Solar energy is incident on the pond surface at 45° at an average rate of 500 W/m² for a period of 4 h. Assuming no convection currents in the water and using the explicit finite difference method with a mesh size of $\Delta x = 0.25$ m and a time step of $\Delta t = 15$ min, determine the temperature distribution in the pond under the most favorable conditions (i.e., no heat losses from the top or bottom surfaces of the pond). The solar energy absorbed by the bottom surface of the pond can be treated as a heat flux to the water at that surface in this case.

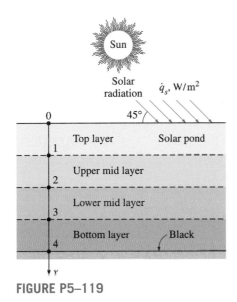

FIGURE P5–119

5–120 Reconsider Prob. 5–119. The absorption of solar radiation in that case can be expressed more accurately as a fourth-degree polynomial as

$$\dot{e}(x) = \dot{q}_s(0.859 - 3.415x + 6.704x^2 - 6.339x^3 + 2.278x^4), \text{ W/m}^3$$

where $\dot{q}_s$ is the solar flux incident on the surface of the pond in W/m² and x is the distance from the free surface of the pond in m. Solve Problem 5–119 using this relation for the absorption of solar radiation.

5–121 A hot surface at 120°C is to be cooled by attaching 8 cm long, 0.8 cm in diameter aluminum pin fins ($k = 237$ W/m · °C and $\alpha = 97.1 \times 10^{-6}$ m²/s) to it with a center-to-center distance of 1.6 cm. The temperature of the surrounding medium is 15°C, and the heat transfer coefficient on the surfaces is 35 W/m² · °C. Initially, the fins are at a uniform temperature of 30°C, and at time $t = 0$, the temperature of the hot surface is raised to 120°C. Assuming one-dimensional heat conduction along the fin and taking the nodal spacing to be

$\Delta x = 2$ cm and a time step to be $\Delta t = 0.5$ s, determine the nodal temperatures after 5 min by using the explicit finite difference method. Also, determine how long it will take for steady conditions to be reached.

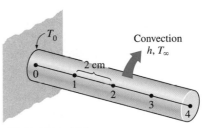

FIGURE P5–121

5–122E Consider a large plane wall of thickness $L = 0.3$ ft and thermal conductivity $k = 1.2$ Btu/h · ft · °F in space. The wall is covered with a material having an emissivity of $\varepsilon = 0.80$ and a solar absorptivity of $\alpha_s = 0.60$. The inner surface of the wall is maintained at 520 R at all times, while the outer surface is exposed to solar radiation that is incident at a rate of $\dot{q}_s = 350$ Btu/h · ft². The outer surface is also losing heat by radiation to deep space at 0 R. Using a uniform nodal spacing of $\Delta x = 0.1$ ft, (*a*) obtain the finite difference formulation for steady one-dimensional heat conduction and (*b*) determine the nodal temperatures by solving those equations.

Answers: (*b*) 528 R, 535 R, 543 R

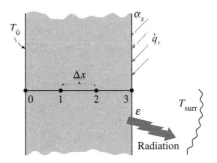

FIGURE P5–122E

5–123 Frozen food items can be defrosted by simply leaving them on the counter, but it takes too long. The process can be speeded up considerably for flat items such as steaks by placing them on a large piece of highly conducting metal, called the defrosting plate, which serves as a fin. The increased surface area enhances heat transfer and thus reduces the defrosting time.

Consider two 1.5-cm-thick frozen steaks at −18°C that resemble a 15-cm-diameter circular object when placed next to each other. The steaks are now placed on a 1-cm-thick black-anodized circular aluminum defrosting plate ($k = $

237 W/m · °C, $\alpha = 97.1 \times 10^{-6}$ m²/s, and $\varepsilon = 0.90$) whose outer diameter is 30 cm. The properties of the frozen steaks are $\rho = 970$ kg/m³, $c_p = 1.55$ kJ/kg · °C, $k = 1.40$ W/m · °C, $\alpha = 0.93 \times 10^{-6}$ m²/s, and $\varepsilon = 0.95$, and the heat of fusion is $h_{if} = 187$ kJ/kg. The steaks can be considered to be defrosted when their average temperature is 0°C and all of the ice in the steaks is melted. Initially, the defrosting plate is at the room temperature of 20°C, and the wooden countertop it is placed on can be treated as insulation. Also, the surrounding surfaces can be taken to be at the same temperature as the ambient air, and the convection heat transfer coefficient for all exposed surfaces can be taken to be 12 W/m² · °C. Heat transfer from the lateral surfaces of the steaks and the defrosting plate can be neglected. Assuming one-dimensional heat conduction in both the steaks and the defrosting plate and using the explicit finite difference method, determine how long it will take to defrost the steaks. Use four nodes with a nodal spacing of $\Delta x = 0.5$ cm for the steaks, and three nodes with a nodal spacing of $\Delta r = 3.75$ cm for the exposed portion of the defrosting plate. Also, use a time step of $\Delta t = 5$ s. *Hint:* First, determine the total amount of heat transfer needed to defrost the steaks, and then determine how long it will take to transfer that much heat.

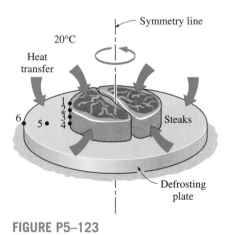

FIGURE P5–123

5–124 Repeat Prob. 5–123 for a copper defrosting plate using a time step of $\Delta t = 3$ s.

Fundamentals of Engineering (FE) Exam Problems

5–125 What is the correct steady-state finite-difference heat conduction equation of node 6 of the rectangular solid shown in Fig. P5–125?

(a) $T_6 = (T_1 + T_3 + T_9 + T_{11})/2$
(b) $T_6 = (T_5 + T_7 + T_2 + T_{10})/2$
(c) $T_6 = (T_1 + T_3 + T_9 + T_{11})/4$
(d) $T_6 = (T_2 + T_5 + T_7 + T_{10})/4$
(e) $T_6 = (T_1 + T_2 + T_9 + T_{10})/4$

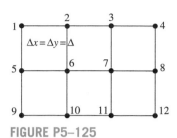

FIGURE P5–125

5–126 Air at T_0 acts on top surface of the rectangular solid shown in Fig. P5–126 with a convection heat transfer coefficient of h. The correct steady-state finite-difference heat conduction equation for node 3 of this solid is

(a) $T_3 = [(k/2\Delta)(T_2 + T_4 + T_7) + hT_0] / [(k/\Delta) + h]$

(b) $T_3 = [(k/2\Delta)(T_2 + T_4 + 2T_7) + hT_0] / [(2k/\Delta) + h]$

(c) $T_3 = [(k/\Delta)(T_2 + T_4) + hT_0] / [(2k/\Delta) + h]$

(d) $T_3 = [(k/\Delta)(T_2 + T_4 + T_7) + hT_0] / [(k/\Delta) + h]$

(e) $T_3 = [(k/\Delta)(2T_2 + 2T_4 + T_7) + hT_0] / [(k/\Delta) + h]$

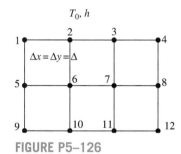

FIGURE P5–126

5–127 What is the correct unsteady forward-difference heat conduction equation of node 6 of the rectangular solid shown in Fig. P5–127 if its temperature at the previous time (Δt) is T_6^*?

(a) $T_6^{i+1} = [k\Delta t / (\rho c_p \Delta^2)](T_5^* + T_2^* + T_7^* + T_{10}^*)$
$+ [1 - 4k\Delta t /(\rho c_p \Delta^2)]T_6^*$

(b) $T_6^{i+1} = [k\Delta t / (\rho c_p \Delta^2)](T_5^* + T_2^* + T_7^* + T_{10}^*)$
$+ [1 - k\Delta t /(\rho c_p \Delta^2)]T_6^*$

(c) $T_6^{i+1} = [k\Delta t / (\rho c_p \Delta^2)](T_5^* + T_2^* + T_7^* + T_{10}^*)$
$+ [2k\Delta t /(\rho c_p \Delta^2)]T_6^*$

(d) $T_6^{i+1} = [2k\Delta t / (\rho c_p \Delta^2)](T_5^* + T_2^* + T_7^* + T_{10}^*)$
$+ [1 - 2k\Delta t /(\rho c_p \Delta^2)]T_6^*$

(e) $T_6^{i+1} = [2k\Delta t / (\rho c_p \Delta^2)](T_5^* + T_2^* + T_7^* + T_{10}^*)$
$+ [1 - 4k\Delta t /(\rho c_p \Delta^2)]T_6^*$

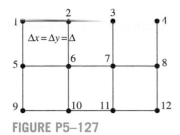

FIGURE P5–127

5–128 The unsteady forward-difference heat conduction for a constant area, A, pin fin with perimeter, p, exposed to air whose temperature is T_0 with a convection heat transfer coefficient of h is

$$T_m^{*+1} = \frac{k}{\rho c_p \Delta x^2}\left[T_{m-1}^* + T_{m+1}^* + \frac{hp\Delta x^2}{A}T_0\right]$$
$$- \left[1 - \frac{2k}{\rho c_p \Delta x^2} - \frac{hp}{\rho c_p A}\right]T_m^*$$

In order for this equation to produce a stable solution, the quantity $\dfrac{2k}{\rho c_p \Delta x^2} + \dfrac{hp}{\rho c_p A}$ must be

(a) negative (b) zero (c) positive
(d) greater than 1 (e) less than 1

5–129 The height of the cells for a finite-difference solution of the temperature in the rectangular solid shown in Fig. P5–129 is one-half the cell width to improve the accuracy of the solution. The correct steady-state finite-difference heat conduction equation for cell 6 is

(a) $T_6 = 0.1(T_5 + T_7) + 0.4(T_2 + T_{10})$
(b) $T_6 = 0.25(T_5 + T_7) + 0.25(T_2 + T_{10})$
(c) $T_6 = 0.5(T_5 + T_7) + 0.5(T_2 + T_{10})$
(d) $T_6 = 0.4(T_5 + T_7) + 0.1(T_2 + T_{10})$
(e) $T_6 = 0.5(T_5 + T_7) + 0.5(T_2 + T_{10})$

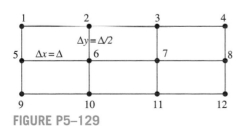

FIGURE P5–129

5–130 The height of the cells for a finite-difference solution of the temperature in the rectangular solid shown in Fig. P5–130 is one-half the cell width to improve the accuracy of the solution. If the left surface is exposed to air at T_0 with a heat transfer coefficient of h, the correct finite-difference heat conduction energy balance for node 5 is

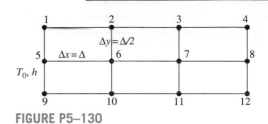

FIGURE P5–130

(a) $2T_1 + 2T_9 + T_6 - T_5 + h\Delta/k\,(T_0 - T_5) = 0$
(b) $2T_1 + 2T_9 + T_6 - 2T_5 + h\Delta/k\,(T_0 - T_5) = 0$
(c) $2T_1 + 2T_9 + T_6 - 3T_5 + h\Delta/k\,(T_0 - T_5) = 0$
(d) $2T_1 + 2T_9 + T_6 - 4T_5 + h\Delta/k\,(T_0 - T_5) = 0$
(e) $2T_1 + 2T_9 + T_6 - 5T_5 + h\Delta/k\,(T_0 - T_5) = 0$

Design and Essay Problems

5–131 Write a two-page essay on the finite element method, and explain why it is used in most commercial engineering software packages. Also explain how it compares to the finite difference method.

5–132 Numerous professional software packages are available in the market for performing heat transfer analysis, and they are widely advertised in professional magazines such as the *Mechanical Engineering* magazine published by the American Society of Mechanical Engineers (ASME). Your company decides to purchase such a software package and asks you to prepare a report on the available packages, their costs, capabilities, ease of use, and compatibility with the available hardware, and other software as well as the reputation of the software company, their history, financial health, customer support, training, and future prospects, among other things. After a preliminary investigation, select the top three packages and prepare a full report on them.

5–133 Design a defrosting plate to speed up defrosting of flat food items such as frozen steaks and packaged vegetables and evaluate its performance using the finite difference method (see Prob. 5–123). Compare your design to the defrosting plates currently available on the market. The plate must perform well, and it must be suitable for purchase and use as a household utensil, durable, easy to clean, easy to manufacture, and affordable. The frozen food is expected to be at an initial temperature of $-18°C$ at the beginning of the thawing process and $0°C$ at the end with all the ice melted. Specify the material, shape, size, and thickness of the proposed plate. Justify your recommendations by calculations. Take the ambient and surrounding surface temperatures to be $20°C$ and the convection heat transfer coefficient to be $15\ W/m^2 \cdot °C$ in your analysis. For a typical case, determine the defrosting time with and without the plate.

5–134 Design a fire-resistant safety box whose outer dimensions are $0.5\ m \times 0.5\ m \times 0.5\ m$ that will protect its combustible contents from fire which may last up to 2 h. Assume the box will be exposed to an environment at an average temperature of 700°C with a combined heat transfer coefficient of 70 W/m² · °C and the temperature inside the box must be below 150°C at the end of 2 h. The cavity of the box must be as large as possible while meeting the design constraints, and the insulation material selected must withstand the high temperatures to which it will be exposed. Cost, durability, and strength are also important considerations in the selection of insulation materials.

FUNDAMENTALS OF CONVECTION

S
o far, we have considered *conduction,* which is the mechanism of heat transfer through a solid or a quiescent fluid. We now consider *convection,* which is the mechanism of heat transfer through a fluid in the presence of bulk fluid motion.

Convection is classified as *natural* (or *free*) and *forced convection,* depending on how the fluid motion is initiated. In forced convection, the fluid is forced to flow over a surface or in a pipe by external means such as a pump or a fan. In natural convection, any fluid motion is caused by natural means such as the buoyancy effect, which manifests itself as the rise of warmer fluid and the fall of the cooler fluid. Convection is also classified as *external* and *internal,* depending on whether the fluid is forced to flow over a surface or in a pipe.

We start this chapter with a general physical description of the convection mechanism. We then discuss the *velocity* and *thermal boundary layers,* and *laminar and turbulent flows.* We continue with the discussion of the dimensionless *Reynolds, Prandtl,* and *Nusselt numbers,* and their physical significance. Next we derive the *convection equations* on the basis of mass, momentum, and energy conservation, and obtain solutions for *flow over a flat plate.* We then nondimensionalize the convection equations, and obtain functional forms of friction and convection coefficients. Finally, we present analogies between momentum and heat transfer.

OBJECTIVES

When you finish studying this chapter, you should be able to:

■ Understand the physical mechanism of convection, and its classification,

■ Visualize the development of velocity and thermal boundary layers during flow over surfaces,

■ Gain a working knowledge of the dimensionless Reynolds, Prandtl, and Nusselt numbers,

■ Distinguish between laminar and turbulent flows, and gain an understanding of the mechanisms of momentum and heat transfer in turbulent flow,

■ Derive the differential equations that govern convection on the basis of mass, momentum, and energy balances, and solve these equations for some simple cases such as laminar flow over a flat plate,

■ Nondimensionalize the convection equations and obtain the functional forms of friction and heat transfer coefficients, and

■ Use analogies between momentum and heat transfer, and determine heat transfer coefficient from knowledge of friction coefficient.

CONTENTS

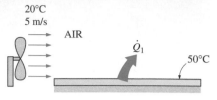

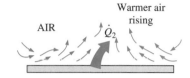

(a) Forced convection

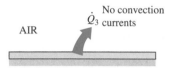

(b) Free convection

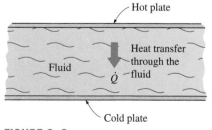

(c) Conduction

FIGURE 6–1

Heat transfer from a hot surface to the surrounding fluid by convection and conduction.

6–1 · PHYSICAL MECHANISM OF CONVECTION

We mentioned in Chapter 1 that there are three basic mechanisms of heat transfer: conduction, convection, and radiation. Conduction and convection are similar in that both mechanisms require the presence of a material medium. But they are different in that convection requires the presence of fluid motion.

Heat transfer through a solid is always by conduction, since the molecules of a solid remain at relatively fixed positions. Heat transfer through a liquid or gas, however, can be by conduction or convection, depending on the presence of any bulk fluid motion. Heat transfer through a fluid is by convection in the presence of bulk fluid motion and by conduction in the absence of it. Therefore, conduction in a fluid can be viewed as the limiting case of convection, corresponding to the case of quiescent fluid (Fig. 6–1).

Convection heat transfer is complicated by the fact that it involves fluid motion as well as heat conduction. The fluid motion enhances heat transfer, since it brings warmer and cooler chunks of fluid into contact, initiating higher rates of conduction at a greater number of sites in a fluid. Therefore, the rate of heat transfer through a fluid is much higher by convection than it is by conduction. In fact, the higher the fluid velocity, the higher the rate of heat transfer.

To clarify this point further, consider steady heat transfer through a fluid contained between two parallel plates maintained at different temperatures, as shown in Figure 6–2. The temperatures of the fluid and the plate are the same at the points of contact because of the continuity of temperature. Assuming no fluid motion, the energy of the hotter fluid molecules near the hot plate is transferred to the adjacent cooler fluid molecules. This energy is then transferred to the next layer of the cooler fluid molecules. This energy is then transferred to the next layer of the cooler fluid, and so on, until it is finally transferred to the other plate. This is what happens during conduction through a fluid. Now let us use a syringe to draw some fluid near the hot plate and inject it next to the cold plate repeatedly. You can imagine that this will speed up the heat transfer process considerably, since some energy is carried to the other side as a result of fluid motion.

Consider the cooling of a hot block with a fan blowing air over its top surface. We know that heat is transferred from the hot block to the surrounding cooler air, and the block eventually cools. We also know that the block cools faster if the fan is switched to a higher speed. Replacing air by water enhances the convection heat transfer even more.

Experience shows that convection heat transfer strongly depends on the fluid properties *dynamic viscosity* μ, *thermal conductivity k, density* ρ, and *specific heat* c_p, as well as the *fluid velocity V*. It also depends on the *geometry* and the *roughness* of the solid surface, in addition to the *type of fluid flow* (such as being streamlined or turbulent). Thus, we expect the convection heat transfer relations to be rather complex because of the dependence of convection on so many variables. This is not surprising, since convection is the most complex mechanism of heat transfer.

Despite the complexity of convection, the rate of convection heat transfer is observed to be proportional to the temperature difference and is conveniently expressed by **Newton's law of cooling** as

$$\dot{q}_{\text{conv}} = h(T_s - T_\infty) \qquad (\text{W/m}^2) \qquad (6\text{–}1)$$

FIGURE 6–2

Heat transfer through a fluid sandwiched between two parallel plates.

or

$$\dot{Q}_{conv} = hA_s(T_s - T_\infty) \qquad (W) \qquad \text{(6–2)}$$

where

    h = convection heat transfer coefficient, W/m² · °C
    A_s = heat transfer surface area, m²
    T_s = temperature of the surface, °C
    T_∞ = temperature of the fluid sufficiently far from the surface, °C

Judging from its units, the **convection heat transfer coefficient** h can be defined as *the rate of heat transfer between a solid surface and a fluid per unit surface area per unit temperature difference.*

You should not be deceived by the simple appearance of this relation, because the convection heat transfer coefficient h depends on the several of the mentioned variables, and thus is difficult to determine.

Fluid flow is often confined by solid surfaces, and it is important to understand how the presence of solid surfaces affects fluid flow. Consider the flow of a fluid in a stationary pipe or over a solid surface that is nonporous (i.e., impermeable to the fluid). All experimental observations indicate that a fluid in motion comes to a complete stop at the surface and assumes a zero velocity relative to the surface. That is, a fluid in direct contact with a solid "sticks" to the surface due to viscous effects, and there is no slip. This is known as the **no-slip condition**.

The photo in Fig. 6–3 obtained from a video clip clearly shows the evolution of a velocity gradient as a result of the fluid sticking to the surface of a blunt nose. The layer that sticks to the surface slows the adjacent fluid layer because of viscous forces between the fluid layers, which slows the next layer, and so on. Therefore, the no-slip condition is responsible for the development of the velocity profile. The flow region adjacent to the wall in which the viscous effects (and thus the velocity gradients) are significant is called the **boundary layer**. The fluid property responsible for the no-slip condition and the development of the boundary layer is *viscosity* and is discussed briefly in Section 6–2.

A fluid layer adjacent to a moving surface has the same velocity as the surface. A consequence of the no-slip condition is that all velocity profiles must have zero values with respect to the surface at the points of contact between a fluid and a solid surface (Fig. 6–4). Another consequence of the no-slip condition is the *surface drag*, which is the force a fluid exerts on a surface in the flow direction.

An implication of the no-slip condition is that heat transfer from the solid surface to the fluid layer adjacent to the surface is by *pure conduction*, since the fluid layer is motionless, and can be expressed as

$$\dot{q}_{conv} = \dot{q}_{cond} = -k_{fluid} \frac{\partial T}{\partial y}\bigg|_{y=0} \qquad (W/m^2) \qquad \text{(6–3)}$$

where T represents the temperature distribution in the fluid and $(\partial T/\partial y)_{y=0}$ is the *temperature gradient* at the surface. Heat is then *convected away* from the surface as a result of fluid motion. Note that convection heat transfer from a solid surface to a fluid is merely the conduction heat transfer from the solid surface to the fluid layer adjacent to the surface. Therefore, we can equate Eqs. 6–1 and 6–3 for the heat flux to obtain

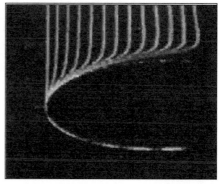

FIGURE 6–3
The development of a velocity profile due to the no-slip condition as a fluid flows over a blunt nose.

"Hunter Rouse: Laminar and Turbulent Flow Film."
Copyright IIHR-Hydroscience & Engineering,
The University of Iowa. Used by permission.

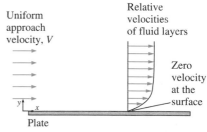

FIGURE 6–4
A fluid flowing over a stationary surface comes to a complete stop at the surface because of the no-slip condition.

$$h = \frac{-k_{\text{fluid}}(\partial T/\partial y)_{y=0}}{T_s - T_\infty} \qquad (\text{W/m}^2 \cdot {}^\circ\text{C}) \tag{6–4}$$

for the determination of the *convection heat transfer coefficient* when the temperature distribution within the fluid is known.

The convection heat transfer coefficient, in general, varies along the flow (or *x*-) direction. The *average* or *mean* convection heat transfer coefficient for a surface in such cases is determined by properly averaging the *local* convection heat transfer coefficients over the entire surface.

Nusselt Number

In convection studies, it is common practice to nondimensionalize the governing equations and combine the variables, which group together into *dimensionless numbers* in order to reduce the number of total variables. It is also common practice to nondimensionalize the heat transfer coefficient h with the Nusselt number, defined as

$$\text{Nu} = \frac{hL_c}{k} \tag{6–5}$$

where k is the thermal conductivity of the fluid and L_c is the *characteristic length*. The Nusselt number is named after Wilhelm Nusselt, who made significant contributions to convective heat transfer in the first half of the twentieth century, and it is viewed as the *dimensionless convection heat transfer coefficient*.

To understand the physical significance of the Nusselt number, consider a fluid layer of thickness L and temperature difference $\Delta T = T_2 - T_1$, as shown in Fig. 6–5. Heat transfer through the fluid layer is by *convection* when the fluid involves some motion and by *conduction* when the fluid layer is motionless. Heat flux (the rate of heat transfer per unit surface area) in either case is

$$\dot{q}_{\text{conv}} = h\Delta T \tag{6–6}$$

and

$$\dot{q}_{\text{cond}} = k\frac{\Delta T}{L} \tag{6–7}$$

Taking their ratio gives

$$\frac{\dot{q}_{\text{conv}}}{\dot{q}_{\text{cond}}} = \frac{h\Delta T}{k\Delta T/L} = \frac{hL}{k} = \text{Nu} \tag{6–8}$$

which is the Nusselt number. Therefore, the Nusselt number represents the enhancement of heat transfer through a fluid layer as a result of convection relative to conduction across the same fluid layer. The larger the Nusselt number, the more effective the convection. A Nusselt number of $\text{Nu} = 1$ for a fluid layer represents heat transfer across the layer by pure conduction.

We use forced convection in daily life more often than you might think (Fig. 6–6). We resort to forced convection whenever we want to increase the rate of heat transfer from a hot object. For example, we turn on the fan on hot

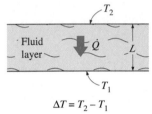

FIGURE 6–5

Heat transfer through a fluid layer of thickness L and temperature difference ΔT.

FIGURE 6–6

We resort to forced convection whenever we need to increase the rate of heat transfer.

summer days to help our body cool more effectively. The higher the fan speed, the better we feel. We *stir* our soup and *blow* on a hot slice of pizza to make them cool faster. The air on windy winter days feels much colder than it actually is. The simplest solution to heating problems in electronics packaging is to use a large enough fan.

6–2 ▪ CLASSIFICATION OF FLUID FLOWS

Convection heat transfer is closely tied with fluid mechanics, which is the science that deals with the behavior of fluids at rest or in motion, and the interaction of fluids with solids or other fluids at the boundaries. There is a wide variety of fluid flow problems encountered in practice, and it is usually convenient to classify them on the basis of some common characteristics to make it feasible to study them in groups. There are many ways to classify fluid flow problems, and here we present some general categories.

Viscous versus Inviscid Regions of Flow

When two fluid layers move relative to each other, a friction force develops between them and the slower layer tries to slow down the faster layer. This internal resistance to flow is quantified by the fluid property *viscosity*, which is a measure of internal stickiness of the fluid. Viscosity is caused by cohesive forces between the molecules in liquids and by molecular collisions in gases. There is no fluid with zero viscosity, and thus all fluid flows involve viscous effects to some degree. Flows in which the frictional effects are significant are called **viscous flows**. However, in many flows of practical interest, there are *regions* (typically regions not close to solid surfaces) where viscous forces are negligibly small compared to inertial or pressure forces. Neglecting the viscous terms in such **inviscid flow regions** greatly simplifies the analysis without much loss in accuracy.

The development of viscous and inviscid regions of flow as a result of inserting a flat plate parallel into a fluid stream of uniform velocity is shown in Fig. 6–7. The fluid sticks to the plate on both sides because of the no-slip condition, and the thin boundary layer in which the viscous effects are significant near the plate surface is the *viscous flow region*. The region of flow on both sides away from the plate and unaffected by the presence of the plate is the *inviscid flow region*.

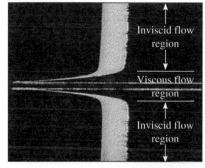

FIGURE 6–7
The flow of an originally uniform fluid stream over a flat plate, and the regions of viscous flow (next to the plate on both sides) and inviscid flow (away from the plate).

Fundamentals of Boundary Layers, National Committee from Fluid Mechanics Films, © Education Development Center.

Internal versus External Flow

A fluid flow is classified as being internal or external, depending on whether the fluid is forced to flow in a confined channel or over a surface. The flow of an unbounded fluid over a surface such as a plate, a wire, or a pipe is **external flow**. The flow in a pipe or duct is **internal flow** if the fluid is completely bounded by solid surfaces. Water flow in a pipe, for example, is internal flow, and airflow over a ball or over an exposed pipe during a windy day is external flow (Fig. 6–8). The flow of liquids in a duct is called *open-channel flow* if the duct is only partially filled with the liquid and there is a free surface. The flows of water in rivers and irrigation ditches are examples of such flows.

Internal flows are dominated by the influence of viscosity throughout the flow field. In external flows the viscous effects are limited to boundary layers near solid surfaces and to wake regions downstream of bodies.

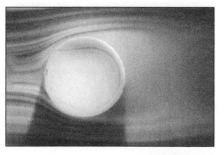

FIGURE 6–8
External flow over a tennis ball, and the turbulent wake region behind.

Courtesy NASA and Cislunar Aerospace, Inc.

Compressible versus Incompressible Flow

A flow is classified as being *compressible* or *incompressible,* depending on the level of variation of density during flow. Incompressibility is an approximation, and a flow is said to be **incompressible** if the density remains nearly constant throughout. Therefore, the volume of every portion of fluid remains unchanged over the course of its motion when the flow (or the fluid) is incompressible.

The densities of liquids are essentially constant, and thus the flow of liquids is typically incompressible. Therefore, liquids are usually referred to as *incompressible substances.* A pressure of 210 atm, for example, causes the density of liquid water at 1 atm to change by just 1 percent. Gases, on the other hand, are highly compressible. A pressure change of just 0.01 atm, for example, causes a change of 1 percent in the density of atmospheric air.

Liquid flows are incompressible to a high level of accuracy, but the level of variation in density in gas flows and the consequent level of approximation made when modeling gas flows as incompressible depends on the Mach number defined as Ma $= V/c$, where c is the **speed of sound** whose value is 346 m/s in air at room temperature at sea level. Gas flows can often be approximated as incompressible if the density changes are under about 5 percent, which is usually the case when Ma < 0.3. Therefore, the compressibility effects of air can be neglected at speeds under about 100 m/s. Note that the flow of a gas is not necessarily a compressible flow.

Small density changes of liquids corresponding to large pressure changes can still have important consequences. The irritating "water hammer" in a water pipe, for example, is caused by the vibrations of the pipe generated by the reflection of pressure waves following the sudden closing of the valves.

Laminar versus Turbulent Flow

Some flows are smooth and orderly while others are rather chaotic. The highly ordered fluid motion characterized by smooth layers of fluid is called **laminar**. The word *laminar* comes from the movement of adjacent fluid particles together in "laminates." The flow of high-viscosity fluids such as oils at low velocities is typically laminar. The highly disordered fluid motion that typically occurs at high velocities and is characterized by velocity fluctuations is called **turbulent** (Fig. 6–9). The flow of low-viscosity fluids such as air at high velocities is typically turbulent. The flow regime greatly influences the required power for pumping. A flow that alternates between being laminar and turbulent is called **transitional**.

Natural (or Unforced) versus Forced Flow

A fluid flow is said to be natural or forced, depending on how the fluid motion is initiated. In **forced flow**, a fluid is forced to flow over a surface or in a pipe by external means such as a pump or a fan. In **natural flows**, any fluid motion is due to natural means such as the buoyancy effect, which manifests itself as the rise of the warmer (and thus lighter) fluid and the fall of cooler (and thus denser) fluid (Fig. 6–10). In solar hot-water systems, for example, the thermosiphoning effect is commonly used to replace pumps by placing the water tank sufficiently above the solar collectors.

Laminar

Transitional

Turbulent

FIGURE 6–9
Laminar, transitional, and turbulent flows.
Courtesy ONERA, photograph by Werlé.

Steady versus Unsteady Flow

The terms *steady* and *uniform* are used frequently in engineering, and thus it is important to have a clear understanding of their meanings. The term **steady** implies *no change at a point with time.* The opposite of steady is **unsteady**. The term **uniform** implies *no change with location* over a specified region. These meanings are consistent with their everyday use (steady girlfriend, uniform distribution, etc.).

The terms *unsteady* and *transient* are often used interchangeably, but these terms are not synonyms. In fluid mechanics, *unsteady* is the most general term that applies to any flow that is not steady, but **transient** is typically used for developing flows. When a rocket engine is fired up, for example, there are transient effects (the pressure builds up inside the rocket engine, the flow accelerates, etc.) until the engine settles down and operates steadily. The term **periodic** refers to the kind of unsteady flow in which the flow oscillates about a steady mean.

Many devices such as turbines, compressors, boilers, condensers, and heat exchangers operate for long periods of time under the same conditions, and they are classified as *steady-flow devices.* (Note that the flow field near the rotating blades of a turbomachine is of course unsteady, but we consider the overall flow field rather than the details at some localities when we classify devices.) During steady flow, the fluid properties can change from point to point within a device, but at any fixed point they remain constant. Therefore, the volume, the mass, and the total energy content of a steady-flow device or flow section remain constant in steady operation.

Steady-flow conditions can be closely approximated by devices that are intended for continuous operation such as turbines, pumps, boilers, condensers, and heat exchangers of power plants or refrigeration systems. Some cyclic devices, such as reciprocating engines or compressors, do not satisfy the steady-flow conditions since the flow at the inlets and the exits is pulsating and not steady. However, the fluid properties vary with time in a periodic manner, and the flow through these devices can still be analyzed as a steady-flow process by using time-averaged values for the properties.

One-, Two-, and Three-Dimensional Flows

A flow field is best characterized by the velocity distribution, and thus a flow is said to be one-, two-, or three-dimensional if the flow velocity varies in one, two, or three primary dimensions, respectively. A typical fluid flow involves a three-dimensional geometry, and the velocity may vary in all three dimensions, rendering the flow three-dimensional [$\vec{V}$ (x, y, z) in rectangular or $\vec{V}$ (r, θ, z) in cylindrical coordinates]. However, the variation of velocity in certain directions can be small relative to the variation in other directions and can be ignored with negligible error. In such cases, the flow can be modeled conveniently as being one- or two-dimensional, which is easier to analyze.

Consider steady flow of a fluid through a circular pipe attached to a large tank. The fluid velocity everywhere on the pipe surface is zero because of the no-slip condition, and the flow is two-dimensional in the entrance region of the pipe since the velocity changes in both the r- and z-directions. The velocity profile develops fully and remains unchanged after some distance from the

FIGURE 6–10

In this schlieren image of a girl, the rise of lighter, warmer air adjacent to her body indicates that humans and warm-blooded animals are surrounded by thermal plumes of rising warm air.

G. S. Settles, Gas Dynamics Lab, Penn State University. Used by permission.

FIGURE 6–11
The development of the velocity profile in a circular pipe. $V = V(r, z)$ and thus the flow is two-dimensional in the entrance region, and becomes one-dimensional downstream when the velocity profile fully develops and remains unchanged in the flow direction, $V = V(r)$.

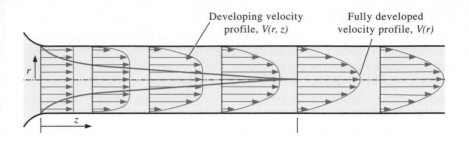

inlet (about 10 pipe diameters in turbulent flow, and less in laminar pipe flow, as in Fig. 6–11), and the flow in this region is said to be *fully developed*. The fully developed flow in a circular pipe is *one-dimensional* since the velocity varies in the radial r-direction but not in the angular θ- or axial z-directions, as shown in Fig. 6–11. That is, the velocity profile is the same at any axial z-location, and it is symmetric about the axis of the pipe.

Note that the dimensionality of the flow also depends on the choice of co-ordinate system and its orientation. The pipe flow discussed, for example, is one-dimensional in cylindrical coordinates, but two-dimensional in Cartesian coordinates—illustrating the importance of choosing the most appropriate co-ordinate system. Also note that even in this simple flow, the velocity cannot be uniform across the cross section of the pipe because of the no-slip condition. However, at a well-rounded entrance to the pipe, the velocity profile may be approximated as being nearly uniform across the pipe, since the velocity is nearly constant at all radii except very close to the pipe wall.

6–3 · VELOCITY BOUNDARY LAYER

Consider the parallel flow of a fluid over a *flat plate,* as shown in Fig. 6–12. Surfaces that are slightly contoured such as turbine blades can also be ap-proximated as flat plates with reasonable accuracy. The x-coordinate is mea-sured along the plate surface from the *leading edge* of the plate in the direction of the flow, and y is measured from the surface in the normal direction. The fluid approaches the plate in the x-direction with a uniform velocity V, which is practically identical to the free-stream velocity over the plate away from the

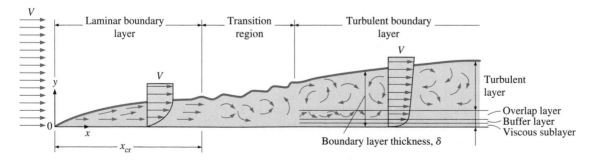

FIGURE 6–12
The development of the boundary layer for flow over a flat plate, and the different flow regimes.

surface (this would not be the case for cross flow over blunt bodies such as a cylinder).

For the sake of discussion, we can consider the fluid to consist of adjacent layers piled on top of each other. The velocity of the particles in the first fluid layer adjacent to the plate becomes zero because of the no-slip condition. This motionless layer slows down the particles of the neighboring fluid layer as a result of friction between the particles of these two adjoining fluid layers at different velocities. This fluid layer then slows down the molecules of the next layer, and so on. Thus, the presence of the plate is felt up to some normal distance δ from the plate beyond which the free-stream velocity remains essentially unchanged. As a result, the x-component of the fluid velocity, u, varies from 0 at $y = 0$ to nearly V at $y = \delta$ (Fig. 6–13).

The region of the flow above the plate bounded by δ in which the effects of the viscous shearing forces caused by fluid viscosity are felt is called the **velocity boundary layer**. The *boundary layer thickness, δ,* is typically defined as the distance y from the surface at which $u = 0.99V$.

The hypothetical line of $u = 0.99V$ divides the flow over a plate into two regions: the **boundary layer region**, in which the viscous effects and the velocity changes are significant, and the **irrotational flow region**, in which the frictional effects are negligible and the velocity remains essentially constant.

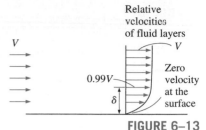

FIGURE 6–13

The development of a boundary layer on a surface is due to the no-slip condition and friction.

Surface Shear Stress

Consider the flow of a fluid over the surface of a plate. The fluid layer in contact with the surface tries to drag the plate along via friction, exerting a *friction force* on it. Likewise, a faster fluid layer tries to drag the adjacent slower layer and exert a friction force because of the friction between the two layers. Friction force per unit area is called **shear stress**, and is denoted by τ. Experimental studies indicate that the shear stress for most fluids is proportional to the *velocity gradient,* and the shear stress at the wall surface is expressed as

$$\tau_s = \mu \frac{\partial u}{\partial y}\bigg|_{y=0} \qquad (N/m^2) \qquad (6\text{–}9)$$

where the constant of proportionality μ is the **dynamic viscosity** of the fluid, whose unit is kg/m · s (or equivalently, N · s/m², or Pa · s, or poise = 0.1 Pa · s).

The fluids that that obey the linear relationship above are called **Newtonian fluids**, after Sir Isaac Newton who expressed it first in 1687. Most common fluids such as water, air, gasoline, and oils are Newtonian fluids. Blood and liquid plastics are examples of non-Newtonian fluids. In this text we consider Newtonian fluids only.

In fluid flow and heat transfer studies, the ratio of dynamic viscosity to density appears frequently. For convenience, this ratio is given the name **kinematic viscosity** v and is expressed as $v = \mu/\rho$. Two common units of kinematic viscosity are m²/s and *stoke* (1 stoke = 1 cm²/s = 0.0001 m²/s).

The viscosity of a fluid is a measure of its *resistance to deformation,* and it is a strong function of temperature. The viscosities of liquids *decrease* with temperature, whereas the viscosities of gases *increase* with temperature (Fig. 6–14). The viscosities of some fluids at 20°C are listed in Table 6–1. Note that the viscosities of different fluids differ by several orders of magnitude.

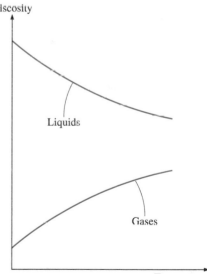

FIGURE 6–14

The viscosity of liquids decreases and the viscosity of gases increases with temperature.

TABLE 6–1

Dynamic viscosities of some fluids at 1 atm and 20°C (unless otherwise stated)

Fluid	Dynamic Viscosity μ, kg/m · s
Glycerin:	
−20°C	134.0
0°C	10.5
20°C	1.52
40°C	0.31
Engine oil:	
SAE 10W	0.10
SAE 10W30	0.17
SAE 30	0.29
SAE 50	0.86
Mercury	0.0015
Ethyl alcohol	0.0012
Water:	
0°C	0.0018
20°C	0.0010
100°C (liquid)	0.00028
100°C (vapor)	0.000012
Blood, 37°C	0.00040
Gasoline	0.00029
Ammonia	0.00015
Air	0.000018
Hydrogen, 0°C	0.0000088

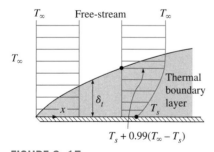

FIGURE 6–15
Thermal boundary layer on a flat plate (the fluid is hotter than the plate surface).

The determination of the surface shear stress τ_s from Eq. 6–9 is not practical since it requires a knowledge of the flow velocity profile. A more practical approach in external flow is to relate τ_s to the upstream velocity V as

$$\tau_s = C_f \frac{\rho V^2}{2} \quad (\text{N/m}^2) \tag{6–10}$$

where C_f is the dimensionless **friction coefficient**, whose value in most cases is determined experimentally, and ρ is the density of the fluid. Note that the friction coefficient, in general, varies with location along the surface. Once the average friction coefficient over a given surface is available, the friction force over the entire surface is determined from

$$F_f = C_f A_s \frac{\rho V^2}{2} \quad (\text{N}) \tag{6–11}$$

where A_s is the surface area.

The friction coefficient is an important parameter in heat transfer studies since it is directly related to the heat transfer coefficient and the power requirements of the pump or fan.

6–4 · THERMAL BOUNDARY LAYER

We have seen that a velocity boundary layer develops when a fluid flows over a surface as a result of the fluid layer adjacent to the surface assuming the surface velocity (i.e., zero velocity relative to the surface). Also, we defined the velocity boundary layer as the region in which the fluid velocity varies from zero to $0.99V$. Likewise, a *thermal boundary layer* develops when a fluid at a specified temperature flows over a surface that is at a different temperature, as shown in Fig. 6–15.

Consider the flow of a fluid at a uniform temperature of T_∞ over an isothermal flat plate at temperature T_s. The fluid particles in the layer adjacent to the surface reach thermal equilibrium with the plate and assume the surface temperature T_s. These fluid particles then exchange energy with the particles in the adjoining-fluid layer, and so on. As a result, a temperature profile develops in the flow field that ranges from T_s at the surface to T_∞ sufficiently far from the surface. The flow region over the surface in which the temperature variation in the direction normal to the surface is significant is the **thermal boundary layer**. The *thickness* of the thermal boundary layer δ_t at any location along the surface is defined as *the distance from the surface at which the temperature difference $T - T_s$ equals $0.99(T_\infty - T_s)$. Note that for the special case of $T_s = 0$, we have $T = 0.99T_\infty$ at the outer edge of the thermal boundary layer, which is analogous to $u = 0.99V$ for the velocity boundary layer.

The thickness of the thermal boundary layer increases in the flow direction, since the effects of heat transfer are felt at greater distances from the surface further down stream.

The convection heat transfer rate anywhere along the surface is directly related to the temperature gradient at that location. Therefore, the shape of the temperature profile in the thermal boundary layer dictates the convection heat transfer between a solid surface and the fluid flowing over it. In flow over a heated (or cooled) surface, both velocity and thermal boundary layers develop simultaneously. Noting that the fluid velocity has a strong influence on the temperature profile, the development of the velocity boundary layer relative to the thermal boundary layer will have a strong effect on the convection heat transfer.

Prandtl Number

The relative thickness of the velocity and the thermal boundary layers is best described by the *dimensionless* parameter **Prandtl number**, defined as

$$\text{Pr} = \frac{\text{Molecular diffusivity of momentum}}{\text{Molecular diffusivity of heat}} = \frac{v}{\alpha} = \frac{\mu c_p}{k} \qquad (6\text{-}12)$$

It is named after Ludwig Prandtl, who introduced the concept of boundary layer in 1904 and made significant contributions to boundary layer theory. The Prandtl numbers of fluids range from less than 0.01 for liquid metals to more than 100,000 for heavy oils (Table 6–2). Note that the Prandtl number is in the order of 10 for water.

The Prandtl numbers of gases are about 1, which indicates that both momentum and heat dissipate through the fluid at about the same rate. Heat diffuses very quickly in liquid metals ($\text{Pr} \ll 1$) and very slowly in oils ($\text{Pr} \gg 1$) relative to momentum. Consequently the thermal boundary layer is much thicker for liquid metals and much thinner for oils relative to the velocity boundary layer.

6–5 · LAMINAR AND TURBULENT FLOWS

If you have been around smokers, you probably noticed that the cigarette smoke rises in a smooth plume for the first few centimeters and then starts fluctuating randomly in all directions as it continues its rise. Other plumes behave similarly (Fig. 6–16). Likewise, a careful inspection of flow in a pipe reveals that the fluid flow is streamlined at low velocities but turns chaotic as the velocity is increased above a critical value, as shown in Figure 6–17. The flow regime in the first case is said to be **laminar**, characterized by *smooth streamlines* and *highly-ordered motion,* and **turbulent** in the second case, where it is characterized by *velocity fluctuations* and *highly-disordered motion.* The **transition** from laminar to turbulent flow does not occur suddenly; rather, it occurs over some region in which the flow fluctuates between laminar and turbulent flows before it becomes fully turbulent. Most flows encountered in practice are turbulent. Laminar flow is encountered when highly viscous fluids such as oils flow in small pipes or narrow passages.

We can verify the existence of these laminar, transitional, and turbulent flow regimes by injecting some dye streak into the flow in a glass tube, as the British scientist Osborn Reynolds (1842–1912) did over a century ago. We observe that the dye streak forms a *straight and smooth line* at low velocities when the flow is laminar (we may see some blurring because of molecular diffusion), has *bursts of fluctuations* in the transitional regime, and *zigzags rapidly and randomly* when the flow becomes fully turbulent. These zigzags and the dispersion of the dye are indicative of the fluctuations in the main flow and the rapid mixing of fluid particles from adjacent layers.

Typical average velocity profiles in laminar and turbulent flow are also given in Fig. 6–12. Note that the velocity profile in turbulent flow is much fuller than that in laminar flow, with a sharp drop near the surface. The turbulent boundary

TABLE 6–2

Typical ranges of Prandtl numbers for common fluids

Fluid	Pr
Liquid metals	0.004–0.030
Gases	0.7–1.0
Water	1.7–13.7
Light organic fluids	5–50
Oils	50–100,000
Glycerin	2000–100,000

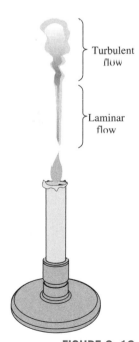

FIGURE 6–16
Laminar and turbulent flow regimes of candle smoke.

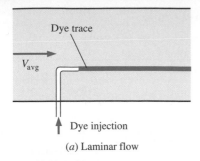

(a) Laminar flow

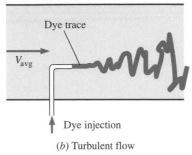

(b) Turbulent flow

FIGURE 6–17
The behavior of colored fluid injected into the flow in laminar and turbulent flows in a pipe.

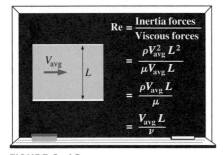

FIGURE 6–18
The Reynolds number can be viewed as the ratio of inertial forces to viscous forces acting on a fluid element.

layer can be considered to consist of four regions, characterized by the distance from the wall. The very thin layer next to the wall where viscous effects are dominant is the **viscous sublayer**. The velocity profile in this layer is very nearly *linear*, and the flow is streamlined. Next to the viscous sublayer is the **buffer layer**, in which turbulent effects are becoming significant, but the flow is still dominated by viscous effects. Above the buffer layer is the **overlap layer**, in which the turbulent effects are much more significant, but still not dominant. Above that is the **turbulent layer** in which turbulent effects dominate over viscous effects.

The *intense mixing* of the fluid in turbulent flow as a result of rapid fluctuations enhances heat and momentum transfer between fluid particles, which increases the friction force on the surface and the convection heat transfer rate. It also causes the boundary layer to enlarge. Both the friction and heat transfer coefficients reach maximum values when the flow becomes *fully turbulent*. So it will come as no surprise that a special effort is made in the design of heat transfer coefficients associated with turbulent flow. The enhancement in heat transfer in turbulent flow does not come for free, however. It may be necessary to use a larger pump to overcome the larger friction forces accompanying the higher heat transfer rate.

Reynolds Number

The transition from laminar to turbulent flow depends on the *surface geometry, surface roughness, flow velocity, surface temperature,* and *type of fluid,* among other things. After exhaustive experiments in the 1880s, Osborn Reynolds discovered that the flow regime depends mainly on the ratio of the *inertia forces* to *viscous forces* in the fluid. This ratio is called the **Reynolds number**, which is a *dimensionless* quantity, and is expressed for external flow as (Fig. 6–18)

$$\text{Re} = \frac{\text{Inertia forces}}{\text{Viscous}} = \frac{VL_c}{v} = \frac{\rho VL_c}{\mu} \qquad (6\text{–}13)$$

where V is the upstream velocity (equivalent to the free-stream velocity for a flat plate), L_c is the characteristic length of the geometry, and $v = \mu/\rho$ is the kinematic viscosity of the fluid. For a flat plate, the characteristic length is the distance x from the leading edge. Note that kinematic viscosity has the unit m²/s, which is identical to the unit of thermal diffusivity, and can be viewed as *viscous diffusivity* or *diffusivity for momentum.*

At *large* Reynolds numbers, the inertia forces, which are proportional to the density and the velocity of the fluid, are large relative to the viscous forces, and thus the viscous forces cannot prevent the random and rapid fluctuations of the fluid. At *small* or *moderate* Reynolds numbers, however, the viscous forces are large enough to suppress these fluctuations and to keep the fluid "in line." Thus the flow is *turbulent* in the first case and *laminar* in the second.

The Reynolds number at which the flow becomes turbulent is called the **critical Reynolds number**. The value of the critical Reynolds number is different for different geometries and flow conditions. For flow over a flat plate, the generally accepted value of the critical Reynolds number is $\text{Re}_{cr} = Vx_{cr}/v = 5 \times 10^5$, where x_{cr} is the distance from the leading edge of the plate at which transition from laminar to turbulent flow occurs. The value of Re_{cr} may change substantially, however, depending on the level of turbulence in the free stream.

6–6 · HEAT AND MOMENTUM TRANSFER IN TURBULENT FLOW

Most flows encountered in engineering practice are turbulent, and thus it is important to understand how turbulence affects wall shear stress and heat transfer. However, turbulent flow is a complex mechanism dominated by fluctuations, and despite tremendous amounts of work done in this area by researchers, the theory of turbulent flow remains largely undeveloped. Therefore, we must rely on experiments and the empirical or semi-empirical correlations developed for various situations.

Turbulent flow is characterized by random and rapid fluctuations of swirling regions of fluid, called **eddies**, throughout the flow. These fluctuations provide an additional mechanism for momentum and energy transfer. In laminar flow, fluid particles flow in an orderly manner along pathlines, and momentum and energy are transferred across streamlines by molecular diffusion. In turbulent flow, the swirling eddies transport mass, momentum, and energy to other regions of flow much more rapidly than molecular diffusion, greatly enhancing mass, momentum, and heat transfer. As a result, turbulent flow is associated with much higher values of friction, heat transfer, and mass transfer coefficients (Fig. 6–19).

Even when the average flow is steady, the eddy motion in turbulent flow causes significant fluctuations in the values of velocity, temperature, pressure, and even density (in compressible flow). Figure 6–20 shows the variation of the instantaneous velocity component u with time at a specified location, as can be measured with a hot-wire anemometer probe or other sensitive device. We observe that the instantaneous values of the velocity fluctuate about an average value, which suggests that the velocity can be expressed as the sum of an *average value* $\bar{u}$ and a *fluctuating component* u',

$$u = \bar{u} + u' \tag{6–14}$$

This is also the case for other properties such as the velocity component v in the y-direction, and thus $v = \bar{v} + v'$, $P = \bar{P} + P'$, and $T = \bar{T} + T'$. The average value of a property at some location is determined by averaging it over a time interval that is sufficiently large so that the time average levels off to a constant. Therefore, the time average of fluctuating components is zero, e.g., $\overline{u'} = 0$. The magnitude of u' is usually just a few percent of $\bar{u}$, but the high frequencies of eddies (in the order of a thousand per second) makes them very effective for the transport of momentum, thermal energy, and mass. In time-averaged *stationary* turbulent flow, the average values of properties (indicated by an overbar) are independent of time. The chaotic fluctuations of fluid particles play a dominant role in pressure drop, and these random motions must be considered in analyses together with the average velocity.

Perhaps the first thought that comes to mind is to determine the shear stress in an analogous manner to laminar flow from $\tau = -\mu\, d\bar{u}/dr$, where $\bar{u}(r)$ is the average velocity profile for turbulent flow. But the experimental studies show that this is not the case, and the shear stress is much larger due to the turbulent fluctuations. Therefore, it is convenient to think of the turbulent shear stress as consisting of two parts: the *laminar component,* which accounts for

(a) Before turbulence   (b) After turbulence

FIGURE 6–19

The intense mixing in turbulent flow brings fluid particles at different temperatures into close contact, and thus enhances heat transfer.

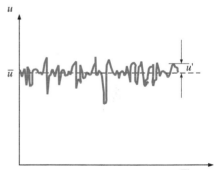

FIGURE 6–20

Fluctuations of the velocity component u with time at a specified location in turbulent flow.

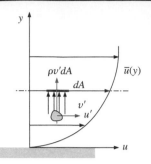

FIGURE 6–21
Fluid particle moving upward through a differential area dA as a result of the velocity fluctuation v'.

the friction between layers in the flow direction (expressed as $\tau_{lam} = -\mu$ $d\overline{u}/dr$), and the *turbulent component,* which accounts for the friction between the fluctuating fluid particles and the fluid body (denoted as τ_{turb} and is related to the fluctuation components of velocity).

Consider turbulent flow in a horizontal pipe, and the upward eddy motion of fluid particles in a layer of lower velocity to an adjacent layer of higher velocity through a differential area dA as a result of the velocity fluctuation v', as shown in Fig. 6–21. The mass flow rate of the fluid particles rising through dA is $\rho v' dA$, and its net effect on the layer above dA is a reduction in its average flow velocity because of momentum transfer to the fluid particles with lower average flow velocity. This momentum transfer causes the horizontal velocity of the fluid particles to increase by u', and thus its momentum in the horizontal direction to increase at a rate of $(\rho v' dA)u'$, which must be equal to the decrease in the momentum of the upper fluid layer.

Noting that force in a given direction is equal to the rate of change of momentum in that direction, the horizontal force acting on a fluid element above dA due to the passing of fluid particles through dA is $\delta F = (\rho v' dA)(-u') = -\rho u'v' dA$. Therefore, the shear force per unit area due to the eddy motion of fluid particles $\delta F/dA = -\rho u'v'$ can be viewed as the instantaneous turbulent shear stress. Then the **turbulent shear stress** can be expressed as $\tau_{turb} = -\rho \overline{u'v'}$ where $\overline{u'v'}$ is the time average of the product of the fluctuating velocity components u' and v'. Similarly, considering that $h = c_p T$ represents the energy of the fluid and T' is the eddy temperature relative to the mean value, the rate of thermal energy transport by turbulent eddies is $\dot{q}_{turb} = \rho c_p \overline{v'T'}$. Note that $\overline{u'v'} \neq 0$ even though $\overline{u'} = 0$ and $\overline{v'} = 0$ (and thus $\overline{u'}\,\overline{v'} = 0$), and experimental results show that $\overline{u'v'}$ is usually a negative quantity. Terms such as $-\rho \overline{u'v'}$ or $-\rho \overline{u'^2}$ are called **Reynolds stresses** or **turbulent stresses**.

The random eddy motion of groups of particles resembles the random motion of molecules in a gas—colliding with each other after traveling a certain distance and exchanging momentum and heat in the process. Therefore, momentum and heat transport by eddies in turbulent boundary layers is analogous to the molecular momentum and heat diffusion. Then turbulent wall shear stress and turbulent heat transfer can be expressed in an analogous manner as

$$\tau_{turb} = -\rho \overline{u'v'} = \mu_t \frac{\partial \overline{u}}{\partial y} \qquad \text{and} \qquad \dot{q}_{turb} = \rho c_p \overline{v'T'} = -k_t \frac{\partial \overline{T}}{\partial y} \qquad \textbf{(6–15)}$$

where μ_t is called the **turbulent** (or **eddy**) **viscosity**, which accounts for momentum transport by turbulent eddies, and k_t is called the **turbulent** (or **eddy**) **thermal conductivity**, which accounts for thermal energy transport by turbulent eddies. Then the total shear stress and total heat flux can be expressed conveniently as

$$\tau_{total} = (\mu + \mu_t) \frac{\partial \overline{u}}{\partial y} = \rho(\nu + \nu_t) \frac{\partial \overline{u}}{\partial y} \qquad \textbf{(6–16)}$$

and

$$\dot{q}_{total} = -(k + k_t) \frac{\partial \overline{T}}{\partial y} = -\rho c_p(\alpha + \alpha_t) \frac{\partial \overline{T}}{\partial y} \qquad \textbf{(6–17)}$$

where $\nu_t = \mu_t/\rho$ is the **kinematic eddy viscosity** (or **eddy diffusivity of momentum**) and $\alpha_t = k_t/\rho c_p$ is the **eddy thermal diffusivity** (or **eddy diffusivity of heat**).

Eddy motion and thus eddy diffusivities are much larger than their molecular counterparts in the core region of a turbulent boundary layer. The eddy motion loses its intensity close to the wall, and diminishes at the wall because of the no-slip condition. Therefore, the velocity and temperature profiles are very slowly changing in the core region of a turbulent boundary layer, but very steep in the thin layer adjacent to the wall, resulting in large velocity and temperature gradients at the wall surface. So it is no surprise that the wall shear stress and wall heat flux are much larger in turbulent flow than they are in laminar flow (Fig. 6–22).

Note that molecular diffusivities ν and α (as well as μ and k) are fluid properties, and their values can be found listed in fluid handbooks. Eddy diffusivities ν_t and α_t (as well as μ_t and k_t), however are *not* fluid properties and their values depend on flow conditions. Eddy diffusivities ν_t and α_t decrease towards the wall, becoming zero at the wall. Their values range from zero at the wall to several thousand times the values of molecular diffusivities in the core region.

6–7 · DERIVATION OF DIFFERENTIAL CONVECTION EQUATIONS*

In this section we derive the governing equations of fluid flow in the boundary layers. To keep the analysis at a manageable level, we assume the flow to be steady and two-dimensional, and the fluid to be Newtonian with constant properties (density, viscosity, thermal conductivity, etc.).

Consider the parallel flow of a fluid over a surface. We take the flow direction along the surface to be x and the direction normal to the surface to be y, and we choose a differential volume element of length dx, height dy, and unit depth in the z-direction (normal to the paper) for analysis (Fig. 6–23). The fluid flows over the surface with a uniform free-stream velocity V, but the velocity within boundary layer is two-dimensional: the x-component of the velocity is u, and the y-component is v. Note that $u = u(x, y)$ and $v = v(x, y)$ in steady two-dimensional flow.

Next we apply three fundamental laws to this fluid element: Conservation of mass, conservation of momentum, and conservation of energy to obtain the continuity, momentum, and energy equations for laminar flow in boundary layers.

The Continuity Equation

The conservation of mass principle is simply a statement that mass cannot be created or destroyed during a process and all the mass must be accounted for during an analysis. In steady flow, the amount of mass within the control volume remains constant, and thus the conservation of mass can be expressed as

---

*This and the upcoming sections of this chapter deal with theoretical aspects of convection, and can be skipped and be used as a reference if desired without a loss in continuity.

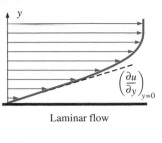

Laminar flow

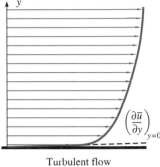

Turbulent flow

FIGURE 6–22
The velocity gradients at the wall, and thus the wall shear stress, are much larger for turbulent flow than they are for laminar flow, even though the turbulent boundary layer is thicker than the laminar one for the same value of free-stream velocity.

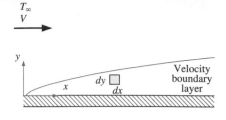

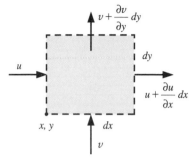

FIGURE 6–23
Differential control volume used in the derivation of mass balance in velocity boundary layer in two-dimensional flow over a surface.

$$\left(\begin{array}{c} \text{Rate of mass flow} \\ \text{into the control volume} \end{array}\right) = \left(\begin{array}{c} \text{Rate of mass flow} \\ \text{out of the control volume} \end{array}\right) \qquad \text{(6–18)}$$

Noting that mass flow rate is equal to the product of density, average velocity, and cross-sectional area normal to flow, the rate at which fluid enters the control volume from the left surface is $\rho u(dy \cdot 1)$. The rate at which the fluid leaves the control volume from the right surface can be expressed as

$$\rho\left(u + \frac{\partial u}{\partial x} dx\right)(dy \cdot 1) \qquad \text{(6–19)}$$

Repeating this for the y direction and substituting the results into Eq. 6–18, we obtain

$$\rho u(dy \cdot 1) + \rho v(dx \cdot 1) = \rho\left(u + \frac{\partial u}{\partial x} dx\right)(dy \cdot 1) + \rho\left(v + \frac{\partial v}{\partial y} dy\right)(dx \cdot 1) \qquad \text{(6–20)}$$

Simplifying and dividing by $dx \cdot dy \cdot 1$ gives

$$\frac{\partial u}{\partial x} + \frac{\partial v}{\partial y} = 0 \qquad \text{(6–21)}$$

This is the *conservation of mass* relation in differential form, which is also known as the **continuity equation** or **mass balance** for steady two-dimensional flow of a fluid with constant density.

The Momentum Equations

The differential forms of the equations of motion in the velocity boundary layer are obtained by applying Newton's second law of motion to a differential control volume element in the boundary layer. Newton's second law is an expression for momentum balance and can be stated as *the net force acting on the control volume is equal to the mass times the acceleration of the fluid element within the control volume, which is also equal to the net rate of momentum outflow from the control volume.*

The forces acting on the control volume consist of *body forces* that act throughout the entire body of the control volume (such as gravity, electric, and magnetic forces) and are proportional to the volume of the body, and *surface forces* that act on the control surface (such as the pressure forces due to hydrostatic pressure and shear stresses due to viscous effects) and are proportional to the surface area. The surface forces appear as the control volume is isolated from its surroundings for analysis, and the effect of the detached body is replaced by a force at that location. Note that pressure represents the compressive force applied on the fluid element by the surrounding fluid, and is always directed to the surface.

We express Newton's second law of motion for the control volume as

$$(\text{Mass})\left(\begin{array}{c} \text{Acceleration} \\ \text{in a specified direction} \end{array}\right) = \left(\begin{array}{c} \text{Net force (body and surface)} \\ \text{acting in that direction} \end{array}\right) \qquad \text{(6–22)}$$

or

$$\delta m \cdot a_x = F_{\text{surface}, x} + F_{\text{body}, x} \qquad \text{(6–23)}$$

where the mass of the fluid element within the control volume is

$$\delta m = \rho(dx \cdot dy \cdot 1) \tag{6-24}$$

Noting that flow is steady and two-dimensional and thus $u = u(x, y)$, the total differential of u is

$$du = \frac{\partial u}{\partial x} dx + \frac{\partial u}{\partial y} dy \tag{6-25}$$

Then the acceleration of the fluid element in the x direction becomes

$$a_x = \frac{du}{dt} = \frac{\partial u}{\partial x}\frac{dx}{dt} + \frac{\partial u}{\partial y}\frac{dy}{dt} = u\frac{\partial u}{\partial x} + v\frac{\partial u}{\partial y} \tag{6-26}$$

You may be tempted to think that acceleration is zero in steady flow since acceleration is the rate of change of velocity with time, and in steady flow there is no change with time. Well, a garden hose nozzle tells us that this understanding is not correct. Even in steady flow and thus constant mass flow rate, water accelerates through the nozzle (Fig. 6–24). *Steady* simply means no change with time at a specified location (and thus $\partial u/\partial t = 0$), but the value of a quantity may change from one location to another (and thus $\partial u/\partial x$ and $\partial u/\partial y$ may be different from zero). In the case of a nozzle, the velocity of water remains constant at a specified point, but it changes from inlet to the exit (water accelerates along the nozzle, which is the reason for attaching a nozzle to the garden hose in the first place).

The forces acting on a surface are due to pressure and viscous effects. In two-dimensional flow, the *viscous stress* at any point on an imaginary surface within the fluid can be resolved into two perpendicular components: one normal to the surface called *normal stress* (which should not be confused with pressure) and another along the surface called *shear stress*. The normal stress is related to the velocity gradients $\partial u/\partial x$ and $\partial v/\partial y$, that are much smaller than $\partial u/\partial y$, to which shear stress is related. Neglecting the normal stresses for simplicity, the surface forces acting on the control volume in the x-direction are as shown in Fig. 6–25. Then the net surface force acting in the x-direction becomes

$$F_{\text{surface}, x} = \left(\frac{\partial \tau}{\partial y} dy\right)(dx \cdot 1) - \left(\frac{\partial P}{\partial x} dx\right)(dy \cdot 1) = \left(\frac{\partial \tau}{\partial y} - \frac{\partial P}{\partial x}\right)(dx \cdot dy \cdot 1)$$

$$= \left(\mu\frac{\partial^2 u}{\partial y^2} - \frac{\partial P}{\partial x}\right)(dx \cdot dy \cdot 1) \tag{6-27}$$

since $\tau = \mu(\partial u/\partial y)$. Substituting Eqs. 6–21, 6–23, and 6–24 into Eq. 6–20 and dividing by $dx \cdot dy \cdot 1$ gives

$$\rho\left(u\frac{\partial u}{\partial x} + v\frac{\partial u}{\partial y}\right) = \mu\frac{\partial^2 u}{\partial y^2} - \frac{\partial P}{\partial x} \tag{6-28}$$

This is the relation for the **momentum balance** in the x-direction, and is known as the **x-momentum equation**. Note that we would obtain the same result if we used momentum flow rates for the left-hand side of this equation instead of mass times acceleration. If there is a body force acting in the x-direction, it can be added to the right side of the equation provided that it is expressed per unit volume of the fluid.

FIGURE 6–24
During steady flow, a fluid may not accelerate in time at a fixed point, but it may accelerate in space.

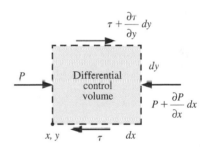

FIGURE 6–25
Differential control volume used in the derivation of x-momentum equation in velocity boundary layer in two-dimensional flow over a surface.

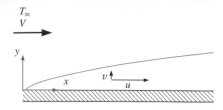

1) Velocity components:
$$v \ll u$$

2) Velocity gradients:
$$\frac{\partial v}{\partial x} \ll 0, \frac{\partial v}{\partial y} \ll 0$$
$$\frac{\partial u}{\partial x} \ll \frac{\partial u}{\partial y}$$

3) Temperature gradients:
$$\frac{\partial T}{\partial x} \ll \frac{\partial T}{\partial y}$$

FIGURE 6–26
Boundary layer approximations.

In a boundary layer, the velocity component in the flow direction is much larger than that in the normal direction, and thus $u \gg v$, and $\partial v/\partial x$ and $\partial v/\partial y$ are negligible. Also, u varies greatly with y in the normal direction from zero at the wall surface to nearly the free-stream value across the relatively thin boundary layer, while the variation of u with x along the flow is typically small. Therefore, $\partial u/\partial y \gg \partial u/\partial x$. Similarly, if the fluid and the wall are at different temperatures and the fluid is heated or cooled during flow, heat conduction occurs primarily in the direction normal to the surface, and thus $\partial T/\partial y \gg \partial T/\partial x$. That is, the velocity and temperature gradients normal to the surface are much greater than those along the surface. These simplifications are known as the **boundary layer approximations**. These approximations greatly simplify the analysis usually with little loss in accuracy, and make it possible to obtain analytical solutions for certain types of flow problems (Fig. 6–26).

When gravity effects and other body forces are negligible and the boundary layer approximations are valid, applying Newton's second law of motion on the volume element in the y-direction gives the y-momentum equation to be

$$\frac{\partial P}{\partial y} = 0 \tag{6–29}$$

That is, *the variation of pressure in the direction normal to the surface is negligible,* and thus $P = P(x)$ and $\partial P/\partial x = dP/dx$. Then it follows that for a given x, the pressure in the boundary layer is equal to the pressure in the free stream, and the pressure determined by a separate analysis of fluid flow in the free stream (which is typically easier because of the absence of viscous effects) can readily be used in the boundary layer analysis.

The velocity components in the free stream region of a flat plate are $u = V$ = constant and $v = 0$. Substituting these into the x-momentum equations (Eq. 6–28) gives $\partial P/\partial x = 0$. Therefore, for flow over a flat plate, the pressure remains constant over the entire plate (both inside and outside the boundary layer).

Conservation of Energy Equation

The energy balance for any system undergoing any process is expressed as $E_{in} - E_{out} = \Delta E_{system}$, which states that the change in the energy content of a system during a process is equal to the difference between the energy input and the energy output. During a *steady-flow process,* the total energy content of a control volume remains constant (and thus $\Delta E_{system} = 0$), and the amount of energy entering a control volume in all forms must be equal to the amount of energy leaving it. Then the rate form of the general energy equation reduces for a steady-flow process to $\dot{E}_{in} - \dot{E}_{out} = 0$.

Noting that energy can be transferred by heat, work, and mass only, the energy balance for a steady-flow control volume can be written explicitly as

$$(\dot{E}_{in} - \dot{E}_{out})_{by\ heat} + (\dot{E}_{in} - \dot{E}_{out})_{by\ work} + (\dot{E}_{in} - \dot{E}_{out})_{by\ mass} = 0 \tag{6–30}$$

The total energy of a flowing fluid stream per unit mass is $e_{stream} = h + ke + pe$ where h is the enthalpy (which is the sum of internal energy and flow energy), $pe = gz$ is the potential energy, and $ke = V^2/2 = (u^2 + v^2)/2$ is the kinetic energy of the fluid per unit mass. The kinetic and potential energies are usually very small relative to enthalpy, and therefore it is common practice to neglect them (besides, it can be shown that if kinetic energy is included in the following analysis, all the terms due to this inclusion cancel each other). We assume the density ρ, specific heat c_p, viscosity μ, and the thermal conductivity k of the fluid to be constant. Then the energy of the fluid per unit mass can be expressed as $e_{\text{stream}} = h = c_p T$.

Energy is a scalar quantity, and thus energy interactions in all directions can be combined in one equation. Noting that mass flow rate of the fluid entering the control volume from the left is $\rho u(dy \cdot 1)$, the rate of energy transfer to the control volume by mass in the x-direction is, from Fig. 6–27,

$$
\begin{aligned}
(\dot{E}_{\text{in}} - \dot{E}_{\text{out}})_{\text{by mass, }x} &= (\dot{m} e_{\text{stream}})_x - \left[(\dot{m} e_{\text{stream}})_x + \frac{\partial (\dot{m} e_{\text{stream}})_x}{\partial x} dx \right] \\
&= -\frac{\partial [\rho u(dy \cdot 1) c_p T]}{\partial x} dx = -\rho c_p \left(u \frac{\partial T}{\partial x} + T \frac{\partial u}{\partial x} \right) dx\, dy \quad \textbf{(6–31)}
\end{aligned}
$$

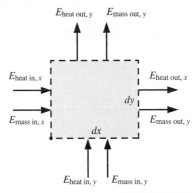

FIGURE 6–27
The energy transfers by heat and mass flow associated with a differential control volume in the thermal boundary layer in steady two-dimensional flow.

Repeating this for the y-direction and adding the results, the net rate of energy transfer to the control volume by mass is determined to be

$$
\begin{aligned}
(\dot{E}_{\text{in}} - \dot{E}_{\text{out}})_{\text{by mass}} &= -\rho c_p \left(u \frac{\partial T}{\partial x} + T \frac{\partial u}{\partial x} \right) dx\, dy - \rho c_p \left(v \frac{\partial T}{\partial y} + T \frac{\partial v}{\partial y} \right) dx\, dy \\
&= -\rho c_p \left(u \frac{\partial T}{\partial x} + v \frac{\partial T}{\partial y} \right) dx\, dy
\end{aligned}
\qquad \textbf{(6–32)}
$$

since $\partial u/\partial x + \partial v/\partial y = 0$ from the continuity equation.

The net rate of heat conduction to the volume element in the x-direction is

$$
\begin{aligned}
(\dot{E}_{\text{in}} - \dot{E}_{\text{out}})_{\text{by heat, }x} &= \dot{Q}_x - \left(\dot{Q}_x + \frac{\partial \dot{Q}_x}{\partial x} dx \right) = -\frac{\partial}{\partial x} \left(-k(dy \cdot 1) \frac{\partial T}{\partial x} \right) dx \\
&= k \frac{\partial^2 T}{\partial x^2} dx\, dy
\end{aligned}
\qquad \textbf{(6–33)}
$$

Repeating this for the y-direction and adding the results, the net rate of energy transfer to the control volume by heat conduction becomes

$$
(\dot{E}_{\text{in}} - \dot{E}_{\text{out}})_{\text{by heat}} = k \frac{\partial^2 T}{\partial x^2} dx\, dy + k \frac{\partial^2 T}{\partial y^2} dx\, dy = k \left(\frac{\partial^2 T}{\partial x^2} + \frac{\partial^2 T}{\partial y^2} \right) dx\, dy \qquad \textbf{(6–34)}
$$

Another mechanism of energy transfer to and from the fluid in the control volume is the work done by the body and surface forces. The work done by a body force is determined by multiplying this force by the velocity in the direction of the force and the volume of the fluid element, and this work needs to be considered only in the presence of significant gravitational, electric, or magnetic effects. The surface forces consist of the forces due to fluid pressure and the viscous shear stresses. The work done by pressure (the flow work) is already accounted for in the analysis above by using enthalpy for the microscopic energy of the fluid instead of internal energy. The shear stresses that result from viscous effects are usually very small, and can be neglected in many cases. This is especially the case for applications that involve low or moderate velocities.

Then the energy equation for the steady two-dimensional flow of a fluid with constant properties and negligible shear stresses is obtained by substituting Eqs. 6–32 and 6–34 into 6–30 to be

$$
\rho c_p \left(u \frac{\partial T}{\partial x} + v \frac{\partial T}{\partial y} \right) = k \left(\frac{\partial^2 T}{\partial x^2} + \frac{\partial^2 T}{\partial y^2} \right)
\qquad \textbf{(6–35)}
$$

which states that *the net energy convected by the fluid out of the control volume is equal to the net energy transferred into the control volume by heat conduction.*

When the viscous shear stresses are not negligible, their effect is accounted for by expressing the energy equation as

$$\rho c_p\left(u\,\frac{\partial T}{\partial x} + v\,\frac{\partial T}{\partial y}\right) = k\left(\frac{\partial^2 T}{\partial x^2} + \frac{\partial^2 T}{\partial y^2}\right) + \mu\Phi \qquad \text{(6–36)}$$

where the *viscous dissipation function* Φ is obtained after a lengthy analysis (see an advanced book such as the one by *Schlichting* for details) to be

$$\Phi = 2\left[\left(\frac{\partial u}{\partial x}\right)^2 + \left(\frac{\partial v}{\partial y}\right)^2\right] + \left(\frac{\partial u}{\partial y} + \frac{\partial v}{\partial x}\right)^2 \qquad \text{(6–37)}$$

Viscous dissipation may play a dominant role in high-speed flows, especially when the viscosity of the fluid is high (like the flow of oil in journal bearings). This manifests itself as a significant rise in fluid temperature due to the conversion of the kinetic energy of the fluid to thermal energy. Viscous dissipation is also significant for high-speed flights of aircraft.

For the special case of a stationary fluid, $u = v = 0$ and the energy equation reduces, as expected, to the steady two-dimensional heat conduction equation,

$$\frac{\partial^2 T}{\partial x^2} + \frac{\partial^2 T}{\partial y^2} = 0 \qquad \text{(6–38)}$$

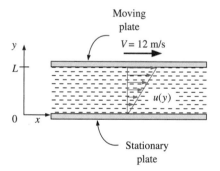

Moving plate

$V = 12$ m/s

$u(y)$

Stationary plate

FIGURE 6–28
Schematic for Example 6–1.

EXAMPLE 6–1   **Temperature Rise of Oil in a Journal Bearing**

The flow of oil in a journal bearing can be approximated as parallel flow between two large plates with one plate moving and the other stationary. Such flows are known as Couette flow.

Consider two large isothermal plates separated by 2-mm-thick oil film. The upper plates moves at a constant velocity of 12 m/s, while the lower plate is stationary. Both plates are maintained at 20°C. (*a*) Obtain relations for the velocity and temperature distributions in the oil. (*b*) Determine the maximum temperature in the oil and the heat flux from the oil to each plate (Fig. 6–28).

SOLUTION   Parallel flow of oil between two plates is considered. The velocity and temperature distributions, the maximum temperature, and the total heat transfer rate are to be determined.

Assumptions   **1** Steady operating conditions exist. **2** Oil is an incompressible substance with constant properties. **3** Body forces such as gravity are negligible. **4** The plates are large so that there is no variation in the *z* direction.

Properties   The properties of oil at 20°C are (Table A–13):

$$k = 0.145 \text{ W/m} \cdot \text{K} \quad \text{and} \quad \mu = 0.8374 \text{ kg/m} \cdot \text{s} = 0.8374 \text{ N} \cdot \text{s/m}^2$$

Analysis   (*a*) We take the *x-axis* to be the flow direction, and *y* to be the normal direction. This is parallel flow between two plates, and thus $v = 0$. Then the continuity equation (Eq. 6–21) reduces to

Continuity: $$\frac{\partial u}{\partial x} + \frac{\partial v}{\partial y} = 0 \;\rightarrow\; \frac{\partial u}{\partial x} = 0 \;\rightarrow\; u = u(y)$$

Therefore, the x-component of velocity does not change in the flow direction (i.e., the velocity profile remains unchanged). Noting that $u = u(y)$, $v = 0$, and $\partial P/\partial x = 0$ (flow is maintained by the motion of the upper plate rather than the pressure gradient), the x-momentum equation (Eq. 6–28) reduces to

x-momentum: $\qquad \rho\left(u\dfrac{\partial u}{\partial x} + v\dfrac{\partial u}{\partial y}\right) = \mu\dfrac{\partial^2 u}{\partial y^2} - \dfrac{\partial P}{\partial x} \qquad \rightarrow \qquad \dfrac{d^2 u}{dy^2} = 0$

This is a second-order ordinary differential equation, and integrating it twice gives

$$u(y) = C_1 y + C_2$$

The fluid velocities at the plate surfaces must be equal to the velocities of the plates because of the no-slip condition. Therefore, the boundary conditions are $u(0) = 0$ and $u(L) = V$, and applying them gives the velocity distribution to be

$$u(y) = \frac{y}{L} V$$

Frictional heating due to viscous dissipation in this case is significant because of the high viscosity of oil and the large plate velocity. The plates are isothermal and there is no change in the flow direction, and thus the temperature depends on y only, $T = T(y)$. Also, $u = u(y)$ and $v = 0$. Then the energy equation with dissipation (Eqs. 6–36 and 6–37) reduce to

Energy: $\qquad 0 = k\dfrac{\partial^2 T}{\partial y^2} + \mu\left(\dfrac{\partial u}{\partial y}\right)^2 \qquad \rightarrow \qquad k\dfrac{d^2 T}{dy^2} = -\mu\left(\dfrac{V}{L}\right)^2$

since $\partial u/\partial y = V/L$. Dividing both sides by k and integrating twice give

$$T(y) = \pm\frac{\mu}{2k}\left(\frac{y}{L}V\right)^2 + C_3 y + C_4$$

Applying the boundary conditions $T(0) = T_0$ and $T(L) = T_0$ gives the temperature distribution to be

$$T(y) = T_0 + \frac{\mu V^2}{2k}\left(\frac{y}{L} - \frac{y^2}{L^2}\right)$$

(b) The temperature gradient is determined by differentiating $T(y)$ with respect to y,

$$\frac{dT}{dy} = \frac{\mu V^2}{2kL}\left(1 - 2\frac{y}{L}\right)$$

The location of maximum temperature is determined by setting $dT/dy = 0$ and solving for y,

$$\frac{dT}{dy} = \frac{\mu V^2}{2kL}\left(1 - 2\frac{y}{L}\right) = 0 \qquad \rightarrow \qquad y = \frac{L}{2}$$

Therefore, maximum temperature occurs at mid plane, which is not surprising since both plates are maintained at the same temperature. The maximum temperature is the value of temperature at $y = L/2$,

$$T_{\max} = T\left(\frac{L}{2}\right) = T_0 + \frac{\mu V^2}{2k}\left(\frac{L/2}{L} - \frac{(L/2)^2}{L^2}\right) = T_0 + \frac{\mu V^2}{8k}$$

$$= 20 + \frac{(0.8374 \text{ N} \cdot \text{s/m}^2)(12 \text{ m/s})^2}{8(0.145 \text{ W/m} \cdot °\text{C})}\left(\frac{1 \text{ W}}{1 \text{ N} \cdot \text{m/s}}\right) = 124°\text{C}$$

Heat flux at the plates is determined from the definition of heat flux,

$$\dot{q}_0 = -k \frac{dT}{dy}\bigg|_{y=0} = -k \frac{\mu V^2}{2kL}\bigg| (1-0) = -\frac{\mu V^2}{2L}$$

$$= -\frac{(0.8374 \text{ N} \cdot \text{s/m}^2)(12 \text{ m/s})^2}{2(0.002 \text{ m})} \left(\frac{1 \text{ kW}}{1000 \text{ N} \cdot \text{m/s}}\right) = -30.1 \text{ kW/m}^2$$

$$\dot{q}_L = -k \frac{dT}{dy}\bigg|_{y=L} = -k \frac{\mu V^2}{2kL} (1-2) = \frac{\mu V^2}{2L} = -\dot{q}_0 = 30.1 \text{ kW/m}^2$$

Therefore, heat fluxes at the two plates are equal in magnitude but opposite in sign.

Discussion A temperature rise of 104°C confirms our suspicion that viscous dissipation is very significant. Also, the heat flux is equivalent to the rate of mechanical energy dissipation. Therefore, mechanical energy is being converted to thermal energy at a rate of 60.2 kW/m² of plate area to overcome friction in the oil. Finally, calculations are done using oil properties at 20°C, but the oil temperature turned out to be much higher. Therefore, knowing the strong dependence of viscosity on temperature, calculations should be repeated using properties at the average temperature of 72°C to improve accuracy.

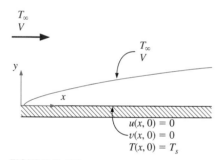

FIGURE 6–29
Boundary conditions for flow over a flat plate.

6–8 · SOLUTIONS OF CONVECTION EQUATIONS FOR A FLAT PLATE

Consider laminar flow of a fluid over a *flat plate,* as shown in Fig. 6–29. Surfaces that are slightly contoured such as turbine blades can also be approximated as flat plates with reasonable accuracy. The x-coordinate is measured along the plate surface from the leading edge of the plate in the direction of the flow, and y is measured from the surface in the normal direction. The fluid approaches the plate in the x-direction with a uniform upstream velocity, which is equivalent to the free stream velocity V.

When viscous dissipation is negligible, the continuity, momentum, and energy equations (Eqs. 6–21, 6–28, and 6–35) reduce for steady, incompressible, laminar flow of a fluid with constant properties over a flat plate to

Continuity:
$$\frac{\partial u}{\partial x} + \frac{\partial v}{\partial y} = 0 \tag{6–39}$$

Momentum:
$$u \frac{\partial u}{\partial x} + v \frac{\partial u}{\partial y} = v \frac{\partial^2 u}{\partial y^2} \tag{6–40}$$

Energy:
$$u \frac{\partial T}{\partial x} + v \frac{\partial T}{\partial y} = \alpha \frac{\partial^2 T}{\partial y^2} \tag{6–41}$$

with the boundary conditions (Fig. 6–26)

At $x = 0$: $u(0, y) = V$, $T(0, y) = T_\infty$
At $y = 0$: $u(x, 0) = 0$, $v(x, 0) = 0, T(x, 0) = T_s$ $\tag{6–42}$
As $y \to \infty$: $u(x, \infty) = V$, $T(x, \infty) = T_\infty$

When fluid properties are assumed to be constant and thus independent of temperature, the first two equations can be solved separately for the velocity components u and v. Once the velocity distribution is available, we can determine the friction coefficient and the boundary layer thickness using their definitions. Also, knowing u and v, the temperature becomes the only unknown in the last equation, and it can be solved for temperature distribution.

The continuity and momentum equations were first solved in 1908 by the German engineer H. Blasius, a student of L. Prandtl. This was done by transforming the two partial differential equations into a single ordinary differential equation by introducing a new independent variable, called the **similarity variable**. The finding of such a variable, assuming it exists, is more of an art than science, and it requires to have a good insight of the problem.

Noticing that the general shape of the velocity profile remains the same along the plate, Blasius reasoned that the nondimensional velocity profile u/V should remain unchanged when plotted against the nondimensional distance y/δ, where δ is the thickness of the local velocity boundary layer at a given x. That is, although both δ and u at a given y vary with x, the velocity u at a fixed y/δ remains constant. Blasius was also aware from the work of Stokes that δ is proportional to $\sqrt{vx/V}$, and thus he defined a *dimensionless similarity variable* as

$$\eta = y\sqrt{\frac{V}{vx}} \tag{6–43}$$

and thus $u/V = $ function(η). He then introduced a *stream function* $\psi(x, y)$ as

$$u = \frac{\partial \psi}{\partial y} \quad \text{and} \quad v = -\frac{\partial \psi}{\partial x} \tag{6–44}$$

so that the continuity equation (Eq. 6–39) is automatically satisfied and thus eliminated (this can be verified easily by direct substitution). Next he defined a function $f(\eta)$ as the dependent variable as

$$f(\eta) = \frac{\psi}{V\sqrt{vx/V}} \tag{6–45}$$

Then the velocity components become

$$u = \frac{\partial \psi}{\partial y} = \frac{\partial \psi}{\partial \eta}\frac{\partial \eta}{\partial y} = V\sqrt{\frac{vx}{V}}\frac{df}{d\eta}\sqrt{\frac{V}{vx}} = V\frac{df}{d\eta} \tag{6–46}$$

$$v = -\frac{\partial \psi}{\partial x} = -V\sqrt{\frac{vx}{V}}\frac{df}{d\eta} - \frac{V}{2}\sqrt{\frac{v}{Vx}}f = \frac{1}{2}\sqrt{\frac{Vv}{x}}\left(\eta\frac{df}{d\eta} - f\right) \tag{6–47}$$

By differentiating these u and v relations, the derivatives of the velocity components can be shown to be

$$\frac{\partial u}{\partial x} = -\frac{V}{2x}\eta\frac{d^2f}{d\eta^2}, \quad \frac{\partial u}{\partial y} = V\sqrt{\frac{V}{vx}}\frac{d^2f}{d\eta^2}, \quad \frac{\partial^2 u}{\partial y^2} = \frac{V^2}{vx}\frac{d^3f}{d\eta^3} \tag{6–48}$$

Substituting these relations into the momentum equation and simplifying, we obtain

$$2\frac{d^3f}{d\eta^3} + f\frac{d^2f}{d\eta^2} = 0 \tag{6–49}$$

which is a third-order nonlinear differential equation. Therefore, the system of two partial differential equations is transformed into a single ordinary differential equation by the use of a similarity variable. Using the definitions

TABLE 6–3

Similarity function f and its derivatives for laminar boundary layer along a flat plate.

η	f	$\dfrac{df}{d\eta} = \dfrac{u}{V}$	$\dfrac{d^2f}{d\eta^2}$
0	0	0	0.332
0.5	0.042	0.166	0.331
1.0	0.166	0.330	0.323
1.5	0.370	0.487	0.303
2.0	0.650	0.630	0.267
2.5	0.996	0.751	0.217
3.0	1.397	0.846	0.161
3.5	1.838	0.913	0.108
4.0	2.306	0.956	0.064
4.5	2.790	0.980	0.034
5.0	3.283	0.992	0.016
5.5	3.781	0.997	0.007
6.0	4.280	0.999	0.002
∞	∞	1	0

of f and η, the boundary conditions in terms of the similarity variables can be expressed as

$$f(0) = 0, \qquad \frac{df}{d\eta}\bigg|_{\eta=0} = 0, \qquad \text{and} \qquad \frac{df}{d\eta}\bigg|_{\eta=\infty} = 1 \qquad \text{(6–50)}$$

The transformed equation with its associated boundary conditions cannot be solved analytically, and thus an alternative solution method is necessary. The problem was first solved by Blasius in 1908 using a power series expansion approach, and this original solution is known as the *Blasius solution*. The problem is later solved more accurately using different numerical approaches, and results from such a solution are given in Table 6–3. The nondimensional velocity profile can be obtained by plotting u/V against η. The results obtained by this simplified analysis are in excellent agreement with experimental results.

Recall that we defined the boundary layer thickness as the distance from the surface for which $u/V = 0.99$. We observe from Table 6–3 that the value of η corresponding to $u/V = 0.99$ is $\eta = 4.91$. Substituting $\eta = 4.91$ and $y = \delta$ into the definition of the similarity variable (Eq. 6–43) gives $4.91 = \delta\sqrt{V/vx}$. Then the velocity boundary layer thickness becomes

$$\delta = \frac{4.91}{\sqrt{V/vx}} = \frac{4.91x}{\sqrt{\text{Re}_x}} \qquad \text{(6–51)}$$

since $\text{Re}_x = Vx/v$, where x is the distance from the leading edge of the plate. Note that the boundary layer thickness increases with increasing kinematic viscosity v and with increasing distance from the leading edge x, but it decreases with increasing free-stream velocity V. Therefore, a large free-stream velocity suppresses the boundary layer and causes it to be thinner.

The shear stress on the wall can be determined from its definition and the $\partial u/\partial y$ relation in Eq. 6–48:

$$\tau_w = \mu \frac{\partial u}{\partial y}\bigg|_{y=0} = \mu V \sqrt{\frac{V}{vx}} \frac{d^2f}{d\eta^2}\bigg|_{\eta=0} \qquad \text{(6–52)}$$

Substituting the value of the second derivative of f at $\eta = 0$ from Table 6–3 gives

$$\tau_w = 0.332V\sqrt{\frac{\rho\mu V}{x}} = \frac{0.332\rho V^2}{\sqrt{\text{Re}_x}} \qquad \text{(6–53)}$$

Then the average local skin friction coefficient becomes

$$C_{f,x} = \frac{\tau_w}{\rho V^2/2} = 0.664\,\text{Re}_x^{-1/2} \qquad \text{(6–54)}$$

Note that unlike the boundary layer thickness, wall shear stress and the skin friction coefficient decrease along the plate as $x^{-1/2}$.

The Energy Equation

Knowing the velocity profile, we are now ready to solve the energy equation for temperature distribution for the case of constant wall temperature T_s. First we introduce the dimensionless temperature θ as

$$\theta(x, y) = \frac{T(x, y) - T_s}{T_\infty - T_s} \tag{6-55}$$

Noting that both T_s and T_∞ are constant, substitution into the energy equation Eq. 6–41 gives

$$u\frac{\partial\theta}{\partial x} + v\frac{\partial\theta}{\partial y} = \alpha\frac{\partial^2\theta}{\partial y^2} \tag{6-56}$$

Temperature profiles for flow over an isothermal flat plate are similar, just like the velocity profiles, and thus we expect a similarity solution for temperature to exist. Further, the thickness of the thermal boundary layer is proportional to $\sqrt{vx/V}$, just like the thickness of the velocity boundary layer, and thus the similarity variable is also η, and $\theta = \theta(\eta)$. Using the chain rule and substituting the u and v expressions from Eqs. 6–46 and 6–47 into the energy equation gives

$$V\frac{df}{d\eta}\frac{d\theta}{d\eta}\frac{\partial\eta}{\partial x} + \frac{1}{2}\sqrt{\frac{Vy}{x}}\left(\eta\frac{df}{d\eta} - f\right)\frac{d\theta}{d\eta}\frac{\partial\eta}{\partial y} = \alpha\frac{d^2\theta}{d\eta^2}\left(\frac{\partial\eta}{\partial y}\right)^2 \tag{6-57}$$

Simplifying and noting that $\mathrm{Pr} = v/\alpha$ gives

$$2\frac{d^2\theta}{d\eta^2} + \mathrm{Pr}\,f\frac{d\theta}{d\eta} = 0 \tag{6-58}$$

with the boundary conditions $\theta(0) = 0$ and $\theta(\infty) = 1$. Obtaining an equation for θ as a function of η alone confirms that the temperature profiles are similar, and thus a similarity solution exists. Again a closed-form solution cannot be obtained for this boundary value problem, and it must be solved numerically.

It is interesting to note that for $\mathrm{Pr} = 1$, this equation reduces to Eq. 6–49 when θ is replaced by $df/d\eta$, which is equivalent to u/V (see Eq. 6–46). The boundary conditions for θ and $df/d\eta$ are also identical. Thus we conclude that the velocity and thermal boundary layers coincide, and the nondimensional velocity and temperature profiles (u/V and θ) are identical for steady, incompressible, laminar flow of a fluid with constant properties and $\mathrm{Pr} = 1$ over an isothermal flat plate (Fig. 6–30). The value of the temperature gradient at the surface ($y = 0$ or $\eta = 0$) in this case is, from Table 6–3, $d\theta/d\eta - d^2f/d\eta^2 = 0.332$.

Equation 6–58 is solved for numerous values of Prandtl numbers. For $\mathrm{Pr} > 0.6$, the nondimensional temperature gradient at the surface is found to be proportional to $\mathrm{Pr}^{1/3}$, and is expressed as

$$\left.\frac{d\theta}{d\eta}\right|_{\eta=0} = 0.332\,\mathrm{Pr}^{1/3} \tag{6-59}$$

The temperature gradient at the surface is

$$\left.\frac{\partial T}{\partial y}\right|_{y=0} = (T_\infty - T_s)\left.\frac{\partial\theta}{\partial y}\right|_{y=0} = (T_\infty - T_s)\left.\frac{d\theta}{d\eta}\right|_{\eta=0}\left.\frac{\partial\eta}{\partial y}\right|_{y=0} \tag{6-60}$$

$$= 0.332\,\mathrm{Pr}^{1/3}(T_\infty - T_s)\sqrt{\frac{V}{vx}}$$

Then the local convection coefficient and Nusselt number become

$$h_x = \frac{\dot{q}_s}{T_s - T_\infty} = \frac{-k(\partial T/\partial y)|_{y=0}}{T_s - T_\infty} = 0.332\,Pr^{1/3}k\sqrt{\frac{V}{vx}} \tag{6-61}$$

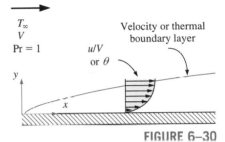

FIGURE 6–30

When $\mathrm{Pr} = 1$, the velocity and thermal boundary layers coincide, and the nondimensional velocity and temperature profiles are identical for steady, incompressible, laminar flow over a flat plate.

and

$$\text{Nu}_x = \frac{h_x x}{k} = 0.332 \, \text{Pr}^{1/3} \text{Re}_x^{1/2} \qquad \text{Pr} > 0.6 \tag{6-62}$$

The Nu_x values obtained from this relation agree well with measured values.

Solving Eq. 6–58 numerically for the temperature profile for different Prandtl numbers, and using the definition of the thermal boundary layer, it is determined that $\delta/\delta_t \cong \text{Pr}^{1/3}$. Then the thermal boundary layer thickness becomes

$$\delta_t = \frac{\delta}{\text{Pr}^{1/3}} = \frac{4.91x}{\text{Pr}^{1/3}\sqrt{\text{Re}_x}} \tag{6-63}$$

Note that these relations are valid only for laminar flow over an isothermal flat plate. Also, the effect of variable properties can be accounted for by evaluating all such properties at the film temperature defined as $T_f = (T_s + T_\infty)/2$.

The Blasius solution gives important insights, but its value is largely historical because of the limitations it involves. Nowadays both laminar and turbulent flows over surfaces are routinely analyzed using numerical methods.

6–9 · NONDIMENSIONALIZED CONVECTION EQUATIONS AND SIMILARITY

When viscous dissipation is negligible, the continuity, momentum, and energy equations for steady, incompressible, laminar flow of a fluid with constant properties are given by Eqs. 6–21, 6–28, and 6–35.

These equations and the boundary conditions can be nondimensionalized by dividing all dependent and independent variables by relevant and meaningful constant quantities: all lengths by a characteristic length L (which is the length for a plate), all velocities by a reference velocity V (which is the free stream velocity for a plate), pressure by ρV^2 (which is twice the free stream dynamic pressure for a plate), and temperature by a suitable temperature difference (which is $T_\infty - T_s$ for a plate). We get

$$x^* = \frac{x}{L}, \quad y^* = \frac{y}{L}, \quad u^* = \frac{u}{V}, \quad v^* = \frac{v}{V}, \quad P^* = \frac{P}{\rho V^2}, \quad \text{and} \quad T^* = \frac{T - T_s}{T_\infty - T_s}$$

where the asterisks are used to denote nondimensional variables. Introducing these variables into Eqs. 6–21, 6–28, and 6–35 and simplifying give

Continuity:
$$\frac{\partial u^*}{\partial x^*} + \frac{\partial v^*}{\partial y^*} = 0 \tag{6-64}$$

Momentum:
$$u^* \frac{\partial u^*}{\partial x^*} + v^* \frac{\partial u^*}{\partial y^*} = \frac{1}{\text{Re}_L} \frac{\partial^2 u^*}{\partial y^{*2}} - \frac{dP^*}{dx^*} \tag{6-65}$$

Energy:
$$u^* \frac{\partial T^*}{\partial x^*} + v^* \frac{\partial T^*}{\partial y^*} = \frac{1}{\text{Re}_L \, \text{Pr}} \frac{\partial^2 T^*}{\partial y^{*2}} \tag{6-66}$$

with the boundary conditions

$$u^*(0, y^*) = 1, \quad u^*(x^*, 0) = 0, \quad u^*(x^*, \infty) = 1, \quad v^*(x^*, 0) = 0, \tag{6-67}$$
$$T^*(0, y^*) = 1, \quad T^*(x^*, 0) = 0, \quad T^*(x^*, \infty) = 1$$

where $\mathrm{Re}_L = VL/\nu$ is the dimensionless Reynolds number and $\mathrm{Pr} = \nu/\alpha$ is the Prandtl number. For a given type of geometry, the solutions of problems with the same Re and Nu numbers are similar, and thus Re and Nu numbers serve as *similarity parameters*. Two physical phenomena are *similar* if they have the same dimensionless forms of governing differential equations and boundary conditions (Fig. 6–31).

A major advantage of nondimensionalizing is the significant reduction in the number of parameters. The original problem involves 6 parameters (L, V, T_∞, T_s, ν, α), but the nondimensionalized problem involves just 2 parameters (Re_L and Pr). For a given geometry, problems that have the same values for the similarity parameters have identical solutions. For example, determining the convection heat transfer coefficient for flow over a given surface requires numerical solutions or experimental investigations for several fluids, with several sets of velocities, surface lengths, wall temperatures, and free stream temperatures. The same information can be obtained with far fewer investigations by grouping data into the dimensionless Re and Pr numbers. Another advantage of similarity parameters is that they enable us to group the results of a large number of experiments and to report them conveniently in terms of such parameters (Fig. 6–32).

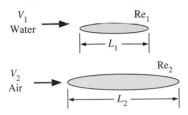

If $\mathrm{Re}_1 = \mathrm{Re}_2$, then $C_{f1} = C_{f2}$

FIGURE 6–31

Two geometrically similar bodies have the same value of friction coefficient at the same Reynolds number.

Parameters before nondimensionalizing

L, V, T_∞, T_s, ν, α

Parameters after nondimensionalizing:

Re, Pr

FIGURE 6–32

The number of parameters is reduced greatly by nondimensionalizing the convection equations.

6–10 ■ FUNCTIONAL FORMS OF FRICTION AND CONVECTION COEFFICIENTS

The three nondimensionalized boundary layer equations (Eqs. 6–64, 6–65, and 6–66) involve three unknown functions u^*, v^*, and T^*, two independent variables x^* and y^*, and two parameters Re_L and Pr. The pressure $P^*(x^*)$ depends on the geometry involved (it is constant for a flat plate), and it has the same value inside and outside the boundary layer at a specified x^*. Therefore, it can be determined separately from the free stream conditions, and dP^*/dx^* in Eq. 6–65 can be treated as a known function of x^*. Note that the boundary conditions do not introduce any new parameters.

For a given geometry, the solution for u^* can be expressed as

$$u^* = f_1(x^*, y^*, \mathrm{Re}_L) \tag{6–68}$$

Then the shear stress at the surface becomes

$$\tau_s = \mu \frac{\partial u}{\partial y}\bigg|_{y=0} = \frac{\mu V}{L} \frac{\partial u^*}{\partial y^*}\bigg|_{y^*=0} = \frac{\mu V}{L} f_2(x^*, \mathrm{Re}_L) \tag{6–69}$$

Substituting into its definition gives the local friction coefficient,

$$C_{f,x} = \frac{\tau_s}{\rho V^2/2} = \frac{\mu V/L}{\rho V^2/2} f_2(x^*, \mathrm{Re}_L) = \frac{2}{\mathrm{Re}_L} f_2(x^*, \mathrm{Re}_L) = f_3(x^*, \mathrm{Re}_L) \tag{6–70}$$

Thus we conclude that the friction coefficient for a given geometry can be expressed in terms of the Reynolds number Re and the dimensionless space variable x^* alone (instead of being expressed in terms of x, L, V, ρ, and μ). This is a very significant finding, and shows the value of nondimensionalized equations.

Similarly, the solution of Eq. 6–66 for the dimensionless temperature T^* for a given geometry can be expressed as

$$T^* = g_1(x^*, y^*, \text{Re}_L, \text{Pr}) \tag{6–71}$$

Using the definition of T^*, the convection heat transfer coefficient becomes

$$h = \frac{-k(\partial T/\partial y)|_{y=0}}{T_s - T_\infty} = \frac{-k(T_\infty - T_s)}{L(T_s - T_\infty)} \frac{\partial T^*}{\partial y^*}\bigg|_{y^*=0} = \frac{k}{L} \frac{\partial T^*}{\partial y^*}\bigg|_{y^*=0} \tag{6–72}$$

Substituting this into the Nusselt number relation gives [or alternately, we can rearrange the relation above in dimensionless form as $hL/k = (\partial T^*/\partial y^*)|_{y^*=0}$ and define the dimensionless group hL/k as the Nusselt number]

$$\text{Nu}_x = \frac{hL}{k} = \frac{\partial T^*}{\partial y^*}\bigg|_{y^*=0} = g_2(x^*, \text{Re}_L, \text{Pr}) \tag{6–73}$$

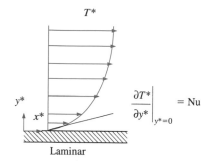

$$\frac{\partial T^*}{\partial y^*}\bigg|_{y^*=0} = \text{Nu}$$

Laminar

FIGURE 6–33
The Nusselt number is equivalent to the dimensionless temperature gradient at the surface.

Note that the Nusselt number is equivalent to the *dimensionless temperature gradient at the surface,* and thus it is properly referred to as the dimensionless heat transfer coefficient (Fig. 6–33). Also, the Nusselt number for a given geometry can be expressed in terms of the Reynolds number Re, the Prandtl number Pr, and the space variable x^*, and such a relation can be used for different fluids flowing at different velocities over similar geometries of different lengths.

The average friction and heat transfer coefficients are determined by integrating $C_{f,x}$ and Nu_x over the surface of the given body with respect to x^* from 0 to 1. Integration removes the dependence on x^*, and the average friction coefficient and Nusselt number can be expressed as

$$C_f = f_4(\text{Re}_L) \qquad \text{and} \qquad \text{Nu} = g_3(\text{Re}_L, \text{Pr}) \tag{6–74}$$

Local Nusselt number:

$$\text{Nu}_x = \text{function } (x^*, \text{Re}_L, \text{Pr})$$

Average Nusselt number:

$$\text{Nu} = \text{function } (\text{Re}_L, \text{Pr})$$

A common form of Nusselt number:

$$\text{Nu} = C \, \text{Re}_L^m \text{Pr}^n$$

FIGURE 6–34
For a given geometry, the average Nusselt number is a function of Reynolds and Prandtl numbers.

These relations are extremely valuable as they state that for a given geometry, the friction coefficient can be expressed as a function of Reynolds number alone, and the Nusselt number as a function of Reynolds and Prandtl numbers alone (Fig. 6–34). Therefore, experimentalists can study a problem with a minimum number of experiments, and report their friction and heat transfer coefficient measurements conveniently in terms of Reynolds and Prandtl numbers. For example, a friction coefficient relation obtained with air for a given surface can also be used for water at the same Reynolds number. But it should be kept in mind that the validity of these relations is limited by the limitations on the boundary layer equations used in the analysis.

The experimental data for heat transfer is often represented with reasonable accuracy by a simple power-law relation of the form

$$\text{Nu} = C \, \text{Re}_L^m \, \text{Pr}^n \tag{6–75}$$

where m and n are constant exponents (usually between 0 and 1), and the value of the constant C depends on geometry. Sometimes more complex relations are used for better accuracy.

6–11 ■ ANALOGIES BETWEEN MOMENTUM AND HEAT TRANSFER

In forced convection analysis, we are primarily interested in the determination of the quantities C_f (to calculate shear stress at the wall) and Nu (to calculate heat transfer rates). Therefore, it is very desirable to have a relation between C_f and Nu so that we can calculate one when the other is available. Such relations are developed on the basis of the similarity between momentum and heat transfers in boundary layers, and are known as *Reynolds analogy* and *Chilton–Colburn analogy*.

Reconsider the nondimensionalized momentum and energy equations for steady, incompressible, laminar flow of a fluid with constant properties and negligible viscous dissipation (Eqs. 6–65 and 6–66). When Pr = 1 (which is approximately the case for gases) and $\partial P^*/\partial x^* = 0$ (which is the case when, $u = V = constant$ in the free stream, as in flow over a flat plate), these equations simplify to

Momentum:
$$u^*\frac{\partial u^*}{\partial x^*} + v^*\frac{\partial u^*}{\partial y^*} = \frac{1}{\mathrm{Re}_L}\frac{\partial^2 u^*}{\partial y^{*2}} \qquad \text{(6–76)}$$

Energy:
$$u^*\frac{\partial T^*}{\partial x^*} + v^*\frac{\partial T^*}{\partial y^*} = \frac{1}{\mathrm{Re}_L}\frac{\partial^2 T^*}{\partial y^{*2}} \qquad \text{(6–77)}$$

which are exactly of the same form for the dimensionless velocity u^* and temperature T^*. The boundary conditions for u^* and T^* are also identical. Therefore, the functions u^* and T^* must be identical, and thus the first derivatives of u^* and T^* at the surface must be equal to each other,

$$\left.\frac{\partial u^*}{\partial y^*}\right|_{y^*=0} = \left.\frac{\partial T^*}{\partial y^*}\right|_{y^*=0} \qquad \text{(6–78)}$$

Then from Eqs. 6–69, 6–70, and 6–73 we have

$$C_{f,x}\frac{\mathrm{Re}_L}{2} = \mathrm{Nu}_x \qquad (\mathrm{Pr} = 1) \qquad \text{(6–79)}$$

which is known as the **Reynolds analogy** (Fig. 6–35). This is an important analogy since it allows us to determine the heat transfer coefficient for fluids with Pr ≈ 1 from a knowledge of friction coefficient which is easier to measure. Reynolds analogy is also expressed alternately as

$$\frac{C_{f,x}}{2} = \mathrm{St}_x \qquad (\mathrm{Pr} = 1) \qquad \text{(6–80)}$$

where

$$\mathrm{St} = \frac{h}{\rho c_p V} = \frac{\mathrm{Nu}}{\mathrm{Re}_L \mathrm{Pr}} \qquad \text{(6–81)}$$

is the **Stanton number**, which is also a dimensionless heat transfer coefficient.

Reynolds analogy is of limited use because of the restrictions Pr = 1 and $\partial P^*/\partial x^* = 0$ on it, and it is desirable to have an analogy that is applicable over a wide range of Pr. This is done by adding a Prandtl number correction.

The friction coefficient and Nusselt number for a flat plate were determined in Section 6–8 to be

$$C_{f,x} = 0.664\,\mathrm{Re}_x^{-1/2} \qquad \text{and} \qquad \mathrm{Nu}_x = 0.332\,\mathrm{Pr}^{1/3}\,\mathrm{Re}_x^{1/2} \qquad \text{(6–82)}$$

Taking their ratio and rearranging give the desired relation, known as the **modified Reynolds analogy** or **Chilton–Colburn analogy**,

$$C_{f,x}\frac{\mathrm{Re}_L}{2} = \mathrm{Nu}_x\,\mathrm{Pr}^{+1/3} \qquad \text{or} \qquad \frac{C_{f,x}}{2} = \frac{h_x}{\rho c_p V}\,\mathrm{Pr}^{2/3} \equiv j_H \qquad \text{(6–83)}$$

Profiles:	$u^* = T^*$
Gradients:	$\left.\dfrac{\partial u^*}{\partial y^*}\right|_{y^*=0} = \left.\dfrac{\partial T^*}{\partial y^*}\right|_{y^*=0}$
Analogy:	$C_{f,x}\dfrac{\mathrm{Re}_L}{2} = \mathrm{Nu}_x$

FIGURE 6–35
When Pr = 1 and $\partial P^*/\partial x^* \approx 0$, the nondimensional velocity and temperature profiles become identical, and Nu is related to C_f by Reynolds analogy.

for $0.6 < \text{Pr} < 60$. Here j_H is called the *Colburn j-factor*. Although this relation is developed using relations for laminar flow over a flat plate (for which $\partial P^*/\partial x^* = 0$), experimental studies show that it is also applicable approximately for turbulent flow over a surface, even in the presence of pressure gradients. For laminar flow, however, the analogy is not applicable unless $\partial P^*/\partial x^* \approx 0$. Therefore, it does not apply to laminar flow in a pipe. Analogies between C_f and Nu that are more accurate are also developed, but they are more complex and beyond the scope of this book. The analogies given above can be used for both local and average quantities.

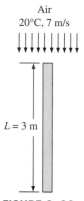

Air
20°C, 7 m/s

$L = 3$ m

FIGURE 6–36
Schematic for Example 6–2.

EXAMPLE 6–2   **Finding Convection Coefficient from Drag Measurement**

A 2-m × 3-m flat plate is suspended in a room, and is subjected to air flow parallel to its surfaces along its 3-m-long side. The free stream temperature and velocity of air are 20°C and 7 m/s. The total drag force acting on the plate is measured to be 0.86 N. Determine the average convection heat transfer coefficient for the plate (Fig. 6–36).

SOLUTION   A flat plate is subjected to air flow, and the drag force acting on it is measured. The average convection coefficient is to be determined.

Assumptions   **1** Steady operating conditions exist. **2** The edge effects are negligible. **3** The local atmospheric pressure is 1 atm.

Properties   The properties of air at 20°C and 1 atm are (Table A–15):

$$\rho = 1.204 \text{ kg/m}^3, \quad c_p = 1.007 \text{ kJ/kg} \cdot \text{K}, \quad \text{Pr} = 0.7309$$

Analysis   The flow is along the 3-m side of the plate, and thus the characteristic length is $L = 3$ m. Both sides of the plate are exposed to air flow, and thus the total surface area is

$$A_s = 2WL = 2(2 \text{ m})(3 \text{ m}) = 12 \text{ m}^2$$

For flat plates, the drag force is equivalent to friction force. The average friction coefficient C_f can be determined from Eq. 6–11,

$$F_f = C_f A_s \frac{\rho V^2}{2}$$

Solving for C_f and substituting,

$$C_f = \frac{F_f}{\rho A_s V^2/2} = \frac{0.86 \text{ N}}{(1.204 \text{ kg/m}^3)(12 \text{ m}^2)(7 \text{ m/s})^2/2} \left(\frac{1 \text{ kg} \cdot \text{m/s}^2}{1 \text{ N}} \right) = 0.00243$$

Then the average heat transfer coefficient can be determined from the modified Reynolds analogy (Eq. 6–83) to be

$$h = \frac{C_f \rho V c_p}{2 \; \text{Pr}^{2/3}} = \frac{0.00243}{2} \frac{(1.204 \text{ kg/m}^3)(7 \text{ m/s})(1007 \text{ J/kg} \cdot {}^\circ\text{C})}{0.7309^{2/3}} = 12.7 \text{ W/m}^2 \cdot {}^\circ\text{C}$$

Discussion   This example shows the great utility of momentum-heat transfer analogies in that the convection heat transfer coefficient can be obtained from a knowledge of friction coefficient, which is easier to determine.

TOPIC OF SPECIAL INTEREST

*Microscale Heat Transfer**

Heat transfer considerations play a crucial role in the design and operation of many modern devices. New approaches and methods of analyses have been developed to understand and modulated (enhance or suppress) such energy interactions. Modulation typically occurs through actively controlling the surface phenomena, or focusing of the volumetric energy. In this section we discuss one such example—microscale heat transfer.

Recent inventions in micro ($\sim 10^{-6}$ m) and nano ($\sim 10^{-9}$ m) scale systems have shown tremendous benefits in fluid flow and heat transfer processes. These devices are extremely tiny and only visible through electron microscopes. The detailed understanding of the governing mechanism of these systems will be at the heart of realizing many future technologies. Examples include chemical and biological sensors, hydrogen storage, space exploration devices, and drug screening. Micro-nanoscale device development also poses several new challenges, however. For example, the classical heat transfer knowledge originates from thermal equilibrium approach and the equations are derived for material continuum. As the length scale of the system becomes minuscule, the heat transfer through these particles in nanoscale systems is no longer an equilibrium process and the continuum based equilibrium approach is no longer valid. Thus, a more general understanding of the concept of heat transfer becomes essential.

Both length and time scales are crucial in micro- and nanoscale heat transfer. The significance of length scale becomes evident from the fact that the surface area per unit volume of an object increases as the length scale of the object shrinks. This means the heat transfer through the surface becomes orders of magnitude more important in microscale than in large everyday objects. Transport of thermal energy in electronic and thermoelectric equipments often occurs at a range of length scales from millimeters to nanometers. For example, in a microelectronic chip (say MOSFET in Fig. 6–37) heat is generated in a nanometer-size drain region and ultimately conducted to the surrounding through substrates whose thickness is of the order of a millimeter. Clearly energy transport and conversion mechanisms in devices involve a wide range of length scales and are quite difficult to model.

Small time scales also play an important role in energy transport mechanisms. For example, ultra-short (pico-second and femto-second) pulse lasers are extremely useful for material processing industry. Here the tiny time scales permit localized laser-material interaction beneficial for high energy deposition and transport.

The applicability of the continuum model is determined by the local value of the non-dimensional Knudsen number (Kn) which is defined as the ratio of the mean free path (mfp) of the heat-carrier medium to the system reference length scale (say thermal diffusion length). Microscale

*This section is contributed by Subrata Roy, Computational Plasma Dynamics Laboratory, Mechanical Engineering, Kettering University, Flint, MI.

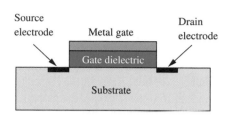

Source electrode   Metal gate   Drain electrode

Gate dielectric

Substrate

FIGURE 6–37
Metal-Oxide Semiconductor Field-Effect Transistor (MOSFET) used in microelectronics.

© Vol. 80/PhotoDisc/Getty Images

effects become important when the mfp becomes comparable to or greater than the reference length of the device, say at Kn > 0.001. As a result, thermophysical properties of materials become dependent on structure, and heat conduction processes are no longer local phenomena, but rather exhibit long-range radiative effects.

The conventional macroscopic Fourier conduction model violates this non-local feature of microscale heat transfer, and alternative approaches are necessary for analysis. The most suitable model to date is the concept of *phonon*. The thermal energy in a uniform solid material can be interpreted as the vibrations of a regular lattice of closely bound atoms inside. These atoms exhibit collective modes of sound waves (phonons) which transports energy at the speed of sound in a material. Following quantum mechanical principles, phonons exhibit particle-like properties of bosons with zero spin (wave-particle duality). Phonons play an important role in many of the physical properties of solids, such as the thermal and the electrical conductivities. In insulating solids, phonons are also the primary mechanism by which heat conduction takes place.

The variation of temperature near the bounding wall continues to be a major determinant of heat transfer though the surface. However, when the continuum approach breaks down, the conventional Newton's law of cooling using wall and bulk fluid temperature needs to be modified. Specifically, unlike in macroscale objects where the wall and adjacent fluid temperatures are equal ($T_w = T_g$), in a micro device there is a temperature slip and the two values are different. One well-known relation for calculating the temperature jump at the wall of a microgeometry was derived by von Smoluchowski in 1898,

$$T_g - T_w = \frac{2 - \sigma_T}{\sigma_T} \left[\frac{2\gamma}{\gamma + 1} \right] \frac{\lambda}{Pr} \left(\frac{\partial T}{\partial y} \right)_w \tag{6–84}$$

where T is the temperature in K, σ_T is the thermal accommodation coefficient and indicates the molecular fraction reflected diffusively from the wall, γ is the specific heat ratio, and Pr is the Prandtl number. Once this value is known, the heat transfer rate may be calculated from:

$$-k \left(\frac{\partial T}{\partial y} \right)_w = \frac{\sigma_T \sqrt{2\pi RT}}{2 - \sigma_T} \left[\frac{\gamma + 1}{2\gamma} \right] \frac{5\rho c_p}{16} (T_w - T_g) \tag{6–85}$$

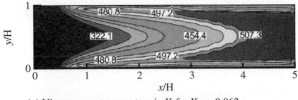

(a) Nitrogen gas temperature in K for Kn = 0.062

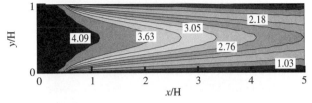

(b) Nitrogen gas velocity relative to the speed of sound (Mach number)

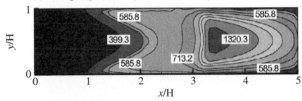

(c) Helium gas temperature in K for Kn = 0.14

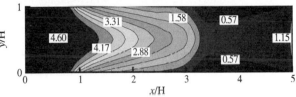

(d) Helium gas velocity relative to the speed of sound (Mach number)

FIGURE 6–38
Fluid-thermal characteristics inside a
microchannel.
(From Raju and Roy, 2005.)

As an example, the temperature distribution and Mach number contours inside a micro-tube of width $H = 1.2 \ \mu$m are plotted in Fig. 6–38 for supersonic flow of nitrogen and helium. For nitrogen gas with an inlet Kn = 0.062, the gas temperature (T_g) adjacent to the wall differs substantially from the fixed wall temperature, as shown in Fig. 6–38a, where T_w is 323 K and T_g is almost 510 K. The effect of this wall heat transfer is to reduce the Mach number, as shown in Figure 6–38b, but the flow remains supersonic. For helium gas with inlet Kn = 0.14 and a lower wall temperature of 298 K, the gas temperature immediately adjacent to the wall is even higher—up to 586 K, as shown in the Fig. 6–38c. This creates very high wall heat flux that is unattainable in macroscale applications. In this case, shown in Fig. 6–38d, heat transfer is large enough to choke the flow.

1. D. G. Cahill, W. K. Ford, K. E. Goodson, *et al.*, "Nanoscale Thermal Transport." *Journal of Applied Physics*, 93, 2 (2003), pp. 793–817.

2. R. Raju and S. Roy, "Hydrodynamic Study of High Speed Flow and Heat Transfer through a Microchannel." *Journal of Thermophysics and Heat Transfer*, 19, 1 (2005), pp. 106–113.

3. S. Roy, R. Raju, H. Chuang, B. Kruden and M. Meyyappan, "Modeling Gas Flow Through Microchannels and Nanopores." *Journal of Applied Physics*, 93, 8 (2003), pp. 4870–79.

4. M. von Smoluchowski, "Ueber Wärmeleitung in Verdünnten Gasen," *Annalen der Physik und Chemi.* 64 (1898), pp. 101–130.

5. C. L. Tien, A. Majumdar, and F. Gerner. *Microscale Energy Tranport.* New York: Taylor & Francis Publishing, 1998.

SUMMARY

Convection heat transfer is expressed by *Newton's law of cooling* as

$$\dot{Q}_{conv} = hA_s(T_s - T_\infty)$$

where h is the convection heat transfer coefficient, T_s is the surface temperature, and T_∞ is the free-stream temperature. The convection coefficient is also expressed as

$$h = \frac{-k_{fluid}(\partial T/\partial y)_{y=0}}{T_s - T_\infty}$$

The *Nusselt number,* which is the dimensionless heat transfer coefficient, is defined as

$$Nu = \frac{hL_c}{k}$$

where k is the thermal conductivity of the fluid and L_c is the characteristic length.

The highly ordered fluid motion characterized by smooth streamlines is called *laminar.* The highly disordered fluid motion that typically occurs at high velocities is characterized by velocity fluctuations is called *turbulent.* The random and rapid fluctuations of groups of fluid particles, called *eddies,* provide an additional mechanism for momentum and heat transfer.

The region of the flow above the plate bounded by δ in which the effects of the viscous shearing forces caused by fluid viscosity are felt is called the *velocity boundary layer.* The *boundary layer thickness,* δ, is defined as the distance from the surface at which $u = 0.99V$. The hypothetical line of $u = 0.99V$ divides the flow over a plate into the *boundary layer region* in which the viscous effects and the velocity changes are significant, and the *irrotational flow region,* in which the frictional effects are negligible.

The friction force per unit area is called *shear stress,* and the shear stress at the wall surface is expressed as

$$\tau_s = \mu \frac{\partial u}{\partial y}\bigg|_{y=0} \qquad \text{or} \qquad \tau_s = C_f \frac{\rho V^2}{2}$$

where μ is the dynamic viscosity, V is the upstream velocity, and C_f is the dimensionless *friction coefficient.* The property $v = \mu/\rho$ is the *kinematic viscosity.* The friction force over the entire surface is determined from

$$F_f = C_f A_s \frac{\rho V^2}{2}$$

The flow region over the surface in which the temperature variation in the direction normal to the surface is significant is the *thermal boundary layer.* The *thickness* of the thermal boundary layer δ_t at any location along the surface is the distance from the surface at which the temperature difference $T - T_s$ equals $0.99(T_\infty - T_s)$. The relative thickness of the velocity and the thermal boundary layers is best described by the dimensionless *Prandtl number,* defined as

$$Pr = \frac{\text{Molecular diffusivity of momentum}}{\text{Molecular diffusivity of heat}} = \frac{v}{\alpha} = \frac{\mu c_p}{k}$$

For external flow, the dimensionless *Reynolds number* is expressed as

$$Re = \frac{\text{Inertia forces}}{\text{Viscous forces}} = \frac{VL_c}{v} = \frac{\rho VL_c}{\mu}$$

For a flat plate, the characteristic length is the distance x from the leading edge. The Reynolds number at which the flow becomes turbulent is called the *critical Reynolds number.* For flow over a flat plate, its value is taken to be $Re_{cr} = Vx_{cr}/v = 5 \times 10^5$.

The continuity, momentum, and energy equations for steady two-dimensional incompressible flow with constant properties are determined from mass, momentum, and energy balances to be

Continuity: $\qquad \dfrac{\partial u}{\partial x} + \dfrac{\partial v}{\partial y} = 0$

x-momentum: $\qquad \rho\left(u \dfrac{\partial u}{\partial x} + v \dfrac{\partial u}{\partial y}\right) = \mu \dfrac{\partial^2 u}{\partial y^2} - \dfrac{\partial P}{\partial x}$

Energy:
$$\rho c_p \left[u \frac{\partial T}{\partial x} + v \frac{\partial T}{\partial y} \right] = k \left(\frac{\partial^2 T}{\partial x^2} + \frac{\partial^2 T}{\partial y^2} \right) + \mu \Phi$$

where the *viscous dissipation function* Φ is

$$\Phi = 2 \left[\left(\frac{\partial u}{\partial x} \right)^2 + \left(\frac{\partial v}{\partial y} \right)^2 \right] + \left(\frac{\partial u}{\partial y} + \frac{\partial v}{\partial x} \right)^2$$

Using the boundary layer approximations and a similarity variable, these equations can be solved for parallel steady incompressible flow over a flat plate, with the following results:

Velocity boundary layer thickness: $\quad \delta = \dfrac{4.91}{\sqrt{V/vx}} = \dfrac{4.91x}{\sqrt{Re_x}}$

Local friction coefficient: $\quad C_{f,x} = \dfrac{\tau_w}{\rho V^2 / 2} = 0.664\, Re_x^{-1/2}$

Local Nusselt number: $\quad Nu_x = \dfrac{h_x x}{k} = 0.332\, Pr^{1/3} Re_x^{1/2}$

Thermal boundary layer thickness: $\quad \delta_t = \dfrac{\delta}{Pr^{1/3}} = \dfrac{4.91x}{Pr^{1/3} \sqrt{Re_x}}$

The average friction coefficient and Nusselt number are expressed in functional form as

$$C_f = f(Re_L) \qquad \text{and} \qquad Nu = g(Re_L, Pr)$$

The Nusselt number can be expressed by a simple power-law relation of the form

$$Nu = C\, Re_L^m\, Pr^n$$

where m and n are constant exponents, and the value of the constant C depends on geometry. The *Reynolds analogy* relates the convection coefficient to the friction coefficient for fluids with $Pr \approx 1$, and is expressed as

$$C_{f,x} \frac{Re_L}{2} = Nu_x \qquad \text{or} \qquad \frac{C_{f,x}}{2} = St_x$$

where

$$St = \frac{h}{\rho c_p V} = \frac{Nu}{Re_L\, Pr}$$

is the *Stanton number.* The analogy is extended to other Prandtl numbers by the *modified Reynolds analogy* or *Chilton–Colburn analogy,* expressed as

$$C_{f,x} \frac{Re_L}{2} = Nu_x Pr^{-1/3}$$

or

$$\frac{C_{f,x}}{2} = \frac{h_x}{\rho c_p V} Pr^{2/3} \equiv j_H \quad (0.6 < Pr < 60)$$

These analogies are also applicable approximately for turbulent flow over a surface, even in the presence of pressure gradients.

REFERENCES AND SUGGESTED READING

1. H. Blasius. "The Boundary Layers in Fluids with Little Friction (in German)." *Z. Math. Phys.*, 56, 1 (1908); pp. 1–37; English translation in National Advisory Committee for Aeronautics Technical Memo No. 1256, February 1950.

2. Y. A. Cengel and J. M. Cimbala. *Fluid Mechanics: Fundamentals and Applications.* New York: McGraw-Hill, 2006.

3. R. W. Fox and A. T. McDonald. *Introduction to Fluid Mechanics.* 5th. ed. New York, Wiley, 1999.

4. W. M. Kays and M. E. Crawford. *Convective Heat and Mass Transfer.* 3rd ed. New York: McGraw-Hill, 1993.

5. O. Reynolds. "On the Experimental Investigation of the Circumstances Which Determine Whether the Motion of Water Shall Be Direct or Sinuous, and the Law of Resistance in Parallel Channels." *Philosophical Transactions of the Royal Society of London* 174 (1883), pp. 935–82.

6. H. Schlichting. *Boundary Layer Theory.* 7th ed. New York: McGraw-Hill, 1979.

7. G. G. Stokes. "On the Effect of the Internal Friction of Fluids on the Motion of Pendulums." *Cambridge Philosophical Transactions,* IX, 8, 1851.

PROBLEMS*

Mechanism and Types of Convection

6–1C What is forced convection? How does it differ from natural convection? Is convection caused by winds forced or natural convection?

6–2C What is external forced convection? How does it differ from internal forced convection? Can a heat transfer system involve both internal and external convection at the same time? Give an example.

6–3C In which mode of heat transfer is the convection heat transfer coefficient usually higher, natural convection or forced convection? Why?

6–4C Consider a hot baked potato. Will the potato cool faster or slower when we blow the warm air coming from our lungs on it instead of letting it cool naturally in the cooler air in the room? Explain.

6–5C What is the physical significance of the Nusselt number? How is it defined?

6–6C When is heat transfer through a fluid conduction and when is it convection? For what case is the rate of heat transfer higher? How does the convection heat transfer coefficient differ from the thermal conductivity of a fluid?

6–7C Define incompressible flow and incompressible fluid. Must the flow of a compressible fluid necessarily be treated as compressible?

6–8 During air cooling of potatoes, the heat transfer coefficient for combined convection, radiation, and evaporation is determined experimentally to be as shown:

Air Velocity, m/s	Heat Transfer Coefficient, $W/m^2 \cdot °C$
0.66	14.0
1.00	19.1
1.36	20.2
1.73	24.4

Consider an 8-cm-diameter potato initially at 20°C with a thermal conductivity of 0.49 W/m · °C. Potatoes are cooled by refrigerated air at 5°C at a velocity of 1 m/s. Determine the initial rate of heat transfer from a potato, and the initial value of the temperature gradient in the potato at the surface.

Answers: 5.8 W, −585°C/m

---

*Problems designated by a "C" are concept questions, and students are encouraged to answer them all. Problems designated by an "E" are in English units, and the SI users can ignore them. Problems with the icon ⊛ are solved using EES, and complete solutions together with parametric studies are included on the enclosed CD. Problems with the icon ▣ are comprehensive in nature, and are intended to be solved with a computer, preferably using the EES software that accompanies this text.

6–9 An average man has a body surface area of 1.8 m² and a skin temperature of 33°C. The convection heat transfer coefficient for a clothed person walking in still air is expressed as $h = 8.6V^{0.53}$ for $0.5 < V < 2$ m/s, where V is the walking velocity in m/s. Assuming the average surface temperature of the clothed person to be 30°C, determine the rate of heat loss from an average man walking in still air at 10°C by convection at a walking velocity of (*a*) 0.5 m/s, (*b*) 1.0 m/s, (*c*) 1.5 m/s, and (*d*) 2.0 m/s.

6–10 The convection heat transfer coefficient for a clothed person standing in moving air is expressed as $h = 14.8V^{0.69}$ for $0.15 < V < 1.5$ m/s, where V is the air velocity. For a person with a body surface area of 1.7 m² and an average surface temperature of 29°C, determine the rate of heat loss from the person in windy air at 10°C by convection for air velocities of (*a*) 0.5 m/s, (*b*) 1.0 m/s, and (*c*) 1.5 m/s.

6–11 During air cooling of oranges, grapefruit, and tangelos, the heat transfer coefficient for combined convection, radiation, and evaporation for air velocities of $0.11 < V < 0.33$ m/s is determined experimentally and is expressed as $h = 5.05 \, k_{air}Re^{1/3}/D$, where the diameter D is the characteristic length. Oranges are cooled by refrigerated air at 5°C and 1 atm at a velocity of 0.3 m/s. Determine (*a*) the initial rate of heat transfer from a 7-cm-diameter orange initially at 15°C with a thermal conductivity of 0.50 W/m · °C, (*b*) the value of the initial temperature gradient inside the orange at the surface, and (*c*) the value of the Nusselt number.

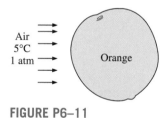

FIGURE P6–11

Velocity and Thermal Boundary Layers

6–12C What is viscosity? What causes viscosity in liquids and in gases? Is dynamic viscosity typically higher for a liquid or for a gas?

6–13C What is Newtonian fluid? Is water a Newtonian fluid?

6–14C What is the no-slip condition? What causes it?

6–15C Consider two identical small glass balls dropped into two identical containers, one filled with water and the other with oil. Which ball will reach the bottom of the container first? Why?

6–16C How does the dynamic viscosity of (*a*) liquids and (*b*) gases vary with temperature?

6–17C What fluid property is responsible for the development of the velocity boundary layer? For what kind of fluids will there be no velocity boundary layer on a flat plate?

6–18C What is the physical significance of the Prandtl number? Does the value of the Prandtl number depend on the type of flow or the flow geometry? Does the Prandtl number of air change with pressure? Does it change with temperature?

6–19C Will a thermal boundary layer develop in flow over a surface even if both the fluid and the surface are at the same temperature?

Laminar and Turbulent Flows

6–20C How does turbulent flow differ from laminar flow? For which flow is the heat transfer coefficient higher?

6–21C What is the physical significance of the Reynolds number? How is it defined for external flow over a plate of length *L*?

6–22C What does the friction coefficient represent in flow over a flat plate? How is it related to the drag force acting on the plate?

6–23C What is the physical mechanism that causes the friction factor to be higher in turbulent flow?

6–24C What is turbulent viscosity? What is it caused by?

6–25C What is turbulent thermal conductivity? What is it caused by?

Convection Equations and Similarity Solutions

6–26C Under what conditions can a curved surface be treated as a flat plate in fluid flow and convection analysis?

6–27C Express continuity equation for steady two-dimensional flow with constant properties, and explain what each term represents.

6–28C Is the acceleration of a fluid particle necessarily zero in steady flow? Explain.

6–29C For steady two-dimensional flow, what are the boundary layer approximations?

6–30C For what types of fluids and flows is the viscous dissipation term in the energy equation likely to be significant?

6–31C For steady two-dimensional flow over an isothermal flat plate in the *x*-direction, express the boundary conditions for the velocity components *u* and *v*, and the temperature *T* at the plate surface and at the edge of the boundary layer.

6–32C What is a similarity variable, and what is it used for? For what kinds of functions can we expect a similarity solution for a set of partial differential equations to exist?

6–33C Consider steady, laminar, two-dimensional flow over an isothermal plate. Does the thickness of the velocity boundary layer increase or decrease with (*a*) distance from the leading edge, (*b*) free-stream velocity, and (*c*) kinematic viscosity?

6–34C Consider steady, laminar, two-dimensional flow over an isothermal plate. Does the wall shear stress increase, decrease, or remain constant with distance from the leading edge?

6–35C What are the advantages of nondimensionalizing the convection equations?

6–36C Consider steady, laminar, two-dimensional, incompressible flow with constant properties and a Prandtl number of unity. For a given geometry, is it correct to say that both the average friction and heat transfer coefficients depend on the Reynolds number only?

6–37 Air at 15°C and 1 atm is flowing over a 0.3-m long plate at 65°C a velocity of 3.0 m/s. Using EES, Excel, or other software, plot the following on a combined graph for the range of $x = 0.0$ m to $x = x_{cr}$.
 (*a*) The hydrodynamic boundary layer as a function of *x*.
 (*b*) The thermal boundary layer as a function of *x*.

6–38 Liquid water at 15°C is flowing over a 0.3-m-wide plate at 65°C a velocity of 3.0 m/s. Using EES, Excel, or other comparable software, plot (*a*) the hydrodynamic boundary layer and (*b*) the thermal boundary layer as a function of *x* on the same graph for the range of $x = 0.0$ m to $x = x_{cr}$. Use a critical Reynolds number of 500,000.

6–39 Oil flow in a journal bearing can be treated as parallel flow between two large isothermal plates with one plate moving at a constant velocity of 12 m/s and the other stationary. Consider such a flow with a uniform spacing of 0.7 mm between the plates. The temperatures of the upper and lower plates are 40°C and 15°C, respectively. By simplifying and solving the continuity, momentum, and energy equations, determine (*a*) the velocity and temperature distributions in the oil, (*b*) the maximum temperature and where it occurs, and (*c*) the heat flux from the oil to each plate.

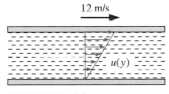

FIGURE P6–39

6–40 Repeat Prob. 6–39 for a spacing of 0.4 mm.

6–41 A 6-cm-diameter shaft rotates at 3000 rpm in a 20-cm-long bearing with a uniform clearance of 0.2 mm. At steady operating conditions, both the bearing and the shaft in the vicinity of the oil gap are at 50°C, and the viscosity and thermal conductivity of lubricating oil are 0.05 N · s/m² and 0.17 W/m · K. By simplifying and solving the continuity, momentum, and energy equations, determine (*a*) the maximum temperature of oil, (b) the rates of heat transfer to the bearing and the shaft, and (c) the mechanical power wasted by the viscous dissipation in the oil. *Answers:* (a) 53.3°C, (b) 419 W, (c) 838 W

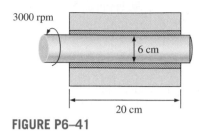

FIGURE P6–41

6–42 Repeat Prob. 6–41 by assuming the shaft to have reached peak temperature and thus heat transfer to the shaft to be negligible, and the bearing surface still to be maintained at 50°C.

6–43 Reconsider Prob. 6–41. Using EES (or other) software, investigate the effect of shaft velocity on the mechanical power wasted by viscous dissipation. Let the shaft rotation vary from 0 rpm to 5000 rpm. Plot the power wasted versus the shaft rpm, and discuss the results.

6–44 Consider a 5-cm-diameter shaft rotating at 4000 rpm in a 25-cm-long bearing with a clearance of 0.5 mm. Determine the power required to rotate the shaft if the fluid in the gap is (a) air, (b) water, and (c) oil at 40°C and 1 atm.

6–45 Consider the flow of fluid between two large parallel isothermal plates separated by a distance L. The upper plate is moving at a constant velocity of V and maintained at temperature T_0 while the lower plate is stationary and insulated. By simplifying and solving the continuity, momentum, and energy equations, obtain relations for the maximum temperature of fluid, the location where it occurs, and heat flux at the upper plate.

6–46 Reconsider Prob. 6–45. Using the results of this problem, obtain a relation for the volumetric heat generation rate $\dot{e}_{gen}$, in W/m³. Then express the convection problem as an equivalent conduction problem in the oil layer. Verify your model by solving the conduction problem and obtaining a relation for the maximum temperature, which should be identical to the one obtained in the convection analysis.

6–47 A 5-cm-diameter shaft rotates at 4500 rpm in a 15-cm-long, 8-cm-outer-diameter cast iron bearing (k = 70 W/m · K) with a uniform clearance of 0.6 mm filled with lubricating oil (μ = 0.03 N · s/m² and k = 0.14 W/m · K). The bearing is cooled externally by a liquid, and its outer surface is maintained at 40°C. Disregarding heat conduction through the shaft and assuming one-dimensional heat transfer, determine (a) the rate of heat transfer to the coolant, (b) the surface temperature of the shaft, and (c) the mechanical power wasted by the viscous dissipation in oil.

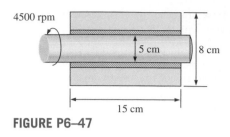

FIGURE P6–47

6–48 Repeat Prob. 6–47 for a clearance of 1 mm.

Momentum and Heat Transfer Analogies

6–49C How is Reynolds analogy expressed? What is the value of it? What are its limitations?

6–50C How is the modified Reynolds analogy expressed? What is the value of it? What are its limitations?

6–51 A 4-m × 4-m flat plate maintained at a constant temperature of 80°C is subjected to parallel flow of air at 1 atm, 20°C, and 10 m/s. The total drag force acting on the upper surface of the plate is measured to be 2.4 N. Using momentum-heat transfer analogy, determine the average convection heat transfer coefficient, and the rate of heat transfer between the upper surface of the plate and the air.

6–52 A metallic airfoil of elliptical cross section has a mass of 50 kg, surface area of 12 m², and a specific heat of 0.50 kJ/kg · °C. The airfoil is subjected to air flow at 1 atm, 25°C, and 5 m/s along its 3-m-long side. The average temperature of the airfoil is observed to drop from 160°C to 150°C within 2 min of cooling. Assuming the surface temperature of the airfoil to be equal to its average temperature and using momentum-heat transfer analogy, determine the average friction coefficient of the airfoil surface. *Answer:* 0.000363

6–53 Repeat Prob. 6–52 for an air-flow velocity of 10 m/s.

6–54 The electrically heated 0.6-m-high and 1.8-m-long windshield of a car is subjected to parallel winds at 1 atm, 0°C, and 80 km/h. The electric power consumption is observed to be 50 W when the exposed surface temperature of the windshield is 4°C. Disregarding radiation and heat transfer from the inner surface and using the momentum-heat transfer analogy, determine drag force the wind exerts on the windshield.

6–55 Consider an airplane cruising at an altitude of 10 km where standard atmospheric conditions are −50°C and 26.5 kPa at a speed of 800 km/h. Each wing of the airplane can be modeled as a 25-m × 3-m flat plate, and the friction coefficient of the wings is 0.0016. Using the momentum-heat transfer analogy, determine the heat transfer coefficient for the wings at cruising conditions. *Answer:* 89.6 W/m² · °C

Special Topic: Microscale Heat Transfer

6–56 Using a cylinder, a sphere, and a cube as examples, show that the rate of heat transfer is inversely proportional to the nominal size of the object. That is, heat transfer per unit area increases as the size of the object decreases.

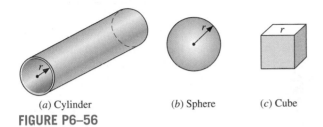

(a) Cylinder   (b) Sphere   (c) Cube

FIGURE P6–56

6–57 Determine the heat flux at the wall of a microchannel of width 1 μm if the wall temperature is 50°C and the average gas temperature near the wall is 100°C for the cases of

(a) $\sigma_T = 1.0$, $\gamma = 1.667$, $k = 0.15$ W/m · K, $\lambda/Pr = 0.5$
(b) $\sigma_T = 0.8$, $\gamma = 2$, $k = 0.1$ W/m · K, $\lambda/Pr = 5$

6–58 If $(\partial T/\partial y)_w = 80$ K/m, calculate the Nusselt number for a microchannel of width 1.2 μm if the wall temperature is 50°C and it is surrounded by (a) ambient air at temperature 30°C, (b) nitrogen gas at temperature −100°C.

Review Problems

6–59 Consider the Couette flow of a fluid with a viscosity of $\mu = 0.8$ N · s/m² and thermal conductivity of $k_f = 0.145$ W/m · K The lower plate is stationary and made of a material of thermal conductivity $k_p = 1.5$ W/m · K and thickness $b = 3$ mm. Its outer surface is maintained at $T_s = 40°C$. The upper plate is insulated and moves with a uniform speed $V = 5$ m/s. The distance between plates is $L = 5$ mm.

(a) Sketch the temperature distribution, $T(y)$, in the fluid and in the stationary plate.
(b) Determine the temperature distribution function, $T(y)$, in the fluid $(0 < y < L)$.
(c) Calculate the maximum temperature of the fluid, as well as the temperature of the fluid at the contact surfaces with the lower and upper plates.

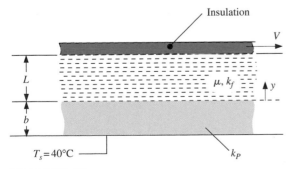

FIGURE P6–59

6–60 Engine oil at 15°C is flowing over a 0.3-m wide plate at 65°C at a velocity of 3.0 m/s. Using EES, Excel, or other comparable software, plot (a) the hydrodynamic boundary layer and (b) the thermal boundary layer as a function of x on the same graph for the range of $x = 0.0$ m to $x = x_{cr}$. Use a critical Reynolds number of 500,000.

Fundamentals of Engineering (FE) Exam Problems

6–61 The _____ number is a significant dimensionless parameter for forced convection and the _____ number is a significant dimensionless parameter for natural convection.
(a) Reynolds, Grashof   (b) Reynolds, Mach
(c) Reynolds, Eckert   (d) Reynolds, Schmidt
(e) Grashof, Sherwood

6–62 For the same initial conditions, one can expect the laminar thermal and momentum boundary layers on a flat plate to have the same thickness when the Prandtl number of the flowing fluid is
(a) Close to zero   (b) Small   (c) Approximately one
(d) Large   (e) Very large

6–63 One can expect the heat transfer coefficient for turbulent flow to be _____ for laminar flow.
(a) less then   (b) same as   (c) greater than

6–64 Most correlations for the convection heat transfer coefficient use the dimensionless Nusselt number, which is defined as
(a) h/k (b) k/h (c) hL_c/k
(d) kL_c/h (e) $k/\rho c_p$

6–65 In any forced or natural convection situation, the velocity of the flowing fluid is zero where the fluid wets any stationary surface. The magnitude of heat flux where the fluid wets a stationary surface is given by

(a) $k(T_{fluid} - T_{wall})$ (b) $k \dfrac{dT}{dy}\bigg|_{wall}$

(c) $k \dfrac{d^2T}{dy^2}\bigg|_{wall}$ (d) $h \dfrac{dT}{dy}\bigg|_{wall}$

(e) None of them

6–66 In turbulent flow, one can estimate the Nusselt number using the analogy between heat and momentum transfer (Colburn analogy). This analogy relates the Nusselt number to the coefficient of friction, C_f, as
(a) $Nu = 0.5 C_f Re Pr^{1/3}$ (b) $Nu = 0.5 C_f Re Pr^{2/3}$
(c) $Nu = C_f Re Pr^{1/3}$ (d) $Nu = C_f Re Pr^{2/3}$
(e) $Nu = C_f Re^{1/2} Pr^{1/3}$

6–67 An electrical water ($k = 0.61$ W/m · K) heater uses natural convection to transfer heat from a 1-cm diameter by 0.65-m long, 110 V electrical resistance heater to the water.

During operation, the surface temperature of this heater is 120°C while the temperature of the water is 35°C, and the Nusselt number (based on the diameter) is 5. Considering only the side surface of the heater (and thus $A = \pi DL$), the current passing through the electrical heating element is

(a) 2.2 A   (b) 2.7 A   (c) 3.6 A   (d) 4.8 A   (e) 5.6 A

6–68   The coefficient of friction C_f for a fluid flowing across a surface in terms of the surface shear stress, τ_s, is given by

(a) $2\rho V^2/\tau_s$ (b) $2\tau_s/\rho V^2$ (c) $2\tau_s/\rho V^2 \Delta T$
(d) $4\tau_s/\rho V^2$ (e) None of them

6–69   The transition from laminar flow to turbulent flow in a forced convection situation is determined by which one of the following dimensionless numbers?

(a) Grasshof   (b) Nusselt   (c) Reynolds
(d) Stanton   (e) Mach

Design and Essay Problems

6–70   Design an experiment to measure the viscosity of liquids using a vertical funnel with a cylindrical reservoir of height h and a narrow flow section of diameter D and length L. Making appropriate assumptions, obtain a relation for viscosity in terms of easily measurable quantities such as density and volume flow rate.

6–71   A facility is equipped with a wind tunnel, and can measure the friction coefficient for flat surfaces and airfoils. Design an experiment to determine the mean heat transfer coefficient for a surface using friction coefficient data.

EXTERNAL FORCED CONVECTION

I n Chapter 6, we considered the general and theoretical aspects of forced convection, with emphasis on differential formulation and analytical solutions. In this chapter we consider the practical aspects of forced convection to or from flat or curved surfaces subjected to *external flow,* characterized by the freely growing boundary layers surrounded by a free flow region that involves no velocity and temperature gradients.

We start this chapter with an overview of external flow, with emphasis on friction and pressure drag, flow separation, and the evaluation of average drag and convection coefficients. We continue with *parallel flow over flat plates.* In Chapter 6, we solved the boundary layer equations for steady, laminar, parallel flow over a flat plate, and obtained relations for the local friction coefficient and the Nusselt number. Using these relations as the starting point, we determine the average friction coefficient and Nusselt number. We then extend the analysis to turbulent flow over flat plates with and without an unheated starting length.

Next we consider *cross flow over cylinders and spheres,* and present graphs and empirical correlations for the drag coefficients and the Nusselt numbers, and discuss their significance. Finally, we consider *cross flow over tube banks* in aligned and staggered configurations, and present correlations for the pressure drop and the average Nusselt number for both configurations.

OBJECTIVES

When you finish studying this chapter, you should be able to:

■   Distinguish between internal and external flow,

■   Develop an intuitive understanding of friction drag and pressure drag, and evaluate the average drag and convection coefficients in external flow,

■   Evaluate the drag and heat transfer associated with flow over a flat plate for both laminar and turbulent flow,

■   Calculate the drag force exerted on cylinders during cross flow, and the average heat transfer coefficient, and

■   Determine the pressure drop and the average heat transfer coefficient associated with flow across a tube bank for both in-line and staggered configurations.

CONTENTS

FIGURE 7–1
Flow over bodies is commonly encountered in practice.

7–1 ▪ DRAG AND HEAT TRANSFER IN EXTERNAL FLOW

Fluid flow over solid bodies frequently occurs in practice, and it is responsible for numerous physical phenomena such as the *drag force* acting on the automobiles, power lines, trees, and underwater pipelines; the *lift* developed by airplane wings; *upward draft* of rain, snow, hail, and dust particles in high winds; and the *cooling* of metal or plastic sheets, steam and hot water pipes, and extruded wires (Fig. 7–1). Therefore, developing a good understanding of external flow and external forced convection is important in the mechanical and thermal design of many engineering systems such as aircraft, automobiles, buildings, electronic components, and turbine blades.

The flow fields and geometries for most external flow problems are too complicated to be solved analytically, and thus we have to rely on correlations based on experimental data. The availability of high-speed computers has made it possible to conduct series of "numerical experimentations" quickly by solving the governing equations numerically, and to resort to the expensive and time-consuming testing and experimentation only in the final stages of design. In this chapter we mostly rely on relations developed experimentally.

The velocity of the fluid relative to an immersed solid body sufficiently far from the body (outside the boundary layer) is called the **free-stream velocity**. It is usually taken to be equal to the **upstream velocity** V, also called the **approach velocity**, which is the velocity of the approaching fluid far ahead of the body. This idealization is nearly exact for very thin bodies, such as a flat plate parallel to flow, but approximate for blunt bodies such as a large cylinder. The fluid velocity ranges from zero at the surface (the no-slip condition) to the free-stream value away from the surface, and the subscript "infinity" serves as a reminder that this is the value at a distance where the presence of the body is not felt. The upstream velocity, in general, may vary with location and time (e.g., the wind blowing past a building). But in the design and analysis, the upstream velocity is usually assumed to be *uniform* and *steady* for convenience, and this is what we will do in this chapter.

Friction and Pressure Drag

It is common experience that a body meets some resistance when it is forced to move through a fluid, especially a liquid. You may have seen high winds knocking down trees, power lines, and even trailers, and have felt the strong "push" the wind exerts on your body. You experience the same feeling when you extend your arm out of the window of a moving car. The force a flowing fluid exerts on a body in the flow direction is called **drag** (Fig. 7–2).

A stationary fluid exerts only normal pressure forces on the surface of a body immersed in it. A moving fluid, however, also exerts tangential shear forces on the surface because of the no-slip condition caused by viscous effects. Both of these forces, in general, have components in the direction of flow, and thus the drag force is due to the combined effects of pressure and wall shear forces in the flow direction. The components of the pressure and wall shear forces in the *normal* direction to flow tend to move the body in that direction, and their sum is called **lift**.

In general, both the skin friction (wall shear) and pressure contribute to the drag and the lift. In the special case of a thin flat plate aligned parallel

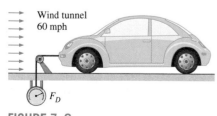

Wind tunnel
60 mph

F_D

FIGURE 7–2
Schematic for measuring the drag force acting on a car in a wind tunnel.

to the flow direction, the drag force depends on the wall shear only and is independent of pressure. When the flat plate is placed normal to the flow direction, however, the drag force depends on the pressure only and is independent of the wall shear since the shear stress in this case acts in the direction normal to flow (Fig. 7–3). For slender bodies such as wings, the shear force acts nearly parallel to the flow direction. The drag force for such slender bodies is mostly due to shear forces (the skin friction).

The drag force F_D depends on the density ρ of the fluid, the upstream velocity V, and the size, shape, and orientation of the body, among other things. The drag characteristics of a body is represented by the dimensionless **drag coefficient** C_D defined as

Drag coefficient:
$$C_D = \frac{F_D}{\frac{1}{2}\rho V^2 A} \qquad (7\text{–}1)$$

where A is the *frontal area* (the area projected on a plane normal to the direction of flow) for blunt bodies—bodies that tends to block the flow. The frontal area of a cylinder of diameter D and length L, for example, is $A = LD$. For parallel flow over flat plates or thin airfoils, A is the surface area. The drag coefficient is primarily a function of the shape of the body, but it may also depend on the Reynolds number and the surface roughness.

The drag force is the net force exerted by a fluid on a body in the direction of flow due to the combined effects of wall shear and pressure forces. The part of drag that is due directly to wall shear stress τ_w is called the **skin friction drag** (or just *friction drag*) since it is caused by frictional effects, and the part that is due directly to pressure P is called the **pressure drag** (also called the *form drag* because of its strong dependence on the form or shape of the body). When the friction and pressure drag coefficients are available, the total drag coefficient is determined by simply adding them,

$$C_D = C_{D,\,\text{friction}} + C_{D,\,\text{pressure}} \qquad (7\text{–}2)$$

The *friction drag* is the component of the wall shear force in the direction of flow, and thus it depends on the orientation of the body as well as the magnitude of the wall shear stress τ_w. The friction drag is *zero* for a surface normal to flow, and *maximum* for a surface parallel to flow since the friction drag in this case equals the total shear force on the surface. Therefore, for parallel flow over a flat plate, the drag coefficient is equal to the *friction drag coefficient*, or simply the *friction coefficient* (Fig. 7–4). That is,

Flat plate:
$$C_D = C_{D,\,\text{friction}} = C_f \qquad (7\text{–}3)$$

Once the average friction coefficient C_f is available, the drag (or friction) force over the surface can be determined from Eq. 7–1. In this case A is the surface area of the plate exposed to fluid flow. When both sides of a thin plate are subjected to flow, A becomes the total area of the top and bottom surfaces. Note that the friction coefficient, in general, varies with location along the surface.

Friction drag is a strong function of viscosity, and an "idealized" fluid with zero viscosity would produce zero friction drag since the wall shear stress would be zero. The pressure drag would also be zero in this case during steady flow regardless of the shape of the body since there are no pressure losses. For flow in the horizontal direction, for example, the pressure along a horizontal line is constant (just like stationary fluids) since the upstream velocity is

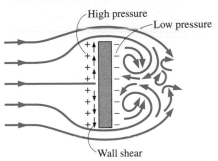

FIGURE 7–3

Drag force acting on a flat plate normal to the flow depends on the pressure only and is independent of the wall shear, which acts normal to the free-stream flow.

$$C_{D,\,\text{pressure}} = 0$$
$$C_D = C_{D,\,\text{friction}} = C_f$$

$$F_{D,\,\text{pressure}} = 0$$
$$F_D = F_{D,\,\text{friction}} = F_f = C_f A \frac{\rho V^2}{2}$$

FIGURE 7–4

For parallel flow over a flat plate, the pressure drag is zero, and thus the drag coefficient is equal to the friction coefficient and the drag force is equal to the friction force.

constant, and thus there is no net pressure force acting on the body in the horizontal direction. Therefore, the total drag is zero for the case of ideal inviscid fluid flow.

At low Reynolds numbers, most drag is due to friction drag. This is especially the case for highly streamlined bodies such as airfoils. The friction drag is also proportional to the surface area. Therefore, bodies with a larger surface area experience a larger friction drag. Large commercial airplanes, for example, reduce their total surface area and thus drag by retracting their wing extensions when they reach the cruising altitudes to save fuel. The friction drag coefficient is independent of *surface roughness* in laminar flow, but is a strong function of surface roughness in turbulent flow due to surface roughness elements protruding further into the boundary layer.

The pressure drag is proportional to the frontal area and to the *difference* between the pressures acting on the front and back of the immersed body. Therefore, the pressure drag is usually dominant for blunt bodies, negligible for streamlined bodies such as airfoils, and zero for thin flat plates parallel to the flow.

When a fluid separates from a body, it forms a separated region between the body and the fluid stream. This low-pressure region behind the body where recirculating and backflows occur is called the **separated region**. The larger the separated region, the larger the pressure drag. The effects of flow separation are felt far downstream in the form of reduced velocity (relative to the upstream velocity). The region of flow trailing the body where the effects of the body on velocity are felt is called the **wake** (Fig. 7–5). The separated region comes to an end when the two flow streams reattach. Therefore, the separated region is an enclosed volume, whereas the wake keeps growing behind the body until the fluid in the wake region regains its velocity and the velocity profile becomes nearly flat again. Viscous and rotational effects are the most significant in the boundary layer, the separated region, and the wake.

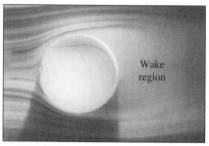

FIGURE 7–5
Separation during flow over a tennis ball and the wake region.
Courtesy of NASA and Cislunar Aerospace, Inc.

Heat Transfer

The phenomena that affect drag force also affect heat transfer, and this effect appears in the Nusselt number. By nondimensionalizing the boundary layer equations, it was shown in Chapter 6 that the local and average Nusselt numbers have the functional form

$$\mathrm{Nu}_x = f_1(x^*, \mathrm{Re}_x, \mathrm{Pr}) \qquad \text{and} \qquad \mathrm{Nu} = f_2(\mathrm{Re}_L, \mathrm{Pr}) \qquad \textbf{(7–4a, b)}$$

The experimental data for heat transfer is often represented conveniently with reasonable accuracy by a simple power-law relation of the form

$$\mathrm{Nu} = C\,\mathrm{Re}_L^m\,\mathrm{Pr}^n \qquad \textbf{(7–5)}$$

where m and n are constant exponents, and the value of the constant C depends on geometry and flow.

The fluid temperature in the thermal boundary layer varies from T_s at the surface to about T_∞ at the outer edge of the boundary. The fluid properties also vary with temperature, and thus with position across the boundary layer. In order to account for the variation of the properties with temperature, the fluid properties are usually evaluated at the so-called **film temperature**, defined as

$$T_f = \frac{T_s + T_\infty}{2} \tag{7–6}$$

which is the *arithmetic average* of the surface and the free-stream temperatures. The fluid properties are then assumed to remain constant at those values during the entire flow. An alternative way of accounting for the variation of properties with temperature is to evaluate all properties at the free stream temperature and to multiply the Nusselt number relation in Eq. 7–5 by $(Pr_\infty/Pr_s)^r$ or $(\mu_\infty/\mu_s)^r$ where r is an experimentally determined constant.

The local drag and convection coefficients vary along the surface as a result of the changes in the velocity boundary layers in the flow direction. We are usually interested in the drag force and the heat transfer rate for the *entire* surface, which can be determined using the *average* friction and convection coefficient. Therefore, we present correlations for both local (identified with the subscript x) and average friction and convection coefficients. When relations for local friction and convection coefficients are available, the *average* friction and convection coefficients for the entire surface can be determined by integration from

$$C_D = \frac{1}{L}\int_0^L C_{D,x}\,dx \tag{7–7}$$

and

$$h = \frac{1}{L}\int_0^L h_x\,dx \tag{7–8}$$

When the average drag and convection coefficients are available, the drag force can be determined from Eq. 7–1 and the rate of heat transfer to or from an isothermal surface can be determined from

$$\dot{Q} = hA_s(T_s - T_\infty) \tag{7–9}$$

where A_s is the surface area.

7–2 · PARALLEL FLOW OVER FLAT PLATES

Consider the parallel flow of a fluid over a flat plate of length L in the flow direction, as shown in Fig. 7–6. The x-coordinate is measured along the plate surface from the leading edge in the direction of the flow. The fluid approaches the plate in the x-direction with a uniform velocity V and temperature T_∞. The flow in the velocity boundary layers starts out as laminar, but if the plate is sufficiently long, the flow becomes turbulent at a distance x_{cr} from the leading edge where the Reynolds number reaches its critical value for transition.

The transition from laminar to turbulent flow depends on the *surface geometry, surface roughness, upstream velocity, surface temperature,* and the *type of fluid,* among other things, and is best characterized by the Reynolds number. The Reynolds number at a distance x from the leading edge of a flat plate

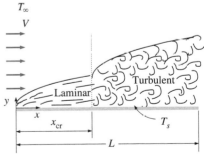

FIGURE 7–6
Laminar and turbulent regions of the boundary layer during flow over a flat plate.

is expressed as

$$\text{Re}_x = \frac{\rho V x}{\mu} = \frac{V x}{v} \qquad (7\text{--}10)$$

Note that the value of the Reynolds number varies for a flat plate along the flow, reaching $\text{Re}_L = VL/v$ at the end of the plate.

For flow over a flat plate, transition from laminar to turbulent begins at about $\text{Re} \cong 1 \times 10^5$, but does not become fully turbulent before the Reynolds number reaches much higher values, typically around 3×10^6. In engineering analysis, a generally accepted value for the critical Reynold number is

$$\text{Re}_{\text{cr}} = \frac{\rho V x_{\text{cr}}}{\mu} = 5 \times 10^5 \qquad (7\text{--}11)$$

The actual value of the engineering critical Reynolds number for a flat plate may vary somewhat from 10^5 to 3×10^6, depending on the surface roughness, the turbulence level, and the variation of pressure along the surface.

Friction Coefficient

Based on analysis, the boundary layer thickness and the local friction coefficient at location x for laminar flow over a flat plate were determined in Chapter 6 to be

Laminar: $\qquad \delta_{v,x} = \dfrac{4.91x}{\text{Re}_x^{1/2}} \qquad$ and $\qquad C_{f,x} = \dfrac{0.664}{\text{Re}_x^{1/2}}, \qquad \text{Re}_x < 5 \times 10^5 \qquad$ **(7–12a, b)**

The corresponding relations for turbulent flow are

Turbulent: $\qquad \delta_{v,x} = \dfrac{0.38x}{\text{Re}_x^{1/5}} \qquad$ and $\qquad C_{f,x} = \dfrac{0.059}{\text{Re}_x^{1/5}}, \qquad 5 \times 10^5 \leq \text{Re}_x \leq 10^7 \qquad$ **(7–13a, b)**

where x is the distance from the leading edge of the plate and $\text{Re}_x = Vx/v$ is the Reynolds number at location x. Note that $C_{f,x}$ is proportional to $\text{Re}_x^{-1/2}$ and thus to $x^{-1/2}$ for laminar flow. Therefore, $C_{f,x}$ is supposedly *infinite* at the leading edge ($x = 0$) and decreases by a factor of $x^{-1/2}$ in the flow direction. The local friction coefficients are higher in turbulent flow than they are in laminar flow because of the intense mixing that occurs in the turbulent boundary layer. Note that $C_{f,x}$ reaches its highest values when the flow becomes fully turbulent, and then decreases by a factor of $x^{-1/5}$ in the flow direction.

The *average* friction coefficient over the entire plate is determined by substituting the relations above into Eq. 7–7 and performing the integrations (Fig. 7–7). We get

Laminar: $\qquad\qquad C_f = \dfrac{1.33}{\text{Re}_L^{1/2}} \qquad \text{Re}_L < 5 \times 10^5 \qquad$ **(7–14)**

Turbulent: $\qquad\qquad C_f = \dfrac{0.074}{\text{Re}_L^{1/5}} \qquad 5 \times 10^5 \leq \text{Re}_L \leq 10^7 \qquad$ **(7–15)**

The first relation gives the average friction coefficient for the entire plate when the flow is *laminar* over the *entire* plate. The second relation gives the average friction coefficient for the entire plate only when the flow is *turbulent* over the *entire* plate, or when the laminar flow region of the plate is too small relative to the turbulent flow region (that is, $x_{\text{cr}} \ll L$).

$$C_f = \frac{1}{L}\int_0^L C_{f,x}\,dx$$

$$= \frac{1}{L}\int_0^L \frac{0.664}{\text{Re}_x^{1/2}}\,dx$$

$$= \frac{0.664}{L}\int_0^L \left(\frac{Vx}{v}\right)^{-1/2} dx$$

$$= \frac{0.664}{L}\left(\frac{V}{v}\right)^{-1/2} \frac{x^{1/2}}{\frac{1}{2}}\bigg|_0^L$$

$$= \frac{2\times 0.664}{L}\left(\frac{VL}{v}\right)^{-1/2}$$

$$= \frac{1.33}{\text{Re}_L^{1/2}}$$

FIGURE 7–7

The average friction coefficient over a surface is determined by integrating the local friction coefficient over the entire surface.

In some cases, a flat plate is sufficiently long for the flow to become turbulent, but not long enough to disregard the laminar flow region. In such cases, the *average* friction coefficient over the entire plate is determined by performing the integration in Eq. 7–7 over two parts: the laminar region $0 \leq x \leq x_{cr}$ and the turbulent region $x_{cr} < x \leq L$ as

$$C_f = \frac{1}{L} \left(\int_0^{x_{cr}} C_{f,x\,\text{laminar}}\, dx + \int_{x_{cr}}^L C_{f,x,\text{turbulent}}\, dx \right) \quad \text{(7–16)}$$

Note that we included the transition region with the turbulent region. Again taking the critical Reynolds number to be $\text{Re}_{cr} = 5 \times 10^5$ and performing the integrations of Eq. 7–16 after substituting the indicated expressions, the *average* friction coefficient over the entire plate is determined to be

$$C_f = \frac{0.074}{\text{Re}_L^{1/5}} - \frac{1742}{\text{Re}_L} \qquad 5 \times 10^5 \leq \text{Re}_L \leq 10^7 \quad \text{(7–17)}$$

The constants in this relation will be different for different critical Reynolds numbers. Also, the surfaces are assumed to be *smooth*, and the free stream to be *turbulent free*. For laminar flow, the friction coefficient depends on only the Reynolds number, and the surface roughness has no effect. For turbulent flow, however, surface roughness causes the friction coefficient to increase severalfold, to the point that in fully turbulent regime the friction coefficient is a function of surface roughness alone, and independent of the Reynolds number (Fig. 7–8). This is also the case in pipe flow.

A curve fit of experimental data for the average friction coefficient in this regime is given by Schlichting as

Rough surface, turbulent: $\qquad C_f = \left(1.89 - 1.62 \log \frac{\varepsilon}{L} \right)^{-2.5} \quad \text{(7–18)}$

were ε is the surface roughness, and L is the length of the plate in the flow direction. In the absence of a better relation, the relation above can be used for turbulent flow on rough surfaces for $\text{Re} > 10^6$, especially when $\varepsilon/L > 10^{-4}$.

Relative roughness, ε/L	Friction coefficient C_f
0.0*	0.0029
1×10^{-5}	0.0032
1×10^{-4}	0.0049
1×10^{-3}	0.0084

*Smooth surface for $\text{Re} = 10^7$. Others calculated from Eq. 7–18.

FIGURE 7–8

For turbulent flow, surface roughness may cause the friction coefficient to increase severalfold.

Heat Transfer Coefficient

The local Nusselt number at a location x for laminar flow over a flat plate was determined in Chapter 6 by solving the differential energy equation to be

Laminar: $\qquad \text{Nu}_x = \frac{h_x x}{k} = 0.332\, \text{Re}_x^{0.5}\, \text{Pr}^{1/3} \qquad \text{Pr} > 0.6 \quad \text{(7–19)}$

The corresponding relation for turbulent flow is

Turbulent: $\qquad \text{Nu}_x = \frac{h_x x}{k} = 0.0296\, \text{Re}_x^{0.8}\, \text{Pr}^{1/3} \qquad \begin{aligned} &0.6 \leq \text{Pr} \leq 60 \\ &5 \times 10^5 \leq \text{Re}_x \leq 10^7 \end{aligned} \quad \text{(7–20)}$

Note that h_x is proportional to $\text{Re}_x^{0.5}$ and thus to $x^{-0.5}$ for laminar flow. Therefore, h_x is *infinite* at the leading edge ($x = 0$) and decreases by a factor of $x^{-0.5}$ in the flow direction. The variation of the boundary layer thickness δ and the friction and heat transfer coefficients along an isothermal flat plate are shown in Fig. 7–9. The local friction and heat transfer coefficients are higher in

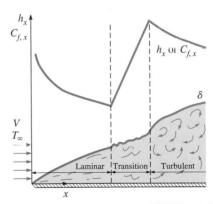

FIGURE 7–9

The variation of the local friction and heat transfer coefficients for flow over a flat plate.

turbulent flow than they are in laminar flow. Also, h_x reaches its highest values when the flow becomes fully turbulent, and then decreases by a factor of $x^{-0.2}$ in the flow direction, as shown in the figure.

The *average* Nusselt number over the entire plate is determined by substituting the relations above into Eq. 7–8 and performing the integrations. We get

$$\textit{Laminar:} \qquad \text{Nu} = \frac{hL}{k} = 0.664\, \text{Re}_L^{0.5}\, \text{Pr}^{1/3} \qquad \text{Re}_L < 5 \times 10^5 \qquad \text{(7–21)}$$

$$\textit{Turbulent:} \qquad \text{Nu} = \frac{hL}{k} = 0.037\, \text{Re}_L^{0.8}\, \text{Pr}^{1/3} \qquad \begin{matrix} 0.6 \le \text{Pr} \le 60 \\ 5 \times 10^5 \le \text{Re}_L \le 10^7 \end{matrix} \qquad \text{(7–22)}$$

The first relation gives the average heat transfer coefficient for the entire plate when the flow is *laminar* over the *entire* plate. The second relation gives the average heat transfer coefficient for the entire plate only when the flow is *turbulent* over the *entire* plate, or when the laminar flow region of the plate is too small relative to the turbulent flow region.

In some cases, a flat plate is sufficiently long for the flow to become turbulent, but not long enough to disregard the laminar flow region. In such cases, the *average* heat transfer coefficient over the entire plate is determined by performing the integration in Eq. 7–8 over two parts as

$$h = \frac{1}{L} \left(\int_0^{x_{cr}} h_{x,\,\text{laminar}}\, dx + \int_{x_{cr}}^L h_{x,\,\text{turbulent}}\, dx \right) \qquad \text{(7–23)}$$

Again taking the critical Reynolds number to be $\text{Re}_{cr} = 5 \times 10^5$ and performing the integrations in Eq. 7–23 after substituting the indicated expressions, the *average* Nusselt number over the *entire* plate is determined to be (Fig. 7–10)

$$\text{Nu} = \frac{hL}{k} = (0.037\, \text{Re}_L^{0.8} - 871)\text{Pr}^{1/3} \qquad \begin{matrix} 0.6 \le \text{Pr} \le 60 \\ 5 \times 10^5 \le \text{Re}_L \le 10^7 \end{matrix} \qquad \text{(7–24)}$$

The constants in this relation will be different for different critical Reynolds numbers.

Liquid metals such as mercury have high thermal conductivities, and are commonly used in applications that require high heat transfer rates. However, they have very small Prandtl numbers, and thus the thermal boundary layer develops much faster than the velocity boundary layer. Then we can assume the velocity in the thermal boundary layer to be constant at the free stream value and solve the energy equation. It gives

$$\text{Nu}_x = 0.565(\text{Re}_x\, \text{Pr})^{1/2} \qquad \text{Pr} < 0.05 \qquad \text{(7–25)}$$

It is desirable to have a single correlation that applies to *all fluids*, including liquid metals. By curve-fitting existing data, Churchill and Ozoe (1973) proposed the following relation which is applicable for *all Prandtl numbers* and is claimed to be accurate to ±1%,

$$\text{Nu}_x = \frac{h_x x}{k} = \frac{0.3387\, \text{Pr}^{1/3}\, \text{Re}_x^{1/2}}{[1 + (0.0468/\text{Pr})^{2/3}]^{1/4}} \qquad \text{(7–26)}$$

These relations have been obtained for the case of *isothermal* surfaces but could also be used approximately for the case of nonisothermal surfaces by assuming the surface temperature to be constant at some average value.

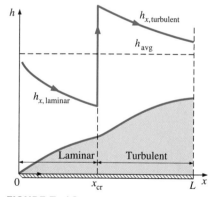

FIGURE 7–10
Graphical representation of the average heat transfer coefficient for a flat plate with combined laminar and turbulent flow.

Also, the surfaces are assumed to be *smooth,* and the free stream to be *turbulent free.* The effect of variable properties can be accounted for by evaluating all properties at the film temperature.

Flat Plate with Unheated Starting Length

So far we have limited our consideration to situations for which the entire plate is heated from the leading edge. But many practical applications involve surfaces with an unheated starting section of length ξ, shown in Fig. 7–11, and thus there is no heat transfer for $0 < x < \xi$. In such cases, the velocity boundary layer starts to develop at the leading edge ($x = 0$), but the thermal boundary layer starts to develop where heating starts ($x = \xi$).

Consider a flat plate whose heated section is maintained at a constant temperature ($T = T_s$ constant for $x > \xi$). Using integral solution methods (see Kays and Crawford, 1994), the local Nusselt numbers for both laminar and turbulent flows are determined to be

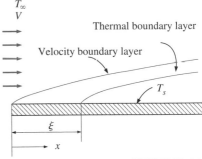

FIGURE 7–11

Flow over a flat plate with an unheated starting length.

Laminar:
$$\text{Nu}_x = \frac{\text{Nu}_{x \,(\text{for } \xi = 0)}}{[1 - (\xi/x)^{3/4}]^{1/3}} = \frac{0.332 \, \text{Re}_x^{0.5} \, \text{Pr}^{1/3}}{[1 - (\xi/x)^{3/4}]^{1/3}} \tag{7–27}$$

Turbulent:
$$\text{Nu}_x = \frac{\text{Nu}_{x \,(\text{for } \xi = 0)}}{[1 - (\xi/x)^{9/10}]^{1/9}} = \frac{0.0296 \, \text{Re}_x^{0.8} \, \text{Pr}^{1/3}}{[1 - (\xi/x)^{9/10}]^{1/9}} \tag{7–28}$$

for $x > \xi$. Note that for $\xi = 0$, these Nu_x relations reduce to $\text{Nu}_{x \,(\text{for } \xi = 0)}$, which is the Nusselt number relation for a flat plate without an unheated starting length. Therefore, the terms in brackets in the denominator serve as correction factors for plates with unheated starting lengths.

The determination of the average Nusselt number for the heated section of a plate requires the integration of the local Nusselt number relations above, which cannot be done analytically. Therefore, integrations must be done numerically. The results of numerical integrations have been correlated for the average convection coefficients [Thomas, (1977)] as

Laminar:
$$h = \frac{2[1 - (\xi/x)^{3/4}]}{1 - \xi/L} \, h_{x=L} \tag{7–29}$$

Turbulent:
$$h = \frac{5[1 - (\xi/x)^{9/10}]}{4(1 - \xi/L)} \, h_{x=L} \tag{7–30}$$

The first relation gives the average convection coefficient for the entire heated section of the plate when the flow is laminar over the entire plate. Note that for $\xi = 0$ it reduces to $h_L = 2h_{x=L}$, as expected. The second relation gives the average convection coefficient for the case of turbulent flow over the entire plate or when the laminar flow region is small relative to the turbulent region.

Uniform Heat Flux

When a flat plate is subjected to *uniform heat flux* instead of uniform temperature, the local Nusselt number is given by

Laminar:
$$\text{Nu}_x = 0.453 \, \text{Re}_x^{0.5} \, \text{Pr}^{1/3} \tag{7–31}$$

Turbulent:
$$\text{Nu}_x = 0.0308 \, \text{Re}_x^{0.8} \, \text{Pr}^{1/3} \tag{7–32}$$

These relations give values that are 36 percent higher for laminar flow and 4 percent higher for turbulent flow relative to the isothermal plate case. When the plate involves an unheated starting length, the relations developed for the uniform surface temperature case can still be used provided that Eqs. 7–31 and 7–32 are used for $Nu_{x(\text{for } \xi = 0)}$ in Eqs. 7–27 and 7–28, respectively.

When heat flux $\dot{q}_s$ is prescribed, the rate of heat transfer to or from the plate and the surface temperature at a distance x are determined from

$$\dot{Q} = \dot{q}_s A_s \tag{7–33}$$

and

$$\dot{q}_s = h_x[T_s(x) - T_\infty] \quad \rightarrow \quad T_s(x) = T_\infty + \frac{\dot{q}_s}{h_x} \tag{7–34}$$

where A_s is the heat transfer surface area.

$T_\infty = 60°C$
$V = 2$ m/s

Oil

A_s $\dot{Q}$ $T_s = 20°C$

$L = 5$ m

FIGURE 7–12
Schematic for Example 7–1.

EXAMPLE 7–1   **Flow of Hot Oil over a Flat Plate**

Engine oil at 60°C flows over the upper surface of a 5-m-long flat plate whose temperature is 20°C with a velocity of 2 m/s (Fig. 7–12). Determine the total drag force and the rate of heat transfer per unit width of the entire plate.

SOLUTION Engine oil flows over a flat plate. The total drag force and the rate of heat transfer per unit width of the plate are to be determined.

Assumptions 1 The flow is steady and incompressible. 2 The critical Reynolds number is $Re_{cr} = 5 \times 10^5$.

Properties The properties of engine oil at the film temperature of $T_f = (T_s + T_\infty)/2 = (20 + 60)/2 = 40°C$ are (Table A–13)

$$\rho = 876 \text{ kg/m}^3 \qquad\qquad Pr = 2962$$
$$k = 0.1444 \text{ W/m} \cdot °C \qquad \nu = 2.485 \times 10^{-4} \text{ m}^2/\text{s}$$

Analysis Noting that $L = 5$ m, the Reynolds number at the end of the plate is

$$Re_L = \frac{VL}{\nu} = \frac{(2 \text{ m/s})(5 \text{ m})}{2.485 \times 10^{-4} \text{ m}^2/\text{s}} = 4.024 \times 10^4$$

which is less than the critical Reynolds number. Thus we have *laminar flow* over the entire plate, and the average friction coefficient is

$$C_f = 1.33 \, Re_L^{-0.5} = 1.33 \times (4.024 \times 10^4)^{-0.5} = 0.00663$$

Noting that the pressure drag is zero and thus $C_D = C_f$ for parallel flow over a flat plate, the drag force acting on the plate per unit width becomes

$$F_D = C_f A \frac{\rho V^2}{2} = 0.00663(5 \times 1 \text{ m}^2)\frac{(876 \text{ kg/m}^3)(2 \text{ m/s})^2}{2}\left(\frac{1 \text{ N}}{1 \text{ kg} \cdot \text{m/s}^2}\right) = 58.1 \text{ N}$$

The total drag force acting on the entire plate can be determined by multiplying the value obtained above by the width of the plate.

This force per unit width corresponds to the weight of a mass of about 6 kg. Therefore, a person who applies an equal and opposite force to the plate to keep il from moving will feel like he or she is using as much force as is necessary to hold a 6-kg mass from dropping.

Similarly, the Nusselt number is determined using the laminar flow relations for a flat plate,

$$\text{Nu} = \frac{hL}{k} = 0.664 \, \text{Re}_L^{0.5} \, \text{Pr}^{1/3} = 0.664 \times (4.024 \times 10^4)^{0.5} \times 2962^{1/3} = 1913$$

Then,

$$h = \frac{k}{L} \text{Nu} = \frac{0.1444 \, \text{W/m} \cdot °\text{C}}{5 \, \text{m}} (1913) = 55.25 \, \text{W/m}^2 \cdot °\text{C}$$

and

$$\dot{Q} = hA_s(T_\infty - T_s) = (55.25 \, \text{W/m}^2 \cdot °\text{C})(5 \times 1 \, \text{m}^2)(60 - 20)°\text{C} = 11{,}050 \, \text{W}$$

Discussion Note that heat transfer is always from the higher-temperature medium to the lower-temperature one. In this case, it is from the oil to the plate. The heat transfer rate is per m width of the plate. The heat transfer for the entire plate can be obtained by multiplying the value obtained by the actual width of the plate.

EXAMPLE 7–2   **Cooling of a Hot Block by Forced Air at High Elevation**

The local atmospheric pressure in Denver, Colorado (elevation 1610 m), is 83.4 kPa. Air at this pressure and 20°C flows with a velocity of 8 m/s over a 1.5 m × 6 m flat plate whose temperature is 140°C (Fig. 7–13). Determine the rate of heat transfer from the plate if the air flows parallel to the (*a*) 6-m-long side and (*b*) the 1.5 m side.

SOLUTION The top surface of a hot block is to be cooled by forced air. The rate of heat transfer is to be determined for two cases.
Assumptions 1 Steady operating conditions exist. 2 The critical Reynolds number is $\text{Re}_{cr} = 5 \times 10^5$. 3 Radiation effects are negligible. 4 Air is an ideal gas.
Properties The properties k, μ, c_p, and Pr of ideal gases are independent of pressure, while the properties ν and α are inversely proportional to density and thus pressure. The properties of air at the film temperature of $T_f = (T_s + T_\infty)/2 = (140 + 20)/2 = 80°\text{C}$ and 1 atm pressure are (Table A–15)

$$k = 0.02953 \, \text{W/m} \cdot °\text{C} \qquad \text{Pr} = 0.7154$$

$$\nu_{@ \, 1 \, \text{atm}} = 2.097 \times 10^{-5} \, \text{m}^2/\text{s}$$

The atmospheric pressure in Denver is $P = (83.4 \, \text{kPa})/(101.325 \, \text{kPa/atm}) = 0.823 \, \text{atm}$. Then the kinematic viscosity of air in Denver becomes

$$\nu = \nu_{@ \, 1 \, \text{atm}}/P = (2.097 \times 10^{-5} \, \text{m}^2/\text{s})/0.823 = 2.548 \times 10^{-5} \, \text{m}^2/\text{s}$$

Analysis (*a*) When air flow is parallel to the long side, we have $L = 6$ m, and the Reynolds number at the end of the plate becomes

$$\text{Re}_L = \frac{VL}{\nu} = \frac{(8 \, \text{m/s})(6 \, \text{m})}{2.548 \times 10^{-5} \, \text{m}^2/\text{s}} = 1.884 \times 10^6$$

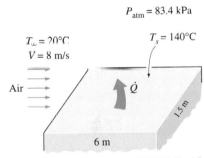

$P_{\text{atm}} = 83.4 \, \text{kPa}$

$T_\infty = 20°\text{C}$
$V = 8$ m/s

$T_s = 140°\text{C}$

Air

$\dot{Q}$

1.5 m

6 m

FIGURE 7–13
Schematic for Example 7–2.

which is greater than the critical Reynolds number. Thus, we have combined laminar and turbulent flow, and the average Nusselt number for the entire plate is determined to be

$$Nu = \frac{hL}{k} = (0.037\ Re_L^{0.8} - 871)Pr^{1/3}$$
$$= [0.037(1.884 \times 10^6)^{0.8} - 871]0.7154^{1/3}$$
$$= 2687$$

Then

$$h = \frac{k}{L}Nu = \frac{0.02953\ \text{W/m} \cdot {}^\circ\text{C}}{6\ \text{m}}(2687) = 13.2\ \text{W/m}^2 \cdot {}^\circ\text{C}$$
$$A_s = wL = (1.5\ \text{m})(6\ \text{m}) = 9\ \text{m}^2$$

and

$$\dot{Q} = hA_s(T_s - T_\infty) = (13.2\ \text{W/m}^2 \cdot {}^\circ\text{C})(9\ \text{m}^2)(140 - 20){}^\circ\text{C} = 1.43 \times 10^4\ \text{W}$$

Note that if we disregarded the laminar region and assumed turbulent flow over the entire plate, we would get Nu = 3466 from Eq. 7–22, which is 29 percent higher than the value calculated above.

(b) When air flow is along the short side, we have L = 1.5 m, and the Reynolds number at the end of the plate becomes

$$Re_L = \frac{VL}{\nu} = \frac{(8\ \text{m/s})(1.5\ \text{m})}{2.548 \times 10^{-5}\ \text{m}^2/\text{s}} = 4.71 \times 10^5$$

which is less than the critical Reynolds number. Thus we have laminar flow over the entire plate, and the average Nusselt number is

$$Nu = \frac{hL}{k} = 0.664\ Re_L^{0.5}\ Pr^{1/3} = 0.664 \times (4.71 \times 10^5)^{0.5} \times 0.7154^{1/3} = 408$$

Then

$$h = \frac{k}{L}Nu = \frac{0.02953\ \text{W/m} \cdot {}^\circ\text{C}}{1.5\ \text{m}}(408) = 8.03\ \text{W/m}^2 \cdot {}^\circ\text{C}$$

and

$$\dot{Q} = hA_s(T_s - T_\infty) = (8.03\ \text{W/m}^2 \cdot {}^\circ\text{C})(9\ \text{m}^2)(140 - 20){}^\circ\text{C} = 8670\ \text{W}$$

which is considerably less than the heat transfer rate determined in case (a).

Discussion Note that the *direction* of fluid flow can have a significant effect on convection heat transfer to or from a surface (Fig. 7–14). In this case, we can increase the heat transfer rate by 65 percent by simply blowing the air along the long side of the rectangular plate instead of the short side.

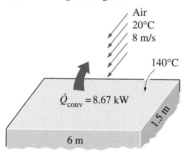

20°C
8 m/s
Air
140°C
$\dot{Q}_{conv} = 14.3\ \text{kW}$
1.5 m
6 m

(a) Flow along the long side

Air
20°C
8 m/s
140°C
$\dot{Q}_{conv} = 8.67\ \text{kW}$
1.5 m
6 m

(b) Flow along the short side

FIGURE 7–14
The direction of fluid flow can have a significant effect on convection heat transfer.

EXAMPLE 7–3   Cooling of Plastic Sheets by Forced Air

The forming section of a plastics plant puts out a continuous sheet of plastic that is 4 ft wide and 0.04 in thick at a velocity of 30 ft/min. The temperature of the plastic sheet is 200°F when it is exposed to the surrounding air, and a 2-ft-long section of the plastic sheet is subjected to air flow at 80°F at a velocity of 10 ft/s on both sides along its surfaces normal to the direction of motion of the sheet, as shown in Fig. 7–15. Determine (a) the rate of heat transfer from the plastic sheet to air by forced convection and radiation and (b) the temperature of the plastic sheet at the end of the cooling section. Take the density, specific heat, and emissivity of the plastic sheet to be $\rho = 75$ lbm/ft³, $c_p = 0.4$ Btu/lbm · °F, and $\varepsilon = 0.9$.

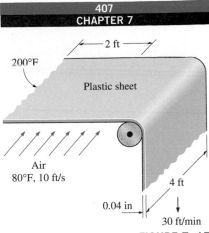

FIGURE 7–15
Schematic for Example 7–3.

SOLUTION Plastic sheets are cooled as they leave the forming section of a plastics plant. The rate of heat loss from the plastic sheet by convection and radiation and the exit temperature of the plastic sheet are to be determined.
Assumptions 1 Steady operating conditions exist. 2 The critical Reynolds number is $Re_{cr} = 5 \times 10^5$. 3 Air is an ideal gas. 4 The local atmospheric pressure is 1 atm. 5 The surrounding surfaces are at the temperature of the room air.
Properties The properties of the plastic sheet are given in the problem statement. The properties of air at the film temperature of $T_f = (T_s + T_\infty)/2 = (200 + 80)/2 = 140°F$ and 1 atm pressure are (Table A–15E)

$$k = 0.01623 \text{ Btu/h} \cdot \text{ft} \cdot °F \qquad Pr = 0.7202$$
$$\nu = 0.204 \times 10^{-3} \text{ ft}^2/\text{s}$$

Analysis (a) We expect the temperature of the plastic sheet to drop somewhat as it flows through the 2-ft-long cooling section, but at this point we do not know the magnitude of that drop. Therefore, we assume the plastic sheet to be isothermal at 200°F to get started. We will repeat the calculations if necessary to account for the temperature drop of the plastic sheet.

Noting that $L = 4$ ft, the Reynolds number at the end of the air flow across the plastic sheet is

$$Re_L = \frac{VL}{\nu} = \frac{(10 \text{ ft/s})(4 \text{ ft})}{0.204 \times 10^{-3} \text{ ft}^2/\text{s}} = 1.961 \times 10^5$$

which is less than the critical Reynolds number. Thus, we have *laminar flow* over the entire sheet, and the Nusselt number is determined from the laminar flow relations for a flat plate to be

$$Nu = \frac{hL}{k} = 0.664 \, Re_L^{0.5} \, Pr^{1/3} = 0.664 \times (1.961 \times 10^5)^{0.5} \times (0.7202)^{1/3} = 263.6$$

Then,

$$h = \frac{k}{L} Nu = \frac{0.01623 \text{ Btu/h} \cdot \text{ft} \cdot °F}{4 \text{ ft}} (263.6) = 1.07 \text{ Btu/h} \cdot \text{ft}^2 \cdot °F$$
$$A_s = (2 \text{ ft})(4 \text{ ft})(2 \text{ sides}) = 16 \text{ ft}^2$$

and

$$\dot{Q}_{conv} = hA_s(T_s - T_\infty)$$
$$= (1.07 \text{ Btu/h} \cdot \text{ft}^2 \cdot °F)(16 \text{ ft}^2)(200 - 80)°F$$
$$= 2054 \text{ Btu/h}$$

$$\dot{Q}_{rad} = \varepsilon\sigma A_s(T_s^4 - T_{surr}^4)$$
$$= (0.9)(0.1714 \times 10^{-8} \text{ Btu/h} \cdot \text{ft}^2 \cdot R^4)(16 \text{ ft}^2)[(660 \text{ R})^4 - (540 \text{ R})^4]$$
$$= 2585 \text{ Btu/h}$$

Therefore, the rate of cooling of the plastic sheet by combined convection and radiation is

$$\dot{Q}_{total} = \dot{Q}_{conv} + \dot{Q}_{rad} = 2054 + 2585 = \textbf{4639 Btu/h}$$

(b) To find the temperature of the plastic sheet at the end of the cooling section, we need to know the mass of the plastic rolling out per unit time (or the mass flow rate), which is determined from

$$\dot{m} = \rho A_c V_{plastic} = (75 \text{ lbm/ft}^3)\left(\frac{4 \times 0.04}{12} \text{ ft}^2\right)\left(\frac{30}{60} \text{ ft/s}\right) = 0.5 \text{ lbm/s}$$

Then, an energy balance on the cooled section of the plastic sheet yields

$$\dot{Q} = \dot{m} c_p (T_2 - T_1) \quad \rightarrow \quad T_2 = T_1 + \frac{\dot{Q}}{\dot{m} c_p}$$

Noting that $\dot{Q}$ is a negative quantity (heat loss) for the plastic sheet and substituting, the temperature of the plastic sheet as it leaves the cooling section is determined to be

$$T_2 = 200°F + \frac{-4639 \text{ Btu/h}}{(0.5 \text{ lbm/s})(0.4 \text{ Btu/lbm} \cdot °F)}\left(\frac{1 \text{ h}}{3600 \text{ s}}\right) = \textbf{193.6°F}$$

Discussion The average temperature of the plastic sheet drops by about 6.4°F as it passes through the cooling section. The calculations now can be repeated by taking the average temperature of the plastic sheet to be 196.8°F instead of 200°F for better accuracy, but the change in the results will be insignificant because of the small change in temperature.

7–3 · FLOW ACROSS CYLINDERS AND SPHERES

Flow across cylinders and spheres is frequently encountered in practice. For example, the tubes in a shell-and-tube heat exchanger involve both *internal flow* through the tubes and *external flow* over the tubes, and both flows must be considered in the analysis of the heat exchanger. Also, many sports such as soccer, tennis, and golf involve flow over spherical balls.

The characteristic length for a circular cylinder or sphere is taken to be the *external diameter D*. Thus, the Reynolds number is defined as $\text{Re} = VD/\nu$ where V is the uniform velocity of the fluid as it approaches the cylinder or sphere. The critical Reynolds number for flow across a circular cylinder or sphere is about $\text{Re}_{cr} \cong 2 \times 10^5$. That is, the boundary layer remains laminar for about $\text{Re} \leqslant 2 \times 10^5$ and becomes turbulent for $\text{Re} \geqslant 2 \times 10^5$.

Cross-flow over a cylinder exhibits complex flow patterns, as shown in Fig. 7–16. The fluid approaching the cylinder branches out and encircles the cylinder, forming a boundary layer that wraps around the cylinder. The fluid particles on the midplane strike the cylinder at the stagnation point, bringing the fluid to a complete stop and thus raising the pressure at that point. The pressure decreases in the flow direction while the fluid velocity increases.

At very low upstream velocities ($\text{Re} \leqslant 1$), the fluid completely wraps around the cylinder and the two arms of the fluid meet on the rear side of the

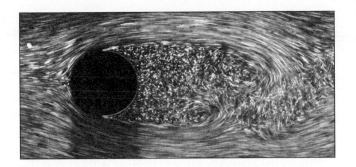

FIGURE 7–16
Laminar boundary layer separation with a turbulent wake; flow over a circular cylinder at Re = 2000.

Courtesy ONERA, photograph by Werlé.

cylinder in an orderly manner. Thus, the fluid follows the curvature of the cylinder. At higher velocities, the fluid still hugs the cylinder on the frontal side, but it is too fast to remain attached to the surface as it approaches the top (or bottom) of the cylinder. As a result, the boundary layer detaches from the surface, forming a separation region behind the cylinder. Flow in the wake region is characterized by periodic vortex formation and pressures much lower than the stagnation point pressure.

The nature of the flow across a cylinder or sphere strongly affects the total drag coefficient C_D. Both the *friction drag* and the *pressure drag* can be significant. The high pressure in the vicinity of the stagnation point and the low pressure on the opposite side in the wake produce a net force on the body in the direction of flow. The drag force is primarily due to friction drag at low Reynolds numbers (Re < 10) and to pressure drag at high Reynolds numbers (Re > 5000). Both effects are significant at intermediate Reynolds numbers.

The average drag coefficients C_D for cross-flow over a smooth single circular cylinder and a sphere are given in Fig. 7–17. The curves exhibit different behaviors in different ranges of Reynolds numbers:

- For Re ≲ 1, we have creeping flow, and the drag coefficient decreases with increasing Reynolds number. For a sphere, it is $C_D = 24/\text{Re}$. There is no flow separation in this regime.

- At about Re = 10, separation starts occurring on the rear of the body with vortex shedding starting at about Re ≅ 90. The region of separation

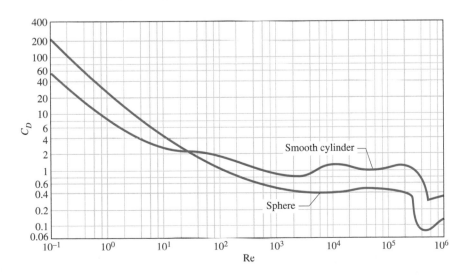

FIGURE 7–17
Average drag coefficient for cross-flow over a smooth circular cylinder and a smooth sphere.

From H. Schlichting, Boundary Layer Theory *7e. Copyright © 1979 The McGraw-Hill Companies, Inc. Used by permission.*

(a)

(b)

FIGURE 7–18

Flow visualization of flow over
(a) a smooth sphere at Re = 15,000,
and *(b)* a sphere at Re = 30,000 with
a trip wire. The delay of boundary
layer separation is clearly seen by
comparing the two photographs.

Courtesy ONERA, photograph by Werlé.

increases with increasing Reynolds number up to about Re = 10^3. At this
point, the drag is mostly (about 95 percent) due to pressure drag. The drag
coefficient continues to decrease with increasing Reynolds number in this
range of $10 < \text{Re} < 10^3$. (A decrease in the drag coefficient does not
necessarily indicate a decrease in drag. The drag force is proportional to
the square of the velocity, and the increase in velocity at higher Reynolds
numbers usually more than offsets the decrease in the drag coefficient.)

• In the moderate range of $10^3 < \text{Re} < 10^5$, the drag coefficient remains
relatively constant. This behavior is characteristic of blunt bodies. The
flow in the boundary layer is laminar in this range, but the flow in the
separated region past the cylinder or sphere is highly turbulent with a
wide turbulent wake.

• There is a sudden drop in the drag coefficient somewhere in the range of
$10^5 < \text{Re} < 10^6$ (usually, at about 2×10^5). This large reduction in C_D is
due to the flow in the boundary layer becoming *turbulent,* which moves the
separation point further on the rear of the body, reducing the size of the
wake and thus the magnitude of the pressure drag. This is in contrast to
streamlined bodies, which experience an increase in the drag coefficient
(mostly due to friction drag) when the boundary layer becomes turbulent.

Flow separation occurs at about $\theta \cong 80°$ (measured from the front stagnation point of a cylinder) when the boundary layer is *laminar* and at about $\theta \cong$
140° when it is *turbulent* (Fig. 7–18). The delay of separation in turbulent flow
is caused by the rapid fluctuations of the fluid in the transverse direction,
which enables the turbulent boundary layer to travel farther along the surface
before separation occurs, resulting in a narrower wake and a smaller pressure
drag. Keep in mind that turbulent flow has a fuller velocity profile as compared to the laminar case, and thus it requires a stronger adverse pressure gradient to overcome the additional momentum close to the wall. In the range of
Reynolds numbers where the flow changes from laminar to turbulent, even the
drag force F_D decreases as the velocity (and thus the Reynolds number) increases. This results in a sudden decrease in drag of a flying body (sometimes
called the *drag crisis*) and instabilities in flight.

Effect of Surface Roughness

We mentioned earlier that *surface roughness,* in general, increases the drag
coefficient in turbulent flow. This is especially the case for streamlined bodies. For blunt bodies such as a circular cylinder or sphere, however, an increase in the surface roughness may actually *decrease* the drag coefficient, as
shown in Fig. 7–19 for a sphere. This is done by tripping the boundary layer
into turbulence at a lower Reynolds number, and thus causing the fluid to
close in behind the body, narrowing the wake and reducing pressure drag considerably. This results in a much smaller drag coefficient and thus drag force
for a rough-surfaced cylinder or sphere in a certain range of Reynolds number
compared to a smooth one of identical size at the same velocity. At Re = 2 ×
10^5, for example, $C_D \cong 0.1$ for a rough sphere with $\varepsilon/D = 0.0015$, whereas
$C_D \cong 0.5$ for a smooth one. Therefore, the drag coefficient in this case is reduced by a factor of 5 by simply roughening the surface. Note, however, that
at Re = 10^6, $C_D \cong 0.4$ for a very rough sphere while $C_D \cong 0.1$ for the smooth

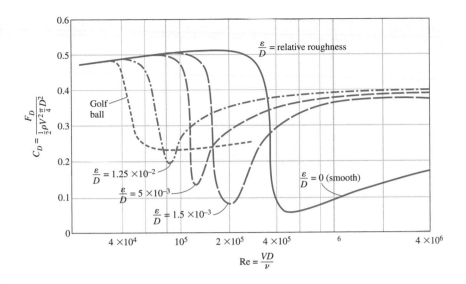

FIGURE 7–19

The effect of surface roughness on the drag coefficient of a sphere.

From Blevins (1984).

one. Obviously, roughening the sphere in this case will increase the drag by a factor of 4 (Fig. 7–20).

The preceding discussion shows that roughening the surface can be used to great advantage in reducing drag, but it can also backfire on us if we are not careful—specifically, if we do not operate in the right range of the Reynolds number. With this consideration, golf balls are intentionally roughened to induce *turbulence* at a lower Reynolds number to take advantage of the sharp *drop* in the drag coefficient at the onset of turbulence in the boundary layer (the typical velocity range of golf balls is 15 to 150 m/s, and the Reynolds number is less than 4×10^5). The critical Reynolds number of dimpled golf balls is about 4×10^4. The occurrence of turbulent flow at this Reynolds number reduces the drag coefficient of a golf ball by about half, as shown in Fig. 7–19. For a given hit, this means a longer distance for the ball. Experienced golfers also give the ball a spin during the hit, which helps the rough ball develop a lift and thus travel higher and farther. A similar argument can be given for a tennis ball. For a table tennis ball, however, the distances are very short, and the balls never reach the speeds in the turbulent range. Therefore, the surfaces of table tennis balls are made smooth.

Once the drag coefficient is available, the drag force acting on a body in cross-flow can be determined from Eq. 7–1 where A is the *frontal area* ($A = LD$ for a cylinder of length L and $A = \pi D^2/4$ for a sphere). It should be kept in mind that free-stream turbulence and disturbances by other bodies in the flow (such as flow over tube bundles) may affect the drag coefficients significantly.

	C_D	
Re	Smooth Surface	Rough Surface, $\varepsilon/D = 0.0015$
2×10^5	0.5	0.1
10^6	0.1	0.4

FIGURE 7–20

Surface roughness may increase or decrease the drag coefficient of a spherical object, depending on the value of the Reynolds number.

EXAMPLE 7–4   Drag Force Acting on a Pipe in a River

A 2.2-cm-outer-diameter pipe is to span across a river at a 30-m-wide section while being completely immersed in water (Fig. 7–21). The average flow velocity of water is 4 m/s and the water temperature is 15°C. Determine the drag force exerted on the pipe by the river.

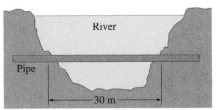

FIGURE 7–21

Schematic for Example 7–4.

SOLUTION A pipe is submerged in a river. The drag force that acts on the pipe is to be determined.

Assumptions **1** The outer surface of the pipe is smooth so that Fig. 7–17 can be used to determine the drag coefficient. **2** Water flow in the river is steady. **3** The direction of water flow is normal to the pipe. **4** Turbulence in river flow is not considered.

Properties The density and dynamic viscosity of water at 15°C are $\rho =$ 999.1 kg/m³ and $\mu = 1.138 \times 10^{-3}$ kg/m · s (Table A–9).

Analysis Noting that $D = 0.022$ m, the Reynolds number is

$$Re = \frac{VD}{\nu} = \frac{\rho VD}{\mu} = \frac{(999.1 \text{ kg/m}^3)(4 \text{ m/s})(0.022 \text{ m})}{1.138 \times 10^{-3} \text{ kg/m} \cdot \text{s}} = 7.73 \times 10^4$$

The drag coefficient corresponding to this value is, from Fig. 7–17, $C_D = 1.0$. Also, the frontal area for flow past a cylinder is $A = LD$. Then the drag force acting on the pipe becomes

$$F_D = C_D A \frac{\rho V^2}{2} = 1.0(30 \times 0.022 \text{ m}^2) \frac{(999.1 \text{ kg/m}^3)(4 \text{ m/s})^2}{2} \left(\frac{1 \text{ N}}{1 \text{ kg} \cdot \text{m/s}^2}\right)$$

$$= 5275 \text{ N} \cong \mathbf{5.30 \text{ kN}}$$

Discussion Note that this force is equivalent to the weight of a mass over 500 kg. Therefore, the drag force the river exerts on the pipe is equivalent to hanging a total of over 500 kg in mass on the pipe supported at its ends 30 m apart. The necessary precautions should be taken if the pipe cannot support this force. If the river were to flow at a faster speed or if turbulent fluctuations in the river were more significant, the drag force would be even larger. *Unsteady* forces on the pipe might then be significant.

Heat Transfer Coefficient

Flows across cylinders and spheres, in general, involve *flow separation,* which is difficult to handle analytically. Therefore, such flows must be studied experimentally or numerically. Indeed, flow across cylinders and spheres has been studied experimentally by numerous investigators, and several empirical correlations have been developed for the heat transfer coefficient.

The complicated flow pattern across a cylinder greatly influences heat transfer. The variation of the local Nusselt number Nu_θ around the periphery of a cylinder subjected to cross flow of air is given in Fig. 7–22. Note that, for all cases, the value of Nu_θ starts out relatively high at the stagnation point ($\theta = 0°$) but decreases with increasing θ as a result of the thickening of the laminar boundary layer. On the two curves at the bottom corresponding to Re = 70,800 and 101,300, Nu_θ reaches a minimum at $\theta \approx 80°$, which is the separation point in laminar flow. Then Nu_θ increases with increasing θ as a result of the intense mixing in the separated flow region (the wake). The curves at the top corresponding to Re = 140,000 to 219,000 differ from the first two curves in that they have *two* minima for Nu_θ. The sharp increase in Nu_θ at about

$\theta \approx 90°$ is due to the transition from laminar to turbulent flow. The later decrease in Nu_θ is again due to the thickening of the boundary layer. Nu_θ reaches its second minimum at about $\theta \approx 140°$, which is the flow separation point in turbulent flow, and increases with θ as a result of the intense mixing in the turbulent wake region.

The discussions above on the local heat transfer coefficients are insightful; however, they are of limited value in heat transfer calculations since the calculation of heat transfer requires the *average* heat transfer coefficient over the entire surface. Of the several such relations available in the literature for the average Nusselt number for cross flow over a cylinder, we present the one proposed by Churchill and Bernstein:

$$Nu_{cyl} = \frac{hD}{k} = 0.3 + \frac{0.62\,Re^{1/2}\,Pr^{1/3}}{[1 + (0.4/Pr)^{2/3}]^{1/4}}\left[1 + \left(\frac{Re}{282,000}\right)^{5/8}\right]^{4/5} \quad (7\text{-}35)$$

This relation is quite comprehensive in that it correlates available data well for $RePr > 0.2$. The fluid properties are evaluated at the *film temperature* $T_f = \frac{1}{2}(T_\infty + T_s)$, which is the average of the free-stream and surface temperatures.

For flow over a *sphere*, Whitaker recommends the following comprehensive correlation:

$$Nu_{sph} = \frac{hD}{k} = 2 + [0.4\,Re^{1/2} + 0.06\,Re^{2/3}]\,Pr^{0.4}\left(\frac{\mu_\infty}{\mu_s}\right)^{1/4} \quad (7\text{-}36)$$

which is valid for $3.5 \le Re \le 80,000$ and $0.7 \le Pr \le 380$. The fluid properties in this case are evaluated at the free-stream temperature T_∞, except for μ_s, which is evaluated at the surface temperature T_s. Although the two relations above are considered to be quite accurate, the results obtained from them can be off by as much as 30 percent.

The average Nusselt number for flow across cylinders can be expressed compactly as

$$Nu_{cyl} = \frac{hD}{k} = C\,Re^m\,Pr^n \quad (7\text{-}37)$$

where $n = \frac{1}{3}$ and the experimentally determined constants C and m are given in Table 7–1 for circular as well as various noncircular cylinders. The characteristic length D for use in the calculation of the Reynolds and the Nusselt numbers for different geometries is as indicated on the figure. All fluid properties are evaluated at the film temperature.

The relations for cylinders above are for *single* cylinders or cylinders oriented such that the flow over them is not affected by the presence of others. Also, they are applicable to *smooth* surfaces. *Surface roughness* and the *free-stream turbulence* may affect the drag and heat transfer coefficients significantly. Eq. 7–37 provides a simpler alternative to Eq. 7–35 for flow over cylinders. However, Eq. 7–35 is more accurate, and thus should be preferred in calculations whenever possible.

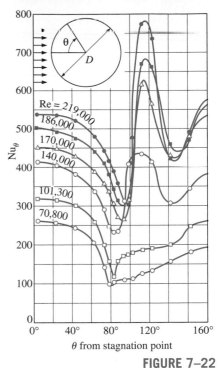

FIGURE 7–22
Variation of the local heat transfer coefficient along the circumference of a circular cylinder in cross flow of air (from Giedt, 1949).

TABLE 7–1

Empirical correlations for the average Nusselt number for forced convection over circular and noncircular cylinders in cross flow (from Zukauskas, 1972 and Jakob, 1949)

Cross-section of the cylinder	Fluid	Range of Re	Nusselt number
Circle D	Gas or liquid	0.4–4 4–40 40–4000 4000–40,000 40,000–400,000	$Nu = 0.989Re^{0.330} Pr^{1/3}$ $Nu = 0.911Re^{0.385} Pr^{1/3}$ $Nu = 0.683Re^{0.466} Pr^{1/3}$ $Nu = 0.193Re^{0.618} Pr^{1/3}$ $Nu = 0.027Re^{0.805} Pr^{1/3}$
Square D	Gas	5000–100,000	$Nu = 0.102Re^{0.675} Pr^{1/3}$
Square (tilted 45°) D	Gas	5000–100,000	$Nu = 0.246Re^{0.588} Pr^{1/3}$
Hexagon D	Gas	5000–100,000	$Nu = 0.153Re^{0.638} Pr^{1/3}$
Hexagon (tilted 45°) D	Gas	5000–19,500 19,500–100,000	$Nu = 0.160Re^{0.638} Pr^{1/3}$ $Nu = 0.0385Re^{0.782} Pr^{1/3}$
Vertical plate D	Gas	4000–15,000	$Nu = 0.228Re^{0.731} Pr^{1/3}$
Ellipse D	Gas	2500–15,000	$Nu = 0.248Re^{0.612} Pr^{1/3}$

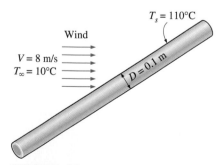

$T_s = 110°C$

Wind

$V = 8$ m/s
$T_\infty = 10°C$
$D = 0.1$ m

FIGURE 7–23
Schematic for Example 7–5.

EXAMPLE 7–5   Heat Loss from a Steam Pipe in Windy Air

A long 10-cm-diameter steam pipe whose external surface temperature is 110°C passes through some open area that is not protected against the winds (Fig. 7–23). Determine the rate of heat loss from the pipe per unit of its length when the air is at 1 atm pressure and 10°C and the wind is blowing across the pipe at a velocity of 8 m/s.

SOLUTION A steam pipe is exposed to windy air. The rate of heat loss from the steam is to be determined.

Assumptions **1** Steady operating conditions exist. **2** Radiation effects are negligible. **3** Air is an ideal gas.

Properties The properties of air at the average film temperature of $T_f = (T_s + T_\infty)/2 = (110 + 10)/2 = 60°C$ and 1 atm pressure are (Table A–15)

$$k = 0.02808 \text{ W/m} \cdot °C \qquad Pr = 0.7202$$
$$\nu = 1.896 \times 10^{-5} \text{ m}^2/\text{s}$$

Analysis The Reynolds number is

$$Re = \frac{VD}{\nu} = \frac{(8 \text{ m/s})(0.1 \text{ m})}{1.896 \times 10^{-5} \text{ m}^2/\text{s}} = 4.219 \times 10^4$$

The Nusselt number can be determined from

$$Nu = \frac{hD}{k} = 0.3 + \frac{0.62 \, Re^{1/2} \, Pr^{1/3}}{[1 + (0.4/Pr)^{2/3}]^{1/4}} \left[1 + \left(\frac{Re}{282,000}\right)^{5/8}\right]^{4/5}$$

$$= 0.3 + \frac{0.62(4.219 \times 10^4)^{1/2} (0.7202)^{1/3}}{[1 + (0.4/0.7202)^{2/3}]^{1/4}} \left[1 + \left(\frac{4.219 \times 10^4}{282,000}\right)^{5/8}\right]^{4/5}$$

$$= 124$$

and

$$h = \frac{k}{D} Nu = \frac{0.02808 \text{ W/m} \cdot °C}{0.1 \text{ m}} (124) = 34.8 \text{ W/m}^2 \cdot °C$$

Then the rate of heat transfer from the pipe per unit of its length becomes

$$A_s = pL = \pi DL = \pi(0.1 \text{ m})(1 \text{ m}) = 0.314 \text{ m}^2$$
$$\dot{Q} = hA_s(T_s - T_\infty) = (34.8 \text{ W/m}^2 \cdot C)(0.314 \text{ m}^2)(110 - 10)°C = \textbf{1093 W}$$

The rate of heat loss from the entire pipe can be obtained by multiplying the value above by the length of the pipe in m.

Discussion The simpler Nusselt number relation in Table 7–1 in this case would give Nu = 128, which is 3 percent higher than the value obtained above using Eq. 7–35.

EXAMPLE 7–6 **Cooling of a Steel Ball by Forced Air**

A 25-cm-diameter stainless steel ball ($\rho = 8055 \text{ kg/m}^3$, $c_p = 480 \text{ J/kg} \cdot °C$) is removed from the oven at a uniform temperature of 300°C (Fig. 7–24). The ball is then subjected to the flow of air at 1 atm pressure and 25°C with a velocity of 3 m/s. The surface temperature of the ball eventually drops to 200°C. Determine the average convection heat transfer coefficient during this cooling process and estimate how long the process will take.

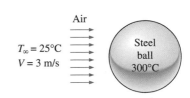

FIGURE 7–24
Schematic for Example 7–6.

SOLUTION A hot stainless steel ball is cooled by forced air. The average convection heat transfer coefficient and the cooling time are to be determined.

Assumptions **1** Steady operating conditions exist. **2** Radiation effects are negligible. **3** Air is an ideal gas. **4** The outer surface temperature of the ball is uniform at all times. **5** The surface temperature of the ball during cooling is changing. Therefore, the convection heat transfer coefficient between the ball and the air will also change. To avoid this complexity, we take the surface temperature of the ball to be constant at the average temperature of $(300 + 200)/2 = 250°C$ in the evaluation of the heat transfer coefficient and use the value obtained for the entire cooling process.

Properties The dynamic viscosity of air at the average surface temperature is $\mu_s = \mu_{@\,250°C} = 2.76 \times 10^{-5}$ kg/m · s. The properties of air at the free-stream temperature of 25°C and 1 atm are (Table A–15)

$$k = 0.02551 \text{ W/m} \cdot °C \qquad \nu = 1.562 \times 10^{-5} \text{ m}^2/\text{s}$$
$$\mu = 1.849 \times 10^{-5} \text{ kg/m} \cdot \text{s} \qquad \text{Pr} = 0.7296$$

Analysis The Reynolds number is determined from

$$\text{Re} = \frac{VD}{\nu} = \frac{(3 \text{ m/s})(0.25 \text{ m})}{1.562 \times 10^{\pm 5} \text{ m}^2/\text{s}} = 4.802 \times 10^4$$

The Nusselt number is

$$\text{Nu} = \frac{hD}{k} = 2 + [0.4 \text{ Re}^{1/2} + 0.06 \text{ Re}^{2/3}] \text{ Pr}^{0.4} \left(\frac{\mu_\infty}{\mu_s}\right)^{1/4}$$
$$= 2 + [0.4(4.802 \times 10^4)^{1/2} + 0.06(4.802 \times 10^4)^{2/3}](0.7296)^{0.4}$$
$$\times \left(\frac{1.849 \times 10^{-5}}{2.76 \times 10^{-5}}\right)^{1/4}$$
$$= 135$$

Then the average convection heat transfer coefficient becomes

$$h = \frac{k}{D} \text{Nu} = \frac{0.02551 \text{ W/m} \cdot °C}{0.25 \text{ m}} (135) = \textbf{13.8 W/m}^2 \cdot \textbf{°C}$$

In order to estimate the time of cooling of the ball from 300°C to 200°C, we determine the *average* rate of heat transfer from Newton's law of cooling by using the *average* surface temperature. That is,

$$A_s = \pi D^2 = \pi(0.25 \text{ m})^2 = 0.1963 \text{ m}^2$$
$$\dot{Q}_{avg} = hA_s(T_{s,\,avg} - T_\infty) = (13.8 \text{ W/m}^2 \cdot °C)(0.1963 \text{ m}^2)(250 - 25)°C = 610 \text{ W}$$

Next we determine the *total* heat transferred from the ball, which is simply the change in the energy of the ball as it cools from 300°C to 200°C:

$$m = \rho V = \rho \tfrac{1}{6}\pi D^3 = (8055 \text{ kg/m}^3) \tfrac{1}{6}\pi(0.25 \text{ m})^3 = 65.9 \text{ kg}$$
$$Q_{total} = mc_p(T_2 - T_1) = (65.9 \text{ kg})(480 \text{ J/kg} \cdot °C)(300 - 200)°C = 3,163,000 \text{ J}$$

In this calculation, we assumed that the entire ball is at 200°C, which is not necessarily true. The inner region of the ball will probably be at a higher temperature than its surface. With this assumption, the time of cooling is determined to be

$$\Delta t \approx \frac{Q}{\dot{Q}_{avg}} = \frac{3{,}163{,}000 \text{ J}}{610 \text{ J/s}} = 5185 \text{ s} = \textbf{1 h 26 min}$$

Discussion The time of cooling could also be determined more accurately using the transient temperature charts or relations introduced in Chapter 4. But the simplifying assumptions we made above can be justified if all we need is a ballpark value. It will be naive to expect the time of cooling to be exactly 1 h 26 min, but, using our engineering judgment, it is realistic to expect the time of cooling to be somewhere between one and two hours.

7–4 ■ FLOW ACROSS TUBE BANKS

Cross-flow over tube banks is commonly encountered in practice in heat transfer equipment such as the condensers and evaporators of power plants, refrigerators, and air conditioners. In such equipment, one fluid moves through the tubes while the other moves over the tubes in a perpendicular direction.

In a heat exchanger that involves a tube bank, the tubes are usually placed in a *shell* (and thus the *name shell-and-tube heat exchanger*), especially when the fluid is a liquid, and the fluid flows through the space between the tubes and the shell. There are numerous types of shell-and-tube heat exchangers, some of which are considered in Chap. 11. In this section we consider the general aspects of flow over a tube bank, and try to develop a better and more intuitive understanding of the performance of heat exchangers involving a tube bank.

Flow *through* the tubes can be analyzed by considering flow through a single tube, and multiplying the results by the number of tubes. This is not the case for flow *over* the tubes, however, since the tubes affect the flow pattern and turbulence level downstream, and thus heat transfer to or from them, as shown in Fig. 7–25. Therefore, when analyzing heat transfer from a tube bank in cross flow, we must consider all the tubes in the bundle at once.

The tubes in a tube bank are usually arranged either *in-line* or *staggered* in the direction of flow, as shown in Fig. 7–26. The outer tube diameter D is taken as the characteristic length. The arrangement of the tubes in the tube bank is characterized by the *transverse pitch* S_T, *longitudinal pitch* S_L, and the *diagonal pitch* S_D between tube centers. The diagonal pitch is determined from

$$S_D = \sqrt{S_L^2 + (S_T/2)^2} \tag{7-38}$$

As the fluid enters the tube bank, the flow area decreases from $A_1 = S_T L$ to $A_T = (S_T - D)L$ between the tubes, and thus flow velocity increases. In staggered arrangement, the velocity may increase further in the diagonal region if the tube rows are very close to each other. In tube banks, the flow characteristics are dominated by the maximum velocitiy V_{max} that occurs within the tube bank rather than the approach velocity V. Therefore, the Reynolds number is defined on the basis of maximum velocity as

$$Re_D = \frac{\rho V_{max} D}{\mu} = \frac{V_{max} D}{\nu} \tag{7-39}$$

Flow
direction

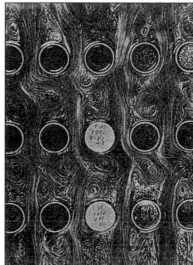

FIGURE 7–25
Flow patterns for staggered and in-line tube banks (photos by R. D. Willis).

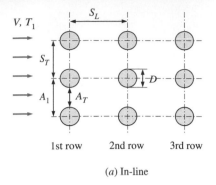

(a) In-line

$A_1 = S_T L$
$A_T = (S_T - D)L$
$A_D = (S_D - D)L$

(b) Staggered

FIGURE 7–26

Arrangement of the tubes in in-line and staggered tube banks (A_1, A_T, and A_D are flow areas at indicated locations, and L is the length of the tubes).

The maximum velocity is determined from the conservation of mass requirement for steady incompressible flow. For *in-line* arrangement, the maximum velocity occurs at the minimum flow area between the tubes, and the conservation of mass can be expressed as (see Fig. 7–26a) $\rho V A_1 = \rho V_{max} A_T$ or $V S_T = V_{max}(S_T - D)$. Then the maximum velocity becomes

$$V_{max} = \frac{S_T}{S_T - D} V \qquad (7\text{–}40)$$

In *staggered* arrangement, the fluid approaching through area A_1 in Fig. 7–26b passes through area A_T and then through area $2A_D$ as it wraps around the pipe in the next row. If $2A_D > A_T$, maximum velocity still occurs at A_T between the tubes and thus the V_{max} relation Eq. 7–40 can also be used for staggered tube banks. But if $2A_D < A_T$ [or, if $2(S_D - D) < (S_T - D)$], maximum velocity occurs at the diagonal cross sections, and the maximum velocity in this case becomes

Staggered and $S_D < (S_T + D)/2$: $\qquad V_{max} = \frac{S_T}{2(S_D - D)} V \qquad (7\text{–}41)$

since $\rho V A_1 = \rho V_{max}(2A_D)$ or $V S_T = 2 V_{max}(S_D - D)$.

The nature of flow around a tube in the first row resembles flow over a single tube discussed in Section 7–3, especially when the tubes are not too close to each other. Therefore, each tube in a tube bank that consists of a single transverse row can be treated as a single tube in cross-flow. The nature of flow around a tube in the second and subsequent rows is very different, however, because of wakes formed and the turbulence caused by the tubes upstream. The level of turbulence, and thus the heat transfer coefficient, increases with row number because of the combined effects of upstream rows. But there is no significant change in turbulence level after the first few rows, and thus the heat transfer coefficient remains constant.

Flow through tube banks is studied experimentally since it is too complex to be treated analytically. We are primarily interested in the average heat transfer coefficient for the entire tube bank, which depends on the number of tube rows along the flow as well as the arrangement and the size of the tubes.

Several correlations, all based on experimental data, have been proposed for the average Nusselt number for cross flow over tube banks. More recently, Zukauskas has proposed correlations whose general form is

$$Nu_D = \frac{hD}{k} = C \, Re_D^m \, Pr^n (Pr/Pr_s)^{0.25} \qquad (7\text{–}42)$$

where the values of the constants C, m, and n depend on Reynolds number. Such correlations are given in Table 7–2 for $0.7 < Pr < 500$ and $0 < Re_D < 2 \times 10^6$. The uncertainty in the values of Nusselt number obtained from these relations is ± 15 percent. Note that all properties except Pr_s are to be evaluated at the arithmetic mean temperature of the fluid determined from

$$T_m = \frac{T_i + T_e}{2} \qquad (7\text{–}43)$$

where T_i and T_e are the fluid temperatures at the inlet and the exit of the tube bank, respectively.

TABLE 7–2

Nusselt number correlations for cross flow over tube banks for $N > 16$ and $0.7 < Pr < 500$ (from Zukauskas, 1987)*

Arrangement	Range of Re_D	Correlation
In-line	0–100	$Nu_D = 0.9\, Re_D^{0.4}Pr^{0.36}(Pr/Pr_s)^{0.25}$
	100–1000	$Nu_D = 0.52\, Re_D^{0.5}Pr^{0.36}(Pr/Pr_s)^{0.25}$
	1000–2×10^5	$Nu_D = 0.27\, Re_D^{0.63}Pr^{0.36}(Pr/Pr_s)^{0.25}$
	2×10^5–2×10^6	$Nu_D = 0.033\, Re_D^{0.8}Pr^{0.4}(Pr/Pr_s)^{0.25}$
Staggered	0–500	$Nu_D = 1.04\, Re_D^{0.4}Pr^{0.36}(Pr/Pr_s)^{0.25}$
	500–1000	$Nu_D = 0.71\, Re_D^{0.5}Pr^{0.36}(Pr/Pr_s)^{0.25}$
	1000–2×10^5	$Nu_D = 0.35(S_T/S_L)^{0.2}\, Re_D^{0.6}Pr^{0.36}(Pr/Pr_s)^{0.25}$
	2×10^5–2×10^6	$Nu_D = 0.031(S_T/S_L)^{0.2}\, Re_D^{0.8}Pr^{0.36}(Pr/Pr_s)^{0.25}$

*All properties except Pr_s are to be evaluated at the arithmetic mean of the inlet and outlet temperatures of the fluid (Pr_s is to be evaluated at T_s).

The average Nusselt number relations in Table 7–2 are for tube banks with 16 or more rows. Those relations can also be used for tube banks with N_L provided that they are modified as

$$Nu_{D, N_L} = F Nu_D \qquad (7\text{–}44)$$

where F is a *correction factor F* whose values are given in Table 7–3. For $Re_D > 1000$, the correction factor is independent of Reynolds number.

Once the Nusselt number and thus the average heat transfer coefficient for the entire tube bank is known, the heat transfer rate can be determined from Newton's law of cooling using a suitable temperature difference ΔT. The first thought that comes to mind is to use $\Delta T = T_s - T_{avg} = T_s - (T_i + T_e)/2$. But this, in general, over predicts the heat transfer rate. We show in the next chapter that the proper temperature difference for internal flow (flow over tube banks is still internal flow through the shell) is the *logarithmic mean temperature difference* ΔT_{ln} defined as

$$\Delta T_{ln} = \frac{(T_s - T_e) - (T_s - T_i)}{\ln[(T_s - T_e)/(T_s - T_i)]} = \frac{\Delta T_e - \Delta T_i}{\ln(\Delta T_e/\Delta T_i)} \qquad (7\text{–}45)$$

We also show that the exit temperature of the fluid T_e can be determined from

$$T_e = T_s - (T_s - T_i) \exp\left(\pm \frac{A_s h}{\dot{m} c_p} \right) \qquad (7\text{–}46)$$

TABLE 7–3

Correction factor F to be used in $Nu_{D, N_L} = F Nu_D$ for $N_L < 16$ and $Re_D > 1000$ (from Zukauskas, 1987)

N_L	1	2	3	4	5	7	10	13
In-line	0.70	0.80	0.86	0.90	0.93	0.96	0.98	0.99
Staggered	0.64	0.76	0.84	0.89	0.93	0.96	0.98	0.99

where $A_s = N\pi DL$ is the heat transfer surface area and $\dot{m} = \rho V(N_T S_T L)$ is the mass flow rate of the fluid. Here N is the total number of tubes in the bank, N_T is the number of tubes in a transverse plane, L is the length of the tubes, and V is the velocity of the fluid just before entering the tube bank. Then the heat transfer rate can be determined from

$$\dot{Q} = hA_s\Delta T_{\ln} = \dot{m}c_p(T_e - T_i) \qquad (7\text{--}47)$$

The second relation is usually more convenient to use since it does not require the calculation of $\Delta T_{\ln}$.

Pressure Drop

Another quantity of interest associated with tube banks is the *pressure drop* ΔP, which is the irreversible pressure loss between the inlet and the exit of the tube bank. It is a measure of the resistance the tubes offer to flow over them, and is expressed as

$$\Delta P = N_L f \chi \frac{\rho V_{\max}^2}{2} \qquad (7\text{--}48)$$

where f is the friction factor and χ is the correction factor, both plotted in Figs. 7–27a and 7–27b against the Reynolds number based on the maximum velocity $V_{\max}$. The friction factor in Fig. 7–27a is for a *square* in-line tube bank ($S_T = S_L$), and the correction factor given in the insert is used to account for the effects of deviation of rectangular in-line arrangements from square arrangement. Similarly, the friction factor in Fig. 7–27b is for an *equilateral* staggered tube bank ($S_T = S_D$), and the correction factor is to account for the effects of deviation from equilateral arrangement. Note that $\chi = 1$ for both square and equilateral triangle arrangements. Also, pressure drop occurs in the flow direction, and thus we used N_L (the number of rows) in the ΔP relation.

The power required to move a fluid through a tube bank is proportional to the pressure drop, and when the pressure drop is available, the pumping power required to overcome flow resistance can be determined from

$$\dot{W}_{\text{pump}} = \dot{V}\Delta P = \frac{\dot{m}\Delta P}{\rho} \qquad (7\text{--}49)$$

where $\dot{V} = V(N_T S_T L)$ is the volume flow rate and $\dot{m} = \rho\dot{V} = \rho V(N_T S_T L)$ is the mass flow rate of the fluid through the tube bank. Note that the power required to keep a fluid flowing through the tube bank (and thus the operating cost) is proportional to the pressure drop. Therefore, the benefits of enhancing heat transfer in a tube bank via rearrangement should be weighed against the cost of additional power requirements.

In this section we limited our consideration to tube banks with base surfaces (no fins). Tube banks with finned surfaces are also commonly used in practice, especially when the fluid is a gas, and heat transfer and pressure drop correlations can be found in the literature for tube banks with pin fins, plate fins, strip fins, etc.

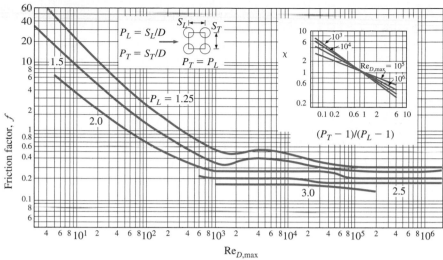

(a) In-line arrangement

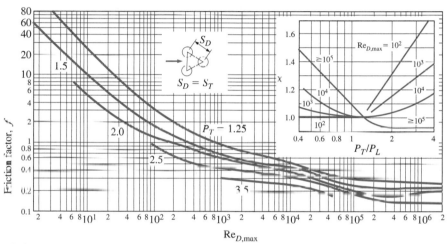

(b) Staggered arrangement

FIGURE 7–27
Friction factor f and correction factor χ for tube banks (from Zukauskas, 1985).

EXAMPLE 7–7   Preheating Air by Geothermal Water in a Tube Bank

In an industrial facility, air is to be preheated before entering a furnace by geothermal water at 120°C flowing through the tubes of a tube bank located in a duct. Air enters the duct at 20°C and 1 atm with a mean velocity of 4.5 m/s, and flows over the tubes in normal direction. The outer diameter of the tubes is 1.5 cm, and the tubes are arranged in-line with longitudinal and transverse pitches of $S_L = S_T = 5$ cm. There are 6 rows in the flow direction with 10 tubes in each row, as shown in Fig. 7–28. Determine the rate of heat transfer per unit length of the tubes, and the pressure drop across the tube bank.

SOLUTION   Air is heated by geothermal water in a tube bank. The rate of heat transfer to air and the pressure drop of air are to be determined.

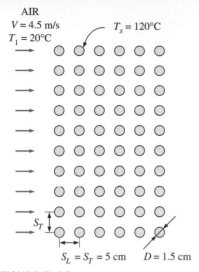

AIR
$V = 4.5$ m/s
$T_1 = 20°C$

$T_s = 120°C$

S_T

$S_L = S_T = 5$ cm $D = 1.5$ cm

FIGURE 7–28
Schematic for Example 7–7.

Assumptions **1** Steady operating conditions exist. **2** The surface temperature of the tubes is equal to the temperature of geothermal water.

Properties The exit temperature of air, and thus the mean temperature, is not known. We evaluate the air properties at the assumed mean temperature of 60°C (will be checked later) and 1 atm (Table A–15):

$$k = 0.02808 \text{ W/m} \cdot \text{K}, \qquad \rho = 1.059 \text{ kg/m}^3$$
$$c_p = 1.007 \text{ kJ/kg} \cdot \text{K}, \qquad \text{Pr} = 0.7202$$
$$\mu = 2.008 \times 10^{-5} \text{ kg/m} \cdot \text{s} \qquad \text{Pr}_s = \text{Pr}_{120°C} = 0.7073$$

Also, the density of air at the inlet temperature of 20°C (for use in the mass flow rate calculation at the inlet) is $\rho_1 = 1.204 \text{ kg/m}^3$.

Analysis It is given that $D = 0.015$ m, $S_L = S_T = 0.05$ m, and $V = 4.5$ m/s. Then the maximum velocity and the Reynolds number based on the maximum velocity become

$$V_{\max} = \frac{S_T}{S_T - D} V = \frac{0.05}{0.05 - 0.015}(4.5 \text{ m/s}) = 6.43 \text{ m/s}$$

$$\text{Re}_D = \frac{\rho V_{\max} D}{\mu} = \frac{(1.059 \text{ kg/m}^3)(6.43 \text{ m/s})(0.015 \text{ m})}{2.008 \times 10^{-5} \text{ kg/m} \cdot \text{s}} = 5086$$

The average Nusselt number is determined using the proper relation from Table 7–2 to be

$$\text{Nu}_D = 0.27 \text{ Re}_D^{0.63} \text{ Pr}^{0.36}(\text{Pr}/\text{Pr}_s)^{0.25}$$
$$= 0.27(5086)^{0.63}(0.7202)^{0.36}(0.7202/0.7073)^{0.25} = 52.1$$

This Nusselt number is applicable to tube banks with $N_L > 16$. In our case, the number of rows is $N_L = 6$, and the corresponding correction factor from Table 7–3 is $F = 0.945$. Then the average Nusselt number and heat transfer coefficient for all the tubes in the tube bank become

$$\text{Nu}_{D, N_L} = F\text{Nu}_D = (0.945)(52.1) = 49.3$$

$$h = \frac{\text{Nu}_{D, N_L} k}{D} = \frac{49.3(0.02808 \text{ W/m} \cdot °C)}{0.015 \text{ m}} = 92.2 \text{ W/m}^2 \cdot °C$$

The total number of tubes is $N = N_L \times N_T = 6 \times 10 = 60$. For a unit tube length ($L = 1$ m), the heat transfer surface area and the mass flow rate of air (evaluated at the inlet) are

$$A_s = N\pi DL = 60\pi(0.015 \text{ m})(1 \text{ m}) = 2.827 \text{ m}^2$$

$$\dot{m} = \dot{m}_1 = \rho_1 V(N_T S_T L)$$
$$= (1.204 \text{ kg/m}^3)(4.5 \text{ m/s})(10)(0.05 \text{ m})(1 \text{ m}) = 2.709 \text{ kg/s}$$

Then the fluid exit temperature, the log mean temperature difference, and the rate of heat transfer become

$$T_e = T_s - (T_s - T_i) \exp\left(-\frac{A_s h}{\dot{m} c_p}\right)$$

$$= 120 - (120 - 20)\exp\left(-\frac{(2.827 \text{ m}^2)(92.2 \text{ W/m}^2 \cdot °C)}{(2.709 \text{ kg/s})(1007 \text{ J/kg} \cdot °C)}\right) = 29.11°C$$

$$\Delta T_{\text{ln}} = \frac{(T_s - T_e) - (T_s - T_i)}{\ln[(T_s - T_e)/(T_s - T_i)]} = \frac{(120 - 29.11) \quad (120 - 20)}{\ln[(120 - 29.11)/(120 - 20)]} = 95.4°C$$

$$\dot{Q} = hA_s\Delta T_{\text{ln}} = (92.2 \text{ W/m}^2 \cdot °C)(2.827 \text{ m}^2)(95.4°C) = 2.49 \times 10^4 \text{ W}$$

The rate of heat transfer can also be determined in a simpler way from

$$\dot{Q} = hA_s\Delta T_{\text{ln}} = \dot{m}c_p(T_e - T_i)$$
$$= (2.709 \text{ kg/s})(1007 \text{ J/kg} \cdot °C)(29.11 - 20)°C = 2.49 \times 10^4 \text{ W}$$

For this square in-line tube bank, the friction coefficient corresponding to $Re_D = 5086$ and $S_L/D = 5/1.5 = 3.33$ is, from Fig. 7–27a, $f = 0.16$. Also, $\chi = 1$ for the square arrangements. Then the pressure drop across the tube bank becomes

$$\Delta P = N_L f\chi \frac{\rho V_{\text{max}}^2}{2}$$

$$= 6(0.16)(1)\frac{(1.059 \text{ kg/m}^3)(6.43 \text{ m/s})^2}{2}\left(\frac{1 \text{ N}}{1 \text{ kg} \cdot \text{m/s}^2}\right) = 21 \text{ Pa}$$

Discussion The arithmetic mean fluid temperature is $(T_i + T_e)/2 = (20 + 29.11)/2 = 24.6°C$, which is not close to the assumed value of 60°C. Repeating calculations for 25°C gives 2.57×10^4 W for the rate of heat transfer and 23.5 Pa for the pressure drop.

TOPIC OF SPECIAL INTEREST*

Reducing Heat Transfer through Surfaces: Thermal Insulation

Thermal insulations are materials or combinations of materials that are used primarily to provide resistance to heat flow (Fig. 7–29). You are probably familiar with several kinds of insulation available in the market. Most insulations are heterogeneous materials made of low thermal conductivity materials, and they involve air pockets. This is not surprising since air has one of the lowest thermal conductivities and is readily available. The *Styrofoam* commonly used as a packaging material for TVs, DVDs, computers, and just about anything because of its light weight is also an excellent insulator.

Temperature difference is the driving force for heat flow, and the greater the temperature difference, the larger the rate of heat transfer. We can slow down the heat flow between two mediums at different temperatures by putting "barriers" on the path of heat flow. Thermal insulations serve as such barriers, and they play a major role in the design and manufacture of all energy-efficient devices or systems, and they are usually the cornerstone of energy conservation projects. A 2001 report by the Alliance to Save Energy revealed that insulation in residential, commercial, and industrial buildings saves the U.S. nearly 4 billion barrels of oil per year, valued at

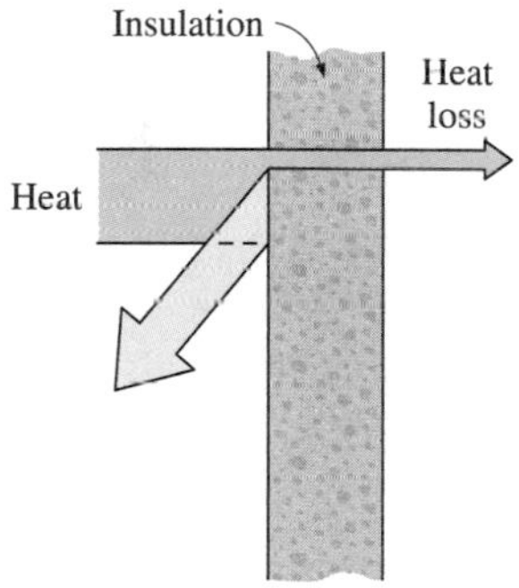

FIGURE 7–29
Thermal insulation retards heat transfer by acting as a barrier in the path of heat flow.

*This section can be skipped without a loss in continuity.

FIGURE 7–30
Insulation also helps the environment by reducing the amount of fuel burned and the air pollutants released.

FIGURE 7–31
In cold weather, we minimize heat loss from our bodies by putting on thick layers of insulation (coats or furs).

$177 billion a year in energy costs, and more can be saved by practicing better insulation techniques and retrofitting the older facilities. It also reduces CO_2 emissions by 1340 million tons a year.

Heat is generated in furnaces or heaters by burning a fuel such as coal, oil, or natural gas or by passing electric current through a resistance heater. Electricity is rarely used for heating purposes since its unit cost is much higher. The heat generated is absorbed by the medium in the furnace and its surfaces, causing a temperature rise above the ambient temperature. This temperature difference drives heat transfer from the hot medium to the ambient, and insulation reduces the amount of heat loss and thus saves fuel and money. Therefore, insulation *pays for itself* from the energy it saves. Insulating properly requires a one-time capital investment, but its effects are dramatic and long term. The payback period of insulation is often less than one year. That is, the money insulation saves during the first year is usually greater than its initial material and installation costs. On a broader perspective, insulation also helps the environment and fights air pollution and the greenhouse effect by reducing the amount of fuel burned and thus the amount of CO_2 and other gases released into the atmosphere (Fig. 7–30).

Saving energy with insulation is not limited to hot surfaces. We can also save energy and money by insulating *cold surfaces* (surfaces whose temperature is below the ambient temperature) such as chilled water lines, cryogenic storage tanks, refrigerated trucks, and air-conditioning ducts. The source of "coldness" is *refrigeration*, which requires energy input, usually electricity. In this case, heat is transferred from the surroundings to the cold surfaces, and the refrigeration unit must now work harder and longer to make up for this heat gain and thus it must consume more electrical energy. A cold canned drink can be kept cold much longer by wrapping it in a blanket. A refrigerator with well-insulated walls consumes much less electricity than a similar refrigerator with little or no insulation. Insulating a house results in reduced cooling load, and thus reduced electricity consumption for air-conditioning.

Whether we realize it or not, we have an *intuitive* understanding and appreciation of thermal insulation. As babies we feel much better in our blankies, and as children we know we should wear a sweater or coat when going outside in cold weather (Fig. 7–31). When getting out of a pool after swimming on a windy day, we quickly wrap in a towel to stop shivering. Similarly, early man used animal furs to keep warm and built shelters using mud bricks and wood. Cork was used as a roof covering for centuries. The need for effective thermal insulation became evident with the development of mechanical refrigeration later in the nineteenth century, and a great deal of work was done at universities and government and private laboratories in the 1910s and 1920s to identify and characterize thermal insulation.

Thermal insulation in the form of *mud, clay, straw, rags,* and *wood strips* was first used in the eighteenth century on steam engines to keep workmen from being burned by hot surfaces. As a result, boiler room temperatures dropped and it was noticed that fuel consumption was also reduced. The realization of improved engine efficiency and energy savings prompted the scarch for materials with improved thermal efficiency. One of the first such materials was *mineral wool* insulation, which, like many materials, was discovered by accident. About 1840, an iron producer in Wales aimed a stream of high-pressure steam at the slag flowing from a blast furnace, and manufactured mineral wool was born. In the early 1860s, this slag wool was a by-product of manufacturing cannons for the Civil War and quickly found its way into many industrial uses. By 1880, builders began installing mineral wool in houses, with one of the most notable applications being General Grant's house. The insulation of this house was described in an article: "it keeps the house cool in summer and warm in winter; it prevents the spread of fire; and it deadens the sound between floors" [Edmunds (1989)]. An article published in 1887 in *Scientific American* detailing the benefits of insulating the entire house gave a major boost to the use of insulation in residential buildings.

The energy crisis of the 1970s had a tremendous impact on the public awareness of energy and limited energy reserves and brought an emphasis on *energy conservation*. We have also seen the development of new and more effective insulation materials since then, and a considerable increase in the use of insulation. Thermal insulation is used in more places than you may be aware of. The walls of your house are probably filled with some kind of insulation, and the roof is likely to have a thick layer of insulation. The "thickness" of the walls of your refrigerator is due to the insulation layer sandwiched between two layers of sheet metal (Fig. 7–32). The walls of your range are also insulated to conserve energy, and your hot water tank contains less water than you think because of the 2- to 4-cm-thick insulation in the walls of the tank. Also, your hot water pipe may look much thicker than the cold water pipe because of insulation.

Reasons for Insulating

If you examine the engine compartment of your car, you will notice that the firewall between the engine and the passenger compartment as well as the inner surface of the hood are insulated. The reason for insulating the hood is not to conserve the waste heat from the engine but to protect people from burning themselves by touching the hood surface, which will be too hot if not insulated. As this example shows, the use of insulation is not limited to energy conservation. Various reasons for using insulation can be summarized as follows:

- **Energy Conservation**   Conserving energy by reducing the rate of heat transfer is the primary reason for insulating surfaces. Insulation materials that performs satisfactorily in the temperature range of −268°C to 1000°C (−450°F to 1800°F) are widely available.

- **Personnel Protection and Comfort**   A surface that is too hot poses a danger to people who are working in that area of accidentally touching the hot surface and burning themselves (Fig. 7–33). To prevent this danger and to comply with the OSHA (Occupational Safety and Health Administration) standards, the temperatures of hot surfaces should be reduced to below 60°C (140°F) by insulating them. Also, the excessive heat coming off the hot surfaces creates an unpleasant environment in

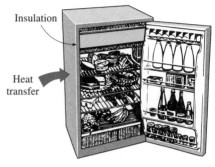

Insulation

Heat transfer

FIGURE 7–32
The insulation layers in the walls of a refrigerator reduce the amount of heat flow into the refrigerator and thus the running time of the refrigerator, saving electricity.

Insulation

FIGURE 7–33
The hood of the engine compartment of a car is insulated to reduce its temperature and to protect people from burning themselves.

which to work, which adversely affects the performance or productivity of the workers, especially in summer months.

- **Maintaining Process Temperature** Some processes in the chemical industry are temperature-sensitive, and it may become necessary to insulate the process tanks and flow sections heavily to maintain a uniform temperature throughout.

- **Reducing Temperature Variation and Fluctuations** The temperature in an enclosure may vary greatly between the midsection and the edges if the enclosure is not insulated. For example, the temperature near the walls of a poorly insulated house is much lower than the temperature at the midsections. Also, the temperature in an uninsulated enclosure follows the temperature changes in the environment closely and fluctuates. Insulation minimizes temperature nonuniformity in an enclosure and slows down fluctuations.

- **Condensation and Corrosion Prevention** Water vapor in air condenses on surfaces whose temperature is below the dew point, and the outer surfaces of the tanks or pipes that contain a cold fluid frequently fall below the dew-point temperature unless they have adequate insulation. The liquid water on exposed surfaces of metal tanks or pipes may promote corrosion as well as algae growth.

- **Fire Protection** Damage during a fire may be minimized by keeping valuable combustibles in a safety box that is well insulated. Insulation may lower the rate of heat transfer to such levels that the temperature in the box never rises to unsafe levels during fire.

- **Freezing Protection** Prolonged exposure to subfreezing temperatures may cause water in pipes or storage vessels to freeze and burst as a result of heat transfer from the water to the cold ambient. The bursting of pipes as a result of freezing can cause considerable damage. Adequate insulation slows down the heat loss from the water and prevents freezing during limited exposure to subfreezing temperatures. For example, covering vegetables during a cold night protects them from freezing, and burying water pipes in the ground at a sufficient depth keeps them from freezing during the entire winter. Wearing thick gloves protects the fingers from possible frostbite. Also, a molten metal or plastic in a container solidifies on the inner surface if the container is not properly insulated.

- **Reducing Noise and Vibration** An added benefit of thermal insulation is its ability to dampen noise and vibrations (Fig. 7–34). The insulation materials differ in their ability to reduce noise and vibration, and the proper kind can be selected if noise reduction is an important consideration.

There are a wide variety of insulation materials available in the market, but most are primarily made of fiberglass, mineral wool, polyethylene, foam, or calcium silicate. They come in various trade names such as granulated bulk mineral wool insulation, cork insulation sheets, foil-faced fiberglass insulation, blended sponge rubber sheeting, and numerous others.

FIGURE 7–34
Insulation materials absorb vibration and sound waves, and are used to minimize sound transmission.

Today various forms of fiberglass insulation are widely used in process industries and heating and air-conditioning applications because of their low cost, light weight, resiliency, and versatility. But they are not suitable for some applications because of their low resistance to moisture and fire and their limited maximum service temperature. Fiberglass insulations come in various forms such as unfaced fiberglass insulation, vinyl-faced fiberglass insulation, foil-faced fiberglass insulation, and fiberglass insulation sheets. The reflective foil-faced fiberglass insulation resists vapor penetration and retards radiation because of the aluminum foil on it and is suitable for use on pipes, ducts, and other surfaces.

Mineral wool is resilient, lightweight, fibrous, wool-like, thermally efficient, and fire resistant up to 1100°C (2000°F), and forms a sound barrier. Mineral wool insulation comes in the form of blankets, rolls, or blocks. *Calcium silicate* is a solid material that is suitable for use at high temperatures, but it is more expensive. Also, it needs to be cut with a saw during installation, and thus it takes longer to install and there is more waste.

Superinsulators

You may be tempted to think that the most effective way to reduce heat transfer is to use insulating materials that are known to have very low thermal conductivities such as urethane or rigid foam ($k = 0.026$ W/m · °C) or fiberglass ($k = 0.035$ W/m · °C). After all, they are widely available, inexpensive, and easy to install. Looking at the thermal conductivities of materials, you may also notice that the thermal conductivity of air at room temperature is 0.026 W/m · °C, which is lower than the conductivities of practically all of the ordinary insulating materials. Thus you may think that a layer of enclosed air space is as effective as any of the common insulating materials of the same thickness. Of course, heat transfer through the air will probably be higher than what a pure conduction analysis alone would indicate because of the natural convection currents that are likely to occur in the air layer. Besides, air is transparent to radiation, and thus heat is also transferred by radiation. The thermal conductivity of air is practically independent of pressure unless the pressure is extremely high or extremely low. Therefore, we can reduce the thermal conductivity of air and thus the conduction heat transfer through the air by evacuating the air space. In the limiting case of absolute vacuum, the thermal conductivity will be zero since there will be no particles in this case to "conduct" heat from one surface to the other, and thus the conduction heat transfer will be zero. Noting that the thermal conductivity cannot be negative, an absolute vacuum must be the ultimate insulator, right? Well, not quite.

The purpose of insulation is to reduce "total" heat transfer from a surface, not just conduction. A vacuum totally eliminates conduction but offers zero resistance to radiation, whose magnitude can be comparable to conduction or natural convection in gases (Fig. 7–35). Thus, a vacuum is no more effective in reducing heat transfer than sealing off one of the lanes of a two-lane road is in reducing the flow of traffic on a one-way road.

Insulation against radiation heat transfer between two surfaces is achieved by placing "barriers" between the two surfaces, which are highly

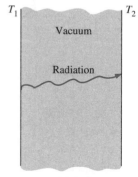

FIGURE 7–35
Evacuating the space between two surfaces completely eliminates heat transfer by conduction or convection but leaves the door wide open for radiation.

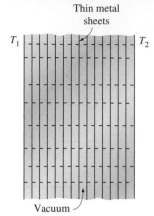

FIGURE 7–36
Superinsulators are built by closely packing layers of highly reflective thin metal sheets and evacuating the space between them.

reflective thin metal sheets. Radiation heat transfer between two surfaces is inversely proportional to the number of such sheets placed between the surfaces. Very effective insulations are obtained by using closely packed layers of highly reflective thin metal sheets such as aluminum foil (usually 25 sheets per cm) separated by fibers made of insulating material such as glass fiber (Fig. 7–36). Further, the space between the layers is evacuated to form a vacuum under 0.000001 atm pressure to minimize conduction or convection heat transfer through the air space between the layers. The result is an insulating material whose apparent thermal conductivity is below 2×10^{-5} W/m · °C, which is one thousand times less than the conductivity of air or any common insulating material. These specially built insulators are called **superinsulators**, and they are commonly used in space applications and cryogenics, which is the branch of heat transfer dealing with temperatures below 100 K (-173°C) such as those encountered in the liquefaction, storage, and transportation of gases, with helium, methane, hydrogen, nitrogen, and oxygen being the most common ones.

The R-value of Insulation

The effectiveness of insulation materials is given by some manufacturers in terms of their **R-value**, which is the *thermal resistance* of the material *per unit surface area*. For *flat insulation* the R-value is obtained by simply dividing the thickness L of the insulation by its thermal conductivity k. That is,

$$R\text{-value} = \frac{L}{k} \qquad \text{(flat insulation)} \tag{7–50}$$

Note that doubling the thickness L doubles the R-value of flat insulation. For *pipe insulation,* the R-value is determined using the thermal resistance relation from

$$R\text{-value} = \frac{r_2}{k} \ln \frac{r_2}{r_1} \qquad \text{(pipe insulation)} \tag{7–51}$$

where r_1 is the inside radius and r_2 is the outside radius of insulation. Once the R-value is available, the rate of heat transfer through the insulation can be determined from

$$\dot{Q} = \frac{\Delta T}{R\text{-value}} \times \text{Area} \tag{7–52}$$

where ΔT is the temperature difference across the insulation and Area is the outer surface area for a cylinder.

In the United States, the R-values of insulation are expressed without any units, such as R-19 and R-30. These R-values are obtained by dividing the thickness of the material in *feet* by its thermal conductivity in the unit Btu/h · ft · °F so that the R-values actually have the unit h · ft^2 · °F/Btu. For example, the R-value of 6-in-thick glass-fiber insulation whose thermal conductivity is 0.025 Btu/h · ft · °F is (Fig. 7–37)

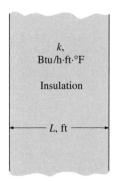

$$R\text{-value} = \frac{L}{k}$$

FIGURE 7–37
The R-value of an insulating material is simply the ratio of the thickness of the material to its thermal conductivity in proper units.

$$R\text{-value} = \frac{L}{k} = \frac{0.5 \text{ ft}}{0.025 \text{ Btu/h} \cdot \text{ft} \cdot {}^\circ\text{F}} = 20 \text{ h} \cdot \text{ft}^2 \cdot {}^\circ\text{F/Btu}$$

Thus, this 6-in-thick glass fiber insulation would be referred to as R-20 insulation by the builders. The unit of R-value is $m^2 \cdot {}^\circ\text{C/W}$ in SI units, with the conversion relation $1 \text{ m}^2 \cdot {}^\circ\text{C/W} = 5.678 \text{ h} \cdot \text{ft}^2 \cdot {}^\circ\text{F/Btu}$. Therefore, a small R-value in SI corresponds to a large R-value in English units.

Optimum Thickness of Insulation

It should be apparent that insulation does not eliminate heat transfer; it merely reduces it. The thicker the insulation, the lower the rate of heat transfer but also the higher the cost of insulation. Therefore, there should be an *optimum* thickness of insulation that corresponds to a minimum combined cost of insulation and heat lost. The determination of the optimum thickness of insulation is illustrated in Fig. 7–38. Notice that the cost of insulation increases roughly linearly with thickness while the cost of heat loss decreases exponentially. The total cost, which is the sum of the insulation cost and the lost heat cost, decreases first, reaches a minimum, and then increases. The thickness corresponding to the minimum total cost is the optimum thickness of insulation, and this is the recommended thickness of insulation to be installed.

If you are mathematically inclined, you can determine the *optimum thickness* by obtaining an expression for the *total cost*, which is the sum of the expressions for the costs of lost heat and insulation as a function of thickness; *differentiating* the total cost expression with respect to thickness; and *setting* it equal to zero. The thickness value satisfying the resulting equation is the optimum thickness. The cost values can be determined from an annualized lifetime analysis or simply from the requirement that the insulation pay for itself within two or three years. Note that the optimum thickness of insulation depends on the fuel cost, and the higher the fuel cost, the larger the optimum thickness of insulation. Considering that insulation will be in service for many years and the fuel prices are likely to escalate, a reasonable increase in fuel prices must be assumed in calculations. Otherwise, what is optimum insulation today may be inadequate insulation in the years to come, and we may have to face the possibility of costly retrofitting projects. This is what happened in the 1970s and 1980s to insulations installed in the 1960s.

The discussion above on optimum thickness is valid when the type and manufacturer of insulation are already selected, and the only thing to be determined is the most economical thickness. But often there are several suitable insulations for a job, and the selection process can be rather confusing since each insulation can have a different thermal conductivity, different installation cost, and different service life. In such cases, a selection can be made by preparing an annualized cost versus thickness chart like Fig. 7–39 for each insulation, and determining the one having the *lowest* minimum cost. The insulation with the lowest annual cost is obviously the most economical insulation, and the insulation thickness corresponding to the *minimum total cost* is the *optimum thickness*. When the optimum thickness falls between two commercially available thicknesses, it is a

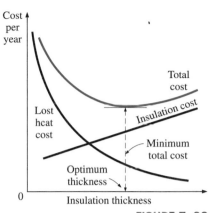

FIGURE 7–38
Determination of the optimum thickness of insulation on the basis of minimum total cost.

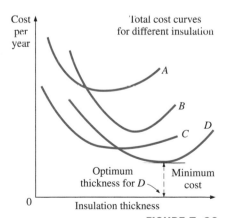

FIGURE 7–39
Determination of the most economical type of insulation and its optimum thickness.

TABLE 7–4

Recommended insulation thicknesses for flat hot surfaces as a function of surface temperature (from TIMA *Energy Savings Guide*)

Surface temperature	Insulation thickness
150°F (66°C)	2″ (5.1 cm)
250°F (121°C)	3″ (7.6 cm)
350°F (177°C)	4″ (10.2 cm)
550°F (288°C)	6″ (15.2 cm)
750°F (400°C)	9″ (22.9 cm)
950°F (510°C)	10″ (25.44 cm)

good practice to be conservative and choose the thicker insulation. The extra thickness provides a little safety cushion for any possible decline in performance over time and helps the environment by reducing the production of greenhouse gases such as CO_2.

The determination of the optimum thickness of insulation requires a heat transfer and economic analysis, which can be tedious and time-consuming. But a selection can be made in a few minutes using the tables and charts prepared by TIMA (Thermal Insulation Manufacturers Association) and member companies. The primary inputs required for using these tables or charts are the operating and ambient temperatures, pipe diameter (in the case of pipe insulation), and the unit fuel cost. Recommended insulation thicknesses for hot surfaces at specified temperatures are given in Table 7–4. Recommended thicknesses of *pipe insulations* as a function of service temperatures are 0.5 to 1 in for 150°F, 1 to 2 in for 250°F, 1.5 to 3 in for 350°F, 2 to 4.5 in for 450°F, 2.5 to 5.5 in for 550°F, and 3 to 6 in for 650°F for nominal pipe diameters of 0.5 to 36 in. The lower recommended insulation thicknesses are for pipes with small diameters, and the larger ones are for pipes with large diameters.

EXAMPLE 7–8   Effect of Insulation on Surface Temperature

Hot water at $T_i = 120°C$ flows in a stainless steel pipe ($k = 15$ W/m · °C) whose inner diameter is 1.6 cm and thickness is 0.2 cm. The pipe is to be covered with adequate insulation so that the temperature of the outer surface of the insulation does not exceed 40°C when the ambient temperature is $T_o = 25°C$. Taking the heat transfer coefficients inside and outside the pipe to be $h_i = 70$ W/m² · °C and $h_o = 20$ W/m² · °C, respectively, determine the thickness of fiberglass insulation ($k = 0.038$ W/m · °C) that needs to be installed on the pipe.

SOLUTION   A steam pipe is to be covered with enough insulation to reduce the exposed surface temperature. The thickness of insulation that needs to be installed is to be determined.

Assumptions   **1** Heat transfer is steady since there is no indication of any change with time. **2** Heat transfer is one-dimensional since there is thermal symmetry about the centerline and no variation in the axial direction. **3** Thermal conductivities are constant.

Properties   The thermal conductivities are given to be $k = 15$ W/m · °C for steel pipe and $k = 0.038$ W/m · °C for fiberglass insulation.

Analysis   The thermal resistance network for this problem involves four resistances in series and is given in Fig. 7–40. The inner radius of the pipe is $r_1 = 0.8$ cm and the outer radius of the pipe and thus the inner radius of the insulation is $r_2 = 1.0$ cm. Letting r_3 represent the outer radius of the insulation, the areas of the surfaces exposed to convection for an $L = 1$-m-long section of the pipe become

$$A_1 = 2\pi r_1 L = 2\pi(0.008 \text{ m})(1 \text{ m}) = 0.0503 \text{ m}^2$$

$$A_3 = 2\pi r_3 L = 2\pi r_3 (1 \text{ m}) = 6.28 r_3 \text{ m}^2$$

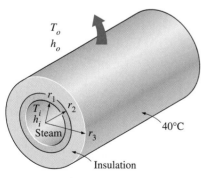

FIGURE 7–40
Schematic for Example 7–8.

Then the individual thermal resistances are determined to be

$$R_i = R_{conv,1} = \frac{1}{h_i A_1} = \frac{1}{(70 \text{ W/m}^2 \cdot {}^\circ\text{C})(0.0503 \text{ m}^2)} = 0.284{}^\circ\text{C/W}$$

$$R_1 = R_{pipe} = \frac{\ln(r_2/r_1)}{2\pi k_1 L} = \frac{\ln(0.01/0.008)}{2\pi(15 \text{ W/m} \cdot {}^\circ\text{C})(1 \text{ m})} = 0.0024{}^\circ\text{C/W}$$

$$R_2 = R_{insulation} = \frac{\ln(r_3/r_2)}{2\pi k_2 L} = \frac{\ln(r_3/0.01)}{2\pi(0.038 \text{ W/m} \cdot {}^\circ\text{C})(1 \text{ m})}$$
$$= 4.188 \ln(r_3/0.01){}^\circ\text{C/W}$$

$$R_o = R_{conv,2} = \frac{1}{h_o A_3} = \frac{1}{(20 \text{ W/m}^2 \cdot {}^\circ\text{C})(6.28 r_3 \text{ m}^2)} = \frac{1}{125.6 r_3}{}^\circ\text{C/W}$$

Noting that all resistances are in series, the total resistance is determined to be

$$R_{total} = R_i + R_1 + R_2 + R_o$$
$$= [0.284 + 0.0024 + 4.188 \ln(r_3/0.01) + 1/(125.6 r_3)]{}^\circ\text{C/W}$$

Then the steady rate of heat loss from the steam becomes

$$\dot{Q} = \frac{T_i - T_o}{R_{total}} = \frac{(120 - 125){}^\circ\text{C}}{[0.284 + 0.0024 + 4.188 \ln(r_3/0.01) + 1/(125.6 r_3)]{}^\circ\text{C/W}}$$

Noting that the outer surface temperature of insulation is specified to be 40°C, the rate of heat loss can also be expressed as

$$\dot{Q} = \frac{T_3 - T_o}{R_o} = \frac{(40 - 25){}^\circ\text{C}}{[1/(125.6 r_3)]{}^\circ\text{C/W}} = 1884 r_3$$

Setting the two relations above equal to each other and solving for r_3 gives $r_3 = 0.0170$ m. Then the minimum thickness of fiberglass insulation required is

$$t = r_3 - r_2 = 0.0170 - 0.0100 = 0.0070 \text{ m} = 0.70 \text{ cm}$$

Discussion Insulating the pipe with at least 0.70-cm-thick fiberglass insulation will ensure that the outer surface temperature of the pipe remains at 40°C or below.

EXAMPLE 7–9   Optimum Thickness of Insulation

During a plant visit, you notice that the outer surface of a cylindrical curing oven is very hot, and your measurements indicate that the average temperature of the exposed surface of the oven is 180°F when the surrounding air temperature is 75°F. You suggest to the plant manager that the oven should be insulated, but the manager does not think it is worth the expense. Then you propose to the manager to pay for the insulation yourself if he lets you keep the savings from the fuel bill for one year. That is, if the fuel bill is $5000/yr before insulation and drops to $2000/yr after insulation, you will get paid $3000. The manager agrees since he has nothing to lose, and a lot to gain. Is this a smart bet on your part?

The oven is 12 ft long and 8 ft in diameter, as shown in Fig. 7–41. The plant operates 16 h a day 365 days a year, and thus 5840 h/yr. The insulation

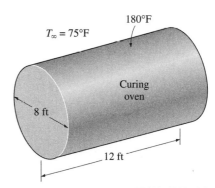

$T_\infty = 75{}^\circ\text{F}$

180°F

Curing oven

8 ft

12 ft

FIGURE 7–41
Schematic for Example 7–9.

to be used is fiberglass (k_{ins} = 0.024 Btu/h · ft · °F), whose cost is \$0.70/ft^2 per inch of thickness for materials, plus \$2.00/ft^2 for labor regardless of thickness. The combined heat transfer coefficient on the outer surface is estimated to be h_o = 3.5 Btu/h · ft^2 · °F. The oven uses natural gas, whose unit cost is \$0.75/therm input, and the efficiency of the oven is 80 percent. Disregarding any inflation or interest, determine how much money you will make out of this venture, if any, and the thickness of insulation (in whole inches) that will maximize your earnings.

SOLUTION A cylindrical oven is to be insulated to reduce heat losses. The optimum thickness of insulation and the potential earnings are to be determined.

Assumptions **1** Steady operating conditions exist. **2** Heat transfer through the insulation is one-dimensional. **3** Thermal conductivities are constant. **4** The thermal contact resistance at the interface is negligible. **5** The surfaces of the cylindrical oven can be treated as plain surfaces since its diameter is large.

Properties The thermal conductivity of insulation is given to be k = 0.024 Btu/h · ft · °F.

Analysis The exposed surface area of the oven is

$$A_s = 2A_{base} + A_{side} = 2\pi r^2 + 2\pi rL = 2\pi(4 \text{ ft})^2 + 2\pi(4 \text{ ft})(12 \text{ ft}) = 402 \text{ ft}^2$$

The rate of heat loss from the oven before the insulation is installed is determined from

$$\dot{Q} = h_o A_s (T_s - T_\infty) = (3.5 \text{ Btu/h} \cdot \text{ft}^2 \cdot °F)(402 \text{ ft}^2)(180 - 75)°F = 147,700 \text{ Btu/h}$$

Noting that the plant operates 5840 h/yr, the total amount of heat loss from the oven per year is

$$Q = \dot{Q}\Delta t = (147,700 \text{ Btu/h})(5840 \text{ h/yr}) = 0.863 \times 10^9 \text{ Btu/yr}$$

The efficiency of the oven is given to be 80 percent. Therefore, to generate this much heat, the oven must consume energy (in the form of natural gas) at a rate of

$$E_{in} = Q/\eta_{oven} = (0.863 \times 10^9 \text{ Btu/yr})/0.80 = 1.079 \times 10^9 \text{ Btu/yr}$$
$$= 10,790 \text{ therms}$$

since 1 therm = 100,000 Btu. Then the annual fuel cost of this oven before insulation becomes

$$\text{Annual cost} = E_{in} \times \text{Unit cost}$$
$$= (10,790 \text{ therm/yr})(\$0.75/\text{therm}) = \$8093/\text{yr}$$

That is, the heat losses from the exposed surfaces of the oven are currently costing the plant over \$8000/yr.

When insulation is installed, the rate of heat transfer from the oven can be determined from

$$\dot{Q}_{ins} = \frac{T_s - T_\infty}{R_{total}} = \frac{T_s - T_\infty}{R_{ins} + R_{conv}} = A_s \frac{T_s - T_\infty}{\dfrac{t_{ins}}{k_{ins}} + \dfrac{1}{h_o}}$$

We expect the surface temperature of the oven to increase and the heat transfer coefficient to decrease somewhat when insulation is installed. We assume these two effects to counteract each other. Then the relation above for 1-in-thick insulation gives the rate of heat loss to be

$$\dot{Q}_{ins} = \frac{A_s(T_s - T_\infty)}{\frac{t_{ins}}{k_{ins}} + \frac{1}{h_o}} = \frac{(402 \text{ ft}^2)(180 - 75)°\text{F}}{\frac{1/12 \text{ ft}}{0.024 \text{ Btu/h} \cdot \text{ft} \cdot °\text{F}} + \frac{1}{3.5 \text{ Btu/h} \cdot \text{ft}^2 \cdot °\text{F}}}$$

$$= 11,230 \text{ Btu/h}$$

Also, the total amount of heat loss from the oven per year and the amount and cost of energy consumption of the oven become

$$Q_{ins} = \dot{Q}_{ins} \Delta t = (11,230 \text{ Btu/h})(5840 \text{ h/yr}) = 0.6558 \times 10^8 \text{ Btu/yr}$$

$$E_{ins} = Q_{ins}/\eta_{oven} = (0.6558 \times 10^8 \text{ Btu/yr})/0.80 = 0.820 \times 10^8 \text{ Btu/yr}$$

$$= 820 \text{ therms}$$

$$\text{Annual cost} = E_{ins} \times \text{Unit cost}$$

$$= (820 \text{ therm/yr})(\$0.75/\text{therm}) = \$615/\text{yr}$$

Therefore, insulating the oven by 1-in-thick fiberglass insulation will reduce the fuel bill by $8093 - $615 = $7362 per year. The unit cost of insulation is given to be $2.70/ft². Then the installation cost of insulation becomes

$$\text{Insulation cost} = (\text{Unit cost})(\text{Surface area}) = (\$2.70/\text{ft}^2)(402 \text{ ft}^2) = \$1085$$

The sum of the insulation and heat loss costs is

$$\text{Total cost} = \text{Insulation cost} + \text{Heat loss cost} = \$1085 + \$615 = \$1700$$

Then the net earnings will be

$$\text{Earnings} = \text{Income} - \text{Expenses} = \$8093 - \$1700 = \$6393$$

To determine the thickness of insulation that maximizes your earnings, we repeat the calculations above for 2-, 3-, 4-, and 5-in-thick insulations, and list the results in Table 7–5. Note that the total cost of insulation decreases first with increasing insulation thickness, reaches a minimum, and then starts to increase.

We observe that the total insulation cost is a minimum at $1687 for the case of **2-in-thick** insulation. The earnings in this case are

$$\text{Maximum earnings} = \text{Income} - \text{Minimum expenses}$$

$$= \$8093 - \$1687 = \$6406$$

TABLE 7–5

The variation of total insulation cost with insulation thickness

Insulation thickness	Heat loss, Btu/h	Lost fuel, therms/yr	Lost fuel cost, $/yr	Insulation cost, $	Total cost, $
1 in	11,230	820	615	1085	1700
2 in	5838	426	320	1367	1687
3 in	3944	288	216	1648	1864
4 in	2978	217	163	1930	2093
5 in	2392	175	131	2211	2342

which is not bad for a day's worth of work. The plant manager is also a big winner in this venture since the heat losses will cost him only $320/yr during the second and consequent years instead of $8093/yr. A thicker insulation could probably be justified in this case if the cost of insulation is annualized over the lifetime of insulation, say 20 years. Several energy conservation measures are marketed as explained above by several power companies and private firms.

SUMMARY

The force a flowing fluid exerts on a body in the flow direction is called *drag*. The part of drag that is due directly to wall shear stress τ_w is called the *skin friction drag* since it is caused by frictional effects, and the part that is due directly to pressure is called the *pressure drag* or *form drag* because of its strong dependence on the form or shape of the body.

The *drag coefficient* C_D is a dimensionless number that represents the drag characteristics of a body, and is defined as

$$C_D = \frac{F_D}{\frac{1}{2}\rho V^2 A}$$

where A is the *frontal area* for blunt bodies, and surface area for parallel flow over flat plates or thin airfoils. For flow over a flat plate, the Reynolds number is

$$\mathrm{Re}_x = \frac{\rho V x}{\mu} = \frac{V x}{\nu}$$

Transition from laminar to turbulent occurs at the *critical Reynolds number* of

$$\mathrm{Re}_{x,\,cr} = \frac{\rho V x_{cr}}{\mu} = 5 \times 10^5$$

For parallel flow over a flat plate, the local friction and convection coefficients are

Laminar: $\quad C_{f,x} = \dfrac{0.664}{\mathrm{Re}_x^{1/2}}, \qquad \mathrm{Re}_x < 5 \times 10^5$

$$\mathrm{Nu}_x = \frac{h_x x}{k} = 0.332\,\mathrm{Re}_x^{0.5}\,\mathrm{Pr}^{1/3}, \qquad \mathrm{Pr} > 0.6$$

Turbulent: $\quad C_{f,x} = \dfrac{0.059}{\mathrm{Re}_x^{1/5}}, \qquad 5 \times 10^5 \leq \mathrm{Re}_x \leq 10^7$

$$\mathrm{Nu}_x = \frac{h_x x}{k} = 0.0296\,\mathrm{Re}_x^{0.8}\,\mathrm{Pr}^{1/3}, \qquad \begin{array}{l} 0.6 \leq \mathrm{Pr} \leq 60 \\ 5 \times 10^5 \leq \mathrm{Re}_x \leq 10^7 \end{array}$$

The *average* friction coefficient relations for flow over a flat plate are:

Laminar: $\quad C_f = \dfrac{1.33}{\mathrm{Re}_L^{1/2}}, \qquad \mathrm{Re}_L < 5 \times 10^5$

Turbulent: $\quad C_f = \dfrac{0.074}{\mathrm{Re}_L^{1/5}}, \qquad 5 \times 10^5 \leq \mathrm{Re}_L \leq 10^7$

Combined: $\quad C_f = \dfrac{0.074}{\mathrm{Re}_L^{1/5}} - \dfrac{1742}{\mathrm{Re}_L}, \qquad 5 \times 10^5 \leq \mathrm{Re}_L \leq 10^7$

Rough surface, turbulent: $\quad C_f = \left(1.89 - 1.62 \log \dfrac{\varepsilon}{L}\right)^{-2.5}$

The average Nusselt number relations for flow over a flat plate are:

Laminar: $\quad \mathrm{Nu} = \dfrac{hL}{k} = 0.664\,\mathrm{Re}_L^{0.5}\,\mathrm{Pr}^{1/3}, \qquad \mathrm{Re}_L < 5 \times 10^5$

Turbulent:

$$\mathrm{Nu} = \frac{hL}{k} = 0.037\,\mathrm{Re}_L^{0.8}\,\mathrm{Pr}^{1/3}, \qquad \begin{array}{l} 0.6 \leq \mathrm{Pr} \leq 60 \\ 5 \times 10^5 \leq \mathrm{Re}_L \leq 10^7 \end{array}$$

Combined:

$$\mathrm{Nu} = \frac{hL}{k} = (0.037\,\mathrm{Re}_L^{0.8} - 871)\,\mathrm{Pr}^{1/3}, \qquad \begin{array}{l} 0.6 \leq \mathrm{Pr} \leq 60 \\ 5 \times 10^5 \leq \mathrm{Re}_L \leq 10^7 \end{array}$$

For isothermal surfaces with an unheated starting section of length ξ, the local Nusselt number and the average convection coefficient relations are

Laminar: $\quad \mathrm{Nu}_x = \dfrac{\mathrm{Nu}_{x\,(\text{for }\xi=0)}}{[1 - (\xi/x)^{3/4}]^{1/3}} = \dfrac{0.332\,\mathrm{Re}_x^{0.5}\,\mathrm{Pr}^{1/3}}{[1 - (\xi/x)^{3/4}]^{1/3}}$

Turbulent: $\quad \mathrm{Nu}_x = \dfrac{\mathrm{Nu}_{x\,(\text{for }\xi=0)}}{[1 - (\xi/x)^{9/10}]^{1/9}} = \dfrac{0.0296\,\mathrm{Re}_x^{0.8}\,\mathrm{Pr}^{1/3}}{[1 - (\xi/x)^{9/10}]^{1/9}}$

Laminar: $\quad h = \dfrac{2[1 - (\xi/x)^{3/4}]}{1 - \xi/L}\,h_{x=L}$

Turbulent: $\quad h = \dfrac{5[1 - (\xi/x)^{9/10}]}{(1 - \xi/L)}\,h_{x=L}$

These relations are for the case of *isothermal* surfaces. When a flat plate is subjected to *uniform heat flux,* the local Nusselt number is given by

Laminar: $\quad\quad\quad$ $Nu_x = 0.453\, Re_x^{0.5}\, Pr^{1/3}$

Turbulent: $\quad\quad\quad$ $Nu_x = 0.0308\, Re_x^{0.8}\, Pr^{1/3}$

The average Nusselt numbers for cross flow over a *cylinder* and *sphere* are

$$Nu_{cyl} = \frac{hD}{k} = 0.3 + \frac{0.62\, Re^{1/2}\, Pr^{1/3}}{[1+(0.4/Pr)^{2/3}]^{1/4}}\left[1+\left(\frac{Re}{282,000}\right)^{5/8}\right]^{4/5}$$

which is valid for Re Pr > 0.2, and

$$Nu_{sph} = \frac{hD}{k} = 2 + [0.4\, Re^{1/2} + 0.06\, Re^{2/3}]Pr^{0.4}\left(\frac{\mu_\infty}{\mu_s}\right)^{1/4}$$

which is valid for $3.5 \le Re \le 80,000$ and $0.7 \le Pr \le 380$. The fluid properties are evaluated at the film temperature $T_f = (T_\infty + T_s)/2$ in the case of a cylinder, and at the free-stream temperature T_∞ (except for μ_s, which is evaluated at the surface temperature T_s) in the case of a sphere.

In tube banks, the Reynolds number is based on the maximum velocity V_{max} that is related to the approach velocity V as

In-line and *Staggered* with $S_D < (S_T + D)/2$:

$$V_{max} = \frac{S_T}{S_T - D}\, V$$

Staggered with $S_D < (S_T + D)/2$:

$$V_{max} = \frac{S_T}{2(S_D - D)}\, V$$

where S_T the transverse pitch and S_D is the diagonal pitch. The average Nusselt number for cross flow over tube banks is expressed as

$$Nu_D = \frac{hD}{k} = C\, Re_D^m\, Pr^n (Pr/Pr_s)^{0.25}$$

where the values of the constants C, m, and n depend on Reynolds number. Such correlations are given in Table 7–2. All properties except Pr_s are to be evaluated at the arithmetic mean of the inlet and exit temperatures of the fluid defined as $T_m = (T_i + T_e)/2$.

The average Nusselt number for tube banks with less than 16 rows is expressed as

$$Nu_{D, N_L} = F\, Nu_D$$

where F is the *correction factor* whose values are given in Table 7–3. The heat transfer rate to or from a tube bank is determined from

$$\dot{Q} = hA_s\Delta T_{ln} = \dot{m}c_p(T_e - T_i)$$

where ΔT_{ln} is the logarithmic mean temperature difference defined as

$$\Delta T_{ln} = \frac{(T_s - T_e) - (T_s - T_i)}{\ln[(T_s - T_e)/(T_s - T_i)]} = \frac{\Delta T_e - \Delta T_i}{\ln(\Delta T_e/\Delta T_i)}$$

and the exit temperature of the fluid T_e is

$$T_e = T_s - (T_s - T_i)\exp\left(-\frac{A_s h}{\dot{m}c_p}\right)$$

where $A_s = N\pi DL$ is the heat transfer surface area and $\dot{m} = \rho V(N_T S_T L)$ is the mass flow rate of the fluid. The pressure drop ΔP for a tube bank is expressed as

$$\Delta P = N_L f\chi \frac{\rho V_{max}^2}{2}$$

where f is the friction factor and χ is the correction factor, both given in Fig. 7–27.

REFERENCES AND SUGGESTED READING

1. R. D. Blevin. *Applied Fluid Dynamics Handbook.* New York: Van Nostrand Reinhold, 1984.

2. S. W. Churchill and M. Bernstein. "A Correlating Equation for Forced Convection from Gases and Liquids to a Circular Cylinder in Cross Flow." *Journal of Heat Transfer* 99 (1977), pp. 300–306.

3. S. W. Churchill and H. Ozoe. "Correlations for Laminar Forced Convection in Flow over an Isothermal Flat Plate and in Developing and Fully Developed Flow in an Isothermal Tube." *Journal of Heat Transfer* 95 (Feb. 1973), pp. 78–84.

4. W. M. Edmunds. "Residential Insulation." *ASTM Standardization News* (Jan. 1989), pp. 36–39.

5. W. H. Giedt. "Investigation of Variation of Point Unit-Heat Transfer Coefficient around a Cylinder Normal to an Air Stream." *Transactions of the ASME* 71 (1949), pp. 375–381.

6. "Green and Clean: The Economic, Energy, and Environmental Benefits of Insulation," *Alliance to Save Energy,* April 2001.

7. M. Jakob. *Heat Transfer.* Vol. 1. New York: John Wiley & Sons, 1949.

8. W. M. Kays and M. E. Crawford. *Convective Heat and Mass Transfer.* 3rd ed. New York: McGraw-Hill, 1993.

9. H. Schlichting. *Boundary Layer Theory,* 7th ed. New York, McGraw-Hill, 1979.

10. W. C. Thomas. "Note on the Heat Transfer Equation for Forced Convection Flow over a Flat Plate with an Unheated Starting Length." *Mechanical Engineering News,* 9, no.1 (1977), p. 361.

11. R. D. Willis. "Photographic Study of Fluid Flow Between Banks of Tubes." *Engineering* (1934), pp. 423–425.

12. A. Zukauskas, "Convection Heat Transfer in Cross Flow." In *Advances in Heat Transfer,* J. P. Hartnett and T. F. Irvine, Jr., Eds. New York: Academic Press, 1972, Vol. 8, pp. 93–106.

13. A. Zukauskas. "Heat Transfer from Tubes in Cross Flow." In *Advances in Heat Transfer,* ed. J. P. Hartnett and T. F. Irvine, Jr. Vol. 8. New York: Academic Press, 1972.

14. A. Zukauskas. "Heat Transfer from Tubes in Cross Flow." In *Handbook of Single Phase Convective Heat Transfer,* Eds. S. Kakac, R. K. Shah, and Win Aung. New York: Wiley Interscience, 1987.

15. A. Zukauskas and R. Ulinskas, "Efficiency Parameters for Heat Transfer in Tube Banks." *Heat Transfer Engineering* no. 2 (1985), pp. 19–25.

PROBLEMS*

Drag Force and Heat Transfer in External Flow

7–1C   What is the difference between the upstream velocity and the free-stream velocity? For what types of flow are these two velocities equal to each other?

7–2C   What is the difference between streamlined and blunt bodies? Is a tennis ball a streamlined or blunt body?

7–3C   What is drag? What causes it? Why do we usually try to minimize it?

7–4C   What is lift? What causes it? Does wall shear contribute to the lift?

7–5C   During flow over a given body, the drag force, the upstream velocity, and the fluid density are measured. Explain how you would determine the drag coefficient. What area would you use in calculations?

7–6C   Define frontal area of a body subjected to external flow. When is it appropriate to use the frontal area in drag and lift calculations?

7–7C   What is the difference between skin friction drag and pressure drag? Which is usually more significant for slender bodies such as airfoils?

7–8C   What is the effect of surface roughness on the friction drag coefficient in laminar and turbulent flows?

7–9C   What is the effect of streamlining on (*a*) friction drag and (*b*) pressure drag? Does the total drag acting on a body necessarily decrease as a result of streamlining? Explain.

7–10C   What is flow separation? What causes it? What is the effect of flow separation on the drag coefficient?

Flow over Flat Plates

7–11C   What does the friction coefficient represent in flow over a flat plate? How is it related to the drag force acting on the plate?

7–12C   Consider laminar flow over a flat plate. Will the friction coefficient change with distance from the leading edge? How about the heat transfer coefficient?

7–13C   How are the average friction and heat transfer coefficients determined in flow over a flat plate?

7–14   Engine oil at 80°C flows over a 10-m-long flat plate whose temperature is 30°C with a velocity of 2.5 m/s. Determine the total drag force and the rate of heat transfer over the entire plate per unit width.

7–15   The local atmospheric pressure in Denver, Colorado (elevation 1610 m), is 83.4 kPa. Air at this pressure and at 30°C flows with a velocity of 6 m/s over a 2.5-m × 8-m flat plate whose temperature is 120°C. Determine the rate of heat transfer from the plate if the air flows parallel to the (*a*) 8-m-long side and (*b*) the 2.5-m side.

7–16   During a cold winter day, wind at 55 km/h is blowing parallel to a 4-m-high and 10-m-long wall of a house. If the air outside is at 5°C and the surface temperature of the wall is

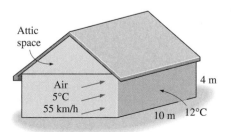

FIGURE P7–16

12°C, determine the rate of heat loss from that wall by convection. What would your answer be if the wind velocity was doubled? *Answers: 9080 W, 16,200 W*

7–17 Reconsider Prob. 7–16. Using EES (or other) software, investigate the effects of wind velocity and outside air temperature on the rate of heat loss from the wall by convection. Let the wind velocity vary from 10 km/h to 80 km/h and the outside air temperature from 0°C to 10°C. Plot the rate of heat loss as a function of the wind velocity and of the outside temperature, and discuss the results.

7–18E Air at 60°F flows over a 10-ft-long flat plate at 7 ft/s. Determine the local friction and heat transfer coefficients at intervals of 1 ft, and plot the results against the distance from the leading edge.

7–19E Reconsider Prob. 7–18E. Using EES (or other) software, evaluate the local friction and heat transfer coefficients along the plate at intervals of 0.1 ft, and plot them against the distance from the leading edge.

7–20 A thin, square flat plate has 0.5 m on each side. Air at 10°C flows over the top and bottom surfaces of the plate in a direction parallel to one edge, with a velocity of 60 m/s. The surface of the plate is maintained at a constant temperature of 54°C. The plate is mounted on a scale that measures a drag force of 1.5 N.

(a) Determine the flow regime (laminar or turbulent).
(b) Determine the total heat transfer rate from the plate to the air.
(c) Assuming a uniform distribution of heat transfer and drag parameters over the plate, estimate the average gradients of the velocity and temperature at the surface, $(\partial u/\partial y)_{y=0}$ and $(\partial T/\partial y)_{y=0}$.

7–21 Water at 43.3°C flows over a large plate at a velocity of 30.0 cm/s. The plate is 1.0 m long (in the flow direction), and its surface is maintained at a uniform temperature of 10.0°C. Calculate the steady rate of heat transfer per unit width of the plate.

7–22 Mercury at 25°C flows over a 3-m-long and 2-m-wide flat plate maintained at 75°C with a velocity of 0.8 m/s. Determine the rate of heat transfer from the entire plate.

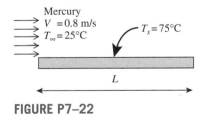

Mercury
$V = 0.8$ m/s
$T_\infty = 25°C$
$T_s = 75°C$
L

FIGURE P7–22

7–23 Parallel plates form a solar collector that covers a roof, as shown in the figure. The plates are maintained at 15°C, while ambient air at 10°C flows over the roof with $V = 2$ m/s.

Determine the rate of convective heat loss from (a) the first plate and (b) the third plate.

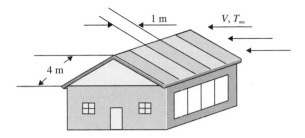

1 m V, T_∞

4 m

FIGURE P7–23

7–24 Consider a hot automotive engine, which can be approximated as a 0.5-m-high, 0.40-m-wide, and 0.8-m-long rectangular block. The bottom surface of the block is at a temperature of 100°C and has an emissivity of 0.95. The ambient air is at 20°C, and the road surface is at 25°C. Determine the rate of heat transfer from the bottom surface of the engine block by convection and radiation as the car travels at a velocity of 80 km/h. Assume the flow to be turbulent over the entire surface because of the constant agitation of the engine block.

7–25 The forming section of a plastics plant puts out a continuous sheet of plastic that is 1.2 m wide and 2 mm thick at a rate of 15 m/min. The temperature of the plastic sheet is 90°C when it is exposed to the surrounding air, and the sheet is subjected to air flow at 30°C at a velocity of 3 m/s on both sides along its surfaces normal to the direction of motion of the sheet. The width of the air cooling section is such that a fixed point on the plastic sheet passes through that section in 2 s. Determine the rate of heat transfer from the plastic sheet to the air.

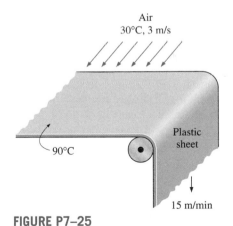

Air
30°C, 3 m/s

Plastic sheet

90°C

15 m/min

FIGURE P7–25

7–26 The top surface of the passenger car of a train moving at a velocity of 70 km/h is 2.8 m wide and 8 m long. The top surface is absorbing solar radiation at a rate of 200 W/m², and the temperature of the ambient air is 30°C. Assuming the roof of the car to be perfectly insulated and the radiation heat exchange with the surroundings to be small relative to convection, determine the equilibrium temperature of the top surface of the car.

Answer: 35.1°C

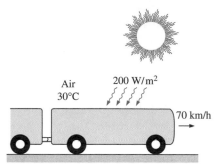

FIGURE P7–26

7–27 [EES] Reconsider Prob. 7–26. Using EES (or other) software, investigate the effects of the train velocity and the rate of absorption of solar radiation on the equilibrium temperature of the top surface of the car. Let the train velocity vary from 10 km/h to 120 km/h and the rate of solar absorption from 100 W/m² to 500 W/m². Plot the equilibrium temperature as functions of train velocity and solar radiation absorption rate, and discuss the results.

7–28 A 15-cm × 15-cm circuit board dissipating 20 W of power uniformly is cooled by air, which approaches the circuit board at 20°C with a velocity of 6 m/s. Disregarding any heat transfer from the back surface of the board, determine the surface temperature of the electronic components (*a*) at the leading edge and (*b*) at the end of the board. Assume the flow to be turbulent since the electronic components are expected to act as turbulators.

7–29 Consider laminar flow of a fluid over a flat plate maintained at a constant temperature. Now the free-stream velocity of the fluid is doubled. Determine the change in the drag force on the plate and rate of heat transfer between the fluid and the plate. Assume the flow to remain laminar.

7–30E Consider a refrigeration truck traveling at 55 mph at a location where the air temperature is 80°F. The refrigerated compartment of the truck can be considered to be a 9-ft-wide, 8-ft-high, and 20-ft-long rectangular box. The refrigeration

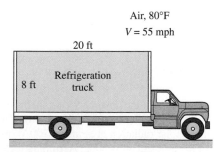

FIGURE P7–30E

system of the truck can provide 3 tons of refrigeration (i.e., it can remove heat at a rate of 600 Btu/min). The outer surface of the truck is coated with a low-emissivity material, and thus radiation heat transfer is very small. Determine the average temperature of the outer surface of the refrigeration compartment of the truck if the refrigeration system is observed to be operating at half the capacity. Assume the air flow over the entire outer surface to be turbulent and the heat transfer coefficient at the front and rear surfaces to be equal to that on side surfaces.

7–31 Solar radiation is incident on the glass cover of a solar collector at a rate of 700 W/m². The glass transmits 88 percent of the incident radiation and has an emissivity of 0.90. The entire hot water needs of a family in summer can be met by two collectors 1.2 m high and 1 m wide. The two collectors are attached to each other on one side so that they appear like a single collector 1.2 m × 2 m in size. The temperature of the glass cover is measured to be 35°C on a day when the surrounding air temperature is 25°C and the wind is blowing at 30 km/h. The effective sky temperature for radiation exchange between the glass cover and the open sky is −40°C. Water enters the tubes attached to the absorber plate at a rate of 1 kg/min. Assuming the back surface of the absorber plate to be heavily insulated and the only heat loss to occur through the glass cover, determine (*a*) the total rate of heat loss from the collector, (*b*) the collector efficiency, which is the ratio of the amount of heat transferred to the water to the solar energy incident on the collector, and (*c*) the temperature rise of water as it flows through the collector.

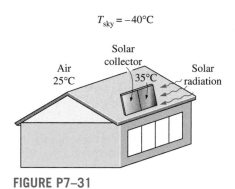

FIGURE P7–31

7–32 A transformer that is 10 cm long, 6.2 cm wide, and 5 cm high is to be cooled by attaching a 10-cm × 6.2-cm wide polished aluminum heat sink (emissivity = 0.03) to its top surface. The heat sink has seven fins, which are 5 mm high, 2 mm thick, and 10 cm long. A fan blows air at 25°C parallel to the passages between the fins. The heat sink is to dissipate 12 W of heat and the base temperature of the heat sink is not to exceed 60°C. Assuming the fins and the base plate to be nearly isothermal and the radiation heat transfer to be negligible, determine the minimum free-stream velocity the fan needs to supply to avoid overheating.

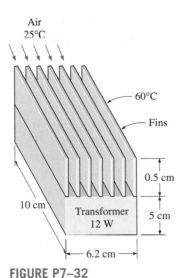

Air
25°C

60°C

Fins

0.5 cm

10 cm

Transformer
12 W

5 cm

6.2 cm

FIGURE P7–32

7–33 Repeat Prob. 7–32 assuming the heat sink to be black-anodized and thus to have an effective emissivity of 0.90. Note that in radiation calculations the base area (10 cm × 6.2 cm) is to be used, not the total surface area.

7–34 An array of power transistors, dissipating 6 W of power each, are to be cooled by mounting them on a 25-cm × 25-cm square aluminum plate and blowing air at 35°C over the plate with a fan at a velocity of 4 m/s. The average temperature of the plate is not to exceed 65°C. Assuming the heat transfer from the back side of the plate to be negligible and disregarding radiation, determine the number of transistors that can be placed on this plate.

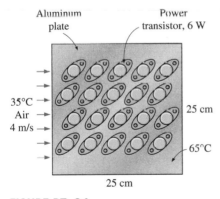

Aluminum
plate

Power
transistor, 6 W

35°C
Air
4 m/s

25 cm

65°C

25 cm

FIGURE P7–34

7–35 Repeat Prob. 7–34 for a location at an elevation of 1610 m where the atmospheric pressure is 83.4 kPa.

Answer: 4

7–36 Air at 25°C and 1 atm is flowing over a long flat plate with a velocity of 8 m/s. Determine the distance from the leading edge of the plate where the flow becomes turbulent, and the thickness of the boundary layer at that location.

7–37 Repeat Prob. 7–36 for water.

7–38 The weight of a thin flat plate 40 cm × 40 cm in size is balanced by a counterweight that has a mass of 2 kg, as shown in the figure. Now a fan is turned on, and air at 1 atm and 25°C flows downward over both surfaces of the plate with a free-stream velocity of 10 m/s. Determine the mass of the counterweight that needs to be added in order to balance the plate in this case.

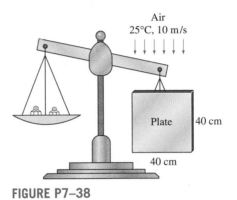

Air
25°C, 10 m/s

Plate
40 cm

40 cm

FIGURE P7–38

Flow across Cylinders and Spheres

7–39C Consider laminar flow of air across a hot circular cylinder. At what point on the cylinder will the heat transfer be highest? What would your answer be if the flow were turbulent?

7–40C In flow over cylinders, why does the drag coefficient suddenly drop when the flow becomes turbulent? Isn't turbulence supposed to increase the drag coefficient instead of decreasing it?

7–41C In flow over blunt bodies such as a cylinder, how does the pressure drag differ from the friction drag?

7–42C Why is flow separation in flow over cylinders delayed in turbulent flow?

7–43 A long 8-cm-diameter steam pipe whose external surface temperature is 90°C passes through some open area that is not protected against the winds. Determine the rate of heat loss from the pipe per unit of its length when the air is at 1 atm pressure and 7°C and the wind is blowing across the pipe at a velocity of 50 km/h.

7–44 In a geothermal power plant, the used geothermal water at 80°C enters a 15-cm-diameter and 400-m-long uninsulated pipe at a rate of 8.5 kg/s and leaves at 70°C before being reinjected back to the ground. Windy air at 15°C flows normal to

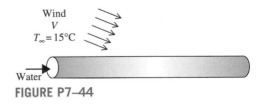

Wind
V
$T_\infty = 15°C$

Water

FIGURE P7–44

the pipe. Disregarding radiation, determine the average wind velocity in km/h.

7–45 A stainless steel ball ($\rho = 8055$ kg/m^3, $c_p = 480$ J/kg · °C) of diameter $D = 15$ cm is removed from the oven at a uniform temperature of 350°C. The ball is then subjected to the flow of air at 1 atm pressure and 30°C with a velocity of 6 m/s. The surface temperature of the ball eventually drops to 250°C. Determine the average convection heat transfer coefficient during this cooling process and estimate how long this process has taken.

7–46 [EES] Reconsider Prob. 7–45. Using EES (or other) software, investigate the effect of air velocity on the average convection heat transfer coefficient and the cooling time. Let the air velocity vary from 1 m/s to 10 m/s. Plot the heat transfer coefficient and the cooling time as a function of air velocity, and discuss the results.

7–47E A person extends his uncovered arms into the windy air outside at 54°F and 20 mph in order to feel nature closely. Initially, the skin temperature of the arm is 86°F. Treating the arm as a 2-ft-long and 3-in-diameter cylinder, determine the rate of heat loss from the arm.

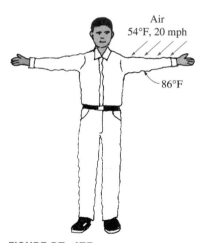

Air
54°F, 20 mph

86°F

FIGURE P7–47E

7–48E [EES] Reconsider Prob. 7–47E. Using EES (or other) software, investigate the effects of air temperature and wind velocity on the rate of heat loss from the arm. Let the air temperature vary from 20°F to 80°F and the wind velocity from 10 mph to 40 mph. Plot the rate of heat loss as a function of air temperature and of wind velocity, and discuss the results.

7–49 An average person generates heat at a rate of 84 W while resting. Assuming one-quarter of this heat is lost from the head and disregarding radiation, determine the average surface temperature of the head when it is not covered and is subjected to winds at 10°C and 25 km/h. The head can be approximated as a 30-cm-diameter sphere. *Answer:* 13.2°C

7–50 Consider the flow of a fluid across a cylinder maintained at a constant temperature. Now the free-stream velocity of the fluid is doubled. Determine the change in the drag force on the cylinder and the rate of heat transfer between the fluid and the cylinder.

7–51 A 6-mm-diameter electrical transmission line carries an electric current of 50 A and has a resistance of 0.002 ohm per meter length. Determine the surface temperature of the wire during a windy day when the air temperature is 10°C and the wind is blowing across the transmission line at 40 km/h.

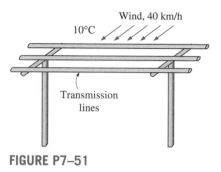

Wind, 40 km/h

10°C

Transmission
lines

FIGURE P7–51

7–52 [EES] Reconsider Prob. 7–51. Using EES (or other) software, investigate the effect of the wind velocity on the surface temperature of the wire. Let the wind velocity vary from 10 km/h to 80 km/h. Plot the surface temperature as a function of wind velocity, and discuss the results.

7–53 A heating system is to be designed to keep the wings of an aircraft cruising at a velocity of 900 km/h above freezing temperatures during flight at 12,200-m altitude where the standard atmospheric conditions are −55.4°C and 18.8 kPa. Approximating the wing as a cylinder of elliptical cross section whose minor axis is 50 cm and disregarding radiation, determine the average convection heat transfer coefficient on the wing surface and the average rate of heat transfer per unit surface area.

7–54 A long aluminum wire of diameter 3 mm is extruded at a temperature of 370°C. The wire is subjected to cross air flow at 30°C at a velocity of 6 m/s. Determine the rate of heat transfer from the wire to the air per meter length when it is first exposed to the air.

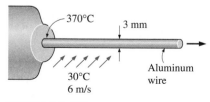

370°C
3 mm

30°C
6 m/s

Aluminum
wire

FIGURE P7–54

7–55E Consider a person who is trying to keep cool on a hot summer day by turning a fan on and exposing his entire body to air flow. The air temperature is 85°F and the fan is blowing air at a velocity of 6 ft/s. If the person is doing light work and generating sensible heat at a rate of 300 Btu/h, determine the average temperature of the outer surface (skin or clothing) of the person. The average human body can be treated as a 1-ft-diameter cylinder with an exposed surface area of 18 ft². Disregard any heat transfer by radiation. What would your answer be if the air velocity were doubled?   *Answers:* 95.1°F, 91.6°F

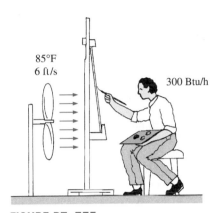

85°F
6 ft/s

300 Btu/h

FIGURE P7–55E

7–56 An incandescent lightbulb is an inexpensive but highly inefficient device that converts electrical energy into light. It converts about 10 percent of the electrical energy it consumes into light while converting the remaining 90 percent into heat. (A fluorescent lightbulb will give the same amount of light while consuming only one-fourth of the electrical energy, and it will last 10 times longer than an incandescent lightbulb.) The glass bulb of the lamp heats up very quickly as a result of absorbing all that heat and dissipating it to the surroundings by convection and radiation.

Consider a 10-cm-diameter 100-W lightbulb cooled by a fan that blows air at 30°C to the bulb at a velocity of 2 m/s. The surrounding surfaces are also at 30°C, and the emissivity of the glass is 0.9. Assuming 10 percent of the energy passes through the glass bulb as light with negligible absorption and the rest of

the energy is absorbed and dissipated by the bulb itself, determine the equilibrium temperature of the glass bulb.

7–57 During a plant visit, it was noticed that a 12-m-long section of a 10-cm-diameter steam pipe is completely exposed to the ambient air. The temperature measurements indicate that the average temperature of the outer surface of the steam pipe is 75°C when the ambient temperature is 5°C. There are also light winds in the area at 10 km/h. The emissivity of the outer surface of the pipe is 0.8, and the average temperature of the surfaces surrounding the pipe, including the sky, is estimated to be 0°C. Determine the amount of heat lost from the steam during a 10-h-long work day.

Steam is supplied by a gas-fired steam generator that has an efficiency of 80 percent, and the plant pays $1.05/therm of natural gas. If the pipe is insulated and 90 percent of the heat loss is saved, determine the amount of money this facility will save a year as a result of insulating the steam pipes. Assume the plant operates every day of the year for 10 h. State your assumptions.

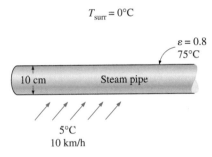

$T_{surr} = 0°C$

$\varepsilon = 0.8$
75°C

10 cm   Steam pipe

5°C
10 km/h

FIGURE P7 57

7–58 Reconsider Prob. 7 57. There seems to be some uncertainty about the average temperature of the surfaces surrounding the pipe used in radiation calculations, and you are asked to determine if it makes any significant difference in overall heat transfer. Repeat the calculations for average surrounding and surface temperatures of −20°C and 25°C, respectively, and determine the change in the values obtained.

7–59E A 12-ft-long, 1.5-kW electrical resistance wire is made of 0.1-in-diameter stainless steel (k = 8.7 Btu/h · ft · °F). The resistance wire operates in an environment at 85°F. Determine the surface temperature of the wire if it is cooled by a fan blowing air at a velocity of 20 ft/s.

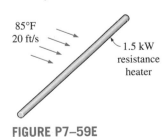

85°F
20 ft/s

1.5 kW
resistance
heater

FIGURE P7–59E

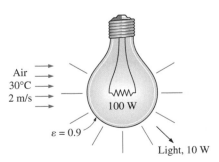

Air
30°C
2 m/s

100 W

$\varepsilon = 0.9$

Light, 10 W

FIGURE P7–56

7–60 The components of an electronic system are located in a 1.5-m-long horizontal duct whose cross section is 20 cm × 20 cm. The components in the duct are not allowed to come into direct contact with cooling air, and thus are cooled by air at 30°C flowing over the duct with a velocity of 200 m/min. If the surface temperature of the duct is not to exceed 65°C, determine the total power rating of the electronic devices that can be mounted into the duct. *Answer:* 640 W

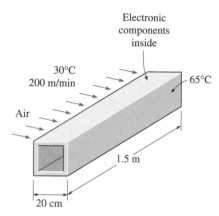

Electronic
components
inside

30°C
200 m/min

65°C

Air

1.5 m

20 cm

FIGURE P7–60

7–61 Repeat Prob. 7–60 for a location at 4000-m altitude where the atmospheric pressure is 61.66 kPa.

7–62 A 0.4-W cylindrical electronic component with diameter 0.3 cm and length 1.8 cm and mounted on a circuit board is cooled by air flowing across it at a velocity of 240 m/min. If the air temperature is 35°C, determine the surface temperature of the component.

7–63 Consider a 50-cm-diameter and 95-cm-long hot water tank. The tank is placed on the roof of a house. The water inside the tank is heated to 80°C by a flat-plate solar collector during the day. The tank is then exposed to windy air at 18°C with an average velocity of 40 km/h during the night. Estimate the temperature of the tank after a 45-min period. Assume the tank surface to be at the same temperature as the water inside, and the heat transfer coefficient on the top and bottom surfaces to be the same as that on the side surface.

7–64 [EES] Reconsider Prob. 7–63. Using EES (or other) software, plot the temperature of the tank as a function of the cooling time as the time varies from 30 min to 5 h, and discuss the results.

7–65 A 1.8-m-diameter spherical tank of negligible thickness contains iced water at 0°C. Air at 25°C flows over the tank with a velocity of 7 m/s. Determine the rate of heat transfer to the tank and the rate at which ice melts. The heat of fusion of water at 0°C is 333.7 kJ/kg.

7–66 A 10-cm-diameter, 30-cm-high cylindrical bottle contains cold water at 3°C. The bottle is placed in windy air at

27°C. The water temperature is measured to be 11°C after 45 min of cooling. Disregarding radiation effects and heat transfer from the top and bottom surfaces, estimate the average wind velocity.

Flow across Tube Banks

7–67C In flow across tube banks, why is the Reynolds number based on the maximum velocity instead of the uniform approach velocity?

7–68C In flow across tube banks, how does the heat transfer coefficient vary with the row number in the flow direction? How does it vary with in the transverse direction for a given row number?

7–69 Combustion air in a manufacturing facility is to be preheated before entering a furnace by hot water at 90°C flowing through the tubes of a tube bank located in a duct. Air enters the duct at 15°C and 1 atm with a mean velocity of 3.8 m/s, and flows over the tubes in normal direction. The outer diameter of the tubes is 2.1 cm, and the tubes are arranged in-line with longitudinal and transverse pitches of $S_L = S_T = 5$ cm. There are eight rows in the flow direction with eight tubes in each row. Determine the rate of heat transfer per unit length of the tubes, and the pressure drop across the tube bank.

7–70 Repeat Prob. 7–69 for staggered arrangement with $S_L = S_T = 6$ cm.

7–71 Air is to be heated by passing it over a bank of 3-m-long tubes inside which steam is condensing at 100°C. Air approaches the tube bank in the normal direction at 20°C and 1 atm with a mean velocity of 5.2 m/s. The outer diameter of the tubes is 1.6 cm, and the tubes are arranged staggered with longitudinal and transverse pitches of $S_L = S_T = 4$ cm. There are 20 rows in the flow direction with 10 tubes in each row. Determine (a) the rate of heat transfer, (b) and pressure drop across the tube bank, and (c) the rate of condensation of steam inside the tubes.

7–72 Repeat Prob. 7–71 for in-line arrangement with $S_L = S_T = 6$ cm.

7–73 Exhaust gases at 1 atm and 300°C are used to preheat water in an industrial facility by passing them over a bank of tubes through which water is flowing at a rate of 6 kg/s. The mean tube wall temperature is 80°C. Exhaust gases approach the tube bank in normal direction at 4.5 m/s. The outer diameter of the tubes is 2.1 cm, and the tubes are arranged in-line with longitudinal and transverse pitches of $S_L = S_T = 8$ cm. There are 16 rows in the flow direction with eight tubes in each row. Using the properties of air for exhaust gases, determine (a) the rate of heat transfer per unit length of tubes, (b) and pressure drop across the tube bank, and (c) the temperature rise of water flowing through the tubes per unit length of tubes.

7–74 Water at 15°C is to be heated to 65°C by passing it over a bundle of 4-m-long 1-cm-diameter resistance heater rods maintained at 90°C. Water approaches the heater rod bundle in normal direction at a mean velocity of 0.8 m/s. The rods are arranged in-line with longitudinal and transverse pitches of $S_L = 4$ cm and $S_T = 3$ cm. Determine the number of tube rows N_L in the flow direction needed to achieve the indicated temperature rise.

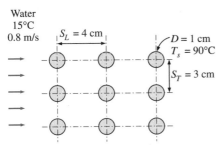

FIGURE P7–74

7–75 Air is to be cooled in the evaporator section of a refrigerator by passing it over a bank of 0.8-cm-outer-diameter and 0.4-m-long tubes inside which the refrigerant is evaporating at $-20°C$. Air approaches the tube bank in the normal direction at $0°C$ and 1 atm with a mean velocity of 4 m/s. The tubes are arranged in-line with longitudinal and transverse pitches of $S_L = S_T = 1.5$ cm. There are 30 rows in the flow direction with 15 tubes in each row. Determine (a) the refrigeration capacity of this system and (b) pressure drop across the tube bank.

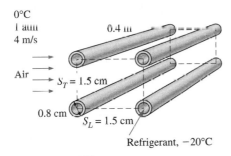

FIGURE P7–75

7–76 Repeat Prob. 7–75 by solving it for staggered arrangement with $S_L = S_T = 1.5$ cm, and compare the performance of the evaporator for the in-line and staggered arrangements.

7–77 A tube bank consists of 300 tubes at a distance of 6 cm between the centerlines of any two adjacent tubes. Air approaches the tube bank in the normal direction at $20°C$ and 1 atm with a mean velocity of 6 m/s. There are 20 rows in the flow direction with 15 tubes in each row with an average surface temperature of $140°C$. For an outer tube diameter of 2 cm, determine the average heat transfer coefficient.

Special Topic: Thermal Insulation

7–78C What is thermal insulation? How does a thermal insulator differ in purpose from an electrical insulator and from a sound insulator?

7–79C Does insulating cold surfaces save energy? Explain.

7–80C What is the R-value of insulation? How is it determined? Will doubling the thickness of flat insulation double its R-value?

7–81C How does the R-value of an insulation differ from its thermal resistance?

7–82C Why is the thermal conductivity of superinsulation orders of magnitude lower than the thermal conductivities of ordinary insulations?

7–83C Someone suggests that one function of hair is to insulate the head. Do you agree with this suggestion?

7–84C Name five different reasons for using insulation in industrial facilities.

7–85C What is optimum thickness of insulation? How is it determined?

7–86 What is the thickness of flat R-8 (in SI units) insulation whose thermal conductivity is 0.04 W/m · °C?

7–87E What is the thickness of flat R-20 (in English units) insulation whose thermal conductivity is 0.04 Btu/h · ft · °F?

7–88 Hot water at $110°C$ flows in a cast iron pipe ($k = 52$ W/m · °C) whose inner radius is 2.0 cm and thickness is 0.3 cm. The pipe is to be covered with adequate insulation so that the temperature of the outer surface of the insulation does not exceed $30°C$ when the ambient temperature is $22°C$. Taking the heat transfer coefficients inside and outside the pipe to be $h_i = 80$ W/m² · °C and $h_o = 22$ W/m² · °C, respectively, determine the thickness of fiberglass insulation ($k = 0.038$ W/m · °C) that needs to be installed on the pipe. *Answer:* 1.32 cm

7–89 Reconsider Prob. 7–88. Using EES (or other) software, plot the thickness of the insulation as a function of the maximum temperature of the outer surface of insulation in the range of $24°C$ to $48°C$. Discuss the results.

7–90 Consider a furnace whose average outer surface temperature is measured to be $90°C$ when the average surrounding air temperature is $27°C$. The furnace is 6 m long and 3 m in diameter. The plant operates 80 h per week for 52 weeks per year. You are to insulate the furnace using fiberglass insulation ($k_{ins} = 0.038$ W/m · °C) whose cost is $10/m² per cm of thickness for materials, plus $30/m² for labor regardless of thickness. The combined heat transfer coefficient on the outer surface is estimated to be $h_o = 30$ W/m² · °C. The furnace uses natural gas whose unit cost is $0.50/therm input (1 therm = 105,500 kJ), and the efficiency of the furnace is 78 percent. The management is willing to authorize the installation of the thickest insulation (in whole cm) that will pay for itself (materials and labor) in one year. That is, the total cost of insulation should be roughly equal to the drop in the fuel cost of the furnace for one year. Determine the thickness of insulation to be used and the money saved per year. Assume the surface temperature of the furnace and the heat transfer coefficient are to remain constant. *Answers:* 14 cm, $12,050/yr

7–91 Repeat Prob. 7–90 for an outer surface temperature of 75°C for the furnace.

7–92E Steam at 300°F is flowing through a steel pipe ($k = 8.7$ Btu/h · ft · °F) whose inner and outer diameters are 3.5 in and 4.0 in, respectively, in an environment at 85°F. The pipe is insulated with 1-in-thick fiberglass insulation ($k = 0.020$ Btu/h · ft · °F), and the heat transfer coefficients on the inside and the outside of the pipe are 30 Btu/h · ft² · °F and 5 Btu/h · ft² · °F, respectively. It is proposed to add another 1-in-thick layer of fiberglass insulation on top of the existing one to reduce the heat losses further and to save energy and money. The total cost of new insulation is $7 per ft length of the pipe, and the net fuel cost of energy in the steam is $0.01 per 1000 Btu (therefore, each 1000 Btu reduction in the heat loss will save the plant $0.01). The policy of the plant is to implement energy conservation measures that pay for themselves within two years. Assuming continuous operation (8760 h/yr), determine if the proposed additional insulation is justified.

7–93 The plumbing system of a plant involves a section of a plastic pipe ($k = 0.16$ W/m · °C) of inner diameter 6 cm and outer diameter 6.6 cm exposed to the ambient air. You are to insulate the pipe with adequate weather-jacketed fiberglass insulation ($k = 0.035$ W/m · °C) to prevent freezing of water in the pipe. The plant is closed for the weekends for a period of 60 h, and the water in the pipe remains still during that period. The ambient temperature in the area gets as low as $-10°C$ in winter, and the high winds can cause heat transfer coefficients as high as 30 W/m² · °C. Also, the water temperature in the pipe can be as cold as 15°C, and water starts freezing when its temperature drops to 0°C. Disregarding the convection resistance inside the pipe, determine the thickness of insulation that will protect the water from freezing under worst conditions.

7–94 Repeat Prob. 7–93 assuming 20 percent of the water in the pipe is allowed to freeze without jeopardizing safety. *Answer:* 27.9 cm

Review Problems

7–95 Consider a house that is maintained at 22°C at all times. The walls of the house have R-3.38 insulation in SI units (i.e., an *L/k* value or a thermal resistance of 3.38 m² · °C/W). During a cold winter night, the outside air temperature is 6°C and wind at 50 km/h is blowing parallel to a 4-m-high and 8-m-long wall of the house. If the heat transfer coefficient on the interior surface of the wall is 8 W/m² · °C, determine the rate of heat loss from that wall of the house. Draw the thermal resistance network and disregard radiation heat transfer. *Answer:* 145 W

7–96 An automotive engine can be approximated as a 0.4-m-high, 0.60-m-wide, and 0.7-m-long rectangular block. The bottom surface of the block is at a temperature of 75°C and has an emissivity of 0.92. The ambient air is at 5°C, and the road surface is at 10°C. Determine the rate of heat transfer from the bottom surface of the engine block by convection and radiation as the car travels at a velocity of 60 km/h. Assume the flow to be turbulent over the entire surface because of the constant agitation of the engine block. How will the heat transfer be affected when a 2-mm-thick gunk ($k = 3$ W/m · °C) has formed at the bottom surface as a result of the dirt and oil collected at that surface over time? Assume the metal temperature under the gunk still to be 75°C.

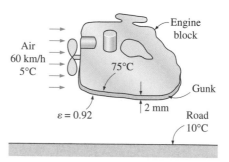

FIGURE P7–96

7–97E The passenger compartment of a minivan traveling at 60 mph can be modeled as a 3.2-ft-high, 6-ft-wide, and 11-ft-long rectangular box whose walls have an insulating value of R-3 (i.e., a wall thickness-to-thermal conductivity ratio of 3 h · ft² · °F/Btu). The interior of a minivan is maintained at an average temperature of 70°F during a trip at night while the outside air temperature is 90°F.

The average heat transfer coefficient on the interior surfaces of the van is 1.2 Btu/h · ft² · °F. The air flow over the exterior surfaces can be assumed to be turbulent because of the intense vibrations involved, and the heat transfer coefficient on the front and back surfaces can be taken to be equal to that on the top surface. Disregarding any heat gain or loss by radiation, determine the rate of heat transfer from the ambient air to the van.

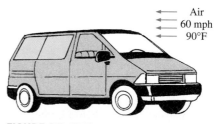

FIGURE P7–97E

7–98 Consider a house that is maintained at a constant temperature of 22°C. One of the walls of the house has three single-pane glass windows that are 1.5 m high and 1.8 m long.

The glass ($k = 0.78$ W/m · °C) is 0.5 cm thick, and the heat transfer coefficient on the inner surface of the glass is 8 W/m² · C. Now winds at 35 km/h start to blow parallel to the surface of this wall. If the air temperature outside is -2°C, determine the rate of heat loss through the windows of this wall. Assume radiation heat transfer to be negligible.

7–99 Consider a person who is trying to keep cool on a hot summer day by turning a fan on and exposing his body to air flow. The air temperature is 32°C, and the fan is blowing air at a velocity of 5 m/s. The surrounding surfaces are at 40°C, and the emissivity of the person can be taken to be 0.9. If the person is doing light work and generating sensible heat at a rate of 90 W, determine the average temperature of the outer surface (skin or clothing) of the person. The average human body can be treated as a 30-cm-diameter cylinder with an exposed surface area of 1.7 m². *Answer: 36.2°C*

7–100 Four power transistors, each dissipating 12 W, are mounted on a thin vertical aluminum plate ($k = 237$ W/m · °C) 22 cm × 22 cm in size. The heat generated by the transistors is to be dissipated by both surfaces of the plate to the surrounding air at 20°C, which is blown over the plate by a fan at a velocity of 250 m/min. The entire plate can be assumed to be nearly isothermal, and the exposed surface area of the transistor can be taken to be equal to its base area. Determine the temperature of the aluminum plate.

7–101 A 3-m-internal-diameter spherical tank made of 1-cm-thick stainless steel ($k = 15$ W/m · °C) is used to store iced water at 0°C. The tank is located outdoors at 30°C and is subjected to winds at 25 km/h. Assuming the entire steel tank to be at 0°C and thus its thermal resistance to be negligible, determine (a) the rate of heat transfer to the iced water in the tank and (b) the amount of ice at 0°C that melts during a 24-h period. The heat of fusion of water at atmospheric pressure is $h_{if} = 333.7$ kJ/kg. Disregard any heat transfer by radiation.

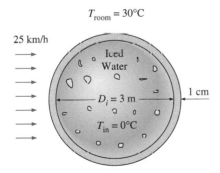

$T_{room} = 30$°C

25 km/h

Iced Water

$D_i = 3$ m

$T_{in} = 0$°C

1 cm

FIGURE P7–101

7–102 Repeat Prob. 7–101, assuming the inner surface of the tank to be at 0°C but by taking the thermal resistance of the tank and heat transfer by radiation into consideration. Assume the average surrounding surface temperature for radiation exchange to be 25°C and the outer surface of the tank to have an emissivity of 0.75. *Answers: (a) 10,530 W, (b) 2727 kg*

7–103E A transistor with a height of 0.25 in and a diameter of 0.22 in is mounted on a circuit board. The transistor is cooled by air flowing over it at a velocity of 500 ft/min. If the air temperature is 120°F and the transistor case temperature is not to exceed 180°F, determine the amount of power this transistor can dissipate safely.

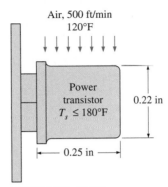

Air, 500 ft/min
120°F

Power transistor
$T_s \leq 180$°F

0.22 in

0.25 in

FIGURE P7–103E

7–104 The roof of a house consists of a 15-cm-thick concrete slab ($k = 2$ W/m · °C) that is 15 m wide and 20 m long. The convection heat transfer coefficient on the inner surface of the roof is 5 W/m² · °C. On a clear winter night, the ambient air is reported to be at 10°C, while the night sky temperature is 100 K. The house and the interior surfaces of the wall are maintained at a constant temperature of 20°C. The emissivity of both surfaces of the concrete roof is 0.9. Considering both radiation and convection heat transfer, determine the rate of heat transfer through the roof when wind at 60 km/h is blowing over the roof.

If the house is heated by a furnace burning natural gas with an efficiency of 85 percent, and the price of natural gas is $1.20/therm, determine the money lost through the roof that night during a 14-h period. *Answers: 28 kW, $18.9*

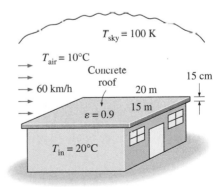

$T_{sky} = 100$ K

$T_{air} = 10$°C

Concrete roof

60 km/h

15 cm

20 m

15 m

$\varepsilon = 0.9$

$T_{in} = 20$°C

FIGURE P7–104

7–105 Steam at 250°C flows in a stainless steel pipe ($k = 15$ W/m · °C) whose inner and outer diameters are 4 cm and 4.6 cm, respectively. The pipe is covered with 3.5-cm-thick glass wool insulation ($k = 0.038$ W/m · °C) whose outer surface has an emissivity of 0.3. Heat is lost to the surrounding air and surfaces at 3°C by convection and radiation. Taking the heat transfer coefficient inside the pipe to be 80 W/m² · °C, determine the rate of heat loss from the steam per unit length of the pipe when air is flowing across the pipe at 4 m/s.

7–106 The boiling temperature of nitrogen at atmospheric pressure at sea level (1 atm pressure) is −196°C. Therefore, nitrogen is commonly used in low-temperature scientific studies, since the temperature of liquid nitrogen in a tank open to the atmosphere will remain constant at −196°C until it is depleted. Any heat transfer to the tank will result in the evaporation of some liquid nitrogen, which has a heat of vaporization of 198 kJ/kg and a density of 810 kg/m³ at 1 atm.

Consider a 4-m-diameter spherical tank that is initially filled with liquid nitrogen at 1 atm and −196°C. The tank is exposed to 20°C ambient air and 40 km/h winds. The temperature of the thin-shelled spherical tank is observed to be almost the same as the temperature of the nitrogen inside. Disregarding any radiation heat exchange, determine the rate of evaporation of the liquid nitrogen in the tank as a result of heat transfer from the ambient air if the tank is (a) not insulated, (b) insulated with 5-cm-thick fiberglass insulation ($k = 0.035$ W/m · °C), and (c) insulated with 2-cm-thick superinsulation that has an effective thermal conductivity of 0.00005 W/m · °C.

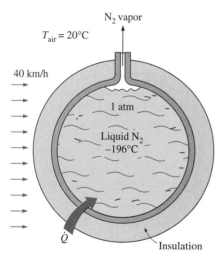

N₂ vapor

$T_{air} = 20$°C

40 km/h

1 atm

Liquid N₂
−196°C

$\dot{Q}$

Insulation

FIGURE P7–106

7–107 Repeat Prob. 7–106 for liquid oxygen, which has a boiling temperature of −183°C, a heat of vaporization of 213 kJ/kg, and a density of 1140 kg/m³ at 1 atm pressure.

7–108 A 0.5-cm-thick, 12-cm-high, and 18-cm-long circuit board houses 80 closely spaced logic chips on one side, each dissipating 0.06 W. The board is impregnated with copper fillings and has an effective thermal conductivity of 16 W/m · °C. All the heat generated in the chips is conducted across the circuit board and is dissipated from the back side of the board to the ambient air at 30°C, which is forced to flow over the surface by a fan at a free-stream velocity of 300 m/min. Determine the temperatures on the two sides of the circuit board.

7–109E It is well known that cold air feels much colder in windy weather than what the thermometer reading indicates because of the "chilling effect" of the wind. This effect is due to the increase in the convection heat transfer coefficient with increasing air velocities. The *equivalent windchill temperature* in °F is given by (1993 *ASHRAE Handbook of Fundamentals,* Atlanta, GA, p. 8.15)

$$T_{equiv} = 91.4 - (91.4 - T_{ambient})(0.475 - 0.0203V + 0.304\sqrt{V})$$

where V is the wind velocity in mph and $T_{ambient}$ is the ambient air temperature in °F in calm air, which is taken to be air with light winds at speeds up to 4 mph. The constant 91.4°F in the above equation is the mean skin temperature of a resting person in a comfortable environment. Windy air at a temperature $T_{ambient}$ and velocity V will feel as cold as calm air at a temperature T_{equiv}. The equation above is valid for winds up to 43 mph. Winds at higher velocities produce little additional chilling effect. Determine the equivalent wind chill temperature of an environment at 10°F at wind speeds of 10, 20, 30, and 40 mph. Exposed flesh can freeze within one minute at a temperature below −25°F in calm weather. Does a person need to be concerned about this possibility in any of the cases above?

Winds
40°F
35 mph

It feels
like 11°F

FIGURE P7–109E

7–110E Reconsider Prob. 7–109E. Using EES (or other) software, plot the equivalent wind chill temperatures in °F as a function of wind velocity in the range of 4 mph to 40 mph for ambient temperatures of 20°F, 40°F and 60°F. Discuss the results.

7–111 Air at 15°C and 1 atm flows over a 0.3 m wide plate at 65°C at a velocity of 3.0 m/s. Compute the following quantities at $x = 0.3$ m and $x = x_{cr}$:

(a) Hydrodynamic boundary layer thickness, m
(b) Local friction coefficient
(c) Average friction coefficient
(d) Local shear stress due to friction, N/m²
(e) Total drag force, N
(f) Thermal boundary layer thickness, m
(g) Local convection heat transfer coefficient, W/m² · °C
(h) Average convection heat transfer coefficient, W/m² · °C
(i) Rate of convective heat transfer, W

7–112 Oil at 60°C flows at a velocity of 20 cm/s over a 5.0 m long and 1.0-m wide flat plate maintained at a constant temperature of 20°C. Determine the rate of heat transfer from the oil to the plate if the average oil properties are: $\rho = 880$ kg/m³, $\mu = 0.005$ kg/m · s, $k = 0.15$ W/m · K, and $c_p = 2.0$ kJ/kg · K.

7–113 A small 2.0-mm diameter sphere of lead is cooled from an average temperature of 200°C to 54°C by dropping it into a tall column filled with air at 27°C and 101.3 kPa. It can be assumed that the terminal velocity (V_t) of the sphere is reached quickly such that the entire fall of the sphere occurs at this constant velocity, which is calculated from:

$$V_t = \left[\frac{2(\rho - \rho_{air})Vg}{C_D \rho_{air} A_p} \right]^{0.5}$$

where, V = volume of sphere, $g = 9.81$ m/s², ρ_{air} = density of air (1.18 kg/m³), C_D = drag coefficient (given as 0.40), and A_p = projected area of sphere ($\pi D^2/4$).

The properties of lead are: $\rho = 11{,}300$ kg/m³, $k = 33$ W/m · K, and $c_p = 0.13$ kJ/kg · K.

(a) Estimate the terminal velocity (V_t) of the sphere.
(b) Calculate the heat transfer coefficient for the lead sphere at its mean temperature.
(c) Calculate the column height for the indicated cooling of the lead sphere.

7–114 Repeat Prob. 7–113 for a 5-mm-diameter sphere.

7–115 Ten square silicon chips of 10 mm on a side are mounted in a single row on an electronic board that is insulated at the bottom side. The top surface is cooled by air flowing parallel to the row of chips with $T_\infty = 24$°C and $V = 30$ m/s. The chips exchange heat by radiation with the surroundings at $T_{surr} = -10$°C. The emissivity of the chips is 0.85. When in use, the same electrical power is dissipated in each chip. The maximum allowable temperature of the chips is 100°C. Assume that the temperature is uniform within each chip, no heat transfer occurs between adjacent chips, and T_∞ is the same throughout the array.

(a) Which chip reaches the highest steady operating temperature? Why?
(b) Determine the maximum electric power that can be dissipated per chip.
(c) Determine the temperature of the 5th chip in the direction of the air flow.
(d) Consider two cooling schemes: one used in parts (a)–(c) with the airflow parallel to the array (solid-line arrows), the other with the flow normal to it (dashed-line arrows). Which scheme is more efficient from a cooling point of view? Why? What other difference(s) between the two schemes would you consider when choosing one for a practical application?

7–116 An array of electrical heating elements is used in an air-duct heater as shown in Fig. P7–116. Each element has a length of 250 mm and a uniform surface temperature of 350°C. Atmospheric air enters the heater with a velocity of 12 m/s and

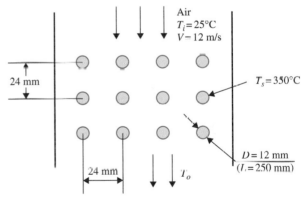

FIGURE P7–116

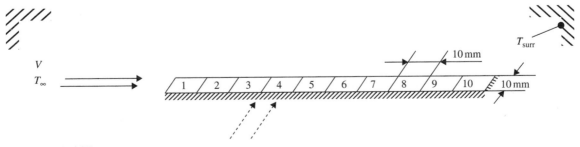

FIGURE P7–115

a temperature of 25°C. Determine the total heat transfer rate and the temperature of the air leaving the heater. Neglect the change in air properties as a result of temperature change across the heater.

Fundamentals of Engineering (FE) Exam Problems

7–117 For laminar flow of a fluid along a flat plate, one would expect the largest local convection heat transfer coefficient for the same Reynolds and Prandtl numbers when

(*a*) The same temperature is maintained on the surface
(*b*) The same heat flux is maintained on the surface
(*c*) The plate has an unheated section
(*d*) The plate surface is polished
(*e*) None of the above

7–118 Air at 20°C flows over a 4-m long and 3-m wide surface of a plate whose temperature is 80°C with a velocity of 5 m/s. The length of the surface for which the flow remains laminar is

(*a*) 1.5 m   (*b*) 1.8 m   (*c*) 2.0 m
(*d*) 2.8 m   (*e*) 4.0 m
(For air, use $k = 0.02735$ W/m · °C, Pr = 0.7228, $\nu = 1.798 \times 10^{-5}$ m^2/s)

7–119 Air at 20°C flows over a 4-m-long and 3-m-wide surface of a plate whose temperature is 80°C with a velocity of 5 m/s. The rate of heat transfer from the laminar flow region of the surface is

(*a*) 950 W   (*b*) 1037 W   (*c*) 2074 W
(*d*) 2640 W   (*e*) 3075 W
(For air, use $k = 0.02735$ W/m · °C, Pr = 0.7228, $\nu = 1.798 \times 10^{-5}$ m^2/s)

7–120 Air at 20°C flows over a 4-m-long and 3-m-wide surface of a plate whose temperature is 80°C with a velocity of 5 m/s. The rate of heat transfer from the surface is

(*a*) 7383 W   (*b*) 8985 W   (*c*) 11,231 W
(*d*) 14,672 W   (*e*) 20,402 W
(For air, use $k = 0.02735$ W/m · °C, Pr = 0.7228, $\nu = 1.798 \times 10^{-5}$ m^2/s)

7–121 Air at 15°C flows over a flat plate subjected to a uniform heat flux of 300 W/m^2 with a velocity of 3.5 m/s. The surface temperature of the plate 6 m from the leading edge is

(*a*) 164°C   (*b*) 68.3°C   (c) 48.1°C
(*d*) 46.8°C   (*e*) 37.5°C
(For air, use $k = 0.02551$ W/m · °C, Pr = 0.7296, $\nu = 1.562 \times 10^{-5}$ m^2/s)

7–122 Water at 75°C flows over a 2-m-long, 2-m-wide surface of a plate whose temperature is 5°C with a velocity of 1.5 m/s. The total drag force acting on the plate is

(*a*) 2.8 N   (*b*) 12.3 N   (*c*) 13.7 N
(*d*) 15.4 N   (*e*) 20.0 N
(For water, use $\nu = 0.658 \times 10^{-6}$ m^2/s, $\rho = 992$ kg/m^3)

7–123 Engine oil at 105°C flows over the surface of a flat plate whose temperature is 15°C with a velocity of 1.5 m/s. The local drag force per unit surface area 0.8 m from the leading edge of the plate is

(*a*) 21.8 N/m^2 (*b*) 14.3 N/m^2 (*c*) 10.9 N/m^2
(*d*) 8.5 N/m^2 (*e*) 5.5 N/m^2
(For oil, use $\nu = 8.565 \times 10^{-5}$ m^2/s, $\rho = 864$ kg/m^3)

7–124 Air at 25°C flows over a 5-cm-diameter, 1.7-m-long pipe with a velocity of 4 m/s. A refrigerant at -15°C flows inside the pipe and the surface temperature of the pipe is essentially the same as the refrigerant temperature inside. Air properties at the average temperature are $k = 0.0240$ W/m · °C, Pr = 0.735, $\nu = 1.382 \times 10^{-5}$ m^2/s. The rate of heat transfer to the pipe is

(*a*) 343 W   (*b*) 419 W   (*c*) 485 W
(*d*) 547 W   (*e*) 610 W

7–125 Air at 25°C flows over a 5-cm-diameter, 1.7-m-long smooth pipe with a velocity of 4 m/s. A refrigerant at -15°C flows inside the pipe and the surface temperature of the pipe is essentially the same as the refrigerant temperature inside. The drag force exerted on the pipe by the air is

(*a*) 0.4 N   (*b*) 1.1 N   (*c*) 8.5 N
(*d*) 13 N   (*e*) 18 N
(For air, use $\nu = 1.382 \times 10^{-5}$ m^2/s, $\rho = 1.269$ kg/m^3)

7–126 Kitchen water at 10°C flows over a 10-cm-diameter pipe with a velocity of 1.1 m/s. Geothermal water enters the pipe at 90°C at a rate of 1.25 kg/s. For calculation purposes, the surface temperature of the pipe may be assumed to be 70°C. If the geothermal water is to leave the pipe at 50°C, the required length of the pipe is

(*a*) 1.1 m   (*b*) 1.8 m   (*c*) 2.5 m
(*d*) 4.3 m   (*e*) 7.6 m
(For both water streams, use $k = 0.631$ W/m · °C, Pr = 4.32, $\nu = 0.658 \times 10^{-6}$ m^2/s, $c_p = 4179$ J/kg · °C)

7–127 Ambient air at 20°C flows over a 30-cm-diameter hot spherical object with a velocity of 2.5 m/s. If the average surface temperature of the object is 200°C, the average convection heat transfer coefficient during this process is

(*a*) 5.0 W/m · °C   (*b*) 6.1 W/m · °C
(*c*) 7.5 W/m · °C   (*d*) 9.3 W/m · °C
(*e*) 11.7 W/m · °C
(For air, use $k = 0.2514$ W/m · °C, Pr = 0.7309, $\nu = 1.516 \times 10^{-5}$ m^2/s, $\mu_\infty = 1.825 \times 10^{-5}$ kg/m · s, $\mu_s = 2.577 \times 10^{-5}$ kg/m · s)

7–128 Wind at 30°C flows over a 0.5-m-diameter spherical tank containing iced water at 0°C with a velocity of 25 km/h. If the tank is thin-shelled with a high thermal conductivity material, the rate at which ice melts is

(*a*) 4.78 kg/h   (*b*) 6.15 kg/h   (*c*) 7.45 kg/h
(*d*) 11.8 kg/h   (*e*) 16.0 kg/h

(Take $h_{if} = 333.7$ kJ/kg, and use the following for air: $k = 0.02588$ W/m · °C, Pr = 0.7282, $v = 1.608 \times 10^{-5}$ m²/s, $\mu_\infty = 1.872 \times 10^{-5}$ kg/m · s, $\mu_s = 1.729 \times 10^{-5}$ kg/m · s)

7–129   Air ($k = 0.028$ W/m · K, Pr = 0.7) at 50°C flows along a 1-m-long flat plate whose temperature is maintained at 20°C with a velocity such that the Reynolds number at the end of the plate is 10,000. The heat transfer per unit width between the plate and air is

(a) 20 W/m   (b) 30 W/m   (c) 40 W/m
(d) 50 W/m   (e) 60 W/m

7–130   Air (Pr = 0.7, $k = 0.026$ W/m · K) at 200°C flows across 2-cm-diameter tubes whose surface temperature is 50°C with a Reynolds number of 8000. The Churchill and Bernstein convective heat transfer correlation for the average Nusselt number in this situation is

$$\text{Nu} = 0.3 + \frac{0.62 \, \text{Re}^{0.5}\text{Pr}^{0.33}}{[1 + (0.4/\text{Pr})^{0.67}]^{0.25}}$$

The average heat flux in this case is

(a) 8.5 kW/m²   (b) 9.7 kW/m²   (c) 10.5 kW/m²
(d) 12.2 kW/m²   (e) 13.9 kW/m²

7–131   Jakob suggests the following correlation be used for square tubes in a liquid cross-flow situation:

$$\text{Nu} = 0.102 \, \text{Re}^{0.625} \, \text{Pr}^{1/3}$$

Water ($k = 0.61$ W/m · K, Pr = 6) flows across a 1-cm square tube with a Reynolds number of 10,000. The convection heat transfer coefficient is

(a) 5.7 kW/m² · K   (b) 8.3 kW/m² · K
(c) 11.2 kW/m² · K   (d) 15.6 kW/m² · K
(e) 18.1 kW/m² · K

7–132   Jakob suggests the following correlation be used for square tubes in a liquid cross-flow situation:

$$\text{Nu} = 0.102 \, \text{Re}^{0.675} \, \text{Pr}^{1/3}$$

Water ($k = 0.61$ W/m · K, Pr = 6) at 50°C flows across a 1-cm square tube with a Reynolds number of 10,000 and surface temperature of 75°C. If the tube is 2 m long, the rate of heat transfer between the tube and water is

(a) 6.0 kW   (b) 8.2 kW   (c) 11.3 kW
(d) 15.7 kW   (e) 18.1 kW

Design and Essay Problems

7–133   On average, superinsulated homes use just 15 percent of the fuel required to heat the same size conventional home built before the energy crisis in the 1970s. Write an essay on superinsulated homes, and identify the features that make them so energy efficient as well as the problems associated with them. Do you think superinsulated homes will be economically attractive in your area?

7–134   Conduct this experiment to determine the heat loss coefficient of your house or apartment in W/°C or Btu/h · °F. First make sure that the conditions in the house are steady and the house is at the set temperature of the thermostat. Use an outdoor thermometer to monitor outdoor temperature. One evening, using a watch or timer, determine how long the heater was on during a 3-h period and the average outdoor temperature during that period. Then using the heat output rating of your heater, determine the amount of heat supplied. Also, estimate the amount of heat generation in the house during that period by noting the number of people, the total wattage of lights that were on, and the heat generated by the appliances and equipment. Using that information, calculate the average rate of heat loss from the house and the heat loss coefficient.

7–135   The decision of whether to invest in an energy-saving measure is made on the basis of the length of time for it to pay for itself in projected energy (and thus cost) savings. The easiest way to reach a decision is to calculate the simple payback period by simply dividing the installed cost of the measure by the annual cost savings and comparing it to the lifetime of the installation. This approach is adequate for short payback periods (less than 5 years) in stable economies with low interest rates (under 10 percent) since the error involved is no larger than the uncertainties. However, if the payback period is long, it may be necessary to consider the *interest rate* if the money is to be borrowed, or the *rate of return* if the money is invested elsewhere instead of the energy conservation measure. For example, a simple payback period of five years corresponds to 5.0, 6.12, 6.64, 7.27, 8.09, 9.919, 10.84, and 13.91 for an interest rate (or return on investment) of 0, 6, 8, 10, 12, 14, 16, and 18 percent, respectively. Finding out the proper relations from engineering economics books, determine the payback periods for the interest rates given above corresponding to simple payback periods of 1 through 10 years.

7–136   Obtain information on frostbite and the conditions under which it occurs. Using the relation in Prob. 7–109E, prepare a table that shows how long people can stay in cold and windy weather for specified temperatures and wind speeds before the exposed flesh is in danger of experiencing frostbite.

7–137   Write an article on forced convection cooling with air, helium, water, and a dielectric liquid. Discuss the advantages and disadvantages of each fluid in heat transfer. Explain the circumstances under which a certain fluid will be most suitable for the cooling job.

INTERNAL FORCED CONVECTION

Liquid or gas flow through pipes or ducts is commonly used in heating and cooling applications. The fluid in such applications is forced to flow by a fan or pump through a flow section that is sufficiently long to accomplish the desired heat transfer. In this chapter we pay particular attention to the determination of the *friction factor* and *convection coefficient* since they are directly related to the *pressure drop* and *heat transfer rate,* respectively. These quantities are then used to determine the pumping power requirement and the required tube length.

There is a fundamental difference between external and internal flows. In *external flow,* considered in Chapter 7, the fluid has a free surface, and thus the boundary layer over the surface is free to grow indefinitely. In *internal flow,* however, the fluid is completely confined by the inner surfaces of the tube, and thus there is a limit on how much the boundary layer can grow.

We start this chapter with a general physical description of internal flow, and the *average velocity* and *average temperature.* We continue with the discussion of the *hydrodynamic* and *thermal entry lengths, developing flow,* and *fully developed flow.* We then obtain the velocity and temperature profiles for fully developed laminar flow, and develop relations for the friction factor and Nusselt number. Finally we present empirical relations for developing and fully developed flows, and demonstrate their use.

OBJECTIVES

When you finish studying this chapter, you should be able to:

- Obtain average velocity from a knowledge of velocity profile, and average temperature from a knowledge of temperature profile in internal flow,

- Have a visual understanding of different flow regions in internal flow, such as the entry and the fully developed flow regions, and calculate hydrodynamic and thermal entry lengths,

- Analyze heating and cooling of a fluid flowing in a tube under constant surface temperature and constant surface heat flux conditions, and work with the logarithmic mean temperature difference,

- Obtain analytic relations for the velocity profile, pressure drop, friction factor, and Nusselt number in fully developed laminar flow, and

- Determine the friction factor and Nusselt number in fully developed turbulent flow using empirical relations, and calculate the pressure drop and heat transfer rate.

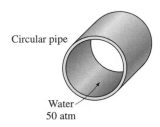

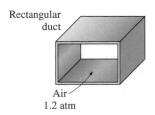

FIGURE 8–1
Circular pipes can withstand large pressure differences between the inside and the outside without undergoing any significant distortion, but noncircular pipes cannot.

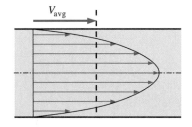

FIGURE 8–2
Average velocity V_{avg} is defined as the average speed through a cross section. For fully developed laminar pipe flow, V_{avg} is half of maximum velocity.

8–1 · INTRODUCTION

The terms *pipe, duct,* and *conduit* are usually used interchangeably for flow sections. In general, flow sections of circular cross section are referred to as *pipes* (especially when the fluid is a liquid), and flow sections of noncircular cross section as *ducts* (especially when the fluid is a gas). Small-diameter pipes are usually referred to as *tubes*. Given this uncertainty, we will use more descriptive phrases (such as *a circular pipe* or *a rectangular duct*) whenever necessary to avoid any misunderstandings.

You have probably noticed that most fluids, especially liquids, are transported in *circular pipes*. This is because pipes with a circular cross section can withstand large pressure differences between the inside and the outside without undergoing significant distortion. *Noncircular pipes* are usually used in applications such as the heating and cooling systems of buildings where the pressure difference is relatively small, the manufacturing and installation costs are lower, and the available space is limited for ductwork (Fig. 8–1). For a fixed surface area, the circular tube gives the most heat transfer for the least pressure drop, which explains the overwhelming popularity of circular tubes in heat transfer equipment.

Although the theory of fluid flow is reasonably well understood, theoretical solutions are obtained only for a few simple cases such as fully developed laminar flow in a circular pipe. Therefore, we must rely on experimental results and empirical relations for most fluid flow problems rather than closed-form analytical solutions. Noting that the experimental results are obtained under carefully controlled laboratory conditions and that no two systems are exactly alike, we must not be so naive as to view the results obtained as "exact." An error of 10 percent (or more) in friction factors calculated using the relations in this chapter is the "norm" rather than the "exception."

The fluid velocity in a pipe changes from *zero* at the surface because of the no-slip condition to a maximum at the pipe center. In fluid flow, it is convenient to work with an *average* velocity V_{avg}, which remains constant in incompressible flow when the cross-sectional area of the pipe is constant (Fig. 8–2). The average velocity in heating and cooling applications may change somewhat because of changes in density with temperature. But, in practice, we evaluate the fluid properties at some average temperature and treat them as constants. The convenience of working with constant properties usually more than justifies the slight loss in accuracy.

Also, the friction between the fluid particles in a pipe does cause a slight rise in fluid temperature as a result of the mechanical energy being converted to sensible thermal energy. But this temperature rise due to *frictional heating* is usually too small to warrant any consideration in calculations and thus is disregarded. For example, in the absence of any heat transfer, no noticeable difference can be detected between the inlet and outlet temperatures of water flowing in a pipe. The primary consequence of friction in fluid flow is pressure drop, and thus any significant temperature change in the fluid is due to heat transfer. But frictional heating must be considered for flows that involve highly viscous fluids with large velocity gradients.

8–2 · AVERAGE VELOCITY AND TEMPERATURE

In external flow, the free-stream velocity served as a convenient reference velocity for use in the evaluation of the Reynolds number and the friction coefficient. In internal flow, there is no free stream and thus we need an alternative. The fluid velocity in a tube changes from zero at the surface because of the no-slip condition, to a maximum at the tube center. Therefore, it is convenient to work with an **average** or **mean velocity** V_{avg}, which remains constant for incompressible flow when the cross sectional area of the tube is constant.

The value of the average velocity V_{avg} at some streamwise cross-section is determined from the requirement that the *conservation of mass* principle be satisfied (Fig. 8–2). That is,

$$\dot{m} = \rho V_{avg} A_c = \int_{A_c} \rho u(r)\, dA_c \qquad (8\text{–}1)$$

where $\dot{m}$ is the mass flow rate, ρ is the density, A_c is the cross-sectional area, and $u(r)$ is the velocity profile. Then the average velocity for incompressible flow in a circular pipe of radius R can be expressed as

$$V_{avg} = \frac{\int_{A_c} \rho u(r)\, dA_c}{\rho A_c} = \frac{\int_0^R \rho u(r) 2\pi r\, dr}{\rho \pi R^2} = \frac{2}{R^2} \int_0^R u(r) r\, dr \qquad (8\text{–}2)$$

Therefore, when we know the flow rate or the velocity profile, the average velocity can be determined easily.

When a fluid is heated or cooled as it flows through a tube, the temperature of the fluid at any cross section changes from T_s at the surface of the wall to some maximum (or minimum in the case of heating) at the tube center. In fluid flow it is convenient to work with an **average** or **mean temperature** T_m, which remains constant at a cross section. Unlike the mean velocity, the mean temperature T_m *changes* in the flow direction whenever the fluid is heated or cooled.

The value of the mean temperature T_m is determined from the requirement that the *conservation of energy* principle be satisfied. That is, the energy transported by the fluid through a cross section in actual flow must be equal to the energy that would be transported through the same cross section if the fluid were at a constant temperature T_m. This can be expressed mathematically as (Fig. 8–3)

$$\dot{E}_{fluid} = \dot{m} c_p T_m = \int_{\dot{m}} c_p T(r)\delta \dot{m} = \int_{A_c} \rho c_p T(r) u(r) V dA_c \qquad (8\text{–}3)$$

where c_p is the specific heat of the fluid. Note that the product $\dot{m} c_p T_m$ at any cross section along the tube represents the *energy flow* with the fluid at that cross section. Then the mean temperature of a fluid with constant density and specific heat flowing in a circular pipe of radius R can be expressed as

$$T_m = \frac{\int_{\dot{m}} c_p T(r)\delta \dot{m}}{\dot{m} c_p} = \frac{\int_0^R c_p T(r)\rho u(r) 2\pi r dr}{\rho V_{avg}(\pi R^2) c_p} = \frac{2}{V_{avg}R^2} \int_0^R T(r) u(r) r\, dr \qquad (8\text{–}4)$$

Note that the mean temperature T_m of a fluid changes during heating or cooling. Also, the fluid properties in internal flow are usually evaluated at the *bulk*

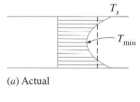

(a) Actual

(b) Idealized

FIGURE 8–3

Actual and idealized temperature profiles for flow in a tube (the rate at which energy is transported with the fluid is the same for both cases).

mean fluid temperature, which is the arithmetic average of the mean temperatures at the inlet and the exit. That is, $T_b = (T_{m, i} + T_{m, e})/2$.

Laminar and Turbulent Flow in Tubes

Flow in a tube can be laminar or turbulent, depending on the flow conditions. Fluid flow is streamlined and thus laminar at low velocities, but turns turbulent as the velocity is increased beyond a critical value. Transition from laminar to turbulent flow does not occur suddenly; rather, it occurs over some range of velocity where the flow fluctuates between laminar and turbulent flows before it becomes fully turbulent. Most pipe flows encountered in practice are turbulent. Laminar flow is encountered when highly viscous fluids such as oils flow in small diameter tubes or narrow passages.

For flow in a circular tube, the Reynolds number is defined as

$$\text{Re} = \frac{\rho V_{\text{avg}} D}{\mu} = \frac{V_{\text{avg}} D}{\nu} \tag{8-5}$$

where V_{avg} is the average flow velocity, D is the diameter of the tube, and $\nu = \mu/\rho$ is the kinematic viscosity of the fluid.

For flow through noncircular tubes, the Reynolds number as well as the Nusselt number, and the friction factor are based on the **hydraulic diameter** D_h defined as (Fig. 8–4)

$$D_h = \frac{4A_c}{p} \tag{8-6}$$

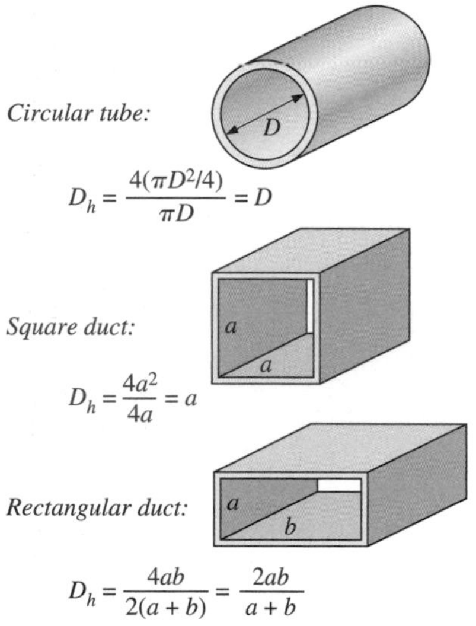

where A_c is the cross sectional area of the tube and p is its perimeter. The hydraulic diameter is defined such that it reduces to ordinary diameter D for circular tubes since

Circular tubes: $\qquad D_h = \dfrac{4A_c}{p} = \dfrac{4\pi D^2/4}{\pi D} = D$

Circular tube:

$$D_h = \frac{4(\pi D^2/4)}{\pi D} = D$$

Square duct:

$$D_h = \frac{4a^2}{4a} = a$$

Rectangular duct:

$$D_h = \frac{4ab}{2(a + b)} = \frac{2ab}{a + b}$$

FIGURE 8–4
The hydraulic diameter $D_h = 4A_c/p$ is defined such that it reduces to ordinary diameter for circular tubes.

It certainly is desirable to have precise values of Reynolds numbers for laminar, transitional, and turbulent flows, but this is not the case in practice. This is because the transition from laminar to turbulent flow also depends on the degree of disturbance of the flow by *surface roughness, pipe vibrations,* and the *fluctuations in the flow.* Under most practical conditions, the flow in a tube is laminar for Re < 2300, fully turbulent for Re > 10,000, and transitional in between. But it should be kept in mind that in many cases the flow becomes fully turbulent for Re > 4000, as discussed in the Topic of Special Interest later in this chapter. When designing piping networks and determining pumping power, a conservative approach is taken and flows with Re > 4000 are assumed to be turbulent.

In transitional flow, the flow switches between laminar and turbulent randomly (Fig. 8–5). It should be kept in mind that laminar flow can be maintained at much higher Reynolds numbers in very smooth pipes by avoiding flow disturbances and tube vibrations. In such carefully controlled experiments, laminar flow has been maintained at Reynolds numbers of up to 100,000.

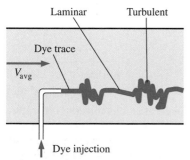

FIGURE 8–5
In the transitional flow region of the flow switches between laminar and turbulent randomly.

8–3 · THE ENTRANCE REGION

Consider a fluid entering a circular pipe at a uniform velocity. Because of the no-slip condition, the fluid particles in the layer in contact with the surface of the pipe come to a complete stop. This layer also causes the fluid particles in the adjacent layers to slow down gradually as a result of friction. To make up for this velocity reduction, the velocity of the fluid at the midsection of the pipe has to increase to keep the mass flow rate through the pipe constant. As a result, a velocity gradient develops along the pipe.

The region of the flow in which the effects of the viscous shearing forces caused by fluid viscosity are felt is called the **velocity boundary layer** or just the **boundary layer**. The hypothetical boundary surface divides the flow in a pipe into two regions: the **boundary layer region**, in which the viscous effects and the velocity changes are significant, and the **irrotational (core) flow region**, in which the frictional effects are negligible and the velocity remains essentially constant in the radial direction.

The thickness of this boundary layer increases in the flow direction until the boundary layer reaches the pipe center and thus fills the entire pipe, as shown in Fig. 8–6. The region from the pipe inlet to the point at which the boundary layer merges at the centerline is called the **hydrodynamic entrance region**, and the length of this region is called the **hydrodynamic entry length** L_h. Flow in the entrance region is called *hydrodynamically developing flow* since this is the region where the velocity profile develops. The region beyond the entrance region in which the velocity profile is fully developed and remains unchanged is called the **hydrodynamically fully developed region**. The velocity profile in the fully developed region is *parabolic* in laminar flow and somewhat *flatter* or fuller in turbulent flow due to eddy motion and more vigorous mixing in radial direction.

Now consider a fluid at a uniform temperature entering a circular tube whose surface is maintained at a different temperature. This time, the fluid particles in the layer in contact with the surface of the tube assume the surface temperature. This initiates convection heat transfer in the tube and the development of a *thermal boundary layer* along the tube. The thickness of this boundary layer also increases in the flow direction until the boundary layer reaches the tube center and thus fills the entire tube, as shown in Fig. 8–7.

The region of flow over which the thermal boundary layer develops and reaches the tube center is called the **thermal entrance region**, and the length of this region is called the **thermal entry length** L_t. Flow in the thermal

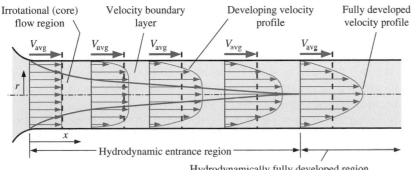

Irrotational (core) flow region   Velocity boundary layer   Developing velocity profile   Fully developed velocity profile

Hydrodynamic entrance region

Hydrodynamically fully developed region

FIGURE 8–6

The development of the velocity boundary layer in a tube. (The developed average velocity profile is parabolic in laminar flow, as shown, but somewhat flatter or fuller in turbulent flow.)

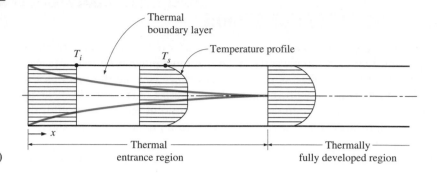

FIGURE 8–7
The development of the
thermal boundary layer in a tube.
(The fluid in the tube is being cooled.)

entrance region is called *thermally developing flow* since this is the region where the temperature profile develops. The region beyond the thermal entrance region in which the dimensionless temperature profile expressed as $(T_s - T)/(T_s - T_m)$ remains unchanged is called the **thermally fully developed region**. The region in which the flow is both hydrodynamically and thermally developed and thus both the velocity and dimensionless temperature profiles remain unchanged is called *fully developed flow.* That is,

Hydrodynamically fully developed: $\qquad \dfrac{\partial u(r, x)}{\partial x} = 0 \quad \longrightarrow \quad u = u(r)$ (8–7)

Thermally fully developed: $\qquad \dfrac{\partial}{\partial x}\left[\dfrac{T_s(x) - T(r, x)}{T_s(x) - T_m(x)}\right] = 0$ (8–8)

The shear stress at the tube wall τ_w is related to the slope of the velocity profile at the surface. Noting that the velocity profile remains unchanged in the hydrodynamically fully developed region, the wall shear stress also remains constant in that region. A similar argument can be given for the heat transfer coefficient in the thermally fully developed region.

In a thermally fully developed region, the derivative of $(T_s - T)/(T_s - T_m)$ with respect to x is zero by definition, and thus $(T_s - T)/(T_s - T_m)$ is independent of x. Then the derivative of $(T_s - T)/(T_s - T_m)$ with respect r must also be independent of x. That is,

$$\frac{\partial}{\partial r}\left(\frac{T_s - T}{T_s - T_m}\right)\bigg|_{r=R} = \frac{-(\partial T/\partial r)|_{r=R}}{T_s - T_m} \neq f(x)$$ (8–9)

Surface heat flux can be expressed as

$$\dot{q}_s = h_x(T_s - T_m) = k\frac{\partial T}{\partial r}\bigg|_{r=R} \quad \longrightarrow \quad h_x = \frac{k(\partial T/\partial r)|_{r=R}}{T_s - T_m}$$ (8–10)

which, from Eq. 8–9, is independent of x. Thus we conclude that *in the thermally fully developed region of a tube, the local convection coefficient is constant* (does not vary with x). Therefore, both *the friction (which is related to wall shear stress) and convection coefficients remain constant in the fully developed region of a tube.*

Note that the *temperature profile* in the thermally fully developed region may vary with x in the flow direction. That is, unlike the velocity profile, the temperature profile can be different at different cross sections of the tube in the developed region, and it usually is. However, the dimensionless temperature profile defined previously remains unchanged in the thermally developed region when the temperature or heat flux at the tube surface remains constant.

During laminar flow in a tube, the magnitude of the dimensionless Prandtl number Pr is a measure of the relative growth of the velocity and thermal boundary layers. For fluids with Pr ≈ 1, such as gases, the two boundary layers essentially coincide with each other. For fluids with Pr ≫ 1, such as oils, the velocity boundary layer outgrows the thermal boundary layer. As a result, the hydrodynamic entry length is smaller than the thermal entry length. The opposite is true for fluids with Pr ≪ 1 such as liquid metals.

Consider a fluid that is being heated (or cooled) in a tube as it flows through it. The wall shear stress and the heat transfer coefficient are *highest* at the tube inlet where the thickness of the boundary layers is smallest, and decrease gradually to the fully developed values, as shown in Fig. 8–8. Therefore, the pressure drop and heat flux are *higher* in the entrance regions of a tube, and the effect of the entrance region is always to *increase* the average friction factor and heat transfer coefficient for the entire tube. This enhancement can be significant for short tubes but negligible for long ones.

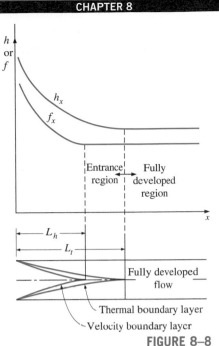

FIGURE 8–8
Variation of the friction factor and the convection heat transfer coefficient in the flow direction for flow in a tube (Pr > 1).

Entry Lengths

The hydrodynamic entry length is usually taken to be the distance from the tube entrance where the wall shear stress (and thus the friction factor) reaches within about 2 percent of the fully developed value. In *laminar flow*, the hydrodynamic and thermal entry lengths are given approximately as [see Kays and Crawford (1993) and Shah and Bhatti (1987)]

$$L_{h,\,laminar} \approx 0.05\,Re\,D \tag{8–11}$$

$$L_{t,\,laminar} \approx 0.05\,Re\,Pr\,D = Pr\,L_{h,\,laminar} \tag{8–12}$$

For Re = 20, the hydrodynamic entry length is about the size of the diameter, but increases linearly with velocity. In the limiting case of Re = 2300, the hydrodynamic entry length is 115D.

In *turbulent flow*, the intense mixing during random fluctuations usually overshadows the effects of molecular diffusion, and therefore the hydrodynamic and thermal entry lengths are of about the same size and independent of the Prandtl number. The hydrodynamic entry length for turbulent flow can be determined from [see Bhatti and Shah (1987) and Zhi-qing (1982)]

$$L_{h,\,turbulent} = 1.359D\,Re^{1/4} \tag{8–13}$$

The entry length is much shorter in turbulent flow, as expected, and its dependence on the Reynolds number is weaker. In many tube flows of practical interest, the entrance effects become insignificant beyond a tube length of 10 diameters, and the hydrodynamic and thermal entry lengths are approximately taken to be

$$L_{h,\,turbulent} \approx L_{t,\,turbulent} \approx 10D \tag{8–14}$$

The variation of local Nusselt number along a tube in turbulent flow for both uniform surface temperature and uniform surface heat flux is given in

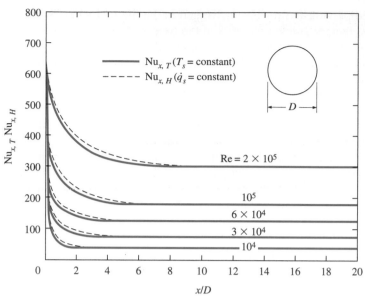

Fig. 8–9 for the range of Reynolds numbers encountered in heat transfer equipment. We make these important observations from this figure:

- The Nusselt numbers and thus the convection heat transfer coefficients are much higher in the entrance region.
- The Nusselt number reaches a constant value at a distance of less than 10 diameters, and thus the flow can be assumed to be fully developed for $x > 10D$.
- The Nusselt numbers for the uniform surface temperature and uniform surface heat flux conditions are identical in the fully developed regions, and nearly identical in the entrance regions. Therefore, Nusselt number is insensitive to the type of thermal boundary condition, and the turbulent flow correlations can be used for either type of boundary condition.

Precise correlations for the friction and heat transfer coefficients for the entrance regions are available in the literature. However, the tubes used in practice in forced convection are usually several times the length of either entrance region, and thus the flow through the tubes is often assumed to be fully developed for the entire length of the tube. This simplistic approach gives *reasonable* results for the rate of heat transfer for long tubes and *conservative* results for short ones.

8–4 ▪ GENERAL THERMAL ANALYSIS

In the absence of any work interactions (such as electric resistance heating), the conservation of energy equation for the steady flow of a fluid in a tube can be expressed as (Fig. 8–10)

$$\dot{Q} = \dot{m}c_p(T_e - T_i) \qquad \text{(W)} \qquad \text{(8–15)}$$

where T_i and T_e are the mean fluid temperatures at the inlet and exit of the tube, respectively, and $\dot{Q}$ is the rate of heat transfer to or from the fluid. Note

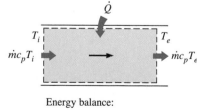

Energy balance:
$$\dot{Q} = \dot{m}c_p(T_e - T_i)$$

FIGURE 8–10
The heat transfer to a fluid flowing in a tube is equal to the increase in the energy of the fluid.

that the temperature of a fluid flowing in a tube remains constant in the absence of any energy interactions through the wall of the tube.

The thermal conditions at the surface can usually be approximated with reasonable accuracy to be *constant surface temperature* (T_s = constant) or *constant surface heat flux* ($\dot{q}_s$ = constant). For example, the constant surface temperature condition is realized when a phase change process such as boiling or condensation occurs at the outer surface of a tube. The constant surface heat flux condition is realized when the tube is subjected to radiation or electric resistance heating uniformly from all directions.

Surface heat flux is expressed as

$$\dot{q}_s = h_x(T_s - T_m) \qquad (\text{W/m}^2) \qquad \textbf{(8–16)}$$

where h_x is the *local* heat transfer coefficient and T_s and T_m are the surface and the mean fluid temperatures at that location. Note that the mean fluid temperature T_m of a fluid flowing in a tube must change during heating or cooling. Therefore, when $h_x = h$ = constant, the surface temperature T_s must change when $\dot{q}_s$ = constant, and the surface heat flux $\dot{q}_s$ must change when T_s = constant. Thus we may have either T_s = constant or $\dot{q}_s$ = constant at the surface of a tube, but not both. Next we consider convection heat transfer for these two common cases.

Constant Surface Heat Flux ($\dot{q}_s$ = constant)

In the case of $\dot{q}_s$ = constant, the rate of heat transfer can also be expressed as

$$\dot{Q} = \dot{q}_s A_s = \dot{m}c_p(T_e - T_i) \qquad (\text{W}) \qquad \textbf{(8–17)}$$

Then the mean fluid temperature at the tube exit becomes

$$T_e = T_i + \frac{\dot{q}_s A_s}{\dot{m}c_p} \qquad \textbf{(8–18)}$$

Note that the mean fluid temperature increases *linearly* in the flow direction in the case of constant surface heat flux, since the surface area increases linearly in the flow direction (A_s is equal to the perimeter, which is constant, times the tube length).

The surface temperature in the case of constant surface heat flux $\dot{q}_s$ can be determined from

$$\dot{q}_s = h(T_s - T_m) \quad \longrightarrow \quad T_s = T_m + \frac{\dot{q}_s}{h} \qquad \textbf{(8–19)}$$

In the fully developed region, the surface temperature T_s will also increase linearly in the flow direction since h is constant and thus $T_s - T_m$ = constant (Fig. 8–11). Of course this is true when the fluid properties remain constant during flow.

The slope of the mean fluid temperature T_m on a T-x diagram can be determined by applying the steady-flow energy balance to a tube slice of thickness dx shown in Fig. 8–12. It gives

$$\dot{m}c_p dT_m = \dot{q}_s(pdx) \quad \longrightarrow \quad \frac{dT_m}{dx} = \frac{\dot{q}_s p}{\dot{m}c_p} = \text{constant} \qquad \textbf{(8–20)}$$

where p is the perimeter of the tube.

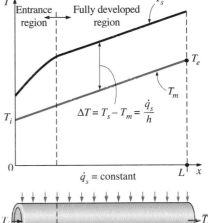

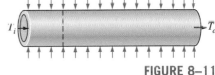

FIGURE 8–11

Variation of the *tube surface* and the *mean fluid* temperatures along the tube for the case of constant surface heat flux.

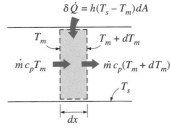

FIGURE 8–12

Energy interactions for a differential control volume in a tube.

Noting that both $\dot{q}_s$ and h are constants, the differentiation of Eq. 8–19 with respect to x gives

$$\frac{dT_m}{dx} = \frac{dT_s}{dx} \tag{8–21}$$

Also, the requirement that the dimensionless temperature profile remains unchanged in the fully developed region gives

$$\frac{\partial}{\partial x}\left(\frac{T_s - T}{T_s - T_m}\right) = 0 \quad \longrightarrow \quad \frac{1}{T_s - T_m}\left(\frac{\partial T_s}{\partial x} - \frac{\partial T}{\partial x}\right) = 0 \quad \longrightarrow \quad \frac{\partial T}{\partial x} = \frac{dT_s}{dx} \tag{8–22}$$

since $T_s - T_m = $ constant. Combining Eqs. 8–20, 8–21, and 8–22 gives

$$\frac{\partial T}{\partial x} = \frac{dT_s}{dx} = \frac{dT_m}{dx} = \frac{\dot{q}_s p}{\dot{m} c_p} = \text{constant} \tag{8–23}$$

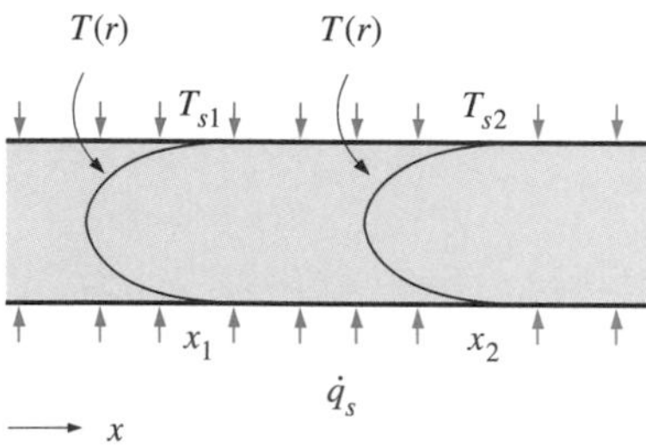

Then we conclude that *in fully developed flow in a tube subjected to constant surface heat flux, the temperature gradient is independent of x and thus the shape of the temperature profile does not change along the tube* (Fig. 8–13).

For a circular tube, $p = 2\pi R$ and $\dot{m} = \rho V_{avg} A_c = \rho V_{avg}(\pi R^2)$, and Eq. 8–23 becomes

Circular tube:
$$\frac{\partial T}{\partial x} = \frac{dT_s}{dx} = \frac{dT_m}{dx} = \frac{2\dot{q}_s}{\rho V_{avg} c_p R} = \text{constant} \tag{8–24}$$

where V_{avg} is the mean velocity of the fluid.

FIGURE 8–13
The shape of the temperature profile remains unchanged in the fully developed region of a tube subjected to constant surface heat flux.

Constant Surface Temperature ($T_s = $ constant)

From Newton's law of cooling, the rate of heat transfer to or from a fluid flowing in a tube can be expressed as

$$\dot{Q} = hA_s \Delta T_{avg} = hA_s(T_s - T_m)_{avg} \qquad \text{(W)} \tag{8–25}$$

where h is the average convection heat transfer coefficient, A_s is the heat transfer surface area (it is equal to πDL for a circular pipe of length L), and ΔT_{avg} is some appropriate *average* temperature difference between the fluid and the surface. Below we discuss two suitable ways of expressing ΔT_{avg}.

In the constant surface temperature ($T_s = $ constant) case, ΔT_{avg} can be expressed *approximately* by the **arithmetic mean temperature difference** ΔT_{am} as

$$\Delta T_{avg} \approx \Delta T_{am} = \frac{\Delta T_i + \Delta T_e}{2} = \frac{(T_s - T_i) + (T_s - T_e)}{2} = T_s - \frac{T_i + T_e}{2}$$

$$= T_s - T_b \tag{8–26}$$

where $T_b = (T_i + T_e)/2$ is the *bulk mean fluid temperature,* which is the *arithmetic average* of the mean fluid temperatures at the inlet and the exit of the tube.

Note that the *arithmetic mean temperature difference* ΔT_{am} is simply the *average* of the *temperature differences* between the surface and the fluid at the inlet and the exit of the tube. Inherent in this definition is the assumption that the mean fluid temperature varies linearly along the tube, which is hardly ever the case when T_s = constant. This simple approximation often gives acceptable results, but not always. Therefore, we need a better way to evaluate ΔT_{avg}.

Consider the heating of a fluid in a tube of constant cross section whose inner surface is maintained at a constant temperature of T_s. We know that the mean temperature of the fluid T_m increases in the flow direction as a result of heat transfer. The energy balance on a differential control volume shown in Fig. 8–12 gives

$$\dot{m}c_p\,dT_m = h(T_s - T_m)dA_s \qquad (8\text{--}27)$$

That is, the increase in the energy of the fluid (represented by an increase in its mean temperature by dT_m) is equal to the heat transferred to the fluid from the tube surface by convection. Noting that the differential surface area is $dA_s = p\,dx$, where p is the perimeter of the tube, and that $dT_m = -d(T_s - T_m)$, since T_s is constant, the relation above can be rearranged as

$$\frac{d(T_s - T_m)}{T_s - T_m} = -\frac{hp}{\dot{m}c_p}\,dx \qquad (8\text{--}28)$$

Integrating from $x = 0$ (tube inlet where $T_m = T_i$) to $x = L$ (tube exit where $T_m = T_e$) gives

$$\ln\frac{T_s - T_e}{T_s - T_i} = -\frac{hA_s}{\dot{m}c_p} \qquad (8\text{--}29)$$

where $A_s = pL$ is the surface area of the tube and h is the constant *average* convection heat transfer coefficient. Taking the exponential of both sides and solving for T_e gives the following relation which is very useful for the determination of the *mean fluid temperature at the tube exit:*

$$T_e = T_s - (T_s - T_i)\exp(-hA_s/\dot{m}c_p) \qquad (8\text{--}30)$$

This relation can also be used to determine the mean fluid temperature $T_m(x)$ at any x by replacing $A_s = pL$ by px.

Note that the temperature difference between the fluid and the surface *decays exponentially* in the flow direction, and the rate of decay depends on the magnitude of the exponent $hA_s/\dot{m}c_p$, as shown in Fig. 8–14. This dimensionless parameter is called the *number of transfer units*, denoted by NTU, and is a measure of the effectiveness of the heat transfer systems. For NTU > 5, the exit temperature of the fluid becomes almost equal to the surface temperature, $T_e \approx T_s$ (Fig. 8–15). Noting that the fluid temperature can approach the surface temperature but cannot cross it, an NTU of about 5 indicates that the limit is reached for heat transfer, and the heat transfer does not increase no matter how much we extend the length of the tube. A small value of NTU, on the other hand, indicates more opportunities for heat transfer, and the heat transfer continues to increase as the tube length is increased. A large NTU and thus a large heat transfer surface area (which means a large tube) may be desirable from a heat transfer point of view, but it may be unacceptable from an economic point of view. The selection of heat transfer equipment usually reflects a compromise between heat transfer performance and cost.

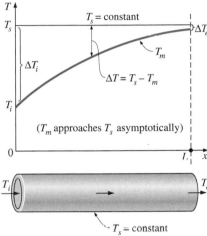

FIGURE 8–14

The variation of the *mean fluid temperature* along the tube for the case of constant temperature.

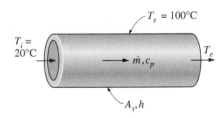

NTU = $hA_s / \dot{m}c_p$	T_e, °C
0.01	20.8
0.05	23.9
0.10	27.6
0.50	51.5
1.00	70.6
5.00	99.5
10.00	100.0

FIGURE 8–15

An NTU greater than 5 indicates that the fluid flowing in a tube will reach the surface temperature at the exit regardless of the inlet temperature.

Solving Eq. 8–29 for $\dot{m}c_p$ gives

$$\dot{m}c_p = -\frac{hA_s}{\ln[(T_s - T_e)/(T_s - T_i)]} \tag{8–31}$$

Substituting this into Eq. 8–15, we obtain

$$\dot{Q} = hA_s\Delta T_{\ln} \tag{8–32}$$

where

$$\Delta T_{\ln} = \frac{T_i - T_e}{\ln[(T_s - T_e)/(T_s - T_i)]} = \frac{\Delta T_e - \Delta T_i}{\ln(\Delta T_e/\Delta T_i)} \tag{8–33}$$

is the **logarithmic mean temperature difference**. Note that $\Delta T_i = T_s - T_i$ and $\Delta T_e = T_s - T_e$ are the temperature differences between the surface and the fluid at the inlet and the exit of the tube, respectively. This $\Delta T_{\ln}$ relation appears to be prone to misuse, but it is practically fail-safe, since using T_i in place of T_e and vice versa in the numerator and/or the denominator will, at most, affect the sign, not the magnitude. Also, it can be used for both heating ($T_s > T_i$ and T_e) and cooling ($T_s < T_i$ and T_e) of a fluid in a tube.

The logarithmic mean temperature difference $\Delta T_{\ln}$ is obtained by tracing the actual temperature profile of the fluid along the tube, and is an *exact* representation of the *average temperature difference* between the fluid and the surface. It truly reflects the exponential decay of the local temperature difference. When ΔT_e differs from ΔT_i by no more than 40 percent, the error in using the arithmetic mean temperature difference is less than 1 percent. But the error increases to undesirable levels when ΔT_e differs from ΔT_i by greater amounts. Therefore, we should always use the logarithmic mean temperature difference when determining the convection heat transfer in a tube whose surface is maintained at a constant temperature T_s.

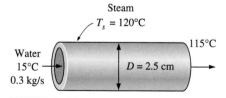

FIGURE 8–16
Schematic for Example 8–1.

EXAMPLE 8–1 **Heating of Water in a Tube by Steam**

Water enters a 2.5-cm-internal-diameter thin copper tube of a heat exchanger at 15°C at a rate of 0.3 kg/s, and is heated by steam condensing outside at 120°C. If the average heat transfer coefficient is 800 W/m² · °C, determine the length of the tube required in order to heat the water to 115°C (Fig. 8–16).

SOLUTION Water is heated by steam in a circular tube. The tube length required to heat the water to a specified temperature is to be determined.

Assumptions **1** Steady operating conditions exist. **2** Fluid properties are constant. **3** The convection heat transfer coefficient is constant. **4** The conduction resistance of copper tube is negligible so that the inner surface temperature of the tube is equal to the condensation temperature of steam.

Properties The specific heat of water at the bulk mean temperature of (15 + 115)/2 = 65°C is 4187 J/kg · °C. The heat of condensation of steam at 120°C is 2203 kJ/kg (Table A–9).

Analysis Knowing the inlet and exit temperatures of water, the rate of heat transfer is determined to be

$$\dot{Q} = \dot{m}c_p(T_e - T_i) = (0.3 \text{ kg/s})(4.187 \text{ kJ/kg} \cdot °C)(115°C - 15°C)$$
$$= 125.6 \text{ kW}$$

The logarithmic mean temperature difference Is

$$\Delta T_e = T_s - T_e = 120°C - 115°C = 5°C$$
$$\Delta T_i = T_s - T_i = 120°C - 15°C = 105°C$$
$$\Delta T_{\ln} = \frac{\Delta T_e - \Delta T_i}{\ln(\Delta T_e / \Delta T_i)} = \frac{5 - 105}{\ln(5/105)} = 32.85°C$$

The heat transfer surface area is

$$\dot{Q} = hA_s\Delta T_{\ln} \longrightarrow A_s = \frac{\dot{Q}}{h\Delta T_{\ln}} = \frac{125.6 \text{ kW}}{(0.8 \text{ kW/m}^2 \cdot °C)(32.85°C)} = 4.78 \text{ m}^2$$

Then the required tube length becomes

$$A_s = \pi DL \longrightarrow L = \frac{A_s}{\pi D} = \frac{4.78 \text{ m}^2}{\pi(0.025 \text{ m})} = 61 \text{ m}$$

Discussion The bulk mean temperature of water during this heating process is 65°C, and thus the *arithmetic* mean temperature difference is $\Delta T_{am} = 120 - 65 = 55°C$. Using ΔT_{am} instead of $\Delta T_{\ln}$ would give $L = 36$ m, which is grossly in error. This shows the importance of using the logarithmic mean temperature in calculations.

8–5 · LAMINAR FLOW IN TUBES

We mentioned in Section 8–2 that flow in tubes is laminar for Re $\lesssim$ 2300, and that the flow is fully developed if the tube is sufficiently long (relative to the entry length) so that the entrance effects are negligible. In this section we consider the steady laminar flow of an incompressible fluid with constant properties in the fully developed region of a straight circular tube. We obtain the momentum equation by applying a force balance to a differential volume element, and obtain the velocity profile by solving it. Then we use it to obtain a relation for the friction factor. An important aspect of the analysis here is that it is one of the few available for viscous flow.

In fully developed laminar flow, each fluid particle moves at a constant axial velocity along a streamline and the velocity profile $u(r)$ remains unchanged in the flow direction. There is no motion in the radial direction, and thus the velocity component in the direction normal to flow is everywhere zero. There is no acceleration since the flow is steady and fully developed.

Now consider a ring-shaped differential volume element of radius r, thickness dr, and length dx oriented coaxially with the tube, as shown in Fig. 8–17. The volume element involves only pressure and viscous effects and thus the pressure and shear forces must balance each other. The pressure force acting on a submerged plane surface is the product of the pressure at the centroid of the surface and the surface area. A force balance on the volume element in the flow direction gives

$$(2\pi r \, dr \, P)_x - (2\pi r \, dr \, P)_{x+dx} + (2\pi r \, dx \, \tau)_r - (2\pi r \, dx \, \tau)_{r+dr} = 0 \quad \textbf{(8–34)}$$

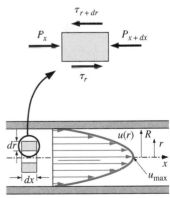

FIGURE 8–17
Free-body diagram of a ring-shaped differential fluid element of radius r, thickness dr, and length dx oriented coaxially with a horizontal tube in fully developed laminar flow.

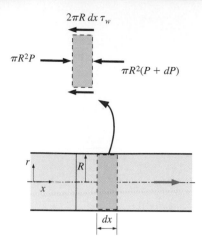

FIGURE 8–18

Free-body diagram of a fluid disk element of radius R and length dx in fully developed laminar flow in a horizontal tube.

which indicates that in fully developed flow in a horizontal tube, the viscous and pressure forces balance each other. Dividing by $2\pi drdx$ and rearranging,

$$r\frac{P_{x+dx} - P_x}{dx} + \frac{(r\tau)_{r+dr} - (r\tau)_r}{dr} = 0 \qquad (8\text{–}35)$$

Taking the limit as $dr, dx \to 0$ gives

$$r\frac{dP}{dx} + \frac{d(r\tau)}{dr} = 0 \qquad (8\text{–}36)$$

Substituting $\tau = -\mu(du/dr)$ and taking $\mu = $ constant gives the desired equation,

$$\frac{\mu}{r}\frac{d}{dr}\left(r\frac{du}{dr}\right) = \frac{dP}{dx} \qquad (8\text{–}37)$$

The quantity du/dr is negative in pipe flow, and the negative sign is included to obtain positive values for τ. (Or, $du/dr = -du/dy$ since $y = R - r$.) The left side of Eq. 8–37 is a function of r, and the right side is a function of x. The equality must hold for any value of r and x, and an equality of the form $f(r) = g(x)$ can be satisfied only if both $f(r)$ and $g(x)$ are equal to the same constant. Thus we conclude that $dP/dx = $ constant. This can be verified by writing a force balance on a volume element of radius R and thickness dx (a slice of the tube), which gives (Fig. 8–18)

$$\frac{dP}{dx} = -\frac{2\tau_w}{R}$$

Here τ_w is constant since the viscosity and the velocity profile are constants in the fully developed region. Therefore, $dP/dx = $ constant.

Equation 8–37 can be solved by rearranging and integrating it twice to give

$$u(r) = \frac{1}{4\mu}\left(\frac{dP}{dx}\right) + C_1 \ln r + C_2 \qquad (8\text{–}38)$$

The velocity profile $u(r)$ is obtained by applying the boundary conditions $\partial u/\partial r = 0$ at $r = 0$ (because of symmetry about the centerline) and $u = 0$ at $r = R$ (the no-slip condition at the tube surface). We get

$$u(r) = -\frac{R^2}{4\mu}\left(\frac{dP}{dx}\right)\left(1 - \frac{r^2}{R^2}\right) \qquad (8\text{–}39)$$

Therefore, the velocity profile in fully developed laminar flow in a tube is *parabolic* with a maximum at the centerline and minimum (zero) at the tube wall. Also, the axial velocity u is positive for any r, and thus the axial pressure gradient dP/dx must be negative (i.e., pressure must decrease in the flow direction because of viscous effects).

The average velocity is determined from its definition by substituting Eq. 8–39 into Eq. 8–2, and performing the integration. It gives

$$V_{avg} = \frac{2}{R^2}\int_0^R u(r)r\,dr = \frac{-2}{R^2}\int_0^R \frac{R^2}{4\mu}\left(\frac{dP}{dx}\right)\left(1 - \frac{r^2}{R^2}\right)r\,dr = -\frac{R^2}{8\mu}\left(\frac{dP}{dx}\right) \qquad (8\text{–}40)$$

Force balance:

$$\pi R^2 P - \pi R^2(P + dP) - 2\pi R\,dx\,\tau_w = 0$$

Simplifying:

$$\frac{dP}{dx} = -\frac{2\tau_w}{R}$$

Combining the last two equations, the velocity profile is rewritten as

$$u(r) = 2V_{avg}\left(1 - \frac{r^2}{R^2}\right) \tag{8-41}$$

This is a convenient form for the velocity profile since V_{avg} can be determined easily from the flow rate information.

The maximum velocity occurs at the centerline and is determined from Eq. 8–41 by substituting $r = 0$,

$$u_{max} = 2V_{avg} \tag{8-42}$$

Therefore, *the average velocity in fully developed laminar pipe flow is one-half of the maximum velocity.*

Pressure Drop

A quantity of interest in the analysis of pipe flow is the *pressure drop* ΔP since it is directly related to the power requirements of the fan or pump to maintain flow. We note that $dP/dx =$ constant, and integrating from $x = x_1$ where the pressure is P_1 to $x = x_1 + L$ where the pressure is P_2 gives

$$\frac{dP}{dx} = \frac{P_2 - P_1}{L} \tag{8-43}$$

Substituting Eq. 8–43 into the V_{avg} expression in Eq. 8–40, the pressure drop can be expressed as

Laminar flow: $\qquad \Delta P = P_1 - P_2 = \frac{8\mu L V_{avg}}{R^2} = \frac{32\mu L V_{avg}}{D^2} \tag{8-44}$

The symbol Δ is typically used to indicate the difference between the final and initial values, like $\Delta y = y_2 - y_1$. But in fluid flow, ΔP is used to designate pressure drop, and thus it is $P_1 - P_2$. A pressure drop due to viscous effects represents an irreversible pressure loss, and it is called **pressure loss** ΔP_L to emphasize that it is a *loss* (just like the head loss h_L, which is proportional to it).

Note from Eq. 8–44 that the pressure drop is proportional to the viscosity μ of the fluid, and ΔP would be zero if there were no friction. Therefore, the drop of pressure from P_1 to P_2 in this case is due entirely to viscous effects, and Eq. 8–44 represents the pressure loss ΔP_L when a fluid of viscosity μ flows through a pipe of constant diameter D and length L at average velocity V_{avg}.

In practice, it is found convenient to express the pressure loss for all types of fully developed internal flows (laminar or turbulent flows, circular or noncircular pipes, smooth or rough surfaces, horizontal or inclined pipes) as (Fig. 8–19)

Pressure loss: $\qquad \Delta P_L = f\frac{L}{D}\frac{\rho V_{avg}^2}{2} \tag{8-45}$

where $\rho V_{avg}^2/2$ is the *dynamic pressure* and f is the **Darcy friction factor**,

$$f = \frac{8\tau_w}{\rho V_{avg}^2}$$

It is also called the **Darcy–Weisbach friction factor**, named after the Frenchman Henry Darcy (1803–1858) and the German Julius Weisbach (1806–1871), the two engineers who provided the greatest contribution in its

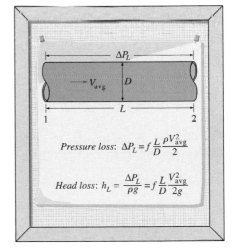

Pressure loss: $\Delta P_L = f\dfrac{L}{D}\dfrac{\rho V_{avg}^2}{2}$

Head loss: $h_L = \dfrac{\Delta P_L}{\rho g} = f\dfrac{L}{D}\dfrac{V_{avg}^2}{2g}$

FIGURE 8–19

The relation for pressure loss (and head loss) is one of the most general relations in fluid mechanics, and it is valid for laminar or turbulent flows, circular or noncircular tubes, and pipes with smooth or rough surfaces.

development. It should not be confused with the *friction coefficient* C_f [also called the *Fanning friction factor*, named after the American engineer John Fanning (1837–1911)], which is defined as $C_f = 2\tau_w/(\rho V_{avg}^2) = f/4$.

Setting Eqs. 8–44 and 8–45 equal to each other and solving for f gives the friction factor for fully developed laminar flow in a circular tube,

Circular tube, laminar:
$$f = \frac{64\mu}{\rho D V_{avg}} = \frac{64}{Re} \tag{8–46}$$

This equation shows that *in laminar flow, the friction factor is a function of the Reynolds number only and is independent of the roughness of the pipe surface.*

In the analysis of piping systems, pressure losses are commonly expressed in terms of the *equivalent fluid column height*, called the **head loss** h_L. Noting from fluid statics that $\Delta P = \rho g h$ and thus a pressure difference of ΔP corresponds to a fluid height of $h = \Delta P/\rho g$, the *pipe head loss* is obtained by dividing ΔP_L by ρg to give

$$h_L = \frac{\Delta P_L}{\rho g} = f \frac{L}{D} \frac{V_{avg}^2}{2g}$$

The head loss h_L represents *the additional height that the fluid needs to be raised by a pump in order to overcome the frictional losses in the pipe.* The head loss is caused by viscosity, and it is directly related to the wall shear stress. Equation 8–45 is valid for both laminar and turbulent flows in both circular and noncircular tubes, but Eq. 8–46 is valid only for fully developed laminar flow in circular pipes.

Once the pressure loss (or head loss) is known, the required pumping power *to overcome the pressure loss* is determined from

$$\dot{W}_{pump, L} = \dot{V} \Delta P_L = \dot{V}\rho g h_L = \dot{m} g h_L \tag{8–47}$$

where $\dot{V}$ is the volume flow rate and $\dot{m}$ is the mass flow rate.

The average velocity for laminar flow in a horizontal tube is, from Eq. 8–44,

Horizontal tube:
$$V_{avg} = \frac{(P_1 - P_2)R^2}{8\mu L} = \frac{(P_1 - P_2)D^2}{32\mu L} = \frac{\Delta P\, D^2}{32\mu L}$$

Then the volume flow rate for laminar flow through a horizontal tube of diameter D and length L becomes

$$\dot{V} = V_{avg} A_c = \frac{(P_1 - P_2)R^2}{8\mu L}\pi R^2 = \frac{(P_1 - P_2)\pi D^4}{128\mu L} = \frac{\Delta P\, \pi D^4}{128\mu L} \tag{8–48}$$

This equation is known as **Poiseuille's law**, and this flow is called *Hagen–Poiseuille flow* in honor of the works of G. Hagen (1797–1884) and J. Poiseuille (1799–1869) on the subject. Note from Eq. 8–48 that *for a specified flow rate, the pressure drop and thus the required pumping power is proportional to the length of the pipe and the viscosity of the fluid, but it is inversely proportional to the fourth power of the radius (or diameter) of the pipe.* Therefore, the pumping power requirement for a piping system can be reduced by a factor of 16 by doubling the tube diameter (Fig. 8–20). Of course the benefits of the reduction in the energy costs must be weighed against the increased cost of construction due to using a larger-diameter tube.

The pressure drop ΔP equals the pressure loss ΔP_L in the case of a horizontal tube, but this is not the case for inclined pipes or pipes with variable cross-sectional area because of the changes in elevation and velocity.

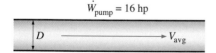

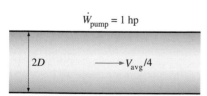

FIGURE 8–20

The pumping power requirement for a laminar flow piping system can be reduced by a factor of 16 by doubling the tube diameter.

Temperature Profile and the Nusselt Number

In the previous analysis, we have obtained the velocity profile for fully devel oped flow in a circular tube from a force balance applied on a volume element, and determined the friction factor and the pressure drop. Below we obtain the energy equation by applying the energy balance on a differential volume element, and solve it to obtain the temperature profile for the constant surface temperature and the constant surface heat flux cases.

Reconsider steady laminar flow of a fluid in a circular tube of radius R. The fluid properties ρ, k, and c_p are constant, and the work done by viscous forces is negligible. The fluid flows along the x-axis with velocity u. The flow is fully developed so that u is independent of x and thus $u = u(r)$. Noting that energy is transferred by mass in the x-direction, and by conduction in the r-direction (heat conduction in the x-direction is assumed to be negligible), the steady-flow energy balance for a cylindrical shell element of thickness dr and length dx can be expressed as (Fig. 8–21)

$$\dot{m}c_pT_x - \dot{m}c_pT_{x+dx} + \dot{Q}_r - \dot{Q}_{r+dr} = 0 \qquad \text{(8–49)}$$

where $\dot{m} = \rho u A_c = \rho u (2\pi r dr)$. Substituting and dividing by $2\pi r dr dx$ gives, after rearranging,

$$\rho c_p u \frac{T_{x+dx} - T_x}{dx} = -\frac{1}{2\pi r dx} \frac{\dot{Q}_{r+dr} - \dot{Q}_r}{dr} \qquad \text{(8–50)}$$

or

$$u \frac{\partial T}{\partial x} = -\frac{1}{2\rho c_p \pi r dx} \frac{\partial \dot{Q}}{\partial r} \qquad \text{(8–51)}$$

But

$$\frac{\partial \dot{Q}}{\partial r} = \frac{\partial}{\partial r}\left(-k2\pi r dx \frac{\partial T}{\partial r}\right) = -2\pi k dx \frac{\partial}{\partial r}\left(r \frac{\partial T}{\partial r}\right) \qquad \text{(8–52)}$$

Substituting and using $\alpha = k/\rho c_p$ gives

$$u\frac{\partial T}{\partial x} = \frac{\alpha}{r}\frac{\partial}{\partial r}\left(r\frac{\partial T}{\partial r}\right) \qquad \text{(8–53)}$$

which states that *the rate of net energy transfer to the control volume by mass flow is equal to the net rate of heat conduction in the radial direction.*

Constant Surface Heat Flux

For fully developed flow in a circular tube subjected to constant surface heat flux, we have, from Eq. 8–24,

$$\frac{\partial T}{\partial x} = \frac{dT_s}{dx} = \frac{dT_m}{dx} = \frac{2\dot{q}_s}{\rho V_{avg}c_p R} = \text{constant} \qquad \text{(8–54)}$$

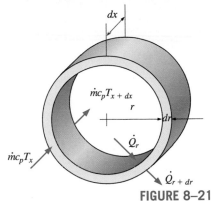

FIGURE 8–21
The differential volume element used in the derivation of energy balance relation.

If heat conduction in the x-direction were considered in the derivation of Eq. 8–53, it would give an additional term $\alpha \partial^2 T/\partial x^2$, which would be equal to zero since $\partial T/\partial x =$ constant and thus $T = T(r)$. Therefore, the assumption that there is no axial heat conduction is satisfied exactly in this case.

Substituting Eq. 8–54 and the relation for velocity profile (Eq. 8–41) into Eq. 8–53 gives

$$\frac{4\dot{q}_s}{kR}\left(1 - \frac{r^2}{R^2}\right) = \frac{1}{r}\frac{d}{dr}\left(r\frac{dT}{dr}\right) \tag{8–55}$$

which is a second-order ordinary differential equation. Its general solution is obtained by separating the variables and integrating twice to be

$$T = \frac{\dot{q}_s}{kR}\left(r^2 - \frac{r^4}{4R^2}\right) + C_1 r + C_2 \tag{8–56}$$

The desired solution to the problem is obtained by applying the boundary conditions $\partial T/\partial x = 0$ at $r = 0$ (because of symmetry) and $T = T_s$ at $r = R$. We get

$$T = T_s - \frac{\dot{q}_s R}{k}\left(\frac{3}{4} - \frac{r^2}{R^2} + \frac{r^4}{4R^4}\right) \tag{8–57}$$

The bulk mean temperature T_m is determined by substituting the velocity and temperature profile relations (Eqs. 8–41 and 8–57) into Eq. 8–4 and performing the integration. It gives

$$T_m = T_s - \frac{11}{24}\frac{\dot{q}_s R}{k} \tag{8–58}$$

Combining this relation with $\dot{q}_s = h(T_s - T_m)$ gives

$$h = \frac{24}{11}\frac{k}{R} = \frac{48}{11}\frac{k}{D} = 4.36\frac{k}{D} \tag{8–59}$$

or

Circular tube, laminar ($\dot{q}_s =$ constant): $\qquad$ $\text{Nu} = \dfrac{hD}{k} = 4.36$ $\qquad$ (8–60)

Therefore, for fully developed laminar flow in a circular tube subjected to constant surface heat flux, the Nusselt number is a constant. There is no dependence on the Reynolds or the Prandtl numbers.

Constant Surface Temperature

A similar analysis can be performed for fully developed laminar flow in a circular tube for the case of constant surface temperature T_s. The solution procedure in this case is more complex as it requires iterations, but the Nusselt number relation obtained is equally simple (Fig. 8–22):

Circular tube, laminar ($T_s =$ constant): $\qquad$ $\text{Nu} = \dfrac{hD}{k} = 3.66$ $\qquad$ (8–61)

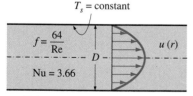

FIGURE 8–22

In laminar flow in a tube with constant surface temperature, both the *friction factor* and the *heat transfer coefficient* remain constant in the fully developed region.

The thermal conductivity k for use in the Nu relations above should be evaluated at the bulk mean fluid temperature, which is the arithmetic average of the mean fluid temperatures at the inlet and the exit of the tube. For laminar flow, the effect of *surface roughness* on the friction factor and the heat transfer coefficient is negligible.

Laminar Flow in Noncircular Tubes

The friction factor f and the Nusselt number relations are given in Table 8–1 for *fully developed laminar flow* in tubes of various cross sections. The Reynolds and Nusselt numbers for flow in these tubes are based on the hydraulic diameter $D_h = 4A_c/p$, where A_c is the cross sectional area of the tube and p is its perimeter. Once the Nusselt number is available, the convection heat transfer coefficient is determined from $h = k\text{Nu}/D_h$.

TABLE 8–1

Nusselt number and friction factor for fully developed laminar flow in tubes of various cross sections ($D_h = 4A_c/p$, $\text{Re} = V_{avg}D_h/\nu$, and $\text{Nu} = hD_h/k$)

Tube Geometry	a/b or $\theta°$	Nusselt Number T_s = Const.	$\dot{q}_s$ = Const.	Friction Factor f
Circle	—	3.66	4.36	64.00/Re
Rectangle	a/b			
	1	2.98	3.61	56.92/Re
	2	3.39	4.12	62.20/Re
	3	3.96	4.79	68.36/Re
	4	4.44	5.33	72.92/Re
	6	5.14	6.05	78.80/Re
	8	5.60	6.49	82.32/Re
	∞	7.54	8.24	96.00/Re
Ellipse	a/b			
	1	3.66	4.36	64.00/Re
	2	3.74	4.56	67.28/Re
	4	3.79	4.88	72.96/Re
	8	3.72	5.09	76.60/Re
	16	3.65	5.18	78.16/Re
Isosceles Triangle	θ			
	10°	1.61	2.45	50.80/Re
	30°	2.26	2.91	52.28/Re
	60°	2.47	3.11	53.32/Re
	90°	2.34	2.98	52.60/Re
	120°	2.00	2.68	50.96/Re

Developing Laminar Flow in the Entrance Region

For a circular tube of length L subjected to constant surface temperature, the average Nusselt number for the *thermal entrance region* can be determined from (Edwards et al., 1979)

Entry region, laminar:
$$\text{Nu} = 3.66 + \frac{0.065\,(D/L)\,\text{Re Pr}}{1 + 0.04[(D/L)\,\text{Re Pr}]^{2/3}} \tag{8-62}$$

Note that the average Nusselt number is larger at the entrance region, as expected, and it approaches asymptotically to the fully developed value of 3.66 as $L \to \infty$. This relation assumes that the flow is hydrodynamically developed when the fluid enters the heating section, but it can also be used approximately for flow developing hydrodynamically.

When the difference between the surface and the fluid temperatures is large, it may be necessary to account for the variation of viscosity with temperature.

The average Nusselt number for developing laminar flow in a circular tube in that case can be determined from [Sieder and Tate (1936)]

$$\text{Nu} = 1.86\left(\frac{\text{Re Pr }D}{L}\right)^{1/3}\left(\frac{\mu_b}{\mu_s}\right)^{0.14} \tag{8-63}$$

All properties are evaluated at the bulk mean fluid temperature, except for μ_s, which is evaluated at the surface temperature.

The average Nusselt number for the thermal entrance region of flow between *isothermal parallel plates* of length L is expressed as (Edwards et al., 1979)

Entry region, laminar:
$$\text{Nu} = 7.54 + \frac{0.03\,(D_h/L)\,\text{Re Pr}}{1 + 0.016[(D_h/L)\,\text{Re Pr}]^{2/3}} \tag{8-64}$$

where D_h is the hydraulic diameter, which is twice the spacing of the plates. This relation can be used for Re $\leq$ 2800.

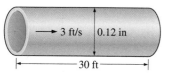

FIGURE 8–23
Schematic for Example 8–2.

EXAMPLE 8–2   Pressure Drop in a Tube

Water at 40°F ($\rho = 62.42$ lbm/ft^3 and $\mu = 1.038 \times 10^{-3}$ lbm/ft · s) is flowing in a 0.12-in-diameter 30-ft-long horizontal tube steadily at an average velocity of 3 ft/s (Fig. 8–23). Determine the pressure drop and the pumping power requirement to overcome this pressure drop.

SOLUTION   The average flow velocity in a tube is given. The pressure drop and the required pumping power are to be determined.

Assumptions   **1** The flow is steady and incompressible. **2** The entrance effects are negligible, and thus the flow is fully developed. **3** The tube involves no components such as bends, valves, and connectors.

Properties   The density and dynamic viscosity of water are given to be $\rho = 62.42$ lbm/ft^3 and $\mu = 1.038 \times 10^{-3}$ lbm/ft · s.

Analysis   First we need to determine the flow regime. The Reynolds number is

$$\text{Re} = \frac{\rho V_{\text{avg}} D}{\mu} = \frac{(62.42\text{ lbm/ft}^3)(3\text{ ft/s})(0.12/12\text{ ft})}{1.038 \times 10^{-3}\text{ lbm/ft} \cdot \text{s}} = 1803$$

which is less than 2300. Therefore, the flow is laminar. Then the friction factor and the pressure drop become

$$f = \frac{64}{Re} = \frac{64}{1803} = 0.0355$$

$$\Delta P = f\frac{L}{D}\frac{\rho V_{avg}^2}{2} = 0.0355\frac{30\text{ ft}}{0.01\text{ ft}}\frac{(62.42\text{ lbm/ft}^3)(3\text{ ft/s})^2}{2}\left(\frac{1\text{ lbf}}{32.174\text{ lbm}\cdot\text{ft/s}^2}\right)$$

$$= 930\text{ lbf/ft}^2 = 6.46\text{ psi}$$

The volume flow rate and the pumping power requirements are

$$\dot{V} = V_{avg}A_c = V_{avg}(\pi D^2/4) = (3\text{ ft/s})[\pi(0.01\text{ ft})^2/4] = 0.000236\text{ ft}^3/\text{s}$$

$$\dot{W}_{pump} = \dot{V}\Delta P = (0.000236\text{ ft}^3/\text{s})(930\text{ lbf/ft}^2)\left(\frac{1\text{ W}}{0.73756\text{ lbf}\cdot\text{ft/s}}\right) = \mathbf{0.30\text{ W}}$$

Therefore, mechanical power input in the amount of 0.30 W is needed to overcome the frictional losses in the flow due to viscosity.

EXAMPLE 8–3   Flow of Oil in a Pipeline through a Lake

Consider the flow of oil at 20°C in a 30-cm-diameter pipeline at an average velocity of 2 m/s (Fig. 8–24). A 200-m-long section of the horizontal pipeline passes through icy waters of a lake at 0°C. Measurements indicate that the surface temperature of the pipe is very nearly 0°C. Disregarding the thermal resistance of the pipe material, determine (a) the temperature of the oil when the pipe leaves the lake, (b) the rate of heat transfer from the oil, and (c) the pumping power required to overcome the pressure losses and to maintain the flow of the oil in the pipe.

SOLUTION   Oil flows in a pipeline that passes through icy waters of a lake at 0°C. The exit temperature of the oil, the rate of heat loss, and the pumping power needed to overcome pressure losses are to be determined.
Assumptions   **1** Steady operating conditions exist. **2** The surface temperature of the pipe is very nearly 0°C. **3** The thermal resistance of the pipe is negligible. **4** The inner surfaces of the pipeline are smooth. **5** The flow is hydrodynamically developed when the pipeline reaches the lake.
Properties   We do not know the exit temperature of the oil, and thus we cannot determine the bulk mean temperature, which is the temperature at which the properties of oil are to be evaluated. The mean temperature of the oil at the inlet is 20°C, and we expect this temperature to drop somewhat as a result of heat loss to the icy waters of the lake. We evaluate the properties of the oil at the inlet temperature, but we will repeat the calculations, if necessary, using properties at the evaluated bulk mean temperature. At 20°C we read (Table A-13)

$$\rho = 888.1\text{ kg/m}^3 \qquad \nu = 9.429\times 10^{-4}\text{ m}^2/\text{s}$$
$$k = 0.145\text{ W/m}\cdot\text{°C} \qquad c_p = 1880\text{ J/kg}\cdot\text{°C} \qquad Pr = 10{,}863$$

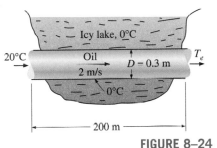

FIGURE 8–24
Schematic for Example 8–3.

Analysis (*a*) The Reynolds number is

$$\text{Re} = \frac{V_{\text{avg}} D}{\nu} = \frac{(2 \text{ m/s})(0.3 \text{ m})}{9.429 \times 10^{-4} \text{ m}^2/\text{s}} = 636$$

which is less than the critical Reynolds number of 2300. Therefore, the flow is laminar, and the thermal entry length in this case is roughly

$$L_t \approx 0.05 \text{ Re Pr } D = 0.05 \times 636 \times 10,863 \times (0.3 \text{ m}) \approx 103,600 \text{ m}$$

which is much greater than the total length of the pipe. This is typical of fluids with high Prandtl numbers. Therefore, we assume thermally developing flow and determine the Nusselt number from

$$\text{Nu} = \frac{hD}{k} = 3.66 + \frac{0.065 \, (D/L) \text{ Re Pr}}{1 + 0.04 \, [(D/L) \text{ Re Pr}]^{2/3}}$$

$$= 3.66 + \frac{0.065(0.3/200) \times 636 \times 10,863}{1 + 0.04[(0.3/200) \times 636 \times 10,863]^{2/3}}$$

$$= 33.7$$

Note that this Nusselt number is considerably higher than the fully developed value of 3.66. Then,

$$h = \frac{k}{D} \text{ Nu} = \frac{0.145 \text{ W/m} \cdot {}^\circ\text{C}}{0.3 \text{ m}} (33.7) = 16.3 \text{ W/m}^2 \cdot {}^\circ\text{C}$$

Also,

$$A_s = \pi DL = \pi(0.3 \text{ m})(200 \text{ m}) = 188.5 \text{ m}^2$$
$$\dot{m} = \rho A_c V_{\text{avg}} = (888.1 \text{ kg/m}^3)[\tfrac{1}{4}\pi(0.3 \text{ m})^2](2 \text{ m/s}) = 125.6 \text{ kg/s}$$

Next we determine the exit temperature of oil,

$$T_e = T_s - (T_s - T_i) \exp(-hA_s/\dot{m}c_p)$$

$$= 0{}^\circ\text{C} - [(0 - 20){}^\circ\text{C}] \exp\left[-\frac{(16.3 \text{ W/m}^2 \cdot {}^\circ\text{C})(188.5 \text{ m}^2)}{(125.6 \text{ kg/s})(1881 \text{ J/kg} \cdot {}^\circ\text{C})}\right]$$

$$= \mathbf{19.74{}^\circ C}$$

Thus, the mean temperature of oil drops by a mere 0.26°C as it crosses the lake. This makes the bulk mean oil temperature 19.87°C, which is practically identical to the inlet temperature of 20°C. Therefore, we do not need to re-evaluate the properties.

(*b*) The logarithmic mean temperature difference and the rate of heat loss from the oil are

$$\Delta T_{\ln} = \frac{T_i - T_e}{\ln \dfrac{T_s - T_e}{T_s - T_i}} = \frac{20 - 19.74}{\ln \dfrac{0 - 19.74}{0 - 20}} = -19.87{}^\circ\text{C}$$

$$\dot{Q} = hA_s \Delta T_{\ln} = (16.3 \text{ W/m}^2 \cdot {}^\circ\text{C})(188.5 \text{ m}^2)(-19.87{}^\circ\text{C}) = \mathbf{-6.11 \times 10^4 \text{ W}}$$

Therefore, the oil will lose heat at a rate of 61.1 kW as it flows through the pipe in the icy waters of the lake. Note that $\Delta T_{\ln}$ is identical to the arithmetic mean temperature in this case, since $\Delta T_i \approx \Delta T_e$.

(c) The laminar flow of oil is hydrodynamically developed. Therefore, the friction factor can be determined from

$$f = \frac{64}{\text{Re}} = \frac{64}{636} = 0.1006$$

Then the pressure drop in the pipe and the required pumping power become

$$\Delta P = f \frac{L}{D} \frac{\rho V_{\text{avg}}^2}{2} = 0.1006 \frac{200 \text{ m}}{0.3 \text{ m}} \frac{(888.1 \text{ kg/m}^3)(2 \text{ m/s})^2}{2} = 1.19 \times 10^5 \text{ N/m}^2$$

$$\dot{W}_{\text{pump}} = \frac{\dot{m} \Delta P}{\rho} = \frac{(125.6 \text{ kg/s})(1.19 \times 10^5 \text{ N/m}^2)}{888.1 \text{ kg/m}^3} = 16.8 \text{ kW}$$

Discussion We need a 16.8-kW pump just to overcome the friction in the pipe as the oil flows in the 200-m-long pipe through the lake.

8–6 · TURBULENT FLOW IN TUBES

We mentioned earlier that flow in smooth tubes is usually fully turbulent for Re > 10,000. Turbulent flow is commonly utilized in practice because of the higher heat transfer coefficients associated with it. Most correlations for the friction and heat transfer coefficients in turbulent flow are based on experimental studies because of the difficulty in dealing with turbulent flow theoretically.

For *smooth* tubes, the friction factor in turbulent flow can be determined from the explicit *first Petukhov equation* [Petukhov (1970)] given as

Smooth tubes: $\quad f = (0.790 \ln \text{Re} - 1.64)^{-2} \qquad 3000 < \text{Re} < 5 \times 10^6 \qquad$ **(8–65)**

The Nusselt number in turbulent flow is related to the friction factor through the *Chilton–Colburn analogy* expressed as

$$\text{Nu} = 0.125 \, f \, \text{Re} \, \text{Pr}^{1/3} \qquad \text{(8–66)}$$

Once the friction factor is available, this equation can be used conveniently to evaluate the Nusselt number for both smooth and rough tubes.

For fully developed turbulent flow in *smooth tubes,* a simple relation for the Nusselt number can be obtained by substituting the simple power law relation $f = 0.184 \, \text{Re}^{-0.2}$ for the friction factor into Eq. 8–66. It gives

$$\text{Nu} = 0.023 \, \text{Re}^{0.8} \, \text{Pr}^{1/3} \qquad \binom{0.7 \leq \text{Pr} \leq 160}{\text{Re} > 10,000} \qquad \text{(8–67)}$$

which is known as the *Colburn equation.* The accuracy of this equation can be improved by modifying it as

$$\text{Nu} = 0.023 \, \text{Re}^{0.8} \, \text{Pr}^n \qquad \text{(8–68)}$$

where $n = 0.4$ for *heating* and 0.3 for *cooling* of the fluid flowing through the tube. This equation is known as the *Dittus–Boelter equation* [Dittus and Boelter (1930)] and it is preferred to the Colburn equation.

The proceeding equations can be used when the temperature difference between the fluid and wall surface is not large by evaluating all fluid properties at the *bulk mean fluid temperature* $T_b = (T_i + T_e)/2$. When the variation in properties is large due to a large temperature difference, the following equation due to Sieder and Tate (1936) can be used:

$$\text{Nu} = 0.027\ \text{Re}^{0.8}\text{Pr}^{1/3}\left(\frac{\mu}{\mu_s}\right)^{0.14} \qquad \left(\begin{matrix}0.7 \le \text{Pr} \le 17{,}600 \\ \text{Re} \ge 10{,}000\end{matrix}\right) \qquad \textbf{(8–69)}$$

Here all properties are evaluated at T_b except μ_s, which is evaluated at T_s.

The Nusselt number relations above are fairly simple, but they may give errors as large as 25 percent. This error can be reduced considerably to less than 10 percent by using more complex but accurate relations such as the *second Petukhov equation* expressed as

$$\text{Nu} = \frac{(f/8)\ \text{Re Pr}}{1.07 + 12.7(f/8)^{0.5}\ (\text{Pr}^{2/3} - 1)} \qquad \left(\begin{matrix}0.5 \le \text{Pr} \le 2000 \\ 10^4 < \text{Re} < 5 \times 10^6\end{matrix}\right) \qquad \textbf{(8–70)}$$

The accuracy of this relation at lower Reynolds numbers is improved by modifying it as [Gnielinski (1976)]

$$\text{Nu} = \frac{(f/8)(\text{Re} - 1000)\ \text{Pr}}{1 + 12.7(f/8)^{0.5}\ (\text{Pr}^{2/3} - 1)} \qquad \left(\begin{matrix}0.5 \le \text{Pr} \le 2000 \\ 3 \times 10^3 < \text{Re} < 5 \times 10^6\end{matrix}\right) \qquad \textbf{(8–71)}$$

where the friction factor f can be determined from an appropriate relation such as the first Petukhov equation. Gnielinski's equation should be preferred in calculations. Again properties should be evaluated at the bulk mean fluid temperature.

The relations above are not very sensitive to the *thermal conditions* at the tube surfaces and can be used for both T_s = constant and $\dot{q}_s$ = constant cases. Despite their simplicity, the correlations already presented give sufficiently accurate results for most engineering purposes. They can also be used to obtain rough estimates of the friction factor and the heat transfer coefficients in the transition region.

The relations given so far do not apply to liquid metals because of their very low Prandtl numbers. For liquid metals ($0.004 < \text{Pr} < 0.01$), the following relations are recommended by Sleicher and Rouse (1975) for $10^4 < \text{Re} < 10^6$:

Liquid metals, T_s = constant: $\qquad \text{Nu} = 4.8 + 0.0156\ \text{Re}^{0.85}\ \text{Pr}_s^{0.93} \qquad \textbf{(8–72)}$

Liquid metals, $\dot{q}_s$ = constant: $\qquad \text{Nu} = 6.3 + 0.0167\ \text{Re}^{0.85}\ \text{Pr}_s^{0.93} \qquad \textbf{(8–73)}$

where the subscript s indicates that the Prandtl number is to be evaluated at the surface temperature.

Rough Surfaces

Any irregularity or roughness on the surface disturbs the laminar sublayer, and affects the flow. Therefore, unlike laminar flow, the friction factor and the convection coefficient in turbulent flow are strong functions of surface roughness.

The friction factor in fully developed turbulent pipe flow depends on the Reynolds number and the **relative roughness** ε/D, which is the ratio of the mean height of roughness of the pipe to the pipe diameter. The functional form of this dependence cannot be obtained from a theoretical analysis, and all available results are obtained from painstaking experiments using artificially roughened surfaces (usually by gluing sand grains of a known size on the inner surfaces of the pipes). Most such experiments were conducted by Prandtl's student J. Nikuradse in 1933, followed by the works of others. The friction factor was calculated from the measurements of the flow rate and the pressure drop.

The experimental results obtained are presented in tabular, graphical, and functional forms obtained by curve-fitting experimental data. In 1939, Cyril F. Colebrook (1910–1997) combined the available data for transition and turbulent flow in smooth as well as rough pipes into the following implicit relation known as the **Colebrook equation**:

$$\frac{1}{\sqrt{f}} = -2.0 \log\left(\frac{\varepsilon/D}{3.7} + \frac{2.51}{\mathrm{Re}\sqrt{f}}\right) \quad \text{(turbulent flow)} \qquad \text{(8–74)}$$

We note that the logarithm in Eq. 8–74 is a base 10 rather than a natural logarithm. In 1942, the American engineer Hunter Rouse (1906–1996) verified Colebrook's equation and produced a graphical plot of f as a function of Re and the product $\mathrm{Re}\sqrt{f}$. He also presented the laminar flow relation and a table of commercial pipe roughness. Two years later, Lewis F. Moody (1880–1953) redrew Rouse's diagram into the form commonly used today. The now famous **Moody chart** is given in the appendix as Fig. A–20. It presents the Darcy friction factor for pipe flow as a function of the Reynolds number and ε/D over a wide range. It is probably one of the most widely accepted and used charts in engineering. Although it is developed for circular pipes, it can also be used for noncircular pipes by replacing the diameter by the hydraulic diameter.

For smooth pipes, the agreement between the Petukhov and Colebrook equations is very good. The friction factor is minimum for a smooth pipe (but still not zero because of the no-slip condition), and increases with roughness (Fig. 8–25).

Commercially available pipes differ from those used in the experiments in that the roughness of pipes in the market is not uniform and it is difficult to give a precise description of it. Equivalent roughness values for some commercial pipes are given in Table 8–3 as well as on the Moody chart. But it should be kept in mind that these values are for new pipes, and the relative roughness of pipes may increase with use as a result of corrosion, scale

Relative Roughness, ε/D	Friction Factor, f
0.0*	0.0119
0.00001	0.0119
0.0001	0.0134
0.0005	0.0172
0.001	0.0199
0.005	0.0305
0.01	0.0380
0.05	0.0716

*Smooth surface. All values are for Re = 10^6, and are calculated from Eq. 8–74.

FIGURE 8–25
The friction factor is minimum for a smooth pipe and increases with roughness.

TABLE 8–2

Standard sizes for Schedule 40 steel pipes

Nominal Size, in	Actual Inside Diameter, in
⅛	0.269
¼	0.364
⅜	0.493
½	0.622
¾	0.824
1	1.049
1½	1.610
2	2.067
2½	2.469
3	3.068
5	5.047
10	10.02

TABLE 8–3

Equivalent roughness values for new commercial pipes*

Material	Roughness, ε	
	ft	mm
Glass, plastic	0 (smooth)	
Concrete	0.003–0.03	0.9–9
Wood stave	0.0016	0.5
Rubber, smoothed	0.000033	0.01
Copper or brass tubing	0.000005	0.0015
Cast iron	0.00085	0.26
Galvanized iron	0.0005	0.15
Wrought iron	0.00015	0.046
Stainless steel	0.000007	0.002
Commercial steel	0.00015	0.045

*The uncertainty in these values can be as much as ±60 percent.

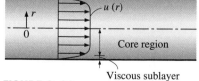

FIGURE 8–26

In turbulent flow, the velocity profile is nearly a straight line in the core region, and any significant velocity gradients occur in the viscous sublayer.

buildup, and precipitation. As a result, the friction factor may increase by a factor of 5 to 10. Actual operating conditions must be considered in the design of piping systems. Also, the Moody chart and its equivalent Colebrook equation involve several uncertainties (the roughness size, experimental error, curve fitting of data, etc.), and thus the results obtained should not be treated as "exact." It is usually considered to be accurate to ±15 percent over the entire range in the figure.

The Colebrook equation is implicit in f, and thus the determination of the friction factor requires some iteration unless an equation solver such as EES is used. An approximate explicit relation for f was given by S. E. Haaland in 1983 as

$$\frac{1}{\sqrt{f}} \cong -1.8 \log \left[\frac{6.9}{\text{Re}} + \left(\frac{\varepsilon/D}{3.7} \right)^{1.11} \right] \qquad \textbf{(8–75)}$$

The results obtained from this relation are within 2 percent of those obtained from the Colebrook equation. If more accurate results are desired, Eq. 8–75 can be used as a good *first guess* in a Newton iteration when using a programmable calculator or a spreadsheet to solve for f with Eq. 8–74.

In turbulent flow, wall roughness increases the heat transfer coefficient h by a factor of 2 or more [Dipprey and Sabersky (1963)]. The convection heat transfer coefficient for rough tubes can be calculated approximately from the Nusselt number relations such as Eq. 8–71 by using the friction factor determined from the Moody chart or the Colebrook equation. However, this approach is not very accurate since there is no further increase in h with f for $f > 4f_{\text{smooth}}$ [Norris (1970)] and correlations developed specifically for rough tubes should be used when more accuracy is desired.

Developing Turbulent Flow in the Entrance Region

The entry lengths for turbulent flow are typically short, often just 10 tube diameters long, and thus the Nusselt number determined for fully developed turbulent flow can be used approximately for the entire tube. This simple approach gives reasonable results for pressure drop and heat transfer for long tubes and conservative results for short ones. Correlations for the friction and heat transfer coefficients for the entrance regions are available in the literature for better accuracy.

Turbulent Flow in Noncircular Tubes

The velocity and temperature profiles in turbulent flow are nearly straight lines in the core region, and any significant velocity and temperature gradients occur in the viscous sublayer (Fig. 8–26). Despite the small thickness of viscous sublayer (usually much less than 1 percent of the pipe diameter), the characteristics of the flow in this layer are very important since they set the stage for flow in the rest of the pipe. Therefore, pressure drop and heat transfer characteristics of turbulent flow in tubes are dominated by the very thin viscous sublayer next to the wall surface, and the shape of the core region is not of much significance. Consequently, the turbulent flow relations given above for circular tubes can also be used for noncircular tubes with reasonable accuracy by replacing the diameter D in the evaluation of the Reynolds number by the hydraulic diameter $D_h = 4A_c/p$.

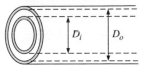

FIGURE 8–27

A double-tube heat exchanger that consists of two concentric tubes.

Flow through Tube Annulus

Some simple heat transfer equipments consist of two concentric tubes, and are properly called *double-tube heat exchangers* (Fig. 8–27). In such devices, one fluid flows through the tube while the other flows through the annular space. The governing differential equations for both flows are identical. Therefore, steady laminar flow through an annulus can be studied analytically by using suitable boundary conditions.

Consider a concentric annulus of inner diameter D_i and outer diameter D_o. The hydraulic diameter of annulus is

$$D_h = \frac{4A_c}{p} = \frac{4\pi(D_o^2 - D_i^2)/4}{\pi(D_o + D_i)} = D_o - D_i$$

Annular flow is associated with two Nusselt numbers—Nu_i on the inner tube surface and Nu_o on the outer tube surface—since it may involve heat transfer on both surfaces. The Nusselt numbers for fully developed laminar flow with one surface isothermal and the other adiabatic are given in Table 8–4. When Nusselt numbers are known, the convection coefficients for the inner and the outer surfaces are determined from

$$Nu_i = \frac{h_i D_h}{k} \quad \text{and} \quad Nu_o = \frac{h_o D_h}{k} \tag{8–76}$$

For fully developed turbulent flow, the inner and outer convection coefficients are approximately equal to each other, and the tube annulus can be treated as a noncircular duct with a hydraulic diameter of $D_h = D_o - D_i$. The Nusselt number in this case can be determined from a suitable turbulent flow relation such as the Gnielinski equation. To improve the accuracy of Nusselt numbers obtained from these relations for annular flow, Petukhov and Roizen (1964) recommend multiplying them by the following correction factors when one of the tube walls is adiabatic and heat transfer is through the other wall:

$$F_i = 0.86 \left(\frac{D_i}{D_o}\right)^{-0.16} \quad \text{(outer wall adiabatic)} \tag{8–77}$$

$$F_o = 0.86 \left(\frac{D_i}{D_o}\right)^{-0.16} \quad \text{(inner wall adiabatic)} \tag{8–78}$$

Heat Transfer Enhancement

Tubes with rough surfaces have much higher heat transfer coefficients than tubes with smooth surfaces. Therefore, tube surfaces are often intentionally *roughened, corrugated,* or *finned* in order to *enhance* the convection heat transfer coefficient and thus the convection heat transfer rate (Fig. 8–28). Heat transfer in turbulent flow in a tube has been increased by as much as 400 percent by roughening the surface. Roughening the surface, of course, also increases the friction factor and thus the power requirement for the pump or the fan.

The convection heat transfer coefficient can also be increased by inducing pulsating flow by pulse generators, by inducing swirl by inserting a twisted tape into the tube, or by inducing secondary flows by coiling the tube.

TABLE 8–4

Nusselt number for fully developed laminar flow in an annulus with one surface isothermal and the other adiabatic (Kays and Perkins, 1972)

D_i/D_o	Nu_i	Nu_o
0	—	3.66
0.05	17.46	4.06
0.10	11.56	4.11
0.25	7.37	4.23
0.50	5.74	4.43
1.00	4.86	4.86

(a) Finned surface

Fin

(b) Roughened surface

Roughness

FIGURE 8–28

Tube surfaces are often *roughened, corrugated,* or *finned* in order to *enhance* convection heat transfer.

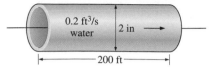

FIGURE 8–29
Schematic for Example 8–4.

EXAMPLE 8–4   Pressure Drop in a Water Tube

Water at 60°F ($\rho = 62.36$ lbm/ft^3 and $\mu = 7.536 \times 10^{-4}$ lbm/ft · s) is flowing steadily in a 2-in-internal-diameter horizontal tube made of stainless steel at a rate of 0.2 ft^3/s (Fig. 8–29). Determine the pressure drop and the required pumping power input for flow through a 200-ft-long section of the tube.

SOLUTION   The flow rate through a specified water tube is given. The pressure drop and the pumping power requirements are to be determined.

Assumptions   **1** The flow is steady and incompressible. **2** The entrance effects are negligible, and thus the flow is fully developed. **3** The tube involves no components such as bends, valves, and connectors. **4** The piping section involves no work devices such as a pump or a turbine.

Properties   The density and dynamic viscosity of water are given to be $\rho = 62.36$ lbm/ft^3 and $\mu = 7.536 \times 10^{-4}$ lbm/ft · s. For stainless steel, $\varepsilon = 0.000007$ ft (Table 8–3).

Analysis   First we calculate the mean velocity and the Reynolds number to determine the flow regime:

$$V = \frac{\dot{V}}{A_c} = \frac{\dot{V}}{\pi D^2/4} = \frac{0.2 \text{ ft}^3/\text{s}}{\pi (2/12 \text{ ft})^2/4} = 9.17 \text{ ft/s}$$

$$\text{Re} = \frac{\rho VD}{\mu} = \frac{(62.36 \text{ lbm/ft}^3)(9.17 \text{ ft/s})(2/12 \text{ ft})}{7.536 \times 10^{-4} \text{ lbm/ft · s}} = 126,400$$

which is greater than 10,000. Therefore, the flow is turbulent. The relative roughness of the tube is

$$\varepsilon/D = \frac{0.000007 \text{ ft}}{2/12 \text{ ft}} = 0.000042$$

The friction factor corresponding to this relative roughness and the Reynolds number can simply be determined from the Moody chart. To avoid the reading error, we determine it from the Colebrook equation:

$$\frac{1}{\sqrt{f}} = -2.0 \log\left(\frac{\varepsilon/D}{3.7} + \frac{2.51}{\text{Re}\sqrt{f}}\right) \rightarrow \frac{1}{\sqrt{f}} = -2.0 \log\left(\frac{0.000042}{3.7} + \frac{2.51}{126,400\sqrt{f}}\right)$$

Using an equation solver or an iterative scheme, the friction factor is determined to be $f = 0.0174$. Then the pressure drop and the required power input become

$$\Delta P = f\frac{L}{D}\frac{\rho V^2}{2} = 0.0174 \frac{200 \text{ ft}}{2/12 \text{ ft}} \frac{(62.36 \text{ lbm/ft}^3)(9.17 \text{ ft/s})^2}{2}\left(\frac{1 \text{ lbf}}{32.174 \text{ lbm · ft/s}^2}\right)$$

$$= 1700 \text{ lbf/ft}^2 = 11.8 \text{ psi}$$

$$\dot{W}_{\text{pump}} = \dot{V}\Delta P = (0.2 \text{ ft}^3/\text{s})(1700 \text{ lbf/ft}^2)\left(\frac{1 \text{ W}}{0.73756 \text{ lbf · ft/s}}\right) = 461 \text{ W}$$

Therefore, power input in the amount of 461 W is needed to overcome the frictional losses in the tube.

Discussion   The friction factor could also be determined easily from the explicit Haaland relation. It would give $f = 0.0172$, which is sufficiently close to 0.0174. Also, the friction factor corresponding to $\varepsilon = 0$ in this case is 0.0170, which indicates that stainless steel tubes can be assumed to be smooth with negligible error.

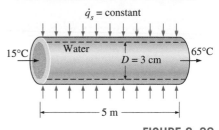

FIGURE 8–30
Schematic for Example 8–5.

EXAMPLE 8–5   Heating of Water by Resistance Heaters in a Tube

Water is to be heated from 15°C to 65°C as it flows through a 3-cm-internal-diameter 5-m-long tube (Fig. 8–30). The tube is equipped with an electric resistance heater that provides uniform heating throughout the surface of the tube. The outer surface of the heater is well insulated, so that in steady operation all the heat generated in the heater is transferred to the water in the tube. If the system is to provide hot water at a rate of 10 L/min, determine the power rating of the resistance heater. Also, estimate the inner surface temperature of the tube at the exit.

SOLUTION   Water is to be heated in a tube equipped with an electric resistance heater on its surface. The power rating of the heater and the inner surface temperature at the exit are to be determined.
Assumptions   **1** Steady flow conditions exist. **2** The surface heat flux is uniform. **3** The inner surfaces of the tube are smooth.
Properties   The properties of water at the bulk mean temperature of $T_b = (T_i + T_e)/2 = (15 + 65)/2 = 40°C$ are (Table A–9)

$$\rho = 992.1 \text{ kg/m}^3 \qquad\qquad c_p = 4179 \text{ J/kg} \cdot °C$$
$$k = 0.631 \text{ W/m} \cdot °C \qquad\qquad Pr = 4.32$$
$$\nu = \mu/\rho = 0.658 \times 10^{-6} \text{ m}^2/\text{s}$$

Analysis   The cross sectional and heat transfer surface areas are

$$A_c = \tfrac{1}{4}\pi D^2 = \tfrac{1}{4}\pi(0.03 \text{ m})^2 = 7.069 \times 10^{-4} \text{ m}^2$$
$$A_s = \pi DL = \pi(0.03 \text{ m})(5 \text{ m}) = 0.471 \text{ m}^2$$

The volume flow rate of water is given as $\dot{V} = 10$ L/min $= 0.01$ m³/min. Then the mass flow rate becomes

$$\dot{m} = \rho\dot{V} = (992.1 \text{ kg/m}^3)(0.01 \text{ m}^3/\text{min}) = 9.921 \text{ kg/min} = 0.1654 \text{ kg/s}$$

To heat the water at this mass flow rate from 15°C to 65°C, heat must be supplied to the water at a rate of

$$\dot{Q} = \dot{m}c_p(T_e - T_i)$$
$$= (0.1654 \text{ kg/s})(4.179 \text{ kJ/kg} \cdot °C)(65 - 15)°C$$
$$= 34.6 \text{ kJ/s} = 34.6 \text{ kW}$$

All of this energy must come from the resistance heater. Therefore, the power rating of the heater must be **34.6 kW**.
   The surface temperature T_s of the tube at any location can be determined from

$$\dot{q}_s = h(T_s - T_m) \quad \rightarrow \quad T_s = T_m + \frac{\dot{q}_s}{h}$$

where h is the heat transfer coefficient and T_m is the mean temperature of the fluid at that location. The surface heat flux is constant in this case, and its value can be determined from

$$\dot{q}_s = \frac{\dot{Q}}{A_s} = \frac{34.6 \text{ kW}}{0.471 \text{ m}^2} = 73.46 \text{ kW/m}^2$$

To determine the heat transfer coefficient, we first need to find the mean velocity of water and the Reynolds number:

$$V_{avg} = \frac{\dot{V}}{A_c} = \frac{0.010 \text{ m}^3/\text{min}}{7.069 \times 10^{-4} \text{ m}^2} = 14.15 \text{ m/min} = 0.236 \text{ m/s}$$

$$\text{Re} = \frac{V_{avg} D}{\nu} = \frac{(0.236 \text{ m/s})(0.03 \text{ m})}{0.658 \times 10^{-6} \text{ m}^2/\text{s}} = 10,760$$

which is greater than 10,000. Therefore, the flow is turbulent and the entry length is roughly

$$L_h \approx L_t \approx 10D = 10 \times 0.03 = 0.3 \text{ m}$$

which is much shorter than the total length of the tube. Therefore, we can assume fully developed turbulent flow in the entire tube and determine the Nusselt number from

$$\text{Nu} = \frac{hD}{k} = 0.023 \text{ Re}^{0.8} \text{ Pr}^{0.4} = 0.023(10,760)^{0.8} (4.32)^{0.4} = 69.4$$

Then,

$$h = \frac{k}{D} \text{Nu} = \frac{0.631 \text{ W/m} \cdot {}^\circ\text{C}}{0.03 \text{ m}} (69.4) = 1460 \text{ W/m}^2 \cdot {}^\circ\text{C}$$

and the surface temperature of the pipe at the exit becomes

$$T_s = T_m + \frac{\dot{q}_s}{h} = 65^\circ\text{C} + \frac{73,460 \text{ W/m}^2}{1460 \text{ W/m}^2 \cdot {}^\circ\text{C}} = \textbf{115}^\circ\textbf{C}$$

Discussion Note that the inner surface temperature of the tube will be 50°C higher than the mean water temperature at the tube exit. This temperature difference of 50°C between the water and the surface will remain constant throughout the fully developed flow region.

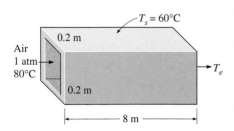

FIGURE 8–31
Schematic for Example 8–6.

EXAMPLE 8–6 Heat Loss from the Ducts of a Heating System

Hot air at atmospheric pressure and 80°C enters an 8-m-long uninsulated square duct of cross section 0.2 m × 0.2 m that passes through the attic of a house at a rate of 0.15 m³/s (Fig. 8–31). The duct is observed to be nearly isothermal at 60°C. Determine the exit temperature of the air and the rate of heat loss from the duct to the attic space.

SOLUTION Heat loss from uninsulated square ducts of a heating system in the attic is considered. The exit temperature and the rate of heat loss are to be determined.

Assumptions **1** Steady operating conditions exist. **2** The inner surfaces of the duct are smooth. **3** Air is an ideal gas.

Properties We do not know the exit temperature of the air in the duct, and thus we cannot determine the bulk mean temperature of air, which is the temperature at which the properties are to be determined. The temperature of air at the inlet is 80°C and we expect this temperature to drop somewhat as a result of heat loss through the duct whose surface is at 60°C. At 80°C and 1 atm we read (Table A–15)

$$\rho = 0.9994 \text{ kg/m}^3 \qquad c_p = 1008 \text{ J/kg} \cdot {}^\circ\text{C}$$
$$k = 0.02953 \text{ W/m} \cdot {}^\circ\text{C} \qquad \text{Pr} = 0.7154$$
$$\nu = 2.097 \times 10^{-5} \text{ m}^2/\text{s}$$

Analysis The characteristic length (which is the hydraulic diameter), the mean velocity, and the Reynolds number in this case are

$$D_h = \frac{4A_c}{p} = \frac{4a^2}{4a} = a = 0.2 \text{ m}$$

$$V_{\text{avg}} = \frac{\dot{V}}{A_c} = \frac{0.15 \text{ m}^3/\text{s}}{(0.2 \text{ m})^2} = 3.75 \text{ m/s}$$

$$\text{Re} = \frac{V_{\text{avg}} D_h}{\nu} = \frac{(3.75 \text{ m/s})(0.2 \text{ m})}{2.097 \times 10^{-5} \text{ m}^2/\text{s}} = 35{,}765$$

which is greater than 10,000. Therefore, the flow is turbulent and the entry lengths in this case are roughly

$$L_h \approx L_t \approx 10D = 10 \times 0.2 \text{ m} = 2 \text{ m}$$

which is much shorter than the total length of the duct. Therefore, we can assume fully developed turbulent flow in the entire duct and determine the Nusselt number from

$$\text{Nu} = \frac{hD_h}{k} = 0.023 \, \text{Re}^{0.8} \, \text{Pr}^{0.3} = 0.023(35{,}765)^{0.8} (0.7154)^{0.3} = 91.4$$

Then,

$$h = \frac{k}{D_h} \text{Nu} = \frac{0.02953 \text{ W/m} \cdot {}^\circ\text{C}}{0.2 \text{ m}} (91.4) = 13.5 \text{ W/m}^2 \cdot {}^\circ\text{C}$$

$$A_s = 4aL = 4 \times (0.2 \text{ m})(8 \text{ m}) = 6.4 \text{ m}^2$$

$$\dot{m} = \rho\dot{V} = (0.9994 \text{ kg/m}^3)(0.15 \text{ m}^3/\text{s}) = 0.150 \text{ kg/s}$$

Next, we determine the exit temperature of air from

$$T_e = T_s - (T_s - T_i) \exp(-hA_s/\dot{m}c_p)$$

$$= 60°\text{C} - [(60 - 80)°\text{C}] \exp\left[-\frac{(13.5 \text{ W/m}^2 \cdot {}^\circ\text{C})(6.4 \text{ m}^2)}{(0.150 \text{ kg/s})(1008 \text{ J/kg} \cdot {}^\circ\text{C})}\right]$$

$$= 71.3°\text{C}$$

Then the logarithmic mean temperature difference and the rate of heat loss from the air become

$$\Delta T_{\ln} = \frac{T_i - T_e}{\ln \dfrac{T_s - T_e}{T_s - T_i}} = \frac{80 - 71.3}{\ln \dfrac{60 - 71.3}{60 - 80}} = -15.2°C$$

$$\dot{Q} = hA_s \, \Delta T_{\ln} = (13.5 \text{ W/m}^2 \cdot °C)(6.4 \text{ m}^2)(-15.2°C) = -1313 \text{ W}$$

Therefore, air will lose heat at a rate of 1313 W as it flows through the duct in the attic.

Discussion The average fluid temperature is $(80 + 71.3)/2 = 75.7°C$, which is sufficiently close to 80°C at which we evaluated the properties of air. Therefore, it is not necessary to re-evaluate the properties at this temperature and to repeat the calculations.

TOPIC OF SPECIAL INTEREST

*Transitional Flow in Tubes**

An important design problem in industrial heat exchangers arises when flow inside the tubes falls into the transition region. In practical engineering design, the usual recommendation is to avoid design and operation in this region; however, this is not always feasible under design constraints. The usually cited transitional Reynolds number range of about 2300 (onset of turbulence) to 10,000 (fully turbulent condition) applies, strictly speaking, to a very steady and uniform entry flow with a rounded entrance. If the flow has a disturbed entrance typical of heat exchangers, in which there is a sudden contraction and possibly even a re-entrant entrance, the transitional Reynolds number range will be much different.

Ghajar and coworkers in a series of papers (listed in the references) have experimentally investigated the inlet configuration effects on the fully developed transitional pressure drop under isothermal and heating conditions; and developing and fully developed transitional forced and mixed convection heat transfer in circular tubes. Based on their experimental data, they have developed practical and easy to use correlations for the friction coefficient and the Nusselt number in the transition region between laminar and turbulent flows. This section provides a brief summary of their work in the transition region.

Pressure Drop in the Transition Region

Pressure drops are measured in circular tubes for fully developed flows in the transition regime for three types of inlet configurations shown in Fig. 8–32: re-entrant (tube extends beyond tubesheet face into head of

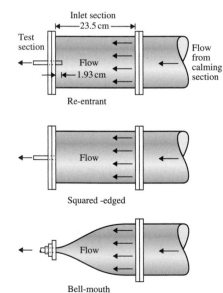

FIGURE 8–32
Schematic of the three differernt inlet configurations.

*This section is contributed by Professor Afshin J. Ghajar of Oklahoma State University.

distributor), square-edged (tube end is flush with tubesheet face), and bell-mouth (a tapered entrance of tube from tubesheet face) under isothermal and heating conditions, respectively. The widely used expressions for the *friction factor f* (also called the *Darcy friction factor*) or the *friction coefficient* C_f (also called the *Fanning friction factor*) in laminar and turbulent flows with heating are

$$f_{\text{lam}} = 4C_{f,\,\text{lam}} = 4\left(\frac{16}{\text{Re}}\right)\left(\frac{\mu_b}{\mu_s}\right)^m \tag{8–79}$$

$$f_{\text{turb}} = 4C_{f,\,\text{turb}} = 4\left(\frac{0.0791}{\text{Re}^{0.25}}\right)\left(\frac{\mu_b}{\mu_s}\right)^m \tag{8–80}$$

where the factors at the end account for the wall temperature effect on viscosity. The exponent m for laminar flows depends on a number of factors while for turbulent flows the most typically quoted value for heating is -0.25. The transition friction factor is given as (Tam and Ghajar, 1997)

$$f_{\text{trans}} = 4C_{f,\,\text{trans}} = 4\left[1 + \left(\frac{\text{Re}}{A}\right)^B\right]^C\left(\frac{\mu_b}{\mu_s}\right)^m \tag{8–81}$$

where

$$m = m_1 - m_2\, \text{Gr}^{m_3}\, \text{Pr}^{m_4} \tag{8–82}$$

and the Grashof number (Gr) which is a dimensionless number representing the ratio of the buoyancy force to the viscous force is defined as $\text{Gr} = g\beta D^3(T_s - T_b)/\nu^2$ (see Chapter 9 for more details). All properties appearing in the dimensionless numbers C_f, f, Re, Pr, and Gr are all evaluated at the bulk fluid temperature T_b. The values of the empirical constants in Eqs. 8–81 and 8–82 are listed in Table 8–5. The range of application of Eq. 8–81 for the transition friction factor is as follows:

Re-entrant: $2700 \leq \text{Re} \leq 5500$, $16 \leq \text{Pr} \leq 35$, $7410 \leq \text{Gr} \leq 158{,}300$, $1.13 \leq \mu_b/\mu_s \leq 2.13$

Square-edged: $3500 \leq \text{Re} \leq 6900$, $12 \leq \text{Pr} \leq 29$, $6800 \leq \text{Gr} \leq 104{,}500$, $1.11 \leq \mu_b/\mu_s \leq 1.89$

Bell-mouth: $5900 \leq \text{Re} \leq 9600$, $8 \leq \text{Pr} \leq 15$, $11{,}900 \leq \text{Gr} \leq 353{,}000$, $1.05 \leq \mu_b/\mu_s \leq 1.47$

TABLE 8–5

Constants for transition friction coefficient correlation

Inlet Geometry	A	B	C	m_1	m_2	m_3	m_4
Re-entrant	5840	−0.0145	−6.23	−1.10	0.460	−0.133	4.10
Square-edged	4230	−0.1600	−6.57	−1.13	0.396	−0.160	5.10
Bell-mouth	5340	−0.0990	−6.32	−2.58	0.420	−0.410	2.46

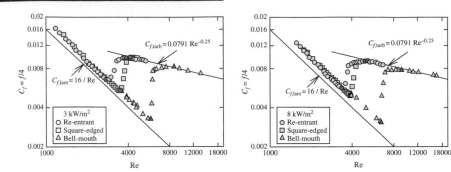
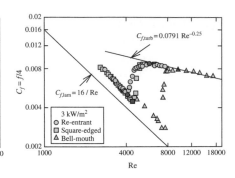

FIGURE 8–33
Fully developed friction coeffficients for three different inlet configurations and heat fluxes (filled symbols designate the start and end of the transition region for each inlet.
(From Tam and Ghajar, 1997.)

These correlations captured about 82% of measured data within an error band of $\pm 10\%$, and 98% of measured data with $\pm 20\%$. For laminar flows with heating, Tam and Ghajar give the following constants for determining the exponent m in Eq. 8–79: $m_1 = 1.65$, $m_2 = 0.013$, $m_3 = 0.170$, and $m_4 = 0.840$, which is applicable over the following range of parameters:

$$1100 \leq Re \leq 7400, \, 6 \leq Pr \leq 36, \, 17{,}100 \leq Gr \leq 95{,}600,$$
$$\text{and } 1.25 \leq \mu_b/\mu_s \leq 2.40.$$

The fully developed friction coefficient results for the three different inlet configurations shown in Fig. 8–33 clearly establish the influence of heating rate on the beginning and end of the transition regions, for each inlet configuration. In the laminar and transition regions, heating seems to have a significant influence on the value of the friction coefficient. However, in the turbulent region, heating did not affect the magnitude of the friction coefficient. The significant influence of heating on the values of friction coefficient in the laminar and transition regions is directly due to the effect of secondary flow.

The isothermal friction coefficients for the three inlet types showed that the range of the Reynolds number values at which transition flow exists is strongly inlet-geometry dependent. Furthermore, heating caused an increase in the laminar and turbulent friction coefficients and an increase in the lower and upper limits of the isothermal transition regime boundaries. The friction coefficient transition Reynolds number ranges for the isothermal and nonisothermal (three different heating rates) and the three different inlets used in their study are summarized in Table 8–6.

TABLE 8–6

Transition Reynolds numbers for friction coefficient

Heat Flux	Re-entrant	Square-Edged	Bell-Mouth
0 kW/m² (isothermal)	2870 < Re < 3500	3100 < Re < 3700	5100 < Re < 6100
3 kW/m²	3060 < Re < 3890	3500 < Re < 4180	5930 < Re < 8730
8 kW/m²	3350 < Re < 4960	3860 < Re < 5200	6480 < Re < 9110
16 kW/m²	4090 < Re < 5940	4450 < Re < 6430	7320 < Re < 9560

Figure 8–34 shows the influence of inlet configuration on the beginning and end of the isothermal fully developed friction coefficients in the transition region.

Note that the isothermal fully developed friction coefficients in the laminar, turbulent, and transition regions can be obtained easily from Eqs. 8–79, 8–80, and 8–81, respectively, by setting the exponent on the viscosity ratio correction to unity (i.e. with $m = 0$).

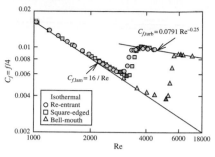

FIGURE 8–34

Influence of different inlet configurations on the isothermal fully developed friction coefficients (filled symbols designate the start and end of the transition region for each inlet).
(From Tam and Ghajar, 1997.)

EXAMPLE 8–7   Nonisothermal Fully Developed Friction Coefficient in the Transition Region

A tube with a bell-mouth inlet configuration is subjected to 8 kW/m² uniform wall heat flux. The tube has an inside diameter of 0.0158 m and a flow rate of 1.32×10^{-4} m³/s. The liquid flowing inside the tube is ethylene glycol-distilled water mixture with a mass fraction of 0.34. The properties of the ethylene glycol-distilled water mixture at the location of interest are Pr = 11.6, $\nu = 1.39 \times 10^{-6}$ m²/s and $\mu_b/\mu_s = 1.14$. Determine the fully developed friction coefficient at a location along the tube where the Grashof number is Gr = 60,800. What would the answer be if a square-edged inlet is used instead?

SOLUTION A liquid mixture flowing in a tube is subjected to uniform wall heat flux. The friction coefficients are to be determined for the bell-mouth and square-edged inlet cases.

Assumptions Steady operating conditions exist.

Properties The properties of the ethylene glycol-distilled water mixture are given to be Pr = 11.6, $\nu - 1.39 \times 10^{-6}$ m²/s and $\mu_b/\mu_s - 1.14$.

Analysis For the calculation of the nonisothermal fully developed friction coefficient, it is necessary to determine the flow regime before making any decision regarding which friction coefficient relation to use. The Reynolds number at the specified location is

$$\text{Re} = \frac{(\dot{V}/A_c)D}{\nu} = \frac{[(1.32 \times 10^{-4}\ \text{m}^3/\text{s})/(1.961 \times 10^{-4}\ \text{m}^2)](0.0158\ \text{m})}{1.39 \times 10^{-6}\ \text{m}^2/\text{s}} = 7651$$

since

$$A_c = \pi D^2/4 = \pi (0.0158\,\text{m})^2/4 = 1.961 \times 10^{-4}\,\text{m}^2$$

From Table 8–6, we see that for a bell-mouth inlet and a heat flux of 8 kW/m² the flow is in the transition region. Therefore, Eq. 8–81 applies. Reading the constants A, B, and C and m_1, m_2, m_3, and m_4 from Table 8–5, the friction coefficient is determined to be

$$C_{f,\text{trans}} = \left[1 + \left(\frac{\text{Re}}{A}\right)^B\right]^C \left(\frac{\mu_b}{\mu_s}\right)^m$$

$$= \left[1 + \left(\frac{7651}{5340}\right)^{-0.099}\right]^{-6.32} (1.14)^{-2.58 - 0.42 \times 60,800^{-0.41} \times 11.6^{2.46}} = \mathbf{0.010}$$

Square-Edged Inlet Case For this inlet shape, the Reynolds number of the flow is the same as that of the bell-mouth inlet (Re = 7651). However, it is necessary to check the type of flow regime for this particular inlet with 8 kW/m^2 of heating. From Table 8–6, the transition Reynolds number range for this case is 3860 < Re < 5200, which means that the flow in this case is turbulent and Eq. 8–80 is the appropriate equation to use. It gives

$$C_{f,\text{turb}} = \left(\frac{0.0791}{\text{Re}^{0.25}}\right)\left(\frac{\mu_b}{\mu_s}\right)^{\mu} = \left(\frac{0.0791}{7651^{0.25}}\right)(1.14)^{-0.25} = \mathbf{0.0082}$$

Discussion Note that the friction factors f can be determined by multiplying the friction coefficient values by 4.

Heat Transfer in the Transition Region

Ghajar and coworkers also experimentally investigated the inlet configuration effects on heat transfer in the transition region between laminar and turbulent flows in tubes for the same three inlet configurations shown in Fig. 8–32. They proposed some prediction methods for this regime to bridge between laminar methods and turbulent methods, applicable to forced and mixed convection in the entrance and fully developed regions for the three types of inlet configurations, which are presented next. The local heat transfer coefficient in transition flow is obtained from the transition Nusselt number, Nu_{trans}, which is calculated as follows at a distance x from the entrance:

$$\text{Nu}_{\text{trans}} = \text{Nu}_{\text{lam}} + \{\exp[(a - \text{Re})/b] + \text{Nu}_{\text{turb}}^c\}^c \qquad \textbf{(8–83)}$$

where Nu_{lam} is the laminar flow Nusselt number for entrance region laminar flows with natural convection effects,

$$\text{Nu}_{\text{lam}} = 1.24\left[\left(\frac{\text{Re}\,\text{Pr}\,D}{x}\right) + 0.025(\text{Gr}\,\text{Pr})^{0.75}\right]^{1/3}\left(\frac{\mu_b}{\mu_s}\right)^{0.14} \qquad \textbf{(8–84)}$$

and Nu_{turb} is the turbulent flow Nusselt number with developing flow effects,

$$\text{Nu}_{\text{turb}} = 0.023\text{Re}^{0.8}\,\text{Pr}^{0.385}\left(\frac{x}{D}\right)^{-0.0054}\left(\frac{\mu_b}{\mu_s}\right)^{0.14} \qquad \textbf{(8–85)}$$

The physical properties appearing in the dimensionless numbers Nu, Re, Pr, and Gr all are evaluated at the bulk fluid temperature T_b. The values of the empirical constants a, b, and c in Eq. 8–83 depend on the inlet configuration and are given in Table 8–7. The viscosity ratio accounts for the temperature effect on the process. The range of application of the heat transfer method based on their database of 1290 points (441 points for re-entrant

TABLE 8–7

Constants for transition heat transfer correlation

Inlet Geometry	a	b	c
Re-entrant	1766	276	−0.955
Square-edged	2617	207	−0.950
Bell-mouth	6628	237	−0.980

inlet, 416 for square-edged inlet and 433 points for bell-mouth inlet) is as follows:

Re-entrant: $3 \leq x/D \leq 192$, $1700 \leq Re \leq 9100$, $5 \leq Pr \leq 51$, $4000 \leq Gr \leq 210{,}000$, $1.2 \leq \mu_b/\mu_s \leq 2.2$

Square-edged: $3 \leq x/D \leq 192$, $1600 \leq Re \leq 10{,}700$, $5 \leq Pr \leq 55$, $4000 \leq Gr \leq 250{,}000$, $1.2 \leq \mu_b/\mu_s \leq 2.6$

Bell-mouth: $3 \leq x/D \leq 192$, $3300 \leq Re \leq 11{,}100$, $13 \leq Pr \leq 77$, $6000 \leq Gr \leq 110{,}000$, $1.2 \leq \mu_b/\mu_s \leq 3.1$

These correlations capture about 70% of measured data within an error band of $\pm 10\%$, and 97% of measured data with $\pm 20\%$, which is remarkable for transition flows. The individual expressions above for Nu_{lam} and Nu_{turb} can be used alone for developing flows in those respective regimes. The lower and upper limits of the heat transfer transition Reynolds number ranges for the three different inlets are summarized in Table 8–8. The results shown in this table indicate that the re-entrant inlet configuration causes the earliest transition from laminar flow into the transition regime (at about 2000) while the bell-mouth entrance retards this regime change (at about 3500). The square-edged entrance falls in between (at about 2400), which is close to the often quoted value of 2300 in most textbooks.

Figure 8–35 clearly shows the influence of inlet configuration on the beginning and end of the heat transfer transition region. This figure plots the local average peripheral heat transfer coefficients in terms of the Colburn j factor ($j_H = St\,Pr^{0.67}$) versus local Reynolds number for all flow regimes at the length-to-diameter ratio of 192, and St is the Stanton number, which is also a dimensionless heat transfer coefficient (see Chapter 6 for more details), defined as $St = Nu/(Re\,Pr)$. The filled symbols in Fig. 8–35 represent the start and end of the heat transfer transition region for each inlet configuration. Note the large influence of natural convection superimposed on the forced convective laminar-flow heat transfer process ($Nu = 4.364$ for a fully developed laminar flow with a uniform heat-flux boundary condition without buoyancy effects), yielding a mixed convection value of about $Nu = 14.5$. Equation 8–84 includes this buoyancy effect through the Grashof number.

In a subsequent study, Tam and Ghajar (1998) experimentally investigated the behaviour of local heat transfer coefficients in the transition region for a tube with a bell-mouth inlet. This type of inlet is used in some heat exchangers mainly to avoid the presence of eddies which are believed to be one of the causes for erosion in the tube inlet region. For the

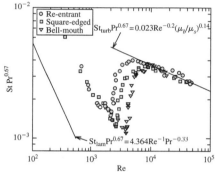

FIGURE 8–35

Influence of different inlets on the heat transfer transition region at $x/D = 192$ (filled symbols designate the start and end of the transition region for each inlet) between limits of Dittus–Boelter correlation ($Nu = 0.023\ Re^{0.8}\ Pr^n$) for fully developed turbulent flow (using $n = 1/3$ for heating) and $Nu = 4.364$ for fully developed laminar flow with a uniform heat flux boundary condition. Note buoyancy effect on the laminar flow data giving the much larger mixed convection heat transfer coefficient.
(From Ghajar and Tam, 1994.)

TABLE 8–8

The lower and upper limits of the heat transfer transition Reynolds numbers

Inlet Geometry	Lower Limit	Upper Limit
Re-entrant	$Re_{lower} = 2157 - 0.65[192 - (x/D)]$	$Re_{upper} = 8475 - 9.28[192 - (x/D)]$
Square-edged	$Re_{lower} = 2524 - 0.82[192 - (x/D)]$	$Re_{upper} = 8791 - 7.69[192 - (x/D)]$
Bell-mouth	$Re_{lower} = 3787 - 1.80[192 - (x/D)]$	$Re_{upper} = 10481 - 5.47[192 - (x/D)]$

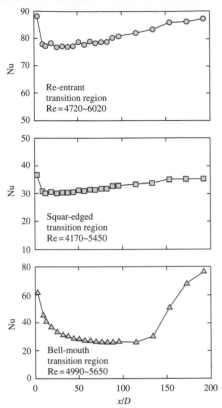

FIGURE 8–36

Variation of local Nusselt number with length for the re-entrant, square-edged, and bell-mouth inlets in the transition region.
(From Tam and Ghajar, 1998.)

bell-mouth inlet, the variation of the local heat transfer coefficient with length in the transition and turbulent flow regions is very unusual. For this inlet geometry, the boundary layer along the tube wall is at first laminar and then changes through a transition to the turbulent condition causing a dip in the Nu versus x/D curve. In their experiments with a fixed inside diameter of 15.84 mm, the length of the dip in the transition region was much longer ($100 < x/D < 175$) than in the turbulent region ($x/D < 25$). The presence of the dip in the transition region causes a significant influence in both the local and the average heat transfer coefficients. This is particularly important for heat transfer calculations in short tube heat exchangers with a bell-mouth inlet. Figure 8–36 shows the variation of local Nusselt number along the tube length in the transition region for the three inlet configurations at comparable Reynolds numbers.

EXAMPLE 8–8   Heat Transfer in the Transition Region

Ethylene glycol-distilled water mixture with a mass fraction of 0.6 and a flow rate of 2.6×10^{-4} m^3/s flows inside a tube with an inside diameter of 0.0158 m subjected to uniform wall heat flux. For this flow, determine the Nusselt number at the location $x/D = 90$ if the inlet configuration of the tube is: (*a*) re-entrant, (*b*) square-edged, and (*c*) bell-mouth. At this location, the local Grashof number is Gr = 51,770. The properties of ethylene glycol-distilled water mixture at the location of interest are Pr = 29.2, $\nu = 3.12 \times 10^{-6}$ m^2/s and $\mu_b/\mu_s = 1.77$.

SOLUTION A liquid mixture flowing in a tube is subjected to uniform wall heat flux. The Nusselt number at a specified location is to be determined for three different tube inlet configurations.

Assumptions Steady operating conditions exist.

Properties The properties of the ethylene glycol-distilled water mixture are given to be Pr = 29.2, $\nu = 3.12 \times 10^{-6}$ m^2/s and $\mu_b/\mu_s = 1.77$.

Analysis For a tube with a known diameter and volume flow rate, the type of flow regime is determined before making any decision regarding which Nusselt number correlation to use. The Reynolds number at the specified location is

$$\text{Re} = \frac{(\dot{V}/A_c)D}{\nu} = \frac{[(2.6 \times 10^{-4}\,\text{m}^3/\text{s})(1.961 \times 10^{-4}\,\text{m}^2)](0.0158\,\text{m})}{3.12 \times 10^{-6}\,\text{m}^2/\text{s}} = 6714$$

since

$$A_c = \pi D^2/4 = \pi(0.0158\,\text{m})^2/4 = 1.961 \times 10^{-4}\,\text{m}^2$$

Therefore, the flow regime is in the transition region for all three inlet configurations (thus use the information given in Table 8–8 with $x/D = 90$) and therefore Eq. 8–83 should be used with the constants a, b, c found in Table 8–7. However, Nu$_{\text{lam}}$ and Nu$_{\text{turb}}$ are the inputs to Eq. 8–83 and they need to be evaluated first from Eqs. 8–84 and 8–85, respectively. It should be mentioned that the correlations for Nu$_{\text{lam}}$ and Nu$_{\text{turb}}$ have no inlet dependency.

From Eq. 8–84:

$$Nu_{lam} = 1.24 \left[\left(\frac{Re\,Pr\,D}{x} \right) + 0.025(Gr\,Pr)^{0.75} \right]^{1/3} \left(\frac{\mu_b}{\mu_s} \right)^{0.14}$$

$$= 1.24 \left[\left(\frac{(6714)(29.2)}{90} \right) + 0.025[(51{,}770)(29.2)]^{0.75} \right]^{1/3} (1.77)^{0.14} = 19.9$$

From Eq. 8–85:

$$Nu_{turb} = 0.023 Re^{0.8} Pr^{0.385} \left(\frac{x}{D} \right)^{-0.0054} \left(\frac{\mu_b}{\mu_s} \right)^{0.14}$$

$$= 0.023(6714)^{0.8}(29.2)^{0.385}(90)^{-0.0054}(1.77)^{0.14} = 102.7$$

Then the transition Nusselt number can be determined from Eq. 8–83,

$$Nu_{trans} = Nu_{lam} + \{\exp[(a - Re)/b] + Nu_{turb}^c\}^c$$

Case 1: For re-entrant inlet:

$$Nu_{trans} = 19.9 + \{\exp[(1766 - 6714)/276] + 102.7^{-0.955}\}^{-0.955} = \mathbf{88.2}$$

Case 2: For square-edged inlet:

$$Nu_{trans} = 19.9 + \{\exp[(2617 - 6714)/207] + 102.7^{-0.950}\}^{-0.950} = \mathbf{85.3}$$

Case 3: For bell-mouth inlet:

$$Nu_{trans} = 19.9 + \{\exp[(6628 - 6714)/237] + 102.7^{-0.980}\}^{-0.980} = \mathbf{21.3}$$

Discussion It is worth mentioning that, for the re-entrant and square-edged inlets, the flow behaves normally. For the bell-mouth inlet, the Nusselt number is low in comparison to the other two inlets. This is because of the unusual behaviour of the bell-mouth inlet noted earlier (see Fig. 8–36); i.e., the boundary layer along the tube wall is at first laminar and then changes through a transition region to the turbulent condition.

REFERENCES

1. A. J. Ghajar and K. F. Madon. "Pressure Drop Measurements in the Transition Region for a Circular Tube with Three Different Inlet Configurations." *Experimental Thermal and Fluid Science*, Vol. 5 (1992), pp. 129–135.

2. A. J. Ghajar and L. M. Tam. "Heat Transfer Measurements and Correlations in the Transition Region for a Circular Tube with Three Different Inlet Configurations." *Experimental Thermal and Fluid Science*, Vol. 8 (1994), pp. 79–90.

3. A. J. Ghajar and L. M. Tam. "Flow Regime Map for a Horizontal Pipe with Uniform Wall Heat Flux and Three Inlet Configurations. *Experimental Thermal and Fluid Science*, Vol. 10 (1995), pp. 287–297.

4. A. J. Ghajar, L. M. Tam, and S. C. Tam. "Improved Heat Transfer Correlation in the Transition Region for a Circular Tube with Three Inlet Configurations Using Artificial Neural Networks." *Heat Transfer Engineering*, Vol. 25, No. 2 (2004), pp. 30–40.

5. L. M. Tam and A. J. Ghajar. "Effect of Inlet Geometry and Heating on the Fully Developed Friction Factor in the Transition Region of a Horizontal Tube." *Experimental Thermal and Fluid Science*, Vol. 15 (1997), pp. 52–64.

6. L. M. Tam and A. J. Ghajar. "The Unusual Behavior of Local Heat Transfer Coefficient in a Circular Tube with a Bell-Mouth Inlet." *Experimental Thermal and Fluid Science*, Vol. 16 (1998), pp. 187–194.

SUMMARY

Internal flow is characterized by the fluid being completely confined by the inner surfaces of the tube. The mean or average velocity and temperature for a circular tube of radius R are expressed as

$$V_{avg} = \frac{2}{R^2} \int_0^R u(r)r\,dr \quad \text{and} \quad T_m = \frac{2}{V_{avg} R^2} \int_0^R u(r)T(r)r\,dr$$

The Reynolds number for internal flow and the hydraulic diameter are defined as

$$\text{Re} = \frac{\rho V_{avg} D}{\mu} = \frac{V_{avg} D}{\nu} \quad \text{and} \quad D_h = \frac{4A_c}{p}$$

The flow in a tube is laminar for $\text{Re} < 2300$, turbulent for about $\text{Re} > 10{,}000$, and transitional in between.

The length of the region from the tube inlet to the point at which the boundary layer merges at the centerline is the *hydrodynamic entry length* L_h. The region beyond the entrance region in which the velocity profile is fully developed is the *hydrodynamically fully developed region*. The length of the region of flow over which the thermal boundary layer develops and reaches the tube center is the *thermal entry length* L_t. The region in which the flow is both hydrodynamically and thermally developed is the *fully developed flow region*. The entry lengths are given by

$$L_{h,\,laminar} \approx 0.05 \text{ Re } D$$
$$L_{t,\,laminar} \approx 0.05 \text{ Re Pr } D = \text{Pr } L_{h,\,laminar}$$
$$L_{h,\,turbulent} \approx L_{t,\,turbulent} = 10D$$

For $\dot{q}_s$ = constant, the rate of heat transfer is expressed as

$$\dot{Q} = \dot{q}_s A_s = \dot{m} c_p (T_e - T_i)$$

For T_s = constant, we have

$$\dot{Q} = hA_s \Delta T_{ln} = \dot{m} c_p (T_e - T_i)$$
$$T_e = T_s - (T_s - T_i)\exp(-hA_s / \dot{m} c_p)$$
$$\Delta T_{ln} = \frac{T_i - T_e}{\ln[(T_s - T_e)/(T_s - T_i)]} = \frac{\Delta T_e - \Delta T_i}{\ln(\Delta T_e / \Delta T_i)}$$

The irreversible pressure loss due to frictional effects and the required pumping power to overcome this loss for a volume flow rate of $\dot{V}$ are

$$\Delta P_L = f \frac{L}{D} \frac{\rho V_{avg}^2}{2} \quad \text{and} \quad \dot{W}_{pump} = \dot{V} \Delta P_L$$

For *fully developed laminar flow* in a circular pipe, we have:

$$u(r) = 2V_{avg}\left(1 - \frac{r^2}{R^2}\right) = u_{max}\left(1 - \frac{r^2}{R^2}\right)$$

$$f = \frac{64\mu}{\rho D V_{avg}} = \frac{64}{\text{Re}}$$

$$\dot{V} = V_{avg} A_c = \frac{\Delta P R^2}{8\mu L} \pi R^2 = \frac{\pi R^4 \Delta P}{8\mu L} = \frac{\pi R^4 \Delta P}{128\mu L}$$

Circular tube, laminar ($\dot{q}_s$ = constant): $\quad \text{Nu} = \dfrac{hD}{k} = 4.36$

Circular tube, laminar (T_s = constant): $\quad \text{Nu} = \dfrac{hD}{k} = 3.66$

For *developing laminar flow* in the entrance region with constant surface temperature, we have

Circular tube: $\quad \text{Nu} = 3.66 + \dfrac{0.065(D/L)\,\text{Re Pr}}{1 + 0.04[(D/L)\,\text{Re Pr}]^{2/3}}$

Circular tube: $\quad \text{Nu} = 1.86 \left(\dfrac{\text{Re Pr } D}{L} \right)^{1/3} \left(\dfrac{\mu_b}{\mu_s} \right)^{0.14}$

Parallel plates: $\quad \text{Nu} = 7.54 + \dfrac{0.03(D_h/L)\,\text{Re Pr}}{1 + 0.016[(D_h/L)\,\text{Re Pr}]^{2/3}}$

For *fully developed turbulent flow with smooth surfaces,* we have

$$f = (0.790 \ln \text{Re} - 1.64)^{-2} \qquad 10^4 < \text{Re} < 10^6$$

$$\text{Nu} = 0.125 f\, \text{Re Pr}^{1/3}$$

$$\text{Nu} = 0.023\, \text{Re}^{0.8}\, \text{Pr}^{1/3} \qquad \left(\begin{matrix} 0.7 \le \text{Pr} \le 160 \\ \text{Re} > 10,000 \end{matrix} \right)$$

$\text{Nu} = 0.023\, \text{Re}^{0.8}\, \text{Pr}^n$ with $n = 0.4$ for *heating* and 0.3 for *cooling* of fluid

$$\text{Nu} = \dfrac{(f/8)(\text{Re} - 1000)\,\text{Pr}}{1 + 12.7(f/8)^{0.5}\,(\text{Pr}^{2/3} - 1)} \left(\begin{matrix} 0.5 \le \text{Pr} \le 2000 \\ 3 \times 10^3 < \text{Re} < 5 \times 10^6 \end{matrix} \right)$$

The fluid properties are evaluated at the *bulk mean fluid temperature* $T_b = (T_i + T_e)/2$. For liquid metal flow in the range of $10^4 < \text{Re} < 10^6$ we have:

$T_s = $ constant: $\qquad \text{Nu} = 4.8 + 0.0156\, \text{Re}^{0.85}\, \text{Pr}_s^{0.93}$

$\dot{q}_s = $ constant: $\qquad \text{Nu} = 6.3 + 0.0167\, \text{Re}^{0.85}\, \text{Pr}_s^{0.93}$

For *fully developed turbulent flow with rough surfaces,* the friction factor f is determined from the Moody chart or

$$\frac{1}{\sqrt{f}} = -2.0 \log \left(\frac{\varepsilon/D}{3.7} + \frac{2.51}{\text{Re}\sqrt{f}} \right) \approx -1.8 \log \left[\frac{6.9}{\text{Re}} + \left(\frac{\varepsilon/D}{3.7} \right)^{1.11} \right]$$

For a *concentric annulus,* the hydraulic diameter is $D_h = D_o - D_i$, and the Nusselt numbers are expressed as

$$\text{Nu}_i = \frac{h_i D_h}{k} \qquad \text{and} \qquad \text{Nu}_o = \frac{h_o D_h}{k}$$

where the values for the Nusselt numbers are given in Table 8–4.

REFERENCES AND SUGGESTED READING

1. M. S. Bhatti and R. K. Shah. "Turbulent and Transition Flow Convective Heat Transfer in Ducts." In *Handbook of Single-Phase Convective Heat Transfer*, ed. S. Kakaç, R. K. Shah, and W. Aung. New York: Wiley Interscience, 1987.

2. Y. A. Cengel and J. M. Cimbala. *Fluid Mechanics: Fundamentals and Applications.* New York: McGraw-Hill, 2005.

3. A. P. Colburn. *Transactions of the AIChE* 26 (1933), p. 174.

4. C. F. Colebrook. "Turbulent flow in Pipes, with Particular Reference to the Transition between the Smooth and Rough Pipe Laws." *Journal of the Institute of Civil Engineers London.* 11 (1939), pp. 133–156.

5. R. G. Deissler. "Analysis of Turbulent Heat Transfer and Flow in the Entrance Regions of Smooth Passages." 1953. Referred to in *Handbook of Single-Phase Convective Heat Transfer*, ed. S. Kakaç, R. K. Shah, and W. Aung. New York: Wiley Interscience, 1987.

6. D. F. Dipprey and D. H. Sabersky. "Heat and Momentum Transfer in Smooth and Rough Tubes at Various Prandtl Numbers." *International Journal of Heat Mass Transfer* 6 (1963), pp. 329–353.

7. F. W. Dittus and L. M. K. Boelter. *University of California Publications on Engineering* 2 (1930), p. 433.

8. D. K. Edwards, V. E. Denny, and A. F. Mills. *Transfer Processes.* 2nd ed. Washington, DC: Hemisphere, 1979.

9. V. Gnielinski. "New Equations for Heat and Mass Transfer in Turbulent Pipe and Channel Flow." *International Chemical Engineering* 16 (1976), pp. 359–368.

10. S. E. Haaland. "Simple and Explicit Formulas for the Friction Factor in Turbulent Pipe Flow." *Journal of Fluids Engineering* (March 1983), pp. 89–90.

11. S. Kakaç, R. K. Shah, and W. Aung, eds. *Handbook of Single-Phase Convective Heat Transfer.* New York: Wiley Interscience, 1987.

12. W. M. Kays and M. E. Crawford. *Convective Heat and Mass Transfer.* 3rd ed. New York: McGraw-Hill, 1993.

13. W. M. Kays and H. C. Perkins. Chapter 7. In *Handbook of Heat Transfer*, ed. W. M. Rohsenow and J. P. Hartnett. New York: McGraw-Hill, 1972.

14. L. F. Moody. "Friction Factors for Pipe Flows." *Transactions of the ASME* 66 (1944), pp. 671–684.

15. M. Molki and E. M. Sparrow. "An Empirical Correlation for the Average Heat Transfer Coefficient in Circular Tubes." *Journal of Heat Transfer* 108 (1986), pp. 482–484.

16. R. H. Norris. "Some Simple Approximate Heat Transfer Correlations for Turbulent Flow in Ducts with Rough Surfaces." In *Augmentation of Convective Heat Transfer*, ed. A. E. Bergles and R. L. Webb. New York: ASME, 1970.

17. B. S. Petukhov. "Heat Transfer and Friction in Turbulent Pipe Flow with Variable Physical Properties." In *Advances*

in Heat Transfer, ed. T. F. Irvine and J. P. Hartnett, Vol. 6. New York: Academic Press, 1970.

18. B. S. Petukhov and L. I. Roizen. "Generalized Relationships for Heat Transfer in a Turbulent Flow of a Gas in Tubes of Annular Section." *High Temperature* (USSR) 2 (1964), pp. 65–68.

19. O. Reynolds. "On the Experimental Investigation of the Circumstances Which Determine Whether the Motion of Water Shall Be Direct or Sinuous, and the Law of Resistance in Parallel Channels." *Philosophical Transactions of the Royal Society of London* 174 (1883), pp. 935–982.

20. H. Schlichting. *Boundary Layer Theory.* 7th ed. New York: McGraw-Hill, 1979.

21. R. K. Shah and M. S. Bhatti. "Laminar Convective Heat Transfer in Ducts." In *Handbook of Single-Phase*

Convective Heat Transfer, ed. S. Kakaç, R. K. Shah, and W. Aung. New York: Wiley Interscience, 1987.

22. E. N. Sieder and G. E. Tate. "Heat Transfer and Pressure Drop of Liquids in Tubes." *Industrial Engineering Chemistry* 28 (1936), pp. 1429–1435.

23. C. A. Sleicher and M. W. Rouse. "A Convenient Correlation for Heat Transfer to Constant and Variable Property Fluids in Turbulent Pipe Flow." *International Journal of Heat Mass Transfer* 18 (1975), pp. 1429–1435.

24. S. Whitaker. "Forced Convection Heat Transfer Correlations for Flow in Pipes, Past Flat Plates, Single Cylinders, and for Flow in Packed Beds and Tube Bundles." *AIChE Journal* 18 (1972), pp. 361–371.

25. W. Zhi-qing. "Study on Correction Coefficients of Laminar and Turbulent Entrance Region Effects in Round Pipes." *Applied Mathematical Mechanics* 3 (1982), p. 433.

PROBLEMS*

General Flow Analysis

8–1C Why are liquids usually transported in circular pipes?

8–2C Show that the Reynolds number for flow in a circular tube of diameter D can be expressed as $Re = 4\dot{m}/(\pi D\mu)$.

8–3C Which fluid at room temperature requires a larger pump to move at a specified velocity in a given tube: water or engine oil? Why?

8–4C What is the generally accepted value of the Reynolds number above which the flow in smooth pipes is turbulent?

8–5C What is hydraulic diameter? How is it defined? What is it equal to for a circular tube of diameter D?

8–6C How is the hydrodynamic entry length defined for flow in a tube? Is the entry length longer in laminar or turbulent flow?

8–7C Consider laminar flow in a circular tube. Will the friction factor be higher near the inlet of the tube or near the exit? Why? What would your response be if the flow were turbulent?

8–8C How does surface roughness affect the pressure drop in a tube if the flow is turbulent? What would your response be if the flow were laminar?

8–9C How does the friction factor f vary along the flow direction in the fully developed region in (a) laminar flow and (b) turbulent flow?

8–10C What fluid property is responsible for the development of the velocity boundary layer? For what kinds of fluids will there be no velocity boundary layer in a pipe?

8–11C What is the physical significance of the number of transfer units NTU $= hA_s/\dot{m}c_p$? What do small and large NTU values tell about a heat transfer system?

8–12C What does the logarithmic mean temperature difference represent for flow in a tube whose surface temperature is constant? Why do we use the logarithmic mean temperature instead of the arithmetic mean temperature?

8–13C How is the thermal entry length defined for flow in a tube? In what region is the flow in a tube fully developed?

8–14C Consider laminar forced convection in a circular tube. Will the heat flux be higher near the inlet of the tube or near the exit? Why?

8–15C Consider turbulent forced convection in a circular tube. Will the heat flux be higher near the inlet of the tube or near the exit? Why?

8–16C In the fully developed region of flow in a circular tube, will the velocity profile change in the flow direction? How about the temperature profile?

---

*Problems designated by a "C" are concept questions, and students are encouraged to answer them all. Problems designated by an "E" are in English units, and the SI users can ignore them. Problems with the icon 🌐 are solved using EES, and complete solutions together with parametric studies are included on the enclosed CD. Problems with the icon 🖥 are comprehensive in nature, and are intended to be solved with a computer, preferably using the EES software that accompanies this text.

8–17C Consider the flow of oil in a tube. How will the hydrodynamic and thermal entry lengths compare if the flow is laminar? How would they compare if the flow were turbulent?

8–18C Consider the flow of mercury (a liquid metal) in a tube. How will the hydrodynamic and thermal entry lengths compare if the flow is laminar? How would they compare if the flow were turbulent?

8–19C What do the average velocity V_{avg} and the mean temperature T_m represent in flow through circular tubes of constant diameter?

8–20C Consider fluid flow in a tube whose surface temperature remains constant. What is the appropriate temperature difference for use in Newton's law of cooling with an average heat transfer coefficient?

8–21 Air enters a 25-cm-diameter 12-m-long underwater duct at 50°C and 1 atm at a mean velocity of 7 m/s, and is cooled by the water outside. If the average heat transfer coefficient is 85 W/m² · °C and the tube temperature is nearly equal to the water temperature of 10°C, determine the exit temperature of air and the rate of heat transfer.

8–22 Cooling water available at 10°C is used to condense steam at 30°C in the condenser of a power plant at a rate of 0.15 kg/s by circulating the cooling water through a bank of 5-m-long 1.2-cm-internal-diameter thin copper tubes. Water enters the tubes at a mean velocity of 4 m/s, and leaves at a temperature of 24°C. The tubes are nearly isothermal at 30°C. Determine the average heat transfer coefficient between the water and the tubes, and the number of tubes needed to achieve the indicated heat transfer rate in the condenser.

8–23 Repeat Prob. 8–22 for steam condensing at a rate of 0.60 kg/s.

8–24 Combustion gases passing through a 3-cm-internal-diameter circular tube are used to vaporize waste water at atmospheric pressure. Hot gases enter the tube at 115 kPa and 250°C at a mean velocity of 5 m/s, and leave at 150°C. If the average heat transfer coefficient is 120 W/m² · °C and the inner surface temperature of the tube is 110°C, determine (a) the tube length and (b) the rate of evaporation of water.

8–25 Repeat Prob. 8–24 for a heat transfer coefficient of 40 W/m² · °C.

Laminar and Turbulent Flow in Tubes

8–26C How is the friction factor for flow in a tube related to the pressure drop? How is the pressure drop related to the pumping power requirement for a given mass flow rate?

8–27C Someone claims that the shear stress at the center of a circular pipe during fully developed laminar flow is zero. Do you agree with this claim? Explain.

8–28C Someone claims that in fully developed turbulent flow in a tube, the shear stress is a maximum at the tube surface. Do you agree with this claim? Explain.

8–29C Consider fully developed flow in a circular pipe with negligible entrance effects. If the length of the pipe is doubled, the pressure drop will (a) double, (b) more than double, (c) less than double, (d) reduce by half, or (e) remain constant.

8–30C Someone claims that the volume flow rate in a circular pipe with laminar flow can be determined by measuring the velocity at the centerline in the fully developed region, multiplying it by the cross sectional area, and dividing the result by 2. Do you agree? Explain.

8–31C Someone claims that the average velocity in a circular pipe in fully developed laminar flow can be determined by simply measuring the velocity at $R/2$ (midway between the wall surface and the centerline). Do you agree? Explain.

8–32C Consider fully developed laminar flow in a circular pipe. If the diameter of the pipe is reduced by half while the flow rate and the pipe length are held constant, the pressure drop will (a) double, (b) triple, (c) quadruple, (d) increase by a factor of 8, or (e) increase by a factor of 16.

8–33C Consider fully developed laminar flow in a circular pipe. If the viscosity of the fluid is reduced by half by heating while the flow rate is held constant, how will the pressure drop change?

8–34C How does surface roughness affect the heat transfer in a tube if the fluid flow is turbulent? What would your response be if the flow in the tube were laminar?

8–35 Water at 15°C ($\rho = 999.1$ kg/m³ and $\mu = 1.138 \times 10^{-3}$ kg/m · s) is flowing in a 4-cm-diameter and 30-m long horizontal pipe made of stainless steel steadily at a rate of 5 L/s. Determine (a) the pressure drop and (b) the pumping power requirement to overcome this pressure drop.

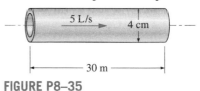

FIGURE P8–35

8–36 In fully developed laminar flow in a circular pipe, the velocity at $R/2$ (midway between the wall surface and the centerline) is measured to be 6 m/s. Determine the velocity at the center of the pipe.   *Answer:* 8 m/s

8–37 The velocity profile in fully developed laminar flow in a circular pipe of inner radius $R = 10$ cm, in m/s, is given by $u(r) = 4(1 - r^2/R^2)$. Determine the mean and maximum velocities in the pipe, and the volume flow rate.

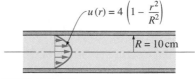

FIGURE P8–37

8–38 Repeat Prob. 8–37 for a pipe of inner radius 5 cm.

8–39 Determine the convection heat transfer coefficient for the flow of (a) air and (b) water at a velocity of 2 m/s in an 8-cm-diameter and 7-m-long tube when the tube is subjected to uniform heat flux from all surfaces. Use fluid properties at 25°C.

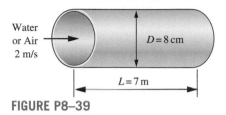

Water or Air 2 m/s

$D = 8\,\text{cm}$

$L = 7\,\text{m}$

FIGURE P8–39

8–40 Air at 10°C enters a 12-cm-diameter and 5-m-long pipe at a rate of 0.065 kg/s. The inner surface of the pipe has a roughness of 0.22 mm and the pipe is nearly isothermal at 50°C. Determine the rate of heat transfer to air using the Nusselt number relation given by (a) Eq. 8–66 and (b) Eq. 8–71.

8–41 An 8-m long, uninsulated square duct of cross section 0.2 m × 0.2 m and relative roughness 10^{-3} passes through the attic space of a house. Hot air enters the duct at 1 atm and 80°C at a volume flow rate of 0.15 m³/s. The duct surface is nearly isothermal at 60°C. Determine the rate of heat loss from the duct to the attic space and the pressure difference between the inlet and outlet sections of the duct.

8–42 A 10-m long and 10-mm inner-diameter pipe made of commercial steel is used to heat a liquid in an industrial process. The liquid enters the pipe with $T_i = 25°C$, $V = 0.8$ m/s. A uniform heat flux is maintained by an electric resistance heater wrapped around the outer surface of the pipe, so that the fluid exits at 75°C. Assuming fully developed flow and taking the average fluid properties to be $\rho = 1000$ kg/m³, $c_p = 4000$ J/kg · K, $\mu = 2 \times 10^{-3}$ kg/m · s, $k = 0.48$ W/m · K, and Pr = 10, determine:

(a) The required surface heat flux $\dot{q}_s$, produced by the heater
(b) The surface temperature at the exit, T_s
(c) The pressure loss through the pipe and the minimum power required to overcome the resistance to flow.

8–43 Water at 10°C ($\rho = 999.7$ kg/m³ and $\mu = 1.307 \times 10^{-3}$ kg/m · s) is flowing in a 0.20-cm-diameter 15-m-long pipe steadily at an average velocity of 1.2 m/s. Determine (a) the pressure drop and (b) the pumping power requirement to overcome this pressure drop.
Answers: (a) 188 kPa, (b) 0.71 W

8–44 Water is to be heated from 10°C to 80°C as it flows through a 2-cm-internal-diameter, 13-m-long tube. The tube is equipped with an electric resistance heater, which provides uniform heating throughout the surface of the tube. The outer surface of the heater is well insulated, so that in steady operation all the heat generated in the heater is transferred to the water in the tube. If the system is to provide hot water at a rate of 5 L/min, determine the power rating of the resistance heater. Also, estimate the inner surface temperature of the pipe at the exit.

8–45 Hot air at atmospheric pressure and 85°C enters a 10-m-long uninsulated square duct of cross section 0.15 m × 0.15 m that passes through the attic of a house at a rate of 0.10 m³/s. The duct is observed to be nearly isothermal at 70°C. Determine the exit temperature of the air and the rate of heat loss from the duct to the air space in the attic.
Answers: 75.7°C, 941 W

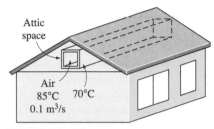

Attic space

Air 85°C 70°C 0.1 m³/s

FIGURE P8–45

8–46 Reconsider Prob. 8–45. Using EES (or other) software, investigate the effect of the volume flow rate of air on the exit temperature of air and the rate of heat loss. Let the flow rate vary from 0.05 m³/s to 0.15 m³/s. Plot the exit temperature and the rate of heat loss as a function of flow rate, and discuss the results.

8–47 Consider an air solar collector that is 1 m wide and 5 m long and has a constant spacing of 3 cm between the glass cover and the collector plate. Air enters the collector at 30°C at a rate of 0.15 m³/s through the 1-m-wide edge and flows along the 5-m-long passage way. If the average temperatures of the glass cover and the collector plate are 20°C and 60°C, respectively, determine (a) the net rate of heat transfer to the air in the collector and (b) the temperature rise of air as it flows through the collector.

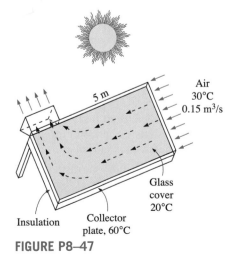

5 m

Air 30°C 0.15 m³/s

Glass cover 20°C

Insulation

Collector plate, 60°C

FIGURE P8–47

8–48 Consider the flow of oil at 10°C in a 40-cm-diameter pipeline at an average velocity of 0.5 m/s. A 1500-m-long section of the pipeline passes through icy waters of a lake at 0°C. Measurements indicate that the surface temperature of the pipe is very nearly 0°C. Disregarding the thermal resistance of the pipe material, determine (*a*) the temperature of the oil when the pipe leaves the lake, (*b*) the rate of heat transfer from the oil, and (*c*) the pumping power required to overcome the pressure losses and to maintain the flow oil in the pipe.

8–49 Consider laminar flow of a fluid through a square channel maintained at a constant temperature. Now the mean velocity of the fluid is doubled. Determine the change in the pressure drop and the change in the rate of heat transfer between the fluid and the walls of the channel. Assume the flow regime remains unchanged. Assume fully developed flow and disregard any changes in ΔT_{ln}.

8–50 Repeat Prob. 8–49 for turbulent flow.

8–51E The hot water needs of a household are to be met by heating water at 55°F to 200°F by a parabolic solar collector at a rate of 4 lbm/s. Water flows through a 1.25-in-diameter thin aluminum tube whose outer surface is blackanodized in order to maximize its solar absorption ability. The centerline of the tube coincides with the focal line of the collector, and a glass sleeve is placed outside the tube to minimize the heat losses. If solar energy is transferred to water at a net rate of 350 Btu/h per ft length of the tube, determine the required length of the parabolic collector to meet the hot water requirements of this house. Also, determine the surface temperature of the tube at the exit.

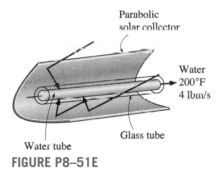

Parabolic
solar collector

Water
200°F
4 lbm/s

Glass tube

Water tube

FIGURE P8–51E

8–52 A 15-cm × 20-cm printed circuit board whose components are not allowed to come into direct contact with air for reliability reasons is to be cooled by passing cool air through a 20-cm-long channel of rectangular cross section 0.2 cm × 14 cm drilled into the board. The heat generated by the electronic components is conducted across the thin layer of the board to the channel, where it is removed by air that enters the channel at 15°C. The heat flux at the top surface of the channel can be considered to be uniform, and heat transfer through other surfaces is negligible. If the velocity of the air at the inlet of the channel is not to exceed 4 m/s and the surface temperature of the channel is to remain under 50°C, determine the maximum total power of the electronic components that can safely be mounted on this circuit board.

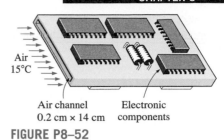

Air
15°C

Air channel   Electronic
0.2 cm × 14 cm   components

FIGURE P8–52

8–53 Repeat Prob. 8–52 by replacing air with helium, which has six times the thermal conductivity of air.

8–54 Reconsider Prob. 8–52. Using EES (or other) software, investigate the effects of air velocity at the inlet of the channel and the maximum surface temperature on the maximum total power dissipation of electronic components. Let the air velocity vary from 1 m/s to 10 m/s and the surface temperature from 30°C to 90°C. Plot the power dissipation as functions of air velocity and surface temperature, and discuss the results.

8–55 Air enters a 7-m-long section of a rectangular duct of cross section 15 cm × 20 cm at 50°C at an average velocity of 7 m/s. If the walls of the duct are maintained at 10°C, determine (*a*) the outlet temperature of the air, (*b*) the rate of heat transfer from the air, and (*c*) the fan power needed to overcome the pressure losses in this section of the duct.

Answers: (*a*) 34.2°C, (*b*) 3775 W, (*c*) 4.7 W

8–56 Reconsider Prob. 8–55. Using EES (or other) software, investigate the effect of air velocity on the exit temperature of air, the rate of heat transfer, and the fan power. Let the air velocity vary from 1 m/s to 10 m/s. Plot the exit temperature, the rate of heat transfer, and the fan power as a function of the air velocity, and discuss the results.

8–57 Hot air at 60°C leaving the furnace of a house enters a 12-m-long section of a sheet metal duct of rectangular cross section 20 cm × 20 cm at an average velocity of 4 m/s. The thermal resistance of the duct is negligible, and the outer surface of the duct, whose emissivity is 0.3, is exposed to the cold air at 10°C in the basement, with a convection heat transfer coefficient of 10 W/m² · °C. Taking the walls of the basement

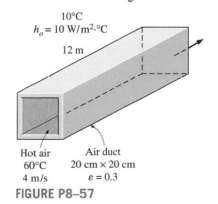

10°C
h_o = 10 W/m²·°C

12 m

Hot air   Air duct
60°C   20 cm × 20 cm
4 m/s ε = 0.3

FIGURE P8–57

to be at 10°C also, determine (a) the temperature at which the hot air will leave the basement and (b) the rate of heat loss from the hot air in the duct to the basement.

8–58 Reconsider Prob. 8–57. Using EES (or other) software, investigate the effects of air velocity and the surface emissivity on the exit temperature of air and the rate of heat loss. Let the air velocity vary from 1 m/s to 10 m/s and the emissivity from 0.1 to 1.0. Plot the exit temperature and the rate of heat loss as functions of air velocity and emissivity, and discuss the results.

8–59 The components of an electronic system dissipating 180 W are located in a 1-m-long horizontal duct whose cross section is 16 cm × 16 cm. The components in the duct are cooled by forced air, which enters at 27°C at a rate of 0.65 m³/min. Assuming 85 percent of the heat generated inside is transferred to air flowing through the duct and the remaining 15 percent is lost through the outer surfaces of the duct, determine (a) the exit temperature of air and (b) the highest component surface temperature in the duct.

8–60 Repeat Prob. 8–59 for a circular horizontal duct of 15-cm diameter.

8–61 Consider a hollow-core printed circuit board 12 cm high and 18 cm long, dissipating a total of 20 W. The width of the air gap in the middle of the PCB is 0.25 cm. The cooling air enters the 12-cm-wide core at 32°C at a rate of 0.8 L/s. Assuming the heat generated to be uniformly distributed over the two side surfaces of the PCB, determine (a) the temperature at which the air leaves the hollow core and (b) the highest temperature on the inner surface of the core.

Answers: (a) 54.0°C, (b) 72.2°C

8–62 Repeat Prob. 8–61 for a hollow-core PCB dissipating 35 W.

8–63E Water at 60°F is heated by passing it through 0.75-in-internal-diameter thin-walled copper tubes. Heat is supplied to the water by steam that condenses outside the copper tubes at 250°F. If water is to be heated to 140°F at a rate of 0.4 lbm/s, determine (a) the length of the copper tube that needs to be used and (b) the pumping power required to overcome pressure losses. Assume the entire copper tube to be at the steam temperature of 250°F.

8–64 A computer cooled by a fan contains eight PCBs, each dissipating 10 W of power. The height of the PCBs is 12 cm and the length is 18 cm. The clearance between the tips of the components on the PCB and the back surface of the adjacent PCB is 0.3 cm. The cooling air is supplied by a 10-W fan mounted at the inlet. If the temperature rise of air as it flows through the case of the computer is not to exceed 10°C, determine (a) the flow rate of the air that the fan needs to deliver, (b) the fraction of the temperature rise of air that is due to the heat generated by the fan and its motor, and (c) the highest allowable inlet air temperature if the surface temperature of the components is not to exceed 70°C anywhere in the system. Use air properties at 25°C.

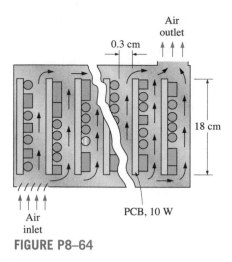

FIGURE P8–64

Special Topic: Transitional Flow

8–65E A tube with a square-edged inlet configuration is subjected to uniform wall heat flux of 8 kW/m². The tube has an inside diameter of 0.622 in and a flow rate of 2.16 gpm. The liquid flowing inside the tube is ethylene glycol-distilled water mixture with a mass fraction of 2.27. Determine the friction coefficient at a location along the tube where the Grashof number is Gr = 35,450. The physical properties of the ethylene glycol-distilled water mixture at the location of interest are Pr = 13.8, $\nu = 18.4 \times 10^{-6}$ ft²/s and $\mu_b/\mu_s = 1.12$. Then recalculate the fully developed friction coefficient if the volume flow rate is increased by 50% while the rest of the parameters remain unchanged. *Answers:* 0.00859, 0.00776

8–66 A tube with a bell-mouth inlet configuration is subjected to uniform wall heat flux of 3 kW/m². The tube has an inside diameter of 0.0158 m (0.622 in.) and a flow rate of 1.43×10^{-4} m³/s (2.27 gpm). The liquid flowing inside the tube is ethylene glycol-distilled water mixture with a mass fraction of 2.27. Determine the fully developed friction coefficient at a location along the tube where the Grashof number is Gr = 16,600. The physical properties of the ethylene glycol-distilled water mixture at the location of interest are Pr = 14.85, $\nu = 1.93 \times 10^{-6}$ m²/s and $\mu_b/\mu_s = 1.07$.

8–67 Reconsider Prob. 8–66. Calculate the fully developed friction coefficient if the volume flow rate is increased by 50 percent while the rest of the parameters remain unchanged.

8–68 Ethylene glycol-distilled water mixture with a mass fraction of 0.72 and a flow rate of 2.05×10^{-4} m³/s flows inside a tube with an inside diameter of 0.0158 m with a uniform wall heat flux boundary condition. For this flow, determine the Nusselt number at the location $x/D = 10$ for the inlet tube configuration of (a) bell-mouth and (b) re-entrant. Compare the results for parts (a) and (b). Assume the Grashof number is Gr = 60,000. The physical properties of ethylene glycol-distilled water mixture are Pr = 33.46, ν = 3.45 × 10^{-6} m²/s and μ_b/μ_s = 2.0.

8–69 Repeat Prob. 8–68 for the location x/D = 90.

Review Problems

8–70 A silicon chip is cooled by passing water through microchannel etched in the back of the chip, as shown in Fig. P8–70. The channels are covered with a silicon cap. Consider a 10-mm × 10-mm square chip in which N = 50 rectangular microchannels, each of width W = 50 μm and height H = 200 μm have been etched. Water enters the microchannels at a temperature T_i = 290 K, and a total flow rate of 0.005 kg/s. The chip and cap are maintained at a uniform temperature of 350 K. Assuming that the flow in the channels is fully developed, all the heat generated by the circuits on the top surface of the chip is transferred to the water, and using circular tube correlations, determine:

(a) The water outlet temperature, T_e
(b) The chip power dissipation, $\dot{W}_e$

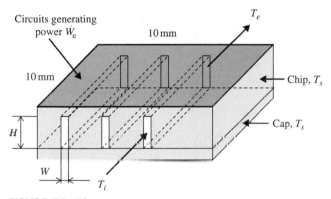

FIGURE P8–70

8–71 Water is heated at a rate of 10 kg/s from a temperature of 15°C to 35°C by passing it through five identical tubes, each 5.0 cm in diameter, whose surface temperature is 60.0°C. Estimate (a) the steady rate of heat transfer and (b) the length of tubes necessary to accomplish this task.

8–72 Repeat Prob. 8–71 for a flow rate of 20 kg/s.

8–73 Water at 1500 kg/h and 10°C enters a 10-mm diameter smooth tube whose wall temperature is maintained at 49°C. Calculate (a) the tube length necessary to heat the water to 40°C, and (b) the water outlet temperature if the tube length is doubled. Assume average water properties to be the same as in (a).

8–74 A geothermal district heating system involves the transport of geothermal water at 110°C from a geothermal well to a city at about the same elevation for a distance of 12 km at a rate of 1.5 m³/s in 60-cm-diameter stainless steel pipes. The fluid pressures at the wellhead and the arrival point in the city are to be the same. The minor losses are negligible because of the large length-to-diameter ratio and the relatively small number of components that cause minor losses. (a) Assuming the pump-motor efficiency to be 65 percent, determine the electric power consumption of the system for pumping. (b) Determine the daily cost of power consumption of the system if the unit cost of electricity is $0.06/kWh. (c) The temperature of geothermal water is estimated to drop 0.5°C during this long flow. Determine if the frictional heating during flow can make up for this drop in temperature.

8–75 Repeat Prob. 8–74 for cast iron pipes of the same diameter.

8–76 The velocity profile in fully developed laminar flow in a circular pipe, in m/s, is given by $u(r) = 6(1 - 100r^2)$ where r is the radial distance from the centerline of the pipe in m. Determine (a) the radius of the pipe, (b) the mean velocity through the pipe, and (c) the maximum velocity in the pipe.

8–77E The velocity profile in fully developed laminar flow of water at 40°F in a 140-ft-long horizontal circular pipe, in ft/s, is given by $u(r) = 0.8(1 - 625r^2)$ where r is the radial distance from the centerline of the pipe in ft. Determine (a) the volume flow rate of water through the pipe, (b) the pressure drop across the pipe, and (c) the useful pumping power required to overcome this pressure drop.

8–78 The compressed air requirements of a manufacturing facility are met by a 150-hp compressor located in a room that is maintained at 20°C. In order to minimize the compressor work, the intake port of the compressor is connected to the outside through an 11-m-long, 20-cm-diameter duct made of thin aluminum sheet. The compressor takes in air at a rate of 0.27 m³/s at the outdoor conditions of 10°C and 95 kPa. Disregarding the thermal resistance of the duct and taking the heat transfer coefficient on the outer surface of the duct to be 10 W/m² · °C, determine (a) the power used by the compressor to overcome the pressure drop in this duct, (b) the rate of heat transfer to the incoming cooler air, and (c) the temperature rise of air as it flows through the duct.

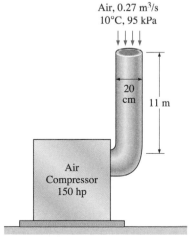

FIGURE P8–78

8–79 A house built on a riverside is to be cooled in summer by utilizing the cool water of the river, which flows at an average temperature of 15°C. A 15-m-long section of a circular duct of 20-cm diameter passes through the water. Air enters the underwater section of the duct at 25°C at a velocity of 3 m/s. Assuming the surface of the duct to be at the temperature of the water, determine the outlet temperature of air as it leaves the underwater portion of the duct. Also, for an overall fan efficiency of 55 percent, determine the fan power input needed to overcome the flow resistance in this section of the duct.

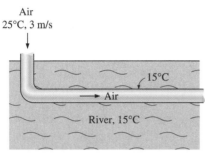

Air
25°C, 3 m/s

15°C

Air

River, 15°C

FIGURE P8–79

8–80 Repeat Prob. 8–79 assuming that a 0.25-mm-thick layer of mineral deposit ($k = 3$ W/m · °C) formed on the inner surface of the pipe.

8–81E The exhaust gases of an automotive engine leave the combustion chamber and enter a 8-ft-long and 3.5-in-diameter thin-walled steel exhaust pipe at 800°F and 15.5 psia at a rate of 0.2 lbm/s. The surrounding ambient air is at a temperature of 80°F, and the heat transfer coefficient on the outer surface of the exhaust pipe is 3 Btu/h · ft² · °F. Assuming the exhaust gases to have the properties of air, determine (a) the velocity of the exhaust gases at the inlet of the exhaust pipe and (b) the temperature at which the exhaust gases will leave the pipe and enter the air.

8–82 Hot water at 90°C enters a 15-m section of a cast iron pipe ($k = 52$ W/m · °C) whose inner and outer diameters are 4 and 4.6 cm, respectively, at an average velocity of 1.2 m/s. The outer surface of the pipe, whose emissivity is 0.7, is exposed to the cold air at 10°C in a basement, with a convection heat transfer coefficient of 12 W/m² · °C. Taking the walls of the basement to be at 10°C also, determine (a) the rate of heat loss from the water and (b) the temperature at which the water leaves the basement.

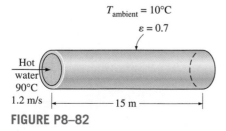

$T_{ambient} = 10°C$

$\varepsilon = 0.7$

Hot water
90°C
1.2 m/s

15 m

FIGURE P8–82

8–83 Repeat Prob. 8–82 for a pipe made of copper ($k = 386$ W/m · °C) instead of cast iron.

8–84 D. B. Tuckerman and R. F. Pease of Stanford University demonstrated in the early 1980s that integrated circuits can be cooled very effectively by fabricating a series of microscopic channels 0.3 mm high and 0.05 mm wide in the back of the substrate and covering them with a plate to confine the fluid flow within the channels. They were able to dissipate 790 W of power generated in a 1-cm² silicon chip at a junction-to-ambient temperature difference of 71°C using water as the coolant flowing at a rate of 0.01 L/s through 100 such channels under a 1-cm × 1-cm silicon chip. Heat is transferred primarily through the base area of the channel, and it was found that the increased surface area and thus the fin effect are of lesser importance. Disregarding the entrance effects and ignoring any heat transfer from the side and cover surfaces, determine (a) the temperature rise of water as it flows through the microchannels and (b) the average surface temperature of the base of the microchannels for a power dissipation of 50 W. Assume the water enters the channels at 20°C.

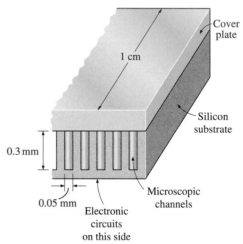

Cover plate

1 cm

Silicon substrate

0.3 mm

0.05 mm

Electronic circuits on this side

Microscopic channels

FIGURE P8–84

8–85 Liquid-cooled systems have high heat transfer coefficients associated with them, but they have the inherent disadvantage that they present potential leakage problems. Therefore, air is proposed to be used as the microchannel coolant. Repeat Prob. 8–84 using air as the cooling fluid instead of water, entering at a rate of 0.5 L/s.

8–86 Hot exhaust gases leaving a stationary diesel engine at 450°C enter a 15-cm-diameter pipe at an average velocity of 4.5 m/s. The surface temperature of the pipe is 180°C. Determine the pipe length if the exhaust gases are to leave the pipe at 250°C after transferring heat to water in a heat recovery unit. Use properties of air for exhaust gases.

8–87 Geothermal steam at 165°C condenses in the shell side of a heat exchanger over the tubes through which water flows.

Water enters the 4-cm-diameter, 14-m-long tubes at 20°C at a rate of 0.8 kg/s. Determine the exit temperature of water and the rate of condensation of geothermal steam.

8–88 Cold air at 5°C enters a 12-cm-diameter 20-m-long isothermal pipe at a velocity of 2.5 m/s and leaves at 19°C. Estimate the surface temperature of the pipe.

8–89 Oil at 15°C is to be heated by saturated steam at 1 atm in a double-pipe heat exchanger to a temperature of 25°C. The inner and outer diameters of the annular space are 3 cm and 5 cm, respectively, and oil enters at with a mean velocity of 0.8 m/s. The inner tube may be assumed to be isothermal at 100°C, and the outer tube is well insulated. Assuming fully developed flow for oil, determine the tube length required to heat the oil to the indicated temperature. In reality, will you need a shorter or longer tube? Explain.

8–90 A liquid hydrocarbon enters a 2.5-cm-diameter tube that is 5.0 m long. The liquid inlet temperature is 20°C and the tube wall temperature is 60°C. Average liquid properties are $c_p = 2.0$ kJ/kg · K, $\mu = 10$ mPa · s, and $\rho = 900$ kg/m³. At a flow rate of 1200 kg/h, the liquid outlet temperature is measured to be 30°C. Estimate the liquid outlet temperature when the flow rate is reduced to 400 kg/h. *Hint*: For heat transfer in tubes, Nu $\propto$ Re$^{1/3}$ in laminar flow and Nu $\propto$ Re$^{4/5}$ in turbulent flow.

8–91 100 kg/s of a crude oil is heated from 20°C to 40°C through the tube side of a multitube heat exchanger. The crude oil flow is divided evenly among all 100 tubes in the tube bundle. The ID of each tube is 10 mm, and the inside tube-wall temperature is maintained at 100°C. Average properties of the crude oil are: $\rho = 950$ kg/m³, $c_p = 1.9$ kJ/kg · K, $k = 0.25$ W/m · K, $\mu = 12$ mPa · s, and $\mu_w = 4$ mPa · s. Estimate the rate of heat transfer and the tube length.

8–92 Crude oil at 22°C enters a 20-cm-diameter pipe with an average velocity of 20 cm/s. The average pipe wall temperature is 2°C. Crude oil-properties are as given below. Calculate the rate of heat transfer and pipe length if the crude oil outlet temperature is 20°C.

T °C	ρ kg/m³	k W/m · K	μ mPa · s	c_p kJ/kg · K
2.0	900	0.145	60.0	1.80
22.0	890	0.145	20.0	1.90

8–93 A heat exchanger with 12 tubes, each 1.0 cm in diameter and 2.0 m in length, is used for heating a liquid stream at the rate of 1.0 kg/s. The tube-wall and liquid-inlet temperatures are 60°C and 20°C, respectively. Average liquid properties are: $\rho = 950$ kg/m³, $\mu = 6$ mPa · s, $\mu_w = 4$ mPa · s, $k = 0.5$ W/m · K, and $c_p = 1.5$ kJ/kg · K. (*a*) Estimate the liquid-outlet temperature and the rate of heat transfer. (*b*) How will the results in part (*a*) change if all but one tubes are plugged (i.e., the entire liquid stream is forced through a single tube)?

Fundamentals of Engineering (FE) Exam Problems

8–94 Internal force flows are said to be fully developed once the _____ at a cross-section no longer changes in the direction of flow.
(*a*) temperature distribution   (*b*) entropy distribution
(*c*) velocity distribution   (*d*) pressure distribution
(*e*) none of the above

8–95 The bulk or mixed temperature of a fluid flowing through a pipe or duct is defined as

$$(a)\ T_b = \frac{1}{A_c} \int_{A_c} T dA_c \qquad (b)\ T_b = \frac{1}{\dot{m}} \int_{A_c} T\rho V dA_c$$

$$(c)\ T_b = \frac{1}{\dot{m}} \int_{A_c} h\rho V dA_c \qquad (d)\ T_b = \frac{1}{A_c} \int_{A_c} h dA_c$$

$$(e)\ T_b = \frac{1}{\dot{V}} \int_{A_c} T\rho V dA_c$$

8–96 Water ($\mu = 9.0 \times 10^{-4}$ kg/m · s, $\rho = 1000$ kg/m³) enters a 2-cm-diameter, and 3-m-long tube whose walls are maintained at 100°C. The water enters this tube with a bulk temperature of 25°C and a volume flow rate of 3 m³/h. The Reynolds number for this internal flow is
(*a*) 59,000   (*b*) 105,000   (*c*) 178,000
(*d*) 236,000   (*e*) 342,000

8–97 Water enters a 2-cm-diameter and 3-m-long tube whose walls are maintained at 100°C with a bulk temperature of 25°C and volume flow rate of 3 m³/h. Neglecting the entrance effects and assuming turbulent flow, the Nusselt number can be determined from Nu = 0.023 Re$^{0.8}$ Pr$^{0.4}$. The convection heat transfer coefficient in this case is
(*a*) 4140 W/m² · K   (*b*) 6160 W/m² · K
(*c*) 8180 W/m² · K   (*d*) 9410 W/m² · K
(*e*) 2870 W/m² · K

(For water, use $k = 0.610$ W/m · °C, Pr = 6.0, $\mu = 9.0 \times 10^{-4}$ kg/m · s, $\rho = 1000$ kg/m³)

8–98 Water enters a circular tube whose walls are maintained at constant temperature at a specified flow rate and temperature. For fully developed turbulent flow, the Nusselt number can be determined from Nu = 0.023 Re$^{0.8}$ Pr$^{0.4}$. The correct temperature difference to use in Newton s law of cooling in this case is
(*a*) The difference between the inlet and outlet water bulk temperature.
(*b*) The difference between the inlet water bulk temperature and the tube wall temperature.
(*c*) The log mean temperature difference.
(*d*) The difference between the average water bulk temperature and the tube temperature.
(*e*) None of the above.

8–99   Water (c_p = 4180 J/kg · K) enters a 4-cm-diameter tube at 15°C at a rate of 0.06 kg/s. The tube is subjected to a uniform heat flux of 2500 W/m² on the surfaces. The length of the tube required in order to heat the water to 45°C is

(a) 6 m   (b) 12 m   (c) 18 m   (d) 24 m   (e) 30 m

8–100   Air (c_p = 1000 J/kg · K) enters a 20-cm-diameter and 19-m-long underwater duct at 50°C and 1 atm at an average velocity of 7 m/s and is cooled by the water outside. If the average heat transfer coefficient is 35 W/m² · °C and the tube temperature is nearly equal to the water temperature of 5 °C, the exit temperature of air is

(a) 8°C   (b) 13°C   (c) 18°C
(d) 28°C   (e) 37°C

8–101   Water (c_p = 4180 J/kg · K) enters a 12-cm-diameter and 8.5-m-long tube at 75°C at a rate of 0.35 kg/s, and is cooled by a refrigerant evaporating outside at −10°C. If the average heat transfer coefficient on the inner surface is 500 W/m² · °C, the exit temperature of water is

(a) 18.4°C   (b) 25.0°C   (c) 33.8°C
(d) 46.5°C   (e) 60.2°C

8–102   Air enters a duct at 20°C at a rate of 0.08 m³/s, and is heated to 150°C by steam condensing outside at 200°C. The error involved in the rate of heat transfer to the air due to using arithmetic mean temperature difference instead of logarithmic mean temperature difference is

(a) 0%   (b) 5.4%   (c) 8.1%   (d) 10.6%   (e) 13.3%

8–103   Engine oil at 60°C (μ = 0.07399 kg/m · s, ρ = 864 kg/m³) flows in a 5-cm-diameter tube with a velocity of 1.3 m/s. The pressure drop along a fully developed 6-m-long section of the tube is

(a) 2.9 kPa   (b) 5.2 kPa   (c) 7.4 kPa
(d) 10.5 kPa   (e) 20.0 kPa

8–104   Engine oil flows in a 15-cm-diameter horizontal tube with a velocity of 1.3 m/s, experiencing a pressure drop of 12 kPa. The pumping power requirement to overcome this pressure drop is

(a) 190 W   (b) 276 W   (c) 407 W
(d) 655 W   (e) 900 W

8–105   Water enters a 5-mm-diameter and 13-m-long tube at 15°C with a velocity of 0.3 m/s, and leaves at 45°C. The tube is subjected to a uniform heat flux of 2000 W/m² on its surface. The temperature of the tube surface at the exit is

(a) 48.7°C   (b) 49.4°C   (c) 51.1°C
(d) 53.7°C   (e) 55.2°C

(For water, use k = 0.615 W/m · °C, Pr = 5.42, v = 0.801 × 10⁻⁶ m²/s.)

8–106   Water enters a 5-mm-diameter and 13-m-long tube at 45°C with a velocity of 0.3 m/s. The tube is maintained at a constant temperature of 5°C. The exit temperature of water is

(a) 7.5°C   (b) 7.0°C   (c) 6.5°C
(d) 6.0°C   (e) 5.5°C

(For water, use k = 0.607 W/m · °C, Pr = 6.14, v = 0.894 × 10⁻⁶ m²/s, c_p = 4180 J/kg · °C, ρ = 997 kg/m³)

8–107   Water enter a 5-mm-diameter and 13-m-long tube at 45°C with a velocity of 0.3 m/s. The tube is maintained at a constant temperature of 5°C. The required length of the tube in order for the water to exit the tube at 25°C is

(a) 1.55 m   (b) 1.72 m   (c) 1.99 m
(d) 2.37 m   (e) 2.96 m

(For water, use k = 0.623 W/m · °C, Pr =4.83, v = 0.724 × 10⁻⁶ m²/s, c_p = 4178 J/kg · °C, ρ = 994 kg/m³.)

8–108   Air at 10°C enters an 18-m-long rectangular duct of cross section 0.15 m × 0.20 m at a velocity of. 4.5 m/s. The duct is subjected to uniform radiation heating throughout the surface at a rate of 400 W/m². The wall temperature at the exit of the duct is

(a) 58.8°C   (b) 61.9°C   (c) 64.6°C
(d) 69.1°C   (e) 75.5°C

(For air, use k = 0.02551 W/m · °C, Pr = 0.7296, v = 1.562 × 10⁻⁵ m²/s, c_p = 1007 J/kg · °C, ρ = 1.184 kg/m³.)

8–109   Air at 110°C enters an 18-cm-diameter and 9-m-long duct at a velocity of 3 m/s. The duct is observed to be nearly isothermal at 85°C. The rate of heat loss from the air in the duct is

(a) 375 W   (b) 510 W   (c) 936 W
(d) 965 W   (e) 987 W

(For air, use k = 0.03095 W/m · °C, Pr = 0.7111, v = 2.306 × 10⁻⁵ m²/s, c_p = 1009 J/kg · °C.)

8–110   Air enters a 7-cm-diameter and 4-m-long tube at 65°C and leaves at 15°C. The tube is observed to be nearly isothermal at 5°C. If the average convection heat transfer coefficient is 20 W/m² · °C, the rate of heat transfer from the air is

(a) 491 W   (b) 616 W   (c) 810 W
(d) 907 W   (e) 975 W

8–111   Air (c_p = 1007 J/kg · °C) enters a 17-cm-diameter and 4-m-long tube at 65°C at a rate of 0.08 kg/s and leaves at 15°C. The tube is observed to be nearly isothermal at 5°C. The average convection heat transfer coefficient is

(a) 24.5 W/m² · °C   (b) 46.2 W/m² · °C
(c) 53.9 W/m² · °C   (d) 67.6 W/m² · °C
(e) 90.7 W/m² · °C

8–112   Air at 40°C (μ = 1.918 × 10⁻⁵ kg/m · s and ρ = 1.127 kg/m³) flows in a 25-cm diameter and 26-m-long horizontal tube at a velocity of 5 m/s. If the roughness of the inner surface of the pipe is 0.2 mm, the required pumping power to overcome the pressure drop is

(a) 0.3 W   (b) 0.9 W   (c) 3.4 W
(d) 5.5 W   (e) 8.0 W

Design and Essay Problems

8–113   Electronic boxes such as computers are commonly cooled by a fan. Write an essay on forced air cooling of electronic boxes and on the selection of the fan for electronic devices.

8–114 Design a heat exchanger to pasteurize milk by steam in a dairy plant. Milk is to flow through a bank of 1.2-cm internal diameter tubes while steam condenses outside the tubes at 1 atm. Milk is to enter the tubes at 4°C, and it is to be heated to 72°C at a rate of 15 L/s. Making reasonable assumptions, you are to specify the tube length and the number of tubes, and the pump for the heat exchanger.

8–115 A desktop computer is to be cooled by a fan. The electronic components of the computer consume 80 W of power under full-load conditions. The computer is to operate in environments at temperatures up to 50°C and at elevations up to 3000 m where the atmospheric pressure is 70.12 kPa. The exit temperature of air is not to exceed 60°C to meet the reliability requirements. Also, the average velocity of air is not to exceed 120 m/min at the exit of the computer case, where the fan is installed to keep the noise level down. Specify the flow rate of the fan that needs to be installed and the diameter of the casing of the fan.

NATURAL CONVECTION

I n Chapters 7 and 8, we considered heat transfer by *forced convection,* where a fluid was *forced* to move over a surface or in a tube by external means such as a pump or a fan. In this chapter, we consider *natural convection,* where any fluid motion occurs by natural means such as buoyancy. The fluid motion in forced convection is quite *noticeable,* since a fan or a pump can transfer enough momentum to the fluid to move it in a certain direction. The fluid motion in natural convection, however, is often not noticeable because of the low velocities involved.

The convection heat transfer coefficient is a strong function of *velocity:* the higher the velocity, the higher the convection heat transfer coefficient. The fluid velocities associated with natural convection are low, typically less than 1 m/s. Therefore, the heat transfer coefficients encountered in natural convection are usually much lower than those encountered in forced convection. Yet several types of heat transfer equipment are designed to operate under natural convection conditions instead of forced convection, because natural convection does not require the use of a fluid mover.

We start this chapter with a discussion of the physical mechanism of *natural convection* and the *Grashof number.* We then present the correlations to evaluate heat transfer by natural convection for various geometries, including finned surfaces and enclosures. Finally, we discuss simultaneous forced and natural convection.

OBJECTIVES

When you finish studying this chapter, you should be able to:

■ Understand the physical mechanism of natural convection,

■ Derive the governing equations of natural convection, and obtain the dimensionless Grashof number by nondimensionalizing them,

■ Evaluate the Nusselt number for natural convection associated with vertical, horizontal, and inclined plates as well as cylinders and spheres,

■ Examine natural convection from finned surfaces, and determine the optimum fin spacing,

■ Analyze natural convection inside enclosures such as double-pane windows, and

■ Consider combined natural and forced convection, and assess the relative importance of each mode.

9–1 ▪ PHYSICAL MECHANISM OF NATURAL CONVECTION

Many familiar heat transfer applications involve natural convection as the primary mechanism of heat transfer. Some examples are cooling of electronic equipment such as power transistors, TVs, and DVDs; heat transfer from electric baseboard heaters or steam radiators; heat transfer from the refrigeration coils and power transmission lines; and heat transfer from the bodies of animals and human beings. Natural convection in gases is usually accompanied by radiation of comparable magnitude except for low-emissivity surfaces.

We know that a hot boiled egg (or a hot baked potato) on a plate eventually cools to the surrounding air temperature (Fig. 9–1). The egg is cooled by transferring heat by convection to the air and by radiation to the surrounding surfaces. Disregarding heat transfer by radiation, the physical mechanism of cooling a hot egg (or any hot object) in a cooler environment can be explained as follows:

As soon as the hot egg is exposed to cooler air, the temperature of the outer surface of the egg shell drops somewhat, and the temperature of the air adjacent to the shell rises as a result of heat conduction from the shell to the air. Consequently, the egg is surrounded by a thin layer of warmer air, and heat is then transferred from this warmer layer to the outer layers of air. The cooling process in this case is rather slow since the egg would always be blanketed by warm air, and it has no direct contact with the cooler air farther away. We may not notice any air motion in the vicinity of the egg, but careful measurements would indicate otherwise.

The temperature of the air adjacent to the egg is higher and thus its density is lower, since at constant pressure the density of a gas is inversely proportional to its temperature. Thus, we have a situation in which some low-density or "light" gas is surrounded by a high-density or "heavy" gas, and the natural laws dictate that the *light gas rise*. This is no different than the oil in a vinegar-and-oil salad dressing rising to the top (since $\rho_{oil} < \rho_{vinegar}$). This phenomenon is characterized incorrectly by the phrase "heat rises," which is understood to mean *heated air rises*. The space vacated by the warmer air in the vicinity of the egg is replaced by the cooler air nearby, and the presence of cooler air in the vicinity of the egg speeds up the cooling process. The rise of warmer air and the flow of cooler air into its place continues until the egg is cooled to the temperature of the surrounding air. The motion that results from the continual replacement of the heated air in the vicinity of the egg by the cooler air nearby is called a **natural convection current**, and the heat transfer that is enhanced as a result of this natural convection current is called **natural convection heat transfer**. Note that in the absence of natural convection currents, heat transfer from the egg to the air surrounding it would be by conduction only, and the rate of heat transfer from the egg would be much lower.

Natural convection is just as effective in the heating of cold surfaces in a warmer environment as it is in the cooling of hot surfaces in a cooler environment, as shown in Fig. 9–2. Note that the direction of fluid motion is reversed in this case.

In a gravitational field, there is a net force that pushes upward a light fluid placed in a heavier fluid. The upward force exerted by a fluid on a body

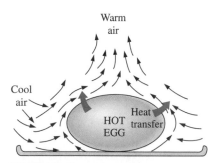

FIGURE 9–1
The cooling of a boiled egg in a cooler environment by natural convection.

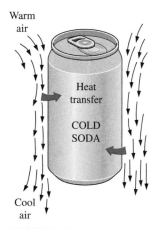

FIGURE 9–2
The warming up of a cold drink in a warmer environment by natural convection.

completely or partially immersed in it is called the **buoyancy force**. The magnitude of the buoyancy force is equal to the weight of the *fluid displaced* by the body. That is,

$$F_{buoyancy} = \rho_{fluid}\, g\, V_{body} \qquad (9\text{-}1)$$

where ρ_{fluid} is the average density of the *fluid* (not the body), g is the gravitational acceleration, and V_{body} is the volume of the portion of the body immersed in the fluid (for bodies completely immersed in the fluid, it is the total volume of the body). In the absence of other forces, the net vertical force acting on a body is the difference between the weight of the body and the buoyancy force. That is,

$$\begin{aligned}
F_{net} &= W - F_{buoyancy} \\
&= \rho_{body}\, g\, V_{body} - \rho_{fluid}\, g\, V_{body} \qquad (9\text{-}2) \\
&= (\rho_{body} - \rho_{fluid})\, g\, V_{body}
\end{aligned}$$

Note that this force is proportional to the difference in the densities of the fluid and the body immersed in it. Thus, a body immersed in a fluid will experience a "weight loss" in an amount equal to the weight of the fluid it displaces. This is known as *Archimedes' principle*.

To have a better understanding of the buoyancy effect, consider an egg dropped into water. If the average density of the egg is greater than the density of water (a sign of freshness), the egg settles at the bottom of the container. Otherwise, it rises to the top. When the density of the egg equals the density of water, the egg settles somewhere in the water while remaining completely immersed, acting like a "weightless object" in space. This occurs when the upward buoyancy force acting on the egg equals the weight of the egg, which acts downward.

The *buoyancy effect* has far-reaching implications in life. For one thing, without buoyancy, heat transfer between a hot (or cold) surface and the fluid surrounding it would be by *conduction* instead of by *natural convection*. The natural convection currents encountered in the oceans, lakes, and the atmosphere owe their existence to buoyancy. Also, light boats as well as heavy warships made of steel float on water because of buoyancy (Fig. 9–3). Ships are designed on the basis of the principle that the entire weight of a ship and its contents is equal to the weight of the water that the submerged volume of the ship can contain. The "chimney effect" that induces the upward flow of hot combustion gases through a chimney is also due to the buoyancy effect, and the upward force acting on the gases in the chimney is proportional to the difference between the densities of the hot gases in the chimney and the cooler air outside. Note that there is *no noticable gravity* in space, and thus there can be no natural convection heat transfer in a spacecraft, even if the spacecraft is filled with atmospheric air.

In heat transfer studies, the primary variable is *temperature,* and it is desirable to express the net buoyancy force (Eq. 9–2) in terms of temperature differences. But this requires expressing the density difference in terms of a temperature difference, which requires a knowledge of a property that represents the *variation of the density of a fluid with temperature at constant pressure. The property that provides that information is the* **volume expansion coefficient** β, defined as (Fig. 9–4)

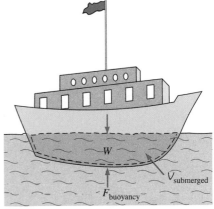

FIGURE 9–3
It is the buoyancy force that keeps the ships afloat in water ($W = F_{buoyancy}$ for floating objects).

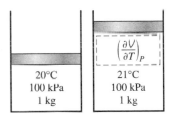

(*a*) A substance with a large β

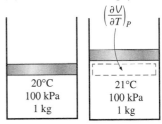

(*b*) A substance with a small β

FIGURE 9–4
The coefficient of volume expansion is a measure of the change in volume of a substance with temperature at constant pressure.

$$\beta = \frac{1}{v}\left(\frac{\partial v}{\partial T}\right)_P = -\frac{1}{\rho}\left(\frac{\partial \rho}{\partial T}\right)_P \qquad (1/K) \qquad \text{(9–3)}$$

In natural convection studies, the condition of the fluid sufficiently far from the hot or cold surface is indicated by the subscript "infinity" to serve as a reminder that this is the value at a distance where the presence of the surface is not felt. In such cases, the volume expansion coefficient can be expressed approximately by replacing differential quantities by differences as

$$\beta \approx -\frac{1}{\rho}\frac{\Delta\rho}{\Delta T} = -\frac{1}{\rho}\frac{\rho_\infty - \rho}{T_\infty - T} \qquad \text{(at constant } P) \qquad \text{(9–4)}$$

or

$$\rho_\infty - \rho = \rho\beta(T - T_\infty) \qquad \text{(at constant } P) \qquad \text{(9–5)}$$

where ρ_∞ is the density and T_∞ is the temperature of the quiescent fluid away from the surface.

We can show easily that the volume expansion coefficient β of an *ideal gas* ($P = \rho RT$) at a temperature T is equivalent to the inverse of the temperature:

$$\beta_{\text{ideal gas}} = \frac{1}{T} \qquad (1/K) \qquad \text{(9–6)}$$

where T is the *thermodynamic* temperature. Note that a large value of β for a fluid means a large change in density with temperature, and that the product $\beta\Delta T$ represents the fraction of volume change of a fluid that corresponds to a temperature change ΔT at constant pressure. Also note that the buoyancy force is proportional to the *density difference,* which is proportional to the *temperature difference* at constant pressure. Therefore, the larger the temperature difference between the fluid adjacent to a hot (or cold) surface and the fluid away from it, the *larger* the buoyancy force and the *stronger* the natural convection currents, and thus the *higher* the heat transfer rate.

The magnitude of the natural convection heat transfer between a surface and a fluid is directly related to the *flow rate* of the fluid. The higher the flow rate, the higher the heat transfer rate. In fact, it is the very high flow rates that increase the heat transfer coefficient by orders of magnitude when forced convection is used. In natural convection, no blowers are used, and therefore the flow rate cannot be controlled externally. The flow rate in this case is established by the dynamic balance of *buoyancy* and *friction.*

As we have discussed earlier, the buoyancy force is caused by the density difference between the heated (or cooled) fluid adjacent to the surface and the fluid surrounding it, and is proportional to this density difference and the volume occupied by the warmer fluid. It is also well known that whenever two bodies in contact (solid–solid, solid–fluid, or fluid–fluid) move relative to each other, a *friction force* develops at the contact surface in the direction opposite to that of the motion. This opposing force slows down the fluid and thus reduces the flow rate of the fluid. Under steady conditions, the airflow rate driven by buoyancy is established at the point where these two effects *balance* each other. The friction force increases as more and more solid surfaces are introduced, seriously disrupting the fluid flow and heat transfer. For that reason, heat sinks with closely spaced fins are not suitable for natural convection cooling.

Most heat transfer correlations in natural convection are based on experimental measurements. The instrument often used in natural convection experiments is the *Mach–Zehnder interferometer*, which gives a plot of isotherms in the fluid in the vicinity of a surface. The operation principle of interferometers is based on the fact that at low pressure, the lines of constant temperature for a gas correspond to the lines of constant density, and that the index of refraction of a gas is a function of its density. Therefore, the degree of refraction of light at some point in a gas is a measure of the temperature gradient at that point. An interferometer produces a map of interference fringes, which can be interpreted as lines of *constant temperature* as shown in Fig. 9–5. The smooth and parallel lines in (*a*) indicate that the flow is *laminar*, whereas the eddies and irregularities in (*b*) indicate that the flow is *turbulent*. Note that the lines are closest near the surface, indicating a *higher temperature gradient.*

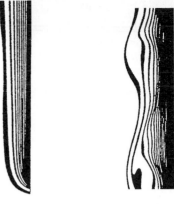

(*a*) Laminar flow   (*b*) Turbulent flow

FIGURE 9–5

Isotherms in natural convection over a hot plate in air.

9–2 ▪ EQUATION OF MOTION AND THE GRASHOF NUMBER

In this section we derive the equation of motion that governs the natural convection flow in laminar boundary layer. The conservation of mass and energy equations derived in Chapter 6 for forced convection are also applicable for natural convection, but the momentum equation needs to be modified to incorporate buoyancy.

Consider a vertical hot flat plate immersed in a quiescent fluid body. We assume the natural convection flow to be steady, laminar, and two-dimensional, and the fluid to be Newtonian with constant properties, including density, with one exception: the density difference $\rho - \rho_\infty$ is to be considered since it is this density difference between the inside and the outside of the boundary layer that gives rise to buoyancy force and sustains flow. (This is known as the *Boussinesq approximation.*) We take the upward direction along the plate to be x, and the direction normal to surface to be y, as shown in Fig. 9–6. Therefore, gravity acts in the $-x$-direction. Noting that the flow is steady and two-dimensional, the x- and y-components of velocity within boundary layer are $u = u(x, y)$ and $v = v(x, y)$, respectively.

The velocity and temperature profiles for natural convection over a vertical hot plate are also shown in Fig. 9–6. Note that as in forced convection, the thickness of the boundary layer increases in the flow direction. Unlike forced convection, however, the fluid velocity is *zero* at the outer edge of the velocity boundary layer as well as at the surface of the plate. This is expected since the fluid beyond the boundary layer is motionless. Thus, the fluid velocity increases with distance from the surface, reaches a maximum, and gradually decreases to zero at a distance sufficiently far from the surface. At the surface, the fluid temperature is equal to the plate temperature, and gradually decreases to the temperature of the surrounding fluid at a distance sufficiently far from the surface, as shown in the figure. In the case of *cold surfaces,* the shape of the velocity and temperature profiles remains the same but their direction is reversed.

Consider a differential volume element of height dx, length dy, and unit depth in the z-direction (normal to the paper) for analysis. The forces acting on this volume element are shown in Fig. 9–7. Newton's second law of motion

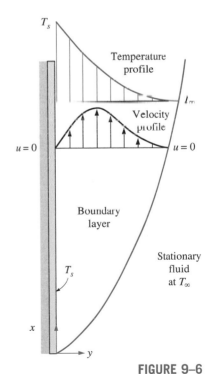

FIGURE 9–6

Typical velocity and temperature profiles for natural convection flow over a hot vertical plate at temperature T_s inserted in a fluid at temperature T_∞.

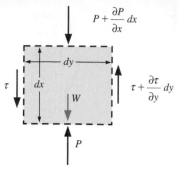

FIGURE 9–7
Forces acting on a differential volume element in the natural convection boundary layer over a vertical flat plate.

for this volume element can be expressed as

$$\delta m \cdot a_x = F_x \tag{9–7}$$

where $\delta m = \rho(dx \cdot dy \cdot 1)$ is the mass of the fluid within the volume element. The acceleration in the x-direction is obtained by taking the total differential of $u(x, y)$, which is $du = (\partial u/\partial x)dx + (\partial u/\partial y)dy$, and dividing it by dt. We get

$$a_x = \frac{du}{dt} = \frac{\partial u}{\partial x}\frac{dx}{dt} + \frac{\partial u}{\partial y}\frac{dy}{dt} = u\frac{\partial u}{\partial x} + v\frac{\partial u}{\partial y} \tag{9–8}$$

The forces acting on the differential volume element in the vertical direction are the pressure forces acting on the top and bottom surfaces, the shear stresses acting on the side surfaces (the normal stresses acting on the top and bottom surfaces are small and are disregarded), and the force of gravity acting on the entire volume element. Then the net surface force acting in the x-direction becomes

$$F_x = \left(\frac{\partial \tau}{\partial y}dy\right)(dx \cdot 1) - \left(\frac{\partial P}{\partial x}dx\right)(dy \cdot 1) - \rho g(dx \cdot dy \cdot 1)$$
$$= \left(\mu\frac{\partial^2 u}{\partial y^2} - \frac{\partial P}{\partial x} - \rho g\right)(dx \cdot dy \cdot 1) \tag{9–9}$$

since $\tau = \mu(\partial u/\partial y)$. Substituting Eqs. 9–8 and 9–9 into Eq. 9–7 and dividing by $\rho \cdot dx \cdot dy \cdot 1$ gives the *conservation of momentum* in the x-direction as

$$\rho\left(u\frac{\partial u}{\partial x} + v\frac{\partial u}{\partial y}\right) = \mu\frac{\partial^2 u}{\partial y^2} - \frac{\partial P}{\partial x} - \rho g \tag{9–10}$$

The x-momentum equation in the quiescent fluid outside the boundary layer can be obtained from the relation above as a special case by setting $u = 0$. It gives

$$\frac{\partial P_\infty}{\partial x} = -\rho_\infty g \tag{9–11}$$

which is simply the relation for the variation of hydrostatic pressure in a quiescent fluid with height, as expected. Also, noting that $v \ll u$ in the boundary layer and thus $\partial v/\partial x \approx \partial v/\partial y \approx 0$, and that there are no body forces (including gravity) in the y-direction, the force balance in that direction gives $\partial P/\partial y = 0$. That is, the variation of pressure in the direction normal to the surface is negligible, and for a given x the pressure in the boundary layer is equal to the pressure in the quiescent fluid. Therefore, $P = P(x) = P_\infty(x)$ and $\partial P/\partial x = \partial P_\infty/\partial x = -\rho_\infty g$. Substituting into Eq. 9–10,

$$\rho\left(u\frac{\partial u}{\partial x} + v\frac{\partial u}{\partial y}\right) = \mu\frac{\partial^2 u}{\partial y^2} + (\rho_\infty - \rho)g \tag{9–12}$$

The last term represents the net upward force per unit volume of the fluid (the difference between the buoyant force and the fluid weight). This is the force that initiates and sustains convection currents.

From Eq. 9–5, we have $\rho_\infty - \rho = \rho\beta(T - T_\infty)$. Substituting it into the last equation and dividing both sides by ρ gives the desired form of the x-momentum equation,

$$u\frac{\partial u}{\partial x} + v\frac{\partial u}{\partial y} = v\frac{\partial^2 u}{\partial y^2} + g\beta(T - T_\infty) \tag{9–13}$$

This is the equation that governs the fluid motion in the boundary layer due to the effect of buoyancy. Note that the momentum equation involves the temperature, and thus the momentum and energy equations must be solved simultaneously.

The set of three partial differential equations (the continuity, momentum, and the energy equations) that govern natural convection flow over vertical isothermal plates can be reduced to a set of two ordinary nonlinear differential equations by the introduction of a similarity variable. But the resulting equations must still be solved numerically [Ostrach (1953)]. Interested readers are referred to advanced books on the topic for detailed discussions [e.g., Kays and Crawford (1993)].

The Grashof Number

The governing equations of natural convection and the boundary conditions can be nondimensionalized by dividing all dependent and independent variables by suitable constant quantities: all lengths by a characteristic length L_c, all velocities by an arbitrary reference velocity V (which, from the definition of Reynolds number, is taken to be $V = \text{Re}_L \, \nu/L_c$), and temperature by a suitable temperature difference (which is taken to be $T_s - T_\infty$) as

$$x^* = \frac{x}{L_c} \qquad y^* = \frac{y}{L_c} \qquad u^* = \frac{u}{V} \qquad v^* = \frac{v}{V} \qquad \text{and} \qquad T^* = \frac{T - T_\infty}{T_s - T_\infty}$$

where asterisks are used to denote nondimensional variables. Substituting them into the momentum equation and simplifying give

$$u^* \frac{\partial u^*}{\partial x^*} + v^* \frac{\partial u^*}{\partial y^*} = \left[\frac{g\beta(T_s - T_\infty)L_c^3}{\nu^2} \right] \frac{T^*}{\text{Re}_L^2} + \frac{1}{\text{Re}_L} \frac{\partial^2 u^*}{\partial y^{*2}} \qquad \textbf{(9–14)}$$

The dimensionless parameter in the brackets represents the natural convection effects, and is called the **Grashof number** Gr_L,

$$\text{Gr}_L = \frac{g\beta(T_s - T_\infty)L_c^3}{\nu^2} \qquad \textbf{(9–15)}$$

where

g = gravitational acceleration, m/s^2
β = coefficient of volume expansion, 1/K ($\beta = 1/T$ for ideal gases)
T_s = temperature of the surface, °C
T_∞ = temperature of the fluid sufficiently far from the surface, °C
L_c = characteristic length of the geometry, m
ν = kinematic viscosity of the fluid, m^2/s

We mentioned in the preceding chapters that the flow regime in forced convection is governed by the dimensionless *Reynolds number,* which represents the ratio of inertial forces to viscous forces acting on the fluid. The flow regime in natural convection is governed by the dimensionless *Grashof number,* which represents the ratio of the *buoyancy force* to the *viscous force* acting on the fluid (Fig. 9–8).

The role played by the Reynolds number in forced convection is played by the Grashof number in natural convection. As such, the Grashof number provides the main criterion in determining whether the fluid flow is laminar or turbulent in natural convection. For vertical plates, for example, the critical

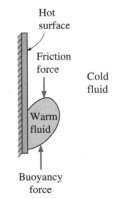

FIGURE 9–8
The Grashof number Gr is a measure of the relative magnitudes of the *buoyancy force* and the opposing *viscous force* acting on the fluid.

Grashof number is observed to be about 10^9. Therefore, the flow regime on a vertical plate becomes *turbulent* at Grashof numbers greater than 10^9.

When a surface is subjected to external flow, the problem involves both natural and forced convection. The relative importance of each mode of heat transfer is determined by the value of the coefficient $\mathrm{Gr}_L/\mathrm{Re}_L^2$: Natural convection effects are negligible if $\mathrm{Gr}_L/\mathrm{Re}_L^2 \ll 1$, free convection dominates and the forced convection effects are negligible if $\mathrm{Gr}_L/\mathrm{Re}_L^2 \gg 1$, and both effects are significant and must be considered if $\mathrm{Gr}_L/\mathrm{Re}_L^2 \approx 1$.

9–3 · NATURAL CONVECTION OVER SURFACES

Natural convection heat transfer on a surface depends on the geometry of the surface as well as its orientation. It also depends on the variation of temperature on the surface and the thermophysical properties of the fluid involved.

Although we understand the mechanism of natural convection well, the complexities of fluid motion make it very difficult to obtain simple analytical relations for heat transfer by solving the governing equations of motion and energy. Some analytical solutions exist for natural convection, but such solutions lack generality since they are obtained for simple geometries under some simplifying assumptions. Therefore, with the exception of some simple cases, heat transfer relations in natural convection are based on experimental studies. Of the numerous such correlations of varying complexity and claimed accuracy available in the literature for any given geometry, we present here the ones that are best known and widely used.

The simple empirical correlations for the average *Nusselt number* Nu in natural convection are of the form (Fig. 9–9)

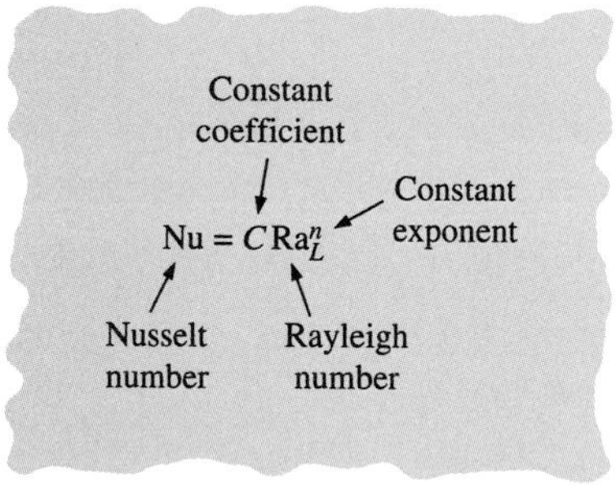

$$\mathrm{Nu} = \frac{hL_c}{k} = C(\mathrm{Gr}_L\,\mathrm{Pr})^n = C\,\mathrm{Ra}_L^n \tag{9–16}$$

where Ra_L is the **Rayleigh number**, which is the product of the Grashof and Prandtl numbers:

$$\mathrm{Ra}_L = \mathrm{Gr}_L\,\mathrm{Pr} = \frac{g\beta(T_s - T_\infty)L_c^3}{\nu^2}\,\mathrm{Pr} \tag{9–17}$$

The values of the constants C and n depend on the *geometry* of the surface and the *flow regime,* which is characterized by the range of the Rayleigh number. The value of n is usually $\frac{1}{4}$ for laminar flow and $\frac{1}{3}$ for turbulent flow. The value of the constant C is normally less than 1.

Simple relations for the average Nusselt number for various geometries are given in Table 9–1, together with sketches of the geometries. Also given in this table are the characteristic lengths of the geometries and the ranges of Rayleigh number in which the relation is applicable. All fluid properties are to be evaluated at the film temperature $T_f = \frac{1}{2}(T_s + T_\infty)$.

When the average Nusselt number and thus the average convection coefficient is known, the rate of heat transfer by natural convection from a solid surface at a uniform temperature T_s to the surrounding fluid is expressed by Newton's law of cooling as

$$\dot{Q}_{\mathrm{conv}} = hA_s(T_s - T_\infty) \quad (\mathrm{W}) \tag{9–18}$$

FIGURE 9–9

Natural convection heat transfer correlations are usually expressed in terms of the Rayleigh number raised to a constant n multiplied by another constant C, both of which are determined experimentally.

TABLE 9–1

Empirical correlations for the average Nusselt number for natural convection over surfaces

Geometry	Characteristic length L_c	Range of Ra	Nu
Vertical plate	L	10^4–10^9 10^{10}–10^{13} Entire range	$Nu = 0.59Ra_L^{1/4}$ **(9–19)** $Nu = 0.1Ra_L^{1/3}$ **(9–20)** $Nu = \left\{ 0.825 + \dfrac{0.387Ra_L^{1/6}}{[1 + (0.492/Pr)^{9/16}]^{8/27}} \right\}^2$ **(9–21)** (complex but more accurate)
Inclined plate	L		Use vertical plate equations for the upper surface of a cold plate and the lower surface of a hot plate Replace g by $g\cos\theta$ for $Ra < 10^9$
Horizontal plate (Surface area A and perimeter p) (a) Upper surface of a hot plate (or lower surface of a cold plate) (b) Lower surface of a hot plate (or upper surface of a cold plate) 	A_s/p	10^4–10^7 10^7–10^{11} 10^5–10^{11}	$Nu = 0.54Ra_L^{1/4}$ **(9–22)** $Nu = 0.15Ra_L^{1/3}$ **(9–23)** $Nu = 0.27Ra_L^{1/4}$ **(9–24)**
Vertical cylinder	L		A vertical cylinder can be treated as a vertical plate when $D > \dfrac{35L}{Gr_L^{1/4}}$
Horizontal cylinder	D	$Ra_D \leq 10^{12}$	$Nu = \left\{ 0.6 + \dfrac{0.387Ra_D^{1/6}}{[1 + (0.559/Pr)^{9/16}]^{8/27}} \right\}^2$ **(9–25)**
Sphere	D	$Ra_D \leq 10^{11}$ $(Pr \geq 0.7)$	$Nu = 2 + \dfrac{0.589Ra_D^{1/4}}{[1 + (0.469/Pr)^{9/16}]^{4/9}}$ **(9–26)**

where A_s is the heat transfer surface area and h is the average heat transfer coefficient on the surface.

Vertical Plates (T_s = constant)

For a vertical flat plate, the characteristic length is the plate height L. In Table 9–1 we give three relations for the average Nusselt number for an isothermal vertical plate. The first two relations are very simple. Despite its complexity, we suggest using the third one (Eq. 9–21) recommended by Churchill and Chu (1975) since it is applicable over the entire range of Rayleigh number. This relation is most accurate in the range of $10^{-1} < \mathrm{Ra}_L < 10^9$.

Vertical Plates ($\dot{q}_s$ = constant)

In the case of constant surface heat flux, the rate of heat transfer is known (it is simply $\dot{Q} = \dot{q}_s A_s$), but the surface temperature T_s is not. In fact, T_s increases with height along the plate. It turns out that the Nusselt number relations for the constant surface temperature and constant surface heat flux cases are nearly identical [Churchill and Chu (1975)]. Therefore, the relations for isothermal plates can also be used for plates subjected to uniform heat flux, provided that the plate midpoint temperature $T_{L/2}$ is used for T_s in the evaluation of the film temperature, Rayleigh number, and the Nusselt number. Noting that $h = \dot{q}_s/(T_{L/2} - T_\infty)$, the average Nusselt number in this case can be expressed as

$$\mathrm{Nu} = \frac{hL}{k} = \frac{\dot{q}_s L}{k(T_{L/2} - T_\infty)} \tag{9–27}$$

The midpoint temperature $T_{L/2}$ is determined by iteration so that the Nusselt numbers determined from Eqs. 9–21 and 9–27 match.

Vertical Cylinders

An outer surface of a vertical cylinder can be treated as a vertical plate when the diameter of the cylinder is sufficiently large so that the curvature effects are negligible. This condition is satisfied if

$$D \geq \frac{35L}{\mathrm{Gr}_L^{1/4}} \tag{9–28}$$

When this criteria is met, the relations for vertical plates can also be used for vertical cylinders. Nusselt number relations for slender cylinders that do not meet this criteria are available in the literature [e.g., Cebeci (1974)].

Inclined Plates

Consider an inclined hot plate that makes an angle θ from the vertical, as shown in Fig. 9–10, in a cooler environment. The net force $F = g(\rho_\infty - \rho)$ (the difference between the buoyancy and gravity) acting on a unit volume of the fluid in the boundary layer is always in the vertical direction. In the case of inclined plate, this force can be resolved into two components: $F_y = F \cos \theta$ parallel to the plate that drives the flow along the plate, and $F_y = F \sin \theta$

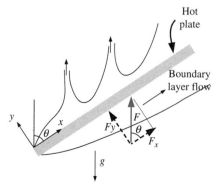

FIGURE 9–10
Natural convection flows on the upper and lower surfaces of an inclined hot plate.

normal to the plate. Noting that the force that drives the motion is reduced, we expect the convection currents to be weaker, and the rate of heat transfer to be lower relative to the vertical plate case.

The experiments confirm what we suspect for the lower surface of a hot plate, but the opposite is observed on the upper surface. The reason for this curious behavior for the upper surface is that the force component F_y initiates upward motion in addition to the parallel motion along the plate, and thus the boundary layer breaks up and forms plumes, as shown in the figure. As a result, the thickness of the boundary layer and thus the resistance to heat transfer decreases, and the rate of heat transfer increases relative to the vertical orientation.

In the case of a cold plate in a warmer environment, the opposite occurs as expected: The boundary layer on the upper surface remains intact with weaker boundary layer flow and thus lower rate of heat transfer, and the boundary layer on the lower surface breaks apart (the colder fluid falls down) and thus enhances heat transfer.

When the boundary layer remains intact (the lower surface of a hot plate or the upper surface of a cold plate), the Nusselt number can be determined from the vertical plate relations provided that g in the Rayleigh number relation is replaced by $g\cos\theta$ for $\theta < 60°$. Nusselt number relations for the other two surfaces (the upper surface of a hot plate or the lower surface of a cold plate) are available in the literature [e.g., Fujiii and Imura (1972)].

Horizontal Plates

The rate of heat transfer to or from a horizontal surface depends on whether the surface is facing upward or downward. For a hot surface in a cooler environment, the net force acts upward, forcing the heated fluid to rise. If the hot surface is facing upward, the heated fluid rises freely, inducing strong natural convection currents and thus effective heat transfer, as shown in Fig. 9–11. But if the hot surface is facing downward, the plate blocks the heated fluid that tends to rise (except near the edges), impeding heat transfer. The opposite is true for a cold plate in a warmer environment since the net force (weight minus buoyancy force) in this case acts downward, and the cooled fluid near the plate tends to descend.

The average Nusselt number for horizontal surfaces can be determined from the simple power-law relations given in Table 9–1. The characteristic length for horizontal surfaces is calculated from

$$L_c = \frac{A_s}{p} \tag{9-29}$$

where A_s is the surface area and p is the perimeter. Note that $L_c = a/4$ for a horizontal square surface of length a, and $D/4$ for a horizontal circular surface of diameter D.

Horizontal Cylinders and Spheres

The boundary layer over a hot horizontal cylinder starts to develop at the bottom, increasing in thickness along the circumference, and forming a rising plume at the top, as shown in Fig. 9–12. Therefore, the local Nusselt number

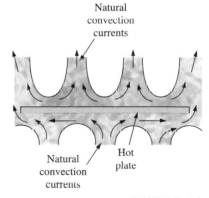

FIGURE 9–11
Natural convection flows on the upper and lower surfaces of a horizontal hot plate.

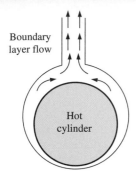

FIGURE 9–12
Natural convection flow over a horizontal hot cylinder.

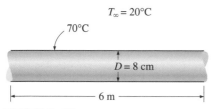

FIGURE 9–13
Schematic for Example 9–1.

is highest at the bottom, and lowest at the top of the cylinder when the boundary layer flow remains laminar. The opposite is true in the case of a cold horizontal cylinder in a warmer medium, and the boundary layer in this case starts to develop at the top of the cylinder and ending with a descending plume at the bottom.

The average Nusselt number over the entire surface can be determined from Eq. 9–25 [Churchill and Chu (1975)] for an isothermal horizontal cylinder, and from Eq. 9–26 for an isothermal sphere [Churchill (1983)] both given in Table 9–1.

EXAMPLE 9–1   Heat Loss from Hot-Water Pipes

A 6-m-long section of an 8-cm-diameter horizontal hot-water pipe shown in Fig. 9–13 passes through a large room whose temperature is 20°C. If the outer surface temperature of the pipe is 70°C, determine the rate of heat loss from the pipe by natural convection.

SOLUTION   A horizontal hot-water pipe passes through a large room. The rate of heat loss from the pipe by natural convection is to be determined.

Assumptions   1 Steady operating conditions exist. 2 Air is an ideal gas. 3 The local atmospheric pressure is 1 atm.

Properties   The properties of air at the film temperature of $T_f = (T_s + T_\infty)/2 = (70 + 20)/2 = 20°C$ and 1 atm are (Table A–15)

$$k = 0.02699 \text{ W/m} \cdot °C \qquad Pr = 0.7241$$

$$\nu = 1.750 \times 10^{-5} \text{ m}^2/\text{s} \qquad \beta = \frac{1}{T_f} = \frac{1}{318 \text{ K}}$$

Analysis   The characteristic length in this case is the outer diameter of the pipe, $L_c = D = 0.08$ m. Then the Rayleigh number becomes

$$Ra_D = \frac{g\beta(T_s - T_\infty)D^3}{\nu^2} Pr$$

$$= \frac{(9.81 \text{ m/s}^2)[1/(318 \text{ K})](70 - 20 \text{ K})(0.08 \text{ m})^3}{(1.750 \times 10^{-5} \text{ m}^2/\text{s})^2} (0.7241) = 1.867 \times 10^6$$

The natural convection Nusselt number in this case can be determined from Eq. 9–25 to be

$$Nu = \left\{ 0.6 + \frac{0.387 Ra_D^{1/6}}{[1 + (0.559/Pr)^{9/16}]^{8/27}} \right\}^2 = \left\{ 0.6 + \frac{0.387(1.867 \times 10^6)^{1/6}}{[1 + (0.559/0.7241)^{9/16}]^{8/27}} \right\}^2$$

$$= 17.39$$

Then,

$$h = \frac{k}{D} Nu = \frac{0.02699 \text{ W/m} \cdot °C}{0.08 \text{ m}} (17.39) = 5.867 \text{ W/m} \cdot °C$$

$$A_s = \pi DL = \pi(0.08 \text{ m})(6 \text{ m}) = 1.508 \text{ m}^2$$

and

$$\dot{Q} = hA_s(T_s - T_\infty) = (5.867 \text{ W/m}^2 \cdot °C)(1.508 \text{ m}^2)(70 - 20)°C = \mathbf{442 \text{ W}}$$

Therefore, the pipe loses heat to the air in the room at a rate of 442 W by natural convection.

Discussion The pipe loses heat to the surroundings by radiation as well as by natural convection. Assuming the outer surface of the pipe to be black (emissivity $\varepsilon = 1$) and the inner surfaces of the walls of the room to be at room temperature, the radiation heat transfer is determined to be (Fig. 9–14)

$$
\begin{aligned}
\dot{Q}_{\text{rad}} &= \varepsilon A_s \sigma (T_s^4 - T_{\text{surr}}^4) \\
&= (1)(1.508 \text{ m}^2)(5.67 \times 10^{-8} \text{ W/m}^2 \cdot \text{K}^4)[(70 + 273 \text{ K})^4 - (20 + 273 \text{ K})^4] \\
&= 553 \text{ W}
\end{aligned}
$$

which is larger than natural convection. The emissivity of a real surface is less than 1, and thus the radiation heat transfer for a real surface will be less. But radiation will still be significant for most systems cooled by natural convection. Therefore, a radiation analysis should normally accompany a natural convection analysis unless the emissivity of the surface is low.

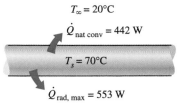

$T_\infty = 20°C$

$\dot{Q}_{\text{nat conv}} = 442 \text{ W}$

$T_s = 70°C$

$\dot{Q}_{\text{rad, max}} = 553 \text{ W}$

FIGURE 9–14

Radiation heat transfer is usually comparable to natural convection in magnitude and should be considered in heat transfer analysis.

EXAMPLE 9–2   Cooling of a Plate in Different Orientations

Consider a 0.6-m × 0.6-m thin square plate in a room at 30°C. One side of the plate is maintained at a temperature of 90°C, while the other side is insulated, as shown in Fig. 9–15. Determine the rate of heat transfer from the plate by natural convection if the plate is (a) vertical, (b) horizontal with hot surface facing up, and (c) horizontal with hot surface facing down.

SOLUTION A hot plate with an insulated back is considered. The rate of heat loss by natural convection is to be determined for different orientations.

Assumptions 1 Steady operating conditions exist. 2 Air is an ideal gas. 3 The local atmospheric pressure is 1 atm.

Properties The properties of air at the film temperature of $T_f = (T_s + T_\infty)/2 = (90 + 30)/2 = 60°C$ and 1 atm are (Table A–15)

$$k = 0.02808 \text{ W/m} \cdot °C \qquad \text{Pr} = 0.7202$$

$$\nu = 1.896 \times 10^{-5} \text{ m}^2/\text{s} \qquad \beta = \frac{1}{T_f} = \frac{1}{333 \text{ K}}$$

Analysis (a) *Vertical.* The characteristic length in this case is the height of the plate, which is $L = 0.6$ m. The Rayleigh number is

$$
\begin{aligned}
\text{Ra}_L &= \frac{g\beta(T_s - T_\infty)L^3}{\nu^2} \text{Pr} \\
&= \frac{(9.81 \text{ m/s}^2)[1/(333 \text{ K})](90 - 30 \text{ K})(0.6 \text{ m})^3}{(1.896 \times 10^{-5} \text{ m}^2/\text{s})^2}(0.7202) = 7.649 \times 10^8
\end{aligned}
$$

Then the natural convection Nusselt number can be determined from Eq. 9–21 to be

$$
\begin{aligned}
\text{Nu} &= \left\{ 0.825 + \frac{0.387 \text{Ra}_L^{1/6}}{[1 + (0.492/\text{Pr})^{9/16}]^{8/27}} \right\}^2 \\
&= \left\{ 0.825 + \frac{0.387(7.649 \times 10^8)^{1/6}}{[1 + (0.492/0.7202)^{9/16}]^{8/27}} \right\}^2 = 113.3
\end{aligned}
$$

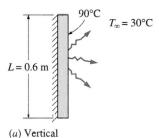

90°C

$T_\infty = 30°C$

$L = 0.6$ m

(a) Vertical

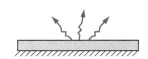

(b) Hot surface facing up

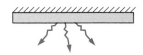

(c) Hot surface facing down

FIGURE 9–15

Schematic for Example 9–2.

Note that the simpler relation Eq. 9–19 would give Nu = 0.59 Ra$_L^{1/4}$ = 98.12, which is 13 percent lower. Then,

$$h = \frac{k}{L} \text{Nu} = \frac{0.02808 \text{ W/m} \cdot °C}{0.6 \text{ m}} (113.3) = 5.302 \text{ W/m}^2 \cdot °C$$

$$A_s = L^2 = (0.6 \text{ m})^2 = 0.36 \text{ m}^2$$

and

$$\dot{Q} = hA_s(T_s - T_\infty) = (5.302 \text{ W/m}^2 \cdot °C)(0.36 \text{ m}^2)(90 - 30)°C = \textbf{115 W}$$

(b) *Horizontal with hot surface facing up.* The characteristic length and the Rayleigh number in this case are

$$L_c = \frac{A_s}{p} = \frac{L^2}{4L} = \frac{L}{4} = \frac{0.6 \text{ m}}{4} = 0.15 \text{ m}$$

$$\text{Ra}_L = \frac{g\beta(T_s - T_\infty)L_c^3}{\nu^2} \text{Pr}$$

$$= \frac{(9.81 \text{ m/s}^2)[1/(333 \text{ K})](90 - 30 \text{ K})(0.15 \text{ m})^3}{(1.896 \times 10^{-5} \text{ m}^2/\text{s})^2} (0.7202) = 1.195 \times 10^7$$

The natural convection Nusselt number can be determined from Eq. 9–22 to be

$$\text{Nu} = 0.54\text{Ra}_L^{1/4} = 0.54(1.195 \times 10^7)^{1/4} = 31.75$$

Then,

$$h = \frac{k}{L_c} \text{Nu} = \frac{0.02808 \text{ W/m} \cdot °C}{0.15 \text{ m}} (31.75) = 5.944 \text{ W/m}^2 \cdot °C$$

and

$$\dot{Q} = hA_s(T_s - T_\infty) = (5.944 \text{ W/m}^2 \cdot °C)(0.36 \text{ m}^2)(90 - 30)°C = \textbf{128 W}$$

(c) *Horizontal with hot surface facing down.* The characteristic length and the Rayleigh number in this case are the same as those determined in (b). But the natural convection Nusselt number is to be determined from Eq. 9–24,

$$\text{Nu} = 0.27\text{Ra}_L^{1/4} = 0.27(1.195 \times 10^7)^{1/4} = 15.87$$

Then,

$$h = \frac{k}{L_c} \text{Nu} = \frac{0.02808 \text{ W/m} \cdot °C}{0.15 \text{ m}} (15.87) = 2.971 \text{ W/m}^2 \cdot °C$$

and

$$\dot{Q} = hA_s(T_s - T_\infty) = (2.971 \text{ W/m}^2 \cdot °C)(0.36 \text{ m}^2)(90 - 30)°C = \textbf{64.2 W}$$

Note that the natural convection heat transfer is the lowest in the case of the hot surface facing down. This is not surprising, since the hot air is "trapped" under the plate in this case and cannot get away from the plate easily. As a result, the cooler air in the vicinity of the plate will have difficulty reaching the plate, which results in a reduced rate of heat transfer.

Discussion The plate will lose heat to the surroundings by radiation as well as by natural convection. Assuming the surface of the plate to be black (emissivity

$\varepsilon = 1$) and the inner surfaces of the walls of the room to be at room temperature, the radiation heat transfer in this case is determined to be

$$\dot{Q}_{rad} = \varepsilon A_s \sigma (T_s^4 - T_{surr}^4)$$
$$= (1)(0.36 \text{ m}^2)(5.67 \times 10^{-8} \text{ W/m}^2 \cdot \text{K}^4)[(90 + 273 \text{ K})^4 - (30 + 273 \text{ K})^4]$$
$$= 182 \text{ W}$$

which is larger than that for natural convection heat transfer for each case. Therefore, radiation can be significant and needs to be considered in surfaces cooled by natural convection.

9–4 · NATURAL CONVECTION FROM FINNED SURFACES AND PCBs

Natural convection flow through a channel formed by two parallel plates as shown in Fig. 9–16 is commonly encountered in practice. When the plates are hot ($T_s > T_\infty$), the ambient fluid at T_∞ enters the channel from the lower end, rises as it is heated under the effect of buoyancy, and the heated fluid leaves the channel from the upper end. The plates could be the fins of a finned heat sink, or the PCBs (printed circuit boards) of an electronic device. The plates can be approximated as being isothermal ($T_s = $ constant) in the first case, and isoflux ($\dot{q}_s = $ constant) in the second case.

Boundary layers start to develop at the lower ends of opposing surfaces, and eventually merge at the midplane if the plates are vertical and sufficiently long. In this case, we will have fully developed channel flow after the merger of the boundary layers, and the natural convection flow is analyzed as channel flow. But when the plates are short or the spacing is large, the boundary layers of opposing surfaces never reach each other, and the natural convection flow on a surface is not affected by the presence of the opposing surface. In that case, the problem should be analyzed as natural convection from two independent plates in a quiescent medium, using the relations given for surfaces, rather than natural convection flow through a channel.

Natural Convection Cooling of Finned Surfaces ($T_s = $ constant)

Finned surfaces of various shapes, called *heat sinks,* are frequently used in the cooling of electronic devices. Energy dissipated by these devices is transferred to the heat sinks by conduction and from the heat sinks to the ambient air by natural or forced convection, depending on the power dissipation requirements. Natural convection is the preferred mode of heat transfer since it involves no moving parts, like the electronic components themselves. However, in the natural convection mode, the components are more likely to run at a higher temperature and thus undermine reliability. A properly selected heat sink may considerably lower the operation temperature of the components and thus reduce the risk of failure.

Natural convection from vertical finned surfaces of rectangular shape has been the subject of numerous studies, mostly experimental. Bar-Cohen and

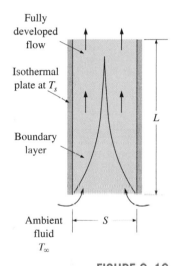

FIGURE 9–16
Natural convection flow through a channel between two isothermal vertical plates.

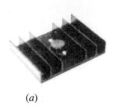

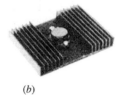

(a)

(b)

FIGURE 9–17
Heat sinks with (a) widely spaced and (b) closely packed fins.

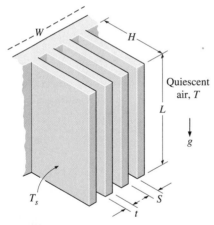

FIGURE 9–18
Various dimensions of a finned surface oriented vertically.

Rohsenow (1984) have compiled the available data under various boundary conditions, and developed correlations for the Nusselt number and optimum spacing. The characteristic length for vertical parallel plates used as fins is usually taken to be the spacing between adjacent fins S, although the fin height L could also be used. The Rayleigh number is expressed as

$$\text{Ra}_S = \frac{g\beta(T_s - T_\infty)S^3}{\nu^2}\text{Pr} \quad \text{and} \quad \text{Ra}_L = \frac{g\beta(T_s - T_\infty)L^3}{\nu^2}\text{Pr} = \text{Ra}_S\frac{L^3}{S^3} \quad \text{(9–30)}$$

The recommended relation for the average Nusselt number for vertical isothermal parallel plates is

$$T_s = \text{constant:} \qquad \text{Nu} = \frac{hS}{k} = \left[\frac{576}{(\text{Ra}_S S/L)^2} + \frac{2.873}{(\text{Ra}_S S/L)^{0.5}}\right]^{-0.5} \quad \text{(9–31)}$$

A question that often arises in the selection of a heat sink is whether to select one with *closely packed* fins or *widely spaced* fins for a given base area (Fig. 9–17). A heat sink with closely packed fins will have greater surface area for heat transfer but a smaller heat transfer coefficient because of the extra resistance the additional fins introduce to fluid flow through the interfin passages. A heat sink with widely spaced fins, on the other hand, will have a higher heat transfer coefficient but a smaller surface area. Therefore, there must be an *optimum spacing* that maximizes the natural convection heat transfer from the heat sink for a given base area WL, where W and L are the width and height of the base of the heat sink, respectively, as shown in Fig. 9–18. When the fins are essentially isothermal and the fin thickness t is small relative to the fin spacing S, the optimum fin spacing for a vertical heat sink is determined by Bar-Cohen and Rohsenow to be

$$T_s = \text{constant:} \qquad S_{\text{opt}} = 2.714\left(\frac{S^3 L}{\text{Ra}_S}\right)^{0.25} = 2.714\frac{L}{\text{Ra}_L^{0.25}} \quad \text{(9–32)}$$

It can be shown by combining the three equations above that when $S = S_{\text{opt}}$, the Nusselt number is a constant and its value is 1.307,

$$S = S_{\text{opt}}: \qquad \text{Nu} = \frac{hS_{\text{opt}}}{k} = 1.307 \quad \text{(9–33)}$$

The rate of heat transfer by natural convection from the fins can be determined from

$$\dot{Q} = h(2nLH)(T_s - T_\infty) \quad \text{(9–34)}$$

where $n = W/(S + t) \approx W/S$ is the number of fins on the heat sink and T_s is the surface temperature of the fins. All fluid properties are to be evaluated at the average temperature $T_{\text{avg}} = (T_s + T_\infty)/2$.

Natural Convection Cooling of Vertical PCBs ($\dot{q}_s$ = constant)

Arrays of printed circuit boards used in electronic systems can often be modeled as parallel plates subjected to uniform heat flux $\dot{q}_s$ (Fig. 9–19). The plate temperature in this case increases with height, reaching a maximum at the upper edge of the board. The modified Rayleigh number for uniform heat flux on both plates is expressed as

$$\text{Ra}_S^* = \frac{g\beta\dot{q}_s S^4}{k\upsilon^2}\text{Pr} \tag{9–35}$$

The Nusselt number at the upper edge of the plate where maximum temperature occurs is determined from [Bar-Cohen and Rohsenow (1984)]

$$\text{Nu}_L = \frac{h_L S}{k} = \left[\frac{48}{\text{Ra}_S^* S/L} + \frac{2.51}{(\text{Ra}_L^* S/L)^{0.4}}\right]^{-0.5} \tag{9–36}$$

The optimum fin spacing for the case of uniform heat flux on both plates is given as

$$\dot{q}_s = \text{constant:} \qquad S_{\text{opt}} = 2.12\left(\frac{S^4 L}{\text{Ra}_S^*}\right)^{0.2} \tag{9–37}$$

The total rate of heat transfer from the plates is

$$\dot{Q} = \dot{q}_s A_s = \dot{q}_s(2nLH) \tag{9–38}$$

where $n = W/(S + t) \approx W/S$ is the number of plates. The critical surface temperature T_L occurs at the upper edge of the plates, and it can be determined from

$$\dot{q}_s = h_L(T_L - T_\infty) \tag{9–39}$$

All fluid properties are to be evaluated at the average temperature $T_{\text{avg}} = (T_L + T_\infty)/2$.

Mass Flow Rate through the Space between Plates

As we mentioned earlier, the magnitude of the natural convection heat transfer is directly related to the mass flow rate of the fluid, which is established by the dynamic balance of two opposing effects: *buoyancy* and *friction*.

The fins of a heat sink introduce both effects: *inducing extra buoyancy* as a result of the elevated temperature of the fin surfaces and *slowing down the fluid* by acting as an added obstacle on the flow path. As a result, increasing the number of fins on a heat sink can either enhance or reduce natural convection, depending on which effect is dominant. The buoyancy-driven fluid flow rate is established at the point where these two effects balance each other. The friction force increases as more and more solid surfaces are introduced, seriously disrupting fluid flow and heat transfer. Under some conditions, the increase in friction may more than offset the increase in buoyancy. This in turn will tend to reduce the flow rate and thus the heat transfer. For that reason, heat sinks with closely spaced fins are not suitable for natural convection cooling.

When the heat sink involves closely spaced fins, the narrow channels formed tend to block or "suffocate" the fluid, especially when the heat sink is long. As a result, the blocking action produced overwhelms the extra buoyancy and downgrades the heat transfer characteristics of the heat sink. Then, at a fixed power setting, the heat sink runs at a higher temperature relative to the no-shroud case. When the heat sink involves widely spaced fins, the

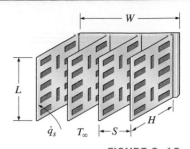

FIGURE 9–19
Arrays of vertical printed circuit boards (PCBs) cooled by natural convection.

shroud does not introduce a significant increase in resistance to flow, and the buoyancy effects dominate. As a result, heat transfer by natural convection may improve, and at a fixed power level the heat sink may run at a lower temperature.

When extended surfaces such as fins are used to enhance natural convection heat transfer between a solid and a fluid, the flow rate of the fluid in the vicinity of the solid adjusts itself to incorporate the changes in buoyancy and friction. It is obvious that this enhancement technique will work to advantage only when the increase in buoyancy is greater than the additional friction introduced. One does not need to be concerned with pressure drop or pumping power when studying natural convection since no pumps or blowers are used in this case. Therefore, an enhancement technique in natural convection is evaluated on heat transfer performance alone.

The failure rate of an electronic component increases almost exponentially with operating temperature. The cooler the electronic device operates, the more reliable it is. A rule of thumb is that the semiconductor failure rate is halved for each 10°C reduction in junction operating temperature. The desire to lower the operating temperature without having to resort to forced convection has motivated researchers to investigate enhancement techniques for natural convection. Sparrow and Prakash have demonstrated that, under certain conditions, the use of discrete plates in lieu of continuous plates of the same surface area increases heat transfer considerably. In other experimental work, using transistors as the heat source, Çengel and Zing (1987) have demonstrated that temperature recorded on the transistor case dropped by as much as 30°C when a shroud was used, as opposed to the corresponding no-shroud case.

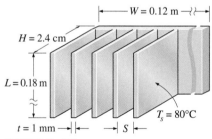

FIGURE 9–20
Schematic for Example 9–3.

EXAMPLE 9–3   Optimum Fin Spacing of a Heat Sink

A 12-cm-wide and 18-cm-high vertical hot surface in 30°C air is to be cooled by a heat sink with equally spaced fins of rectangular profile (Fig. 9–20). The fins are 0.1 cm thick and 18 cm long in the vertical direction and have a height of 2.4 cm from the base. Determine the optimum fin spacing and the rate of heat transfer by natural convection from the heat sink if the base temperature is 80°C.

SOLUTION   A heat sink with equally spaced rectangular fins is to be used to cool a hot surface. The optimum fin spacing and the rate of heat transfer are to be determined.

Assumptions   **1** Steady operating conditions exist. **2** Air is an ideal gas. **3** The atmospheric pressure at that location is 1 atm. **4** The thickness t of the fins is very small relative to the fin spacing S so that Eq. 9–32 for optimum fin spacing is applicable. **5** All fin surfaces are isothermal at base temperature.

Properties   The properties of air at the film temperature of $T_f = (T_s + T_\infty)/2 = (80 + 30)/2 = 55°C$ and 1 atm pressure are (Table A–15)

$$k = 0.02772 \text{ W/m} \cdot °C \qquad \text{Pr} = 0.7215$$
$$\nu = 1.847 \times 10^{-5} \text{ m}^2/\text{s} \qquad \beta = 1/T_f = 1/328 \text{ K}$$

Analysis   We take the characteristic length to be the length of the fins in the vertical direction (since we do not know the fin spacing). Then the Rayleigh number becomes

$$Ra_L = \frac{g\beta(T_s - T_\infty)L^3}{v^2}\,Pr$$
$$= \frac{(9.81 \text{ m/s}^2)[1/(328 \text{ K})](80 - 30 \text{ K})(0.18 \text{ m})^3}{(1.847 \times 10^{-5} \text{ m}^2/\text{s})^2}(0.7215) = 1.845 \times 10^7$$

The optimum fin spacing is determined from Eq. 9–32 to be

$$S_{opt} = 2.714\frac{L}{Ra_L^{0.25}} = 2.714\frac{0.18 \text{ m}}{(1.845 \times 10^7)^{0.25}} = 7.45 \times 10^{-3} \text{ m} = \mathbf{7.45 \text{ mm}}$$

which is about seven times the thickness of the fins. Therefore, the assumption of negligible fin thickness in this case is acceptable. The number of fins and the heat transfer coefficient for this optimum fin spacing case are

$$n = \frac{W}{S + t} = \frac{0.12 \text{ m}}{(0.00745 + 0.001) \text{ m}} \approx 14 \text{ fins}$$

The convection coefficient for this optimum fin spacing case is, from Eq. 9–33,

$$h = Nu_{opt}\frac{k}{S_{opt}} = 1.307\frac{0.02772 \text{ W/m} \cdot ^\circ\text{C}}{0.00745 \text{ m}} = 4.863 \text{ W/m}^2 \cdot ^\circ\text{C}$$

Then the rate of natural convection heat transfer becomes

$$\dot{Q} = hA_s(T_s - T_\infty) = h(2nLH)(T_s - T_\infty)$$
$$= (4.863 \text{ W/m}^2 \cdot ^\circ\text{C})[2 \times 14(0.18 \text{ m})(0.024 \text{ m})](80 - 30)^\circ\text{C} = \mathbf{29.4 \text{ W}}$$

Therefore, this heat sink can dissipate heat by natural convection at a rate of 29.4 W.

9–5 • NATURAL CONVECTION INSIDE ENCLOSURES

A considerable portion of heat loss from a typical residence occurs through the windows. We certainly would insulate the windows, if we could, in order to conserve energy. The problem is finding an insulating material that is transparent. An examination of the thermal conductivities of the insulating materials reveals that *air* is a *better insulator* than most common insulating materials. Besides, it is transparent. Therefore, it makes sense to insulate the windows with a layer of air. Of course, we need to use another sheet of glass to trap the air. The result is an *enclosure,* which is known as a *double-pane window* in this case. Other examples of enclosures include wall cavities, solar collectors, and cryogenic chambers involving concentric cylinders or spheres.

Enclosures are frequently encountered in practice, and heat transfer through them is of practical interest. Heat transfer in enclosed spaces is complicated by the fact that the fluid in the enclosure, in general, does not remain stationary. In a vertical enclosure, the fluid adjacent to the hotter surface rises and the fluid adjacent to the cooler one falls, setting off a rotationary motion within the enclosure that enhances heat transfer through the enclosure. Typical flow patterns in vertical and horizontal rectangular enclosures are shown in Figs. 9–21 and 9–22.

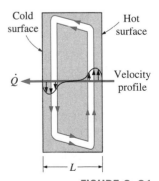

FIGURE 9–21
Convective currents in a vertical rectangular enclosure.

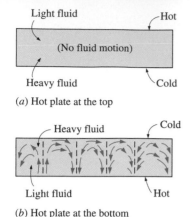

FIGURE 9–22
Convective currents in a horizontal enclosure with (*a*) hot plate at the top and (*b*) hot plate at the bottom.

The characteristics of heat transfer through a horizontal enclosure depend on whether the hotter plate is at the top or at the bottom, as shown in Fig. 9–22. When the hotter plate is at the *top,* no convection currents develop in the enclosure, since the lighter fluid is always on top of the heavier fluid. Heat transfer in this case is by *pure conduction,* and we have Nu = 1. When the hotter plate is at the *bottom,* the heavier fluid will be on top of the lighter fluid, and there will be a tendency for the lighter fluid to topple the heavier fluid and rise to the top, where it comes in contact with the cooler plate and cools down. Until that happens, however, heat transfer is still by *pure conduction* and Nu = 1. When Ra > 1708, the buoyant force overcomes the fluid resistance and initiates natural convection currents, which are observed to be in the form of hexagonal cells called *Bénard cells.* For Ra > 3 × 10⁵, the cells break down and the fluid motion becomes turbulent.

The Rayleigh number for an enclosure is determined from

$$\text{Ra}_L = \frac{g\beta(T_1 - T_2)L_c^3}{\nu^2}\text{Pr} \tag{9–40}$$

where the characteristic length L_c is the distance between the hot and cold surfaces, and T_1 and T_2 are the temperatures of the hot and cold surfaces, respectively. All fluid properties are to be evaluated at the average fluid temperature $T_{avg} = (T_1 + T_2)/2$.

Effective Thermal Conductivity

When the Nusselt number is known, the rate of heat transfer through the enclosure can be determined from

$$\dot{Q} = hA_s(T_1 - T_2) = k\text{Nu}A_s\frac{T_1 - T_2}{L_c} \tag{9–41}$$

since $h = k\text{Nu}/L$. The rate of steady heat conduction across a layer of thickness L_c, area A_s, and thermal conductivity k is expressed as

$$\dot{Q}_{cond} = kA_s\frac{T_1 - T_2}{L_c} \tag{9–42}$$

where T_1 and T_2 are the temperatures on the two sides of the layer. A comparison of this relation with Eq. 9–41 reveals that the convection heat transfer in an enclosure is analogous to heat conduction across the fluid layer in the enclosure provided that the thermal conductivity k is replaced by $k\text{Nu}$. That is, *the fluid in an enclosure behaves like a fluid whose thermal conductivity is $k\text{Nu}$ as a result of convection currents.* Therefore, the quantity $k\text{Nu}$ is called the **effective thermal conductivity** of the enclosure. That is,

$$k_{eff} = k\text{Nu} \tag{9–43}$$

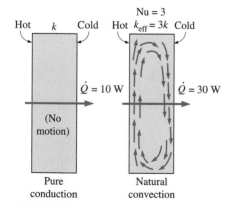

FIGURE 9–23
A Nusselt number of 3 for an enclosure indicates that heat transfer through the enclosure by *natural convection* is three times that by *pure conduction.*

Note that for the special case of Nu = 1, the effective thermal conductivity of the enclosure becomes equal to the conductivity of the fluid. This is expected since this case corresponds to pure conduction (Fig. 9–23).

Natural convection heat transfer in enclosed spaces has been the subject of many experimental and numerical studies, and numerous correlations for the Nusselt number exist. Simple power-law type relations in the form of

$\mathrm{Nu} = C\mathrm{Ra}_L^n$, where C and n are constants, are sufficiently accurate, but they are usually applicable to a narrow range of Prandtl and Rayleigh numbers and aspect ratios. The relations that are more comprehensive are naturally more complex. Next we present some widely used relations for various types of enclosures.

Horizontal Rectangular Enclosures

We need no Nusselt number relations for the case of the hotter plate being at the top, since there are no convection currents in this case and heat transfer is downward by conduction ($\mathrm{Nu} - 1$). When the hotter plate is at the bottom, however, significant convection currents set in for $\mathrm{Ra}_L > 1708$, and the rate of heat transfer increases (Fig. 9–24).

For horizontal enclosures that contain air, Jakob (1949) recommends the following simple correlations

$$\mathrm{Nu} = 0.195\mathrm{Ra}_L^{1/4} \qquad 10^4 < \mathrm{Ra}_L < 4 \times 10^5 \tag{9–44}$$

$$\mathrm{Nu} = 0.068\mathrm{Ra}_L^{1/3} \qquad 4 \times 10^5 < \mathrm{Ra}_L < 10^7 \tag{9–45}$$

These relations can also be used for other gases with $0.5 < \mathrm{Pr} < 2$. Using water, silicone oil, and mercury in their experiments, Globe and Dropkin (1959) obtained this correlation for horizontal enclosures heated from below,

$$\mathrm{Nu} = 0.069\mathrm{Ra}_L^{1/3}\,\mathrm{Pr}^{0.074} \qquad 3 \times 10^5 < \mathrm{Ra}_L < 7 \times 10^9 \tag{9–46}$$

Based on experiments with air, Hollands et al. (1976) recommend this correlation for horizontal enclosures,

$$\mathrm{Nu} = 1 + 1.44\left[1 - \frac{1708}{\mathrm{Ra}_L}\right]^+ + \left[\frac{\mathrm{Ra}_L^{1/3}}{18} - 1\right]^+ \qquad \mathrm{Ra}_L < 10^8 \tag{9–47}$$

The notation $[\ ]^+$ indicates that if the quantity in the bracket is negative, it should be set equal to zero. This relation also correlates data well for liquids with moderate Prandtl numbers for $\mathrm{Ra}_L < 10^5$, and thus it can also be used for water.

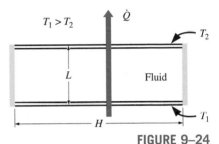

$T_1 > T_2$

FIGURE 9–24
A horizontal rectangular enclosure with isothermal surfaces.

Inclined Rectangular Enclosures

Air spaces between two inclined parallel plates are commonly encountered in flat-plate solar collectors (between the glass cover and the absorber plate) and the double-pane skylights on inclined roofs. Heat transfer through an inclined enclosure depends on the **aspect ratio** H/L as well as the tilt angle θ from the horizontal (Fig. 9–25).

For large aspect ratios ($H/L \geq 12$), this equation [Hollands et al., 1976] correlates experimental data extremely well for tilt angles up to 70°,

$$\mathrm{Nu} = 1 + 1.44\left[1 - \frac{1708}{\mathrm{Ra}_L\cos\theta}\right]^+\left(1 - \frac{1708(\sin 1.8\theta)^{1.6}}{\mathrm{Ra}_L\cos\theta}\right) + \left[\frac{(\mathrm{Ra}_L\cos\theta)^{1/3}}{18} - 1\right]^+ \tag{9–48}$$

for $\mathrm{Ra}_L < 10^5$, $0 < \theta < 70°$, and $H/L \geq 12$. Again any quantity in $[\ ]^+$ should be set equal to zero if it is negative. This is to ensure that $\mathrm{Nu} = 1$ for $\mathrm{Ra}_L \cos\theta < 1708$. Note that this relation reduces to Eq. 9–47 for horizontal enclosures for $\theta = 0°$, as expected.

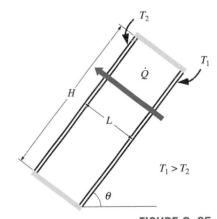

FIGURE 9–25
An inclined rectangular enclosure with isothermal surfaces.

TABLE 9–2

Critical angles for inclined rectangular enclosures

Aspect ratio, H/L	Critical angle, θ_{cr}
1	25°
3	53°
6	60°
12	67°
>12	70°

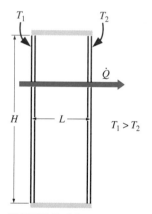

FIGURE 9–26
A vertical rectangular enclosure with isothermal surfaces.

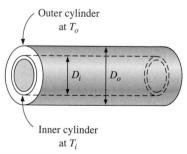

Outer cylinder at T_o

D_i D_o

Inner cylinder at T_i

FIGURE 9–27
Two concentric horizontal isothermal cylinders.

For enclosures with smaller aspect ratios ($H/L < 12$), the next correlation can be used provided that the tilt angle is less than the critical value θ_{cr} listed in Table 9–2 [Catton (1978)]

$$\text{Nu} = \text{Nu}_{\theta=0°}\left(\frac{\text{Nu}_{\theta=90°}}{\text{Nu}_{\theta=0°}}\right)^{\theta/\theta_{cr}}(\sin\theta_{cr})^{\theta/(4\theta_{cr})} \qquad 0° < \theta < \theta_{cr} \qquad (9\text{–}49)$$

For tilt angles greater than the critical value ($\theta_{cr} < \theta < 90°$), the Nusselt number can be obtained by multiplying the Nusselt number for a vertical enclosure by $(\sin\theta)^{1/4}$ [Ayyaswamy and Catton (1973)],

$$\text{Nu} = \text{Nu}_{\theta=90°}(\sin\theta)^{1/4} \qquad \theta_{cr} < \theta < 90°, \text{ any } H/L \qquad (9\text{–}50)$$

For enclosures tilted more than 90°, the recommended relation is [Arnold et al. (1974)]

$$\text{Nu} = 1 + (\text{Nu}_{\theta=90°} - 1)\sin\theta \qquad 90° < \theta < 180°, \text{ any } H/L \qquad (9\text{–}51)$$

More recent but more complex correlations are also available in the literature [e.g., and ElSherbiny et al. (1982)].

Vertical Rectangular Enclosures

For vertical enclosures (Fig. 9–26), Catton (1978) recommends these two correlations due to Berkovsky and Polevikov (1977),

$$\text{Nu} = 0.18\left(\frac{\text{Pr}}{0.2+\text{Pr}}\text{Ra}_L\right)^{0.29} \qquad \begin{array}{l} 1 < H/L < 2 \\ \text{any Prandtl number} \\ \text{Ra}_L\,\text{Pr}/(0.2+\text{Pr}) > 10^3 \end{array} \qquad (9\text{–}52)$$

$$\text{Nu} = 0.22\left(\frac{\text{Pr}}{0.2+\text{Pr}}\text{Ra}_L\right)^{0.28}\left(\frac{H}{L}\right)^{-1/4} \qquad \begin{array}{l} 2 < H/L < 10 \\ \text{any Prandtl number} \\ \text{Ra}_L < 10^{10} \end{array} \qquad (9\text{–}53)$$

For vertical enclosures with larger aspect ratios, the following correlations can be used [MacGregor and Emery (1969)]

$$\text{Nu} = 0.42\text{Ra}_L^{1/4}\text{Pr}^{0.012}\left(\frac{H}{L}\right)^{-0.3} \qquad \begin{array}{l} 10 < H/L < 40 \\ 1 < \text{Pr} < 2\times10^4 \\ 10^4 < \text{Ra}_L < 10^7 \end{array} \qquad (9\text{–}54)$$

$$\text{Nu} = 0.46\text{Ra}_L^{1/3} \qquad \begin{array}{l} 1 < H/L < 40 \\ 1 < \text{Pr} < 20 \\ 10^6 < \text{Ra}_L < 10^9 \end{array} \qquad (9\text{–}55)$$

Again all fluid properties are to be evaluated at the average temperature $(T_1 + T_2)/2$.

Concentric Cylinders

Consider two long concentric horizontal cylinders maintained at uniform but different temperatures of T_i and T_o, as shown in Fig. 9–27. The diameters of the inner and outer cylinders are D_i and D_o, respectively, and the characteristic length is the spacing between the cylinders, $L_c = (D_o - D_i)/2$. The rate of heat transfer through the annular space between the cylinders by natural convection per unit length is expressed as

$$\dot{Q} = \frac{2\pi k_{eff}}{\ln(D_o/D_i)}(T_i - T_o) \qquad \text{(W/m)} \qquad \text{(9–56)}$$

The recommended relation for effective thermal conductivity is [Raithby and Hollands (1975)]

$$\frac{k_{eff}}{k} = 0.386\left(\frac{Pr}{0.861 + Pr}\right)^{1/4}(F_{cyl}Ra_L)^{1/4} \qquad \text{(9–57)}$$

where the geometric factor for concentric cylinders F_{cyl} is

$$F_{cyl} = \frac{[\ln(D_o/D_i)]^4}{L_c^3(D_i^{-3/5} + D_o^{-3/5})^5} \qquad \text{(9–58)}$$

The k_{eff} relation in Eq. 9–57 is applicable for $0.70 \le Pr \le 6000$ and $10^2 \le F_{cyl}Ra_L \le 10^7$. For $F_{cyl}Ra_L < 100$, natural convection currents are negligible and thus $k_{eff} = k$. Note that k_{eff} cannot be less than k, and thus we should set $k_{eff} = k$ if $k_{eff}/k < 1$. The fluid properties are evaluated at the average temperature of $(T_i + T_o)/2$.

Concentric Spheres

For concentric isothermal spheres, the rate of heat transfer through the gap between the spheres by natural convection is expressed as (Fig. 9–28)

$$\dot{Q} = k_{eff}\frac{\pi D_i D_o}{L_c}(T_i - T_o) \qquad \text{(W)} \qquad \text{(9–59)}$$

where $L_c = (D_o - D_i)/2$ is the characteristic length. The recommended relation for effective thermal conductivity is [Raithby and Hollands (1975)]

$$\frac{k_{eff}}{k} = 0.74\left(\frac{Pr}{0.861 + Pr}\right)^{1/4}(F_{sph}Ra_L)^{1/4} \qquad \text{(9–60)}$$

where the geometric factor for concentric spheres F_{sph} is

$$F_{sph} = \frac{L_c}{(D_iD_o)^4(D_i^{-7/5} + D_o^{-7/5})^5} \qquad \text{(9–61)}$$

The k_{eff} relation in Eq. 9–60 is applicable for $0.70 \le Pr \le 4200$ and $10^2 \le F_{sph}Ra_L \le 10^4$. If $k_{eff}/k < 1$, we should set $k_{eff} = k$.

Combined Natural Convection and Radiation

Gases are nearly transparent to radiation, and thus heat transfer through a gas layer is by simultaneous convection (or conduction, if the gas is quiescent) and radiation. Natural convection heat transfer coefficients are typically very low compared to those for forced convection. Therefore, radiation is usually disregarded in forced convection problems, but it must be considered in natural convection problems that involve a gas. This is especially the case for surfaces with high emissivities. For example, about half of the heat transfer through the air space of a double-pane window is by radiation. The total rate of heat transfer is determined by adding the convection and radiation components,

$$\dot{Q}_{total} = \dot{Q}_{conv} + \dot{Q}_{rad} \qquad \text{(9–62)}$$

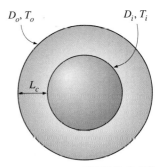

D_o, T_o D_i, T_i

L_c

FIGURE 9–28

Two concentric isothermal spheres

Radiation heat transfer from a surface at temperature T_s surrounded by surfaces at a temperature T_{surr} (both in K) is determined from

$$\dot{Q}_{rad} = \varepsilon\sigma A_s(T_s^4 - T_{surr}^4) \qquad \text{(W)} \tag{9–63}$$

where ε is the emissivity of the surface, A_s is the surface area, and $\sigma = 5.67 \times 10^{-8} \text{ W/m}^2 \cdot \text{K}^4$ is the Stefan–Boltzmann constant.

When the end effects are negligible, radiation heat transfer between two large parallel plates at temperatures T_1 and T_2 is expressed as (see Chapter 13 for details)

$$\dot{Q}_{rad} = \frac{\pi A_s(T_1^4 - T_2^4)}{1/\varepsilon_1 + 1/\varepsilon_2 - 1} = \varepsilon_{effective}\,\sigma A_s(T_1^4 - T_2^4) \qquad \text{(W)} \tag{9–64}$$

where ε_1 and ε_2 are the emissivities of the plates, and $\varepsilon_{effective}$ is the *effective emissivity* defined as

$$\varepsilon_{effective} = \frac{1}{1/\varepsilon_1 + 1/\varepsilon_2 - 1} \tag{9–65}$$

The emissivity of an ordinary glass surface, for example, is 0.84. Therefore, the effective emissivity of two parallel glass surfaces facing each other is 0.72. Radiation heat transfer between concentric cylinders and spheres is discussed in Chapter 13.

Note that in some cases the temperature of the surrounding medium may be below the surface temperature ($T_\infty < T_s$), while the temperature of the surrounding surfaces is above the surface temperature ($T_{surr} > T_s$). In such cases, convection and radiation heat transfers are subtracted from each other instead of being added since they are in opposite directions. Also, for a metal surface, the radiation effect can be reduced to negligible levels by polishing the surface and thus lowering the surface emissivity to a value near zero.

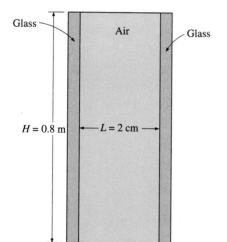

Glass — Air — Glass

$H = 0.8 \text{ m}$ ← $L = 2 \text{ cm}$ →

FIGURE 9–29
Schematic for Example 9–4.

EXAMPLE 9–4   Heat Loss through a Double-Pane Window

The vertical 0.8-m-high, 2-m-wide double-pane window shown in Fig. 9–29 consists of two sheets of glass separated by a 2-cm air gap at atmospheric pressure. If the glass surface temperatures across the air gap are measured to be 12°C and 2°C, determine the rate of heat transfer through the window.

SOLUTION   Two glasses of a double-pane window are maintained at specified temperatures. The rate of heat transfer through the window is to be determined.
Assumptions   1 Steady operating conditions exist. 2 Air is an ideal gas. 3 Radiation heat transfer is not considered.
Properties   The properties of air at the average temperature of $T_{avg} = (T_1 + T_2)/2 = (12 + 2)/2 = 7°C$ and 1 atm pressure are (Table A–15)

$$k = 0.02416 \text{ W/m} \cdot °C \qquad \text{Pr} = 0.7344$$

$$\nu = 1.400 \times 10^{-5} \text{ m}^2/\text{s} \qquad \beta = \frac{1}{T_{avg}} = \frac{1}{280 \text{ K}}$$

Analysis   We have a rectangular enclosure filled with air. The characteristic length in this case is the distance between the two glasses, $L_c = L = 0.02$ m. Then the Rayleigh number becomes

$$Ra_L = \frac{g\beta(T_1 - T_2)L_c^3}{\nu^2}Pr$$

$$= \frac{(9.81 \text{ m/s}^2)[1/(280 \text{ K})](12 - 2 \text{ K})(0.02 \text{ m})^3}{(1.400 \times 10^{-5} \text{ m}^2/\text{s})^2}(0.7344) = 1.050 \times 10^4$$

The aspect ratio of the geometry is $H/L = 0.8/0.02 = 40$. Then the Nusselt number in this case can be determined from Eq. 9–54 to be

$$Nu = 0.42Ra_L^{1/4} Pr^{0.012}\left(\frac{H}{L}\right)^{-0.3}$$

$$= 0.42(1.050 \times 10^4)^{1/4}(0.7344)^{0.012}\left(\frac{0.8}{0.02}\right)^{-0.3} = 1.40$$

Then,

$$A_s = H \times W = (0.8 \text{ m})(2 \text{ m}) = 1.6 \text{ m}^2$$

and

$$\dot{Q} = hA_s(T_1 - T_2) = kNuA_s\frac{T_1 - T_2}{L}$$

$$= (0.02416 \text{ W/m} \cdot °\text{C})(1.40)(1.6 \text{ m}^2)\frac{(12 - 2)°\text{C}}{0.02 \text{ m}} = \textbf{27.1 W}$$

Therefore, heat is lost through the window at a rate of 27.1 W.

Discussion Recall that a Nusselt number of Nu = 1 for an enclosure corresponds to pure conduction heat transfer through the enclosure. The air in the enclosure in this case remains still, and no natural convection currents occur in the enclosure. The Nusselt number in our case is 1.40, which indicates that heat transfer through the enclosure is 1.40 times that by pure conduction. The increase in heat transfer is due to the natural convection currents that develop in the enclosure.

EXAMPLE 9–5   Heat Transfer through a Spherical Enclosure

The two concentric spheres of diameters $D_i = 20$ cm and $D_o = 30$ cm shown in Fig. 9–30 are separated by air at 1 atm pressure. The surface temperatures of the two spheres enclosing the air are $T_i = 320$ K and $T_o = 280$ K, respectively. Determine the rate of heat transfer from the inner sphere to the outer sphere by natural convection.

SOLUTION Two surfaces of a spherical enclosure are maintained at specified temperatures. The rate of heat transfer through the enclosure is to be determined.
Assumptions **1** Steady operating conditions exist. **2** Air is an ideal gas. **3** Radiation heat transfer is not considered.
Properties The properties of air at the average temperature of $T_{avg} = (T_i + T_o)/2 = (320 + 280)/2 = 300$ K $= 27°$C and 1 atm pressure are (Table A–15)

$$k = 0.02566 \text{ W/m} \cdot °\text{C} \qquad Pr = 0.7290$$

$$\nu = 1.580 \times 10^{-5} \text{ m}^2/\text{s} \qquad \beta = \frac{1}{T_{avg}} = \frac{1}{300 \text{ K}}$$

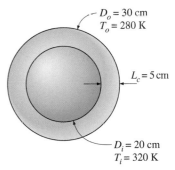

$D_o = 30$ cm
$T_o = 280$ K

$L_c = 5$ cm

$D_i = 20$ cm
$T_i = 320$ K

FIGURE 9–30
Schematic for Example 9–5.

Analysis We have a spherical enclosure filled with air. The characteristic length in this case is the distance between the two spheres,

$$L_c = (D_o - D_i)/2 = (0.3 - 0.2)/2 = 0.05 \text{ m}$$

The Rayleigh number is

$$\text{Ra}_L = \frac{g\beta(T_i - T_o)L_c^3}{\nu^2} \text{Pr}$$

$$= \frac{(9.81 \text{ m/s}^2)[1/(300 \text{ K})](320 - 280 \text{ K})(0.05 \text{ m})^3}{(1.58 \times 10^{-5} \text{ m}^2/\text{s})^2}(0.729) = 4.775 \times 10^5$$

The effective thermal conductivity is

$$F_{\text{sph}} = \frac{L_c}{(D_i D_o)^4 (D_i^{-7/5} + D_o^{-7/5})^5}$$

$$= \frac{0.05 \text{ m}}{[(0.2 \text{ m})(0.3 \text{ m})]^4 [(0.2 \text{ m})^{-7/5} + (0.3 \text{ m})^{-7/5}]^5} = 0.005229$$

$$k_{\text{eff}} = 0.74k \left(\frac{\text{Pr}}{0.861 + \text{Pr}}\right)^{1/4} (F_{\text{sph}}\text{Ra}_L)^{1/4}$$

$$= 0.74(0.02566 \text{ W/m} \cdot °\text{C})\left(\frac{0.729}{0.861 + 0.729}\right)^{1/4}(0.005229 \times 4.775 \times 10^5)^{1/4}$$

$$= 0.1105 \text{ W/m} \cdot °\text{C}$$

Then the rate of heat transfer between the spheres becomes

$$\dot{Q} = k_{\text{eff}}\frac{\pi D_i D_o}{L_c}(T_i - T_o)$$

$$= (0.1105 \text{ W/m} \cdot °\text{C})\frac{\pi(0.2 \text{ m})(0.3 \text{ m})}{0.05 \text{ m}}(320 - 280) \text{ K} = \mathbf{16.7 \text{ W}}$$

Therefore, heat will be lost from the inner sphere to the outer one at a rate of 16.7 W.

Discussion Note that the air in the spherical enclosure acts like a stationary fluid whose thermal conductivity is $k_{\text{eff}}/k = 0.1105/0.02566 = 4.3$ times that of air as a result of natural convection currents. Also, radiation heat transfer between spheres is usually significant, and should be considered in a complete analysis.

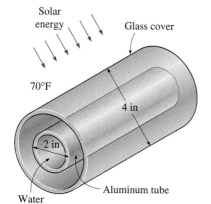

Solar energy

Glass cover

70°F

4 in

2 in

Aluminum tube

Water

FIGURE 9–31
Schematic for Example 9–6.

EXAMPLE 9–6   Heating Water in a Tube by Solar Energy

A solar collector consists of a horizontal aluminum tube having an outer diameter of 2 in enclosed in a concentric thin glass tube of 4-in-diameter (Fig. 9–31). Water is heated as it flows through the tube, and the annular space between the aluminum and the glass tubes is filled with air at 1 atm pressure. The pump circulating the water fails during a clear day, and the water temperature in the tube starts rising. The aluminum tube absorbs solar radiation at a rate of 30 Btu/h per foot length, and the temperature of the ambient air outside is 70°F. Disregarding any heat loss by radiation, determine the temperature of the aluminum tube when steady operation is established (i.e., when the rate of heat loss from the tube equals the amount of solar energy gained by the tube).

SOLUTION The circulating pump of a solar collector that consists of a horizontal tube and its glass cover fails. The equilibrium temperature of the tube is to be determined.

Assumptions 1 Steady operating conditions exist. 2 The tube and its cover are isothermal. 3 Air is an ideal gas. 4 Heat loss by radiation is negligible.

Properties The properties of air should be evaluated at the average temperature. But we do not know the exit temperature of the air in the duct, and thus we cannot determine the bulk fluid and glass cover temperatures at this point, and we cannot evaluate the average temperatures. Therefore, we assume the glass temperature to be 110°F, and use properties at an anticipated average temperature of $(70 + 110)/2 = 90°F$ (Table A–15E),

$$k = 0.01505 \text{ Btu/h} \cdot \text{ft} \cdot °F \qquad Pr = 0.7275$$

$$\nu = 1.753 \times 10^{-4} \text{ ft}^2/\text{s} \qquad \beta = \frac{1}{T_{avg}} = \frac{1}{550 \text{ R}}$$

Analysis We have a horizontal cylindrical enclosure filled with air at 1 atm pressure. The problem involves heat transfer from the aluminum tube to the glass cover and from the outer surface of the glass cover to the surrounding ambient air. When steady operation is reached, these two heat transfer rates must equal the rate of heat gain. That is,

$$\dot{Q}_{\text{tube-glass}} = \dot{Q}_{\text{glass-ambient}} = \dot{Q}_{\text{solar gain}} = 30 \text{ Btu/h} \qquad \text{(per foot of tube)}$$

The heat transfer surface area of the glass cover is

$$A_o = A_{\text{glass}} = (\pi D_o L) = \pi(4/12 \text{ ft})(1 \text{ ft}) = 1.047 \text{ ft}^2 \qquad \text{(per foot of tube)}$$

To determine the Rayleigh number, we need to know the surface temperature of the glass, which is not available. Therefore, it is clear that the solution will require a trial-and-error approach. Assuming the glass cover temperature to be 110°F, the Rayleigh number, the Nusselt number, the convection heat transfer coefficient, and the rate of natural convection heat transfer from the glass cover to the ambient air are determined to be

$$Ra_{D_o} = \frac{g\beta(T_s - T_\infty)D_o^3}{\nu^2} Pr$$

$$= \frac{(32.2 \text{ ft/s}^2)[1/(550 \text{ R})](110 - 70 \text{ R})(4/12 \text{ ft})^3}{(1.753 \times 10^{-4} \text{ ft}^2/\text{s})^2}(0.7275) = 2.054 \times 10^6$$

$$Nu = \left\{0.6 + \frac{0.387 Ra_D^{1/6}}{[1 + (0.559/Pr)^{9/16}]^{8/27}}\right\}^2 = \left\{0.6 + \frac{0.387(2.054 \times 10^6)^{1/6}}{[1 + (0.559/0.7275)^{9/16}]^{8/27}}\right\}^2$$

$$= 17.89$$

$$h_o = \frac{k}{D_0} Nu = \frac{0.01505 \text{ Btu/h} \cdot \text{ft} \cdot °F}{4/12 \text{ ft}}(17.89) = 0.8077 \text{ Btu/h} \cdot \text{ft}^2 \cdot °F$$

$$\dot{Q}_o = h_o A_o (T_o - T_\infty) = (0.8077 \text{ Btu/h} \cdot \text{ft}^2 \cdot °F)(1.047 \text{ ft}^2)(110 - 70)°F$$

$$= 33.8 \text{ Btu/h}$$

which is more than 30 Btu/h. Therefore, the assumed temperature of 110°F for the glass cover is high. Repeating the calculations with lower temperatures, the glass cover temperature corresponding to 30 Btu/h is determined to be 106°F.

The temperature of the aluminum tube is determined in a similar manner using the natural convection relations for two horizontal concentric cylinders. The characteristic length in this case is the distance between the two cylinders, which is

$$L_c = (D_o - D_i)/2 = (4 - 2)/2 = 1 \text{ in} = 1/12 \text{ ft}$$

We start the calculations by assuming the tube temperature to be 200°F, and thus an average temperature of $(106 + 200)/2 = 153°F = 613$ R. Using air properties at this temperature gives

$$\text{Ra}_L = \frac{g\beta(T_i - T_o)L_c^3}{\nu^2} \text{Pr}$$

$$= \frac{(32.2 \text{ ft/s}^2)[1/613 \text{ R}](200 - 106 \text{ R})(1/12 \text{ ft})^3}{(2.117 \times 10^{-4} \text{ ft}^2\text{/s})^2}(0.7184) = 4.580 \times 10^4$$

The effective thermal conductivity is

$$F_{\text{cyl}} = \frac{[\ln(D_o/D_i)]^4}{L_c^3(D_i^{-3/5} + D_o^{-3/5})^5}$$

$$= \frac{[\ln(4/2)]^4}{(1/12 \text{ ft})^3[(2/12 \text{ ft})^{-3/5} + (4/12 \text{ ft})^{-3/5}]^5} = 0.1466$$

$$k_{\text{eff}} = 0.386k\left(\frac{\text{Pr}}{0.861 + \text{Pr}}\right)^{1/4}(F_{\text{cyl}}\text{Ra}_L)^{1/4}$$

$$= 0.386(0.01653 \text{ Btu/h} \cdot \text{ft} \cdot °\text{F})\left(\frac{0.7184}{0.861 + 0.7184}\right)^{1/4}$$

$$\times (0.1466 \times 4.580 \times 10^4)^{1/4}$$

$$= 0.04743 \text{ Btu/h} \cdot \text{ft} \cdot °\text{F}$$

Then the rate of heat transfer between the cylinders becomes

$$\dot{Q} = \frac{2\pi k_{\text{eff}}}{\ln(D_o/D_i)}(T_i - T_o)$$

$$= \frac{2\pi(0.04743 \text{ Btu/h} \cdot \text{ft} \cdot °\text{F})}{\ln(4/2)}(200 - 106)°\text{F} = 40.4 \text{ Btu/h}$$

which is more than 30 Btu/h. Therefore, the assumed temperature of 200°F for the tube is high. By trying other values, the tube temperature corresponding to 30 Btu/h is determined to be **180°F**. Therefore, the tube will reach an equilibrium temperature of 180°F when the pump fails.

Discussion   Note that we have not considered heat loss by radiation in the calculations, and thus the tube temperature determined is probably too high. This problem is considered in Chapter 13 by accounting for the effect of radiation heat transfer.

9–6 · COMBINED NATURAL AND FORCED CONVECTION

The presence of a temperature gradient in a fluid in a gravity field always gives rise to natural convection currents, and thus heat transfer by natural convection. Therefore, forced convection is always accompanied by natural convection.

We mentioned earlier that the convection heat transfer coefficient, natural or forced, is a strong function of the fluid velocity. Heat transfer coefficients encountered in forced convection are typically much higher than those encountered in natural convection because of the higher fluid velocities associated with forced convection. As a result, we tend to ignore natural convection in heat transfer analyses that involve forced convection, although we recognize that natural convection always accompanies forced convection. The error involved in ignoring natural convection is negligible at high velocities but may be considerable at low velocities. Therefore, it is desirable to have a criterion to assess the relative magnitude of natural convection in the presence of forced convection.

For a given fluid, it is observed that the parameter Gr/Re^2 represents the importance of natural convection relative to forced convection. This is not surprising since the convection heat transfer coefficient is a strong function of the Reynolds number Re in forced convection and the Grashof number Gr in natural convection.

A plot of the nondimensionalized heat transfer coefficient for combined natural and forced convection on a vertical plate is given in Fig. 9–32 for different fluids. We note from this figure that natural convection is negligible when $Gr/Re^2 < 0.1$, forced convection is negligible when $Gr/Re^2 > 10$, and neither is negligible when $0.1 < Gr/Re^2 < 10$. Therefore, both natural and forced convection must be considered in heat transfer calculations when the Gr and Re^2 are of the same order of magnitude (one is within a factor of 10 times the other). Note that forced convection is small relative to natural convection only in the rare case of extremely low forced flow velocities.

Natural convection may *help* or *hurt* forced convection heat transfer, depending on the relative directions of *buoyancy-induced* and the *forced convection* motions (Fig. 9–33):

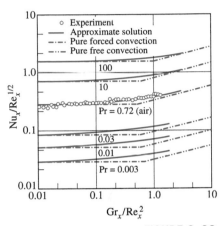

FIGURE 9–32

Variation of the local Nusselt number Nu_x for combined natural and forced convection from a hot isothermal vertical plate.

(From Lloyd and Sparrow, 1970.)

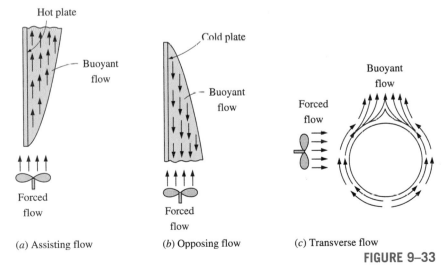

(a) Assisting flow   (b) Opposing flow   (c) Transverse flow

FIGURE 9–33

Natural convection can *enhance* or *inhibit* heat transfer, depending on the relative directions of *buoyancy-induced motion* and the *forced convection motion.*

1. In *assisting flow,* the buoyant motion is in the *same* direction as the forced motion. Therefore, natural convection assists forced convection and *enhances* heat transfer. An example is upward forced flow over a hot surface.

2. In *opposing flow,* the buoyant motion is in the *opposite* direction to the forced motion. Therefore, natural convection resists forced convection and *decreases* heat transfer. An example is upward forced flow over a cold surface.

3. In *transverse flow,* the buoyant motion is *perpendicular* to the forced motion. Transverse flow enhances fluid mixing and thus *enhances* heat transfer. An example is horizontal forced flow over a hot or cold cylinder or sphere.

When determining heat transfer under combined natural and forced convection conditions, it is tempting to add the contributions of natural and forced convection in assisting flows and to subtract them in opposing flows. However, the evidence indicates differently. A review of experimental data suggests a correlation of the form

$$\text{Nu}_{\text{combined}} = (\text{Nu}_{\text{forced}}^{n} \pm \text{Nu}_{\text{natural}}^{n})^{1/n} \tag{9-66}$$

where $\text{Nu}_{\text{forced}}$ and $\text{Nu}_{\text{natural}}$ are determined from the correlations for *pure forced* and *pure natural convection,* respectively. The plus sign is for *assisting* and *transverse* flows and the minus sign is for *opposing* flows. The value of the exponent n varies between 3 and 4, depending on the geometry involved. It is observed that $n = 3$ correlates experimental data for vertical surfaces well. Larger values of n are better suited for horizontal surfaces.

A question that frequently arises in the cooling of heat-generating equipment such as electronic components is whether to use a fan (or a pump if the cooling medium is a liquid)—that is, whether to utilize *natural* or *forced* convection in the cooling of the equipment. The answer depends on the maximum allowable operating temperature. Recall that the convection heat transfer rate from a surface at temperature T_s in a medium at T_∞ is given by

$$\dot{Q}_{\text{conv}} = hA_s(T_s - T_\infty)$$

where h is the convection heat transfer coefficient and A_s is the surface area. Note that for a fixed value of power dissipation and surface area, h and T_s are *inversely proportional.* Therefore, the device operates at a *higher* temperature when h is low (typical of natural convection) and at a *lower* temperature when h is high (typical of forced convection).

Natural convection is the preferred mode of heat transfer since no blowers or pumps are needed and thus all the problems associated with these, such as noise, vibration, power consumption, and malfunctioning, are avoided. Natural convection is adequate for cooling *low-power-output* devices, especially when they are attached to extended surfaces such as heat sinks. For *high-power-output* devices, however, we have no choice but to use a blower or a pump to keep the operating temperature below the maximum allowable level. For *very-high-power-output* devices, even forced convection may not be sufficient to keep the surface temperature at the desirable levels. In such cases, we may have to use *boiling* and *condensation* to take advantage of the very high heat transfer coefficients associated with phase-change processes.

Heat Transfer through Windows

Windows are *glazed apertures* in the building envelope that typically consist of single or multiple glazing (glass or plastic), framing, and shading. In a building envelope, windows offer the least resistance to heat transfer. In a typical house, about one-third of the total heat loss in winter occurs through the windows. Also, most air infiltration occurs at the edges of the windows. The solar heat gain through the windows is responsible for much of the cooling load in summer. The net effect of a window on the heat balance of a building depends on the characteristics and orientation of the window as well as the solar and weather data. Workmanship is very important in the construction and installation of windows to provide effective sealing around the edges while allowing them to be opened and closed easily.

Despite being so undesirable from an energy conservation point of view, windows are an essential part of any building envelope since they enhance the appearance of the building, allow daylight and solar heat to come in, and allow people to view and observe outside without leaving their home. For low-rise buildings, windows also provide easy exit areas during emergencies such as fire. Important considerations in the selection of windows are *thermal comfort* and *energy conservation*. A window should have a good light transmittance while providing effective resistance to heat transfer. The lighting requirements of a building can be minimized by maximizing the use of natural daylight. Heat loss in winter through the windows can be minimized by using airtight double- or triple-pane windows with spectrally selective films or coatings, and letting in as much solar radiation as possible. Heat gain and thus cooling load in summer can be minimized by using effective internal or external shading on the windows.

Even in the absence of solar radiation and air infiltration, heat transfer through the windows is more complicated than it appears to be. This is because the structure and properties of the frame are quite different than the glazing. As a result, heat transfer through the frame and the edge section of the glazing adjacent to the frame is two-dimensional. Therefore, it is customary to consider the window in three regions when analyzing heat transfer through it: (1) the *center-of-glass*, (2) the *edge-of-glass*, and (3) the *frame* regions, as shown in Fig. 9–34. Then the total rate of heat transfer through the window is determined by adding the heat transfer through each region as

$$\dot{Q}_{window} = \dot{Q}_{center} + \dot{Q}_{edge} + \dot{Q}_{frame}$$
$$= U_{window} A_{window} (T_{indoors} - T_{outdoors}) \qquad (9\text{–}67)$$

where

$$U_{window} = (U_{center} A_{center} + U_{edge} A_{edge} + U_{frame} A_{frame})/A_{window} \qquad (9\text{–}68)$$

is the *U*-factor or the **overall heat transfer coefficient** of the window; A_{window} is the window area; A_{center}, A_{edge}, and A_{frame} are the areas of the

---

*This section can be skipped without a loss of continuity.

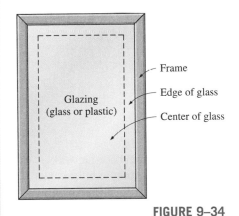

FIGURE 9–34
The three regions of a window considered in heat transfer analysis.

center, edge, and frame sections of the window, respectively; and U_{center}, U_{edge}, and U_{frame} are the heat transfer coefficients for the center, edge, and frame sections of the window. Note that $A_{window} = A_{center} + A_{edge} + A_{frame}$, and the overall U-factor of the window is determined from the area-weighed U-factors of each region of the window. Also, the inverse of the U-factor is the R-value, which is the unit thermal resistance of the window (thermal resistance for a unit area).

Consider steady one-dimensional heat transfer through a single-pane glass of thickness L and thermal conductivity k. The thermal resistance network of this problem consists of surface resistances on the inner and outer surfaces and the conduction resistance of the glass in series, as shown in Fig. 9–35, and the total resistance on a unit area basis can be expressed as

$$R_{total} = R_{inside} + R_{glass} + R_{outside} = \frac{1}{h_i} + \frac{L_{glass}}{k_{glass}} + \frac{1}{h_o} \qquad (9\text{-}69)$$

Using common values of 3 mm for the thickness and 0.92 W/m · °C for the thermal conductivity of the glass and the winter design values of 8.29 and 34.0 W/m² · °C for the inner and outer surface heat transfer coefficients, the thermal resistance of the glass is determined to be

$$R_{total} = \frac{1}{8.29 \text{ W/m}^2 \cdot \text{°C}} + \frac{0.003 \text{ m}}{0.92 \text{ W/m} \cdot \text{°C}} + \frac{1}{34.0 \text{ W/m}^2 \cdot \text{°C}}$$
$$= 0.121 + 0.003 + 0.029 = 0.153 \text{ m}^2 \cdot \text{°C/W}$$

Note that the ratio of the glass resistance to the total resistance is

$$\frac{R_{glass}}{R_{total}} = \frac{0.003 \text{ m}^2 \cdot \text{°C/W}}{0.153 \text{ m}^2 \cdot \text{°C/W}} = 2.0\%$$

That is, the glass layer itself contributes about 2 percent of the total thermal resistance of the window, which is negligible. The situation would not be much different if we used acrylic, whose thermal conductivity is 0.19 W/m · °C, instead of glass. Therefore, we cannot reduce the heat transfer through the window effectively by simply increasing the thickness of the glass. But we can reduce it by trapping still air between two layers of glass. The result is a **double-pane window**, which has become the norm in window construction.

The thermal conductivity of air at room temperature is $k_{air} = 0.025$ W/m · °C, which is one-thirtieth that of glass. Therefore, the thermal resistance of 1-cm-thick still air is equivalent to the thermal resistance of a 30-cm-thick glass layer. Disregarding the thermal resistances of glass layers, the thermal resistance and U-factor of a double-pane window can be expressed as (Fig. 9–36)

$$\frac{1}{U_{double\text{-}pane \text{ (center region)}}} \cong \frac{1}{h_i} + \frac{1}{h_{space}} + \frac{1}{h_o} \qquad (9\text{-}70)$$

where $h_{space} = h_{rad, space} + h_{conv, space}$ is the combined radiation and convection heat transfer coefficient of the space trapped between the two glass layers.

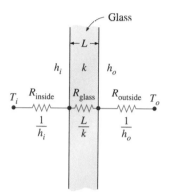

FIGURE 9–35
The thermal resistance network for heat transfer through a single glass.

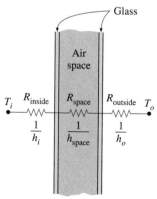

FIGURE 9–36
The thermal resistance network for heat transfer through the center section of a double-pane window (the resistances of the glasses are neglected).

Roughly half of the heat transfer through the air space of a double-pane window is by radiation and the other half is by conduction (or convection, if there is any air motion). Therefore, there are two ways to minimize h_{space} and thus the rate of heat transfer through a double-pane window:

1. *Minimize radiation heat transfer through the air space.* This can be done by reducing the emissivity of glass surfaces by coating them with low-emissivity (or "low-e" for short) material. Recall that the *effective emissivity* of two parallel plates of emissivities ε_1 and ε_2 is given by

$$\varepsilon_{effective} = \frac{1}{1/\varepsilon_1 + 1/\varepsilon_2 - 1} \tag{9-71}$$

The emissivity of an ordinary glass surface is 0.84. Therefore, the effective emissivity of two parallel glass surfaces facing each other is 0.72. But when the glass surfaces are coated with a film that has an emissivity of 0.1, the effective emissivity reduces to 0.05, which is one-fourteenth of 0.72. Then for the same surface temperatures, radiation heat transfer will also go down by a factor of 14. Even if only one of the surfaces is coated, the overall emissivity reduces to 0.1, which is the emissivity of the coating. Thus it is no surprise that about one-fourth of all windows sold for residences have a low-e coating. The heat transfer coefficient h_{space} for the air space trapped between the two vertical parallel glass layers is given in Table 9–3 for 13-mm- ($\frac{1}{2}$-in) and 6-mm- ($\frac{1}{4}$-in) thick air spaces for various effective emissivities and temperature differences.

It can be shown that coating just one of the two parallel surfaces facing each other by a material of emissivity ε reduces the effective emissivity nearly to ε.

TABLE 9–3

The heat transfer coefficient h_{space} for the air space trapped between the two vertical parallel glass layers for 13-mm- and 6-mm-thick air spaces (from Building Materials and Structures, Report 151, U.S. Dept. of Commerce).

(a) Air space thickness = 13 mm

T_{avg}, °C	ΔT, °C	h_{space}, W/m² · °C* $\varepsilon_{effective}$ 0.72	0.4	0.2	0.1
0	5	5.3	3.8	2.9	2.4
0	15	5.3	3.8	2.9	2.4
0	30	5.5	4.0	3.1	2.6
10	5	5.7	4.1	3.0	2.5
10	15	5.7	4.1	3.1	2.5
10	30	6.0	4.3	3.3	2.7
30	5	5.7	4.6	3.4	2.7
30	15	5.7	4.7	3.4	2.8
30	30	6.0	4.9	3.6	3.0

(b) Air space thickness = 6 mm

T_{avg}, °C	ΔT, °C	h_{space}, W/m² · °C* $\varepsilon_{effective}$ 0.72	0.4	0.2	0.1
0	5	7.2	5.7	4.8	4.3
0	50	7.2	5.7	4.8	4.3
10	5	7.7	6.0	5.0	4.5
10	50	7.7	6.1	5.0	4.5
30	5	8.8	6.8	5.5	4.9
30	50	8.8	6.8	5.5	4.9
50	5	10.0	7.5	6.0	5.2
50	50	10.0	7.5	6.0	5.2

*Multiply by 0.176 to convert to Btu/h · ft² · °F.

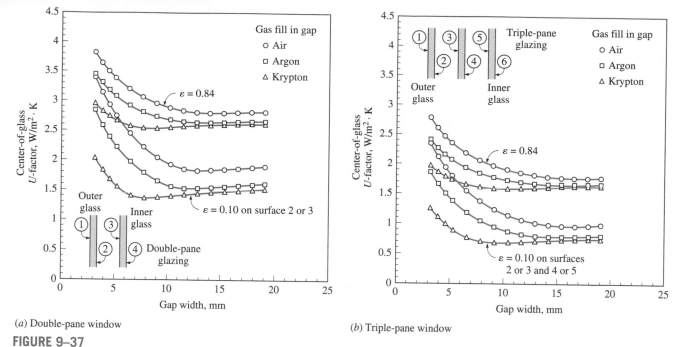

(a) Double-pane window

(b) Triple-pane window

FIGURE 9–37

The variation of the U-factor for the center section of double- and triple-pane windows with uniform spacing between the panes (from ASHRAE *Handbook of Fundamentals,* Chap. 27, Fig. 1).

Therefore, it is usually more economical to coat only one of the facing surfaces. Note from Fig. 9–37 that coating one of the interior surfaces of a double-pane window with a material having an emissivity of 0.1 reduces the rate of heat transfer through the center section of the window by half.

2. *Minimize conduction heat transfer through air space.* This can be done by *increasing* the distance d between the two glasses. However, this cannot be done indefinitely since increasing the spacing beyond a critical value initiates convection currents in the enclosed air space, which increases the heat transfer coefficient and thus defeats the purpose. Besides, increasing the spacing also increases the thickness of the necessary framing and the cost of the window. Experimental studies have shown that when the spacing d is less than about 13 mm, there is no convection, and heat transfer through the air is by conduction. But as the spacing is increased further, convection currents appear in the air space, and the increase in heat transfer coefficient offsets any benefit obtained by the thicker air layer. As a result, the heat transfer coefficient remains nearly constant, as shown in Fig. 9–37. Therefore, it makes no sense to use an air space thicker than 13 mm in a double-pane window unless a thin polyester film is used to divide the air space into two halves to suppress convection currents. The film provides added insulation without adding much to the weight or cost of the double-pane window. The thermal resistance of the window can be increased further by using triple- or quadruple-pane windows whenever it is economical to do so. Note that using a triple-pane window instead of a double-pane reduces the rate of heat transfer through the center section of the window by about one-third.

Another way of reducing conduction heat transfer through a double-pane window is to use a *less-conducting fluid* such as argon or krypton to fill the gap between the glasses instead of air. The gap in this case needs to be well sealed to prevent the gas from leaking outside. Of course, another alternative is to evacuate the gap between the glasses completely, but it is not practical to do so.

Edge-of-Glass *U*-Factor of a Window

The glasses in double- and triple-pane windows are kept apart from each other at a uniform distance by **spacers** made of metals or insulators like aluminum, fiberglass, wood, and butyl. Continuous spacer strips are placed around the glass perimeter to provide an edge seal as well as uniform spacing. However, the spacers also serve as undesirable "thermal bridges" between the glasses, which are at different temperatures, and this short circuiting may increase heat transfer through the window considerably. Heat transfer in the edge region of a window is two-dimensional, and lab measurements indicate that the edge effects are limited to a 6.5-cm-wide band around the perimeter of the glass.

The *U*-factor for the edge region of a window is given in Fig. 9–38 relative to the *U*-factor for the center region of the window. The curve would be a straight diagonal line if the two *U*-values were equal to each other. Note that this is almost the case for insulating spacers such as wood and fiberglass. But the *U*-factor for the edge region can be twice that of the center region for conducting spacers such as those made of aluminum. Values for steel spacers fall between the two curves for metallic and insulating spacers. The edge effect is not applicable to single-pane windows.

Frame *U*-Factor

The framing of a window consists of the entire window except the glazing. Heat transfer through the framing is difficult to determine because of the different window configurations, different sizes, different constructions, and different combination of materials used in the frame construction. The type of glazing such as single pane, double pane, and triple pane affects the thickness of the framing and thus heat transfer through the frame. Most frames are made of *wood, aluminum, vinyl,* or *fiberglass.* However, using a combination of these materials (such as aluminum-clad wood and vinyl-clad aluminum) is also common to improve appearance and durability.

Aluminum is a popular framing material because it is inexpensive, durable, and easy to manufacture, and does not rot or absorb water like wood. However, from a heat transfer point of view, it is the least desirable framing material because of its high thermal conductivity. It will come as no surprise that the *U*-factor of solid aluminum frames is the highest, and thus a window with aluminum framing will lose much more heat than a comparable window with wood or vinyl framing. Heat transfer through the aluminum framing members can be reduced by using plastic inserts between

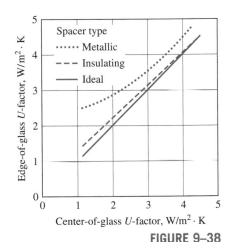

FIGURE 9–38

The edge-of-glass *U*-factor relative to the center-of-glass *U*-factor for windows with various spacers. (From ASHRAE *Handbook of Fundamentals,* Chap. 27, Fig. 2)

TABLE 9–4

Representative frame U-factors for fixed vertical windows (from ASHRAE *Handbook of Fundamentals,* Chap. 27, Table 2.)

Frame material	U-factor, W/m² · °C*
Aluminum:	
Single glazing (3 mm)	10.1
Double glazing (18 mm)	10.1
Triple glazing (33 mm)	10.1
Wood or vinyl:	
Single glazing (3 mm)	2.9
Double glazing (18 mm)	2.8
Triple glazing (33 mm)	2.7

*Multiply by 0.176 to convert to Btu/h · ft² · °F

TABLE 9–5

Combined convection and radiation heat transfer coefficient h_i at the inner surface of a vertical glass under still air conditions (in W/m² · °C)*

T_i (°C)	T_g (°C)	Glass emissivity, ε_g		
		0.05	0.20	0.84
20	17	2.6	3.5	7.1
20	15	2.9	3.8	7.3
20	10	3.4	4.2	7.7
20	5	3.7	4.5	7.9
20	0	4.0	4.8	8.1
20	−5	4.2	5.0	8.2
20	−10	4.4	5.1	8.3

*Multiply by 0.176 to convert to Btu/h · ft² · °F.

components to serve as thermal barriers. The thickness of these inserts greatly affects heat transfer through the frame. For aluminum frames without the plastic strips, the primary resistance to heat transfer is due to the interior surface heat transfer coefficient. The U-factors for various frames are listed in Table 9–4 as a function of spacer materials and the glazing unit thicknesses. Note that the U-factor of metal framing and thus the rate of heat transfer through a metal window frame is more than three times that of a wood or vinyl window frame.

Interior and Exterior Surface Heat Transfer Coefficients

Heat transfer through a window is also affected by the convection and radiation heat transfer coefficients between the glass surfaces and surroundings. The effects of convection and radiation on the inner and outer surfaces of glazings are usually combined into the combined convection and radiation heat transfer coefficients h_i and h_o, respectively. Under still air conditions, the combined heat transfer coefficient at the inner surface of a vertical window can be determined from

$$h_i = h_{\text{conv}} + h_{\text{rad}} = 1.77(T_g - T_i)^{0.25} + \frac{\varepsilon_g \sigma (T_g^4 - T_i^4)}{T_g - T_i} \quad (\text{W/m}^2 \cdot \text{°C}) \tag{9–72}$$

where T_g = glass temperature in K, T_i = indoor air temperature in K, ε_g = emissivity of the inner surface of the glass exposed to the room (taken to be 0.84 for uncoated glass), and $\sigma = 5.67 \times 10^{-8}$ W/m² · K⁴ is the Stefan–Boltzmann constant. Here the temperature of the interior surfaces facing the window is assumed to be equal to the indoor air temperature. This assumption is reasonable when the window faces mostly interior walls, but it becomes questionable when the window is exposed to heated or cooled surfaces or to other windows. The commonly used value of h_i for peak load calculation is

$$h_i = 8.29 \text{ W/m}^2 \cdot \text{°C} = 1.46 \text{ Btu/h} \cdot \text{ft}^2 \cdot \text{°F} \quad (\text{winter and summer})$$

which corresponds to the winter design conditions of $T_i = 22$°C and $T_g = -7$°C for uncoated glass with $\varepsilon_g = 0.84$. But the same value of h_i can also be used for summer design conditions as it corresponds to summer conditions of $T_i = 24$°C and $T_g = 32$°C. The values of h_i for various temperatures and glass emissivities are given in Table 9–5. The commonly used values of h_o for peak load calculations are the same as those used for outer wall surfaces (34.0 W/m² · °C for winter and 22.7 W/m² · °C for summer).

Overall U-Factor of Windows

The overall U-factors for various kinds of windows and skylights are evaluated using computer simulations and laboratory testing for winter design conditions; representative values are given in Table 9–6. Test data may provide more accurate information for specific products and should be preferred when available. However, the values listed in the table can be used to obtain satisfactory results under various conditions in the absence of product-specific data. The U-factor of a fenestration product that differs

TABLE 9–6

Overall U-factors (heat transfer coefficients) for various windows and skylights in W/m$^2 \cdot$ °C
(from ASHRAE *Handbook of Fundamentals*, Chap. 27, Table 5)

Type →	Glass section (glazing) only			Aluminum frame (without thermal break)			Wood or vinyl frame					
	Center-of-glass	Edge-of-glass		Fixed	Double door	Sloped skylight	Fixed		Double door		Sloped skylight	
Frame width →	(Not applicable)			32 mm ($1\frac{1}{4}$ in)	53 mm (2 in)	19 mm ($\frac{3}{4}$ in)	41 mm ($1\frac{5}{8}$ in)		88 mm ($3\frac{7}{18}$ in)		23 mm ($\frac{7}{8}$ in)	
Spacer type →	—	Metal	Insul.	All	All	All	Metal	Insul.	Metal	Insul.	Metal	Insul.
Glazing Type												
Single Glazing												
3 mm ($\frac{1}{8}$ in) glass	6.30	6.30	—	6.63	7.16	9.88	5.93	—	5.57	—	7.57	—
6.4 mm ($\frac{1}{4}$ in) acrylic	5.28	5.28	—	5.69	6.27	8.86	5.02	—	4.77	—	6.57	—
3 mm ($\frac{1}{8}$ in) acrylic	5.79	5.79	—	6.16	6.71	9.94	5.48	—	5.17	—	7.63	—
Double Glazing (no coating)												
6.4 mm air space	3.24	3.71	3.34	3.90	4.55	6.70	3.26	3.16	3.20	3.09	4.37	4.22
12.7 mm air space	2.78	3.40	2.91	3.51	4.18	6.65	2.88	2.76	2.86	2.74	4.32	4.17
6.4 mm argon space	2.95	3.52	3.07	3.66	4.32	6.47	3.03	2.91	2.98	2.87	4.14	3.97
12.7 mm argon space	2.61	3.28	2.76	3.36	4.04	6.47	2.74	2.61	2.73	2.60	4.14	3.97
Double Glazing [$\varepsilon = 0.1$, coating on one of the surfaces of air space (surface 2 or 3, counting from the outside toward inside)]												
6.4 mm air space	2.44	3.16	2.60	3.21	3.89	6.04	2.59	2.46	2.60	2.47	3.73	3.53
12.7 mm air space	1.82	2.71	2.06	2.67	3.37	6.04	2.06	1.92	2.13	1.99	3.73	3.53
6.4 mm argon space	1.99	2.83	2.21	2.82	3.52	5.62	2.21	2.07	2.26	2.12	3.32	3.09
12.7 mm argon space	1.63	2.49	1.83	2.42	3.14	5.71	1.82	1.67	1.91	1.78	3.41	3.19
Triple Glazing (no coating)												
6.4 mm air space	2.16	2.96	2.35	2.97	3.66	5.81	2.34	2.18	2.36	2.21	3.48	3.24
12.7 mm air space	1.76	2.67	2.02	2.62	3.33	5.67	2.01	1.84	2.07	1.91	3.34	3.09
6.4 mm argon space	1.93	2.79	2.16	2.77	3.47	5.57	2.15	1.99	2.19	2.04	3.25	3.00
12.7 mm argon space	1.65	2.58	1.92	2.52	3.23	5.53	1.91	1.74	1.98	1.82	3.20	2.95
Triple Glazing [$\varepsilon = 0.1$, coating on one of the surfaces of air spaces (surfaces 3 and 5, counting from the outside toward inside)]												
6.4 mm air space	1.53	2.49	1.83	2.42	3.14	5.24	1.81	1.64	1.89	1.73	2.92	2.66
12.7 mm air space	0.97	2.05	1.38	1.92	2.66	5.10	1.33	1.15	1.46	1.30	2.78	2.52
6.4 mm argon space	1.19	2.23	1.56	2.12	2.85	4.90	1.52	1.35	1.64	1.47	2.59	2.33
12.7 mm argon space	0.80	1.92	1.25	1.77	2.51	4.86	1.18	1.01	1.33	1.17	2.55	2.28

Notes:

(1) Multiply by 0.176 to obtain U-factors in Btu/h $\cdot$ ft$^2 \cdot$ °F.

(2) The U-factors in this table include the effects of surface heat transfer coefficients and are based on winter conditions of -18°C outdoor air and 21°C indoor air temperature, with 24 km/h (15 mph) winds outdoors and zero solar flux. Small changes in indoor and outdoor temperatures will not affect the overall U-factors much. Windows are assumed to be vertical, and the skylights are tilted 20° from the horizontal with upward heat flow. Insulation spacers are wood, fiberglass, or butyl. Edge-of-glass effects are assumed to extend the 65-mm band around perimeter of each glazing. The product sizes are 1.2 m × 1.8 m for fixed windows, 1.8 m × 2.0 m for double-door windows, and 1.2 m × 0.6 m for the skylights, but the values given can also be used for products of similar sizes. All data are based on 3-mm ($\frac{1}{8}$-in) glass unless noted otherwise.

considerably from the ones in the table can be determined by (1) determining the fractions of the area that are frame, center-of-glass, and edge-of-glass (assuming a 65-mm-wide band around the perimeter of each glazing), (2) determining the U-factors for each section (the center-of-glass and edge-of-glass U-factors can be taken from the first two columns of Table 9–6 and the frame U-factor can be taken from Table 9–5 or other sources), and (3) multiplying the area fractions and the U-factors for each section and adding them up.

Glazed wall systems can be treated as fixed windows. Also, the data for double-door windows can be used for single-glass doors. Several observations can be made from the data in the table:

1. Skylight U-factors are considerably greater than those of vertical windows. This is because the skylight area, including the curb, can be 13 to 240 percent greater than the rough opening area. The slope of the skylight also has some effect.

2. The U-factor of multiple-glazed units can be reduced considerably by filling cavities with argon gas instead of dry air. The performance of CO_2-filled units is similar to those filled with argon. The U-factor can be reduced even further by filling the glazing cavities with krypton gas.

3. Coating the glazing surfaces with low-e (low-emissivity) films reduces the U-factor significantly. For multiple-glazed units, it is adequate to coat one of the two surfaces facing each other.

4. The thicker the air space in multiple-glazed units, the lower the U-factor, for a thickness of up to 13 mm ($\frac{1}{2}$ in) of air space. For a specified number of glazings, the window with thicker air layers will have a lower U-factor. For a specified overall thickness of glazing, the higher the number of glazings, the lower the U-factor. Therefore, a triple-pane window with air spaces of 6.4 mm (two such air spaces) will have a lower U-value than a double-pane window with an air space of 12.7 mm.

5. Wood or vinyl frame windows have a considerably lower U-value than comparable metal-frame windows. Therefore, wood or vinyl frame windows are called for in energy-efficient designs.

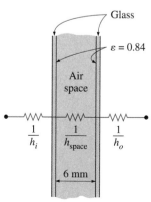

FIGURE 9–39
Schematic of Example 9–7.

EXAMPLE 9–7 **U-Factor for Center-of-Glass Section of Windows**

Determine the U-factor for the center-of-glass section of a double-pane window with a 6-mm air space for winter design conditions (Fig. 9–39). The glazings are made of clear glass that has an emissivity of 0.84. Take the average air space temperature at design conditions to be 0°C.

SOLUTION The U-factor for the center-of-glass section of a double-pane window is to be determined.
Assumptions 1 Steady operating conditions exist. 2 Heat transfer through the window is one-dimensional. 3 The thermal resistance of glass sheets is negligible.
Properties The emissivity of clear glass is 0.84.

Analysis Disregarding the thermal resistance of glass sheets, which are small, the U-factor for the center region of a double-pane window is determined from

$$\frac{1}{U_{\text{center}}} \cong \frac{1}{h_i} + \frac{1}{h_{\text{space}}} + \frac{1}{h_o}$$

where h_i, h_{space}, and h_o are the heat transfer coefficients at the inner surface of the window, the air space between the glass layers, and the outer surface of the window, respectively. The values of h_i and h_o for winter design conditions were given earlier to be $h_i = 8.29 \text{ W/m}^2 \cdot °C$ and $h_o = 34.0 \text{ W/m}^2 \cdot °C$. The effective emissivity of the air space of the double-pane window is

$$\varepsilon_{\text{effective}} = \frac{1}{1/\varepsilon_1 + 1/\varepsilon_2 - 1} = \frac{1}{1/0.84 + 1/0.84 - 1} = 0.72$$

For this value of emissivity and an average air space temperature of 0°C, we read $h_{\text{space}} = 7.2 \text{ W/m}^2 \cdot °C$ from Table 9–3 for 6-mm-thick air space. Therefore,

$$\frac{1}{U_{\text{center}}} = \frac{1}{8.29} + \frac{1}{7.2} + \frac{1}{34.0} \quad \rightarrow \quad U_{\text{center}} = 3.46 \text{ W/m}^2 \cdot °C$$

Discussion The center-of-glass U-factor value of 3.24 W/m² · °C in Table 9–6 (fourth row and second column) is obtained by using a standard value of $h_o = 29 \text{ W/m}^2 \cdot °C$ (instead of 34.0 W/m² · °C) and $h_{\text{space}} = 6.5 \text{ W/m}^2 \cdot °C$ at an average air space temperature of $-15°C$.

EXAMPLE 9–8   Heat Loss through Aluminum Framed Windows

A fixed aluminum-framed window with glass glazing is being considered for an opening that is 4 ft high and 6 ft wide in the wall of a house that is maintained at 72°F (Fig. 9–40). Determine the rate of heat loss through the window and the inner surface temperature of the window glass facing the room when the outdoor air temperature is 15°F if the window is selected to be (a) $\frac{1}{8}$-in single glazing, (b) double glazing with an air space of $\frac{1}{2}$ in, and (c) low-e-coated triple glazing with an air space of $\frac{1}{2}$ in.

SOLUTION The rate of heat loss through an aluminum framed window and the inner surface temperature are to be determined from the cases of single-pane, double-pane, and low-e triple-pane windows.

Assumptions 1 Steady operating conditions exist. 2 Heat transfer through the window is one-dimensional. 3 Thermal properties of the windows and the heat transfer coefficients are constant.

Properties The U-factors of the windows are given in Table 9–6.

Analysis The rate of heat transfer through the window can be determined from

$$\dot{Q}_{\text{window}} = U_{\text{overall}} A_{\text{window}} (T_i - T_o)$$

where T_i and T_o are the indoor and outdoor air temperatures, respectively; U_{overall} is the U-factor (the overall heat transfer coefficient) of the window; and A_{window} is the window area, which is determined to be

$$A_{\text{window}} = \text{Height} \times \text{Width} = (4 \text{ ft})(6 \text{ ft}) = 24 \text{ ft}^2$$

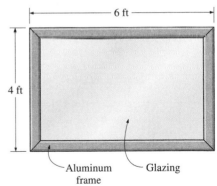

FIGURE 9–40
Schematic for Example 9–8.

The *U*-factors for the three cases can be determined directly from Table 9–6 to be 6.63, 3.51, and 1.92 W/m² · °C, respectively, to be multiplied by the factor 0.176 to convert them to Btu/h · ft² · °F. Also, the inner surface temperature of the window glass can be determined from

$$\dot{Q}_{\text{window}} = h_i A_{\text{window}} (T_i - T_{\text{glass}}) \quad \rightarrow \quad T_{\text{glass}} = T_i - \frac{\dot{Q}_{\text{window}}}{h_i A_{\text{window}}}$$

where h_i is the heat transfer coefficient on the inner surface of the window, which is determined from Table 9–5 to be $h_i = 8.3$ W/m² · °C = 1.46 Btu/h · ft² · °F. Then the rate of heat loss and the interior glass temperature for each case are determined as follows:

(*a*) Single glazing:

$$\dot{Q}_{\text{window}} = (6.63 \times 0.176 \text{ Btu/h} \cdot \text{ft}^2 \cdot °\text{F})(24 \text{ ft}^2)(72 - 15)°\text{F} = \mathbf{1596 \text{ Btu/h}}$$

$$T_{\text{glass}} = T_i - \frac{\dot{Q}_{\text{window}}}{h_i A_{\text{window}}} = 72°\text{F} - \frac{1596 \text{ Btu/h}}{(1.46 \text{ Btu/h} \cdot \text{ft}^2 \cdot °\text{F})(24 \text{ ft}^2)} = \mathbf{26.5°F}$$

(*b*) Double glazing ($\frac{1}{2}$ in air space):

$$\dot{Q}_{\text{window}} = (3.51 \times 0.176 \text{ Btu/h} \cdot \text{ft}^2 \cdot °\text{F})(24 \text{ ft}^2)(72 - 15)°\text{F} = \mathbf{845 \text{ Btu/h}}$$

$$T_{\text{glass}} = T_i - \frac{\dot{Q}_{\text{window}}}{h_i A_{\text{window}}} = 72°\text{F} - \frac{845 \text{ Btu/h}}{(1.46 \text{ Btu/h} \cdot \text{ft}^2 \cdot °\text{F})(24 \text{ ft}^2)} = \mathbf{47.9°F}$$

(*c*) Triple glazing ($\frac{1}{2}$ in air space, low-e coated):

$$\dot{Q}_{\text{window}} = (1.92 \times 0.176 \text{ Btu/h} \cdot \text{ft}^2 \cdot °\text{F})(24 \text{ ft}^2)(72 - 15)°\text{F} = \mathbf{462 \text{ Btu/h}}$$

$$T_{\text{glass}} = T_i - \frac{\dot{Q}_{\text{window}}}{h_i A_{\text{window}}} = 72°\text{F} - \frac{462 \text{ Btu/h}}{(1.46 \text{ Btu/h} \cdot \text{ft}^2 \cdot °\text{F})(24 \text{ ft}^2)} = \mathbf{58.8°F}$$

Therefore, heat loss through the window will be reduced by 47 percent in the case of double glazing and by 71 percent in the case of triple glazing relative to the single-glazing case. Also, in the case of single glazing, the low inner-glass surface temperature will cause considerable discomfort in the occupants because of the excessive heat loss from the body by radiation. It is raised from 26.5°F, which is below freezing, to 47.9°F in the case of double glazing and to 58.8°F in the case of triple glazing.

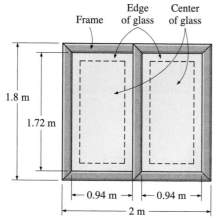

FIGURE 9–41
Schematic for Example 9–9.

EXAMPLE 9–9   *U*-Factor of a Double-Door Window

Determine the overall *U*-factor for a double-door-type, wood-framed double-pane window with metal spacers, and compare your result to the value listed in Table 9–6. The overall dimensions of the window are 1.80 m × 2.00 m, and the dimensions of each glazing are 1.72 m × 0.94 m (Fig. 9–41).

SOLUTION   The overall *U*-factor for a double-door type window is to be determined, and the result is to be compared to the tabulated value.
Assumptions   **1** Steady operating conditions exist. **2** Heat transfer through the window is one-dimensional.

Properties The U-factors for the various sections of windows are given in Tables 9–4 and 9–6.

Analysis The areas of the window, the glazing, and the frame are

$$A_{\text{window}} = \text{Height} \times \text{Width} = (1.8 \text{ m})(2.0 \text{ m}) = 3.60 \text{ m}^2$$

$$A_{\text{glazing}} = 2 \times (\text{Height} \times \text{Width}) = 2(1.72 \text{ m})(0.94 \text{ m}) = 3.23 \text{ m}^2$$
$$A_{\text{frame}} = A_{\text{window}} - A_{\text{glazing}} = 3.60 - 3.23 = 0.37 \text{ m}^2$$

The edge-of-glass region consists of a 6.5-cm-wide band around the perimeter of the glazings, and the areas of the center and edge sections of the glazing are determined to be

$$A_{\text{center}} = 2 \times (\text{Height} \times \text{Width}) = 2(1.72 - 0.13 \text{ m})(0.94 - 0.13 \text{ m}) = 2.58 \text{ m}^2$$
$$A_{\text{edge}} = A_{\text{glazing}} - A_{\text{center}} = 3.23 - 2.58 = 0.65 \text{ m}^2$$

The U-factor for the frame section is determined from Table 9–4 to be $U_{\text{frame}} = 2.8 \text{ W/m}^2 \cdot {}^\circ\text{C}$. The U-factors for the center and edge sections are determined from Table 9–6 (fifth row, second and third columns) to be $U_{\text{center}} = 3.24 \text{ W/m}^2 \cdot {}^\circ\text{C}$ and $U_{\text{edge}} = 3.71 \text{ W/m}^2 \cdot {}^\circ\text{C}$. Then the overall U-factor of the entire window becomes

$$U_{\text{window}} = (U_{\text{center}} A_{\text{center}} + U_{\text{edge}} A_{\text{edge}} + U_{\text{frame}} A_{\text{frame}})/A_{\text{window}}$$
$$= (3.24 \times 2.58 + 3.71 \times 0.65 + 2.8 \times 0.37)/3.60$$
$$= \mathbf{3.28 \text{ W/m}^2 \cdot {}^\circ\text{C}}$$

The overall U-factor listed in Table 9–6 for the specified type of window is 3.20 W/m$^2 \cdot$ °C, which is sufficiently close to the value obtained above.

SUMMARY

In this chapter, we have considered natural convection heat transfer where any fluid motion occurs by natural means such as buoyancy. The volume expansion coefficient of a substance represents the variation of the density of that substance with temperature at constant pressure, and for an ideal gas, it is expressed as $\beta = 1/T$, where T is the absolute temperature in K or R.

The flow regime in natural convection is governed by a dimensionless number called the *Grashof number,* which represents the ratio of the buoyancy force to the viscous force acting on the fluid and is expressed as

$$\text{Gr}_L = \frac{g\beta(T_s - T_\infty)L_c^3}{\nu^2}$$

where L_c is the *characteristic length,* which is the height L for a vertical plate and the diameter D for a horizontal cylinder.

The correlations for the Nusselt number $\text{Nu} = hL_c/k$ in natural convection are expressed in terms of the *Rayleigh number* defined as

$$\text{Ra}_L = \text{Gr}_L \text{Pr} = \frac{g\beta(T_s - T_\infty)L_c^3}{\nu^2} \text{Pr}$$

Nusselt number relations for various surfaces are given in Table 9–1. All fluid properties are evaluated at the film temperature of $T_f = \frac{1}{2}(T_s + T_\infty)$. The outer surface of a vertical cylinder can be treated as a vertical plate when the curvature effects are negligible. The characteristic length for a horizontal surface is $L_c = A_s/p$, where A_s is the surface area and p is the perimeter.

The average Nusselt number for vertical isothermal *parallel plates* of spacing S and height L is given as

$$\text{Nu} = \frac{hS}{k} = \left[\frac{576}{(\text{Ra}_S S/L)^2} + \frac{2.873}{(\text{Ra}_S S/L)^{0.5}}\right]^{-0.5}$$

The optimum fin spacing for a vertical heat sink and the Nusselt number for optimally spaced fins is

$$S_{opt} = 2.714 \left(\frac{S^3 L}{Ra_S} \right)^{0.25} = 2.714 \frac{L}{Ra_L^{0.25}} \text{ and } Nu = \frac{hS_{opt}}{k} = 1.307$$

In a *horizontal rectangular enclosure* with the hotter plate at the top, heat transfer is by pure conduction and $Nu = 1$. When the hotter plate is at the bottom, the Nusselt number is

$$Nu = 1 + 1.44 \left[1 - \frac{1708}{Ra_L} \right]^+ + \left[\frac{Ra_L^{1/3}}{18} - 1 \right]^+ \qquad Ra_L < 10^8$$

The notation $[\ ]^+$ indicates that if the quantity in the bracket is negative, it should be set equal to zero. For *vertical horizontal enclosures,* the Nusselt number can be determined from

$$Nu = 0.18 \left(\frac{Pr}{0.2 + Pr} Ra_L \right)^{0.29} \quad \begin{array}{l} 1 < H/L < 2 \\ \text{any Prandtl number} \\ Ra_L Pr/(0.2 + Pr) > 10^3 \end{array}$$

$$Nu = 0.22 \left(\frac{Pr}{0.2 + Pr} Ra_L \right)^{0.28} \left(\frac{H}{L} \right)^{-1/4} \quad \begin{array}{l} 2 < H/L < 10 \\ \text{any Prandtl number} \\ Ra_L < 10^{10} \end{array}$$

For aspect ratios greater than 10, Eqs. 9–54 and 9–55 should be used. For inclined enclosures, Eqs. 9–48 through 9–51 should be used.

For *concentric horizontal cylinders,* the rate of heat transfer through the annular space between the cylinders by natural convection per unit length is

$$\dot{Q} = \frac{2\pi k_{eff}}{\ln(D_o/D_i)} (T_i - T_o)$$

where

$$\frac{k_{eff}}{k} = 0.386 \left(\frac{Pr}{0.861 + Pr} \right)^{1/4} (F_{cyl} Ra_L)^{1/4}$$

and

$$F_{cyl} = \frac{[\ln(D_o/D_i)]^4}{L_c^3 (D_i^{-3/5} + D_o^{-3/5})^5}$$

For a *spherical enclosure,* the rate of heat transfer through the space between the spheres by natural convection is expressed as

$$\dot{Q} = k_{eff} \frac{\pi D_i D_o}{L_c} (T_i - T_o)$$

where

$$\frac{k_{eff}}{k} = 0.74 \left(\frac{Pr}{0.861 + Pr} \right)^{1/4} (F_{sph} Ra_L)^{1/4}$$

$$L_c = (D_o - D_i)/2$$

$$F_{sph} = \frac{L_c}{(D_i D_o)^4 (D_i^{-7/5} + D_o^{-7/5})^5}$$

The quantity kNu is called the *effective thermal conductivity* of the enclosure, since a fluid in an enclosure behaves like a quiescent fluid whose thermal conductivity is kNu as a result of convection currents. The fluid properties are evaluated at the average temperature of $(T_i + T_o)/2$.

For a given fluid, the parameter Gr/Re^2 represents the importance of natural convection relative to forced convection. Natural convection is negligible when $Gr/Re^2 < 0.1$, forced convection is negligible when $Gr/Re^2 > 10$, and neither is negligible when $0.1 < Gr/Re^2 < 10$.

REFERENCES AND SUGGESTED READINGS

1. American Society of Heating, Refrigeration, and Air Conditioning Engineers. *Handbook of Fundamentals.* Atlanta: ASHRAE, 1993.

2. J. N Arnold, I. Catton, and D. K. Edwards. "Experimental Investigation of Natural Convection in Inclined Rectangular Region of Differing Aspects Ratios." ASME Paper No. 75-HT-62, 1975.

3. P. S. Ayyaswamy and I. Catton. "The Boundary-Layer Regime for Natural Convection in a Differently Heated

Tilted Rectangular Cavity." *Journal of Heat Transfer* 95 (1973), p. 543.

4. A. Bar-Cohen. "Fin Thickness for an Optimized Natural Convection Array of Rectangular Fins." *Journal of Heat Transfer* 101 (1979), pp. 564–566.

5. A. Bar-Cohen and W. M. Rohsenow. "Thermally Optimum Spacing of Vertical Natural Convection Cooled Parallel Plates." *Journal of Heat Transfer* 106 (1984), p. 116.

6. B. M. Berkovsky and V. K. Polevikov. "Numerical Study of Problems on High-Intensive Free Convection." In *Heat Transfer and Turbulent Buoyant Convection,* eds. D. B. Spalding and N. Afgan, pp. 443–445. Washington, DC: Hemisphere, 1977.

7. I. Catton. "Natural Convection in Enclosures." *Proceedings of Sixth International Heat Transfer Conference.* Toronto: Canada, 1978, Vol. 6, pp. 13–31.

8. T. Cebeci. "Laminar Free Convection Heat Transfer from the Outer Surface of a Vertical Slender Circular Cylinder." *Proceedings of Fifth International Heat Transfer Conference* paper NCI.4, 1974 pp. 15–19.

9. Y. A. Çengel and P. T. L. Zing. "Enhancement of Natural Convection Heat Transfer from Heat Sinks by Shrouding." *Proceedings of ASME/JSME Thermal Engineering Conference.* Honolulu: HA, 1987, Vol. 3, pp. 451–475.

10. S. W. Churchill. "A Comprehensive Correlating Equation for Laminar Assisting Forced and Free Convection." *AIChE Journal* 23 (1977), pp. 10–16.

11. S. W. Churchill. "Free Convection around Immersed Bodies." In *Heat Exchanger Design Handbook,* ed. E. U. Schlünder, Section 2.5.7. New York: Hemisphere, 1983.

12. S. W. Churchill. "Combined Free and Forced Convection around Immersed Bodies." In *Heat Exchanger Design Handbook,* Section 2.5.9. New York: Hemisphere Publishing, 1986.

13. S. W. Churchill and H. H. S. Chu. "Correlating Equations for Laminar and Turbulent Free Convection from a Horizontal Cylinder." *International Journal of Heat Mass Transfer* 18 (1975), p. 1049.

14. S. W. Churchill and H. H. S. Chu. "Correlating Equations for Laminar and Turbulent Free Convection from a Vertical Plate." *International Journal of Heat Mass Transfer* 18 (1975), p. 1323.

15. E. R. G. Eckert and E. Soehngen. "Studies on Heat Transfer in Laminar Free Convection with Zehnder–Mach Interferometer." USAF Technical Report 5747, December 1948.

16. E. R. G. Eckert and E. Soehngen. "Interferometric Studies on the Stability and Transition to Turbulence of a Free Convection Boundary Layer." *Proceedings of General Discussion, Heat Transfer ASME-IME,* London, 1951.

17. S. M. ElSherbiny, G. D. Raithby, and K. G. T. Hollands. "Heat Transfer by Natural Convection Across Vertical and Inclined Air Layers. *Journal of Heat Transfer* 104 (1982), pp. 96–102.

18. T. Fujiii and H. Imura. "Natural Convection Heat Transfer from a Plate with Arbitrary Inclination." *International Journal of Heat Mass Transfer* 15 (1972), p. 755.

19. K. G. T. Hollands, T. E. Unny, G. D. Raithby, and L. Konicek. "Free Convective Heat Transfer Across Inclined Air Layers." *Journal of Heat Transfer* 98 (1976), pp. 182–193.

20. M. Jakob. *Heat Transfer.* New York: Wiley, 1949.

21. W. M. Kays and M. E. Crawford. *Convective Heat and Mass Transfer.* 3rd ed. New York: McGraw-Hill, 1993.

22. Reprinted from J. R. Lloyd and E. M. Sparrows. "Combined Force and Free Convection Flow on Vertical Surfaces." *International Journal of Heat Mass Transfer* 13 copyright 1970, with permission from Elsevier.

23. R. K. MacGregor and A. P. Emery. "Free Convection Through Vertical Plane Layers: Moderate and High Prandtl Number Fluids." *Journal of Heat Transfer* 91 (1969), p. 391.

24. S. Ostrach. "An Analysis of Laminar Free Convection Flow and Heat Transfer About a Flat Plate Parallel to the Direction of the Generating Body Force." National Advisory Committee for Aeronautics, Report 1111, 1953.

25. G. D. Raithby and K. G. T. Hollands. "A General Method of Obtaining Approximate Solutions to Laminar and Turbulent Free Convection Problems." In *Advances in Heat Transfer,* ed. F. Irvine and J. P. Hartnett, Vol. II, pp. 265–315. New York: Academic Press, 1975.

26. E. M. Sparrow and J. L. Gregg. "Laminar Free Convection from a Vertical Flat Plate." *Transactions of the ASME* 78 (1956), p. 438.

27. E. M. Sparrow and J. L. Gregg. "Laminar Free Convection Heat Transfer from the Outer Surface of a Vertical Circular Cylinder." *ASME* 78 (1956), p. 1823.

28. E. M. Sparrow and C. Prakash. "Enhancement of Natural Convection Heat Transfer by a Staggered Array of Vertical Plates." *Journal of Heat Transfer* 102 (1980), pp. 215–220.

29. E. M. Sparrow and S. B. Vemuri. "Natural Convection/Radiation Heat Transfer from Highly Populated Pin Fin Arrays." *Journal of Heat Transfer* 107 (1985), pp. 190–197.

PROBLEMS*

Physical Mechanism of Natural Convection

9–1C What is natural convection? How does it differ from forced convection? What force causes natural convection currents?

9–2C In which mode of heat transfer is the convection heat transfer coefficient usually higher, natural convection or forced convection? Why?

9–3C Consider a hot boiled egg in a spacecraft that is filled with air at atmospheric pressure and temperature at all times. Will the egg cool faster or slower when the spacecraft is in space instead of on the ground? Explain.

9–4C What is buoyancy force? Compare the relative magnitudes of the buoyancy force acting on a body immersed in these mediums: (*a*) air, (*b*) water, (*c*) mercury, and (*d*) an evacuated chamber.

9–5C When will the hull of a ship sink in water deeper: when the ship is sailing in fresh water or in seawater? Why?

9–6C A person weighs himself on a waterproof spring scale placed at the bottom of a 1-m-deep swimming pool. Will the person weigh more or less in water? Why?

9–7C Consider two fluids, one with a large coefficient of volume expansion and the other with a small one. In what fluid will a hot surface initiate stronger natural convection currents? Why? Assume the viscosity of the fluids to be the same.

9–8C Consider a fluid whose volume does not change with temperature at constant pressure. What can you say about natural convection heat transfer in this medium?

9–9C What do the lines on an interferometer photograph represent? What do closely packed lines on the same photograph represent?

9–10C Physically, what does the Grashof number represent? How does the Grashof number differ from the Reynolds number?

9–11 Show that the volume expansion coefficient of an ideal gas is $\beta = 1/T$, where T is the absolute temperature.

Natural Convection over Surfaces

9–12C How does the Rayleigh number differ from the Grashof number?

9–13C Under what conditions can the outer surface of a vertical cylinder be treated as a vertical plate in natural convection calculations?

9–14C Will a hot horizontal plate whose back side is insulated cool faster or slower when its hot surface is facing down instead of up?

9–15C Consider laminar natural convection from a vertical hot-plate. Will the heat flux be higher at the top or at the bottom of the plate? Why?

9–16 Consider a thin 16-cm-long and 20-cm-wide horizontal plate suspended in air at 20°C. The plate is equipped with electric resistance heating elements with a rating of 20 W. Now the heater is turned on and the plate temperature rises. Determine the temperature of the plate when steady operating conditions are reached. The plate has an emissivity of 0.90 and the surrounding surfaces are at 17°C.

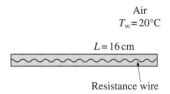

Air
$T_\infty = 20°C$

$L = 16$ cm

Resistance wire

FIGURE P9–16

9–17 Flue gases from an incinerator are released to atmosphere using a stack that is 0.6 m in diameter and 10.0 m high. The outer surface of the stack is at 40°C and the surrounding air is at 10°C. Determine the rate of heat transfer from the stack assuming (*a*) there is no wind and (*b*) the stack is exposed to 20 km/h winds.

9–18 Thermal energy generated by the electrical resistance of a 5-mm-diameter and 4-m-long bare cable is dissipated to the surrounding air at 20°C. The voltage drop and the electric current across the cable in steady operation are measured to be 60 V and 1.5 A, respectively. Disregarding radiation, estimate the surface temperature of the cable.

9–19 A 10-m-long section of a 6-cm-diameter horizontal hot-water pipe passes through a large room whose temperature is 27°C. If the temperature and the emissivity of the outer surface of the pipe are 73°C and 0.8, respectively, determine the rate of heat loss from the pipe by (*a*) natural convection and (*b*) radiation.

9–20 Consider a wall-mounted power transistor that dissipates 0.18 W of power in an environment at 35°C. The transistor is 0.45 cm long and has a diameter of 0.4 cm. The emissivity of the outer surface of the transistor is 0.1, and the average temperature of the surrounding surfaces is 25°C. Disregarding any heat transfer from the base surface, determine

---

*Problems designated by a "C" are concept questions, and students are encouraged to answer them all. Problems designated by an "E" are in English units, and the SI users can ignore them. Problems with the icon ⊛ are solved using EES, and complete solutions together with parametric studies are included on the enclosed CD. Problems with the icon ▣ are comprehensive in nature, and are intended to be solved with a computer, preferably using the EES software that accompanies this text.

the surface temperature of the transistor. Use air properties at 100°C. *Answer:* 183°C

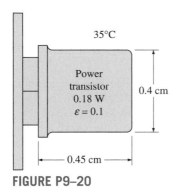

35°C

Power transistor
0.18 W
$\varepsilon = 0.1$

0.4 cm

0.45 cm

FIGURE P9–20

9–21 Reconsider Prob. 9–20. Using EES (or other) software, investigate the effect of ambient temperature on the surface temperature of the transistor. Let the environment temperature vary from 10°C to 40°C and assume that the surrounding surfaces are 10°C colder than the environment temperature. Plot the surface temperature of the transistor versus the environment temperature, and discuss the results.

9–22E Consider a 2-ft × 2-ft thin square plate in a room at 75°F. One side of the plate is maintained at a temperature of 130°F, while the other side is insulated. Determine the rate of heat transfer from the plate by natural convection if the plate is (a) vertical; (b) horizontal with hot surface facing up; and (c) horizontal with hot surface facing down.

9–23E Reconsider Prob. 9–22E. Using EES (or other) software, plot the rate of natural convection heat transfer for different orientations of the plate as a function of the plate temperature as the temperature varies from 80°F to 180°F, and discuss the results.

9–24 A 300-W cylindrical resistance heater is 0.75 m long and 0.5 cm in diameter. The resistance wire is placed horizontally in a fluid at 20°C. Determine the outer surface temperature of the resistance wire in steady operation if the fluid is (a) air and (b) water. Ignore any heat transfer by radiation. Use properties at 500°C for air and 40°C for water.

9–25 Water is boiling in a 12-cm-deep pan with an outer diameter of 25 cm that is placed on top of a stove. The ambient air and the surrounding surfaces are at a temperature of 25°C, and the emissivity of the outer surface of the pan is 0.80. Assuming the entire pan to be at an average temperature of 98°C, determine the rate of heat loss from the cylindrical side surface of the pan to the surroundings by (a) natural convection and (b) radiation. (c) If water is boiling at a rate of 1.5 kg/h at 100°C, determine the ratio of the heat lost from the side surfaces of the pan to that by the evaporation of water. The enthalpy of vaporization of water at 100°C is 2257 kJ/kg. *Answers:* 46.2 W, 47.3 W, 0.099

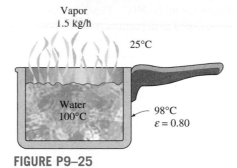

Vapor
1.5 kg/h

25°C

Water
100°C

98°C
$\varepsilon = 0.80$

FIGURE P9–25

9–26 Repeat Prob. 9–25 for a pan whose outer surface is polished and has an emissivity of 0.1.

9–27 In a plant that manufactures canned aerosol paints, the cans are temperature-tested in water baths at 55°C before they are shipped to ensure that they withstand temperatures up to 55°C during transportation and shelving. The cans, moving on a conveyor, enter the open hot water bath, which is 0.5 m deep, 1 m wide, and 3.5 m long, and move slowly in the hot water toward the other end. Some of the cans fail the test and explode in the water bath. The water container is made of sheet metal, and the entire container is at about the same temperature as the hot water. The emissivity of the outer surface of the container is 0.7. If the temperature of the surrounding air and surfaces is 20°C, determine the rate of heat loss from the four side surfaces of the container (disregard the top surface, which is open).

The water is heated electrically by resistance heaters, and the cost of electricity is $0.085/kWh. If the plant operates 24 h a day 365 days a year and thus 8760 h a year, determine the annual cost of the heat losses from the container for this facility.

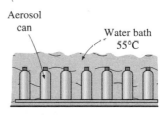

Aerosol
can

Water bath
55°C

FIGURE P9–27

9–28 Reconsider Prob. 9–27. In order to reduce the heating cost of the hot water, it is proposed to insulate the side and bottom surfaces of the container with 5-cm-thick fiberglass insulation ($k = 0.035$ W/m · °C) and to wrap the insulation with aluminum foil ($\varepsilon = 0.1$) in order to minimize the heat loss by radiation. An estimate is obtained from a local insulation contractor, who proposes to do the insulation job for $350, including materials and labor. Would you support this proposal? How long will it take for the insulation to pay for itself from the energy it saves?

9–29 Consider a 15-cm × 20-cm printed circuit board (PCB) that has electronic components on one side. The board is placed in a room at 20°C. The heat loss from the back surface of the board is negligible. If the circuit board is dissipating 8 W of power in steady operation, determine the average temperature of the hot surface of the board, assuming the board is (*a*) vertical; (*b*) horizontal with hot surface facing up; and (*c*) horizontal with hot surface facing down. Take the emissivity of the surface of the board to be 0.8 and assume the surrounding surfaces to be at the same temperature as the air in the room. *Answers:* (*a*) 46.6°C, (*b*) 42.6°C, (*c*) 50.3°C

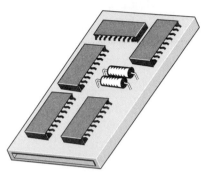

FIGURE P9–29

9–30 [EES] Reconsider Prob. 9–29. Using EES (or other) software, investigate the effects of the room temperature and the emissivity of the board on the temperature of the hot surface of the board for different orientations of the board. Let the room temperature vary from 5°C to 35°C and the emissivity from 0.1 to 1.0. Plot the hot surface temperature for different orientations of the board as the functions of the room temperature and the emissivity, and discuss the results.

9–31 [CD/EES] A manufacturer makes absorber plates that are 1.2 m × 0.8 m in size for use in solar collectors. The back side of the plate is heavily insulated, while its front surface is coated with black chrome, which has an absorptivity of 0.87 for solar radiation and an emissivity of 0.09. Consider such a plate placed horizontally outdoors in calm air at 25°C. Solar radiation is incident on the plate at a rate of 700 W/m². Taking the effective sky temperature to be 10°C, determine the equilibrium temperature of the absorber plate.

What would your answer be if the absorber plate is made of ordinary aluminum plate that has a solar absorptivity of 0.28 and an emissivity of 0.07?

9–32 Repeat Prob. 9–31 for an aluminum plate painted flat black (solar absorptivity 0.98 and emissivity 0.98) and also for a plate painted white (solar absorptivity 0.26 and emissivity 0.90).

9–33 The following experiment is conducted to determine the natural convection heat transfer coefficient for a horizontal cylinder that is 80 cm long and 2 cm in diameter. A 80-cm-long resistance heater is placed along the centerline of the cylinder, and the surfaces of the cylinder are polished to minimize the radiation effect. The two circular side surfaces of the cylinder are well insulated. The resistance heater is turned on, and the power dissipation is maintained constant at 60 W. If the average surface temperature of the cylinder is measured to be 120°C in the 20°C room air when steady operation is reached, determine the natural convection heat transfer coefficient. If the emissivity of the outer surface of the cylinder is 0.1 and a 5 percent error is acceptable, do you think we need to do any correction for the radiation effect? Assume the surrounding surfaces to be at 20°C also.

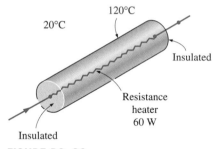

FIGURE P9–33

9–34 Thick fluids such as asphalt and waxes and the pipes in which they flow are often heated in order to reduce the viscosity of the fluids and thus to reduce the pumping costs. Consider the flow of such a fluid through a 100-m-long pipe of outer diameter 30 cm in calm ambient air at 0°C. The pipe is heated electrically, and a thermostat keeps the outer surface temperature of the pipe constant at 25°C. The emissivity of the outer surface of the

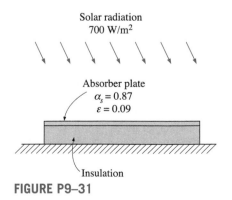

Solar radiation
700 W/m²

Absorber plate
$\alpha_s = 0.87$
$\varepsilon = 0.09$

Insulation

FIGURE P9–31

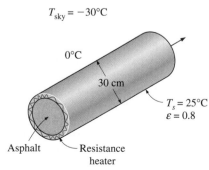

$T_{sky} = -30°C$

0°C

30 cm

$T_s = 25°C$
$\varepsilon = 0.8$

Asphalt   Resistance heater

FIGURE P9–34

pipe is 0.8, and the effective sky temperature is −30°C. Determine the power rating of the electric resistance heater, in kW, that needs to be used. Also, determine the cost of electricity associated with heating the pipe during a 10-h period under the above conditions if the price of electricity is $0.09/kWh.

Answers: 29.1 kW, $26.2

9–35 Reconsider Prob. 9–34. To reduce the heating cost of the pipe, it is proposed to insulate it with sufficiently thick fiberglass insulation ($k = 0.035$ W/m · °C) wrapped with aluminum foil ($\varepsilon = 0.1$) to cut down the heat losses by 85 percent. Assuming the pipe temperature to remain constant at 25°C, determine the thickness of the insulation that needs to be used. How much money will the insulation save during this 10-h period? *Answers:* 1.3 cm, $22.3

9–36E Consider an industrial furnace that resembles a 13-ft-long horizontal cylindrical enclosure 8 ft in diameter whose end surfaces are well insulated. The furnace burns natural gas at a rate of 48 therms/h. The combustion efficiency of the furnace is 82 percent (i.e., 18 percent of the chemical energy of the fuel is lost through the flue gases as a result of incomplete combustion and the flue gases leaving the furnace at high temperature). If the heat loss from the outer surfaces of the furnace by natural convection and radiation is not to exceed 1 percent of the heat generated inside, determine the highest allowable surface temperature of the furnace. Assume the air and wall surface temperature of the room to be 75°F, and take the emissivity of the outer surface of the furnace to be 0.85. If the cost of natural gas is $1.15/therm and the furnace operates 2800 h per year, determine the annual cost of this heat loss to the plant.

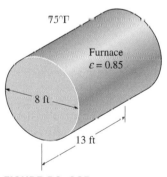

75°F

Furnace
$\varepsilon = 0.85$

8 ft

13 ft

FIGURE P9–36E

9–37 Consider a 1.2-m-high and 2-m-wide glass window with a thickness of 6 mm, thermal conductivity $k = 0.78$ W/m · °C, and emissivity $\varepsilon = 0.9$. The room and the walls that face the window are maintained at 25°C, and the average temperature of the inner surface of the window is measured to be 5°C. If the temperature of the outdoors is −5°C, determine (*a*) the convection heat transfer coefficient on the inner surface of the window, (*b*) the rate of total heat transfer through the window, and (*c*) the combined natural convection and radiation heat transfer coefficient on the outer

surface of the window. Is it reasonable to neglect the thermal resistance of the glass in this case?

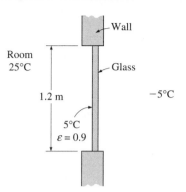

Room
25°C

Wall

Glass

1.2 m

−5°C

5°C
$\varepsilon = 0.9$

FIGURE P9–37

9–38 A 3-mm-diameter and 12-m-long electric wire is tightly wrapped with a 1.5-mm-thick plastic cover whose thermal conductivity and emissivity are $k = 0.20$ W/m · °C and $\varepsilon = 0.9$. Electrical measurements indicate that a current of 10 A passes through the wire and there is a voltage drop of 7 V along the wire. If the insulated wire is exposed to calm atmospheric air at $T_\infty = 30$°C, determine the temperature at the interface of the wire and the plastic cover in steady operation. Take the surrounding surfaces to be at about the same temperature as the air.

9–39 During a visit to a plastic sheeting plant, it was observed that a 60-m-long section of a 2-in nominal (6.03-cm-outer-diameter) steam pipe extended from one end of the plant to the other with no insulation on it. The temperature measurements at several locations revealed that the average temperature of the exposed surfaces of the steam pipe was 170°C, while the temperature of the surrounding air was 20°C. The outer surface of the pipe appeared to be oxidized, and its emissivity can be taken to be 0.7. Taking the temperature of the surrounding surfaces to be 20°C also, determine the rate of heat loss from the steam pipe.

Steam is generated in a gas furnace that has an efficiency of 78 percent, and the plant pays $1.10 per therm (1 therm = 105,500 kJ) of natural gas. The plant operates 24 h a day 365 days a year, and thus 8760 h a year. Determine the annual cost of the heat losses from the steam pipe for this facility.

170°C
$\varepsilon = 0.7$

20°C

6.03 cm

60 m

Steam

FIGURE P9–39

9–40 Reconsider Prob. 9–39. Using EES (or other) software, investigate the effect of the surface temperature of the steam pipe on the rate of heat loss from the pipe and the annual cost of this heat loss. Let the surface temperature vary from 100°C to 200°C. Plot the rate of heat loss and the annual cost as a function of the surface temperature, and discuss the results.

9–41 Reconsider Prob. 9–39. In order to reduce heat losses, it is proposed to insulate the steam pipe with 5-cm-thick fiberglass insulation ($k = 0.038$ W/m · °C) and to wrap it with aluminum foil ($\varepsilon = 0.1$) in order to minimize the radiation losses. Also, an estimate is obtained from a local insulation contractor, who proposed to do the insulation job for $750, including materials and labor. Would you support this proposal? How long will it take for the insulation to pay for itself from the energy it saves? Assume the temperature of the steam pipe to remain constant at 170°C.

9–42 A 50-cm × 50-cm circuit board that contains 121 square chips on one side is to be cooled by combined natural convection and radiation by mounting it on a vertical surface in a room at 25°C. Each chip dissipates 0.18 W of power, and the emissivity of the chip surfaces is 0.7. Assuming the heat transfer from the back side of the circuit board to be negligible, and the temperature of the surrounding surfaces to be the same as the air temperature of the room, determine the surface temperature of the chips.   *Answer:* 36.2°C

9–43 Repeat Prob. 9–42 assuming the circuit board to be positioned horizontally with (*a*) chips facing up and (*b*) chips facing down.

9–44 The side surfaces of a 2-m-high cubic industrial furnace burning natural gas are not insulated, and the temperature at the outer surface of this section is measured to be 110°C. The temperature of the furnace room, including its surfaces, is 30°C, and the emissivity of the outer surface of the furnace is 0.7. It is proposed that this section of the furnace wall be insulated with glass wool insulation ($k = 0.038$ W/m · °C) wrapped by a reflective sheet ($\varepsilon = 0.2$) in order to reduce the

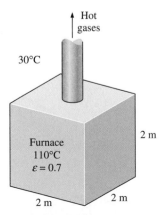

FIGURE P9–44

heat loss by 90 percent. Assuming the outer surface temperature of the metal section still remains at about 110°C, determine the thickness of the insulation that needs to be used.

The furnace operates continuously throughout the year and has an efficiency of 78 percent. The price of the natural gas is $0.55/therm (1 therm = 105,500 kJ of energy content). If the installation of the insulation will cost $550 for materials and labor, determine how long it will take for the insulation to pay for itself from the energy it saves.

9–45 A 1.5-m-diameter, 4-m-long cylindrical propane tank is initially filled with liquid propane, whose density is 581 kg/m³. The tank is exposed to the ambient air at 25°C in calm weather. The outer surface of the tank is polished so that the radiation heat transfer is negligible. Now a crack develops at the top of the tank, and the pressure inside drops to 1 atm while the temperature drops to −42°C, which is the boiling temperature of propane at 1 atm. The heat of vaporization of propane at 1 atm is 425 kJ/kg. The propane is slowly vaporized as a result of the heat transfer from the ambient air into the tank, and the propane vapor escapes the tank at −42°C through the crack. Assuming the propane tank to be at about the same temperature as the propane inside at all times, determine how long it will take for the tank to empty if it is not insulated.

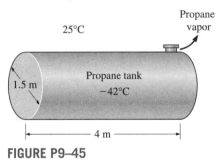

FIGURE P9–45

9–46E An average person generates heat at a rate of 240 Btu/h while resting in a room at 70°F. Assuming one-quarter of this heat is lost from the head and taking the emissivity of the skin to be 0.9, determine the average surface temperature of the head when it is not covered. The head can be approximated as a 12-in-diameter sphere, and the interior surfaces of the room can be assumed to be at the room temperature.

9–47 An incandescent lightbulb is an inexpensive but highly inefficient device that converts electrical energy into light. It converts about 10 percent of the electrical energy it consumes into light while converting the remaining 90 percent into heat. The glass bulb of the lamp heats up very quickly as a result of absorbing all that heat and dissipating it to the surroundings by convection and radiation. Consider an 8-cm-diameter 60-W lightbulb in a room at 25°C. The emissivity of the glass is 0.9. Assuming that 10 percent of the energy passes through the glass bulb as light with negligible absorption and the rest of the energy is absorbed and dissipated by the bulb itself by natural convection and radiation, determine the equilibrium temperature of the glass bulb. Assume the interior surfaces of the room to be at room temperature. *Answer:* 169°C

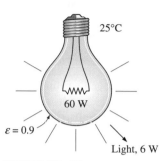

25°C

60 W

$\varepsilon = 0.9$

Light, 6 W

FIGURE P9–47

9–48 A 40-cm-diameter, 110-cm-high cylindrical hot-water tank is located in the bathroom of a house maintained at 20°C. The surface temperature of the tank is measured to be 44°C and its emissivity is 0.4. Taking the surrounding surface temperature to be also 20°C, determine the rate of heat loss from all surfaces of the tank by natural convection and radiation.

9–49 A 28-cm-high, 18-cm-long, and 18-cm-wide rectangular container suspended in a room at 24°C is initially filled with cold water at 2°C. The surface temperature of the container is observed to be nearly the same as the water temperature inside. The emissivity of the container surface is 0.6, and the temperature of the surrounding surfaces is about the same as the air temperature. Determine the water temperature in the container after 3 h, and the average rate of heat transfer to the water. Assume the heat transfer coefficient on the top and bottom surfaces to be the same as that on the side surfaces.

9–50 Reconsider Prob. 9–49. Using EES (or other) software, plot the water temperature in the container as a function of the heating time as the time varies from 30 min to 10 h, and discuss the results.

9–51 A room is to be heated by a coal-burning stove, which is a cylindrical cavity with an outer diameter of 32 cm and a height of 70 cm. The rate of heat loss from the room is estimated to be 1.5 kW when the air temperature in the room is maintained constant at 24°C. The emissivity of the stove surface is 0.85 and the average temperature of the surrounding wall surfaces is 14°C. Determine the surface temperature of the stove. Neglect the transfer from the bottom surface and take the heat transfer coefficient at the top surface to be the same as that on the side surface.

The heating value of the coal is 30,000 kJ/kg, and the combustion efficiency is 65 percent. Determine the amount of coal burned a day if the stove operates 14 h a day.

9–52 The water in a 40-L tank is to be heated from 15°C to 45°C by a 6-cm-diameter spherical heater whose surface temperature is maintained at 85°C. Determine how long the heater should be kept on.

Natural Convection from Finned Surfaces and PCBs

9–53C Why are finned surfaces frequently used in practice? Why are the finned surfaces referred to as heat sinks in the electronics industry?

9–54C Why are heat sinks with closely packed fins not suitable for natural convection heat transfer, although they increase the heat transfer surface area more?

9–55C Consider a heat sink with optimum fin spacing. Explain how heat transfer from this heat sink will be affected by (*a*) removing some of the fins on the heat sink and (*b*) doubling the number of fins on the heat sink by reducing the fin spacing. The base area of the heat sink remains unchanged at all times.

9–56 Aluminum heat sinks of rectangular profile are commonly used to cool electronic components. Consider a 7.62-cm-long and 9.68-cm-wide commercially available heat sink whose cross section and dimensions are as shown in Fig. P9–56. The heat sink is oriented vertically and is used to cool a power transistor that can dissipate up to 125 W of power. The back surface of the heat sink is insulated. The surfaces of the heat sink are untreated, and thus they have a low emissivity (under 0.1). Therefore, radiation heat transfer from the heat sink can be neglected. During an experiment conducted in room air at 22°C, the base temperature of the heat sink was measured to be 120°C when the power dissipation of the transistor was 15 W. Assuming the entire heat sink to be at the base temperature, determine the average natural convection heat transfer coefficient for this case. *Answer:* 7.13 W/m² · °C

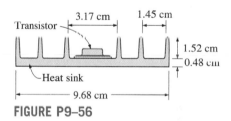

Transistor

3.17 cm   1.45 cm

1.52 cm
0.48 cm

Heat sink

9.68 cm

FIGURE P9–56

9–57 Reconsider the heat sink in Prob. 9–56. In order to enhance heat transfer, a shroud (a thin rectangular metal plate) whose surface area is equal to the base area of the heat sink is

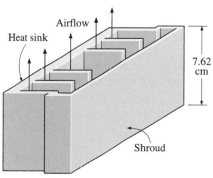

Airflow

Heat sink

7.62 cm

Shroud

FIGURE P9–57

placed very close to the tips of the fins such that the interfin spaces are converted into rectangular channels. The base temperature of the heat sink in this case was measured to be 108°C. Noting that the shroud loses heat to the ambient air from both sides, determine the average natural convection heat transfer coefficient in this shrouded case. (For complete details, see Çengel and Zing, 1987.)

9–58E A 6-in-wide and 8-in-high vertical hot surface in 78°F air is to be cooled by a heat sink with equally spaced fins of rectangular profile. The fins are 0.08 in thick and 8 in long in the vertical direction and have a height of 1.2 in from the base. Determine the optimum fin spacing and the rate of heat transfer by natural convection from the heat sink if the base temperature is 180°F.

9–59E [EES] Reconsider Prob. 9–58E. Using EES (or other) software, investigate the effect of the length of the fins in the vertical direction on the optimum fin spacing and the rate of heat transfer by natural convection. Let the fin length vary from 2 in to 10 in. Plot the optimum fin spacing and the rate of convection heat transfer as a function of the fin length, and discuss the results.

9–60 A 15-cm-wide and 18-cm-high vertical hot surface in 25°C air is to be cooled by a heat sink with equally spaced fins of rectangular profile. The fins are 0.1 cm thick and 18 cm long in the vertical direction. Determine the optimum fin height and the rate of heat transfer by natural convection from the heat sink if the base temperature is 85°C.

The criteria for optimum fin height H in the literature is given by $H = \sqrt{hA_c/pk}$. Take the thermal conductivity of fin material to be 177 W/m · °C

Natural Convection inside Enclosures

9–61C The upper and lower compartments of a well-insulated container are separated by two parallel sheets of glass with an air space between them. One of the compartments is to be filled with a hot fluid and the other with a cold fluid. If it is desired that heat transfer between the two compartments be minimal, would you recommend putting the hot fluid into the upper or the lower compartment of the container? Why?

9–62C Someone claims that the air space in a double-pane window enhances the heat transfer from a house because of the natural convection currents that occur in the air space and recommends that the double-pane window be replaced by a single sheet of glass whose thickness is equal to the sum of the thicknesses of the two glasses of the double-pane window to save energy. Do you agree with this claim?

9–63C Consider a double-pane window consisting of two glass sheets separated by a 1-cm-wide air space. Someone suggests inserting a thin vinyl sheet in the middle of the two glasses to form two 0.5-cm-wide compartments in the window in order to reduce natural convection heat transfer through the window. From a heat transfer point of view, would you be in favor of this idea to reduce heat losses through the window?

9–64C What does the effective conductivity of an enclosure represent? How is the ratio of the effective conductivity to thermal conductivity related to the Nusselt number?

9–65 Show that the thermal resistance of a rectangular enclosure can be expressed as $R = L_c/(Ak \, \mathrm{Nu})$, where k is the thermal conductivity of the fluid in the enclosure.

9–66 Determine the U-factors for the center-of-glass section of a double-pane window and a triple-pane window. The heat transfer coefficients on the inside and outside surfaces are 6 and 25 W/m² · °C, respectively. The thickness of the air layer is 1.5 cm and there are two such air layers in triple-pane window. The Nusselt number across an air layer is estimated to be 1.2. Take the thermal conductivity of air to be 0.025 W/m · °C and neglect the thermal resistance of glass sheets. Also, assume that the effect of radiation through the air space is of the same magnitude as the convection.

Considering that about 70 percent of total heat transfer through a window is due to center-of-glass section, estimate the percentage decrease in total heat transfer when triple-pane window is used in place of double-pane window.

9–67 A vertical 1.5-m-high and 3.0-m-wide enclosure consists of two surfaces separated by a 0.4-m air gap at atmospheric pressure. If the surface temperatures across the air gap are measured to be 280 K and 336 K and the surface emissivities to be 0.15 and 0.90, determine the fraction of heat transferred through the enclosure by radiation. *Answer:* 0.30

9–68E A vertical 4-ft-high and 6-ft-wide double-pane window consists of two sheets of glass separated by a 1-in air gap at atmospheric pressure. If the glass surface temperatures across the air gap are measured to be 65°F and 40°F, determine the rate of heat transfer through the window by (*a*) natural convection and (*b*) radiation. Also, determine the R-value of insulation of this window such that multiplying the inverse of the R-value by the surface area and the temperature difference gives the total rate of heat transfer through the window. The effective emissivity for use in radiation calculations between two large parallel glass plates can be taken to be 0.82.

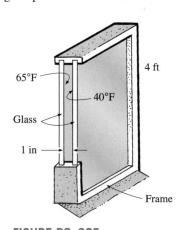

65°F
40°F
4 ft
Glass
1 in
Frame

FIGURE P9–68E

9–69E Reconsider Prob. 9–68E. Using EES (or other) software, investigate the effect of the air gap thickness on the rates of heat transfer by natural convection and radiation, and the R-value of insulation. Let the air gap thickness vary from 0.2 in to 2.0 in. Plot the rates of heat transfer by natural convection and radiation, and the R-value of insulation as a function of the air gap thickness, and discuss the results.

9–70 Two concentric spheres of diameters 15 cm and 25 cm are separated by air at 1 atm pressure. The surface temperatures of the two spheres enclosing the air are $T_1 = 350$ K and $T_2 = 275$ K, respectively. Determine the rate of heat transfer from the inner sphere to the outer sphere by natural convection.

9–71 Reconsider Prob. 9–70. Using EES (or other) software, plot the rate of natural convection heat transfer as a function of the hot surface temperature of the sphere as the temperature varies from 300 K to 500 K, and discuss the results.

9–72 Flat-plate solar collectors are often tilted up toward the sun in order to intercept a greater amount of direct solar radiation. The tilt angle from the horizontal also affects the rate of heat loss from the collector. Consider a 1.5-m-high and 3-m-wide solar collector that is tilted at an angle θ from the horizontal. The back side of the absorber is heavily insulated. The absorber plate and the glass cover, which are spaced 2.5 cm from each other, are maintained at temperatures of 80°C and 40°C, respectively. Determine the rate of heat loss from the absorber plate by natural convection for $\theta = 0°$, 30°, and 90°.

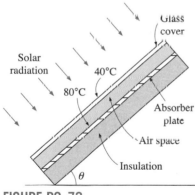

FIGURE P9–72

9–73 A simple solar collector is built by placing a 5-cm-diameter clear plastic tube around a garden hose whose outer diameter is 1.6 cm. The hose is painted black to maximize solar absorption, and some plastic rings are used to keep the spacing between the hose and the clear plastic cover constant. During a clear day, the temperature of the hose is measured to be 65°C, while the ambient air temperature is 26°C. Determine the rate of heat loss from the water in the hose per meter of its length by natural convection. Also, discuss how the performance of this solar collector can be improved.   *Answer: 8.2 W*

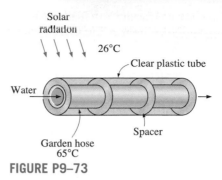

FIGURE P9–73

9–74 Reconsider Prob. 9–73. Using EES (or other) software, plot the rate of heat loss from the water by natural convection as a function of the ambient air temperature as the temperature varies from 4°C to 40°C, and discuss the results.

9–75 A vertical 1.3-m-high, 2.8-m-wide double-pane window consists of two layers of glass separated by a 2.2-cm air gap at atmospheric pressure. The room temperature is 26°C while the inner glass temperature is 18°C. Disregarding radiation heat transfer, determine the temperature of the outer glass layer and the rate of heat loss through the window by natural convection.

9–76 Consider two concentric horizontal cylinders of diameters 55 cm and 65 cm, and length 125 cm. The surfaces of the inner and outer cylinders are maintained at 54°C and 106°C, respectively. Determine the rate of heat transfer between the cylinders by natural convection if the annular space is filled with (*a*) water and (*b*) air.

Combined Natural and Forced Convection

9–77C When is natural convection negligible and when is it not negligible in forced convection heat transfer?

9–78C Under what conditions does natural convection enhance forced convection, and under what conditions does it hurt forced convection?

9–79C When neither natural nor forced convection is negligible, is it correct to calculate each independently and add them to determine the total convection heat transfer?

9–80 Consider a 5-m-long vertical plate at 85°C in air at 30°C. Determine the forced motion velocity above which natural convection heat transfer from this plate is negligible.
  Answer: 9.04 m/s

9–81 Reconsider Prob. 9–80. Using EES (or other) software, plot the forced motion velocity above which natural convection heat transfer is negligible as a function of the plate temperature as the temperature varies from 50°C to 150°C, and discuss the results.

9–82 Consider a 5-m-long vertical plate at 60°C in water at 25°C. Determine the forced motion velocity above which natural convection heat transfer from this plate is negligible.

9–83 In a production facility, thin square plates 2 m × 2 m in size coming out of the oven at 270°C are cooled by blowing ambient air at 18°C horizontally parallel to their surfaces. Determine the air velocity above which the natural convection effects on heat transfer are less than 10 percent and thus are negligible.

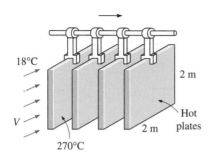

FIGURE P9–83

9–84 A 12-cm-high and 20-cm-wide circuit board houses 100 closely spaced logic chips on its surface, each dissipating 0.05 W. The board is cooled by a fan that blows air over the hot surface of the board at 35°C at a velocity of 0.5 m/s. The heat transfer from the back surface of the board is negligible. Determine the average temperature on the surface of the circuit board assuming the air flows vertically upward along the 12-cm-long side by (a) ignoring natural convection and (b) considering the contribution of natural convection. Disregard any heat transfer by radiation.

Special Topic: Heat Transfer through Windows

9–85C Why are the windows considered in three regions when analyzing heat transfer through them? Name those regions and explain how the overall U-value of the window is determined when the heat transfer coefficients for all three regions are known.

9–86C Consider three similar double-pane windows with air gap widths of 5, 10, and 20 mm. For which case will the heat transfer through the window will be a minimum?

9–87C In an ordinary double-pane window, about half of the heat transfer is by radiation. Describe a practical way of reducing the radiation component of heat transfer.

9–88C Consider a double-pane window whose air space width is 20 mm. Now a thin polyester film is used to divide the air space into two 10-mm-wide layers. How will the film affect (a) convection and (b) radiation heat transfer through the window?

9–89C Consider a double-pane window whose air space is flashed and filled with argon gas. How will replacing the air in the gap by argon affect (a) convection and (b) radiation heat transfer through the window?

9–90C Is the heat transfer rate through the glazing of a double-pane window higher at the center or edge section of the glass area? Explain.

9–91C How do the relative magnitudes of U-factors of windows with aluminum, wood, and vinyl frames compare? Assume the windows are identical except for the frames.

9–92 Determine the U-factor for the center-of-glass section of a double-pane window with a 13-mm air space for winter design conditions. The glazings are made of clear glass having an emissivity of 0.84. Take the average air space temperature at design conditions to be 10°C and the temperature difference across the air space to be 15°C.

9–93 A double-door wood-framed window with glass glazing and metal spacers is being considered for an opening that is 1.2 m high and 1.8 m wide in the wall of a house maintained at 20°C. Determine the rate of heat loss through the window and the inner surface temperature of the window glass facing the room when the outdoor air temperature is −8°C if the window is selected to be (a) 3-mm single glazing, (b) double glazing with an air space of 13 mm, and (c) low-e-coated triple glazing with an air space of 13 mm.

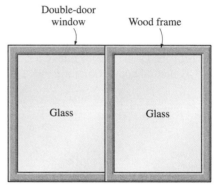

FIGURE P9–93

9–94 Determine the overall U-factor for a double-door-type wood-framed double-pane window with 13-mm air space and metal spacers, and compare your result to the value listed in Table 9–6. The overall dimensions of the window are 2.00 m × 2.40 m, and the dimensions of each glazing are 1.92 m × 1.14 m.

9–95 Consider a house in Atlanta, Georgia, that is maintained at 22°C and has a total of 14 m² of window area. The windows are double-door-type with wood frames and metal spacers. The glazing consists of two layers of glass with 12.7 mm of air space with one of the inner surfaces coated with reflective film. The winter average temperature of Atlanta is 11.3°C. Determine the average rate of heat loss through the windows in winter.
Answer: 319 W

9–96E Consider an ordinary house with R-13 walls (walls that have an R-value of 13 h · ft² · °F/Btu). Compare this to the

R-value of the common double-door windows that are double pane with $\frac{1}{4}$ in of air space and have aluminum frames. If the windows occupy only 20 percent of the wall area, determine if more heat is lost through the windows or through the remaining 80 percent of the wall area. Disregard infiltration losses.

9–97 The overall U-factor of a fixed wood-framed window with double glazing is given by the manufacturer to be $U = 2.76$ W/m$^2 \cdot$ °C under the conditions of still air inside and winds of 12 km/h outside. What will the U-factor be when the wind velocity outside is doubled? *Answer:* 2.88 W/m$^2 \cdot$ °C

9–98 The owner of an older house in Wichita, Kansas, is considering replacing the existing double-door type wood-framed single-pane windows with vinyl-framed double-pane windows with an air space of 6.4 mm. The new windows are of double-door type with metal spacers. The house is maintained at 22°C at all times, but heating is needed only when the outdoor temperature drops below 18°C because of the internal heat gain from people, lights, appliances, and the sun. The average winter temperature of Wichita is 7.1°C, and the house is heated by electric resistance heaters. If the unit cost of electricity is $0.085/kWh and the total window area of the house is 17 m^2, determine how much money the new windows will save the home owner per month in winter.

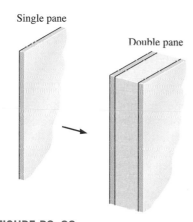

Single pane

Double pane

FIGURE P9–98

Review Problems

9–99 A 10-cm-diameter and 10-m-long cylinder with a surface temperature of 10°C is placed horizontally in air at 40°C. Calculate the steady rate of heat transfer for the cases of (a) free-stream air velocity of 10 m/s due to normal winds and (b) no winds and thus a free stream velocity of zero.

9–100 A spherical vessel, with 30.0-cm outside diameter, is used as a reactor for a slow endothermic reaction. The vessel is completely submerged in a large water-filled tank, held at a constant temperature of 30°C. The outside surface temperature of the vessel is 20°C. Calculate the rate of heat transfer in

steady operation for the following cases: (a) the water in the tank is still, (b) the water in the tank is still (as in a part a), however, the buoyancy force caused by the difference in water density is assumed to be negligible, and (c) the water in the tank is circulated at an average velocity of 20 cm/s.

9–101 A vertical cylindrical pressure vessel is 1.0 m in diameter and 3.0 m in height. Its outside average wall temperature is 60°C, while the surrounding air is at 0°C. Calculate the rate of heat loss from the vessel's cylindrical surface when there is (a) no wind and (b) a crosswind of 20 km/h.

9–102 Consider a solid sphere, 50 cm in diameter embedded with electrical heating elements such that its surface temperature is always maintained constant at 60°C. The sphere is placed in a large pool of oil held at a constant temperature of 20°C. Using the oil properties tabulated below, calculate the rate of heat transfer in steady operation for each of the following scenarios.

(a) Heat flow in the oil is assumed to occur only by conduction.
(b) The oil is circulated across the sphere at an average velocity of 1.50 m/s.
(c) The pump causing the oil circulation in part (b) has broken down.

T, °C	k, W/m·K	ρ, kg/m^3	c_p, J/kg·K	μ, mPa·s	β, K^{-1}
20.0	0.22	888.0	1880	10.0	0.00070
40.0	0.21	876.0	1965	7.0	0.00070
60.0	0.20	864.0	2050	4.0	0.00070

9–103E A 0.1-W small cylindrical resistor mounted on a lower part of a vertical circuit board is 0.3 in long and has a diameter of 0.2 in. The view of the resistor is largely blocked by another circuit board facing it, and the heat transfer through the connecting wires is negligible. The air is free to flow through the large parallel flow passages between the boards as a result of natural convection currents. If the air temperature at the vicinity of the resistor is 120°F, determine the approximate surface temperature of the resistor. *Answer:* 211°F

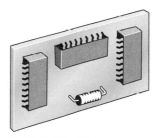

FIGURE P9–103E

9–104 An ice chest whose outer dimensions are 30 cm $\times$ 40 cm $\times$ 40 cm is made of 3-cm-thick Styrofoam ($k = 0.033$ W/m · °C). Initially, the chest is filled with 30 kg of ice at 0°C, and the inner surface temperature of the ice chest can be taken to be 0°C at all times. The heat of fusion of water at 0°C is 333.7 kJ/kg, and the surrounding ambient air is at 20°C. Disregarding any heat transfer from the 40 cm × 40 cm base of the ice chest, determine how long it will take for the ice in the chest to melt completely if the ice chest is subjected to (a) calm air and (b) winds at 50 km/h. Assume the heat transfer coefficient on the front, back, and top surfaces to be the same as that on the side surfaces.

9–105 An electronic box that consumes 200 W of power is cooled by a fan blowing air into the box enclosure. The dimensions of the electronic box are 15 cm × 50 cm × 50 cm, and all surfaces of the box are exposed to the ambient except the base surface. Temperature measurements indicate that the box is at an average temperature of 32°C when the ambient temperature and the temperature of the surrounding walls are 25°C. If the emissivity of the outer surface of the box is 0.75, determine the fraction of the heat lost from the outer surfaces of the electronic box.

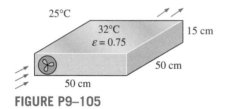

25°C

32°C
$\varepsilon = 0.75$

15 cm

50 cm

50 cm

FIGURE P9–105

9–106 A 6-m-internal-diameter spherical tank made of 1.5-cm-thick stainless steel ($k = 15$ W/m · °C) is used to store iced water at 0°C in a room at 20°C. The walls of the room are also at 20°C. The outer surface of the tank is black (emissivity $\varepsilon = 1$), and heat transfer between the outer surface of the tank and the surroundings is by natural convection and radiation. Assuming the entire steel tank to be at 0°C and thus the thermal resistance of the tank to be negligible, determine (a) the rate of heat transfer to the iced water in the tank and (b) the amount of ice at 0°C that melts during a 24-h period. The heat of fusion of water is 333.7 kJ/kg
Answers: (a) 15.4 kW, (b) 3988 kg

9–107 Consider a 1.2-m-high and 2-m-wide double-pane window consisting of two 3-mm-thick layers of glass ($k = 0.78$ W/m · °C) separated by a 3-cm-wide air space. Determine the steady rate of heat transfer through this window and the temperature of its inner surface for a day during which the room is maintained at 20°C while the temperature of the outdoors is 0°C. Take the heat transfer coefficients on the inner and outer surfaces of the window to be $h_1 = 10$ W/m² · °C and $h_2 = 25$ W/m² · °C and disregard any heat transfer by radiation.

9–108 An electric resistance space heater is designed such that it resembles a rectangular box 50 cm high, 80 cm long, and 15 cm wide filled with 45 kg of oil. The heater is to be placed against a wall, and thus heat transfer from its back surface is

negligible. The surface temperature of the heater is not to exceed 75°C in a room at 25°C for safety considerations. Disregarding heat transfer from the bottom and top surfaces of the heater in anticipation that the top surface will be used as a shelf, determine the power rating of the heater in W. Take the emissivity of the outer surface of the heater to be 0.8 and the average temperature of the ceiling and wall surfaces to be the same as the room air temperature.

Also, determine how long it will take for the heater to reach steady operation when it is first turned on (i.e., for the oil temperature to rise from 25°C to 75°C). State your assumptions in the calculations.

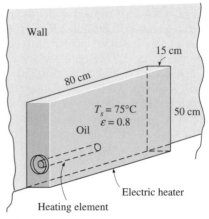

Wall

15 cm

80 cm

$T_s = 75°C$
$\varepsilon = 0.8$

50 cm

Oil

Electric heater

Heating element

FIGURE P9–108

9–109 Skylights or "roof windows" are commonly used in homes and manufacturing facilities since they let natural light in during day time and thus reduce the lighting costs. However, they offer little resistance to heat transfer, and large amounts of energy are lost through them in winter unless they are equipped with a motorized insulating cover that can be used in cold weather and at nights to reduce heat losses. Consider a 1-m-wide and 2.5-m-long horizontal skylight on the roof of a house that is kept at 20°C. The glazing of the skylight is made of a single layer of 0.5-cm-thick glass ($k = 0.78$ W/m · °C and $\varepsilon = 0.9$). Determine the rate of

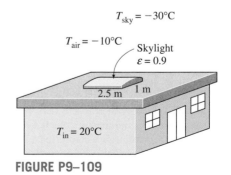

$T_{sky} = -30°C$

$T_{air} = -10°C$
Skylight
$\varepsilon = 0.9$

2.5 m   1 m

$T_{in} = 20°C$

FIGURE P9–109

heat loss through the skylight when the air temperature outside is $-10°C$ and the effective sky temperature is $-30°C$. Compare your result with the rate of heat loss through an equivalent surface area of the roof that has a common R-5.34 construction in SI units (i.e., a thickness-to-effective-thermal-conductivity ratio of 5.34 m² · °C/W).

9–110 A solar collector consists of a horizontal copper tube of outer diameter 5 cm enclosed in a concentric thin glass tube of 9 cm diameter. Water is heated as it flows through the tube, and the annular space between the copper and glass tube is filled with air at 1 atm pressure. During a clear day, the temperatures of the tube surface and the glass cover are measured to be 60°C and 32°C, respectively. Determine the rate of heat loss from the collector by natural convection per meter length of the tube. *Answer:* 17.4 W

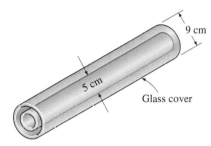

FIGURE P9–110

9–111 A solar collector consists of a horizontal aluminum tube of outer diameter 5 cm enclosed in a concentric thin glass tube of 7 cm diameter. Water is heated as it flows through the aluminum tube, and the annular space between the aluminum and glass tubes is filled with air at 1 atm pressure. The pump circulating the water fails during a clear day, and the water temperature in the tube starts rising. The aluminum tube absorbs solar radiation at a rate of 20 W per meter length, and the temperature of the ambient air outside is 30°C. Approximating the surfaces of the tube and the glass cover as being black (emissivity $\varepsilon = 1$) in radiation calculations and taking the effective sky temperature to be 20°C, determine the temperature of the aluminum tube when equilibrium is established (i.e., when the net heat loss from the tube by convection and radiation equals the amount of solar energy absorbed by the tube).

9–112E The components of an electronic system dissipating 180 W are located in a 4-ft-long horizontal duct whose cross section is 6 in × 6 in. The components in the duct are cooled by forced air, which enters at 85°F at a rate of 22 cfm and leaves at 100°F. The surfaces of the sheet metal duct are not painted, and thus radiation heat transfer from the outer surfaces is negligible. If the ambient air temperature is 80°F, determine (a) the heat transfer from the outer surfaces of the duct to the ambient air by natural convection and (b) the average temperature of the duct.

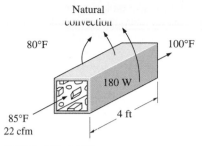

FIGURE P9–112E

9–113E Repeat Prob. 9–112E for a circular horizontal duct of diameter 4 in.

9–114E Repeat Prob. 9–112E assuming the fan fails and thus the entire heat generated inside the duct must be rejected to the ambient air by natural convection through the outer surfaces of the duct.

9–115 Consider a cold aluminum canned drink that is initially at a uniform temperature of 5°C. The can is 12.5 cm high and has a diameter of 6 cm. The emissivity of the outer surface of the can is 0.6. Disregarding any heat transfer from the bottom surface of the can, determine how long it will take for the average temperature of the drink to rise to 7°C if the surrounding air and surfaces are at 25°C. *Answer:* 12.1 min

9–116 Consider a 2-m-high electric hot-water heater that has a diameter of 40 cm and maintains the hot water at 60°C. The tank is located in a small room at 20°C whose walls and the ceiling are at about the same temperature. The tank is placed in a 44-cm-diameter sheet metal shell of negligible thickness, and the space between the tank and the shell is filled with foam insulation. The average temperature and emissivity of the outer surface of the shell are 40°C and 0.7, respectively. The price of electricity is $0.08/kWh. Hot-water tank insulation kits large

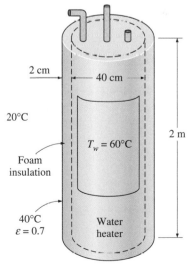

FIGURE P9–116

enough to wrap the entire tank are available on the market for about $60. If such an insulation is installed on this water tank by the home owner himself, how long will it take for this additional insulation to pay for itself? Disregard any heat loss from the top and bottom surfaces, and assume the insulation to reduce the heat losses by 80 percent.

9–117 During a plant visit, it was observed that a 1.5-m-high and 1-m-wide section of the vertical front section of a natural gas furnace wall was too hot to touch. The temperature measurements on the surface revealed that the average temperature of the exposed hot surface was 110°C, while the temperature of the surrounding air was 25°C. The surface appeared to be oxidized, and its emissivity can be taken to be 0.7. Taking the temperature of the surrounding surfaces to be 25°C also, determine the rate of heat loss from this furnace.

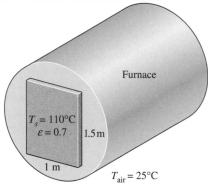

FIGURE P9–117

The furnace has an efficiency of 79 percent, and the plant pays $1.20 per therm of natural gas. If the plant operates 10 h a day, 310 days a year, and thus 3100 h a year, determine the annual cost of the heat loss from this vertical hot surface on the front section of the furnace wall.

9–118 A group of 25 power transistors, dissipating 1.5 W each, are to be cooled by attaching them to a black-anodized

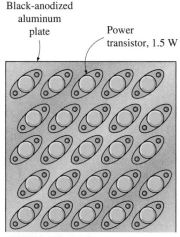

FIGURE P9–118

square aluminum plate and mounting the plate on the wall of a room at 30°C. The emissivity of the transistor and the plate surfaces is 0.9. Assuming the heat transfer from the back side of the plate to be negligible and the temperature of the surrounding surfaces to be the same as the air temperature of the room, determine the size of the plate if the average surface temperature of the plate is not to exceed 50°C.   *Answer:* 43 cm × 43 cm

9–119 Repeat Prob. 9–118 assuming the plate to be positioned horizontally with (*a*) transistors facing up and (*b*) transistors facing down.

9–120E Hot water is flowing at an average velocity of 4 ft/s through a cast iron pipe ($k = 30$ Btu/h · ft · °F) whose inner and outer diameters are 1.0 in and 1.2 in, respectively. The pipe passes through a 50-ft-long section of a basement whose temperature is 60°F. The emissivity of the outer surface of the pipe is 0.5, and the walls of the basement are also at about 60°F. If the inlet temperature of the water is 150°F and the heat transfer coefficient on the inner surface of the pipe is 30 Btu/h · ft² · °F, determine the temperature drop of water as it passes through the basement.

9–121 Consider a flat-plate solar collector placed horizontally on the flat roof of a house. The collector is 1.5 m wide and 6 m long, and the average temperature of the exposed surface of the collector is 42°C. Determine the rate of heat loss from the collector by natural convection during a calm day when the ambient air temperature is 8°C. Also, determine the heat loss by radiation by taking the emissivity of the collector surface to be 0.9 and the effective sky temperature to be −15°C.   *Answers:* 1750 W, 2490 W

9–122 Solar radiation is incident on the glass cover of a solar collector at a rate of 650 W/m². The glass transmits 88 percent of the incident radiation and has an emissivity of 0.90. The hot water needs of a family in summer can be met completely by a collector 1.5 m high and 2 m wide, and tilted 40° from the horizontal. The temperature of the glass cover is measured to be 40°C on a calm day when the surrounding air temperature is 20°C. The effective sky temperature for radiation exchange

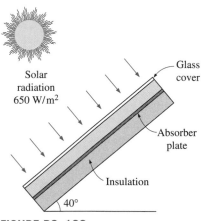

FIGURE P9–122

between the glass cover and the open sky is $-40°C$. Water enters the tubes attached to the absorber plate at a rate of 1 kg/min. Assuming the back surface of the absorber plate to be heavily insulated and the only heat loss occurs through the glass cover, determine (a) the total rate of heat loss from the collector; (b) the collector efficiency, which is the ratio of the amount of heat transferred to the water to the solar energy incident on the collector; and (c) the temperature rise of water as it flows through the collector.

Fundamentals of Engineering (FE) Exam Problems

9–123 Consider a hot, boiled egg in a spacecraft that is filled with air at atmospheric pressure and temperature at all times. Disregarding any radiation effect, will the egg cool faster or slower when the spacecraft is in space instead of on the ground?
(a) faster   (b) no difference   (c) slower
(d) insufficient information

9–124 A hot object suspended by a string is to be cooled by natural convection in fluids whose volume changes differently with temperature at constant pressure. In which fluid will the rate of cooling be lowest? With increasing temperature, a fluid whose volume (a) increases a lot, (b) increases slightly, (c) does not change, (d) decreases slightly, or (e) decreases a lot.

9–125 The primary driving force for natural convection is
(a) shear stress forces   (b) buoyancy forces
(c) pressure forces   (d) surface tension forces
(e) none of them

9–126 A spherical block of dry ice at $-79°C$ is exposed to atmospheric air at $30°C$. The general direction in which the air moves in this situation is
(a) horizontal   (b) up   (c) down
(d) recirculation around the sphere
(e) no motion

9–127 Consider a horizontal 0.7-m-wide and 0.85-m-long plate in a room at $30°C$. Top side of the plate is insulated while the bottom side is maintained at $0°C$. The rate of heat transfer from the room air to the plate by natural convection is
(a) 36.8 W   (b) 43.7 W   (c) 128.5 W
(d) 92.7 W   (e) 69.7 W
(For air, use $k = 0.02476$ W/m · °C, Pr = 0.7323, $v = 1.470 \times 10^{-5}$ m²/s)

9–128 Consider a 0.3-m-diameter and 1.8-m-long horizontal cylinder in a room at $20°C$. If the outer surface temperature of the cylinder is $40°C$, the natural convection heat transfer coefficient is
(a) 3.0 W/m² · °C   (b) 3.5 W/m² · °C   (c) 3.9 W/m² · °C
(d) 4.6 W/m² · °C   (e) 5.7 W/m² · °C

(For air, use $k = 0.02588$ W/m · °C, Pr = 0.7282, $v = 1.608 \times 10^{-5}$ m²/s)

9–129 A 4-m-diameter spherical tank contains iced water at $0°C$. The tank is thin-shelled and thus its outer surface temperature may be assumed to be same as the temperature of the iced water inside. Now the tank is placed in a large lake at $20°C$. The rate at which the ice melts is
(a) 0.42 kg/s   (b) 0.58 kg/s   (c) 0.70 kg/s
(d) 0.83 kg/s   (e) 0.98 kg/s
(For lake water, use $k = 0.580$ W/m · °C, Pr = 9.45, $v = 0.1307 \times 10^{-5}$ m²/s, $\beta = 0.138 \times 10^{-3}$ K⁻¹)

9–130 A 4-m-long section of a 5-cm-diameter horizontal pipe in which a refrigerant flows passes through a room at $20°C$. The pipe is not well insulated and the outer surface temperature of the pipe is observed to be $-10°C$. The emissivity of the pipe surface is 0.85 and the surrounding surfaces are at $15°C$. The fraction of heat transferred to the pipe by radiation is
(a) 0.24   (b) 0.30   (c) 0.37
(d) 0.48   (e) 0.58
(For air, use $k = 0.02401$ W/m · °C, Pr = 0.735, $v = 1.382 \times 10^{-5}$ m²/s)

9–131 A vertical 0.9-m-high and 1.8-m-wide double-pane window consists of two sheets of glass separated by a 2.2-cm air gap at atmospheric pressure. If the glass surface temperatures across the air gap are measured to be $20°C$ and $30°C$, the rate of heat transfer through the window is
(a) 19.8 W   (b) 26.1 W   (c) 30.5 W
(d) 34.7 W   (e) 55.0 W
(For air, use $k = 0.02551$ W/m · °C, Pr = 0.7296, $v = 1.562 \times 10^{-5}$ m²/s. Also, the applicable correlation is Nu $= 0.42$Ra$^{1/4}$ Pr$^{0.012}$ $(H/L)^{-0.3}$)

9–132 A horizontal 1.5-m-wide, 4.5-m-long double-pane window consists of two sheets of glass separated by a 3.5-cm gap filled with water. If the glass surface temperatures at the bottom and the top are measured to be $60°C$ and $40°C$, respectively, the rate of heat transfer through the window is
(a) 27.6 kW   (b) 39.4 kW   (c) 59.6 kW
(d) 66.4 kW   (e) 75.5 kW
(For water, use $k = 0.644$ W/m · °C, Pr = 3.55, $v = 0.554 \times 10^{-6}$ m²/s, $\beta = 0.451 \times 10^{-3}$ K⁻¹. Also, the applicable correlation is Nu $= 0.069$Ra$^{1/3}$ Pr$^{0.074}$).

9–133 Two concentric cylinders of diameters $D_i = 30$ cm and $D_o = 40$ cm and length $L = 5$ m are separated by air at 1 atm pressure. Heat is generated within the inner cylinder uniformly at a rate of 1100 W/m³ and the inner surface temperature of the outer cylinder is 300 K. The steady–state outer surface temperature of the inner cylinder is

(*a*) 402 K   (*b*) 415 K   (*c*) 429 K
(*d*) 442 K   (*e*) 456 K

(For air, use $k = 0.03095$ W/m · °C, Pr $= 0.7111$, $v = 2.306 \times 10^{-5}$ m²/s)

9–134 A vertical double-pane window consists of two sheets of glass separated by a 1.5-cm air gap at atmospheric pressure. The glass surface temperatures across the air gap are measured to be 278 K and 288 K. If it is estimated that the heat transfer by convection through the enclosure is 1.5 times that by pure conduction and that the rate of heat transfer by radiation through the enclosure is about the same magnitude as the convection, the effective emissivity of the two glass surfaces is

(*a*) 0.47   (*b*) 0.53   (*c*) 0.61
(*d*) 0.65   (*e*) 0.72

Design and Essay Problems

9–135 Write a computer program to evaluate the variation of temperature with time of thin square metal plates that are removed from an oven at a specified temperature and placed vertically in a large room. The thickness, the size, the initial temperature, the emissivity, and the thermophysical properties of the plate as well as the room temperature are to be specified by the user. The program should evaluate the temperature of the plate at specified intervals and tabulate the results against time. The computer should list the assumptions made during calculations before printing the results.

For each step or time interval, assume the surface temperature to be constant and evaluate the heat loss during that time interval and the temperature drop of the plate as a result of this heat loss. This gives the temperature of the plate at the end of a time interval, which is to serve as the initial temperature of the plate for the beginning of the next time interval.

Try your program for 0.2-cm-thick vertical copper plates of 40 cm × 40 cm in size initially at 300°C cooled in a room at 25°C. Take the surface emissivity to be 0.9. Use a time interval of 1 s in calculations, but print the results at 10-s intervals for a total cooling period of 15 min.

9–136 Write a computer program to optimize the spacing between the two glasses of a double-pane window. Assume the spacing is filled with dry air at atmospheric pressure. The program should evaluate the recommended practical value of the spacing to minimize the heat losses and list it when the size of the window (the height and the width) and the temperatures of the two glasses are specified.

9–137 Contact a manufacturer of aluminum heat sinks and obtain their product catalog for cooling electronic components by natural convection and radiation. Write an essay on how to select a suitable heat sink for an electronic component when its maximum power dissipation and maximum allowable surface temperature are specified.

9–138 The top surfaces of practically all flat-plate solar collectors are covered with glass in order to reduce the heat losses from the absorber plate underneath. Although the glass cover reflects or absorbs about 15 percent of the incident solar radiation, it saves much more from the potential heat losses from the absorber plate, and thus it is considered to be an essential part of a well-designed solar collector. Inspired by the energy efficiency of double-pane windows, someone proposes to use double glazing on solar collectors instead of a single glass. Investigate if this is a good idea for the town in which you live. Use local weather data and base your conclusion on heat transfer analysis and economic considerations.

BOILING AND CONDENSATION

We know from thermodynamics that when the temperature of a liquid at a specified pressure is raised to the saturation temperature T_{sat} at that pressure, *boiling* occurs. Likewise, when the temperature of a vapor is lowered to T_{sat}, *condensation* occurs. In this chapter we study the rates of heat transfer during such liquid-to-vapor and vapor-to-liquid phase transformations.

Although boiling and condensation exhibit some unique features, they are considered to be forms of *convection* heat transfer since they involve fluid motion (such as the rise of the bubbles to the top and the flow of condensate to the bottom). Boiling and condensation differ from other forms of convection in that they depend on the *latent heat of vaporization* h_{fg} of the fluid and the *surface tension* σ at the liquid–vapor interface, in addition to the properties of the fluid in each phase. Noting that under equilibrium conditions the temperature remains constant during a phase-change process at a fixed pressure, large amounts of heat (due to the large latent heat of vaporization released or absorbed) can be transferred during boiling and condensation essentially at constant temperature. In practice, however, it is necessary to maintain some difference between the surface temperature T_s and T_{sat} for effective heat transfer. Heat transfer coefficients h associated with boiling and condensation are typically much higher than those encountered in other forms of convection processes that involve a single phase.

We start this chapter with a discussion of the *boiling curve* and the modes of pool boiling such as *free convection boiling, nucleate boiling,* and *film boiling.* We then discuss boiling in the presence of forced convection. In the second part of this chapter, we describe the physical mechanism of *film condensation* and discuss condensation heat transfer in several geometrical arrangements and orientations. Finally, we introduce *dropwise condensation* and discuss ways of maintaining it.

OBJECTIVES

When you finish studying this chapter, you should be able to:

■ Differentiate between evaporation and boiling, and gain familiarity with different types of boiling,

■ Develop a good understanding of the boiling curve, and the different boiling regimes corresponding to different regions of the boiling curve,

■ Calculate the heat flux and its critical value associated with nucleate boiling, and examine the methods of boiling heat transfer enhancement,

■ Derive a relation for the heat transfer coefficient in laminar film condensation over a vertical plate,

■ Calculate the heat flux associated with condensation on inclined and horizontal plates, vertical and horizontal cylinders or spheres, and tube bundles,

■ Examine dropwise condensation and understand the uncertainties associated with them.

CONTENTS

FIGURE 10–1

A liquid-to-vapor phase change process is called *evaporation* if it originates at a liquid–vapor interface and *boiling* if it occurs at a solid–liquid interface.

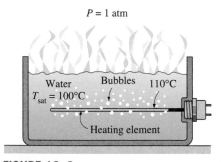

FIGURE 10–2

Boiling occurs when a liquid is brought into contact with a surface at a temperature above the saturation temperature of the liquid.

10–1 · BOILING HEAT TRANSFER

Many familiar engineering applications involve condensation and boiling heat transfer. In a household refrigerator, for example, the refrigerant absorbs heat from the refrigerated space by boiling in the *evaporator* section and rejects heat to the kitchen air by condensing in the *condenser* section (the long coils behind or under the refrigerator). Also, in steam power plants, heat is transferred to the steam in the *boiler* where water is vaporized, and the waste heat is rejected from the steam in the *condenser* where the steam is condensed. Some electronic components are cooled by boiling by immersing them in a fluid with an appropriate boiling temperature.

Boiling is a liquid-to-vapor phase change process just like evaporation, but there are significant differences between the two. **Evaporation** occurs at the *liquid–vapor interface* when the vapor pressure is less than the saturation pressure of the liquid at a given temperature. Water in a lake at 20°C, for example, evaporates to air at 20°C and 60 percent relative humidity since the saturation pressure of water at 20°C is 2.3 kPa and the vapor pressure of air at 20°C and 60 percent relative humidity is 1.4 kPa (evaporation rates are determined in Chapter 14). Other examples of evaporation are the drying of clothes, fruits, and vegetables; the evaporation of sweat to cool the human body; and the rejection of waste heat in wet cooling towers. Note that evaporation involves no bubble formation or bubble motion (Fig. 10–1).

Boiling, on the other hand, occurs at the *solid–liquid interface* when a liquid is brought into contact with a surface maintained at a temperature T_s sufficiently above the saturation temperature T_{sat} of the liquid (Fig. 10–2). At 1 atm, for example, liquid water in contact with a solid surface at 110°C boils since the saturation temperature of water at 1 atm is 100°C. The boiling process is characterized by the rapid formation of *vapor bubbles* at the solid–liquid interface that detach from the surface when they reach a certain size and attempt to rise to the free surface of the liquid. When cooking, we do not say water is boiling until we see the bubbles rising to the top. Boiling is a complicated phenomenon because of the large number of variables involved in the process and the complex fluid motion patterns caused by the bubble formation and growth.

As a form of convection heat transfer, the *boiling heat flux* from a solid surface to the fluid is expressed from Newton's law of cooling as

$$\dot{q}_{boiling} = h(T_s - T_{sat}) = h\Delta T_{excess} \quad (\text{W/m}^2) \qquad (10\text{–}1)$$

where $\Delta T_{excess} = T_s - T_{sat}$ is called the *excess temperature,* which represents the temperature excess of the surface above the saturation temperature of the fluid.

In the preceding chapters we considered forced and free convection heat transfer involving a single phase of a fluid. The analysis of such convection processes involves the thermophysical properties ρ, μ, k, and c_p of the fluid. The analysis of boiling heat transfer involves these properties of the liquid (indicated by the subscript l) or vapor (indicated by the subscript v) as well as the properties h_{fg} (the latent heat of vaporization) and σ (the surface tension). The h_{fg} represents the energy absorbed as a unit mass of liquid vaporizes at a specified temperature or pressure and is the primary quantity of energy transferred during boiling heat transfer. The h_{fg} values of water at various temperatures are given in Table A–9.

Bubbles owe their existence to the *surface-tension* σ at the liquid–vapor interface due to the attraction force on molecules at the interface toward the liquid phase. The surface tension decreases with increasing temperature and becomes zero at the critical temperature. This explains why no bubbles are formed during boiling at supercritical pressures and temperatures. Surface tension has the unit N/m.

The boiling processes in practice do not occur under *equilibrium* conditions, and normally the bubbles are not in thermodynamic equilibrium with the surrounding liquid. That is, the temperature and pressure of the vapor in a bubble are usually different than those of the liquid. The pressure difference between the liquid and the vapor is balanced by the surface tension at the interface. The temperature difference between the vapor in a bubble and the surrounding liquid is the driving force for heat transfer between the two phases. When the liquid is at a *lower temperature* than the bubble, heat is transferred from the bubble into the liquid, causing some of the vapor inside the bubble to condense and the bubble eventually to collapse. When the liquid is at a *higher temperature* than the bubble, heat is transferred from the liquid to the bubble, causing the bubble to grow and rise to the top under the influence of buoyancy.

Boiling is classified as *pool boiling* or *flow boiling,* depending on the presence of bulk fluid motion (Fig. 10–3). Boiling is called **pool boiling** in the absence of bulk fluid flow and **flow boiling** (or *forced convection boiling*) in the presence of it. In pool boiling, the fluid body is stationary, and any motion of the fluid is due to natural convection currents and the motion of the bubbles under the influence of buoyancy. The boiling of water in a pan on top of a stove is an example of pool boiling. Pool boiling of a fluid can also be achieved by placing a heating coil in the fluid. In flow boiling, the fluid is forced to move in a heated pipe or over a surface by external means such as a pump. Therefore, flow boiling is always accompanied by other convection effects.

Pool and flow boiling are further classified as *subcooled boiling* or *saturated boiling,* depending on the bulk liquid temperature (Fig. 10–4). Boiling is said to be **subcooled** (or *local*) when the temperature of the main body of the liquid is below the saturation temperature T_{sat} (i.e., the bulk of the liquid is subcooled) and **saturated** (or *bulk*) when the temperature of the liquid is equal to T_{sat} (i.e., the bulk of the liquid is saturated). At the early stages of boiling, the bubbles are confined to a narrow region near the hot surface. This is because the liquid adjacent to the hot surface vaporizes as a result of being heated above its saturation temperature. But these bubbles disappear soon after they move away from the hot surface as a result of heat transfer from the bubbles to the cooler liquid surrounding them. This happens when the bulk of the liquid is at a lower temperature than the saturation temperature. The bubbles serve as "energy movers" from the hot surface into the liquid body by absorbing heat from the hot surface and releasing it into the liquid as they condense and collapse. Boiling in this case is confined to a region in the locality of the hot surface and is appropriately called *local* or *subcooled* boiling. When the entire liquid body reaches the saturation temperature, the bubbles

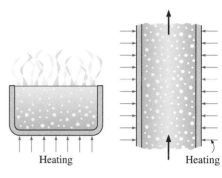

(a) Pool boiling (b) Flow boiling

FIGURE 10–3
Classification of boiling on the basis of the presence of bulk fluid motion.

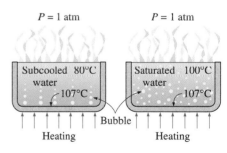

(a) Subcooled boiling (b) Saturated boiling

FIGURE 10–4
Classification of boiling on the basis of the presence of bulk liquid temperature.

start rising to the top. We can see bubbles throughout the bulk of the liquid, and boiling in this case is called the *bulk* or *saturated* boiling. Next, we consider different boiling regimes in detail.

10–2 ▪ POOL BOILING

So far we presented some general discussions on boiling. Now we turn our attention to the physical mechanisms involved in *pool boiling,* that is, the boiling of stationary fluids. In pool boiling, the fluid is not forced to flow by a mover such as a pump, and any motion of the fluid is due to natural convection currents and the motion of the bubbles under the influence of buoyancy.

As a familiar example of pool boiling, consider the boiling of tap water in a pan on top of a stove. The water is initially at about 15°C, far below the saturation temperature of 100°C at standard atmospheric pressure. At the early stages of boiling, you will not notice anything significant except some bubbles that stick to the surface of the pan. These bubbles are caused by the release of air molecules dissolved in liquid water and should not be confused with vapor bubbles. As the water temperature rises, you will notice chunks of liquid water rolling up and down as a result of natural convection currents, followed by the first vapor bubbles forming at the bottom surface of the pan. These bubbles get smaller as they detach from the surface and start rising, and eventually collapse in the cooler water above. This is *subcooled boiling* since the bulk of the liquid water has not reached saturation temperature yet. The intensity of bubble formation increases as the water temperature rises further, and you will notice waves of vapor bubbles coming from the bottom and rising to the top when the water temperature reaches the saturation temperature (100°C at standard atmospheric conditions). This full scale boiling is the *saturated boiling*.

Boiling Regimes and the Boiling Curve

Boiling is probably the most familiar form of heat transfer, yet it remains to be the least understood form. After hundreds of papers written on the subject, we still do not fully understand the process of bubble formation and we must still rely on empirical or semi-empirical relations to predict the rate of boiling heat transfer.

The pioneering work on boiling was done in 1934 by S. Nukiyama, who used electrically heated nichrome and platinum wires immersed in liquids in his experiments. Nukiyama noticed that boiling takes different forms, depending on the value of the excess temperature ΔT_{excess}. Four different boiling regimes are observed: *natural convection boiling, nucleate boiling, transition boiling,* and *film boiling* (Fig. 10–5). These regimes are illustrated on the **boiling curve** in Fig. 10–6, which is a plot of boiling heat flux versus the excess temperature. Although the boiling curve given in this figure is for water, the general shape of the boiling curve remains the same for different fluids. The specific shape of the curve depends on the fluid–heating surface material combination and the fluid pressure, but it is practically independent of the geometry of the heating surface. We now describe each boiling regime in detail.

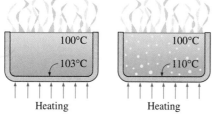

(a) Natural convection boiling

(b) Nucleate boiling

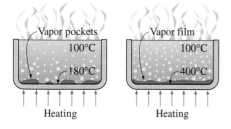

(c) Transition boiling

(d) Film boiling

FIGURE 10–5
Different boiling regimes in pool boiling.

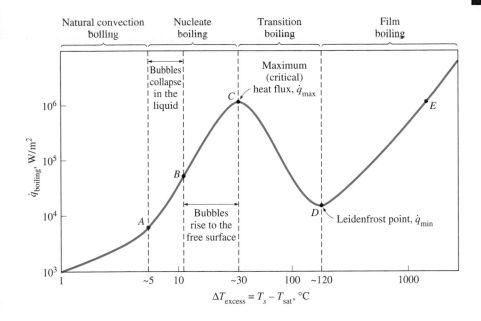

FIGURE 10–6
Typical boiling curve for water
at 1 atm pressure.

Natural Convection Boiling (to Point *A* on the Boiling Curve)

We know from thermodynamics that a pure substance at a specified pressure starts boiling when it reaches the saturation temperature at that pressure. But in practice we do not see any bubbles forming on the heating surface until the liquid is heated a few degrees above the saturation temperature (about 2 to 6°C for water). Therefore, the liquid is slightly *superheated* in this case (a *metastable* condition) and evaporates when it rises to the free surface. The fluid motion in this mode of boiling is governed by natural convection currents, and heat transfer from the heating surface to the fluid is by natural convection.

Nucleate Boiling (between Points *A* and *C*)

The first bubbles start forming at point *A* of the boiling curve at various preferential sites on the heating surface. The bubbles form at an increasing rate at an increasing number of nucleation sites as we move along the boiling curve toward point *C*.

The nucleate boiling regime can be separated into two distinct regions. In region *A–B*, *isolated bubbles* are formed at various preferential nucleation sites on the heated surface. But these bubbles are dissipated in the liquid shortly after they separate from the surface. The space vacated by the rising bubbles is filled by the liquid in the vicinity of the heater surface, and the process is repeated. The stirring and agitation caused by the entrainment of the liquid to the heater surface is primarily responsible for the increased heat transfer coefficient and heat flux in this region of nucleate boiling.

In region *B–C*, the heater temperature is further increased, and bubbles form at such great rates at such a large number of nucleation sites that they form numerous *continuous columns of vapor* in the liquid. These bubbles move all the way up to the free surface, where they break up and release their vapor content. The large heat fluxes obtainable in this region are caused by the combined effect of liquid entrainment and evaporation.

At large values of ΔT_{excess}, the rate of evaporation at the heater surface reaches such high values that a large fraction of the heater surface is covered by bubbles, making it difficult for the liquid to reach the heater surface and wet it. Consequently, the heat flux increases at a lower rate with increasing ΔT_{excess}, and reaches a maximum at point C. The heat flux at this point is

(a)

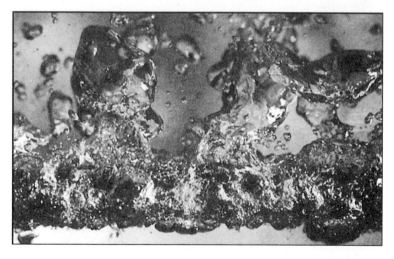

(b)

FIGURE 10–7
Various boiling regimes during boiling of methanol on a horizontal 1-cm-diameter steam-heated copper tube: (a) nucleate boiling, (b) transition boiling, and (c) film boiling.
(From J. W. Westwater and J. G. Santangelo, University of Illinois at Champaign-Urbana.)

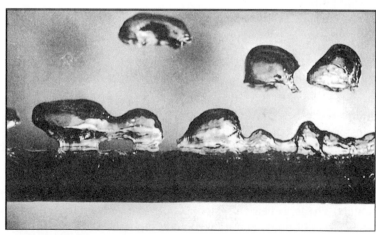

(c)

called the **critical** (or **maximum**) **heat flux**, $\dot{q}_{max}$. For water, the critical heat flux exceeds 1 MW/m^2.

Nucleate boiling is the most desirable boiling regime in practice because high heat transfer rates can be achieved in this regime with relatively small values of ΔT_{excess}, typically under 30°C for water. The photographs in Fig. 10–7 show the nature of bubble formation and bubble motion associated with nucleate, transition, and film boiling.

Transition Boiling (between Points C and D)

As the heater temperature and thus the ΔT_{excess} is increased past point C, the heat flux decreases, as shown in Fig. 10–6. This is because a large fraction of the heater surface is covered by a vapor film, which acts as an insulation due to the low thermal conductivity of the vapor relative to that of the liquid. In the transition boiling regime, both nucleate and film boiling partially occur. Nucleate boiling at point C is completely replaced by film boiling at point D. Operation in the transition boiling regime, which is also called the *unstable film boiling regime,* is avoided in practice. For water, transition boiling occurs over the excess temperature range from about 30°C to about 120°C.

Film Boiling (beyond Point D)

In this region the heater surface is completely covered by a continuous stable vapor film. Point D, where the heat flux reaches a minimum, is called the **Leidenfrost point**, in honor of J. C. Leidenfrost, who observed in 1756 that liquid droplets on a very hot surface jump around and slowly boil away. The presence of a vapor film between the heater surface and the liquid is responsible for the low heat transfer rates in the film boiling region. The heat transfer rate increases with increasing excess temperature as a result of heat transfer from the heated surface to the liquid through the vapor film by radiation, which becomes significant at high temperatures.

A typical boiling process does not follow the boiling curve beyond point C, as Nukiyama has observed during his experiments. Nukiyama noticed, with surprise, that when the power applied to the nichrome wire immersed in water exceeded $\dot{q}_{max}$ even slightly, the wire temperature increased suddenly to the melting point of the wire and *burnout* occurred beyond his control. When he repeated the experiments with platinum wire, which has a much higher melting point, he was able to avoid burnout and maintain heat fluxes higher than $\dot{q}_{max}$. When he gradually reduced power, he obtained the cooling curve shown in Fig. 10–8 with a sudden drop in excess temperature when $\dot{q}_{min}$ is reached. Note that the boiling process cannot follow the transition boiling part of the boiling curve past point C unless the power applied is reduced suddenly.

The *burnout phenomenon* in boiling can be explained as follows: In order to move beyond point C where $\dot{q}_{max}$ occurs, we must increase the heater surface temperature T_s. To increase T_s, however, we must increase the heat flux. But the fluid cannot receive this increased energy at an excess temperature just beyond point C. Therefore, the heater surface ends up absorbing the increased energy, causing the heater surface temperature T_s to rise. But the fluid can receive even less energy at this increased excess temperature, causing the heater surface temperature T_s to rise even further. This continues until the surface

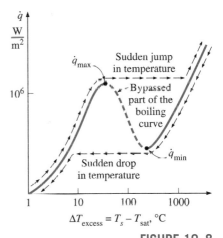

FIGURE 10–8

The actual boiling curve obtained with heated platinum wire in water as the heat flux is increased and then decreased.

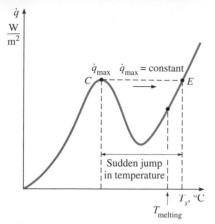

FIGURE 10–9
An attempt to increase the boiling heat flux beyond the critical value often causes the temperature of the heating element to jump suddenly to a value that is above the melting point, resulting in *burnout*.

TABLE 10–1

Surface tension of liquid–vapor interface for water

T, °C	σ, N/m*
0	0.0757
20	0.0727
40	0.0696
60	0.0662
80	0.0627
100	0.0589
120	0.0550
140	0.0509
160	0.0466
180	0.0422
200	0,0377
220	0.0331
240	0.0284
260	0.0237
280	0.0190
300	0.0144
320	0.0099
340	0.0056
360	0.0019
374	0.0

*Multiply by 0.06852 to convert to lbf/ft or by 2.2046 to convert to lbm/s².

temperature reaches a point at which it no longer rises and the heat supplied can be transferred to the fluid steadily. This is point E on the boiling curve, which corresponds to very high surface temperatures. Therefore, any attempt to increase the heat flux beyond $\dot{q}_{max}$ will cause the operation point on the boiling curve to jump suddenly from point C to point E. However, surface temperature that corresponds to point E is beyond the melting point of most heater materials, and *burnout* occurs. Therefore, point C on the boiling curve is also called the **burnout point**, and the heat flux at this point the **burnout heat flux** (Fig. 10–9).

Most boiling heat transfer equipment in practice operate slightly below $\dot{q}_{max}$ to avoid any disastrous burnout. However, in cryogenic applications involving fluids with very low boiling points such as oxygen and nitrogen, point E usually falls below the melting point of the heater materials, and steady film boiling can be used in those cases without any danger of burnout.

Heat Transfer Correlations in Pool Boiling

Boiling regimes discussed above differ considerably in their character, and thus different heat transfer relations need to be used for different boiling regimes. In the *natural convection boiling* regime, boiling is governed by natural convection currents, and heat transfer rates in this case can be determined accurately using natural convection relations presented in Chapter 9.

Nucleate Boiling

In the *nucleate boiling* regime, the rate of heat transfer strongly depends on the nature of nucleation (the number of active nucleation sites on the surface, the rate of bubble formation at each site, etc.), which is difficult to predict. The type and the condition of the heated surface also affect the heat transfer. These complications made it difficult to develop theoretical relations for heat transfer in the nucleate boiling regime, and we had to rely on relations based on experimental data. The most widely used correlation for the rate of heat transfer in the nucleate boiling regime was proposed in 1952 by Rohsenow, and expressed as

$$\dot{q}_{\text{nucleate}} = \mu_l h_{fg} \left[\frac{g(\rho_l - \rho_v)}{\sigma} \right]^{1/2} \left[\frac{c_{pl}(T_s - T_{\text{sat}})}{C_{sf} h_{fg} \, \text{Pr}_l^n} \right]^3 \tag{10–2}$$

where

$\dot{q}_{\text{nucleate}}$ = nucleate boiling heat flux, W/m²
μ_l = viscosity of the liquid, kg/m · s
h_{fg} = enthalpy of vaporization, J/kg
g = gravitational acceleration, m/s²
ρ_l = density of the liquid, kg/m³
ρ_v = density of the vapor, kg/m³
σ = surface tension of liquid–vapor interface, N/m
c_{pl} = specific heat of the liquid, J/kg · °C
T_s = surface temperature of the heater, °C
T_{sat} = saturation temperature of the fluid, °C
C_{sf} = experimental constant that depends on surface–fluid combination
Pr_l = Prandtl number of the liquid
n = experimental constant that depends on the fluid

It can be shown easily that using property values in the specified units in the Rohsenow equation produces the desired unit W/m² for the boiling heat flux, thus saving one from having to go through tedious unit manipulations (Fig. 10–10).

The surface tension at the vapor–liquid interface is given in Table 10–1 for water, and Table 10–2 for some other fluids. Experimentally determined values of the constant C_{sf} are given in Table 10–3 for various fluid–surface combinations. These values can be used for *any geometry* since it is found that the rate of heat transfer during nucleate boiling is essentially independent of the geometry and orientation of the heated surface. The fluid properties in Eq. 10–2 are to be evaluated at the saturation temperature T_{sat}.

The *condition* of the heater surface greatly affects heat transfer, and the Rohsenow equation given above is applicable to *clean* and relatively *smooth* surfaces. The results obtained using the Rohsenow equation can be in error by $\pm 100\%$ for the heat transfer rate for a given excess temperature and by $\pm 30\%$ for the excess temperature for a given heat transfer rate. Therefore, care should be exercised in the interpretation of the results.

Recall from thermodynamics that the enthalpy of vaporization h_{fg} of a pure substance decreases with increasing pressure (or temperature) and reaches zero at the critical point. Noting that h_{fg} appears in the denominator of the Rohsenow equation, we should see a significant rise in the rate of heat transfer at *high pressures* during nucleate boiling.

$$\dot{q} = \left(\frac{kg}{m \cdot s}\right)\left(\frac{J}{kg}\right)$$

$$\times \left(\frac{\dfrac{m}{s^2}\dfrac{kg}{m^3}}{\dfrac{N}{m}}\right)^{1/2}\left(\frac{\dfrac{J}{kg \cdot {}^\circ C}{}^\circ C}{\dfrac{J}{kg}}\right)^3$$

$$= \frac{W}{m}\left(\frac{1}{m^2}\right)^{1/2}(1)^3$$

$$= W/m^2$$

FIGURE 10–10
Equation 10–2 gives the boiling heat flux in W/m² when the quantities are expressed in the units specified in their descriptions.

Peak Heat Flux

In the design of boiling heat transfer equipment, it is extremely important for the designer to have a knowledge of the maximum heat flux in order to avoid the danger of burnout. The *maximum* (or *critical*) *heat flux* in nucleate pool boiling was determined theoretically by S. S. Kutateladze in Russia in 1948 and N. Zuber in the United States in 1958 using quite different approaches, and is expressed as (Fig. 10–11)

$$\dot{q}_{max} = C_{cr}\, h_{fg}[\sigma g\rho_v^2\,(\rho_l - \rho_v)]^{1/4} \tag{10–3}$$

where C_{cr} is a constant whose value depends on the heater geometry. Exhaustive experimental studies by Lienhard and his coworkers indicated that the value of C_{cr} is about 0.15. Specific values of C_{cr} for different heater geometries are listed in Table 10–4. Note that the heaters are classified as being large or small based on the value of the parameter L^*.

Equation 10–3 will give the maximum heat flux in W/m² if the properties are used *in the units specified* earlier in their descriptions following Eq. 10–2. The maximum heat flux is independent of the fluid–heating surface combination, as well as the viscosity, thermal conductivity, and the specific heat of the liquid.

Note that ρ_v increases but σ and h_{fg} decrease with increasing pressure, and thus the change in $\dot{q}_{max}$ with pressure depends on which effect dominates. The experimental studies of Cichelli and Bonilla indicate that $\dot{q}_{max}$ increases with pressure up to about one-third of the critical pressure, and then starts to decrease and becomes zero at the critical pressure. Also note that $\dot{q}_{max}$ is proportional to h_{fg}, and large maximum heat fluxes can be obtained using fluids with a large enthalpy of vaporization, such as water.

TABLE 10–2

Surface tension of some fluids (from Suryanarayana, originally based on data from Jasper)

Substance and Temp. Range	Surface Tension, σ, N/m* (T in °C)
Ammonia, −75 to −40°C:	$0.0264 + 0.000223T$
Benzene, 10 to 80°C:	$0.0315 - 0.000129T$
Butane, −70 to −20°C:	$0.0149 - 0.000121T$
Carbon dioxide, −30 to −20°C:	$0.0043 - 0.000160T$
Ethyl alcohol, 10 to 70°C:	$0.0241 - 0.000083T$
Mercury, 5 to 200°C:	$0.4906 - 0.000205T$
Methyl alcohol, 10 to 60°C:	$0.0240 - 0.000077T$
Pentane, 10 to 30°C:	$0.0183 - 0.000110T$
Propane, −90 to −10°C:	$0.0092 - 0.000087T$

*Multiply by 0.06852 to convert to lbf/ft or by 2.2046 to convert to lbm/s².

TABLE 10–3

Values of the coefficient C_{sf} and n for various fluid–surface combinations

Fluid–Heating Surface Combination	C_{sf}	n
Water–copper (polished)	0.0130	1.0
Water–copper (scored)	0.0068	1.0
Water–stainless steel (mechanically polished)	0.0130	1.0
Water–stainless steel (ground and polished)	0.0060	1.0
Water–stainless steel (teflon pitted)	0.0058	1.0
Water–stainless steel (chemically etched)	0.0130	1.0
Water–brass	0.0060	1.0
Water–nickel	0.0060	1.0
Water–platinum	0.0130	1.0
n-Pentane–copper (polished)	0.0154	1.7
n-Pentane–chromium	0.0150	1.7
Benzene–chromium	0.1010	1.7
Ethyl alcohol–chromium	0.0027	1.7
Carbon tetrachloride–copper	0.0130	1.7
Isopropanol–copper	0.0025	1.7

TABLE 10–4

Values of the coefficient C_{cr} for use in Eq. 10–3 for maximum heat flux (dimensionless parameter $L^* = L[g(\rho_l - \rho_v)/\sigma]^{1/2}$)

Heater Geometry	C_{cr}	Charac. Dimension of Heater, L	Range of L^*
Large horizontal flat heater	0.149	Width or diameter	$L^* > 27$
Small horizontal flat heater[1]	$18.9K_1$	Width or diameter	$9 < L^* < 20$
Large horizontal cylinder	0.12	Radius	$L^* > 1.2$
Small horizontal cylinder	$0.12L^{*-0.25}$	Radius	$0.15 < L^* < 1.2$
Large sphere	0.11	Radius	$L^* > 4.26$
Small sphere	$0.227L^{*-0.5}$	Radius	$0.15 < L^* < 4.26$

[1] $K_1 = \sigma/[g(\rho_l - \rho_v)A_{\text{heater}}]$

FIGURE 10–11
Different relations are used to determine the heat flux in different boiling regimes.

Minimum Heat Flux

Minimum heat flux, which occurs at the Leidenfrost point, is of practical interest since it represents the lower limit for the heat flux in the film boiling regime. Using the stability theory, Zuber derived the following expression for the minimum heat flux for a *large horizontal plate,*

$$\dot{q}_{\min} = 0.09\rho_v\, h_{fg} \left[\frac{\sigma g(\rho_l - \rho_v)}{(\rho_l + \rho_v)^2}\right]^{1/4} \tag{10–4}$$

where the constant 0.09 was determined by Berensen in 1961. He replaced the theoretically determined value of $\frac{\pi}{24}$ by 0.09 to match the experimental data better. Still, the relation above can be in error by 50 percent or more.

Film Boiling

Using an analysis similar to Nusselt's theory on filmwise condensation presented in the next section, Bromley developed a theory for the prediction of heat flux for stable film boiling on the outside of a horizontal cylinder. The heat flux for film boiling on a *horizontal cylinder* or *sphere* of diameter D is given by

$$\dot{q}_{film} = C_{film} \left[\frac{gk_v^3 \rho_v (\rho_l - \rho_v)[h_{fg} + 0.4c_{pv}(T_s - T_{sat})]}{\mu_v D(T_s - T_{sat})} \right]^{1/4} (T_s - T_{sat}) \quad (10\text{-}5)$$

where k_v is the thermal conductivity of the vapor in W/m · °C and

$$C_{film} = \begin{cases} 0.62 \text{ for horizontal cylinders} \\ 0.67 \text{ for spheres} \end{cases}$$

Other properties are as listed before in connection with Eq. 10–2. We used a modified latent heat of vaporization in Eq. 10–5 to account for the heat transfer associated with the superheating of the vapor.

The *vapor* properties are to be evaluated at the *film temperature* $T_f = (T_s + T_{sat})/2$, which is the *average temperature* of the vapor film. The liquid properties and h_{fg} are to be evaluated at the saturation temperature at the specified pressure. Again, this relation gives the film boiling heat flux in W/m² if the properties are used *in the units specified* earlier in their descriptions following Eq. 10–2.

At high surface temperatures (typically above 300°C), heat transfer across the vapor film by *radiation* becomes significant and needs to be considered (Fig. 10–12). Treating the vapor film as a transparent medium sandwiched between two large parallel plates and approximating the liquid as a blackbody, *radiation heat transfer* can be determined from

$$\dot{q}_{rad} = \varepsilon\sigma (T_s^4 - T_{sat}^4) \quad (10\text{-}6)$$

where ε is the emissivity of the heating surface and $\sigma = 5.67 \times 10^{-8}$ W/m² · K⁴ is the Stefan–Boltzman constant. Note that the temperature in this case must be expressed in K, not °C, and that surface tension and the Stefan–Boltzman constant share the same symbol.

You may be tempted to simply add the convection and radiation heat transfers to determine the total heat transfer during film boiling. However, these two mechanisms of heat transfer adversely affect each other, causing the total heat transfer to be less than their sum. For example, the radiation heat transfer from the surface to the liquid enhances the rate of evaporation, and thus the thickness of the vapor film, which impedes convection heat transfer. For $\dot{q}_{rad} < \dot{q}_{film}$, Bromley determined that the relation

$$\dot{q}_{total} = \dot{q}_{film} + \frac{3}{4}\dot{q}_{rad} \quad (10\text{-}7)$$

correlates experimental data well.

Operation in the *transition boiling* regime is normally avoided in the design of heat transfer equipment, and thus no major attempt has been made to develop general correlations for boiling heat transfer in this regime.

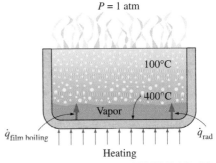

$P = 1$ atm

100°C

400°C

Vapor

$\dot{q}_{film\ boiling}$

$\dot{q}_{rad}$

Heating

FIGURE 10–12

At high heater surface temperatures, radiation heat transfer becomes significant during film boiling.

Note that the gravitational acceleration g, whose value is approximately 9.81 m/s² at sea level, appears in all of the relations above for boiling heat transfer. The effects of low and high gravity (as encountered in aerospace applications and turbomachinery) are studied experimentally. The studies confirm that the critical heat flux and heat flux in film boiling are proportional to $g^{1/4}$. However, they indicate that heat flux in nucleate boiling is practically independent of gravity g, instead of being proportional to $g^{1/2}$, as dictated by Eq. 10–2.

Enhancement of Heat Transfer in Pool Boiling

The pool boiling heat transfer relations given above apply to smooth surfaces. Below we discuss some methods to enhance heat transfer in pool boiling.

We pointed out earlier that the rate of heat transfer in the nucleate boiling regime strongly depends on the number of active nucleation sites on the surface, and the rate of bubble formation at each site. Therefore, any modification that enhances *nucleation* on the heating surface will also enhance *heat transfer* in nucleate boiling. It is observed that *irregularities* on the heating surface, including roughness and dirt, serve as additional nucleation sites during boiling, as shown in Fig. 10–13. For example, the first bubbles in a pan filled with water are most likely to form at the *scratches* at the bottom surface. These scratches act like "nests" for the bubbles to form and thus increase the rate of bubble formation. Berensen has shown that heat flux in the nucleate boiling regime can be increased by a factor of 10 by *roughening* the heating surface. However, these high heat transfer rates cannot be sustained for long since the effect of surface roughness is observed to decay with time, and the heat flux to drop eventually to values encountered on smooth surfaces. The effect of surface roughness is negligible on the critical heat flux and the heat flux in film boiling.

Surfaces that provide enhanced heat transfer in nucleate boiling *permanently* are being manufactured and are available in the market. Enhancement in nucleation and thus heat transfer in such special surfaces is achieved either by *coating* the surface with a thin layer (much less than 1 mm) of very porous material or by *forming cavities* on the surface mechanically to facilitate continuous vapor formation. Such surfaces are reported to enhance heat transfer

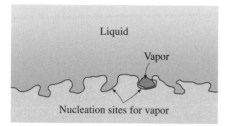

FIGURE 10–13

The cavities on a rough surface act as nucleation sites and enhance boiling heat transfer.

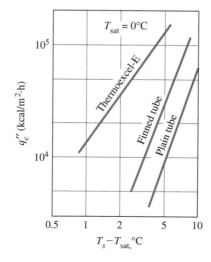

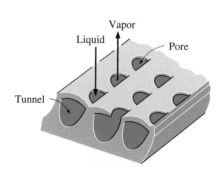

FIGURE 10–14

The enhancement of boiling heat transfer in Freon-12 by a mechanically roughened surface, thermoexcel-E.

in the nucleate boiling regime by a factor of up to 10, and the critical heat flux by a factor of 3. The enhancement provided by one such material prepared by machine roughening, the thermoexcel-E, is shown in Fig. 10–14. The use of finned surfaces is also known to enhance nucleate boiling heat transfer and the critical heat flux.

Boiling heat transfer can also be enhanced by other techniques such as *mechanical agitation* and *surface vibration*. These techniques are not practical, however, because of the complications involved.

EXAMPLE 10–1   Nucleate Boiling of Water in a Pan

Water is to be boiled at atmospheric pressure in a mechanically polished stainless steel pan placed on top of a heating unit, as shown in Fig. 10–15. The inner surface of the bottom of the pan is maintained at 108°C. If the diameter of the bottom of the pan is 30 cm, determine (a) the rate of heat transfer to the water and (b) the rate of evaporation of water.

SOLUTION   Water is boiled at 1 atm pressure on a stainless steel surface. The rate of heat transfer to the water and the rate of evaporation of water are to be determined.

Assumptions   **1** Steady operating conditions exist. **2** Heat losses from the heater and the pan are negligible.

Properties   The properties of water at the saturation temperature of 100°C are $\sigma = 0.0589$ N/m (Table 10–1) and, from Table A–9,

$$\rho_l = 957.9 \text{ kg/m}^3 \qquad h_{fg} = 2257 \times 10^3 \text{ J/kg}$$
$$\rho_v = 0.6 \text{ kg/m}^3 \qquad \mu_l = 0.282 \times 10^{-3} \text{ kg/m} \cdot \text{s}$$
$$\text{Pr}_l = 1.75 \qquad c_{pl} = 4217 \text{ J/kg} \cdot °\text{C}$$

Also, $C_{sf} = 0.0130$ and $n = 1.0$ for the boiling of water on a mechanically polished stainless steel surface (Table 10–3). Note that we expressed the properties in units specified under Eq. 10–2 in connection with their definitions in order to avoid unit manipulations.

Analysis   (a) The excess temperature in this case is $\Delta T = T_s - T_{sat} = 108 - 100 = 8°\text{C}$ which is relatively low (less than 30°C). Therefore, nucleate boiling will occur. The heat flux in this case can be determined from the Rohsenow relation to be

$$\dot{q}_{nucleate} = \mu_l h_{fg} \left[\frac{g(\rho_l - \rho_v)}{\sigma} \right]^{1/2} \left[\frac{c_{pl}(T_s - T_{sat})}{C_{sf} h_{fg} \text{Pr}_l^n} \right]^3$$

$$= (0.282 \times 10^{-3})(2257 \times 10^3) \left[\frac{9.81 \times (957.9 - 0.6)}{0.0589} \right]^{1/2}$$

$$\times \left(\frac{4217(108 - 100)}{0.0130(2257 \times 10^3)1.75} \right)^3$$

$$= 7.21 \times 10^4 \text{ W/m}^2$$

The surface area of the bottom of the pan is

$$A = \pi D^2 / 4 = \pi (0.3 \text{ m})^2 / 4 = 0.07069 \text{ m}^2$$

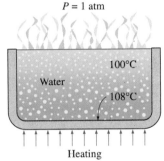

$P = 1$ atm

Water   100°C

108°C

Heating

FIGURE 10–15
Schematic for Example 10–1.

Then the rate of heat transfer during nucleate boiling becomes

$$\dot{Q}_{boiling} = A\dot{q}_{nucleate} = (0.07069 \text{ m}^2)(7.21 \times 10^4 \text{ W/m}^2) = \mathbf{5097 \text{ W}}$$

(b) The rate of evaporation of water is determined from

$$\dot{m}_{evaporation} = \frac{\dot{Q}_{boiling}}{h_{fg}} = \frac{5097 \text{ J/s}}{2257 \times 10^3 \text{ J/kg}} = \mathbf{2.26 \times 10^{-3} \text{ kg/s}}$$

That is, water in the pan will boil at a rate of more than 2 grams per second.

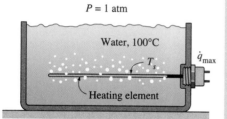

FIGURE 10–16
Schematic for Example 10–2.

EXAMPLE 10–2   **Peak Heat Flux in Nucleate Boiling**

Water in a tank is to be boiled at sea level by a 1-cm-diameter nickel plated steel heating element equipped with electrical resistance wires inside, as shown in Fig. 10–16. Determine the maximum heat flux that can be attained in the nucleate boiling regime and the surface temperature of the heater in that case.

SOLUTION Water is boiled at 1 atm pressure on a nickel plated steel surface. The maximum heat flux and the surface temperature are to be determined.

Assumptions **1** Steady operating conditions exist. **2** Heat losses from the boiler are negligible.

Properties The properties of water at the saturation temperature of 100°C are $\sigma = 0.0589$ N/m (Table 10–1) and, from Table A–9,

$$\rho_l = 957.9 \text{ kg/m}^3 \qquad h_{fg} = 2257 \times 10^3 \text{ J/kg}$$
$$\rho_v = 0.6 \text{ kg/m}^3 \qquad \mu_l = 0.282 \times 10^{-3} \text{ kg/m} \cdot \text{s}$$
$$\text{Pr}_l = 1.75 \qquad c_{pl} = 4217 \text{ J/kg} \cdot °\text{C}$$

Also, $C_{sf} = 0.0060$ and $n = 1.0$ for the boiling of water on a nickel plated surface (Table 10–3). Note that we expressed the properties in units specified under Eqs. 10–2 and 10–3 in connection with their definitions in order to avoid unit manipulations.

Analysis The heating element in this case can be considered to be a short cylinder whose characteristic dimension is its radius. That is, $L = r = 0.005$ m. The dimensionless parameter L^* and the constant C_{cr} are determined from Table 10–4 to be

$$L^* = L\left(\frac{g(\rho_l - \rho_v)}{\sigma}\right)^{1/2} = (0.005)\left(\frac{(9.81)(957.9 - 0.6)}{0.0589}\right)^{1/2} = 2.00 > 1.2$$

which corresponds to $C_{cr} = 0.12$.

Then the maximum or critical heat flux is determined from Eq. 10–3 to be

$$\dot{q}_{max} = C_{cr} h_{fg} [\sigma g \rho_v^2 (\rho_l - \rho_v)]^{1/4}$$
$$= 0.12(2257 \times 10^3)[0.0589 \times 9.81 \times (0.6)^2 (957.9 - 0.6)]^{1/4}$$
$$= \mathbf{1.017 \times 10^6 \text{ W/m}^2}$$

The Rohsenow relation, which gives the nucleate boiling heat flux for a specified surface temperature, can also be used to determine the surface temperature when the heat flux is given. Substituting the maximum heat flux into Eq. 10–2 together with other properties gives

$$q_{nucleate} = \mu_l \, h_{fg} \left[\frac{g(\rho_l - \rho_v)}{\sigma} \right]^{1/2} \left[\frac{c_{pl} \, (T_s - T_{sat})}{C_{sf} \, h_{fg} \, Pr_l^n} \right]^3$$

$$1.017 \times 10^6 = (0.282 \times 10^{-3})(2257 \times 10^3) \left[\frac{9.81(957.9 - 0.6)}{0.0589} \right]^{1/2}$$

$$\times \left[\frac{4217(T_s - 100)}{0.0130(2257 \times 10^3) \, 1.75} \right]^3$$

$$T_s = 119°C$$

Discussion Note that heat fluxes on the order of 1 MW/m² can be obtained in nucleate boiling with a temperature difference of less than 20°C.

EXAMPLE 10–3   Film Boiling of Water on a Heating Element

Water is boiled at atmospheric pressure by a horizontal polished copper heating element of diameter $D = 5$ mm and emissivity $\varepsilon = 0.05$ immersed in water, as shown in Fig. 10–17. If the surface temperature of the heating wire is 350°C, determine the rate of heat transfer from the wire to the water per unit length of the wire.

SOLUTION Water is boiled at 1 atm by a horizontal polished copper heating element. The rate of heat transfer to the water per unit length of the heater is to be determined.

Assumptions 1 Steady operating conditions exist. 2 Heat losses from the boiler are negligible.

Properties The properties of water at the saturation temperature of 100°C are $h_{fg} = 2257 \times 10^3$ J/kg and $\rho_l = 957.9$ kg/m³ (Table A–9). The properties of vapor at the film temperature of $T_f = (T_{sat} + T_s)/2 = (100 + 350)/2 = 225°C$ are, from Table A–16,

$$\rho_v = 0.444 \text{ kg/m}^3 \qquad c_{pv} = 1951 \text{ J/kg} \cdot °C$$
$$\mu_v = 1.75 \times 10^{-5} \text{ kg/m} \cdot \text{s} \qquad k_v = 0.0358 \text{ W/m} \cdot °C$$

Note that we expressed the properties in units that cancel each other in boiling heat transfer relations. Also note that we used vapor properties at 1 atm pressure from Table A–16 instead of the properties of saturated vapor from Table A–9 at 225°C since the latter are at the saturation pressure of 2.55 MPa.

Analysis The excess temperature in this case is $\Delta T = T_s - T_{sat} = 350 - 100 = 250°C$, which is much larger than 30°C for water. Therefore, film boiling will occur. The film boiling heat flux in this case can be determined from Eq. 10–5 to be

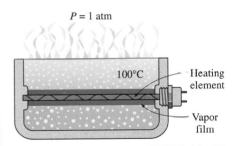

$P = 1$ atm

100°C — Heating element

Vapor film

FIGURE 10–17
Schematic for Example 10–3.

$$\dot{q}_{film} = 0.62 \left[\frac{gk_v^3 \rho_v (\rho_l - \rho_v)[h_{fg} + 0.4c_{pv}(T_s - T_{sat})]}{\mu_v D(T_s - T_{sat})} \right]^{1/4} (T_s - T_{sat})$$

$$= 0.62 \left[\frac{9.81(0.0358)^3 (0.444)(957.9 - 0.441)}{(1.75 \times 10^{-5})(5 \times 10^{-3})(250)} \right]^{1/4} \times 250$$

$$= 5.93 \times 10^4 \ \text{W/m}^2$$

The radiation heat flux is determined from Eq. 10–6 to be

$$\dot{q}_{rad} = \varepsilon \sigma (T_s^4 - T_{sat}^4)$$
$$= (0.05)(5.67 \times 10^{-8} \ \text{W/m}^2 \cdot \text{K}^4)[(350 + 273 \ \text{K})^4 - (100 + 273 \ \text{K})^4]$$
$$= 372 \ \text{W/m}^2$$

Note that heat transfer by radiation is negligible in this case because of the low emissivity of the surface and the relatively low surface temperature of the heating element. Then the total heat flux becomes (Eq. 10–7)

$$\dot{q}_{total} = \dot{q}_{film} + \frac{3}{4} \dot{q}_{rad} = 5.93 \times 10^4 + \frac{3}{4} \times 372 = 5.96 \times 10^4 \ \text{W/m}^2$$

Finally, the rate of heat transfer from the heating element to the water is determined by multiplying the heat flux by the heat transfer surface area,

$$\dot{Q}_{total} = A\dot{q}_{total} = (\pi DL)\dot{q}_{total}$$
$$= (\pi \times 0.005 \ \text{m} \times 1 \ \text{m})(5.96 \times 10^4 \ \text{W/m}^2)$$
$$= \textbf{936 W}$$

Discussion Note that the 5-mm-diameter copper heating element consumes about 1 kW of electric power per unit length in steady operation in the film boiling regime. This energy is transferred to the water through the vapor film that forms around the wire.

10–3 · FLOW BOILING

The pool boiling we considered so far involves a pool of seemingly motionless liquid, with vapor bubbles rising to the top as a result of buoyancy effects. In **flow boiling**, the fluid is forced to move by an external source such as a pump as it undergoes a phase-change process. The boiling in this case exhibits the combined effects of convection and pool boiling. The flow boiling is also classified as either *external* and *internal flow boiling* depending on whether the fluid is forced to flow over a heated surface or inside a heated tube.

External flow boiling over a plate or cylinder is similar to pool boiling, but the added motion increases both the nucleate boiling heat flux and the critical heat flux considerably, as shown in Fig. 10–18. Note that the higher the velocity, the higher the nucleate boiling heat flux and the critical heat flux. In experiments with water, critical heat flux values as high as 35 MW/m² have been obtained (compare this to the pool boiling value of 1.02 MW/m² at 1 atm pressure) by increasing the fluid velocity.

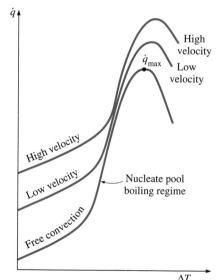

FIGURE 10–18
The effect of forced convection on external flow boiling for different flow velocities.

Internal flow boiling is much more complicated in nature because there is no free surface for the vapor to escape, and thus both the liquid and the vapor are forced to flow together. The two-phase flow in a tube exhibits different flow boiling regimes, depending on the relative amounts of the liquid and the vapor phases. This complicates the analysis even further.

The different stages encountered in flow boiling in a heated tube are illustrated in Fig. 10–19 together with the variation of the heat transfer coefficient along the tube. Initially, the liquid is subcooled and heat transfer to the liquid is by *forced convection.* Then bubbles start forming on the inner surfaces of the tube, and the detached bubbles are drafted into the mainstream. This gives the fluid flow a bubbly appearance, and thus the name *bubbly flow regime.* As the fluid is heated further, the bubbles grow in size and eventually coalesce into slugs of vapor. Up to half of the volume in the tube in this *slug-flow regime* is occupied by vapor. After a while the core of the flow consists of vapor only, and the liquid is confined only in the annular space between the vapor core and the tube walls. This is the *annular-flow regime,* and very high heat transfer coefficients are realized in this regime. As the heating continues, the annular liquid layer gets thinner and thinner, and eventually dry spots start to appear on the inner surfaces of the tube. The appearance of dry spots is accompanied by a sharp decrease in the heat transfer coefficient. This *transition regime* continues until the inner surface of the tube is completely dry. Any liquid at this moment is in the form of droplets suspended in the vapor core, which resembles a mist, and we have a *mist-flow regime* until all the liquid droplets are vaporized. At the end of the mist-flow regime we have saturated vapor, which becomes superheated with any further heat transfer.

Note that the tube contains a liquid before the bubbly flow regime and a vapor after the mist-flow regime. Heat transfer in those two cases can be determined using the appropriate relations for single-phase convection heat transfer. Many correlations are proposed for the determination of heat transfer

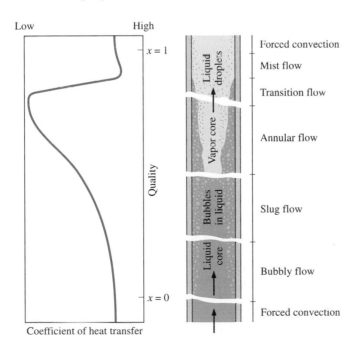

FIGURE 10–19
Different flow regimes encountered in flow boiling in a tube under forced convection.

in the two-phase flow (bubbly flow, slug-flow, annular-flow, and mist-flow) cases, but they are beyond the scope of this introductory text. A crude estimate for heat flux in flow boiling can be obtained by simply adding the forced convection and pool boiling heat fluxes.

10–4 · CONDENSATION HEAT TRANSFER

Condensation occurs when the temperature of a vapor is reduced *below* its saturation temperature T_{sat}. This is usually done by bringing the vapor into contact with a solid surface whose temperature T_s is below the saturation temperature T_{sat} of the vapor. But condensation can also occur on the free surface of a liquid or even in a gas when the temperature of the liquid or the gas to which the vapor is exposed is below T_{sat}. In the latter case, the liquid droplets suspended in the gas form a fog. In this chapter, we consider condensation on solid surfaces only.

Two distinct forms of condensation are observed: *film condensation* and *dropwise condensation*. In **film condensation**, the condensate wets the surface and forms a liquid film on the surface that slides down under the influence of gravity. The thickness of the liquid film increases in the flow direction as more vapor condenses on the film. This is how condensation normally occurs in practice. In **dropwise condensation**, the condensed vapor forms droplets on the surface instead of a continuous film, and the surface is covered by countless droplets of varying diameters (Fig. 10–20).

In film condensation, the surface is blanketed by a liquid film of increasing thickness, and this "liquid wall" between solid surface and the vapor serves as a *resistance* to heat transfer. The heat of vaporization h_{fg} released as the vapor condenses must pass through this resistance before it can reach the solid surface and be transferred to the medium on the other side. In dropwise condensation, however, the droplets slide down when they reach a certain size, clearing the surface and exposing it to vapor. There is no liquid film in this case to resist heat transfer. As a result, heat transfer rates that are more than 10 times larger than those associated with film condensation can be achieved with dropwise condensation. Therefore, dropwise condensation is the preferred mode of condensation in heat transfer applications, and people have long tried to achieve sustained dropwise condensation by using various vapor additives and surface coatings. These attempts have not been very successful, however, since the dropwise condensation achieved did not last long and converted to film condensation after some time. Therefore, it is common practice to be conservative and assume film condensation in the design of heat transfer equipment.

10–5 · FILM CONDENSATION

We now consider film condensation on a vertical plate, as shown in Fig. 10–21. The liquid film starts forming at the top of the plate and flows downward under the influence of gravity. The thickness of the film δ *increases* in the flow direction x because of continued condensation at the liquid–vapor interface. Heat in the amount h_{fg} (the latent heat of vaporization) is released during condensation and is *transferred* through the film to the plate surface at temperature T_s. Note that T_s must be below the saturation temperature T_{sat} of the vapor for condensation to occur.

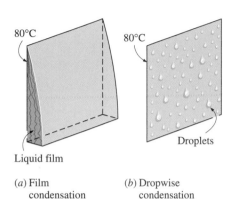

(a) Film condensation   (b) Dropwise condensation

FIGURE 10–20

When a vapor is exposed to a surface at a temperature below T_{sat}, condensation in the form of a liquid film or individual droplets occurs on the surface.

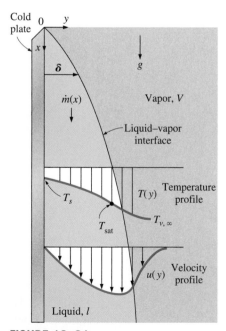

FIGURE 10–21

Film condensation on a vertical plate.

Typical velocity and temperature profiles of the condensate are also given in Fig. 10–21. Note that the *velocity* of the condensate at the wall is zero because of the "no-slip" condition and reaches a *maximum* at the liquid–vapor interface. The *temperature* of the condensate is T_{sat} at the interface and decreases gradually to T_s at the wall.

As was the case in forced convection involving a single phase, heat transfer in condensation also depends on whether the condensate flow is *laminar* or *turbulent*. Again the criterion for the flow regime is provided by the Reynolds number, which is defined as

$$\text{Re} = \frac{D_h \rho_l V_l}{\mu_l} = \frac{4 A_c \rho_l V_l}{p \mu_l} = \frac{4 \rho_l V_l \delta}{\mu_l} = \frac{4 \dot{m}}{p \mu_l} \tag{10–8}$$

where

$D_h = 4A_c/p = 4\delta$ = hydraulic diameter of the condensate flow, m

p = wetted perimeter of the condensate, m

$A_c = p\delta$ = wetted perimeter × film thickness, m², cross-sectional area of the condensate flow at the lowest part of the flow

ρ_l = density of the liquid, kg/m³

μ_l = viscosity of the liquid, kg/m · s

V_l = average velocity of the condensate at the lowest part of the flow, m/s

$\dot{m} = \rho_l V_l A_c$ = mass flow rate of the condensate at the lowest part, kg/s

The evaluation of the hydraulic diameter D_h for some common geometries is illustrated in Fig. 10–22. Note that the hydraulic diameter is again defined such that it reduces to the ordinary diameter for flow in a circular tube, as was done in Chapter 8 for internal flow, and it is equivalent to 4 times the thickness of the condensate film at the location where the hydraulic diameter is evaluated. That is, $D_h = 4\delta$.

The latent heat of vaporization h_{fg} is the heat released as a unit mass of vapor condenses, and it normally represents the heat transfer per unit mass of condensate formed during condensation. However, the condensate in an actual

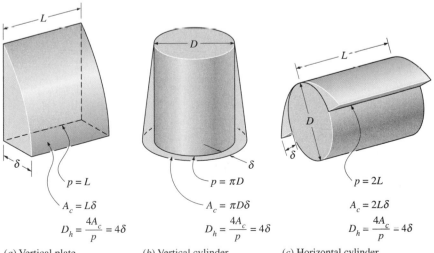

(a) Vertical plate
$p = L$
$A_c = L\delta$
$D_h = \dfrac{4A_c}{p} = 4\delta$

(b) Vertical cylinder
$p = \pi D$
$A_c = \pi D\delta$
$D_h = \dfrac{4A_c}{p} = 4\delta$

(c) Horizontal cylinder
$p = 2L$
$A_c = 2L\delta$
$D_h = \dfrac{4A_c}{p} = 4\delta$

FIGURE 10–22

The wetted perimeter p, the condensate cross-sectional area A_c, and the hydraulic diameter D_h for some common geometries.

condensation process is cooled further to some average temperature between T_{sat} and T_s, releasing more heat in the process. Therefore, the actual heat transfer will be larger. Rohsenow showed in 1956 that the cooling of the liquid below the saturation temperature can be accounted for by replacing h_{fg} by the **modified latent heat of vaporization** h_{fg}^*, defined as

$$h_{fg}^* = h_{fg} + 0.68c_{pl}(T_{sat} - T_s) \tag{10–9a}$$

where c_{pl} is the specific heat of the liquid at the average film temperature.

We can have a similar argument for vapor that enters the condenser as **superheated vapor** at a temperature T_v instead of as saturated vapor. In this case the vapor must be cooled first to T_{sat} before it can condense, and this heat must be transferred to the wall as well. The amount of heat released as a unit mass of superheated vapor at a temperature T_v is cooled to T_{sat} is simply $c_{pv}(T_v - T_{sat})$, where c_{pv} is the specific heat of the vapor at the average temperature of $(T_v + T_{sat})/2$. The modified latent heat of vaporization in this case becomes

$$h_{fg}^* = h_{fg} + 0.68c_{pl}(T_{sat} - T_s) + c_{pv}(T_v - T_{sat}) \tag{10–9b}$$

With these considerations, the rate of heat transfer can be expressed as

$$\dot{Q}_{conden} = hA_s(T_{sat} - T_s) = \dot{m}h_{fg}^* \tag{10–10}$$

where A_s is the heat transfer area (the surface area on which condensation occurs). Solving for $\dot{m}$ from the equation above and substituting it into Eq. 10–8 gives yet another relation for the Reynolds number,

$$Re = \frac{4\dot{Q}_{conden}}{p\mu_l h_{fg}^*} = \frac{4A_s h(T_{sat} - T_s)}{p\mu_l h_{fg}^*} \tag{10–11}$$

This relation is convenient to use to determine the Reynolds number when the condensation heat transfer coefficient or the rate of heat transfer is known.

The temperature of the liquid film varies from T_{sat} on the liquid–vapor interface to T_s at the wall surface. Therefore, the properties of the liquid should be evaluated at the *film temperature* $T_f = (T_{sat} + T_s)/2$, which is approximately the *average* temperature of the liquid. The h_{fg}, however, should be evaluated at T_{sat} since it is not affected by the subcooling of the liquid.

Flow Regimes

The Reynolds number for condensation on the outer surfaces of vertical tubes or plates increases in the flow direction due to the increase of the liquid film thickness δ. The flow of liquid film exhibits *different regimes*, depending on the value of the Reynolds number. It is observed that the outer surface of the liquid film remains *smooth* and *wave-free* for about Re $\leq$ 30, as shown in Fig. 10–23, and thus the flow is clearly *laminar*. Ripples or waves appear on the free surface of the condensate flow as the Reynolds number increases, and the condensate flow becomes fully *turbulent* at about Re $\approx$ 1800. The condensate flow is called *wavy-laminar* in the range of 450 < Re < 1800 and *turbulent* for Re > 1800. However, some disagreement exists about the value of Re at which the flow becomes wavy-laminar or turbulent.

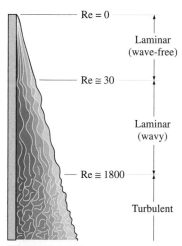

Re = 0

Laminar
(wave-free)

Re ≅ 30

Laminar
(wavy)

Re ≅ 1800

Turbulent

FIGURE 10–23
Flow regimes during film condensation on a vertical plate.

Heat Transfer Correlations for Film Condensation

Below we discuss relations for the average heat transfer coefficient h for the case of *laminar* film condensation for various geometries.

1 Vertical Plates

Consider a vertical plate of height L and width b maintained at a constant temperature T_s that is exposed to vapor at the saturation temperature T_{sat}. The downward direction is taken as the positive x-direction with the origin placed at the top of the plate where condensation initiates, as shown in Fig. 10–24. The surface temperature is below the saturation temperature ($T_s < T_{sat}$) and thus the vapor condenses on the surface. The liquid film flows downward under the influence of gravity. The film thickness δ and thus the mass flow rate of the condensate increases with x as a result of continued condensation on the existing film. Then heat transfer from the vapor to the plate must occur through the film, which offers resistance to heat transfer. Obviously the thicker the film, the larger its thermal resistance and thus the lower the rate of heat transfer.

The analytical relation for the heat transfer coefficient in film condensation on a vertical plate described above was first developed by Nusselt in 1916 under the following simplifying assumptions:

1. Both the plate and the vapor are maintained at *constant temperatures* of T_s and T_{sat}, respectively, and the temperature across the liquid film varies *linearly*.
2. Heat transfer across the liquid film is by pure *conduction* (no convection currents in the liquid film).
3. The velocity of the vapor is low (or zero) so that it exerts *no drag* on the condensate (no viscous shear on the liquid–vapor interface).
4. The flow of the condensate is *laminar* and the properties of the liquid are constant.
5. The acceleration of the condensate layer is negligible.

Then Newton's second law of motion for the volume element shown in Fig. 10–24 in the vertical x-direction can be written as

$$\sum F_x = ma_x = 0$$

since the acceleration of the fluid is zero. Noting that the only force acting downward is the weight of the liquid element, and the forces acting upward are the viscous shear (or fluid friction) force at the left and the buoyancy force, the force balance on the volume element becomes

$$F_{\text{downward}} \downarrow = F_{\text{upward}} \uparrow$$
$$\text{Weight} = \text{Viscous shear force} + \text{Buoyancy force}$$
$$\rho_l g(\delta - y)(b\,dx) = \mu_l \frac{du}{dy}(b\,dx) + \rho_v g(\delta - y)(b\,dx)$$

Canceling the plate width b and solving for du/dy gives

$$\frac{du}{dy} = \frac{g(\rho_l - \rho_v)g(\rho - y)}{\mu_l}$$

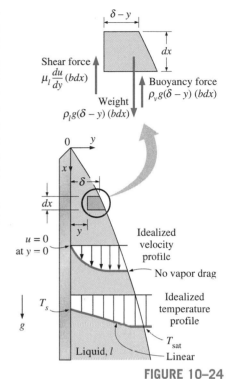

FIGURE 10–24

The volume element of condensate on a vertical plate considered in Nusselt's analysis.

Integrating from $y = 0$ where $u = 0$ (because of the no-slip boundary condition) to $y = y$ where $u = u(y)$ gives

$$u(y) = \frac{g(\rho_l - \rho_v)g}{\mu_l} \left(y\delta - \frac{y^2}{2} \right) \tag{10-12}$$

The mass flow rate of the condensate at a location x, where the boundary layer thickness is δ, is determined from

$$\dot{m}(x) = \int_A \rho_l u(y) dA = \int_{y=0}^{\delta} \rho_l u(y) b dy \tag{10-13}$$

Substituting the $u(y)$ relation from Equation 10–12 into Eq. 10–13 gives

$$\dot{m}(x) = \frac{gb\rho_l(\rho_l - \rho_v)\delta^3}{3\mu_l} \tag{10-14}$$

whose derivative with respect to x is

$$\frac{d\dot{m}}{dx} = \frac{gb\rho_l(\rho_l - \rho_v)\delta^2}{\mu_l} \frac{d\delta}{dx} \tag{10-15}$$

which represents the rate of condensation of vapor over a vertical distance dx. The rate of heat transfer from the vapor to the plate through the liquid film is simply equal to the heat released as the vapor is condensed and is expressed as

$$d\dot{Q} = h_{fg} d\dot{m} = k_l(b dx) \frac{T_{\text{sat}} - T_s}{\delta} \quad \rightarrow \quad \frac{d\dot{m}}{dx} = \frac{k_l \, b}{h_{fg}} \frac{T_{\text{sat}} - T_s}{\delta} \tag{10-16}$$

Equating Eqs. 10–15 and 10–16 for $d\dot{m}/dx$ to each other and separating the variables give

$$\delta^3 d\delta = \frac{\mu_l k_l (T_{\text{sat}} - T_s)}{g\rho_l(\rho_l - \rho_v)h_{fg}} dx \tag{10-17}$$

Integrating from $x = 0$ where $\delta = 0$ (the top of the plate) to $x = x$ where $\delta = \delta(x)$, the liquid film thickness at any location x is determined to be

$$\delta(x) = \left[\frac{4\mu_l k_l (T_{\text{sat}} - T_s)x}{g\rho_l(\rho_l - \rho_v)h_{fg}} \right]^{1/4} \tag{10-18}$$

The heat transfer rate from the vapor to the plate at a location x can be expressed as

$$\dot{q}_x = h_x(T_{\text{sat}} - T_s) = k_l \frac{T_{\text{sat}} - T_s}{\delta} \quad \rightarrow \quad h_x = \frac{k_l}{\delta(x)} \tag{10-19}$$

Substituting the $\delta(x)$ expression from Eq. 10–18, the local heat transfer coefficient h_x is determined to be

$$h_x = \left[\frac{g\rho_l(\rho_l - \rho_v)h_{fg} k_l^3}{4\mu_l (T_{\text{sat}} - T_s)x} \right]^{1/4} \tag{10-20}$$

The average heat transfer coefficient over the entire plate is determined from its definition by substituting the h_x relation and performing the integration. It gives

$$h = h_{\text{avg}} = \frac{1}{L} \int_0^L h_x \, dx = \frac{4}{3} h_{x=L} = 0.943 \left[\frac{g\rho_l (\rho_l - \rho_v) h_{fg} k_l^3}{\mu_l (T_{\text{sat}} - T_s) L} \right]^{1/4} \quad \textbf{(10–21)}$$

Equation 10–21, which is obtained with the simplifying assumptions stated earlier, provides good insight on the functional dependence of the condensation heat transfer coefficient. However, it is observed to underpredict heat transfer because it does not take into account the effects of the nonlinear temperature profile in the liquid film and the cooling of the liquid below the saturation temperature. Both of these effects can be accounted for by replacing h_{fg} by h_{fg}^* given by Eq. 10–9. With this modification, the *average heat transfer coefficient* for laminar film condensation over a vertical flat plate of height L is determined to be

$$h_{\text{vert}} = 0.943 \left[\frac{g\rho_l (\rho_l - \rho_v) h_{fg}^* k_l^3}{\mu_l (T_{\text{sat}} - T_s) L} \right]^{1/4} \quad (\text{W/m}^2 \cdot °\text{C}), \quad 0 < \text{Re} < 30 \quad \textbf{(10–22)}$$

where

g = gravitational acceleration, m/s^2

ρ_l, ρ_v = densities of the liquid and vapor, respectively, kg/m^3

μ_l = viscosity of the liquid, kg/m · s

$h_{fg}^* = h_{fg} + 0.68 c_{pl} (T_{\text{sat}} - T_s)$ = modified latent heat of vaporization, J/kg

k_l = thermal conductivity of the liquid, W/m · °C

L = height of the vertical plate, m

T_s = surface temperature of the plate, °C

T_{sat} = saturation temperature of the condensing fluid, °C

At a given temperature, $\rho_v \ll \rho_l$ and thus $\rho_l - \rho_v \approx \rho_l$ except near the critical point of the substance. Using this approximation and substituting Eqs. 10–14 and 10–18 at $x = L$ into Eq. 10–8 by noting that $\delta_{x=L} = k_l / h_{x=L}$ and $h_{\text{vert}} = \frac{4}{3} h_{x=L}$ (Eqs. 10–19 and 10–21) give

$$\text{Re} \cong \frac{4g\rho_l (\rho_l - \rho_v)\delta^3}{3\mu_l^2} = \frac{4g\rho_l^2}{3\mu_l^2} \left(\frac{k_l}{h_{x=L}} \right)^3 = \frac{4g}{3v_l^2} \left(\frac{k_l}{3h_{\text{vert}}/4} \right)^3 \quad \textbf{(10–23)}$$

Then the heat transfer coefficient h_{vert} in terms of Re becomes

$$h_{\text{vert}} \cong 1.47 k_l \, \text{Re}^{-1/3} \left(\frac{g}{v_l^2} \right)^{1/3}, \qquad \begin{matrix} 0 < \text{Re} < 30 \\ \rho_v \ll \rho_l \end{matrix} \quad \textbf{(10–24)}$$

The results obtained from the theoretical relations above are in excellent agreement with the experimental results. It can be shown easily that using property values in Eqs. 10–22 and 10–24 in the *specified units* gives the condensation heat transfer coefficient in W/m^2 · °C, thus saving one from having

$$h_{vert} = \left(\dfrac{\dfrac{m}{s^2} \dfrac{kg}{m^3} \dfrac{kg}{m^3} \dfrac{J}{kg} \left(\dfrac{W}{m \cdot {}^\circ C} \right)^3}{\dfrac{kg}{m \cdot s} \cdot {}^\circ C \cdot m} \right)^{1/4}$$

$$= \left[\dfrac{m}{s} \dfrac{1}{m^6} \dfrac{W^3}{m^3 \cdot {}^\circ C^3} \dfrac{J}{{}^\circ C} \right]$$

$$= \left(\dfrac{W^4}{m^8 \cdot {}^\circ C^4} \right)^{1/4}$$

$$= W/m^2 \cdot {}^\circ C$$

FIGURE 10–25
Equation 10–22 gives the condensation heat transfer coefficient in W/m² · °C when the quantities are expressed in the units specified in their descriptions.

to go through tedious unit manipulations each time (Fig. 10–25). This is also true for the equations below. All properties of the liquid are to be evaluated at the film temperature $T_f = (T_{sat} + T_s)/2$. The h_{fg} and ρ_v are to be evaluated at the saturation temperature T_{sat}.

Wavy Laminar Flow on Vertical Plates

At Reynolds numbers greater than about 30, it is observed that waves form at the liquid–vapor interface although the flow in liquid film remains laminar. The flow in this case is said to be *wavy laminar*. The waves at the liquid–vapor interface tend to increase heat transfer. But the waves also complicate the analysis and make it very difficult to obtain analytical solutions. Therefore, we have to rely on experimental studies. The increase in heat transfer due to the wave effect is, on average, about 20 percent, but it can exceed 50 percent. The exact amount of enhancement depends on the Reynolds number. Based on his experimental studies, Kutateladze (1963) recommended the following relation for the average heat transfer coefficient in wavy laminar condensate flow for $\rho_v \ll \rho_l$ and $30 < Re < 1800$,

$$h_{vert, wavy} = \dfrac{Re\, k_l}{1.08\, Re^{1.22} - 5.2} \left(\dfrac{g}{v_l^2} \right)^{1/3}, \qquad \begin{matrix} 30 < Re < 1800 \\ \rho_v \ll \rho_l \end{matrix} \qquad \text{(10–25)}$$

A simpler alternative to the relation above proposed by Kutateladze (1963) is

$$h_{vert, wavy} = 0.8\, Re^{0.11} h_{vert\,(smooth)} \qquad \text{(10–26)}$$

which relates the heat transfer coefficient in wavy laminar flow to that in wave-free laminar flow. McAdams (1954) went even further and suggested accounting for the increase in heat transfer in the wavy region by simply increasing the heat transfer coefficient determined from Eq. 10–22 for the laminar case by 20 percent. It is also suggested using Eq. 10–22 for the wavy region also, with the understanding that this is a conservative approach that provides a safety margin in thermal design. In this book we use Eq. 10–25.

A relation for the Reynolds number in the wavy laminar region can be determined by substituting the h relation in Eq. 10–25 into the Re relation in Eq. 10–11 and simplifying. It yields

$$Re_{vert, wavy} = \left[4.81 + \dfrac{3.70\, L k_l (T_{sat} - T_s)}{\mu_l\, h_{fg}^*} \left(\dfrac{g}{v_l^2} \right)^{1/3} \right]^{0.820}, \quad \rho_v \ll \rho_l \qquad \text{(10–27)}$$

Turbulent Flow on Vertical Plates

At a Reynolds number of about 1800, the condensate flow becomes turbulent. Several empirical relations of varying degrees of complexity are proposed for the heat transfer coefficient for turbulent flow. Again assuming $\rho_v \ll \rho_l$ for simplicity, Labuntsov (1957) proposed the following relation for the turbulent flow of condensate on *vertical plates:*

$$h_{vert, turbulent} = \dfrac{Re\, k_l}{8750 + 58\, Pr^{-0.5} (Re^{0.75} - 253)} \left(\dfrac{g}{v_l^2} \right)^{1/3}, \quad \begin{matrix} Re > 1800 \\ \rho_v \ll \rho_l \end{matrix} \qquad \text{(10–28)}$$

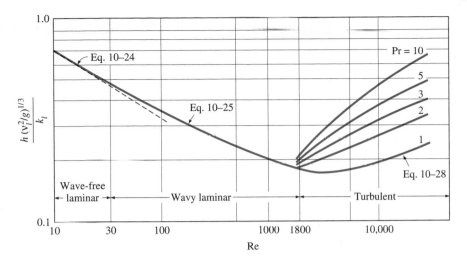

FIGURE 10–26
Nondimensionalized heat transfer coefficients for the wave-free laminar, wavy laminar, and turbulent flow of condensate on vertical plates.

The physical properties of the condensate are again to be evaluated at the film temperature $T_f = (T_{sat} + T_s)/2$. The Re relation in this case is obtained by substituting the h relation above into the Re relation in Eq. 10–11, which gives

$$\text{Re}_{\text{vert, turbulent}} = \left[\frac{0.0690 \, Lk_l \, \text{Pr}^{0.5} \, (T_{sat} - T_s)}{\mu_l \, h_{fg}^*} \left(\frac{g}{v_l^2} \right)^{1/3} - 151 \, \text{Pr}^{0.5} + 253 \right]^{4/3} \quad \text{(10–29)}$$

Nondimensionalized heat transfer coefficients for the wave-free laminar, wavy laminar, and turbulent flow of condensate on vertical plates are plotted in Fig. 10–26.

2 Inclined Plates

Equation 10–22 was developed for vertical plates, but it can also be used for laminar film condensation on the upper surfaces of plates that are *inclined* by an angle θ from the *vertical,* by replacing g in that equation by $g \cos \theta$ (Fig. 10–27). This approximation gives satisfactory results especially for $\theta \le 60°$. Note that the condensation heat transfer coefficients on vertical and inclined plates are related to each other by

$$h_{\text{inclined}} = h_{\text{vert}} (\cos \theta)^{1/4} \qquad \text{(laminar)} \qquad \text{(10–30)}$$

Equation 10–30 is developed for laminar flow of condensate, but it can also be used for wavy laminar flows as an approximation.

3 Vertical Tubes

Equation 10–22 for vertical plates can also be used to calculate the average heat transfer coefficient for laminar film condensation on the outer surfaces of vertical tubes provided that the tube diameter is large relative to the thickness of the liquid film.

4 Horizontal Tubes and Spheres

Nusselt's analysis of film condensation on vertical plates can also be extended to horizontal tubes and spheres. The average heat transfer coefficient for film condensation on the outer surfaces of a *horizontal tube* is determined to be

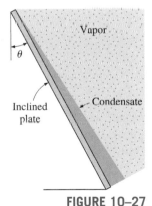

FIGURE 10–27
Film condensation on an inclined plate.

$$h_{\text{horiz}} = 0.729 \left[\frac{g\rho_l (\rho_l - \rho_v)\, h_{fg}^* k_l^3}{\mu_l (T_{\text{sat}} - T_s)D} \right]^{1/4} \qquad (\text{W/m}^2 \cdot {}^\circ\text{C}) \qquad (10\text{--}31)$$

where D is the diameter of the horizontal tube. Equation 10–31 can easily be modified for a *sphere* by replacing the constant 0.729 by 0.815.

A comparison of the heat transfer coefficient relations for a vertical tube of height L and a horizontal tube of diameter D yields

$$\frac{h_{\text{vert}}}{h_{\text{horiz}}} = 1.29 \left(\frac{D}{L} \right)^{1/4} \qquad (10\text{--}32)$$

Setting $h_{\text{vertical}} = h_{\text{horizontal}}$ gives $L = 1.29^4 D = 2.77D$, which implies that for a tube whose length is 2.77 times its diameter, the average heat transfer coefficient for laminar film condensation will be the *same* whether the tube is positioned horizontally or vertically. For $L > 2.77D$, the heat transfer coefficient is higher in the horizontal position. Considering that the length of a tube in any practical application is several times its diameter, it is common practice to place the tubes in a condenser *horizontally* to maximize the condensation heat transfer coefficient on the outer surfaces of the tubes.

5 Horizontal Tube Banks

Horizontal tubes stacked on top of each other as shown in Fig. 10–28 are commonly used in condenser design. The average thickness of the liquid film at the lower tubes is much larger as a result of condensate falling on top of them from the tubes directly above. Therefore, the average heat transfer coefficient at the lower tubes in such arrangements is smaller. Assuming the condensate from the tubes above to the ones below drain smoothly, the average film condensation heat transfer coefficient for all tubes in a vertical tier can be expressed as

$$h_{\text{horiz}, N \text{ tubes}} = 0.729 \left[\frac{g\rho_l (\rho_l - \rho_v)\, h_{fg}^* k_l^3}{\mu_l (T_{\text{sat}} - T_s)\, ND} \right]^{1/4} = \frac{1}{N^{1/4}}\, h_{\text{horiz, 1 tube}} \qquad (10\text{--}33)$$

Note that Eq. 10–33 can be obtained from the heat transfer coefficient relation for a horizontal tube by replacing D in that relation by ND. This relation does not account for the increase in heat transfer due to the ripple formation and turbulence caused during drainage, and thus generally yields conservative results.

Effect of Vapor Velocity

In the analysis above we assumed the vapor velocity to be small and thus the vapor drag exerted on the liquid film to be negligible, which is usually the case. However, when the vapor velocity is high, the vapor will "pull" the liquid at the interface along since the vapor velocity at the interface must drop to the value of the liquid velocity. If the vapor flows downward (i.e., in the same direction as the liquid), this additional force will increase the average velocity of the liquid and thus decrease the film thickness. This, in turn, will decrease the thermal resistance of the liquid film and thus increase heat transfer. Upward vapor flow has the opposite effects: the vapor exerts a force on the

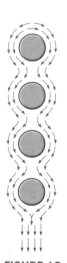

FIGURE 10–28

Film condensation on a vertical tier of horizontal tubes.

liquid in the opposite direction to flow, thickens the liquid film, and thus decreases heat transfer. Condensation in the presence of high vapor flow is studied [e.g., Shekriladze and Gomelauri (1966)] and heat transfer relations are obtained, but a detailed analysis of this topic is beyond the scope of this introductory text.

The Presence of Noncondensable Gases in Condensers

Most condensers used in steam power plants operate at pressures well below the atmospheric pressure (usually under 0.1 atm) to maximize cycle thermal efficiency, and operation at such low pressures raises the possibility of air (a noncondensable gas) leaking into the condensers. Experimental studies show that the presence of noncondensable gases in the vapor has a detrimental effect on condensation heat transfer. Even small amounts of a noncondensable gas in the vapor cause significant drops in heat transfer coefficient during condensation. For example, the presence of less than 1 percent (by mass) of air in steam can reduce the condensation heat transfer coefficient by more than half. Therefore, it is common practice to periodically vent out the noncondensable gases that accumulate in the condensers to ensure proper operation.

The drastic reduction in the condensation heat transfer coefficient in the presence of a noncondensable gas can be explained as follows: When the vapor mixed with a noncondensable gas condenses, only the noncondensable gas remains in the vicinity of the surface (Fig. 10–29). This gas layer acts as a *barrier* between the vapor and the surface, and makes it difficult for the vapor to reach the surface. The vapor now must diffuse through the noncondensable gas first before reaching the surface, and this reduces the effectiveness of the condensation process.

Experimental studies show that heat transfer in the presence of a noncondensable gas strongly depends on the nature of the vapor flow and the flow velocity. As you would expect, a *high flow velocity* is more likely to remove the stagnant noncondensable gas from the vicinity of the surface, and thus *improve* heat transfer.

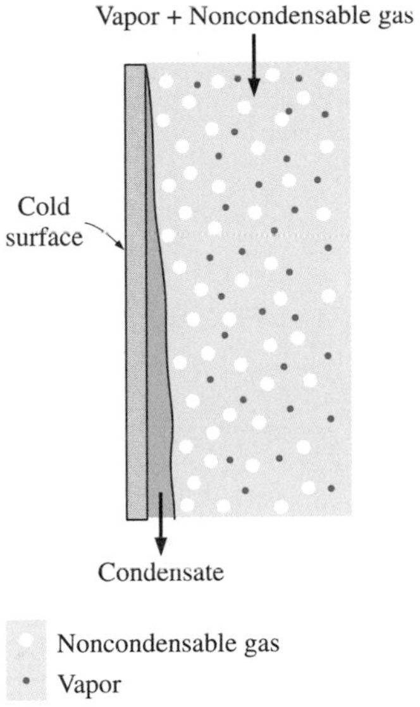

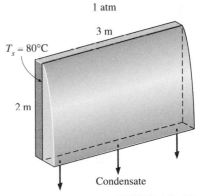

Vapor + Noncondensable gas

Cold surface

Condensate

○ Noncondensable gas
• Vapor

FIGURE 10–29
The presence of a noncondensable gas in a vapor prevents the vapor molecules from reaching the cold surface easily, and thus impedes condensation heat transfer.

EXAMPLE 10–4   Condensation of Steam on a Vertical Plate

Saturated steam at atmospheric pressure condenses on a 2-m-high and 3-m-wide vertical plate that is maintained at 80°C by circulating cooling water through the other side (Fig. 10–30). Determine (a) the rate of heat transfer by condensation to the plate and (b) the rate at which the condensate drips off the plate at the bottom.

SOLUTION   Saturated steam at 1 atm condenses on a vertical plate. The rates of heat transfer and condensation are to be determined.

Assumptions   **1** Steady operating conditions exist. **2** The plate is isothermal. **3** The condensate flow is wavy-laminar over the entire plate (will be verified). **4** The density of vapor is much smaller than the density of liquid, $\rho_v \ll \rho_l$.

Properties   The properties of water at the saturation temperature of 100°C are $h_{fg} = 2257 \times 10^3$ J/kg and $\rho_v = 0.60$ kg/m³. The properties of liquid water at the film temperature of $T_f = (T_{sat} + T_s)/2 = (100 + 80)/2 = 90°C$ are (Table A–9)

1 atm

3 m

$T_s = 80°C$

2 m

Condensate

FIGURE 10–30
Schematic for Example 10–4.

$$\rho_l = 965.3 \text{ kg/m}^3 \qquad\qquad c_{pl} = 4206 \text{ J/kg} \cdot \text{°C}$$
$$\mu_l = 0.315 \times 10^{-3} \text{ kg/m} \cdot \text{s} \qquad k_l = 0.675 \text{ W/m} \cdot \text{°C}$$
$$v_l = \mu_l/\rho_l = 0.326 \times 10^{-6} \text{ m}^2/\text{s}$$

Analysis (a) The modified latent heat of vaporization is

$$h_{fg}^* = h_{fg} + 0.68c_{pl}(T_{sat} - T_s)$$
$$= 2257 \times 10^3 \text{ J/kg} + 0.68 \times (4206 \text{ J/kg} \cdot \text{°C})(100 - 80)\text{°C}$$
$$= 2314 \times 10^3 \text{ J/kg}$$

For wavy-laminar flow, the Reynolds number is determined from Eq. 10–27 to be

$$\text{Re} = \text{Re}_{vertical,\, wavy} = \left[4.81 + \frac{3.70\, Lk_l(T_{sat} - T_s)}{\mu_l\, h_{fg}^*}\left(\frac{g}{v_l^2}\right)^{1/3}\right]^{0.820}$$

$$= \left[4.81 + \frac{3.70(2 \text{ m})(0.675 \text{ W/m} \cdot \text{°C})(100 - 80)\text{°C}}{(0.315 \times 10^{-3} \text{ kg/m} \cdot \text{s})(2314 \times 10^3 \text{ J/kg})}\right.$$
$$\left. \times \left(\frac{9.81 \text{ m/s}^2}{(0.326 \times 10^{-6} \text{ m}^2/\text{s})^2}\right)^{1/3}\right]^{0.820}$$

$$= 1287$$

which is between 30 and 1800, and thus our assumption of wavy laminar flow is verified. Then the condensation heat transfer coefficient is determined from Eq. 10–25 to be

$$h = h_{vertical,\, wavy} = \frac{\text{Re}\, k_l}{1.08\, \text{Re}^{1.22} - 5.2}\left(\frac{g}{v_l^2}\right)^{1/3}$$

$$= \frac{1287 \times (0.675 \text{ W/m} \cdot \text{°C})}{1.08(1287)^{1.22} - 5.2}\left(\frac{9.81 \text{ m/s}^2}{(0.326 \times 10^{-6} \text{ m}^2/\text{s})^2}\right)^{1/3} = 5850 \text{ W/m}^2 \cdot \text{°C}$$

The heat transfer surface area of the plate is $A_s = W \times L = (3 \text{ m})(2 \text{ m}) = 6 \text{ m}^2$. Then the rate of heat transfer during this condensation process becomes

$$\dot{Q} = hA_s(T_{sat} - T_s) = (5850 \text{ W/m}^2 \cdot \text{°C})(6 \text{ m}^2)(100 - 80)\text{°C} = \textbf{7.02} \times \textbf{10}^5 \text{ W}$$

(b) The rate of condensation of steam is determined from

$$\dot{m}_{condensation} = \frac{\dot{Q}}{h_{fg}^*} = \frac{7.02 \times 10^5 \text{ J/s}}{2314 \times 10^3 \text{ J/kg}} = \textbf{0.303 kg/s}$$

That is, steam will condense on the surface at a rate of 303 grams per second.

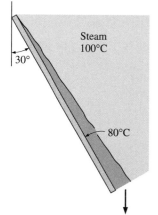

FIGURE 10–31
Schematic for Example 10–5.

EXAMPLE 10–5   **Condensation of Steam on a Tilted Plate**

What would your answer be to the preceding example problem if the plate were tilted 30° from the vertical, as shown in Fig. 10–31?

SOLUTION (a) The heat transfer coefficient in this case can be determined from the vertical plate relation by replacing g by $g \cos \theta$. But we will use Eq. 10–30 instead since we already know the value for the vertical plate from the preceding example:

$$h = h_{\text{inclined}} = h_{\text{vert}} (\cos \theta)^{1/4} = (5850 \text{ W/m}^2 \cdot °C)(\cos 30°)^{1/4} = 5643 \text{ W/m}^2 \cdot °C$$

The heat transfer surface area of the plate is still 6 m². Then the rate of condensation heat transfer in the tilted plate case becomes

$$\dot{Q} = hA_s(T_{\text{sat}} - T_s) = (5643 \text{ W/m}^2 \cdot °C)(6 \text{ m}^2)(100 - 80)°C = \mathbf{6.77 \times 10^5 \ W}$$

(b) The rate of condensation of steam is again determined from

$$\dot{m}_{\text{condensation}} = \frac{\dot{Q}}{h_{fg}^*} = \frac{6.77 \times 10^5 \text{ J/s}}{2314 \times 10^3 \text{ J/kg}} = \mathbf{0.293 \ kg/s}$$

Discussion Note that the rate of condensation decreased by about 3.3 percent when the plate is tilted.

EXAMPLE 10–6 **Condensation of Steam on Horizontal Tubes**

The condenser of a steam power plant operates at a pressure of 7.38 kPa. Steam at this pressure condenses on the outer surfaces of horizontal tubes through which cooling water circulates. The outer diameter of the pipes is 3 cm, and the outer surfaces of the tubes are maintained at 30°C (Fig. 10–32). Determine (a) the rate of heat transfer to the cooling water circulating in the tubes and (b) the rate of condensation of steam per unit length of a horizontal tube.

SOLUTION Saturated steam at a pressure of 7.38 kPa condenses on a horizontal tube at 30°C. The rates of heat transfer and condensation are to be determined.
Assumptions 1 Steady operating conditions exist. 2 The tube is isothermal.
Properties The properties of water at the saturation temperature of 40°C corresponding to 7.38 kPa are $h_{fg} = 2407 \times 10^3$ J/kg and $\rho_v = 0.05$ kg/m³. The properties of liquid water at the film temperature of $T_f = (T_{\text{sat}} + T_s)/2 = (40 + 30)/2 = 35°C$ are (Table A–9)

$$\rho_l = 994 \text{ kg/m}^3 \qquad\qquad c_{pl} = 4178 \text{ J/kg} \cdot °C$$
$$\mu_l = 0.720 \times 10^{-3} \text{ kg/m} \cdot \text{s} \qquad k_l = 0.623 \text{ W/m} \cdot °C$$

Analysis (a) The modified latent heat of vaporization is

$$h_{fg}^* = h_{fg} + 0.68c_{pl}(T_{\text{sat}} - T_s)$$
$$= 2407 \times 10^3 \text{ J/kg} + 0.68 \times (4178 \text{ J/kg} \cdot °C)(40 - 30)°C$$
$$= 2435 \times 10^3 \text{ J/kg}$$

Noting that $\rho_v \ll \rho_l$ (since $0.05 \ll 994$), the heat transfer coefficient for condensation on a single horizontal tube is determined from Eq. 10–31 to be

Steam, 40°C   30°C

Cooling water

FIGURE 10–32
Schematic for Example 10–6.

$$h = h_{horiz} = 0.729 \left[\frac{g\rho_l(\rho_l - \rho_v)\, h_{fg}^*\, k_l^3}{\mu(T_{sat} - T_s)\, D} \right]^{1/4} \cong 0.729 \left[\frac{g\rho_l^2\, h_{fg}^*\, k_l^3}{\mu_l\, (T_{sat} - T_s)\, D} \right]^{1/4}$$

$$= 0.729 \left[\frac{(9.81\ m/s^2)(994\ kg/m^3)^2\, (2435 \times 10^3\ J/kg)(0.623\ W/m \cdot °C)^3}{(0.720 \times 10^{-3}\ kg/m \cdot s)(40 - 30)°C(0.03\ m)} \right]^{1/4}$$

$$= 9294\ W/m^2 \cdot °C$$

The heat transfer surface area of the pipe per unit of its length is $A_s = \pi DL = \pi(0.03\ m)(1\ m) = 0.09425\ m^2$. Then the rate of heat transfer during this condensation process becomes

$$\dot{Q} = hA_s(T_{sat} - T_s) = (9292\ W/m^2 \cdot °C)(0.09425\ m^2)(40 - 30)°C = \textbf{8760 W}$$

(b) The rate of condensation of steam is

$$\dot{m}_{condensation} = \frac{\dot{Q}}{h_{fg}^*} = \frac{8760\ J/s}{2435 \times 10^3\ J/kg} = \textbf{0.00360 kg/s}$$

Therefore, steam will condense on the horizontal tube at a rate of 3.6 g/s or 13.0 kg/h per meter of its length.

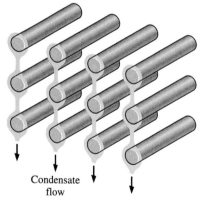

Condensate
flow

FIGURE 10–33
Schematic for Example 10–7.

EXAMPLE 10–7   **Condensation of Steam on Horizontal Tube Banks**

Repeat the preceding example problem for the case of 12 horizontal tubes arranged in a rectangular array of 3 tubes high and 4 tubes wide, as shown in Fig. 10–33.

SOLUTION   (a) Condensation heat transfer on a tube is not influenced by the presence of other tubes in its neighborhood unless the condensate from other tubes drips on it. In our case, the horizontal tubes are arranged in four vertical tiers, each tier consisting of 3 tubes. The average heat transfer coefficient for a vertical tier of N horizontal tubes is related to the one for a single horizontal tube by Eq. 10–33 and is determined to be

$$h_{horiz,\ N\ tubes} = \frac{1}{N^{1/4}}\, h_{horiz,\ 1\ tube} = \frac{1}{3^{1/4}}\, (9294\ W/m^2 \cdot °C) = 7062\ W/m^2 \cdot °C$$

Each vertical tier consists of 3 tubes, and thus the heat transfer coefficient determined above is valid for each of the four tiers. In other words, this value can be taken to be the average heat transfer coefficient for all 12 tubes.
   The surface area for all 12 tubes per unit length of the tubes is

$$A_s = N_{total}\, \pi DL = 12\pi(0.03\ m)(1\ m) = 1.1310\ m^2$$

Then the rate of heat transfer during this condensation process becomes

$$\dot{Q} = hA_s(T_{sat} - T_s) = (7062\ W/m^2 \cdot °C)(1.131\ m^2)(40 - 30)°C = \textbf{79,870 W}$$

(b) The rate of condensation of steam is again determined from

$$\dot{m}_{\text{condensation}} = \frac{\dot{Q}}{h_{fg}^*} = \frac{79{,}870 \text{ J/s}}{2435 \times 10^3 \text{ J/kg}} = 0.0328 \text{ kg/s}$$

Therefore, steam will condense on the horizontal pipes at a rate of 32.8 g/s per meter length of the tubes.

10–6 · FILM CONDENSATION INSIDE HORIZONTAL TUBES

So far we have discussed film condensation on the *outer surfaces* of tubes and other geometries, which is characterized by negligible vapor velocity and the unrestricted flow of the condensate. Most condensation processes encountered in refrigeration and air-conditioning applications, however, involve condensation on the *inner surfaces* of horizontal or vertical tubes. Heat transfer analysis of condensation inside tubes is complicated by the fact that it is strongly influenced by the vapor velocity and the rate of liquid accumulation on the walls of the tubes (Fig. 10–34).

For *low vapor velocities,* Chato (1962) recommends this expression for condensation

$$h_{\text{internal}} = 0.555 \left[\frac{g\rho_l(\rho_l - \rho_v) k_l^3}{\mu_l(T_{\text{sat}} - T_s)} \left(h_{fg} + \frac{3}{8} c_{pl}(T_{\text{sat}} - T_s) \right) \right]^{1/4} \qquad (10\text{–}34)$$

for

$$\text{Re}_{\text{vapor}} = \left(\frac{\rho_v V_v D}{\mu_v} \right)_{\text{inlet}} < 35{,}000 \qquad (10\text{–}35)$$

where the Reynolds number of the vapor is to be evaluated at the tube *inlet* conditions using the internal tube diameter as the characteristic length. Heat transfer coefficient correlations for higher vapor velocities are given by Rohsenow.

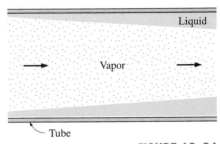

FIGURE 10–34

Condensate flow in a horizontal tube with large vapor velocities.

10–7 · DROPWISE CONDENSATION

Dropwise condensation, characterized by countless droplets of varying diameters on the condensing surface instead of a continuous liquid film, is one of the most effective mechanisms of heat transfer, and extremely large heat transfer coefficients can be achieved with this mechanism (Fig. 10–35).

In dropwise condensation, the small droplets that form at the nucleation sites on the surface grow as a result of continued condensation, coalesce into large droplets, and slide down when they reach a certain size, clearing the surface and exposing it to vapor. There is no liquid film in this case to resist heat transfer. As a result, with dropwise condensation, heat transfer coefficients can be achieved that are more than 10 times larger than those associated with film condensation. Large heat transfer coefficients enable designers to achieve a specified heat transfer rate with a smaller surface area, and thus a smaller (and less expensive) condenser. Therefore, dropwise condensation is the preferred mode of condensation in heat transfer applications.

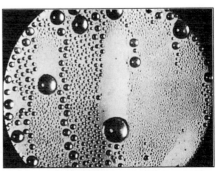

FIGURE 10–35

Dropwise condensation of steam on a vertical surface.
(From Hampson and Özişik.)

The challenge in dropwise condensation is not to achieve it, but rather, to *sustain* it for prolonged periods of time. Dropwise condensation is achieved by *adding* a promoting chemical into the vapor, *treating* the surface with a promoter chemical, or *coating* the surface with a polymer such as teflon or a noble metal such as gold, silver, rhodium, palladium, or platinum. The *promoters* used include various waxes and fatty acids such as oleic, stearic, and linoic acids. They lose their effectiveness after a while, however, because of fouling, oxidation, and the removal of the promoter from the surface. It is possible to sustain dropwise condensation for over a year by the combined effects of surface coating and periodic injection of the promoter into the vapor. However, any gain in heat transfer must be weighed against the cost associated with sustaining dropwise condensation.

Dropwise condensation has been studied experimentally for a number of surface–fluid combinations. Of these, the studies on the condensation of steam on copper surfaces has attracted the most attention because of their widespread use in steam power plants. P. Griffith (1983) recommends these simple correlations for dropwise condensation of *steam* on *copper surfaces:*

$$h_{\text{dropwise}} = \begin{cases} 51{,}104 + 2044 T_{\text{sat}}, & 22°C < T_{\text{sat}} < 100°C \qquad \textbf{(10–36)} \\ 255{,}310, & T_{\text{sat}} > 100°C \qquad \textbf{(10–37)} \end{cases}$$

where T_{sat} is in °C and the heat transfer coefficient h_{dropwise} is in W/m² · °C.

The very high heat transfer coefficients achievable with dropwise condensation are of little significance if the material of the condensing surface is not a good conductor like copper or if the thermal resistance on the other side of the surface is too large. In steady operation, heat transfer from one medium to another depends on the sum of the thermal resistances on the path of heat flow, and a large thermal resistance may overshadow all others and dominate the heat transfer process. In such cases, improving the accuracy of a small resistance (such as one due to condensation or boiling) makes hardly any difference in overall heat transfer calculations.

TOPIC OF SPECIAL INTEREST*

Heat Pipes

A **heat pipe** is a simple device with no moving parts that can transfer large quantities of heat over fairly large distances essentially at a constant temperature without requiring any power input. A heat pipe is basically a sealed slender tube containing a wick structure lined on the inner surface and a small amount of fluid such as water at the saturated state, as shown in Fig. 10–36. It is composed of three sections: the *evaporator* section at one end, where heat is absorbed and the fluid is vaporized; a *condenser* section at the other end, where the vapor is condensed and heat is rejected; and the *adiabatic* section in between, where the vapor and the liquid phases of the fluid flow in opposite directions through the core and the wick,

*This section can be skipped without a loss in continuity.

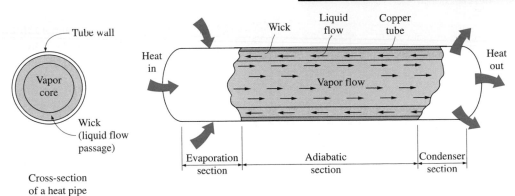

FIGURE 10–36
Schematic and operation of a heat pipe.

respectively, to complete the cycle with no significant heat transfer between the fluid and the surrounding medium.

The *type of fluid* and the *operating pressure* inside the heat pipe depend on the *operating temperature* of the heat pipe. For example, the triple- and critical-point temperatures of water are 0.01°C and 374.1°C, respectively. Therefore, water can undergo a liquid-to-vapor or vapor-to-liquid phase change process in this temperature range only, and thus it is not a suitable fluid for applications involving temperatures beyond this range. Furthermore, water undergoes a phase-change process at a specified temperature only if its pressure equals the saturation pressure at that temperature. For example, if a heat pipe with water as the working fluid is designed to remove heat at 70°C, the pressure inside the heat pipe must be maintained at 31.2 kPa, which is the boiling pressure of water at this temperature. Note that this value is well below the atmospheric pressure of 101 kPa, and thus the heat pipe operates in a vacuum environment in this case. If the pressure inside is maintained at the local atmospheric pressure instead, heat transfer would result in an increase in the temperature of the water instead of evaporation.

Although water is a suitable fluid to use in the moderate temperature range encountered in electronic equipment, several other fluids can be used in the construction of heat pipes to enable them to be used in cryogenic as well as high-temperature applications. The suitable temperature ranges for some common heat pipe fluids are given in Table 10–5. Note that the overall temperature range extends from almost absolute zero for cryogenic fluids such as helium to over 1600°C for liquid metals such as lithium. The ultimate temperature limits for a fluid are the *triple-* and *critical-point* temperatures. However, a narrower temperature range is used in practice to avoid the extreme pressures and low heats of vaporization that occur near the critical point. Other desirable characteristics of the candidate fluids are having a high surface tension to enhance the capillary effect and being compatible with the wick material, as well as being readily available, chemically stable, nontoxic, and inexpensive.

The concept of heat pipe was originally conceived by R. S. Gaugler of the General Motors Corporation, who filed a patent application for it in

TABLE 10–5

Suitable temperature ranges for some fluids used in heat pipes

Fluid	Temperature Range, °C
Helium	−271 to −268
Hydrogen	−259 to −240
Neon	−248 to −230
Nitrogen	−210 to −150
Methane	−182 to −82
Ammonia	−78 to −130
Water	5 to 230
Mercury	200 to 500
Cesium	400 to 1000
Sodium	500 to 1200
Lithium	850 to 1600

1942. However, it did not receive much attention until 1962, when it was suggested for use in space applications. Since then, heat pipes have found a wide range of applications, including the cooling of electronic equipment.

The Operation of a Heat Pipe

The operation of a heat pipe is based on the following physical principles:

- At a specified pressure, a liquid vaporizes or a vapor condenses at a certain temperature, called the *saturation temperature*. Thus, fixing the pressure insides a heat pipe fixes the temperature at which phase change occurs.

- At a specified pressure or temperature, the amount of heat *absorbed* as a unit mass of liquid vaporizes is equal to the amount of heat *rejected* as that vapor condenses.

- The capillary pressure developed in a wick moves a liquid in the wick even *against* the gravitational field as a result of the capillary effect.

- A fluid in a channel flows in the direction of *decreasing pressure.*

Initially, the *wick* of the heat pipe is saturated with liquid and the *core section* is filled with vapor. When the evaporator end of the heat pipe is brought into contact with a hot surface or is placed into a hot environment, heat transfers into the heat pipe. Being at a saturated state, the liquid in the evaporator end of the heat pipe *vaporizes* as a result of this heat transfer, causing the vapor pressure there to rise. This resulting pressure difference drives the vapor through the core of the heat pipe from the evaporator toward the condenser section. The condenser end of the heat pipe is in a cooler environment, and thus its surface is slightly cooler. The vapor that comes into contact with this cooler surface *condenses,* releasing the heat a vaporization, which is rejected to the surrounding medium. The liquid then returns to the evaporator end of the heat pipe through the wick as a result of *capillary action* in the wick, completing the cycle. As a result, heat is absorbed at one end of the heat pipe and is rejected at the other end, with the fluid inside serving as a transport medium for heat.

The boiling and condensation processes are associated with extremely high heat transfer coefficients, and thus it is natural to expect the heat pipe to be a very effective heat transfer device, since its operation is based on alternative boiling and condensation of the working fluid. Indeed, heat pipes have effective conductivities *several hundred times* that of copper or silver. That is, replacing a copper bar between two mediums at different temperatures by a heat pipe of equal size can increase the rate of heat transfer between those two mediums by several hundred times. A simple heat pipe with water as the working fluid has an effective thermal conductivity of the order of 100,000 W/m · °C compared with about 400 W/m · °C for copper. For a heat pipe, it is not unusual to have an effective conductivity of 400,000 W/m · °C, which is 1000 times that of copper. A 15-cm-long, 0.6-cm-diameter horizontal cylindrical heat pipe with water inside, for example, can transfer heat at a rate of 300 W. Therefore, heat pipes are preferred in some critical applications, despite their high initial cost.

There is a small pressure difference between the evaporator and condenser ends, and thus a small temperature difference between the two ends of the heat pipe. This temperature difference is usually between 1°C and 5°C.

The Construction of a Heat Pipe

The wick of a heat pipe provides the means for the return of the liquid to the evaporator. Therefore, the structure of the wick has a strong effect on the performance of a heat pipe, and the design and construction of the wick are the most critical aspects of the manufacturing process.

The wicks are often made of porous ceramic or woven stainless wire mesh. They can also be made together with the tube by extruding axial grooves along its inner surface, but this approach presents manufacturing difficulties.

The performance of a wick depends on its structure. The characteristics of a wick can be changed by changing the *size* and the *number* of the pores per unit volume and the *continuity* of the passageway. Liquid motion in the wick depends on the dynamic balance between two opposing effects: the *capillary pressure,* which creates the suction effect to draw the liquid, and the *internal resistance to flow* as a result of friction between the mesh surfaces and the liquid. A small pore size increases the capillary action, since the capillary pressure is inversely proportional to the effective capillary radius of the mesh. But decreasing the pore size and thus the capillary radius also increases the friction force opposing the motion. Therefore, the core size of the mesh should be reduced so long as the increase in capillary force is greater than the increase in the friction force.

Note that the *optimum pore size* is different for different fluids and different orientations of the heat pipe. An improperly designed wick results in an inadequate liquid supply and eventual failure of the heat pipe.

Capillary action permits the heat pipe to operate in any orientation in a gravity field. However, the performance of a heat pipe is best when the capillary and gravity forces act in the same direction (evaporator end down) and is worst when these two forces act in opposite directions (evaporator end up). Gravity does not affect the capillary force when the heat pipe is in the horizontal position. The heat removal capacity of a horizontal heat pipe can be *doubled* by installing it vertically with evaporator end down so that gravity helps the capillary action. In the opposite case, vertical orientation with evaporator end up, the performance declines considerably relative the horizontal case since the capillary force in this case must work against the gravity force.

Most heat pipes are cylindrical in shape. However, they can be manufactured in a variety of shapes involving 90° bends, S-turns, or spirals. They can also be made as a flat layer with a thickness of about 0.3 cm. Flat heat pipes are very suitable for cooling high-power-output (say, 50 W or greater) PCBs. In this case, flat heat pipes are attached directly to the back surface of the PCB, and they absorb and transfer the heat to the edges. Cooling fins are usually attached to the condenser end of the heat pipe to improve its effectiveness and to eliminate a bottleneck in the path of heat flow from the components to the environment when the ultimate heat sink is the ambient air.

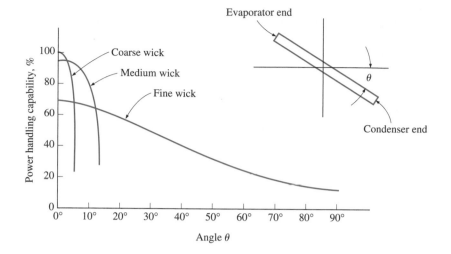

FIGURE 10–37
Variation of the heat removal capacity of a heat pipe with tilt angle from the horizontal when the liquid flows in the wick against gravity (from Steinberg).

TABLE 10–6

Typical heat removal capacity of various heat pipes

Outside Diameter, cm (in)	Length, cm (in)	Heat Removal Rate, W
$0.64(\frac{1}{4})$	15.2(6)	300
	30.5(12)	175
	45.7(18)	150
$0.95(\frac{3}{8})$	15.2(6)	500
	30.5(12)	375
	45.7(18)	350
$1.27(\frac{1}{2})$	15.2(6)	700
	30.5(12)	575
	45.7(18)	550

The decline in the performance of a 122-cm-long water heat pipe with the tilt angle from the horizontal is shown in Fig. 10–37 for heat pipes with coarse, medium, and fine wicks. Note that for the horizontal case, the heat pipe with a coarse wick performs best, but the performance drops off sharply as the evaporator end is raised from the horizontal. The heat pipe with a fine wick does not perform as well in the horizontal position but maintains its level of performance greatly at tilted positions. It is clear from this figure that heat pipes that work against gravity must be equipped with *fine* wicks. The heat removal capacities of various heat pipes are given in Table 10–6.

A major concern about the performance of a heat pipe is degradation with time. Some heat pipes have failed within just a few months after they are put into operation. The major cause of degradation appears to be *contamination* that occurs during the sealing of the ends of the heat pipe tube and affects the vapor pressure. This form of contamination has been minimized by electron beam welding in clean rooms. Contamination of the wick prior to installation in the tube is another cause of degradation. Cleanliness of the wick is essential for its reliable operation for a long time. Heat pipes usually undergo extensive testing and quality control process before they are put into actual use.

An important consideration in the design of heat pipes is the compatibility of the materials used for the tube, wick, and fluid. Otherwise, reaction between the incompatible materials produces noncondensable gases, which degrade the performance of the heat pipe. For example, the reaction between stainless steel and water in some early heat pipes generated hydrogen gas, which destroyed the heat pipe.

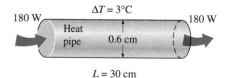

FIGURE 10–38
Schematic for Example 10–8.

EXAMPLE 10–8   **Replacing a Heat Pipe by a Copper Rod**

A 30-cm-long cylindrical heat pipe having a diameter of 0.6 cm is dissipating heat at a rate of 180 W, with a temperature difference of 3°C across the heat pipe, as shown in Fig. 10–38. If we were to use a 30-cm-long copper rod

Instead to remove heat at the same rate, determine the diameter and the mass of the copper rod that needs to be installed.

SOLUTION A cylindrical heat pipe dissipates heat at a specified rate. The diameter and mass of a copper rod that can conduct heat at the same rate are to be determined.

Assumptions Steady operating conditions exist.

Properties The properties of copper at room temperature are $\rho = 8933 \text{ kg/m}^3$ and $k = 401 \text{ W/m} \cdot °C$ (Table A–3).

Analysis The rate of heat transfer through the copper rod can be expressed as

$$\dot{Q} = kA \frac{\Delta T}{L}$$

where k is the thermal conductivity, L is the length, and ΔT is the temperature difference across the copper bar. Solving for the cross-sectional area A and substituting the specified values gives

$$A = \frac{L}{k\Delta T} \dot{Q} = \frac{0.3 \text{ m}}{(401 \text{ W/m} \cdot °C)(3°C)} (180 \text{ W}) = 0.04489 \text{ m}^2 = 448.9 \text{ cm}^2$$

Then the diameter and the mass of the copper rod become

$$A = \frac{1}{4}\pi D^2 \longrightarrow D = \sqrt{4 A/\pi} = \sqrt{4(466.3 \text{ cm}^2)/\pi} = \textbf{24.4 cm}$$

$$m = \rho V = \rho A L = (8533 \text{ kg/m}^3)(0.04489 \text{ m}^2)(0.3 \text{ m}) = \textbf{120 kg}$$

Therefore, the diameter of the copper rod needs to be almost 40 times that of the heat pipe to transfer heat at the same rate. Also, the rod would have a mass of 120 kg, which is impossible for an average person to lift

SUMMARY

Boiling occurs when a liquid is in contact with a surface maintained at a temperature T_s sufficiently above the saturation temperature T_{sat} of the liquid. Boiling is classified as *pool boiling* or *flow boiling* depending on the presence of bulk fluid motion. Boiling is called *pool boiling* in the absence of bulk fluid flow and *flow boiling* (or *forced convection boiling*) in its presence. Pool and flow boiling are further classified as *subcooled boiling* and *saturated boiling* depending on the bulk liquid temperature. Boiling is said to be *subcooled* (or *local*) when the temperature of the main body of the liquid is below the saturation temperature T_{sat} and *saturated* (or *bulk*) when the temperature of the liquid is equal to T_{sat}. Boiling exhibits different regimes depending on the value of the excess temperature ΔT_{excess}. Four different boiling regimes are observed: natural convection boiling, nucleate boiling, transition boiling, and film boiling.

These regimes are illustrated on the *boiling curve*. The rate of evaporation and the rate of heat transfer in nucleate boiling increase with increasing ΔT_{excess} and reach a maximum at some point. The heat flux at this point is called the *critical* (or *maximum*) *heat flux*, $\dot{q}_{max}$. The rate of heat transfer in nucleate pool boiling is determined from

$$\dot{q}_{nucleate} = \mu_l h_{fg} \left[\frac{g(\rho_l - \rho_v)}{\sigma} \right]^{1/2} \left[\frac{c_{pl}(T_s - T_{sat})}{C_{sf} h_{fg} \text{Pr}_l^n} \right]^3$$

The *maximum* (or *critical*) *heat flux* in nucleate pool boiling is determined from

$$\dot{q}_{max} = C_{cr} h_{fg} [\sigma g \rho_v^2 (\rho_l - \rho_v)]^{1/4}$$

where the value of the constant C_{cr} is about 0.15. The minimum heat flux is given by

$$\dot{q}_{min} = 0.09 \rho_v h_{fg} \left[\frac{\sigma g(\rho_l - \rho_v)}{(\rho_l + \rho_v)^2} \right]^{1/4}$$

The heat flux for stable *film boiling* on the outside of a *horizontal cylinder* or *sphere* of diameter D is given by

$$\dot{q}_{film} = C_{film} \left[\frac{g k_v^3 \rho_v(\rho_l - \rho_v)[h_{fg} + 0.4 c_{pv}(T_s - T_{sat})]}{\mu_v D(T_s - T_{sat})} \right]^{1/4}$$
$$\times (T_s - T_{sat})$$

where the constant $C_{film} = 0.62$ for horizontal cylinders and 0.67 for spheres. The *vapor* properties are to be evaluated at the *film temperature* $T_f = (T_{sat} + T_s)/2$, which is the *average temperature* of the vapor film. The liquid properties and h_{fg} are to be evaluated at the saturation temperature at the specified pressure.

Two distinct forms of condensation are observed in nature: film condensation and dropwise condensation. In *film condensation,* the condensate wets the surface and forms a liquid film on the surface that slides down under the influence of gravity. In *dropwise condensation,* the condensed vapor forms countless droplets of varying diameters on the surface instead of a continuous film.

The Reynolds number for the condensate flow is defined as

$$Re = \frac{D_h \rho_l V_l}{\mu_l} = \frac{4 A_c \rho_l V_l}{p \mu_l} = \frac{4 \dot{m}}{p \mu_l}$$

and

$$Re = \frac{4 \dot{Q}_{conden}}{p \mu_l h_{fg}^*} = \frac{4 A_s h(T_{sat} - T_s)}{p \mu_l h_{fg}^*}$$

where h_{fg}^* is the *modified latent heat of vaporization,* defined as

$$h_{fg}^* = h_{fg} + 0.68 c_{pl}(T_{sat} - T_s)$$

and represents heat transfer during condensation per unit mass of condensate.

Using some simplifying assumptions, the *average heat transfer coefficient* for film condensation on a vertical plate of height L is determined to be

$$h_{vert} = 0.943 \left[\frac{g \rho_l(\rho_l - \rho_v) h_{fg}^* k_l^3}{\mu_l(T_s - T_{sat})L} \right]^{1/4}$$

All properties of the *liquid* are to be evaluated at the film temperature $T_f = (T_{sat} + T_s)/2$. The h_{fg} and ρ_v are to be evaluated at T_{sat}. Condensate flow is *smooth* and *wave-free laminar* for

about $Re \leq 30$, *wavy-laminar* in the range of $30 < Re < 1800$, and *turbulent* for $Re > 1800$. Heat transfer coefficients in the wavy-laminar and turbulent flow regions are determined from

$$h_{vert,\ wavy} = \frac{Re\ k_l}{1.08\ Re^{1.22} - 5.2} \left(\frac{g}{v_l^2} \right)^{1/3}, \quad \begin{array}{l} 30 < Re < 1800 \\ \rho_v \ll \rho_l \end{array}$$

$$h_{vert,\ turbulent} = \frac{Re\ k_l}{8750 + 58\ Pr^{-0.5}\ (Re^{0.75} - 253)} \left(\frac{g}{v_l^2} \right)^{1/3},$$

$$\begin{array}{l} Re > 1800 \\ \rho_v \ll \rho_l \end{array}$$

Equations for vertical plates can also be used for laminar film condensation on the upper surfaces of the plates that are inclined by an angle θ from the vertical, by replacing g in that equation by $g \cos \theta$. Vertical plate equations can also be used to calculate the average heat transfer coefficient for laminar film condensation on the outer surfaces of vertical tubes provided that the tube diameter is large relative to the thickness of the liquid film.

The average heat transfer coefficient for film condensation on the outer surfaces of a *horizontal tube* is determined to be

$$h_{horiz} = 0.729 \left[\frac{g \rho_l(\rho_l - \rho_v) h_{fg}^* k_l^3}{\mu_l(T_s - T_{sat})D} \right]^{1/4}$$

where D is the diameter of the horizontal tube. This relation can easily be modified for a *sphere* by replacing the constant 0.729 by 0.815. It can also be used for *N horizontal tubes* stacked on top of each other by replacing D in the denominator by ND.

For low vapor velocities, film condensation heat transfer *inside horizontal tubes* can be determined from

$$h_{internal} = 0.555 \left[\frac{g \rho_l(\rho_l - \rho_v) k_l^3}{\mu_l(T_{sat} - T_s)} \left(h_{fg} + \frac{3}{8} c_{pl}(T_{sat} - T_s) \right) \right]^{1/4}$$

$$Re_{vapor} = \left(\frac{\rho_v V_v D}{\mu_v} \right)_{inlet} < 35,000$$

where the Reynolds number of the vapor is to be evaluated at the tube inlet conditions using the internal tube diameter as the characteristic length. Finally, the heat transfer coefficient for *dropwise condensation* of steam on copper surfaces is given by

$$h_{dropwise} = \begin{cases} 51,104 + 2044 T_{sat}, & 22°C < T_{sat} < 100°C \\ 255,310 & T_{sat} > 100°C \end{cases}$$

where T_{sat} is in °C and the heat transfer coefficient $h_{dropwise}$ is in $W/m^2 \cdot °C$ or its equivalent $W/m^2 \cdot K$.

REFERENCES AND SUGGESTED READING

1. N. Arai, T. Fukushima, A. Arai, T. Nakajima, K. Fujie, and Y. Nakayama. "Heat Transfer Tubes Enhancing Boiling and Condensation in Heat Exchangers of a Refrigeration Machine." *ASHRAE Journal 83* (1977), p. 58.

2. P. J. Berensen. "Film Boiling Heat Transfer for a Horizontal Surface." *Journal of Heat Transfer* 83 (1961), pp. 351–358.

3. P. J. Berensen. "Experiments in Pool Boiling Heat Transfer." *International Journal of Heat Mass Transfer* 5 (1962), pp. 985–999.

4. L. A. Bromley. "Heat Transfer in Stable Film Boiling." *Chemical Engineering Prog.* 46 (1950), pp. 221–227.

5. J. C. Chato. "Laminar Condensation inside Horizontal and Inclined Tubes." *ASHRAE Journal* 4 (1962), p. 52.

6. S. W. Chi. *Heat Theory and Practice.* Washington, D.C.: Hemisphere, 1976.

7. M. T. Cichelli and C. F. Bonilla. "Heat Transfer to Liquids Boiling under Pressure." *Transactions of AIChE* 41 (1945), pp. 755–787.

8. R. A. Colclaser, D. A. Neaman, and C. F. Hawkins. *Electronic Circuit Analysis.* New York: John Wiley & Sons, 1984.

9. J. W. Dally. *Packaging of Electronic Systems.* New York: McGraw-Hill, 1960.

10. P. Griffith. "Dropwise Condensation." In *Heat Exchanger Design Handbook,* ed. E. U. Schlunder, Vol 2, Ch. 2.6.5. New York: Hemisphere, 1983

11. H. Hampson and N. Özişik. "An Investigation into the Condensation of Steam." *Proceedings of the Institute of Mechanical Engineers,* London 1B (1952), pp. 282–294.

12. J. J. Jasper. "The Surface Tension of Pure Liquid Compounds." *Journal of Physical and Chemical Reference Data* 1, No. 4 (1972), pp. 841–1009.

13. R. Kemp. "The Heat Pipe—A New Tune on an Old Pipe." *Electronics and Power* (August 9, 1973), p. 326.

14. S. S. Kutateladze. *Fundamentals of Heat Transfer.* New York: Academic Press, 1963.

15. S. S. Kutateladze. "On the Transition to Film Boiling under Natural Convection." *Kotloturbostroenie* 3 (1948), p. 48.

16. D. A. Labuntsov. "Heat Transfer in Film Condensation of Pure Steam on Vertical Surfaces and Horizontal Tubes." *Teploenergetika* 4 (1957), pp. 72–80.

17. J. H. Lienhard and V. K. Dhir. "Extended Hydrodynamic Theory of the Peak and Minimum Pool Boiling Heat Fluxes." NASA Report, NASA-CR-2270, July 1973.

18. J. H. Lienhard and V. K. Dhir. "Hydrodynamic Prediction of Peak Pool Boiling Heat Fluxes from Finite Bodies." *Journal of Heat Transfer* 95 (1973), pp. 152–158.

19. W. H. McAdams. *Heat Transmission.* 3rd ed. New York: McGraw-Hill, 1954.

20. W. M. Rohsenow. "A Method of Correlating Heat Transfer Data for Surface Boiling of Liquids." *ASME Transactions* 74 (1952), pp. 969–975.

21. D. S. Steinberg. *Cooling Techniques for Electronic Equipment.* New York: John Wiley & Sons, 1980.

22. W. M. Rohsenow. "Film Condensation." In *Handbook of Heat Transfer,* ed. W. M. Rohsenow and J. P. Hartnett, Ch. 12A. New York: McGraw-Hill, 1973.

23. I. G. Shekriladze, I. G. Gomelauri, and V. I. Gomelauri. "Theoretical Study of Laminar Film Condensation of Flowing Vapor." *International Journal of Heat Mass Transfer* 9 (1966), pp. 591–592.

24. N. V. Suryanarayana. *Engineering Heat Transfer.* St. Paul, MN: West Publishing, 1995.

25. J. W. Westwater and J. G. Santangelo. *Industrial Engineering Chemistry* 47 (1955), p. 1605.

26. N. Zuber. "On the Stability of Boiling Heat Transfer." *ASME Transactions* 80 (1958), pp. 711–720.

PROBLEMS*

Boiling Heat Transfer

10–1C What is boiling? What mechanisms are responsible for the very high heat transfer coefficients in nucleate boiling?

10–2C Does the amount of heat absorbed as 1 kg of saturated liquid water boils at 100°C have to be equal to the amount of heat released as 1 kg of saturated water vapor condenses at 100°C?

*Problems designated by a "C" are concept questions, and students are encouraged to answer them all. Problems designated by an "E" are in English units, and the SI users can ignore them. Problems with the icon ⊛ are solved using EES, and complete solutions together with parametric studies are included on the enclosed CD. Problems with the icon ▤ are comprehensive in nature, and are intended to be solved with a computer, preferably using the EES software that accompanies this text.

10–3C What is the difference between evaporation and boiling?

10–4C What is the difference between pool boiling and flow boiling?

10–5C What is the difference between subcooled and saturated boiling?

10–6C Draw the boiling curve and identify the different boiling regimes. Also, explain the characteristics of each regime.

10–7C How does film boiling differ from nucleate boiling? Is the boiling heat flux necessarily higher in the stable film boiling regime than it is in the nucleate boiling regime?

10–8C Draw the boiling curve and identify the burnout point on the curve. Explain how burnout is caused. Why is the burnout point avoided in the design of boilers?

10–9C Discuss some methods of enhancing pool boiling heat transfer permanently.

10–10C Name the different boiling regimes in the order they occur in a vertical tube during flow boiling.

10–11 Water is boiled at 120°C in a mechanically polished stainless-steel pressure cooker placed on top of a heating unit. The inner surface of the bottom of the cooker is maintained at 130°C. Determine the heat flux on the surface.

Answer: 228.4 kW/m²

10–12 Water is boiled at 90°C by a horizontal brass heating element of diameter 7 mm. Determine the maximum heat flux that can be attained in the nucleate boiling regime.

10–13 Water is boiled at 90°C by a horizontal brass heating element of diameter 7 mm. Determine the surface temperature of the heater for the minimum heat flux case.

10–14 Water is to be boiled at atmospheric pressure in a mechanically polished steel pan placed on top of a heating unit. The inner surface of the bottom of the pan is maintained at 110°C. If the diameter of the bottom of the pan is 30 cm, determine (*a*) the rate of heat transfer to the water and (*b*) the rate of evaporation.

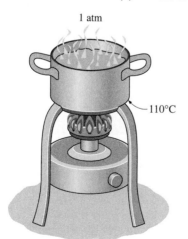

1 atm

110°C

FIGURE P10–14

10–15 Water is to be boiled at atmospheric pressure on a 3-cm-diameter mechanically polished steel heater. Determine the maximum heat flux that can be attained in the nucleate boiling regime and the surface temperature of the heater surface in that case.

10–16 Reconsider Prob. 10–15. Using EES (or other) software, investigate the effect of local atmospheric pressure on the maximum heat flux and the temperature difference $T_s - T_{sat}$. Let the atmospheric pressure vary from 70 kPa to 101.3 kPa. Plot the maximum heat flux and the temperature difference as a function of the atmospheric pressure, and discuss the results.

10–17E Water is boiled at atmospheric pressure by a horizontal polished copper heating element of diameter $D = 0.5$ in and emissivity $\varepsilon = 0.05$ immersed in water. If the surface temperature of the heating element is 788°F, determine the rate of heat transfer to the water per unit length of the heating element.

Answer: 2453 Btu/h

10–18E Repeat Prob. 10–17E for a heating element temperature of 988°F.

10–19 Water is to be boiled at sea level in a 30-cm-diameter mechanically polished AISI 304 stainless steel pan placed on top of a 3-kW electric burner. If 60 percent of the heat generated by the burner is transferred to the water during boiling, determine the temperature of the inner surface of the bottom of the pan. Also, determine the temperature difference between the inner and outer surfaces of the bottom of the pan if it is 6-mm thick.

1 atm

3 kW

FIGURE P10–19

10–20 Repeat Prob. 10–19 for a location at an elevation of 1500 m where the atmospheric pressure is 84.5 kPa and thus the boiling temperature of water is 95°C.

Answers: 100.9°C, 10.3°C

10–21 Water is boiled at sea level in a coffee maker equipped with a 20-cm-long 0.4-cm-diameter immersion-type electric heating element made of mechanically polished stainless steel. The coffee maker initially contains 1 L of water at 14°C. Once boiling starts, it is observed that half of the water in the coffee maker evaporates in 25 min. Determine the power rating of the electric heating element immersed in water and the surface temperature of the heating element. Also determine how long it will take for this heater to raise the temperature of 1 L of cold water from 14°C to the boiling temperature.

Coffee maker

1 atm

1 L

FIGURE P10–21

10–22 Repeat Prob. 10–21 for a copper heating element.

10–23 A 65-cm-long, 2-cm-diameter brass heating element is to be used to boil water at 120°C. If the surface temperature of the heating element is not to exceed 125°C, determine the highest rate of steam production in the boiler, in kg/h.

Answer: 19.4 kg/h

10–24 To understand the burnout phenomenon, boiling experiments are conducted in water at atmospheric pressure using an electrically heated 30-cm-long, 3-mm-diameter nickel-plated horizontal wire. Determine (*a*) the critical heat flux and (*b*) the increase in the temperature of the wire as the operating point jumps from the nucleate boiling to the film boiling regime at the critical heat flux. Take the emissivity of the wire to be 0.5.

10–25 Reconsider Prob. 10–24. Using EES (or other) software, investigate the effects of the local atmospheric pressure and the emissivity of the wire on the critical heat flux and the temperature rise of wire. Let the atmospheric pressure vary from 70 kPa to 101.3 kPa and the emissivity from 0.1 to 1.0. Plot the critical heat flux and the temperature rise as functions of the atmospheric pressure and the emissivity, and discuss the results.

10–26 Water is boiled at 1 atm pressure in a 20-cm-internal-diameter teflon-pitted stainless-steel pan on an electric range. If it is observed that the water level in the pan drops by 10 cm in 15 min, determine the inner surface temperature of the pan.

Answer: 105.3°C

10–27 Repeat Prob. 10–26 for a polished copper pan.

10–28 In a gas-fired boiler, water is boiled at 150°C by hot gases flowing through 50-m-long, 5-cm-outer-diameter mechanically polished stainless-steel pipes submerged in water. If the outer surface temperature of the pipes is 165°C, determine (*a*) the rate of heat transfer from the hot gases to water, (*b*) the

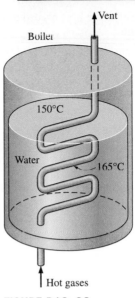

Vent

Boiler

150°C

Water

165°C

Hot gases

FIGURE P10–28

rate of evaporation, (*c*) the ratio of the critical heat flux to the present heat flux, and (*d*) the surface temperature of the pipe at which critical heat flux occurs.

Answers: (*a*) 10,865 kW, (*b*) 5.139 kg/s, (*c*) 1.34, (*d*) 166.5°C

10–29 Repeat Prob. 10–28 for a boiling temperature of 160°C.

10–30E Water is boiled at 250°F by a 2-ft-long and 0.5-in diameter nickel-plated electric heating element maintained at 280°F. Determine (*a*) the boiling heat transfer coefficient, (*b*) the electric power consumed by the heating element, and (*c*) the rate of evaporation of water.

10–31E Repeat Prob. 10–30E for a platinum-plated heating element.

10–32E Reconsider Prob. 10–30E. Using EES (or other) software, investigate the effect of surface temperature of the heating element on the boiling heat transfer coefficient, the electric power, and the rate of evaporation of water. Let the surface temperature vary from 260°F to 300°F. Plot the boiling heat transfer coefficient, the electric power consumption, and the rate of evaporation of water as a function of the surface temperature, and discuss the results.

10–33 Cold water enters a steam generator at 15°C and leaves as saturated steam at 200°C. Determine the fraction of heat used to preheat the liquid water from 15°C to the saturation temperature of 200°C in the steam generator.

Answer: 28.7 percent

10–34 Cold water enters a steam generator at 20°C and leaves as saturated steam at the boiler pressure. At what pressure will the amount of heat needed to preheat the water to saturation temperature be equal to the heat needed to vaporize the liquid at the boiler pressure?

10–35 Reconsider Prob. 10–34. Using EES (or other) software, plot the boiler pressure as a function of the cold water temperature as the temperature varies from 0°C to 30°C, and discuss the results.

10–36 A 50-cm-long, 2-mm-diameter electric resistance wire submerged in water is used to determine the boiling heat transfer coefficient in water at 1 atm experimentally. The wire temperature is measured to be 130°C when a wattmeter indicates the electric power consumed to be 3.8 kW. Using Newton's law of cooling, determine the boiling heat transfer coefficient.

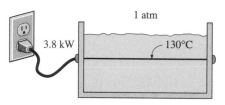

FIGURE P10–36

10–37 Water is boiled at 120°C in a mechanically polished stainless-steel pressure cooker placed on top of a heating unit. If the inner surface of the bottom of the cooker is maintained at 132°C, determine the boiling heat transfer coefficient.
Answer: 32.9 kW/m² · °C

10–38 Water is boiled at 100°C by a spherical platinum heating element of diameter 15 cm and emissivity 0.10 immersed in the water. If the surface temperature of the heating element is 350°C, determine the rate of heat transfer from the heating element to the water.

Condensation Heat Transfer

10–39C What is condensation? How does it occur?

10–40C What is the difference between film and dropwise condensation? Which is a more effective mechanism of heat transfer?

10–41C In condensate flow, how is the wetted perimeter defined? How does wetted perimeter differ from ordinary perimeter?

10–42C What is the modified latent heat of vaporization? For what is it used? How does it differ from the ordinary latent heat of vaporization?

10–43C Consider film condensation on a vertical plate. Will the heat flux be higher at the top or at the bottom of the plate? Why?

10–44C Consider film condensation on the outer surfaces of a tube whose length is 10 times its diameter. For which orientation of the tube will the heat transfer rate be the highest:

horizontal or vertical? Explain. Disregard the base and top surfaces of the tube.

10–45C Consider film condensation on the outer surfaces of four long tubes. For which orientation of the tubes will the condensation heat transfer coefficient be the highest: (*a*) vertical, (*b*) horizontal side by side, (*c*) horizontal but in a vertical tier (directly on top of each other), or (*d*) a horizontal stack of two tubes high and two tubes wide?

10–46C How does the presence of a noncondensable gas in a vapor influence the condensation heat transfer?

10–47 The Reynolds number for condensate flow is defined as $Re = 4\dot{m}/p\mu_l$, where p is the wetted perimeter. Obtain simplified relations for the Reynolds number by expressing p and $\dot{m}$ by their equivalence for the following geometries: (*a*) a vertical plate of height L and width w, (*b*) a tilted plate of height L and width W inclined at an angle θ from the vertical, (*c*) a vertical cylinder of length L and diameter D, (*d*) a horizontal cylinder of length L and diameter D, and (*e*) a sphere of diameter D.

10–48 Consider film condensation on the outer surfaces of N horizontal tubes arranged in a vertical tier. For what value of N will the average heat transfer coefficient for the entire stack of tubes be equal to half of what it is for a single horizontal tube?

10–49 Saturated steam at 1 atm condenses on a 3-m-high and 8-m-wide vertical plate that is maintained at 90°C by circulating cooling water through the other side. Determine (*a*) the rate of heat transfer by condensation to the plate, and (*b*) the rate at which the condensate drips off the plate at the bottom.
Answers: (*a*) 1507 kW, (*b*) 0.659 kg/s

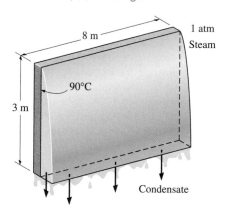

FIGURE P10–49

10–50 Repeat Prob. 10–49 for the case of the plate being tilted 60° from the vertical.

10–51 Saturated steam at 30°C condenses on the outside of a 4-cm-outer-diameter, 2-m-long vertical tube. The temperature of the tube is maintained at 20°C by the cooling water. Determine (*a*) the rate of heat transfer from the steam to the cooling water, (b) the rate of condensation of steam, and (c) the approximate thickness of the liquid film at the bottom of the tube.

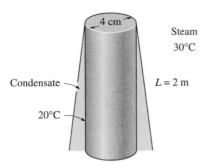

FIGURE P10–51

10–52E Saturated steam at 95°F is condensed on the outer surfaces of an array of horizontal pipes through which cooling water circulates. The outer diameter of the pipes is 1 in and the outer surfaces of the pipes are maintained at 65°F. Determine (a) the rate of heat transfer to the cooling water circulating in the pipes and (b) the rate of condensation of steam per unit length of a single horizontal pipe.

10–53E Repeat Prob. 10–52E for the case of 32 horizontal pipes arranged in a rectangular array of 4 pipes high and 8 pipes wide.

10–54 Saturated steam at 55°C is to be condensed at a rate of 10 kg/h on the outside of a 3-cm-outer-diameter vertical tube whose surface is maintained at 45°C by the cooling water. Determine the required tube length.

10–55 Repeat Prob. 10–54 for a horizontal tube.
Answer: 0.70 m

10–56 Saturated steam at 100°C condenses on a 2-m × 2-m plate that is tilted 40° from the vertical. The plate is maintained at 80°C by cooling it from the other side. Determine (a) the average heat transfer coefficient over the entire plate and (b) the rate at which the condensate drips off the plate at the bottom.

10–57 Reconsider Prob. 10–56. Using EES (or other) software, investigate the effects of plate temperature and the angle of the plate from the vertical on the average heat transfer coefficient and the rate at which the condensate drips off. Let the plate temperature vary from 40°C to 90°C and the plate angle from 0° to 60°. Plot the heat transfer coefficient and the rate at which the condensate drips off as functions of the plate temperature and the tilt angle, and discuss the results.

10–58 Saturated ammonia vapor at 10°C condenses on the outside of a 4-cm-outer-diameter, 15-m-long horizontal tube whose outer surface is maintained at −10°C. Determine (a) the rate of heat transfer from the ammonia and (b) the rate of condensation of ammonia.

10–59 The condenser of a steam power plant operates at a pressure of 4.25 kPa. The condenser consists of 100 horizontal tubes arranged in a 10 × 10 square array. The tubes are 8 m long and have an outer diameter of 3 cm. If the tube surfaces are at 20°C, determine (a) the rate of heat transfer from the steam to the cooling water and (b) the rate of condensation of steam in the condenser. *Answers:* (a) 3678 kW, (b) 1.496 kg/s

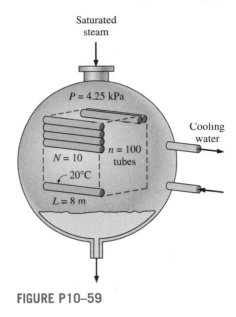

FIGURE P10–59

10–60 Reconsider Prob. 10–59. Using EES (or other) software, investigate the effect of the condenser pressure on the rate of heat transfer and the rate of condensation of the steam. Let the condenser pressure vary from 3 kPa to 15 kPa. Plot the rate of heat transfer and the rate of condensation of the steam as a function of the condenser pressure, and discuss the results.

10–61 A large heat exchanger has several columns of tubes, with 33 tubes in each column. The outer diameter of the tubes is 1.5 cm. Saturated steam at 50°C condenses on the outer surfaces of the tubes, which are maintained at 20°C. Determine (a) the average heat transfer coefficient and (b) the rate of condensation of steam per m length of a column.

10–62 Saturated refrigerant-134a vapor at 30°C is to be condensed in a 5-m-long, 1-cm-diameter horizontal tube that is maintained at a temperature of 20°C. If the refrigerant enters the tube at a rate of 2.5 kg/min, determine the fraction of the refrigerant that is condensed at the end of the tube.

10–63 Repeat Prob. 10–62 for a tube length of 8 m.
Answer: 17.2 percent

10–64 Reconsider Prob. 10–62. Using EES (or other) software, plot the fraction of the refrigerant condensed at the end of the tube as a function of the temperature of the saturated R-134a vapor as the temperature varies from 25°C to 50°C, and discuss the results.

10–65 A horizontal condenser uses a 4 × 4 array of tubes that have an outer diameter of 5.0 cm and length 2.0 m. Saturated steam at 101.3 kPa condenses on the outside tube surface held at a temperature of 80°C. Calculate the steady rate of steam condensation in kg/h.

10–66 Saturated ammonia vapor at 30°C is passed over 20 vertical flat plates, each of which is 10 cm high and 15 cm wide. The average surface temperature of the plates is 10°C. Estimate the average heat transfer coefficient and the rate of ammonia condensation.

Special Topic: Heat Pipes

10–67C What is a heat pipe? How does it operate? Does it have any moving parts?

10–68C A heat pipe with water as the working fluid is said to have an effective thermal conductivity of 100,000 W/m · °C, which is more than 100,000 times the conductivity of water. How can this happen?

10–69C What is the effect of a small amount of noncondensable gas such as air on the performance of a heat pipe?

10–70C Why do water-based heat pipes used in the cooling of electronic equipment operate below atmospheric pressure?

10–71C What happens when the wick of a heat pipe is too coarse or too fine?

10–72C Does the orientation of a heat pipe affect its performance? Does it matter if the evaporator end of the heat pipe is up or down? Explain.

10–73C How can the liquid in a heat pipe move up against gravity without a pump? For heat pipes that work against gravity, is it better to have coarse or fine wicks? Why?

10–74C What are the important considerations in the design and manufacture of heat pipes?

10–75C What is the major cause for the premature degradation of the performance of some heat pipes?

10–76 A 40-cm-long cylindrical heat pipe having a diameter of 0.5 cm is dissipating heat at a rate of 150 W, with a temperature difference of 4°C across the heat pipe. If we were to use a 40-cm-long copper rod ($k = 401$ W/m · °C and $\rho = 8933$ kg/m³) instead to remove heat at the same rate, determine the diameter and the mass of the copper rod that needs to be installed.

10–77 Repeat Prob. 10–76 for an aluminum rod instead of copper.

10–78E A plate that supports 10 power transistors, each dissipating 45 W, is to be cooled with 1.5-ft-long heat pipes having a diameter of $\frac{1}{4}$ in. Using Table 10–6, determine how many pipes need to be attached to this plate. *Answer:* 3

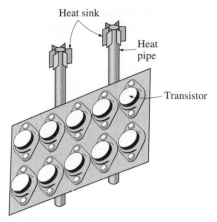

FIGURE P10–78E

Review Problems

10–79 Water is boiled at 100°C by a spherical platinum heating element of diameter 15 cm and emissivity 0.10 immersed in the water. If the surface temperature of the heating element is 350°C, determine the convective boiling heat transfer coefficient.

10–80 Water is boiled at 120°C in a mechanically polished stainless-steel pressure cooker placed on top of a heating unit. The inner surface of the bottom of the cooker is maintained at 130°C. The cooker that has a diameter of 20 cm and a height of 30 cm is half filled with water. Determine the time it will take for the tank to empty.
Answer: 22.8 min

10–81 Saturated ammonia vapor at 25°C condenses on the outside surface of 16 thin-walled tubes, 2.5 cm in diameter, arranged horizontally in a 4 × 4 square array. Cooling water enters the tubes at 14°C at an average velocity of 2 m/s and exits at 17°C. Calculate (*a*) the rate of NH₃ condensation, (*b*) the overall heat transfer coefficient, and (*c*) the tube length.

10–82 Steam at 40°C condenses on the outside of a 3-cm diameter thin horizontal copper tube by cooling water that enters the tube at 25°C at an average velocity of 2 m/s and leaves at 35°C. Determine the rate of condensation of steam, the average overall heat transfer coefficient between the steam and the cooling water, and the tube length.

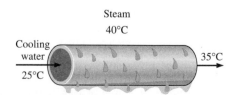

FIGURE P10–82

10–83 Saturated ammonia vapor at 25°C condenses on the outside of a 2-m-long, 3.2-cm-outer-diameter vertical tube maintained at 15°C. Determine (*a*) the average heat transfer co-efficient, (*b*) the rate of heat transfer, and (*c*) the rate of condensation of ammonia.

10–84 Saturated isobutane vapor in a binary geothermal power plant is to be condensed outside an array of eight horizontal tubes. Determine the ratio of the condensation rate for the cases of the tubes being arranged in a horizontal tier versus in a vertical tier of horizontal tubes.

10–85E The condenser of a steam power plant operates at a pressure of 0.95 psia. The condenser consists of 144 horizontal tubes arranged in a 12×12 square array. The tubes are 15 ft long and have an outer diameter of 1.2 in. If the outer surfaces of the tubes are maintained at 80°F, determine (*a*) the rate of heat transfer from the steam to the cooling water and (*b*) the rate of condensation of steam in the condenser.

10–86E Repeat Prob. 10–85E for a tube diameter of 2 in.

10–87 Water is boiled at 100°C electrically by an 80-cm-long, 2-mm-diameter horizontal resistance wire made of chemically etched stainless steel. Determine (*a*) the rate of heat transfer to the water and the rate of evaporation of water if the temperature of the wire is 115°C and (*b*) the maximum rate of evaporation in the nucleate boiling regime.
Answers: (*a*) 2387 W, 3.81 kg/h, (*b*) 1280 kW/m²

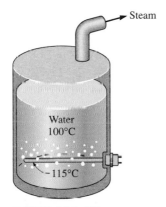

FIGURE P10–87

10–88E Saturated steam at 100°F is condensed on a 4-ft-high vertical plate that is maintained at 80°F. Determine the rate of heat transfer from the steam to the plate and the rate of condensation per foot width of the plate.

10–89 Saturated refrigerant-134a vapor at 35°C is to be condensed on the outer surface of a 7-m-long, 1.5-cm-diameter horizontal tube that is maintained at a temperature of 25°C. Determine the rate at which the refrigerant will condense, in kg/min.

10–90 Repeat Prob. 10–89 for a tube diameter of 3 cm.

10–91 Saturated steam at 270.1 kPa condenses inside a horizontal, 10-m-long, 2.5-cm-internal-diameter pipe whose sur-face is maintained at 110°C. Assuming low vapor velocity, determine the average heat transfer coefficient and the rate of condensation of the steam inside the pipe.
Answers: 3345 W/m² · °C, 0.0242 kg/s

10–92 A 1.5-cm-diameter silver sphere initially at 30°C is suspended in a room filled with saturated steam at 100°C. Using the lumped system analysis, determine how long it will take for the temperature of the ball to rise to 50°C. Also, determine the amount of steam that condenses during this process and verify that the lumped system analysis is applicable.

10–93 Repeat Prob. 10–92 for a 3-cm-diameter copper ball.

10–94 You have probably noticed that water vapor that condenses on a canned drink slides down, clearing the surface for further condensation. Therefore, condensation in this case can be considered to be dropwise. Determine the condensation heat transfer coefficient on a cold canned drink at 2°C that is placed in a large container filled with saturated steam at 95°C.

Steam 95°C

2°C

FIGURE P10–94

10–95 A resistance heater made of 2-mm-diameter nickel wire is used to heat water at 1 atm pressure. Determine the highest temperature at which this heater can operate safely without the danger of burning out. *Answer:* 109.6°C

10–96 Atmospheric-pressure steam is to be generated in the shell side of a horizontal heat exchanger. There are 100 tubes, each with 5.0 cm in outer diameter and 2.0 m in length. The heat transfer coefficient on the tube surface can be expressed in W/m² · K as $h = 5.56(T_s - T_{sat})^3$, where T_s is the tube surface temperature and T_{sat} is the boiling temperature. Estimate the tube surface temperature for producing 50 kg/min of steam.

10–97 An electrical heating rod, 1.0 cm in diameter and 30.0 cm in length, is rated at 1.5 kW. It is immersed horizontally in a vessel filled with water at 101.3 kPa. The heat transfer coefficient on the heater surface can be expressed in W/m² · K as $h = 5.56(T_s - T_{sat})^3$, where T_s is the heater surface temperature and T_{sat} is the boiling temperature. Calculate the heater surface temperature and the rate of steam generation after the water starts boiling.

10–98 Shown in Fig. P10–98 is the tube layout for a horizontal condenser that is used for liquefying 900 kg/h of saturated ammonia vapor at 37°C. There are 14 copper tubes, each with inner diameter $D_i = 3.0$ cm and outer diameter $D_o = 3.8$ cm. A coolant flows through the tubes at an average temperature of 20°C such that it yields a heat transfer coefficient of 4.0 kW/m² · K. For this condenser, estimate (a) the average value of the overall heat transfer coefficient and (b) the tube length.

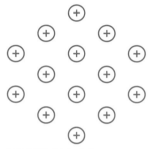

FIGURE P10–98

Fundamentals of Engineering (FE) Exam Problems

10–99 Saturated water vapor at 40°C is to be condensed as it flows through a tube at a rate of 0.2 kg/s. The condensate leaves the tube as a saturated liquid at 40°C. The rate of heat transfer from the tube is
(a) 34 kJ/s   (b) 268 kJ/s   (c) 453 kJ/s
(d) 481 kJ/s   (e) 515 kJ/s

10–100 Heat transfer coefficients for a vapor condensing on a surface can be increased by promoting
(a) film condensation   (b) dropwise condensation
(c) rolling action   (d) none of them

10–101 At a distance x down a vertical, isothermal flat plate on which a saturated vapor is condensing in a continuous film, the thickness of the liquid condensate layer is δ. The heat transfer coefficient at this location on the plate is given by
(a) k_l/δ (b) δh_f (c) δh_{fg}
(d) δh_g (e) none of them

10–102 When a saturated vapor condenses on a vertical, isothermal flat plate in a continuous film, the rate of heat transfer is proportional to
(a) $(T_s - T_{sat})^{1/4}$ (b) $(T_s - T_{sat})^{1/2}$ (c) $(T_s - T_{sat})^{3/4}$
(d) $(T_s - T_{sat})$ (e) $(T_s - T_{sat})^{2/3}$

10–103 Saturated water vapor is condensing on a 0.5 m² vertical flat plate in a continuous film with an average heat transfer coefficient of 7 kW/m² · K. The temperature of the water is 80°C ($h_{fg} = 2309$ kJ/kg) and the temperature of the plate is 60°C. The rate at which condensate is being formed is
(a) 0.03 kg/s   (b) 0.07 kg/s   (c) 0.15 kg/s
(d) 0.24 kg/s   (e) 0.28 kg/s

10–104 An air conditioner condenser in an automobile consists of 2 m² of tubular heat exchange area whose surface temperature is 30°C. Saturated refrigerant-134a vapor at 50°C ($h_{fg} = 152$ kJ/kg) condenses on these tubes. What heat transfer coefficient must exist between the tube surface and condensing vapor to produce 1.5 kg/min of condensate?
(a) 95 W/m² · K   (b) 640 W/m² · K
(c) 727 W/m² · K   (d) 799 W/m² · K
(e) 960 W/m² · K

10–105 When boiling a saturated liquid, one must be careful while increasing the heat flux to avoid burnout. Burnout occurs when the boiling transitions from _____ boiling.
(a) convection to nucleate   (b) convection to film
(c) film to nucleate   (d) nucleate to film
(e) none of them

10–106 Steam condenses at 50°C on a 0.8-m-high and 2.4-m-wide vertical plate that is maintained at 30°C. The condensation heat transfer coefficient is
(a) 3975 W/m² · °C   (b) 5150 W/m² · °C
(c) 8060 W/m² · °C   (d) 11,300 W/m² · °C
(e) 14,810 W/m² · °C

(For water, use $\rho_l = 992.1$ kg/m³, $\mu_l = 0.653 \times 10^{-3}$ kg/m · s, $k_l = 0.631$ W/m · °C, $c_{pl} = 4179$ J/kg · °C, $h_{fg@T_{sat}} = 2383$ kJ/kg)

10–107 Steam condenses at 50°C on the outer surface of a horizontal tube with an outer diameter of 6 cm. The outer surface of the tube is maintained at 30°C. The condensation heat transfer coefficient is
(a) 5493 W/m² · °C   (b) 5921 W/m² · °C
(c) 6796 W/m² · °C   (d) 7040 W/m² · °C
(e) 7350 W/m² · °C

(For water, use $\rho_l = 992.1$ kg/m³, $\mu_l = 0.653 \times 10^{-3}$ kg/m · s, $k_l = 0.631$ W/m · °C, $c_{pl} = 4179$ J/kg · °C, $h_{fg@T_{sat}} = 2383$ kJ/kg)

10–108 Steam condenses at 50°C on the tube bank consisting of 20 tubes arranged in a rectangular array of 4 tubes high and 5 tubes wide. Each tube has a diameter of 6 cm and a length of 3 m and the outer surfaces of the tubes are maintained at 30°C. The rate of condensation of steam is
(a) 0.054 kg/s   (b) 0.076 kg/s   (c) 0.115 kg/s
(d) 0.284 kg/s   (e) 0.446 kg/s

(For water, use $\rho_l = 992.1$ kg/m³, $\mu_l = 0.653 \times 10^{-3}$ kg/m · s, $k_l = 0.631$ W/m · °C, $c_{pl} = 4179$ J/kg · °C, $h_{fg@T_{sat}} = 2383$ kJ/kg)

Design and Essay Problems

10–109 Design the condenser of a steam power plant that has a thermal efficiency of 40 percent and generates 10 MW of net electric power. Steam enters the condenser as saturated vapor at 10 kPa, and it is to be condensed outside horizontal tubes through which cooling water from a nearby river flows. The temperature rise of the cooling water is limited to 8°C, and the velocity of the cooling water in the pipes is limited to 6 m/s to keep the pressure drop at an acceptable level. Specify the pipe diameter, total pipe length, and the arrangement of the pipes to minimize the condenser volume.

10–110 The refrigerant in a household refrigerator is condensed as it flows through the coil that is typically placed behind the refrigerator. Heat transfer from the outer surface of the coil to the surroundings is by natural convection and radiation. Obtaining information about the operating conditions of the refrigerator, including the pressures and temperatures of the refrigerant at the inlet and the exit of the coil, show that the coil is selected properly, and determine the safety margin in the selection.

10–111 Water-cooled steam condensers are commonly used in steam power plants. Obtain information about water-cooled steam condensers by doing a literature search on the topic and also by contacting some condenser manufacturers. In a report, describe the various types, the way they are designed, the limitation on each type, and the selection criteria.

10–112 Steam boilers have long been used to provide process heat as well as to generate power. Write an essay on the history of steam boilers and the evolution of modern supercritical steam power plants. What was the role of the American Society of Mechanical Engineers in this development?

10–113 The technology for power generation using geothermal energy is well established, and numerous geothermal power plants throughout the world are currently generating electricity economically. Binary geothermal plants utilize a volatile secondary fluid such as isobutane, n-pentane, and R-114 in a closed loop. Consider a binary geothermal plant with R-114 as the working fluid that is flowing at a rate of 600 kg/s. The R-114 is vaporized in a boiler at 115°C by the geothermal fluid that enters at 165°C and is condensed at 30°C outside the tubes by cooling water that enters the tubes at 18°C. Design the condenser of this binary plant.

Specify (a) the length, diameter, and number of tubes and their arrangement in the condenser, (b) the mass flow rate of cooling water, and (c) the flow rate of make-up water needed if a cooling tower is used to reject the waste heat from the cooling water. The liquid velocity is to remain under 6 m/s and the length of the tubes is limited to 8 m.

10–114 A manufacturing facility requires saturated steam at 120°C at a rate of 1.2 kg/min. Design an electric steam boiler for this purpose under these constraints:

- The boiler will be in cylindrical shape with a height-to-diameter ratio of 1.5. The boiler can be horizontal or vertical.

- The boiler will operate in the nucleate boiling regime, and the design heat flux will not exceed 60 percent of the critical heat flux to provide an adequate safety margin.

- A commercially available plug-in type electrical heating element made of mechanically polished stainless steel will be used. The diameter of the heater cannot be between 0.5 cm and 3 cm.

- Half of the volume of the boiler should be occupied by steam, and the boiler should be large enough to hold enough water for 2 h supply of steam. Also, the boiler will be well insulated.

You are to specify the following: (a) The height and inner diameter of the tank, (b) the length, diameter, power rating, and surface temperature of the electric heating element, (c) the maximum rate of steam production during short periods of overload conditions, and how it can be accomplished.

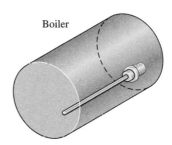

Boiler

FIGURE P10–114

10–115 Repeat Prob. 10–114 for a boiler that produces steam at 150°C at a rate of 2.5 kg/min.

10–116 Conduct this experiment to determine the boiling heat transfer coefficient. You will need a portable immersion-type electric heating element, an indoor-outdoor thermometer, and metal glue (all can be purchased for about $15 in a hardware store). You will also need a piece of string and a ruler to calculate the surface area of the heater. First, boil water in a pan using the heating element and measure the temperature of the boiling water away from the heating element. Based on your reading, estimate the elevation of your location, and compare it to the actual value. Then glue the tip of the thermocouple wire of the thermometer to the midsection of the heater surface. The temperature reading in this case will give the surface temperature of the heater. Assuming the rated power of the heater to be the actual power consumption during heating (you can check this by measuring the electric current and voltage), calculate the heat transfer coefficients from Newton's law of cooling.

FIGURE P10–116

HEAT EXCHANGERS

Heat exchangers are devices that facilitate the *exchange of heat* between two fluids that are at different temperatures while keeping them from mixing with each other. Heat exchangers are commonly used in practice in a wide range of applications, from heating and air-conditioning systems in a household, to chemical processing and power production in large plants. Heat exchangers differ from mixing chambers in that they do not allow the two fluids involved to mix.

Heat transfer in a heat exchanger usually involves *convection* in each fluid and *conduction* through the wall separating the two fluids. In the analysis of heat exchangers, it is convenient to work with an *overall heat transfer coefficient U* that accounts for the contribution of all these effects on heat transfer. The rate of heat transfer between the two fluids at a location in a heat exchanger depends on the magnitude of the temperature difference at that location, which varies along the heat exchanger.

Heat exchangers are manufactured in a variety of types, and thus we start this chapter with the *classification* of heat exchangers. We then discuss the determination of the overall heat transfer coefficient in heat exchangers, and the *logarithmic mean temperature difference* (LMTD) for some configurations. We then introduce the *correction factor F* to account for the deviation of the mean temperature difference from the LMTD in complex configurations. Next we discuss the effectiveness–NTU method, which enables us to analyze heat exchangers when the outlet temperatures of the fluids are not known. Finally, we discuss the selection of heat exchangers.

CONTENTS

OBJECTIVES

When you finish studying this chapter, you should be able to:

- Recognize numerous types of heat exchangers, and classify them,
- Develop an awareness of fouling on surfaces, and determine the overall heat transfer coefficient for a heat exchanger,
- Perform a general energy analysis on heat exchangers,
- Obtain a relation for the logarithmic mean temperature difference for use in the LMTD method, and modify it for different types of heat exchangers using the correction factor,
- Develop relations for effectiveness, and analyze heat exchangers when outlet temperatures are not known using the effectiveness-NTU method,
- Know the primary considerations in the selection of heat exchangers.

11–1 · TYPES OF HEAT EXCHANGERS

Different heat transfer applications require different types of hardware and different configurations of heat transfer equipment. The attempt to match the heat transfer hardware to the heat transfer requirements within the specified constraints has resulted in numerous types of innovative heat exchanger designs.

The simplest type of heat exchanger consists of two concentric pipes of different diameters, as shown in Fig. 11–1, called the **double-pipe** heat exchanger. One fluid in a double-pipe heat exchanger flows through the smaller pipe while the other fluid flows through the annular space between the two pipes. Two types of flow arrangement are possible in a double-pipe heat exchanger: in **parallel flow**, both the hot and cold fluids enter the heat exchanger at the same end and move in the *same* direction. In **counter flow**, on the other hand, the hot and cold fluids enter the heat exchanger at opposite ends and flow in *opposite* directions.

Another type of heat exchanger, which is specifically designed to realize a large heat transfer surface area per unit volume, is the **compact** heat exchanger. The ratio of the heat transfer surface area of a heat exchanger to its volume is called the *area density* β. A heat exchanger with $\beta > 700$ m^2/m^3 (or 200 ft^2/ft^3) is classified as being compact. Examples of compact heat exchangers are car radiators ($\beta \approx 1000$ m^2/m^3), glass-ceramic gas turbine heat exchangers ($\beta \approx 6000$ m^2/m^3), the regenerator of a Stirling engine ($\beta \approx 15{,}000$ m^2/m^3), and the human lung ($\beta \approx 20{,}000$ m^2/m^3). Compact heat

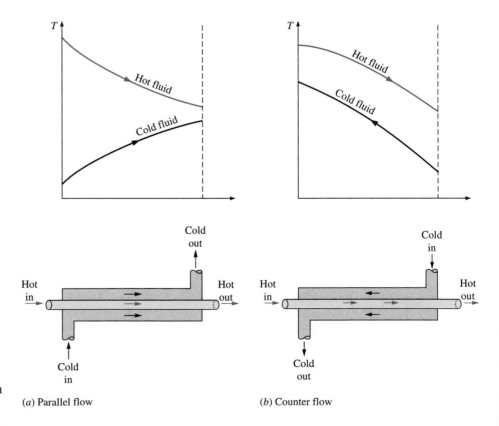

FIGURE 11–1

Different flow regimes and associated temperature profiles in a double-pipe heat exchanger.

(*a*) Parallel flow

(*b*) Counter flow

exchangers enable us to achieve high heat transfer rates between two fluids in a small volume, and they are commonly used in applications with strict limitations on the weight and volume of heat exchangers (Fig. 11–2).

The large surface area in compact heat exchangers is obtained by attaching closely spaced *thin plate* or *corrugated fins* to the walls separating the two fluids. Compact heat exchangers are commonly used in gas-to-gas and gas-to-liquid (or liquid-to-gas) heat exchangers to counteract the low heat transfer coefficient associated with gas flow with increased surface area. In a car radiator, which is a water-to-air compact heat exchanger, for example, it is no surprise that fins are attached to the air side of the tube surface.

In compact heat exchangers, the two fluids usually move *perpendicular* to each other, and such flow configuration is called **cross-flow**. The cross-flow is further classified as *unmixed* and *mixed flow,* depending on the flow configuration, as shown in Fig. 11–3. In (*a*) the cross-flow is said to be *unmixed* since the plate fins force the fluid to flow through a particular interfin spacing and prevent it from moving in the transverse direction (i.e., parallel to the tubes). The cross-flow in (*b*) is said to be *mixed* since the fluid now is free to move in the transverse direction. Both fluids are unmixed in a car radiator. The presence of mixing in the fluid can have a significant effect on the heat transfer characteristics of the heat exchanger.

Perhaps the most common type of heat exchanger in industrial applications is the **shell-and-tube** heat exchanger, shown in Fig. 11–4. Shell-and-tube heat exchangers contain a large number of tubes (sometimes several hundred) packed in a shell with their axes parallel to that of the shell. Heat transfer takes place as one fluid flows inside the tubes while the other fluid flows outside the tubes through the shell. *Baffles* are commonly placed in the shell to force the shell-side fluid to flow across the shell to enhance heat transfer and to maintain uniform spacing between the tubes. Despite their widespread use, shell-and-tube heat exchangers are not suitable for use in automotive and aircraft applications because of their relatively large size and weight. Note that the tubes in a shell-and-tube heat exchanger open to some large flow areas called *headers* at both ends of the shell, where the tube-side fluid accumulates before entering the tubes and after leaving them.

Shell-and-tube heat exchangers are further classified according to the number of shell and tube passes involved. Heat exchangers in which all the tubes make one U-turn in the shell, for example, are called *one-shell-pass and*

FIGURE 11–2
A gas-to-liquid compact heat exchanger for a residential air-conditioning system.
(© Yunus Çengel)

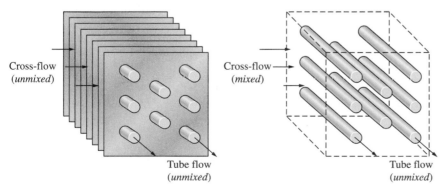

Cross-flow
(*unmixed*)

Tube flow
(*unmixed*)

(*a*) Both fluids unmixed

Cross-flow
(*mixed*)

Tube flow
(*unmixed*)

(*b*) One fluid mixed, one fluid unmixed

FIGURE 11–3
Different flow configurations in cross-flow heat exchangers.

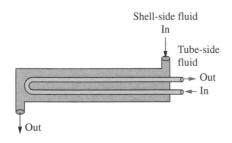

FIGURE 11–4

The schematic of a shell-and-tube heat exchanger (one-shell pass and one-tube pass).

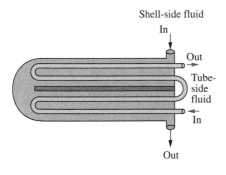

(a) One-shell pass and two-tube passes

(b) Two-shell passes and four-tube passes

FIGURE 11–5

Multipass flow arrangements in shell-and-tube heat exchangers.

two-tube-passes heat exchangers. Likewise, a heat exchanger that involves two passes in the shell and four passes in the tubes is called a *two-shell-passes and four-tube-passes* heat exchanger (Fig. 11–5).

An innovative type of heat exchanger that has found widespread use is the **plate and frame** (or just plate) heat exchanger, which consists of a series of plates with corrugated flat flow passages (Fig. 11–6). The hot and cold fluids flow in alternate passages, and thus each cold fluid stream is surrounded by two hot fluid streams, resulting in very effective heat transfer. Also, plate heat exchangers can grow with increasing demand for heat transfer by simply mounting more plates. They are well suited for liquid-to-liquid heat exchange applications, provided that the hot and cold fluid streams are at about the same pressure.

Another type of heat exchanger that involves the alternate passage of the hot and cold fluid streams through the same flow area is the **regenerative** heat exchanger. The *static*-type regenerative heat exchanger is basically a porous mass that has a large heat storage capacity, such as a ceramic wire mesh. Hot and cold fluids flow through this porous mass alternatively. Heat is transferred from the hot fluid to the matrix of the regenerator during the flow of the hot fluid, and from the matrix to the cold fluid during the flow of the cold fluid. Thus, the matrix serves as a temporary heat storage medium.

The *dynamic*-type regenerator involves a rotating drum and continuous flow of the hot and cold fluid through different portions of the drum so that any portion of the drum passes periodically through the hot stream, storing heat, and then through the cold stream, rejecting this stored heat. Again the drum serves as the medium to transport the heat from the hot to the cold fluid stream.

Heat exchangers are often given specific names to reflect the specific application for which they are used. For example, a *condenser* is a heat exchanger in which one of the fluids is cooled and condenses as it flows through the heat exchanger. A *boiler* is another heat exchanger in which one of the fluids absorbs heat and vaporizes. A *space radiator* is a heat exchanger that transfers heat from the hot fluid to the surrounding space by radiation.

11–2 · THE OVERALL HEAT TRANSFER COEFFICIENT

A heat exchanger typically involves two flowing fluids separated by a solid wall. Heat is first transferred from the hot fluid to the wall by *convection,*

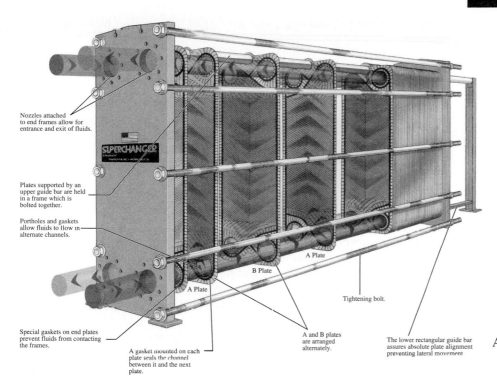

FIGURE 11–6
A plate-and-frame liquid-to-liquid
heat exchanger.
(Courtesy of Tranter PHE, Inc.)

through the wall by *conduction,* and from the wall to the cold fluid again by *convection.* Any radiation effects are usually included in the convection heat transfer coefficients.

The thermal resistance network associated with this heat transfer process involves two convection and one conduction resistances, as shown in Fig. 11–7. Here the subscripts i and o represent the inner and outer surfaces of the inner tube. For a double-pipe heat exchanger, the *thermal resistance* of the tube wall is

$$R_{wall} = \frac{\ln (D_o/D_i)}{2\pi kL} \qquad (11\text{–}1)$$

where k is the thermal conductivity of the wall material and L is the length of the tube. Then the *total thermal resistance* becomes

$$R = R_{total} = R_i + R_{wall} + R_o = \frac{1}{h_i A_i} + \frac{\ln (D_o/D_i)}{2\pi kL} + \frac{1}{h_o A_o} \qquad (11\text{–}2)$$

The A_i is the area of the inner surface of the wall that separates the two fluids, and A_o is the area of the outer surface of the wall. In other words, A_i and A_o are surface areas of the separating wall wetted by the inner and the outer fluids, respectively. When one fluid flows inside a circular tube and the other outside of it, we have $A_i = \pi D_i L$ and $A_o = \pi D_o L$ (Fig. 11–8).

In the analysis of heat exchangers, it is convenient to combine all the thermal resistances in the path of heat flow from the hot fluid to the cold one into

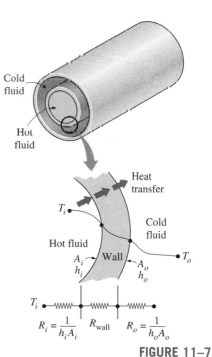

FIGURE 11–7
Thermal resistance network
associated with heat transfer
in a double-pipe heat exchanger.

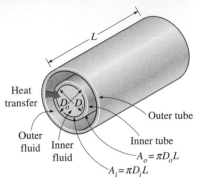

FIGURE 11–8
The two heat transfer surface areas associated with a double-pipe heat exchanger (for thin tubes, $D_i \approx D_o$ and thus $A_i \approx A_o$).

a single resistance R, and to express the rate of heat transfer between the two fluids as

$$\dot{Q} = \frac{\Delta T}{R} = UA\Delta T = U_i A_i \Delta T = U_o A_o \Delta T \qquad (11\text{–}3)$$

where U is the **overall heat transfer coefficient**, whose unit is $W/m^2 \cdot {}^\circ C$, which is identical to the unit of the ordinary convection coefficient h. Canceling ΔT, Eq. 11–3 reduces to

$$\frac{1}{UA_s} = \frac{1}{U_i A_i} = \frac{1}{U_o A_o} = R = \frac{1}{h_i A_i} + R_{wall} + \frac{1}{h_o A_o} \qquad (11\text{–}4)$$

Perhaps you are wondering why we have two overall heat transfer coefficients U_i and U_o for a heat exchanger. The reason is that every heat exchanger has two heat transfer surface areas A_i and A_o, which, in general, are not equal to each other.

Note that $U_i A_i = U_o A_o$, but $U_i \neq U_o$ unless $A_i = A_o$. Therefore, the overall heat transfer coefficient U of a heat exchanger is meaningless unless the area on which it is based is specified. This is especially the case when one side of the tube wall is finned and the other side is not, since the surface area of the finned side is several times that of the unfinned side.

When the wall thickness of the tube is small and the thermal conductivity of the tube material is high, as is usually the case, the thermal resistance of the tube is negligible ($R_{wall} \approx 0$) and the inner and outer surfaces of the tube are almost identical ($A_i \approx A_o \approx A_s$). Then Eq. 11–4 for the overall heat transfer coefficient simplifies to

$$\frac{1}{U} \approx \frac{1}{h_i} + \frac{1}{h_o} \qquad (11\text{–}5)$$

where $U \approx U_i \approx U_o$. The individual convection heat transfer coefficients inside and outside the tube, h_i and h_o, are determined using the convection relations discussed in earlier chapters.

The overall heat transfer coefficient U in Eq. 11–5 is dominated by the *smaller* convection coefficient, since the inverse of a large number is small. When one of the convection coefficients is *much smaller* than the other (say, $h_i \ll h_o$), we have $1/h_i \gg 1/h_o$, and thus $U \approx h_i$. Therefore, the smaller heat transfer coefficient creates a *bottleneck* on the path of heat transfer and seriously impedes heat transfer. This situation arises frequently when one of the fluids is a gas and the other is a liquid. In such cases, fins are commonly used on the gas side to enhance the product UA and thus the heat transfer on that side.

Representative values of the overall heat transfer coefficient U are given in Table 11–1. Note that the overall heat transfer coefficient ranges from about 10 $W/m^2 \cdot {}^\circ C$ for gas-to-gas heat exchangers to about 10,000 $W/m^2 \cdot {}^\circ C$ for heat exchangers that involve phase changes. This is not surprising, since gases have very low thermal conductivities, and phase-change processes involve very high heat transfer coefficients.

TABLE 11-1

Representative values of the overall heat transfer coefficients in heat exchangers

Type of heat exchanger	U, W/m$^2 \cdot$ °C*
Water-to-water	850–1700
Water-to-oil	100–350
Water-to-gasoline or kerosene	300–1000
Feedwater heaters	1000–8500
Steam-to-light fuel oil	200–400
Steam-to-heavy fuel oil	50–200
Steam condenser	1000–6000
Freon condenser (water cooled)	300–1000
Ammonia condenser (water cooled)	800–1400
Alcohol condensers (water cooled)	250–700
Gas-to-gas	10–40
Water-to-air in finned tubes (water in tubes)	30–60[†]
	400–850[†]
Steam-to-air in finned tubes (steam in tubes)	30–300[†]
	400–4000[‡]

*Multiply the listed values by 0.176 to convert them to Btu/h · ft^2 · °F.

[†]Based on air-side surface area.

[‡]Based on water- or steam-side surface area.

When the tube is *finned* on one side to enhance heat transfer, the total heat transfer surface area on the finned side becomes

$$A_s = A_{total} = A_{fin} + A_{unfinned} \qquad (11\text{-}6)$$

where A_{fin} is the surface area of the fins and $A_{unfinned}$ is the area of the unfinned portion of the tube surface. For short fins of high thermal conductivity, we can use this total area in the convection resistance relation $R_{conv} = 1/hA_s$, since the fins in this case will be very nearly isothermal. Otherwise, we should determine the effective surface area A from

$$A_s = A_{unfinned} + \eta_{fin} A_{fin} \qquad (11\text{-}7)$$

where η_{fin} is the fin efficiency. This way, the temperature drop along the fins is accounted for. Note that $\eta_{fin} = 1$ for isothermal fins, and thus Eq. 11–7 reduces to Eq. 11–6 in that case.

Fouling Factor

The performance of heat exchangers usually deteriorates with time as a result of accumulation of *deposits* on heat transfer surfaces. The layer of deposits represents *additional resistance* to heat transfer and causes the rate of heat transfer in a heat exchanger to decrease. The net effect of these accumulations on heat transfer is represented by a **fouling factor** R_f, which is a measure of the *thermal resistance* introduced by fouling.

The most common type of fouling is the *precipitation* of solid deposits in a fluid on the heat transfer surfaces. You can observe this type of fouling even in your house. If you check the inner surfaces of your teapot after prolonged use, you will probably notice a layer of calcium-based deposits on the surfaces at which boiling occurs. This is especially the case in areas where the water is hard. The scales of such deposits come off by scratching, and the surfaces can be cleaned of such deposits by chemical treatment. Now imagine those mineral deposits forming on the inner surfaces of fine tubes in a heat exchanger (Fig. 11–9) and the detrimental effect it may have on the flow passage area and the heat transfer. To avoid this potential problem, water in power and process plants is extensively treated and its solid contents are removed before it is allowed to circulate through the system. The solid ash particles in the flue gases accumulating on the surfaces of air preheaters create similar problems.

Another form of fouling, which is common in the chemical process industry, is *corrosion* and other *chemical fouling*. In this case, the surfaces are fouled by the accumulation of the products of chemical reactions on the surfaces. This form of fouling can be avoided by coating metal pipes with glass or using plastic pipes instead of metal ones. Heat exchangers may also be fouled by the growth of algae in warm fluids. This type of fouling is called *biological fouling* and can be prevented by chemical treatment.

In applications where it is likely to occur, fouling should be considered in the design and selection of heat exchangers. In such applications, it may be necessary to select a larger and thus more expensive heat exchanger to ensure that it meets the design heat transfer requirements even after fouling occurs. The periodic cleaning of heat exchangers and the resulting down time are additional penalties associated with fouling.

The fouling factor is obviously zero for a new heat exchanger and increases with time as the solid deposits build up on the heat exchanger surface. The fouling factor depends on the *operating temperature* and the *velocity* of the fluids, as well as the length of service. Fouling increases with *increasing temperature* and *decreasing velocity*.

FIGURE 11–9

Precipitation fouling of ash particles on superheater tubes.
(From Steam: Its Generation, and Use, Babcock and Wilcox Co., 1978. Reprinted by permission.)

The overall heat transfer coefficient relation given above is valid for clean surfaces and needs to be modified to account for the effects of fouling on both the inner and the outer surfaces of the tube. For an unfinned shell-and-tube heat exchanger, it can be expressed as

$$\frac{1}{UA_s} = \frac{1}{U_i A_i} = \frac{1}{U_o A_o} = R = \frac{1}{h_i A_i} + \frac{R_{f,i}}{A_i} + \frac{\ln(D_o/D_i)}{2\pi k L} + \frac{R_{f,o}}{A_o} + \frac{1}{h_o A_o} \quad (11\text{–}8)$$

where $R_{f,i}$ and $R_{f,o}$ are the fouling factors at those surfaces.

Representative values of fouling factors are given in Table 11–2. More comprehensive tables of fouling factors are available in handbooks. As you would expect, considerable uncertainty exists in these values, and they should be used as a guide in the selection and evaluation of heat exchangers to account for the effects of anticipated fouling on heat transfer. Note that most fouling factors in the table are of the order of 10^{-4} m² · °C/W, which is equivalent to the thermal resistance of a 0.2-mm-thick limestone layer ($k = 2.9$ W/m · °C) per unit surface area. Therefore, in the absence of specific data, we can assume the surfaces to be coated with 0.2 mm of limestone as a starting point to account for the effects of fouling.

TABLE 11–2

Representative fouling factors (thermal resistance due to fouling for a unit surface area)

Fluid	R_f, m² · °C/W
Distilled water, sea-water, river water, boiler feedwater:	
Below 50°C	0.0001
Above 50°C	0.0002
Fuel oil	0.0009
Steam (oil-free)	0.0001
Refrigerants (liquid)	0.0002
Refrigerants (vapor)	0.0004
Alcohol vapors	0.0001
Air	0.0004

(*Source:* Tubular Exchange Manufacturers Association.)

EXAMPLE 11–1   Overall Heat Transfer Coefficient of a Heat Exchanger

Hot oil is to be cooled in a double-tube counter-flow heat exchanger. The copper inner tubes have a diameter of 2 cm and negligible thickness. The inner diameter of the outer tube (the shell) is 3 cm. Water flows through the tube at a rate of 0.5 kg/s, and the oil through the shell at a rate of 0.8 kg/s. Taking the average temperatures of the water and the oil to be 45°C and 80°C, respectively, determine the overall heat transfer coefficient of this heat exchanger.

SOLUTION Hot oil is cooled by water in a double-tube counter-flow heat exchanger. The overall heat transfer coefficient is to be determined.

Assumptions 1 The thermal resistance of the inner tube is negligible since the tube material is highly conductive and its thickness is negligible. 2 Both the oil and water flow are fully developed. 3 Properties of the oil and water are constant.

Properties The properties of water at 45°C are (Table A–9)

$$\rho = 990.1 \text{ kg/m}^3 \qquad \text{Pr} = 3.91$$

$$k = 0.637 \text{ W/m} \cdot \text{°C} \qquad \nu = \mu/\rho = 0.602 \times 10^{-6} \text{ m}^2/\text{s}$$

The properties of oil at 80°C are (Table A–13)

$$\rho = 852 \text{ kg/m}^3 \qquad \text{Pr} = 499.3$$

$$k = 0.138 \text{ W/m} \cdot \text{°C} \qquad \nu = 3.794 \times 10^{-5} \text{ m}^2/\text{s}$$

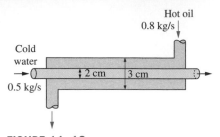

FIGURE 11–10
Schematic for Example 11–1.

Analysis The schematic of the heat exchanger is given in Fig. 11–10. The overall heat transfer coefficient U can be determined from Eq. 11–5:

$$\frac{1}{U} \approx \frac{1}{h_i} + \frac{1}{h_o}$$

where h_i and h_o are the convection heat transfer coefficients inside and outside the tube, respectively, which are to be determined using the forced convection relations.

The hydraulic diameter for a circular tube is the diameter of the tube itself, $D_h = D = 0.02$ m. The average velocity of water in the tube and the Reynolds number are

$$V = \frac{\dot{m}}{\rho A_c} = \frac{\dot{m}}{\rho(\frac{1}{4}\pi D^2)} = \frac{0.5 \text{ kg/s}}{(990.1 \text{ kg/m}^3)[\frac{1}{4}\pi (0.02 \text{ m})^2]} = 1.61 \text{ m/s}$$

and

$$\text{Re} = \frac{VD}{\nu} = \frac{(1.61 \text{ m/s})(0.02 \text{ m})}{0.602 \times 10^{-6} \text{ m}^2/\text{s}} = 53{,}490$$

which is greater than 10,000. Therefore, the flow of water is turbulent. Assuming the flow to be fully developed, the Nusselt number can be determined from

$$\text{Nu} = \frac{hD}{k} = 0.023 \, \text{Re}^{0.8} \text{Pr}^{0.4} = 0.023(53{,}490)^{0.8}(3.91)^{0.4} = 240.6$$

Then,

$$h = \frac{k}{D} \text{Nu} = \frac{0.637 \text{ W/m} \cdot \text{°C}}{0.02 \text{ m}}(240.6) = 7663 \text{ W/m}^2 \cdot \text{°C}$$

Now we repeat the analysis above for oil. The properties of oil at 80°C are

$$\rho = 852 \text{ kg/m}^3 \qquad \nu = 37.5 \times 10^{-6} \text{ m}^2/\text{s}$$
$$k = 0.138 \text{ W/m} \cdot \text{°C} \qquad \text{Pr} = 490$$

The hydraulic diameter for the annular space is

$$D_h = D_o - D_i = 0.03 - 0.02 = 0.01 \text{ m}$$

The average velocity and the Reynolds number in this case are

$$V = \frac{\dot{m}}{\rho A_c} = \frac{\dot{m}}{\rho[\frac{1}{4}\pi(D_o^2 - D_i^2)]} = \frac{0.8 \text{ kg/s}}{(852 \text{ kg/m}^3)[\frac{1}{4}\pi(0.03^2 - 0.02^2)] \text{ m}^2} = 2.39 \text{ m/s}$$

and

$$\text{Re} = \frac{VD}{\nu} = \frac{(2.39 \text{ m/s})(0.01 \text{ m})}{3.794 \times 10^{-5} \text{ m}^2/\text{s}} = 630$$

which is less than 2300. Therefore, the flow of oil is laminar. Assuming fully developed flow, the Nusselt number on the tube side of the annular space Nu_i corresponding to $D_i/D_o = 0.02/0.03 = 0.667$ can be determined from Table 11–3 by interpolation to be

$$\text{Nu} = 5.45$$

TABLE 11–3

Nusselt number for fully developed laminar flow in a circular annulus with one surface insulated and the other isothermal (Kays and Perkins)

D_i/D_o	Nu_i	Nu_o
0.00	—	3.66
0.05	17.46	4.06
0.10	11.56	4.11
0.25	7.37	4.23
0.50	5.74	4.43
1.00	4.86	4.86

and

$$h_o = \frac{k}{D_h} \text{Nu} = \frac{0.138 \text{ W/m} \cdot °\text{C}}{0.01 \text{ m}} (5.45) = 75.2 \text{ W/m}^2 \cdot °\text{C}$$

Then the overall heat transfer coefficient for this heat exchanger becomes

$$U = \frac{1}{\dfrac{1}{h_i} + \dfrac{1}{h_o}} = \frac{1}{\dfrac{1}{7663 \text{ W/m}^2 \cdot °\text{C}} + \dfrac{1}{75.2 \text{ W/m}^2 \cdot °\text{C}}} = \mathbf{74.5 \text{ W/m}^2 \cdot °\text{C}}$$

Discussion Note that $U \approx h_o$ in this case, since $h_i \gg h_o$. This confirms our earlier statement that the overall heat transfer coefficient in a heat exchanger is dominated by the smaller heat transfer coefficient when the difference between the two values is large.

To improve the overall heat transfer coefficient and thus the heat transfer in this heat exchanger, we must use some enhancement techniques on the oil side, such as a finned surface.

EXAMPLE 11–2   Effect of Fouling on the Overall Heat Transfer Coefficient

A double-pipe (shell-and-tube) heat exchanger is constructed of a stainless steel ($k = 15.1$ W/m · °C) inner tube of inner diameter $D_i = 1.5$ cm and outer diameter $D_o = 1.9$ cm and an outer shell of inner diameter 3.2 cm. The convection heat transfer coefficient is given to be $h_i = 800$ W/m² · °C on the inner surface of the tube and $h_o = 1200$ W/m² · °C on the outer surface. For a fouling factor of $R_{f,i} = 0.0004$ m² · °C/W on the tube side and $R_{f,o} = 0.0001$ m² · °C/W on the shell side, determine (*a*) the thermal resistance of the heat exchanger per unit length and (*b*) the overall heat transfer coefficients, U_i and U_o based on the inner and outer surface areas of the tube, respectively.

SOLUTION The heat transfer coefficients and the fouling factors on the tube and shell sides of a heat exchanger are given. The thermal resistance and the overall heat transfer coefficients based on the inner and outer areas are to be determined.

Assumptions The heat transfer coefficients and the fouling factors are constant and uniform.

Analysis (*a*) The schematic of the heat exchanger is given in Fig. 11–11. The thermal resistance for an unfinned shell-and-tube heat exchanger with fouling on both heat transfer surfaces is given by Eq. 11–8 as

$$R = \frac{1}{UA_s} = \frac{1}{U_i A_i} = \frac{1}{U_o A_o} = \frac{1}{h_i A_i} + \frac{R_{f,i}}{A_i} + \frac{\ln (D_o/D_i)}{2\pi kL} + \frac{R_{f,o}}{A_o} + \frac{1}{h_o A_o}$$

where

$$A_i = \pi D_i L = \pi(0.015 \text{ m})(1 \text{ m}) = 0.0471 \text{ m}^2$$
$$A_o = \pi D_o L = \pi(0.019 \text{ m})(1 \text{ m}) = 0.0597 \text{ m}^2$$

Substituting, the total thermal resistance is determined to be

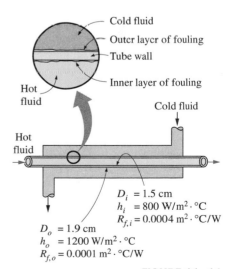

Cold fluid
Outer layer of fouling
Tube wall
Inner layer of fouling
Hot fluid
Cold fluid
Hot fluid

$D_i = 1.5$ cm
$h_i = 800$ W/m² · °C
$R_{f,i} = 0.0004$ m² · °C/W
$D_o = 1.9$ cm
$h_o = 1200$ W/m² · °C
$R_{f,o} = 0.0001$ m² · °C/W

FIGURE 11–11
Schematic for Example 11–2.

$$R = \frac{1}{(800 \text{ W/m}^2 \cdot {}^\circ\text{C})(0.0471 \text{ m}^2)} + \frac{0.0004 \text{ m}^2 \cdot {}^\circ\text{C/W}}{0.0471 \text{ m}^2}$$

$$+ \frac{\ln{(0.019/0.015)}}{2\pi(15.1 \text{ W/m} \cdot {}^\circ\text{C})(1 \text{ m})}$$

$$+ \frac{0.0001 \text{ m}^2 \cdot {}^\circ\text{C/W}}{0.0597 \text{ m}^2} + \frac{1}{(1200 \text{ W/m}^2 \cdot {}^\circ\text{C})(0.0597 \text{ m}^2)}$$

$$= (0.02654 + 0.00849 + 0.0025 + 0.00168 + 0.01396){}^\circ\text{C/W}$$

$$= \mathbf{0.0532 {}^\circ\text{C/W}}$$

Note that about 19 percent of the total thermal resistance in this case is due to fouling and about 5 percent of it is due to the steel tube separating the two fluids. The rest (76 percent) is due to the convection resistances.

(*b*) Knowing the total thermal resistance and the heat transfer surface areas, the overall heat transfer coefficients based on the inner and outer surfaces of the tube are

$$U_i = \frac{1}{RA_i} = \frac{1}{(0.0532 \, {}^\circ\text{C/W})(0.0471 \text{ m}^2)} = \mathbf{399 \text{ W/m}^2 \cdot {}^\circ\text{C}}$$

and

$$U_o = \frac{1}{RA_o} = \frac{1}{(0.0532 \, {}^\circ\text{C/W})(0.0597 \text{ m}^2)} = \mathbf{315 \text{ W/m}^2 \cdot {}^\circ\text{C}}$$

Discussion Note that the two overall heat transfer coefficients differ significantly (by 27 percent) in this case because of the considerable difference between the heat transfer surface areas on the inner and the outer sides of the tube. For tubes of negligible thickness, the difference between the two overall heat transfer coefficients would be negligible.

11–3 · ANALYSIS OF HEAT EXCHANGERS

Heat exchangers are commonly used in practice, and an engineer often finds himself or herself in a position to *select a heat exchanger* that will achieve a *specified temperature change* in a fluid stream of known mass flow rate, or to *predict the outlet temperatures* of the hot and cold fluid streams in a *specified heat exchanger.*

In upcoming sections, we discuss the two methods used in the analysis of heat exchangers. Of these, the *log mean temperature difference* (or LMTD) method is best suited for the first task and the *effectiveness–NTU* method for the second task. But first we present some general considerations.

Heat exchangers usually operate for long periods of time with no change in their operating conditions. Therefore, they can be modeled as *steady-flow* devices. As such, the mass flow rate of each fluid remains constant, and the fluid properties such as temperature and velocity at any inlet or outlet remain the same. Also, the fluid streams experience little or no change in their velocities and elevations, and thus the kinetic and potential energy changes are negligible. The specific heat of a fluid, in general, changes with temperature. But, in a specified temperature range, it can be treated as a constant at some average value with little loss in accuracy. Axial heat conduction along the tube is usually insignificant and can be considered negligible. Finally, the outer surface of the heat exchanger is assumed to be *perfectly insulated,* so that there is no heat loss to the surrounding medium, and any heat transfer occurs between the two fluids only.

The idealizations stated above are closely approximated in practice, and they greatly simplify the analysis of a heat exchanger with little sacrifice from accuracy. Therefore, they are commonly used. Under these assumptions, the *first law of thermodynamics* requires that the rate of heat transfer from the hot fluid be equal to the rate of heat transfer to the cold one. That is,

$$\dot{Q} = \dot{m}_c c_{pc}(T_{c,\,out} - T_{c,\,in}) \tag{11-9}$$

and

$$\dot{Q} = \dot{m}_h c_{ph}(T_{h,\,in} - T_{h,\,out}) \tag{11-10}$$

where the subscripts c and h stand for *cold* and *hot* fluids, respectively, and

$$\dot{m}_c, \dot{m}_h = \text{mass flow rates}$$
$$c_{pc}, c_{ph} = \text{specific heats}$$
$$T_{c,\,out}, T_{h,\,out} = \text{outlet temperatures}$$
$$T_{c,\,in}, T_{h,\,in} = \text{inlet temperatures}$$

Note that the heat transfer rate $\dot{Q}$ is taken to be a positive quantity, and its direction is understood to be from the hot fluid to the cold one in accordance with the second law of thermodynamics.

In heat exchanger analysis, it is often convenient to combine the product of the mass flow rate and the specific heat of a fluid into a single quantity. This quantity is called the **heat capacity rate** and is defined for the hot and cold fluid streams as

$$C_h = \dot{m}_h c_{ph} \quad \text{and} \quad C_c = \dot{m}_c c_{pc} \tag{11-11}$$

The heat capacity rate of a fluid stream represents the rate of heat transfer needed to change the temperature of the fluid stream by 1°C as it flows through a heat exchanger. Note that in a heat exchanger, the fluid with a *large* heat capacity rate experiences a *small* temperature change, and the fluid with a *small* heat capacity rate experiences a *large* temperature change. Therefore, *doubling* the mass flow rate of a fluid while leaving everything else unchanged will *halve* the temperature change of that fluid.

With the definition of the heat capacity rate above, Eqs. 11–9 and 11–10 can also be expressed as

$$\dot{Q} = C_c(T_{c,\,out} - T_{c,\,in}) \tag{11-12}$$

and

$$\dot{Q} = C_h(T_{h,\,in} - T_{h,\,out}) \tag{11-13}$$

That is, the heat transfer rate in a heat exchanger is equal to the heat capacity rate of either fluid multiplied by the temperature change of that fluid. Note that *the only time the temperature rise of a cold fluid is equal to the temperature drop of the hot fluid is when the heat capacity rates of the two fluids are equal to each other* (Fig. 11–12).

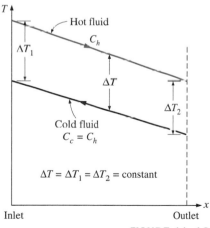

FIGURE 11–12

Two fluid streams that have the same capacity rates experience the same temperature change in a well-insulated heat exchanger.

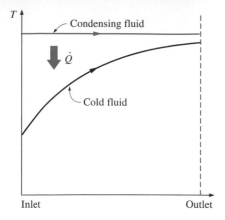

(a) Condenser $(C_h \rightarrow \infty)$

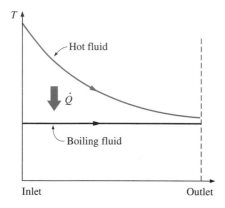

(b) Boiler $(C_c \rightarrow \infty)$

FIGURE 11–13

Variation of fluid temperatures in a heat exchanger when one of the fluids condenses or boils.

Two special types of heat exchangers commonly used in practice are *condensers* and *boilers*. One of the fluids in a condenser or a boiler undergoes a phase-change process, and the rate of heat transfer is expressed as

$$\dot{Q} = \dot{m} h_{fg} \tag{11–14}$$

where $\dot{m}$ is the rate of evaporation or condensation of the fluid and h_{fg} is the enthalpy of vaporization of the fluid at the specified temperature or pressure.

An ordinary fluid absorbs or releases a large amount of heat essentially at constant temperature during a phase-change process, as shown in Fig. 11–13. The heat capacity rate of a fluid during a phase-change process must approach infinity since the temperature change is practically zero. That is, $C = \dot{m} c_p \rightarrow \infty$ when $\Delta T \rightarrow 0$, so that the heat transfer rate $\dot{Q} = \dot{m} c_p \, \Delta T$ is a finite quantity. Therefore, in heat exchanger analysis, a condensing or boiling fluid is conveniently modeled as a fluid whose heat capacity rate is *infinity*.

The rate of heat transfer in a heat exchanger can also be expressed in an analogous manner to Newton's law of cooling as

$$\dot{Q} = U A_s \, \Delta T_m \tag{11–15}$$

where U is the overall heat transfer coefficient, A_s is the heat transfer area, and ΔT_m is an appropriate average temperature difference between the two fluids. Here the surface area A_s can be determined precisely using the dimensions of the heat exchanger. However, the overall heat transfer coefficient U and the temperature difference ΔT between the hot and cold fluids, in general, may vary along the heat exchanger.

The average value of the overall heat transfer coefficient can be determined as described in the preceding section by using the average convection coefficients for each fluid. It turns out that the appropriate form of the average temperature difference between the two fluids is *logarithmic* in nature, and its determination is presented in Section 11–4.

11–4 ▪ THE LOG MEAN TEMPERATURE DIFFERENCE METHOD

Earlier, we mentioned that the temperature difference between the hot and cold fluids varies along the heat exchanger, and it is convenient to have a *mean temperature difference* ΔT_m for use in the relation $\dot{Q} = U A_s \, \Delta T_m$.

In order to develop a relation for the equivalent average temperature difference between the two fluids, consider the *parallel-flow double-pipe* heat exchanger shown in Fig. 11–14. Note that the temperature difference ΔT between the hot and cold fluids is large at the inlet of the heat exchanger but decreases exponentially toward the outlet. As you would expect, the temperature of the hot fluid decreases and the temperature of the cold fluid increases along the heat exchanger, but the temperature of the cold fluid can never exceed that of the hot fluid no matter how long the heat exchanger is.

Assuming the outer surface of the heat exchanger to be well insulated so that any heat transfer occurs between the two fluids, and disregarding any changes in kinetic and potential energy, an energy balance on each fluid in a differential section of the heat exchanger can be expressed as

$$\delta \dot{Q} = -\dot{m}_h c_{ph} \, dT_h \qquad (11\text{--}16)$$

and

$$\delta \dot{Q} = \dot{m}_c c_{pc} \, dT_c \qquad (11\text{--}17)$$

That is, the rate of heat loss from the hot fluid at any section of a heat exchanger is equal to the rate of heat gain by the cold fluid in that section. The temperature change of the hot fluid is a *negative* quantity, and so a *negative sign* is added to Eq. 11–16 to make the heat transfer rate $\dot{Q}$ a positive quantity. Solving the equations above for dT_h and dT_c gives

$$dT_h = -\frac{\delta \dot{Q}}{\dot{m}_h c_{ph}} \qquad (11\text{--}18)$$

and

$$dT_c = \frac{\delta \dot{Q}}{\dot{m}_c c_{pc}} \qquad (11\text{--}19)$$

Taking their difference, we get

$$dT_h - dT_c = d(T_h - T_c) = -\delta \dot{Q} \left(\frac{1}{\dot{m}_h c_{ph}} + \frac{1}{\dot{m}_c c_{pc}} \right) \qquad (11\text{--}20)$$

The rate of heat transfer in the differential section of the heat exchanger can also be expressed as

$$\delta \dot{Q} = U(T_h - T_c) \, dA_s \qquad (11\text{--}21)$$

Substituting this equation into Eq. 11–20 and rearranging give

$$\frac{d(T_h - T_c)}{T_h - T_c} = -U \, dA_s \left(\frac{1}{\dot{m}_h c_{ph}} + \frac{1}{\dot{m}_c c_{pc}} \right) \qquad (11\text{--}22)$$

Integrating from the inlet of the heat exchanger to its outlet, we obtain

$$\ln \frac{T_{h,\,out} - T_{c,\,out}}{T_{h,\,in} - T_{c,\,in}} = -UA_s \left(\frac{1}{\dot{m}_h c_{ph}} + \frac{1}{\dot{m}_c c_{pc}} \right) \qquad (11\text{--}23)$$

Finally, solving Eqs. 11–9 and 11–10 for $\dot{m}_c c_{pc}$ and $\dot{m}_h c_{ph}$ and substituting into Eq. 11–23 give, after some rearrangement,

$$\dot{Q} = UA_s \, \Delta T_{lm} \qquad (11\text{--}24)$$

where

$$\Delta T_{lm} = \frac{\Delta T_1 - \Delta T_2}{\ln (\Delta T_1/\Delta T_2)} \qquad (11\text{--}25)$$

is the **log mean temperature difference**, which is the suitable form of the average temperature difference for use in the analysis of heat exchangers. Here ΔT_1 and ΔT_2 represent the temperature difference between the two fluids

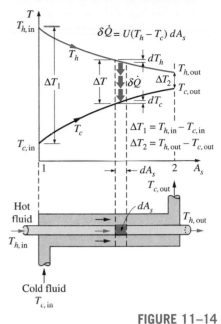

FIGURE 11–14

Variation of the fluid temperatures in a parallel-flow double-pipe heat exchanger.

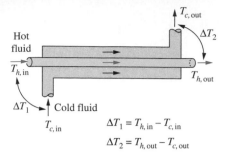

$$\Delta T_1 = T_{h,\,in} - T_{c,\,in}$$
$$\Delta T_2 = T_{h,\,out} - T_{c,\,out}$$

(a) Parallel-flow heat exchangers

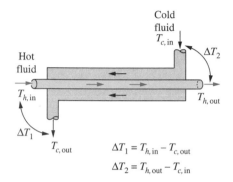

$$\Delta T_1 = T_{h,\,in} - T_{c,\,out}$$
$$\Delta T_2 = T_{h,\,out} - T_{c,\,in}$$

(b) Counter-flow heat exchangers

FIGURE 11–15

The ΔT_1 and ΔT_2 expressions in parallel-flow and counter-flow heat exchangers.

at the two ends (inlet and outlet) of the heat exchanger. It makes no difference which end of the heat exchanger is designated as the inlet or the outlet (Fig. 11–15).

The temperature difference between the two fluids decreases from ΔT_1 at the inlet to ΔT_2 at the outlet. Thus, it is tempting to use the arithmetic mean temperature $\Delta T_{am} = \frac{1}{2}(\Delta T_1 + \Delta T_2)$ as the average temperature difference. The logarithmic mean temperature difference ΔT_{lm} is obtained by tracing the actual temperature profile of the fluids along the heat exchanger and is an *exact* representation of the *average temperature difference* between the hot and cold fluids. It truly reflects the exponential decay of the local temperature difference.

Note that ΔT_{lm} is always less than ΔT_{am}. Therefore, using ΔT_{am} in calculations instead of ΔT_{lm} will overestimate the rate of heat transfer in a heat exchanger between the two fluids. When ΔT_1 differs from ΔT_2 by no more than 40 percent, the error in using the arithmetic mean temperature difference is less than 1 percent. But the error increases to undesirable levels when ΔT_1 differs from ΔT_2 by greater amounts. Therefore, we should always use the *logarithmic mean temperature difference* when determining the rate of heat transfer in a heat exchanger.

Counter-Flow Heat Exchangers

The variation of temperatures of hot and cold fluids in a counter-flow heat exchanger is given in Fig. 11–16. Note that the hot and cold fluids enter the heat exchanger from opposite ends, and the outlet temperature of the *cold fluid* in this case may exceed the outlet temperature of the *hot fluid*. In the limiting case, the cold fluid will be heated to the inlet temperature of the hot fluid. However, the outlet temperature of the cold fluid can *never* exceed the inlet temperature of the hot fluid, since this would be a violation of the second law of thermodynamics.

The relation already given for the log mean temperature difference is developed using a parallel-flow heat exchanger, but we can show by repeating the analysis for a counter-flow heat exchanger that is also applicable to counter-flow heat exchangers. But this time, ΔT_1 and ΔT_2 are expressed as shown in Fig. 11–15.

For specified inlet and outlet temperatures, the log mean temperature difference for a counter-flow heat exchanger is always greater than that for a parallel-flow heat exchanger. That is, $\Delta T_{lm,\,CF} > \Delta T_{lm,\,PF}$, and thus a smaller surface area (and thus a smaller heat exchanger) is needed to achieve a specified heat transfer rate in a counter-flow heat exchanger. Therefore, it is common practice to use counter-flow arrangements in heat exchangers.

In a counter-flow heat exchanger, the temperature difference between the hot and the cold fluids remains constant along the heat exchanger when the *heat capacity rates* of the two fluids are *equal* (that is, $\Delta T = $ constant when $C_h = C_c$ or $\dot{m}_h c_{ph} = \dot{m}_c c_{pc}$). Then we have $\Delta T_1 = \Delta T_2$, and the log mean temperature difference relation gives $\Delta T_{lm} = \frac{0}{0}$, which is indeterminate. It can be shown by the application of l'Hôpital's rule that in this case we have $\Delta T_{lm} = \Delta T_1 = \Delta T_2$, as expected.

A *condenser* or a *boiler* can be considered to be either a parallel- or counter-flow heat exchanger since both approaches give the same result.

CHAPTER 11

Multipass and Cross-Flow Heat Exchangers: Use of a Correction Factor

The log mean temperature difference ΔT_{lm} relation developed earlier is limited to parallel-flow and counter-flow heat exchangers only. Similar relations are also developed for *cross-flow* and *multipass shell-and-tube* heat exchangers, but the resulting expressions are too complicated because of the complex flow conditions.

In such cases, it is convenient to relate the equivalent temperature difference to the log mean temperature difference relation for the counter-flow case as

$$\Delta T_{lm} = F \Delta T_{lm, CF} \qquad (11-26)$$

where F is the **correction factor**, which depends on the *geometry* of the heat exchanger and the inlet and outlet temperatures of the hot and cold fluid streams. The $\Delta T_{lm, CF}$ is the log mean temperature difference for the case of a *counter-flow* heat exchanger with the same inlet and outlet temperatures and is determined from Eq. 11–25 by taking $\Delta T_1 = T_{h, in} - T_{c, out}$ and $\Delta T_2 = T_{h, out} - T_{c, in}$ (Fig. 11–17).

The correction factor is less than unity for a cross-flow and multipass shell-and-tube heat exchanger. That is, $F \le 1$. The limiting value of $F = 1$ corresponds to the counter-flow heat exchanger. Thus, the correction factor F for a heat exchanger is *a measure of deviation of the ΔT_{lm} from the corresponding values for the counter-flow case.*

The correction factor F for common cross-flow and shell-and-tube heat exchanger configurations is given in Fig. 11–18 versus two temperature ratios P and R defined as

$$P = \frac{t_2 - t_1}{T_1 - t_1} \qquad (11-27)$$

and

$$R = \frac{T_1 - T_2}{t_2 - t_1} = \frac{(\dot{m}c_p)_{tube\ side}}{(\dot{m}c_p)_{shell\ side}} \qquad (11-28)$$

where the subscripts 1 and 2 represent the *inlet* and *outlet,* respectively. Note that for a shell-and-tube heat exchanger, T and t represent the *shell-* and *tube-side* temperatures, respectively, as shown in the correction factor charts. It makes no difference whether the hot or the cold fluid flows through the shell or the tube. The determination of the correction factor F requires the availability of the inlet and the outlet temperatures for both the cold and hot fluids.

Note that the value of P ranges from 0 to 1. The value of R, on the other hand, ranges from 0 to infinity, with $R = 0$ corresponding to the phase-change (condensation or boiling) on the shell-side and $R \to \infty$ to phase-change on the tube side. The correction factor is $F = 1$ for both of these limiting cases. Therefore, the correction factor for a condenser or boiler is $F = 1$, regardless of the configuration of the heat exchanger.

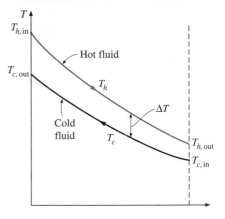

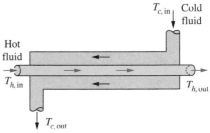

FIGURE 11–16

The variation of the fluid temperatures in a counter-flow double-pipe heat exchanger.

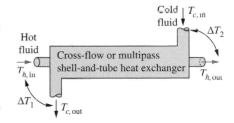

Heat transfer rate:
$$\dot{Q} = UA_s F \Delta T_{lm, CF}$$

where
$$\Delta T_{lm, CF} = \frac{\Delta T_1 - \Delta T_2}{\ln(\Delta T_1 / \Delta T_2)}$$

$$\Delta T_1 = T_{h, in} - T_{c, out}$$

$$\Delta T_2 = T_{h, out} - T_{c, in}$$

and
$$F = \dots \text{ (Fig. 11–18)}$$

FIGURE 11–17

The determination of the heat transfer rate for cross-flow and multipass shell-and-tube heat exchangers using the correction factor.

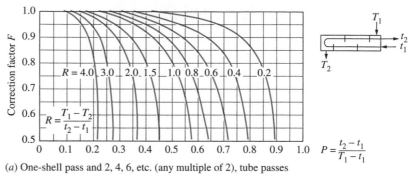

(a) One-shell pass and 2, 4, 6, etc. (any multiple of 2), tube passes

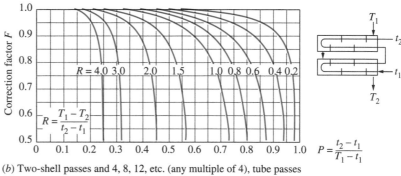

(b) Two-shell passes and 4, 8, 12, etc. (any multiple of 4), tube passes

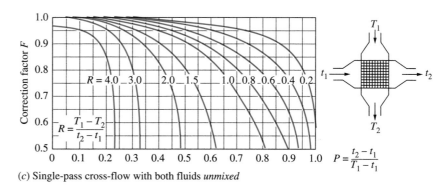

(c) Single-pass cross-flow with both fluids *unmixed*

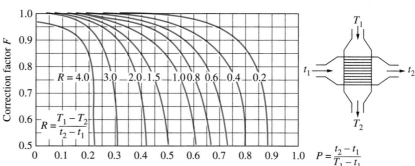

(d) Single-pass cross-flow with one fluid *mixed* and the other *unmixed*

FIGURE 11–18

Correction factor F charts
for common shell-and-tube and
cross-flow heat exchangers.
(From Bowman, Mueller, and Nagle)

EXAMPLE 11–3   The Condensation of Steam in a Condenser

Steam in the condenser of a power plant is to be condensed at a temperature of 30°C with cooling water from a nearby lake, which enters the tubes of the condenser at 14°C and leaves at 22°C. The surface area of the tubes is 45 m², and the overall heat transfer coefficient is 2100 W/m² · °C. Determine the mass flow rate of the cooling water needed and the rate of condensation of the steam in the condenser.

SOLUTION   Steam is condensed by cooling water in the condenser of a power plant. The mass flow rate of the cooling water and the rate of condensation are to be determined.

Assumptions   **1** Steady operating conditions exist. **2** The heat exchanger is well insulated so that heat loss to the surroundings is negligible. **3** Changes in the kinetic and potential energies of fluid streams are negligible. **4** There is no fouling. **5** Fluid properties are constant.

Properties   The heat of vaporization of water at 30°C is h_{fg} = 2431 kJ/kg and the specific heat of cold water at the average temperature of 18°C is c_p = 4184 J/kg · °C (Table A–9).

Analysis   The schematic of the condenser is given in Fig. 11–19. The condenser can be treated as a counter-flow heat exchanger since the temperature of one of the fluids (the steam) remains constant.

The temperature difference between the steam and the cooling water at the two ends of the condenser is

$$\Delta T_1 = T_{h, in} - T_{c, out} = (30 - 22)°C = 8°C$$

$$\Delta T_2 = T_{h, out} - T_{c, in} = (30 - 14)°C = 16°C$$

That is, the temperature difference between the two fluids varies from 8°C at one end to 16°C at the other. The proper average temperature difference between the two fluids is the *logarithmic mean temperature difference* (not the arithmetic), which is determined from

$$\Delta T_{lm} = \frac{\Delta T_1 - \Delta T_2}{\ln(\Delta T_1 / \Delta T_2)} = \frac{8 - 16}{\ln(8/16)} = 11.5°C$$

This is a little less than the arithmetic mean temperature difference of $\frac{1}{2}(8 + 16) = 12°C$. Then the heat transfer rate in the condenser is determined from

$$\dot{Q} = UA_s \, \Delta T_{lm} = (2100 \text{ W/m}^2 \cdot °C)(45 \text{ m}^2)(11.5°C) = 1.087 \times 10^6 \text{ W} = 1087 \text{ kW}$$

Therefore, steam will lose heat at a rate of 1087 kW as it flows through the condenser, and the cooling water will gain practically all of it, since the condenser is well insulated.

The mass flow rate of the cooling water and the rate of the condensation of the steam are determined from $\dot{Q} = [\dot{m}c_p(T_{out} - T_{in})]_{cooling\ water} = (\dot{m}h_{fg})_{steam}$ to be

$$\dot{m}_{cooling\ water} = \frac{\dot{Q}}{c_p(T_{out} - T_{in})} = \frac{1087 \text{ kJ/s}}{(4.184 \text{ kJ/kg} \cdot °C)(22 - 14)°C} = 32.5 \text{ kg/s}$$

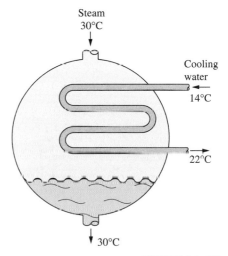

Steam
30°C

Cooling
water
14°C

22°C

30°C

FIGURE 11–19
Schematic for Example 11–3.

and

$$\dot{m}_{\text{steam}} = \frac{\dot{Q}}{h_{fg}} = \frac{1087 \text{ kJ/s}}{2431 \text{ kJ/kg}} = 0.45 \text{ kg/s}$$

Therefore, we need to circulate about 72 kg of cooling water for each 1 kg of steam condensing to remove the heat released during the condensation process.

EXAMPLE 11–4 **Heating Water in a Counter-Flow Heat Exchanger**

A counter-flow double-pipe heat exchanger is to heat water from 20°C to 80°C at a rate of 1.2 kg/s. The heating is to be accomplished by geothermal water available at 160°C at a mass flow rate of 2 kg/s. The inner tube is thin-walled and has a diameter of 1.5 cm. If the overall heat transfer coefficient of the heat exchanger is 640 W/m² · °C, determine the length of the heat exchanger required to achieve the desired heating.

SOLUTION Water is heated in a counter-flow double-pipe heat exchanger by geothermal water. The required length of the heat exchanger is to be determined.

Assumptions **1** Steady operating conditions exist. **2** The heat exchanger is well insulated so that heat loss to the surroundings is negligible. **3** Changes in the kinetic and potential energies of fluid streams are negligible. **4** There is no fouling. **5** Fluid properties are constant.

Properties We take the specific heats of water and geothermal fluid to be 4.18 and 4.31 kJ/kg · °C, respectively.

Analysis The schematic of the heat exchanger is given in Fig. 11–20. The rate of heat transfer in the heat exchanger can be determined from

$$\dot{Q} = [\dot{m}c_p(T_{\text{out}} - T_{\text{in}})]_{\text{water}} = (1.2 \text{ kg/s})(4.18 \text{ kJ/kg} \cdot °C)(80 - 20)°C = 301 \text{ kW}$$

Noting that all of this heat is supplied by the geothermal water, the outlet temperature of the geothermal water is determined to be

$$\dot{Q} = [\dot{m}c_p(T_{\text{in}} - T_{\text{out}})]_{\text{geothermal}} \longrightarrow T_{\text{out}} = T_{\text{in}} - \frac{\dot{Q}}{\dot{m}c_p}$$

$$= 160°C - \frac{301 \text{ kW}}{(2 \text{ kg/s})(4.31 \text{ kJ/kg} \cdot °C)}$$

$$= 125°C$$

Knowing the inlet and outlet temperatures of both fluids, the logarithmic mean temperature difference for this counter-flow heat exchanger becomes

$$\Delta T_1 = T_{h, \text{ in}} - T_{c, \text{ out}} = (160 - 80)°C = 80°C$$
$$\Delta T_2 = T_{h, \text{ out}} - T_{c, \text{ in}} = (125 - 20)°C = 105°C$$

and

$$\Delta T_{\text{lm}} = \frac{\Delta T_1 - \Delta T_2}{\ln (\Delta T_1/\Delta T_2)} = \frac{80 - 105}{\ln (80/105)} = 91.9°C$$

Then the surface area of the heat exchanger is determined to be

$$\dot{Q} = UA_s \, \Delta T_{\text{lm}} \longrightarrow A_s = \frac{\dot{Q}}{U \, \Delta T_{\text{lm}}} = \frac{301,000 \text{ W}}{(640 \text{ W/m}^2 \cdot °C)(91.9°C)} = 5.12 \text{ m}^2$$

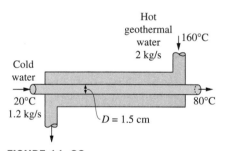

Hot geothermal water 160°C 2 kg/s

Cold water

20°C 1.2 kg/s

80°C

D = 1.5 cm

FIGURE 11–20
Schematic for Example 11–4.

To provide this much heat transfer surface area, the length of the tube must be

$$A_s = \pi DL \longrightarrow L = \frac{A_s}{\pi D} = \frac{5.12 \text{ m}^2}{\pi (0.015 \text{ m})} = 109 \text{ m}$$

Discussion The inner tube of this counter-flow heat exchanger (and thus the heat exchanger itself) needs to be over 100 m long to achieve the desired heat transfer, which is impractical. In cases like this, we need to use a plate heat exchanger or a multipass shell-and-tube heat exchanger with multiple passes of tube bundles.

EXAMPLE 11–5   **Heating of Glycerin in a Multipass Heat Exchanger**

A 2-shell passes and 4-tube passes heat exchanger is used to heat glycerin from 20°C to 50°C by hot water, which enters the thin-walled 2-cm-diameter tubes at 80°C and leaves at 40°C (Fig. 11–21). The total length of the tubes in the heat exchanger is 60 m. The convection heat transfer coefficient is 25 W/m² · °C on the glycerin (shell) side and 160 W/m² · °C on the water (tube) side. Determine the rate of heat transfer in the heat exchanger (*a*) before any fouling and (*b*) after fouling with a fouling factor of 0.0006 m² · °C/W occurs on the outer surfaces of the tubes.

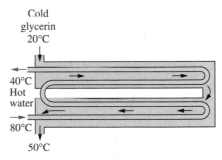

FIGURE 11–21
Schematic for Example 11–5.

SOLUTION Glycerin is heated in a 2-shell passes and 4-tube passes heat exchanger by hot water. The rate of heat transfer for the cases of fouling and no fouling are to be determined.

Assumptions **1** Steady operating conditions exist. **2** The heat exchanger is well insulated so that heat loss to the surroundings is negligible **3** Changes in the kinetic and potential energies of fluid streams are negligible. **4** Heat transfer coefficients and fouling factors are constant and uniform. **5** The thermal resistance of the inner tube is negligible since the tube is thin-walled and highly conductive.

Analysis The tubes are said to be thin-walled, and thus it is reasonable to assume the inner and outer surface areas of the tubes to be equal. Then the heat transfer surface area becomes

$$A_s = \pi DL = \pi (0.02 \text{ m})(60 \text{ m}) = 3.77 \text{ m}^2$$

The rate of heat transfer in this heat exchanger can be determined from

$$\dot{Q} = UA_s F \, \Delta T_{\text{lm, CF}}$$

where F is the correction factor and $\Delta T_{\text{lm, CF}}$ is the log mean temperature difference for the counter-flow arrangement. These two quantities are determined from

$$\Delta T_1 = T_{h,\text{ in}} - T_{c,\text{ out}} = (80 - 50)\text{°C} = 30\text{°C}$$

$$\Delta T_2 = T_{h,\text{ out}} - T_{c,\text{ in}} = (40 - 20)\text{°C} = 20\text{°C}$$

$$\Delta T_{\text{lm, CF}} = \frac{\Delta T_1 - \Delta T_2}{\ln (\Delta T_1/\Delta T_2)} = \frac{30 - 20}{\ln (30/20)} = 24.7\text{°C}$$

and

$$P = \frac{t_2 - t_1}{T_1 - t_1} = \frac{40 - 80}{20 - 80} = 0.67 \left.\right\}$$

$$R = \frac{T_1 - T_2}{t_2 - t_1} = \frac{20 - 50}{40 - 80} = 0.75 \left.\right\} \quad F = 0.91 \qquad \text{(Fig. 11–18}b\text{)}$$

(a) In the case of no fouling, the overall heat transfer coefficient U is

$$U = \frac{1}{\dfrac{1}{h_i} + \dfrac{1}{h_o}} = \frac{1}{\dfrac{1}{160 \text{ W/m}^2 \cdot {}^\circ\text{C}} + \dfrac{1}{25 \text{ W/m}^2 \cdot {}^\circ\text{C}}} = 21.6 \text{ W/m}^2 \cdot {}^\circ\text{C}$$

Then the rate of heat transfer becomes

$$\dot{Q} = UA_s F \, \Delta T_{\text{lm, CF}} = (21.6 \text{ W/m}^2 \cdot {}^\circ\text{C})(3.77 \text{ m}^2)(0.91)(24.7{}^\circ\text{C}) = 1830 \text{ W}$$

(b) When there is fouling on one of the surfaces, we have

$$U = \frac{1}{\dfrac{1}{h_i} + \dfrac{1}{h_o} + R_f} = \frac{1}{\dfrac{1}{160 \text{ W/m}^2 \cdot {}^\circ\text{C}} + \dfrac{1}{25 \text{ W/m}^2 \cdot {}^\circ\text{C}} + 0.0006 \text{ m}^2 \cdot {}^\circ\text{C/ W}}$$

$$= 21.3 \text{ W/m}^2 \cdot {}^\circ\text{C}$$

and

$$\dot{Q} = UA_s F \, \Delta T_{\text{lm, CF}} = (21.3 \text{ W/m}^2 \cdot {}^\circ\text{C})(3.77 \text{ m}^2)(0.91)(24.7{}^\circ\text{C}) = 1805 \text{ W}$$

Discussion Note that the rate of heat transfer decreases as a result of fouling, as expected. The decrease is not dramatic, however, because of the relatively low convection heat transfer coefficients involved.

90°C

Air flow
(*unmixed*)
20°C

40°C

65°C
Water flow
(*unmixed*)

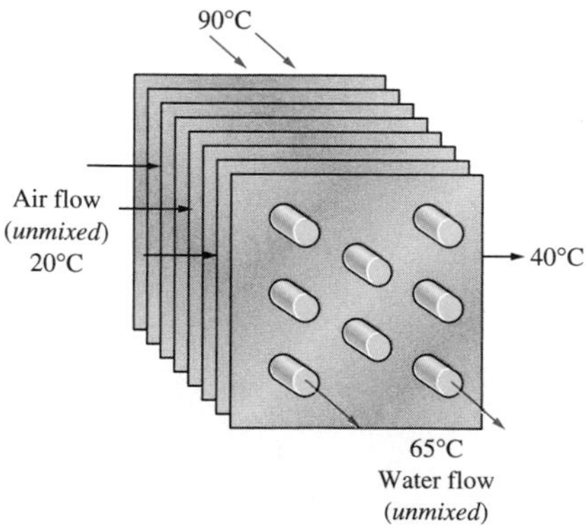

FIGURE 11–22
Schematic for Example 11–6.

EXAMPLE 11–6   Cooling of Water in an Automotive Radiator

A test is conducted to determine the overall heat transfer coefficient in an automotive radiator that is a compact cross-flow water-to-air heat exchanger with both fluids (air and water) unmixed (Fig. 11–22). The radiator has 40 tubes of internal diameter 0.5 cm and length 65 cm in a closely spaced plate-finned matrix. Hot water enters the tubes at 90°C at a rate of 0.6 kg/s and leaves at 65°C. Air flows across the radiator through the interfin spaces and is heated from 20°C to 40°C. Determine the overall heat transfer coefficient U_i of this radiator based on the inner surface area of the tubes.

SOLUTION During an experiment involving an automotive radiator, the inlet and exit temperatures of water and air and the mass flow rate of water are measured. The overall heat transfer coefficient based on the inner surface area is to be determined.

Assumptions **1** Steady operating conditions exist. **2** Changes in the kinetic and potential energies of fluid streams are negligible. **3** Fluid properties are constant.

Properties The specific heat of water at the average temperature of $(90 + 65)/2 = 77.5°C$ is 4.195 kJ/kg · °C (Table A–9).

Analysis The rate of heat transfer in this radiator from the hot water to the air is determined from an energy balance on water flow,

$$\dot{Q} = [\dot{m}c_p(T_{in} - T_{out})]_{water} = (0.6 \text{ kg/s})(4.195 \text{ kJ/kg} \cdot °C)(90 - 65)°C$$

$$= 62.93 \text{ kW}$$

The tube-side heat transfer area is the total surface area of the tubes, and is determined from

$$A_i = n\pi D_i L = (40)\pi(0.005 \text{ m})(0.65 \text{ m}) = 0.408 \text{ m}^2$$

Knowing the rate of heat transfer and the surface area, the overall heat transfer coefficient can be determined from

$$\dot{Q} = U_i A_i F \Delta T_{lm, CF} \longrightarrow U_i = \frac{\dot{Q}}{A_i F \Delta T_{lm, CF}}$$

where F is the correction factor and $\Delta T_{lm, CF}$ is the log mean temperature difference for the counter-flow arrangement. These two quantities are found to be

$$\Delta T_1 = T_{h, in} - T_{c, out} = (90 - 40)°C = 50°C$$

$$\Delta T_2 = T_{h, out} - T_{c, in} = (65 - 20)°C = 45°C$$

$$\Delta T_{lm, CF} = \frac{\Delta T_1 - \Delta T_2}{\ln(\Delta T_1/\Delta T_2)} = \frac{50 - 45}{\ln(50/45)} = 47.5°C$$

and

$$\left.\begin{array}{l} P = \dfrac{t_2 - t_1}{T_1 - t_1} = \dfrac{65 - 90}{20 - 90} = 0.36 \\[2ex] R = \dfrac{T_1 - T_2}{t_2 - t_1} = \dfrac{20 - 40}{65 - 90} = 0.80 \end{array}\right\} \quad F = 0.97 \qquad \text{(Fig. 11–18c)}$$

Substituting, the overall heat transfer coefficient U_i is determined to be

$$U_i = \frac{\dot{Q}}{A_i F \Delta T_{lm, CF}} = \frac{62,930 \text{ W}}{(0.408 \text{ m}^2)(0.97)(47.5°C)} = 3347 \text{ W/m}^2 \cdot °C$$

Discussion Note that the overall heat transfer coefficient on the air side will be much lower because of the large surface area involved on that side.

11–5 · THE EFFECTIVENESS–NTU METHOD

The log mean temperature difference (LMTD) method discussed in Section 11–4 is easy to use in heat exchanger analysis when the inlet and the outlet temperatures of the hot and cold fluids are known or can be determined from an energy balance. Once ΔT_{lm}, the mass flow rates, and the overall heat transfer coefficient are available, the heat transfer surface area of the heat exchanger can be determined from

$$\dot{Q} = UA_s\,\Delta T_{lm}$$

Therefore, the LMTD method is very suitable for determining the *size* of a heat exchanger to realize prescribed outlet temperatures when the mass flow rates and the inlet and outlet temperatures of the hot and cold fluids are specified.

With the LMTD method, the task is to *select* a heat exchanger that will meet the prescribed heat transfer requirements. The procedure to be followed by the selection process is:

1. Select the type of heat exchanger suitable for the application.
2. Determine any unknown inlet or outlet temperature and the heat transfer rate using an energy balance.
3. Calculate the log mean temperature difference ΔT_{lm} and the correction factor F, if necessary.
4. Obtain (select or calculate) the value of the overall heat transfer co-efficient U.
5. Calculate the heat transfer surface area A_s.

The task is completed by selecting a heat exchanger that has a heat transfer surface area equal to or larger than A_s.

A second kind of problem encountered in heat exchanger analysis is the determination of the *heat transfer rate* and the *outlet temperatures* of the hot and cold fluids for prescribed fluid mass flow rates and inlet temperatures when the *type* and *size* of the heat exchanger are specified. The heat transfer surface area of the heat exchanger in this case is known, but the *outlet temperatures* are not. Here the task is to determine the heat transfer performance of a specified heat exchanger or to determine if a heat exchanger available in storage will do the job.

The LMTD method could still be used for this alternative problem, but the procedure would require tedious iterations, and thus it is not practical. In an attempt to eliminate the iterations from the solution of such problems, Kays and London came up with a method in 1955 called the **effectiveness–NTU method**, which greatly simplified heat exchanger analysis.

This method is based on a dimensionless parameter called the **heat transfer effectiveness** ε, defined as

$$\varepsilon = \frac{\dot{Q}}{Q_{max}} = \frac{\text{Actual heat transfer rate}}{\text{Maximum possible heat transfer rate}} \tag{11–29}$$

The *actual* heat transfer rate in a heat exchanger can be determined from an energy balance on the hot or cold fluids and can be expressed as

$$\dot{Q} = C_c(T_{c,\,out} - T_{c,\,in}) = C_h(T_{h,\,in} - T_{h,\,out}) \tag{11–30}$$

where $C_c = \dot{m}_c c_{pc}$ and $C_h = \dot{m}_c c_{ph}$ are the heat capacity rates of the cold and hot fluids, respectively.

To determine the maximum possible heat transfer rate in a heat exchanger, we first recognize that the *maximum temperature difference* in a heat exchanger is the difference between the *inlet* temperatures of the hot and cold fluids. That is,

$$\Delta T_{max} = T_{h,\,in} - T_{c,\,in} \qquad (11\text{–}31)$$

The heat transfer in a heat exchanger will reach its maximum value when (1) the cold fluid is heated to the inlet temperature of the hot fluid or (2) the hot fluid is cooled to the inlet temperature of the cold fluid. These two limiting conditions will not be reached simultaneously unless the heat capacity rates of the hot and cold fluids are identical (i.e., $C_c = C_h$). When $C_c \ne C_h$, which is usually the case, the fluid with the *smaller* heat capacity rate will experience a larger temperature change, and thus it will be the first to experience the maximum temperature, at which point the heat transfer will come to a halt. Therefore, the maximum possible heat transfer rate in a heat exchanger is (Fig. 11–23)

$$\dot{Q}_{max} = C_{min}(T_{h,\,in} - T_{c,\,in}) \qquad (11\text{–}32)$$

where C_{min} is the smaller of C_h and C_c. This is further clarified by Example 11–7.

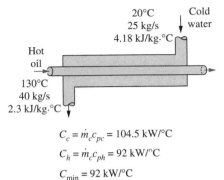

$C_c = \dot{m}_c c_{pc} = 104.5 \text{ kW/°C}$

$C_h = \dot{m}_c c_{ph} = 92 \text{ kW/°C}$

$C_{min} = 92 \text{ kW/°C}$

$\Delta T_{max} = T_{h,in} - T_{c,in} = 110°C$

$\dot{Q}_{max} = C_{min}\Delta T_{max} = 10{,}120 \text{ kW}$

FIGURE 11–23
The determination of the maximum rate of heat transfer in a heat exchanger.

EXAMPLE 11–7   Upper Limit for Heat Transfer in a Heat Exchanger

Cold water enters a counter-flow heat exchanger at 10°C at a rate of 8 kg/s, where it is heated by a hot-water stream that enters the heat exchanger at 70°C at a rate of 2 kg/s. Assuming the specific heat of water to remain constant at $c_p = 4.18$ kJ/kg · °C, determine the maximum heat transfer rate and the outlet temperatures of the cold- and the hot-water streams for this limiting case.

SOLUTION   Cold- and hot-water streams enter a heat exchanger at specified temperatures and flow rates. The maximum rate of heat transfer in the heat exchanger and the outlet temperatures are to be determined.
Assumptions   **1** Steady operating conditions exist. **2** The heat exchanger is well insulated so that heat loss to the surroundings is negligible. **3** Changes in the kinetic and potential energies of fluid streams are negligible. **4** Fluid properties are constant.
Properties   The specific heat of water is given to be $c_p = 4.18$ kJ/kg · °C.
Analysis   A schematic of the heat exchanger is given in Fig. 11–24. The heat capacity rates of the hot and cold fluids are

$$C_h = \dot{m}_h c_{ph} = (2 \text{ kg/s})(4.18 \text{ kJ/kg} \cdot °\text{C}) = 8.36 \text{ kW/°C}$$

and

$$C_c = \dot{m}_c c_{pc} = (8 \text{ kg/s})(4.18 \text{ kJ/kg} \cdot °\text{C}) = 33.4 \text{ kW/°C}$$

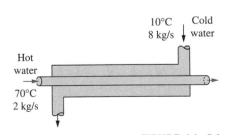

FIGURE 11–24
Schematic for Example 11–7.

Therefore,

$$C_{min} = C_h = 8.36 \text{ kW/°C}$$

which is the smaller of the two heat capacity rates. Then the maximum heat transfer rate is determined from Eq. 11–32 to be

$$\dot{Q}_{max} = C_{min}(T_{h,\text{ in}} - T_{c,\text{ in}})$$
$$= (8.36 \text{ kW/°C})(70 - 10)\text{°C}$$
$$= 502 \text{ kW}$$

That is, the maximum possible heat transfer rate in this heat exchanger is 502 kW. This value would be approached in a counter-flow heat exchanger with a *very large* heat transfer surface area.

The maximum temperature difference in this heat exchanger is $\Delta T_{max} = T_{h,\text{ in}} - T_{c,\text{ in}} = (70 - 10)\text{°C} = 60\text{°C}$. Therefore, the hot water cannot be cooled by more than 60°C (to 10°C) in this heat exchanger, and the cold water cannot be heated by more than 60°C (to 70°C), no matter what we do. The outlet temperatures of the cold and the hot streams in this limiting case are determined to be

$$\dot{Q} = C_c(T_{c,\text{ out}} - T_{c,\text{ in}}) \longrightarrow T_{c,\text{ out}} = T_{c,\text{ in}} + \frac{\dot{Q}}{C_c} = 10\text{°C} + \frac{502 \text{ kW}}{33.4 \text{ kW/°C}} = 25\text{°C}$$

$$\dot{Q} = C_h(T_{h,\text{ in}} - T_{h,\text{ out}}) \longrightarrow T_{h,\text{ out}} = T_{h,\text{ in}} - \frac{\dot{Q}}{C_h} = 70\text{°C} - \frac{502 \text{ kW}}{8.38 \text{ kW/°C}} = 10\text{°C}$$

Discussion Note that the hot water is cooled to the limit of 10°C (the inlet temperature of the cold-water stream), but the cold water is heated to 25°C only when maximum heat transfer occurs in the heat exchanger. This is not surprising, since the mass flow rate of the hot water is only one-fourth that of the cold water, and, as a result, the temperature of the cold water increases by 0.25°C for each 1°C drop in the temperature of the hot water.

You may be tempted to think that the cold water should be heated to 70°C in the limiting case of maximum heat transfer. But this will require the temperature of the hot water to drop to −170°C (below 10°C), which is impossible. Therefore, heat transfer in a heat exchanger reaches its maximum value when the fluid with the smaller heat capacity rate (or the smaller mass flow rate when both fluids have the same specific heat value) experiences the maximum temperature change. This example explains why we use C_{min} in the evaluation of $\dot{Q}_{max}$ instead of C_{max}.

We can show that the hot water will leave at the inlet temperature of the cold water and vice versa in the limiting case of maximum heat transfer when the mass flow rates of the hot- and cold-water streams are identical (Fig. 11–25). We can also show that the outlet temperature of the cold water will reach the 70°C limit when the mass flow rate of the hot water is greater than that of the cold water.

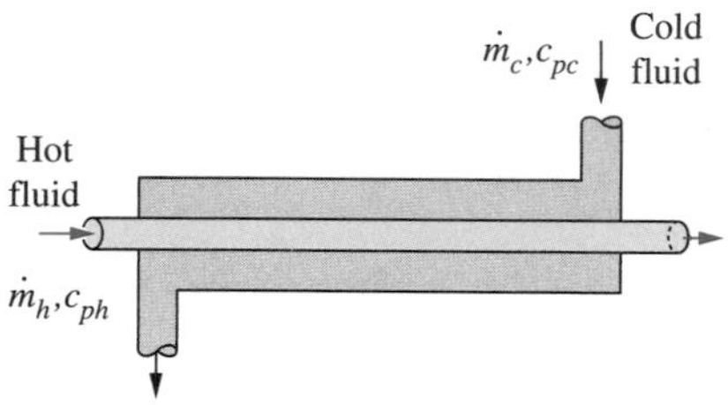

$$\dot{Q} = \dot{m}_h c_{ph} \Delta T_h$$
$$= \dot{m}_c c_{pc} \Delta T_c$$

If $\dot{m}_c c_{pc} = \dot{m}_h c_{ph}$

then $\Delta T_h = \Delta T_c$

FIGURE 11–25
The temperature rise of the cold fluid in a heat exchanger will be equal to the temperature drop of the hot fluid when the heat capcity rates of the hot and cold fluids are identical.

The determination of $\dot{Q}_{max}$ requires the availability of the *inlet temperature* of the hot and cold fluids and their *mass flow rates,* which are usually specified. Then, once the effectiveness of the heat exchanger is known, the actual heat transfer rate $\dot{Q}$ can be determined from

$$\dot{Q} = \varepsilon \dot{Q}_{max} = \varepsilon C_{min}(T_{h,\text{ in}} - T_{c,\text{ in}}) \tag{11–33}$$

Therefore, the effectiveness of a heat exchanger enables us to determine the heat transfer rate without knowing the *outlet temperatures* of the fluids.

The effectiveness of a heat exchanger depends on the *geometry* of the heat exchanger as well as the *flow arrangement*. Therefore, different types of heat exchangers have different effectiveness relations. Below we illustrate the development of the effectiveness ε relation for the double-pipe *parallel-flow* heat exchanger.

Equation 11–23 developed in Section 11–4 for a parallel-flow heat exchanger can be rearranged as

$$\ln \frac{T_{h,\,out} - T_{c,\,out}}{T_{h,\,in} - T_{c,\,in}} = -\frac{UA_s}{C_c}\left(1 + \frac{C_c}{C_h}\right) \tag{11–34}$$

Also, solving Eq. 11–30 for $T_{h,\,out}$ gives

$$T_{h,\,out} = T_{h,\,in} - \frac{C_c}{C_h}(T_{c,\,out} - T_{c,\,in}) \tag{11–35}$$

Substituting this relation into Eq. 11–34 after adding and subtracting $T_{c,\,in}$ gives

$$\ln \frac{T_{h,\,in} - T_{c,\,in} + T_{c,\,in} - T_{c,\,out} - \dfrac{C_c}{C_h}(T_{c,\,out} - T_{c,\,in})}{T_{h,\,in} - T_{c,\,in}} = -\frac{UA_s}{C_c}\left(1 + \frac{C_c}{C_h}\right)$$

which simplifies to

$$\ln\left[1 - \left(1 + \frac{C_c}{C_h}\right)\frac{T_{c,\,out} - T_{c,\,in}}{T_{h,\,in} - T_{c,\,in}}\right] = -\frac{UA_s}{C_c}\left(1 + \frac{C_c}{C_h}\right) \tag{11–36}$$

We now manipulate the definition of effectiveness to obtain

$$\varepsilon = \frac{\dot{Q}}{\dot{Q}_{max}} = \frac{C_c(T_{c,\,out} - T_{c,\,in})}{C_{min}(T_{h,\,in} - T_{c,\,in})} \quad \longrightarrow \quad \frac{T_{c,\,out} - T_{c,\,in}}{T_{h,\,in} - T_{c,\,in}} = \varepsilon\frac{C_{min}}{C_c}$$

Substituting this result into Eq. 11–36 and solving for ε gives the following relation for the effectiveness of a *parallel-flow* heat exchanger:

$$\varepsilon_{\text{parallel flow}} = \frac{1 - \exp\left[-\dfrac{UA_s}{C_c}\left(1 + \dfrac{C_c}{C_h}\right)\right]}{\left(1 + \dfrac{C_c}{C_h}\right)\dfrac{C_{min}}{C_c}} \tag{11–37}$$

Taking either C_c or C_h to be C_{min} (both approaches give the same result), the relation above can be expressed more conveniently as

$$\varepsilon_{\text{parallel flow}} = \frac{1 - \exp\left[-\dfrac{UA_s}{C_{min}}\left(1 + \dfrac{C_{min}}{C_{max}}\right)\right]}{1 + \dfrac{C_{min}}{C_{max}}} \tag{11–38}$$

Again C_{min} is the *smaller* heat capacity ratio and C_{max} is the larger one, and it makes no difference whether C_{min} belongs to the hot or cold fluid.

Effectiveness relations of the heat exchangers typically involve the *dimensionless* group UA_s/C_{min}. This quantity is called the **number of transfer units NTU** and is expressed as

$$\text{NTU} = \frac{UA_s}{C_{min}} = \frac{UA_s}{(\dot{m}c_p)_{min}} \tag{11–39}$$

where U is the overall heat transfer coefficient and A_s is the heat transfer surface area of the heat exchanger. Note that NTU is proportional to A_s. Therefore, for specified values of U and C_{min}, the value of NTU *is a measure of the heat transfer surface area* A_s. Thus, the larger the NTU, the larger the heat exchanger.

In heat exchanger analysis, it is also convenient to define another dimensionless quantity called the **capacity ratio c** as

$$c = \frac{C_{min}}{C_{max}} \tag{11–40}$$

It can be shown that the effectiveness of a heat exchanger is a function of the number of transfer units NTU and the capacity ratio c. That is,

$$\varepsilon = \text{function} \,(UA_s/C_{min}, \, C_{min}/C_{max}) = \text{function} \,(\text{NTU}, \, c)$$

Effectiveness relations have been developed for a large number of heat exchangers, and the results are given in Table 11–4. The effectivenesses of some common types of heat exchangers are also plotted in Fig. 11–26.

TABLE 11–4

Effectiveness relations for heat exchangers: NTU $= UA_s/C_{min}$ and $c = C_{min}/C_{max} = (\dot{m}c_p)_{min}/(\dot{m}c_p)_{max}$

Heat exchanger type	Effectiveness relation
1 *Double pipe:*	
Parallel-flow	$\varepsilon = \dfrac{1 - \exp\,[-\text{NTU}(1 + c)]}{1 + c}$
Counter-flow	$\varepsilon = \dfrac{1 - \exp\,[-\text{NTU}(1 - c)]}{1 - c\,\exp\,[-\text{NTU}(1 - c)]}$
2 *Shell-and-tube:* One-shell pass 2, 4, . . . tube passes	$\varepsilon = 2\left\{1 + c + \sqrt{1 + c^2}\,\dfrac{1 + \exp\,[-\text{NTU}\sqrt{1 + c^2}]}{1 - \exp\,[-\text{NTU}\sqrt{1 + c^2}]}\right\}^{-1}$
3 *Cross-flow (single-pass)*	
Both fluids unmixed	$\varepsilon = 1 - \exp\left\{\dfrac{\text{NTU}^{0.22}}{c}\,[\exp\,(-c\,\text{NTU}^{0.78}) - 1]\right\}$
C_{max} mixed, C_{min} unmixed	$\varepsilon = \dfrac{1}{c}(1 - \exp\,\{1 - c[1 - \exp\,(-\text{NTU})]\})$
C_{min} mixed, C_{max} unmixed	$\varepsilon = 1 - \exp\left\{-\dfrac{1}{c}[1 - \exp\,(-c\,\text{NTU})]\right\}$
4 *All heat exchangers with c = 0*	$\varepsilon = 1 - \exp(-\text{NTU})$

From W. M. Kays and A. L. London. *Compact Heat Exchangers, 3/e.* McGraw-Hill, 1984. Reprinted by permission of William M. Kays.

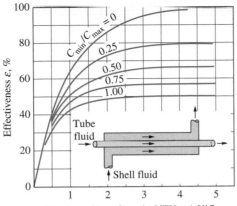

(a) Parallel-flow

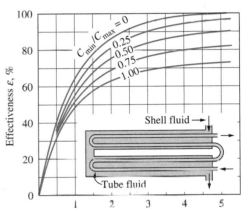

(b) Counter-flow

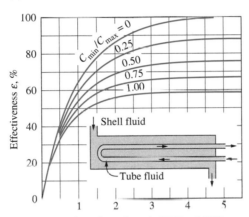

(c) One-shell pass and 2, 4, 6, ... tube passes

(d) Two-shell passes and 4, 8, 12, ... tube passes

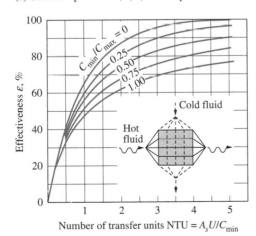

(e) Cross-flow with both fluids unmixed

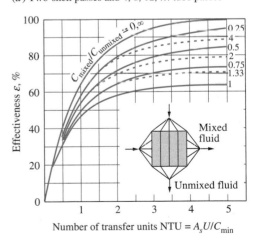

(f) Cross-flow with one fluid mixed and the other unmixed

FIGURE 11–26

Effectiveness for heat exchangers.
(From Kays and London)

More extensive effectiveness charts and relations are available in the literature. The dashed lines in Fig. 11–26f are for the case of C_{min} unmixed and C_{max} mixed and the solid lines are for the opposite case. The analytic relations for the effectiveness give more accurate results than the charts, since reading errors in charts are unavoidable, and the relations are very suitable for computerized analysis of heat exchangers.

We make these observations from the effectiveness relations and charts already given:

1. The value of the effectiveness ranges from 0 to 1. It increases rapidly with NTU for small values (up to about NTU = 1.5) but rather slowly for larger values. Therefore, the use of a heat exchanger with a large NTU (usually larger than 3) and thus a large size cannot be justified economically, since a large increase in NTU in this case corresponds to a small increase in effectiveness. Thus, a heat exchanger with a very high effectiveness may be desirable from a heat transfer point of view but undesirable from an economical point of view.

2. For a given NTU and capacity ratio $c = C_{min}/C_{max}$, the *counter-flow* heat exchanger has the *highest* effectiveness, followed closely by the cross-flow heat exchangers with both fluids unmixed. As you might expect, the lowest effectiveness values are encountered in parallel-flow heat exchangers (Fig. 11–27).

3. The effectiveness of a heat exchanger is independent of the capacity ratio c for NTU values of less than about 0.3.

4. The value of the capacity ratio c ranges between 0 and 1. For a given NTU, the effectiveness becomes a *maximum* for $c = 0$ and a *minimum* for $c = 1$. The case $c = C_{min}/C_{max} \rightarrow 0$ corresponds to $C_{max} \rightarrow \infty$, which is realized during a phase-change process in a *condenser* or *boiler*. All effectiveness relations in this case reduce to

$$\varepsilon = \varepsilon_{max} = 1 - \exp(-NTU) \qquad (11\text{–}41)$$

regardless of the type of heat exchanger (Fig. 11–28). Note that the temperature of the condensing or boiling fluid remains constant in this case. The effectiveness is the *lowest* in the other limiting case of $c = C_{min}/C_{max} = 1$, which is realized when the heat capacity rates of the two fluids are equal.

Once the quantities $c = C_{min}/C_{max}$ and NTU $= UA_s/C_{min}$ have been evaluated, the effectiveness ε can be determined from either the charts or the effectiveness relation for the specified type of heat exchanger. Then the rate of heat transfer $\dot{Q}$ and the outlet temperatures $T_{h,\,out}$ and $T_{c,\,out}$ can be determined from Eqs. 11–33 and 11–30, respectively. Note that the analysis of heat exchangers with unknown outlet temperatures is a straightforward matter with the effectiveness–NTU method but requires rather tedious iterations with the LMTD method.

We mentioned earlier that when all the inlet and outlet temperatures are specified, the *size* of the heat exchanger can easily be determined using the LMTD method. Alternatively, it can also be determined from the effectiveness–NTU method by first evaluating the effectiveness ε from its definition (Eq. 11–29) and then the NTU from the appropriate NTU relation in Table 11–5.

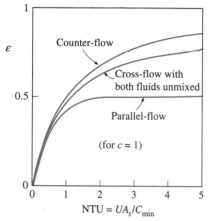

FIGURE 11–27
For a specified NTU and capacity ratio c, the counter-flow heat exchanger has the highest effectiveness and the parallel-flow the lowest.

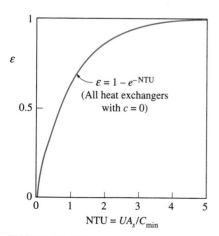

FIGURE 11–28
The effectiveness relation reduces to $\varepsilon = \varepsilon_{max} = 1 - \exp(-NTU)$ for all heat exchangers when the capacity ratio $c = 0$.

TABLE 11–5

NTU relations for heat exchangers: $\text{NTU} = UA_s/C_{\min}$ and $c = C_{\min}/C_{\max} = (\dot{m}c_p)_{\min}/(\dot{m}c_p)_{\max}$

Heat exchanger type	NTU relation
1 *Double-pipe:* Parallel-flow	$\text{NTU} = -\dfrac{\ln\,[1 - \varepsilon(1 + c)]}{1 + c}$
Counter-flow	$\text{NTU} = \dfrac{1}{c - 1}\ln\left(\dfrac{\varepsilon - 1}{\varepsilon c - 1}\right)$
2 *Shell and tube:* One-shell pass 2, 4, . . . tube passes	$\text{NTU} = -\dfrac{1}{\sqrt{1 + c^2}}\ln\left(\dfrac{2/\varepsilon - 1 - c - \sqrt{1 + c^2}}{2/\varepsilon - 1 - c + \sqrt{1 + c^2}}\right)$
3 *Cross-flow (single-pass):* $C_{\max}$ mixed, $C_{\min}$ unmixed	$\text{NTU} = -\ln\left[1 + \dfrac{\ln\,(1 - \varepsilon c)}{c}\right]$
$C_{\min}$ mixed, $C_{\max}$ unmixed	$\text{NTU} = -\dfrac{\ln\,[c\,\ln\,(1 - \varepsilon) + 1]}{c}$
4 *All heat exchangers* *with $c = 0$*	$\text{NTU} = -\ln(1 - \varepsilon)$

From W. M. Kays and A. L. London. *Compact Heat Exchangers, 3/e.* McGraw-Hill, 1984. Reprinted by permission of William M. Kays.

Note that the relations in Table 11–5 are equivalent to those in Table 11–4. Both sets of relations are given for convenience. The relations in Table 11–4 give the effectiveness directly when NTU is known, and the relations in Table 11–5 give the NTU directly when the effectiveness ε is known.

EXAMPLE 11–8   Using the Effectiveness–NTU Method

Repeat Example 11–4, which was solved with the LMTD method, using the effectiveness–NTU method.

SOLUTION The schematic of the heat exchanger is redrawn in Fig. 11–29, and the same assumptions are utilized.

Analysis In the effectiveness–NTU method, we first determine the heat capacity rates of the hot and cold fluids and identify the smaller one:

$$C_h = \dot{m}_h c_{ph} = (2\ \text{kg/s})(4.31\ \text{kJ/kg} \cdot {}^\circ\text{C}) = 8.62\ \text{kW/}{}^\circ\text{C}$$

$$C_c = \dot{m}_c c_{pc} = (1.2\ \text{kg/s})(4.18\ \text{kJ/kg} \cdot {}^\circ\text{C}) = 5.02\ \text{kW/}{}^\circ\text{C}$$

Therefore,

$$C_{\min} = C_c = 5.02\ \text{kW/}{}^\circ\text{C}$$

and

$$c = C_{\min}/C_{\max} = 5.02/8.62 = 0.582$$

Then the maximum heat transfer rate is determined from Eq. 11–32 to be

$$\dot{Q}_{\max} = C_{\min}(T_{h,\ \text{in}} - T_{c,\ \text{in}})$$
$$= (5.02\ \text{kW/}{}^\circ\text{C})(160 - 20){}^\circ\text{C}$$
$$= 702.8\ \text{kW}$$

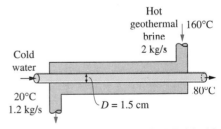

FIGURE 11–29

Schematic for Example 11–8.

That is, the maximum possible heat transfer rate in this heat exchanger is 702.8 kW. The actual rate of heat transfer is

$$\dot{Q} = [\dot{m}c_p(T_{out} - T_{in})]_{water} = (1.2 \text{ kg/s})(4.18 \text{ kJ/kg} \cdot {}^\circ\text{C})(80 - 20){}^\circ\text{C} = 301.0 \text{ kW}$$

Thus, the effectiveness of the heat exchanger is

$$\varepsilon = \frac{\dot{Q}}{\dot{Q}_{max}} = \frac{301.0 \text{ kW}}{702.8 \text{ kW}} = 0.428$$

Knowing the effectiveness, the NTU of this counter-flow heat exchanger can be determined from Fig. 11–26b or the appropriate relation from Table 11–5. We choose the latter approach for greater accuracy:

$$\text{NTU} = \frac{1}{c - 1} \ln\left(\frac{\varepsilon - 1}{\varepsilon c - 1}\right) = \frac{1}{0.582 - 1} \ln\left(\frac{0.428 - 1}{0.428 \times 0.582 - 1}\right) = 0.651$$

Then the heat transfer surface area becomes

$$\text{NTU} = \frac{UA_s}{C_{min}} \longrightarrow A_s = \frac{\text{NTU } C_{min}}{U} = \frac{(0.651)(5020 \text{ W/}^\circ\text{C})}{640 \text{ W/m}^2 \cdot {}^\circ\text{C}} = 5.11 \text{ m}^2$$

To provide this much heat transfer surface area, the length of the tube must be

$$A_s = \pi D L \longrightarrow L = \frac{A_s}{\pi D} = \frac{5.11 \text{ m}^2}{\pi(0.015 \text{ m})} = 108 \text{ m}$$

Discussion Note that we obtained practically the same result with the effectiveness–NTU method in a systematic and straightforward manner.

EXAMPLE 11–9   **Cooling Hot Oil by Water in a Multipass Heat Exchanger**

Hot oil is to be cooled by water in a 1-shell-pass and 8-tube-passes heat exchanger. The tubes are thin-walled and are made of copper with an internal diameter of 1.4 cm. The length of each tube pass in the heat exchanger is 5 m, and the overall heat transfer coefficient is 310 W/m² · °C. Water flows through the tubes at a rate of 0.2 kg/s, and the oil through the shell at a rate of 0.3 kg/s. The water and the oil enter at temperatures of 20°C and 150°C, respectively. Determine the rate of heat transfer in the heat exchanger and the outlet temperatures of the water and the oil.

SOLUTION Hot oil is to be cooled by water in a heat exchanger. The mass flow rates and the inlet temperatures are given. The rate of heat transfer and the outlet temperatures are to be determined.

Assumptions 1 Steady operating conditions exist. 2 The heat exchanger is well insulated so that heat loss to the surroundings is negligible. 3 The thickness of the tube is negligible since it is thin-walled. 4 Changes in the kinetic and potential energies of fluid streams are negligible. 5 The overall heat transfer coefficient is constant and uniform.

Properties We take the specific heats of water and oil to be 4.18 and 2.13 kJ/kg · °C, respectively.

Analysis The schematic of the heat exchanger is given in Fig. 11–30. The outlet temperatures are not specified, and they cannot be determined from an energy balance. The use of the LMTD method in this case will involve tedious iterations, and thus the ε–NTU method is indicated. The first step in the ε–NTU method is to determine the heat capacity rates of the hot and cold fluids and identify the smaller one:

$$C_h = \dot{m}_h c_{ph} = (0.3 \text{ kg/s})(2.13 \text{ kJ/kg} \cdot °C) = 0.639 \text{ kW/°C}$$

$$C_c = \dot{m}_c c_{pc} = (0.2 \text{ kg/s})(4.18 \text{ kJ/kg} \cdot °C) = 0.836 \text{ kW/°C}$$

Therefore,

$$C_{\min} = C_h = 0.639 \text{ kW/°C} \quad \text{and} \quad c = \frac{C_{\min}}{C_{\max}} = \frac{0.639}{0.836} = 0.764$$

Then the maximum heat transfer rate is determined from Eq. 11–32 to be

$$\dot{Q}_{\max} = C_{\min}(T_{h,\text{ in}} - T_{c,\text{ in}}) = (0.639 \text{ kW/°C})(150 - 20)°C = 83.1 \text{ kW}$$

That is, the maximum possible heat transfer rate in this heat exchanger is 83.1 kW. The heat transfer surface area is

$$A_s = n(\pi DL) = 8\pi(0.014 \text{ m})(5 \text{ m}) = 1.76 \text{ m}^2$$

Then the NTU of this heat exchanger becomes

$$\text{NTU} = \frac{UA_s}{C_{\min}} = \frac{(310 \text{ W/m}^2 \cdot °C)(1.76 \text{ m}^2)}{639 \text{ W/°C}} = 0.854$$

The effectiveness of this heat exchanger corresponding to $c = 0.764$ and NTU $= 0.854$ is determined from Fig. 11–26c to be

$$\varepsilon = 0.47$$

We could also determine the effectiveness from the third relation in Table 11–4 more accurately but with more labor. Then the actual rate of heat transfer becomes

$$\dot{Q} = \varepsilon \dot{Q}_{\max} - (0.47)(83.1 \text{ kW}) = \mathbf{39.1 \text{ kW}}$$

Finally, the outlet temperatures of the cold and the hot fluid streams are determined to be

$$\dot{Q} = C_c(T_{c,\text{ out}} - T_{c,\text{ in}}) \quad \longrightarrow \quad T_{c,\text{ out}} = T_{c,\text{ in}} + \frac{\dot{Q}}{C_c}$$

$$= 20°C + \frac{39.1 \text{ kW}}{0.836 \text{ kW/°C}} = \mathbf{66.8°C}$$

$$\dot{Q} = C_h(T_{h,\text{ in}} - T_{h,\text{ out}}) \quad \longrightarrow \quad T_{h,\text{ out}} = T_{h,\text{ in}} - \frac{\dot{Q}}{C_h}$$

$$= 150°C - \frac{39.1 \text{ kW}}{0.639 \text{ kW/°C}} = \mathbf{88.8°C}$$

Therefore, the temperature of the cooling water will rise from 20°C to 66.8°C as it cools the hot oil from 150°C to 88.8°C in this heat exchanger.

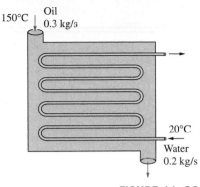

FIGURE 11–30
Schematic for Example 11–9.

11–6 · SELECTION OF HEAT EXCHANGERS

Heat exchangers are complicated devices, and the results obtained with the simplified approaches presented above should be used with care. For example, we assumed that the overall heat transfer coefficient U is constant throughout the heat exchanger and that the convection heat transfer coefficients can be predicted using the convection correlations. However, it should be kept in mind that the uncertainty in the predicted value of U can exceed 30 percent. Thus, it is natural to tend to overdesign the heat exchangers in order to avoid unpleasant surprises.

Heat transfer enhancement in heat exchangers is usually accompanied by *increased pressure drop,* and thus *higher pumping power.* Therefore, any gain from the enhancement in heat transfer should be weighed against the cost of the accompanying pressure drop. Also, some thought should be given to which fluid should pass through the tube side and which through the shell side. Usually, the *more viscous fluid is more suitable for the shell side* (larger passage area and thus lower pressure drop) and *the fluid with the higher pressure for the tube side.*

Engineers in industry often find themselves in a position to select heat exchangers to accomplish certain heat transfer tasks. Usually, the goal is to heat or cool a certain fluid at a known mass flow rate and temperature to a desired temperature. Thus, the *rate of heat transfer* in the prospective heat exchanger is

$$\dot{Q}_{max} = \dot{m}c_p(T_{in} - T_{out})$$

which gives the heat transfer requirement of the heat exchanger before having any idea about the heat exchanger itself.

An engineer going through catalogs of heat exchanger manufacturers will be overwhelmed by the type and number of readily available off-the-shelf heat exchangers. The proper selection depends on several factors.

Heat Transfer Rate

This is the most important quantity in the selection of a heat exchanger. A heat exchanger should be capable of transferring heat at the specified rate in order to achieve the desired temperature change of the fluid at the specified mass flow rate.

Cost

Budgetary limitations usually play an important role in the selection of heat exchangers, except for some specialized cases where "money is no object." An off-the-shelf heat exchanger has a definite cost advantage over those made to order. However, in some cases, none of the existing heat exchangers will do, and it may be necessary to undertake the expensive and time-consuming task of designing and manufacturing a heat exchanger from scratch to suit the needs. This is often the case when the heat exchanger is an integral part of the overall device to be manufactured.

The operation and maintenance costs of the heat exchanger are also important considerations in assessing the overall cost.

Pumping Power

In a heat exchanger, both fluids are usually forced to flow by pumps or fans that consume electrical power. The annual cost of electricity associated with the operation of the pumps and fans can be determined from

$$\text{Operating cost} = (\text{Pumping power, kW}) \times (\text{Hours of operation, h})$$
$$\times (\text{Unit cost of electricity, \$/kWh})$$

where the pumping power is the total electrical power consumed by the motors of the pumps and fans. For example, a heat exchanger that involves a 1-hp pump and a $\frac{1}{3}$-hp fan (1 hp = 0.746 kW) operating at full load 8 h a day and 5 days a week will consume 2069 kWh of electricity per year, which will cost \$166 at an electricity cost of 8 cents/kWh.

Minimizing the pressure drop and the mass flow rate of the fluids will minimize the operating cost of the heat exchanger, but it will maximize the size of the heat exchanger and thus the initial cost. As a rule of thumb, doubling the mass flow rate will reduce the initial cost by *half* but will increase the pumping power requirements by a factor of roughly *eight*.

Typically, fluid velocities encountered in heat exchangers range between 0.7 and 7 m/s for liquids and between 3 and 30 m/s for gases. Low velocities are helpful in avoiding erosion, tube vibrations, and noise as well as pressure drop.

Size and Weight

Normally, the *smaller* and the *lighter* the heat exchanger, the better it is. This is especially the case in the *automotive* and *aerospace* industries, where size and weight requirements are most stringent. Also, a larger heat exchanger normally carries a higher price tag. The space available for the heat exchanger in some cases limits the length of the tubes that can be used.

Type

The type of heat exchanger to be selected depends primarily on the type of *fluids* involved, the *size* and *weight* limitations, and the presence of any *phase-change* processes. For example, a heat exchanger is suitable to cool a liquid by a gas if the surface area on the gas side is many times that on the liquid side. On the other hand, a plate or shell-and-tube heat exchanger is very suitable for cooling a liquid by another liquid.

Materials

The materials used in the construction of the heat exchanger may be an important consideration in the selection of heat exchangers. For example, the thermal and structural *stress effects* need not be considered at pressures below 15 atm or temperatures below 150°C. But these effects are major considerations above 70 atm or 550°C and seriously limit the acceptable materials of the heat exchanger.

A temperature difference of 50°C or more between the tubes and the shell will probably pose *differential thermal expansion* problems and needs to be considered. In the case of corrosive fluids, we may have to select expensive *corrosion-resistant* materials such as stainless steel or even titanium if we are not willing to replace low-cost heat exchangers frequently.

Other Considerations

There are other considerations in the selection of heat exchangers that may or may not be important, depending on the application. For example, being *leak-tight* is an important consideration when *toxic* or *expensive* fluids are involved. Ease of servicing, low maintenance cost, and safety and reliability are some other important considerations in the selection process. Quietness is one of the primary considerations in the selection of liquid-to-air heat exchangers used in heating and air-conditioning applications.

EXAMPLE 11–10   Installing a Heat Exchanger to Save Energy and Money

In a dairy plant, milk is pasteurized by hot water supplied by a natural gas furnace. The hot water is then discharged to an open floor drain at 80°C at a rate of 15 kg/min. The plant operates 24 h a day and 365 days a year. The furnace has an efficiency of 80 percent, and the cost of the natural gas is $1.10 per therm (1 therm = 105,500 kJ). The average temperature of the cold water entering the furnace throughout the year is 15°C. The drained hot water cannot be returned to the furnace and recirculated, because it is contaminated during the process.

In order to save energy, installation of a water-to-water heat exchanger to preheat the incoming cold water by the drained hot water is proposed. Assuming that the heat exchanger will recover 75 percent of the available heat in the hot water, determine the heat transfer rating of the heat exchanger that needs to be purchased and suggest a suitable type. Also, determine the amount of money this heat exchanger will save the company per year from natural gas savings.

SOLUTION A water-to-water heat exchanger is to be installed to transfer energy from drained hot water to the incoming cold water to preheat it. The rate of heat transfer in the heat exchanger and the amounts of energy and money saved per year are to be determined.

Assumptions **1** Steady operating conditions exist. **2** The effectiveness of the heat exchanger remains constant.

Properties We use the specific heat of water at room temperature, c_p = 4.18 kJ/kg · °C, and treat it as a constant.

Analysis A schematic of the prospective heat exchanger is given in Fig. 11–31. The heat recovery from the hot water will be a maximum when it leaves the heat exchanger at the inlet temperature of the cold water. Therefore,

$$\dot{Q}_{max} = \dot{m}_h c_p (T_{h,\,in} - T_{c,\,in})$$

$$= \left(\frac{15}{60}\,\text{kg/s}\right)(4.18\,\text{kJ/kg} \cdot \text{°C})(80 - 15)\text{°C}$$

$$= 67.9\,\text{kJ/s}$$

That is, the existing hot-water stream has the potential to supply heat at a rate of 67.9 kJ/s to the incoming cold water. This value would be approached in a counter-flow heat exchanger with a *very large* heat transfer surface area. A heat exchanger of reasonable size and cost can capture 75 percent of this heat

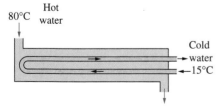

80°C   Hot water

Cold water
15°C

FIGURE 11–31
Schematic for Example 11–10.

transfer potential. Thus, the heat transfer rating of the prospective heat exchanger must be

$$\dot{Q} = \varepsilon \dot{Q}_{max} = (0.75)(67.9 \text{ kJ/s}) = 50.9 \text{ kJ/s}$$

That is, the heat exchanger should be able to deliver heat at a rate of 50.9 kJ/s from the hot to the cold water. An ordinary plate or *shell-and-tube* heat exchanger should be adequate for this purpose, since both sides of the heat exchanger involve the same fluid at comparable flow rates and thus comparable heat transfer coefficients. (Note that if we were heating air with hot water, we would have to specify a heat exchanger that has a large surface area on the air side.)

The heat exchanger will operate 24 h a day and 365 days a year. Therefore, the annual operating hours are

$$\text{Operating hours} = (24 \text{ h/day})(365 \text{ days/year}) = 8760 \text{ h/year}$$

Noting that this heat exchanger saves 50.9 kJ of energy per second, the energy saved during an entire year will be

$$\begin{aligned}
\text{Energy saved} &= (\text{Heat transfer rate})(\text{Operation time}) \\
&= (50.9 \text{ kJ/s})(8760 \text{ h/year})(3600 \text{ s/h}) \\
&= 1.605 \times 10^9 \text{ kJ/year}
\end{aligned}$$

The furnace is said to be 80 percent efficient. That is, for each 80 units of heat supplied by the furnace, natural gas with an energy content of 100 units must be supplied to the furnace. Therefore, the energy savings determined above result in fuel savings in the amount of

$$\begin{aligned}
\text{Fuel saved} &= \frac{\text{Energy saved}}{\text{Furnace efficiency}} = \frac{1.605 \times 10^9 \text{ kJ/year}}{0.80}\left(\frac{1 \text{ therm}}{105{,}500 \text{ kJ}}\right) \\
&= 19{,}020 \text{ therms/year}
\end{aligned}$$

Noting that the price of natural gas is $1.10 per therm, the amount of money saved becomes

$$\begin{aligned}
\text{Money saved} &= (\text{Fuel saved}) \times (\text{Price of fuel}) \\
&= (19{,}020 \text{ therms/year})(\$1.10/\text{therm}) \\
&= \$20{,}920/\text{year}
\end{aligned}$$

Therefore, the installation of the proposed heat exchanger will save the company $20,920 a year, and the installation cost of the heat exchanger will probably be paid from the fuel savings in a short time.

SUMMARY

Heat exchangers are devices that allow the exchange of heat between two fluids without allowing them to mix with each other. Heat exchangers are manufactured in a variety of types, the simplest being the *double-pipe* heat exchanger. In a *parallel-flow* type, both the hot and cold fluids enter the heat exchanger at the same end and move in the same direction, whereas in a *counter-flow* type, the hot and cold fluids enter the heat exchanger at opposite ends and flow in opposite directions. In *compact* heat exchangers, the two fluids move perpendicular to each other, and such a flow configuration is called *cross-flow*. Other common types of heat exchangers in industrial applications are the *plate* and the *shell-and-tube* heat exchangers.

Heat transfer in a heat exchanger usually involves convection in each fluid and conduction through the wall separating the two fluids. In the analysis of heat exchangers, it is convenient to work with an *overall heat transfer coefficient U* or a *total thermal resistance R*, expressed as

$$\frac{1}{UA_s} = \frac{1}{U_i A_i} = \frac{1}{U_o A_o} = R = \frac{1}{h_i A_i} + R_{\text{wall}} + \frac{1}{h_o A_o}$$

where the subscripts i and o stand for the inner and outer surfaces of the wall that separates the two fluids, respectively. When the wall thickness of the tube is small and the thermal conductivity of the tube material is high, the relation simplifies to

$$\frac{1}{U} \approx \frac{1}{h_i} + \frac{1}{h_o}$$

where $U \approx U_i \approx U_o$. The effects of fouling on both the inner and the outer surfaces of the tubes of a heat exchanger can be accounted for by

$$\frac{1}{UA_s} = \frac{1}{U_i A_i} = \frac{1}{U_o A_o} = R$$
$$= \frac{1}{h_i A_i} + \frac{R_{f,i}}{A_i} + \frac{\ln (D_o/D_i)}{2\pi kL} + \frac{R_{f,o}}{A_o} + \frac{1}{h_o A_o}$$

where $A_i = \pi D_i L$ and $A_o = \pi D_o L$ are the areas of the inner and outer surfaces and $R_{f,i}$ and $R_{f,o}$ are the fouling factors at those surfaces.

In a well-insulated heat exchanger, the rate of heat transfer from the hot fluid is equal to the rate of heat transfer to the cold one. That is,

$$\dot{Q} = \dot{m}_c c_{pc}(T_{c,\text{out}} - T_{c,\text{in}}) = C_c(T_{c,\text{out}} - T_{c,\text{in}})$$

and

$$\dot{Q} = \dot{m}_h c_{ph}(T_{h,\text{in}} - T_{h,\text{out}}) = C_h(T_{h,\text{in}} - T_{h,\text{out}})$$

where the subscripts c and h stand for the cold and hot fluids, respectively, and the product of the mass flow rate and the specific heat of a fluid $\dot{m}c_p$ is called the *heat capacity rate.*

Of the two methods used in the analysis of heat exchangers, the *log mean temperature difference* (or LMTD) method is best suited for determining the size of a heat exchanger when all the inlet and the outlet temperatures are known. The *effectiveness–NTU* method is best suited to predict the outlet temperatures of the hot and cold fluid streams in a specified heat exchanger. In the LMTD method, the rate of heat transfer is determined from

$$\dot{Q} = UA_s \, \Delta T_{\text{lm}}$$

where

$$\Delta T_{\text{lm}} = \frac{\Delta T_1 - \Delta T_2}{\ln (\Delta T_1/\Delta T_2)}$$

is the *log mean temperature difference,* which is the suitable form of the average temperature difference for use in the analysis of heat exchangers. Here ΔT_1 and ΔT_2 represent the temperature differences between the two fluids at the two ends (inlet and outlet) of the heat exchanger. For cross-flow and multipass shell-and-tube heat exchangers, the logarithmic mean temperature difference is related to the counter-flow one $\Delta T_{\text{lm, CF}}$ as

$$\Delta T_{\text{lm}} = F \, \Delta T_{\text{lm, CF}}$$

where F is the *correction factor,* which depends on the geometry of the heat exchanger and the inlet and outlet temperatures of the hot and cold fluid streams.

The *effectiveness* of a heat exchanger is defined as

$$\varepsilon = \frac{\dot{Q}}{Q_{\text{max}}} = \frac{\text{Actual heat transfer rate}}{\text{Maximum possible heat transfer rate}}$$

where

$$\dot{Q}_{\text{max}} = C_{\text{min}}(T_{h,\text{in}} - T_{c,\text{in}})$$

and C_{min} is the smaller of $C_h = \dot{m}_h c_{ph}$ and $C_c = \dot{m}_c c_{pc}$. The effectiveness of heat exchangers can be determined from effectiveness relations or charts.

The selection or design of a heat exchanger depends on several factors such as the heat transfer rate, cost, pressure drop, size, weight, construction type, materials, and operating environment.

REFERENCES AND SUGGESTED READINGS

1. N. Afgan and E. U. Schlunder. *Heat Exchanger: Design and Theory Sourcebook.* Washington, DC: McGraw-Hill/Scripta, 1974.

2. R. A. Bowman, A. C. Mueller, and W. M. Nagle. "Mean Temperature Difference in Design." *Trans. ASME* 62, 1940, p. 283. Reprined with permission of ASME International.

3. A. P. Fraas. *Heat Exchanger Design.* 2d ed. New York: John Wiley & Sons, 1989.

4. K. A. Gardner. "Variable Heat Transfer Rate Correction in Multipass Exchangers, Shell Side Film Controlling." *Transactions of the ASME* 67 (1945), pp. 31–38.

5. W. M. Kays and A. L. London. *Compact Heat Exchangers.* 3rd ed. New York: McGraw-Hill, 1984. Reprinted by permission of William M. Kays.

6. W. M. Kays and H. C. Perkins. In *Handbook of Heat Transfer,* ed. W. M. Rohsenow and J. P. Hartnett. New York: McGraw-Hill, 1972, Chap. 7.

7. A. C. Mueller. "Heat Exchangers." In *Handbook of Heat Transfer,* ed. W. M. Rohsenow and J. P. Hartnett. New York: McGraw-Hill, 1972, Chap. 18.

8. M. N. Özişik. *Heat Transfer—A Basic Approach.* New York: McGraw-Hill, 1985.

9. E. U. Schlunder. *Heat Exchanger Design Handbook.* Washington, DC: Hemisphere, 1982.

10. *Standards of Tubular Exchanger Manufacturers Association.* New York: Tubular Exchanger Manufacturers Association, latest ed.

11. R. A. Stevens, J. Fernandes, and J. R. Woolf. "Mean Temperature Difference in One, Two, and Three Pass Crossflow Heat Exchangers." *Transactions of the ASME* 79 (1957), pp. 287–297.

12. J. Taborek, G. F. Hewitt, and N. Afgan. *Heat Exchangers: Theory and Practice.* New York: Hemisphere, 1983.

13. G. Walker. *Industrial Heat Exchangers.* Washington, DC: Hemisphere, 1982.

PROBLEMS*

Types of Heat Exchangers

11–1C Classify heat exchangers according to flow type and explain the characteristics of each type.

11–2C Classify heat exchangers according to construction type and explain the characteristics of each type.

11–3C When is a heat exchanger classified as being compact? Do you think a double-pipe heat exchanger can be classified as a compact heat exchanger?

11–4C How does a cross-flow heat exchanger differ from a counter-flow one? What is the difference between mixed and unmixed fluids in cross-flow?

11–5C What is the role of the baffles in a shell-and-tube exchanger? How does the presence of baffles affect the heat transfer and the pumping power requirements? Explain.

11–6C Draw a 1-shell-pass and 6-tube-passes shell-and-tube heat exchanger. What are the advantages and disadvantages of using 6 tube passes instead of just 2 of the same diameter?

11–7C Draw a 2-shell-passes and 8-tube-passes shell-and-tube heat exchanger. What is the primary reason for using so many tube passes?

11–8C What is a regenerative heat exchanger? How does a static type of regenerative heat exchanger differ from a dynamic type?

The Overall Heat Transfer Coefficient

11–9C What are the heat transfer mechanisms involved during heat transfer from the hot to the cold fluid?

11–10C Under what conditions is the thermal resistance of the tube in a heat exchanger negligible?

11–11C Consider a double-pipe parallel-flow heat exchanger of length L. The inner and outer diameters of the inner tube are D_1 and D_2, respectively, and the inner diameter of the outer tube is D_3. Explain how you would determine the two heat transfer surface areas A_i and A_o. When is it reasonable to assume $A_i \approx A_o \approx A_s$?

11–12C Is the approximation $h_i \approx h_o \approx h$ for the convection heat transfer coefficient in a heat exchanger a reasonable one when the thickness of the tube wall is negligible?

11–13C Under what conditions can the overall heat transfer coefficient of a heat exchanger be determined from $U = (1/h_i + 1/h_o)^{-1}$?

11–14C What are the restrictions on the relation $UA_s = U_i A_i = U_o A_o$ for a heat exchanger? Here A_s is the heat transfer surface area and U is the overall heat transfer coefficient.

11–15C In a thin-walled double-pipe heat exchanger, when is the approximation $U = h_i$ a reasonable one? Here U is the overall heat transfer coefficient and h_i is the convection heat transfer coefficient inside the tube.

11–16C What are the common causes of fouling in a heat exchanger? How does fouling affect heat transfer and pressure drop?

11–17C How is the thermal resistance due to fouling in a heat exchanger accounted for? How do the fluid velocity and temperature affect fouling?

11–18 A double-pipe heat exchanger is constructed of a copper ($k = 380$ W/m · °C) inner tube of internal diameter $D_i = 1.2$ cm and external diameter $D_o = 1.6$ cm and an outer tube of diameter 3.0 cm. The convection heat transfer coefficient is reported to be $h_i = 700$ W/m² · °C on the inner surface of the tube and $h_o = 1400$ W/m² · °C on its outer surface. For a fouling factor $R_{f, i} = 0.0005$ m² · °C/W on the tube side and $R_{f, o} = 0.0002$ m² · °C/W on the shell side, determine (a) the thermal resistance of the heat exchanger per unit length and

---

*Problems designated by a "C" are concept questions, and students are encouraged to answer them all. Problems designated by an "E" are in English units, and the SI users can ignore them. Problems with the icon ⊛ are solved using EES, and complete solutions together with parametric studies are included on the enclosed CD. Problems with the icon ▨ are comprehensive in nature, and are intended to be solved with a computer, preferably using the EES software that accompanies this text.

(b) the overall heat transfer coefficients U_i and U_o based on the inner and outer surface areas of the tube, respectively.

11–19 [EES] Reconsider Prob. 11–18. Using EES (or other) software, investigate the effects of pipe conductivity and heat transfer coefficients on the thermal resistance of the heat exchanger. Let the thermal conductivity vary from 10 W/m · °C to 400 W/m · °C, the convection heat transfer coefficient from 500 W/m² · °C to 1500 W/m² · °C on the inner surface, and from 1000 W/m² · °C to 2000 W/m² · °C on the outer surface. Plot the thermal resistance of the heat exchanger as functions of thermal conductivity and heat transfer coefficients, and discuss the results.

11–20 A jacketted-agitated vessel, fitted with a turbine agitator, is used for heating a water stream from 10°C to 54°C. The average heat transfer coefficient for water at the vessel's inner-wall can be estimated from $Nu = 0.76 Re^{2/3} Pr^{1/3}$. Saturated steam at 100°C condenses in the jacket, for which the average heat transfer coefficient in kW/m² · K is: $h_o = 13.1(T_g - T_w)^{-0.25}$. The vessel dimensions are: $D_t = 0.6$ m, $H = 0.6$ m and $D_a = 0.2$ m. The agitator speed is 60 rpm. Calculate the mass rate of water that can be heated in this agitated vessel steadily.

11–21 Water at an average temperature of 110°C and an average velocity of 3.5 m/s flows through a 5-m-long stainless steel tube ($k = 14.2$ W/m · °C) in a boiler. The inner and outer diameters of the tube are $D_i = 1.0$ cm and $D_o = 1.4$ cm, respectively. If the convection heat transfer coefficient at the outer surface of the tube where boiling is taking place is $h_o = 8400$ W/m² · °C, determine the overall heat transfer coefficient U_i of this boiler based on the inner surface area of the tube.

11–22 Repeat Prob. 11–21, assuming a fouling factor $R_{f, i} = 0.0005$ m² · °C/W on the inner surface of the tube.

11–23 [EES] Reconsider Prob. 11–22. Using EES (or other) software, plot the overall heat transfer coefficient based on the inner surface as a function of fouling factor as it varies from 0.0001 m² · °C/W to 0.0008 m² · °C/W, and discuss the results.

11–24 A long thin-walled double-pipe heat exchanger with tube and shell diameters of 1.0 cm and 2.5 cm, respectively, is used to condense refrigerant-134a by water at 20°C. The refrigerant flows through the tube, with a convection heat transfer coefficient of $h_i = 5000$ W/m² · °C. Water flows through the shell at a rate of 0.3 kg/s. Determine the overall heat transfer coefficient of this heat exchanger. *Answer:* 2020 W/m² · °C

11–25 Repeat Prob. 11–24 by assuming a 2-mm-thick layer of limestone ($k = 1.3$ W/m · °C) forms on the outer surface of the inner tube.

11–26 [EES] Reconsider Prob. 11–25. Using EES (or other) software, plot the overall heat transfer coefficient as a function of the limestone thickness as it varies from 1 mm to 3 mm, and discuss the results.

11–27E Water at an average temperature of 180°F and an average velocity of 4 ft/s flows through a thin-walled $\frac{3}{4}$-in-diameter tube. The water is cooled by air that flows across the tube with a velocity of 12 ft/s at an average temperature of 80°F. Determine the overall heat transfer coefficient.

Analysis of Heat Exchangers

11–28C What are the common approximations made in the analysis of heat exchangers?

11–29C Under what conditions is the heat transfer relation

$$\dot{Q} = \dot{m}_c c_{pc}(T_{c, \text{out}} - T_{c, \text{in}}) = \dot{m}_h c_{ph}(T_{h, \text{in}} - T_{h, \text{out}})$$

valid for a heat exchanger?

11–30C What is the heat capacity rate? What can you say about the temperature changes of the hot and cold fluids in a heat exchanger if both fluids have the same capacity rate? What does a heat capacity of infinity for a fluid in a heat exchanger mean?

11–31C Consider a condenser in which steam at a specified temperature is condensed by rejecting heat to the cooling water. If the heat transfer rate in the condenser and the temperature rise of the cooling water is known, explain how the rate of condensation of the steam and the mass flow rate of the cooling water can be determined. Also, explain how the total thermal resistance R of this condenser can be evaluated in this case.

11–32C Under what conditions will the temperature rise of the cold fluid in a heat exchanger be equal to the temperature drop of the hot fluid?

The Log Mean Temperature Difference Method

11–33C In the heat transfer relation $\dot{Q} = UA_s \Delta T_{\text{lm}}$ for a heat exchanger, what is ΔT_{lm} called? How is it calculated for a parallel-flow and counter-flow heat exchanger?

11–34C How does the log mean temperature difference for a heat exchanger differ from the arithmetic mean temperature difference? For specified inlet and outlet temperatures, which one of these two quantities is larger?

11–35C The temperature difference between the hot and cold fluids in a heat exchanger is given to be ΔT_1 at one end and ΔT_2 at the other end. Can the logarithmic temperature difference ΔT_{lm} of this heat exchanger be greater than both ΔT_1 and ΔT_2? Explain.

11–36C Can the logarithmic mean temperature difference ΔT_{lm} of a heat exchanger be a negative quantity? Explain.

11–37C Can the outlet temperature of the cold fluid in a heat exchanger be higher than the outlet temperature of the hot fluid in a parallel-flow heat exchanger? How about in a counter-flow heat exchanger? Explain.

11–38C For specified inlet and outlet temperatures, for what kind of heat exchanger will the ΔT_{lm} be greatest: double-pipe parallel-flow, double-pipe counter-flow, cross-flow, or multipass shell-and-tube heat exchanger?

11–39C In the heat transfer relation $\dot{Q} = UA_s F\Delta T_{lm}$ for a heat exchanger, what is the quantity F called? What does it represent? Can F be greater than one?

11–40C When the outlet temperatures of the fluids in a heat exchanger are not known, is it still practical to use the LMTD method? Explain.

11–41C Explain how the LMTD method can be used to determine the heat transfer surface area of a multipass shell-and-tube heat exchanger when all the necessary information, including the outlet temperatures, is given.

11–42 Ethylene glycol is heated from 20°C to 40°C at a rate of 1.0 kg/s in a horizontal copper tube ($k = 386$ W/m · K) with an inner diameter of 2.0 cm and an outer diameter of 2.5 cm. A saturated vapor ($T_g = 110$°C) condenses on the outside-tube surface with the heat transfer coefficient (in kW/m² · K) given by $9.2/(T_g - T_w)^{0.25}$, where T_w is the average outside-tube wall temperature. What tube length must be used? Take the properties of ethylene glycol to be $\rho = 1109$ kg/m³, $c_p = 2428$ kj/kg · K, $k = 0.253$ W/m · °C, $\mu = 0.01545$ kg/m · s, and Pr = 148.5.

11–43 A double-pipe parallel-flow heat exchanger is used to heat cold tap water with hot water. Hot water ($c_p = 4.25$ kJ/kg · °C) enters the tube at 85°C at a rate of 1.4 kg/s and leaves at 50°C. The heat exchanger is not well insulated, and it is estimated that 3 percent of the heat given up by the hot fluid is lost from the heat exchanger. If the overall heat transfer coefficient and the surface area of the heat exchanger are 1150 W/m² · °C and 4 m², respectively, determine the rate of heat transfer to the cold water and the log mean temperature difference for this heat exchanger.

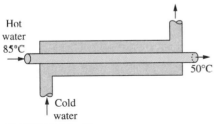

Hot
water
85°C

50°C

Cold
water

FIGURE P11–43

11–44 A stream of hydrocarbon ($c_p = 2.2$ kJ/kg · K) is cooled at a rate of 720 kg/h from 150°C to 40°C in the tube side of a double-pipe counter-flow heat exchanger. Water ($c_p = 4.18$ kJ/kg · K) enters the heat exchanger at 10°C at a rate of 540 kg/h. The outside diameter of the inner tube is 2.5 cm, and its length is 6.0 m. Calculate the overall heat transfer coefficient.

11–45 A shell-and-tube heat exchanger is used for heating 10 kg/s of oil ($c_p = 2.0$ kJ/kg · K) from 25°C to 46°C. The heat exchanger has 1-shell pass and 6-tube passes. Water enters the shell side at 80°C and leaves at 60°C. The overall heat transfer coefficient is estimated to be 1000 W/m² · K. Calculate the rate of heat transfer and the heat transfer area.

11–46 Steam in the condenser of a steam power plant is to be condensed at a temperature of 50°C ($h_{fg} = 2383$ kJ/kg) with cooling water ($c_p = 4180$ J/kg · °C) from a nearby lake, which enters the tubes of the condenser at 18°C and leaves at 27°C. The surface area of the tubes is 42 m², and the overall heat transfer coefficient is 2400 W/m² · °C. Determine the mass flow rate of the cooling water needed and the rate of condensation of the steam in the condenser. *Answers: 73.1 kg/s, 1.15 kg/s*

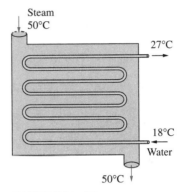

Steam
50°C

27°C

18°C
Water

50°C

FIGURE P11–46

11–47 A double-pipe parallel-flow heat exchanger is to heat water ($c_p = 4180$ J/kg · °C) from 25°C to 60°C at a rate of 0.2 kg/s. The heating is to be accomplished by geothermal water ($c_p = 4310$ J/kg · °C) available at 140°C at a mass flow rate of 0.3 kg/s. The inner tube is thin-walled and has a diameter of 0.8 cm. If the overall heat transfer coefficient of the heat exchanger is 550 W/m² · °C, determine the length of the tube required to achieve the desired heating.

11–48 Reconsider Prob. 11–47. Using EES (or other) software, investigate the effects of temperature and mass flow rate of geothermal water on the length of the tube. Let the temperature vary from 100°C to 200°C, and the mass flow rate from 0.1 kg/s to 0.5 kg/s. Plot the length of the tube as functions of temperature and mass flow rate, and discuss the results.

11–49E A 1-shell-pass and 8-tube-passes heat exchanger is used to heat glycerin ($c_p = 0.60$ Btu/lbm · °F) from 65°F to 140°F by hot water ($c_p = 1.0$ Btu/lbm · °F) that enters the thin-walled 0.5-in-diameter tubes at 175°F and leaves at 120°F. The total length of the tubes in the heat exchanger is 500 ft. The convection heat transfer coefficient is 4 Btu/h · ft² · °F on the glycerin (shell) side and 50 Btu/h · ft² · °F on the water (tube) side. Determine the rate of heat transfer in the heat exchanger (*a*) before any fouling occurs and (*b*) after fouling with a fouling factor of 0.002 h · ft² · °F/Btu on the outer surfaces of the tubes.

11–50 A test is conducted to determine the overall heat transfer coefficient in a shell-and-tube oil-to-water heat exchanger that has 24 tubes of internal diameter 1.2 cm and length 2 m in a single shell. Cold water (c_p = 4180 J/kg · °C) enters the tubes at 20°C at a rate of 3 kg/s and leaves at 55°C. Oil (c_p = 2150 J/kg · °C) flows through the shell and is cooled from 120°C to 45°C. Determine the overall heat transfer coefficient U_i of this heat exchanger based on the inner surface area of the tubes. *Answer:* 8.31 kW/m² · °C

11–51 A double-pipe counter-flow heat exchanger is to cool ethylene glycol (c_p = 2560 J/kg · °C) flowing at a rate of 3.5 kg/s from 80°C to 40°C by water (c_p = 4180 J/kg · °C) that enters at 20°C and leaves at 55°C. The overall heat transfer coefficient based on the inner surface area of the tube is 250 W/m² · °C. Determine (*a*) the rate of heat transfer, (*b*) the mass flow rate of water, and (*c*) the heat transfer surface area on the inner side of the tube.

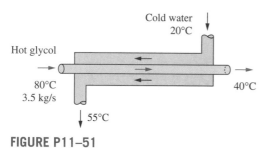

FIGURE P11–51

11–52 Water (c_p = 4180 J/kg · °C) enters the 2.5-cm-internal-diameter tube of a double-pipe counter-flow heat exchanger at 17°C at a rate of 3 kg/s. It is heated by steam condensing at 120°C (h_{fg} = 2203 kJ/kg) in the shell. If the overall heat transfer coefficient of the heat exchanger is 1500 W/m² · °C, determine the length of the tube required in order to heat the water to 80°C.

11–53 A thin-walled double-pipe counter-flow heat exchanger is to be used to cool oil (c_p = 2200 J/kg · °C) from 150°C to 40°C at a rate of 2 kg/s by water (c_p = 4180 J/kg · °C) that enters at 22°C at a rate of 1.5 kg/s. The diameter of the tube is 2.5 cm, and its length is 6 m. Determine the overall heat transfer coefficient of this heat exchanger.

11–54 Reconsider Prob. 11–53. Using EES (or other) software, investigate the effects of oil exit temperature and water inlet temperature on the overall heat transfer coefficient of the heat exchanger. Let the oil exit temperature vary from 30°C to 70°C and the water inlet temperature from 5°C to 25°C. Plot the overall heat transfer coefficient as functions of the two temperatures, and discuss the results.

11–55 Consider a water-to-water double-pipe heat exchanger whose flow arrangement is not known. The temperature measurements indicate that the cold water enters at 20°C and leaves at 50°C, while the hot water enters at 80°C and leaves at 45°C.

Do you think this is a parallel-flow or counter-flow heat exchanger? Explain.

11–56 Cold water (c_p = 4180 J/kg · °C) leading to a shower enters a thin-walled double-pipe counter-flow heat exchanger at 15°C at a rate of 1.25 kg/s and is heated to 45°C by hot water (c_p = 4190 J/kg · °C) that enters at 100°C at a rate of 3 kg/s. If the overall heat transfer coefficient is 880 W/m² · °C, determine the rate of heat transfer and the heat transfer surface area of the heat exchanger.

11–57 Engine oil (c_p = 2100 J/kg · °C) is to be heated from 20°C to 60°C at a rate of 0.3 kg/s in a 2-cm-diameter thin-walled copper tube by condensing steam outside at a temperature of 130°C (h_{fg} = 2174 kJ/kg). For an overall heat transfer coefficient of 650 W/m² · °C, determine the rate of heat transfer and the length of the tube required to achieve it.
Answers: 25.2 kW, 7.0 m

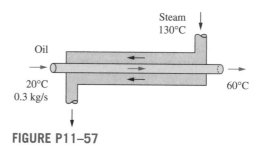

FIGURE P11–57

11–58E Geothermal water (c_p = 1.03 Btu/lbm · °F) is to be used as the heat source to supply heat to the hydronic heating system of a house at a rate of 40 Btu/s in a double-pipe counter-flow heat exchanger. Water (c_p = 1.0 Btu/lbm · °F) is heated from 140°F to 200°F in the heat exchanger as the geothermal water is cooled from 270°F to 180°F. Determine the mass flow rate of each fluid and the total thermal resistance of this heat exchanger.

11–59 Glycerin (c_p = 2400 J/kg · °C) at 20°C and 0.3 kg/s is to be heated by ethylene glycol (c_p = 2500 J/kg · °C) at 60°C in a thin-walled double-pipe parallel-flow heat exchanger. The temperature difference between the two fluids is 15°C at the outlet of the heat exchanger. If the overall heat transfer coefficient is 240 W/m² · °C and the heat transfer surface area is 3.2 m², determine (*a*) the rate of heat transfer, (*b*) the outlet temperature of the glycerin, and (*c*) the mass flow rate of the ethylene glycol.

11–60 Air (c_p = 1005 J/kg · °C) is to be preheated by hot exhaust gases in a cross-flow heat exchanger before it enters the furnace. Air enters the heat exchanger at 95 kPa and 20°C at a rate of 0.8 m³/s. The combustion gases (c_p = 1100 J/kg · °C) enter at 180°C at a rate of 1.1 kg/s and leave at 95°C. The product of the overall heat transfer coefficient and the heat transfer surface area is UA_s = 1200 W/°C. Assuming both fluids to be unmixed, determine the rate of heat transfer and the outlet temperature of the air.

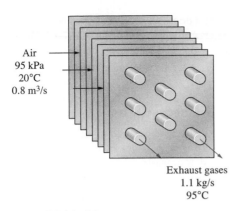

FIGURE P11–60

11–61 A shell-and-tube heat exchanger with 2-shell passes and 12-tube passes is used to heat water (c_p = 4180 J/kg · °C) in the tubes from 20°C to 70°C at a rate of 4.5 kg/s. Heat is supplied by hot oil (c_p = 2300 J/kg · °C) that enters the shell side at 170°C at a rate of 10 kg/s. For a tube-side overall heat transfer coefficient of 350 W/m² · °C, determine the heat transfer surface area on the tube side. *Answer:* 25.7 m²

11–62 Repeat Prob. 11–61 for a mass flow rate of 2 kg/s for water.

11–63 A shell-and-tube heat exchanger with 2-shell passes and 8-tube passes is used to heat ethyl alcohol (c_p = 2670 J/kg · °C) in the tubes from 25°C to 70°C at a rate of 2.1 kg/s. The heating is to be done by water (c_p = 4190 J/kg · °C) that enters the shell side at 95°C and leaves at 45°C. If the overall heat transfer coefficient is 950 W/m² · °C, determine the heat transfer surface area of the heat exchanger.

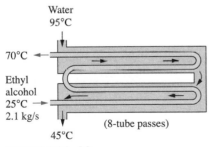

FIGURE P11–63

11–64 A shell-and-tube heat exchanger with 2-shell passes and 12-tube passes is used to heat water (c_p = 4180 J/kg · °C) with ethylene glycol (c_p = 2680 J/kg · °C). Water enters the tubes at 22°C at a rate of 0.8 kg/s and leaves at 70°C. Ethylene glycol enters the shell at 110°C and leaves at 60°C. If the overall heat transfer coefficient based on the tube side is 280 W/m² · °C, determine the rate of heat transfer and the heat transfer surface area on the tube side.

11–65 Reconsider Prob. 11–64. Using EES (or other) software, investigate the effect of the mass flow rate of water on the rate of heat transfer and the tube-side surface area. Let the mass flow rate vary from 0.4 kg/s to 2.2 kg/s. Plot the rate of heat transfer and the surface area as a function of the mass flow rate, and discuss the results.

11–66E Steam is to be condensed on the shell side of a 1-shell-pass and 8-tube-passes condenser, with 50 tubes in each pass at 90°F (h_{fg} = 1043 Btu/lbm). Cooling water (c_p = 1.0 Btu/lbm · °F) enters the tubes at 60°F and leaves at 73°F. The tubes are thin walled and have a diameter of 3/4 in and length of 5 ft per pass. If the overall heat transfer coefficient is 600 Btu/h · ft² · °F, determine (a) the rate of heat transfer, (b) the rate of condensation of steam, and (c) the mass flow rate of cold water.

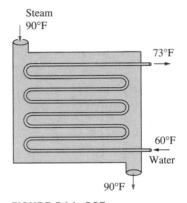

FIGURE P11–66E

11–67E Reconsider Prob. 11–66E. Using EES (or other) software, investigate the effect of the condensing steam temperature on the rate of heat transfer, the rate of condensation of steam, and the mass flow rate of cold water. Let the steam temperature vary from 80°F to 120°F. Plot the rate of heat transfer, the condensation rate of steam, and the mass flow rate of cold water as a function of steam temperature, and discuss the results.

11–68 A shell-and-tube heat exchanger with 1-shell pass and 20–tube passes is used to heat glycerin (c_p = 2480 J/kg · °C) in the shell, with hot water in the tubes. The tubes are thin-walled and have a diameter of 4 cm and length of 2 m per pass. The water enters the tubes at 100°C at a rate of 0.5 kg/s and leaves at 55°C. The glycerin enters the shell at 15°C and leaves at 55°C. Determine the mass flow rate of the glycerin and the overall heat transfer coefficient of the heat exchanger.

11–69 In a binary geothermal power plant, the working fluid isobutane is to be condensed by air in a condenser at 75°C (h_{fg} = 255.7 kJ/kg) at a rate of 2.7 kg/s. Air enters the condenser at 21°C and leaves at 28°C (see Fig. P11–69 on the next page). The heat transfer surface area based on the isobutane side is 24 m². Determine the mass flow rate of air and the overall heat transfer coefficient.

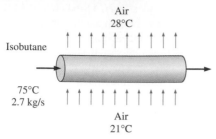

FIGURE P11–69

11–70 Hot exhaust gases of a stationary diesel engine are to be used to generate steam in an evaporator. Exhaust gases ($c_p = 1051$ J/kg · °C) enter the heat exchanger at 550°C at a rate of 0.25 kg/s while water enters as saturated liquid and evaporates at 200°C ($h_{fg} = 1941$ kJ/kg). The heat transfer surface area of the heat exchanger based on water side is 0.5 m² and overall heat transfer coefficient is 1780 W/m² · °C. Determine the rate of heat transfer, the exit temperature of exhaust gases, and the rate of evaporation of water.

11–71 [EES] Reconsider Prob. 11–70. Using EES (or other) software, investigate the effect of the exhaust gas inlet temperature on the rate of heat transfer, the exit temperature of exhaust gases, and the rate of evaporation of water. Let the temperature of exhaust gases vary from 300°C to 600°C. Plot the rate of heat transfer, the exit temperature of exhaust gases, and the rate of evaporation of water as a function of the temperature of the exhaust gases, and discuss the results.

11–72 In a textile manufacturing plant, the waste dyeing water ($c_p = 4295$ J/kg · °C) at 75°C is to be used to preheat fresh water ($c_p = 4180$ J/kg · °C) at 15°C at the same flow rate in a double-pipe counter-flow heat exchanger. The heat transfer surface area of the heat exchanger is 1.65 m² and the overall heat transfer coefficient is 625 W/m² · °C. If the rate of heat transfer in the heat exchanger is 35 kW, determine the outlet temperature and the mass flow rate of each fluid stream.

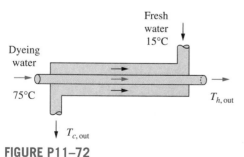

FIGURE P11–72

The Effectiveness–NTU Method

11–73C Under what conditions is the effectiveness–NTU method definitely preferred over the LMTD method in heat exchanger analysis?

11–74C What does the effectiveness of a heat exchanger represent? Can effectiveness be greater than one? On what factors does the effectiveness of a heat exchanger depend?

11–75C For a specified fluid pair, inlet temperatures, and mass flow rates, what kind of heat exchanger will have the highest effectiveness: double-pipe parallel-flow, double-pipe counter-flow, cross-flow, or multipass shell-and-tube heat exchanger?

11–76C Explain how you can evaluate the outlet temperatures of the cold and hot fluids in a heat exchanger after its effectiveness is determined.

11–77C Can the temperature of the hot fluid drop below the inlet temperature of the cold fluid at any location in a heat exchanger? Explain.

11–78C Can the temperature of the cold fluid rise above the inlet temperature of the hot fluid at any location in a heat exchanger? Explain.

11–79C Consider a heat exchanger in which both fluids have the same specific heats but different mass flow rates. Which fluid will experience a larger temperature change: the one with the lower or higher mass flow rate?

11–80C Explain how the maximum possible heat transfer rate $\dot{Q}_{max}$ in a heat exchanger can be determined when the mass flow rates, specific heats, and the inlet temperatures of the two fluids are specified. Does the value of $\dot{Q}_{max}$ depend on the type of the heat exchanger?

11–81C Consider two double-pipe counter-flow heat exchangers that are identical except that one is twice as long as the other one. Which heat exchanger is more likely to have a higher effectiveness?

11–82C Consider a double-pipe counter-flow heat exchanger. In order to enhance heat transfer, the length of the heat exchanger is now doubled. Do you think its effectiveness will also double?

11–83C Consider a shell-and-tube water-to-water heat exchanger with identical mass flow rates for both the hot- and cold-water streams. Now the mass flow rate of the cold water is reduced by half. Will the effectiveness of this heat exchanger increase, decrease, or remain the same as a result of this modification? Explain. Assume the overall heat transfer coefficient and the inlet temperatures remain the same.

11–84C Under what conditions can a counter-flow heat exchanger have an effectiveness of one? What would your answer be for a parallel-flow heat exchanger?

11–85C How is the NTU of a heat exchanger defined? What does it represent? Is a heat exchanger with a very large NTU (say, 10) necessarily a good one to buy?

11–86C Consider a heat exchanger that has an NTU of 4. Someone proposes to double the size of the heat exchanger and thus double the NTU to 8 in order to increase the effectiveness of the heat exchanger and thus save energy. Would you support this proposal?

11–87C Consider a heat exchanger that has an NTU of 0.1. Someone proposes to triple the size of the heat exchanger and thus triple the NTU to 0.3 in order to increase the effectiveness of the heat exchanger and thus save energy. Would you support this proposal?

11–88 The radiator in an automobile is a cross-flow heat exchanger ($UA_s = 10$ kW/K) that uses air ($c_p = 1.00$ kJ/kg · K) to cool the engine-coolant fluid ($c_p = 4.00$ kJ/kg · K). The engine fan draws 30°C air through this radiator at a rate of 10 kg/s while the coolant pump circulates the engine coolant at a rate of 5 kg/s. The coolant enters this radiator at 80°C. Under these conditions, the effectiveness of the radiator is 0.4. Determine (a) the outlet temperature of the air and (b) the rate of heat transfer between the two fluids.

11–89 During an experiment, a shell-and-tube heat exchanger that is used to transfer heat from a hot-water stream to a cold-water stream is tested, and the following measurements are taken:

	Hot-Water Stream	Cold-Water Stream
Inlet temperature, °C	71.5	19.7
Outlet temperature, °C	58.2	27.8
Volume flow rate, L/min	1.05	1.55

The heat transfer area is calculated to be 0.0200 m².
  (a) Calculate the rate of heat transfer to the cold water.
  (b) Calculate the overall heat transfer coefficient.
  (c) Determine if the heat exchanger is truly adiabatic. If not, determine the fraction of heat loss and calculate the heat transfer efficiency.
  (d) Determine the effectiveness and the NTU values of the heat exchanger.

Also, discuss if the measured values are reasonable.

11–90 Cold water ($c_p = 4.18$ kJ/kg · °C) enters a cross-flow heat exchanger at 14°C at a rate of 0.35 kg/s where it is heated by hot air ($c_p = 1.0$ kJ/kg · °C) that enters the heat exchanger at 65°C at a rate of 0.8 kg/s and leaves at 25°C. Determine the maximum outlet temperature of the cold water and the effectiveness of this heat exchanger.

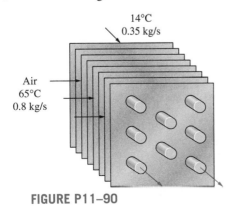

14°C
0.35 kg/s

Air
65°C
0.8 kg/s

FIGURE P11–90

11–91 Water from a lake is used as the cooling agent in a power plant. To achieve condensation of 2.5 kg/s of steam exiting the turbine, a shell-and-tube heat exchanger is used, which has a single shell and 300 thin-walled, 25-mm-diameter tubes, each tube making two passes. Steam flows through the shell, while cooling water flows through the tubes. Steam enters as saturated vapor at 60°C and leaves as saturated liquid. Cooling water at 20°C is available at a rate of 200 kg/s. The convection coefficient at the outer surface of the tubes is 8500 W/m² · K. Determine (a) the temperature of the cooling water leaving the condenser and (b) the required tube length per pass. (Use the following average properties for water: $c_p = 4180$ J/kg · K, $\mu = 8 \times 10^{-4}$ N · s/m², $k = 0.6$ W/m · K, Pr = 6).

11–92 Air ($c_p = 1005$ J/kg · °C) enters a cross-flow heat exchanger at 20°C at a rate of 3 kg/s, where it is heated by a hot water stream ($c_p = 4190$ J/kg · °C) that enters the heat exchanger at 70°C at a rate of 1 kg/s. Determine the maximum heat transfer rate and the outlet temperatures of both fluids for that case.

11–93 Hot oil ($c_p = 2200$ J/kg · °C) is to be cooled by water ($c_p = 4180$ J/kg · °C) in a 2-shell-passes and 12-tube-passes heat exchanger. The tubes are thin-walled and are made of copper with a diameter of 1.8 cm. The length of each tube pass in the heat exchanger is 3 m, and the overall heat transfer coefficient is 340 W/m² · °C. Water flows through the tubes at a total rate of 0.1 kg/s, and the oil through the shell at a rate of 0.2 kg/s. The water and the oil enter at temperatures 18°C and 160°C, respectively. Determine the rate of heat transfer in the heat exchanger and the outlet temperatures of the water and the oil. *Answers.* 36.2 kW, 104.6°C, 77.7°C

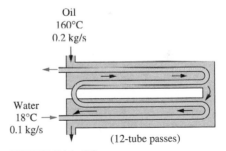

Oil
160°C
0.2 kg/s

Water
18°C
0.1 kg/s

(12-tube passes)

FIGURE P11–93

11–94 Consider an oil-to-oil double-pipe heat exchanger whose flow arrangement is not known. The temperature measurements indicate that the cold oil enters at 20°C and leaves at 55°C, while the hot oil enters at 80°C and leaves at 45°C. Do you think this is a parallel-flow or counter-flow heat exchanger? Why? Assuming the mass flow rates of both fluids to be identical, determine the effectiveness of this heat exchanger.

11–95E Hot water enters a double-pipe counter-flow water-to-oil heat exchanger at 190°F and leaves at 100°F. Oil enters at 70°F and leaves at 130°F. Determine which fluid has the smaller heat capacity rate and calculate the effectiveness of this heat exchanger.

11–96 A thin-walled double-pipe parallel-flow heat exchanger is used to heat a chemical whose specific heat is 1800 J/kg · °C with hot water (c_p = 4180 J/kg · °C). The chemical enters at 20°C at a rate of 3 kg/s, while the water enters at 110°C at a rate of 2 kg/s. The heat transfer surface area of the heat exchanger is 7 m² and the overall heat transfer coefficient is 1200 W/m² · °C. Determine the outlet temperatures of the chemical and the water.

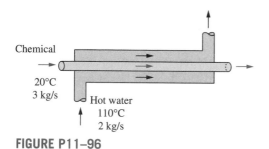

FIGURE P11–96

11–97 Reconsider Prob. 11–96. Using EES (or other) software, investigate the effects of the inlet temperatures of the chemical and the water on their outlet temperatures. Let the inlet temperature vary from 10°C to 50°C for the chemical and from 80°C to 150°C for water. Plot the outlet temperature of each fluid as a function of the inlet temperature of that fluid, and discuss the results.

11–98 A cross-flow air-to-water heat exchanger with an effectiveness of 0.65 is used to heat water (c_p = 4180 J/kg · °C) with hot air (c_p = 1010 J/kg · °C). Water enters the heat exchanger at 20°C at a rate of 4 kg/s, while air enters at 100°C at a rate of 9 kg/s. If the overall heat transfer coefficient based on the water side is 260 W/m² · °C, determine the heat transfer surface area of the heat exchanger on the water side. Assume both fluids are unmixed. *Answer:* 52.4 m²

11–99 Water (c_p = 4180 J/kg · °C) enters the 2.5-cm-internal-diameter tube of a double-pipe counter-flow heat exchanger at 17°C at a rate of 1.8 kg/s. Water is heated by steam condensing at 120°C (h_{fg} = 2203 kJ/kg) in the shell. If the overall heat transfer coefficient of the heat exchanger is 700 W/m² · °C, determine the length of the tube required in order to heat the water to 80°C using (*a*) the LMTD method and (*b*) the ε–NTU method.

11–100 Ethanol is vaporized at 78°C (h_{fg} = 846 kJ/kg) in a double-pipe parallel-flow heat exchanger at a rate of 0.03 kg/s by hot oil (c_p = 2200 J/kg · °C) that enters at 120°C. If the heat transfer surface area and the overall heat transfer coefficients are 6.2 m² and 320 W/m² · °C, respectively, determine the outlet temperature and the mass flow rate of oil using (*a*) the LMTD method and (*b*) the ε–NTU method.

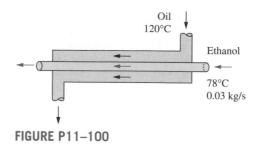

FIGURE P11–100

11–101 Water (c_p = 4180 J/kg · °C) is to be heated by solar-heated hot air (c_p = 1010 J/kg · °C) in a double-pipe counter-flow heat exchanger. Air enters the heat exchanger at 90°C at a rate of 0.3 kg/s, while water enters at 22°C at a rate of 0.1 kg/s. The overall heat transfer coefficient based on the inner side of the tube is given to be 80 W/m² · °C. The length of the tube is 12 m and the internal diameter of the tube is 1.2 cm. Determine the outlet temperatures of the water and the air.

11–102 Reconsider Prob. 11–101. Using EES (or other) software, investigate the effects of the mass flow rate of water and the tube length on the outlet temperatures of water and air. Let the mass flow rate vary from 0.05 kg/s to 1.0 kg/s and the tube length from 5 m to 25 m. Plot the outlet temperatures of the water and the air as functions of the mass flow rate and the tube length, and discuss the results.

11–103E A thin-walled double-pipe, counter-flow heat exchanger is to be used to cool oil (c_p = 0.525 Btu/lbm · °F) from 300°F to 105°F at a rate of 5 lbm/s by water (c_p = 1.0 Btu/lbm · °F) that enters at 70°F at a rate of 3 lbm/s. The diameter of the tube is 5 in and its length is 200 ft. Determine the overall heat transfer coefficient of this heat exchanger using (*a*) the LMTD method and (*b*) the ε–NTU method.

11–104 Cold water (c_p = 4180 J/kg · °C) leading to a shower enters a thin-walled double-pipe counter-flow heat exchanger at 15°C at a rate of 0.25 kg/s and is heated to 45°C by hot water (c_p = 4190 J/kg · °C) that enters at 100°C at a rate of 3 kg/s. If the overall heat transfer coefficient is 950 W/m² · °C, determine the rate of heat transfer and the heat transfer surface area of the heat exchanger using the ε–NTU method. *Answers:* 31.35 kW, 0.482 m²

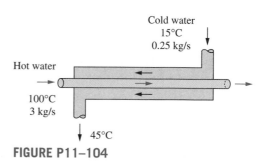

FIGURE P11–104

11–105 Reconsider Prob. 11–104. Using EES (or other) software, investigate the effects of the inlet temperature of hot water and the heat transfer coefficient on the rate of heat transfer and the surface area. Let the inlet temperature vary from 60°C to 120°C and the overall heat transfer coefficient from 750 W/m² · °C to 1250 W/m² · °C. Plot the rate of heat transfer and surface area as functions of the inlet temperature and the heat transfer coefficient, and discuss the results.

11–106 Glycerin (c_p = 2400 J/kg · °C) at 20°C and 0.3 kg/s is to be heated by ethylene glycol (c_p = 2500 J/kg · °C) at 60°C and the same mass flow rate in a thin-walled double-pipe parallel-flow heat exchanger. If the overall heat transfer coefficient is 380 W/m² · °C and the heat transfer surface area is 5.3 m², determine (*a*) the rate of heat transfer and (*b*) the outlet temperatures of the glycerin and the glycol.

11–107 A cross-flow heat exchanger consists of 80 thin-walled tubes of 3-cm diameter located in a duct of 1 m × 1 m cross section. There are no fins attached to the tubes. Cold water (c_p = 4180 J/kg · °C) enters the tubes at 18°C with an average velocity of 3 m/s, while hot air (c_p = 1010 J/kg · °C) enters the channel at 130°C and 105 kPa at an average velocity of 12 m/s. If the overall heat transfer coefficient is 130 W/m² · °C, determine the outlet temperatures of both fluids and the rate of heat transfer.

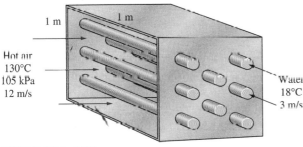

Hot air
130°C
105 kPa
12 m/s

Water
18°C
3 m/s

1 m   1 m

FIGURE P11–107

11–108 A shell-and-tube heat exchanger with 2-shell passes and 8-tube passes is used to heat ethyl alcohol (c_p = 2670 J/kg · °C) in the tubes from 25°C to 70°C at a rate of 2.1 kg/s. The heating is to be done by water (c_p = 4190 J/kg · °C) that enters the shell at 95°C and leaves at 60°C. If the overall heat transfer coefficient is 800 W/m² · °C, determine the heat transfer surface area of the heat exchanger using (*a*) the LMTD method and (*b*) the ε–NTU method.
Answer: (*a*) 11.4 m²

11–109 Steam is to be condensed on the shell side of a 1-shell-pass and 8-tube-passes condenser, with 50 tubes in each pass, at 30°C (h_{fg} = 2431 kJ/kg). Cooling water (c_p = 4180 J/kg · °C) enters the tubes at 15°C at a rate of 1800 kg/h. The tubes are thin-walled, and have a diameter of 1.5 cm and length of 2 m per pass. If the overall heat transfer coefficient is 3000 W/m² · °C, determine (*a*) the rate of heat transfer and (*b*) the rate of condensation of steam.

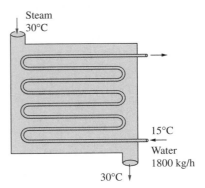

Steam
30°C

15°C
Water
1800 kg/h

30°C

FIGURE P11–109

11–110 Reconsider Prob. 11–109. Using EES (or other) software, investigate the effects of the condensing steam temperature and the tube diameter on the rate of heat transfer and the rate of condensation of steam. Let the steam temperature vary from 20°C to 70°C and the tube diameter from 1.0 cm to 2.0 cm. Plot the rate of heat transfer and the rate of condensation as functions of steam temperature and tube diameter, and discuss the results.

11–111 Cold water (c_p = 4180 J/kg · °C) enters the tubes of a heat exchanger with 2-shell passes and 23-tube passes at 14°C at a rate of 3 kg/s, while hot oil (c_p = 2200 J/kg · °C) enters the shell at 200°C at the same mass flow rate. The overall heat transfer coefficient based on the outer surface of the tube is 300 W/m² · °C and the heat transfer surface area on that side is 20 m². Determine the rate of heat transfer using (*a*) the LMTD method and (*b*) the ε–NTU method.

Selection of Heat Exchangers

11–112C A heat exchanger is to be selected to cool a hot liquid chemical at a specified rate to a specified temperature. Explain the steps involved in the selection process.

11–113C There are two heat exchangers that can meet the heat transfer requirements of a facility. One is smaller and cheaper but requires a larger pump, while the other is larger and more expensive but has a smaller pressure drop and thus requires a smaller pump. Both heat exchangers have the same life expectancy and meet all other requirements. Explain which heat exchanger you would choose under what conditions.

11–114C There are two heat exchangers that can meet the heat transfer requirements of a facility. Both have the same pumping power requirements, the same useful life, and the same price tag. But one is heavier and larger in size. Under what conditions would you choose the smaller one?

11–115 A heat exchanger is to cool oil (c_p = 2200 J/kg · °C) at a rate of 13 kg/s from 120°C to 50°C by air. Determine the heat transfer rating of the heat exchanger and propose a suitable type.

11–116 A shell-and-tube process heater is to be selected to heat water (c_p = 4190 J/kg · °C) from 20°C to 90°C by steam flowing on the shell side. The heat transfer load of the heater is 600 kW. If the inner diameter of the tubes is 1 cm and the velocity of water is not to exceed 3 m/s, determine how many tubes need to be used in the heat exchanger.

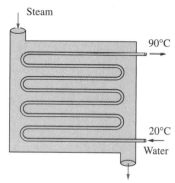

FIGURE P11–116

11–117 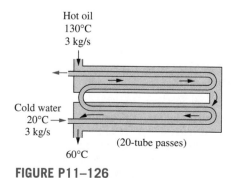 Reconsider Prob. 11–116. Using EES (or other) software, plot the number of tube passes as a function of water velocity as it varies from 1 m/s to 8 m/s, and discuss the results.

11–118 The condenser of a large power plant is to remove 500 MW of heat from steam condensing at 30°C (h_{fg} = 2431 kJ/kg). The cooling is to be accomplished by cooling water (c_p = 4180 J/kg · °C) from a nearby river, which enters the tubes at 18°C and leaves at 26°C. The tubes of the heat exchanger have an internal diameter of 2 cm, and the overall heat transfer coefficient is 3500 W/m² · °C. Determine the total length of the tubes required in the condenser. What type of heat exchanger is suitable for this task? *Answer: 312.3 km*

11–119 Repeat Prob. 11–118 for a heat transfer load of 50 MW.

Review Problems

11–120 The mass flow rate, specific heat, and inlet temperature of the tube-side stream in a double-pipe, parallel-flow heat exchanger are 2700 kg/h, 2.0 kJ/kg · K, and 120°C, respectively. The mass flow rate, specific heat, and inlet temperature of the other stream are 1800 kg/h, 4.2 kJ/kg · K, and 20°C, respectively. The heat transfer area and overall heat transfer coefficient are 0.50 m² and 2.0 kW/m² · K, respectively. Find the outlet temperatures of both streams in steady operation using (*a*) the LMTD method and (*b*) the effectiveness–NTU method.

11–121 A shell-and-tube heat exchanger is used for cooling 47 kg/s of a process stream flowing through the tubes from 160°C to 100°C. This heat exchanger has a total of 100 identical tubes, each with an inside diameter of 2.5 cm and negligible wall thickness. The average properties of the process stream are: ρ = 950 kg/m³, k = 0.50 W/m · K, c_p = 3.5 kJ/kg · K and μ = 2.0 mPa · s. The coolant stream is water (c_p = 4.18 kJ/kg · K) at a flow rate of 66 kg/s and an inlet temperature of 10°C, which yields an average shell-side heat transfer coefficient of 4.0 kW/m² · K. Calculate the tube length if the heat exchanger has (*a*) a 1-shell pass and a 1-tube pass and (*b*) a 1-shell pass and 4-tube passes.

11–122 A 2-shell passes and 4-tube passes heat exchanger is used for heating a hydrocarbon stream (c_p = 2.0 kJ/kg · K) steadily from 20°C to 50°C. A water stream enters the shell-side at 80°C and leaves at 40°C. There are 160 thin-walled tubes, each with a diameter of 2.0 cm and length of 1.5 m. The tube-side and shell-side heat transfer coefficients are 1.6 and 2.5 kW/m² · K, respectively. (*a*) Calculate the rate of heat transfer and the mass rates of water and hydrocarbon streams. (*b*) With usage, the outlet hydrocarbon-stream temperature was found to decrease by 5°C due to the deposition of solids on the tube surface. Estimate the magnitude of fouling factor.

11–123 Hot water at 60°C is cooled to 36°C through the tube side of a 1–shell pass and 2-tube passes heat exchanger. The coolant is also a water stream, for which the inlet and outlet temperatures are 7°C and 31°C, respectively. The overall heat transfer coefficient and the heat transfer area are 950 W/m² · K and 15 m², respectively. Calculate the mass flow rates of hot and cold water streams in steady operation.

11–124 Hot oil is to be cooled in a multipass shell-and-tube heat exchanger by water. The oil flows through the shell, with a heat transfer coefficient of h_o = 35 W/m² · °C, and the water flows through the tube with an average velocity of 3 m/s. The tube is made of brass (k = 110 W/m · °C) with internal and external diameters of 1.3 cm and 1.5 cm, respectively. Using water properties at 25°C, determine the overall heat transfer coefficient of this heat exchanger based on the inner surface.

11–125 Repeat Prob. 11–124 by assuming a fouling factor $R_{f,o}$ = 0.0004 m² · °C/W on the outer surface of the tube.

11–126 Cold water (c_p = 4180 J/kg · °C) enters the tubes of a heat exchanger with 2-shell passes and 20-tube passes at 20°C at a rate of 3 kg/s, while hot oil (c_p = 2200 J/kg · °C)

FIGURE P11–126

enters the shell at 130°C at the same mass flow rate and leaves at 60°C. If the overall heat transfer coefficient based on the outer surface of the tube is 220 W/m^2 · °C, determine (a) the rate of heat transfer and (b) the heat transfer surface area on the outer side of the tube. *Answers: (a) 462 kW, (b) 39.8 m^2*

11–127E Water (c_p = 1.0 Btu/lbm · °F) is to be heated by solar-heated hot air (c_p = 0.24 Btu/lbm · °F) in a double-pipe counter-flow heat exchanger. Air enters the heat exchanger at 190°F at a rate of 0.7 lbm/s and leaves at 135°F. Water enters at 70°F at a rate of 0.35 lbm/s. The overall heat transfer coefficient based on the inner side of the tube is given to be 20 Btu/h · ft^2 · °F. Determine the length of the tube required for a tube internal diameter of 0.5 in.

11–128 By taking the limit as $\Delta T_2 \rightarrow \Delta T_1$, show that when $\Delta T_1 = \Delta T_2$ for a heat exchanger, the ΔT_{lm} relation reduces to $\Delta T_{lm} = \Delta T_1 = \Delta T_2$.

11–129 The condenser of a room air conditioner is designed to reject heat at a rate of 15,000 kJ/h from refrigerant-134a as the refrigerant is condensed at a temperature of 40°C. Air (c_p = 1005 J/kg · °C) flows across the finned condenser coils, entering at 25°C and leaving at 35°C. If the overall heat transfer coefficient based on the refrigerant side is 150 W/m^2 · °C, determine the heat transfer area on the refrigerant side. *Answer: 3.05 m^2*

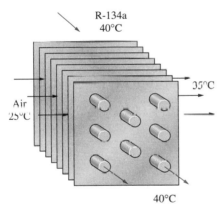

R-134a
40°C

35°C

Air
25°C

40°C

FIGURE P11–129

11–130 Air (c_p = 1005 J/kg · °C) is to be preheated by hot exhaust gases in a cross-flow heat exchanger before it enters the furnace. Air enters the heat exchanger at 95 kPa and 20°C at a rate of 0.4 m^3/s. The combustion gases (c_p = 1100 J/kg · °C) enter at 180°C at a rate of 0.65 kg/s and leave at 95°C. The product of the overall heat transfer coefficient and the heat transfer surface area is UA_s = 1620 W/°C. Assuming both fluids to be unmixed, determine the rate of heat transfer.

11–131 In a chemical plant, a certain chemical is heated by hot water supplied by a natural gas furnace. The hot water (c_p = 4180 J/kg · °C) is then discharged at 60°C at a rate of 8 kg/min. The plant operates 8 h a day, 5 days a week, 52 weeks a year. The furnace has an efficiency of 78 percent, and the cost of the

natural gas is $1.00 per therm (1 therm = 105,500 kJ). The average temperature of the cold water entering the furnace throughout the year is 14°C. In order to save energy, it is proposed to install a water-to-water heat exchanger to preheat the incoming cold water by the drained hot water. Assuming that the heat exchanger will recover 72 percent of the available heat in the hot water, determine the heat transfer rating of the heat exchanger that needs to be purchased and suggest a suitable type. Also, determine the amount of money this heat exchanger will save the company per year from natural gas savings.

11–132 A shell-and-tube heat exchanger with 1-shell pass and 14-tube passes is used to heat water in the tubes with geothermal steam condensing at 120°C (h_{fg} = 2203 kJ/kg) on the shell side. The tubes are thin-walled and have a diameter of 2.4 cm and length of 3.2 m per pass. Water (c_p = 4180 J/kg · °C) enters the tubes at 22°C at a rate of 3.9 kg/s. If the temperature difference between the two fluids at the exit is 46°C, determine (a) the rate of heat transfer, (b) the rate of condensation of steam, and (c) the overall heat transfer coefficient.

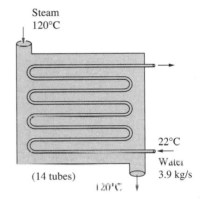

Steam
120°C

22°C

Water
3.9 kg/s

(14 tubes)

120°C

FIGURE P11–132

11–133 Geothermal water (c_p = 4250 J/kg · °C) at 75°C is to be used to heat fresh water (c_p = 4180 J/kg · °C) at 17°C at a rate of 1.2 kg/s in a double-pipe counter-flow heat exchanger. The heat transfer surface area is 25 m^2, the overall heat transfer coefficient is 480 W/m^2 · °C, and the mass flow rate of geothermal water is larger than that of fresh water. If the effectiveness of the heat exchanger is desired to be 0.823, determine the mass flow rate of geothermal water and the outlet temperatures of both fluids.

11–134 Air at 18°C (c_p = 1006 J/kg · °C) is to be heated to 58°C by hot oil at 80°C (c_p = 2150 J/kg · °C) in a cross-flow heat exchanger with air mixed and oil unmixed. The product of heat transfer surface area and the overall heat transfer coefficient is 750 W/°C and the mass flow rate of air is twice that of oil. Determine (a) the effectiveness of the heat exchanger, (b) the mass flow rate of air, and (c) the rate of heat transfer.

11–135 Consider a water-to-water counter-flow heat exchanger with these specifications. Hot water enters at 95°C while cold water enters at 20°C. The exit temperature of hot water is 15°C greater than that of cold water, and the mass flow rate of hot water is 50 percent greater than that of cold water. The product of heat transfer surface area and the overall heat transfer coefficient is 1400 W/°C. Taking the specific heat of both cold and hot water to be c_p = 4180 J/kg · °C, determine (a) the outlet temperature of the cold water, (b) the effectiveness of the heat exchanger, (c) the mass flow rate of the cold water, and (d) the heat transfer rate.

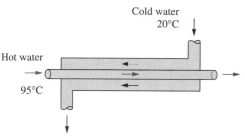

FIGURE P11–135

11–136 A shell-and-tube heat exchanger with 2-shell passes and 4-tube passes is used for cooling oil (c_p = 2.0 kJ/kg · K) from 125°C to 55°C. The coolant is water, which enters the shell side at 25°C and leaves at 46°C. The overall heat transfer coefficient is 900 W/m² · K. For an oil flow rate of 10 kg/s, calculate the cooling water flow rate and the heat transfer area.

11–137 A polymer solution (c_p = 2.0 kJ/kg · K) at 20°C and 0.3 kg/s is heated by ethylene glycol (c_p = 2.5 kJ/kg · K) at 60°C in a thin-walled double-pipe parallel-flow heat exchanger. The temperature difference between the two outlet fluids is 15°C. The overall heat transfer coefficient is 240 W/m² · K and the heat transfer area is 0.8 m². Calculate (a) the rate of heat transfer, (b) the outlet temperature of polymer solution, and (c) the mass flow rate of ethylene glycol.

11–138 During an experiment, a plate heat exchanger that is used to transfer heat from a hot-water stream to a cold-water stream is tested, and the following measurements are taken:

	Hot Water Stream	Cold Water Stream
Inlet temperature, °C	38.9	14.3
Outlet temperature, °C	27.0	19.8
Volume flow rate, L/min	2.5	4.5

The heat transfer area is calculated to be 0.0400 m².

(a) Calculate the rate of heat transfer to the cold water.
(b) Calculate the overall heat transfer coefficient.
(c) Determine if the heat exchanger is truly adiabatic. If not, determine the fraction of heat loss and calculate the heat transfer efficiency.
(d) Determine the effectiveness and the NTU values of the heat exchanger.

Also, discuss if the measured values are reasonable.

Fundamentals of Engineering (FE) Exam Problems

11–139 Hot water coming from the engine is to be cooled by ambient air in a car radiator. The aluminum tubes in which the water flows have a diameter of 4 cm and negligible thickness. Fins are attached on the outer surface of the tubes in order to increase the heat transfer surface area on the air side. The heat transfer coefficients on the inner and outer surfaces are 2000 and 150 W/m² · °C, respectively. If the effective surface area on the finned side is 10 times the inner surface area, the overall heat transfer coefficient of this heat exchanger based on the inner surface area is

(a) 150 W/m² · °C   (b) 857 W/m² · °C
(c) 1075 W/m² · °C   (d) 2000 W/m² · °C
(e) 2150 W/m² · °C

11–140 A double-pipe heat exchanger is used to heat cold tap water with hot geothermal brine. Hot geothermal brine (c_p = 4.25 kJ/kg · °C) enters the tube at 95°C at a rate of 2.8 kg/s and leaves at 60°C. The heat exchanger is not well insulated, and it is estimated that 5 percent of the heat given up by the hot fluid is lost from the heat exchanger. If the total thermal resistance of the heat exchanger is calculated to be 0.12°C/kW, the temperature difference between the hot and cold fluid is

(a) 32.5°C   (b) 35.0°C   (c) 45.0°C
(d) 47.5°C   (e) 50.0°C

11–141 Consider a double-pipe heat exchanger with a tube diameter of 10 cm and negligible tube thickness. The total thermal resistance of the heat exchanger was calculated to be 0.025°C/W when it was first constructed. After some prolonged use, fouling occurs at both the inner and outer surfaces with the fouling factors 0.00045 m² · °C/W and 0.00015 m² · °C/W, respectively. The percentage decrease in the rate of heat transfer in this heat exchanger due to fouling is

(a) 2.3%   (b) 6.8%   (c) 7.1%
(d) 7.6%   (e) 8.5%

11–142 Saturated water vapor at 40°C is to be condensed as it flows through the tubes of an air-cooled condenser at a rate of 0.2 kg/s. The condensate leaves the tubes as a saturated liquid at 40°C. The rate of heat transfer to air is

(a) 34 kJ/s   (b) 268 kJ/s   (c) 453 kJ/s
(d) 481 kJ/s   (e) 515 kJ/s

11–143 A heat exchanger is used to condense steam coming off the turbine of a steam power plant by cold water from a nearby lake. The cold water (c_p = 4.18 kJ/kg · °C) enters the condenser at 16°C at a rate of 20 kg/s and leaves at 25°C, while the steam condenses at 45°C. The condenser is not insulated, and it is estimated that heat at a rate of 8 kW is lost from the condenser to the surrounding air. The rate at which the steam condenses is

(a) 0.282 kg/s   (b) 0.290 kg/s   (c) 0.305 kg/s
(d) 0.314 kg/s   (e) 0.318 kg/s

11–144 A counter-flow heat exchanger is used to cool oil (c_p = 2.20 kJ/kg · °C) from 110°C to 85°C at a rate of 0.75 kg/s by cold water (c_p = 4.18 kJ/kg · °C) that enters the heat exchanger at 20°C at a rate of 0.6 kg/s. If the overall heat transfer coefficient is 800 W/m² · °C, the heat transfer area of the heat exchanger is

(a) 0.745 m²   (b) 0.760 m²   (c) 0.775 m²
(d) 0.790 m²   (e) 0.805 m²

11–145   In a parallel-flow, liquid-to-liquid heat exchanger, the inlet and outlet temperatures of the hot fluid are 150°C and 90°C while that of the cold fluid are 30°C and 70°C, respectively. For the same overall heat transfer coefficient, the percentage decrease in the surface area of the heat exchanger if counter-flow arrangement is used is

(a) 3.9%   (b) 9.7%   (c) 14.5%
(d) 19.7%   (e) 24.6%

11–146   A heat exchanger is used to heat cold water entering at 8°C at a rate of 1.2 kg/s by hot air entering at 90°C at rate of 2.5 kg/s. The highest rate of heat transfer in the heat exchanger is

(a) 205 kW   (b) 411 kW   (c) 311 kW
(d) 114 kW   (e) 78 kW

11–147   Cold water (c_p = 4.18 kJ/kg · °C) enters a heat exchanger at 15°C at a rate of 0.5 kg/s, where it is heated by hot air (c_p = 1.0 kJ/kg · °C) that enters the heat exchanger at 50°C at a rate of 1.8 kg/s. The maximum possible heat transfer rate in this heat exchanger is

(a) 51.1 kW   (b) 63.0 kW   (c) 66.8 kW
(d) 73.2 kW   (e) 80.0 kW

11–148   Cold water (c_p = 4.18 kJ/kg · °C) enters a counter-flow heat exchanger at 10°C at a rate of 0.35 kg/s, where it is heated by hot air (c_p = 1.0 kJ/kg · °C) that enters the heat exchanger at 50°C at a rate of 1.9 kg/s and leaves at 25°C. The effectiveness of this heat exchanger is

(a) 0.50   (b) 0.63   (c) 0.72   (d) 0.81   (e) 0.89

11–149   Hot oil (c_p = 2.1 kJ/kg · °C) at 110°C and 8 kg/s is to be cooled in a heat exchanger by cold water (c_p = 4.18 kJ/kg · °C) entering at 10°C and at a rate of 2 kg/s. The lowest temperature that oil can be cooled in this heat exchanger is

(a) 10.0°C   (b) 33.5°C   (c) 46.1°C
(d) 60.2°C   (e) 71.4°C

11–150   Cold water (c_p = 4.18 kJ/kg · °C) enters a counter-flow heat exchanger at 18°C at a rate of 0.7 kg/s where it is heated by hot air (c_p = 1.0 kJ/kg · °C) that enters the heat exchanger at 50°C at a rate of 1.6 kg/s and leaves at 25°C. The maximum possible outlet temperature of the cold water is

(a) 25.0°C   (b) 32.0°C   (c) 35.5°C
(d) 39.7°C   (e) 50.0°C

11–151   Steam is to be condensed on the shell side of a 2-shell-passes and 8-tube-passes condenser, with 20 tubes in each pass. Cooling water enters the tubes a rate of 2 kg/s. If the heat transfer area is 14 m² and the overall heat transfer

coefficient is 1800 W/m² · °C, the effectiveness of this condenser is

(a) 0.70   (b) 0.80   (c) 0.90   (d) 0.95   (e) 1.0

11–152   Water is boiled at 150°C in a boiler by hot exhaust gases (c_p = 1.05 kJ/kg · °C) that enter the boiler at 400°C at a rate of 0.4 kg/s and leaves at 200°C. The surface area of the heat exchanger is 0.64 m². The overall heat transfer coefficient of this heat exchanger is

(a) 940 W/m² · °C   (b) 1056 W/m² · °C
(c) 1145 W/m² · °C   (d) 1230 W/m² · °C
(e) 1393 W/m² · °C

11–153   In a parallel-flow, water-to-water heat exchanger, the hot water enters at 75°C at a rate of 1.2 kg/s and cold water enters at 20°C at a rate of 0.9 kg/s. The overall heat transfer coefficient and the surface area for this heat exchanger are 750 W/m² · °C and 6.4 m², respectively. The specific heat for both the hot and cold fluid may be taken to be 4.18 kJ/kg · °C. For the same overall heat transfer coefficient and the surface area, the increase in the effectiveness of this heat exchanger if counter-flow arrangement is used is

(a) 0.09   (b) 0.11   (c) 0.14   (d) 0.17   (e) 0.19

11–154   In a parallel-flow heat exchanger, the NTU is calculated to be 2.5. The lowest possible effectiveness for this heat exchanger is

(a) 10%   (b) 27%   (c) 41%   (d) 50%   (e) 92%

11–155   In a parallel-flow, air-to-air heat exchanger, hot air (c_p = 1.05 kJ/kg · °C) enters at 400°C at a rate of 0.06 kg/s and cold air (c_p = 1.0 kJ/kg · °C) enters at 25°C. The overall heat transfer coefficient and the surface area for this heat exchanger are 500 W/m² · °C and 0.12 m², respectively. The lowest possible heat transfer rate in this heat exchanger is

(a) 3.8 kW   (b) 7.9 kW   (c) 10.1 kW
(d) 14.5 kW   (e) 23.6 kW

11–156   Steam is to be condensed on the shell side of a 1-shell-pass and 4-tube-passes condenser, with 30 tubes in each pass, at 30°C (h_{fg} = 2431 kJ/kg). Cooling water (c_p = 4.18 kJ/kg · °C) enters the tubes at 12°C at a rate of 2 kg/s. If the heat transfer area is 14 m² and the overall heat transfer coefficient is 1800 W/m² · °C, the rate of heat transfer in this condenser is

(a) 112 kW   (b) 94 kW   (c) 166 kW
(d) 151 kW   (e) 143 kW

11–157   An air-cooled condenser is used to condense isobutane in a binary geothermal power plant. The isobutane is condensed at 85°C by air (c_p = 1.0 kJ/kg · °C) that enters at 22°C at a rate of 18 kg/s. The overall heat transfer coefficient and the surface area for this heat exchanger are 2.4 kW/m² · °C and 1.25 m², respectively. The outlet temperature of air is

(a) 45.4°C   (b) 40.9°C   (c) 37.5°C
(d) 34.2°C   (e) 31.7°C

11–158   An air handler is a large unmixed heat exchanger used for comfort control in large buildings. In one such application, chilled water (c_p = 4.2 kJ/kg · K) enters an air handler at 5°C and leaves at 12°C with a flow rate of 1000 kg/h. This cold water cools 5000 kg/h of air ($c_p = 1.0$ kJ/kg · K) which enters the air handler at 25°C. If these streams are in counterflow and the water-stream conditions remain fixed, the minimum temperature at the air outlet is

(a) 5°C   (b) 12°C   (c) 19°C   (d) 22°C   (e) 25°C

11–159 An air handler is a large unmixed heat exchanger used for comfort control in large buildings. In one such application, chilled water ($c_p = 4.2$ kJ/kg · K) enters an air handler at 5°C and leaves at 12°C with a flow rate of 1000 kg/h. This cold water cools air ($c_p = 1.0$ kJ/kg · K) from 25°C to 15°C. The rate of heat transfer between the two streams is

(a) 8.2 kW   (b) 23.7 kW   (c) 33.8 kW
(d) 44.8 kW   (e) 52.8 kW

11–160 The radiator in an automobile is a cross-flow heat exchanger ($UA_s = 10$ kW/K) that uses air ($c_p = 1.00$ kJ/kg · K) to cool the engine coolant fluid ($c_p = 4.00$ kJ/kg · K). The engine fan draws 30°C air through this radiator at a rate of 10 kg/s while the coolant pump circulates the engine coolant at a rate of 5 kg/s. The coolant enters this radiator at 80°C. Under these conditions, what is the number of transfer units (NTU) of this radiator?

(a) 1   (b) 2   (c) 3   (d) 4   (e) 5

Design and Essay Problems

11–161 Write an interactive computer program that will give the effectiveness of a heat exchanger and the outlet temperatures of both the hot and cold fluids when the type of fluids, the inlet temperatures, the mass flow rates, the heat transfer surface area, the overall heat transfer coefficient, and the type of heat exchanger are specified. The program should allow the user to select from the fluids water, engine oil, glycerin, ethyl alcohol, and ammonia. Assume constant specific heats at about room temperature.

11–162 Water flows through a shower head steadily at a rate of 8 kg/min. The water is heated in an electric water heater from 15°C to 45°C. In an attempt to conserve energy, it is proposed to pass the drained warm water at a temperature of 38°C through a heat exchanger to preheat the incoming cold water. Design a heat exchanger that is suitable for this task, and discuss the potential savings in energy and money for your area.

11–163 Open the engine compartment of your car and search for heat exchangers. How many do you have? What type are they? Why do you think those specific types are selected? If you were redesigning the car, would you use different kinds? Explain.

11–164 Write an essay on the static and dynamic types of regenerative heat exchangers and compile information about the manufacturers of such heat exchangers. Choose a few models by different manufacturers and compare their costs and performance.

11–165 Design a hydrocooling unit that can cool fruits and vegetables from 30°C to 5°C at a rate of 20,000 kg/h under the following conditions:

The unit will be of flood type that will cool the products as they are conveyed into the channel filled with water. The products will be dropped into the channel filled with water at one end and picked up at the other end. The channel can be as wide as 3 m and as high as 90 cm. The water is to be circulated and cooled by the evaporator section of a refrigeration system. The refrigerant temperature inside the coils is to be –2°C, and the water temperature is not to drop below 1°C and not to exceed 6°C.

Assuming reasonable values for the average product density, specific heat, and porosity (the fraction of air volume in a box), recommend reasonable values for the quantities related to the thermal aspects of the hydrocooler, including (a) how long the fruits and vegetables need to remain in the channel, (b) the length of the channel, (c) the water velocity through the channel, (d) the velocity of the conveyor and thus the fruits and vegetables through the channel, (e) the refrigeration capacity of the refrigeration system, and (f) the type of heat exchanger for the evaporator and the surface area on the water side.

11–166 Design a scalding unit for slaughtered chicken to loosen their feathers before they are routed to feather-picking machines with a capacity of 1200 chickens per h under the following conditions:

The unit will be of immersion type filled with hot water at an average temperature of 53°C at all times. Chickens with an average mass of 2.2 kg and an average temperature of 36°C will be dipped into the tank, held in the water for 1.5 min, and taken out by a slow-moving conveyor. Each chicken is expected to leave the tank 15 percent heavier as a result of the water that sticks to its surface. The center-to-center distance between chickens in any direction will be at least 30 cm. The tank can be as wide as 3 m and as high as 60 cm. The water is to be circulated through and heated by a natural gas furnace, but the temperature rise of water will not exceed 5°C as it passes through the furnace. The water loss is to be made up by the city water at an average temperature of 16°C. The ambient air temperature can be taken to be 20°C. The walls and the floor of the tank are to be insulated with a 2.5-cm-thick urethane layer. The unit operates 24 h a day and 6 days a week.

Assuming reasonable values for the average properties, recommend reasonable values for the quantities related to the thermal aspects of the scalding tank, including (a) the mass flow rate of the make-up water that must be supplied to the tank, (b) the length of the tank, (c) the rate of heat transfer from the water to the chicken, in kW, (d) the velocity of the conveyor and thus the chickens through the tank, (e) the rate of heat loss from the exposed surfaces of the tank and if it is significant, (f) the size of the heating system in kJ/h, (g) the type of heat exchanger for heating the water with flue gases of the furnace and the surface area on the water side, and (h) the operating cost of the scalding unit per month for a unit cost of $0.90 therm of natural gas.

11–167 A company owns a refrigeration system whose refrigeration capacity is 200 tons (1 ton of refrigeration = 211 kJ/min), and you are to design a forced-air cooling system for fruits whose diameters do not exceed 7 cm under the following conditions:

The fruits are to be cooled from 28°C to an average temperature of 8°C. The air temperature is to remain above –2°C and below 10°C at all times, and the velocity of air approaching the fruits must remain under 2 m/s. The cooling section can be as wide as 3.5 m and as high as 2 m.

Assuming reasonable values for the average fruit density, specific heat, and porosity (the fraction of air volume in a box), recommend reasonable values for the quantities related to the thermal aspects of the forced-air cooling, including (*a*) how long the fruits need to remain in the cooling section, (*b*) the length of the cooling section, (*c*) the air velocity approaching the cooling section, (*d*) the product cooling capacity of the system, in kg · fruit/h, (*e*) the volume flow rate of air, and (*f*) the type of heat exchanger for the evaporator and the surface area on the air side.

11–168 A counterflow double-pipe heat exchanger with $A_s = 9.0$ m^2 is used for cooling a liquid stream ($c_p = 3.15$ kJ/kg · K) at a rate of 10.0 kg/s with an inlet temperature of 90°C. The coolant ($c_p = 4.2$ kJ/kg · K) enters the heat exchanger at a rate of 8.0 kg/s with an inlet temperature of 10°C. The plant data gave the following equation for the overall heat transfer coefficient in W/m^2 · K: $U = 600/(1/\dot{m}_c^{0.8} + 2/\dot{m}_h^{0.8})$, where $\dot{m}_c$ and $\dot{m}_h$ are the cold-and hot-stream flow rates in kg/s, respectively. (*a*) Calculate the rate of heat transfer and the outlet stream temperatures for this unit. (*b*) The existing unit is to be replaced. A vendor is offering a very attractive discount on two identical heat exchangers that are presently stocked in its warehouse, each with $A_s = 5$ m^2. Because the tube diameters in the existing and new units are the same, the above heat transfer coefficient equation is expected to be valid for the new units as well. The vendor is proposing that the two new units could be operated in parallel, such that each unit would process exactly one-half the flow rate of each of the hot and cold streams in a counterflow manner; hence, they together would meet (or exceed) the present plant heat duty. Give your recommendation, with supporting calculations, on this replacement proposal.

FUNDAMENTALS OF THERMAL RADIATION

So far, we have considered the conduction and convection modes of heat transfer, which are related to the nature of the media involved and the presence of fluid motion, among other things. We now turn our attention to the third mechanism of heat transfer: *radiation,* which is characteristically different from the other two.

We start this chapter with a discussion of *electromagnetic waves* and the *electromagnetic spectrum,* with particular emphasis on *thermal radiation.* Then we introduce the idealized *blackbody, blackbody radiation,* and *blackbody radiation function,* together with the *Stefan–Boltzmann law, Planck's law,* and *Wien's displacement law.*

Radiation is emitted by every point on a plane surface in all directions into the hemisphere above the surface. The quantity that describes the magnitude of radiation emitted or incident in a specified direction in space is the *radiation intensity.* Various radiation fluxes such as *emissive power, irradiation,* and *radiosity* are expressed in terms of intensity. This is followed by a discussion of radiative properties of materials such as *emissivity, absorptivity, reflectivity,* and *transmissivity* and their dependence on wavelength, direction, and temperature. The *greenhouse effect* is presented as an example of the consequences of the wavelength dependence of radiation properties. We end this chapter with a discussion of *atmospheric* and *solar radiation.*

OBJECTIVES

When you finish studying this chapter, you should be able to.

- Classify electromagnetic radiation, and identify thermal radiation,

- Understand the idealized blackbody, and calculate the total and spectral blackbody emissive power,

- Calculate the fraction of radiation emitted in a specified wavelength band using the blackbody radiation functions,

- Understand the concept of radiation intensity, and define spectral directional quantities using intensity,

- Develop a clear understanding of the properties emissivity, absorptivity, relfectivity, and transmissivity on spectral, directional, and total basis,

- Apply Kirchhoff law's law to determine the absorptivity of a surface when its emissivity is known,

- Model the atmospheric radiation by the use of an effective sky temperature, and appreciate the importance of greenhouse effect.

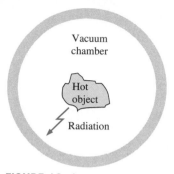

FIGURE 12–1

A hot object in a vacuum chamber loses heat by radiation only.

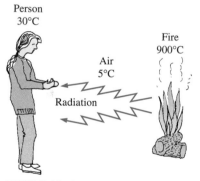

FIGURE 12–2

Unlike conduction and convection, heat transfer by radiation can occur between two bodies, even when they are separated by a medium colder than both.

12–1 · INTRODUCTION

Consider a hot object that is suspended in an evacuated chamber whose walls are at room temperature (Fig. 12–1). The hot object will eventually cool down and reach thermal equilibrium with its surroundings. That is, it will lose heat until its temperature reaches the temperature of the walls of the chamber. Heat transfer between the object and the chamber could not have taken place by conduction or convection, because these two mechanisms cannot occur in a vacuum. Therefore, heat transfer must have occurred through another mechanism that involves the emission of the internal energy of the object. This mechanism is *radiation.*

Radiation differs from the other two heat transfer mechanisms in that it does not require the presence of a material medium to take place. In fact, energy transfer by radiation is fastest (at the speed of light) and it suffers no attenuation in a *vacuum.* Also, radiation transfer occurs in solids as well as liquids and gases. In most practical applications, all three modes of heat transfer occur concurrently at varying degrees. But heat transfer through an evacuated space can occur only by radiation. For example, the energy of the sun reaches the earth by radiation.

You will recall that heat transfer by conduction or convection takes place in the direction of decreasing temperature; that is, from a high-temperature medium to a lower-temperature one. It is interesting that radiation heat transfer can occur between two bodies separated by a medium colder than both bodies (Fig. 12–2). For example, solar radiation reaches the surface of the earth after passing through cold air layers at high altitudes. Also, the radiation-absorbing surfaces inside a greenhouse reach high temperatures even when its plastic or glass cover remains relatively cool.

The theoretical foundation of radiation was established in 1864 by physicist James Clerk Maxwell, who postulated that accelerated charges or changing electric currents give rise to electric and magnetic fields. These rapidly moving fields are called **electromagnetic waves** or **electromagnetic radiation**, and they represent the energy emitted by matter as a result of the changes in the electronic configurations of the atoms or molecules. In 1887, Heinrich Hertz experimentally demonstrated the existence of such waves. Electromagnetic waves transport energy just like other waves, and all electromagnetic waves travel at the *speed of light* in a vacuum, which is $c_0 = 2.9979 \times 10^8$ m/s. Electromagnetic waves are characterized by their *frequency ν* or *wavelength λ*. These two properties in a medium are related by

$$\lambda = \frac{c}{\nu} \tag{12–1}$$

where c is the speed of propagation of a wave in that medium. The speed of propagation in a medium is related to the speed of light in a vacuum by $c = c_0/n$, where n is the *index of refraction* of that medium. The refractive index is essentially unity for air and most gases, about 1.5 for glass, and 1.33 for water. The commonly used unit of wavelength is the *micrometer* (μm) or micron, where 1 μm $= 10^{-6}$ m. Unlike the wavelength and the speed of propagation, the frequency of an electromagnetic wave depends only on the source and is independent of the medium through which the wave travels. The *frequency* (the number of oscillations per second) of an electromagnetic wave can range from less than a million Hz to a septillion Hz or higher, depending on the source. Note from Eq. 12–1 that the wavelength and the frequency of electromagnetic radiation are inversely proportional.

It has proven useful to view electromagnetic radiation as the propagation of a collection of discrete packets of energy called **photons** or **quanta**, as proposed by Max Planck in 1900 in conjunction with his *quantum theory*. In this view, each photon of frequency ν is considered to have an energy of

$$e = h\nu = \frac{hc}{\lambda} \tag{12–2}$$

where $h = 6.626069 \times 10^{-34}$ J · s is *Planck's constant*. Note from the second part of Eq. 12–2 that the energy of a photon is inversely proportional to its wavelength. Therefore, shorter-wavelength radiation possesses larger photon energies. It is no wonder that we try to avoid very-short-wavelength radiation such as gamma rays and X-rays since they are highly destructive.

12–2 ▪ THERMAL RADIATION

Although all electromagnetic waves have the same general features, waves of different wavelength differ significantly in their behavior. The electromagnetic radiation encountered in practice covers a wide range of wavelengths, varying from less than 10^{-10} μm for cosmic rays to more than 10^{10} μm for electrical power waves. The **electromagnetic spectrum** also includes gamma rays, X-rays, ultraviolet radiation, visible light, infrared radiation, thermal radiation, microwaves, and radio waves, as shown in Fig. 12–3.

Different types of electromagnetic radiation are produced through various mechanisms. For example, *gamma rays* are produced by nuclear reactions, *X-rays* by the bombardment of metals with high-energy electrons, *microwaves* by special types of electron tubes such as klystrons and magnetrons, and *radio waves* by the excitation of some crystals or by the flow of alternating current through electric conductors.

The short-wavelength gamma rays and X-rays are primarily of concern to nuclear engineers, while the long-wavelength microwaves and radio waves are of concern to electrical engineers. The type of electromagnetic radiation that is pertinent to heat transfer is the **thermal radiation** emitted as a result of energy transitions of molecules, atoms, and electrons of a substance. Temperature is a measure of the strength of these activities at the microscopic level, and the rate of thermal radiation emission increases with increasing temperature. Thermal radiation is continuously emitted by all matter whose temperature is above absolute zero. That is, everything around us such as walls, furniture, and our friends constantly emits (and absorbs) radiation (Fig. 12–4). Thermal radiation is also defined as the portion of the electromagnetic spectrum that extends from about 0.1 to 100 μm, since the radiation emitted by bodies due to their temperature falls almost entirely into this wavelength range. Thus, thermal radiation includes the entire visible and infrared (IR) radiation as well as a portion of the ultraviolet (UV) radiation.

What we call **light** is simply the *visible* portion of the electromagnetic spectrum that lies between 0.40 and 0.76 μm. Light is characteristically no different than other electromagnetic radiation, except that it happens to trigger the

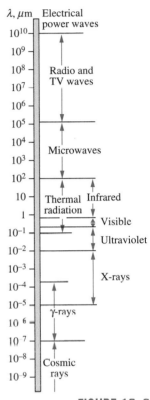

FIGURE 12–3
The electromagnetic wave spectrum.

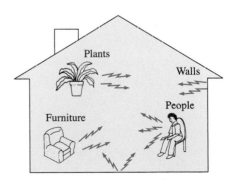

FIGURE 12–4
Everything around us constantly emits thermal radiation.

TABLE 12–1

The wavelength ranges of different colors

Color	Wavelength band
Violet	0.40–0.44 μm
Blue	0.44–0.49 μm
Green	0.49–0.54 μm
Yellow	0.54–0.60 μm
Orange	0.60–0.67 μm
Red	0.63–0.76 μm

FIGURE 12–5

Food is heated or cooked in a microwave oven by absorbing the electromagnetic radiation energy generated by the magnetron of the oven.

sensation of seeing in the human eye. Light, or the visible spectrum, consists of narrow bands of color from violet (0.40–0.44 μm) to red (0.63–0.76 μm), as shown in Table 12–1.

A body that emits some radiation in the visible range is called a light source. The sun is obviously our primary light source. The electromagnetic radiation emitted by the sun is known as **solar radiation**, and nearly all of it falls into the wavelength band 0.3–3 μm. Almost *half* of solar radiation is light (i.e., it falls into the visible range), with the remaining being ultraviolet and infrared.

The radiation emitted by bodies at room temperature falls into the **infrared** region of the spectrum, which extends from 0.76 to 100 μm. Bodies start emitting noticeable visible radiation at temperatures above 800 K. The tungsten filament of a lightbulb must be heated to temperatures above 2000 K before it can emit any significant amount of radiation in the visible range.

The **ultraviolet** radiation includes the low-wavelength end of the thermal radiation spectrum and lies between the wavelengths 0.01 and 0.40 μm. Ultraviolet rays are to be avoided since they can kill microorganisms and cause serious damage to humans and other living beings. About 12 percent of solar radiation is in the ultraviolet range, and it would be devastating if it were to reach the surface of the earth. Fortunately, the ozone (O_3) layer in the atmosphere acts as a protective blanket and absorbs most of this ultraviolet radiation. The ultraviolet rays that remain in sunlight are still sufficient to cause serious sunburns to sun worshippers, and prolonged exposure to direct sunlight is the leading cause of skin cancer, which can be lethal. Recent discoveries of "holes" in the ozone layer have prompted the international community to ban the use of ozone-destroying chemicals such as the refrigerant Freon-12 in order to save the earth. Ultraviolet radiation is also produced artificially in fluorescent lamps for use in medicine as a bacteria killer and in tanning parlors as an artificial tanner.

Microwave ovens utilize electromagnetic radiation in the **microwave** region of the spectrum generated by microwave tubes called *magnetrons*. Microwaves in the range of 10^2–10^5 μm are very suitable for use in cooking since they are *reflected* by metals, *transmitted* by glass and plastics, and *absorbed* by food (especially water) molecules. Thus, the electric energy converted to radiation in a microwave oven eventually becomes part of the internal energy of the food. The fast and efficient cooking of microwave ovens has made them one of the essential appliances in modern kitchens (Fig. 12–5).

Radars and cordless telephones also use electromagnetic radiation in the microwave region. The wavelength of the electromagnetic waves used in radio and TV broadcasting usually ranges between 1 and 1000 m in the **radio wave** region of the spectrum.

In heat transfer studies, we are interested in the energy emitted by bodies because of their temperature only. Therefore, we limit our consideration to *thermal radiation,* which we simply call *radiation*. The relations developed below are restricted to thermal radiation only and may not be applicable to other forms of electromagnetic radiation.

The electrons, atoms, and molecules of all solids, liquids, and gases above absolute zero temperature are constantly in motion, and thus radiation is constantly emitted, as well as being absorbed or transmitted throughout the entire volume of matter. That is, radiation is a **volumetric phenomenon**. However, for opaque (nontransparent) solids such as metals, wood, and rocks, radiation is considered to be a **surface phenomenon**, since the radiation emitted by the interior regions can never reach the surface, and the radiation incident on such bodies is usually absorbed within a few microns from the surface (Fig. 12–6). Note that the radiation characteristics of surfaces can be changed completely by applying thin layers of coatings on them.

12–3 · BLACKBODY RADIATION

A body at a thermodynamic (or absolute) temperature above zero emits radiation in all directions over a wide range of wavelengths. The amount of radiation energy emitted from a surface at a given wavelength depends on the material of the body and the condition of its surface as well as the surface temperature. Therefore, different bodies may emit different amounts of radiation per unit surface area, even when they are at the same temperature. Thus, it is natural to be curious about the *maximum* amount of radiation that can be emitted by a surface at a given temperature. Satisfying this curiosity requires the definition of an idealized body, called a *blackbody,* to serve as a standard against which the radiative properties of real surfaces may be compared.

A **blackbody** is defined as *a perfect emitter and absorber of radiation.* At a specified temperature and wavelength, no surface can emit more energy than a blackbody. A blackbody absorbs *all* incident radiation, regardless of wavelength and direction. Also, a blackbody emits radiation energy uniformly in all directions per unit area normal to direction of emission (Fig. 12–7). That is, a blackbody is a *diffuse* emitter. The term *diffuse* means "independent of direction."

The radiation energy emitted by a blackbody per unit time and per unit surface area was determined experimentally by Joseph Stefan in 1879 and expressed as

$$E_b(T) = \sigma T^4 \qquad (\text{W/m}^2) \qquad (12\text{–}3)$$

where $\sigma = 5.670 \times 10^{-8}$ W/m² · K⁴ is the *Stefan–Boltzmann constant* and T is the absolute temperature of the surface in K. This relation was theoretically verified in 1884 by Ludwig Boltzmann. Equation 12–3 is known as the **Stefan–Boltzmann law** and E_b is called the **blackbody emissive power**. Note that the emission of thermal radiation is proportional to the *fourth power* of the absolute temperature.

Although a blackbody would appear *black* to the eye, a distinction should be made between the idealized blackbody and an ordinary black surface. Any surface that absorbs light (the visible portion of radiation) would appear black to the eye, and a surface that reflects it completely would appear white. Considering that visible radiation occupies a very narrow band of the spectrum from 0.4 to 0.76 μm, we cannot make any judgments about the blackness of a surface on the basis of visual observations. For example, snow and white paint reflect light and thus appear white. But they are essentially black for infrared radiation since they strongly absorb long-wavelength radiation. Surfaces coated with lampblack paint approach idealized blackbody behavior.

Another type of body that closely resembles a blackbody is a *large cavity with a small opening,* as shown in Fig. 12–8. Radiation coming in through the opening of area A undergoes multiple reflections, and thus it has several chances to be absorbed by the interior surfaces of the cavity before any part of

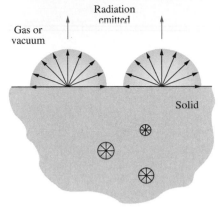

FIGURE 12–6
Radiation in opaque solids is considered a surface phenomenon since the radiation emitted only by the molecules at the surface can escape the solid.

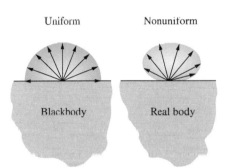

FIGURE 12–7
A blackbody is said to be a *diffuse* emitter since it emits radiation energy uniformly in all directions.

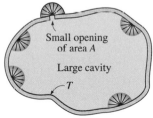

FIGURE 12–8
A large isothermal cavity at temperature T with a small opening of area A closely resembles a blackbody of surface area A at the same temperature.

it can possibly escape. Also, if the surface of the cavity is isothermal at temperature T, the radiation emitted by the interior surfaces streams through the opening after undergoing multiple reflections, and thus it has a diffuse nature. Therefore, the cavity acts as a perfect absorber and perfect emitter, and the opening will resembles a blackbody of surface area A at temperature T, regardless of the actual radiative properties of the cavity.

The Stefan–Boltzmann law in Eq. 12–3 gives the *total* blackbody emissive power E_b, which is the sum of the radiation emitted over all wavelengths. Sometimes we need to know the **spectral blackbody emissive power**, which is *the amount of radiation energy emitted by a blackbody at a thermodynamic temperature T per unit time, per unit surface area, and per unit wavelength about the wavelength λ.* For example, we are more interested in the amount of radiation an incandescent lightbulb emits in the visible wavelength spectrum than we are in the total amount emitted.

The relation for the spectral blackbody emissive power $E_{b\lambda}$ was developed by Max Planck in 1901 in conjunction with his famous quantum theory. This relation is known as **Planck's law** and is expressed as

$$E_{b\lambda}(\lambda, T) = \frac{C_1}{\lambda^5[\exp(C_2/\lambda T) - 1]} \qquad (\text{W/m}^2 \cdot \mu\text{m}) \qquad (12\text{–}4)$$

where

$$C_1 = 2\pi h c_0^2 = 3.74177 \times 10^8 \text{ W} \cdot \mu\text{m}^4/\text{m}^2$$
$$C_2 = h c_0/k = 1.43878 \times 10^4 \ \mu\text{m} \cdot \text{K}$$

Also, T is the absolute temperature of the surface, λ is the wavelength of the radiation emitted, and $k = 1.38065 \times 10^{-23}$ J/K is *Boltzmann's constant.* This relation is valid for a surface in a *vacuum* or a *gas.* For other mediums, it needs to be modified by replacing C_1 by C_1/n^2, where n is the index of refraction of the medium. Note that the term *spectral* indicates dependence on wavelength.

The variation of the spectral blackbody emissive power with wavelength is plotted in Fig. 12–9 for selected temperatures. Several observations can be made from this figure:

1. The emitted radiation is a continuous function of *wavelength.* At any specified temperature, it increases with wavelength, reaches a peak, and then decreases with increasing wavelength.
2. At any wavelength, the amount of emitted radiation *increases* with increasing temperature.
3. As temperature increases, the curves shift to the left to the shorter-wavelength region. Consequently, a larger fraction of the radiation is emitted at *shorter wavelengths* at higher temperatures.
4. The radiation emitted by the *sun,* which is considered to be a blackbody at 5780 K (or roughly at 5800 K), reaches its peak in the visible region of the spectrum. Therefore, the sun is in tune with our eyes. On the other hand, surfaces at $T \leq 800$ K emit almost entirely in the infrared region and thus are not visible to the eye unless they reflect light coming from other sources.

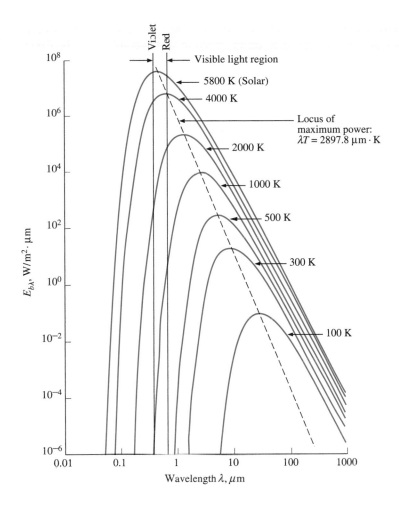

FIGURE 12–9
The variation of the blackbody emissive power with wavelength for several temperatures.

As the temperature increases, the peak of the curve in Fig. 12–9 shifts toward shorter wavelengths. The wavelength at which the peak occurs for a specified temperature is given by **Wien's displacement law** as

$$(\lambda T)_{\text{max power}} = 2897.8 \ \mu\text{m} \cdot \text{K} \tag{12–5}$$

This relation was originally developed by Willy Wien in 1894 using classical thermodynamics, but it can also be obtained by differentiating Eq. 12–4 with respect to λ while holding T constant and setting the result equal to zero. A plot of Wien's displacement law, which is the locus of the peaks of the radiation emission curves, is also given in Fig. 12–9.

The peak of the solar radiation, for example, occurs at $\lambda = 2897.8/5780 = 0.50 \ \mu\text{m}$, which is near the middle of the visible range. The peak of the radiation emitted by a surface at room temperature ($T = 298 \ \text{K}$) occurs at $9.72 \ \mu\text{m}$, which is well into the infrared region of the spectrum.

An electrical resistance heater starts radiating heat soon after it is plugged in, and we can feel the emitted radiation energy by holding our hands against the heater. But this radiation is entirely in the infrared region and thus cannot be sensed by our eyes. The heater would appear dull red when its temperature reaches about 1000 K, since it starts emitting a detectable amount (about 1 W/m^2 · μm) of visible red radiation at that temperature. As the temperature rises even more, the heater appears bright red and is said to be *red hot*. When the temperature reaches about 1500 K, the heater emits enough radiation in the entire visible range of the spectrum to appear almost *white* to the eye, and it is called *white hot*.

Although it cannot be sensed directly by the human eye, infrared radiation can be detected by infrared cameras, which transmit the information to microprocessors to display visual images of objects at night. *Rattlesnakes* can sense the infrared radiation or the "body heat" coming off warm-blooded animals, and thus they can see at night without using any instruments. Similarly, honeybees are sensitive to ultraviolet radiation. A surface that reflects all of the light appears *white*, while a surface that absorbs all of the light incident on it appears black. (Then how do we see a black surface?)

It should be clear from this discussion that the color of an object is not due to emission, which is primarily in the infrared region, unless the surface temperature of the object exceeds about 1000 K. Instead, the color of a surface depends on the absorption and reflection characteristics of the surface and is due to selective absorption and reflection of the incident visible radiation coming from a light source such as the sun or an incandescent lightbulb. A piece of clothing containing a pigment that reflects red while absorbing the remaining parts of the incident light appears "red" to the eye (Fig. 12–10). Leaves appear "green" because their cells contain the pigment chlorophyll, which strongly reflects green while absorbing other colors.

It is left as an exercise to show that integration of the *spectral* blackbody emissive power $E_{b\lambda}$ over the entire wavelength spectrum gives the *total* blackbody emissive power E_b:

$$E_b(T) = \int_0^\infty E_{b\lambda}(\lambda, T)\, d\lambda = \sigma T^4 \qquad (\text{W/m}^2) \qquad \textbf{(12–6)}$$

Thus, we obtained the Stefan–Boltzmann law (Eq. 12–3) by integrating Planck's law (Eq. 12–4) over all wavelengths. Note that on an $E_{b\lambda}$–λ chart, $E_{b\lambda}$ corresponds to any value on the curve, whereas E_b corresponds to the area under the entire curve for a specified temperature (Fig. 12–11). Also, the term *total* means "integrated over all wavelengths."

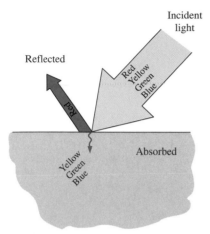

FIGURE 12–10
A surface that reflects red while absorbing the remaining parts of the incident light appears red to the eye.

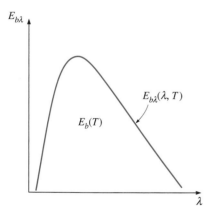

FIGURE 12–11
On an $E_{b\lambda}$–λ chart, the area under a curve for a given temperature represents the total radiation energy emitted by a blackbody at that temperature.

EXAMPLE 12–1   Radiation Emission from a Black Ball

Consider a 20-cm-diameter spherical ball at 800 K suspended in air as shown in Fig. 12–12. Assuming the ball closely approximates a blackbody, determine (*a*) the total blackbody emissive power, (*b*) the total amount of radiation emitted by the ball in 5 min, and (*c*) the spectral blackbody emissive power at a wavelength of 3 μm.

SOLUTION An isothermal sphere is suspended in air. The total blackbody emissive power, the total radiation emitted in 5 min, and the spectral blackbody emissive power at 3 μm are to be determined.
Assumptions The ball behaves as a blackbody.

Analysis (*a*) The total blackbody emissive power is determined from the Stefan–Boltzmann law to be

$$E_b = \sigma T^4 = (5.67 \times 10^{-8} \text{ W/m}^2 \cdot \text{K}^4)(800 \text{ K})^4 = \textbf{23.2 kW/m}^2$$

That is, the ball emits 23.2 kJ of energy in the form of electromagnetic radiation per second per m^2 of the surface area of the ball.

(*b*) The total amount of radiation energy emitted from the entire ball in 5 min is determined by multiplying the blackbody emissive power obtained above by the total surface area of the ball and the given time interval:

$$A_s = \pi D^2 = \pi (0.2 \text{ m})^2 = 0.1257 \text{ m}^2$$

$$\Delta t = (5 \text{ min})\left(\frac{60 \text{ s}}{1 \text{ min}}\right) = 300 \text{ s}$$

$$Q_{rad} = E_b A_s \, \Delta t = (23.2 \text{ kW/m}^2)(0.1257 \text{ m}^2)(300 \text{ s})\left(\frac{1 \text{ kJ}}{1 \text{ kW} \cdot \text{s}}\right)$$
$$= \textbf{875 kJ}$$

That is, the ball loses 875 kJ of its internal energy in the form of electromagnetic waves to the surroundings in 5 min, which is enough energy to heat 20 kg of water from 0°C to 100°C. Note that the surface temperature of the ball cannot remain constant at 800 K unless there is an equal amount of energy flow to the surface from the surroundings or from the interior regions of the ball through some mechanisms such as chemical or nuclear reactions.

(*c*) The spectral blackbody emissive power at a wavelength of 3 μm is determined from Planck's distribution law to be

$$E_{b\lambda} = \frac{C_1}{\lambda^5 \left[\exp\left(\dfrac{C_2}{\lambda T}\right) - 1\right]} = \frac{3.74177 \times 10^8 \text{ W} \cdot \mu\text{m}^4/\text{m}^2}{(3 \, \mu\text{m})^5 \left[\exp\left(\dfrac{1.43878 \times 10^4 \, \mu\text{m} \cdot \text{K}}{(3 \, \mu\text{m})(800 \text{ K})}\right) - 1\right]}$$
$$= \textbf{3846 W/m}^2 \cdot \mu\textbf{m}$$

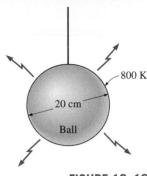

FIGURE 12–12
The spherical ball considered in Example 12–1.

The Stefan–Boltzmann law $E_b(T) = \sigma T^4$ gives the *total* radiation emitted by a blackbody at all wavelengths from $\lambda = 0$ to $\lambda = \infty$. But we are often interested in the amount of radiation emitted over *some wavelength band*. For example, an incandescent lightbulb is judged on the basis of the radiation it emits in the visible range rather than the radiation it emits at all wavelengths.

The radiation energy emitted by a blackbody per unit area over a wavelength band from $\lambda = 0$ to λ is determined from (Fig. 12–13)

$$E_{b, 0-\lambda}(T) = \int_0^\lambda E_{b\lambda}(\lambda, T) \, d\lambda \qquad \text{(W/m}^2) \qquad \text{(12–7)}$$

It looks like we can determine $E_{b, 0-\lambda}$ by substituting the $E_{b\lambda}$ relation from Eq. 12–4 and performing this integration. But it turns out that this integration does not have a simple closed-form solution, and performing a numerical integration each time we need a value of $E_{b,0-\lambda}$ is not practical. Therefore, we define a dimensionless quantity f_λ called the **blackbody radiation function** as

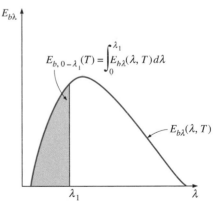

FIGURE 12–13
On an $E_{b\lambda}$–λ chart, the area under the curve to the left of the $\lambda = \lambda_1$ line represents the radiation energy emitted by a blackbody in the wavelength range 0–λ_1 for the given temperature.

TABLE 12–2

Blackbody radiation functions f_λ

λT, $\mu m \cdot K$	f_λ	λT, $\mu m \cdot K$	f_λ
200	0.000000	6200	0.754140
400	0.000000	6400	0.769234
600	0.000000	6600	0.783199
800	0.000016	6800	0.796129
1000	0.000321	7000	0.808109
1200	0.002134	7200	0.819217
1400	0.007790	7400	0.829527
1600	0.019718	7600	0.839102
1800	0.039341	7800	0.848005
2000	0.066728	8000	0.856288
2200	0.100888	8500	0.874608
2400	0.140256	9000	0.890029
2600	0.183120	9500	0.903085
2800	0.227897	10,000	0.914199
3000	0.273232	10,500	0.923710
3200	0.318102	11,000	0.931890
3400	0.361735	11,500	0.939959
3600	0.403607	12,000	0.945098
3800	0.443382	13,000	0.955139
4000	0.480877	14,000	0.962898
4200	0.516014	15,000	0.969981
4400	0.548796	16,000	0.973814
4600	0.579280	18,000	0.980860
4800	0.607559	20,000	0.985602
5000	0.633747	25,000	0.992215
5200	0.658970	30,000	0.995340
5400	0.680360	40,000	0.997967
5600	0.701046	50,000	0.998953
5800	0.720158	75,000	0.999713
6000	0.737818	100,000	0.999905

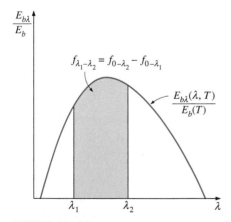

FIGURE 12–14
Graphical representation of the fraction of radiation emitted in the wavelength band from λ_1 to λ_2.

$$f_\lambda(T) = \frac{\int_0^\lambda E_{b\lambda}(\lambda, T) \, d\lambda}{\sigma T^4} \qquad (12\text{–}8)$$

The function f_λ represents *the fraction of radiation emitted from a blackbody at temperature T in the wavelength band from $\lambda = 0$ to λ*. The values of f_λ are listed in Table 12–2 as a function of λT, where λ is in μm and T is in K.

The fraction of radiation energy emitted by a blackbody at temperature T over a finite wavelength band from $\lambda = \lambda_1$ to $\lambda = \lambda_2$ is determined from (Fig. 12–14)

$$f_{\lambda_1-\lambda_2}(T) = f_{\lambda_2}(T) - f_{\lambda_1}(T) \qquad (12\text{–}9)$$

where $f_{\lambda_1}(T)$ and $f_{\lambda_2}(T)$ are blackbody radiation functions corresponding to $\lambda_1 T$ and $\lambda_2 T$, respectively.

EXAMPLE 12–2   Emission of Radiation from a Lightbulb

The temperature of the filament of an incandescent lightbulb is 2500 K. Assuming the filament to be a blackbody, determine the fraction of the radiant energy emitted by the filament that falls in the visible range. Also, determine the wavelength at which the emission of radiation from the filament peaks.

SOLUTION   The temperature of the filament of an incandescent lightbulb is given. The fraction of visible radiation emitted by the filament and the wavelength at which the emission peaks are to be determined.
Assumptions   The filament behaves as a blackbody.
Analysis   The visible range of the electromagnetic spectrum extends from $\lambda_1 = 0.4\ \mu m$ to $\lambda_2 = 0.76\ \mu m$. Noting that $T = 2500$ K, the blackbody radiation functions corresponding to $\lambda_1 T$ and $\lambda_2 T$ are determined from Table 12–2 to be

$$\lambda_1 T = (0.40\ \mu m)(2500\ K) = 1000\ \mu m \cdot K \longrightarrow f_{\lambda_1} = 0.000321$$
$$\lambda_2 T = (0.76\ \mu m)(2500\ K) = 1900\ \mu m \cdot K \longrightarrow f_{\lambda_2} = 0.053035$$

That is, 0.03 percent of the radiation is emitted at wavelengths less than 0.4 μm and 5.3 percent at wavelengths less than 0.76 μm. Then the fraction of radiation emitted between these two wavelengths is (Fig. 12–15)

$$f_{\lambda_1-\lambda_2} = f_{\lambda_2} - f_{\lambda_1} = 0.053035 - 0.000321 = \mathbf{0.052714}$$

Therefore, only about 5 percent of the radiation emitted by the filament of the lightbulb falls in the visible range. The remaining 95 percent of the radiation appears in the infrared region in the form of radiant heat or "invisible light," as it used to be called. This is certainly not a very efficient way of converting electrical energy to light and explains why fluorescent tubes are a wiser choice for lighting.
   The wavelength at which the emission of radiation from the filament peaks is easily determined from Wien's displacement law to be

$$(\lambda T)_{\text{max power}} = 2897.8\ \mu m \cdot K \quad \rightarrow \quad \lambda_{\text{max power}} = \frac{2897.8\ \mu m \cdot K}{2500\ K} - \mathbf{1.16\ \mu m}$$

Discussion   Note that the radiation emitted from the filament peaks in the infrared region.

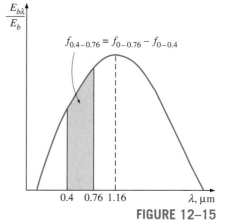

$$f_{0.4-0.76} = f_{0-0.76} - f_{0-0.4}$$

0.4   0.76 1.16 $\lambda,\ \mu m$

FIGURE 12–15
Graphical representation of the fraction of radiation emitted in the visible range in Example 12–2.

12–4 · RADIATION INTENSITY

Radiation is emitted by all parts of a plane surface in all directions into the hemisphere above the surface, and the directional distribution of emitted (or incident) radiation is usually not uniform. Therefore, we need a quantity that describes the magnitude of radiation emitted (or incident) in a specified direction in space. This quantity is *radiation intensity*, denoted by I. Before we can describe a directional quantity, we need to specify direction in space. The direction of radiation passing through a point is best described in spherical coordinates in terms of the zenith angle θ and the azimuth angle ϕ, as shown in

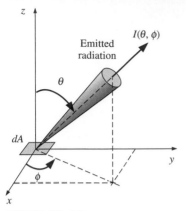

FIGURE 12–16
Radiation intensity is used to describe the variation of radiation energy with direction.

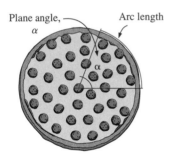

A slice of pizza of plane angle α

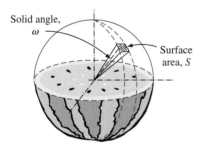

A slice of watermelon of solid angle ω

FIGURE 12–17
Describing the size of a slice of pizza by a plane angle, and the size of a watermelon slice by a solid angle.

Fig. 12–16. Radiation intensity is used to describe how the emitted radiation varies with the zenith and azimuth angles.

If all surfaces emitted radiation uniformly in all directions, the *emissive power* would be sufficient to quantify radiation, and we would not need to deal with intensity. The radiation emitted by a blackbody per unit normal area is the same in all directions, and thus there is no directional dependence. But this is not the case for real surfaces. Before we define intensity, we need to quantify the size of an opening in space.

Solid Angle

Let us try to quantify the size of a slice of pizza. One way of doing that is to specify the arc length of the outer edge of the slice, and to form the slice by connecting the endpoints of the arc to the center. A more general approach is to specify the angle of the slice at the center, as shown in Fig. 12–17. An angle of 90° (or $\pi/2$ radians), for example, always represents a quarter pizza, no matter what the radius is. For a circle of unit radius, the length of an arc is equivalent in magnitude to the *plane angle* it subtends (both are 2π for a complete circle of radius $r = 1$).

Now consider a watermelon, and let us attempt to quantify the size of a slice. Again we can do it by specifying the outer surface area of the slice (the green part), or by working with angles for generality. Connecting all points at the edges of the slice to the center in this case will form a three-dimensional body (like a cone whose tip is at the center), and thus the angle at the center in this case is properly called the **solid angle**. The solid angle is denoted by ω, and its unit is the *steradian* (sr). In analogy to plane angle, we can say that *the area of a surface on a sphere of unit radius is equivalent in magnitude to the solid angle it subtends* (both are 4π for a sphere of radius $r = 1$).

This can be shown easily by considering a differential surface area on a sphere $dS = r^2 \sin\theta \, d\theta \, d\phi$, as shown in Fig. 12–18, and integrating it from $\theta = 0$ to $\theta = \pi$, and from $\phi = 0$ to $\phi = 2\pi$. We get

$$S = \int_{\text{sphere}} dS = \int_{\phi=0}^{2\pi}\int_{\theta=0}^{\pi} r^2 \sin\theta \, d\theta\phi = 2\pi r^2 \int_{\theta=0}^{\pi}\sin\theta \, d\theta = 4\pi r^2 \quad \text{(12–10)}$$

which is the formula for the area of a sphere. For $r = 1$ it reduces to $S = 4\pi$, and thus the solid angle associated with a sphere is $\omega = 4\pi$ sr. For a hemisphere, which is more relevant to radiation emitted or received by a surface, it is $\omega = 2\pi$ sr.

The differential solid angle $d\omega$ subtended by a differential area dS on a sphere of radius r can be expressed as

$$d\omega = \frac{dS}{r^2} = \sin\theta \, d\theta \, d\phi \quad \text{(12–11)}$$

Note that the area dS is normal to the direction of viewing since dS is viewed from the center of the sphere. In general, the differential solid angle $d\omega$ subtended by a differential surface area dA when viewed from a point at a distance r from dA is expressed as

$$d\omega = \frac{dA_n}{r^2} = \frac{dA \cos\alpha}{r^2} \quad \text{(12–12)}$$

where α is the angle between the normal of the surface and the direction of viewing, and thus $dA_n = dA \cos \alpha$ is the normal (or projected) area to the direction of viewing.

Small surfaces viewed from relatively large distances can approximately be treated as differential areas in solid angle calculations. For example, the solid angle subtended by a 5 cm² plane surface when viewed from a point at a distance of 80 cm along the normal of the surface is

$$\omega \cong \frac{A_n}{r^2} = \frac{5 \text{ cm}^2}{(80 \text{ cm})^2} = 7.81 \times 10^{-4} \text{ sr}$$

If the surface is tilted so that the normal of the surface makes an angle of $\alpha = 60°$ with the line connecting the point of viewing to the center of the surface, the projected area would be $dA_n = dA \cos \alpha = (5 \text{ cm}^2)\cos 60° = 2.5 \text{ cm}^2$, and the solid angle in this case would be half of the value just determined.

Intensity of Emitted Radiation

Consider the emission of radiation by a differential area element dA of a surface, as shown in Fig. 12–18. Radiation is emitted in all directions into the hemispherical space, and the radiation streaming though the surface area dS is proportional to the solid angle $d\omega$ subtended by dS. It is also proportional to the radiating area dA as seen by an observer on dS, which varies from a maximum of dA when dS is at the top directly above dA ($\theta = 0°$) to a minimum of zero when dS is at the bottom ($\theta = 90°$). Therefore, the effective area of dA for emission in the direction of θ is the projection of dA on a plane normal to θ, which is $dA \cos \theta$. Radiation intensity in a given direction is based on a unit area *normal* to that direction to provide a common basis for the comparison of radiation emitted in different directions.

The **radiation intensity** for emitted radiation $I_e(\theta, \phi)$ is defined as *the rate at which radiation energy dQ_e is emitted in the $(0, \phi)$ direction per unit area normal to this direction and per unit solid angle about this direction.* That is,

$$I_e(0, \phi) = \frac{d\dot{Q}_e}{dA \cos \theta \cdot d\omega} = \frac{d\dot{Q}_e}{dA \cos \theta \sin \theta \, d\theta \, d\phi} \quad \text{(W/m}^2 \cdot \text{sr)} \quad \textbf{(12–13)}$$

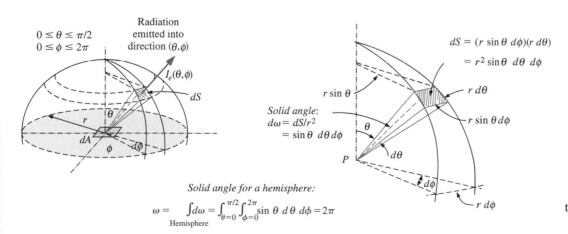

$0 \le \theta \le \pi/2$
$0 \le \phi \le 2\pi$

Radiation emitted into direction (θ, ϕ)

$I_e(\theta, \phi)$

dS

r

θ

dA

ϕ

$d\phi$

Solid angle for a hemisphere:

$$\omega = \int_{\text{Hemisphere}} d\omega = \int_{\theta=0}^{\pi/2} \int_{\phi=0}^{2\pi} \sin\theta \, d\theta \, d\phi = 2\pi$$

$dS = (r \sin\theta \, d\phi)(r \, d\theta)$
$= r^2 \sin\theta \, d\theta \, d\phi$

$r \sin\theta$

Solid angle:
$d\omega = dS/r^2$
$= \sin\theta \, d\theta \, d\phi$

θ

P

$d\theta$

$d\phi$

$r \, d\theta$

$r \sin\theta \, d\phi$

$r \, d\phi$

FIGURE 12–18
The emission of radiation from a differential surface element into the surrounding hemispherical space through a differential solid angle.

The *radiation flux* for emitted radiation is the **emissive power** E (the rate at which radiation energy is emitted per unit area of the emitting surface), which can be expressed in differential form as

$$dE = \frac{d\dot{Q}_e}{dA} = I_e(\theta, \phi) \cos\theta \sin\theta \, d\theta \, d\phi \qquad \text{(12–14)}$$

Noting that the hemisphere above the surface intercepts all the radiation rays emitted by the surface, the emissive power from the surface into the hemisphere surrounding it can be determined by integration as

$$E = \int_{\text{hemisphere}} dE = \int_{\phi=0}^{2\pi} \int_{\theta=0}^{\pi/2} I_e(\theta, \phi) \cos\theta \sin\theta \, d\theta \, d\phi \qquad (\text{W/m}^2) \qquad \text{(12–15)}$$

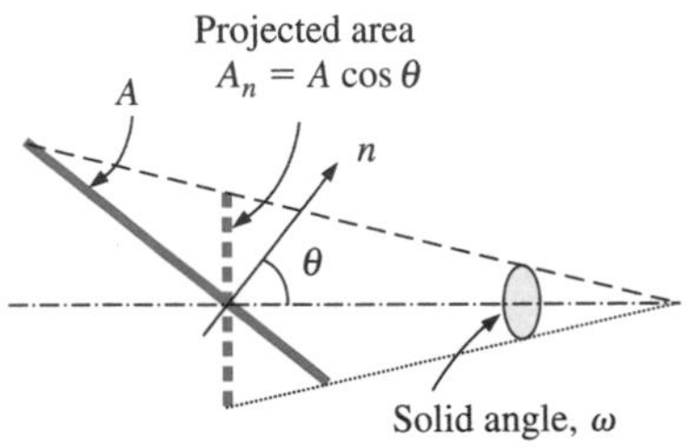

The intensity of radiation emitted by a surface, in general, varies with direction (especially with the zenith angle θ). But many surfaces in practice can be approximated as being diffuse. For a *diffusely emitting* surface, the intensity of the emitted radiation is independent of direction and thus $I_e = $ constant. Noting that $\int_{\phi=0}^{2\pi} \int_{\theta=0}^{\pi/2} \cos\theta \sin\theta \, d\theta \, d\phi = \pi$, the emissive power relation in Eq. 12–15 reduces in this case to

Diffusely emitting surface: $\qquad E = \pi I_e \qquad (\text{W/m}^2) \qquad \text{(12–16)}$

Note that the factor in Eq. 12–16 is π. You might have expected it to be 2π since intensity is radiation energy per unit solid angle, and the solid angle associated with a hemisphere is 2π. The reason for the factor being π is that the emissive power is based on the *actual* surface area whereas the intensity is based on the *projected* area (and thus the factor $\cos\theta$ that accompanies it), as shown in Fig. 12–19.

For a *blackbody,* which is a diffuse emitter, Eq. 12–16 can be expressed as

Blackbody: $\qquad\qquad\qquad E_b = \pi I_b \qquad\qquad\qquad \text{(12–17)}$

where $E_b = \sigma T^4$ is the blackbody emissive power. Therefore, the intensity of the radiation emitted by a blackbody at absolute temperature T is

Blackbody: $\qquad I_b(T) = \dfrac{E_b(T)}{\pi} = \dfrac{\sigma T^4}{\pi} \qquad (\text{W/m}^2 \cdot \text{sr}) \qquad \text{(12–18)}$

Incident Radiation

All surfaces emit radiation, but they also receive radiation emitted or reflected by other surfaces. The intensity of incident radiation $I_i(\theta, \phi)$ is defined as *the rate at which radiation energy dG is incident from the (θ, ϕ) direction per unit area of the receiving surface normal to this direction and per unit solid angle about this direction* (Fig. 12–20). Here θ is the angle between the direction of incident radiation and the normal of the surface.

The radiation flux incident on a surface from *all directions* is called **irradiation** G, and is expressed as

$$G = \int_{\text{hemisphere}} dG = \int_{\phi=0}^{2\pi} \int_{\theta=0}^{\pi/2} I_i(\theta, \phi) \cos\theta \sin\theta \, d\theta \, d\phi \qquad (\text{W/m}^2) \qquad \text{(12–19)}$$

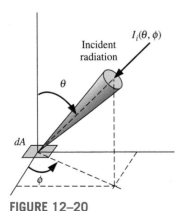

FIGURE 12–19
Radiation intensity is based on projected area, and thus the calculation of radiation emission from a surface involves the projection of the surface.

FIGURE 12–20
Radiation incident on a surface in the direction (θ, ϕ).

Therefore irradiation represents the rate at which radiation energy is incident on a surface per unit area of the surface. When the incident radiation is diffuse and thus $I_i =$ constant, Eq. 12–19 reduces to

Diffusely incident radiation: $G = \pi I_i$ (W/m^2)   (12–20)

Again note that irradiation is based on the *actual* surface area (and thus the factor cos θ), whereas the intensity of incident radiation is based on the *projected* area.

Radiosity

Surfaces emit radiation as well as reflecting it, and thus the radiation leaving a surface consists of emitted and reflected components, as shown in Fig. 12–21. The calculation of radiation heat transfer between surfaces involves the *total* radiation energy streaming away from a surface, with no regard for its origin. Thus, we need to define a quantity that represents *the rate at which radiation energy leaves a unit area of a surface in all directions.* This quantity is called the **radiosity** *J,* and is expressed as

$$J = \int_{\phi=0}^{2\pi} \int_{\theta=0}^{\pi/2} I_{e+r}(\theta, \phi) \cos \theta \sin \theta \, d\theta \, d\phi \quad (\text{W/m}^2) \quad (12\text{–}21)$$

where I_{e+r} is the sum of the emitted and reflected intensities. For a surface that is both a diffuse emitter and a diffuse reflector, $I_{e+r} =$ constant, and the radiosity relation reduces to

Diffuse emitter and reflector: $J = \pi I_{e+r}$ (W/m^2)   (12–22)

For a blackbody, radiosity J is equivalent to the emissive power E_b since a blackbody absorbs the entire radiation incident on it and there is no reflected component in radiosity.

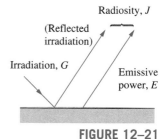

Radiosity, J
(Reflected irradiation)
Irradiation, G
Emissive power, E

FIGURE 12–21
The three kinds of radiation flux (in W/m^2): emissive power, irradiation, and radiosity.

Spectral Quantities

So far we considered *total* radiation quantities (quantities integrated over all wavelengths), and made no reference to wavelength dependence. This lumped approach is adequate for many radiation problems encountered in practice. But sometimes it is necessary to consider the variation of radiation with wavelength as well as direction, and to express quantities at a certain wavelength λ or per unit wavelength interval about λ. Such quantities are referred to as *spectral* quantities to draw attention to wavelength dependence. The modifier "spectral" is used to indicate "at a given wavelength."

The *spectral radiation intensity* $I_\lambda(\lambda, \theta, \phi)$, for example, is simply the total radiation intensity $I(\theta, \phi)$ per unit wavelength interval about λ. The **spectral intensity** for emitted radiation $I_{\lambda, e}(\lambda, \theta, \phi)$ can be defined as *the rate at which radiation energy $d\dot{Q}_e$ is emitted at the wavelength λ in the (θ, ϕ) direction per unit area normal to this direction, per unit solid angle about this direction,* and it can be expressed as

$$I_{\lambda, e}(\lambda, \theta, \phi) = \frac{d\dot{Q}_e}{dA \cos \theta \cdot d\omega \cdot d\lambda} \quad (\text{W/m}^2 \cdot \text{sr} \cdot \mu\text{m}) \quad (12\text{–}23)$$

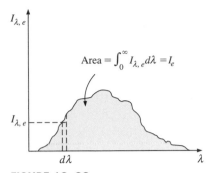

FIGURE 12–22
Integration of a "spectral" quantity for all wavelengths gives the "total" quantity.

Then the *spectral emissive power* becomes

$$E_\lambda = \int_{\phi=0}^{2\pi} \int_{\theta=0}^{\pi/2} I_{\lambda,e}(\lambda, \theta, \phi)\cos\theta \sin\theta \, d\theta \, d\phi \qquad (\text{W/m}^2) \qquad (12\text{–}24)$$

Similar relations can be obtained for spectral irradiation G_λ, and spectral radiosity J_λ by replacing $I_{\lambda,e}$ in this equation by $I_{\lambda,i}$ and $I_{\lambda,e+r}$, respectively.

When the variation of spectral radiation intensity I_λ with wavelength λ is known, the total radiation intensity I for emitted, incident, and emitted + reflected radiation can be determined by integration over the entire wavelength spectrum as (Fig. 12–22)

$$I_e = \int_0^\infty I_{\lambda,e} \, d\lambda, \qquad I_i = \int_0^\infty I_{\lambda,i} \, d\lambda, \qquad \text{and} \qquad I_{e+r} = \int_0^\infty I_{\lambda,e+r} \, d\lambda \qquad (12\text{–}25)$$

These intensities can then be used in Eqs. 12–15, 12–19, and 12–21 to determine the emissive power E, irradiation G, and radiosity J, respectively.

Similarly, when the variations of spectral radiation fluxes E_λ, G_λ, and J_λ with wavelength λ are known, the total radiation fluxes can be determined by integration over the entire wavelength spectrum as

$$E = \int_0^\infty E_\lambda \, d\lambda, \qquad G = \int_0^\infty G_\lambda \, d\lambda, \qquad \text{and} \qquad J = \int_0^\infty J_\lambda \, d\lambda \qquad (12\text{–}26)$$

When the surfaces and the incident radiation are *diffuse,* the spectral radiation fluxes are related to spectral intensities as

$$E_\lambda = \pi I_{\lambda,e}, \qquad G_\lambda = \pi I_{\lambda,i}, \qquad \text{and} \qquad J_\lambda = \pi I_{\lambda,e+r} \qquad (12\text{–}27)$$

Note that the relations for spectral and total radiation quantities are of the same form.

The spectral intensity of radiation emitted by a blackbody at a thermodynamic temperature T at a wavelength λ has been determined by Max Planck, and is expressed as

$$I_{b\lambda}(\lambda, T) = \frac{2hc_0^2}{\lambda^5[\exp(hc_0/\lambda kT) - 1]} \qquad (\text{W/m}^2 \cdot \text{sr} \cdot \mu\text{m}) \qquad (12\text{–}28)$$

where $h = 6.626069 \times 10^{-34}$ J $\cdot$ s is the Planck constant, $k = 1.38065 \times 10^{-23}$ J/K is the Boltzmann constant, and $c_0 = 2.9979 \times 10^8$ m/s is the speed of light in a vacuum. Then the spectral blackbody emissive power is, from Eq. 12–27,

$$E_{b\lambda}(\lambda, T) = \pi I_{b\lambda}(\lambda, T) \qquad (12\text{–}29)$$

A simplified relation for $E_{b\lambda}$ is given by Eq. 12–4.

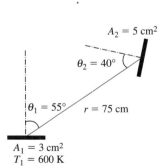

FIGURE 12–23
Schematic for Example 12–3.

EXAMPLE 12–3   **Radiation Incident on a Small Surface**

A small surface of area $A_1 = 3$ cm² emits radiation as a blackbody at $T_1 = 600$ K. Part of the radiation emitted by A_1 strikes another small surface of area $A_2 = 5$ cm² oriented as shown in Fig. 12–23. Determine the solid angle subtended by A_2 when viewed from A_1, and the rate at which radiation emitted by A_1 strikes A_2.

SOLUTION A surface Is subjected to radiation emitted by another surface. The solid angle subtended and the rate at which emitted radiation is received are to be determined.

Assumptions **1** Surface A_1 emits diffusely as a blackbody. **2** Both A_1 and A_2 can be approximated as differential surfaces since both are very small compared to the square of the distance between them.

Analysis Approximating both A_1 and A_2 as differential surfaces, the solid angle subtended by A_2 when viewed from A_1 can be determined from Eq. 12–12 to be

$$\omega_{2-1} \cong \frac{A_{n,2}}{r^2} = \frac{A_2 \cos \theta_2}{r^2} = \frac{(5\ \text{cm}^2) \cos 40°}{(75\ \text{cm})^2} = 6.81 \times 10^{-4}\ \text{sr}$$

since the normal of A_2 makes 40° with the direction of viewing. Note that solid angle subtended by A_2 would be maximum if A_2 were positioned normal to the direction of viewing. Also, the point of viewing on A_1 is taken to be a point in the middle, but it can be any point since A_1 is assumed to be very small.

The radiation emitted by A_1 that strikes A_2 is equivalent to the radiation emitted by A_1 through the solid angle ω_{2-1}. The intensity of the radiation emitted by A_1 is

$$I_1 = \frac{E_b(T_1)}{\pi} = \frac{\sigma T_1^4}{\pi} = \frac{(5.67 \times 10^{-8}\ \text{W/m}^2 \cdot \text{K}^4)(600\ \text{K})^4}{\pi} = 2339\ \text{W/m}^2 \cdot \text{sr}$$

This value of intensity is the same in all directions since a blackbody is a diffuse emitter. Intensity represents the rate of radiation emission per unit area normal to the direction of emission per unit solid angle. Therefore, the rate of radiation energy emitted by A_1 in the direction of θ_1 through the solid angle ω_{2-1} is determined by multiplying I_1 by the area of A_1 normal to θ_1 and the solid angle ω_{2-1}. That is,

$$\begin{aligned}
\dot{Q}_{1-2} &= I_1(A_1 \cos \theta_1)\omega_{2-1} \\
&= (2339\ \text{W/m}^2 \cdot \text{sr})(3 \times 10^{-4} \cos 55°\ \text{m}^2)(6.81 \times 10^{-4}\ \text{sr}) \\
&= 2.74 \times 10^{-4}\ \text{W}
\end{aligned}$$

Therefore, the radiation emitted from surface A_1 will strike surface A_2 at a rate of 2.74×10^{-4} W.

Discussion The total rate of radiation emission from surface A_1 is $\dot{Q}_e = A_1 \sigma T_1^4 = 2.204$ W. Therefore, the fraction of emitted radiation that strikes A_2 is $2.74 \times 10^{-4}/2.204 = 0.00012$ (or 0.012 percent). Noting that the solid angle associated with a hemisphere is 2π, the fraction of the solid angle subtended by A_2 is $6.81 \times 10^{-4}/(2\pi) = 0.000108$ (or 0.0108 percent), which is 0.9 times the fraction of emitted radiation. Therefore, the fraction of the solid angle a surface occupies does not represent the fraction of radiation energy the surface will receive even when the intensity of emitted radiation is constant. This is because radiation energy emitted by a surface in a given direction is proportional to the *projected area* of the surface in that direction, and reduces from a maximum at $\theta = 0°$ (the direction normal to surface) to zero at $\theta = 90°$ (the direction parallel to surface).

12–5 ▪ RADIATIVE PROPERTIES

Most materials encountered in practice, such as metals, wood, and bricks, are *opaque* to thermal radiation, and radiation is considered to be a *surface phenomenon* for such materials. That is, thermal radiation is emitted or absorbed within the first few microns of the surface, and thus we speak of radiative properties of *surfaces* for opaque materials.

Some other materials, such as glass and water, allow visible radiation to penetrate to considerable depths before any significant absorption takes place. Radiation through such *semitransparent* materials obviously cannot be considered to be a surface phenomenon since the entire volume of the material interacts with radiation. On the other hand, both glass and water are practically opaque to infrared radiation. Therefore, materials can exhibit different behavior at different wavelengths, and the dependence on wavelength is an important consideration in the study of radiative properties such as emissivity, absorptivity, reflectivity, and transmissivity of materials.

In the preceding section, we defined a *blackbody* as a perfect emitter and absorber of radiation and said that no body can emit more radiation than a blackbody at the same temperature. Therefore, a blackbody can serve as a convenient *reference* in describing the emission and absorption characteristics of real surfaces.

Emissivity

The **emissivity** of a surface represents *the ratio of the radiation emitted by the surface at a given temperature to the radiation emitted by a blackbody at the same temperature.* The emissivity of a surface is denoted by ε, and it varies between zero and one, $0 \leq \varepsilon \leq 1$. Emissivity is a measure of how closely a surface approximates a blackbody, for which $\varepsilon = 1$.

The emissivity of a real surface is not a constant. Rather, it varies with the *temperature* of the surface as well as the *wavelength* and the *direction* of the emitted radiation. Therefore, different emissivities can be defined for a surface, depending on the effects considered. The most elemental emissivity of a surface at a given temperature is the **spectral directional emissivity**, which is defined as the ratio of the intensity of radiation emitted by the surface at a specified wavelength in a specified direction to the intensity of radiation emitted by a blackbody at the same temperature at the same wavelength. That is,

$$\varepsilon_{\lambda,\theta}(\lambda, \theta, \phi, T) = \frac{I_{\lambda,e}(\lambda, \theta, \phi, T)}{I_{b\lambda}(\lambda, T)} \qquad (12\text{--}30)$$

where the subscripts λ and θ are used to designate *spectral* and *directional* quantities, respectively. Note that blackbody radiation intensity is independent of direction, and thus it has no functional dependence on θ and ϕ.

The **total directional emissivity** is defined in a like manner by using total intensities (intensities integrated over all wavelengths) as

$$\varepsilon_{\theta}(\theta, \phi, T) = \frac{I_e(\theta, \phi, T)}{I_b(T)} \qquad (12\text{--}31)$$

In practice, it is usually more convenient to work with radiation properties averaged over all directions, called *hemispherical properties*. Noting that the integral of the rate of radiation energy emitted at a specified wavelength per unit surface area over the entire hemisphere is *spectral emissive power*, the **spectral hemispherical emissivity** can be expressed as

$$\varepsilon_\lambda(\lambda, T) = \frac{E_\lambda(\lambda, T)}{E_{b\lambda}(\lambda, T)} \qquad (12\text{--}32)$$

Note that the emissivity of a surface at a given wavelength can be different at different temperatures since the spectral distribution of emitted radiation (and thus the amount of radiation emitted at a given wavelength) changes with temperature.

Finally, the **total hemispherical emissivity** is defined in terms of the radiation energy emitted over all wavelengths in all directions as

$$\varepsilon(T) = \frac{E(T)}{E_b(T)} \qquad (12\text{--}33)$$

Therefore, the total hemispherical emissivity (or simply the "average emissivity") of a surface at a given temperature represents the ratio of the total radiation energy emitted by the surface to the radiation emitted by a blackbody of the same surface area at the same temperature.

Noting from Eqs. 12–26 and 12–32 that $E = \int_0^\infty E_\lambda d\lambda$ and $E_\lambda(\lambda, T) = \varepsilon_\lambda(\lambda, T)E_{b\lambda}(\lambda, T)$, and the total hemispherical emissivity can also be expressed as

$$\varepsilon(T) = \frac{E(T)}{E_b(T)} = \frac{\int_0^\infty \varepsilon_\lambda(\lambda, T)E_{b\lambda}(\lambda, T)d\lambda}{\sigma T^4} \qquad (12\text{--}34)$$

since $E_b(T) = \sigma T^4$. To perform this integration, we need to know the variation of spectral emissivity with wavelength at the specified temperature. The integrand is usually a complicated function, and the integration has to be performed numerically. However, the integration can be performed quite easily by dividing the spectrum into a sufficient number of *wavelength bands* and assuming the emissivity to remain constant over each band; that is, by expressing the function $\varepsilon_\lambda(\lambda, T)$ as a step function. This simplification offers great convenience for little sacrifice of accuracy, since it allows us to transform the integration into a summation in terms of blackbody emission functions.

As an example, consider the emissivity function plotted in Fig. 12–24. It seems like this function can be approximated reasonably well by a step function of the form

$$\varepsilon_\lambda = \begin{cases} \varepsilon_1 = \text{constant}, & 0 \le \lambda < \lambda_1 \\ \varepsilon_2 = \text{constant}, & \lambda_1 \le \lambda < \lambda_2 \\ \varepsilon_3 = \text{constant}, & \lambda_2 \le \lambda < \infty \end{cases} \qquad (12\text{--}35)$$

Then the average emissivity can be determined from Eq. 12–34 by breaking the integral into three parts and utilizing the definition of the blackbody radiation function as

$$\varepsilon(T) = \frac{\varepsilon_1 \int_0^{\lambda_1} E_{b\lambda}\, d\lambda}{E_b} + \frac{\varepsilon_2 \int_{\lambda_1}^{\lambda_2} E_{b\lambda}\, d\lambda}{E_b} + \frac{\varepsilon_3 \int_{\lambda_2}^{\infty} E_{b\lambda}\, d\lambda}{E_b}$$

$$= \varepsilon_1 f_{0-\lambda_1}(T) + \varepsilon_2 f_{\lambda_1-\lambda_2}(T) + \varepsilon_3 f_{\lambda_2-\infty}(T) \qquad (12\text{--}36)$$

Radiation is a complex phenomenon as it is, and the consideration of the wavelength and direction dependence of properties, assuming sufficient data

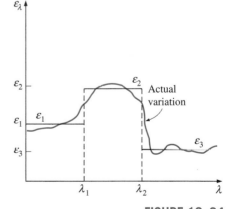

FIGURE 12–24
Approximating the actual variation of emissivity with wavelength by a step function.

Real surface:
$\varepsilon_\theta \ne$ constant
$\varepsilon_\lambda \ne$ constant

Diffuse surface:
$\varepsilon_\theta =$ constant

Gray surface:
$\varepsilon_\lambda =$ constant

Diffuse, gray surface:
$\varepsilon = \varepsilon_\lambda = \varepsilon_\theta =$ constant

FIGURE 12–25

The effect of diffuse and gray approximations on the emissivity of a surface.

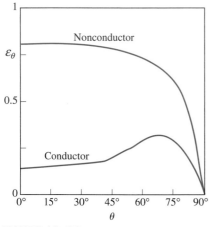

FIGURE 12–26

Typical variations of emissivity with direction for electrical conductors and nonconductors.

exist, makes it even more complicated. Therefore, the *gray* and *diffuse* approximations are often utilized in radiation calculations. A surface is said to be *diffuse* if its properties are *independent of direction,* and *gray* if its properties are *independent of wavelength.* Therefore, the emissivity of a gray, diffuse surface is simply the total hemispherical emissivity of that surface because of independence of direction and wavelength (Fig. 12–25).

A few comments about the validity of the diffuse approximation are in order. Although real surfaces do not emit radiation in a perfectly diffuse manner as a blackbody does, they often come close. The variation of emissivity with direction for both electrical conductors and nonconductors is given in Fig. 12–26. Here θ is the angle measured from the normal of the surface, and thus $\theta = 0$ for radiation emitted in a direction normal to the surface. Note that ε_θ remains nearly constant for about $\theta < 40°$ for conductors such as metals and for $\theta < 70°$ for nonconductors such as plastics. Therefore, the directional emissivity of a surface in the normal direction is representative of the hemispherical emissivity of the surface. In radiation analysis, it is common practice to assume the surfaces to be diffuse emitters with an emissivity equal to the value in the normal ($\theta = 0$) direction.

The effect of the gray approximation on emissivity and emissive power of a real surface is illustrated in Fig. 12–27. Note that the radiation emission from a real surface, in general, differs from the Planck distribution, and the emission curve may have several peaks and valleys. A gray surface should emit as much radiation as the real surface it represents at the same temperature. Therefore, the areas under the emission curves of the real and gray surfaces must be equal.

The emissivities of common materials are listed in Table A–9 in the appendix, and the variation of emissivity with wavelength and temperature is illustrated in Fig. 12–28. Typical ranges of emissivity of various materials are given in Fig. 12–29. Note that metals generally have low emissivities, as low as 0.02 for polished surfaces, and nonmetals such as ceramics and organic materials have high ones. The emissivity of metals increases with temperature. Also, oxidation causes significant increases in the emissivity of metals. Heavily oxidized metals can have emissivities comparable to those of nonmetals.

FIGURE 12–27

Comparison of the emissivity (*a*) and emissive power (*b*) of a real surface with those of a gray surface and a blackbody at the same temperature.

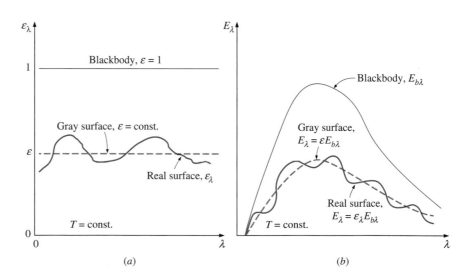

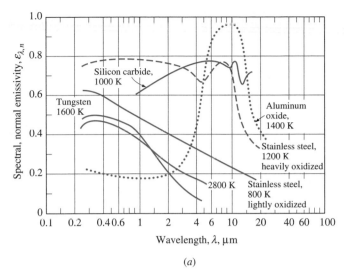

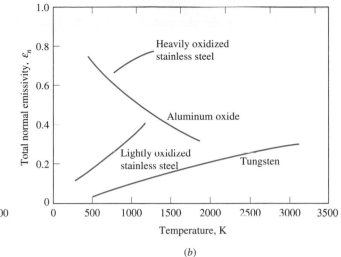

(a)

(b)

FIGURE 12–28

The variation of normal emissivity with (a) wavelength and (b) temperature for various materials.

Care should be exercised in the use and interpretation of radiation property data reported in the literature, since the properties strongly depend on the surface conditions such as oxidation, roughness, type of finish, and cleanliness. Consequently, there is considerable discrepancy and uncertainty in the reported values. This uncertainty is largely due to the difficulty in characterizing and describing the surface conditions precisely.

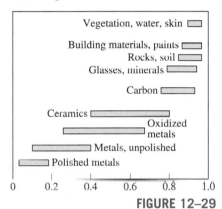

FIGURE 12–29

Typical ranges of emissivity for various materials.

EXAMPLE 12–4   Emissivity of a Surface and Emissive Power

The spectral emissivity function of an opaque surface at 800 K is approximated as (Fig. 12–30)

$$\varepsilon_\lambda = \begin{cases} \varepsilon_1 = 0.3, & 0 \le \lambda < 3 \ \mu m \\ \varepsilon_2 = 0.8, & 3 \ \mu m \le \lambda < 7 \ \mu m \\ \varepsilon_3 = 0.1, & 7 \ \mu m \le \lambda < \infty \end{cases}$$

Determine the average emissivity of the surface and its emissive power.

SOLUTION The variation of emissivity of a surface at a specified temperature with wavelength is given. The average emissivity of the surface and its emissive power are to be determined.

Analysis The variation of the emissivity with wavelength is given as a step function. Therefore, the average emissivity of the surface can be determined from Eq. 12–34 by breaking the integral into three parts,

$$\varepsilon(T) = \frac{\varepsilon_1 \int_0^{\lambda_1} E_{b\lambda} \, d\lambda}{\sigma T^4} + \frac{\varepsilon_2 \int_{\lambda_1}^{\lambda_2} E_{b\lambda} \, d\lambda}{\sigma T^4} + \frac{\varepsilon_3 \int_{\lambda_2}^{\infty} E_{b\lambda} \, d\lambda}{\sigma T^4}$$

$$= \varepsilon_1 f_{0-\lambda_1}(T) + \varepsilon_2 f_{\lambda_1-\lambda_2}(T) + \varepsilon_3 f_{\lambda_2-\infty}(T)$$

$$= \varepsilon_1 f_{\lambda_1} + \varepsilon_2 (f_{\lambda_2} - f_{\lambda_1}) + \varepsilon_3 (1 - f_{\lambda_2})$$

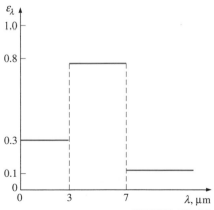

FIGURE 12–30

The spectral emissivity of the surface considered in Example 12–4 where f_{λ_1} and f_{λ_2} are blackbody radiation functions and are determined from Table 12–2 to be

$$\lambda_1 T = (3 \ \mu m)(800 \ K) = 2400 \ \mu m \cdot K \quad \rightarrow \quad f_{\lambda_1} = 0.140256$$
$$\lambda_2 T = (7 \ \mu m)(800 \ K) = 5600 \ \mu m \cdot K \quad \rightarrow \quad f_{\lambda_2} = 0.701046$$

Note that $f_{0-\lambda_1} = f_{\lambda_1} - f_0 = f_{\lambda_1}$ since $f_0 = 0$, and $f_{\lambda_2-\infty} = f_\infty - f_{\lambda_2} = 1 - f_{\lambda_2}$ since $f_\infty = 1$. Substituting,

$$\varepsilon = 0.3 \times 0.140256 + 0.8(0.701046 - 0.140256) + 0.1(1 - 0.701046)$$
$$= \mathbf{0.521}$$

That is, the surface will emit as much radiation energy at 800 K as a gray surface having a constant emissivity of $\varepsilon = 0.521$. The emissive power of the surface is

$$E = \varepsilon \sigma T^4 = 0.521(5.67 \times 10^{-8} \ W/m^2 \cdot K^4)(800 \ K)^4 = \mathbf{12,100 \ W/m^2}$$

Discussion   Note that the surface emits 12.1 kJ of radiation energy per second per m^2 area of the surface.

Absorptivity, Reflectivity, and Transmissivity

Everything around us constantly emits radiation, and the emissivity represents the emission characteristics of those bodies. This means that every body, including our own, is constantly bombarded by radiation coming from all directions over a range of wavelengths. Recall that radiation flux *incident on a surface* is called **irradiation** and is denoted by G.

When radiation strikes a surface, part of it is absorbed, part of it is reflected, and the remaining part, if any, is transmitted, as illustrated in Fig. 12–31. *The fraction of irradiation absorbed by the surface* is called the **absorptivity** α, *the fraction reflected by the surface* is called the **reflectivity** ρ, and *the fraction transmitted* is called the **transmissivity** τ. That is,

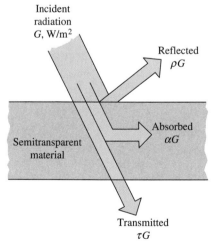

Incident
radiation
G, W/m²

Reflected
ρG

Semitransparent
material

Absorbed
αG

Transmitted
τG

FIGURE 12–31
The absorption, reflection, and transmission of incident radiation by a semitransparent material.

Absorptivity: $\alpha = \dfrac{\text{Absorbed radiation}}{\text{Incident radiation}} = \dfrac{G_{abs}}{G}, \qquad 0 \le \alpha \le 1$ (12–37)

Reflectivity: $\rho = \dfrac{\text{Reflected radiation}}{\text{Incident radiation}} = \dfrac{G_{ref}}{G}, \qquad 0 \le \rho \le 1$ (12–38)

Transmissivity: $\tau = \dfrac{\text{Transmitted radiation}}{\text{Incident radiation}} = \dfrac{G_{tr}}{G}, \qquad 0 \le \tau \le 1$ (12–39)

where G is the radiation flux incident on the surface, and G_{abs}, G_{ref}, and G_{tr} are the absorbed, reflected, and transmitted portions of it, respectively. The first law of thermodynamics requires that the sum of the absorbed, reflected, and transmitted radiation be equal to the incident radiation. That is,

$$G_{abs} + G_{ref} + G_{tr} = G \qquad (12\text{–}40)$$

Dividing each term of this relation by G yields

$$\alpha + \rho + \tau = 1 \qquad (12\text{–}41)$$

For opaque surfaces, $\tau = 0$, and thus

$$\alpha + \rho = 1 \qquad (12\text{–}42)$$

This is an important property relation since it enables us to determine both the absorptivity and reflectivity of an opaque surface by measuring either of these properties.

These definitions are for *total hemispherical* properties, since G represents the radiation flux incident on the surface from all directions over the hemispherical space and over all wavelengths. Thus, α, ρ, and τ are the *average* properties of a medium for all directions and all wavelengths. However, like emissivity, these properties can also be defined for a specific wavelength and/or direction. For example, the **spectral directional absorptivity** and **spectral directional reflectivity** of a surface are defined, respectively, as the absorbed and reflected fractions of the intensity of radiation incident at a specified wavelength in a specified direction as

$$\alpha_{\lambda, \theta}(\lambda, \theta, \phi) = \frac{I_{\lambda, \text{abs}}(\lambda, \theta, \phi)}{I_{\lambda, i}(\lambda, \theta, \phi)} \quad \text{and} \quad \rho_{\lambda, \theta}(\lambda, \theta, \phi) = \frac{I_{\lambda, \text{ref}}(\lambda, \theta, \phi)}{I_{\lambda, i}(\lambda, \theta, \phi)} \tag{12–43}$$

Likewise, the **spectral hemispherical absorptivity** and **spectral hemispherical reflectivity** of a surface are defined as

$$\alpha_{\lambda}(\lambda) = \frac{G_{\lambda, \text{abs}}(\lambda)}{G_{\lambda}(\lambda)} \quad \text{and} \quad \rho_{\lambda}(\lambda) = \frac{G_{\lambda, \text{ref}}(\lambda)}{G_{\lambda}(\lambda)} \tag{12–44}$$

where G_{λ} is the spectral irradiation (in $W/m^2 \cdot \mu m$) incident on the surface, and $G_{\lambda, \text{abs}}$ and $G_{\lambda, \text{ref}}$ are the reflected and absorbed portions of it, respectively.

Similar quantities can be defined for the transmissivity of semitransparent materials. For example, the **spectral hemispherical transmissivity** of a medium can be expressed as

$$\tau_{\lambda}(\lambda) = \frac{G_{\lambda, \text{tr}}(\lambda)}{G_{\lambda}(\lambda)} \tag{12–45}$$

The average absorptivity, reflectivity, and transmissivity of a surface can also be defined in terms of their spectral counterparts as

$$\alpha = \frac{\int_0^\infty \alpha_{\lambda} G_{\lambda} \, d\lambda}{\int_0^\infty G_{\lambda} \, d\lambda}, \quad \rho = \frac{\int_0^\infty \rho_{\lambda} G_{\lambda} \, d\lambda}{\int_0^\infty G_{\lambda} \, d\lambda}, \quad \tau = \frac{\int_0^\infty \tau_{\lambda} G_{\lambda} \, d\lambda}{\int_0^\infty G_{\lambda} \, d\lambda} \tag{12–46}$$

The reflectivity differs somewhat from the other properties in that it is *bidirectional* in nature. That is, the value of the reflectivity of a surface depends not only on the direction of the incident radiation but also the direction of reflection. Therefore, the reflected rays of a radiation beam incident on a real surface in a specified direction forms an irregular shape, as shown in Fig. 12–32. Such detailed reflectivity data do not exist for most surfaces, and even if they did, they would be of little value in radiation calculations since this would usually add more complication to the analysis.

In practice, for simplicity, surfaces are assumed to reflect in a perfectly *specular* or *diffuse* manner. In **specular** (or *mirrorlike*) **reflection**, *the angle of reflection equals the angle of incidence of the radiation beam*. In **diffuse reflection**, *radiation is reflected equally in all directions*, as shown in Fig. 12–32. Reflection

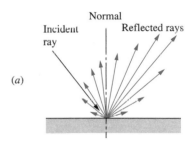

(a)

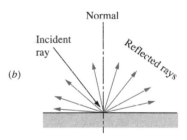

(b)

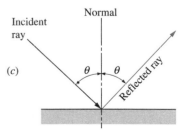

(c)

FIGURE 12–32

Different types of reflection from a surface: (*a*) actual or irregular, (*b*) diffuse, and (*c*) specular or mirrorlike.

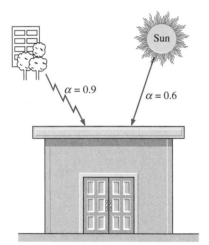

FIGURE 12–33

Variation of absorptivity with the temperature of the source of irradiation for various common materials at room temperature.

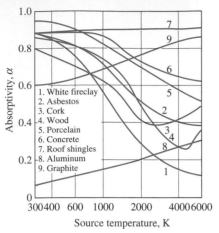

FIGURE 12–34

The absorptivity of a material may be quite different for radiation originating from sources at different temperatures.

from smooth and polished surfaces approximates specular reflection, whereas reflection from rough surfaces approximates diffuse reflection. In radiation analysis, smoothness is defined relative to wavelength. A surface is said to be *smooth* if the height of the surface roughness is much smaller than the wavelength of the incident radiation.

Unlike emissivity, the absorptivity of a material is practically independent of surface temperature. However, the absorptivity depends strongly on the temperature of the source at which the incident radiation is originating. This is also evident from Fig. 12–33, which shows the absorptivities of various materials at room temperature as functions of the temperature of the radiation source. For example, the absorptivity of the concrete roof of a house is about 0.6 for solar radiation (source temperature: 5780 K) and 0.9 for radiation originating from the surrounding trees and buildings (source temperature: 300 K), as illustrated in Fig. 12–34.

Notice that the absorptivity of aluminum increases with the source temperature, a characteristic for metals, and the absorptivity of electric nonconductors, in general, decreases with temperature. This decrease is most pronounced for surfaces that appear white to the eye. For example, the absorptivity of a white painted surface is low for solar radiation, but it is rather high for infrared radiation.

Kirchhoff's Law

Consider a small body of surface area A_s, emissivity ε, and absorptivity α at temperature T contained in a large isothermal enclosure at the same temperature, as shown in Fig. 12–35. Recall that a large isothermal enclosure forms a blackbody cavity regardless of the radiative properties of the enclosure surface, and the body in the enclosure is too small to interfere with the blackbody nature of the cavity. Therefore, the radiation incident on any part of the surface of the small body is equal to the radiation emitted by a blackbody at temperature T. That is, $G = E_b(T) = \sigma T^4$, and the radiation absorbed by the small body per unit of its surface area is

$$G_{\text{abs}} = \alpha G = \alpha \sigma T^4$$

The radiation emitted by the small body is

$$E_{\text{emit}} = \varepsilon \sigma T^4$$

Considering that the small body is in thermal equilibrium with the enclosure, the net rate of heat transfer to the body must be zero. Therefore, the radiation emitted by the body must be equal to the radiation absorbed by it:

$$A_s \varepsilon \sigma T^4 = A_s \alpha \sigma T^4$$

Thus, we conclude that

$$\varepsilon(T) = \alpha(T) \tag{12–47}$$

That is, *the total hemispherical emissivity of a surface at temperature T is equal to its total hemispherical absorptivity for radiation coming from a blackbody at the same temperature.* This relation, which greatly simplifies the radiation analysis, was first developed by Gustav Kirchhoff in 1860 and is now called **Kirchhoff's law**. Note that this relation is derived under the condition that the surface temperature is equal to the temperature of the source of irradiation, and the reader is cautioned against using it when considerable difference (more than a few hundred degrees) exists between the surface temperature and the temperature of the source of irradiation.

The derivation above can also be repeated for radiation at a specified wavelength to obtain the *spectral* form of Kirchhoff's law:

$$\varepsilon_\lambda(T) = \alpha_\lambda(T) \qquad (12\text{–}48)$$

This relation is valid when the irradiation or the emitted radiation is independent of direction. The form of Kirchhoff's law that involves no restrictions is the *spectral directional* form expressed as $\varepsilon_{\lambda,\,\theta}(T) = \alpha_{\lambda,\,\theta}(T)$. That is, the emissivity of a surface at a specified wavelength, direction, and temperature is always equal to its absorptivity at the same wavelength, direction, and temperature.

It is very tempting to use Kirchhoff's law in radiation analysis since the relation $\varepsilon = \alpha$ together with $\rho = 1 - \alpha$ enables us to determine all three properties of an opaque surface from a knowledge of only *one* property. Although Eq. 12–47 gives acceptable results in most cases, in practice, care should be exercised when there is considerable difference between the surface temperature and the temperature of the source of incident radiation.

The Greenhouse Effect

You have probably noticed that when you leave your car under direct sunlight on a sunny day, the interior of the car gets much warmer than the air outside, and you may have wondered why the car acts like a *heat trap*. The answer lies in the spectral transmissivity curve of the *glass*, which resembles an inverted U, as shown in Fig. 12–36. We observe from this figure that glass at thicknesses encountered in practice transmits over 90 percent of radiation in the visible range and is practically opaque (nontransparent) to radiation in the longer-wavelength infrared regions of the electromagnetic spectrum (roughly $\lambda > 3\ \mu m$). Therefore, glass has a transparent window in the wavelength range $0.3\ \mu m < \lambda < 3\ \mu m$ in which over 90 percent of solar radiation is emitted. On the other hand, the entire radiation emitted by surfaces at room temperature falls in the infrared region. Consequently, glass allows the solar radiation to enter but does not allow the infrared radiation from the interior surfaces to escape. This causes a rise in the interior temperature as a result of the energy buildup in the car. This heating effect, which is due to the nongray characteristic of glass (or clear plastics), is known as the **greenhouse effect**, since it is utilized extensively in greenhouses (Fig. 12–37).

The greenhouse effect is also experienced on a larger scale on earth. The surface of the earth, which warms up during the day as a result of the absorption of solar energy, cools down at night by radiating its energy into deep space as infrared radiation. The combustion gases such as CO_2 and water vapor in the atmosphere transmit the bulk of the solar radiation but absorb the infrared radiation emitted by the surface of the earth. Thus, there is concern that the energy trapped on earth will eventually cause global warming and thus drastic changes in weather patterns.

In *humid* places such as coastal areas, there is not a large change between the daytime and nighttime temperatures, because the humidity acts as a barrier

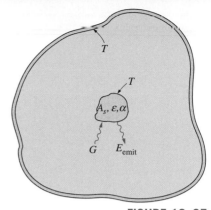

FIGURE 12–35
The small body contained in a large isothermal enclosure used in the development of Kirchhoff's law.

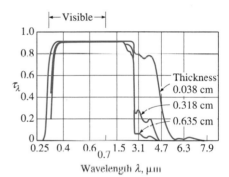

FIGURE 12–36
The spectral transmissivity of low-iron glass at room temperature for different thicknesses.

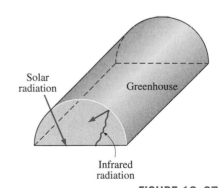

FIGURE 12–37
A greenhouse traps energy by allowing the solar radiation to come in but not allowing the infrared radiation to go out.

on the path of the infrared radiation coming from the earth, and thus slows down the cooling process at night. In areas with clear skies such as deserts, there is a large swing between the daytime and nighttime temperatures because of the absence of such barriers for infrared radiation.

12–6 · ATMOSPHERIC AND SOLAR RADIATION

The sun is our primary source of energy. The energy coming off the sun, called *solar energy*, reaches us in the form of electromagnetic waves after experiencing considerable interactions with the atmosphere. The radiation energy emitted or reflected by the constituents of the atmosphere form the *atmospheric radiation*. Here we give an overview of the solar and atmospheric radiation because of their importance and relevance to daily life. Also, our familiarity with solar energy makes it an effective tool in developing a better understanding for some of the new concepts introduced earlier. Detailed treatment of this exciting subject can be found in numerous books devoted to this topic.

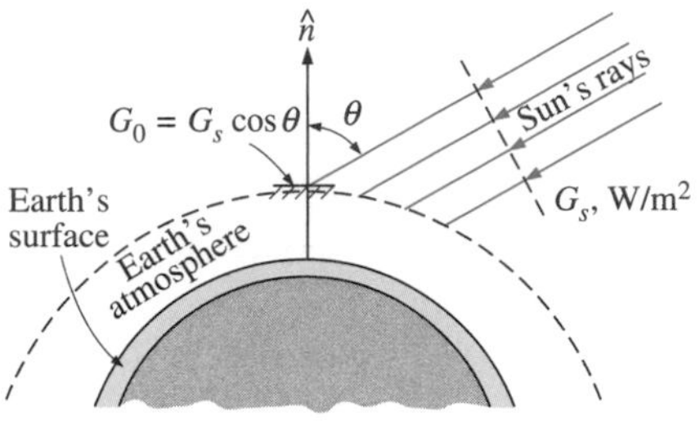

The *sun* is a nearly spherical body that has a diameter of $D \approx 1.39 \times 10^9$ m and a mass of $m \approx 2 \times 10^{30}$ kg and is located at a mean distance of $L = 1.50 \times 10^{11}$ m from the earth. It emits radiation energy continuously at a rate of $E_{\text{sun}} \approx 3.8 \times 10^{26}$ W. Less than a billionth of this energy (about 1.7×10^{17} W) strikes the earth, which is sufficient to keep the earth warm and to maintain life through the photosynthesis process. The energy of the sun is due to the continuous *fusion* reaction during which two hydrogen atoms fuse to form one atom of helium. Therefore, the sun is essentially a *nuclear reactor*, with temperatures as high as 40,000,000 K in its core region. The temperature drops to about 5800 K in the outer region of the sun, called the convective zone, as a result of the dissipation of this energy by radiation.

The solar energy reaching the earth's atmosphere is called the **total solar irradiance** G_s, whose value is

$$G_s = 1373 \text{ W/m}^2 \tag{12–49}$$

The total solar irradiance (also called the **solar constant**) represents *the rate at which solar energy is incident on a surface normal to the sun's rays at the outer edge of the atmosphere when the earth is at its mean distance from the sun* (Fig. 12–38).

The value of the total solar irradiance can be used to estimate the effective surface temperature of the sun from the requirement that

$$(4\pi L^2)G_s = (4\pi r^2)\,\sigma T_{\text{sun}}^4 \tag{12–50}$$

where L is the mean distance between the sun's center and the earth and r is the radius of the sun. The left-hand side of this equation represents the total solar energy passing through a spherical surface whose radius is the mean earth–sun distance, and the right-hand side represents the total energy that leaves the sun's outer surface. The conservation of energy principle requires that these two quantities be equal to each other, since the solar energy experiences no attenuation (or enhancement) on its way through the vacuum (Fig. 12–39). The **effective surface temperature** of the sun is determined from Eq. 12–50 to be $T_{\text{sun}} = 5780$ K. That is, the sun can be treated as a

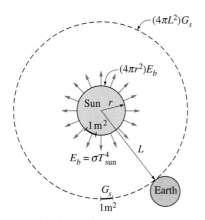

FIGURE 12–38
Solar radiation reaching the earth's atmosphere and the total solar irradiance.

FIGURE 12–39
The total solar energy passing through concentric spheres remains constant, but the energy falling per unit area decreases with increasing radius.

blackbody at a temperature of 5780 K. This is also confirmed by the measurements of the spectral distribution of the solar radiation just outside the atmosphere plotted in Fig. 12–40, which shows only small deviations from the idealized blackbody behavior.

The spectral distribution of solar radiation on the ground plotted in Fig. 12–40 shows that the solar radiation undergoes considerable *attenuation* as it passes through the atmosphere as a result of absorption and scattering. About 99 percent of the atmosphere is contained within a distance of 30 km from the earth's surface. The several dips on the spectral distribution of radiation on the earth's surface are due to *absorption* by the gases O_2, O_3 (ozone), H_2O, and CO_2. Absorption by *oxygen* occurs in a narrow band about $\lambda = 0.76$ μm. The *ozone* absorbs *ultraviolet* radiation at wavelengths below 0.3 μm almost completely, and radiation in the range 0.3–0.4 μm considerably. Thus, the ozone layer in the upper regions of the atmosphere protects biological systems on earth from harmful ultraviolet radiation. In turn, we must protect the ozone layer from the destructive chemicals commonly used as refrigerants, cleaning agents, and propellants in aerosol cans. The use of these chemicals is now banned. The ozone gas also absorbs some radiation in the visible range. Absorption in the infrared region is dominated by *water vapor* and *carbon dioxide*. The dust particles and other pollutants in the atmosphere also absorb radiation at various wavelengths.

As a result of these absorptions, the solar energy reaching the earth's surface is weakened considerably, to about 950 W/m^2 on a clear day and much less on cloudy or smoggy days. Also, practically all of the solar radiation reaching the earth's surface falls in the wavelength band from 0.3 to 2.5 μm.

Another mechanism that attenuates solar radiation as it passes through the atmosphere is *scattering* or *reflection* by air molecules and the many other kinds of particles such as dust, smog, and water droplets suspended in the atmosphere. Scattering is mainly governed by the size of the particle relative to the wavelength of radiation. The oxygen and nitrogen molecules primarily scatter radiation at very short wavelengths, comparable to the size of the molecules themselves. Therefore, radiation at wavelengths corresponding to violet and blue colors is scattered the most. This molecular scattering in all directions is what gives the sky its bluish color. The same phenomenon is responsible for red sunrises and sunsets. Early in the morning and late in the afternoon, the sun's rays pass through a greater thickness of the atmosphere than they do at midday, when the sun is at the top. Therefore, the violet and blue colors of the light encounter a greater number of molecules by the time they reach the earth's surface, and thus a greater fraction of them are scattered (Fig. 12–41). Consequently, the light that reaches the earth's surface consists primarily of colors corresponding to longer wavelengths such as red, orange, and yellow. The clouds appear in reddish-orange color during sunrise and sunset because the light they reflect is reddish-orange at those times. For the same reason, a red traffic light is visible from a longer distance than is a green light under the same circumstances.

The solar energy incident on a surface on earth is considered to consist of *direct* and *diffuse* parts. The part of solar radiation that reaches the earth's surface without being scattered or absorbed by the atmosphere is called **direct solar radiation** G_D. The scattered radiation is assumed to reach the earth's surface uniformly from all directions and is called **diffuse solar radiation** G_d.

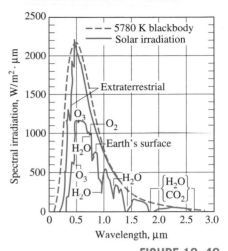

FIGURE 12–40

Spectral distribution of solar radiation just outside the atmosphere, at the surface of the earth on a typical day, and comparison with blackbody radiation at 5780 K.

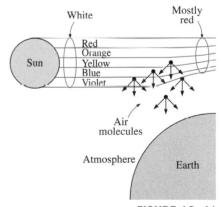

FIGURE 12–41

Air molecules scatter blue light much more than they do red light. At sunset, light travels through a thicker layer of atmosphere, which removes much of the blue from the natural light, allowing the red to dominate.

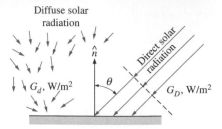

FIGURE 12–42
The direct and diffuse radiation incident on a horizontal surface on earth's surface.

Then the *total* solar energy incident on the unit area of a *horizontal surface* on the ground is (Fig. 12–42)

$$G_{\text{solar}} = G_D \cos \theta + G_d \qquad (\text{W/m}^2) \qquad (12\text{–}51)$$

where θ is the angle of incidence of direct solar radiation (the angle that the sun's rays make with the normal of the surface). The diffuse radiation varies from about 10 percent of the total radiation on a clear day to nearly 100 percent on a totally cloudy day.

The gas molecules and the suspended particles in the atmosphere emit radiation as well as absorbing it. The atmospheric emission is primarily due to the CO_2 and H_2O molecules and is concentrated in the regions from 5 to 8 μm and above 13 μm. Although this emission is far from resembling the distribution of radiation from a blackbody, it is found convenient in radiation calculations to treat the atmosphere as a blackbody at some lower fictitious temperature that emits an equivalent amount of radiation energy. This fictitious temperature is called the **effective sky temperature** T_{sky}. Then the radiation emission from the atmosphere to the earth's surface is expressed as

$$G_{\text{sky}} = \sigma T_{\text{sky}}^4 \qquad (\text{W/m}^2) \qquad (12\text{–}52)$$

The value of T_{sky} depends on the atmospheric conditions. It ranges from about 230 K for cold, clear-sky conditions to about 285 K for warm, cloudy-sky conditions.

Note that the effective sky temperature does not deviate much from the room temperature. Thus, in the light of Kirchhoff's law, we can take the absorptivity of a surface to be equal to its emissivity at room temperature, $\alpha = \varepsilon$. Then the sky radiation absorbed by a surface can be expressed as

$$E_{\text{sky, absorbed}} = \alpha G_{\text{sky}} = \alpha \sigma T_{\text{sky}}^4 = \varepsilon \sigma T_{\text{sky}}^4 \qquad (\text{W/m}^2) \qquad (12\text{–}53)$$

The net rate of radiation heat transfer to a surface exposed to solar and atmospheric radiation is determined from an energy balance (Fig. 12–43):

$$\begin{aligned}
\dot{q}_{\text{net, rad}} &= \Sigma E_{\text{absorbed}} - \Sigma E_{\text{emitted}} \\
&= E_{\text{solar, absorbed}} + E_{\text{sky, absorbed}} - E_{\text{emitted}} \\
&= \alpha_s G_{\text{solar}} + \varepsilon \sigma T_{\text{sky}}^4 - \varepsilon \sigma T_s^4 \\
&= \alpha_s G_{\text{solar}} + \varepsilon \sigma (T_{\text{sky}}^4 - T_s^4) \qquad (\text{W/m}^2) \qquad (12\text{–}54)
\end{aligned}$$

where T_s is the temperature of the surface in K and ε is its emissivity at room temperature. A positive result for $\dot{q}_{\text{net, rad}}$ indicates a radiation heat gain by the surface and a negative result indicates a heat loss.

The absorption and emission of radiation by the *elementary gases* such as H_2, O_2, and N_2 at moderate temperatures are negligible, and a medium filled with these gases can be treated as a *vacuum* in radiation analysis. The absorption and emission of gases with *larger molecules* such as H_2O and CO_2, however, can be *significant* and may need to be considered when considerable amounts of such gases are present in a medium. For example, a 1-m-thick layer of water vapor at 1 atm pressure and 100°C emits more than 50 percent of the energy that a blackbody would emit at the same temperature.

In solar energy applications, the spectral distribution of incident solar radiation is very different than the spectral distribution of emitted radiation by the

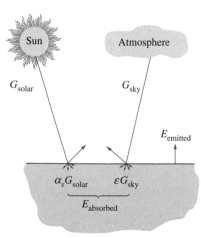

FIGURE 12–43
Radiation interactions of a surface exposed to solar and atmospheric radiation.

surfaces, since the former is concentrated in the short-wavelength region and the latter in the infrared region. Therefore, the radiation properties of surfaces are quite different for the incident and emitted radiation, and the surfaces cannot be assumed to be gray. Instead, the surfaces are assumed to have two sets of properties: one for solar radiation and another for infrared radiation at room temperature. Table 12–3 lists the *emissivity* ε and the *solar absorptivity* α_s of some common materials. Surfaces that are intended to *collect solar energy,* such as the absorber surfaces of solar collectors, are desired to have high α_s but low ε values to maximize the absorption of solar radiation and to minimize the emission of radiation. Surfaces that are intended to *remain cool* under the sun, such as the outer surfaces of fuel tanks and refrigerator trucks, are desired to have just the opposite properties. Surfaces are often given the desired properties by coating them with thin layers of *selective* materials. A surface can be kept cool, for example, by simply painting it white.

We close this section by pointing out that what we call *renewable energy* is usually nothing more than the manifestation of solar energy in different forms. Such energy sources include wind energy, hydroelectric power, ocean thermal energy, ocean wave energy, and wood. For example, no hydroelectric power plant can generate electricity year after year unless the water evaporates by absorbing solar energy and comes back as a rainfall to replenish the water source (Fig. 12–44). Although solar energy is sufficient to meet the entire energy needs of the world, currently it is not economical to do so because of the low concentration of solar energy on earth and the high capital cost of harnessing it.

TABLE 12–3

Comparison of the solar absorptivity α_s of some surfaces with their emissivity ε at room temperature

Surface	α_s	ε
Aluminum		
Polished	0.09	0.03
Anodized	0.14	0.84
Foil	0.15	0.05
Copper		
Polished	0.18	0.03
Tarnished	0.65	0.75
Stainless steel		
Polished	0.37	0.60
Dull	0.50	0.21
Plated metals		
Black nickel oxide	0.92	0.08
Black chrome	0.87	0.09
Concrete	0.60	0.88
White marble	0.46	0.95
Red brick	0.63	0.93
Asphalt	0.90	0.90
Black paint	0.97	0.97
White paint	0.14	0.93
Snow	0.28	0.97
Human skin		
(Caucasian)	0.62	0.97

EXAMPLE 12–5   Selective Absorber and Reflective Surfaces

Consider a surface exposed to solar radiation. At a given time, the direct and diffuse components of solar radiation are $G_D = 400$ and $G_d = 300$ W/m², and the direct radiation makes a 20° angle with the normal of the surface. The surface temperature is observed to be 320 K at that time. Assuming an effective sky temperature of 260 K, determine the net rate of radiation heat transfer for these cases (Fig. 12–45):

(a) $\alpha_s = 0.9$ and $\varepsilon = 0.9$ (gray absorber surface)
(b) $\alpha_s = 0.1$ and $\varepsilon = 0.1$ (gray reflector surface)
(c) $\alpha_s = 0.9$ and $\varepsilon = 0.1$ (selective absorber surface)
(d) $\alpha_s = 0.1$ and $\varepsilon = 0.9$ (selective reflector surface)

SOLUTION A surface is exposed to solar and sky radiation. The net rate of radiation heat transfer is to be determined for four different combinations of emissivities and solar absorptivities.

Analysis The total solar energy incident on the surface is

$$G_{\text{solar}} = G_D \cos \theta + G_d$$
$$= (400 \text{ W/m}^2) \cos 20° + (300 \text{ W/m}^2)$$
$$= 676 \text{ W/m}^2$$

Then the net rate of radiation heat transfer for each of the four cases is determined from:

$$\dot{q}_{\text{net, rad}} = \alpha_s G_{\text{solar}} + \varepsilon \sigma (T_{\text{sky}}^4 - T_s^4)$$

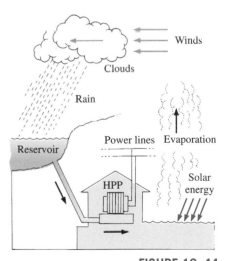

FIGURE 12–44
The cycle that water undergoes in a hydroelectric power plant.

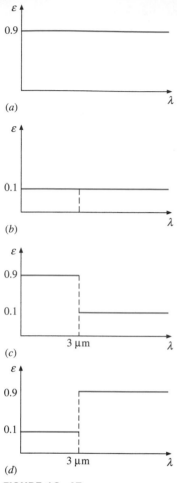

FIGURE 12–45

Graphical representation of the spectral emissivities of the four surfaces considered in Example 12–5.

(a) $\alpha_s = 0.9$ and $\varepsilon = 0.9$ (gray absorber surface):

$$\dot{q}_{net,\,rad} = 0.9(676\ \text{W/m}^2) + 0.9(5.67 \times 10^{-8}\ \text{W/m}^2 \cdot \text{K}^4)[(260\ \text{K})^4 - (320\ \text{K})^4]$$
$$= 307\ \text{W/m}^2$$

(b) $\alpha_s = 0.1$ and $\varepsilon = 0.1$ (gray reflector surface):

$$\dot{q}_{net,\,rad} = 0.1(676\ \text{W/m}^2) + 0.1(5.67 \times 10^{-8}\ \text{W/m}^2 \cdot \text{K}^4)[(260\ \text{K})^4 - (320\ \text{K})^4]$$
$$= 34\ \text{W/m}^2$$

(c) $\alpha_s = 0.9$ and $\varepsilon = 0.1$ (selective absorber surface):

$$\dot{q}_{net,\,rad} = 0.9(676\ \text{W/m}^2) + 0.1(5.67 \times 10^{-8}\ \text{W/m}^2 \cdot \text{K}^4)[(260\ \text{K})^4 - (320\ \text{K})^4]$$
$$= 575\ \text{W/m}^2$$

(d) $\alpha_s = 0.1$ and $\varepsilon = 0.9$ (selective reflector surface):

$$\dot{q}_{net,\,rad} = 0.1(676\ \text{W/m}^2) + 0.9(5.67 \times 10^{-8}\ \text{W/m}^2 \cdot \text{K}^4)[(260\ \text{K})^4 - (320\ \text{K})^4]$$
$$= -234\ \text{W/m}^2$$

Discussion Note that the surface of an ordinary gray material of high absorptivity gains heat at a rate of 307 W/m². The amount of heat gain increases to 575 W/m² when the surface is coated with a selective material that has the same absorptivity for solar radiation but a low emissivity for infrared radiation. Also note that the surface of an ordinary gray material of high reflectivity still gains heat at a rate of 34 W/m². When the surface is coated with a selective material that has the same reflectivity for solar radiation but a high emissivity for infrared radiation, the surface loses 234 W/m² instead. Therefore, the temperature of the surface will decrease when a selective reflector surface is used.

TOPIC OF SPECIAL INTEREST*

Solar Heat Gain Through Windows

The sun is the primary heat source of the earth, and the solar irradiance on a surface normal to the sun's rays beyond the earth's atmosphere at the mean earth–sun distance of 149.5 million km is called the total solar irradiance or solar constant. The accepted value of the solar constant is 1373 W/m² (435.4 Btu/h · ft²), but its value changes by 3.5 percent from a maximum of 1418 W/m² on January 3 when the earth is closest to the sun, to a minimum of 1325 W/m² on July 4 when the earth is farthest away from the sun. The spectral distribution of solar radiation beyond the earth's atmosphere resembles the energy emitted by a blackbody at 5780°C, with about 9 percent of the energy contained in the ultraviolet region (at wavelengths between 0.29 to 0.4 μm), 39 percent in the visible region (0.4 to 0.7 μm), and the remaining 52 percent in the near-infrared region (0.7 to

*This section can be skipped without a loss in continuity.

3.5 μm). The peak radiation occurs at a wavelength of about 0.48 μm, which corresponds to the green color portion of the visible spectrum. Obviously a glazing material that transmits the visible part of the spectrum while absorbing the infrared portion is ideally suited for an application that calls for maximum daylight and minimum solar heat gain. Surprisingly, the ordinary window glass approximates this behavior remarkably well (Fig. 12–46).

Part of the solar radiation entering the earth's atmosphere is scattered and absorbed by air and water vapor molecules, dust particles, and water droplets in the clouds, and thus the solar radiation incident on earth's surface is less than the solar constant. The extent of the attenuation of solar radiation depends on the length of the path of the rays through the atmosphere as well as the composition of the atmosphere (the clouds, dust, humidity, and smog) along the path. Most ultraviolet radiation is absorbed by the ozone in the upper atmosphere. At a solar altitude of 41.8°, the total energy of direct solar radiation incident at sea level on a clear day consists of about 3 percent ultraviolet, 38 percent visible, and 59 percent infrared radiation.

The part of solar radiation that reaches the earth's surface without being scattered or absorbed is the *direct radiation*. Solar radiation that is scattered or reemitted by the constituents of the atmosphere is the *diffuse radiation*. Direct radiation comes directly from the sun following a straight path, whereas diffuse radiation comes from all directions in the sky. The entire radiation reaching the ground on an overcast day is diffuse radiation. The radiation reaching a surface, in general, consists of three components: direct radiation, diffuse radiation, and radiation reflected onto the surface from surrounding surfaces (Fig. 12–47). Common surfaces such as grass, trees, rocks, and concrete reflect about 20 percent of the radiation while absorbing the rest. Snow-covered surfaces, however, reflect 70 percent of the incident radiation. Radiation incident on a surface that does not have a direct view of the sun consists of diffuse and reflected radiation. Therefore, at solar noon, solar radiations incident on the east, west, and north surfaces of a south-facing house are identical since they all consist of diffuse and reflected components. The difference between the radiations incident on the south and north walls in this case gives the magnitude of direct radiation incident on the south wall.

When solar radiation strikes a glass surface, part of it (about 8 percent for uncoated clear glass) is reflected back to outdoors, part of it (5 to 50 percent, depending on composition and thickness) is absorbed within the glass, and the remainder is transmitted indoors, as shown in Fig. 12–48. The conservation of energy principle requires that the sum of the transmitted, reflected, and absorbed solar radiations be equal to the incident solar radiation. That is,

$$\tau_s + \rho_s + \alpha_s = 1$$

where τ_s is the transmissivity, ρ_s is the reflectivity, and α_s is the absorptivity of the glass for solar energy, which are the fractions of incident solar radiation transmitted, reflected, and absorbed, respectively. The standard 3-mm-($\frac{1}{8}$-in)-thick single-pane double-strength clear window glass transmits

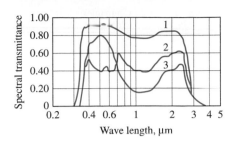

1. 3 mm regular sheet
2. 6 mm gray heat-absorbing plate/float
3. 6 mm green heat-absorbing plate/float

FIGURE 12–46
The variation of the transmittance of typical architectural glass with wavelength (from ASHRAE *Handbook of Fundamentals*, Chap. 27, Fig. 11).

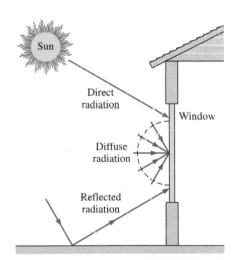

FIGURE 12–47
Direct, diffuse, and reflected components of solar radiation incident on a window.

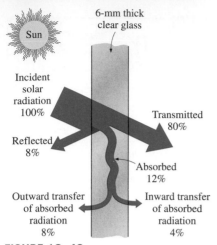

FIGURE 12–48
Distribution of solar radiation incident on a clear glass.

86 percent, reflects 8 percent, and absorbs 6 percent of the solar energy incident on it. The radiation properties of materials are usually given for normal incidence, but can also be used for radiation incident at other angles since the transmissivity, reflectivity, and absorptivity of the glazing materials remain essentially constant for incidence angles up to about 60° from the normal.

The hourly variation of solar radiation incident on the walls and windows of a house is given in Table 12–4. Solar radiation that is transmitted indoors is partially absorbed and partially reflected each time it strikes a surface, but all of it is eventually absorbed as sensible heat by the furniture, walls, people, and so forth. Therefore, the solar energy transmitted inside a building represents a heat gain for the building. Also, the solar radiation absorbed by the glass is subsequently transferred to the indoors and outdoors by convection and radiation. The sum of the *transmitted* solar radiation and the portion of the *absorbed* radiation that flows indoors constitutes the **solar heat gain** of the building.

The fraction of incident solar radiation that enters through the glazing is called the **solar heat gain coefficient** (SHGC) and is expressed as

$$\text{SHGC} = \frac{\text{Solar heat gain through the window}}{\text{Solar radiation incident on the window}}$$

$$= \frac{\dot{q}_{\text{solar, gain}}}{\dot{q}_{\text{solar, incident}}} = \tau_s + f_i \alpha_s \tag{12–55}$$

where α_s is the solar absorptivity of the glass and f_i is the inward flowing fraction of the solar radiation absorbed by the glass. Therefore, the dimensionless quantity SHGC is the sum of the fractions of the directly transmitted (τ_s) and the absorbed and reemitted ($f_i \alpha_s$) portions of solar radiation incident on the window. The value of SHGC ranges from 0 to 1, with 1 corresponding to an opening in the wall (or the ceiling) with no glazing. When the SHGC of a window is known, the total solar heat gain through that window is determined from

$$\dot{Q}_{\text{solar, gain}} = \text{SHGC} \times A_{\text{glazing}} \times \dot{q}_{\text{solar, incident}} \quad (\text{W}) \tag{12–56}$$

where A_{glazing} is the glazing area of the window and $\dot{q}_{\text{solar, incident}}$ is the solar heat flux incident on the outer surface of the window, in W/m².

Another way of characterizing the solar transmission characteristics of different kinds of glazing and shading devices is to compare them to a well-known glazing material that can serve as a base case. This is done by taking the standard 3-mm-($\frac{1}{8}$-in)-thick double-strength clear window glass sheet whose SHGC is 0.87 as the *reference glazing* and defining a **shading coefficient** SC as

$$\text{SC} = \frac{\text{Solar heat gain of product}}{\text{Solar heat gain of reference glazing}}$$

$$= \frac{\text{SHGC}}{\text{SHGC}_{\text{ref}}} = \frac{\text{SHGC}}{0.87} = 1.15 \times \text{SHGC} \tag{12–57}$$

Therefore, the shading coefficient of a single-pane clear glass window is SC = 1.0. The shading coefficients of other commonly used fenestration

TABLE 12–4

Hourly variation of solar radiation incident on various surfaces and the daily totals throughout the year at 40° latitude (from ASHRAE *Handbook of Fundamentals*, Chap. 27, Table 15)

		Solar Radiation Incident on the Surface,* W/m²															
		Solar Time															
Date	Direction of Surface	5	6	7	8	9	10	11	12 noon	13	14	15	16	17	18	19	Daily Total
Jan.	N	0	0	0	20	43	66	68	71	68	66	43	20	0	0	0	446
	NE	0	0	0	63	47	66	68	71	68	59	43	20	0	0	0	489
	E	0	0	0	402	557	448	222	76	68	59	43	20	0	0	0	1863
	SE	0	0	0	483	811	875	803	647	428	185	48	20	0	0	0	4266
	S	0	0	0	271	579	771	884	922	884	771	579	271	0	0	0	5897
	SW	0	0	0	20	48	185	428	647	803	875	811	483	0	0	0	4266
	W	0	0	0	20	43	59	68	76	222	448	557	402	0	0	0	1863
	NW	0	0	0	20	43	59	68	71	68	66	47	63	0	0	0	489
	Horizontal	0	0	0	51	198	348	448	482	448	348	198	51	0	0	0	2568
	Direct	0	0	0	446	753	865	912	926	912	865	753	446	0	0	0	—
Apr.	N	0	41	57	79	97	110	120	122	120	110	97	79	57	41	0	1117
	NE	0	262	508	462	291	134	123	122	120	110	97	77	52	17	0	2347
	E	0	321	728	810	732	552	293	131	120	110	97	77	52	17	0	4006
	SE	0	189	518	682	736	699	582	392	187	116	97	77	52	17	0	4323
	S	0	18	59	149	333	437	528	559	528	437	333	149	59	18	0	3536
	SW	0	17	52	77	97	116	187	392	582	699	736	682	518	189	0	4323
	W	0	17	52	77	97	110	120	392	293	552	732	810	728	321	0	4006
	NW	0	17	52	77	97	110	120	122	123	134	291	462	508	262	0	2347
	Horizontal	0	39	222	447	640	786	880	911	880	786	640	447	222	39	0	6938
	Direct	0	282	651	794	864	901	919	925	919	901	864	794	651	282	0	—
July	N	3	133	109	103	117	126	134	138	134	126	117	103	109	133	3	1621
	NE	8	454	590	540	383	203	144	138	134	126	114	95	71	39	0	3068
	E	7	498	739	782	701	531	294	149	134	126	114	95	71	39	0	4313
	SE	2	248	460	580	617	576	460	291	155	131	114	95	71	39	0	3849
	S	0	39	76	108	190	292	369	395	369	292	190	108	76	39	0	2552
	SW	0	39	71	95	114	131	155	291	460	576	617	580	460	248	2	3849
	W	0	39	71	95	114	126	134	149	294	531	701	782	739	498	7	4313
	NW	0	39	71	95	114	126	134	138	144	203	383	540	590	454	8	3068
	Horizontal	1	115	320	528	702	838	922	949	922	838	702	528	320	115	1	3902
	Direct	7	434	656	762	818	850	866	871	866	850	818	762	656	434	7	—
Oct.	N	0	0	7	40	62	77	87	90	87	77	62	40	7	0	0	453
	NE	0	0	74	178	84	80	87	90	87	87	62	40	7	0	0	869
	E	0	0	163	626	652	505	256	97	87	87	62	40	7	0	0	2578
	SE	0	0	152	680	853	864	770	599	364	137	66	40	7	0	0	4543
	S	0	0	44	321	547	711	813	847	813	711	547	321	44	0	0	5731
	SW	0	0	7	40	66	137	364	599	770	864	853	680	152	0	0	4543
	W	0	0	7	40	62	87	87	97	256	505	652	626	163	0	0	2578
	NW	0	0	7	40	62	87	87	90	87	80	84	178	74	0	0	869
	Horizontal	0	0	14	156	351	509	608	640	608	509	351	156	14	0	0	3917
	Direct	0	0	152	643	811	884	917	927	917	884	811	643	152	0	0	—

*Multiply by 0.3171 to convert to Btu/h · ft².

Values given are for the 21st of the month for average days with no clouds. The values can be up to 15 percent higher at high elevations under very clear skies and up to 30 percent lower at very humid locations with very dusty industrial atmospheres. Daily totals are obtained using Simpson's rule for integration with 10-min time intervals. Solar reflectance of the ground is assumed to be 0.2, which is valid for old concrete, crushed rock, and bright green grass. For a specified location, use solar radiation data obtained for that location. The direction of a surface indicates the direction a vertical surface is facing. For example, W represents the solar radiation incident on a west-facing wall per unit area of the wall.

Solar time may deviate from the local time. Solar noon at a location is the time when the sun is at the highest location (and thus when the shadows are shortest). Solar radiation data are symmetric about the solar noon: the value on a west wall before the solar noon is equal to the value on an east wall two hours after the solar noon.

TABLE 12–5

Shading coefficient SC and solar transmissivity τ_{solar} for some common glass types for summer design conditions (from ASHRAE *Handbook of Fundamentals*, Chap. 27, Table 11).

Type of Glazing	Nominal Thickness		τ_{solar}	SC*
	mm	in		
(a) Single Glazing				
Clear	3	$\frac{1}{8}$	0.86	1.0
	6	$\frac{1}{4}$	0.78	0.95
	10	$\frac{3}{8}$	0.72	0.92
	13	$\frac{1}{2}$	0.67	0.88
Heat absorbing	3	$\frac{1}{8}$	0.64	0.85
	6	$\frac{1}{4}$	0.46	0.73
	10	$\frac{3}{8}$	0.33	0.64
	13	$\frac{1}{2}$	0.24	0.58
(b) Double Glazing				
Clear in,	3[a]	$\frac{1}{8}$	0.71[b]	0.88
clear out	6	$\frac{1}{4}$	0.61	0.82
Clear in, heat absorbing out[c]	6	$\frac{1}{4}$	0.36	0.58

*Multiply by 0.87 to obtain SHGC.

[a]The thickness of each pane of glass.

[b]Combined transmittance for assembled unit.

[c]Refers to gray-, bronze-, and green-tinted heat-absorbing float glass.

products are given in Table 12–5 for summer design conditions. The values for winter design conditions may be slightly lower because of the higher heat transfer coefficients on the outer surface due to high winds and thus higher rate of outward flow of solar heat absorbed by the glazing, but the difference is small.

Note that the larger the shading coefficient, the smaller the shading effect, and thus the larger the amount of solar heat gain. A glazing material with a large shading coefficient allows a large fraction of solar radiation to come in.

Shading devices are classified as *internal shading* and *external shading*, depending on whether the shading device is placed *inside* or *outside*. External shading devices are more effective in reducing the solar heat gain since they intercept the sun's rays before they reach the glazing. The solar heat gain through a window can be reduced by as much as 80 percent by exterior shading. Roof overhangs have long been used for exterior shading of windows. The sun is high in the horizon in summer and low in winter. A properly sized roof overhang or a horizontal projection blocks off the sun's rays completely in summer while letting in most of them in winter, as shown in Figure 12–49. Such shading structures can reduce the solar heat gain on the south, southeast, and southwest windows in the northern hemisphere considerably. A window can also be shaded from outside by vertical or horizontal architectural projections, insect or shading screens, and sun screens. To be effective, air must be able to move freely around the exterior device to carry away the heat absorbed by the shading and the glazing materials.

Some type of internal shading is used in most windows to provide privacy and aesthetic effects as well as some control over solar heat gain. Internal shading devices reduce solar heat gain by reflecting transmitted solar radiation back through the glazing before it can be absorbed and converted into heat in the building.

Draperies reduce the annual heating and cooling loads of a building by 5 to 20 percent, depending on the type and the user habits. In summer, they reduce heat gain primarily by reflecting back direct solar radiation (Fig. 12–50). The semiclosed air space formed by the draperies serves as an additional barrier against heat transfer, resulting in a lower U-factor for the window and thus a lower rate of heat transfer in summer and winter. The solar optical properties of draperies can be measured accurately, or they can be obtained directly from the manufacturers. The shading coefficient of draperies depends on the openness factor, which is the ratio of the open area between the fibers that permits the sun's rays to pass freely, to the total area of the fabric. Tightly woven fabrics allow little direct radiation to pass through, and thus they have a small openness factor. The *reflectance* of the surface of the drapery facing the glazing has a major effect on the amount of solar heat gain. *Light-colored* draperies made of closed or tightly woven fabrics maximize the back reflection and thus minimize the solar gain. *Dark-colored* draperies made of open or semi-open woven fabrics, on the other hand, minimize the back reflection and thus maximize the solar gain.

The shading coefficients of drapes also depend on the way they are hung. Usually, the width of drapery used is twice the width of the draped area to allow folding of the drapes and to give them their characteristic "full" or "wavy" appearance. A flat drape behaves like an ordinary window shade. A flat drape has a higher reflectance and thus a lower shading coefficient than a full drape.

External shading devices such as overhangs and tinted glazings do not require operation, and provide reliable service over a long time without significant degradation during their service life. Their operation does not depend on a person or an automated system, and these passive shading devices are considered fully effective when determining the peak cooling load and the annual energy use. The effectiveness of manually operated shading devices, on the other hand, varies greatly depending on the user habits, and this variation should be considered when evaluating performance.

The primary function of an indoor shading device is to provide *thermal comfort* for the occupants. An unshaded window glass allows most of the incident solar radiation in, and also dissipates part of the solar energy it absorbs by emitting infrared radiation to the room. The emitted radiation and the transmitted direct sunlight may bother the occupants near the window. In winter, the temperature of the glass is lower than the room air temperature, causing excessive heat loss by radiation from the occupants. A shading device allows the control of direct solar and infrared radiation while providing various degrees of privacy and outward vision. The shading device is also at a higher temperature than the glass in winter, and thus reduces radiation loss from occupants. *Glare* from draperies can be minimized by using off-white colors. Indoor shading devices, especially draperies made of a closed-weave fabric, are effective in reducing *sounds* that originate in the room, but they are not as effective against the sounds coming from outside.

The type of climate in an area usually dictates the type of windows to be used in buildings. In *cold climates* where the heating load is much larger than the cooling load, the windows should have the highest transmissivity for the entire solar spectrum, and a high reflectivity (or low emissivity) for the far infrared radiation emitted by the walls and furnishings of the room. Low-e windows are well suited for such heating-dominated buildings. Properly designed and operated windows allow more heat into the building over a heating season than it loses, making them energy contributors rather then energy losers. In *warm climates* where the cooling load is much larger than the heating load, the windows should allow the visible solar radiation (light) in, but should block off the infrared solar radiation. Such windows can reduce the solar heat gain by 60 percent with no appreciable loss in daylighting. This behavior is approximated by window glazings that are coated with a heat-absorbing film outside and a low-e film inside (Fig. 12–51). Properly selected windows can reduce the cooling load by 15 to 30 percent compared to windows with clear glass.

Note that radiation heat transfer between a room and its windows is proportional to the emissivity of the glass surface facing the room, ε_{glass}, and can be expressed as

$$\dot{Q}_{\text{rad, room-window}} = \varepsilon_{glass} A_{glass} \sigma (T_{room}^4 - T_{glass}^4) \qquad (12\text{–}58)$$

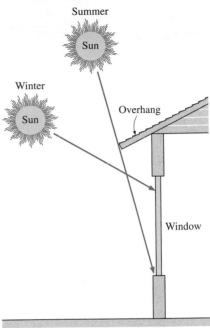

FIGURE 12–49
A properly sized overhang blocks off the sun's rays completely in summer while letting them in in winter.

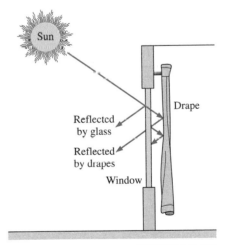

FIGURE 12–50
Draperies reduce heat gain in summer by reflecting back solar radiation, and reduce heat loss in winter by forming an air space before the window.

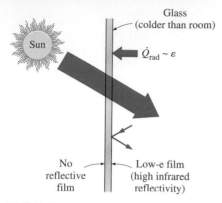

(a) Cold climates

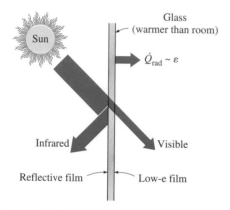

(b) Warm climates

FIGURE 12–51
Radiation heat transfer between a room and its windows is proportional to the emissivity of the glass surface, and low-e coatings on the inner surface of the windows reduce heat loss in winter and heat gain in summer.

Therefore, a low-e interior glass will reduce the heat loss by radiation in winter ($T_{\text{glass}} < T_{\text{room}}$) and heat gain by radiation in summer ($T_{\text{glass}} > T_{\text{room}}$).

Tinted glass and glass coated with reflective films reduce solar heat gain in summer and heat loss in winter. The conductive heat gains or losses can be minimized by using multiple-pane windows. Double-pane windows are usually called for in climates where the winter design temperature is less than 7°C (45°F). Double-pane windows with tinted or reflective films are commonly used in buildings with large window areas. Clear glass is preferred for showrooms since it affords maximum visibility from outside, but bronze-, gray-, and green-colored glass are preferred in office buildings since they provide considerable privacy while reducing glare.

EXAMPLE 12–6   Installing Reflective Films on Windows

A manufacturing facility located at 40° N latitude has a glazing area of 40 m² that consists of double-pane windows made of clear glass (SHGC = 0.766). To reduce the solar heat gain in summer, a reflective film that reduces the SHGC to 0.261 is considered. The cooling season consists of June, July, August, and September, and the heating season October through April. The average daily solar heat fluxes incident on the west side at this latitude are 1.86, 2.66, 3.43, 4.00, 4.36, 5.13, 4.31, 3.93, 3.28, 2.80, 1.84, and 1.54 kWh/day · m² for January through December, respectively. Also, the unit cost of electricity and natural gas are \$0.08/kWh and \$0.50/therm, respectively. If the coefficient of performance of the cooling system is 2.5 and efficiency of the furnace is 0.8, determine the net annual cost savings due to installing reflective coating on the windows. Also, determine the simple payback period if the installation cost of reflective film is \$20/m² (Fig. 11–52).

SOLUTION   The net annual cost savings due to installing reflective film on the west windows of a building and the simple payback period are to be determined.

Assumptions   **1** The calculations given below are for an average year. **2** The unit costs of electricity and natural gas remain constant.

Analysis   Using the daily averages for each month and noting the number of days of each month, the total solar heat flux incident on the glazing during summer and winter months are determined to be

$$Q_{\text{solar, summer}} = 5.13 \times 30 + 4.31 \times 31 + 3.93 \times 31 + 3.28 \times 30 = 508 \text{ kWh/year}$$

$$\begin{aligned} Q_{\text{solar, winter}} &= 2.80 \times 31 + 1.84 \times 30 + 1.54 \times 31 + 1.86 \times 31 \\ &\quad + 2.66 \times 28 + 3.43 \times 31 + 4.00 \times 30 \\ &= 548 \text{ kWh/year} \end{aligned}$$

Then the decrease in the annual cooling load and the increase in the annual heating load due to the reflective film become

$$\begin{aligned} \text{Cooling load decrease} &= Q_{\text{solar, summer}} A_{\text{glazing}} (\text{SHGC}_{\text{without film}} - \text{SHGC}_{\text{with film}}) \\ &= (508 \text{ kWh/year})(40 \text{ m}^2)(0.766 - 0.261) \\ &= 10{,}262 \text{ kWh/year} \end{aligned}$$

Heating load increase $= Q_{\text{solar, winter}} A_{\text{glazing}} (\text{SHGC}_{\text{without film}} - \text{SHGC}_{\text{with film}})$
$= (548 \text{ kWh/year})(40 \text{ m}^2)(0.766 - 0.261)$
$= 11{,}070 \text{ kWh/year} = 377.7 \text{ therms/year}$

since 1 therm $= 29.31$ kWh. The corresponding decrease in cooling costs and the increase in heating costs are

Decrease in cooling costs $=$ (Cooling load decrease)(Unit cost of electricity)/COP
$= (10{,}262 \text{ kWh/year})(\$0.08/\text{kWh})/2.5 = \$328/\text{year}$

Increase in heating costs $=$ (Heating load increase)(Unit cost of fuel)/Efficiency
$= (377.7 \text{ therms/year})(\$0.50/\text{therm})/0.80 = \$236/\text{year}$

Then the net annual cost savings due to the reflective film become

Cost savings $=$ Decrease in cooling costs $-$ Increase in heating costs
$= \$328 - \$236 = \mathbf{\$92/year}$

The implementation cost of installing films is

Implementation cost $= (\$20/\text{m}^2)(40 \text{ m}^2) = \$800$

This gives a simple payback period of

Simple payback period $= \dfrac{\text{Implementation cost}}{\text{Annual cost savings}} = \dfrac{\$800}{\$92/\text{year}} = \mathbf{8.7 \text{ years}}$

Discussion The reflective film will pay for itself in this case in about nine years. This may be unacceptable to most manufacturers since they are not usually interested in any energy conservation measure that does not pay for itself within three years. But the enhancement in thermal comfort and thus the resulting increase in productivity often makes it worthwhile to install reflective film.

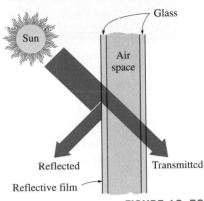

FIGURE 12–52
Schematic for Example 12–6.

SUMMARY

Radiation propagates in the form of electromagnetic waves. The *frequency ν* and *wavelength λ* of electromagnetic waves in a medium are related by $\lambda = c/\nu$, where c is the speed of propagation in that medium. All matter continuously emits *thermal radiation* as a result of vibrational and rotational motions of molecules, atoms, and electrons of a substance.

A *blackbody* is defined as a perfect emitter and absorber of radiation. At a specified temperature and wavelength, no surface can emit more energy than a blackbody. A blackbody absorbs *all* incident radiation, regardless of wavelength and direction. The radiation energy emitted by a blackbody per unit time and per unit surface area is called the *blackbody emissive power E_b* and is expressed by the *Stefan–Boltzmann law* as

$$E_b(T) = \sigma T^4$$

where $\sigma = 5.670 \times 10^{-8} \text{ W/m}^2 \cdot \text{K}^4$ is the *Stefan–Boltzmann constant* and T is the absolute temperature of the surface in K. At any specified temperature, the spectral blackbody emissive power $E_{b\lambda}$ increases with wavelength, reaches a peak, and then decreases with increasing wavelength. The wavelength at which the peak occurs for a specified temperature is given by *Wien's displacement law* as

$$(\lambda T)_{\text{max power}} = 2897.8 \ \mu\text{m} \cdot \text{K}$$

The *blackbody radiation function* f_λ represents the fraction of radiation emitted by a blackbody at temperature T in the wavelength band from $\lambda = 0$ to λ. The fraction of radiation energy emitted by a blackbody at temperature T over a finite wavelength band from $\lambda = \lambda_1$ to $\lambda = \lambda_2$ is determined from

$$f_{\lambda_1-\lambda_2}(T) = f_{\lambda_2}(T) - f_{\lambda_1}(T)$$

where $f_{\lambda_1}(T)$ and $f_{\lambda_2}(T)$ are the blackbody radiation functions corresponding to $\lambda_1 T$ and $\lambda_2 T$, respectively.

The magnitude of a viewing angle in space is described by solid angle expressed as $d\omega = dA_n/r^2$. The *radiation intensity* for emitted radiation $I_e(\theta, \phi)$ is defined as the rate at which radiation energy is emitted in the (θ, ϕ) direction per unit area normal to this direction and per unit solid angle about this direction. The *radiation flux* for emitted radiation is the *emissive power E*, and is expressed as

$$E = \int_{\text{hemisphere}} dE = \int_{\phi=0}^{2\pi} \int_{\theta=0}^{\pi/2} I_e(\theta, \phi) \cos\theta \sin\theta \, d\theta \, d\phi$$

For a *diffusely emitting* surface, intensity is independent of direction and thus

$$E = \pi I_e$$

For a blackbody, we have

$$E_b = \pi I_b \quad \text{and} \quad I_b(T) = \frac{E_b(T)}{\pi} = \frac{\sigma T^4}{\pi}$$

The radiation flux incident on a surface from all directions is *irradiation G,* and for diffusely incident radiation of intensity I_i it is expressed as

$$G = \pi I_i$$

The rate at which radiation energy leaves a unit area of a surface in all directions is *radiosity J,* and for a surface that is both a diffuse emitter and a diffuse reflector it is expressed as

$$J = \pi I_{e+r}$$

where I_{e+r} is the sum of the emitted and reflected intensities. The spectral emitted quantities are related to total quantities as

$$I_e = \int_0^\infty I_{\lambda,e} \, d\lambda \quad \text{and} \quad E = \int_0^\infty E_\lambda \, d\lambda$$

They reduce for a diffusely emitting surface and for a blackbody to

$$E_\lambda = \pi I_{\lambda,e} \quad \text{and} \quad E_{b\lambda}(\lambda, T) = \pi I_{b\lambda}(\lambda, T)$$

The *emissivity* of a surface represents the ratio of the radiation emitted by the surface at a given temperature to the radiation emitted by a blackbody at the same temperature. Different emissivities are defined as

Spectral directional emissivity:

$$\varepsilon_{\lambda,\theta}(\lambda, \theta, \phi, T) = \frac{I_{\lambda,e}(\lambda, \theta, \phi, T)}{I_{b\lambda}(\lambda, T)}$$

Total directional emissivity:

$$\varepsilon_\theta(\theta, \phi, T) = \frac{I_e(\theta, \phi, T)}{I_b(T)}$$

Spectral hemispherical emissivity:

$$\varepsilon_\lambda(\lambda, T) = \frac{E_\lambda(\lambda, T)}{E_{b\lambda}(\lambda, T)}$$

Total hemispherical emissivity:

$$\varepsilon(T) = \frac{E(T)}{E_b(T)} = \frac{\int_0^\infty \varepsilon_\lambda(\lambda, T)E_{b\lambda}(\lambda, T) \, d\lambda}{\sigma T^4}$$

Emissivity can also be expressed as a step function by dividing the spectrum into a sufficient number of *wavelength bands* of constant emissivity as, for example,

$$\varepsilon(T) = \varepsilon_1 f_{0-\lambda_1}(T) + \varepsilon_2 f_{\lambda_1-\lambda_2}(T) + \varepsilon_3 f_{\lambda_2-\infty}(T)$$

The *total hemispherical emissivity* ε of a surface is the average emissivity over all directions and wavelengths.

When radiation strikes a surface, part of it is absorbed, part of it is reflected, and the remaining part, if any, is transmitted. The fraction of incident radiation (intensity I_i or irradiation G) absorbed by the surface is called the *absorptivity,* the fraction reflected by the surface is called the *reflectivity,* and the fraction transmitted is called the *transmissivity.* Various absorptivities, reflectivities, and transmissivities for a medium are expressed as

$$\alpha_{\lambda,\theta}(\lambda, \theta, \phi) = \frac{I_{\lambda,\text{abs}}(\lambda, \theta, \phi)}{I_{\lambda,i}(\lambda, \theta, \phi)} \text{ and } \rho_{\lambda,\theta}(\lambda, \theta, \phi) = \frac{I_{\lambda,\text{ref}}(\lambda, \theta, \phi)}{I_{\lambda,i}(\lambda, \theta, \phi)}$$

$$\alpha_\lambda(\lambda) = \frac{G_{\lambda,\text{abs}}(\lambda)}{G_\lambda(\lambda)}, \quad \rho_\lambda(\lambda) = \frac{G_{\lambda,\text{ref}}(\lambda)}{G_\lambda(\lambda)}, \quad \text{and} \quad \tau_\lambda(\lambda) = \frac{G_{\lambda,\text{tr}}(\lambda)}{G_\lambda(\lambda)}$$

$$\alpha = \frac{G_{\text{abs}}}{G}, \quad \rho = \frac{G_{\text{ref}}}{G}, \quad \text{and} \quad \tau = \frac{G_{\text{tr}}}{G}$$

The consideration of wavelength and direction dependence of properties makes radiation calculations very complicated. Therefore, the *gray* and *diffuse* approximations are commonly utilized in radiation calculations. A surface is said to be *diffuse* if its properties are independent of direction and *gray* if its properties are independent of wavelength.

The sum of the absorbed, reflected, and transmitted fractions of radiation energy must be equal to unity,

$$\alpha + \rho + \tau = 1$$

For *opaque* surfaces, $\tau = 0$, and thus

$$\alpha + \rho = 1$$

Surfaces are usually assumed to reflect in a perfectly *specular* or *diffuse* manner for simplicity. In *specular* (or *mirrorlike*) *reflection,* the angle of reflection equals the angle of incidence of the radiation beam. In *diffuse reflection,* radiation is reflected equally in all directions. Reflection from smooth and polished surfaces approximates specular reflection, whereas reflection from rough surfaces approximates diffuse reflection. *Kirchhoff's law* of radiation is expressed as

$$\varepsilon_{\lambda, \theta}(T) = \alpha_{\lambda, \theta}(T), \quad \varepsilon_\lambda(T) = \alpha_\lambda(T), \quad \text{and} \quad \varepsilon(T) = \alpha(T)$$

Gas molecules and the suspended particles in the atmosphere emit radiation as well as absorbing it. The atmosphere can be treated as a blackbody at some lower fictitious temperature, called the *effective sky temperature* T_{sky} that emits an equivalent amount of radiation energy,

$$G_{\text{sky}} = \sigma T_{\text{sky}}^4$$

The net rate of radiation heat transfer to a surface exposed to solar and atmospheric radiation is determined from an energy balance expressed as

$$\dot{q}_{\text{net, rad}} = \alpha_s G_{\text{solar}} + \varepsilon\sigma(T_{\text{sky}}^4 - T_s^4)$$

where T_s is the surface temperature in K, and ε is the surface emissivity at room temperature.

REFERENCES AND SUGGESTED READINGS

1. American Society of Heating, Refrigeration, and Air Conditioning Engineers, *Handbook of Fundamentals,* Atlanta, ASHRAE, 1993.

2. A. G. H. Dietz. "Diathermanous Materials and Properties of Surfaces." In *Space Heating with Solar Energy,* ed. R. W. Hamilton. Cambridge, MA: MIT Press, 1954.

3. J. A. Duffy and W. A. Beckman. *Solar Energy Thermal Process.* New York: John Wiley & Sons, 1974.

4. H. C. Hottel. "Radiant Heat Transmission." In *Heat Transmission.* 3rd ed., ed. W. H. McAdams. New York: McGraw-Hill, 1954.

5. M. F. Modest. *Radiative Heat Transfer.* New York: McGraw-Hill, 1993.

6. M. Planck. *The Theory of Heat Radiation.* New York: Dover, 1959.

7. W. Sieber. *Zeitschrift für Technische Physics* 22 (1941), pp. 130–135.

8. R. Siegel and J. R. Howell. *Thermal Radiation Heat Transfer.* 3rd ed. Washington, DC: Hemisphere, 1992.

9. Y. S. Touloukain and D. P. DeWitt. "Nonmetallic Solids." In *Thermal Radiative Properties.* Vol. 8. New York: IFI/Plenum, 1970.

10. Y. S. Touloukian and D. P. DeWitt. "Metallic Elements and Alloys." In *Thermal Radiative Properties,* Vol. 7. New York: IFI/Plenum, 1970.

PROBLEMS*

Electromagnetic and Thermal Radiation

12–1C What is an electromagnetic wave? How does it differ from a sound wave?

*Problems designated by a "C" are concept questions, and students are encouraged to answer them all. Problems designated by an "E" are in English units, and the SI users can ignore them. Problems with the icon ⊕ are solved using EES, and complete solutions together with parametric studies are included on the enclosed CD. Problems with the icon ▨ are comprehensive in nature, and are intended to be solved with a computer, preferably using the EES software that accompanies this text.

12–2C By what properties is an electromagnetic wave characterized? How are these properties related to each other?

12–3C What is visible light? How does it differ from the other forms of electromagnetic radiation?

12–4C How do ultraviolet and infrared radiation differ? Do you think your body emits any radiation in the ultraviolet range? Explain.

12–5C What is thermal radiation? How does it differ from the other forms of electromagnetic radiation?

12–6C What is the cause of color? Why do some objects appear blue to the eye while others appear red? Is the color of a surface at room temperature related to the radiation it emits?

12–7C Why is radiation usually treated as a surface phenomenon?

12–8C Why do skiers get sunburned so easily?

12–9C How does microwave cooking differ from conventional cooking?

12–10 The speed of light in vacuum is given to be 3.0×10^8 m/s. Determine the speed of light in air ($n = 1$), in water ($n = 1.33$), and in glass ($n = 1.5$).

12–11 Electricity is generated and transmitted in power lines at a frequency of 60 Hz (1 Hz = 1 cycle per second). Determine the wavelength of the electromagnetic waves generated by the passage of electricity in power lines.

12–12 A microwave oven is designed to operate at a frequency of 2.2×10^9 Hz. Determine the wavelength of these microwaves and the energy of each microwave.

12–13 A radio station is broadcasting radio waves at a wavelength of 200 m. Determine the frequency of these waves.

Answer: 1.5×10^6 Hz

12–14 A cordless telephone is designed to operate at a frequency of 8.5×10^8 Hz. Determine the wavelength of these telephone waves.

Blackbody Radiation

12–15C What is a blackbody? Does a blackbody actually exist?

12–16C Define the total and spectral blackbody emissive powers. How are they related to each other? How do they differ?

12–17C Why did we define the blackbody radiation function? What does it represent? For what is it used?

12–18C Consider two identical bodies, one at 1000 K and the other at 1500 K. Which body emits more radiation in the shorter-wavelength region? Which body emits more radiation at a wavelength of 20 μm?

12–19 Consider a surface at a uniform temperature of 800 K. Determine the maximum rate of thermal radiation that can be emitted by this surface, in W/m^2.

12–20 Consider a 20-cm $\times$ 20-cm $\times$ 20-cm cubical body at 750 K suspended in the air. Assuming the body closely approximates a blackbody, determine (a) the rate at which the cube emits radiation energy, in W and (b) the spectral blackbody emissive power at a wavelength of 4 μm.

12–21E The sun can be treated as a blackbody at an effective surface temperature of 10,400 R. Determine the rate at which infrared radiation energy ($\lambda = 0.76$–100 μm) is emitted by the sun, in Btu/h · ft^2.

12–22 [EES] The sun can be treated as a blackbody at 5780 K. Using EES (or other) software, calculate and plot the spectral blackbody emissive power $E_{b\lambda}$ of the sun versus wavelength in the range of 0.01 μm to 1000 μm. Discuss the results.

12–23 The temperature of the filament of an incandescent lightbulb is 3200 K. Treating the filament as a blackbody, determine the fraction of the radiant energy emitted by the filament that falls in the visible range. Also, determine the wavelength at which the emission of radiation from the filament peaks.

12–24 [EES] Reconsider Prob. 12–23. Using EES (or other) software, investigate the effect of temperature on the fraction of radiation emitted in the visible range. Let the surface temperature vary from 1000 K to 4000 K, and plot fraction of radiation emitted in the visible range versus the surface temperature.

12–25 An incandescent lightbulb is desired to emit at least 15 percent of its energy at wavelengths shorter than 0.8 μm. Determine the minimum temperature to which the filament of the lightbulb must be heated.

12–26 It is desired that the radiation energy emitted by a light source reach a maximum in the blue range ($\lambda = 0.47$ μm). Determine the temperature of this light source and the fraction of radiation it emits in the visible range ($\lambda = 0.40$–0.76 μm).

12–27 A 3-mm-thick glass window transmits 90 percent of the radiation between $\lambda = 0.3$ and 3.0 μm and is essentially opaque for radiation at other wavelengths. Determine the rate of radiation transmitted through a 2-m $\times$ 2-m glass window from blackbody sources at (a) 5800 K and (b) 1000 K.

Answers: (a) 218,400 kW, (b) 55.8 kW

Radiation Intensity

12–28C What does a solid angle represent, and how does it differ from a plane angle? What is the value of a solid angle associated with a sphere?

12–29C How is the intensity of emitted radiation defined? For a diffusely emitting surface, how is the emissive power related to the intensity of emitted radiation?

12–30C For a surface, how is irradiation defined? For diffusely incident radiation, how is irradiation on a surface related to the intensity of incident radiation?

12–31C For a surface, how is radiosity defined? For diffusely emitting and reflecting surfaces, how is radiosity related to the intensities of emitted and reflected radiation?

12–32C When the variation of spectral radiation quantity with wavelength is known, how is the corresponding total quantity determined?

12–33 A small surface of area $A_1 = 8$ cm^2 emits radiation as a blackbody at $T_1 = 800$ K. Part of the radiation emitted by A_1

strikes another small surface of area $A_2 = 8$ cm² oriented as shown in the figure. Determine the solid angle subtended by A_2 when viewed from A_1, and the rate at which radiation emitted by A_1 strikes A_2 directly. What would your answer be if A_2 were directly above A_1 at a distance of 80 cm?

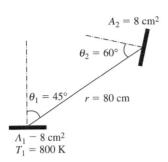

FIGURE P12–33

12–34 A small circular surface of area $A_1 = 2$ cm² located at the center of a 2-m-diameter sphere emits radiation as a blackbody at $T_1 = 1000$ K. Determine the rate at which radiation energy is streaming through a $D_2 = 1$-cm-diameter hole located (a) on top of the sphere directly above A_1 and (b) on the side of sphere such that the line that connects the centers of A_1 and A_2 makes 45° with surface A_1.

12–35 Repeat Prob. 12–34 for a 4-m-diameter sphere.

12–36 A small surface of area $A = 1$ cm² emits radiation as a blackbody at 1800 K. Determine the rate at which radiation energy is emitted through a band defined by $0 \le \phi \le 2\pi$ and $45 \le \theta \le 60°$, where θ is the angle a radiation beam makes with the normal of the surface and ϕ is the azimuth angle.

12–37 A small surface of area $A = 1$ cm² is subjected to incident radiation of constant intensity $I_i = 2.2 \times 10^4$ W/m² · sr over the entire hemisphere. Determine the rate at which radiation energy is incident on the surface through (a) $0 \le \theta \le 45°$ and (b) $45 \le \theta \le 90°$, where θ is the angle a radiation beam makes with the normal of the surface.

Radiation Properties

12–38C Define the properties emissivity and absorptivity. When are these two properties equal to each other?

12–39C Define the properties reflectivity and transmissivity and discuss the different forms of reflection.

12–40C What is a graybody? How does it differ from a blackbody? What is a diffuse gray surface?

12–41C What is the greenhouse effect? Why is it a matter of great concern among atmospheric scientists?

12–42C We can see the inside of a microwave oven during operation through its glass door, which indicates that visible radiation is escaping the oven. Do you think that the harmful microwave radiation might also be escaping?

12–43 The spectral emissivity function of an opaque surface at 1000 K is approximated as

$$\varepsilon_\lambda = \begin{cases} \varepsilon_1 = 0.4, & 0 \le \lambda < 2 \ \mu m \\ \varepsilon_2 = 0.7, & 2 \ \mu m \le \lambda < 6 \ \mu m \\ \varepsilon_3 = 0.3, & 6 \ \mu m \le \lambda < \infty \end{cases}$$

Determine the average emissivity of the surface and the rate of radiation emission from the surface, in W/m².
Answers: 0.575, 32.6 kW/m²

12–44 The reflectivity of aluminum coated with lead sulfate is 0.35 for radiation at wavelengths less than 3 μm and 0.95 for radiation greater than 3 μm. Determine the average reflectivity of this surface for solar radiation ($T \approx 5800$ K) and radiation coming from surfaces at room temperature ($T \approx 300$ K). Also, determine the emissivity and absorptivity of this surface at both temperatures. Do you think this material is suitable for use in solar collectors?

12–45 A furnace that has a 40-cm × 40-cm glass window can be considered to be a blackbody at 1200 K. If the transmissivity of the glass is 0.7 for radiation at wavelengths less than 3 μm and zero for radiation at wavelengths greater than 3 μm, determine the fraction and the rate of radiation coming from the furnace and transmitted through the window.

12–46 The emissivity of a tungsten filament can be approximated to be 0.5 for radiation at wavelengths less than 1 μm and 0.15 for radiation at greater than 1 μm. Determine the average emissivity of the filament at (a) 2000 K and (b) 3000 K. Also, determine the absorptivity and reflectivity of the filament at both temperatures.

12–47 The variations of the spectral emissivity of two surfaces are as given in Fig. P12–47. Determine the average emissivity of each surface at $T = 3000$ K. Also, determine the average absorptivity and reflectivity of each surface for radiation coming from a source at 3000 K. Which surface is more suitable to serve as a solar absorber?

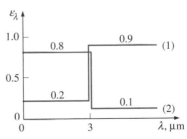

FIGURE P12–47

12–48 The emissivity of a surface coated with aluminum oxide can be approximated to be 0.15 for radiation at wavelengths less than 5 μm and 0.9 for radiation at wavelengths greater than 5 μm. Determine the average emissivity of this surface at (a) 5800 K and (b) 300 K. What can you say about the absorptivity of this surface for radiation coming from sources at 5800 K and 300 K?   *Answers:* (a) 0.154, (b) 0.89

12–49   The variation of the spectral absorptivity of a surface is as given in Fig. P12–49. Determine the average absorptivity and reflectivity of the surface for radiation that originates from a source at $T = 2500$ K. Also, determine the average emissivity of this surface at 3000 K.

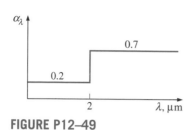

FIGURE P12–49

12–50E   A 5-in-diameter spherical ball is known to emit radiation at a rate of 550 Btu/h when its surface temperature is 950 R. Determine the average emissivity of the ball at this temperature.

12–51   The variation of the spectral transmissivity of a 0.6-cm-thick glass window is as given in Fig. P12–51. Determine the average transmissivity of this window for solar radiation ($T \approx 5800$ K) and radiation coming from surfaces at room temperature ($T \approx 300$ K). Also, determine the amount of solar radiation transmitted through the window for incident solar radiation of 650 W/m².
   Answers: 0.870, 0.00016, 566 W/m²

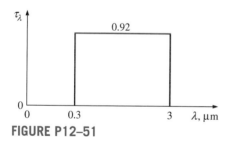

FIGURE P12–51

Atmospheric and Solar Radiation

12–52C   What is the solar constant? How is it used to determine the effective surface temperature of the sun? How would the value of the solar constant change if the distance between the earth and the sun doubled?

12–53C   What changes would you notice if the sun emitted radiation at an effective temperature of 2000 K instead of 5762 K?

12–54C   Explain why the sky is blue and the sunset is yellow-orange.

12–55C   When the earth is closest to the sun, we have winter in the northern hemisphere. Explain why. Also explain why we have summer in the northern hemisphere when the earth is farthest away from the sun.

12–56C   What is the effective sky temperature?

12–57C   You have probably noticed warning signs on the highways stating that bridges may be icy even when the roads are not. Explain how this can happen.

12–58C   Unless you live in a warm southern state, you have probably had to scrape ice from the windshield and windows of your car many mornings. You may have noticed, with frustration, that the thickest layer of ice always forms on the windshield instead of the side windows. Explain why this is the case.

12–59C   Explain why surfaces usually have quite different absorptivities for solar radiation and for radiation originating from the surrounding bodies.

12–60   A surface has an absorptivity of $\alpha_s = 0.85$ for solar radiation and an emissivity of $\varepsilon = 0.5$ at room temperature. The surface temperature is observed to be 350 K when the direct and the diffuse components of solar radiation are $G_D = 350$ and $G_d = 400$ W/m², respectively, and the direct radiation makes a 30° angle with the normal of the surface. Taking the effective sky temperature to be 280 K, determine the net rate of radiation heat transfer to the surface at that time.

12–61E   Solar radiation is incident on the outer surface of a spaceship at a rate of 400 Btu/h · ft². The surface has an absorptivity of $\alpha_s = 0.10$ for solar radiation and an emissivity of $\varepsilon = 0.6$ at room temperature. The outer surface radiates heat into space at 0 R. If there is no net heat transfer into the spaceship, determine the equilibrium temperature of the surface.
   Answer: 444 R

12–62   The air temperature on a clear night is observed to remain at about 4°C. Yet water is reported to have frozen that night due to radiation effect. Taking the convection heat transfer coefficient to be 18 W/m² · °C, determine the value of the maximum effective sky temperature that night.

12–63   The absorber surface of a solar collector is made of aluminum coated with black chrome ($\alpha_s = 0.87$ and $\varepsilon = 0.09$). Solar radiation is incident on the surface at a rate of 600 W/m². The air and the effective sky temperatures are 25°C and 15°C, respectively, and the convection heat transfer coefficient is 10 W/m² · °C. For an absorber surface temperature of 70°C, determine the net rate of solar energy delivered by the absorber plate to the water circulating behind it.

12–64   Reconsider Prob. 12–63. Using EES (or other) software, plot the net rate of solar energy transferred to water as a function of the absorptivity of the absorber plate. Let the absorptivity vary from 0.5 to 1.0, and discuss the results.

12–65   Determine the equilibrium temperature of the absorber surface in Prob. 12–63 if the back side of the absorber is insulated.

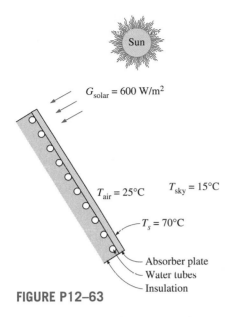

FIGURE P12–63

Special Topic: Solar Heat Gain through Windows

12–66C What fraction of the solar energy is in the visible range (a) outside the earth's atmosphere and (b) at sea level on earth? Answer the same question for infrared radiation.

12–67C Describe the solar radiation properties of a window that is ideally suited for minimizing the air-conditioning load.

12–68C Define the SHGC (solar heat gain coefficient), and explain how it differs from the SC (shading coefficient). What are the values of the SHGC and SC of a single-pane clear-glass window?

12–69C What does the SC (shading coefficient) of a device represent? How do the SCs of clear glass and heat-absorbing glass compare?

12–70C What is a shading device? Is an internal or external shading device more effective in reducing the solar heat gain through a window? How does the color of the surface of a shading device facing outside affect the solar heat gain?

12–71C What is the effect of a low-e coating on the inner surface of a window glass on the (a) heat loss in winter and (b) heat gain in summer through the window?

12–72C What is the effect of a reflective coating on the outer surface of a window glass on the (a) heat loss in winter and (b) heat gain in summer through the window?

12–73 A manufacturing facility located at 32° N latitude has a glazing area of 60 m^2 facing west that consists of double-pane windows made of clear glass (SHGC = 0.766). To reduce the solar heat gain in summer, a reflective film that will reduce the SHGC to 0.35 is considered. The cooling season consists of June, July, August, and September, and the heating season, October through April. The average daily solar heat fluxes incident on the west side at this latitude are 2.35, 3.03, 3.62, 4.00, 4.20, 4.24, 4.16, 3.93, 3.48, 2.94, 2.33, and 2.07

kWh/day · m^2 for January through December, respectively. Also, the unit costs of electricity and natural gas are $0.09/kWh and $0.45/therm, respectively. If the coefficient of performance of the cooling system is 3.2 and the efficiency of the furnace is 0.90, determine the net annual cost savings due to installing reflective coating on the windows. Also, determine the simple payback period if the installation cost of reflective film is $20/$m^2$. *Answers:* $76, 16 years

12–74 A house located in Boulder, Colorado (40° N latitude), has ordinary double-pane windows with 6-mm-thick glasses and the total window areas are 8, 6, 6, and 4 m^2 on the south, west, east, and north walls, respectively. Determine the total solar heat gain of the house at 9:00, 12:00, and 15:00 solar time in July. Also, determine the total amount of solar heat gain per day for an average day in January.

12–75 Repeat Prob. 11–74 for double-pane windows that are gray-tinted.

12–76 Consider a building in New York (40° N latitude) that has 130 m^2 of window area on its south wall. The windows are double-pane heat-absorbing type, and are equipped with light-colored venetian blinds with a shading coefficient of SC = 0.30. Determine the total solar heat gain of the building through the south windows at solar noon in April. What would your answer be if there were no blinds at the windows?

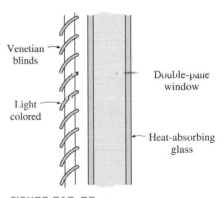

FIGURE P12–76

12–77 A typical winter day in Reno, Nevada (39° N latitude), is cold but sunny, and thus the solar heat gain through the windows can be more than the heat loss through them during daytime. Consider a house with double-door-type windows that are double paned with 3-mm-thick glasses and 6.4 mm of air space and have aluminum frames and spacers. The house is maintained at 22°C at all times. Determine if the house is losing more or less heat than it is gaining from the sun through an east window on a typical day in January for a 24-h period if the average outdoor temperature is 10°C. *Answer:* less

12–78 Repeat Prob. 12–77 for a south window.

12–79E Determine the rate of net heat gain (or loss) through a 9-ft-high, 15-ft-wide, fixed $\frac{1}{8}$-in single-glass window with

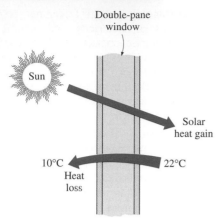

FIGURE P12–77

aluminum frames on the west wall at 3 PM solar time during a typical day in January at a location near 40° N latitude when the indoor and outdoor temperatures are 70°F and 20°F, respectively. *Answer: 12,890 Btu/h gain*

12–80 Consider a building located near 40° N latitude that has equal window areas on all four sides. The building owner is considering coating the south-facing windows with reflective film to reduce the solar heat gain and thus the cooling load. But someone suggests that the owner will reduce the cooling load even more if she coats the west-facing windows instead. What do you think?

Review Problems

12–81 The spectral emissivity of an opaque surface at 1500 K is approximated as

$$\varepsilon_1 = 0 \quad \text{for} \quad \lambda < 2 \ \mu\text{m}$$
$$\varepsilon_2 = 0.85 \quad \text{for} \quad 2 \leq \lambda \leq 6 \ \mu\text{m}$$
$$\varepsilon_3 = 0 \quad \text{for} \quad \lambda > 6 \ \mu\text{m}$$

Determine the total emissivity and the emissive flux of the surface.

12–82 The spectral transmissivity of a 3-mm-thick regular glass can be expressed as

$$\tau_1 = 0 \quad \text{for} \quad \lambda < 0.35 \ \mu\text{m}$$
$$\tau_2 = 0.85 \quad \text{for} \quad 0.35 < \lambda < 2.5 \ \mu\text{m}$$
$$\tau_3 = 0 \quad \text{for} \quad \lambda > 2.5 \ \mu\text{m}$$

Determine the transmissivity of this glass for solar radiation. What is the transmissivity of this glass for light?

12–83 A 1-m-diameter spherical cavity is maintained at a uniform temperature of 600 K. Now a 5-mm-diameter hole is drilled. Determine the maximum rate of radiation energy streaming through the hole. What would your answer be if the diameter of the cavity were 3 m?

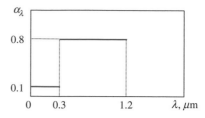

FIGURE P12–84

12–84 The spectral absorptivity of an opaque surface is as shown on the graph. Determine the absorptivity of the surface for radiation emitted by a source at (*a*) 1000 K and (*b*) 3000 K.

12–85 The surface in Prob. 12–84 receives solar radiation at a rate of 470 W/m². Determine the solar absorptivity of the surface and the rate of absorption of solar radiation.

12–86 The spectral transmissivity of a glass cover used in a solar collector is given as

$$\tau_1 = 0 \quad \text{for} \quad \lambda < 0.3 \ \mu\text{m}$$
$$\tau_2 = 0.9 \quad \text{for} \quad 0.3 < \lambda < 3 \ \mu\text{m}$$
$$\tau_3 = 0 \quad \text{for} \quad \lambda > 3 \ \mu\text{m}$$

Solar radiation is incident at a rate of 950 W/m², and the absorber plate, which can be considered to be black, is maintained at 340 K by the cooling water. Determine (*a*) the solar flux incident on the absorber plate, (*b*) the transmissivity of the glass cover for radiation emitted by the absorber plate, and (*c*) the rate of heat transfer to the cooling water if the glass cover temperature is also 340 K.

12–87 Consider a small black surface of area $A = 3.5$ cm² maintained at 600 K. Determine the rate at which radiation energy is emitted by the surface through a ring-shaped opening defined by $0 \leq \phi \leq 2\pi$ and $40 \leq \theta \leq 50°$, where ϕ is the azimuth angle and θ is the angle a radiation beam makes with the normal of the surface.

12–88 Solar radiation is incident on the front surface of a thin plate with direct and diffuse components of 300 and 250 W/m², respectively. The direct radiation makes a 30° angle with the normal of the surface. The plate surfaces have a solar absorptivity of 0.63 and an emissivity of 0.93. The air temperature is 5°C and the convection heat transfer coefficient is 20 W/m² · °C.

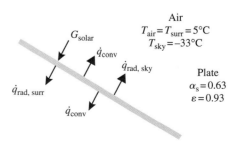

FIGURE P12–88

The effective sky temperature for the front surface is $-33°C$ while the surrounding surfaces are at $5°C$ for the back surface. Determine the equilibrium temperature of the plate.

Fundamentals of Engineering (FE) Exam Problems

12–89 Consider a surface at $-5°C$ in an environment at $25°C$. The maximum rate of heat that can be emitted from this surface by radiation is

(a) 0 W/m^2 (b) 155 W/m^2 (c) 293 W/m^2
(d) 354 W/m^2 (e) 567 W/m^2

12–90 The wavelength at which the blackbody emissive power reaches its maximum value at 300 K is

(a) 5.1 μm   (b) 9.7 μm   (c) 15.5 μm
(d) 38.0 μm   (e) 73.1 μm

12–91 Consider a surface at 500 K. The spectral blackbody emissive power at a wavelength of 50 μm is

(a) 1.54 W/m^2 · μm   (b) 26.3 W/m^2 · μm
(c) 108.4 W/m^2 · μm   (d) 2750 W/m^2 · μm
(e) 8392 W/m^2 · μm

12–92 A surface absorbs 10 percent of radiation at wavelengths less than 3 μm and 50 percent of radiation at wavelengths greater than 3 μm. The average absorptivity of this surface for radiation emitted by a source at 3000 K is

(a) 0.14 (b) 0.22 (c) 0.30 (d) 0.38 (e) 0.42

12–93 Consider a 4-cm-diameter and 6-cm-long cylindrical rod at 1000 K. If the emissivity of the rod surface is 0.75, the total amount of radiation emitted by all surfaces of the rod in 20 min is

(a) 43 kJ   (b) 385 kJ   (c) 434 kJ
(d) 513 kJ   (e) 684 kJ

12 94 Solar radiation is incident on a semi-transparent body at a rate of 500 W/m^2. If 150 W/m^2 of this incident radiation is reflected back and 225 W/m^2 is transmitted across the body, the absorptivity of the body is

(a) 0 (b) 0.25 (c) 0.30 (d) 0.45 (e) 1

12–95 Solar radiations is incident on an opaque surface at a rate of 400 W/m^2. The emissivity of the surface is 0.65 and the absorptivity to solar radiation is 0.85. The convection coefficient between the surface and the environment at 25°C is 6 W/m^2 · °C. If the surface is exposed to atmosphere with an effective sky temperature of 250 K, the equilibrium temperature of the surface is

(a) 281 K   (b) 298 K   (c) 303 K
(d) 317 K   (e) 339 K

12–96 A surface is exposed to solar radiation. The direct and diffuse components of solar radiation are 350 and 250 W/m^2, and the direct radiation makes a 35° angle with the normal of the surface. The solar absorptivity and the emissivity of the surface are 0.24 and 0.41, respectively. If the surface is observed to be at 315 K and the effective sky temperature is 256 K, the net rate of radiation heat transfer to the surface is

(a) -129 W/m^2 (b) -44 W/m^2 (c) 0 W/m^2
(d) 129 W/m^2 (e) 537 W/m^2

12–97 A surface at 300°C has an emissivity of 0.7 in the wavelength range of 0–4.4 μm and 0.3 over the rest of the wavelength range. At a temperature of 300°C, 19 percent of the blackbody emissive power is in wavelength range up to 4.4 μm. The total emissivity of this surface is

(a) 0.300 (b) 0.376 (c) 0.624
(d) 0.70 (e) 0.50

Design and Essay Problems

12–98 Write an essay on the radiation properties of selective surfaces used on the absorber plates of solar collectors. Find out about the various kinds of such surfaces, and discuss the performance and cost of each type. Recommend a selective surface that optimizes cost and performance.

12–99 According to an Atomic Energy Commission report, a hydrogen bomb can be approximated as a large fireball at a temperature of 7200 K. You are to assess the impact if such a bomb exploded 5 km above a city. Assume the diameter of the fireball to be 1 km, and the blast to last 15 s. Investigate the level of radiation energy people, plants, and houses will be exposed to, and how adversely they will be affected by the blast.

RADIATION HEAT TRANSFER

In Chapter 12, we considered the fundamental aspects of radiation and the radiation properties of surfaces. We are now in a position to consider radiation exchange between two or more surfaces, which is the primary quantity of interest in most radiation problems.

We start this chapter with a discussion of view factors and the rules associated with them. View factor expressions and charts for some common configurations are given, and the crossed-strings method is presented. We then discuss radiation heat transfer, first between black surfaces and then between nonblack surfaces using the radiation network approach. We continue with radiation shields and discuss the radiation effect on temperature measurements and comfort. Finally, we consider gas radiation, and discuss the effective emissivities and absorptivities of gas bodies of various shapes. We also discuss radiation exchange between the walls of combustion chambers and the high-temperature emitting and absorbing combustion gases inside.

OBJECTIVES

When you finish studying this chapter, you should be able to:

- Define view factor, and understand its importance in radiation heat transfer calculations,
- Develop view factor relations, and calculate the unknown view factors in an enclosure by using these relations,
- Calculate radiation heat transfer between black surfaces,
- Determine radiation heat transfer between diffuse and gray surfaces in an enclosure using the concept of radiosity,
- Obtain relations for net rate of radiation heat transfer between the surfaces of a two-zone enclosure, including two large parallel plates, two long concentric cylinders, and two concentric spheres,
- Quantify the effect of radiation shields on the reduction of radiation heat transfer between two surfaces, and become aware of the importance of radiation effect in temperature measurements.

CONTENTS

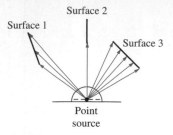

FIGURE 13–1

Radiation heat exchange between surfaces depends on the *orientation* of the surfaces relative to each other, and this dependence on orientation is accounted for by the *view factor*.

13–1 ▪ THE VIEW FACTOR

Radiation heat transfer between surfaces depends on the *orientation* of the surfaces relative to each other as well as their radiation properties and temperatures, as illustrated in Fig. 13–1. For example, a camper can make the most use of a campfire on a cold night by standing as close to the fire as possible and by blocking as much of the radiation coming from the fire by turning his or her front to the fire instead of the side. Likewise, a person can maximize the amount of solar radiation incident on him or her and take a sunbath by lying down on his or her back instead of standing.

To account for the effects of orientation on radiation heat transfer between two surfaces, we define a new parameter called the *view factor*, which is a purely geometric quantity and is independent of the surface properties and temperature. It is also called the *shape factor, configuration factor,* and *angle factor*. The view factor based on the assumption that the surfaces are diffuse emitters and diffuse reflectors is called the *diffuse view factor,* and the view factor based on the assumption that the surfaces are diffuse emitters but specular reflectors is called the *specular view factor*. In this book, we consider radiation exchange between diffuse surfaces only, and thus the term *view factor* simply means *diffuse view factor*.

The view factor from a surface i to a surface j is denoted by $F_{i \to j}$ or just F_{ij}, and is defined as

F_{ij} = *the fraction of the radiation leaving surface i that strikes surface j directly*

The notation $F_{i \to j}$ is instructive for beginners, since it emphasizes that the view factor is for radiation that travels from surface i to surface j. However, this notation becomes rather awkward when it has to be used many times in a problem. In such cases, it is convenient to replace it by its *shorthand* version F_{ij}.

The view factor F_{12} represents the fraction of radiation leaving surface 1 that strikes surface 2 directly, and F_{21} represents the fraction of radiation leaving surface 2 that strikes surface 1 directly. Note that the radiation that strikes a surface does not need to be absorbed by that surface. Also, radiation that strikes a surface after being reflected by other surfaces is not considered in the evaluation of view factors.

To develop a general expression for the view factor, consider two differential surfaces dA_1 and dA_2 on two arbitrarily oriented surfaces A_1 and A_2, respectively, as shown in Fig. 13–2. The distance between dA_1 and dA_2 is r, and the angles between the normals of the surfaces and the line that connects dA_1 and dA_2 are θ_1 and θ_2, respectively. Surface 1 emits and reflects radiation diffusely in all directions with a constant intensity of I_1, and the solid angle subtended by dA_2 when viewed by dA_1 is $d\omega_{21}$.

The rate at which radiation leaves dA_1 in the direction of θ_1 is $I_1 \cos \theta_1 dA_1$. Noting that $d\omega_{21} = dA_2 \cos \theta_2 / r^2$, the portion of this radiation that strikes dA_2 is

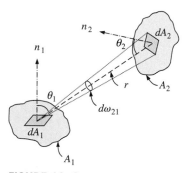

FIGURE 13–2

Geometry for the determination of the view factor between two surfaces.

$$\dot{Q}_{dA_1 \to dA_2} = I_1 \cos \theta_1 \, dA_1 \, d\omega_{21} = I_1 \cos \theta_1 \, dA_1 \frac{dA_2 \cos \theta_2}{r^2} \qquad \textbf{(13–1)}$$

The total rate at which radiation leaves dA_1 (via emission and reflection) in all directions is the radiosity (which is $J_1 = \pi I_1$) times the surface area,

$$\dot{Q}_{dA_1} = J_1 \, dA_1 = \pi I_1 \, dA_1 \tag{13-2}$$

Then the *differential view factor* $dF_{dA_1 \rightarrow dA_2}$, which is the fraction of radiation leaving dA_1 that strikes dA_2 directly, becomes

$$dF_{dA_1 \rightarrow dA_2} = \frac{\dot{Q}_{dA_1 \rightarrow dA_2}}{\dot{Q}_{dA_1}} = \frac{\cos\theta_1 \cos\theta_2}{\pi r^2} \, dA_2 \tag{13-3}$$

The differential view factor $dF_{dA_2 \rightarrow dA_1}$ can be determined from Eq. 13–3 by interchanging the subscripts 1 and 2.

The view factor from a differential area dA_1 to a finite area A_2 can be determined from the fact that the fraction of radiation leaving dA_1 that strikes A_2 is the sum of the fractions of radiation striking the differential areas dA_2. Therefore, the view factor $F_{dA_1 \rightarrow A_2}$ is determined by integrating $dF_{dA_1 \rightarrow dA_2}$ over A_2,

$$F_{dA_1 \rightarrow A_2} = \int_{A_2} \frac{\cos\theta_1 \cos\theta_2}{\pi r^2} \, dA_2 \tag{13-4}$$

The total rate at which radiation leaves the entire A_1 (via emission and reflection) in all directions is

$$\dot{Q}_{A_1} = J_1 A_1 = \pi I_1 A_1 \tag{13-5}$$

The portion of this radiation that strikes dA_2 is determined by considering the radiation that leaves dA_1 and strikes dA_2 (given by Eq. 13–1), and integrating it over A_1,

$$\dot{Q}_{A_1 \rightarrow dA_2} = \int_{A_1} \dot{Q}_{dA_1 \rightarrow dA_2} = \int_{A_1} \frac{I_1 \cos\theta_1 \cos\theta_2 \, dA_2}{r^2} \, dA_1 \tag{13-6}$$

Integration of this relation over A_2 gives the radiation that strikes the entire A_2,

$$\dot{Q}_{A_1 \rightarrow A_2} = \int_{A_2} \dot{Q}_{A_1 \rightarrow dA_2} = \int_{A_2} \int_{A_1} \frac{I_1 \cos\theta_1 \cos\theta_2}{r^2} \, dA_1 \, dA_2 \tag{13-7}$$

Dividing this by the total radiation leaving A_1 (from Eq. 13–5) gives the fraction of radiation leaving A_1 that strikes A_2, which is the view factor $F_{A_1 \rightarrow A_2}$ (or F_{12} for short),

$$F_{12} = F_{A_1 \rightarrow A_2} = \frac{\dot{Q}_{A_1 \rightarrow A_2}}{\dot{Q}_{A_1}} = \frac{1}{A_1} \int_{A_2} \int_{A_1} \frac{\cos\theta_1 \cos\theta_2}{\pi r^2} \, dA_1 \, dA_2 \tag{13-8}$$

The view factor $F_{A_2 \rightarrow A_1}$ is readily determined from Eq. 13–8 by interchanging the subscripts 1 and 2,

$$F_{21} = F_{A_2 \rightarrow A_1} = \frac{\dot{Q}_{A_2 \rightarrow A_1}}{\dot{Q}_{A_2}} = \frac{1}{A_2} \int_{A_2} \int_{A_1} \frac{\cos\theta_1 \cos\theta_2}{\pi r^2} \, dA_1 \, dA_2 \tag{13-9}$$

$F_{2\to2} = 0$

(b) Convex surface

$F_{1\to1} = 0$

(a) Plane surface

$F_{3\to3} \neq 0$

(c) Concave surface

FIGURE 13–3

The view factor from a surface to itself is *zero* for *plane* or *convex* surfaces and *nonzero* for *concave* surfaces.

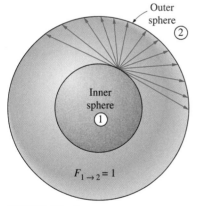

$F_{1\to2} = 1$

FIGURE 13–4

In a geometry that consists of two concentric spheres, the view factor $F_{1\to2} = 1$ since the entire radiation leaving the surface of the smaller sphere is intercepted by the larger sphere.

Note that I_1 is constant but r, θ_1, and θ_2 are variables. Also, integrations can be performed in any order since the integration limits are constants. These relations confirm that the view factor between two surfaces depends on their relative orientation and the distance between them.

Combining Eqs. 13–8 and 13–9 after multiplying the former by A_1 and the latter by A_2 gives

$$A_1 F_{12} = A_2 F_{21} \tag{13–10}$$

which is known as the **reciprocity relation** for view factors. It allows the calculation of a view factor from a knowledge of the other.

The view factor relations developed above are applicable to any two surfaces i and j provided that the surfaces are diffuse emitters and diffuse reflectors (so that the assumption of constant intensity is valid). For the special case of $j = i$, we have

$F_{i\to i}$ = the fraction of radiation leaving surface i that strikes itself directly

Noting that in the absence of strong electromagnetic fields radiation beams travel in straight paths, the view factor from a surface to itself is zero unless the surface "sees" itself. Therefore, $F_{i\to i} = 0$ for *plane* or *convex* surfaces and $F_{i\to i} \neq 0$ for concave surfaces, as illustrated in Fig. 13–3.

The value of the view factor ranges between *zero* and *one*. The limiting case $F_{i\to j} = 0$ indicates that the two surfaces do not have a direct view of each other, and thus radiation leaving surface i cannot strike surface j directly. The other limiting case $F_{i\to j} = 1$ indicates that surface j completely surrounds surface i, so that the entire radiation leaving surface i is intercepted by surface j. For example, in a geometry consisting of two concentric spheres, the entire radiation leaving the surface of the smaller sphere (surface 1) strikes the larger sphere (surface 2), and thus $F_{1\to2} = 1$, as illustrated in Fig. 13–4.

The view factor has proven to be very useful in radiation analysis because it allows us to express the *fraction of radiation* leaving a surface that strikes another surface in terms of the orientation of these two surfaces relative to each other. The underlying assumption in this process is that the radiation a surface receives from a source is directly proportional to the angle the surface subtends when viewed from the source. This would be the case only if the radiation coming off the source is *uniform* in all directions throughout its surface and the medium between the surfaces does not *absorb, emit,* or *scatter* radiation. That is, it is the case when the surfaces are *isothermal* and *diffuse* emitters and reflectors and the surfaces are separated by a *nonparticipating* medium such as a vacuum or air.

The view factor $F_{1\to2}$ between two surfaces A_1 and A_2 can be determined in a systematic manner first by expressing the view factor between two differential areas dA_1 and dA_2 in terms of the spatial variables and then by performing the necessary integrations. However, this approach is not practical, since, even for simple geometries, the resulting integrations are usually very complex and difficult to perform.

View factors for hundreds of common geometries are evaluated and the results are given in analytical, graphical, and tabular form in several publications. View factors for selected geometries are given in Tables 13–1 and 13–2 in *analytical* form and in Figs. 13–5 to 13–8 in *graphical* form. The view

TABLE 13–1

View factor expressions for some common geometries of finite size (3-D)

Geometry	Relation
Aligned parallel rectangles	$\overline{X} = X/L$ and $\overline{Y} = Y/L$ $F_{i \to j} = \dfrac{2}{\pi \overline{X}\,\overline{Y}} \left\{ \ln \left[\dfrac{(1 + \overline{X}^2)(1 + \overline{Y}^2)}{1 + \overline{X}^2 + \overline{Y}^2} \right]^{1/2} + \overline{X}(1 + \overline{Y}^2)^{1/2} \tan^{-1} \dfrac{\overline{X}}{(1 + \overline{Y}^2)^{1/2}} \right.$ $\left. + \overline{Y}(1 + \overline{X}^2)^{1/2} \tan^{-1} \dfrac{\overline{Y}}{(1 + \overline{X}^2)^{1/2}} - \overline{X} \tan^{-1} \overline{X} - \overline{Y} \tan^{-1} \overline{Y} \right\}$
Coaxial parallel disks	$R_i = r_i/L$ and $R_j = r_j/L$ $S = 1 + \dfrac{1 + R_j^2}{R_i^2}$ $F_{i \to j} = \dfrac{1}{2} \left\{ S - \left[S^2 - 4\left(\dfrac{r_j}{r_i} \right)^2 \right]^{1/2} \right\}$
Perpendicular rectangles with a common edge	$H = Z/X$ and $W = Y/X$ $F_{i \to j} = \dfrac{1}{\pi W} \left(W \tan^{-1} \dfrac{1}{W} + H \tan^{-1} \dfrac{1}{H} - (H^2 + W^2)^{1/2} \tan^{-1} \dfrac{1}{(H^2 + W^2)^{1/2}} \right.$ $+ \dfrac{1}{4} \ln \left\{ \dfrac{(1 + W^2)(1 + H^2)}{1 + W^2 + H^2} \left[\dfrac{W^2(1 + W^2 + H^2)}{(1 + W^2)(W^2 + H^2)} \right]^{W^2} \right.$ $\left. \left. \times \left[\dfrac{H^2(1 + H^2 + W^2)}{(1 + H^2)(H^2 + W^2)} \right]^{H^2} \right\} \right)$

factors in Table 13–1 are for three-dimensional geometries. The view factors in Table 13–2, on the other hand, are for geometries that are *infinitely long* in the direction perpendicular to the plane of the paper and are therefore two-dimensional.

13–2 ▪ VIEW FACTOR RELATIONS

Radiation analysis on an enclosure consisting of N surfaces requires the evaluation of N^2 view factors, and this evaluation process is probably the most time-consuming part of a radiation analysis. However, it is neither practical nor necessary to evaluate all of the view factors directly. Once a sufficient number of view factors are available, the rest of them can be determined by utilizing some fundamental relations for view factors, as discussed next.

TABLE 13–2

View factor expressions for some infinitely long (2-D) geometries

Geometry	Relation
Parallel plates with midlines connected by perpendicular line	$W_i = w_i/L$ and $W_j = w_j/L$ $F_{i \rightarrow j} = \dfrac{[(W_i + W_j)^2 + 4]^{1/2} - (W_j - W_i)^2 + 4]^{1/2}}{2W_i}$
Inclined plates of equal width and with a common edge	$F_{i \rightarrow j} = 1 - \sin \dfrac{1}{2}\alpha$
Perpendicular plates with a common edge	$F_{i \rightarrow j} = \dfrac{1}{2} \left\{ 1 + \dfrac{w_j}{w_i} - \left[1 + \left(\dfrac{w_j}{w_i} \right)^2 \right]^{1/2} \right\}$
Three-sided enclosure	$F_{i \rightarrow j} = \dfrac{w_i + w_j - w_k}{2w_i}$
Infinite plane and row of cylinders	$F_{i \rightarrow j} = 1 - \left[1 - \left(\dfrac{D}{s} \right)^2 \right]^{1/2}$ $+ \dfrac{D}{s} \tan^{-1} \left(\dfrac{s^2 - D^2}{D^2} \right)^{1/2}$

1   The Reciprocity Relation

The view factors $F_{i \rightarrow j}$ and $F_{j \rightarrow i}$ are *not* equal to each other unless the areas of the two surfaces are. That is,

$$F_{j \rightarrow i} = F_{i \rightarrow j} \qquad \text{when} \qquad A_i = A_j$$
$$F_{j \rightarrow i} \neq F_{i \rightarrow j} \qquad \text{when} \qquad A_i \neq A_j$$

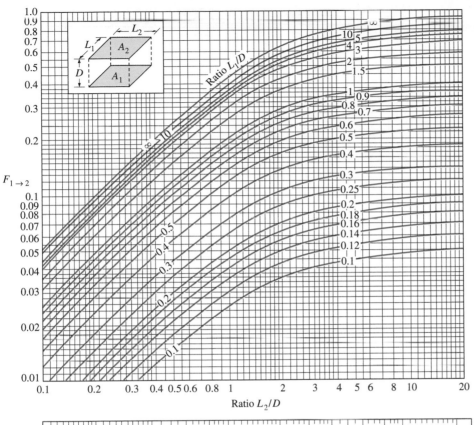

FIGURE 13–5
View factor between two
aligned parallel rectangles of
equal size.

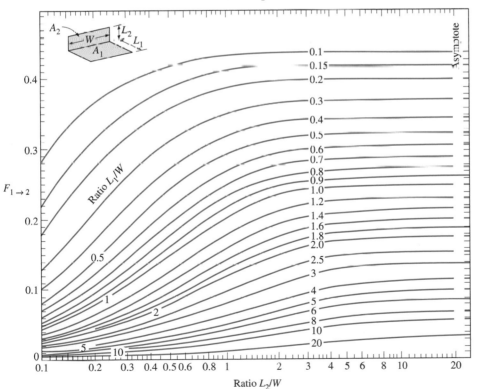

FIGURE 13–6
View factor between two
perpendicular rectangles with
a common edge.

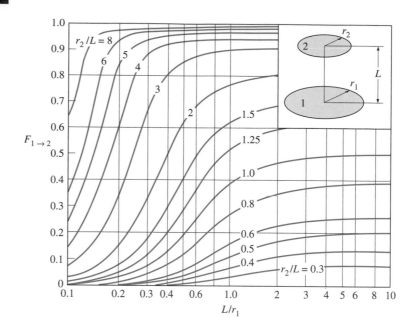

FIGURE 13–7
View factor between two coaxial parallel disks.

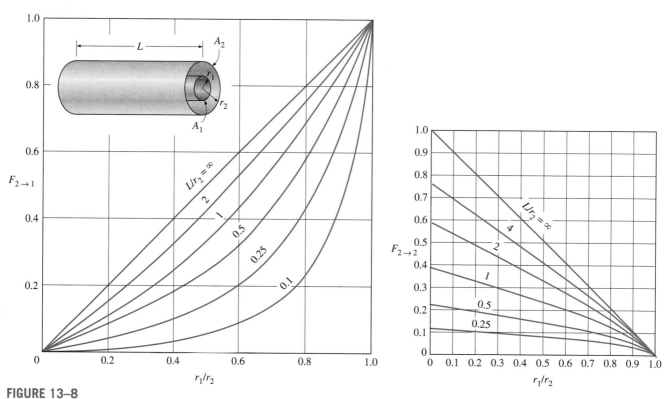

FIGURE 13–8
View factors for two concentric cylinders of finite length: (*a*) outer cylinder to inner cylinder; (*b*) outer cylinder to itself.

We have shown earlier that the pair of view factors $F_{i \rightarrow j}$ and $F_{j \rightarrow i}$ are related to each other by

$$A_i F_{i \rightarrow j} = A_j F_{j \rightarrow i} \tag{13–11}$$

This relation is referred to as the **reciprocity relation** or the **reciprocity rule**, and it enables us to determine the counterpart of a view factor from a knowledge of the view factor itself and the areas of the two surfaces. When determining the pair of view factors $F_{i \rightarrow j}$ and $F_{j \rightarrow i}$, it makes sense to evaluate first the easier one directly and then the more difficult one by applying the reciprocity relation.

2   The Summation Rule

The radiation analysis of a surface normally requires the consideration of the radiation coming in or going out in all directions. Therefore, most radiation problems encountered in practice involve enclosed spaces. When formulating a radiation problem, we usually form an *enclosure* consisting of the surfaces interacting radiatively. Even openings are treated as imaginary surfaces with radiation properties equivalent to those of the opening.

The conservation of energy principle requires that the entire radiation leaving any surface i of an enclosure be intercepted by the surfaces of the enclosure. Therefore, *the sum of the view factors from surface i of an enclosure to all surfaces of the enclosure, including to itself, must equal unity.* This is known as the **summation rule** for an enclosure and is expressed as (Fig. 13–9)

$$\sum_{j-1}^{N} F_{i \rightarrow j} = 1 \tag{13–12}$$

where N is the number of surfaces of the enclosure. For example, applying the summation rule to surface 1 of a three-surface enclosure yields

$$\sum_{j=1}^{3} F_{1 \rightarrow j} = F_{1 \rightarrow 1} + F_{1 \rightarrow 2} + F_{1 \rightarrow 3} = 1$$

The summation rule can be applied to each surface of an enclosure by varying i from 1 to N. Therefore, the summation rule applied to each of the N surfaces of an enclosure gives N relations for the determination of the view factors. Also, the reciprocity rule gives $\frac{1}{2}N(N - 1)$ additional relations. Then the total number of view factors that need to be evaluated directly for an N-surface enclosure becomes

$$N^2 - [N + \tfrac{1}{2}N(N - 1)] = \tfrac{1}{2}N(N - 1)$$

For example, for a six-surface enclosure, we need to determine only $\frac{1}{2} \times 6(6 - 1) = 15$ of the $6^2 = 36$ view factors directly. The remaining 21 view factors can be determined from the 21 equations that are obtained by applying the reciprocity and the summation rules.

Surface i

FIGURE 13–9
Radiation leaving any surface i of an enclosure must be intercepted completely by the surfaces of the enclosure. Therefore, the sum of the view factors from surface i to each one of the surfaces of the enclosure must be unity.

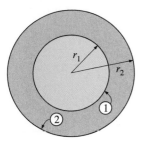

FIGURE 13–10
The geometry considered
in Example 13–1.

EXAMPLE 13–1 **View Factors Associated with Two Concentric Spheres**

Determine the view factors associated with an enclosure formed by two concentric spheres, shown in Fig. 13–10.

SOLUTION The view factors associated with two concentric spheres are to be determined.

Assumptions The surfaces are diffuse emitters and reflectors.

Analysis The outer surface of the smaller sphere (surface 1) and inner surface of the larger sphere (surface 2) form a two-surface enclosure. Therefore, $N = 2$ and this enclosure involves $N^2 = 2^2 = 4$ view factors, which are F_{11}, F_{12}, F_{21}, and F_{22}. In this two-surface enclosure, we need to determine only

$$\tfrac{1}{2}N(N-1) = \tfrac{1}{2} \times 2(2-1) = 1$$

view factor directly. The remaining three view factors can be determined by the application of the summation and reciprocity rules. But it turns out that we can determine not only one but *two* view factors directly in this case by a simple *inspection*:

$F_{11} = 0,$ since no radiation leaving surface 1 strikes itself

$F_{12} = 1,$ since all radiation leaving surface 1 strikes surface 2

Actually it would be sufficient to determine only one of these view factors by inspection, since we could always determine the other one from the summation rule applied to surface 1 as $F_{11} + F_{12} = 1$.

The view factor F_{21} is determined by applying the reciprocity relation to surfaces 1 and 2:

$$A_1 F_{12} = A_2 F_{21}$$

which yields

$$F_{21} = \frac{A_1}{A_2} F_{12} = \frac{4\pi r_1^2}{4\pi r_2^2} \times 1 = \left(\frac{r_1}{r_2}\right)^2$$

Finally, the view factor F_{22} is determined by applying the summation rule to surface 2:

$$F_{21} + F_{22} = 1$$

and thus

$$F_{22} = 1 - F_{21} = 1 - \left(\frac{r_1}{r_2}\right)^2$$

Discussion Note that when the outer sphere is much larger than the inner sphere ($r_2 \gg r_1$), F_{22} approaches one. This is expected, since the fraction of radiation leaving the outer sphere that is intercepted by the inner sphere will be negligible in that case. Also note that the two spheres considered above do not need to be concentric. However, the radiation analysis will be most accurate for the case of concentric spheres, since the radiation is most likely to be uniform on the surfaces in that case.

3 The Superposition Rule

Sometimes the view factor associated with a given geometry is not available in standard tables and charts. In such cases, it is desirable to express the given geometry as the sum or difference of some geometries with known view factors, and then to apply the **superposition rule**, which can be expressed as *the view factor from a surface i to a surface j is equal to the sum of the view factors from surface i to the parts of surface j*. Note that the reverse of this is not true. That is, the view factor from a surface j to a surface i is *not* equal to the sum of the view factors from the parts of surface j to surface i.

Consider the geometry in Fig. 13–11, which is infinitely long in the direction perpendicular to the plane of the paper. The radiation that leaves surface 1 and strikes the combined surfaces 2 and 3 is equal to the sum of the radiation that strikes surfaces 2 and 3. Therefore, the view factor from surface 1 to the combined surfaces of 2 and 3 is

$$F_{1 \to (2, 3)} = F_{1 \to 2} + F_{1 \to 3} \qquad \text{(13–13)}$$

Suppose we need to find the view factor $F_{1 \to 3}$. A quick check of the view factor expressions and charts in this section reveals that such a view factor cannot be evaluated directly. However, the view factor $F_{1 \to 3}$ can be determined from Eq. 13–13 after determining both $F_{1 \to 2}$ and $F_{1 \to (2, 3)}$ from the chart in Table 13–2. Therefore, it may be possible to determine some difficult view factors with relative ease by expressing one or both of the areas as the sum or differences of areas and then applying the superposition rule.

To obtain a relation for the view factor $F_{(2, 3) \to 1}$, we multiply Eq. 13–13 by A_1,

$$A_1 F_{1 \to (2, 3)} = A_1 F_{1 \to 2} + A_1 F_{1 \to 3}$$

and apply the reciprocity relation to each term to get

$$(A_2 + A_3)F_{(2, 3) \to 1} = A_2 F_{2 \to 1} + A_3 F_{3 \to 1}$$

or

$$F_{(2, 3) \to 1} = \frac{A_2 F_{2 \to 1} + A_3 F_{3 \to 1}}{A_2 + A_3} \qquad \text{(13–14)}$$

Areas that are expressed as the sum of more than two parts can be handled in a similar manner.

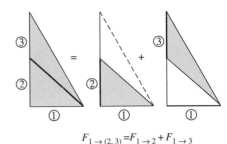

$$F_{1 \to (2, 3)} = F_{1 \to 2} + F_{1 \to 3}$$

FIGURE 13–11

The view factor from a surface to a composite surface is equal to the sum of the view factors from the surface to the parts of the composite surface.

> **EXAMPLE 13–2   Fraction of Radiation Leaving through an Opening**
>
> Determine the fraction of the radiation leaving the base of the cylindrical enclosure shown in Fig. 13–12 that escapes through a coaxial ring opening at its top surface. The radius and the length of the enclosure are $r_1 = 10$ cm and $L = 10$ cm, while the inner and outer radii of the ring are $r_2 = 5$ cm and $r_3 = 8$ cm, respectively.

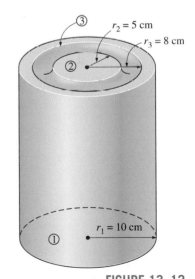

FIGURE 13–12

The cylindrical enclosure considered in Example 13–2.

SOLUTION The fraction of radiation leaving the base of a cylindrical enclosure through a coaxial ring opening at its top surface is to be determined.

Assumptions The base surface is a diffuse emitter and reflector.

Analysis We are asked to determine the fraction of the radiation leaving the base of the enclosure that escapes through an opening at the top surface. Actually, what we are asked to determine is simply the *view factor* $F_{1 \rightarrow \text{ring}}$ from the base of the enclosure to the ring-shaped surface at the top.

We do not have an analytical expression or chart for view factors between a circular area and a coaxial ring, and so we cannot determine $F_{1 \rightarrow \text{ring}}$ directly. However, we do have a chart for view factors between two coaxial parallel disks, and we can always express a ring in terms of disks.

Let the base surface of radius $r_1 = 10$ cm be surface 1, the circular area of $r_2 = 5$ cm at the top be surface 2, and the circular area of $r_3 = 8$ cm be surface 3. Using the superposition rule, the view factor from surface 1 to surface 3 can be expressed as

$$F_{1 \rightarrow 3} = F_{1 \rightarrow 2} + F_{1 \rightarrow \text{ring}}$$

since surface 3 is the sum of surface 2 and the ring area. The view factors $F_{1 \rightarrow 2}$ and $F_{1 \rightarrow 3}$ are determined from the chart in Fig. 13–7.

$$\frac{L}{r_1} = \frac{10 \text{ cm}}{10 \text{ cm}} = 1 \quad \text{and} \quad \frac{r_2}{L} = \frac{5 \text{ cm}}{10 \text{ cm}} = 0.5 \xrightarrow{\text{(Fig. 13–7)}} F_{1 \rightarrow 2} = 0.11$$

$$\frac{L}{r_1} = \frac{10 \text{ cm}}{10 \text{ cm}} = 1 \quad \text{and} \quad \frac{r_3}{L} = \frac{8 \text{ cm}}{10 \text{ cm}} = 0.8 \xrightarrow{\text{(Fig. 13–7)}} F_{1 \rightarrow 3} = 0.28$$

Therefore,

$$F_{1 \rightarrow \text{ring}} = F_{1 \rightarrow 3} - F_{1 \rightarrow 2} = 0.28 - 0.11 = \mathbf{0.17}$$

which is the desired result. Note that $F_{1 \rightarrow 2}$ and $F_{1 \rightarrow 3}$ represent the fractions of radiation leaving the base that strike the circular surfaces 2 and 3, respectively, and their difference gives the fraction that strikes the ring area.

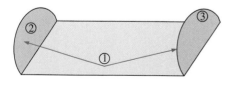

$F_{1 \rightarrow 2} = F_{1 \rightarrow 3}$

(Also, $F_{2 \rightarrow 1} = F_{3 \rightarrow 1}$)

FIGURE 13–13

Two surfaces that are symmetric about a third surface will have the same view factor from the third surface.

4 The Symmetry Rule

The determination of the view factors in a problem can be simplified further if the geometry involved possesses some sort of symmetry. Therefore, it is good practice to check for the presence of any *symmetry* in a problem before attempting to determine the view factors directly. The presence of symmetry can be determined *by inspection*, keeping the definition of the view factor in mind. Identical surfaces that are oriented in an identical manner with respect to another surface will intercept identical amounts of radiation leaving that surface. Therefore, the **symmetry rule** can be expressed as *two (or more) surfaces that possess symmetry about a third surface will have identical view factors from that surface* (Fig. 13–13).

The symmetry rule can also be expressed as *if the surfaces j and k are symmetric about the surface i then $F_{i \rightarrow j} = F_{i \rightarrow k}$*. Using the reciprocity rule, we can show that the relation $F_{j \rightarrow i} = F_{k \rightarrow i}$ is also true in this case.

EXAMPLE 13–3   View Factors Associated with a Tetragon

Determine the view factors from the base of the pyramid shown in Fig. 13–14 to each of its four side surfaces. The base of the pyramid is a square, and its side surfaces are isosceles triangles.

SOLUTION The view factors from the base of a pyramid to each of its four side surfaces for the case of a square base are to be determined.
Assumptions The surfaces are diffuse emitters and reflectors.
Analysis The base of the pyramid (surface 1) and its four side surfaces (surfaces 2, 3, 4, and 5) form a five-surface enclosure. The first thing we notice about this enclosure is its symmetry. The four side surfaces are symmetric about the base surface. Then, from the *symmetry rule,* we have

$$F_{12} = F_{13} = F_{14} = F_{15}$$

Also, the *summation rule* applied to surface 1 yields

$$\sum_{j=1}^{5} F_{1j} = F_{11} + F_{12} + F_{13} + F_{14} + F_{15} = 1$$

However, $F_{11} = 0$, since the base is a *flat* surface. Then the two relations above yield

$$F_{12} = F_{13} = F_{14} = F_{15} = \mathbf{0.25}$$

Discussion Note that each of the four side surfaces of the pyramid receive one-fourth of the entire radiation leaving the base surface, as expected. Also note that the presence of symmetry greatly simplified the determination of the view factors.

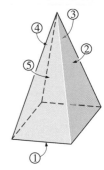

FIGURE 13–14
The pyramid considered in Example 13–3.

EXAMPLE 13–4   View Factors Associated with a Triangular Duct

Determine the view factor from any one side to any other side of the infinitely long triangular duct whose cross section is given in Fig. 13–15.

SOLUTION The view factors associated with an infinitely long triangular duct are to be determined.
Assumptions The surfaces are diffuse emitters and reflectors.
Analysis The widths of the sides of the triangular cross section of the duct are L_1, L_2, and L_3, and the surface areas corresponding to them are A_1, A_2, and A_3, respectively. Since the duct is infinitely long, the fraction of radiation leaving any surface that escapes through the ends of the duct is negligible. Therefore, the infinitely long duct can be considered to be a three-surface enclosure, $N = 3$.

This enclosure involves $N^2 = 3^2 = 9$ view factors, and we need to determine

$$\tfrac{1}{2}N(N-1) = \tfrac{1}{2} \times 3(3-1) = 3$$

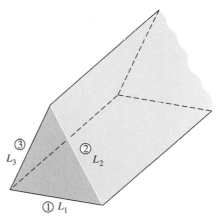

FIGURE 13–15
The infinitely long triangular duct considered in Example 13–4.

of these view factors directly. Fortunately, we can determine all three of them by inspection to be

$$F_{11} = F_{22} = F_{33} = 0$$

since all three surfaces are flat. The remaining six view factors can be determined by the application of the summation and reciprocity rules.

Applying the summation rule to each of the three surfaces gives

$$F_{11} + F_{12} + F_{13} = 1$$
$$F_{21} + F_{22} + F_{23} = 1$$
$$F_{31} + F_{32} + F_{33} = 1$$

Noting that $F_{11} = F_{22} = F_{33} = 0$ and multiplying the first equation by A_1, the second by A_2, and the third by A_3 gives

$$A_1 F_{12} + A_1 F_{13} = A_1$$
$$A_2 F_{21} + A_2 F_{23} = A_2$$
$$A_3 F_{31} + A_3 F_{32} = A_3$$

Finally, applying the three reciprocity relations $A_1 F_{12} = A_2 F_{21}$, $A_1 F_{13} = A_3 F_{31}$, and $A_2 F_{23} = A_3 F_{32}$ gives

$$A_1 F_{12} + A_1 F_{13} = A_1$$
$$A_1 F_{12} + A_2 F_{23} = A_2$$
$$A_1 F_{13} + A_2 F_{23} = A_3$$

This is a set of three algebraic equations with three unknowns, which can be solved to obtain

$$F_{12} = \frac{A_1 + A_2 - A_3}{2A_1} = \frac{L_1 + L_2 - L_3}{2L_1}$$

$$F_{13} = \frac{A_1 + A_3 - A_2}{2A_1} = \frac{L_1 + L_3 - L_2}{2L_1}$$

$$F_{23} = \frac{A_2 + A_3 - A_1}{2A_2} = \frac{L_2 + L_3 - L_1}{2L_2} \qquad \text{(13–15)}$$

Discussion Note that we have replaced the areas of the side surfaces by their corresponding widths for simplicity, since $A = Ls$ and the length s can be factored out and canceled. We can generalize this result as *the view factor from a surface of a very long triangular duct to another surface is equal to the sum of the widths of these two surfaces minus the width of the third surface, divided by twice the width of the first surface.*

View Factors between Infinitely Long Surfaces: The Crossed-Strings Method

Many problems encountered in practice involve geometries of constant cross section such as channels and ducts that are *very long* in one direction relative

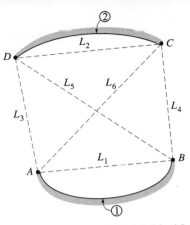

FIGURE 13–16
Determination of the view factor
$F_{1 \to 2}$ by the application of
the crossed-strings method.

to the other directions. Such geometries can conveniently be considered to be *two-dimensional,* since any radiation interaction through their end surfaces is negligible. These geometries can subsequently be modeled as being *infinitely long,* and the view factor between their surfaces can be determined by the amazingly simple *crossed-strings method* developed by H. C. Hottel in the 1950s. The surfaces of the geometry do not need to be flat; they can be convex, concave, or any irregular shape.

To demonstrate this method, consider the geometry shown in Fig. 13–16, and let us try to find the view factor $F_{1 \to 2}$ between surfaces 1 and 2. The first thing we do is identify the endpoints of the surfaces (the points A, B, C, and D) and connect them to each other with tightly stretched strings, which are indicated by dashed lines. Hottel has shown that the view factor $F_{1 \to 2}$ can be expressed in terms of the lengths of these stretched strings, which are straight lines, as

$$F_{1 \to 2} = \frac{(L_5 + L_6) - (L_3 + L_4)}{2L_1} \tag{13–16}$$

Note that $L_5 + L_6$ is the sum of the lengths of the *crossed strings,* and $L_3 + L_4$ is the sum of the lengths of the *uncrossed strings* attached to the endpoints. Therefore, Hottel's crossed-strings method can be expressed verbally as

$$F_{i \to j} = \frac{\Sigma (\text{Crossed strings}) - \Sigma (\text{Uncrossed strings})}{2 \times (\text{String on surface } i)} \tag{13–17}$$

The crossed-strings method is applicable even when the two surfaces considered share a common edge, as in a triangle. In such cases, the common edge can be treated as an imaginary string of zero length. The method can also be applied to surfaces that are partially blocked by other surfaces by allowing the strings to bend around the blocking surfaces.

EXAMPLE 13–5   The Crossed-Strings Method for View Factors

Two infinitely long parallel plates of widths $a = 12$ cm and $b = 5$ cm are located a distance $c = 6$ cm apart, as shown in Fig. 13–17. (a) Determine the view factor $F_{1 \to 2}$ from surface 1 to surface 2 by using the crossed-strings method. (b) Derive the crossed-strings formula by forming triangles on the given geometry and using Eq. 13–15 for view factors between the sides of triangles.

SOLUTION The view factors between two infinitely long parallel plates are to be determined using the crossed-strings method, and the formula for the view factor is to be derived.
Assumptions The surfaces are diffuse emitters and reflectors.
Analysis (a) First we label the endpoints of both surfaces and draw straight dashed lines between the endpoints, as shown in Fig. 13–17. Then we identify the crossed and uncrossed strings and apply the crossed-strings method (Eq. 13–17) to determine the view factor $F_{1 \to 2}$:

$$F_{1 \to 2} = \frac{\Sigma (\text{Crossed strings}) - \Sigma (\text{Uncrossed strings})}{2 \times (\text{String on surface 1})} = \frac{(L_5 + L_6) - (L_3 + L_4)}{2L_1}$$

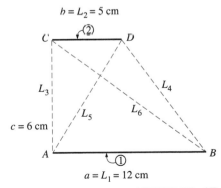

FIGURE 13–17
The two infinitely long parallel plates considered in Example 13–5.

where

$$L_1 = a = 12 \text{ cm} \qquad L_4 = \sqrt{7^2 + 6^2} = 9.22 \text{ cm}$$
$$L_2 = b = 5 \text{ cm} \qquad L_5 = \sqrt{5^2 + 6^2} = 7.81 \text{ cm}$$
$$L_3 = c = 6 \text{ cm} \qquad L_6 = \sqrt{12^2 + 6^2} = 13.42 \text{ cm}$$

Substituting,

$$F_{1 \rightarrow 2} = \frac{[(7.81 + 13.42) - (6 + 9.22)] \text{ cm}}{2 \times 12 \text{ cm}} = 0.250$$

(b) The geometry is infinitely long in the direction perpendicular to the plane of the paper, and thus the two plates (surfaces 1 and 2) and the two openings (imaginary surfaces 3 and 4) form a four-surface enclosure. Then applying the summation rule to surface 1 yields

$$F_{11} + F_{12} + F_{13} + F_{14} = 1$$

But $F_{11} = 0$ since it is a flat surface. Therefore,

$$F_{12} = 1 - F_{13} - F_{14}$$

where the view factors F_{13} and F_{14} can be determined by considering the triangles *ABC* and *ABD,* respectively, and applying Eq. 13–15 for view factors between the sides of triangles. We obtain

$$F_{13} = \frac{L_1 + L_3 - L_6}{2L_1}, \qquad F_{14} = \frac{L_1 + L_4 - L_5}{2L_1}$$

Substituting,

$$F_{12} = 1 - \frac{L_1 + L_3 - L_6}{2L_1} - \frac{L_1 + L_4 - L_5}{2L_1}$$
$$= \frac{(L_5 + L_6) - (L_3 + L_4)}{2L_1}$$

which is the desired result. This is also a miniproof of the crossed-strings method for the case of two infinitely long plain parallel surfaces.

13–3 · RADIATION HEAT TRANSFER: BLACK SURFACES

So far, we have considered the nature of radiation, the radiation properties of materials, and the view factors, and we are now in a position to consider the rate of heat transfer between surfaces by radiation. The analysis of radiation exchange between surfaces, in general, is complicated because of reflection: a radiation beam leaving a surface may be reflected several times, with partial reflection occurring at each surface, before it is completely absorbed. The analysis is simplified greatly when the surfaces involved can be approximated as blackbodies because of the absence of reflection. In this section, we consider radiation exchange between *black surfaces* only; we extend the analysis to reflecting surfaces in the next section.

Consider two black surfaces of arbitrary shape maintained at uniform temperatures T_1 and T_2, as shown in Fig. 13–18. Recognizing that radiation leaves a black surface at a rate of $E_b = \sigma T^4$ per unit surface area and that the view factor $F_{1 \to 2}$ represents the fraction of radiation leaving surface 1 that strikes surface 2, the *net* rate of radiation heat transfer from surface 1 to surface 2 can be expressed as

$$\dot{Q}_{1 \to 2} = \begin{pmatrix} \text{Radiation leaving} \\ \text{the entire surface 1} \\ \text{that strikes surface 2} \end{pmatrix} - \begin{pmatrix} \text{Radiation leaving} \\ \text{the entire surface 2} \\ \text{that strikes surface 1} \end{pmatrix}$$

$$= A_1 E_{b1} F_{1 \to 2} - A_2 E_{b2} F_{2 \to 1} \quad \text{(W)} \quad \text{(13–18)}$$

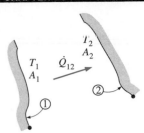

FIGURE 13–18
Two general black surfaces maintained at uniform temperatures T_1 and T_2.

Applying the reciprocity relation $A_1 F_{1 \to 2} = A_2 F_{2 \to 1}$ yields

$$\dot{Q}_{1 \to 2} = A_1 F_{1 \to 2} \, \sigma(T_1^4 - T_2^4) \quad \text{(W)} \quad \text{(13–19)}$$

which is the desired relation. A negative value for $\dot{Q}_{1 \to 2}$ indicates that net radiation heat transfer is from surface 2 to surface 1.

Now consider an *enclosure* consisting of N *black* surfaces maintained at specified temperatures. The *net* radiation heat transfer *from* any surface i of this enclosure is determined by adding up the net radiation heat transfers from surface i to each of the surfaces of the enclosure:

$$\dot{Q}_i = \sum_{j-1}^{N} \dot{Q}_{i \to j} = \sum_{j-1}^{N} A_i F_{i \to j} \sigma(T_i^4 - T_j^4) \quad \text{(W)} \quad \text{(13–20)}$$

Again a negative value for $\dot{Q}$ indicates that net radiation heat transfer is *to* surface i (i.e., surface i *gains* radiation energy instead of losing). Also, the net heat transfer from a surface to itself is zero, regardless of the shape of the surface.

EXAMPLE 13–6   **Radiation Heat Transfer in a Black Furnace**

Consider the 5-m × 5-m × 5-m cubical furnace shown in Fig. 13–19, whose surfaces closely approximate black surfaces. The base, top, and side surfaces of the furnace are maintained at uniform temperatures of 800 K, 1500 K, and 500 K, respectively. Determine (*a*) the net rate of radiation heat transfer between the base and the side surfaces, (*b*) the net rate of radiation heat transfer between the base and the top surface, and (*c*) the net radiation heat transfer from the base surface.

SOLUTION   The surfaces of a cubical furnace are black and are maintained at uniform temperatures. The net rate of radiation heat transfer between the base and side surfaces, between the base and the top surface, and from the base surface are to be determined.

Assumptions   The surfaces are black and isothermal.

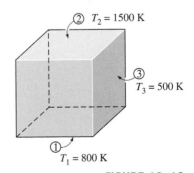

FIGURE 13–19
The cubical furnace of black surfaces considered in Example 13–6.

Analysis (*a*) The geometry involves six surfaces, and thus we may be tempted at first to treat the furnace as a six-surface enclosure. However, the four side surfaces possess the same properties, and thus we can treat them as a single side surface in radiation analysis. We consider the base surface to be surface 1, the top surface to be surface 2, and the side surfaces to be surface 3. Then the problem reduces to determining $\dot{Q}_{1\to3}$, $\dot{Q}_{1\to2}$, and $\dot{Q}_1$.

The net rate of radiation heat transfer $\dot{Q}_{1\to3}$ from surface 1 to surface 3 can be determined from Eq. 13–19, since both surfaces involved are black, by replacing the subscript 2 by 3:

$$\dot{Q}_{1\to3} = A_1 F_{1\to3}\sigma(T_1^4 - T_3^4)$$

But first we need to evaluate the view factor $F_{1\to3}$. After checking the view factor charts and tables, we realize that we cannot determine this view factor directly. However, we can determine the view factor $F_{1\to2}$ from Fig. 13–5 to be $F_{1\to2} = 0.2$, and we know that $F_{1\to1} = 0$ since surface 1 is a plane. Then applying the summation rule to surface 1 yields

$$F_{1\to1} + F_{1\to2} + F_{1\to3} = 1$$

or

$$F_{1\to3} = 1 - F_{1\to1} - F_{1\to2} = 1 - 0 - 0.2 = 0.8$$

Substituting,

$$\dot{Q}_{1\to3} = (25\ \text{m}^2)(0.8)(5.67 \times 10^{-8}\ \text{W/m}^2 \cdot \text{K}^4)[(800\ \text{K})^4 - (500\ \text{K})^4]$$
$$= \mathbf{394\ kW}$$

(*b*) The net rate of radiation heat transfer $\dot{Q}_{1\to2}$ from surface 1 to surface 2 is determined in a similar manner from Eq. 13–19 to be

$$\dot{Q}_{1\to2} = A_1 F_{1\to2}\sigma(T_1^4 - T_2^4)$$
$$= (25\ \text{m}^2)(0.2)(5.67 \times 10^{-8}\ \text{W/m}^2 \cdot \text{K}^4)[(800\ \text{K})^4 - (1500\ \text{K})^4]$$
$$= \mathbf{-1319\ kW}$$

The negative sign indicates that net radiation heat transfer is from surface 2 to surface 1.

(*c*) The net radiation heat transfer from the base surface $\dot{Q}_1$ is determined from Eq. 13–20 by replacing the subscript *i* by 1 and taking $N = 3$:

$$\dot{Q}_1 = \sum_{j=1}^{3} \dot{Q}_{1\to j} = \dot{Q}_{1\to1} + \dot{Q}_{1\to2} + \dot{Q}_{1\to3}$$
$$= 0 + (-1319\ \text{kW}) + (394\ \text{kW})$$
$$= \mathbf{-925\ kW}$$

Again the negative sign indicates that net radiation heat transfer is *to* surface 1. That is, the base of the furnace is gaining net radiation at a rate of 925 kW.

13–4 ▪ RADIATION HEAT TRANSFER: DIFFUSE, GRAY SURFACES

The analysis of radiation transfer in enclosures consisting of black surfaces is relatively easy, as we have seen, but most enclosures encountered in practice involve nonblack surfaces, which allow multiple reflections to occur. Radiation analysis of such enclosures becomes very complicated unless some simplifying assumptions are made.

To make a simple radiation analysis possible, it is common to assume the surfaces of an enclosure to be *opaque, diffuse,* and *gray.* That is, the surfaces are nontransparent, they are diffuse emitters and diffuse reflectors, and their radiation properties are independent of wavelength. Also, each surface of the enclosure is *isothermal,* and both the incoming and outgoing radiation are *uniform* over each surface. But first we review the concept of radiosity introduced in Chap. 12.

Radiosity

Surfaces emit radiation as well as reflect it, and thus the radiation leaving a surface consists of emitted and reflected parts. The calculation of radiation heat transfer between surfaces involves the *total* radiation energy streaming away from a surface, with no regard for its origin. The *total radiation energy leaving a surface per unit time and per unit area* is the **radiosity** and is denoted by J (Fig. 13–20).

For a surface i that is *gray* and *opaque* ($\varepsilon_i = \alpha_i$ and $\alpha_i + \rho_i = 1$), the radiosity can be expressed as

$$
\begin{aligned}
J_i &= \left(\begin{array}{c}\text{Radiation emitted}\\\text{by surface } i\end{array}\right) + \left(\begin{array}{c}\text{Radiation reflected}\\\text{by surface } i\end{array}\right)\\
&= \varepsilon_i E_{bi} + \rho_i G_i\\
&= \varepsilon_i E_{bi} + (1 - \varepsilon_i) G_i \qquad (\text{W/m}^2)
\end{aligned}
\qquad (13\text{–}21)
$$

where $E_{bi} = \sigma T_i^4$ is the blackbody emissive power of surface i and G_i is irradiation (i.e., the radiation energy incident on surface i per unit time per unit area).

For a surface that can be approximated as a *blackbody* ($\varepsilon_i = 1$), the radiosity relation reduces to

$$
J_i = E_{bi} = \sigma T_i^4 \qquad (\text{blackbody}) \qquad (13\text{–}22)
$$

That is, *the radiosity of a blackbody is equal to its emissive power.* This is expected, since a blackbody does not reflect any radiation, and thus radiation coming from a blackbody is due to emission only.

Net Radiation Heat Transfer to or from a Surface

During a radiation interaction, a surface *loses* energy by emitting radiation and *gains* energy by absorbing radiation emitted by other surfaces. A surface experiences a net gain or a net loss of energy, depending on which quantity is larger. The *net rate of radiation heat transfer from a surface i of surface area A_i is denoted by $\dot{Q}_i$ and is expressed as

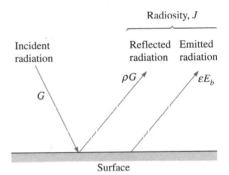

FIGURE 13–20
Radiosity represents the sum of the radiation energy emitted and reflected by a surface.

$$\dot{Q}_i = \begin{pmatrix} \text{Radiation leaving} \\ \text{entire surface } i \end{pmatrix} - \begin{pmatrix} \text{Radiation incident} \\ \text{on entire surface } i \end{pmatrix}$$
$$= A_i(J_i - G_i) \qquad \text{(W)} \qquad \qquad \text{(13–23)}$$

Solving for G_i from Eq. 13–21 and substituting into Eq. 13–23 yields

$$\dot{Q}_i = A_i\left(J_i - \frac{J_i - \varepsilon_i E_{bi}}{1 - \varepsilon_i}\right) = \frac{A_i \varepsilon_i}{1 - \varepsilon_i}(E_{bi} - J_i) \qquad \text{(W)} \qquad \text{(13–24)}$$

In an electrical analogy to Ohm's law, this equation can be rearranged as

$$\dot{Q}_i = \frac{E_{bi} - J_i}{R_i} \qquad \text{(W)} \qquad \qquad \text{(13–25)}$$

where

$$R_i = \frac{1 - \varepsilon_i}{A_i \varepsilon_i} \qquad \qquad \text{(13–26)}$$

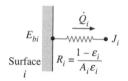

E_{bi} ●───wwww───● J_i

Surface $R_i = \dfrac{1 - \varepsilon_i}{A_i \varepsilon_i}$
i

FIGURE 13–21
Electrical analogy of surface resistance to radiation.

is the **surface resistance** to radiation. The quantity $E_{bi} - J_i$ corresponds to a *potential difference* and the net rate of radiation heat transfer corresponds to *current* in the electrical analogy, as illustrated in Fig. 13–21.

The direction of the net radiation heat transfer depends on the relative magnitudes of J_i (the radiosity) and E_{bi} (the emissive power of a blackbody at the temperature of the surface). It is *from* the surface if $E_{bi} > J_i$ and *to* the surface if $J_i > E_{bi}$. A negative value for $\dot{Q}_i$ indicates that heat transfer is *to* the surface. All of this radiation energy gained must be removed from the other side of the surface through some mechanism if the surface temperature is to remain constant.

The surface resistance to radiation for a *blackbody* is *zero* since $\varepsilon_i = 1$ and $J_i = E_{bi}$. The net rate of radiation heat transfer in this case is determined directly from Eq. 13–23.

Some surfaces encountered in numerous practical heat transfer applications are modeled as being *adiabatic* since their back sides are well insulated and the net heat transfer through them is zero. When the convection effects on the front (heat transfer) side of such a surface is negligible and steady-state conditions are reached, the surface must lose as much radiation energy as it gains, and thus $\dot{Q}_i = 0$. In such cases, the surface is said to *reradiate* all the radiation energy it receives, and such a surface is called a **reradiating surface**. Setting $\dot{Q}_i = 0$ in Eq. 13–25 yields

$$J_i = E_{bi} = \sigma T_i^4 \qquad \text{(W/m}^2) \qquad \qquad \text{(13–27)}$$

Therefore, the *temperature* of a reradiating surface under steady conditions can easily be determined from the equation above once its radiosity is known. Note that the temperature of a reradiating surface is *independent of its emissivity*. In radiation analysis, the surface resistance of a reradiating surface is disregarded since there is no net heat transfer through it. (This is like the fact that there is no need to consider a resistance in an electrical network if no current is flowing through it.)

Net Radiation Heat Transfer between Any Two Surfaces

Consider two diffuse, gray, and opaque surfaces of arbitrary shape maintained at uniform temperatures, as shown in Fig. 13–22. Recognizing that the radiosity J represents the rate of radiation leaving a surface per unit surface area and that the view factor $F_{i \rightarrow j}$ represents the fraction of radiation leaving surface i that strikes surface j, the *net* rate of radiation heat transfer from surface i to surface j can be expressed as

$$\dot{Q}_{i \rightarrow j} = \begin{pmatrix} \text{Radiation leaving} \\ \text{the entire surface } i \\ \text{that strikes surface } j \end{pmatrix} - \begin{pmatrix} \text{Radiation leaving} \\ \text{the entire surface } j \\ \text{that strikes surface } i \end{pmatrix} \quad \textbf{(13–28)}$$

$$= A_i J_i F_{i \rightarrow j} - A_j J_j F_{j \rightarrow i} \quad \text{(W)}$$

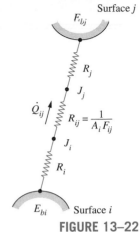

Applying the reciprocity relation $A_i F_{i \rightarrow j} = A_j F_{j \rightarrow i}$ yields

$$\dot{Q}_{i \rightarrow j} = A_i F_{i \rightarrow j} (J_i - J_j) \quad \text{(W)} \quad \textbf{(13–29)}$$

Again in analogy to Ohm's law, this equation can be rearranged as

$$\dot{Q}_{i \rightarrow j} = \frac{J_i - J_j}{R_{i \rightarrow j}} \quad \text{(W)} \quad \textbf{(13–30)}$$

where

$$R_{i \rightarrow j} = \frac{1}{A_i F_{i \rightarrow j}} \quad \textbf{(13–31)}$$

FIGURE 13–22
Electrical analogy of space resistance to radiation.

is the **space resistance** to radiation. Again the quantity $J_i - J_j$ corresponds to a *potential difference,* and the net rate of heat transfer between two surfaces corresponds to *current* in the electrical analogy, as illustrated in Fig. 13–22.

The direction of the net radiation heat transfer between two surfaces depends on the relative magnitudes of J_i and J_j. A positive value for $\dot{Q}_{i \rightarrow j}$ indicates that net heat transfer is *from* surface i *to* surface j. A negative value indicates the opposite.

In an N-surface enclosure, the conservation of energy principle requires that the net heat transfer from surface i be equal to the sum of the net heat transfers from surface i to each of the N surfaces of the enclosure. That is,

$$\dot{Q}_i = \sum_{j=1}^{N} \dot{Q}_{i \rightarrow j} = \sum_{j=1}^{N} A_i F_{i \rightarrow j} (J_i - J_j) = \sum_{j=1}^{N} \frac{J_i - J_j}{R_{i \rightarrow j}} \quad \text{(W)} \quad \textbf{(13–32)}$$

The network representation of net radiation heat transfer from surface i to the remaining surfaces of an N-surface enclosure is given in Fig. 13–23. Note that $\dot{Q}_{i \rightarrow i}$ (the net rate of heat transfer from a surface to itself) is zero regardless of the shape of the surface. Combining Eqs. 13–25 and 13–32 gives

$$\frac{E_{bi} - J_i}{R_i} = \sum_{j=1}^{N} \frac{J_i - J_j}{R_{i \rightarrow j}} \quad \text{(W)} \quad \textbf{(13–33)}$$

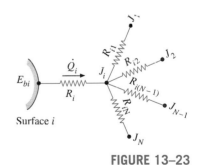

FIGURE 13–23
Network representation of net radiation heat transfer from surface i to the remaining surfaces of an N-surface enclosure.

which has the electrical analogy interpretation that *the net radiation flow from a surface through its surface resistance is equal to the sum of the radiation flows from that surface to all other surfaces through the corresponding space resistances.*

Methods of Solving Radiation Problems

In the radiation analysis of an enclosure, either the temperature or the net rate of heat transfer must be given for each of the surfaces to obtain a unique solution for the unknown surface temperatures and heat transfer rates. There are two methods commonly used to solve radiation problems. In the first method, Eqs. 13–32 (for surfaces with specified heat transfer rates) and 13–33 (for surfaces with specified temperatures) are simplified and re-arranged as

Surfaces with specified net heat transfer rate $\dot{Q}$

$$\dot{Q}_i = A_i \sum_{j=1}^{N} F_{i \to j}(J_i - J_j) \tag{13–34}$$

Surfaces with specified temperature T_i

$$\sigma T_i^4 = J_i + \frac{1 - \varepsilon_i}{\varepsilon_i} \sum_{j=1}^{N} F_{i \to j}(J_i - J_j) \tag{13–35}$$

Note that $\dot{Q}_i = 0$ for insulated (or reradiating) surfaces, and $\sigma T_i^4 = J_i$ for black surfaces since $\varepsilon_i = 1$ in that case. Also, the term corresponding to $j = i$ drops out from either relation since $J_i - J_j = J_i - J_i = 0$ in that case.

The equations above give N linear algebraic equations for the determination of the N unknown radiosities for an N-surface enclosure. Once the radiosities $J_1, J_2, \ldots, J_N$ are available, the unknown heat transfer rates can be determined from Eq. 13–34 while the unknown surface temperatures can be determined from Eq. 13–35. The temperatures of insulated or reradiating surfaces can be determined from $\sigma T_i^4 = J_i$. A positive value for $\dot{Q}_i$ indicates net radiation heat transfer *from* surface i to other surfaces in the enclosure while a negative value indicates net radiation heat transfer *to* the surface.

The systematic approach described above for solving radiation heat transfer problems is very suitable for use with today's popular equation solvers such as EES, Mathcad, and Matlab, especially when there are a large number of surfaces, and is known as the **direct method** (formerly, the *matrix method,* since it resulted in matrices and the solution required a knowledge of linear algebra). The second method described below, called the **network method**, is based on the electrical network analogy.

The network method was first introduced by A. K. Oppenheim in the 1950s and found widespread acceptance because of its simplicity and emphasis on the physics of the problem. The application of the method is straightforward: draw a surface resistance associated with each surface of an enclosure and connect them with space resistances. Then solve the radiation problem by treating it as an electrical network problem where the radiation heat transfer replaces the current and radiosity replaces the potential.

The network method is not practical for enclosures with more than three or four surfaces, however, because of the increased complexity of the network. Next we apply the method to solve radiation problems in two- and three-surface enclosures.

Radiation Heat Transfer in Two-Surface Enclosures

Consider an enclosure consisting of two opaque surfaces at specified temperatures T_1 and T_2, as shown in Fig. 13–24, and try to determine the net rate of radiation heat transfer between the two surfaces with the network method. Surfaces 1 and 2 have emissivities ε_1 and ε_2 and surface areas A_1 and A_2 and are maintained at uniform temperatures T_1 and T_2, respectively. There are only two surfaces in the enclosure, and thus we can write

$$\dot{Q}_{12} = \dot{Q}_1 = -\dot{Q}_2$$

That is, the net rate of radiation heat transfer from surface 1 to surface 2 must equal the net rate of radiation heat transfer *from* surface 1 and the net rate of radiation heat transfer *to* surface 2.

The radiation network of this two-surface enclosure consists of two surface resistances and one space resistance, as shown in Fig. 13–24. In an electrical network, the electric current flowing through these resistances connected in series would be determined by dividing the potential difference between points A and B by the total resistance between the same two points. The net rate of radiation transfer is determined in the same manner and is expressed as

$$\dot{Q}_{12} - \frac{E_{b1} - E_{b2}}{R_1 + R_{12} + R_2} = \dot{Q}_1 = -\dot{Q}_2$$

or

$$\dot{Q}_{12} = \frac{\sigma(T_1^4 - T_2^4)}{\dfrac{1 - \varepsilon_1}{A_1 \varepsilon_1} + \dfrac{1}{A_1 F_{12}} + \dfrac{1 - \varepsilon_2}{A_2 \varepsilon_2}} \quad \text{(W)} \qquad (13\text{–}36)$$

This important result is applicable to any two gray, diffuse, and opaque surfaces that form an enclosure. The view factor F_{12} depends on the geometry and must be determined first. Simplified forms of Eq. 13–36 for some familiar arrangements that form a two-surface enclosure are given in Table 13–3. Note that $F_{12} = 1$ for all of these special cases.

$$E_{b1} \xrightarrow{\dot{Q}_1} J_1 \xrightarrow{\dot{Q}_{12}} J_2 \xrightarrow{\dot{Q}_2} E_{b2}$$

$$R_1 = \frac{1 - \varepsilon_1}{A_1 \varepsilon_1} \qquad R_{12} = \frac{1}{A_1 F_{12}} \qquad R_2 = \frac{1 - \varepsilon_2}{A_2 \varepsilon_2}$$

FIGURE 13–24
Schematic of a two-surface enclosure and the radiation network associated with it.

EXAMPLE 13–7   Radiation Heat Transfer between Parallel Plates

Two very large parallel plates are maintained at uniform temperatures $T_1 = 800$ K and $T_2 = 500$ K and have emissivities $\varepsilon_1 = 0.2$ and $\varepsilon_2 = 0.7$, respectively, as shown in Fig. 13–25. Determine the net rate of radiation heat transfer between the two surfaces per unit surface area of the plates.

SOLUTION   Two large parallel plates are maintained at uniform temperatures. The net rate of radiation heat transfer between the plates is to be determined.
Assumptions   Both surfaces are opaque, diffuse, and gray.
Analysis   The net rate of radiation heat transfer between the two plates per unit area is readily determined from Eq. 13–38 to be

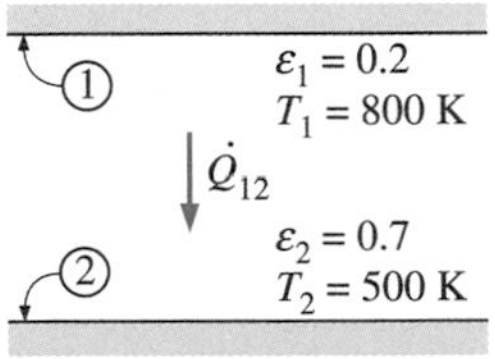

$\varepsilon_1 = 0.2$
$T_1 = 800$ K

$\dot{Q}_{12}$

$\varepsilon_2 = 0.7$
$T_2 = 500$ K

FIGURE 13–25
The two parallel plates considered in Example 13–7.

TABLE 13–3

Small object in a large cavity 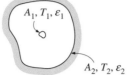	$\dfrac{A_1}{A_2} \approx 0$ $F_{12} = 1$	$\dot{Q}_{12} = A_1 \sigma \varepsilon_1 (T_1^4 - T_2^4)$ **(13–37)**
Infinitely large parallel plates A_1, T_1, ε_1 —————————— A_2, T_2, ε_2 ——————————	$A_1 = A_2 = A$ $F_{12} = 1$	$\dot{Q}_{12} = \dfrac{A\sigma(T_1^4 - T_2^4)}{\dfrac{1}{\varepsilon_1} + \dfrac{1}{\varepsilon_2} - 1}$ **(13–38)**
Infinitely long concentric cylinders 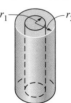	$\dfrac{A_1}{A_2} = \dfrac{r_1}{r_2}$ $F_{12} = 1$	$\dot{Q}_{12} = \dfrac{A_1\sigma(T_1^4 - T_2^4)}{\dfrac{1}{\varepsilon_1} + \dfrac{1 - \varepsilon_2}{\varepsilon_2}\left(\dfrac{r_1}{r_2}\right)}$ **(13–39)**
Concentric spheres	$\dfrac{A_1}{A_2} = \left(\dfrac{r_1}{r_2}\right)^2$ $F_{12} = 1$	$\dot{Q}_{12} = \dfrac{A_1\sigma(T_1^4 - T_2^4)}{\dfrac{1}{\varepsilon_1} + \dfrac{1 - \varepsilon_2}{\varepsilon_2}\left(\dfrac{r_1}{r_2}\right)^2}$ **(13–40)**

$$\dot{q}_{12} = \frac{\dot{Q}_{12}}{A} = \frac{\sigma(T_1^4 - T_2^4)}{\dfrac{1}{\varepsilon_1} + \dfrac{1}{\varepsilon_2} - 1} = \frac{(5.67 \times 10^{-8}\ \text{W/m}^2 \cdot \text{K}^4)[(800\ \text{K})^4 - (500\ \text{K})^4]}{\dfrac{1}{0.2} + \dfrac{1}{0.7} - 1}$$

$$= 3625\ \text{W/m}^2$$

Discussion Note that heat at a net rate of 3625 W is transferred from plate 1 to plate 2 by radiation per unit surface area of either plate.

Radiation Heat Transfer
in Three-Surface Enclosures

We now consider an enclosure consisting of three opaque, diffuse, and gray surfaces, as shown in Fig. 13–26. Surfaces 1, 2, and 3 have surface areas A_1, A_2, and A_3; emissivities ε_1, ε_2, and ε_3; and uniform temperatures T_1, T_2, and T_3, respectively. The radiation network of this geometry is constructed by following the standard procedure: draw a surface resistance associated with each of the three surfaces and connect these surface resistances with space resistances, as shown in the figure. Relations for the surface and space resistances are given by Eqs. 13–26 and 13–31. The three endpoint potentials E_{b1}, E_{b2}, and E_{b3} are considered known, since the surface temperatures are specified. Then all we need to find are the radiosities J_1, J_2, and J_3. The three equations for the determination of these three unknowns are obtained from the requirement that *the algebraic sum of the currents (net radiation heat transfer) at each node must equal zero.* That is,

$$\frac{E_{b1} - J_1}{R_1} + \frac{J_2 - J_1}{R_{12}} + \frac{J_3 - J_1}{R_{13}} = 0$$

$$\frac{J_1 - J_2}{R_{12}} + \frac{E_{b2} - J_2}{R_2} + \frac{J_3 - J_2}{R_{23}} = 0$$

$$\frac{J_1 - J_3}{R_{13}} + \frac{J_2 - J_3}{R_{23}} + \frac{E_{b3} - J_3}{R_3} = 0 \qquad \textbf{(13–41)}$$

Once the radiosities J_1, J_2, and J_3 are available, the net rate of radiation heat transfers at each surface can be determined from Eq. 13–32.

The set of equations above simplify further if one or more surfaces are "special" in some way. For example, $J_i = E_{bi} = \sigma T_i^4$ for a *black* or *reradiating* surface. Also, $\dot{Q}_i = 0$ for a reradiating surface. Finally, when the net rate of radiation heat transfer $\dot{Q}_i$ is specified at surface i instead of the temperature, the term $(E_{bi} - J_i)/R_i$ should be replaced by the specified $\dot{Q}_i$.

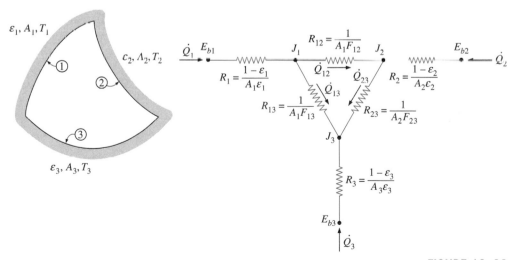

FIGURE 13–26

Schematic of a three-surface enclosure and the radiation network associated with it.

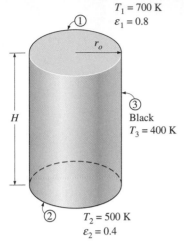

$T_1 = 700$ K
$\varepsilon_1 = 0.8$

r_o

①

H

③
Black
$T_3 = 400$ K

②
$T_2 = 500$ K
$\varepsilon_2 = 0.4$

FIGURE 13–27
The cylindrical furnace
considered in Example 13–8.

EXAMPLE 13–8   Radiation Heat Transfer in a Cylindrical Furnace

Consider a cylindrical furnace with $r_o = H = 1$ m, as shown in Fig. 13–27. The top (surface 1) and the base (surface 2) of the furnace have emissivities $\varepsilon_1 = 0.8$ and $\varepsilon_2 = 0.4$, respectively, and are maintained at uniform temperatures $T_1 = 700$ K and $T_2 = 500$ K. The side surface closely approximates a blackbody and is maintained at a temperature of $T_3 = 400$ K. Determine the net rate of radiation heat transfer at each surface during steady operation and explain how these surfaces can be maintained at specified temperatures.

SOLUTION   The surfaces of a cylindrical furnace are maintained at uniform temperatures. The net rate of radiation heat transfer at each surface during steady operation is to be determined.
Assumptions   **1** Steady operating conditions exist. **2** The surfaces are opaque, diffuse, and gray. **3** Convection heat transfer is not considered.
Analysis   We will solve this problem systematically using the direct method to demonstrate its use. The cylindrical furnace can be considered to be a three-surface enclosure with surface areas of

$$A_1 = A_2 = \pi r_o^2 = \pi (1 \text{ m})^2 = 3.14 \text{ m}^2$$
$$A_3 = 2\pi r_o H = 2\pi (1 \text{ m})(1 \text{ m}) = 6.28 \text{ m}^2$$

The view factor from the base to the top surface is, from Fig. 13–7, $F_{12} = 0.38$. Then the view factor from the base to the side surface is determined by applying the summation rule to be

$$F_{11} + F_{12} + F_{13} = 1 \quad \rightarrow \quad F_{13} = 1 - F_{11} - F_{12} = 1 - 0 - 0.38 = 0.62$$

since the base surface is flat and thus $F_{11} = 0$. Noting that the top and bottom surfaces are symmetric about the side surface, $F_{21} = F_{12} = 0.38$ and $F_{23} = F_{13} = 0.62$. The view factor F_{31} is determined from the reciprocity relation,

$$A_1 F_{13} = A_3 F_{31} \quad \rightarrow \quad F_{31} = F_{13}(A_1/A_3) = (0.62)(0.314/0.628) = 0.31$$

Also, $F_{32} = F_{31} = 0.31$ because of symmetry. Now that all the view factors are available, we apply Eq. 13–35 to each surface to determine the radiosities:

Top surface (i = 1): $\sigma T_1^4 = J_1 + \dfrac{1 - \varepsilon_1}{\varepsilon_1} [F_{12}(J_1 - J_2) + F_{13}(J_1 - J_3)]$

Bottom surface (i = 2): $\sigma T_2^4 = J_2 + \dfrac{1 - \varepsilon_2}{\varepsilon_2} [F_{21}(J_2 - J_1) + F_{23}(J_2 - J_3)]$

Side surface (i = 3): $\sigma T_3^4 = J_3 + 0$ (since surface 3 is black and thus $\varepsilon_3 = 1$)

Substituting the known quantities,

$$(5.67 \times 10^{-8} \text{ W/m}^2 \cdot \text{K}^4)(700 \text{ K})^4 = J_1 + \frac{1 - 0.8}{0.8} [0.38(J_1 - J_2) + 0.62(J_1 - J_3)]$$

$$(5.67 \times 10^{-8} \text{ W/m}^2 \cdot \text{K}^4)(500 \text{ K})^4 = J_2 + \frac{1 - 0.4}{0.4} [0.38(J_2 - J_1) + 0.62(J_2 - J_3)]$$

$$(5.67 \times 10^{-8} \text{ W/m}^2 \cdot \text{K}^4)(400 \text{ K})^4 = J_3$$

Solving these equations for J_1, J_2, and J_3 gives

$$J_1 = 11{,}418 \text{ W/m}^2, \quad J_2 = 4562 \text{ W/m}^2, \quad \text{and} \quad J_3 = 1452 \text{ W/m}^2$$

Then the net rates of radiation heat transfer at the three surfaces are determined from Eq. 13–34 to be

$$
\begin{aligned}
\dot{Q}_1 &= A_1[F_{1 \rightarrow 2}(J_1 - J_2) + F_{1 \rightarrow 3}(J_1 - J_3)] \\
&= (3.14 \text{ m}^2)[0.38(11{,}418 - 4562) + 0.62(11{,}418 - 1452)] \text{ W/m}^2 \\
&= 27.6 \text{ kW}
\end{aligned}
$$

$$
\begin{aligned}
\dot{Q}_2 &= A_2[F_{2 \rightarrow 1}(J_2 - J_1) + F_{2 \rightarrow 3}(J_2 - J_3)] \\
&= (3.14 \text{ m}^2)[0.38(4562 - 11{,}418) + 0.62(4562 - 1452)] \text{ W/m}^2 \\
&= -2.13 \text{ kW}
\end{aligned}
$$

$$
\begin{aligned}
\dot{Q}_3 &= A_3[F_{3 \rightarrow 1}(J_3 - J_1) + F_{3 \rightarrow 2}(J_3 - J_2)] \\
&= (6.28 \text{ m}^2)[0.31(1452 - 11{,}418) + 0.31(1452 - 4562)] \text{ W/m}^2 \\
&= -25.5 \text{ kW}
\end{aligned}
$$

Note that the direction of net radiation heat transfer is *from* the top surface *to* the base and side surfaces, and the algebraic sum of these three quantities must be equal to zero. That is,

$$\dot{Q}_1 + \dot{Q}_2 + \dot{Q}_3 = 27.6 + (-2.13) + (-25.5) \cong 0$$

Discussion To maintain the surfaces at the specified temperatures, we must supply heat to the top surface continuously at a rate of 27.6 kW while removing 2.13 kW from the base and 25.5 kW from the side surfaces.

The direct method presented here is straightforward, and it does not require the evaluation of radiation resistances. Also, it can be applied to enclosures with any number of surfaces in the same manner.

EXAMPLE 13–9   Radiation Heat Transfer in a Triangular Furnace

A furnace is shaped like a long equilateral triangular duct, as shown in Fig. 13–28. The width of each side is 1 m. The base surface has an emissivity of 0.7 and is maintained at a uniform temperature of 600 K. The heated left-side surface closely approximates a blackbody at 1000 K. The right-side surface is well insulated. Determine the rate at which heat must be supplied to the heated side externally per unit length of the duct in order to maintain these operating conditions.

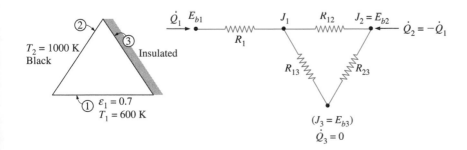

FIGURE 13–28
The triangular furnace
considered in Example 13–9.

SOLUTION Two of the surfaces of a long equilateral triangular furnace are maintained at uniform temperatures while the third surface is insulated. The external rate of heat transfer to the heated side per unit length of the duct during steady operation is to be determined.

Assumptions **1** Steady operating conditions exist. **2** The surfaces are opaque, diffuse, and gray. **3** Convection heat transfer is not considered.

Analysis The furnace can be considered to be a three-surface enclosure with a radiation network as shown in the figure, since the duct is very long and thus the end effects are negligible. We observe that the view factor from any surface to any other surface in the enclosure is 0.5 because of symmetry. Surface 3 is a reradiating surface since the net rate of heat transfer at that surface is zero. Then we must have $\dot{Q}_1 = -\dot{Q}_2$, since the entire heat lost by surface 1 must be gained by surface 2. The radiation network in this case is a simple series–parallel connection, and we can determine $\dot{Q}_1$ directly from

$$\dot{Q}_1 = \frac{E_{b1} - E_{b2}}{R_1 + \left(\dfrac{1}{R_{12}} + \dfrac{1}{R_{13} + R_{23}}\right)^{-1}} = \frac{E_{b1} - E_{b2}}{\dfrac{1 - \varepsilon_1}{A_1 \varepsilon_1} + \left(A_1 F_{12} + \dfrac{1}{1/A_1 F_{13} + 1/A_2 F_{23}}\right)^{-1}}$$

where

$$A_1 = A_2 = A_3 = wL = 1 \text{ m} \times 1 \text{ m} = 1 \text{ m}^2 \qquad \text{(per unit length of the duct)}$$
$$F_{12} = F_{13} = F_{23} = 0.5 \qquad \text{(symmetry)}$$
$$E_{b1} = \sigma T_1^4 = (5.67 \times 10^{-8} \text{ W/m}^2 \cdot \text{K}^4)(600 \text{ K})^4 = 7348 \text{ W/m}^2$$
$$E_{b2} = \sigma T_2^4 = (5.67 \times 10^{-8} \text{ W/m}^2 \cdot \text{K}^4)(1000 \text{ K})^4 = 56{,}700 \text{ W/m}^2$$

Substituting,

$$\dot{Q}_1 = \frac{(56{,}700 - 7348) \text{ W/m}^2}{\dfrac{1 - 0.7}{0.7 \times 1 \text{ m}^2} + \left[(0.5 \times 1 \text{ m}^2) + \dfrac{1}{1/(0.5 \times 1 \text{ m}^2) + 1/(0.5 \times 1 \text{ m}^2)}\right]^{-1}}$$
$$= \textbf{28.0 kW}$$

Therefore, heat at a rate of 28 kW must be supplied to the heated surface per unit length of the duct to maintain steady operation in the furnace.

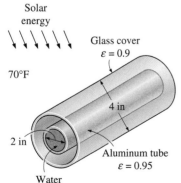

FIGURE 13–29
Schematic for Example 13–10.

EXAMPLE 13–10   **Heat Transfer through a Tubular Solar Collector**

A solar collector consists of a horizontal aluminum tube having an outer diameter of 2 in enclosed in a concentric thin glass tube of 4-in diameter, as shown in Fig. 13–29. Water is heated as it flows through the tube, and the space between the aluminum and the glass tubes is filled with air at 1 atm pressure. The pump circulating the water fails during a clear day, and the water temperature in the tube starts rising. The aluminum tube absorbs solar radiation at a rate of 30 Btu/h per foot length, and the temperature of the ambient air outside is 70°F. The emissivities of the tube and the glass cover are 0.95 and 0.9, respectively. Taking the effective sky temperature to be 50°F, determine the temperature of the aluminum tube when steady operating conditions are established (i.e., when the rate of heat loss from the tube equals the amount of solar energy gained by the tube).

SOLUTION The circulating pump of a solar collector that consists of a horizontal tube and its glass cover fails. The equilibrium temperature of the tube is to be determined.

Assumptions **1** Steady operating conditions exist. **2** The tube and its cover are isothermal. **3** Air is an ideal gas. **4** The surfaces are opaque, diffuse, and gray for infrared radiation. **5** The glass cover is transparent to solar radiation.

Properties The properties of air should be evaluated at the average temperature. But we do not know the exit temperature of the air in the duct, and thus we cannot determine the bulk fluid and glass cover temperatures at this point, and thus we cannot evaluate the average temperatures. Therefore, we assume the glass temperature to be 110°F, and use properties at an anticipated average temperature of $(70 + 110)/2 = 90°F$ (Table A–15E),

$$k = 0.01505 \text{ Btu/h} \cdot \text{ft} \cdot °F \qquad\qquad Pr = 0.7275$$

$$\nu = 1.753 \times 10^{-4} \text{ ft}^2/\text{s} \qquad\qquad \beta = \frac{1}{T_{ave}} = \frac{1}{550 \text{ R}}$$

Analysis This problem was solved in Chap. 9 by disregarding radiation heat transfer. Now we repeat the solution by considering natural convection and radiation occurring simultaneously.

We have a horizontal cylindrical enclosure filled with air at 1 atm pressure. The problem involves heat transfer from the aluminum tube to the glass cover and from the outer surface of the glass cover to the surrounding ambient air. When steady operation is reached, these two heat transfer rates must equal the rate of heat gain. That is,

$$\dot{Q}_{\text{tube-glass}} = \dot{Q}_{\text{glass ambient}} = \dot{Q}_{\text{solar gain}} = 30 \text{ Btu/h} \qquad \text{(per foot of tube)}$$

The heat transfer surface area of the glass cover is

$$A_o = A_{\text{glass}} = (\pi D_o L) = \pi(4/12 \text{ ft})(1 \text{ ft}) = 1.047 \text{ ft}^2 \qquad \text{(per foot of tube)}$$

To determine the Rayleigh number, we need to know the surface temperature of the glass, which is not available. Therefore, it is clear that the solution requires a trial-and-error approach unless we use an equation solver such as EES. Assuming the glass cover temperature to be 110°F, the Rayleigh number, the Nusselt number, the convection heat transfer coefficient, and the rate of natural convection heat transfer from the glass cover to the ambient air are determined to be

$$Ra_{D_o} = \frac{g\beta(T_o - T_\infty) D_o^3}{\nu^2} Pr$$

$$= \frac{(32.2 \text{ ft/s}^2)[1/(550 \text{ R})](110 - 70 \text{ R})(4/12 \text{ ft})^3}{(1.753 \times 10^{-4} \text{ ft}^2/\text{s})^2} (0.7275) = 2.053 \times 10^6$$

$$Nu = \left\{ 0.6 + \frac{0.387 \, Ra_{D_o}^{1/6}}{[1 + (0.559/Pr)^{9/16}]^{8/27}} \right\}^2 = \left\{ 0.6 + \frac{0.387(2.053 \times 10^6)^{1/6}}{[1 + (0.559/0.7275)^{9/16}]^{8/27}} \right\}^2$$

$$= 17.88$$

$$h_o = \frac{k}{D_o} \text{Nu} = \frac{0.01505 \text{ Btu/h} \cdot \text{ft} \cdot °\text{F}}{4/12 \text{ ft}} (17.88) = 0.8073 \text{ Btu/h} \cdot \text{ft}^2 \cdot °\text{F}$$

$$\dot{Q}_{o,\text{conv}} = h_o A_o (T_o - T_\infty) = (0.8073 \text{ Btu/h} \cdot \text{ft}^2 \cdot °\text{F})(1.047 \text{ ft}^2)(110 - 70)°\text{F}$$
$$= 33.8 \text{ Btu/h}$$

Also,

$$\dot{Q}_{o,\text{rad}} = \varepsilon_o \sigma A_o (T_o^4 - T_{\text{sky}}^4)$$
$$= (0.9)(0.1714 \times 10^{-8} \text{ Btu/h} \cdot \text{ft}^2 \cdot \text{R}^4)(1.047 \text{ ft}^2)[(570 \text{ R})^4 - (510 \text{ R})^4]$$
$$= 61.2 \text{ Btu/h}$$

Then the total rate of heat loss from the glass cover becomes

$$\dot{Q}_{o,\text{total}} = \dot{Q}_{o,\text{conv}} + \dot{Q}_{o,\text{rad}} = 33.8 + 61.2 = 95.0 \text{ Btu/h}$$

which is much larger than 30 Btu/h. Therefore, the assumed temperature of 110°F for the glass cover is high. Repeating the calculations with lower temperatures (including the evaluation of properties), the glass cover temperature corresponding to 30 Btu/h is determined to be 78°F (it would be 106°F if radiation were ignored).

The temperature of the aluminum tube is determined in a similar manner using the natural convection and radiation relations for two horizontal concentric cylinders. The characteristic length in this case is the distance between the two cylinders, which is

$$L_c = (D_o - D_i)/2 = (4 - 2)/2 = 1 \text{ in} = 1/12 \text{ ft}$$

Also,

$$A_i = A_{\text{tube}} = (\pi D_i L) = \pi(2/12 \text{ ft})(1 \text{ ft}) = 0.5236 \text{ ft}^2 \qquad \text{(per foot of tube)}$$

We start the calculations by assuming the tube temperature to be 122°F, and thus an average temperature of $(78 + 122)/2 = 100°\text{F} = 560 \text{ R}$. Using properties at 100°F,

$$\text{Ra}_L = \frac{g\beta(T_i - T_o)L_c^3}{\nu^2} \text{Pr}$$

$$= \frac{(32.2 \text{ ft/s}^2)[1/(560 \text{ R})](122 - 78 \text{ R})(1/12 \text{ ft})^3}{(1.809 \times 10^{-4} \text{ ft}^2/\text{s})^2} (0.726) = 3.246 \times 10^4$$

The effective thermal conductivity is

$$F_{\text{cyl}} = \frac{[\ln(D_o/D_i)]^4}{L_c^3 (D_i^{-3/5} + D_o^{-3/5})^5} = \frac{[\ln(4/2)]^4}{(1/12 \text{ ft})^3 [(2/12 \text{ ft})^{-3/5} + (4/12 \text{ ft})^{-3/5}]^5} = 0.1466$$

$$k_{\text{eff}} = 0.386k \left(\frac{\text{Pr}}{0.861 + \text{Pr}}\right)^{1/4} (F_{\text{cyl}} \text{Ra}_L)^{1/4}$$

$$= 0.386(0.01529 \text{ Btu/h} \cdot \text{ft} \cdot °\text{F})\left(\frac{0.726}{0.861 + 0.726}\right)^{1/4} (0.1466 \times 3.248 \times 10^4)^{1/4}$$

$$= 0.04032 \text{ Btu/h} \cdot \text{ft} \cdot °\text{F}$$

Then the rate of heat transfer between the cylinders by convection becomes

$$\dot{Q}_{i,\,\text{conv}} = \frac{2\pi k_{\text{eff}}}{\ln(D_o/D_i)}\,(T_i - T_o)$$

$$= \frac{2\pi(0.04032 \text{ Btu/h} \cdot \text{ft }^\circ\text{F})}{\ln(4/2)}\,(122 - 78)^\circ\text{F} = 16.1 \text{ Btu/h}$$

Also,

$$\dot{Q}_{i,\,\text{rad}} = \frac{\sigma A_i\,(T_i^4 - T_o^4)}{\dfrac{1}{\varepsilon_i} + \dfrac{1 - \varepsilon_o}{\varepsilon_o}\left(\dfrac{D_i}{D_o}\right)}$$

$$= \frac{(0.1714 \times 10^{-8} \text{ Btu/h} \cdot \text{ft}^2 \cdot \text{R}^4)(0.5236 \text{ ft}^2)[(582 \text{ R})^4 - (538 \text{ R})^4]}{\dfrac{1}{0.95} + \dfrac{1 - 0.9}{0.9}\left(\dfrac{2 \text{ in}}{4 \text{ in}}\right)}$$

$$= 25.1 \text{ Btu/h}$$

Then the total rate of heat loss from the glass cover becomes

$$\dot{Q}_{i,\,\text{total}} = \dot{Q}_{i,\,\text{conv}} + \dot{Q}_{i,\,\text{rad}} = 16.1 + 25.1 = 41.2 \text{ Btu/h}$$

which is larger than 30 Btu/h. Therefore, the assumed temperature of 122°F for the tube is high. By trying other values, the tube temperature corresponding to 30 Btu/h is determined to be **112°F** (it would be 180°F if radiation were ignored). Therefore, the tube will reach an equilibrium temperature of 112°F when the pump fails.

Discussion It is clear from the results obtained that radiation should always be considered in systems that are heated or cooled by natural convection, unless the surfaces involved are polished and thus have very low emissivities.

13–5 · RADIATION SHIELDS AND THE RADIATION EFFECTS

Radiation heat transfer between two surfaces can be reduced greatly by inserting a thin, high-reflectivity (low-emissivity) sheet of material between the two surfaces. Such highly reflective thin plates or shells are called **radiation shields**. Multilayer radiation shields constructed of about 20 sheets per cm thickness separated by evacuated space are commonly used in cryogenic and space applications. Radiation shields are also used in temperature measurements of fluids to reduce the error caused by the radiation effect when the temperature sensor is exposed to surfaces that are much hotter or colder than the fluid itself. The role of the radiation shield is to reduce the rate of radiation heat transfer by placing additional resistances in the path of radiation heat flow. The lower the emissivity of the shield, the higher the resistance.

Radiation heat transfer between two large parallel plates of emissivities ε_1 and ε_2 maintained at uniform temperatures T_1 and T_2 is given by Eq. 13–38:

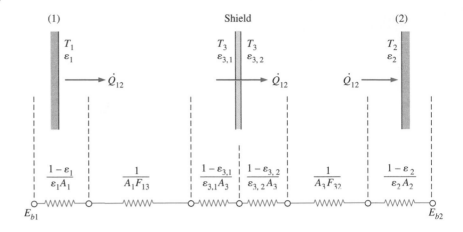

FIGURE 13–30

The radiation shield placed between two parallel plates and the radiation network associated with it.

$$\dot{Q}_{12,\text{ no shield}} = \frac{A\sigma(T_1^4 - T_2^4)}{\dfrac{1}{\varepsilon_1} + \dfrac{1}{\varepsilon_2} - 1}$$

Now consider a radiation shield placed between these two plates, as shown in Fig. 13–30. Let the emissivities of the shield facing plates 1 and 2 be $\varepsilon_{3,1}$ and $\varepsilon_{3,2}$, respectively. Note that the emissivity of different surfaces of the shield may be different. The radiation network of this geometry is constructed, as usual, by drawing a surface resistance associated with each surface and connecting these surface resistances with space resistances, as shown in the figure. The resistances are connected in series, and thus the rate of radiation heat transfer is

$$\dot{Q}_{12,\text{ one shield}} = \frac{E_{b1} - E_{b2}}{\dfrac{1-\varepsilon_1}{A_1\,\varepsilon_1} + \dfrac{1}{A_1\,F_{12}} + \dfrac{1-\varepsilon_{3,1}}{A_3\,\varepsilon_{3,1}} + \dfrac{1-\varepsilon_{3,2}}{A_3\,\varepsilon_{3,2}} + \dfrac{1}{A_3\,F_{32}} + \dfrac{1-\varepsilon_2}{A_2\,\varepsilon_2}} \qquad \textbf{(13-42)}$$

Noting that $F_{13} = F_{23} = 1$ and $A_1 = A_2 = A_3 = A$ for infinite parallel plates, Eq. 13–42 simplifies to

$$\dot{Q}_{12,\text{ one shield}} = \frac{A\sigma(T_1^4 - T_2^4)}{\left(\dfrac{1}{\varepsilon_1} + \dfrac{1}{\varepsilon_2} - 1\right) + \left(\dfrac{1}{\varepsilon_{3,1}} + \dfrac{1}{\varepsilon_{3,2}} - 1\right)} \qquad \textbf{(13-43)}$$

where the terms in the second set of parentheses in the denominator represent the additional resistance to radiation introduced by the shield. The appearance of the equation above suggests that parallel plates involving multiple radiation shields can be handled by adding a group of terms like those in the second set of parentheses to the denominator for each radiation shield. Then the radiation heat transfer through large parallel plates separated by N radiation shields becomes

$$\dot{Q}_{12,\,N\text{ shields}} = \frac{A\sigma(T_1^4 - T_2^4)}{\left(\dfrac{1}{\varepsilon_1} + \dfrac{1}{\varepsilon_2} - 1\right) + \left(\dfrac{1}{\varepsilon_{3,1}} + \dfrac{1}{\varepsilon_{3,2}} - 1\right) + \cdots + \left(\dfrac{1}{\varepsilon_{N,1}} + \dfrac{1}{\varepsilon_{N,2}} - 1\right)}$$

$$\textbf{(13-44)}$$

If the emissivities of all surfaces are equal, Eq. 13–44 reduces to

$$\dot{Q}_{12,\,N\text{ shields}} = \frac{A\sigma(T_1^4 - T_2^4)}{(N + 1)\left(\dfrac{1}{\varepsilon} + \dfrac{1}{\varepsilon} - 1\right)} = \frac{1}{N + 1}\dot{Q}_{12,\text{ no shield}} \qquad \textbf{(13-45)}$$

Therefore, when all emissivities are equal, 1 shield reduces the rate of radiation heat transfer to one-half, 9 shields reduce it to one-tenth, and 19 shields reduce it to one-twentieth (or 5 percent) of what it was when there were no shields.

The equilibrium temperature of the radiation shield T_3 in Figure 13–30 can be determined by expressing Eq. 13–43 for $\dot{Q}_{13}$ or $\dot{Q}_{23}$ (which involves T_3) after evaluating $\dot{Q}_{12}$ from Eq. 13–43 and noting that $\dot{Q}_{12} = \dot{Q}_{13} = \dot{Q}_{23}$ when steady conditions are reached.

Radiation shields used to reduce the rate of radiation heat transfer between concentric cylinders and spheres can be handled in a similar manner. In case of one shield, Eq. 13–42 can be used by taking $F_{13} = F_{23} = 1$ for both cases and by replacing the A's by the proper area relations.

Radiation Effect on Temperature Measurements

A temperature measuring device indicates the temperature of its *sensor,* which is supposed to be, but is not necessarily, the temperature of the medium that the sensor is in contact with. When a thermometer (or any other temperature measuring device such as a thermocouple) is placed in a medium, heat transfer takes place between the sensor of the thermometer and the medium by convection until the sensor reaches the temperature of the medium. But when the sensor is surrounded by surfaces that are at a different temperature than the fluid, radiation exchange also takes place between the sensor and the surrounding surfaces. When the heat transfers by convection and radiation balance each other, the sensor indicates a temperature that falls between the fluid and surface temperatures. Below we develop a procedure to account for the radiation effect and to determine the actual fluid temperature.

Consider a thermometer that is used to measure the temperature of a fluid flowing through a large channel whose walls are at a lower temperature than the fluid (Fig. 13–31). Equilibrium will be established and the reading of the thermometer will stabilize when heat gain by convection, as measured by the sensor, equals heat loss by radiation (or vice versa). That is, on a unit-area basis,

$$\dot{q}_{\text{conv, to sensor}} = \dot{q}_{\text{rad, from sensor}}$$

$$h(T_f - T_{\text{th}}) = \varepsilon\sigma(T_{\text{th}}^4 - T_w^4)$$

or

$$T_f = T_{\text{th}} + \frac{\varepsilon\sigma(T_{\text{th}}^4 - T_w^4)}{h} \qquad \text{(K)} \qquad \textbf{(13-46)}$$

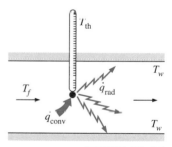

FIGURE 13–31
A thermometer used to measure the temperature of a fluid in a channel.

where

T_f = actual temperature of the fluid, K

T_{th} = temperature value measured by the thermometer, K

T_w = temperature of the surrounding surfaces, K

h = convection heat transfer coefficient, W/m² · K

ε = emissivity of the sensor of the thermometer

The last term in Eq. 13–46 is due to the *radiation effect* and represents the *radiation correction*. Note that the radiation correction term is most significant when the convection heat transfer coefficient is small and the emissivity of the surface of the sensor is large. Therefore, the sensor should be coated with a material of high reflectivity (low emissivity) to reduce the radiation effect.

Placing the sensor in a radiation shield without interfering with the fluid flow also reduces the radiation effect. The sensors of temperature measurement devices used outdoors must be protected from direct sunlight since the radiation effect in that case is sure to reach unacceptable levels.

The radiation effect is also a significant factor in *human comfort* in heating and air-conditioning applications. A person who feels fine in a room at a specified temperature may feel chilly in another room at the same temperature as a result of the radiation effect if the walls of the second room are at a considerably lower temperature. For example, most people feel comfortable in a room at 22°C if the walls of the room are also roughly at that temperature. When the wall temperature drops to 5°C for some reason, the interior temperature of the room must be raised to at least 27°C to maintain the same level of comfort. Therefore, well-insulated buildings conserve energy not only by reducing the heat loss or heat gain, but also by allowing the thermostats to be set at a lower temperature in winter and at a higher temperature in summer without compromising the comfort level.

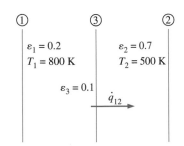

FIGURE 13–32
Schematic for Example 13–11.

EXAMPLE 13–11   Radiation Shields

A thin aluminum sheet with an emissivity of 0.1 on both sides is placed between two very large parallel plates that are maintained at uniform temperatures $T_1 = 800$ K and $T_2 = 500$ K and have emissivities $\varepsilon_1 = 0.2$ and $\varepsilon_2 = 0.7$, respectively, as shown in Fig. 13–32. Determine the net rate of radiation heat transfer between the two plates per unit surface area of the plates and compare the result to that without the shield.

SOLUTION   A thin aluminum sheet is placed between two large parallel plates maintained at uniform temperatures. The net rates of radiation heat transfer between the two plates with and without the radiation shield are to be determined.

Assumptions   The surfaces are opaque, diffuse, and gray.

Analysis   The net rate of radiation heat transfer between these two plates without the shield was determined in Example 13–7 to be 3625 W/m². Heat transfer in the presence of one shield is determined from Eq. 13–43 to be

$$\dot{q}_{12, \text{ one shield}} = \frac{\dot{Q}_{12, \text{ one shield}}}{A} = \frac{\sigma(T_1^4 - T_2^4)}{\left(\dfrac{1}{\varepsilon_1} + \dfrac{1}{\varepsilon_2} - 1\right) + \left(\dfrac{1}{\varepsilon_{3,1}} + \dfrac{1}{\varepsilon_{3,2}} - 1\right)}$$

$$= \frac{(5.67 \times 10^{\pm 8} \text{ W/m}^2 \cdot \text{K}^4)[(800 \text{ K})^4 - (500 \text{ K})^4]}{\left(\dfrac{1}{0.2} + \dfrac{1}{0.7} - 1\right) + \left(\dfrac{1}{0.1} + \dfrac{1}{0.1} - 1\right)}$$

$$= 806 \text{ W/m}^2$$

Discussion Note that the rate of radiation heat transfer reduces to about one-fourth of what it was as a result of placing a radiation shield between the two parallel plates.

EXAMPLE 13–12 **Radiation Effect on Temperature Measurements**

A thermocouple used to measure the temperature of hot air flowing in a duct whose walls are maintained at $T_w = 400$ K shows a temperature reading of $T_{th} = 650$ K (Fig. 13–33). Assuming the emissivity of the thermocouple junction to be $\varepsilon = 0.6$ and the convection heat transfer coefficient to be $h = 80$ W/m² · K, determine the actual temperature of the air.

SOLUTION The temperature of air in a duct is measured. Accounting for the radiation effect, and the actual air temperature is to be determined.
Assumptions The surfaces are opaque, diffuse, and gray.
Analysis The walls of the duct are at a considerably lower temperature than the air in it, and thus we expect the thermocouple to show a reading lower than the actual air temperature as a result of the radiation effect. The actual air temperature is determined from Eq. 13–46 to be

$$T_f = T_{th} + \frac{\varepsilon\sigma(T_{th}^4 - T_w^4)}{h}$$

$$= (650 \text{ K}) + \frac{0.6 \times (5.67 \times 10^{-8} \text{ W/m}^2 \cdot \text{K}^4)[(650 \text{ K})^4 - (400 \text{ K})^4]}{80 \text{ W/m}^2 \cdot \text{K}}$$

$$= 715 \text{ K}$$

Note that the radiation effect causes a difference of 65°C (or 65 K since °C ≡ K for temperature differences) in temperature reading in this case.

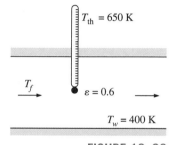

FIGURE 13–33
Schematic for Example 13–12.

13–6 · RADIATION EXCHANGE WITH EMITTING AND ABSORBING GASES

So far we considered radiation heat transfer between surfaces separated by a medium that does not emit, absorb, or scatter radiation—a nonparticipating medium that is completely transparent to thermal radiation. A vacuum satisfies this condition perfectly, and air at ordinary temperatures and pressures comes very close. Gases that consist of monatomic molecules such as Ar and He and symmetric diatomic molecules such as N_2 and O_2 are essentially transparent to radiation, except at extremely high temperatures at which ionization occurs. Therefore, atmospheric air can be considered to be a nonparticipating medium in radiation calculations.

Gases with asymmetric molecules such as H_2O, CO_2, CO, SO_2, and hydrocarbons H_mC_n may participate in the radiation process by absorption at moderate temperatures, and by absorption and emission at high temperatures such as those encountered in combustion chambers. Therefore, air or any other medium that contains such gases with asymmetric molecules at sufficient concentrations must be treated as a participating medium in radiation calculations. Combustion gases in a furnace or a combustion chamber, for example, contain sufficient amounts of H_2O and CO_2, and thus the emission and absorption of gases in furnaces must be taken into consideration.

The presence of a participating medium complicates the radiation analysis considerably for several reasons:

- A participating medium emits and absorbs radiation throughout its entire volume. That is, gaseous radiation is a *volumetric phenomena,* and thus it depends on the size and shape of the body. This is the case even if the temperature is uniform throughout the medium.

- Gases emit and absorb radiation at a number of narrow wavelength bands. This is in contrast to solids, which emit and absorb radiation over the entire spectrum. Therefore, the gray assumption may not always be appropriate for a gas even when the surrounding surfaces are gray.

- The emission and absorption characteristics of the constituents of a gas mixture also depends on the temperature, pressure, and composition of the gas mixture. Therefore, the presence of other participating gases affects the radiation characteristics of a particular gas.

The propagation of radiation through a medium can be complicated further by presence of *aerosols* such as dust, ice particles, liquid droplets, and soot (unburned carbon) particles that *scatter* radiation. Scattering refers to the change of direction of radiation due to reflection, refraction, and diffraction. Scattering caused by gas molecules themselves is known as the *Rayleigh scattering,* and it has negligible effect on heat transfer. Radiation transfer in scattering media is considered in advanced books such as the ones by Modest (1993) and Siegel and Howell (1992).

The participating medium can also be semitransparent liquids or solids such as water, glass, and plastics. To keep complexities to a manageable level, we limit our consideration to gases that emit and absorb radiation. In particular, we consider the emission and absorption of radiation by H_2O and CO_2 only since they are the participating gases most commonly encountered in practice (combustion products in furnaces and combustion chambers burning hydrocarbon fuels contain both gases at high concentrations), and they are sufficient to demonstrate the basic principles involved.

Radiation Properties of a Participating Medium

Consider a participating medium of thickness L. A spectral radiation beam of intensity $I_{\lambda, 0}$ is incident on the medium, which is attenuated as it propagates due to absorption. The decrease in the intensity of radiation as it passes through a layer of thickness dx is proportional to the intensity itself and the thickness dx. This is known as **Beer's law**, and is expressed as (Fig. 13–34)

$$dI_\lambda(x) = -\kappa_\lambda I_\lambda(x)dx \qquad (13\text{–}47)$$

where the constant of proportionality κ_λ is the **spectral absorption coefficient** of the medium whose unit is m^{-1} (from the requirement of dimensional homogeneity). This is just like the amount of interest earned by a bank account during a time interval being proportional to the amount of money in the account and the time interval, with the interest rate being the constant of proportionality.

Separating the variables and integrating from $x = 0$ to $x = L$ gives

$$\frac{I_{\lambda, L}}{I_{\lambda, 0}} = e^{-\kappa_\lambda L} \qquad (13\text{–}48)$$

where we have assumed the absorptivity of the medium to be independent of x. Note that radiation intensity decays exponentially in accordance with Beer's law.

The **spectral transmissivity** of a medium can be defined as the ratio of the intensity of radiation leaving the medium to that entering the medium. That is,

$$\tau_\lambda = \frac{I_{\lambda, L}}{I_{\lambda, 0}} = e^{-\kappa_\lambda L} \qquad (13\text{–}49)$$

Note that $\tau_\lambda = 1$ when no radiation is absorbed and thus radiation intensity remains constant. Also, the spectral transmissivity of a medium represents the fraction of radiation transmitted by the medium at a given wavelength.

Radiation passing through a nonscattering (and thus nonreflecting) medium is either absorbed or transmitted. Therefore $\alpha_\lambda + \tau_\lambda = 1$, and the **spectral absorptivity** of a medium of thickness L is

$$\alpha_\lambda = 1 - \tau_\lambda = 1 - e^{-\kappa_\lambda L} \qquad (13\text{–}50)$$

From Kirchoff's law, the **spectral emissivity** of the medium is

$$\varepsilon_\lambda = \alpha_\lambda = 1 - e^{-\kappa_\lambda L} \qquad (13\text{–}51)$$

Note that the spectral absorptivity, transmissivity, and emissivity of a medium are dimensionless quantities, with values less than or equal to 1. The spectral absorption coefficient of a medium (and thus ε_λ, α_λ, and τ_λ), in general, vary with wavelength, temperature, pressure, and composition.

For an *optically thick* medium (a medium with a large value of $\kappa_\lambda L$), Eq. 13–51 gives $\varepsilon_\lambda \approx \alpha_\lambda \approx 1$. For $\kappa_\lambda L = 5$, for example, $\varepsilon_\lambda = \alpha_\lambda = 0.993$. Therefore, an optically thick medium emits like a blackbody at the given wavelength. As a result, an optically thick absorbing-emitting medium with no significant scattering at a given temperature T_g can be viewed as a "black surface" at T_g since it will absorb essentially all the radiation passing through it, and it will emit the maximum possible radiation that can be emitted by a surface at T_g, which is $E_{b\lambda}(T_g)$.

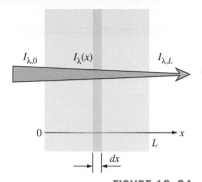

FIGURE 13–34
The attenuation of a radiation beam while passing through an absorbing medium of thickness L.

Emissivity and Absorptivity of Gases and Gas Mixtures

The spectral absorptivity of CO_2 is given in Figure 13–35 as a function of wavelength. The various peaks and dips in the figure together with discontinuities show clearly the band nature of absorption and the strong nongray characteristics. The shape and the width of these absorption bands vary with temperature and pressure, but the magnitude of absorptivity also varies with the thickness of the gas layer. Therefore, absorptivity values without specified thickness and pressure are meaningless.

The nongray nature of properties should be considered in radiation calculations for high accuracy. This can be done using a band model, and thus performing calculations for each absorption band. However, satisfactory results can be obtained by assuming the gas to be gray, and using an effective total absorptivity and emissivity determined by some averaging process. Charts for the total emissivities of gases are first presented by Hottel (1954), and they have been widely used in radiation calculations with reasonable accuracy. Alternative emissivity charts and calculation procedures have been developed more recently by Edwards and Matavosian (1984). Here we present the Hottel approach because of its simplicity.

Even with gray assumption, the total emissivity and absorptivity of a gas depends on the geometry of the gas body as well as the temperature, pressure, and composition. Gases that participate in radiation exchange such as CO_2 and H_2O typically coexist with nonparticipating gases such as N_2 and O_2, and thus radiation properties of an absorbing and emitting gas are usually reported for a mixture of the gas with nonparticipating gases rather than the pure gas. The emissivity and absorptivity of a gas component in a mixture depends primarily on its density, which is a function of temperature and partial pressure of the gas.

The emissivity of H_2O vapor in a mixture of nonparticipating gases is plotted in Figure 13–36a for a total pressure of $P = 1$ atm as a function of gas temperature T_g for a range of values for P_wL, where P_w is the partial pressure of water vapor and L is the mean distance traveled by the radiation beam.

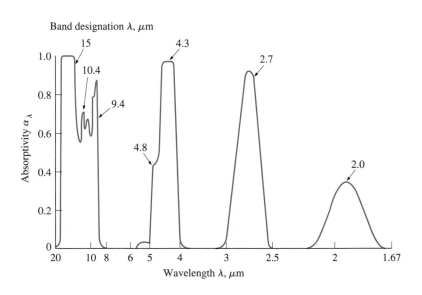

FIGURE 13–35

Spectral absorptivity of CO_2 at 830 K and 10 atm for a path length of 38.8 cm
(from Siegel and Howell, 1992).

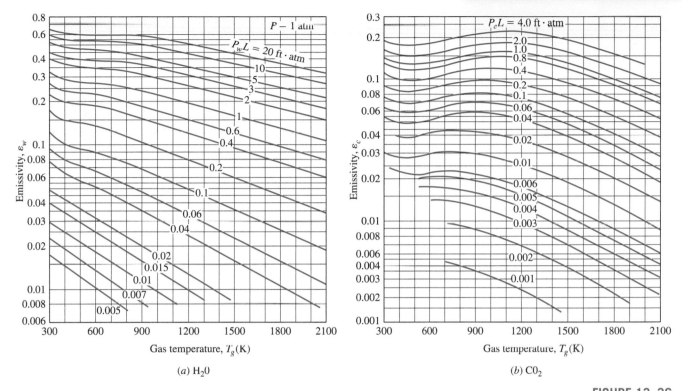

FIGURE 13–36

Emissivities of H_2O and CO_2 gases in a mixture of nonparticipating gases at a total pressure of 1 atm for a mean beam length of L (1 m · atm = 3.28 ft · atm)

(from Hottel, 1954).

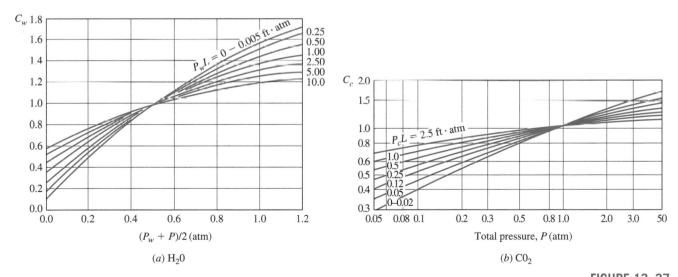

FIGURE 13–37

Correction factors for the emissivities of H_2O and CO_2 gases at pressures other than 1 atm for use in the relations $\varepsilon_w = C_w \varepsilon_{w, \, 1 \, atm}$ and $\varepsilon_c = C_c \varepsilon_{c, \, 1 \, atm}$ (1 m · atm = 3.28 ft · atm)

(from Hottel, 1954).

Emissivity at a total pressure P other than $P = 1$ atm is determined by multiplying the emissivity value at 1 atm by a **pressure correction factor** C_w obtained from Figure 13–37a for water vapor. That is,

$$\varepsilon_w = C_w \varepsilon_{w,\,1\,\text{atm}} \qquad (13\text{–}52)$$

Note that $C_w = 1$ for $P = 1$ atm and thus $(P_w + P)/2 \cong 0.5$ (a very low concentration of water vapor is used in the preparation of the emissivity chart in Fig. 13–36a and thus P_w is very low). Emissivity values are presented in a similar manner for a mixture of CO_2 and nonparticipating gases in Fig. 13–36b and 13–37b.

Now the question that comes to mind is what will happen if the CO_2 and H_2O gases exist *together* in a mixture with nonparticipating gases. The emissivity of each participating gas can still be determined as explained above using its partial pressure, but the effective emissivity of the mixture cannot be determined by simply adding the emissivities of individual gases (although this would be the case if different gases emitted at different wavelengths). Instead, it should be determined from

$$\varepsilon_g = \varepsilon_c + \varepsilon_w - \Delta\varepsilon$$
$$= C_c \varepsilon_{c,\,1\,\text{atm}} + C_w \varepsilon_{w,\,1\,\text{atm}} - \Delta\varepsilon \qquad (13\text{–}53)$$

where $\Delta\varepsilon$ is the **emissivity correction factor,** which accounts for the overlap of emission bands. For a gas mixture that contains both CO_2 and H_2O gases, $\Delta\varepsilon$ is plotted in Figure 13–38.

The emissivity of a gas also depends on the *mean length* an emitted radiation beam travels in the gas before reaching a bounding surface, and thus the shape and the size of the gas body involved. During their experiments in the 1930s, Hottel and his coworkers considered the emission of radiation from a hemispherical gas body to a small surface element located at the center of the base of the hemisphere. Therefore, the given charts represent emissivity

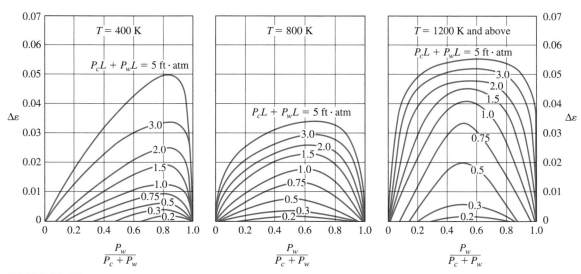

FIGURE 13–38

Emissivity correction $\Delta\varepsilon$ for use in $\varepsilon_g = \varepsilon_w + \varepsilon_c - \Delta\varepsilon$ when both CO_2 and H_2O vapor are present in a gas mixture ($1\,\text{m} \cdot \text{atm} = 3.28\,\text{ft} \cdot \text{atm}$)

(from Hottel, 1954).

data for the emission of radiation from a hemispherical gas body of radius L toward the center of the base of the hemisphere. It is certainly desirable to extend the reported emissivity data to gas bodies of other geometries, and this is done by introducing the concept of **mean beam length** L, which represents the radius of an equivalent hemisphere. The mean beam lengths for various gas geometries are listed in Table 13–4. More extensive lists are available in the literature [such as Hottel (1954), and Siegel and Howell, (1992)]. The emissivities associated with these geometries can be determined from Figures 13–36 through 13–38 by using the appropriate mean beam length.

Following a procedure recommended by Hottel, the absorptivity of a gas that contains CO_2 and H_2O gases for radiation emitted by a source at temperature T_s can be determined similarly from

$$\alpha_g = \alpha_c + \alpha_w - \Delta\alpha \tag{13-54}$$

where $\Delta\alpha = \Delta\varepsilon$ and is determined from Figure 13–38 at the source temperature T_s. The absorptivities of CO_2 and H_2O can be determined from the emissivity charts (Figs. 12–36 and 12–37) as

CO_2:
$$\alpha_c = C_c \times (T_g/T_s)^{0.65} \times \varepsilon_c(T_s, P_cLT_s/T_g) \tag{13-55}$$

and

H_2O:
$$\alpha_w = C_w \times (T_g/T_s)^{0.45} \times \varepsilon_w(T_s, P_wLT_s/T_g) \tag{13-56}$$

The notation indicates that the emissivities should be evaluated using T_s instead of T_g (both in K or R), P_cLT_s/T_g instead of P_cL, and P_wLT_s/T_g instead of P_wL. Note that the absorptivity of the gas depends on the source temperature T_s as well as the gas temperature T_g. Also, $\alpha = \varepsilon$ when $T_s = T_g$, as expected. The pressure correction factors C_c and C_w are evaluated using P_cL and P_wL, as in emissivity calculations.

TABLE 13–4

Mean beam length L for various gas volume shapes

Gas Volume Geometry	L
Hemisphere of radius R radiating to the center of its base	R
Sphere of diameter D radiating to its surface	$0.65D$
Infinite circular cylinder of diameter D radiating to curved surface	$0.95D$
Semi-infinite circular cylinder of diameter D radiating to its base	$0.65D$
Semi-infinite circular cylinder of diameter D radiating to center of its base	$0.90D$
Infinite semicircular cylinder of radius R radiating to center of its base	$1.26R$
Circular cylinder of height equal to diameter D radiating to entire surface	$0.60D$
Circular cylinder of height equal to diameter D radiating to center of its base	$0.71D$
Infinite slab of thickness D radiating to either bounding plane	$1.80D$
Cube of side length L radiating to any face	$0.66L$
Arbitrary shape of volume V and surface area A_s radiating to surface	$3.6V/A_s$

When the total emissivity of a gas ε_g at temperature T_g is known, the emissive power of the gas (radiation emitted by the gas per unit surface area) can be expressed as $E_g = \varepsilon_g \sigma T_g^4$. Then the rate of radiation energy emitted by a gas to a bounding surface of area A_s becomes

$$\dot{Q}_{g,e} = \varepsilon_g A_s \sigma T_g^4 \qquad (13\text{--}57)$$

If the bounding surface is black at temperature T_s, the surface will emit radiation to the gas at a rate of $A_s \sigma T_s^4$ without reflecting any, and the gas will absorb this radiation at a rate of $\alpha_g A_s \sigma T_s^4$, where α_g is the absorptivity of the gas. Then the net rate of radiation heat transfer between the gas and a black surface surrounding it becomes

Black enclosure: $\qquad \dot{Q}_{net} = A_s \sigma (\varepsilon_g T_g^4 - \alpha_g T_s^4) \qquad (13\text{--}58)$

If the surface is not black, the analysis becomes more complicated because of the radiation reflected by the surface. But for surfaces that are nearly black with an emissivity $\varepsilon_s > 0.7$, Hottel (1954), recommends this modification,

$$\dot{Q}_{net,\,gray} = \frac{\varepsilon_s + 1}{2} \dot{Q}_{net,\,black} = \frac{\varepsilon_s + 1}{2} A_s \sigma (\varepsilon_g T_g^4 - \alpha_g T_s^4) \qquad (13\text{--}59)$$

The emissivity of wall surfaces of furnaces and combustion chambers are typically greater than 0.7, and thus the relation above provides great convenience for preliminary radiation heat transfer calculations.

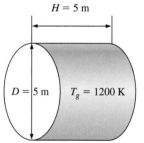

$H = 5$ m

$D = 5$ m $T_g = 1200$ K

FIGURE 13–39
Schematic for Example 13–13.

EXAMPLE 13–13   **Effective Emissivity of Combustion Gases**

A cylindrical furnace whose height and diameter are 5 m contains combustion gases at 1200 K and a total pressure of 2 atm. The composition of the combustion gases is determined by volumetric analysis to be 80 percent N_2, 8 percent H_2O, 7 percent O_2, and 5 percent CO_2. Determine the effective emissivity of the combustion gases (Fig. 13–39).

SOLUTION   The temperature, pressure, and composition of a gas mixture is given. The emissivity of the mixture is to be determined.
Assumptions   **1** All the gases in the mixture are ideal gases. **2** The emissivity determined is the mean emissivity for radiation emitted to all surfaces of the cylindrical enclosure.
Analysis   The volumetric analysis of a gas mixture gives the mole fractions y_i of the components, which are equivalent to pressure fractions for an ideal gas mixture. Therefore, the partial pressures of CO_2 and H_2O are

$$P_c = y_{CO_2} P = 0.05(2 \text{ atm}) = 0.10 \text{ atm}$$
$$P_w = y_{H_2O} P = 0.08(2 \text{ atm}) = 0.16 \text{ atm}$$

The mean beam length for a cylinder of equal diameter and height for radiation emitted to all surfaces is, from Table 13–4,

$$L = 0.60D = 0.60(5 \text{ m}) = 3 \text{ m}$$

Then,

$$P_c L = (0.10 \text{ atm})(3 \text{ m}) = 0.30 \text{ m} \cdot \text{atm} = 0.98 \text{ ft} \cdot \text{atm}$$
$$P_w L = (0.16 \text{ atm})(3 \text{ m}) = 0.48 \text{ m} \cdot \text{atm} = 1.57 \text{ ft} \cdot \text{atm}$$

The emissivities of CO_2 and H_2O corresponding to these values at the gas temperature of $T_g = 1200$ K and 1 atm are, from Figure 13–36,

$$\varepsilon_{c, 1 \text{ atm}} = 0.16 \quad \text{and} \quad \varepsilon_{w, 1 \text{ atm}} = 0.23$$

These are the base emissivity values at 1 atm, and they need to be corrected for the 2 atm total pressure. Noting that $(P_w + P)/2 = (0.16 + 2)/2 = 1.08$ atm, the pressure correction factors are, from Figure 13–37,

$$C_c = 1.1 \quad \text{and} \quad C_w = 1.4$$

Both CO_2 and H_2O are present in the same mixture, and we need to correct for the overlap of emission bands. The emissivity correction factor at $T = T_g = 1200$ K is, from Figure 13–38,

$$\left. \begin{array}{l} P_c L + P_w L = 0.98 + 1.57 = 2.55 \\[2mm] \dfrac{P_w}{P_w + P_c} = \dfrac{0.16}{0.16 + 0.10} = 0.615 \end{array} \right\} \quad \Delta\varepsilon = 0.048$$

Then the effective emissivity of the combustion gases becomes

$$\varepsilon_g = C_c \varepsilon_{c, 1 \text{ atm}} + C_w \varepsilon_{w, 1 \text{ atm}} - \Delta\varepsilon = 1.1 \times 0.16 + 1.4 \times 0.23 - 0.048 = \mathbf{0.45}$$

Discussion This is the average emissivity for radiation emitted to all surfaces of the cylindrical enclosure. For radiation emitted towards the center of the base, the mean beam length is $0.71D$ instead of $0.60D$, and the emissivity value would be different.

EXAMPLE 13–14 Radiation Heat Transfer in a Cylindrical Furnace

Reconsider the cylindrical furnace discussed in Example 13–13. For a wall temperature of 600 K, determine the absorptivity of the combustion gases and the rate of radiation heat transfer from the combustion gases to the furnace walls (Fig. 13–40).

SOLUTION The temperatures for the wall surfaces and the combustion gases are given for a cylindrical furnace. The absorptivity of the gas mixture and the rate of radiation heat transfer are to be determined.

Assumptions **1** All the gases in the mixture are ideal gases. **2** All interior surfaces of furnace walls are black. **3** Scattering by soot and other particles is negligible.

Analysis The average emissivity of the combustion gases at the gas temperature of $T_g = 1200$ K was determined in the preceding example to be $\varepsilon_g = 0.45$.

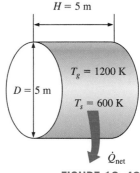

FIGURE 13–40
Schematic for Example 13–14.

For a source temperature of $T_s = 600$ K, the absorptivity of the gas is again determined using the emissivity charts as

$$P_c L \frac{T_s}{T_g} = (0.10 \text{ atm})(3 \text{ m}) \frac{600 \text{ K}}{1200 \text{ K}} = 0.15 \text{ m} \cdot \text{atm} = 0.49 \text{ ft} \cdot \text{atm}$$

$$P_w L \frac{T_s}{T_g} = (0.16 \text{ atm})(3 \text{ m}) \frac{600 \text{ K}}{1200 \text{ K}} = 0.24 \text{ m} \cdot \text{atm} = 0.79 \text{ ft} \cdot \text{atm}$$

The emissivities of CO_2 and H_2O corresponding to these values at a temperature of $T_s = 600$ K and 1 atm are, from Figure 13–36,

$$\varepsilon_{c, 1 \text{ atm}} = 0.11 \qquad \text{and} \qquad \varepsilon_{w, 1 \text{ atm}} = 0.25$$

The pressure correction factors were determined in the preceding example to be $C_c = 1.1$ and $C_w = 1.4$, and they do not change with surface temperature. Then the absorptivities of CO_2 and H_2O become

$$\alpha_c = C_c \left(\frac{T_g}{T_s}\right)^{0.65} \varepsilon_{c, 1 \text{ atm}} = (1.1)\left(\frac{1200 \text{ K}}{600 \text{ K}}\right)^{0.65} (0.11) = 0.19$$

$$\alpha_w = C_w \left(\frac{T_g}{T_s}\right)^{0.45} \varepsilon_{w, 1 \text{ atm}} = (1.4)\left(\frac{1200 \text{ K}}{600 \text{ K}}\right)^{0.45} (0.25) = 0.48$$

Also $\Delta \alpha = \Delta \varepsilon$, but the emissivity correction factor is to be evaluated from Figure 13–38 at $T = T_s = 600$ K instead of $T_g = 1200$ K. There is no chart for 600 K in the figure, but we can read $\Delta \varepsilon$ values at 400 K and 800 K, and take their average. At $P_w / (P_w + P_c) = 0.615$ and $P_c L + P_w L = 2.55$ we read $\Delta \varepsilon = 0.027$. Then the absorptivity of the combustion gases becomes

$$\alpha_g = \alpha_c + \alpha_w - \Delta \alpha = 0.19 + 0.48 - 0.027 = \mathbf{0.64}$$

The surface area of the cylindrical surface is

$$A_s = \pi D H + 2 \frac{\pi D^2}{4} = \pi (5 \text{ m})(5 \text{ m}) + 2 \frac{\pi (5 \text{ m})^2}{4} = 118 \text{ m}^2$$

Then the net rate of radiation heat transfer from the combustion gases to the walls of the furnace becomes

$$\dot{Q}_{\text{net}} = A_s \sigma (\varepsilon_g T_g^4 - \alpha_g T_s^4)$$
$$= (118 \text{ m}^2)(5.67 \times 10^{-8} \text{ W/m}^2 \cdot \text{K}^4)[0.45(1200 \text{ K})^4 - 0.64(600 \text{ K})^4]$$
$$= \mathbf{5.69 \times 10^6 \text{ W}}$$

Discussion The heat transfer rate determined above is for the case of black wall surfaces. If the surfaces are not black but the surface emissivity ε_s is greater than 0.7, the heat transfer rate can be determined by multiplying the rate of heat transfer already determined by $(\varepsilon_s + 1)/2$.

Heat Transfer from the Human Body

The metabolic heat generated in the body is dissipated to the environment through the skin and the lungs by convection and radiation as *sensible heat* and by evaporation as *latent heat* (Fig. 13–41). Latent heat represents the heat of vaporization of water as it evaporates in the lungs and on the skin by absorbing body heat, and latent heat is released as the moisture condenses on cold surfaces. The warming of the inhaled air represents sensible heat transfer in the lungs and is proportional to the temperature rise of inhaled air. The total rate of heat loss from the body can be expressed as

$$
\begin{aligned}
\dot{Q}_{\text{body, total}} &= \dot{Q}_{\text{skin}} + \dot{Q}_{\text{lungs}} \\
&= (\dot{Q}_{\text{sensible}} + \dot{Q}_{\text{latent}})_{\text{skin}} + (\dot{Q}_{\text{sensible}} + \dot{Q}_{\text{latent}})_{\text{lungs}} \\
&= (\dot{Q}_{\text{convection}} + \dot{Q}_{\text{radiation}} + \dot{Q}_{\text{latent}})_{\text{skin}} + (\dot{Q}_{\text{convection}} + \dot{Q}_{\text{latent}})_{\text{lungs}} \quad \textbf{(13–60)}
\end{aligned}
$$

Therefore, the determination of heat transfer from the body by analysis alone is difficult. Clothing further complicates the heat transfer from the body, and thus we must rely on experimental data. Under steady conditions, the total rate of heat transfer from the body is equal to the rate of metabolic heat generation in the body, which varies from about 100 W for light office work to roughly 1000 W during heavy physical work.

Sensible heat loss from the skin depends on the temperatures of the skin, the environment, and the surrounding surfaces as well as the air motion. The *latent heat loss*, on the other hand, depends on the skin wettedness and the relative humidity of the environment as well. *Clothing* serves as insulation and reduces both the sensible and latent forms of heat loss. The heat transfer from the lungs through respiration obviously depends on the frequency of breathing and the volume of the lungs as well as the environmental factors that affect heat transfer from the skin.

Sensible heat from the clothed skin is first transferred to the clothing and then from the clothing to the environment. The convection and radiation heat losses from the outer surface of a clothed body can be expressed as

$$
\dot{Q}_{\text{conv}} = h_{\text{conv}} A_{\text{clothing}}(T_{\text{clothing}} - T_{\text{ambient}}) \quad \text{(W)} \quad \textbf{(13–61)}
$$
$$
\dot{Q}_{\text{rad}} = h_{\text{rad}} A_{\text{clothing}}(T_{\text{clothing}} - T_{\text{surr}}) \quad \textbf{(13–62)}
$$

where

h_{conv} = convection heat transfer coefficient, as given in Table 13–5

h_{rad} = radiation heat transfer coefficient, 4.7 W/m² · °C for typical indoor conditions; the emissivity is assumed to be 0.95, which is typical

A_{clothing} = outer surface area of a clothed person

T_{clothing} = average temperature of exposed skin and clothing

T_{ambient} = ambient air temperature

T_{surr} = average temperature of the surrounding surfaces

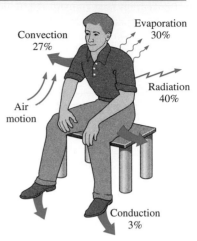

Convection
27%

Evaporation
30%

Air
motion

Radiation
40%

Conduction
3%

Floor

FIGURE 13–41
Mechanisms of heat loss from the human body and relative magnitudes for a resting person.

TABLE 13–5

Convection heat transfer coefficients for a clothed body at 1 atm (V is in m/s) (compiled from various sources)

Activity	h_{conv},* W/m² · °C
Seated in air moving at	
$0 < V < 0.2$ m/s	3.1
$0.2 < V < 4$ m/s	$8.3V^{0.6}$
Walking in still air at	
$0.5 < V < 2$ m/s	$8.6V^{0.53}$
Walking on treadmill in still air at	
$0.5 < V < 2$ m/s	$6.5V^{0.39}$
Standing in moving air at	
$0 < V < 0.15$ m/s	4.0
$0.15 < V < 1.5$ m/s	$14.8V^{0.69}$

*At pressures other than 1 atm, multiply by $P^{0.55}$, where P is in atm.

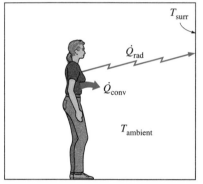

(a) Convection and radiation, separate

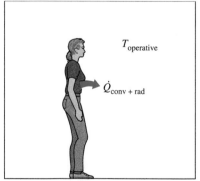

(b) Convection and radiation, combined

FIGURE 13–42

Heat loss by convection and radiation from the body can be combined into a single term by defining an equivalent operative temperature.

The convection heat transfer coefficients at 1 atm pressure are given in Table 13–5. Convection coefficients at pressures P other than 1 atm are obtained by multiplying the values at atmospheric pressure by $P^{0.55}$ where P is in atm. Also, it is recognized that the temperatures of different surfaces surrounding a person are probably different, and T_{surr} represents the **mean radiation temperature**, which is the temperature of an imaginary isothermal enclosure in which radiation heat exchange with the human body equals the radiation heat exchange with the actual enclosure. Noting that most clothing and building materials are very nearly black, the *mean radiation temperature* of an enclosure that consists of N surfaces at different temperatures can be determined from

$$T_{surr} \cong F_{person-1} T_1 + F_{person-2} T_2 + \cdots + F_{person-N} T_N \qquad \text{(13–63)}$$

where T_i is the *temperature of the surface i* and $F_{person-i}$ is the *view factor* between the person and surface i.

Total sensible heat loss can also be expressed conveniently by combining the convection and radiation heat losses as

$$\dot{Q}_{conv+rad} = h_{combined} A_{clothing} (T_{clothing} - T_{operative}) \qquad \text{(13–64)}$$
$$= (h_{conv} + h_{rad}) A_{clothing} (T_{clothing} - T_{operative}) \quad \text{(W)} \qquad \text{(13–65)}$$

where the **operative temperature** $T_{operative}$ is the average of the mean radiant and ambient temperatures weighed by their respective convection and radiation heat transfer coefficients and is expressed as (Fig. 13–42)

$$T_{operative} = \frac{h_{conv} T_{ambient} + h_{rad} T_{surr}}{h_{conv} + h_{rad}} \cong \frac{T_{ambient} + T_{surr}}{2} \qquad \text{(13–66)}$$

Note that the operative temperature will be the arithmetic average of the ambient and surrounding surface temperatures when the convection and radiation heat transfer coefficients are equal to each other. Another environmental index used in thermal comfort analysis is the **effective temperature**, which combines the effects of temperature and humidity. Two environments with the same effective temperature evokes the same thermal response in people even though they are at different temperatures and humidities.

Heat transfer through the *clothing* can be expressed as

$$\dot{Q}_{conv+rad} = \frac{A_{clothing} (T_{skin} - T_{clothing})}{R_{clothing}} \qquad \text{(13–67)}$$

where $R_{clothing}$ is the **unit thermal resistance of clothing** in m² · °C/W, which involves the combined effects of conduction, convection, and radiation between the skin and the outer surface of clothing. The thermal resistance of clothing is usually expressed in the unit **clo** where 1 clo = 0.155 m² · °C/W = 0.880 ft² · °F · h/Btu. The thermal resistance of trousers, long-sleeve shirt, long-sleeve sweater, and T-shirt is 1.0 clo, or 0.155 m² · °C/W. Summer clothing such as light slacks and short-sleeved shirt has an insulation value of 0.5 clo, whereas winter clothing such as heavy slacks, long-sleeve shirt, and a sweater or jacket has an insulation value of 0.9 clo.

Then the total sensible heat loss can be expressed in terms of the skin temperature instead of the inconvenient clothing temperature as (Fig. 13–43)

$$\dot{Q}_{\text{conv + rad}} = \frac{A_{\text{clothing}}\,(T_{\text{skin}} - T_{\text{operative}})}{R_{\text{clothing}} + \dfrac{1}{h_{\text{combined}}}} \qquad (13\text{–}68)$$

At a state of thermal comfort, the average skin temperature of the body is observed to be 33°C (91.5°F). No discomfort is experienced as the skin temperature fluctuates by ±1.5°C (2.5°F). This is the case whether the body is clothed or unclothed.

Evaporative or **latent heat loss** from the skin is proportional to the difference between the water vapor pressure at the skin and the ambient air, and the skin wettedness, which is a measure of the amount of moisture on the skin. It is due to the combined effects of the *evaporation of sweat* and the *diffusion* of water through the skin, and can be expressed as

$$\dot{Q}_{\text{latent}} = \dot{m}_{\text{vapor}}\,h_{fg} \qquad (13\text{–}69)$$

where

$\dot{m}_{\text{vapor}}$ = the rate of evaporation from the body, kg/s

h_{fg} = the enthalpy of vaporization of water = 2430 kJ/kg at 30°C

Heat loss by evaporation is maximum when the skin is completely wetted. Also, clothing offers resistance to evaporation, and the rate of evaporation in clothed bodies depends on the moisture permeability of the clothes. The maximum evaporation rate for an average man is about 1 L/h (0.3 g/s), which represents an upper limit of 730 W for the evaporative cooling rate. A person can lose as much as 2 kg of water per hour during a workout on a hot day, but any excess sweat slides off the skin surface without evaporating (Fig. 13–44).

During *respiration*, the inhaled air enters at ambient conditions and exhaled air leaves nearly saturated at a temperature close to the deep body temperature (Fig. 13–45). Therefore, the body loses both sensible heat by convection and latent heat by evaporation from the lungs, and these can be expressed as

$$\dot{Q}_{\text{conv, lungs}} = \dot{m}_{\text{air, lungs}}\,c_{p,\,\text{air}}(T_{\text{exhale}} - T_{\text{ambient}}) \qquad (13\text{–}70)$$

$$\dot{Q}_{\text{latent, lungs}} = \dot{m}_{\text{vapor, lungs}}\,h_{fg} = \dot{m}_{\text{air, lungs}}\,(\omega_{\text{exhale}} - \omega_{\text{ambient}})h_{fg} \qquad (13\text{–}71)$$

where

$\dot{m}_{\text{air, lungs}}$ = rate of air intake to the lungs, kg/s

$c_{p,\,\text{air}}$ = specific heat of air = 1.0 kJ/kg · °C

T_{exhale} = temperature of exhaled air

ω = humidity ratio (the mass of moisture per unit mass of dry air)

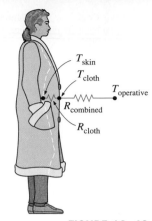

FIGURE 13–43
Simplified thermal resistance network for heat transfer from a clothed person.

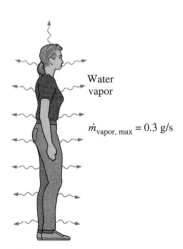

$\dot{Q}_{\text{latent, max}} = \dot{m}_{\text{latent, max}}\,h_{fg\,@\,30°C}$
$= (0.3\ \text{g/s})(2430\ \text{kJ/kg})$
$= 729\ \text{W}$

FIGURE 13–44
An average person can lose heat at a rate of up to 730 W by evaporation.

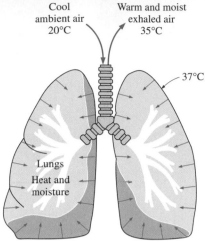

FIGURE 13–45

Part of the metabolic heat generated in the body is rejected to the air from the lungs during respiration.

The rate of air intake to the lungs is directly proportional to the metabolic rate $\dot{Q}_{met}$. The rate of total heat loss from the lungs through respiration can be expressed approximately as

$$\dot{Q}_{conv + latent, lungs} = 0.0014\dot{Q}_{met}(34 - T_{ambient}) + 0.0173\dot{Q}_{met}(5.87 - P_{v, ambient})$$

(13–72)

where $P_{v, ambient}$ is the vapor pressure of ambient air in kPa.

The fraction of sensible heat varies from about 40 percent in the case of heavy work to about 70 percent during light work. The rest of the energy is rejected from the body by perspiration in the form of latent heat.

EXAMPLE 13–15   **Effect of Clothing on Thermal Comfort**

It is well established that a clothed or unclothed person feels comfortable when the skin temperature is about 33°C. Consider an average man wearing summer clothes whose thermal resistance is 0.6 clo. The man feels very comfortable while standing in a room maintained at 22°C. The air motion in the room is negligible, and the interior surface temperature of the room is about the same as the air temperature. If this man were to stand in that room unclothed, determine the temperature at which the room must be maintained for him to feel thermally comfortable.

SOLUTION A man wearing summer clothes feels comfortable in a room at 22°C. The room temperature at which this man would feel thermally comfortable when unclothed is to be determined.

Assumptions **1** Steady conditions exist. **2** The latent heat loss from the person remains the same. **3** The heat transfer coefficients remain the same.

Analysis The body loses heat in sensible and latent forms, and the sensible heat consists of convection and radiation heat transfer. At low air velocities, the convection heat transfer coefficient for a standing man is given in Table 13–5 to be 4.0 W/m² · °C. The radiation heat transfer coefficient at typical indoor conditions is 4.7 W/m² · °C. Therefore, the surface heat transfer coefficient for a standing person for combined convection and radiation is

$$h_{combined} = h_{conv} + h_{rad} = 4.0 + 4.7 = 8.7 \text{ W/m}^2 \cdot °C$$

The thermal resistance of the clothing is given to be

$$R_{clothing} = 0.6 \text{ clo} = 0.6 \times 0.155 \text{ m}^2 \cdot °C/W = 0.093 \text{ m}^2 \cdot °C/W$$

Noting that the surface area of an average man is 1.8 m², the sensible heat loss from this person when clothed is determined to be (Fig. 13–46)

$$\dot{Q}_{sensible, clothed} = \frac{A_s(T_{skin} - T_{ambient})}{R_{clothing} + \dfrac{1}{h_{combined}}} = \frac{(1.8 \text{ m}^2)(33 - 22)°C}{0.093 \text{ m}^2 \cdot °C/W + \dfrac{1}{8.7 \text{ W/m}^2 \cdot °C}}$$

$$= 95.2 \text{ W}$$

From a heat transfer point of view, taking the clothes off is equivalent to removing the clothing insulation or setting $R_{clothing} = 0$. The heat transfer in this case can be expressed as

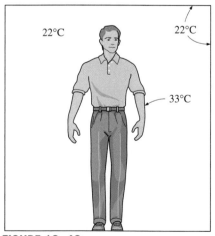

FIGURE 13–46

Schematic for Example 13–15.

$$\dot{Q}_{\text{sensible, unclothed}} = \frac{A_s(T_{\text{skin}} - T_{\text{ambient}})}{\dfrac{1}{h_{\text{combined}}}} = \frac{(1.8 \text{ m}^2)(33 - T_{\text{ambient}})°\text{C}}{\dfrac{1}{8.7 \text{ W/m}^2 \cdot °\text{C}}}$$

To maintain thermal comfort after taking the clothes off, the skin temperature of the person and the rate of heat transfer from him must remain the same. Then setting the equation above equal to 95.2 W gives

$$T_{\text{ambient}} = \mathbf{26.9°C}$$

Therefore, the air temperature needs to be raised from 22 to 26.9°C to ensure that the person feels comfortable in the room after he takes his clothes off (Fig. 13–47).

Discussion Note that the effect of clothing on latent heat is assumed to be negligible in the solution above. We also assumed the surface area of the clothed and unclothed person to be the same for simplicity, and these two effects should counteract each other.

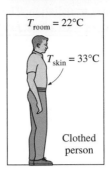
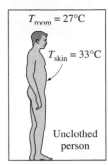

FIGURE 13–47
Clothing serves as insulation, and the room temperature needs to be raised when a person is unclothed to maintain the same comfort level.

SUMMARY

Radiaton heat transfer between surfaces depends on the orientation of the surfaces relative to each other. In a radiation analysis, this effect is accounted for by the geometric parameter *view factor*. The *view factor* from a surface i to a surface j is denoted by $F_{i \to j}$ or F_{ij}, and is defined as the fraction of the radiation leaving surface i that strikes surface j directly. The view factors between differential and finite surfaces are expressed as

$$dF_{dA_1 \to dA_2} = \frac{\dot{Q}_{dA_1 \to dA_2}}{\dot{Q}_{dA_1}} = \frac{\cos\theta_1 \cos\theta_2}{\pi r^2} dA_2$$

$$F_{dA_1 \to A_2} = \int_{A_2} \frac{\cos\theta_1 \cos\theta_2}{\pi r^2} dA_2$$

$$F_{12} = F_{A_1 \to A_2} = \frac{\dot{Q}_{A_1 \to A_2}}{\dot{Q}_{A_1}} = \frac{1}{A_1} \int_{A_2} \int_{A_1} \frac{\cos\theta_1 \cos\theta_2}{\pi r^2} dA_1\, dA_2$$

where r is the distance between dA_1 and dA_2, and θ_1 and θ_2 are the angles between the normals of the surfaces and the line that connects dA_1 and dA_2.

The view factor $F_{i \to i}$ represents the fraction of the radiation leaving surface i that strikes itself directly; $F_{i \to i} = 0$ for *plane* or *convex* surfaces and $F_{i \to i} \neq 0$ for *concave* surfaces. For view factors, the *reciprocity rule* is expressed as

$$A_i F_{i \to j} = A_j F_{j \to i}$$

The sum of the view factors from surface i of an enclosure to all surfaces of the enclosure, including to itself, must equal unity. This is known as the *summation rule* for an enclosure. The *superposition rule* is expressed as the view factor from a surface i to a surface j is equal to the sum of the view factors from surface i to the parts of surface j. The symmetry rule is expressed as if the surfaces j and k are symmetric about the surface i then $F_{i \to j} = F_{i \to k}$.

The rate of net radiation heat transfer between two *black* surfaces is determined from

$$\dot{Q}_{1 \to 2} = A_1 F_{1 \to 2} \sigma(T_1^4 - T_2^4)$$

The *net* radiation heat transfer from any surface i of a *black* enclosure is determined by adding up the net radiation heat transfers from surface i to each of the surfaces of the enclosure:

$$\dot{Q}_i = \sum_{j=1}^{N} \dot{Q}_{i \to j} = \sum_{j=1}^{N} A_i F_{i \to j} \sigma(T_i^4 - T_j^4)$$

The total radiation energy leaving a surface per unit time and per unit area is called the *radiosity* and is denoted by J. The *net* rate of radiation heat transfer from a surface i of surface area A_i is expressed as

$$\dot{Q}_i = \frac{E_{bi} - J_i}{R_i}$$

where

$$R_i = \frac{1 - \varepsilon_i}{A_i \varepsilon_i}$$

is the *surface resistance* to radiation. The *net* rate of radiation heat transfer from surface i to surface j can be expressed as

$$\dot{Q}_{i \to j} = \frac{J_i - J_j}{R_{i \to j}}$$

where

$$R_{i \to j} = \frac{1}{A_i F_{i \to j}}$$

is the *space resistance* to radiation. The *network method* is applied to radiation enclosure problems by drawing a surface resistance associated with each surface of an enclosure and connecting them with space resistances. Then the problem is solved by treating it as an electrical network problem where the radiation heat transfer replaces the current and the radiosity replaces the potential. The *direct method* is based on the following two equations:

Surfaces with specified net heat transfer rate $\dot{Q}_i$
$$\dot{Q}_i = A_i \sum_{j=1}^{N} F_{i \to j}(J_i - J_j)$$

Surfaces with specified temperature T_i
$$\sigma T_i^4 = J_i + \frac{1 - \varepsilon_i}{\varepsilon_i} \sum_{j=1}^{N} F_{i \to j}(J_i - J_j)$$

The first and the second groups of equations give N linear algebraic equations for the determination of the N unknown radiosities for an N-surface enclosure. Once the radiosities $J_1, J_2, \ldots, J_N$ are available, the unknown surface temperatures and heat transfer rates can be determined from the equations just shown.

The net rate of radiation transfer between any two gray, diffuse, opaque surfaces that form an enclosure is given by

$$\dot{Q}_{12} = \frac{\sigma(T_1^4 - T_2^4)}{\dfrac{1 - \varepsilon_1}{A_1 \varepsilon_1} + \dfrac{1}{A_1 F_{12}} + \dfrac{1 - \varepsilon_2}{A_2 \varepsilon_2}}$$

Radiation heat transfer between two surfaces can be reduced greatly by inserting between the two surfaces thin, high-reflectivity (low-emissivity) sheets of material called *radiation shields*. Radiation heat transfer between two large parallel plates separated by N radiation shields is

$$\dot{Q}_{12,\, N \text{ shields}} = \frac{A\sigma(T_1^4 - T_2^4)}{\left(\dfrac{1}{\varepsilon_1} + \dfrac{1}{\varepsilon_2} - 1\right) + \ldots + \left(\dfrac{1}{\varepsilon_{N,1}} + \dfrac{1}{\varepsilon_{N,2}} - 1\right)}$$

The radiation effect in temperature measurements can be properly accounted for by

$$T_f = T_{\text{th}} + \frac{\varepsilon\sigma(T_{\text{th}}^4 - T_w^4)}{h}$$

where T_f is the actual fluid temperature, T_{th} is the temperature value measured by the thermometer, and T_w is the temperature of the surrounding walls, all in K.

Gases with asymmetric molecules such as H_2O, CO_2 CO, SO_2, and hydrocarbons H_nC_m participate in the radiation process by absorption and emission. The *spectral transmissivity, absorptivity,* and *emissivity* of a medium are expressed as

$$\tau_\lambda = e^{-\kappa_\lambda L}, \qquad \alpha_\lambda = 1 - \tau_\lambda = 1 - e^{-\kappa_\lambda L}, \qquad \text{and}$$
$$\varepsilon_\lambda = \alpha_\lambda = 1 - e^{-\kappa_\lambda L}$$

where κ_λ is the *spectral absorption coefficient* of the medium.

The emissivities of H_2O and CO_2 gases are given in Figure 13–36 for a total pressure of $P = 1$ atm. Emissivities at other pressures are determined from

$$\varepsilon_w = C_w \varepsilon_{w,\, 1\, \text{atm}} \qquad \text{and} \qquad \varepsilon_c = C_c \varepsilon_{c,\, 1\, \text{atm}}$$

where C_w and C_c are the *pressure correction factors*. For gas mixtures that contain both of H_2O and CO_2, the emissivity is determined from

$$\varepsilon_g = \varepsilon_c + \varepsilon_w - \Delta\varepsilon = C_c\varepsilon_{c,\, 1\, \text{atm}} + C_w\varepsilon_{w,\, 1\, \text{atm}} - \Delta\varepsilon$$

where $\Delta\varepsilon$ is the *emissivity correction factor,* which accounts for the overlap of emission bands. The gas absorptivities for radiation emitted by a source at temperature T_s are determined similarly from

$$\alpha_g = \alpha_c + \alpha_w - \Delta\alpha$$

where $\Delta\alpha = \Delta\varepsilon$ at the source temperature T_s and

CO_2: $\qquad \alpha_c = C_c \times (T_g/T_s)^{0.65} \times \varepsilon_c(T_s, P_cLT_s/T_g)$

H_2O: $\qquad \alpha_w = C_w \times (T_g/T_s)^{0.45} \times \varepsilon_w(T_s, P_wLT_s/T_g)$

The rate of radiation heat transfer between a gas and a surrounding surface is

Black enclosure: $\qquad \dot{Q}_{\text{net}} = A_s\sigma(\varepsilon_g T_g^4 - \alpha_g T_s^4)$

Gray enclosure, with $\varepsilon_s > 0.7$: $\qquad \dot{Q}_{\text{net, gray}} = \dfrac{\varepsilon_s + 1}{2} A_s\sigma(\varepsilon_g T_g^4 - \alpha_g T_s^4)$

REFERENCES AND SUGGESTED READING

1. D. K. Edwards. *Radiation Heat Transfer Notes.* Washington, D.C.: Hemisphere, 1981.

2. D. K. Edwards and R. Matavosian. "Scaling Rules for Total Absorptivity and Emissivity of Gases." *Journal of Heat Transfer* 106 (1984), pp. 684–689.

3. D. K. Edwards and R. Matavosian. "Emissivity Data for Gases." Section 5.5.5, in *Hemisphere Handbook of Heat Exchanger Design,* ed. G. F. Hewitt. New York: Hemisphere, 1990.

4. D. C. Hamilton and W. R. Morgan. "Radiation Interchange Configuration Factors." National Advisory Committee for Aeronautics, Technical Note 2836, 1952.

5. H. C. Hottel. "Radiant Heat Transmission." In *Heat Transmission,* ed. W. H. McAdams. 3rd ed. New York: McGraw-Hill, 1954.

6. H. C. Hottel. "Heat Transmission by Radiation from Non-luminous Gases," *Transaction of the AIChE* (1927), pp. 173–205.

7. H. C. Hottel and R. B. Egbert. "Radiant Heat Transmission from Water Vapor." *Transactions of the AIChE* 38 (1942), pp. 531–565.

8. J. R. Howell. *A Catalog of Radiation Configuration Factors.* New York: McGraw-Hill, 1982.

9. M. F. Modest. *Radiative Heat Transfer.* New York: McGraw-Hill, 1993.

10. A. K. Oppenheim. "Radiation Analysis by the Network Method." *Transactions of the ASME* 78 (1956), pp. 725–735.

11. R. Siegel and J. R. Howell. *Thermal Radiation Heat Transfer.* 3rd ed. Washington, D.C.: Hemisphere, 1992.

PROBLEMS*

The View Factor

13–1C What does the view factor represent? When is the view factor from a surface to itself not zero?

13–2C How can you determine the view factor F_{12} when the view factor F_{21} and the surface areas are available?

13–3C What are the summation rule and the superposition rule for view factors?

13–4C What is the crossed-strings method? For what kind of geometries is the crossed-strings method applicable?

13–5 Consider an enclosure consisting of eight surfaces. How many view factors does this geometry involve? How many of these view factors can be determined by the application of the reciprocity and the summation rules?

13–6 Consider an enclosure consisting of five surfaces. How many view factors does this geometry involve? How many of these view factors can be determined by the application of the reciprocity and summation rules?

13–7 Consider an enclosure consisting of 12 surfaces. How many view factors does this geometry involve? How many of these view factors can be determined by the application of the reciprocity and the summation rules?   *Answers: 144, 78*

13–8 Determine the view factors F_{13} and F_{23} between the rectangular surfaces shown in Fig. P13–8.

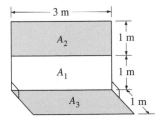

FIGURE P13–8

13–9 Consider a cylindrical enclosure whose height is twice the diameter of its base. Determine the view factor from the side surface of this cylindrical enclosure to its base surface.

---

*Problems designated by a "C" are concept questions, and students are encouraged to answer them all. Problems designated by an "E" are in English units, and the SI users can ignore them. Problems with the icon 🖐 are solved using EES, and complete solutions together with parametric studies are included on the enclosed CD. Problems with the icon 📓 are comprehensive in nature, and are intended to be solved with a computer, preferably using the EES software that accompanies this text.

13–10 Consider a hemispherical furnace with a flat circular base of diameter D. Determine the view factor from the dome of this furnace to its base. *Answer:* 0.5

13–11 Determine the view factors F_{12} and F_{21} for the very long ducts shown in Fig. P13–11 without using any view factor tables or charts. Neglect end effects.

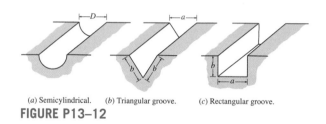

(a)   (b)   (c)

FIGURE P13–11

13–12 Determine the view factors from the very long grooves shown in Fig. P13–12 to the surroundings without using any view factor tables or charts. Neglect end effects.

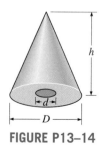

(a) Semicylindrical.   (b) Triangular groove.   (c) Rectangular groove.

FIGURE P13–12

13–13 Determine the view factors from the base of a cube to each of the other five surfaces.

13–14 Consider a conical enclosure of height h and base diameter D. Determine the view factor from the conical side surface to a hole of diameter d located at the center of the base.

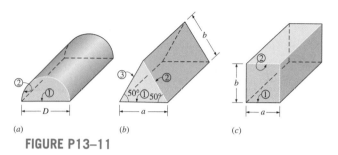

FIGURE P13–14

13–15 Determine the four view factors associated with an enclosure formed by two very long concentric cylinders of radii r_1 and r_2. Neglect the end effects.

13–16 Determine the view factor F_{12} between the rectangular surfaces shown in Fig. P13–16.

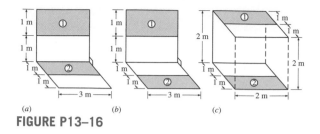

(a)   (b)   (c)

FIGURE P13–16

13–17 Two infinitely long parallel cylinders of diameter D are located a distance s apart from each other. Determine the view factor F_{12} between these two cylinders.

13–18 Three infinitely long parallel cylinders of diameter D are located a distance s apart from each other. Determine the view factor between the cylinder in the middle and the surroundings.

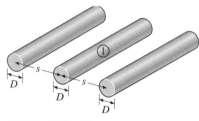

FIGURE P13–18

Radiation Heat Transfer between Surfaces

13–19C Why is the radiation analysis of enclosures that consist of black surfaces relatively easy? How is the rate of radiation heat transfer between two surfaces expressed in this case?

13–20C How does radiosity for a surface differ from the emitted energy? For what kind of surfaces are these two quantities identical?

13–21C What are the radiation surface and space resistances? How are they expressed? For what kind of surfaces is the radiation surface resistance zero?

13–22C What are the two methods used in radiation analysis? How do these two methods differ?

13–23C What is a reradiating surface? What simplifications does a reradiating surface offer in the radiation analysis?

13–24 A solid sphere of 1 m diameter at 500 K is kept in an evacuated, long, equilateral triangular enclosure whose sides are 2 m long. The emissivity of the sphere is 0.45 and the temperature of the enclosure is 380 K. If heat is generated uniformly within the sphere at a rate of 3100 W, determine (a) the view factor from the enclosure to the sphere and (b) the emissivity of the enclosure.

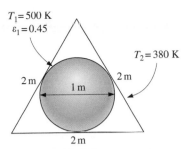

$T_1 = 500$ K
$\varepsilon_1 = 0.45$

$T_2 = 380$ K

2 m

2 m

1 m

2 m

FIGURE P13–24

13–25 This question deals with steady–state radiation heat transfer between a sphere ($r_1 = 30$ cm) and a circular disk ($r_2 = 120$ cm), which are separated by a center-to-center distance $h = 60$ cm. When the normal to the center of disk passes through the center of sphere, the radiation view factor is given by

$$F_{12} = 0.5 \left\{ 1 - \left[1 + \left(\frac{r_2}{h} \right)^2 \right]^{-0.5} \right\}$$

Surface temperatures of the sphere and the disk are 600°C and 200°C, respectively; and their emissivities are 0.9 and 0.5, respectively.

(a) Calculate the view factors F_{12} and F_{21}.

(b) Calculate the net rate of radiation heat exchange between the sphere and the disk.

(c) For the given radii and temperatures of the sphere and the disk, the following four possible modifications could increase the net rate of radiation heat exchange: paint each of the two surfaces to alter their emissivities, adjust the distance between them, and provide an (refractory) enclosure. Calculate the net rate of radiation heat exchange between the two bodies if the *best values* are selected for each of the above modifications.

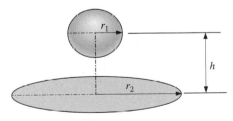

r_1

r_2

h

FIGURE P13–25

13–26E Consider a 10-ft $\times$ 10-ft $\times$ 10-ft cubical furnace whose top and side surfaces closely approximate black surfaces and whose base surface has an emissivity $\varepsilon = 0.7$. The base, top, and side surfaces of the furnace are maintained at uniform temperatures of 800 R, 1600 R, and 2400 R, respectively. Determine the net rate of radiation heat transfer between (a) the base and the side surfaces and (b) the base and the top surfaces. Also, determine the net rate of radiation heat transfer to the base surface.

13–27E Reconsider Prob. 13–26E. Using EES (or other) software, investigate the effect of base surface emissivity on the net rates of radiation heat transfer between the base and the side surfaces, between the base and top surfaces, and to the base surface. Let the emissivity vary from 0.1 to 0.9. Plot the rates of heat transfer as a function of emissivity, and discuss the results.

13–28 Two very large parallel plates are maintained at uniform temperatures of $T_1 = 600$ K and $T_2 = 400$ K and have emissivities $\varepsilon_1 = 0.5$ and $\varepsilon_2 = 0.9$, respectively. Determine the net rate of radiation heat transfer between the two surfaces per unit area of the plates.

13–29 Reconsider Prob. 13–28. Using EES (or other) software, investigate the effects of the temperature and the emissivity of the hot plate on the net rate of radiation heat transfer between the plates. Let the temperature vary from 500 K to 1000 K and the emissivity from 0.1 to 0.9. Plot the net rate of radiation heat transfer as functions of temperature and emissivity, and discuss the results.

13–30 A furnace is of cylindrical shape with $R = H = 2$ m. The base, top, and side surfaces of the furnace are all black and are maintained at uniform temperatures of 500, 700, and 1400 K, respectively. Determine the net rate of radiation heat transfer to or from the top surface during steady operation.

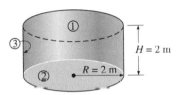

③

①

$H = 2$ m

②

$R = 2$ m

FIGURE P13–30

13–31 Consider a hemispherical furnace of diameter $D = 5$ m with a flat base. The dome of the furnace is black, and the base has an emissivity of 0.7. The base and the dome of the furnace are maintained at uniform temperatures of 400 and 1000 K, respectively. Determine the net rate of radiation heat transfer from the dome to the base surface during steady operation. *Answer:* 759 kW

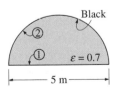

Black

②

①

$\varepsilon = 0.7$

5 m

FIGURE P13–31

13–32 Two very long concentric cylinders of diameters $D_1 = 0.35$ m and $D_2 = 0.5$ m are maintained at uniform temperatures of $T_1 = 950$ K and $T_2 = 500$ K and have emissivities $\varepsilon_1 = 1$ and $\varepsilon_2 = 0.55$, respectively. Determine the net rate of radiation heat transfer between the two cylinders per unit length of the cylinders.

13–33 This experiment is conducted to determine the emissivity of a certain material. A long cylindrical rod of diameter $D_1 = 0.01$ m is coated with this new material and is placed in an evacuated long cylindrical enclosure of diameter $D_2 = 0.1$ m and emissivity $\varepsilon_2 = 0.95$, which is cooled externally and maintained at a temperature of 200 K at all times. The rod is heated by passing electric current through it. When steady operating conditions are reached, it is observed that the rod is dissipating electric power at a rate of 8 W per unit of its length and its surface temperature is 500 K. Based on these measurements, determine the emissivity of the coating on the rod.

13–34E A furnace is shaped like a long semicylindrical duct of diameter $D = 15$ ft. The base and the dome of the furnace have emissivities of 0.5 and 0.9 and are maintained at uniform temperatures of 550 and 1800 R, respectively. Determine the net rate of radiation heat transfer from the dome to the base surface per unit length during steady operation.

FIGURE P13–34E

13–35 Two parallel disks of diameter $D = 0.6$ m separated by $L = 0.4$ m are located directly on top of each other. Both disks are black and are maintained at a temperature of 450 K. The back sides of the disks are insulated, and the environment that the disks are in can be considered to be a blackbody at 300 K. Determine the net rate of radiation heat transfer from the disks to the environment. *Answer: 781 W*

13–36 A furnace is shaped like a long equilateral-triangular duct where the width of each side is 2 m. Heat is supplied from the base surface, whose emissivity is $\varepsilon_1 = 0.8$, at a rate of 800 W/m² while the side surfaces, whose emissivities are 0.5, are maintained at 500 K. Neglecting the end effects, determine the temperature of the base surface. Can you treat this geometry as a two-surface enclosure?

13–37 Reconsider Prob. 13–36. Using EES (or other) software, investigate the effects of the rate of the heat transfer at the base surface and the temperature of the side surfaces on the temperature of the base surface. Let the rate of heat transfer vary from 500 W/m² to 1000 W/m² and the temperature from 300 K to 700 K. Plot the temperature of the base surface as functions of the rate of heat transfer and the temperature of the side surfaces, and discuss the results.

13–38 Consider a 4-m × 4-m × 4-m cubical furnace whose floor and ceiling are black and whose side surfaces are

reradiating. The floor and the ceiling of the furnace are maintained at temperatures of 550 K and 1100 K, respectively. Determine the net rate of radiation heat transfer between the floor and the ceiling of the furnace.

13–39 Two concentric spheres of diameters $D_1 = 0.3$ m and $D_2 = 0.4$ m are maintained at uniform temperatures $T_1 = 700$ K and $T_2 = 500$ K and have emissivities $\varepsilon_1 = 0.5$ and $\varepsilon_2 = 0.7$, respectively. Determine the net rate of radiation heat transfer between the two spheres. Also, determine the convection heat transfer coefficient at the outer surface if both the surrounding medium and the surrounding surfaces are at 30°C. Assume the emissivity of the outer surface is 0.35.

13–40 A spherical tank of diameter $D = 2$ m that is filled with liquid nitrogen at 100 K is kept in an evacuated cubic enclosure whose sides are 3 m long. The emissivities of the spherical tank and the enclosure are $\varepsilon_1 = 0.1$ and $\varepsilon_2 = 0.8$, respectively. If the temperature of the cubic enclosure is measured to be 240 K, determine the net rate of radiation heat transfer to the liquid nitrogen. *Answer: 228 W*

Liquid N₂

FIGURE P13–40

13–41 Repeat Prob. 13–40 by replacing the cubic enclosure by a spherical enclosure whose diameter is 3 m.

13–42 Reconsider Prob. 13–40. Using EES (or other) software, investigate the effects of the side length and the emissivity of the cubic enclosure, and the emissivity of the spherical tank on the net rate of radiation heat transfer. Let the side length vary from 2.5 m to 5.0 m and both emissivities from 0.1 to 0.9. Plot the net rate of radiation heat transfer as functions of side length and emissivities, and discuss the results.

13–43 Consider a circular grill whose diameter is 0.3 m. The bottom of the grill is covered with hot coal bricks at 950 K, while the wire mesh on top of the grill is covered with steaks initially at 5°C. The distance between the coal bricks and the steaks is 0.20 m. Treating both the steaks and the coal bricks as blackbodies, determine the initial rate of radiation heat transfer from the coal bricks to the steaks. Also, determine the initial rate of radiation heat transfer to the steaks if the side opening of the grill is covered by aluminum foil, which can be approximated as a reradiating surface. *Answers.* 928 W, 2085 W

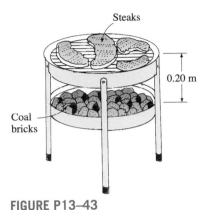

FIGURE P13–43

13–44E A 9-ft-high room with a base area of 12 ft × 12 ft is to be heated by electric resistance heaters placed on the ceiling, which is maintained at a uniform temperature of 90°F at all times. The floor of the room is at 65°F and has an emissivity of 0.8. The side surfaces are well insulated. Treating the ceiling as a blackbody, determine the rate of heat loss from the room through the floor.

13–45 Consider two rectangular surfaces perpendicular to each other with a common edge which is 1.6 m long. The horizontal surface is 0.8 m wide and the vertical surface is 1.2 m high. The horizontal surface has an emissivity of 0.75 and is maintained at 400 K. The vertical surface is black and is maintained at 550 K. The back sides of the surfaces are insulated. The surrounding surfaces are at 290 K, and can be considered to have an emissivity of 0.85. Determine the net rate of radiation heat transfers between the two surfaces, and between the horizontal surface and the surroundings.

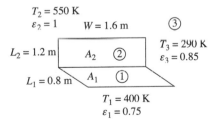

FIGURE P13–45

13–46 Two long parallel 20-cm-diameter cylinders are located 30 cm apart from each other. Both cylinders are black, and are maintained at temperatures 425 K and 275 K. The surroundings can be treated as a blackbody at 300 K. For a 1-m-long section of the cylinders, determine the rates of radiation heat transfer between the cylinders and between the hot cylinder and the surroundings.

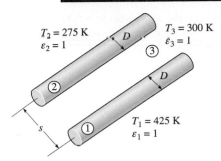

FIGURE P13–46

13–47 Consider a long semicylindrical duct of diameter 1.0 m. Heat is supplied from the base surface, which is black, at a rate of 1200 W/m², while the side surface with an emissivity of 0.4 are is maintained at 650 K. Neglecting the end effects, determine the temperature of the base surface.

13–48 Consider a 20-cm-diameter hemispherical enclosure. The dome is maintained at 600 K and heat is supplied from the dome at a rate of 50 W while the base surface with an emissivity is 0.55 is maintained at 400 K. Determine the emissivity of the dome.

Radiation Shields and the Radiation Effect

13–49C What is a radiation shield? Why is it used?

13–50C What is the radiation effect? How does it influence the temperature measurements?

13–51C Give examples of radiation effects that affect human comfort.

13–52 Consider a person whose exposed surface area is 1.9 m², emissivity is 0.85, and surface temperature is 30°C. Determine the rate of heat loss from that person by radiation in a large room whose walls are at a temperature of (*a*) 300 K and (*b*) 280 K.

13–53 A thin aluminum sheet with an emissivity of 0.15 on both sides is placed between two very large parallel plates, which are maintained at uniform temperatures T_1 = 900 K and T_2 = 650 K and have emissivities ε_1 = 0.5 and ε_2 = 0.8, respectively. Determine the net rate of radiation heat transfer between the two plates per unit surface area of the plates and compare the result with that without the shield.

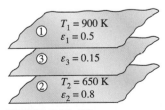

FIGURE P13–53

13–54 Reconsider Prob. 13–53. Using EES (or other) software, plot the net rate of radiation heat transfer between the two plates as a function of the emissivity of the aluminum sheet as the emissivity varies from 0.05 to 0.25, and discuss the results.

13–55 Two very large parallel plates are maintained at uniform temperatures of $T_1 = 1000$ K and $T_2 = 800$ K and have emissivities of $\varepsilon_1 = \varepsilon_2 = 0.5$, respectively. It is desired to reduce the net rate of radiation heat transfer between the two plates to one-fifth by placing thin aluminum sheets with an emissivity of 0.1 on both sides between the plates. Determine the number of sheets that need to be inserted.

13–56 Five identical thin aluminum sheets with emissivities of 0.1 on both sides are placed between two very large parallel plates, which are maintained at uniform temperatures of $T_1 = 800$ K and $T_2 = 450$ K and have emissivities of $\varepsilon_1 = \varepsilon_2 = 0.1$, respectively. Determine the net rate of radiation heat transfer between the two plates per unit surface area of the plates and compare the result to that without the shield.

13–57 Reconsider Prob. 13–56. Using EES (or other) software, investigate the effects of the number of the aluminum sheets and the emissivities of the plates on the net rate of radiation heat transfer between the two plates. Let the number of sheets vary from 1 to 10 and the emissivities of the plates from 0.1 to 0.9. Plot the rate of radiation heat transfer as functions of the number of sheets and the emissivities of the plates, and discuss the results.

13–58E Two parallel disks of diameter $D = 3$ ft separated by $L = 2$ ft are located directly on top of each other. The disks are separated by a radiation shield whose emissivity is 0.15. Both disks are black and are maintained at temperatures of 1200 R and 700 R, respectively. The environment that the disks are in can be considered to be a blackbody at 540 R. Determine the net rate of radiation heat transfer through the shield under steady conditions.   *Answer:* 872 Btu/h

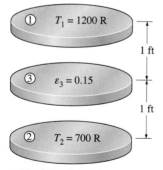

$T_1 = 1200$ R

①

③ $\varepsilon_3 = 0.15$

② $T_2 = 700$ R

1 ft

1 ft

FIGURE P13–58E

13–59 A radiation shield that has the same emissivity ε_3 on both sides is placed between two large parallel plates, which are maintained at uniform temperatures of $T_1 = 650$ K and $T_2 = 400$ K and have emissivities of $\varepsilon_1 = 0.6$ and $\varepsilon_2 = 0.9$, respectively. Determine the emissivity of the radiation shield if

the radiation heat transfer between the plates is to be reduced to 15 percent of that without the radiation shield.

13–60 Reconsider Prob. 13–59. Using EES (or other) software, investigate the effect of the percent reduction in the net rate of radiation heat transfer between the plates on the emissivity of the radiation shields. Let the percent reduction vary from 40 to 95 percent. Plot the emissivity versus the percent reduction in heat transfer, and discuss the results.

13–61 Two coaxial cylinders of diameters $D_1 = 0.10$ m and $D_2 = 0.30$ m and emissivities $\varepsilon_1 = 0.7$ and $\varepsilon_2 = 0.4$ are maintained at uniform temperatures of $T_1 = 750$ K and $T_2 = 500$ K, respectively. Now a coaxial radiation shield of diameter $D_3 = 0.20$ m and emissivity $\varepsilon_3 = 0.2$ is placed between the two cylinders. Determine the net rate of radiation heat transfer between the two cylinders per unit length of the cylinders and compare the result with that without the shield.

13–62 Reconsider Prob. 13–61. Using EES (or other) software, investigate the effects of the diameter of the outer cylinder and the emissivity of the radiation shield on the net rate of radiation heat transfer between the two cylinders. Let the diameter vary from 0.25 m to 0.50 m and the emissivity from 0.05 to 0.35. Plot the rate of radiation heat transfer as functions of the diameter and the emissivity, and discuss the results.

Radiation Exchange with Absorbing and Emitting Gases

13–63C How does radiation transfer through a participating medium differ from that through a nonparticipating medium?

13–64C Define spectral transmissivity of a medium of thickness L in terms of (a) spectral intensities and (b) the spectral absorption coefficient.

13–65C Define spectral emissivity of a medium of thickness L in terms of the spectral absorption coefficient.

13–66C How does the wavelength distribution of radiation emitted by a gas differ from that of a surface at the same temperature?

13–67 Consider an equimolar mixture of CO_2 and O_2 gases at 800 K and a total pressure of 0.5 atm. For a path length of 1.2 m, determine the emissivity of the gas.

13–68 A cubic furnace whose side length is 6 m contains combustion gases at 1000 K and a total pressure of 1 atm. The composition of the combustion gases is 75 percent N_2, 9 percent H_2O, 6 percent O_2, and 10 percent CO_2. Determine the effective emissivity of the combustion gases.

13–69 A cylindrical container whose height and diameter are 8 m is filled with a mixture of CO_2 and N_2 gases at 600 K and 1 atm. The partial pressure of CO_2 in the mixture is 0.15 atm. If the walls are black at a temperature of 450 K, determine the rate of radiation heat transfer between the gas and the container walls.

13–70 Repeat Prob. 13–69 by replacing CO_2 by the H_2O gas.

13–71 A 3-m-diameter spherical furnace contains a mixture of CO_2 and N_2 gases at 1200 K and 1 atm. The mole fraction of CO_2 in the mixture is 0.15. If the furnace wall is black and its temperature is to be maintained at 600 K, determine the net rate of radiation heat transfer between the gas mixture and the furnace walls.

13–72 A flow-through combustion chamber consists of 15-cm diameter long tubes immersed in water. Compressed air is routed to the tube, and fuel is sprayed into the compressed air. The combustion gases consist of 70 percent N_2, 9 percent H_2O, 15 percent O_2, and 6 percent CO_2, and are maintained at 1 atm and 1500 K. The tube surfaces are near black, with an emissivity of 0.9. If the tubes are to be maintained at a temperature of 600 K, determine the rate of heat transfer from combustion gases to tube wall by radiation per m length of tube.

13–73 Repeat Prob. 13–72 for a total pressure of 3 atm.

13–74 In a cogeneration plant, combustion gases at 1 atm and 800 K are used to preheat water by passing them through 6-m-long 10-cm-diameter tubes. The inner surface of the tube is black, and the partial pressures of CO_2 and H_2O in combustion gases are 0.12 atm and 0.18 atm, respectively. If the tube temperature is 500 K, determine the rate of radiation heat transfer from the gases to the tube.

13–75 A gas at 1200 K and 1 atm consists of 10 percent CO_2, 10 percent H_2O, 10 percent O_2, and 70 percent N_2 by volume. The gas flows between two large parallel black plates maintained at 600 K. If the plates are 20 cm apart, determine the rate of heat transfer from the gas to each plate per unit surface area.

Special Topic: Heat Transfer from the Human Body

13–76C Consider a person who is resting or doing light work. Is it fair to say that roughly one-third of the metabolic heat generated in the body is dissipated to the environment by convection, one-third by evaporation, and the remaining one-third by radiation?

13–77C What is sensible heat? How is the sensible heat loss from a human body affected by (a) skin temperature, (b) environment temperature, and (c) air motion?

13–78C What is latent heat? How is the latent heat loss from the human body affected by (a) skin wettedness and (b) relative humidity of the environment? How is the rate of evaporation from the body related to the rate of latent heat loss?

13–79C How is the insulating effect of clothing expressed? How does clothing affect heat loss from the body by convection, radiation, and evaporation? How does clothing affect heat gain from the sun?

13–80C Explain all the different mechanisms of heat transfer from the human body (a) through the skin and (b) through the lungs.

13–81C What is operative temperature? How is it related to the mean ambient and radiant temperatures? How does it differ from effective temperature?

13–82 The convection heat transfer coefficient for a clothed person while walking in still air at a velocity of 0.5 to 2 m/s is given by $h = 8.6V^{0.53}$, where V is in m/s and h is in W/m$^2 \cdot$ °C. Plot the convection coefficient against the walking velocity, and compare the convection coefficients in that range to the average radiation coefficient of about 5 W/m$^2 \cdot$ °C.

13–83 A clothed or unclothed person feels comfortable when the skin temperature is about 33°C. Consider an average man wearing summer clothes whose thermal resistance is 1.1 clo. The man feels very comfortable while standing in a room maintained at 20°C. If this man were to stand in that room unclothed, determine the temperature at which the room must be maintained for him to feel thermally comfortable. Assume the latent heat loss from the person to remain the same.
Answer: 27.8°C

13–84E An average person produces 0.50 lbm of moisture while taking a shower and 0.12 lbm while bathing in a tub. Consider a family of four who shower once a day in a bathroom that is not ventilated. Taking the heat of vaporization of water to be 1050 Btu/lbm, determine the contribution of showers to the latent heat load of the air conditioner in summer per day.

13–85 An average (1.82 kg or 4.0 lbm) chicken has a basal metabolic rate of 5.47 W and an average metabolic rate of 10.2 W (3.78 W sensible and 6.42 W latent) during normal activity. If there are 100 chickens in a breeding room, determine the rate of total heat generation and the rate of moisture production in the room. Take the heat of vaporization of water to be 2430 kJ/kg.

13–86 Consider a large classroom with 90 students on a hot summer day. All the lights with 2.0 kW of rated power are kept on. The room has no external walls, and thus heat gain through the walls and the roof is negligible. Chilled air is available at 15°C, and the temperature of the return air is not to exceed 25°C. The average rate of metabolic heat generation by a person sitting or doing light work is 115 W (70 W sensible and 45 W latent). Determine the required flow rate of air that needs to be supplied to the room. *Answer: 0.83 kg/s*

13–87 A person feels very comfortable in his house in light clothing when the thermostat is set at 22°C and the mean radiation temperature (the average temperature of the surrounding surfaces) is also 22°C. During a cold day, the average mean radiation temperature drops to 18°C. To what level must the indoor air temperature be raised to maintain the same level of comfort in the same clothing?

13–88 Repeat Prob. 13–87 for a mean radiation temperature of 12°C.

13–89 A car mechanic is working in a shop whose interior space is not heated. Comfort for the mechanic is provided by

FIGURE P13–87

two radiant heaters that radiate heat at a total rate of 4 kJ/s. About 5 percent of this heat strikes the mechanic directly. The shop and its surfaces can be assumed to be at the ambient temperature, and the emissivity and absorptivity of the mechanic can be taken to be 0.95 and the surface area to be 1.8 m². The mechanic is generating heat at a rate of 350 W, half of which is latent, and is wearing medium clothing with a thermal resistance of 0.7 clo. Determine the lowest ambient temperature in which the mechanic can work comfortably.

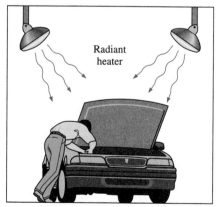

FIGURE P13–89

Review Problems

13–90 A thermocouple used to measure the temperature of hot air flowing in a duct whose walls are maintained at $T_w = 500$ K shows a temperature reading of $T_{th} = 850$ K. Assuming the emissivity of the thermocouple junction to be $\varepsilon = 0.6$ and the convection heat transfer coefficient to be $h = 60$ W/m² · °C, determine the actual temperature of air.
Answer: 1111 K

13–91 Consider the two parallel coaxial disks of diameters a and b, shown in Fig. P13–91. For this geometry, the view factor from the smaller disk to the larger disk can be calculated from

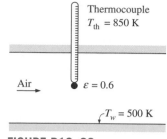

FIGURE P13–90

$$F_{ij} = 0.5\left(\frac{B}{A}\right)^2\left\{C - \left[C^2 - 4\left(\frac{A}{B}\right)^2\right]^{0.5}\right\}$$

where, $A = a/2L$, $B = b/2L$, and $C = 1 + [(1 + A^2)/B^2]$. The diameter, emissivity and temperature are 20 cm, 0.60, and 600°C, respectively, for disk a, and 40 cm, 0.80 and 200°C for disk b. The distance between the two disks is $L = 10$ cm.

(a) Calculate F_{ab} and F_{ba}.

(b) Calculate the net rate of radiation heat exchange between disks a and b in steady operation.

(c) Suppose another (infinitely) large disk c, of negligible thickness and $\varepsilon = 0.7$, is inserted between disks a and b such that it is parallel and equidistant to both disks. Calculate the net rate of radiation heat exchange between disks a and c and disks c and b in steady operation.

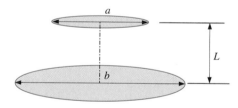

FIGURE P13–91

13–92 A large number of long tubes, each of diameter D, are placed parallel to each other and at a center-to-center distance of s. Since all of the tubes are geometrically similar and at the same temperature, these could be treated collectively as one surface (A_j) for radiation heat transfer calculations. As shown in Fig. P13–92, the tube-bank (A_j) is placed opposite a large flat wall (A_i) such that the tube-bank is parallel to the wall. The radiation view factor, F_{ij}, for this arrangement is given by

$$F_{ij} = 1 - \left[1 - \left(\frac{D}{s}\right)^2\right]^{0.5} + \frac{D}{s}\left\{\tan^{-1}\left[\left(\frac{s}{D}\right)^2 - 1\right]^{0.5}\right\}$$

(a) Calculate the view factors F_{ij} and F_{ji} for $s = 3.0$ cm and $D = 1.5$ cm.

(b) Calculate the net rate of radiation heat transfer between the wall and the tube-bank per unit area of the wall when $T_i = 900$°C, $T_j = 60$°C, $\varepsilon_i = 0.8$, and $\varepsilon_j = 0.9$.

(c) A fluid flows through the tubes at an average temperature of 40°C, resulting in a heat transfer coefficient of 2.0 kW/m² · K. Assuming $T_i = 900$°C, $\varepsilon_i = 0.8$ and $\varepsilon_j = 0.9$ (as above) and neglecting the tube wall thickness and convection from the outer surface, calculate the temperature of the tube surface in steady operation.

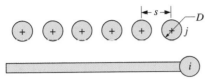

FIGURE P13–92

13–93 A thermocouple shielded by aluminum foil of emissivity 0.15 is used to measure the temperature of hot gases flowing in a duct whose walls are maintained at $T_w = 380$ K. The thermometer shows a temperature reading of $T_{th} = 530$ K. Assuming the emissivity of the thermocouple junction to be $\varepsilon = 0.7$ and the convection heat transfer coefficient to be $h = 120$ W/m² · °C, determine the actual temperature of the gas. What would the thermometer reading be if no radiation shield was used?

13–94E Consider a sealed 8-in-high electronic box whose base dimensions are 12 in × 12 in placed in a vacuum chamber. The emissivity of the outer surface of the box is 0.95. If the electronic components in the box dissipate a total of 90 W of power and the outer surface temperature of the box is not to exceed 130°F, determine the highest temperature at which the surrounding surfaces must be kept if this box is to be cooled by radiation alone. Assume the heat transfer from the bottom surface of the box to the stand to be negligible. *Answer:* 54°F

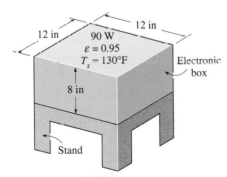

FIGURE P13–94E

13–95 A 2-m-internal-diameter double-walled spherical tank is used to store iced water at 0°C. Each wall is 0.5 cm thick, and the 1.5-cm-thick air space between the two walls of the tank is evacuated in order to minimize heat transfer. The surfaces surrounding the evacuated space are polished so that each surface has an emissivity of 0.15. The temperature of the outer

wall of the tank is measured to be 20°C. Assuming the inner wall of the steel tank to be at 0°C, determine (a) the rate of heat transfer to the iced water in the tank and (b) the amount of ice at 0°C that melts during a 24-h period.

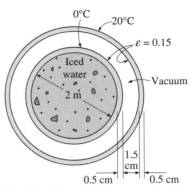

FIGURE P13–95

13–96 Two concentric spheres of diameters $D_1 = 15$ cm and $D_2 = 25$ cm are separated by air at 1 atm pressure. The surface temperatures of the two spheres enclosing the air are $T_1 = 350$ K and $T_2 = 275$ K, respectively, and their emissivities are 0.75. Determine the rate of heat transfer from the inner sphere to the outer sphere by (a) natural convection and (b) radiation.

13–97 Consider a 1.5-m-high and 3-m-wide solar collector that is tilted at an angle 20° from the horizontal. The distance between the glass cover and the absorber plate is 3 cm, and the back side of the absorber is heavily insulated. The absorber plate and the glass cover are maintained at temperatures of 80°C and 32°C, respectively. The emissivity of the glass surface is 0.9 and that of the absorber plate is 0.8. Determine the rate of heat loss from the absorber plate by natural convection and radiation. *Answers:* 750 W, 1289 W

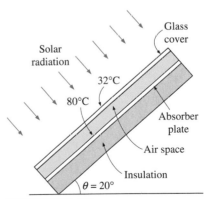

FIGURE P13–97

13–98E A solar collector consists of a horizontal aluminum tube having an outer diameter of

2.5 in enclosed in a concentric thin glass tube of diameter 5 in. Water is heated as it flows through the tube, and the annular space between the aluminum and the glass tube is filled with air at 0.5 atm pressure. The pump circulating the water fails during a clear day, and the water temperature in the tube starts rising. The aluminum tube absorbs solar radiation at a rate of 30 Btu/h per foot length, and the temperature of the ambient air outside is 75°F. The emissivities of the tube and the glass cover are 0.9. Taking the effective sky temperature to be 60°F, determine the temperature of the aluminum tube when thermal equilibrium is established (i.e., when the rate of heat loss from the tube equals the amount of solar energy gained by the tube).

13–99 A vertical 2-m-high and 5-m-wide double-pane window consists of two sheets of glass separated by a 3-cm-thick air gap. In order to reduce heat transfer through the window, the air space between the two glasses is partially evacuated to 0.3 atm pressure. The emissivities of the glass surfaces are 0.9. Taking the glass surface temperatures across the air gap to be 15°C and 5°C, determine the rate of heat transfer through the window by natural convection and radiation.

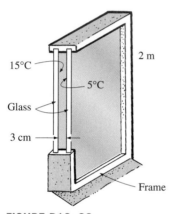

FIGURE P13–99

13–100 A simple solar collector is built by placing a 6-cm-diameter clear plastic tube around a garden hose whose outer diameter is 2 cm. The hose is painted black to maximize solar absorption, and some plastic rings are used to keep the spacing between the hose and the clear

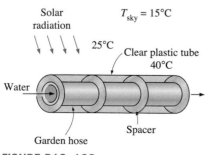

FIGURE P13–100

plastic cover constant. The emissivities of the hose surface and the glass cover are 0.9, and the effective sky temperature is estimated to be 15°C. The temperature of the plastic tube is measured to be 40°C, while the ambient air temperature is 25°C. Determine the rate of heat loss from the water in the hose by natural convection and radiation per meter of its length under steady conditions.   *Answers:* 5.2 W, 26.1 W

13–101 A solar collector consists of a horizontal copper tube of outer diameter 5 cm enclosed in a concentric thin glass tube of diameter 12 cm. Water is heated as it flows through the tube, and the annular space between the copper and the glass tubes is filled with air at 1 atm pressure. The emissivities of the tube surface and the glass cover are 0.85 and 0.9, respectively. During a clear day, the temperatures of the tube surface and the glass cover are measured to be 60°C and 40°C, respectively. Determine the rate of heat loss from the collector by natural convection and radiation per meter length of the tube.

13–102 A furnace is of cylindrical shape with a diameter of 1.2 m and a length of 1.2 m. The top surface has an emissivity of 0.70 and is maintained at 500 K. The bottom surface has an emissivity of 0.50 and is maintained at 650 K. The side surface has an emissivity of 0.40. Heat is supplied from the base surface at a net rate of 1400 W. Determine the temperature of the side surface and the net rates of heat transfer between the top and the bottom surfaces, and between the bottom and side surfaces.

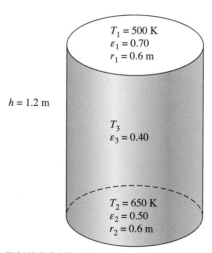

FIGURE P13–102

13–103 Consider a cubical furnace with a side length of 3 m. The top surface is maintained at 700 K. The base surface has an emissivity of 0.90 and is maintained at 950 K. The side surface is black and is maintained at 450 K. Heat is supplied from the base surface at a rate of 340 kW. Determine the emissivity of the top surface and the net rates of heat transfer between the top and the bottom surfaces, and between the bottom and side surfaces.

13–104 A thin aluminum sheet with an emissivity of 0.12 on both sides is placed between two very large parallel plates maintained at uniform temperatures of $T_1 = 750$ K and $T_2 = 400$ K. The emissivities of the plates are $\varepsilon_1 = 0.8$ and $\varepsilon_2 = 0.7$. Determine the net rate of radiation heat transfer between the two plates per unit surface area of the plates, and the temperature of the radiation shield in steady operation.

13–105 Two thin radiation shields with emissivities of $\varepsilon_3 = 0.10$ and $\varepsilon_4 = 0.15$ on both sides are placed between two very large parallel plates, which are maintained at uniform temperatures $T_1 = 600$ K and $T_2 = 300$ K and have emissivities $\varepsilon_1 = 0.6$ and $\varepsilon_2 = 0.7$, respectively. Determine the net rates of radiation heat transfer between the two plates with and without the shields per unit surface area of the plates, and the temperatures of the radiation shields in steady operation.

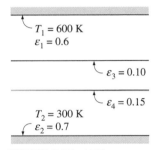

$T_1 = 600$ K
$\varepsilon_1 = 0.6$

$\varepsilon_3 = 0.10$

$\varepsilon_4 = 0.15$

$T_2 = 300$ K
$\varepsilon_2 = 0.7$

FIGURE P13–105

13–106 Two square plates, with the sides a and b (and $b > a$), are coaxial and parallel to each other, as shown in Fig. P13–106, and they are separated by a center-to-center distance of L. The radiation view factor from the smaller to the larger plate, F_{ab}, is given by

$$F_{ab} = \frac{1}{2A}\left\{[(B + A)^2 + 4]^{0.5} - [(B - A)^2 + 4]^{0.5}\right\}$$

where, $A = a/L$ and $B = b/L$.
 (a) Calculate the view factors F_{ab} and F_{ba} for $a = 20$ cm, $b = 60$ cm, and $L = 40$ cm.
 (b) Calculate the net rate of radiation heat exchange between the two plates described above if $T_a = 800°C$, $T_b = 200°C$, $\varepsilon_a = 0.8$, and $\varepsilon_b = 0.4$.

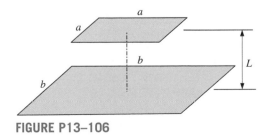

a

a

b

b

L

FIGURE P13–106

 (c) A large square plate (with the side $c = 2.0$ m, $\varepsilon_c = 0.1$, and negligible thickness) is inserted symmetrically between the two plates such that it is parallel to and equidistant from them. For the data given above, calculate the temperature of this third plate when steady operating conditions are established.

13–107 Two parallel concentric disks, 20 cm and 40 cm in diameter, are separated by a distance of 10 cm. The smaller disk ($\varepsilon = 0.80$) is at a temperature of 300°C. The larger disk ($\varepsilon = 0.60$) is at a temperature of 800°C.
 (a) Calculate the radiation view factors.
 (b) Determine the rate of radiation heat exchange between the two disks.
 (c) Suppose that the space between the two disks is completely surrounded by a reflective surface. Estimate the rate of radiation heat exchange between the two disks.

13–108 In a natural-gas fired boiler, combustion gases pass through 6-m-long 15-cm-diameter tubes immersed in water at 1 atm pressure. The tube temperature is measured to be 105°C, and the emissivity of the inner surfaces of the tubes is estimated to be 0.9. Combustion gases enter the tube at 1 atm and 1200 K at a mean velocity of 3 m/s. The mole fractions of CO_2 and H_2O in combustion gases are 8 percent and 16 percent, respectively. Assuming fully developed flow and using properties of air for combustion gases, determine (a) the rates of heat transfer by convection and by radiation from the combustion gases to the tube wall and (b) the rate of evaporation of water.

13–109 Repeat Prob. 13–108 for a total pressure of 3 atm for the combustion gases.

Fundamentals of Engineering (FE) Exam Problems

13–110 Consider two concentric spheres with diameters 12 cm and 18 cm forming an enclosure. The view factor from the inner surface of the outer sphere to the inner sphere is
 (a) 0 (b) 0.18 (c) 0.44 (d) 0.56 (e) 0.67

13–111 Consider an infinitely long three-sided enclosure with side lengths 2 cm, 3 cm, and 4 cm. The view factor from the 2 cm side to the 4 cm side is
 (a) 0.25 (b) 0.50 (c) 0.64 (d) 0.75 (e) 0.87

13–112 Consider a 15-cm-diameter sphere placed within a cubical enclosure with a side length of 15 cm. The view factor from any of the square-cube surfaces to the sphere is
 (a) 0.09 (b) 0.26 (c) 0.52 (d) 0.78 (e) 1

13–113 The number of view factors that need to be evaluated directly for a 10-surface enclosure is
 (a) 1 (b) 10 (c) 22 (d) 34 (e) 45

13–114 A 70-cm-diameter flat black disk is placed in the center of the top surface of a 1-m × 1-m × 1-m black box. The view factor from the entire interior surface of the box to the interior surface of the disk is
 (a) 0.077 (b) 0.144 (c) 0.356
 (d) 0.220 (e) 1.0

13–115 Consider two concentric spheres forming an enclosure with diameters of 12 cm and 18 cm and surface temperatures 300 K and 500 K, respectively. Assuming that the surfaces are black, the net radiation exchange between the two spheres is

(a) 21 W   (b) 140 W   (c) 160 W
(d) 1275 W   (e) 3084 W

13–116 The base surface of a cubical furnace with a side length of 3 m has an emissivity of 0.80 and is maintained at 500 K. If the top and side surfaces also have an emissivity of 0.80 and are maintained at 900 K, the net rate of radiation heat transfer from the top and side surfaces to the bottom surface is

(a) 194 kW   (b) 233 kW   (c) 288 kW
(d) 312 kW   (e) 242 kW

13–117 Consider a vertical 2-m-diameter cylindrical furnace whose surfaces closely approximate black surfaces. The base, top, and side surfaces of the furnace are maintained at 400 K, 600 K, and 900 K, respectively. If the view factor from the base surface to the top surface is 0.2, the net radiation heat transfer between the base and the side surfaces is

(a) 22.5 kW   (b) 38.6 kW   (c) 60.7 kW
(d) 89.8 kW   (e) 151 kW

13–118 Consider a vertical 2-m-diameter cylindrical furnace whose surfaces closely approximate black surfaces. The base, top, and side surfaces of the furnace are maintained at 400 K, 600 K, and 900 K, respectively. If the view factor from the base surface to the top surface is 0.2, the net radiation heat transfer from the bottom surface is

(a) −93.6 kW   (b) −86.1 kW   (c) 0 kW
(d) 86.1 kW   (e) 93.6 kW

13–119 Consider a surface at 0°C that may be assumed to be a blackbody in an environment at 25°C. If 300 W/m² of radiation is incident on the surface, the radiosity of this black surface is

(a) 0 W/m²   (b) 15 W/m²   (c) 132 W/m²
(d) 300 W/m²   (e) 315 W/m²

13–120 Consider a gray and opaque surface at 0°C in an environment at 25°C. The surface has an emissivity of 0.8. If 300 W/m² of radiation is incident on the surface, the radiosity of the surface is

(a) 60 W/m²   (b) 132 W/m²   (c) 300 W/m²
(d) 312 W/m²   (e) 315 W/m²

13–121 Consider a two-surface enclosure with T_1 = 550 K, A_1 = 0.25 m², ε_1 = 0.65, T_2 = 350 K, A_2 = 0.40 m², and ε_2 = 1. If the view factor F_{21} is 0.55, the net rate of radiation heat transfer between the surfaces is

(a) 460 W   (b) 539 W   (c) 648 W
(d) 772 W   (e) 828 W

13–122 Consider two infinitely long concentric cylinders with diameters 20 and 25 cm. The inner surface is maintained at 700 K and has an emissivity of 0.40, while the outer surface is black. If the rate of radiation heat transfer from the inner surface to the outer surface is 2400 W per unit area of the inner surface, the temperature of the outer surface is

(a) 605 K   (b) 538 K   (c) 517 K
(d) 451 K   (e) 415 K

13–123 Two concentric spheres are maintained at uniform temperatures T_1 = 45°C and T_2 = 280°C and have emissivities ε_1 = 0.25 and ε_2 = 0.7, respectively. If the ratio of the diameters is D_1/D_2 = 0.30, the net rate of radiation heat transfer between the two spheres per unit surface area of the inner sphere is

(a) 86 W/m²   (b) 1169 W/m²   (c) 1181 W/m²
(d) 2510 W/m²   (e) 3306 W/m²

13–124 Consider a 3-m × 3-m × 3-m cubical furnace. The base surface of the furnace is black and has a temperature of 400 K. The radiosities for the top and side surfaces are calculated to be 7500 W/m² and 3200 W/m², respectively. The net rate of radiation heat transfer to the bottom surface is

(a) 2.61 kW   (b) 8.27 kW   (c) 14.7 kW
(d) 23.5 kW   (e) 141 kW

13–125 Consider a 3-m × 3-m × 3-m cubical furnace. The base surface is black and has a temperature of 400 K. The radiosities for the top and side surfaces are calculated to be 7500 W/m² and 3200 W/m², respectively. If the temperature of the side surfaces is 485 K, the emissivity of the side surfaces is

(a) 0.37   (b) 0.55   (c) 0.63   (d) 0.80   (e) 0.89

13–126 Two very large parallel plates are maintained at uniform temperatures T_1 = 750 K and T_2 = 500 K and have emissivities ε_1 = 0.85 and ε_2 = 0.7, respectively. If a thin aluminum sheet with the same emissivity on both sides is to be placed between the plates in order to reduce the net rate of radiation heat transfer between the plates by 90 percent, the emissivity of the aluminum sheet must be

(a) 0.07   (b) 0.10   (c) 0.13   (d) 0.16   (e) 0.19

13–127 A 70-cm-diameter flat black disk is placed in the center of the top surface of a 1-m × 1-m × 1-m black box. If the temperature of the box is 427°C and the temperature of the disk is 27°C, the rate of heat transfer by radiation between the interior of the box and the disk is

(a) 2 kW   (b) 3 kW   (c) 4 kW
(d) 5 kW   (e) 6 kW

13–128 A 70-cm-diameter flat disk is placed in the center of the top of a 1-m × 1-m × 1-m black box. If the temperature of the box is 427 °C, the temperature of the disk is 27°C, and emissivity of the interior surface of the disk is 0.3, the rate of heat transfer by radiation between the interior of the box and the disk is

(a) 1.0 kW   (b) 1.5 kW   (c) 2.0 kW
(d) 2.5 kW   (e) 3.2 kW

13–129 Two grey surfaces that form an enclosure exchange heat with one another by thermal radiation. Surface 1 has a temperature of 400 K, an area of 0.2 m², and a total emissivity of 0.4. Surface 2 has a temperature of 600 K, an area of 0.3 m², and a total emissivity of 0.6. If the view factor F_{12} is 0.3, the rate of radiation heat transfer between the two surfaces is

(a) 135 W   (b) 223 W   (c) 296 W
(d) 342 W   (e) 422 W

13–130   The surfaces of a two-surface enclosure exchange heat with one another by thermal radiation. Surface 1 has a temperature of 400 K, an area of 0.2 m², and a total emissivity of 0.4. Surface 2 is black, has a temperature of 600 K, and an area of 0.3 m². If the view factor F_{12} is 0.3, the rate of radiation heat transfer between the two surfaces is

(a) 87 W   (b) 135 W   (c) 244 W
(d) 342 W   (e) 386 W

13–131   A solar flux of 1400 W/m² directly strikes a space-vehicle surface which has a solar absortivity of 0.4 and thermal emissivity of 0.6. The equilibrium temperature of this surface in space at 0 K is

(a) 300 K   (b) 360 K   (c) 410 K
(d) 467 K   (e) 510 K

Design and Essay Problems

13–132   Consider an enclosure consisting of N diffuse and gray surfaces. The emissivity and temperature of each surface as well as the view factors between the surfaces are specified. Write a program to determine the net rate of radiation heat transfer for each surface.

13–133   Radiation shields are commonly used in the design of superinsulations for use in space and cryogenics applications. Write an essay on superinsulations and how they are used in different applications.

13–134   Thermal comfort in a house is strongly affected by the so-called radiation effect, which is due to radiation heat transfer between the person and surrounding surface. A person feels much colder in the morning, for example, becaus of the lower surface temperature of the walls at that time, although the thermostat setting of the house is fixed. Write an essay on thc radiation effect, how it affect human comfort, and how it is accounted for in heating and air-conditioning applications.

MASS TRANSFER

To this point we have restricted our attention to heat transfer problems that did not involve any mass transfer. However, many significant heat transfer problems encountered in practice involve mass transfer. For example, about one-third of the heat loss from a resting person is due to evaporation. It turns out that mass transfer is analogous to heat transfer in many respects, and there is close resemblance between heat and mass transfer relations. In this chapter we discuss the mass transfer mechanisms and develop relations for mass transfer rate for situations commonly encountered in practice.

Distinction should be made between *mass transfer* and the *bulk fluid motion* (or *fluid flow*) that occurs on a macroscopic level as a fluid is transported from one location to another. Mass transfer requires the presence of two regions at different chemical compositions, and mass transfer refers to the movement of a chemical species from a high concentration region toward a lower concentration one. The primary driving force for fluid flow is the *pressure difference,* whereas for mass transfer it is the *concentration difference.*

We begin this chapter by pointing out numerous analogies between heat and mass transfer and draw several parallels between them. We then discuss boundary conditions associated with mass transfer and one-dimensional steady and transient mass diffusion, followed by a discussion of mass transfer in a moving medium. Finally, we consider convection mass transfer and simultaneous heat and mass transfer.

OBJECTIVES

When you finish studying this chapter, you should be able to:

■ Understand the concenration gradient and the physical mechanism of mass transfer,

■ Reccognize the analogy between heat and mass transfer,

■ Describe the concenration at a location on mass or mole basis, and relate the rate of diffusion to the concentration gradient by Fick's law,

■ Calculate the rate of mass diffusion through a plain layer under steady conditions,

■ Predict the migration of water vapor in buildings,

■ Perform a transient mass diffusion analysis in large mediums,

■ Calculate mass transfer by convection, and

■ Analyze simultaneous heat and mass transfer.

14–1 · INTRODUCTION

It is a common observation that whenever there is an imbalance of a commodity in a medium, nature tends to redistribute it until a "balance" or "equality" is established. This tendency is often referred to as the *driving force*, which is the mechanism behind many naturally occurring transport phenomena.

If we define the amount of a commodity per unit volume as the **concentration** of that commodity, we can say that the flow of a commodity is always in the direction of decreasing concentration; that is, from the region of high concentration to the region of low concentration (Fig. 14–1). The commodity simply creeps away during redistribution, and thus the flow is a *diffusion process*. The rate of flow of the commodity is proportional to the *concentration gradient dC/dx*, which is the change in the concentration C per unit length in the flow direction x, and the area A normal to flow direction, and is expressed as

Flow rate ∝ (Normal area)(Concentration gradient)

or

$$\dot{Q} = -k_{\text{diff}} A \frac{dC}{dx} \qquad (14\text{–}1)$$

Here the proportionality constant k_{diff} is the *diffusion coefficient* of the medium, which is a measure of how fast a commodity diffuses in the medium, and the negative sign is to make the flow in the positive direction a positive quantity (note that dC/dx is a negative quantity since concentration decreases in the flow direction). You may recall that *Fourier's law of heat conduction, Ohm's law of electrical conduction,* and *Newton's law of viscosity* are all in the form of Eq. 14–1.

To understand the diffusion process better, consider a tank that is divided into two equal parts by a partition. Initially, the left half of the tank contains nitrogen N_2 gas while the right half contains air (about 21 percent O_2 and 79 percent N_2) at the same temperature and pressure. The O_2 and N_2 molecules are indicated by dark and light circles, respectively. When the partition is removed, we know that the N_2 molecules will start diffusing into the air while the O_2 molecules diffuse into the N_2, as shown in Fig. 14–2. If we wait long enough, we will have a homogeneous mixture of N_2 and O_2 in the tank. This mass diffusion process can be explained by considering an imaginary plane indicated by the dashed line in the figure as: Gas molecules move randomly, and thus the probability of a molecule moving to the right or to the left is the same. Consequently, half of the molecules on one side of the dashed line at any given moment will move to the other side. Since the concentration of N_2 is greater on the left side than it is on the right side, more N_2 molecules will move toward the right than toward the left, resulting in a net flow of N_2 toward the right. As a result, N_2 is said to be *transferred* to the right. A similar argument can be given for O_2 being transferred to the left. The process continues until uniform concentrations of N_2 and O_2 are established throughout the tank so that the number of N_2 (or O_2) molecules moving to the right equals

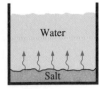

(*a*) Before   (*b*) After

FIGURE 14–1
Whenever there is concentration difference of a physical quantity in a medium, nature tends to equalize things by forcing a flow from the high to the low concentration region.

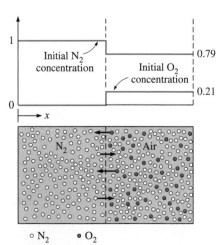

FIGURE 14–2
A tank that contains N_2 and air in its two compartments, and the diffusion of N_2 into the air (and the diffusion of O_2 into N_2) when the partition is removed.

the number moving to the left, resulting in zero net transfer of N_2 or O_2 across an imaginary plane.

The molecules in a gas mixture continually collide with each other, and the diffusion process is strongly influenced by this collision process. The collision of like molecules is of little consequence since both molecules are identical and it makes no difference which molecule crosses a certain plane. The collisions of unlike molecules, however, influence the rate of diffusion since unlike molecules may have different masses and thus different momentums, and thus the diffusion process is dominated by the heavier molecules. The diffusion coefficients and thus diffusion rates of gases depend strongly on *temperature* since the temperature is a measure of the average velocity of gas molecules. Therefore, the diffusion rates are higher at higher temperatures.

Mass transfer can also occur in liquids and solids as well as in gases. For example, a cup of water left in a room eventually evaporates as a result of water molecules diffusing into the air *(liquid-to-gas mass transfer)*. A piece of solid CO_2 (dry ice) also gets smaller and smaller in time as the CO_2 molecules diffuse into the air *(solid-to-gas mass transfer)*. A spoon of sugar in a cup of coffee eventually moves up and sweetens the coffee although the sugar molecules are much heavier than the water molecules, and the molecules of a colored pencil inserted into a glass of water diffuses into the water as evidenced by the gradual spread of color in the water *(solid-to-liquid mass transfer)*. Of course, mass transfer can also occur from a gas to a liquid or solid if the concentration of the species is higher in the gas phase. For example, a small fraction of O_2 in the air diffuses into the water and meets the oxygen needs of marine animals. The diffusion of carbon into iron during case-hardening, doping of semiconductors for transistors, and the migration of doped molecules in semiconductors at high temperature are examples of solid-to-solid diffusion processes (Fig. 14–3).

Another factor that influences the diffusion process is the *molecular spacing*. The larger the spacing, in general, the higher the diffusion rate. Therefore, the diffusion rates are typically much higher in gases than they are in liquids and much higher in liquids than in solids. Diffusion coefficients in gas mixtures are a few orders of magnitude larger than those of liquid or solid solutions.

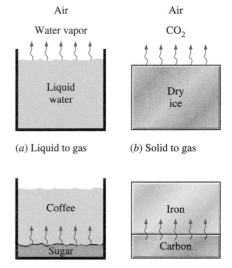

(a) Liquid to gas (b) Solid to gas

(c) Solid to liquid (d) Solid to solid

FIGURE 14–3

Some examples of mass transfer that involve a liquid and/or a solid.

14–2 · ANALOGY BETWEEN HEAT AND MASS TRANSFER

We have spent a considerable amount of time studying heat transfer, and we could spend just as much time studying mass transfer. However, the mechanisms of heat and mass transfer are analogous to each other, and thus we can develop an understanding of mass transfer in a short time with little effort by simply drawing *parallels* between heat and mass transfer. Establishing those "bridges" between the two seemingly unrelated areas will make it possible to use our heat transfer knowledge to solve mass transfer problems. Alternately, gaining a working knowledge of mass transfer will help us to better understand the heat transfer processes by thinking of heat as a massless substance as they did in the nineteenth century. The short-lived caloric theory of heat is the origin of most heat transfer terminology used today and served its purpose well until it was replaced by the kinetic theory. Mass is, in essence, energy

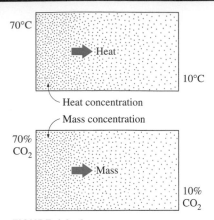

FIGURE 14–4
Analogy between
heat and mass transfer.

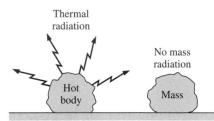

FIGURE 14–5
Unlike heat radiation, there is
no such thing as mass radiation.

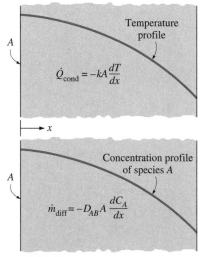

FIGURE 14–6
Analogy between heat conduction
and mass diffusion.

since mass and energy can be converted to each other according to Einstein's formula $E = mc^2$, where c is the speed of light. Therefore, we can look at mass and heat as two different forms of energy and exploit this to advantage without going overboard.

Temperature

The driving force for heat transfer is the *temperature difference*. In contrast, the driving force for mass transfer is the *concentration difference*. We can view temperature as a measure of "heat concentration," and thus a high temperature region as one that has a high heat concentration (Fig. 14–4). Therefore, both heat and mass are transferred from the more concentrated regions to the less concentrated ones. If there is no temperature difference between two regions, then there is no heat transfer. Likewise, if there is no difference between the concentrations of a species at different parts of a medium, there will be no mass transfer.

Conduction

You will recall that heat is transferred by conduction, convection, and radiation. Mass, however, is transferred by *conduction* (called diffusion) and *convection* only, and there is no such thing as "mass radiation" (unless there is something Scotty knows that we don't when he "beams" people to anywhere in space at the speed of light) (Fig. 14–5). The rate of heat conduction in a direction x is proportional to the temperature gradient dT/dx in that direction and is expressed by **Fourier's law of heat conduction** as

$$\dot{Q}_{\text{cond}} = -kA \frac{dT}{dx} \qquad (14\text{–}2)$$

where k is the thermal conductivity of the medium and A is the area normal to the direction of heat transfer. Likewise, the rate of mass diffusion $\dot{m}_{\text{diff}}$ of a chemical species A in a stationary medium in the direction x is proportional to the concentration gradient dC/dx in that direction and is expressed by **Fick's law of diffusion** by (Fig. 14–6)

$$\dot{m}_{\text{diff}} = -D_{AB} A \frac{dC_A}{dx} \qquad (14\text{–}3)$$

where D_{AB} is the **diffusion coefficient** (or *mass diffusivity*) of the species in the mixture and C_A is the concentration of the species in the mixture at that location.

It can be shown that the differential equations for both heat conduction and mass diffusion are of the same form. Therefore, the solutions of mass diffusion equations can be obtained from the solutions of corresponding heat conduction equations for the same type of boundary conditions by simply switching the corresponding coefficients and variables.

Heat Generation

Heat generation refers to the conversion of some form of energy such as electrical, chemical, or nuclear energy into *sensible thermal* energy in the medium. Heat generation occurs throughout the medium and exhibits itself as a rise in temperature. Similarly, some mass transfer problems involve chemical reactions that occur within the medium and result in the *generation of a species* throughout. Therefore, species generation is a *volumetric phenomenon,* and the rate of generation may vary from point to point in the medium. Such reactions that occur within the medium are called **homogeneous reactions** and are analogous to internal heat generation. In contrast, some chemical reactions result in the generation of a species *at the surface* as a result of chemical reactions occurring at the surface due to contact between the medium and the surroundings. This is a *surface phenomenon,* and as such it needs to be treated as a boundary condition. In mass transfer studies, such reactions are called **heterogeneous reactions** and are analogous to *specified surface heat flux.*

Convection

You may recall that *heat convection* is the heat transfer mechanism that involves both *heat conduction* (molecular diffusion) and *bulk fluid motion.* Fluid motion enhances heat transfer considerably by removing the heated fluid near the surface and replacing it by the cooler fluid further away. In the limiting case of no bulk fluid motion, convection reduces to conduction. Likewise, **mass convection** (or *convective mass transfer*) is the mass transfer mechanism between a surface and a moving fluid that involves both *mass diffusion* and *bulk fluid motion.* Fluid motion also enhances mass transfer considerably by removing the high concentration fluid near the surface and replacing it by the lower concentration fluid further away. In mass convection, we define a *concentration boundary layer* in an analogous manner to the thermal boundary layer and define new dimensionless numbers that are counterparts of the Nusselt and Prandtl numbers.

The rate of heat convection for external flow was expressed conveniently by *Newton's law of cooling* as

$$\dot{Q}_{\text{conv}} = h_{\text{conv}} A_s (T_s - T_\infty) \tag{14-4}$$

where h_{conv} is the heat transfer coefficient, A_s is the surface area, and $T_s - T_\infty$ is the temperature difference across the thermal boundary layer. Likewise, the rate of mass convection can be expressed as (Fig. 14–7)

$$\dot{m}_{\text{conv}} = h_{\text{mass}} A_s (C_s - C_\infty) \tag{14-5}$$

where h_{mass} is the *mass transfer coefficient,* A_s is the surface area, and $C_s - C_\infty$ is a suitable concentration difference across the concentration boundary layer.

Various aspects of the analogy between heat and mass convection are explored in Section 14–9. The analogy is valid for *low mass transfer rate* cases in which the flow rate of species undergoing mass flow is low (under 10 percent) relative to the total flow rate of the liquid or gas mixture.

14–3 · MASS DIFFUSION

Fick's law of diffusion, proposed in 1855, states that the rate of diffusion of a chemical species at a location in a gas mixture (or liquid or solid solution) is proportional to the *concentration gradient* of that species at that location.

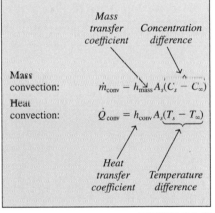

Mass convection:
$$\dot{m}_{\text{conv}} = h_{\text{mass}} A_s (C_s - C_\infty)$$

Heat convection:
$$\dot{Q}_{\text{conv}} = h_{\text{conv}} A_s (T_s - T_\infty)$$

FIGURE 14–7

Analogy between convection heat transfer and convection mass transfer.

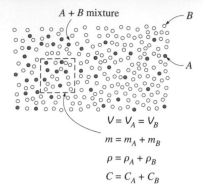

$$V = V_A = V_B$$
$$m = m_A + m_B$$
$$\rho = \rho_A + \rho_B$$
$$C = C_A + C_B$$

Mass basis:

$$\rho_A = \frac{m_A}{V}, \quad \rho = \frac{m}{V}, \quad w_A = \frac{\rho_A}{\rho}$$

Mole basis:

$$C_A = \frac{N_A}{V}, \quad C = \frac{N}{V}, \quad y_A = \frac{C_A}{C}$$

Relation between them:

$$C_A = \frac{\rho_A}{M_A}, \quad w_A = y_A \frac{M_A}{M}$$

FIGURE 14–8

Different ways of expressing the concentration of species A of a binary mixture A and B.

Although a higher concentration for a species means more molecules of that species per unit volume, the concentration of a species can be expressed in several ways. Next we describe two common ways.

1 Mass Basis

On a *mass basis,* concentration is expressed in terms of **density** (or *mass concentration*), which is mass per unit volume. Considering a small volume V at a location within the mixture, the densities of a species (subscript i) and of the mixture (no subscript) at that location are given by (Fig. 14–8)

Partial density of species i: $\qquad\qquad \rho_i = m_i/V \qquad\qquad\qquad$ (kg/m³)

Total density of mixture: $\qquad\qquad \rho = m/V = \sum m_i/V = \sum \rho_i$

Therefore, the *density of a mixture* at a location is equal to the sum of the *densities of its constituents* at that location. Mass concentration can also be expressed in dimensionless form in terms of **mass fraction** w as

Mass fraction of species i: $\qquad\qquad w_i = \frac{m_i}{m} = \frac{m_i/V}{m/V} = \frac{\rho_i}{\rho} \qquad$ (14–6)

Note that the mass fraction of a species ranges between 0 and 1, and the conservation of mass requires that the sum of the mass fractions of the constituents of a mixture be equal to 1. That is, $\Sigma w_i = 1$. Also note that the density and mass fraction of a constituent in a mixture, in general, vary with location unless the concentration gradients are zero.

2 Mole Basis

On a *mole basis,* concentration is expressed in terms of **molar concentration** (or *molar density*), which is the amount of matter in kmol per unit volume. Again considering a small volume V at a location within the mixture, the molar concentrations of a species (subscript i) and of the mixture (no subscript) at that location are given by

Partial molar concentration of species i: $\quad C_i = N_i/V \qquad\qquad$ (kmol/m³)

Total molar concentration of mixture: $\qquad C = N/V = \sum N_i/V = \sum C_i$

Therefore, the molar concentration of a mixture at a location is equal to the sum of the molar concentrations of its constituents at that location. Molar concentration can also be expressed in dimensionless form in terms of **mole fraction** y as

Mole fraction of species i: $\qquad\qquad y_i = \frac{N_i}{N} = \frac{N_i/V}{N/V} = \frac{C_i}{C} \qquad$ (14–7)

Again the mole fraction of a species ranges between 0 and 1, and the sum of the mole fractions of the constituents of a mixture is unity, $\Sigma y_i = 1$.

The mass m and mole number N of a substance are related to each other by $m = NM$ (or, for a unit volume, $\rho = CM$) where M is the *molar mass* (also called the *molecular weight*) of the substance. This is expected since the mass of 1 kmol of the substance is M kg, and thus the mass of N kmol is NM kg. Therefore, the mass and molar concentrations are related to each other by

$$C_i = \frac{\rho_i}{M_i} \quad \text{(for species } i) \quad \text{and} \quad C = \frac{\rho}{M} \quad \text{(for the mixture)} \qquad (14\text{–}8)$$

where M is the molar mass of the mixture which can be determined from

$$M = \frac{m}{N} = \frac{\sum N_i M_i}{N} = \sum \frac{N_i}{N} M_i = \sum y_i M_i \qquad (14\text{–}9)$$

The mass and mole fractions of species i of a mixture are related to each other by

$$w_i = \frac{\rho_i}{\rho} = \frac{C_i M_i}{CM} = y_i \frac{M_i}{M} \qquad (14\text{–}10)$$

Two different approaches are presented above for the description of concentration at a location, and you may be wondering which approach is better to use. Well, the answer depends on the situation on hand. Both approaches are equivalent, and the better approach for a given problem is the one that yields the desired solution more easily.

Special Case: Ideal Gas Mixtures

At low pressures, a gas or gas mixture can conveniently be approximated as an ideal gas with negligible error. For example, a mixture of dry air and water vapor at atmospheric conditions can be treated as an ideal gas with an error much less than 1 percent. The total pressure of a gas mixture P is equal to the sum of the partial pressures P_i of the individual gases in the mixture and is expressed as $P = \sum P_i$. Here P_i is called the **partial pressure** of species i, which is the pressure species i would exert if it existed alone at the mixture temperature and volume. This is known as **Dalton's law of additive pressures**. Then using the ideal gas relation $PV = NR_uT$ where R_u is the universal gas constant for both the species i and the mixture, the **pressure fraction** of species i can be expressed as (Fig. 14–9)

$$\frac{P_i}{P} = \frac{N_i R_u T/V}{NR_u T/V} = \frac{N_i}{N} = y_i \qquad (14\text{–}11)$$

Therefore, the *pressure fraction* of species i of an ideal gas mixture is equivalent to the *mole fraction* of that species and can be used in place of it in mass transfer analysis.

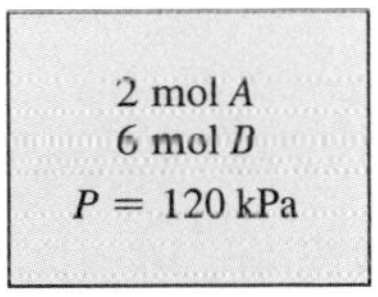

| 2 mol A |
| 6 mol B |
| $P = 120$ kPa |

A mixture of two ideal gases A and B

$$y_A = \frac{N_A}{N} = \frac{2}{2 + 6} = 0.25$$
$$P_A = y_A P = 0.25 \times 120 = 30 \text{ kPa}$$

FIGURE 14–9
For ideal gas mixtures, pressure fraction of a gas is equal to its mole fraction.

Fick's Law of Diffusion: Stationary Medium Consisting of Two Species

We mentioned earlier that the rate of mass diffusion of a chemical species in a stagnant medium in a specified direction is proportional to the local concentration gradient in that direction. This linear relationship between the rate of diffusion and the concentration gradient proposed by Fick in 1855 is known as **Fick's law of diffusion** and can be expressed as

Mass flux = Constant of proportionality $\times$ Concentration gradient

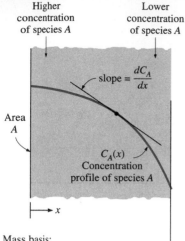

Higher concentration of species A

Lower concentration of species A

slope = $\dfrac{dC_A}{dx}$

Area A

$C_A(x)$
Concentration profile of species A

x

Mass basis:

$$\dot{m}_{\text{diff}} = -\rho A D_{AB} \frac{dw_A}{dx}$$

$$= -\rho A D_{AB} \frac{d(\rho_A/\rho)}{dx}$$

$$= -A D_{AB} \frac{d\rho_A}{dx} \quad (\text{if } \rho = \text{constant})$$

Mole basis:

$$\dot{N}_{\text{diff},A} = -C A D_{AB} \frac{dy_A}{dx}$$

$$= -C A D_{AB} \frac{d(C_A/C)}{dx}$$

$$= -A D_{AB} \frac{dC_A}{dx} \quad (\text{if } C = \text{constant})$$

FIGURE 14–10
Various expressions of Fick's law of diffusion for a binary mixture.

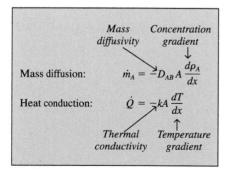

	Mass diffusivity	Concentration gradient
Mass diffusion:		$\dot{m}_A = -D_{AB} A \dfrac{d\rho_A}{dx}$
Heat conduction:		$\dot{Q} = -kA \dfrac{dT}{dx}$
	Thermal conductivity	Temperature gradient

FIGURE 14–11
Analogy between Fourier's law of heat conduction and Fick's law of mass diffusion.

But the concentration of a species in a gas mixture or liquid or solid solution can be defined in several ways such as density, mass fraction, molar concentration, and mole fraction, as already discussed, and thus Fick's law can be expressed mathematically in many ways. It turns out that it is best to express the concentration gradient in terms of the mass or mole fraction, and the most appropriate formulation of *Fick's law* for the diffusion of a species A in a stationary binary mixture of species A and B in a specified direction x is given by (Fig. 14–10)

Mass basis: $\quad j_{\text{diff},A} = \dfrac{\dot{m}_{\text{diff},A}}{A} = -\rho D_{AB} \dfrac{d(\rho_A/\rho)}{dx} = -\rho D_{AB} \dfrac{dw_A}{dx} \quad (\text{kg/s} \cdot \text{m}^2)$

Mole basis: $\quad \bar{j}_{\text{diff},A} = \dfrac{\dot{N}_{\text{diff},A}}{A} = -CD_{AB} \dfrac{d(C_A/C)}{dx} = -CD_{AB} \dfrac{dy_A}{dx} \quad (\text{kmol/s} \cdot \text{m}^2)$

$$(14\text{–}12)$$

Here $j_{\text{diff},A}$ is the **(diffusive) mass flux** of species A (mass transfer by diffusion per unit time and per unit area normal to the direction of mass transfer, in kg/s · m²) and $\bar{j}_{\text{diff},A}$ is the **(diffusive) molar flux** (in kmol/s · m²). The mass flux of a species at a location is proportional to the density of the mixture at that location. Note that $\rho = \rho_A + \rho_B$ is the density and $C = C_A + C_B$ is the molar concentration of the binary mixture, and in general, they may vary throughout the mixture. Therefore, $\rho d(\rho_A/\rho) \neq d\rho_A$ or $Cd(C_A/C) \neq dC_A$. But in the special case of constant mixture density ρ or constant molar concentration C, the relations above simplify to

Mass basis (ρ = constant): $\qquad j_{\text{diff},A} = -D_{AB} \dfrac{d\rho_A}{dx} \qquad (\text{kg/s} \cdot \text{m}^2)$

Mole basis (C = constant): $\qquad \bar{j}_{\text{diff},A} = -D_{AB} \dfrac{dC_A}{dx} \qquad (\text{kmol/s} \cdot \text{m}^2) \qquad (14\text{–}13)$

The constant density or constant molar concentration assumption is usually appropriate for *solid* and *dilute liquid* solutions, but often this is not the case for gas mixtures or concentrated liquid solutions. Therefore, Eq. 14–12 should be used in the latter case. In this introductory treatment we limit our consideration to one-dimensional mass diffusion. For two- or three-dimensional cases, Fick's law can conveniently be expressed in vector form by simply replacing the derivatives in the above relations by the corresponding gradients (such as $\mathbf{j}_A = -\rho D_{AB} \nabla w_A$).

Remember that the constant of proportionality in Fourier's law was defined as the transport property *thermal conductivity*. Similarly, the constant of proportionality in Fick's law is defined as another transport property called the **binary diffusion coefficient** or **mass diffusivity**, D_{AB}. The unit of mass diffusivity is m²/s, which is the same as the units of *thermal diffusivity* or *momentum diffusivity* (also called *kinematic viscosity*) (Fig. 14–11).

Because of the complex nature of mass diffusion, the diffusion coefficients are usually determined experimentally. The kinetic theory of gases indicates that the diffusion coefficient for dilute gases at ordinary pressures is essentially independent of mixture composition and tends to increase with temperature while decreasing with pressure as

$$D_{AB} \propto \frac{T^{3/2}}{P} \qquad \text{or} \qquad \frac{D_{AB,1}}{D_{AB,2}} = \frac{P_2}{P_1}\left(\frac{T_1}{T_2}\right)^{3/2} \qquad (14\text{–}14)$$

This relation is useful in determining the diffusion coefficient for gases at different temperatures and pressures from a knowledge of the diffusion coefficient at a specified temperature and pressure. More general but complicated relations that account for the effects of molecular collisions are also available. The diffusion coefficients of some gases in air at 1 atm pressure are given in Table 14–1 at various temperatures.

The diffusion coefficients of solids and liquids also tend to increase with temperature while exhibiting a strong dependence on the composition. The diffusion process in solids and liquids is a great deal more complicated than that in gases, and the diffusion coefficients in this case are almost exclusively determined experimentally.

The binary diffusion coefficients for several binary gas mixtures and solid and liquid solutions are given in Tables 14–2 and 14–3. We make two observations from these tables:

1. The diffusion coefficients, in general, are *highest in gases* and *lowest in solids*. The diffusion coefficients of gases are several orders of magnitude greater than those of liquids.
2. Diffusion coefficients *increase with temperature*. The diffusion coefficient (and thus the mass diffusion rate) of carbon through iron during a hardening process, for example, increases by 6000 times as the temperature is raised from 500°C to 1000°C.

TABLE 14–1

Binary diffusion coefficients of some gases in air at 1 atm pressure (from Mills, 1995; Table A.17a, p. 869)

	Binary Diffusion Coefficient,* $m^2/s \times 10^5$			
T, K	O_2	CO_2	H_2	NO
200	0.95	0.74	3.75	0.88
300	1.88	1.57	7.77	1.80
400	5.25	2.63	12.5	3.03
500	4.75	3.85	17.1	4.43
600	6.46	5.37	24.4	6.03
700	8.38	6.84	31.7	7.82
800	10.5	8.57	39.3	9.78
900	12.6	10.5	47.7	11.8
1000	15.2	12.4	56.9	14.1
1200	20.6	16.9	77.7	19.2
1400	26.6	21.7	99.0	24.5
1600	33.2	27.5	125	30.4
1800	40.3	32.8	152	37.0
2000	48.0	39.4	180	44.8

*Multiply by 10.76 to convert to ft^2/s.

TABLE 14–2

Binary diffusion coefficients of dilute gas mixtures at 1 atm
(from Barrer, 1941; Geankoplis, 1972; Perry, 1963; and Reid et al., 1977)

Substance A	Substance B	T, K	D_{AB} or D_{BA}, m^2/s	Substance A	Substance B	T, K	D_{AB} or D_{BA}, m^2/s
Air	Acetone	273	1.1×10^{-5}	Argon, Ar	Nitrogen, N_2	293	1.9×10^{-5}
Air	Ammonia, NH_3	298	2.6×10^{-5}	Carbon dioxide, CO_2	Benzene	318	0.72×10^{-5}
Air	Benzene	298	0.88×10^{-5}	Carbon dioxide, CO_2	Hydrogen, H_2	273	5.5×10^{-5}
Air	Carbon dioxide	298	1.6×10^{-5}	Carbon dioxide, CO_2	Nitrogen, N_2	293	1.6×10^{-5}
Air	Chlorine	273	1.2×10^{-5}	Carbon dioxide, CO_2	Oxygen, O_2	273	1.4×10^{-5}
Air	Ethyl alcohol	298	1.2×10^{-5}	Carbon dioxide, CO_2	Water vapor	298	1.6×10^{-5}
Air	Ethyl ether	298	0.93×10^{-5}	Hydrogen, H_2	Nitrogen, N_2	273	6.8×10^{-5}
Air	Helium, He	298	7.2×10^{-5}	Hydrogen, H_2	Oxygen, O_2	273	7.0×10^{-5}
Air	Hydrogen, H_2	298	7.2×10^{-5}	Oxygen, O_2	Ammonia	293	2.5×10^{-5}
Air	Iodine, I_2	298	0.83×10^{-5}	Oxygen, O_2	Benzene	296	0.39×10^{-5}
Air	Methanol	298	1.6×10^{-5}	Oxygen, O_2	Nitrogen, N_2	273	1.8×10^{-5}
Air	Mercury	614	4.7×10^{-5}	Oxygen, O_2	Water vapor	298	2.5×10^{-5}
Air	Napthalene	300	0.62×10^{-5}	Water vapor	Argon, Ar	298	2.4×10^{-5}
Air	Oxygen, O_2	298	2.1×10^{-5}	Water vapor	Helium, He	298	9.2×10^{-5}
Air	Water vapor	298	2.5×10^{-5}	Water vapor	Nitrogen, N_2	298	2.5×10^{-5}

Note: The effect of pressure and temperature on D_{AB} can be accounted for through $D_{AB} \sim T^{3/2}/P$. Also, multiply D_{AB} values by 10.76 to convert them to ft^2/s.

TABLE 14–3

Binary diffusion coefficients of dilute liquid solutions and solid solutions at 1 atm
(from Barrer, 1941; Reid et al., 1977; Thomas, 1991; and van Black, 1980)

(a) Diffusion through Liquids				(b) Diffusion through Solids			
Substance A (Solute)	Substance B (Solvent)	T, K	D_{AB}, m²/s	Substance A (Solute)	Substance B (Solvent)	T, K	D_{AB}, m²/s
Ammonia	Water	285	1.6×10^{-9}	Carbon dioxide	Natural rubber	298	1.1×10^{-10}
Benzene	Water	293	1.0×10^{-9}	Nitrogen	Natural rubber	298	1.5×10^{-10}
Carbon dioxide	Water	298	2.0×10^{-9}	Oxygen	Natural rubber	298	2.1×10^{-10}
Chlorine	Water	285	1.4×10^{-9}	Helium	Pyrex	773	2.0×10^{-12}
Ethanol	Water	283	0.84×10^{-9}	Helium	Pyrex	293	4.5×10^{-15}
Ethanol	Water	288	1.0×10^{-9}	Helium	Silicon dioxide	298	4.0×10^{-14}
Ethanol	Water	298	1.2×10^{-9}	Hydrogen	Iron	298	2.6×10^{-13}
Glucose	Water	298	0.69×10^{-9}	Hydrogen	Nickel	358	1.2×10^{-12}
Hydrogen	Water	298	6.3×10^{-9}	Hydrogen	Nickel	438	1.0×10^{-11}
Methane	Water	275	0.85×10^{-9}	Cadmium	Copper	293	2.7×10^{-19}
Methane	Water	293	1.5×10^{-9}	Zinc	Copper	773	4.0×10^{-18}
Methane	Water	333	3.6×10^{-9}	Zinc	Copper	1273	5.0×10^{-13}
Methanol	Water	288	1.3×10^{-9}	Antimony	Silver	293	3.5×10^{-25}
Nitrogen	Water	298	2.6×10^{-9}	Bismuth	Lead	293	1.1×10^{-20}
Oxygen	Water	298	2.4×10^{-9}	Mercury	Lead	293	2.5×10^{-19}
Water	Ethanol	298	1.2×10^{-9}	Copper	Aluminum	773	4.0×10^{-14}
Water	Ethylene glycol	298	0.18×10^{-9}	Copper	Aluminum	1273	1.0×10^{-10}
Water	Methanol	298	1.8×10^{-9}	Carbon	Iron (fcc)	773	5.0×10^{-15}
Chloroform	Methanol	288	2.1×10^{-9}	Carbon	Iron (fcc)	1273	3.0×10^{-11}

TABLE 14–4

In a binary ideal gas mixture of species A and B, the diffusion coefficient of A in B is equal to the diffusion coefficient of B in A, and both increase with temperature

T, °C	D_{H_2O-Air} or D_{Air-H_2O} at 1 atm, in m²/s (from Eq. 14–15)
0	2.09×10^{-5}
5	2.17×10^{-5}
10	2.25×10^{-5}
15	2.33×10^{-5}
20	2.42×10^{-5}
25	2.50×10^{-5}
30	2.59×10^{-5}
35	2.68×10^{-5}
40	2.77×10^{-5}
50	2.96×10^{-5}
100	3.99×10^{-5}
150	5.18×10^{-5}

Due to its practical importance, the diffusion of *water vapor* in *air* has been the topic of several studies, and some empirical formulas have been developed for the diffusion coefficient D_{H_2O-air}. Marrero and Mason (1972) proposed this popular formula (Table 14–4):

$$D_{H_2O-Air} = 1.87 \times 10^{-10} \frac{T^{2.072}}{P} \quad (m^2/s), \quad 280\,K < T < 450\,K \quad (14\text{–}15)$$

where P is total pressure in atm and T is the temperature in K.

The primary driving mechanism of mass diffusion is the concentration gradient, and mass diffusion due to a concentration gradient is known as the **ordinary diffusion**. However, diffusion may also be caused by other effects. Temperature gradients in a medium can cause **thermal diffusion** (also called the **soret effect**), and pressure gradients may result in **pressure diffusion**. Both of these effects are usually negligible, however, unless the gradients are very large. In centrifuges, the pressure gradient generated by the centrifugal effect is used to separate liquid solutions and gaseous isotopes. An external force field such as an electric or magnetic field applied on a mixture or solution can be used successfully to separate electrically charged or magnetized molecules (as in an electrolyte or ionized gas) from the mixture. This is called **forced diffusion**. Also, when the pores of a porous solid such as silica-gel are smaller than the mean free path of the gas molecules, the molecular collisions may be negligible and a free molecule flow may be initiated. This is known as **Knudsen diffusion**. When the size of the gas molecules is comparable to the pore size, adsorbed molecules move along the pore walls. This is known as **surface diffusion**. Finally, particles whose diameter is under 0.1 μm such as mist and soot particles act like large molecules, and the diffusion process of such particles due to the concentration gradient is called **Brownian motion**. Large particles (those whose diameter is greater than 1 μm) are not affected by diffusion as the motion of such particles is governed by Newton's laws. In our elementary treatment of mass diffusion, we assume these additional effects to be nonexistent or negligible, as is usually the case, and refer the interested reader to advanced books on these topics.

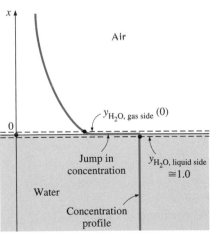

AIR

78.1% N_2
20.9% O_2
1.0% Ar

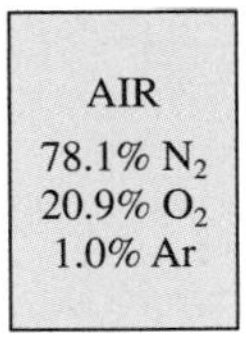

FIGURE 14–12
Schematic for Example 14–1.

EXAMPLE 14–1   **Determining Mass Fractions from Mole Fractions**

The composition of dry standard atmosphere is given on a molar basis to be 78.1 percent N_2, 20.9 percent O_2, and 1.0 percent Ar and small amounts of other constituents (Fig. 14–12). Treating other constituents as Ar, determine the mass fractions of the constituents of air.

SOLUTION   The molar fractions of the constituents of air are given. The mass fractions are to be determined.
Assumptions   The small amounts of other gases in air are treated as argon.
Properties   The molar masses of N_2, O_2, and Ar are 28.0, 32.0, and 39.9 kg/kmol, respectively (Table A–1).
Analysis   The molar mass of air is determined to be

$$M = \Sigma\, y_i M_i = 0.781 \times 28.0 + 0.209 \times 32.0 + 0.01 \times 39.9 = 29.0 \text{ kg/kmol}$$

Then the mass fractions of constituent gases are determined from Eq. 14–10 to be

N_2:
$$w_{N_2} = y_{N_2}\,\frac{M_{N_2}}{M} = (0.781)\,\frac{28.0}{29.0} = 0.754$$

O_2:
$$w_{O_2} = y_{O_2}\,\frac{M_{O_2}}{M} = (0.209)\,\frac{32.0}{29.0} = 0.231$$

Ar:
$$w_{Ar} = y_{Ar}\,\frac{M_{Ar}}{M} = (0.01)\,\frac{39.9}{29.0} = 0.014$$

Therefore, the mass fractions of N_2, O_2, and Ar in dry standard atmosphere are 75.4 percent, 23.1 percent, and 1.4 percent, respectively.

14–4 · BOUNDARY CONDITIONS

We mentioned earlier that the mass diffusion equation is analogous to the heat diffusion (conduction) equation, and thus we need comparable boundary conditions to determine the species concentration distribution in a medium. Two common types of boundary conditions are the (1) *specified species concentration,* which corresponds to specified temperature, and (2) *specified species flux,* which corresponds to specified heat flux.

Despite their apparent similarity, an important difference exists between temperature and concentration: temperature is necessarily a *continuous* function, but concentration, in general, is not. The wall and air temperatures at a wall surface, for example, are always the same. The concentrations of air on the two sides of a water–air interface, however, are obviously very different (in fact, the

FIGURE 14–13
Unlike temperature, the concentration of species on the two sides of a liquid–gas (or solid–gas or solid–liquid) interface are usually not the same.

concentration of air in water is close to zero). Likewise, the concentrations of water on the two sides of a water–air interface are also different even when air is saturated (Fig. 14–13). Therefore, when specifying a boundary condition, specifying the location is not enough. We also need to specify the side of the boundary. To do this, we consider two imaginary surfaces on the two sides of the interface that are infinitesimally close to the interface. Whenever there is a doubt, we indicate the desired side of the interface by specifying its phase as a subscript. For example, the water (liquid or vapor) concentration at the liquid and gas sides of a water–air interface at $x = 0$ can be expressed on a molar basis is

$$y_{\text{H}_2\text{O, liquid side}} (0) = y_1 \quad \text{and} \quad y_{\text{H}_2\text{O, gas side}} (0) = y_2 \tag{14–16}$$

Using Fick's law, the constant species flux boundary condition for a diffusing species A at a boundary at $x = 0$ is expressed, in the absence of any blowing or suction, as

$$-CD_{AB} \frac{dy_A}{dx}\bigg|_{x=0} = \bar{j}_{A,\,0} \quad \text{or} \quad -\rho D_{AB} \frac{dw_A}{dx}\bigg|_{x=0} = j_{A,\,0} \tag{14–17}$$

where $\bar{j}_{A,\,0}$ and $j_{A,\,0}$ are the specified mole and mass fluxes of species A at the boundary, respectively. The special case of zero mass flux ($\bar{j}_{A,\,0} = j_{A,\,0} = 0$) corresponds to an **impermeable surface** for which $dy_A(0)/dx = dw_A(0)/dx = 0$ (Fig. 14–14).

To apply the *specified concentration* boundary condition, we must know the concentration of a species at the boundary. This information is usually obtained from the requirement that *thermodynamic equilibrium* must exist at the interface of two phases of a species. In the case of air–water interface, the concentration values of water vapor in the air are easily determined from saturation data, as shown in Example 14–2.

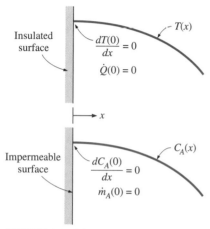

FIGURE 14–14
An impermeable surface in mass transfer is analogous to an insulated surface in heat transfer.

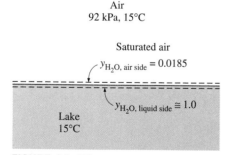

FIGURE 14–15
Schematic for Example 14–2.

EXAMPLE 14–2 **Mole Fraction of Water Vapor at the Surface of a Lake**

Determine the mole fraction of the water vapor at the surface of a lake whose temperature is 15°C and compare it to the mole fraction of water in the lake (Fig. 14–15). Take the atmospheric pressure at lake level to be 92 kPa.

SOLUTION The mole fraction of the water vapor at the surface of a lake and the mole fraction of water in the lake are to be determined and compared.
Assumptions **1** Both the air and water vapor are ideal gases. **2** The mole fraction of dissolved air in water is negligible.
Properties The saturation pressure of water at 15°C is 1.705 kPa (Table A–9).
Analysis The air at the water surface is saturated. Therefore, the partial pressure of water vapor in the air at the lake surface simply is the saturation pressure of water at 15°C,

$$P_{\text{vapor}} = P_{\text{sat @ 15°C}} = 1.705 \text{ kPa}$$

Assuming both the air and vapor to be ideal gases, the mole fraction of water vapor in the air at the surface of the lake is determined from Eq. 14–11 to be

$$y_{\text{vapor}} = \frac{P_{\text{vapor}}}{P} = \frac{1.705 \text{ kPa}}{92 \text{ kPa}} = \mathbf{0.0185} \quad \text{(or 1.85 percent)}$$

Water contains some dissolved air, but the amount is negligible. Therefore, we can assume the entire lake to be liquid water. Then its mole fraction becomes

$$y_{water,\ liquid\ side} \cong 1.0 \ \text{(or 100 percent)}$$

Discussion Note that the concentration of water on a molar basis is 100 percent just beneath the air–water interface and 1.85 percent just above it, even though the air is assumed to be saturated (so this is the highest value at 15°C). Therefore, huge discontinuities can occur in the concentrations of a species across phase boundaries.

The situation is similar at *solid–liquid* interfaces. Again, at a given temperature, only a certain amount of solid can be dissolved in a liquid, and the solubility of the solid in the liquid is determined from the requirement that thermodynamic equilibrium exists between the solid and the solution at the interface. The **solubility** represents *the maximum amount of solid that can be dissolved in a liquid at a specified temperature* and is widely available in chemistry handbooks. In Table 14–5 we present sample solubility data for sodium chloride (NaCl) and calcium bicarbonate [Ca(HCO₃)₂] at various temperatures. For example, the solubility of salt (NaCl) in water at 310 K is 36.5 kg per 100 kg of water. Therefore, the mass fraction of salt in the brine at the interface is simply

$$w_{salt,\ liquid\ side} = \frac{m_{salt}}{m} = \frac{36.5\ \text{kg}}{(100 + 36.5)\ \text{kg}} = 0.267 \quad \text{(or 26.7 percent)}$$

whereas the mass fraction of salt in the pure solid salt is $w = 1.0$. Note that water becomes *saturated* with salt when 36.5 kg of salt are dissolved in 100 kg of water at 310 K.

Many processes involve the absorption of a gas into a liquid. Most gases are weakly soluble in liquids (such as air in water), and for such dilute solutions the mole fractions of a species i in the gas and liquid phases at the interface are observed to be proportional to each other. That is, $y_{i,\ gas\ side} \propto y_{i,\ liquid\ side}$ or $P_{i,\ gas\ side} \propto P\ y_{i,\ liquid\ side}$ since $y_{i,\ gas\ side} = P_{i,\ gas\ side}/P$ for ideal gas mixtures. This is known as **Henry's law** and is expressed as

$$y_{i,\ liquid\ side} = \frac{P_{i,\ gas\ side}}{H} \ \text{(at interface)} \tag{14–18}$$

where *H* is **Henry's constant**, which is the product of the total pressure of the gas mixture and the proportionality constant. For a given species, it is a function of temperature only and is practically independent of pressure for pressures under about 5 atm. Values of Henry's constant for a number of aqueous solutions are given in Table 14–6 for various temperatures. From this table and the equation above we make the following observations:

1. The concentration of a gas dissolved in a liquid is inversely proportional to Henry's constant. Therefore, the larger the Henry's constant, the smaller the concentration of dissolved gases in the liquid.

TABLE 14–5

Solubility of two inorganic compounds in water at various temperatures, in kg, in 100 kg of water [from *Handbook of Chemistry* (New York: McGraw-Hill, 1961)]

Tempera-ture, K	Solute	
	Salt, NaCl	Calcium Bicarbonate, Ca(HCO₃)₂
273.15	35.7	16.15
280	35.8	16.30
290	35.9	16.53
300	36.2	16.75
310	36.5	16.98
320	36.9	17.20
330	37.2	17.43
340	37.6	17.65
350	38.2	17.88
360	38.8	18.10
370	39.5	18.33
373.15	39.8	18.40

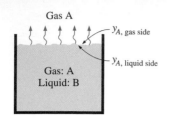

$y_{A, \text{ gas side}} \propto y_{A, \text{ liquid side}}$

or

$\dfrac{P_{A, \text{ gas side}}}{P} \propto y_{A, \text{ liquid side}}$

or

$P_{A, \text{ gas side}} = H y_{A, \text{ liquid side}}$

FIGURE 14–16
Dissolved gases in a liquid can be driven off by heating the liquid.

TABLE 14–6

Henry's constant H (in bars) for selected gases in water at low to moderate pressures (for gas i, $H = P_{i, \text{ gas side}}/y_{i, \text{ water side}}$)
(from Mills, 1995; Table A.21)

Solute	290 K	300 K	310 K	320 K	330 K	340 K
H_2S	440	560	700	830	980	1140
CO_2	1280	1710	2170	2720	3220	—
O_2	38,000	45,000	52,000	57,000	61,000	65,000
H_2	67,000	72,000	75,000	76,000	77,000	76,000
CO	51,000	60,000	67,000	74,000	80,000	84,000
Air	62,000	74,000	84,000	92,000	99,000	104,000
N_2	76,000	89,000	101,000	110,000	118,000	124,000

2. Henry's constant increases (and thus the fraction of a dissolved gas in the liquid decreases) with increasing temperature. Therefore, the dissolved gases in a liquid can be driven off by heating the liquid (Fig. 14–16).

3. The concentration of a gas dissolved in a liquid is proportional to the partial pressure of the gas. Therefore, the amount of gas dissolved in a liquid can be increased by increasing the pressure of the gas. This can be used to advantage in the carbonation of soft drinks with CO_2 gas.

Strictly speaking, the result obtained from Eq. 14–18 for the mole fraction of dissolved gas is valid for the liquid layer just beneath the interface and not necessarily the entire liquid. The latter will be the case only when thermodynamic phase equilibrium is established throughout the entire liquid body.

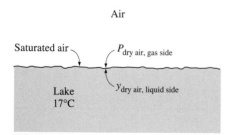

FIGURE 14–17
Schematic for Example 14–3.

EXAMPLE 14–3   **Mole Fraction of Dissolved Air in Water**

Determine the mole fraction of air dissolved in water at the surface of a lake whose temperature is 17°C (Fig. 14–17). Take the atmospheric pressure at lake level to be 92 kPa.

SOLUTION The mole fraction of air dissolved in water at the surface of a lake is to be determined.

Assumptions **1** Both the air and water vapor are ideal gases. **2** Air is weakly soluble in water so that Henry's law is applicable.

Properties The saturation pressure of water at 17°C is 1.96 kPa (Table A–9). Henry's constant for air dissolved in water at 290 K is $H = 62,000$ bar (Table 14–6).

Analysis This example is similar to the previous example. Again the air at the water surface is saturated, and thus the partial pressure of water vapor in the air at the lake surface is the saturation pressure of water at 17°C,

$$P_{\text{vapor}} = P_{\text{sat @ 17°C}} = 1.96 \text{ kPa}$$

Assuming both the air and vapor to be ideal gases, the partial pressure of dry air is determined to be

$$P_{\text{dry air}} = P - P_{\text{vapor}} = 92 - 1.96 = 90.04 \text{ kPa} = 0.9004 \text{ bar}$$

Note that with little loss in accuracy (an error of about 2 percent), we could have ignored the vapor pressure since the amount of vapor in air is so small. Then the mole fraction of air in the water becomes

$$y_{\text{dry air, liquid state}} = \frac{P_{\text{dry air, gas side}}}{H} = \frac{0.9004 \text{ bar}}{62,900 \text{ bar}} = 1.45 \times 10^{-5}$$

which is very small, as expected. Therefore, the concentration of air in water just below the air–water interface is 1.45 moles per 100,0000 moles. But obviously this is enough oxygen for fish and other creatures in the lake. Note that the amount of air dissolved in water decreases with increasing depth.

We mentioned earlier that the use of Henry's law is limited to dilute gas–liquid solutions; that is, a liquid with a small amount of gas dissolved in it. Then the question that arises naturally is, what do we do when the gas is highly soluble in the liquid (or solid), such as ammonia in water? In this case the linear relationship of Henry's law does not apply, and the mole fraction of a gas dissolved in the liquid (or solid) is usually expressed as a function of the partial pressure of the gas in the gas phase and the temperature. An approximate relation in this case for the *mole fractions* of a species on the *liquid* and *gas* sides of the interface is given by **Raoult's law** as

$$P_{i, \text{gas side}} = y_{i, \text{gas side}} P = y_{i, \text{liquid side}} P_{i, \text{sat}}(T) \tag{14–19}$$

where $P_{i, \text{sat}}(T)$ is the *saturation pressure* of the species i at the interface temperature and P is the *total pressure* on the gas phase side. Tabular data are available in chemical handbooks for common solutions such as the ammonia–water solution that is widely used in absorption-refrigeration systems.

Gases may also dissolve in *solids*, but the diffusion process in this case can be very complicated. The dissolution of a gas may be independent of the structure of the solid, or it may depend strongly on its porosity. Some dissolution processes (such as the dissolution of hydrogen in titanium, similar to the dissolution of CO_2 in water) are *reversible*, and thus maintaining the gas content in the solid requires constant contact of the solid with a reservoir of that gas. Some other dissolution processes are *irreversible*. For example, oxygen gas dissolving in titanium forms TiO_2 on the surface, and the process does not reverse itself.

The concentration of the gas species i in the solid at the interface $C_{i, \text{solid side}}$ is proportional to the *partial pressure* of the species i in the gas $P_{i, \text{gas side}}$ on the gas side of the interface and is expressed as

$$C_{i, \text{solid side}} = \mathcal{S} \times P_{i, \text{gas side}} \quad (\text{kmol/m}^3) \tag{14–20}$$

where $\mathcal{S}$ is the **solubility**. Expressing the pressure in bars and noting that the unit of molar concentration is kmol of species i per m^3, the unit of solubility is kmol/m$^3 \cdot$ bar. Solubility data for selected gas–solid combinations are given in Table 14–7. The product of the *solubility* of a gas and the *diffusion coefficient* of the gas in a solid is referred to as the **permeability** $\mathcal{P}$, which is a measure of the ability of the gas to penetrate a solid. That is, $\mathcal{P} = \mathcal{S}D_{AB}$ where D_{AB} is

TABLE 14–7

Solubility of selected gases and solids
(for gas i, $\mathcal{S} = C_{i, \text{solid side}}/P_{i, \text{gas side}}$)
(from Barrer, 1941)

Gas	Solid	T, K	$\mathcal{S}$ kmol/m$^3 \cdot$ bar
O_2	Rubber	298	0.00312
N_2	Rubber	298	0.00156
CO_2	Rubber	298	0.04015
He	SiO_2	293	0.00045
H_2	Ni	358	0.00901

the diffusivity of the gas in the solid. Permeability is inversely proportional to thickness and has the unit kmol/s · bar.

Finally, if a process involves the *sublimation* of a pure solid (such as ice or solid CO_2) or the *evaporation* of a pure liquid (such as water) in a different medium such as air, the mole (or mass) fraction of the substance in the liquid or solid phase is simply taken to be 1.0, and the partial pressure and thus the mole fraction of the substance in the gas phase can readily be determined from the saturation data of the substance at the specified temperature. Also, the assumption of thermodynamic equilibrium at the interface is very reasonable for pure solids, pure liquids, and solutions, except when chemical reactions are occurring at the interface.

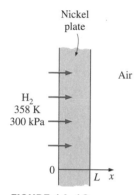

FIGURE 14–18
Schematic for Example 14–4.

EXAMPLE 14–4   Diffusion of Hydrogen Gas into a Nickel Plate

Consider a nickel plate that is in contact with hydrogen gas at 358 K and 300 kPa. Determine the molar and mass density of hydrogen in the nickel at the interface (Fig. 14–18).

SOLUTION A nickel plate is exposed to hydrogen. The molar and mass density of hydrogen in the nickel at the interface is to be determined.
Assumptions Nickel and hydrogen are in thermodynamic equilibrium at the interface.
Properties The molar mass of hydrogen is $M = 2$ kg/kmol (Table A–1). The solubility of hydrogen in nickel at 358 K is 0.00901 kmol/m³ · bar (Table 14–7).
Analysis Noting that 300 kPa = 3 bar, the molar density of hydrogen in the nickel at the interface is determined from Eq. 14–20 to be

$$C_{H_2, \text{ solid side}} = \mathcal{S} \times P_{H_2, \text{ gas side}}$$
$$= (0.00901 \text{ kmol/m}^3 \cdot \text{bar})(3 \text{ bar}) = \mathbf{0.027 \text{ kmol/m}^3}$$

It corresponds to a mass density of

$$\rho_{H_2, \text{ solid side}} = C_{H_2, \text{ solid side}} M_{H_2}$$
$$= (0.027 \text{ kmol/m}^3)(2) = \mathbf{0.054 \text{ kg/m}^3}$$

That is, there will be 0.027 kmol (or 0.054 kg) of H_2 gas in each m³ volume of nickel adjacent to the interface.

14–5 · STEADY MASS DIFFUSION THROUGH A WALL

Many practical mass transfer problems involve the diffusion of a species through a plane-parallel medium that does not involve any homogeneous chemical reactions under *one-dimensional steady* conditions. Such mass transfer problems are analogous to the steady one-dimensional heat conduction problems in a plane wall with no heat generation and can be analyzed similarly. In fact, many of the relations developed in Chapter 3 can be used for mass transfer by replacing temperature by mass (or molar) fraction, thermal conductivity by ρD_{AB} (or CD_{AB}), and heat flux by mass (or molar) flux (Table 14–8).

TABLE 14–8

Analogy between heat conduction and mass diffusion in a stationary medium

	Mass Diffusion	
Heat	Mass	Molar
Conduction	Basis	Basis
T	w_i	y_i
k	ρD_{AB}	CD_{AB}
$\dot{q}$	j_i	$\bar{j}_i$
α	D_{AB}	D_{AB}
L	L	L

Consider a solid plane wall (medium B) of area A, thickness L, and density ρ. The wall is subjected on both sides to different concentrations of a species A to which it is permeable. The boundary surfaces at $x = 0$ and $x = L$ are located within the solid adjacent to the interfaces, and the mass fractions of A at those surfaces are maintained at $w_{A,1}$ and $w_{A,2}$, respectively, at all times (Fig. 14–19). The mass fraction of species A in the wall varies in the x-direction only and can be expressed as $w_A(x)$. Therefore, mass transfer through the wall in this case can be modeled as *steady* and *one-dimensional*. Here we determine the rate of mass diffusion of species A through the wall using a similar approach to that used in Chapter 3 for heat conduction.

The concentration of species A at any point does not change with time since operation is steady, and there is no production or destruction of species A since no chemical reactions are occurring in the medium. Then the **conservation of mass** principle for species A can be expressed as *the mass flow rate of species A through the wall at any cross section is the same.* That is

$$\dot{m}_{\text{diff},A} = j_A A = \text{constant} \qquad \text{(kg/s)}$$

Then Fick's law of diffusion becomes

$$j_A = \frac{\dot{m}_{\text{diff},A}}{A} = -\rho D_{AB} \frac{dw_A}{dx} = \text{constant}$$

Separating the variables in this equation and integrating across the wall from $x = 0$, where $w(0) = w_{A,1}$, to $x = L$, where $w(L) = w_{A,2}$, we get

$$\frac{\dot{m}_{\text{diff},A}}{A} \int_0^L dx = -\int_{w_{A,1}}^{w_{A,2}} \rho D_{AB}\, dw_A \qquad \textbf{(14–21)}$$

where the mass transfer rate $\dot{m}_{\text{diff},A}$ and the wall area A are taken out of the integral sign since both are constants. If the density ρ and the mass diffusion coefficient D_{AB} vary little along the wall, they can be assumed to be constant. The integration can be performed in that case to yield

$$\dot{m}_{\text{diff},A,\text{wall}} = \rho D_{AB} A \frac{w_{A,1} - w_{A,2}}{L} = D_{AB} A \frac{\rho_{A,1} - \rho_{A,2}}{L} \qquad \text{(kg/s)} \quad \textbf{(14–22)}$$

This relation can be rearranged as

$$\dot{m}_{\text{diff},A,\text{wall}} = \frac{w_{A,1} - w_{A,2}}{L/\rho D_{AB} A} = \frac{w_{A,1} - w_{A,2}}{R_{\text{diff},\text{wall}}} \qquad \textbf{(14–23)}$$

where

$$R_{\text{diff},\text{wall}} = \frac{L}{\rho D_{AB} A}$$

is the **diffusion resistance** of the wall, in s/kg, which is analogous to the electrical or conduction resistance of a plane wall of thickness L and area A (Fig. 14–20). Thus, we conclude that *the rate of mass diffusion through a plane wall is proportional to the average density, the wall area, and the concentration difference across the wall, but is inversely proportional to the wall thickness.* Also, once the rate of mass diffusion is determined, the mass fraction $w_A(x)$ at any location x can be determined by replacing $w_{A,2}$ in Eq. 14–22 by $w_A(x)$ and L by x.

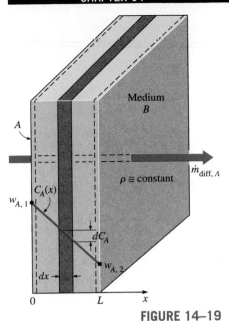

FIGURE 14–19
Schematic for steady one-dimensional mass diffusion of species A through a plane wall.

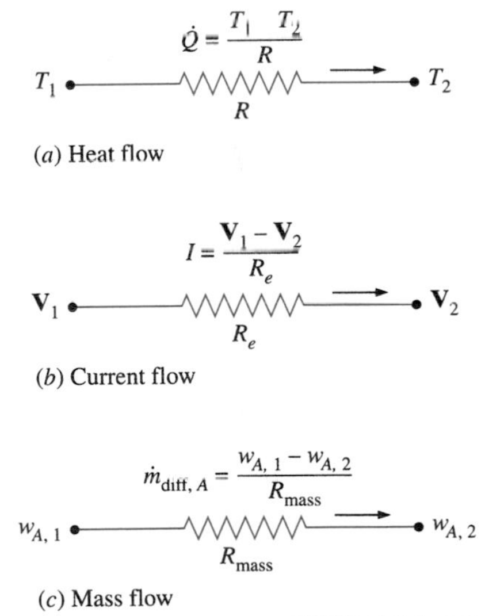

$$\dot{Q} = \frac{T_1 - T_2}{R}$$

$T_1 \bullet\!\!-\!\!\mathsf{WWW}\!\!-\!\!\rightarrow\!\bullet T_2$
$$R$$

(a) Heat flow

$$I = \frac{V_1 - V_2}{R_e}$$

$V_1 \bullet\!\!-\!\!\mathsf{WWW}\!\!-\!\!\rightarrow\!\bullet V_2$
$$R_e$$

(b) Current flow

$$\dot{m}_{\text{diff},A} = \frac{w_{A,1} - w_{A,2}}{R_{\text{mass}}}$$

$w_{A,1} \bullet\!\!-\!\!\mathsf{WWW}\!\!-\!\!\rightarrow\!\bullet w_{A,2}$
$$R_{\text{mass}}$$

(c) Mass flow

FIGURE 14–20
Analogy between thermal, electrical, and mass diffusion resistance concepts.

The preceding analysis can be repeated on a molar basis with this result,

$$\dot{N}_{\text{diff, }A,\text{ wall}} = CD_{AB}A\frac{y_{A,1} - y_{A,2}}{L} = D_{AB}A\frac{C_{A,1} - C_{A,2}}{L} = \frac{y_{A,1} - y_{A,2}}{\overline{R}_{\text{diff, wall}}} \quad (14\text{--}24)$$

where $\overline{R}_{\text{diff, wall}} = L/CD_{AB}A$ is the **molar diffusion resistance** of the wall in s/kmol. Note that mole fractions are accompanied by molar concentrations and mass fractions are accompanied by density. Either relation can be used to determine the diffusion rate of species A across the wall, depending on whether the mass or molar fractions of species A are known at the boundaries. Also, the concentration gradients on both sides of an interface are different, and thus diffusion resistance networks cannot be constructed in an analogous manner to thermal resistance networks.

In developing these relations, we assumed the density and the diffusion coefficient of the wall to be nearly constant. This assumption is reasonable when a small amount of species A diffuses through the wall and thus *the concentration of A is small*. The species A can be a gas, a liquid, or a solid. Also, the wall can be a plane layer of a liquid or gas provided that it is *stationary*.

The analogy between heat and mass transfer also applies to *cylindrical* and *spherical* geometries. Repeating the approach outlined in Chapter 3 for heat conduction, we obtain the following analogous relations for steady one-dimensional mass transfer through nonreacting cylindrical and spherical layers (Fig. 14–21)

$$\dot{m}_{\text{diff, }A,\text{ cyl}} = 2\pi L\rho D_{AB}\frac{w_{A,1} - w_{A,2}}{\ln(r_2/r_1)} = 2\pi LD_{AB}\frac{\rho_{A,1} - \rho_{A,2}}{\ln(r_2/r_1)} \quad (14\text{--}25)$$

$$\dot{m}_{\text{diff, }A,\text{ sph}} = 4\pi r_1 r_2\rho D_{AB}\frac{w_{A,1} - w_{A,2}}{r_2 - r_1} = 4\pi r_1 r_2 D_{AB}\frac{\rho_{A,1} - \rho_{A,2}}{r_2 - r_1} \quad (14\text{--}26)$$

or, on a molar basis,

$$\dot{N}_{\text{diff, }A,\text{ cyl}} = 2\pi LCD_{AB}\frac{y_{A,1} - y_{A,2}}{\ln(r_2/r_1)} = 2\pi LD_{AB}\frac{C_{A,1} - C_{A,2}}{\ln(r_2/r_1)} \quad (14\text{--}27)$$

$$\dot{N}_{\text{diff, }A,\text{ sph}} = 4\pi r_1 r_2 CD_{Ab}\frac{y_{A,1} - y_{A,2}}{r_2 - r_1} = 4\pi r_1 r_2 D_{AB}\frac{C_{A,1} - C_{A,2}}{r_2 - r_1} \quad (14\text{--}28)$$

Here, L is the length of the cylinder, r_1 is the inner radius, and r_2 is the outer radius for the cylinder or the sphere. Again, the boundary surfaces at $r = r_1$ and $r = r_2$ are located within the solid adjacent to the interfaces, and the mass fractions of A at those surfaces are maintained at $w_{A,1}$ and $w_{A,2}$, respectively, at all times. (We could make similar statements for the density, molar concentration, and mole fraction of species A at the boundaries.)

We mentioned earlier that the concentration of the gas species in a solid at the interface is proportional to the partial pressure of the adjacent gas and was expressed as $C_{A,\text{ solid side}} = \mathcal{S}_{AB}P_{A,\text{ gas side}}$ where $\mathcal{S}_{AB}$ is the *solubility* (in kmol/m³ · bar) of the gas A in the solid B. We also mentioned that the product of solubility and the diffusion coefficient is called the *permeability*, $\mathcal{P}_{Ab} = \mathcal{S}_{AB}D_{AB}$ (in kmol/m · s · bar). Then the molar flow rate of a gas through a solid under steady one-dimensional conditions can be expressed in terms of the partial pressures of the adjacent gas on the two sides of the solid by replacing C_A in these relations by $\mathcal{S}_{AB}P_A$ or $\mathcal{P}_{AB}P_A/D_{AB}$. In the case of a *plane wall*, for example, it gives (Fig. 14–22)

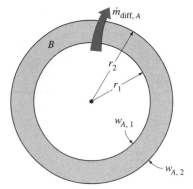

FIGURE 14–21
One-dimensional mass diffusion through a cylindrical or spherical shell.

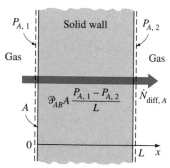

FIGURE 14–22
The diffusion rate of a gas species through a solid can be determined from a knowledge of the partial pressures of the gas on both sides and the permeability of the solid to that gas.

$$\dot{N}_{\text{diff}, A, \text{wall}} = D_{AB}\mathcal{S}_{AB}A\,\frac{P_{A,1} - P_{A,2}}{L} = \mathcal{P}_{AB}A\,\frac{P_{A,1} - P_{A,2}}{L} \qquad \text{(kmol/s)} \qquad \textbf{(14–29)}$$

where $P_{A,1}$ and $P_{A,2}$ are the *partial pressures* of gas A on the two sides of the wall. Similar relations can be obtained for cylindrical and spherical walls by following the same procedure. Also, if the permeability is given on a mass basis (in kg/m · s · bar), then Eq. 14–29 gives the diffusion mass flow rate.

Noting that 1 kmol of an ideal gas at the standard conditions of 0°C and 1 atm occupies a volume of 22.414 m³, the volume flow rate of the gas through the wall by diffusion can be determined from

$$\dot{V}_{\text{diff}, A} = 22.414\dot{N}_{\text{diff}, A} \qquad \text{(standard m}^3\text{/s, at 0°C and 1 atm)}$$

The volume flow rate at other conditions can be determined from the ideal gas relation $P_A\dot{V} = \dot{N}_A R_u T$.

EXAMPLE 14–5   **Diffusion of Hydrogen through a Spherical Container**

Pressurized hydrogen gas is stored at 358 K in a 4.8-m-outer-diameter spherical container made of nickel (Fig. 14–23). The shell of the container is 6 cm thick. The molar concentration of hydrogen in the nickel at the inner surface is determined to be 0.087 kmol/m³. The concentration of hydrogen in the nickel at the outer surface is negligible. Determine the mass flow rate of hydrogen by diffusion through the nickel container.

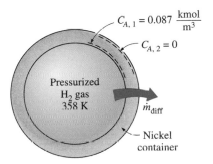

FIGURE 14–23
Schematic for Example 14–5.

SOLUTION Pressurized hydrogen gas is stored in a spherical container. The diffusion rate of hydrogen through the container is to be determined.
Assumptions **1** Mass diffusion is *steady* and *one-dimensional* since the hydrogen concentration in the tank and thus at the inner surface of the container is practically constant, and the hydrogen concentration in the atmosphere and thus at the outer surface is practically zero. Also, there is thermal symmetry about the center. **2** There are no chemical reactions in the nickel shell that result in the generation or depletion of hydrogen.
Properties The binary diffusion coefficient for hydrogen in the nickel at the specified temperature is 1.2×10^{-12} m²/s (Table 14–3*b*).
Analysis We can consider the total molar concentration to be constant ($C = C_A + C_B \cong C_B$ = constant), and the container to be a *stationary* medium since there is no diffusion of nickel molecules ($\dot{N}_B = 0$) and the concentration of the hydrogen in the container is extremely low ($C_A \ll 1$). Then the molar flow rate of hydrogen through this spherical shell by diffusion can readily be determined from Eq. 14–28 to be

$$\dot{N}_{\text{diff}} = 4\pi r_1 r_2 D_{AB}\,\frac{C_{A,1} - C_{A,2}}{r_2 - r_1}$$

$$= 4\pi(2.34\ \text{m})(2.40\ \text{m})(1.2 \times 10^{-12}\ \text{m}^2\text{/s})\,\frac{(0.087 - 0)\ \text{kmol/m}^3}{(2.40 - 2.34)\ \text{m}}$$

$$= 1.228 \times 10^{-10}\ \text{kmol/s}$$

The mass flow rate is determined by multiplying the molar flow rate by the molar mass of hydrogen, which is $M = 2$ kg/kmol,

$$\dot{m}_{diff} = M\dot{N}_{diff} = (2 \text{ kg/kmol})(1.228 \times 10^{-10} \text{ kmol/s}) = 2.46 \times 10^{-10} \text{ kg/s}$$

Therefore, hydrogen will leak out through the shell of the container by diffusion at a rate of 2.46×10^{-10} kg/s or 7.8 g/year. Note that the concentration of hydrogen in the nickel at the inner surface depends on the temperature and pressure of the hydrogen in the tank and can be determined as explained in Example 14–4. Also, the assumption of zero hydrogen concentration in nickel at the outer surface is reasonable since there is only a trace amount of hydrogen in the atmosphere (0.5 part per million by mole numbers).

14–6 · WATER VAPOR MIGRATION IN BUILDINGS

Moisture greatly influences the performance and durability of building materials, and thus moisture transmission is an important consideration in the construction and maintenance of buildings.

The *dimensions* of wood and other hygroscopic substances change with moisture content. For example, a variation of 4.5 percent in moisture content causes the volume of white oak wood to change by 2.5 percent. Such cyclic changes of dimensions weaken the joints and can jeopardize the structural integrity of building components, causing "squeaking" at the minimum. Excess moisture can also cause changes in the *appearance* and *physical properties* of materials: *corrosion* and *rusting* in metals, *rotting* in woods, and *peeling of paint* on the interior and exterior wall surfaces. Soaked wood with a water content of 24 to 31 percent is observed to decay rapidly at temperatures 10 to 38°C. Also, *molds* grow on wood surfaces at relative humidities above 85 percent. The expansion of water during freezing may damage the cell structure of porous materials.

Moisture content also affects the *effective conductivity* of porous mediums such as soils, building materials, and insulations, and thus heat transfer through them. Several studies have indicated that heat transfer increases almost linearly with moisture content, at a rate of 3 to 5 percent for each percent increase in moisture content by volume. Insulation with 5 percent moisture content by volume, for example, increases heat transfer by 15 to 25 percent relative to dry insulation (ASHRAE *Handbook of Fundamentals*, 1993, Chap. 20) (Fig. 14–24). Moisture migration may also serve as a transfer mechanism for latent heat by alternate evaporation and condensation. During a hot and humid day, for example, water vapor may migrate through a wall and condense on the inner side, releasing the heat of vaporization, with the process reversing during a cool night. Moisture content also affects the *specific heat* and thus the heat storage characteristics of building materials.

Moisture migration in the walls, floors, or ceilings of buildings and in other applications is controlled by either **vapor barriers** or **vapor retarders**. *Vapor barriers* are materials that are impermeable to moisture, such as sheet metals,

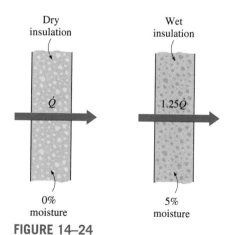

Dry insulation

Wet insulation

$\dot{Q}$

$1.25\dot{Q}$

0% moisture

5% moisture

FIGURE 14–24

A 5 percent moisture content can increase heat transfer through wall insulation by 25 percent.

heavy metal foils, and thick plastic layers, and they effectively bar the vapor from migrating. *Vapor retarders,* on the other hand, *retard* or *slow down* the flow of moisture through the structures but do not totally eliminate it. Vapor retarders are available as solid, flexible, or coating materials, but they usually consist of a thin sheet or coating. Common forms of vapor retarders are *reinforced plastics or metals, thin foils, plastic films, treated papers, coated felts,* and *polymeric or asphaltic paint coatings.* In applications such as the building of walls where vapor penetration is unavoidable because of numerous openings such as electrical boxes, telephone lines, and plumbing passages, vapor retarders are used instead of vapor barriers to allow the vapor that somehow leaks in to exit to the outside instead of trapping it in. Vapor retarders with a permeance of 57.4×10^{-9} kg/s $\cdot$ m² are commonly used in residential buildings.

The insulation on *chilled water lines* and other impermeable surfaces that are always cold must be wrapped with a *vapor barrier jacket,* or such cold surfaces must be insulated with a material that is impermeable to moisture. This is because moisture that migrates through the insulation to the cold surface condenses and remains there indefinitely with no possibility of vaporizing and moving back to the outside. The accumulation of moisture in such cases may render the insulation useless, resulting in excessive energy consumption.

Atmospheric air can be viewed as a mixture of dry air and water vapor, and the atmospheric pressure is the sum of the pressure of dry air and the pressure of water vapor, which is called the **vapor pressure** P_v. Air can hold a certain amount of moisture only, and the ratio of the actual amount of moisture in the air at a given temperature to the maximum amount air can hold at that temperature is called the **relative humidity** ϕ. The relative humidity ranges from 0 for dry air to 100 percent for *saturated* air (air that cannot hold any more moisture). The partial pressure of water vapor in saturated air is called the **saturation pressure** P_{sat}. Table 14–9 lists the saturation pressure at various temperatures.

The amount of moisture in the air is completely specified by the temperature and the relative humidity, and the vapor pressure is related to relative humidity ϕ by

$$P_v = \phi P_{sat} \tag{14–30}$$

where P_{sat} is the saturation (or boiling) pressure of water at the specified temperature. Then the mass flow rate of moisture through a plain layer of thickness L and normal area A can be expressed as

$$\dot{m}_v = \mathcal{P}A \frac{P_{v,1} - P_{v,2}}{L} = \mathcal{P}A \frac{\phi_1 P_{sat,1} - \phi_2 P_{sat,2}}{L} \quad \text{(kg/s)} \tag{14–31}$$

where $\mathcal{P}$ is the vapor permeability of the material, which is usually expressed on a mass basis in the unit ng/s $\cdot$ m $\cdot$ Pa, where ng $= 10^{-12}$ kg and 1 Pa $= 10^{-5}$ bar. Note that vapor migrates or diffuses from a region of higher vapor pressure toward a region of lower vapor pressure.

TABLE 14–9

Saturation pressure of water at various temperatures

Temperature, °C	Saturation Pressure, Pa
−40	13
−36	20
−32	31
−28	47
−24	70
−20	104
−16	151
−12	218
−8	310
−4	438
0	611
5	872
10	1228
15	1705
20	2339
25	3169
30	4246
35	5628
40	7384
50	12,350
100	101,330
200	1.55×10^6
300	8.58×10^6

The permeability of most construction materials is usually expressed for a given thickness instead of per unit thickness. It is called the **permeance** $\mathcal{M}$, which is the ratio of the permeability of the material to its thickness. That is,

$$\text{Permeance} = \frac{\text{Permeability}}{\text{Thickness}}$$

$$\mathcal{M} = \frac{\mathcal{P}}{L} \qquad (\text{kg/s} \cdot \text{m}^2 \cdot \text{Pa}) \qquad (14\text{--}32)$$

The reciprocal of permeance is called (unit) **vapor resistance** and is expressed as

$$\text{Vapor resistance} = \frac{1}{\text{Permeance}}$$

$$R_\nu = \frac{1}{\mathcal{M}} = \frac{L}{\mathcal{P}} \qquad (\text{s} \cdot \text{m}^2 \cdot \text{Pa/kg}) \qquad (14\text{--}33)$$

Note that vapor resistance represents the resistance of a material to water vapor transmission.

It should be pointed out that the amount of moisture that enters or leaves a building by *diffusion* is usually negligible compared to the amount that enters with *infiltrating air* or leaves with *exfiltrating air*. The primary cause of interest in the moisture diffusion is its impact on the performance and longevity of building materials.

The overall vapor resistance of a *composite* building structure that consists of several layers in series is the sum of the resistances of the individual layers and is expressed as

$$R_{\nu,\,\text{total}} = R_{\nu,\,1} + R_{\nu,\,2} + \cdots + R_{\nu,\,n} = \sum R_{\nu,\,i} \qquad (14\text{--}34)$$

Then the rate of vapor transmission through a composite structure can be determined in an analogous manner to heat transfer from

$$\dot{m}_\nu = A \frac{\Delta P_\nu}{R_{\nu,\,\text{total}}} \qquad (\text{kg/s}) \qquad (14\text{--}35)$$

Vapor permeance of common building materials is given in Table 14–10.

TABLE 14–10

Typical vapor permeance of common building materials (from ASHRAE, 1993, Chap. 22, Table 9)*

Materials and Its Thickness	Permeance ng/s $\cdot$ m^2 $\cdot$ Pa
Concrete (1:2:4 mix, 1 m)	4.7
Brick, masonry, 100 mm	46
Plaster on metal lath, 19 mm	860
Plaster on wood lath, 19 mm	630
Gypsum wall board, 9.5 mm	2860
Plywood, 6.4 mm	40–109
Still air, 1 m	174
Mineral wool insulation (unprotected), 1 m	245
Expanded polyurethane insulation board, 1 m	0.58–2.3
Aluminum foil, 0.025 mm	0.0
Aluminum foil, 0.009 mm	2.9
Polyethylene, 0.051 mm	9.1
Polyethylene, 0.2 mm	2.3
Polyester, 0.19 mm	4.6
Vapor retarder latex paint, 0.070 mm	26
Exterior acrylic house and trim paint, 0.040 mm	313
Building paper, unit mass of 0.16–0.68 kg/m^2	0.1–2400

*Data vary greatly. Consult manufacturer for more accurate data. Multiply by 1.41×10^{-6} to convert to lbm/s $\cdot$ ft^2 $\cdot$ psi. Also, 1 ng $= 10^{-12}$ kg.

EXAMPLE 14–6   Condensation and Freezing of Moisture in Walls

The condensation and even freezing of moisture in walls without effective vapor retarders is a real concern in cold climates, and it undermines the effectiveness of insulations. Consider a wood frame wall that is built around 38 mm $\times$ 90 mm (2 $\times$ 4 nominal) wood studs. The 90-mm-wide cavity between the studs is filled with glass fiber insulation. The inside is finished with 13-mm gypsum wallboard and the outside with 13-mm wood fiberboard and 13-mm $\times$ 200-mm wood bevel lapped siding. Using manufacturer's data, the thermal and vapor resistances of various components for a unit wall area are determined to be

Construction	R-Value, m² · °C/W	R_v-Value, s · m² · Pa/ng
1. Outside surface, 24 km/h wind	0.030	—
2. Painted wood bevel lapped siding	0.14	0.019
3. Wood fiberboard sheeting, 13 mm	0.23	0.0138
4. Glass fiber insulation, 90 mm	2.45	0.0004
5. Painted gypsum wallboard, 13 mm	0.079	0.012
6. Inside surface, still air	0.12	—
TOTAL	3.05	0.0452

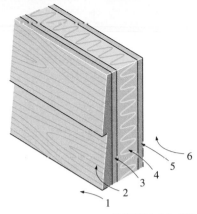

FIGURE 14–25
Schematic for Example 14–6.

The indoor conditions are 20°C and 60 percent relative humidity while the outside conditions are −16°C and 70 percent relative humidity. Determine if condensation or freezing of moisture will occur in the insulation.

SOLUTION The thermal and vapor resistances of different layers of a wall are given. The possibility of condensation or freezing of moisture in the wall is to be investigated.

Assumptions **1** Steady operating conditions exist. **2** Heat transfer through the wall is one-dimensional. **3** Thermal and vapor resistances of different layers of the wall and the heat transfer coefficients are constant.

Properties The thermal and vapor resistances are as given in the problem statement. The saturation pressures of water at 20°C and −16°C are 2339 Pa and 151 Pa, respectively (Table 14–9).

Analysis The schematic of the wall as well as the different elements used in its construction are shown in Figure 14–25. Condensation is most likely to occur at the coldest part of insulation, which is the part adjacent to the exterior sheathing. Noting that the total thermal resistance of the wall is 3.05 m² · °C/W, the rate of heat transfer through a unit area $A = 1$ m² of the wall is

$$\dot{Q}_{wall} = A \frac{T_i - T_o}{R_{total}} = (1 \text{ m}^2) \frac{[20 - (-16)°C]}{3.05 \text{ m}^2 \cdot °C/W} = 11.8 \text{ W}$$

The thermal resistance of the exterior part of the wall beyond the insulation is $0.03 + 0.14 + 0.23 = 0.40$ m² · °C/W. Then the temperature of the insulation–outer sheathing interface is

$$T_I = T_o + \dot{Q}_{wall} R_{ext} = -16°C + (11.8 \text{ W})(0.40°C/W) = -11.3°C$$

The saturation pressure of water at −11.3°C is 234 Pa, as shown in Table 14–9, and if there is condensation or freezing, the vapor pressure at the insulation–outer sheathing interface will have to be this value. The vapor pressure at the indoors and the outdoors is

$$P_{v,1} = \phi_1 P_{sat,1} = 0.60 \times (2340 \text{ Pa}) = 1404 \text{ Pa}$$
$$P_{v,2} = \phi_2 P_{sat,2} = 0.70 \times (151 \text{ Pa}) = 106 \text{ Pa}$$

Then the rate of moisture flow through the interior and exterior parts of the wall becomes

$$\dot{m}_{v, \text{interior}} = A\left(\frac{\Delta P}{R_v}\right)_{\text{interior}} = A\frac{P_{v, I} - P_{v, l}}{R_{v, \text{interior}}}$$

$$= (1 \text{ m}^2)\frac{(1404 - 234) \text{ Pa}}{(0.012 + 0.0004) \text{ Pa} \cdot \text{m}^2 \cdot \text{s/ng}} = 94{,}355 \text{ ng/s} = 94.4 \text{ μg/s}$$

$$\dot{m}_{v, \text{exterior}} = A\left(\frac{\Delta P}{R_v}\right)_{\text{exterior}} = A\frac{P_{v, I} - P_{v, 2}}{R_{v, \text{exterior}}}$$

$$= (1 \text{ m}^2)\frac{(234 - 106) \text{ Pa}}{(0.019 + 0.0138) \text{ Pa} \cdot \text{m}^2 \cdot \text{s/ng}} = 3902 \text{ ng/s} = 3.9 \text{ μg/s}$$

That is, moisture is flowing toward the interface at a rate of 94.4 μg/s but flowing from the interface to the outdoors at a rate of only 3.9 μg/s. Noting that the interface pressure cannot exceed 234 Pa, these results indicate that moisture is freezing in the insulation at a rate of

$$\dot{m}_{v, \text{freezing}} = \dot{m}_{v, \text{interior}} - \dot{m}_{v, \text{exterior}} = 94.4 - 3.9 = 90.5 \text{ μg/s}$$

Discussion This result corresponds to 7.82 g during a 24-h period, which can be absorbed by the insulation or sheathing, and then flows out when the conditions improve. However, excessive condensation (or frosting at temperatures below 0°C) of moisture in the walls during long cold spells can cause serious problems. This problem can be avoided or minimized by installing vapor barriers on the interior side of the wall, which will limit the moisture flow rate to 3.9 μg/s. Note that if there were no condensation or freezing, the flow rate of moisture through a 1 m² section of the wall would be 28.7 μg/s (can you verify this?).

14–7 · TRANSIENT MASS DIFFUSION

The steady analysis discussed earlier is useful when determining the leakage rate of a species through a stationary layer. But sometimes we are interested in the diffusion of a species into a body during a limited time before steady operating conditions are established. Such problems are studied using **transient analysis**. For example, the surface of a mild steel component is commonly hardened by packing the component in a carbonaceous material in a furnace at high temperature. During the short time period in the furnace, the carbon molecules diffuse through the surface of the steel component, but they penetrate to a depth of only a few millimeters. The carbon concentration decreases exponentially from the surface to the inner parts, and the result is a steel component with a very hard surface and a relatively soft core region (Fig. 14–26).

The same process is used in the gem industry to color clear stones. For example, a clear sapphire is given a brilliant blue color by packing it in titanium and iron oxide powders and baking it in an oven at about 2000°C for about a month. The titanium and iron molecules penetrate less than 0.5 mm in the sapphire during this process. Diffusion in solids is usually done at high temperatures to take advantage of the high diffusion coefficients at high temperatures and thus to keep the diffusion time at a reasonable level. Such diffusion or "doping" is also commonly practiced in the production of n- or p-type semiconductor materials used in the manufacture of electronic components. Drying processes such as the drying of coal, timber, food, and textiles constitute another major application area of transient mass diffusion.

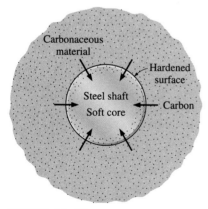

FIGURE 14–26
The surface hardening of a mild steel component by the diffusion of carbon molecules is a transient mass diffusion process.

Transient mass diffusion in a stationary medium is analogous to transient heat transfer provided that the solution is dilute and thus the density of the medium ρ is constant. In Chapter 4 we presented analytical and graphical solutions for one-dimensional transient heat conduction problems in solids with constant properties, no heat generation, and uniform initial temperature. The analogous one-dimensional transient mass diffusion problems satisfy these requirements:

1. The *diffusion coefficient is constant.* This is valid for an isothermal medium since D_{AB} varies with temperature (corresponds to constant thermal diffusivity).
2. There are *no homogeneous reactions* in the medium that generate or deplete the diffusing species A (corresponds to no heat generation).
3. Initially ($t = 0$) the concentration of species A is *constant* throughout the medium (corresponds to uniform initial temperature).

Then the solution of a mass diffusion problem can be obtained directly from the analytical or graphical solution of the corresponding heat conduction problem given in Chapter 4. The analogous quantities between heat and mass transfer are summarized in Table 14–11 for easy reference. For the case of a semi-infinite medium with constant surface concentration, for example, the solution can be expressed in an analogous manner to Eq. 4–45 as

$$\frac{C_A(x, t) - C_{A,i}}{C_{A,s} - C_{A,i}} = \text{erfc}\left(\frac{x}{2\sqrt{D_{AB}t}}\right) \tag{14-36}$$

where $C_{A,i}$ is the initial concentration of species A at time $t = 0$ and $C_{A,s}$ is the concentration at the inner side of the exposed surface of the medium. By using the definitions of molar fraction, mass fraction, and density, it can be shown that for dilute solutions,

$$\frac{C_A(x, t) - C_{A,i}}{C_{A,s} - C_{A,i}} = \frac{\rho_A(x, t) - \rho_{A,i}}{\rho_{A,s} - \rho_{A,i}} = \frac{w_A(x, t) - w_{A,i}}{w_{A,s} - w_{A,i}} = \frac{y_A(x, t) - y_{A,i}}{y_{A,s} - y_{A,i}} \tag{14-37}$$

since the total density or total molar concentration of dilute solutions is usually constant (ρ = constant or C = constant). Therefore, other measures of concentration can be used in Eq. 14–36.

A quantity of interest in mass diffusion processes is the depth of diffusion at a given time. This is usually characterized by the **penetration depth** defined as *the location x where the tangent to the concentration profile at the surface* ($x = 0$) *intercepts the* $C_A = C_{A,i}$ *line,* as shown in Figure 14–27. Obtaining the concentration gradient at $x = 0$ by differentiating Eq. 14–36, the penetration depth is determined to be

$$\delta_{\text{diff}} = \frac{C_{A,s} - C_{A,i}}{-(dC_A/dx)_{x=0}} = \frac{C_{A,s} - C_{A,i}}{(C_{A,s} - C_{A,i})/\sqrt{\pi D_{AB}t}} = \sqrt{\pi D_{AB}t} \tag{14-38}$$

Therefore, the penetration depth is proportional to the square root of both the diffusion coefficient and time. The diffusion coefficient of zinc in copper at 1000°C, for example, is 5.0×10^{-13} m²/s (Table 14–3). Then the penetration depth of zinc in copper in 10 h is

TABLE 14–11

Analogy between the quantities that appear in the formulation and solution of transient heat conduction and transient mass diffusion in a stationary medium

Heat Conduction	Mass Diffusion
T	C, y, ρ or w
α	D_{AB}
$\theta = \dfrac{T(x, t) - T_\infty}{T_i - T_\infty}$	$\theta_{\text{mass}} = \dfrac{w_A(X, t) - w_{A,\infty}}{w_{A,i} - w_{A,\infty}}$
$\dfrac{T(x, t) - T_s}{T_i - T_s}$	$\dfrac{w_A(x, t) - w_A}{w_{A,i} - w_A}$
$\xi = \dfrac{x}{2\sqrt{\alpha t}}$	$\xi_{\text{mass}} = \dfrac{x}{2\sqrt{D_{AB}t}}$
$\text{Bi} = \dfrac{h_{\text{conv}} L}{k}$	$\text{Bi}_{\text{mass}} = \dfrac{h_{\text{mass}} L}{D_{AB}}$
$\tau = \dfrac{\alpha t}{L^2}$	$\tau = \dfrac{D_{AB}t}{L^2}$

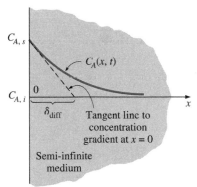

Slope of tangent line	$\dfrac{dC_A}{dx}\bigg|_{x=0} = -\dfrac{C_{A,s} - C_{A,i}}{\delta_{\text{diff}}}$

FIGURE 14–27

The concentration profile of species A in a semi-infinite medium during transient mass diffusion and the penetration depth.

$$\delta_{\text{diff}} = \sqrt{\pi D_{AB}t} = \sqrt{\pi(5.0 \times 10^{-13} \text{ m}^2/\text{s})(10 \times 3600 \text{ s})}$$
$$= 0.00024 \text{ m} = 0.24 \text{ mm}$$

That is, zinc will penetrate to a depth of about 0.24 mm in an appreciable amount in 10 h, and there will hardly be any zinc in the copper block beyond a depth of 0.24 mm.

The diffusion coefficients in solids are typically very low (on the order of 10^{-9} to 10^{-15} m²/s), and thus the diffusion process usually affects a thin layer at the surface. A solid can conveniently be treated as a semi-infinite medium during transient mass diffusion regardless of its size and shape when the penetration depth is small relative to the thickness of the solid. When this is not the case, solutions for one-dimensional transient mass diffusion through a plane wall, cylinder, and sphere can be obtained from the solutions of analogous heat conduction problems using the Heisler charts or one-term solutions presented in Chapter 4.

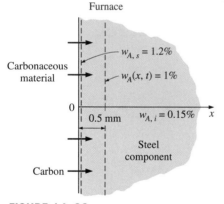

FIGURE 14–28
Schematic for Example 14–7.

EXAMPLE 14–7 **Hardening of Steel by the Diffusion of Carbon**

The surface of a mild steel component is commonly hardened by packing the component in a carbonaceous material in a furnace at a high temperature for a predetermined time. Consider such a component with a uniform initial carbon concentration of 0.15 percent by mass. The component is now packed in a carbonaceous material and is placed in a high-temperature furnace. The diffusion coefficient of carbon in steel at the furnace temperature is given to be 4.8×10^{-10} m²/s, and the equilibrium concentration of carbon in the iron at the interface is determined from equilibrium data to be 1.2 percent by mass. Determine how long the component should be kept in the furnace for the mass concentration of carbon 0.5 mm below the surface to reach 1 percent (Fig. 14–28).

SOLUTION A steel component is to be surface hardened by packing it in a carbonaceous material in a furnace. The length of time the component should be kept in the furnace is to be determined.

Assumptions Carbon penetrates into a very thin layer beneath the surface of the component, and thus the component can be modeled as a semi-infinite medium regardless of its thickness or shape.

Properties The relevant properties are given in the problem statement.

Analysis This problem is analogous to the one-dimensional transient heat conduction problem in a semi-infinite medium with specified surface temperature, and thus can be solved accordingly. Using mass fraction for concentration since the data are given in that form, the solution can be expressed as

$$\frac{w_A(x, t) - w_{A,i}}{w_{A,s} - w_{A,i}} = \text{erfc}\left(\frac{x}{2\sqrt{D_{AB}t}}\right)$$

Substituting the specified quantities gives

$$\frac{0.01 - 0.0015}{0.012 - 0.0015} = 0.81 = \text{erfc}\left(\frac{x}{2\sqrt{D_{AB}t}}\right)$$

The argument whose complementary error function is 0.81 is determined from Table 4–4 to be 0.17. That is,

$$\frac{x}{2\sqrt{D_{AB}t}} = 0.17$$

Then solving for the time t gives

$$t = \frac{x^2}{4D_{AB}(0.17)^2} = \frac{(0.0005 \text{ m})^2}{4 \times (4.8 \times 10^{-10} \text{ m}^2/\text{s})(0.17)^2} = 4505 \text{ s} = 1 \text{ h } 15 \text{ min}$$

Discussion The steel component in this case must be held in the furnace for 1 h and 15 min to achieve the desired level of hardening. The diffusion coefficient of carbon in steel increases exponentially with temperature, and thus this process is commonly done at high temperatures to keep the diffusion time at a reasonable level.

14–8 · DIFFUSION IN A MOVING MEDIUM

To this point we have limited our consideration to mass diffusion in a *stationary medium*, and thus the only motion involved was the creeping motion of molecules in the direction of decreasing concentration, and there was no motion of the mixture as a whole. Many practical problems, such as the evaporation of water from a lake under the influence of the wind or the mixing of two fluids as they flow in a pipe, involve diffusion in a **moving medium** where the bulk motion is caused by an external force. Mass diffusion in such cases is complicated by the fact that chemical species are transported both by *diffusion* and by the bulk motion of the medium (i.e., *convection*). The velocities and mass flow rates of species in a moving medium consist of two components: one due to *molecular diffusion* and one due to *convection* (Fig. 14–29).

Diffusion in a moving medium, in general, is difficult to analyze since various species can move at different velocities in different directions. Turbulence complicates the things even more. To gain a firm understanding of the physical mechanism while keeping the mathematical complexities to a minimum, we limit our consideration to systems that involve only *two components* (species A and B) in *one-dimensional flow* (velocity and other properties change in one direction only, say the x-direction). We also assume the total density (or molar concentration) of the medium remains constant. That is, $\rho = \rho_A + \rho_B = \text{constant}$ (or $C = C_A + C_B = \text{constant}$) but the densities of species A and B may vary in the x-direction.

Several possibilities are summarized in Figure 14–30. In the trivial case (case *a*) of a *stationary homogeneous mixture*, there will be no mass transfer by molecular diffusion or convection since there is no concentration gradient or bulk motion. The next case (case *b*) corresponds to the *flow of a well-mixed fluid mixture* through a pipe. Note that there is no concentration gradients and thus molecular diffusion in this case, and all species move at the bulk flow velocity of *V*. The mixture in the third case (case *c*) is *stationary* ($V = 0$) and thus it corresponds to ordinary molecular diffusion in stationary mediums, which we discussed before. Note that the velocity of a species at a location in this

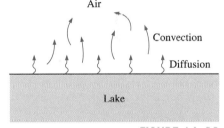

FIGURE 14–29

In a moving medium, mass transfer is due to both diffusion and convection.

	Species	Density	Velocity	Mass flow rate
(a) Homogeneous mixture without bulk motion (no concentration gradients and thus no diffusion) $V = 0$	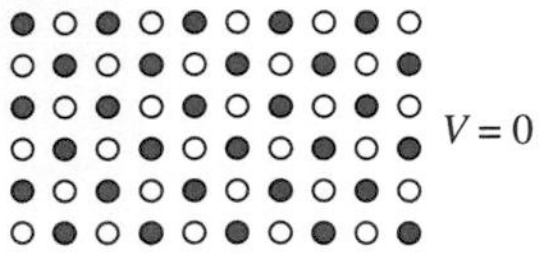 Species A	ρ_A = constant	$V_A = 0$	$\dot{m}_A = 0$
	Species B	ρ_B = constant	$V_B = 0$	$\dot{m}_B = 0$
	Mixture of A and B	$\rho = \rho_A + \rho_B$ = constant	$V = 0$	$\dot{m} = 0$
(b) Homogeneous mixture with bulk motion (no concentration gradients and thus no diffusion) $\rightarrow V$	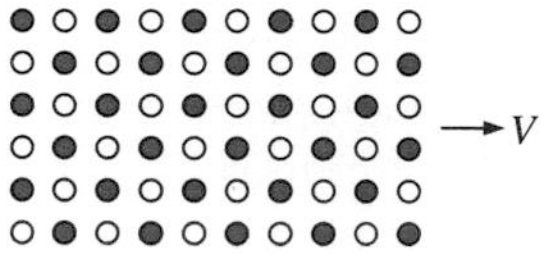 Species A	ρ_A = constant	$V_A = V$	$\dot{m}_A = \rho_A V_A A$
	Species B	ρ_B = constant	$V_B = V$	$\dot{m}_B = \rho_B V_B A$
	Mixture of A and B	$\rho = \rho_A + \rho_B$ = constant	$V = V$	$\dot{m} = \rho V A$ $= \dot{m}_A + \dot{m}_B$
(c) Nonhomogeneous mixture without bulk motion (stationary medium with concentration gradients) $V = 0$ $V_{\text{diff}, A} \rightarrow$ $\leftarrow V_{\text{diff}, B}$	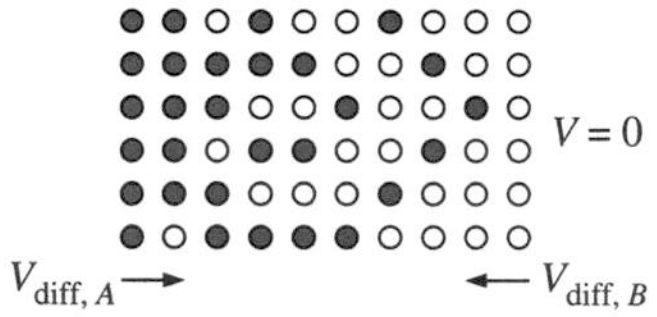 Species A	$\rho_A \neq$ constant	$V_A = V_{\text{diff}, A}$	$\dot{m}_A = \rho_A V_{\text{diff}, A} A$
	Species B	$\rho_B \neq$ constant	$V_B = V_{\text{diff}, B}$	$\dot{m}_B = \rho_B V_{\text{diff}, B} A$
	Mixture of A and B	$\rho = \rho_A + \rho_B$ = constant	$V = 0$	$\dot{m} = \rho V A = 0$ (thus $\dot{m}_A = -\dot{m}_B$)
(d) Nonhomogeneous mixture with bulk motion (moving medium with concentration gradients) $\rightarrow V$ $V_{\text{diff}, A} \rightarrow$ $\leftarrow V_{\text{diff}, B}$	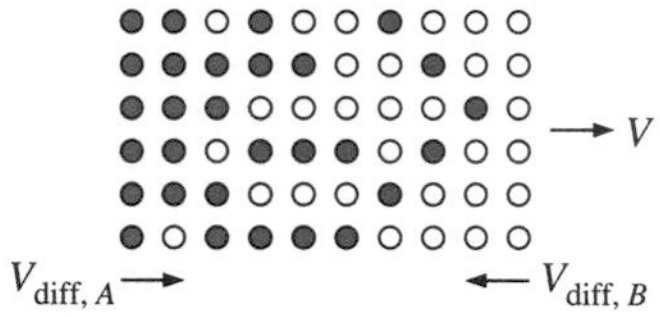 Species A	$\rho_A \neq$ constant	$V_A = V + V_{\text{diff}, A}$	$\dot{m}_A = \rho_A V_{\text{diff}, A} A$
	Species B	$\rho_B \neq$ constant	$V_B = V + V_{\text{diff}, B}$	$\dot{m}_B = \rho_B V_{\text{diff}, B} A$
	Mixture of A and B	$\rho = \rho_A + \rho_B$ = constant	$V = V$	$\dot{m} = \rho V A$ $= \dot{m}_A + \dot{m}_B$

FIGURE 14–30
Various quantities associated with a mixture of two species A and B at a location x under one-dimensional flow or no-flow conditions. (The density of the mixture $\rho = \rho_A + \rho_B$ is assumed to remain constant.)

case is simply the **diffusion velocity**, which is the average velocity of a group of molecules at that location moving under the influence of concentration gradient. Finally, the last case (case *d*) involves both *molecular diffusion* and *convection,* and the velocity of a species in this case is equal to the sum of the bulk flow velocity and the diffusion velocity. Note that the flow and the diffusion velocities can be in the same or opposite directions, depending on the direction of the concentration gradient. The diffusion velocity of a species is *negative* when the bulk flow is in the positive *x*-direction and the concentration gradient is positive (i.e., the concentration of the species increases in the *x*-direction).

Noting that the mass flow rate at any flow section is expressed as $\dot{m} = \rho V A$ where ρ is the density, V is the velocity, and A is the cross-sectional area, the conservation of mass relation for the flow of a mixture that involves two species A and B can be expressed as

$$\dot{m} = \dot{m}_A + \dot{m}_B$$

or

$$\rho V A = \rho_A V_A A + \rho_B V_B A$$

Canceling A and solving for V gives

$$V = \frac{\rho_A V_A + \rho_B V_B}{\rho} = \frac{\rho_A}{\rho} V_A + \frac{\rho_B}{\rho} V_B = w_A V_A + w_B V_B \qquad \text{(14–39)}$$

where V is called the **mass-average velocity** of the flow, which is the velocity that would be measured by a velocity sensor such as a pitot tube, a turbine device, or a hot wire anemometer inserted into the flow.

The special case $V = 0$ corresponds to a **stationary medium**, which can now be defined more precisely as *a medium whose mass-average velocity is zero*. Therefore, mass transport in a stationary medium is by diffusion only, and zero mass-average velocity indicates that there is no bulk fluid motion.

When there is *no concentration gradient* (and thus no molecular mass diffusion) in the fluid, the velocity of all species will be equal to the *mass-average velocity of the flow*. That is, $V = V_A = V_B$. But when there is a concentration gradient, there will also be a simultaneous flow of species in the direction of decreasing concentration at a diffusion velocity of V_{diff}. Then the average velocity of the species A and B can be determined by superimposing the average flow velocity and the diffusion velocity as (Fig. 14–31)

$$V_A = V + V_{\text{diff}, A}$$
$$V_B = V + V_{\text{diff}, B} \qquad \text{(14–40)}$$

Similarly, we apply the superposition principle to the species mass flow rates to get

$$\dot{m}_A = \rho_A V_A A = \rho_A (V + V_{\text{diff}, A})A = \rho_A VA + \rho_A V_{\text{diff}, A} A = \dot{m}_{\text{conv}, A} + \dot{m}_{\text{diff}, A}$$
$$\dot{m}_B = \rho_B V_B A = \rho_B (V + V_{\text{diff}, B})A = \rho_B VA + \rho_B V_{\text{diff}, B} A = \dot{m}_{\text{conv}, B} + \dot{m}_{\text{diff}, B} \qquad \text{(14–41)}$$

Using Fick's law of diffusion, the total mass fluxes $j = \dot{m}/A$ can be expressed as

$$j_A = \rho_A V + \rho_A V_{\text{diff}, A} = \frac{\rho_A}{\rho} \rho V - \rho D_{AB} \frac{dw_A}{dx} = w_A (j_A + j_B) - \rho D_{AB} \frac{dw_A}{dx}$$
$$j_B = \rho_B V + \rho_B V_{\text{diff}, B} = \frac{\rho_B}{\rho} \rho V - \rho D_{BA} \frac{dw_B}{dx} = w_B (j_A + j_B) - \rho D_{BA} \frac{dw_B}{dx} \qquad \text{(14–42)}$$

Note that the diffusion velocity of a species is negative when the molecular diffusion occurs in the negative x-direction (opposite to flow direction). The mass diffusion rates of the species A and B at a specified location x can be expressed as

$$\dot{m}_{\text{diff}, A} = \rho_A V_{\text{diff}, A} A = \rho_A (V_A - V)A$$
$$\dot{m}_{\text{diff}, B} = \rho_B V_{\text{diff}, B} A = \rho_B (V_B - V)A \qquad \text{(14–43)}$$

By substituting the V relation from Eq. 14–39 into Eq. 11–43, it can be shown that at any cross section

$$\dot{m}_{\text{diff}, A} + \dot{m}_{\text{diff}, B} = 0 \quad \rightarrow \quad \dot{m}_{\text{diff}, A} = -\dot{m}_{\text{diff}, B} \quad \rightarrow \quad -\rho A D_{AB} \frac{dw_A}{dx} = \rho A D_{BA} \frac{dw_B}{dx} \qquad \text{(14–44)}$$

which indicates that the rates of diffusion of species A and B must be equal in magnitude but opposite in sign. This is a consequence of the assumption

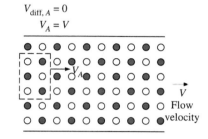

$V_{\text{diff}, A} = 0$
$V_A = V$

(a) No concentration gradient

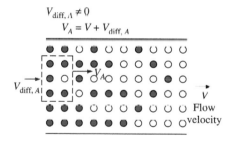

$V_{\text{diff}, A} \neq 0$
$V_A = V + V_{\text{diff}, A}$

(b) Mass concentration gradient and thus mass diffusion

FIGURE 14–31
The velocity of a species at a point is equal to the sum of the bulk flow velocity and the diffusion velocity of that species at that point.

$\rho = \rho_A + \rho_B$ = constant, and it indicates that *anytime the species A diffuses in one direction, an equal amount of species B must diffuse in the opposite direction to maintain the density (or the molar concentration) constant.* This behavior is closely approximated by dilute gas mixtures and dilute liquid or solid solutions. For example, when a small amount of gas diffuses into a liquid, it is reasonable to assume the density of the liquid to remain constant.

Note that for a binary mixture, $w_A + w_B = 1$ at any location x. Taking the derivative with respect to x gives

$$\frac{dw_A}{dx} = -\frac{dw_B}{dx} \tag{14–45}$$

Thus we conclude from Eq. 14–44 that (Fig. 14–32)

$$D_{AB} = D_{BA} \tag{14–46}$$

That is, in the case of constant total concentration, the diffusion coefficient of species A into B is equal to the diffusion coefficient of species B into A.

We now repeat the analysis presented above with molar concentration C and the molar flow rate $\dot{N}$. The conservation of matter in this case is expressed as

$$\dot{N} = \dot{N}_A + \dot{N}_B$$

or

$$\rho \overline{V} A = \rho_A \overline{V}_A A + \rho_B \overline{V}_B A \tag{14–47}$$

Canceling A and solving for $\overline{V}$ gives

$$\overline{V} = \frac{C_A \overline{V}_A + C_B \overline{V}_B}{C} = \frac{C_A}{C} \overline{V}_A + \frac{C_B}{C} \overline{V}_B = y_A \overline{V}_A + y_B \overline{V}_B \tag{14–48}$$

where $\overline{V}$ is called the **molar-average velocity** of the flow. Note that $\overline{V} \neq V$ unless the mass and molar fractions are the same. The molar flow rates of species are determined similarly to be

$$\dot{N}_A = C_A V_A A = C_A(\overline{V} + \overline{V}_{\text{diff}, A})A = C_A \overline{V} A + C_A \overline{V}_{\text{diff}, A}A = \dot{N}_{\text{conv}, A} + \dot{N}_{\text{diff}, A}$$
$$\dot{N}_B = C_B V_B A = C_B(\overline{V} + \overline{V}_{\text{diff}, B})A = C_B \overline{V} A + C_B \overline{V}_{\text{diff}, B}A = \dot{N}_{\text{conv}, B} + \dot{N}_{\text{diff}, B} \tag{14–49}$$

Using Fick's law of diffusion, the total molar fluxes $\bar{j} = \dot{N}/A$ and diffusion molar flow rates $\dot{N}_{\text{diff}}$ can be expressed as

$$\bar{j}_A = C_A \overline{V} + C_A \overline{V}_{\text{diff}, A} = \frac{C_A}{C} C\overline{V} - CD_{AB} \frac{dy_A}{dx} = y_A(\bar{j}_A + \bar{j}_B) - CD_{AB} \frac{dy_A}{dx}$$
$$\bar{j}_B = C_B \overline{V} + C_B \overline{V}_{\text{diff}, B} = \frac{C_B}{C} C\overline{V} - CD_{BA} \frac{dy_B}{dx} = y_B(\bar{j}_A + \bar{j}_B) - CD_{BA} \frac{dy_B}{dx} \tag{14–50}$$

and

$$\dot{N}_{\text{diff}, A} = C_A \overline{V}_{\text{diff}, A} A = C_A(V_A - \overline{V})A$$
$$\dot{N}_{\text{diff}, B} = C_B \overline{V}_{\text{diff}, B} A = C_B(V_B - \overline{V})A \tag{14–51}$$

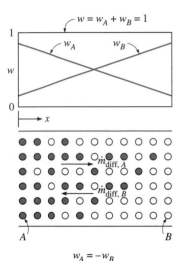

FIGURE 14–32

In a binary mixture of species A and B with $\rho = \rho_A + \rho_B$ = constant, the rates of mass diffusion of species A and B are equal magnitude and opposite in direction.

By substituting the $\overline{V}$ relation from Eq. 14–48 into these two equations, it can be shown that

$$\dot{N}_{\text{diff}, A} + \dot{N}_{\text{diff}, B} = 0 \quad \rightarrow \quad \dot{N}_{\text{diff}, A} = -\dot{N}_{\text{diff}, B} \qquad \text{(14–52)}$$

which again indicates that the rates of diffusion of species A and B must be equal in magnitude but opposite in sign.

It is important to note that when working with molar units, a medium is said to be stationary when the *molar-average velocity* is zero. The average velocity of the molecules will be zero in this case, but the apparent velocity of the mixture as measured by a velocimeter placed in the flow will not necessarily be zero because of the different masses of different molecules. In a *mass-based stationary medium,* for each unit mass of species A moving in one direction, a unit mass of species B moves in the opposite direction. In a *mole-based stationary medium,* however, for each mole of species A moving in one direction, one mole of species B moves in the opposite direction. But this may result in a net mass flow rate in one direction that can be measured by a velocimeter since the masses of different molecules are different.

You may be wondering whether to use the mass analysis or molar analysis in a problem. The two approaches are equivalent, and either approach can be used in mass transfer analysis. But sometimes it may be easier to use one of the approaches, depending on what is given. When mass-average velocity is known or can easily be obtained, obviously it is more convenient to use the mass-based formulation. When the *total pressure* and *temperature* of a mixture are constant, however, it is more convenient to use the molar formulation, as explained next.

Special Case: Gas Mixtures at Constant Pressure and Temperature

Consider a gas mixture whose total pressure and temperature are constant throughout. When the mixture is homogeneous, the mass density ρ, the molar density (or concentration) C, the gas constant R, and the molar mass M of the mixture are the same throughout the mixture. But when the concentration of one or more gases in the mixture is not constant, setting the stage for mass diffusion, then the mole fractions y_i of the species will vary throughout the mixture. As a result, the gas constant R, the molar mass M, and the mass density ρ of the mixture will also vary since, assuming ideal gas behavior,

$$M = \sum y_i M_i, \qquad R = \frac{R_u}{M}, \qquad \text{and} \qquad \rho = \frac{P}{RT}$$

where $R_u = 8.314 \text{ kJ/kmol} \cdot \text{K}$ is the universal gas constant. Therefore, the assumption of *constant mixture density* ($\rho = $ constant) in such cases will not be accurate unless the gas or gases with variable concentrations constitute a very small fraction of the mixture. However, the *molar density* C of a mixture *remains constant* when the mixture pressure P and temperature T are constant since

$$P = \rho RT = \rho \frac{R_u}{M} T = C R_u T \qquad \text{(14–53)}$$

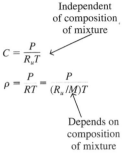

Gas
mixture
T = constant
P = constant

Independent
of composition
of mixture

$$C = \frac{P}{R_u T}$$

$$\rho = \frac{P}{RT} = \frac{P}{(R_u/M)T}$$

Depends on
composition
of mixture

FIGURE 14–33
When the total pressure P and
temperature T of a binary mixture
of ideal gases is held constant, then
the molar concentration C of the
mixture remains constant.

The condition C = constant offers considerable simplification in mass
transfer analysis, and thus it is more convenient to use the molar formulation
when dealing with gas mixtures at constant total pressure and temperature
(Fig. 14–33).

Diffusion of Vapor through a Stationary Gas: Stefan Flow

Many engineering applications such as heat pipes, cooling ponds, and the familiar perspiration involve condensation, evaporation, and transpiration in the
presence of a noncondensable gas, and thus the *diffusion* of a vapor through a
stationary (or stagnant) gas. To understand and analyze such processes, consider a liquid layer of species A in a tank surrounded by a gas of species B,
such as a layer of liquid water in a tank open to the atmospheric air (Fig.
14–34), at constant pressure P and temperature T. Equilibrium exists between
the liquid and vapor phases at the interface ($x = 0$), and thus the vapor pressure at the interface must equal the saturation pressure of species A at the
specified temperature. We assume the gas to be insoluble in the liquid, and
both the gas and the vapor to behave as ideal gases.

If the surrounding gas at the top of the tank ($x = L$) is not saturated, the
vapor pressure at the interface will be greater than the vapor pressure at the top
of the tank ($P_{A,0} > P_{A,L}$ and thus $y_{A,0} > y_{A,L}$ since $y_A = P_A/P$), and this pressure (or concentration) difference will drive the vapor upward from the
air–water interface into the stagnant gas. The upward flow of vapor will be
sustained by the evaporation of water at the interface. Under steady conditions, the molar (or mass) flow rate of vapor throughout the stagnant gas column remains constant. That is,

$$\bar{j}_A = \dot{N}_A/A = \text{constant} \qquad (\text{or } j_A = \dot{m}_A/A = \text{constant})$$

The pressure and temperature of the gas–vapor mixture are said to be constant, and thus the molar density of the mixture must be constant throughout
the mixture, as shown earlier. That is, $C = C_A + C_B = \text{constant}$, and it is more
convenient to work with mole fractions or molar concentrations in this case
instead of mass fractions or densities since $\rho \neq \text{constant}$.

Noting that $y_A + y_B = 1$ and that $y_{A,0} > y_{A,L}$, we must have $y_{B,0} < y_{B,L}$. That
is, the mole fraction of the gas must be decreasing downward by the same
amount that the mole fraction of the vapor is increasing. Therefore, gas must
be diffusing from the top of the column toward the liquid interface. However,
the gas is said to be *insoluble* in the liquid, and thus there can be no net mass
flow of the gas downward. Then under steady conditions, there must be an
upward bulk fluid motion with an average velocity V that is just large enough
to balance the diffusion of air downward so that the net molar (or mass) flow
rate of the gas at any point is zero. In other words, the upward bulk motion offsets the downward diffusion, and for each air molecule that moves downward,
there is another air molecule that moves upward. As a result, the air appears to
be *stagnant* (it does not move). That is,

$$\bar{j}_B = \dot{N}_B/A = 0 \qquad (\text{or } j_B = \dot{m}_B/A = 0)$$

The diffusion medium is no longer stationary because of the bulk motion. The
implication of the bulk motion of the gas is that it transports vapor as well as

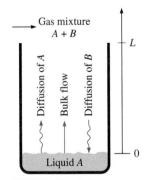

Gas mixture
$A + B$

Diffusion of A

Bulk flow

Diffusion of B

L

0

Liquid A

FIGURE 14–34
Diffusion of a vapor A
through a stagnant gas B.

the gas upward with a velocity of V, which results in *additional* mass flow of vapor upward. Therefore, the molar flux of the vapor can be expressed as

$$\bar{j}_A = \dot{N}_A/A = \bar{j}_{A,\,conv} + \bar{j}_{A,\,diff} = y_A(\bar{j}_A + \bar{j}_B) - CD_{AB}\frac{dy_A}{dx} \qquad \textbf{(14-54)}$$

Noting that $\bar{j}_B = 0$, it simplifies to

$$\bar{j}_A = y_A\bar{j}_A - CD_{AB}\frac{dy_A}{dx} \qquad \textbf{(14-55)}$$

Solving for $\bar{j}_A$ gives

$$\bar{j}_A = -\frac{CD_{AB}}{1-y_A}\frac{dy_A}{dx} \quad \rightarrow \quad -\frac{1}{1-y_A}\frac{dy_A}{dx} = \frac{\bar{j}_A}{CD_{AB}} = \text{constant} \qquad \textbf{(14-56)}$$

since $\bar{j}_A$ = constant, C = constant, and D_{AB} = constant. Separating the variables and integrating from $x = 0$, where $y_A(0) = y_{A,\,0}$, to $x = L$, where $y_A(L) = y_{A,\,L}$ gives

$$-\int_{A,0}^{y_{A,L}}\frac{dy_A}{1-y_A} = \int_0^L \frac{\bar{j}_A}{CD_{AB}}\,dx \qquad \textbf{(14-57)}$$

Performing the integrations,

$$\ln\frac{1-y_{A,L}}{1-y_{A,0}} = \frac{\bar{j}_A}{CD_{AB}}L \qquad \textbf{(14-58)}$$

Then the molar flux of vapor A, which is the *evaporation rate of species A per unit interface area*, becomes

$$\bar{j}_A = \dot{N}_A/A = \frac{CD_{AB}}{L}\ln\frac{1-y_{A,L}}{1-y_{A,0}} \qquad (\text{kmol/s}\cdot\text{m}^2) \qquad \textbf{(14-59)}$$

This relation is known as **Stefan's law**, and the *induced convective flow* described that enhances mass diffusion is called the **Stefan flow**. Noting that $y_A = P_A/P$ and $C = P/R_uT$ for an ideal gas mixture, the evaporation rate of species A can also be expressed as

$$\dot{N}_A = \frac{D_{AB}P}{LR_uT}\ln\frac{P-P_{A,L}}{P-P_{A,0}} \qquad (\text{kmol/s}) \qquad \textbf{(14-60)}$$

An expression for the variation of the mole fraction of A with x can be determined by performing the integration in Eq. 14–57 to the upper limit of x where $y_A(x) = y_A$ (instead of to L where $y_A(L) = y_{A,L}$). It yields

$$\ln\frac{1-y_A}{1-y_{A,0}} = \frac{\bar{j}_A}{CD_{AB}}x$$

Substituting the $\bar{j}_A$ expression from Eq. 14–59 into this relation and rearranging gives

$$\frac{1-y_A}{1-y_{A,0}} = \left(\frac{1-y_{A,L}}{1-y_{A,0}}\right)^{x/L} \quad \text{and} \quad \frac{y_B}{y_{B,0}} = \left(\frac{y_{B,L}}{y_{B,0}}\right)^{x/L} \qquad \textbf{(14-61)}$$

The second relation for the variation of the mole fraction of the stationary gas B is obtained from the first one by substituting $1 - y_A = y_B$ since $y_A + y_B = 1$.

To maintain isothermal conditions in the tank during evaporation, heat must be supplied to the tank at a rate of

$$\dot{Q} = \dot{m}_A h_{fg,A} = j_A A_s h_{fg,A} = (\bar{j}_A M_A) A_s h_{fg,A} \qquad \text{(kJ/s)} \qquad \textbf{(14–62)}$$

where A_s is the surface area of the liquid–vapor interface, $h_{fg,A}$ is the latent heat of vaporization, and M_A is the molar mass of species A.

Equimolar Counterdiffusion

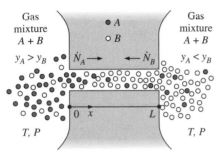

FIGURE 14–35
Equimolar isothermal counterdiffusion of two gases A and B.

Consider two large reservoirs connected by a channel of length L, as shown in Figure 14–35. The entire system contains a binary mixture of gases A and B at a uniform temperature T and pressure P throughout. The concentrations of species are maintained constant in each of the reservoirs such that $y_{A,0} > y_{A,L}$ and $y_{B,0} < y_{B,L}$. The resulting concentration gradients will cause the species A to diffuse in the positive x-direction and the species B in the opposite direction. Assuming the gases to behave as ideal gases and thus $P = CR_uT$, the total molar concentration of the mixture C will remain constant throughout the mixture since P and T are constant. That is,

$$C = C_A + C_B = \text{constant} \qquad \text{(kmol/m}^3\text{)}$$

This requires that for each molecule of A that moves to the right, a molecule of B moves to the left, and thus the molar flow rates of species A and B must be equal in magnitude and opposite in sign. That is,

$$\dot{N}_A = -\dot{N}_B \qquad \text{or} \qquad \dot{N}_A + \dot{N}_B = 0 \qquad \text{(kmol/s)}$$

This process is called **equimolar counterdiffusion** for obvious reasons. The net molar flow rate of the mixture for such a process, and thus the molar-average velocity, is zero since

$$\dot{N} = \dot{N}_A + \dot{N}_B = 0 \qquad \rightarrow \qquad C A \overline{V} = 0 \qquad \rightarrow \qquad \overline{V} = 0$$

Therefore, the mixture is *stationary* on a molar basis and thus mass transfer is by diffusion only (there is no mass transfer by convection) so that

$$\bar{j}_A = \dot{N}_A/A = -CD_{AB}\frac{dy_A}{dx} \qquad \text{and} \qquad \bar{j}_B = \dot{N}_B/A = -CD_{BA}\frac{dy_B}{dx} \qquad \textbf{(14–63)}$$

Under steady conditions, the molar flow rates of species A and B can be determined directly from Eq. 14–24 developed earlier for one-dimensional steady diffusion in a stationary medium, noting that $P = CR_uT$ and thus $C = P/R_uT$ for each constituent gas and the mixture. For one-dimensional flow through a channel of uniform cross sectional area A with no homogeneous chemical reactions, they are expressed as

$$\dot{N}_{\text{diff},A} = CD_{AB}A\frac{y_{A,1} - y_{A,2}}{L} = D_{AB}A\frac{C_{A,1} - C_{A,2}}{L} = \frac{D_{AB}}{R_uT}A\frac{P_{A,0} - P_{A,L}}{L}$$

$$\dot{N}_{\text{diff},B} = CD_{BA}A\frac{y_{B,1} - y_{B,2}}{L} = D_{BA}A\frac{C_{B,1} - C_{B,2}}{L} = \frac{D_{BA}}{R_uT}A\frac{P_{B,0} - P_{B,L}}{L} \qquad \textbf{(14–64)}$$

These relations imply that the mole fraction, molar concentration, and the partial pressure of either gas vary linearly during equimolar counterdiffusion.

It is interesting to note that the mixture is *stationary* on a molar basis, but it is not stationary on a mass basis unless the molar masses of A and B are equal. Although the net molar flow rate through the channel is zero, the net mass flow rate of the mixture through the channel is not zero and can be determined from

$$\dot{m} = \dot{m}_A + \dot{m}_B = \dot{N}_A M_A + \dot{N}_B M_B = \dot{N}_A (M_A - M_B) \qquad (14\text{--}65)$$

since $\dot{N}_B = -\dot{N}_A$. Note that the direction of net mass flow rate is the flow direction of the gas with the larger molar mass. A velocity measurement device such as an anemometer placed in the channel will indicate a velocity of $V = \dot{m}/\rho A$ where ρ is the total density of the mixture at the site of measurement.

EXAMPLE 14–8 Venting of Helium into the Atmosphere by Diffusion

The pressure in a pipeline that transports helium gas at a rate of 2 kg/s is maintained at 1 atm by venting helium to the atmosphere through a 5-mm-internal-diameter tube that extends 15 m into the air, as shown in Figure 14–36. Assuming both the helium and the atmospheric air to be at 25°C, determine (*a*) the mass flow rate of helium lost to the atmosphere through the tube, (*b*) the mass flow rate of air that infiltrates into the pipeline, and (*c*) the flow velocity at the bottom of the tube where it is attached to the pipeline that will be measured by an anemometer in steady operation.

SOLUTION The pressure in a helium pipeline is maintained constant by venting to the atmosphere through a long tube. The mass flow rates of helium and air through the tube and the net flow velocity at the bottom are to be determined.

Assumptions 1 Steady operating conditions exist. 2 Helium and atmospheric air are ideal gases. 3 No chemical reactions occur in the tube. 4 Air concentration in the pipeline and helium concentration in the atmosphere are negligible so that the mole fraction of the helium is 1 in the pipeline and 0 in the atmosphere (we will check this assumption later).

Properties The diffusion coefficient of helium in air (or air in helium) at normal atmospheric conditions is $D_{AB} = 7.2 \times 10^{-5}$ m²/s (Table 14–2). The molar masses of air and helium are 29 and 4 kg/kmol, respectively (Table A–1).

Analysis This is a typical equimolar counterdiffusion process since the problem involves two large reservoirs of ideal gas mixtures connected to each other by a channel, and the concentrations of species in each reservoir (the pipeline and the atmosphere) remain constant.

(*a*) The flow area, which is the cross sectional area of the tube, is

$$A = \pi D^2/4 = \pi (0.005 \text{ m})^2/4 = 1.963 \times 10^{-5} \text{ m}^2$$

Noting that the pressure of helium is 1 atm at the bottom of the tube ($x = 0$) and 0 at the top ($x = L$), its molar flow rate is determined from Eq. 14–64 to be

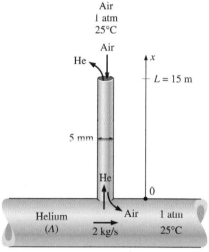

FIGURE 14–36
Schematic for Example 14–8.

$$\dot{N}_{\text{helium}} = \dot{N}_{\text{diff}, A} = \frac{D_{AB} A}{R_u T} \frac{P_{A,0} - P_{A,L}}{L}$$

$$= \frac{(7.20 \times 10^{-5} \text{ m}^2/\text{s})(1.963 \times 10^{-5} \text{ m}^2)}{(8.314 \text{ kPa} \cdot \text{m}^3/\text{kmol} \cdot \text{K})(298 \text{ K})} \left(\frac{1 \text{ atm} - 0}{15 \text{ m}} \right) \left(\frac{101.3 \text{ kPa}}{1 \text{ atm}} \right)$$

$$= 3.85 \times 10^{-12} \text{ kmol/s}$$

Therefore,

$$\dot{m}_{\text{helium}} = (\dot{N} M)_{\text{helium}} = (3.85 \times 10^{-12} \text{ kmol/s})(4 \text{ kg/kmol}) = \mathbf{1.54 \times 10^{-11} \text{ kg/s}}$$

which corresponds to about 0.5 g per year.

(b) Noting that $\dot{N}_B = -\dot{N}_A$ during an equimolar counterdiffusion process, the molar flow rate of air into the helium pipeline is equal to the molar flow rate of helium. The mass flow rate of air into the pipeline is

$$\dot{m}_{\text{air}} = (\dot{N} M)_{\text{air}} = (-3.85 \times 10^{-12} \text{ kmol/s})(29 \text{ kg/kmol}) = -112 \times 10^{-12} \text{ kg/s}$$

The mass fraction of air in the helium pipeline is

$$w_{\text{air}} = \frac{|\dot{m}_{\text{air}}|}{\dot{m}_{\text{total}}} = \frac{112 \times 10^{-12} \text{ kg/s}}{(2 + 112 \times 10^{-12} - 1.54 \times 10^{-11}) \text{ kg/s}} = 5.6 \times 10^{-11} \approx 0$$

which validates our original assumption of negligible air in the pipeline.

(c) The net mass flow rate through the tube is

$$\dot{m}_{\text{net}} = \dot{m}_{\text{helium}} + \dot{m}_{\text{air}} = 1.54 \times 10^{-11} - 112 \times 10^{-12} = -9.66 \times 10^{-11} \text{ kg/s}$$

The mass fraction of air at the bottom of the tube is very small, as shown above, and thus the density of the mixture at $x = 0$ can simply be taken to be the density of helium, which is

$$\rho \cong \rho_{\text{helium}} = \frac{P}{RT} = \frac{101.325 \text{ kPa}}{(2.0769 \text{ kPa} \cdot \text{m}^3/\text{kg} \cdot \text{K})(298 \text{ K})} = 0.1637 \text{ kg/m}^3$$

Then the average flow velocity at the bottom part of the tube becomes

$$V = \frac{\dot{m}}{\rho A} = \frac{-9.66 \times 10^{-11} \text{ kg/s}}{(0.1637 \text{ kg/m}^3)(1.963 \times 10^{-5} \text{ m}^2)} = \mathbf{-3.01 \times 10^{-5} \text{ m/s}}$$

which is difficult to measure by even the most sensitive velocity measurement devices. The negative sign indicates flow in the negative x-direction (toward the pipeline).

EXAMPLE 14–9   **Measuring Diffusion Coefficient by the Stefan Tube**

A 3-cm-diameter Stefan tube is used to measure the binary diffusion coefficient of water vapor in air at 20°C at an elevation of 1600 m where the atmospheric pressure is 83.5 kPa. The tube is partially filled with water, and the distance from the water surface to the open end of the tube is 40 cm (Fig. 14–37). Dry air is blown over the open end of the tube so that water vapor rising to the top is removed immediately and the concentration of vapor at the top of the tube is zero. In 15 days of continuous operation at constant pressure and temperature, the amount of water that has evaporated is measured to be 1.23 g. Determine the diffusion coefficient of water vapor in air at 20°C and 83.5 kPa.

SOLUTION The amount of water that evaporates from a Stefan tube at a specified temperature and pressure over a specified time period is measured. The diffusion coefficient of water vapor in air is to be determined.

Assumptions **1** Water vapor and atmospheric air are ideal gases. **2** The amount of air dissolved in liquid water is negligible. **3** Heat is transferred to the water from the surroundings to make up for the latent heat of vaporization so that the temperature of water remains constant at 20°C.

Properties The saturation pressure of water at 20°C is 2.34 kPa (Table A–9).

Analysis The vapor pressure at the air–water interface is the saturation pressure of water at 20°C, and the mole fraction of water vapor (species A) at the interface is determined from

$$y_{vapor, 0} = y_{A, 0} = \frac{P_{vapor, 0}}{P} = \frac{2.34 \text{ kPa}}{83.5 \text{ kPa}} = 0.0280$$

Dry air is blown on top of the tube and, thus, $y_{vapor, L} = y_{A, L} = 0$. Also, the total molar density throughout the tube remains constant because of the constant temperature and pressure conditions and is determined to be

$$C = \frac{P}{R_u T} = \frac{83.5 \text{ kPa}}{(8.314 \text{ kPa} \cdot \text{m}^3/\text{kmol} \cdot \text{K})(293 \text{ K})} = 0.0343 \text{ kmol/m}^3$$

The cross-sectional area of the tube is

$$A = \pi D^2/4 = \pi (0.03 \text{ m})^2/4 = 7.069 \times 10^{-4} \text{ m}^2$$

The evaporation rate is given to be 1.23 g per 15 days. Then the molar flow rate of vapor is determined to be

$$\dot{N}_A = \dot{N}_{vapor} = \frac{\dot{m}_{vapor}}{M_{vapor}} = \frac{1.23 \times 10^{-3} \text{ kg}}{(15 \times 24 \times 3600 \text{ s})(18 \text{ kg/kmol})}$$

$$= 5.27 \times 10^{-11} \text{ kmol/s}$$

Finally, substituting the information above into Eq. 14–59 we get

$$\frac{5.27 \times 10^{-11} \text{ kmol/s}}{7.069 \times 10^{-4} \text{ m}^2} = \frac{(0.0343 \text{ kmol/m}^3)D_{AB}}{0.4 \text{ m}} \ln \frac{1-0}{1-0.028}$$

which gives

$$D_{AB} = 3.06 \times 10^{-5} \text{ m}^2/\text{s}$$

for the binary diffusion coefficient of water vapor in air at 20°C and 83.5 kPa.

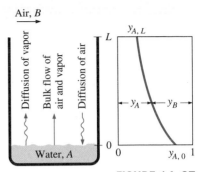

FIGURE 14–37
Schematic for Example 14–9.

14–9 · MASS CONVECTION

So far we have considered *mass diffusion,* which is the transfer of mass due to a concentration gradient. Now we consider *mass convection* (or *convective mass transfer*), which is the transfer of mass between a surface and a moving fluid due to both *mass diffusion* and *bulk fluid motion.* We mentioned earlier that fluid motion enhances heat transfer considerably by removing the heated fluid near the surface and replacing it by the cooler fluid further away. Likewise, fluid motion enhances mass transfer considerably by removing the high-concentration fluid near the surface and replacing it by the lower-concentration fluid further away. In the limiting case of no bulk fluid motion, mass convection reduces to mass diffusion, just as convection reduces to conduction. The analogy between heat and mass convection holds for both *forced* and *natural* convection, *laminar* and *turbulent* flow, and *internal* and *external* flow.

Like heat convection, mass convection is also complicated because of the complications associated with fluid flow such as the *surface geometry, flow regime, flow velocity,* and the *variation of the fluid properties* and *composition.* Therefore, we have to rely on experimental relations to determine mass transfer. Also, mass convection is usually analyzed on a *mass basis* rather than on a molar basis. Therefore, we will present formulations in terms of mass concentration (density ρ or mass fraction w) instead of molar concentration (molar density C or mole fraction y). But the formulations on a molar basis can be obtained using the relation $C = \rho/M$ where M is the molar mass. Also, for simplicity, we will restrict our attention to convection in fluids that are (or can be treated as) *binary mixtures.*

Consider the flow of air over the free surface of a water body such as a lake under isothermal conditions. If the air is not saturated, the concentration of water vapor will vary from a maximum at the water surface where the air is always saturated to the free steam value far from the surface. In heat convection, we defined the region in which temperature gradients exist as the *thermal boundary layer.* Similarly, in mass convection, we define the region of the fluid in which concentration gradients exist as the **concentration boundary layer,** as shown in Figure 14–38. In **external flow,** the thickness of the concentration boundary layer δ_c for a species A at a specified location on the surface is defined as the normal distance y from the surface at which

$$\frac{\rho_{A,s} - \rho_A}{\rho_{A,s} - \rho_{A,\infty}} = 0.99$$

where $\rho_{A,s}$ and $\rho_{A,\infty}$ are the densities of species A at the surface (on the fluid side) and the free stream, respectively.

In **internal flow,** we have a **concentration entrance region** where the concentration profile develops, in addition to the hydrodynamic and thermal entry regions (Fig. 14–39). The concentration boundary layer continues to develop in the flow direction until its thickness reaches the tube center and the boundary layers merge. The distance from the tube inlet to the location where this merging occurs is called the **concentration entry length** L_c, and the region beyond that point is called the **fully developed region,** which is characterized by

$$\frac{\partial}{\partial x}\left(\frac{\rho_{A,s} - \rho_A}{\rho_{A,s} - \rho_{A,b}}\right) = 0 \qquad \textbf{(14–66)}$$

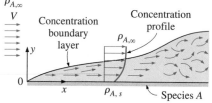

FIGURE 14–38

The development of a concentration boundary layer for species A during external flow on a flat surface.

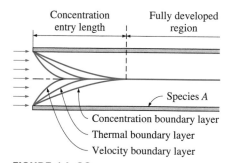

FIGURE 14–39

The development of the velocity, thermal, and concentration boundary layers in internal flow.

where $\rho_{A,b}$ is the *bulk mean density* of species A defined as

$$\rho_{A,b} = \frac{1}{A_c V_{avg}} \int_{A_c} \rho_A u \, dA_c \qquad \text{(14-67)}$$

Therefore, the nondimensionalized concentration difference profile as well as the mass transfer coefficient remain constant in the fully developed region. This is analogous to the friction and heat transfer coefficients remaining constant in the fully developed region.

In heat convection, the relative magnitudes of momentum and heat diffusion in the velocity and thermal boundary layers are expressed by the dimensionless *Prandtl number,* defined as (Fig. 14–40)

Prandtl number:
$$\text{Pr} = \frac{\nu}{\alpha} = \frac{\text{Momentum diffusivity}}{\text{Thermal diffusivity}} \qquad \text{(14-68)}$$

The corresponding quantity in mass convection is the dimensionless **Schmidt number**, defined as

Schmidt number:
$$\text{Sc} = \frac{\nu}{D_{AB}} = \frac{\text{Momentum diffusivity}}{\text{Mass diffusivity}} \qquad \text{(14-69)}$$

which represents the relative magnitudes of molecular momentum and mass diffusion in the velocity and concentration boundary layers, respectively.

The relative growth of the velocity and thermal boundary layers in laminar flow is governed by the Prandtl number, whereas the relative growth of the velocity and concentration boundary layers is governed by the Schmidt number. A Prandtl number of near unity ($\text{Pr} \approx 1$) indicates that momentum and heat transfer by diffusion are comparable, and velocity and thermal boundary layers almost coincide with each other. *A Schmidt number of near unity ($\text{Sc} \approx 1$) indicates that momentum and mass transfer by diffusion are comparable, and velocity and concentration boundary layers almost coincide with each other.*

It seems like we need one more dimensionless number to represent the relative magnitudes of heat and mass diffusion in the thermal and concentration boundary layers. That is the **Lewis number**, defined as (Fig. 14–41)

Lewis number:
$$\text{Le} = \frac{\text{Sc}}{\text{Pr}} = \frac{\alpha}{D_{AB}} = \frac{\text{Thermal diffusivity}}{\text{Mass diffusivity}} \qquad \text{(14-70)}$$

The relative thicknesses of velocity, thermal, and concentration boundary layers in laminar flow are expressed as

$$\frac{\delta_{\text{velocity}}}{\delta_{\text{thermal}}} = \text{Pr}^n, \qquad \frac{\delta_{\text{velocity}}}{\delta_{\text{concentration}}} = \text{Sc}^n, \qquad \text{and} \qquad \frac{\delta_{\text{thermal}}}{\delta_{\text{concentration}}} = \text{Le}^n \qquad \text{(14-71)}$$

where $n = \frac{1}{3}$ for most applications in all three relations. These relations, in general, are not applicable to turbulent boundary layers since turbulent mixing in this case may dominate the diffusion processes.

Note that species transfer at the surface ($y = 0$) is by diffusion only because of the *no-slip boundary condition,* and mass flux of species A at the surface can be expressed by Fick's law as (Fig. 14–42)

$$
\begin{array}{ll}
\text{Heat transfer:} & \text{Pr} = \dfrac{\nu}{\alpha} \\[2mm]
\text{Mass transfer:} & \text{Sc} = \dfrac{\nu}{D_{AB}}
\end{array}
$$

FIGURE 14–40
In mass transfer, the Schmidt number plays the role of the Prandtl number in heat transfer.

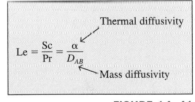

$$\text{Le} = \frac{\text{Sc}}{\text{Pr}} = \frac{\alpha}{D_{AB}}$$

FIGURE 14–41
Lewis number is a measure of heat diffusion relative to mass diffusion.

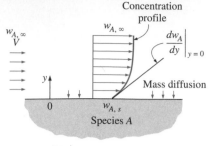

$$-D_{AB}\frac{\partial C_A}{\partial y}\Big|_{y=0} = h_{\text{mass}}(w_{A,s} - w_{A,\infty})$$

FIGURE 14–42

Mass transfer at a surface occurs by diffusion because of the no-slip boundary condition, just like heat transfer occurring by conduction.

$$j_A = \dot{m}_A/A_s = -\rho D_{AB}\frac{\partial w_A}{\partial y}\Big|_{y=0} \qquad \text{(kg/s} \cdot \text{m}^2) \qquad \textbf{(14–72)}$$

This is analogous to heat transfer at the surface being by conduction only and expressing it by Fourier's law.

The rate of heat convection for external flow was expressed conveniently by *Newton's law of cooling* as

$$\dot{Q}_{\text{conv}} = h_{\text{conv}}A_s(T_s - T_\infty)$$

where h_{conv} is the average heat transfer coefficient, A_s is the surface area, and $T_s - T_\infty$ is the temperature difference across the thermal boundary layer. Likewise, the rate of mass convection can be expressed as

$$\dot{m}_{\text{conv}} = h_{\text{mass}}A_s(\rho_{A,s} - \rho_{A,\infty}) = h_{\text{mass}}\rho A_s(w_{A,s} - w_{A,\infty}) \qquad \text{(kg/s)} \qquad \textbf{(14–73)}$$

where h_{mass} is the average **mass transfer coefficient,** in m/s; A_s is the surface area; $\rho_{A,s} - \rho_{A,\infty}$ is the mass concentration difference of species A across the concentration boundary layer; and ρ is the average density of the fluid in the boundary layer. The product $h_{\text{mass}}\rho$, whose unit is kg/m$^2 \cdot$ s, is called the *mass transfer conductance.* For internal flow we have

$$\dot{m}_{\text{conv}} = h_{\text{mass}}A_s\frac{\Delta\rho_{A,e} - \Delta\rho_{A,i}}{\ln(\Delta\rho_{A,e}/\Delta\rho_{A,i})} \qquad \textbf{(14–74)}$$

where $\Delta\rho_{A,e} = \rho_{A,s} - \rho_{A,e}$ and $\Delta\rho_{A,i} = \rho_{A,s} - \rho_{A,i}$. If the local mass transfer coefficient varies in the flow direction, the *average mass transfer coefficient* can be determined from

$$h_{\text{mass, avg}} = \frac{1}{A_s}\int_{A_s} h_{\text{mass}}dA_s$$

In heat convection analysis, it is often convenient to express the heat transfer coefficient in a nondimensionalized form in terms of the dimensionless *Nusselt number,* defined as

Nusselt number: $\qquad\qquad\qquad\qquad \text{Nu} = \dfrac{h_{\text{conv}}L_c}{k} \qquad \textbf{(14–75)}$

where L_c is the characteristic length and k is the thermal conductivity of the fluid. The corresponding quantity in mass convection is the dimensionless **Sherwood number,** defined as (Fig. 14–43)

Sherwood number: $\qquad\qquad\qquad\qquad \text{Sh} = \dfrac{h_{\text{mass}}L_c}{D_{AB}} \qquad \textbf{(14–76)}$

where h_{mass} is the mass transfer coefficient and D_{AB} is the mass diffusivity. The Nusselt and Sherwood numbers represent the effectiveness of heat and mass convection at the surface, respectively.

Sometimes it is more convenient to express the heat and mass transfer coefficients in terms of the dimensionless **Stanton number** as

Heat transfer Stanton number: $\qquad\quad \text{St} = \dfrac{h_{\text{conv}}}{\rho V c_p} = \text{Nu}\dfrac{1}{\text{Re Pr}} \qquad \textbf{(14–77)}$

Heat transfer:	$\text{Nu} = \dfrac{h_{\text{conv}}L_c}{k}$
Mass transfer:	$\text{Sh} = \dfrac{h_{\text{mass}}L_c}{D_{AB}}$

FIGURE 14–43

In mass transfer, the Sherwood number plays the role the Nusselt number plays in heat transfer.

and

Mass transfer Stanton number:
$$\text{St}_{\text{mass}} = \frac{h_{\text{mass}}}{V} = \text{Sh} \frac{1}{\text{Re Sc}} \tag{14–78}$$

where V is the free steam velocity in external flow and the bulk mean fluid velocity in internal flow.

For a given geometry, the average Nusselt number in forced convection depends on the Reynolds and Prandtl numbers, whereas the average Sherwood number depends on the Reynolds and Schmidt numbers. That is,

Nusselt number: $\text{Nu} = f(\text{Re}, \text{Pr})$
Sherwood number: $\text{Sh} = f(\text{Re}, \text{Sc})$

where the functional form of f is the same for both the Nusselt and Sherwood numbers in a given geometry, provided that the thermal and concentration boundary conditions are of the same type. Therefore, *the Sherwood number can be obtained from the Nusselt number expression by simply replacing the Prandtl number by the Schmidt number.* This shows what a powerful tool analogy can be in the study of natural phenomena (Table 14–12).

In *natural convection mass transfer,* the analogy between the Nusselt and Sherwood numbers still holds, and thus $\text{Sh} = f(\text{Gr}, \text{Sc})$. But the Grashof number in this case should be determined directly from

$$\text{Gr} = \frac{g(\rho_\infty - \rho_s) L_c^3}{\rho\nu^2} = \frac{g(\Delta\rho/\rho) L_c^3}{\nu^2} \tag{14–79}$$

which is applicable to both temperature- and/or concentration-driven natural convection flows. Note that in *homogeneous* fluids (i.e., fluids with no concentration gradients), density differences are due to temperature differences only, and thus we can replace $\Delta\rho/\rho$ by $\beta\Delta T$ for convenience, as we did in natural convection heat transfer. However, in *nonhomogeneous* fluids, density differences are due to the *combined effects* of *temperature* and *concentration differences,* and $\Delta\rho/\rho$ cannot be replaced by $\beta\Delta T$ in such cases even when all we care about is heat transfer and we have no interest in mass transfer. For example, hot water at the bottom of a pond rises to the top. But when salt is placed at the bottom, as it is done in solar ponds, the salty water (brine) at the bottom will not rise because it is now heavier than the fresh water at the top (Fig. 14–44).

Concentration-driven natural convection flows are based on the densities of different species in a mixture being different. Therefore, at isothermal conditions, there will be no natural convection in a gas mixture that is composed of gases with identical molar masses. Also, the case of a hot surface facing up corresponds to diffusing fluid having a lower density than the mixture (and thus rising under the influence of buoyancy), and the case of a hot surface facing down corresponds to the diffusing fluid having a higher density. For example, the evaporation of water into air corresponds to a hot surface facing up since water vapor is lighter than the air and it tends to rise. But this is not the case for gasoline unless the temperature of the gasoline–air mixture at the gasoline surface is so high that thermal expansion overwhelms the density differential due to higher gasoline concentration near the surface.

TABLE 14–12

Analogy between the quantities that appear in the formulation and solution of heat convection and mass convection

Heat Convection	Mass Convection
T	C, y, ρ, or w
h_{conv}	h_{mass}
δ_{thermal}	$\delta_{\text{concentration}}$
$\text{Re} = \dfrac{VL_c}{\nu}$	$\text{Re} = \dfrac{VL_c}{\nu}$
$\text{Gr} = \dfrac{g\beta(T_s - T_\infty) L_c^3}{\nu^2}$,	$\text{Gr} = \dfrac{g(\rho_\infty - \rho_s) L_c^3}{\rho\nu^2}$
$\text{Pr} = \dfrac{\nu}{\alpha}$	$\text{Sc} = \dfrac{\nu}{D_{AB}}$
$\text{St} = \dfrac{h_{\text{conv}}}{\rho V c_p}$	$\text{St}_{\text{mass}} = \dfrac{h_{\text{mass}}}{V}$
$\text{Nu} = \dfrac{h_{\text{conv}} L_c}{k}$	$\text{Sh} = \dfrac{h_{\text{mass}} L_c}{D_{AB}}$
$\text{Nu} = f(\text{Re}, \text{Pr})$	$\text{Sh} = f(\text{Re}, \text{Sc})$
$\text{Nu} = f(\text{Gr}, \text{Pr})$	$\text{Sh} = f(\text{Gr}, \text{Sc})$

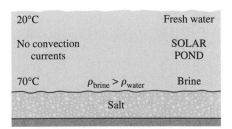

FIGURE 14–44

A hot fluid at the bottom will rise and initiate natural convection currents only if its density is lower.

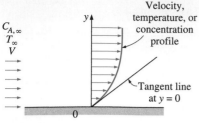

FIGURE 14–45

The friction, heat, and mass transfer coefficients for flow over a surface are proportional to the slope of the tangent line of the velocity, temperature, and concentration profiles, respectively, at the surface.

Analogy between Friction, Heat Transfer, and Mass Transfer Coefficients

Consider the flow of a fluid over a flat plate of length L with free steam conditions of T_∞, V, and $w_{A,\infty}$ (Fig. 14–45). Noting that convection at the surface ($y = 0$) is equal to diffusion because of the no-slip condition, the friction, heat transfer, and mass transfer conditions at the surface can be expressed as

Wall friction:
$$\tau_s = \mu \frac{\partial u}{\partial y}\Big|_{y=0} = \frac{f}{2}\rho V^2 \tag{14–80}$$

Heat transfer:
$$\dot{q}_s = -k \frac{\partial T}{\partial y}\Big|_{y=0} = h_{\text{heat}}(T_s - T_\infty) \tag{14–81}$$

Mass transfer:
$$j_{A,s} = -D_{AB} \frac{\partial w_A}{\partial y}\Big|_{y=0} = h_{\text{mass}}(w_{A,s} - w_{A,\infty}) \tag{14–82}$$

These relations can be rewritten for internal flow by using *bulk mean properties* instead of free stream properties. After some simple mathematical manipulations, the three relations above can be rearranged as

Wall friction:
$$\frac{d(u/V)}{d(y/L_c)}\Big|_{y=0} = \frac{f}{2}\frac{\rho V L_c}{\mu} = \frac{f}{2}\text{Re} \tag{14–83}$$

Heat transfer:
$$\frac{d[(T - T_s)/(T_\infty - T_s)]}{d(y/L_c)}\Big|_{y=0} = \frac{h_{\text{heat}} L_c}{k} = \text{Nu} \tag{14–84}$$

Mass transfer:
$$\frac{d[(w_A - w_{A,s})/(w_{A,\infty} - w_{A,s})]}{d(y/L_c)}\Big|_{y=0} = \frac{h_{\text{mass}} L_c}{D_{AB}} = \text{Sh} \tag{14–85}$$

The left sides of these three relations are the slopes of the normalized velocity, temperature, and concentration profiles at the surface, and the right sides are the dimensionless numbers discussed earlier.

Special Case: Pr ≈ Sc ≈ 1 (Reynolds Analogy)

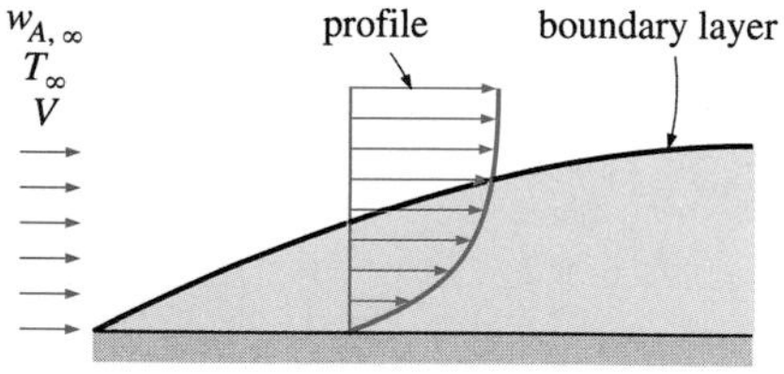

Now consider the hypothetical case in which the molecular diffusivities of momentum, heat, and mass are identical. That is, $\nu = \alpha = D_{AB}$ and thus Pr = Sc = Le = 1. In this case the normalized velocity, temperature, and concentration profiles will coincide, and thus the slope of these three curves at the surface (the left sides of Eqs. 14–83 through 14–85) will be identical (Fig. 14–46). Then we can set the right sides of those three equations equal to each other and obtain

$$\frac{f}{2}\text{Re} = \text{Nu} = \text{Sh} \qquad \text{or} \qquad \frac{f}{2}\frac{V L_c}{\nu} = \frac{h_{\text{heat}} L_c}{k} = \frac{h_{\text{mass}} L_c}{D_{AB}} \tag{14–86}$$

Noting that Pr = Sc = 1, we can also write this equation as

$$\frac{f}{2} = \frac{\text{Nu}}{\text{Re Pr}} = \frac{\text{Sh}}{\text{Re Sc}} \qquad \text{or} \qquad \frac{f}{2} = \text{St} = \text{St}_{\text{mass}} \tag{14–87}$$

This relation is known as the **Reynolds analogy**, and it enables us to determine the seemingly unrelated friction, heat transfer, and mass transfer coefficients when only one of them is known or measured. (Actually the original

FIGURE 14–46

When the molecular diffusivities of momentum, heat, and mass are equal to each other, the velocity, temperature, and concentration boundary layers coincide.

Reynolds analogy proposed by O. Reynolds in 1874 is $\mathrm{St} = f/2$, which is then extended to include mass transfer.) However, it should always be remembered that the analogy is restricted to situations for which $\mathrm{Pr} \approx \mathrm{Sc} \approx 1$. Of course the first part of the analogy between friction and heat transfer coefficients can always be used for gases since their Prandtl number is close to unity.

General Case: Pr ≠ Sc ≠ 1 (Chilton–Colburn Analogy)

The Reynolds analogy is a very useful relation, and it is certainly desirable to extend it to a wider range of Pr and Sc numbers. Several attempts have been made in this regard, but the simplest and the best known is the one suggested by Chilton and Colburn in 1934 as

$$\frac{f}{2} = \mathrm{St}\,\mathrm{Pr}^{2/3} = \mathrm{St}_{\mathrm{mass}}\mathrm{Sc}^{2/3} \tag{14–88}$$

for $0.6 < \mathrm{Pr} < 60$ and $0.6 < \mathrm{Sc} < 3000$. This equation is known as the **Chilton–Colburn analogy.** Using the definition of heat and mass Stanton numbers, the analogy between heat and mass transfer can be expressed more conveniently as (Fig. 14–47)

or

$$\frac{\mathrm{St}}{\mathrm{St}_{\mathrm{mass}}} = \left(\frac{\mathrm{Sc}}{\mathrm{Pr}}\right)^{2/3}$$

$$\frac{h_{\mathrm{heat}}}{h_{\mathrm{mass}}} = \rho c_p \left(\frac{\mathrm{Sc}}{\mathrm{Pr}}\right)^{2/3} = \rho c_p \left(\frac{\alpha}{D_{AB}}\right)^{2/3} = \rho c_p\,\mathrm{Le}^{2/3} \tag{14–89}$$

For air–water vapor mixtures at 298 K, the mass and thermal diffusivities are $D_{AB} = 2.5 \times 10^{-5}\ \mathrm{m^2/s}$ and $\alpha = 2.18 \times 10^{-5}\ \mathrm{m^2/s}$ and thus the Lewis number is $\mathrm{Le} = \alpha/D_{AB} = 0.872$. (We simply use the α value of dry air instead of the moist air since the fraction of vapor in the air at atmospheric conditions is low.) Then $(\alpha/D_{AB})^{2/3} = 0.872^{2/3} = 0.913$, which is close to unity. Also, the Lewis number is relatively insensitive to variations in temperature. Therefore, for air–water vapor mixtures, the relation between heat and mass transfer coefficients can be expressed with a good accuracy as

$$h_{\mathrm{heat}} \cong \rho c_p h_{\mathrm{mass}} \qquad \text{(air–water vapor mixtures)} \tag{14–90}$$

where ρ and c_p are the density and specific heat of air at average conditions (or ρc_p is the specific heat of air per unit volume). Equation 14–90 is known as the **Lewis relation** and is commonly used in air-conditioning applications. Another important consequence of $\mathrm{Le} \cong 1$ is that the *wet-bulb* and *adiabatic saturation temperatures* of moist air are nearly identical. In *turbulent flow*, the Lewis relation can be used even when the Lewis number is not 1 since eddy mixing in turbulent flow overwhelms any molecular diffusion, and heat and mass are transported at the same rate.

The Chilton–Colburn analogy has been observed to hold quite well in laminar or turbulent flow over plane surfaces. But this is not always the case for internal flow and flow over irregular geometries, and in such cases specific relations developed should be used. When dealing with flow over blunt bodies, it is important to note that f in these relations is the *skin friction coefficient,* not the total drag coefficient, which also includes the pressure drag.

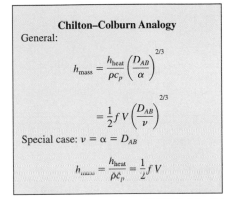

Chilton–Colburn Analogy

General:

$$h_{\mathrm{mass}} = \frac{h_{\mathrm{heat}}}{\rho c_p}\left(\frac{D_{AB}}{\alpha}\right)^{2/3}$$

$$= \frac{1}{2} f\, V \left(\frac{D_{AB}}{\nu}\right)^{2/3}$$

Special case: $\nu = \alpha = D_{AB}$

$$h_{\mathrm{mass}} = \frac{h_{\mathrm{heat}}}{\rho c_p} = \frac{1}{2} f\, V$$

FIGURE 14–47

When the friction or heat transfer coefficient is known, the mass transfer coefficient can be determined directly from the Chilton–Colburn analogy.

Limitation on the Heat–Mass Convection Analogy

Caution should be exercised when using the analogy in Eq. 14–88 since there are a few factors that put some shadow on the accuracy of that relation. For one thing, the Nusselt numbers are usually evaluated for smooth surfaces, but many mass transfer problems involve wavy or roughened surfaces. Also, many Nusselt relations are obtained for constant surface temperature situations, but the concentration may not be constant over the entire surface because of the possible surface dryout. The blowing or suction at the surface during mass transfer may also cause some deviation, especially during high speed blowing or suction.

Finally, the heat–mass convection analogy is valid for **low mass flux** cases in which the flow rate of species undergoing mass flow is low relative to the total flow rate of the liquid or gas mixture so that the mass transfer between the fluid and the surface does not affect the *flow velocity*. (Note that convection relations are based on *zero* fluid velocity at the surface, which is true only when there is no net mass transfer at the surface.) Therefore, the heat–mass convection analogy is not applicable when the rate of mass transfer of a species is high relative to the flow rate of that species.

Consider, for example, the evaporation and transfer of water vapor into air in an air washer, an evaporative cooler, a wet cooling tower, or just at the free surface of a lake or river (Fig. 14–48). Even at a temperature of 40°C, the vapor pressure at the water surface is the saturation pressure of 7.4 kPa, which corresponds to a mole fraction of 0.074 or a mass fraction of $w_{A, s} = 0.047$ for the vapor. Then the mass fraction difference across the boundary layer will be, at most, $\Delta w = w_{A, s} - w_{A, \infty} = 0.047 - 0 = 0.047$. For the evaporation of water into air, the error involved in the low mass flux approximation is roughly $\Delta w/2$, which is 2.5 percent in the worst case considered above. Therefore, in processes that involve the evaporation of water into air, we can use the heat–mass convection analogy with confidence. However, the mass fraction of vapor approaches 1 as the water temperature approaches the saturation temperature, and thus the low mass flux approximation is not applicable to mass transfer in boilers, condensers, and the evaporation of fuel droplets in combustion chambers. In this chapter, we limit our consideration to low mass flux applications.

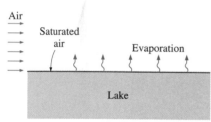

FIGURE 14–48

Evaporation from the free surface of water into air.

Mass Convection Relations

Under low mass flux conditions, the mass convection coefficients can be determined by either (1) determining the friction or heat transfer coefficient and then using the Chilton–Colburn analogy or (2) picking the appropriate Nusselt number relation for the given geometry and analogous boundary conditions, replacing the Nusselt number by the Sherwood number and the Prandtl number by the Schmidt number, as shown in Table 14–13 for some representative cases. The first approach is obviously more convenient when the friction or heat transfer coefficient is already known. Otherwise, the second approach should be preferred since it is generally more accurate, and the Chilton–Colburn analogy offers no significant advantage in this case. Relations for convection mass transfer in other geometries can be written similarly using the corresponding heat transfer relation in Chapters 6 through 9.

TABLE 14-13

Sherwood number relations in mass convection for specified concentration at the surface corresponding to the Nusselt number relations in heat convection for specified surface temperature

Convective Heat Transfer	Convective Mass Transfer
1. *Forced Convection over a Flat Plate*	
(a) Laminar flow (Re $< 5 \times 10^5$)	
$\text{Nu} = 0.664\,\text{Re}_L^{0.5}\,\text{Pr}^{1/3}$, Pr > 0.6	$\text{Sh} = 0.664\,\text{Re}_L^{0.5}\,\text{Sc}^{1/3}$, Sc > 0.5
(b) Turbulent flow ($5 \times 10^5 <$ Re $< 10^7$)	
$\text{Nu} = 0.037\,\text{Re}_L^{0.8}\,\text{Pr}^{1/3}$, Pr > 0.6	$\text{Sh} = 0.037\,\text{Re}_L^{0.8}\,\text{Sc}^{1/3}$, Sc > 0.5
2. *Fully Developed Flow in Smooth Circular Pipes*	
(a) Laminar flow (Re < 2300)	
$\text{Nu} = 3.66$	$\text{Sh} = 3.66$
(b) Turbulent flow (Re $> 10{,}000$)	
$\text{Nu} = 0.023\,\text{Re}^{0.8}\,\text{Pr}^{0.4}$, $0.7 <$ Pr < 160	$\text{Sh} = 0.023\,\text{Re}^{0.8}\,\text{Sc}^{0.4}$, $0.7 <$ Sc 160
3. *Natural Convection over Surfaces*	
(a) Vertical plate	
$\text{Nu} = 0.59(\text{Gr Pr})^{1/4}$, $10^5 <$ Gr Pr $< 10^9$	$\text{Sh} = 0.59(\text{Gr Sc})^{1/4}$, $10^5 <$ Gr Sc $< 10^9$
$\text{Nu} = 0.1(\text{Gr Pr})^{1/3}$, $10^9 <$ Gr Pr $< 10^{13}$	$\text{Sh} = 0.1(\text{Gr Sc})^{1/3}$, $10^9 <$ Gr Sc $< 10^{13}$
(b) Upper surface of a horizontal plate	
Surface is hot ($T_s > T_\infty$)	Fluid near the surface is light ($\rho_s < \rho_\infty$)
$\text{Nu} = 0.54(\text{Gr Pr})^{1/4}$, $10^4 <$ Gr Pr $< 10^7$	$\text{Sh} = 0.54(\text{Gr Sc})^{1/4}$, $10^4 <$ Gr Sc $< 10^7$
$\text{Nu} = 0.15(\text{Gr Pr})^{1/3}$, $10^7 <$ Gr Pr $< 10^{11}$	$\text{Sh} = 0.15(\text{Gr Sc})^{1/3}$, $10^7 <$ Gr Sc $< 10^{11}$
(c) Lower surface of a horizontal plate	
Surface is hot ($T_s > T_\infty$)	Fluid near the surface is light ($\rho_s < \rho_\infty$)
$\text{Nu} = 0.27(\text{Gr Pr})^{1/4}$, $10^5 <$ Gr Pr $< 10^{11}$	$\text{Sh} = 0.27(\text{Gr Sc})^{1/4}$, $10^5 <$ Gr Sc $< 10^{11}$

EXAMPLE 14–10   Mass Convection inside a Circular Pipe

Consider a circular pipe of inner diameter $D = 0.015$ m whose inner surface is covered with a layer of liquid water as a result of condensation (Fig. 14–49). In order to dry the pipe, air at 300 K and 1 atm is forced to flow through it with an average velocity of 1.2 m/s. Using the analogy between heat and mass transfer, determine the mass transfer coefficient inside the pipe for fully developed flow.

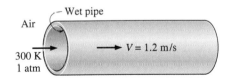

FIGURE 14–49
Schematic for Example 14–10.

SOLUTION The liquid layer on the inner surface of a circular pipe is dried by blowing air through it. The mass transfer coefficient is to be determined.

Assumptions **1** The low mass flux model and thus the analogy between heat and mass transfer is applicable since the mass fraction of vapor in the air is low (about 2 percent for saturated air at 300 K). **2** The flow is fully developed.

Properties Because of low mass flux conditions, we can use dry air properties for the mixture at the specified temperature of 300 K and 1 atm, for which $\nu = 1.58 \times 10^{-5}$ m²/s (Table A–15). The mass diffusivity of water vapor in the air at 300 K is determined from Eq. 14–15 to be

$$D_{AB} - D_{\text{H}_2\text{O-air}} = 1.87 \times 10^{-10}\,\frac{T^{2.072}}{P} = 1.87 \times 10^{-10}\,\frac{300^{2.072}}{1} = 2.54 \times 10^{-5} \text{ m}^2/\text{s}$$

Analysis The Reynolds number for this internal flow is

$$Re = \frac{VD}{\nu} = \frac{(1.2 \text{ m/s})(0.015 \text{ m})}{1.58 \times 10^{-5} \text{ m}^2/\text{s}} = 1139$$

which is less than 2300 and thus the flow is laminar. Therefore, based on the analogy between heat and mass transfer, the Nusselt and the Sherwood numbers in this case are Nu = Sh = 3.66. Using the definition of Sherwood number, the mass transfer coefficient is determined to be

$$h_{\text{mass}} = \frac{Sh D_{AB}}{D} = \frac{(3.66)(2.54 \times 10^{-5} \text{ m}^2/\text{s})}{0.015 \text{ m}} = 0.00620 \text{ m/s}$$

The mass transfer rate (or the evaporation rate) in this case can be determined by defining the logarithmic mean concentration difference in an analogous manner to the logarithmic mean temperature difference.

EXAMPLE 14–11   Analogy between Heat and Mass Transfer

Heat transfer coefficients in complex geometries with complicated boundary conditions can be determined by mass transfer measurements on similar geometries under similar flow conditions using volatile solids such as naphthalene and dichlorobenzene and utilizing the Chilton–Colburn analogy between heat and mass transfer at low mass flux conditions. The amount of mass transfer during a specified time period is determined by weighing the model or measuring the surface recession.

During a certain experiment involving the flow of dry air at 25°C and 1 atm at a free stream velocity of 2 m/s over a body covered with a layer of naphthalene, it is observed that 12 g of naphthalene has sublimated in 15 min (Fig. 14–50). The surface area of the body is 0.3 m². Both the body and the air were kept at 25°C during the study. The vapor pressure of naphthalene at 25°C is 11 Pa and the mass diffusivity of naphthalene in air at 25°C is $D_{AB} = 0.61 \times 10^{-5}$ m²/s. Determine the heat transfer coefficient under the same flow conditions over the same geometry.

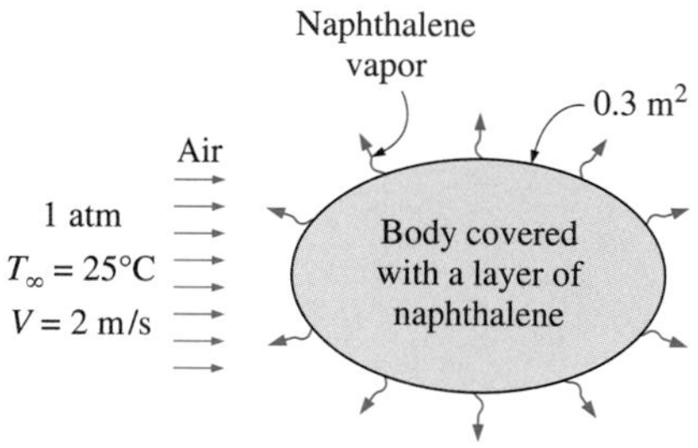

FIGURE 14–50
Schematic for Example 14–11.

SOLUTION Air is blown over a body covered with a layer of naphthalene, and the rate of sublimation is measured. The heat transfer coefficient under the same flow conditions over the same geometry is to be determined.
Assumptions **1** The low mass flux conditions exist so that the Chilton–Colburn analogy between heat and mass transfer is applicable (will be verified). **2** Both air and naphthalene vapor are ideal gases.
Properties The molar mass of naphthalene is 128.2 kg/kmol. Because of low mass flux conditions, we can use dry air properties for the mixture at the specified temperature of 25°C and 1 atm, at which $\rho = 1.184$ kg/m³, $c_p = 1007$ J/kg · K, and $\alpha = 2.141 \times 10^{-5}$ m²/s (Table A–15).
Analysis The incoming air is free of naphthalene, and thus the mass fraction of naphthalene at free stream conditions is zero, $w_{A,\infty} = 0$. Noting that the vapor pressure of naphthalene at the surface is 11 Pa, its mass fraction at the surface is determined to be

$$w_{A,s} = \frac{P_{A,s}}{P}\left(\frac{M_A}{M_{air}}\right) = \frac{11 \text{ Pa}}{101,325 \text{ Pa}}\left(\frac{128.2 \text{ kg/kmol}}{29 \text{ kg/kmol}}\right) = 4.8 \times 10^{-4}$$

which confirms that the low mass flux approximation is valid. The rate of evaporation of naphthalene in this case is

$$\dot{m}_{evap} = \frac{m}{\Delta t} = \frac{0.012 \text{ kg}}{(15 \times 60 \text{ s})} = 1.33 \times 10^{-5} \text{ kg/s}$$

Then the mass convection coefficient becomes

$$h_{mass} = \frac{\dot{m}}{\rho A_s(w_{A,s} - w_{A,\infty})} = \frac{1.33 \times 10^{-5} \text{ kg/s}}{(1.184 \text{ kg/m}^3)(0.3 \text{ m}^2)(4.8 \times 10^{-4} - 0)}$$

$$= 0.0780 \text{ m/s}$$

Using the analogy between heat and mass transfer, the average heat transfer coefficient is determined from Eq. 14–89 to be

$$h_{heat} = \rho c_p h_{mass}\left(\frac{\alpha}{D_{AB}}\right)^{2/3}$$

$$= (1.184 \text{ kg/m}^3)(1007 \text{ J/kg} \cdot {}^\circ\text{C})(0.0780 \text{ m/s})\left(\frac{2.141 \times 10^{-5} \text{ m}^2/\text{s}}{0.61 \times 10^{-5} \text{ m}^2/\text{s}}\right)^{2/3}$$

$$= 215 \text{ W/m}^2 \cdot {}^\circ\text{C}$$

Discussion Because of the convenience it offers, naphthalene has been used in numerous heat transfer studies to determine convection heat transfer coefficients.

14–10 · SIMULTANEOUS HEAT AND MASS TRANSFER

Many mass transfer processes encountered in practice occur isothermally, and thus they do not involve any heat transfer. But some engineering applications involve the vaporization of a liquid and the diffusion of this vapor into the surrounding gas. Such processes require the transfer of the latent heat of vaporization h_{fg} to the liquid in order to vaporize it, and thus such problems involve simultaneous heat and mass transfer. To generalize, any mass transfer problem involving *phase change* (evaporation, sublimation, condensation, melting, etc.) must also involve *heat transfer*, and the solution of such problems needs to be analyzed by considering *simultaneous heat and mass transfer*. Some examples of simultaneous heat and mass problems are drying, evaporative cooling, transpiration (or sweat) cooling, cooling by dry ice, combustion of fuel droplets, and ablation cooling of space vehicles during reentry, and even ordinary events like rain, snow, and hail. In warmer locations, for example, the snow melts and the rain evaporates before reaching the ground (Fig. 14–51).

To understand the mechanism of simultaneous heat and mass transfer, consider the *evaporation of water* from a swimming pool into air. Let us assume that the water and the air are initially at the same temperature. If the air is saturated (a relative humidity of $\phi = 100$ percent), there will be no heat or mass transfer as long as the isothermal conditions remain. But if the air is not

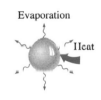

(a) Ablation

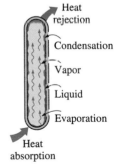

(b) Evaporation of rain droplet

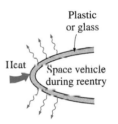

(c) Drying of clothes

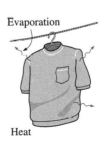

(d) Heat pipes

FIGURE 14–51
Many problems encountered in practice involve simultaneous heat and mass transfer.

saturated ($\phi < 100$ percent), there will be a difference between the concentration of water vapor at the water–air interface (which is always saturated) and some distance above the interface (the concentration boundary layer). Concentration difference is the driving force for mass transfer, and thus this concentration difference drives the water into the air. But the water must vaporize first, and it must absorb the latent heat of vaporization in order to vaporize. Initially, the entire heat of vaporization comes from the water near the interface since there is no temperature difference between the water and the surroundings and thus there cannot be any heat transfer. The temperature of water near the surface must drop as a result of the sensible heat loss, which also drops the saturation pressure and thus vapor concentration at the interface.

This temperature drop creates temperature differences within the water at the top as well as between the water and the surrounding air. These temperature differences drive heat transfer toward the water surface from both the air and the deeper parts of the water, as shown in Figure 14–52. If the evaporation rate is high and thus the demand for the heat of vaporization is higher than the amount of heat that can be supplied from the lower parts of the water body and the surroundings, the deficit is made up from the sensible heat of the water at the surface, and thus the temperature of water at the surface drops further. The process continues until the latent heat of vaporization equals the heat transfer to the water at the surface. Once the steady operation conditions are reached and the interface temperature stabilizes, the energy balance on a thin layer of liquid at the surface can be expressed as

$$\dot{Q}_{\text{sensible, transferred}} = \dot{Q}_{\text{latent, absorbed}} \qquad \text{or} \qquad \dot{Q} = \dot{m}_v \, h_{fg} \qquad \text{(14–91)}$$

where $\dot{m}_v$ is the rate of evaporation and h_{fg} is the latent heat of vaporization of water at the surface temperature. Various expressions for $\dot{m}_v$ under various approximations are given in Table 14–14. The mixture properties such as the specific heat c_p and molar mass M should normally be evaluated at the *mean film composition* and *mean film temperature*. However, when dealing with air–water vapor mixtures at atmospheric conditions or other low mass flux situations, we can simply use the properties of the gas with reasonable accuracy.

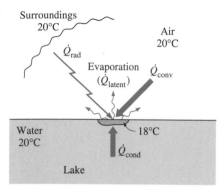

FIGURE 14–52

Various mechanisms of heat transfer involved during the evaporation of water from the surface of a lake.

TABLE 14–14

Various expressions for evaporation rate of a liquid into a gas through an interface area A_s under various approximations (subscript v stands for vapor, s for liquid–gas interface, and ∞ away from surface)

Assumption	Evaporation Rate
General	$\dot{m}_v = h_{\text{mass}} A_s (\rho_{v,\,s} - \rho_{v,\,\infty})$
Assuming vapor to be an ideal gas, $P_v = \rho_v R_v T$	$\dot{m}_v = \dfrac{h_{\text{mass}} A_s}{R_v} \left(\dfrac{P_{v,\,s}}{T_s} - \dfrac{P_{v,\,\infty}}{T_\infty} \right)$
Using Chilton–Colburn analogy, $h_{\text{heat}} = \rho c_p h_{\text{mass}} \text{Le}^{2/3}$	$\dot{m}_v = \dfrac{h_{\text{mass}} A_s}{\rho c_p \text{Le}^{2/3} R_v} \left(\dfrac{P_{v,\,s}}{T_s} - \dfrac{P_{v,\,\infty}}{T_\infty} \right)$
Using $\dfrac{1}{T_s} - \dfrac{1}{T_\infty} \approx \dfrac{1}{T}$, where $T = \dfrac{T_s + T_\infty}{2}$ and $P = \rho R T = \rho (R_u/M) T$	$\dot{m}_v = \dfrac{h_{\text{mass}} A_s}{\rho c_p \text{Le}^{2/3}} \dfrac{M_v}{M} \dfrac{P_{v,\,s} - P_{v,\,\infty}}{P}$

The $\dot{Q}$ in Eq. 14–91 represents all forms of heat from all sources transferred to the surface, including convection and radiation from the surroundings and conduction from the deeper parts of the water due to the sensible energy of the water itself or due to heating the water body by a resistance heater, heating coil, or even chemical reactions in the water. If heat transfer from the water body to the surface as well as radiation from the surroundings is negligible, which is often the case, then the heat loss by evaporation must equal heat gain by convection. That is,

$$\dot{Q}_{conv} = \dot{m}_v\,h_{fg} \quad \text{or} \quad h_{conv}A_s(T_\infty - T_s) = \frac{h_{conv}A_s\,h_{fg}}{c_p\,Le^{2/3}}\frac{M_v}{M}\frac{P_{v,s} - P_{v,\infty}}{P}$$

Canceling $h_{conv}A_s$ from both sides of the second equation gives

$$T_s = T_\infty - \frac{h_{fg}}{c_p\,Le^{2/3}}\frac{M_v}{M}\frac{P_{v,s} - P_{v,\infty}}{P} \tag{14–92}$$

which is a relation for the temperature of the liquid under steady conditions.

EXAMPLE 14–12   Evaporative Cooling of a Canned Drink

During a hot summer day, a canned drink is to be cooled by wrapping it in a cloth that is kept wet continually, and blowing air to it by a fan (Fig. 14–53). If the environment conditions are 1 atm, 30°C, and 40 percent relative humidity, determine the temperature of the drink when steady conditions are reached.

SOLUTION Air is blown over a canned drink wrapped in a wet cloth to cool it by simultaneous heat and mass transfer. The temperature of the drink when steady conditions are reached is to be determined.

Assumptions 1 The low mass flux conditions exist so that the Chilton–Colburn analogy between heat and mass transfer is applicable since the mass fraction of vapor in the air is low (about 2 percent for saturated air at 25°C). 2 Both air and water vapor at specified conditions are ideal gases (the error involved in this assumption is less than 1 percent). 3 Radiation effects are negligible.

Properties Because of low mass flux conditions, we can use dry air properties for the mixture at the average temperature of $(T_\infty + T_s)/2$ which cannot be determined at this point because of the unknown surface temperature T_s. We know that $T_s < T_\infty$ and, for the purpose of property evaluation, we take T_s to be 20°C. Then the properties of water at 20°C and the properties of dry air at the average temperature of 25°C and 1 atm are (Tables A–9 and A–15)

Water: $h_{fg} = 2454$ kJ/kg, $P_v = 2.34$ kPa; also, $P_v = 4.25$ kPa at 30°C
Dry air: $c_p = 1.007$ kJ/kg · °C, $\alpha = 2.141 \times 10^{-5}$ m²/s

The molar masses of water and air are 18 and 29 kg/kmol, respectively (Table A–1). Also, the mass diffusivity of water vapor in air at 25°C is $D_{H_2O\text{-air}} = 2.50 \times 10^{-5}$ m²/s (Table 14–4).

Analysis Utilizing the Chilton–Colburn analogy, the surface temperature of the drink can be determined from Eq. 14–92,

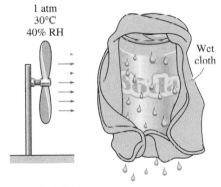

1 atm
30°C
40% RH

Wet cloth

FIGURE 14–53
Schematic for Example 14–12.

$$T_s = T_\infty - \frac{h_{fg}}{c_p \, \text{Le}^{2/3}} \frac{M_v}{M} \frac{P_{v,s} - P_{v,\infty}}{P}$$

where the Lewis number is

$$\text{Le} = \frac{\alpha}{D_{AB}} = \frac{2.141 \times 10^{-5} \text{ m}^2/\text{s}}{2.5 \times 10^{-5} \text{ m}^2/\text{s}} = 0.856$$

Note that we could take the Lewis number to be 1 for simplicity, but we chose to incorporate it for better accuracy.

The air at the surface is saturated, and thus the vapor pressure at the surface is simply the saturation pressure of water at the surface temperature (2.34 kPa). The vapor pressure of air away from the surface is

$$P_{v,\infty} = \phi P_{\text{sat} @ T_\infty} = (0.40) P_{\text{sat} @ 30°\text{C}} = (0.40)(4.25 \text{ kPa}) = 1.70 \text{ kPa}$$

Noting that the atmospheric pressure is 1 atm = 101.3 kPa, substituting gives

$$T_s = 30°\text{C} - \frac{2454 \text{ kJ/kg}}{(1.007 \text{ kJ/kg} \cdot °\text{C})(0.856)^{2/3}} \frac{18 \text{ kg/kmol}}{29 \text{ kg/kmol}} \frac{(2.34 - 1.70) \text{ kPa}}{101.3 \text{ kPa}}$$

$$= \mathbf{19.4°\text{C}}$$

Therefore, the temperature of the drink can be lowered to 19.4°C by this process.

EXAMPLE 14–13   Heat Loss from Uncovered Hot Water Baths

Hot water baths with open tops are commonly used in manufacturing facilities for various reasons. In a plant that manufactures spray paints, the pressurized paint cans are temperature tested by submerging them in hot water at 50°C in a 40-cm-deep rectangular bath and keeping them there until the cans are heated to 50°C to ensure that the cans can withstand temperatures up to 50°C during transportation and storage (Fig. 14–54). The water bath is 1 m wide and 3.5 m long, and its top surface is open to ambient air to facilitate easy observation for the workers. If the average conditions in the plant are 92 kPa, 25°C, and 52 percent relative humidity, determine the rate of heat loss from the top surface of the water bath by (*a*) radiation, (*b*) natural convection, and (*c*) evaporation. Assume the water is well agitated and maintained at a uniform temperature of 50°C at all times by a heater, and take the average temperature of the surrounding surfaces to be 20°C.

SOLUTION   Spray paint cans are temperature tested by submerging them in an uncovered hot water bath. The rates of heat loss from the top surface of the bath by radiation, natural convection, and evaporation are to be determined.

Assumptions   **1** The low mass flux conditions exist so that the Chilton–Colburn analogy between heat and mass transfer is applicable since the mass fraction of vapor in the air is low (about 2 percent for saturated air at 300 K). **2** Both air and water vapor at specified conditions are ideal gases (the error involved in this assumption is less than 1 percent). **3** Water is maintained at a uniform temperature of 50°C.

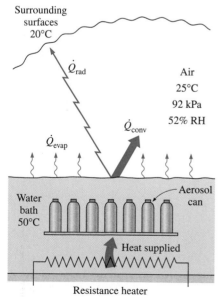

FIGURE 14–54
Schematic for Example 14–13.

Properties Relevant properties for each mode of heat transfer are determined below in respective sections.

Analysis (*a*) The emissivity of liquid water is given in Table A–18 to be 0.95. Then the radiation heat loss from the water to the surrounding surfaces becomes

$$\dot{Q}_{rad} = \varepsilon A_s \sigma (T_s^4 - T_{surr}^4)$$
$$= (0.95)(3.5 \text{ m}^2)(5.67 \times 10^{-8} \text{ W/m}^2 \cdot \text{K}^4)[(323 \text{ K})^4 - (293 \text{ K})^4]$$
$$= 663 \text{ W}$$

(*b*) The air–water vapor mixture is dilute and thus we can use dry air properties for the mixture at the average temperature of $(T_\infty + T_s)/2 = (25 + 50)/2 = 37.5°C$. Noting that the total atmospheric pressure is $92/101.3 = 0.9080$ atm, the properties of dry air at $37.5°C$ and 0.9080 atm are (Table A–15)

$$k = 0.02644 \text{ W/m} \cdot °C, \quad Pr = 0.7262 \text{ (independent of pressure)}$$
$$\alpha = (2.312 \times 10^{-5} \text{ m}^2/\text{s})/0.9080 = 2.546 \times 10^{-5} \text{ m}^2/\text{s}$$
$$\nu = (1.679 \times 10^{-5} \text{ m}^2/\text{s})/0.9080 = 1.849 \times 10^{-5} \text{ m}^2/\text{s}$$

The properties of water at $50°C$ are

$$h_{fg} = 2383 \text{ kJ/kg} \quad \text{and} \quad P_v = 12.35 \text{ kPa}$$

The air at the surface is saturated, and thus the vapor pressure at the surface is simply the saturation pressure of water at the surface temperature. The vapor pressure of air far from the water surface is

$$P_{v,\infty} = \phi P_{sat @ T_\infty} = (0.52)P_{sat @ 25°C} = (0.52)(3.17 \text{ kPa}) = 1.65 \text{ kPa}$$

Treating the water vapor and the air as ideal gases and noting that the total atmospheric pressure is the sum of the vapor and dry air pressures, the densities of the water vapor, dry air, and their mixture at the water–air interface and far from the surface are determined to be

At the surface:
$$\rho_{v,s} = \frac{P_{v,s}}{R_v T_s} = \frac{12.35 \text{ kPa}}{(0.4615 \text{ kPa} \cdot \text{m}^3/\text{kg} \cdot \text{K})(323 \text{ K})} = 0.0829 \text{ kg/m}^3$$

$$\rho_{a,s} = \frac{P_{a,s}}{R_a T_s} = \frac{(92 - 12.35) \text{ kPa}}{(0.287 \text{ kPa} \cdot \text{m}^3/\text{kg} \cdot \text{K})(323 \text{ K})} = 0.8592 \text{ kg/m}^3$$

$$\rho_s = \rho_{v,s} + \rho_{a,s} = 0.0829 + 0.8592 = 0.9421 \text{ kg/m}^3$$

Away from the surface:
$$\rho_{v,\infty} = \frac{P_{v,\infty}}{R_v T_\infty} = \frac{1.65 \text{ kPa}}{(0.4615 \text{ kPa} \cdot \text{m}^3/\text{kg} \cdot \text{K})(298 \text{ K})} = 0.0120 \text{ kg/m}^3$$

$$\rho_{a,\infty} = \frac{P_{a,\infty}}{R_a T_\infty} = \frac{(92 - 1.65) \text{ kPa}}{(0.287 \text{ kPa} \cdot \text{m}^3/\text{kg} \cdot \text{K})(298 \text{ K})} = 1.0564 \text{ kg/m}^3$$

$$\rho_\infty = \rho_{v,\infty} + \rho_{a,\infty} = 0.0120 + 1.0564 = 1.0684 \text{ kg/m}^3$$

The area of the top surface of the water bath is $A_s = (3.5 \text{ m})(1 \text{ m}) = 3.5 \text{ m}^2$ and its perimeter is $p = 2(3.5 + 1) = 9$ m. Therefore, the characteristic length is

$$L_c = \frac{A_s}{p} = \frac{3.5 \text{ m}^2}{9 \text{ m}} = 0.3889 \text{ m}$$

Then using densities (instead of temperatures) since the mixture is not homogeneous, the Grashof number is

$$\text{Gr} = \frac{g(\rho_\infty - \rho_S)L_c^3}{\rho\nu^2}$$

$$= \frac{(9.81 \text{ m/s}^2)(1.0684 - 0.9421 \text{ kg/m}^3)(0.3889 \text{ m})^3}{[(0.9421 + 1.0684)/2 \text{ kg/m}^3](1.849 \times 10^{-5} \text{ m}^2/\text{s})^2}$$

$$= 2.121 \times 10^8$$

Recognizing that this is a natural convection problem with hot horizontal surface facing up, the Nusselt number and the convection heat transfer coefficients are determined to be

$$\text{Nu} = 0.15(\text{Gr Pr})^{1/3} = 0.15(2.121 \times 10^8 \times 0.7262)^{1/3} = 80.41$$

$$h_{\text{conv}} = \frac{\text{Nu}k}{L_c} = \frac{(80.41)(0.02644 \text{ W/m} \cdot {}^\circ\text{C})}{0.3889 \text{ m}} = 5.47 \text{ W/m}^2 \cdot {}^\circ\text{C}$$

Then the natural convection heat transfer rate becomes

$$\dot{Q}_{\text{conv}} = h_{\text{conv}}A_s(T_s - T_\infty)$$
$$= (5.47 \text{ W/m}^2 \cdot {}^\circ\text{C})(3.5 \text{ m}^2)(50 - 25){}^\circ\text{C} = \textbf{479 W}$$

Note that the magnitude of natural convection heat transfer is comparable to that of radiation, as expected.

(c) Utilizing the analogy between heat and mass convection, the mass transfer coefficient is determined the same way by replacing Pr by Sc. The mass diffusivity of water vapor in air at the average temperature of 310.5 K is determined from Eq. 14–15 to be

$$D_{AB} = D_{\text{H}_2\text{O-air}} = 1.87 \times 10^{-10}\frac{T^{2.072}}{P} = 1.87 \times 10^{-10}\frac{310.5^{2.072}}{0.908}$$

$$= 3.00 \times 10^{-5} \text{ m}^2/\text{s}$$

The Schmidt number is

$$\text{Sc} = \frac{\nu}{D_{AB}} = \frac{1.849 \times 10^{\pm 5} \text{ m}^2/\text{s}}{3.00 \times 10^{\pm 5} \text{ m}^2/\text{s}} = 0.616$$

The Sherwood number and the mass transfer coefficients are determined to be

$$\text{Sh} = 0.15(\text{Gr Sc})^{1/3} = 0.15(2.121 \times 10^8 \times 0.616)^{1/3} = 76.1$$

$$h_{\text{mass}} = \frac{\text{Sh}D_{AB}}{L_c} = \frac{(76.1)(3.00 \times 10^{-5} \text{ m}^2/\text{s})}{0.3889 \text{ m}} = 0.00587 \text{ m/s}$$

Then the evaporation rate and the rate of heat transfer by evaporation become

$$\dot{m}_v = h_{mass} A_s (\rho_{v,s} - \rho_{v,\infty})$$
$$= (0.00587 \text{ m/s})(3.5 \text{ m}^2)(0.0829 - 0.0120)\text{kg/m}^3$$
$$= 0.00146 \text{ kg/s} = 5.24 \text{ kg/h}$$

$$\dot{Q}_{evap} = \dot{m}_v\, h_{fg} = (0.00146 \text{ kg/s})(2383 \text{ kJ/kg}) = 3.479 \text{ kW} = \mathbf{3479 \text{ W}}$$

which is more than seven times the rate of heat transfer by natural convection.

Finally, noting that the direction of heat transfer is always from high to low temperature, all forms of heat transfer determined above are in the same direction, and the total rate of heat loss from the water to the surrounding air and surfaces is

$$\dot{Q}_{total} = \dot{Q}_{rad} + \dot{Q}_{conv} + \dot{Q}_{evap} = 663 + 479 + 3479 = \mathbf{4621 \text{ W}}$$

Discussion Note that if the water bath is heated electrically, a 4.6 kW resistance heater will be needed just to make up for the heat loss from the top surface. The total heater size will have to be larger to account for the heat losses from the side and bottom surfaces of the bath as well as the heat absorbed by the spray paint cans as they are heated to 50°C. Also note that water needs to be supplied to the bath at a rate of 5.24 kg/h to make up for the water loss by evaporation. Also, in reality, the surface temperature will probably be a little lower than the bulk water temperature, and thus the heat transfer rates will be somewhat lower than indicated here.

SUMMARY

Mass transfer is the movement of a chemical species from a high concentration region toward a lower concentration one relative to the other chemical species present in the medium. Heat and mass transfer are analogous to each other, and several parallels can be drawn between them. The driving forces are the *temperature difference* in heat transfer and the *concentration difference* in mass transfer. Fick's law of mass diffusion is of the same form as Fourier's law of heat conduction. The species generation in a medium due to *homogeneous reactions* is analogous to heat generation. Also, mass convection due to bulk fluid motion is analogous to heat convection. Constant surface temperature corresponds to constant concentration at the surface, and an adiabatic wall corresponds to an imperme-able wall. However, concentration is usually not a continuous function at a phase interface.

The concentration of a species A can be expressed in terms of density ρ_A or molar concentration C_A. It can also be expressed in dimensionless form in terms of *mass* or *molar fraction* as

Mass fraction of species A: $\quad w_A = \dfrac{m_A}{m} = \dfrac{m_A/V}{m/V} = \dfrac{\rho_A}{\rho}$

Mole fraction of species A: $\quad y_A = \dfrac{N_A}{N} = \dfrac{N_A/V}{N/V} = \dfrac{C_A}{C}$

In the case of an ideal gas mixture, the mole fraction of a gas is equal to its pressure fraction. *Fick's law* for the diffusion of a species A in a stationary binary mixture of species A and B in a specified direction x is expressed as

Mass basis: $\quad j_{diff,\,A} = \dfrac{\dot{m}_{diff,\,A}}{A} = -\rho D_{AB}\dfrac{d(\rho_A/\rho)}{dx}$
$$= -\rho D_{AB}\dfrac{dw_A}{dx}$$

Mole basis: $\quad \bar{j}_{diff,\,A} = \dfrac{\dot{N}_{diff,\,A}}{A} = -C D_{AB}\dfrac{d(C_A/C)}{dx}$
$$= -C D_{AB}\dfrac{dy_A}{dx}$$

where D_{AB} is the *diffusion coefficient* (or *mass diffusivity*) of the species in the mixture, $j_{\text{diff},A}$ is the diffusive *mass flux* of species A, and $\bar{j}_{\text{diff},A}$ is the *molar flux*.

The mole fractions of a species i in the gas and liquid phases at the interface of a dilute mixture are proportional to each other and are expressed by *Henry's law* as

$$y_{i,\text{ liquid side}} = \frac{P_{i,\text{ gas side}}}{H}$$

where H is *Henry's constant*. When the mixture is not dilute, an approximate relation for the mole fractions of a species on the liquid and gas sides of the interface are expressed approximately by *Raoult's law* as

$$P_{i,\text{ gas side}} = y_{i,\text{ gas side}} P = y_{i,\text{ liquid side}} P_{i,\text{ sat}}(T)$$

where $P_{i,\text{ sat}}(T)$ is the saturation pressure of the species i at the interface temperature and P is the *total pressure* on the gas phase side.

The concentration of the gas species i in the solid at the interface $C_{i,\text{ solid side}}$ is proportional to the *partial pressure* of the species i in the gas $P_{i,\text{ gas side}}$ on the gas side of the interface and is expressed as

$$C_{i,\text{ solid side}} = \mathscr{S} \times P_{i,\text{ gas side}}$$

where $\mathscr{S}$ is the *solubility*. The product of the *solubility* of a gas and the *diffusion coefficient* of the gas in a solid is referred to as the *permeability* $\mathscr{P}$, which is a measure of the ability of the gas to penetrate a solid.

In the absence of any chemical reactions, the mass transfer rates $\dot{m}_{\text{diff},A}$ through a plane wall of area A and thickness L and cylindrical and spherical shells of inner and outer radii r_1 and r_2 under one-dimensional steady conditions are expressed as

$$\dot{m}_{\text{diff},A,\text{ wall}} = \rho D_{AB} A \frac{w_{A,1} - w_{A,2}}{L} = D_{AB} A \frac{\rho_{A,1} - \rho_{A,2}}{L}$$

$$\dot{m}_{\text{diff},A,\text{ cyl}} = 2\pi L \rho D_{AB} \frac{w_{A,1} - w_{A,2}}{\ln(r_2/r_1)} = 2\pi L D_{AB} \frac{\rho_{A,1} - \rho_{A,2}}{\ln(r_2/r_1)}$$

$$\dot{m}_{\text{diff},A,\text{ sph}} = 4\pi r_1 r_2 \rho D_{AB} \frac{w_{A,1} - w_{A,2}}{r_2 - r_1}$$

$$= 4\pi r_1 r_2 D_{AB} \frac{\rho_{A,1} - \rho_{A,2}}{r_2 - r_1}$$

The flow rate of a gas through a solid plane wall under steady one-dimensional conditions can also be expressed in terms of the partial pressures of the adjacent gas on the two sides of the solid as

$$\dot{N}_{\text{diff},A,\text{ wall}} = D_{AB} \mathscr{S}_{AB} A \frac{P_{A,1} - P_{A,2}}{L}$$

$$= \mathscr{P}_{AB} A \frac{P_{A,1} - P_{A,2}}{L}$$

where $P_{A,1}$ and $P_{A,2}$ are the partial pressures of gas A on the two sides of the wall.

During mass transfer in a *moving medium*, chemical species are transported both by molecular diffusion and by the bulk fluid motion, and the velocities of the species are expressed as

$$V_A = V + V_{\text{diff},A}$$

$$V_B = V + V_{\text{diff},B}$$

where V is the *mass-average velocity* of the flow. It is the velocity that would be measured by a velocity sensor and is expressed as

$$V = w_A V_A + w_B V_B$$

The special case $V = 0$ corresponds to a *stationary medium*. Using Fick's law of diffusion, the total mass fluxes $j = \dot{m}/A$ in a moving medium are expressed as

$$j_A = \rho_A V + \rho_A V_{\text{diff},A} = w_A(j_A + j_B) - \rho D_{AB} \frac{dw_A}{dx}$$

$$j_B = \rho_B V + \rho_B V_{\text{diff},B} = w_B(j_A + j_B) - \rho D_{BA} \frac{dw_B}{dx}$$

The *rate of mass convection* of species A in a binary mixture is expressed in an analogous manner to Newton's law of cooling as

$$\dot{m}_{\text{conv}} = h_{\text{mass}} A_s(\rho_{A,s} - \rho_{A,\infty}) = h_{\text{mass}} \rho A_s(w_{A,s} - w_{A,\infty})$$

where h_{mass} is the average *mass transfer coefficient*, in m/s.

The counterparts of the Prandtl and Nusselt numbers in mass convection are the *Schmidt number* Sc and the *Sherwood number* Sh, defined as

$$\text{Sc} = \frac{\nu}{D_{AB}} = \frac{\text{Momentum diffusivity}}{\text{Mass diffusivity}} \quad \text{and} \quad \text{Sh} = \frac{h_{\text{mass}} L_c}{D_{AB}}$$

The relative magnitudes of heat and mass diffusion in the thermal and concentration boundary layers are represented by the *Lewis number*, defined as

$$\text{Le} = \frac{\text{Sc}}{\text{Pr}} = \frac{\alpha}{D_{AB}} = \frac{\text{Thermal diffusivity}}{\text{Mass diffusivity}}$$

Heat and mass transfer coefficients are sometimes expressed in terms of the dimensionless *Stanton number,* defined as

$$St = \frac{h_{conv}}{\rho V c_p} = Nu \frac{1}{Re\ Pr} \quad \text{and} \quad St_{mass} = \frac{h_{mass}}{V} = Sh \frac{1}{Re\ Sc}$$

where V is the free-stream velocity in external flow and the bulk mean fluid velocity in internal flow. For a given geometry and boundary conditions, the Sherwood number in natural or forced convection can be determined from the corresponding Nusselt number expression by simply replacing the Prandtl number by the Schmidt number. But in natural convection, the Grashof number should be expressed in terms of density difference instead of temperature difference.

When the molecular diffusivities of momentum, heat, and mass are identical, we have $\nu = \alpha = D_{AB}$, and thus $Pr = Sc = Le = 1$. The similarity between momentum, heat, and mass transfer in this case is given by the *Reynolds analogy,* expressed as

$$\frac{f}{2} Re = Nu = Sh \quad \text{or}$$

$$\frac{f}{2} \frac{V_\infty L}{\nu} = \frac{h_{heat} L}{k} = \frac{h_{mass} L}{D_{AB}} \quad \text{or} \quad \frac{f}{2} = St = St_{mass}$$

For the general case of $Pr \neq Sc \neq 1$, it is modified as

$$\frac{f}{2} = St\ Pr^{2/3} = St_{mass}Sc^{2/3}$$

which is known as the *Chilton–Colburn analogy.* The analogy between heat and mass transfer is expressed more conveniently as

$$h_{heat} = \rho c_p Le^{2/3} h_{mass} = \rho c_p (\alpha/D_{AB})^{2/3} h_{mass}$$

For air–water vapor mixtures, $Le \cong 1$, and thus this relation simplifies further. The heat–mass convection analogy is limited to *low mass flux* cases in which the flow rate of species undergoing mass flow is low relative to the total flow rate of the liquid or gas mixture. The mass transfer problems that involve phase change (evaporation, sublimation, condensation, melting, etc.) also involve heat transfer, and such problems are analyzed by considering heat and mass transfer simultaneously.

REFERENCES AND SUGGESTED READING

1. American Society of Heating, Refrigeration, and Air Conditioning Engineers. *Handbook of Fundamentals.* Atlanta: ASHRAE, 1993.

2. R. M. Barrer. *Diffusion In and through Solids.* New York: Macmillan, 1941.

3. R. B. Bird. "Theory of Diffusion." *Advances in Chemical Engineering* 1 (1956), p. 170.

4. R. B. Bird, W. E. Stewart, and E. N. Lightfoot. *Transport Phenomena.* New York: John Wiley & Sons, 1960.

5. C. J. Geankoplis. *Mass Transport Phenomena.* New York: Holt, Rinehart, and Winston, 1972.

6. *Handbook of Chemistry and Physics* 56th ed. Cleveland, OH: Chemical Rubber Publishing Co., 1976.

7. J. O. Hirshfelder, F. Curtis, and R. B. Bird. *Molecular Theory of Gases and Liquids.* New York: John Wiley & Sons, 1954.

8. *International Critical Tables.* Vol. 3. New York: McGraw-Hill, 1928.

9. W. M. Kays and M. E. Crawford. *Convective Heat and Mass Transfer.* 2nd ed. New York: McGraw-Hill, 1980.

10. T. R. Marrero and E. A. Mason. "Gaseous Diffusion Coefficients." *Journal of Phys. Chem. Ref. Data* 1 (1972), pp. 3–118.

11. A. F. Mills. *Basic Heat and Mass Transfer.* Burr Ridge, IL: Richard D. Irwin, 1995.

12. J. H. Perry, ed. *Chemical Engineer's Handbook.* 4th ed. New York: McGraw-Hill, 1963.

13. R. D. Reid, J. M. Prausnitz, and T. K. Sherwood. *The Properties of Gases and Liquids.* 3rd ed. New York: McGraw-Hill, 1977.

14. A. H. P. Skelland. *Diffusional Mass Transfer.* New York: John Wiley & Sons, 1974.

15. D. B. Spalding. *Convective Mass Transfer.* New York: McGraw-Hill, 1963.

16. W. F. Stoecker and J. W. Jones. *Refrigeration and Air Conditioning.* New York: McGraw-Hill, 1982.

17. L. C. Thomas. *Mass Transfer Supplement—Heat Transfer.* Englewood Cliffs, NJ: Prentice Hall, 1991.

18. L. Van Black. *Elements of Material Science and Engineering.* Reading, MA: Addison-Wesley, 1980.

PROBLEMS*

Analogy between Heat and Mass Transfer

14–1C How does mass transfer differ from bulk fluid flow? Can mass transfer occur in a homogeneous medium?

14–2C How is the concentration of a commodity defined? How is the concentration gradient defined? How is the diffusion rate of a commodity related to the concentration gradient?

14–3C Give examples for (a) liquid-to-gas, (b) solid-to-liquid, (c) solid-to-gas, and (d) gas-to-liquid mass transfer.

14–4C Someone suggests that thermal (or heat) radiation can also be viewed as mass radiation since, according to Einstein's formula, an energy transfer in the amount of E corresponds to a mass transfer in the amount of $m = E/c^2$. What do you think?

14–5C What is the driving force for (a) heat transfer, (b) electric current flow, (c) fluid flow, and (d) mass transfer?

14–6C What do (a) homogeneous reactions and (b) heterogeneous reactions represent in mass transfer? To what do they correspond in heat transfer?

Mass Diffusion

14–7C Both Fourier's law of heat conduction and Fick's law of mass diffusion can be expressed as $\dot{Q} = -kA(dT/dx)$. What do the quantities $\dot{Q}$, k, A, and T represent in (a) heat conduction and (b) mass diffusion?

14–8C Mark these statements as being True or False for a binary mixture of substances A and B.

____(a) The density of a mixture is always equal to the sum of the densities of its constituents.

____(b) The ratio of the density of component A to the density of component B is equal to the mass fraction of component A.

____(c) If the mass fraction of component A is greater than 0.5, then at least half of the moles of the mixture are component A.

____(d) If the molar masses of A and B are equal to each other, then the mass fraction of A will be equal to the mole fraction of A.

____(e) If the mass fractions of A and B are both 0.5, then the molar mass of the mixture is simply the arithmetic average of the molar masses of A and B.

*Problems designated by a "C" are concept questions, and students are encouraged to answer them all. Problems designated by an "E" are in English units, and the SI users can ignore them. Problems with the icon ⊛ are solved using EES, and complete solutions together with parametric studies are included on the enclosed CD. Problems with the icon 📱 are comprehensive in nature, and are intended to be solved with a computer, preferably using the EES software that accompanies this text.

14–9C Mark these statements as being True or False for a binary mixture of substances A and B.

____(a) The molar concentration of a mixture is always equal to the sum of the molar concentrations of its constituents.

____(b) The ratio of the molar concentration of A to the molar concentration of B is equal to the mole fraction of component A.

____(c) If the mole fraction of component A is greater than 0.5, then at least half of the mass of the mixture is component A.

____(d) If both A and B are ideal gases, then the pressure fraction of A is equal to its mole fraction.

____(e) If the mole fractions of A and B are both 0.5, then the molar mass of the mixture is simply the arithmetic average of the molar masses of A and B.

14–10C Fick's law of diffusion is expressed on the mass and mole basis as $\dot{m}_{diff,\,A} = -\rho A D_{AB}(dw_A/dx)$ and $\dot{N}_{diff,\,A} = -C A D_{AB}(dy_A/dx)$, respectively. Are the diffusion coefficients D_{AB} in the two relations the same or different?

14–11C How does the mass diffusivity of a gas mixture change with (a) temperature and (b) pressure?

14–12C At a given temperature and pressure, do you think the mass diffusivity of air in water vapor will be equal to the mass diffusivity of water vapor in air? Explain.

14–13C At a given temperature and pressure, do you think the mass diffusivity of copper in aluminum will be equal to the mass diffusivity of aluminum in copper? Explain.

14–14C In a mass production facility, steel components are to be hardened by carbon diffusion. Would you carry out the hardening process at room temperature or in a furnace at a high temperature, say 900°C? Why?

14–15C Someone claims that the mass and the mole fractions for a mixture of CO_2 and N_2O gases are identical. Do you agree? Explain.

14–16 Determine the maximum mass fraction of calcium bicarbonate [$Ca(HCO_3)_2$)] in water at 350 K.
Answer: 0.152

14–17 The composition of moist air is given on a molar basis to be 78 percent N_2, 20 percent O_2, and 2 percent water vapor. Determine the mass fractions of the constituents of air.
Answers: 76.4 percent N_2, 22.4 percent O_2, 1.2 percent H_2O

14–18E A gas mixture consists of 7 lbm of O_2, 8 lbm of N_2, and 10 lbm of CO_2. Determine (a) the mass fraction of each component, (b) the mole fraction of each component, and (c) the average molar mass of the mixture.

14–19 A gas mixture consists of 8 kmol of H_2 and 2 kmol of N_2. Determine the mass of each gas and the apparent gas constant of the mixture.

14–20 The molar analysis of a gas mixture at 290 K and 250 kPa is 65 percent N_2, 20 percent O_2, and 15 percent CO_2. Determine the mass fraction and partial pressure of each gas.

14–21 Determine the binary diffusion coefficient of CO_2 in air at (a) 200 K and 1 atm, (b) 400 K and 0.5 atm, and (c) 600 K and 5 atm.

14–22 Repeat Prob. 14–21 for O_2 in N_2.

14–23E The relative humidity of air at 80°F and 14.7 psia is increased from 30 percent to 90 percent during a humidification process at constant temperature and pressure. Determine the percent error involved in assuming the density of air to have remained constant. *Answer:* 2.1 percent

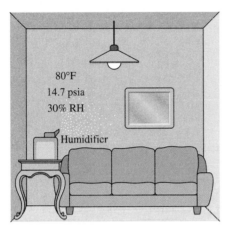

80°F
14.7 psia
30% RH

Humidifier

FIGURE P14–23E

14–24 The diffusion coefficient of hydrogen in steel is given as a function of temperature as

$$D_{AB} = 1.65 \times 10^{-6} \exp(-4630/T) \quad (m^2/s)$$

where T is in K. Determine the diffusion coefficients at 200 K, 500 K, 1000 K, and 1500 K.

14–25 Reconsider Prob. 14–24. Using EES (or other) software, plot the diffusion coefficient as a function of the temperature in the range of 200 K to 1200 K.

Boundary Conditions

14–26C Write three boundary conditions for mass transfer (on a mass basis) for species A at $x = 0$ that correspond to specified temperature, specified heat flux, and convection boundary conditions in heat transfer.

14–27C What is an impermeable surface in mass transfer? How is it expressed mathematically (on a mass basis)? To what does it correspond in heat transfer?

14–28C Consider the free surface of a lake exposed to the atmosphere. If the air at the lake surface is saturated, will the mole fraction of water vapor in air at the lake surface be the same as the mole fraction of water in the lake (which is nearly 1)?

14–29C When prescribing a boundary condition for mass transfer at a solid–gas interface, why do we need to specify the side of the surface (whether the solid or the gas side)? Why did we not do it in heat transfer?

14–30C Using properties of saturated water, explain how you would determine the mole fraction of water vapor at the surface of a lake when the temperature of the lake surface and the atmospheric pressure are specified.

14–31C Using solubility data of a solid in a specified liquid, explain how you would determine the mass fraction of the solid in the liquid at the interface at a specified temperature.

14–32C Using solubility data of a gas in a solid, explain how you would determine the molar concentration of the gas in the solid at the solid–gas interface at a specified temperature.

14–33C Using Henry's constant data for a gas dissolved in a liquid, explain how you would determine the mole fraction of the gas dissolved in the liquid at the interface at a specified temperature.

14–34C What is permeability? How is the permeability of a gas in a solid related to the solubility of the gas in that solid?

14–35 Determine the mole fraction of carbon dioxide (CO_2) dissolved in water at the surface of water at 300 K. The mole fraction of CO_2 in air is 0.005, and the local atmosphere pressure is 100 kPa.

14–36E Determine the mole fraction of the water vapor at the surface of a lake whose temperature at the surface is 70°F, and compare it to the mole fraction of water in the lake. Take the atmospheric pressure at lake level to be 13.8 psia.

14–37 Determine the mole fraction of dry air at the surface of a lake whose temperature is 15°C. Take the atmospheric pressure at lake level to be 100 kPa. *Answer:* 98.3 percent

14–38 Reconsider Prob. 14–37. Using EES (or other) software, plot the mole fraction of dry air at the surface of the lake as a function of the lake temperature as the temperature varies from 5°C to 25°C, and discuss the results.

14–39 Consider a rubber plate that is in contact with nitrogen gas at 298 K and 250 kPa. Determine the molar and mass densities of nitrogen in the rubber at the interface. *Answers:* 0.0039 kmol/m³, 0.1092 kg/m³

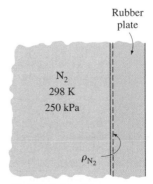

Rubber plate

N_2
298 K
250 kPa

ρ_{N_2}

FIGURE P14–39

14–40 A wall made of natural rubber separates O_2 and N_2 gases at 25°C and 750 kPa. Determine the molar concentrations of O_2 and N_2 in the wall.

14–41 Consider a glass of water in a room at 20°C and 97 kPa. If the relative humidity in the room is 100 percent and the water and the air are in thermal and phase equilibrium, determine (a) the mole fraction of the water vapor in the air and (b) the mole fraction of air in the water.

14–42E Water is sprayed into air at 80°F and 14.3 psia, and the falling water droplets are collected in a container on the floor. Determine the mass and mole fractions of air dissolved in the water.

14–43 Consider a carbonated drink in a bottle at 37°C and 130 kPa. Assuming the gas space above the liquid consists of a saturated mixture of CO_2 and water vapor and treating the drink as water, determine (a) the mole fraction of the water vapor in the CO_2 gas and (b) the mass of dissolved CO_2 in a 200-ml drink. *Answers:* (a) 4.9 percent, (b) 0.28 g

CO₂
H₂O

37°C
130 kPa

FIGURE P14–43

Steady Mass Diffusion through a Wall

14–44C Write down the relations for steady one-dimensional heat conduction and mass diffusion through a plane wall, and identify the quantities in the two equations that correspond to each other.

14–45C Consider steady one-dimensional mass diffusion through a wall. Mark these statements as being True or False.

____(a) Other things being equal, the higher the density of the wall, the higher the rate of mass transfer.
____(b) Other things being equal, doubling the thickness of the wall will double the rate of mass transfer.
____(c) Other things being equal, the higher the temperature, the higher the rate of mass transfer.

____(d) Other things being equal, doubling the mass fraction of the diffusing species at the high concentration side will double the rate of mass transfer.

14–46C Consider one-dimensional mass diffusion of species A through a plane wall of thickness L. Under what conditions will the concentration profile of species A in the wall be a straight line?

14–47C Consider one-dimensional mass diffusion of species A through a plane wall. Does the species A content of the wall change during steady mass diffusion? How about during transient mass diffusion?

14–48 Helium gas is stored at 293 K in a 3-m-outer-diameter spherical container made of 5-cm-thick Pyrex. The molar concentration of helium in the Pyrex is 0.00073 $kmol/m^3$ at the inner surface and negligible at the outer surface. Determine the mass flow rate of helium by diffusion through the Pyrex container. *Answer:* 7.2×10^{-15} kg/s

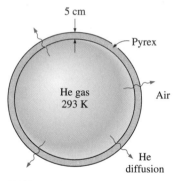

5 cm

Pyrex

He gas
293 K

Air

He
diffusion

FIGURE P14–48

14–49 A thin plastic membrane separates hydrogen from air. The molar concentrations of hydrogen in the membrane at the inner and outer surfaces are determined to be 0.045 and 0.002 $kmol/m^3$, respectively. The binary diffusion coefficient of hydrogen in plastic at the operation temperature is 5.3×10^{-10} m^2/s. Determine the mass flow rate of hydrogen by diffusion through the membrane under steady conditions if the thickness of the membrane is (a) 2 mm and (b) 0.5 mm.

14–50 The solubility of hydrogen gas in steel in terms of its mass fraction is given as $w_{H_2} = 2.09 \times 10^{-4} \exp(-3950/T) P_{H_2}^{0.5}$ where P_{H_2} is the partial pressure of hydrogen in bars and T is the temperature in K. If natural gas is transported in a 1-cm-thick, 3-m-internal-diameter steel pipe at 500 kPa pressure and the mole fraction of hydrogen in the natural gas is 8 percent, determine the highest rate of hydrogen loss through a 100-m-long section of the pipe at steady conditions at a temperature of 293 K if the pipe is exposed to air. Take the diffusivity of hydrogen in steel to be 2.9×10^{-13} m^2/s.
Answer: 3.98×10^{-14} kg/s

14–51 Reconsider Prob. 14–50. Using EES (or other) software, plot the highest rate of hydrogen loss as a function of the mole fraction of hydrogen in natural gas as the mole fraction varies from 5 to 15 percent, and discuss the results.

14–52 Helium gas is stored at 293 K and 500 kPa in a 1-cm-thick, 2-m-inner-diameter spherical tank made of fused silica (SiO_2). The area where the container is located is well ventilated. Determine (a) the mass flow rate of helium by diffusion through the tank and (b) the pressure drop in the tank in one week as a result of the loss of helium gas.

14–53 You probably have noticed that balloons inflated with helium gas rise in the air the first day during a party but they fall down the next day and act like ordinary balloons filled with air. This is because the helium in the balloon slowly leaks out through the wall while air leaks in by diffusion.

Consider a balloon that is made of 0.1-mm-thick soft rubber and has a diameter of 15 cm when inflated. The pressure and temperature inside the balloon are initially 110 kPa and 25°C. The permeability of rubber to helium, oxygen, and nitrogen at 25°C are 9.4×10^{-13}, 7.05×10^{-13}, and 2.6×10^{-13} kmol/m · s · bar, respectively. Determine the initial rates of diffusion of helium, oxygen, and nitrogen through the balloon wall and the mass fraction of helium that escapes the balloon during the first 5 h assuming the helium pressure inside the balloon remains nearly constant. Assume air to be 21 percent oxygen and 79 percent nitrogen by mole numbers and take the room conditions to be 100 kPa and 25°C.

FIGURE P14–53

14–54 Reconsider the balloon discussed in Prob. 14–53. Assuming the volume to remain constant and disregarding the diffusion of air into the balloon, obtain a relation for the variation of pressure in the balloon with time. Using the results obtained and the numerical values given in the problem, determine how long it will take for the pressure inside the balloon to drop to 100 kPa.

14–55 Pure N_2 gas at 1 atm and 25°C is flowing through a 10-m-long, 3-cm-inner diameter pipe made of 2-mm-thick rubber. Determine the rate at which N_2 leaks out of the pipe if the medium surrounding the pipe is (a) a vacuum and (b) atmospheric air at 1 atm and 25°C with 21 percent O_2 and 79 percent N_2.

Answers: (a) 2.28×10^{-10} kmol/s, (b) 4.78×10^{-11} kmol/s

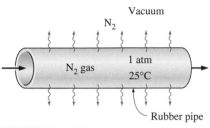

FIGURE P14–55

Water Vapor Migration in Buildings

14–56C Consider a tank that contains moist air at 3 atm and whose walls are permeable to water vapor. The surrounding air at 1 atm pressure also contains some moisture. Is it possible for the water vapor to flow into the tank from surroundings? Explain.

14–57C Express the mass flow rate of water vapor through a wall of thickness L in terms of the partial pressure of water vapor on both sides of the wall and the permeability of the wall to the water vapor.

14–58C How does the condensation or freezing of water vapor in the wall affect the effectiveness of the insulation in the wall? How does the moisture content affect the effective thermal conductivity of soil?

14–59C Moisture migration in the walls, floors, and ceilings of buildings is controlled by vapor barriers or vapor retarders. Explain the difference between the two, and discuss which is more suitable for use in the walls of residential buildings.

14–60C What are the adverse effects of excess moisture on the wood and metal components of a house and the paint on the walls?

14–61C Why are the insulations on the chilled water lines always wrapped with vapor barrier jackets?

14–62C Explain how vapor pressure of the ambient air is determined when the temperature, total pressure, and relative humidity of the air are given.

14–63 Consider a 20-cm-thick brick wall of a house. The indoor conditions are 25°C and 50 percent relative humidity while the outside conditions are 40°C and 50 percent relative humidity. Assuming that there is no condensation or freezing within the wall, determine the amount of moisture flowing through a unit surface area of the wall during a 24-h period.

14–64 The diffusion of water vapor through plaster boards and its condensation in the wall insulation in cold weather are of concern since they reduce the effectiveness of insulation. Consider a house that is maintained at 20°C and 60 percent relative humidity at a location where the atmospheric pressure is 97 kPa. The inside of the walls is finished with 9.5-mm-thick gypsum wallboard. Taking the vapor pressure at the outer side of the wallboard to be zero, determine the maximum amount of water vapor that will diffuse through a 3-m × 8-m section of a wall during a 24-h period. The permeance of the 9.5-mm-thick gypsum wallboard to water vapor is 2.86×10^{-9} kg/s · m² · Pa.

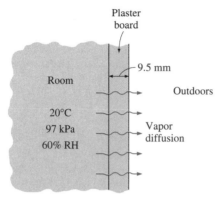

Plaster
board

9.5 mm

Room

Outdoors

20°C
97 kPa
60% RH

Vapor
diffusion

FIGURE P14–64

14–65 Reconsider Prob. 14–64. In order to reduce the migration of water vapor through the wall, it is proposed to use a 0.051-mm-thick polyethylene film with a permeance of 9.1×10^{-12} kg/s · m² · Pa. Determine the amount of water vapor that will diffuse through the wall in this case during a 24-h period.

Answer: 26.4 g

14–66 The roof of a house is 15 m × 8 m and is made of a 20-cm-thick concrete layer. The interior of the house is maintained at 25°C and 50 percent relative humidity and the local atmospheric pressure is 100 kPa. Determine the amount of water vapor that will migrate through the roof in 24 h if the average outside conditions during that period are 3°C and 30 percent relative humidity. The permeability of concrete to water vapor is 24.7×10^{-12} kg/s · m · Pa.

14–67 Reconsider Prob. 14–66. Using EES (or other) software, investigate the effects of temperature and relative humidity of air inside the house on the amount of water vapor that will migrate through the roof. Let the temperature vary from 15°C to 30°C and the relative humidity from 30 to 70 percent. Plot the amount of water vapor that will migrate as functions of the temperature and the relative humidity, and discuss the results.

14–68 Reconsider Prob. 14–66. In order to reduce the migration of water vapor, the inner surface of the wall is painted with vapor retarder latex paint whose permeance is 26×10^{-12} kg/s · m² · Pa. Determine the amount of water vapor that will diffuse through the roof in this case during a 24-h period.

14–69 A glass of milk left on top of a counter in the kitchen at 15°C, 88 kPa, and 50 percent relative humidity is tightly sealed by a sheet of 0.009-mm-thick aluminum foil whose permeance is 2.9×10^{-12} kg/s · m² · Pa. The inner diameter of the glass is 12 cm. Assuming the air in the glass to be saturated at all times, determine how much the level of the milk in the glass will recede in 12 h. *Answer:* 0.00011 mm

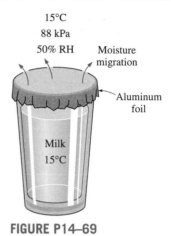

15°C
88 kPa
50% RH

Moisture
migration

Aluminum
foil

Milk
15°C

FIGURE P14–69

Transient Mass Diffusion

14–70C In transient mass diffusion analysis, can we treat the diffusion of a solid into another solid of finite thickness (such as the diffusion of carbon into an ordinary steel component) as a diffusion process in a semi-infinite medium? Explain.

14–71C Define the penetration depth for mass transfer, and explain how it can be determined at a specified time when the diffusion coefficient is known.

14–72C When the density of a species A in a semi-infinite medium is known at the beginning and at the surface, explain how you would determine the concentration of the species A at a specified location and time.

14–73 A steel part whose initial carbon content is 0.12 percent by mass is to be case-hardened in a furnace at 1150 K by exposing it to a carburizing gas. The diffusion coefficient of carbon in steel is strongly temperature dependent, and at the

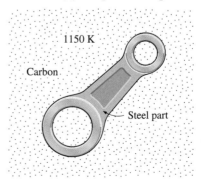

1150 K

Carbon

Steel part

FIGURE P14–73

furnace temperature it is given to be $D_{AB} = 7.2 \times 10^{-12}$ m²/s. Also, the mass fraction of carbon at the exposed surface of the steel part is maintained at 0.011 by the carbon-rich environment in the furnace. If the hardening process is to continue until the mass fraction of carbon at a depth of 0.7 mm is raised to 0.32 percent, determine how long the part should be held in the furnace.   *Answer:* 5.9 h

14–74 Repeat Prob. 14–73 for a furnace temperature of 500 K at which the diffusion coefficient of carbon in steel is $D_{AB} = 2.1 \times 10^{-20}$ m²/s.

14–75 A pond with an initial oxygen content of zero is to be oxygenated by forming a tent over the water surface and filling the tent with oxygen gas at 25°C and 130 kPa. Determine the mole fraction of oxygen at a depth of 1 cm from the surface after 24 h.

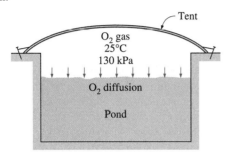

O₂ gas
25°C
130 kPa

O₂ diffusion

Pond

Tent

FIGURE P14–75

14–76 A long nickel bar with a diameter of 5 cm has been stored in a hydrogen-rich environment at 358 K and 300 kPa for a long time, and thus it contains hydrogen gas throughout uniformly. Now the bar is taken into a well-ventilated area so that the hydrogen concentration at the outer surface remains at almost zero at all times. Determine how long it will take for the hydrogen concentration at the center of the bar to drop by half. The diffusion coefficient of hydrogen in the nickel bar at the room temperature of 298 K can be taken to be $D_{AB} = 1.2 \times 10^{-12}$ m²/s.   *Answer:* 3.3 years

Diffusion in a Moving Medium

14–77C Define the following terms: mass-average velocity, diffusion velocity, stationary medium, and moving medium.

14–78C What is diffusion velocity? How does it affect the mass-average velocity? Can the velocity of a species in a moving medium relative to a fixed reference point be zero in a moving medium? Explain.

14–79C What is the difference between mass-average velocity and mole-average velocity during mass transfer in a moving medium? If one of these velocities is zero, will the other also necessarily be zero? Under what conditions will these two velocities be the same for a binary mixture?

14–80C Consider one-dimensional mass transfer in a moving medium that consists of species A and B with $\rho = \rho_A + \rho_B =$ constant. Mark these statements as being True or False.

___(*a*) The rates of mass diffusion of species A and B are equal in magnitude and opposite in direction.

___(*b*) $D_{AB} = D_{BA}$.

___(*c*) During equimolar counterdiffusion through a tube, equal numbers of moles of A and B move in opposite directions, and thus a velocity measurement device placed in the tube will read zero.

___(*d*) The lid of a tank containing propane gas (which is heavier than air) is left open. If the surrounding air and the propane in the tank are at the same temperature and pressure, no propane will escape the tank and no air will enter.

14–81C What is Stefan flow? Write the expression for Stefan's law and indicate what each variable represents.

14–82E The pressure in a pipeline that transports helium gas at a rate of 5 lbm/s is maintained at 14.5 psia by venting helium to the atmosphere through a $\frac{1}{4}$-in internal diameter tube that extends 30 ft into the air. Assuming both the helium and the atmospheric air to be at 80°F, determine (*a*) the mass flow rate of helium lost to the atmosphere through the tube, (*b*) the mass flow rate of air that infiltrates into the pipeline, and (*c*) the flow velocity at the bottom of the tube where it is attached to the pipeline that will be measured by an anemometer in steady operation.

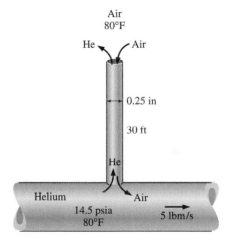

Air
80°F

He   Air

0.25 in

30 ft

He

Helium
14.5 psia
80°F

Air

5 lbm/s

FIGURE P14–82E

14–83E Repeat Prob. 14–82E for a pipeline that transports carbon dioxide instead of helium.

14–84 A tank with a 2-cm-thick shell contains hydrogen gas at the atmospheric conditions of 25°C and 90 kPa. The charging valve of the tank has an internal diameter of 3 cm and extends 8 cm above the tank. If the lid of the tank is left open so that hydrogen and air can undergo equimolar counterdiffusion through the 10-cm-long passageway, determine the mass flow rate of hydrogen lost to the atmosphere through the valve at the initial stages of the process.   *Answer:* 4.20 × 10⁻⁸ kg/s

14–85 Reconsider Prob. 14–84. Using EES (or other) software, plot the mass flow rate of hydrogen lost as a function of the diameter of the charging valve as the diameter varies from 1 cm to 5 cm, and discuss the results.

14–86E A 1-in-diameter Stefan tube is used to measure the binary diffusion coefficient of water vapor in air at 80°F and 13.8 psia. The tube is partially filled with water with a distance from the water surface to the open end of the tube of 10 in. Dry air is blown over the open end of the tube so that water vapor rising to the top is removed immediately and the concentration of vapor at the top of the tube is zero. During 10 days of continuous operation at constant pressure and temperature, the amount of water that has evaporated is measured to be 0.0025 lbm. Determine the diffusion coefficient of water vapor in air at 80°F and 13.8 psia.

14–87 An 8-cm-internal-diameter, 30-cm-high pitcher half filled with water is left in a dry room at 15°C and 87 kPa with its top open. If the water is maintained at 15°C at all times also, determine how long it will take for the water to evaporate completely. *Answer:* 1125 days

Room
15°C
87 kPa
Water vapor

Water
15°C

FIGURE P14–87

14–88 A large tank containing ammonia at 1 atm and 25°C is vented to the atmosphere through a 2-m-long tube whose internal diameter is 1.5 cm. Determine the rate of loss of ammonia and the rate of infiltration of air into the tank.

Mass Convection

14–89C Heat convection is expressed by Newton's law of cooling as $\dot{Q} = hA_s(T_s - T_\infty)$. Express mass convection in an analogous manner on a mass basis, and identify all the quantities in the expression and state their units.

14–90C What is a concentration boundary layer? How is it defined for flow over a plate?

14–91C What is the physical significance of the Schmidt number? How is it defined? To what dimensionless number does it correspond in heat transfer? What does a Schmidt number of 1 indicate?

14–92C What is the physical significance of the Sherwood number? How is it defined? To what dimensionless number does it correspond in heat transfer? What does a Sherwood number of 1 indicate for a plain fluid layer?

14–93C What is the physical significance of the Lewis number? How is it defined? What does a Lewis number of 1 indicate?

14–94C In natural convection mass transfer, the Grashof number is evaluated using density difference instead of temperature difference. Can the Grashof number evaluated this way be used in heat transfer calculations also?

14–95C Using the analogy between heat and mass transfer, explain how the mass transfer coefficient can be determined from the relations for the heat transfer coefficient.

14–96C It is well known that warm air in a cooler environment rises. Now consider a warm mixture of air and gasoline (C_8H_{18}) on top of an open gasoline can. Do you think this gas mixture will rise in a cooler environment?

14–97C Consider two identical cups of coffee, one with no sugar and the other with plenty of sugar at the bottom. Initially, both cups are at the same temperature. If left unattended, which cup of coffee will cool faster?

14–98C Under what conditions will the normalized velocity, thermal, and concentration boundary layers coincide during flow over a flat plate?

14–99C What is the relation ($f/2$) Re = Nu = Sh known as? Under what conditions is it valid? What is the practical importance of it?

14–100C What is the name of the relation $f/2 = St\ Pr^{2/3} = St_{mass}Sc^{2/3}$ and what are the names of the variables in it? Under what conditions is it valid? What is the importance of it in engineering?

14–101C What is the relation $h_{heat} = \rho c_p h_{mass}$ known as? For what kind of mixtures is it valid? What is the practical importance of it?

14–102C What is the low mass flux approximation in mass transfer analysis? Can the evaporation of water from a lake be treated as a low mass flux process?

14–103 Air at 40°C and 1 atm flows over a 5-m-long wet plate with an average velocity of 2.5 m/s in order to dry the surface. Using the analogy between heat and mass transfer, determine the mass transfer coefficient on the plate.

14–104E Consider a circular pipe of inner diameter $D = 0.7$ in. whose inner surface is covered with a thin layer of liquid water as a result of condensation. In order to dry the pipe, air at

540 R and 1 atm is forced to flow through it with an average velocity of 6 ft/s. Using the analogy between heat and mass transfer, determine the mass transfer coefficient inside the pipe for fully developed flow.   *Answer:* 0.017 ft/s

14–105   The average heat transfer coefficient for air flow over an odd-shaped body is to be determined by mass transfer measurements and using the Chilton–Colburn analogy between heat and mass transfer. The experiment is conducted by blowing dry air at 1 atm at a free stream velocity of 2 m/s over a body covered with a layer of naphthalene. The surface area of the body is 0.75 m², and it is observed that 100 g of naphthalene has sublimated in 45 min. During the experiment, both the body and the air were kept at 25°C, at which the vapor pressure and mass diffusivity of naphthalene are 11 Pa and $D_{AB} = 0.61 \times 10^{-5}$ m²/s, respectively. Determine the heat transfer coefficient under the same flow conditions over the same geometry.

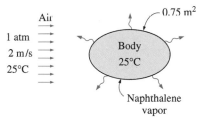

FIGURE P14–105

14–106   Consider a 15-cm-internal-diameter, 10-m-long circular duct whose interior surface is wet. The duct is to be dried by forcing dry air at 1 atm and 15°C through it at an average velocity of 3 m/s. The duct passes through a chilled room, and it remains at an average temperature of 15°C at all times. Determine the mass transfer coefficient in the duct.

14–107   Reconsider Prob. 14–106. Using EES (or other) software, plot the mass transfer coefficient as a function of the air velocity as the velocity varies from 1 m/s to 8 m/s, and discuss the results.

14–108   Dry air at 15°C and 85 kPa flows over a 2-m-long wet surface with a free stream velocity of 3 m/s. Determine the average mass transfer coefficient.   *Answer:* 0.00463 m/s

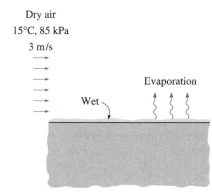

FIGURE P14–108

14–109   Consider a 5-m × 5-m wet concrete patio with an average water film thickness of 0.3 mm. Now wind at 50 km/h is blowing over the surface. If the air is at 1 atm, 15°C, and 35 percent relative humidity, determine how long it will take for the patio to dry completely.   *Answer:* 18.6 min

14–110E   A 2-in-diameter spherical naphthalene ball is suspended in a room at 1 atm and 80°F. Determine the average mass transfer coefficient between the naphthalene and the air if air is forced to flow over naphthalene with a free stream velocity of 15 ft/s. The Schmidt number of naphthalene in air at room temperature is 2.35.   *Answer:* 0.0524 ft/s

14–111   Consider a 3-mm-diameter raindrop that is falling freely in atmospheric air at 25°C. Taking the temperature of the raindrop to be 9°C, determine the terminal velocity of the raindrop at which the drag force equals the weight of the drop and the average mass transfer coefficient at that time.

14–112   In a manufacturing facility, 40 cm × 40 cm wet brass plates coming out of a water bath are to be dried by passing them through a section where dry air at 1 atm and 25°C is blown parallel to their surfaces at 4 m/s. If the plates are at 15°C and there are no dry spots, determine the rate of evaporation from both sides of a plate.

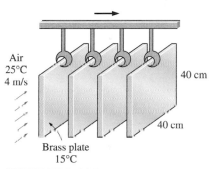

FIGURE P14–112

14–113E   Air at 80°F, 1 atm, and 30 percent relative humidity is blown over the surface of a 15-in × 15-in square pan filled with water at a free stream velocity of 10 ft/s. If the water is maintained at a uniform temperature of 80°F, determine the rate of evaporation of water and the amount of heat that needs to be supplied to the water to maintain its temperature constant.

14–114E   Repeat Prob. 14–113E for temperature of 60°F for both the air and water.

Simultaneous Heat and Mass Transfer

14–115C   Does a mass transfer process have to involve heat transfer? Describe a process that involves both heat and mass transfer.

14–116C   Consider a shallow body of water. Is it possible for this water to freeze during a cold and dry night even when the ambient air and surrounding surface temperatures never drop to 0°C? Explain.

14–117C During evaporation from a water body to air, under what conditions will the latent heat of vaporization be equal to convection heat transfer from the air?

14–118 Jugs made of porous clay were commonly used to cool water in the past. A small amount of water that leaks out keeps the outer surface of the jug wet at all times, and hot and relatively dry air flowing over the jug causes this water to evaporate. Part of the latent heat of evaporation comes from the water in the jug, and the water is cooled as a result. If the environment conditions are 1 atm, 30°C, and 35 percent relative humidity, determine the temperature of the water when steady conditions are reached.

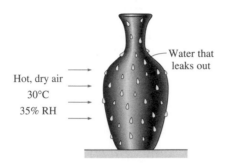

Hot, dry air
30°C
35% RH

Water that leaks out

FIGURE P14–118

14–119 Reconsider Prob. 14–118. Using EES (or other) software, plot the water temperature as a function of the relative humidity of air as the relative humidity varies from 10 to 100 percent, and discuss the results.

14–120E During a hot summer day, a 2-L bottle drink is to be cooled by wrapping it in a cloth kept wet continually and blowing air to it with a fan. If the environment conditions are 1 atm, 80°F, and 30 percent relative humidity, determine the temperature of the drink when steady conditions are reached.

14–121 A glass bottle washing facility uses a well agitated hot water bath at 55°C with an open top that is placed on the ground. The bathtub is 1 m high, 2 m wide, and 4 m long and is made of sheet metal so that the outer side surfaces are also at about 55°C. The bottles enter at a rate of 800 per minute at ambient temperature and leave at the water temperature. Each bottle has a mass of 150 g and removes 0.6 g of water as it leaves the bath wet. Makeup water is supplied at 15°C. If the average conditions in the plant are 1 atm, 25°C, and 50 percent relative humidity, and the average temperature of the surrounding surfaces is 15°C, determine (a) the amount of heat and water removed by the bottles themselves per second, (b) the rate of heat loss from the top surface of the water bath by radiation, natural convection, and evaporation, (c) the rate of heat loss from the side surfaces by natural convection and radiation, and (d) the rate at which heat and water must be supplied to maintain steady operating conditions. Disregard heat loss through the bottom surface of the bath and take

the emissivities of sheet metal and water to be 0.61 and 0.95, respectively.
Answers: (a) 61.3 kW, 28.8 kg/h, (b) 14.2 kW, (c) 3.22 kW, (d) 80.9 kW, 45.1 kg/h

14–122 Repeat Prob. 14–121 for a water bath temperature of 50°C.

14–123 One way of increasing heat transfer from the head on a hot summer day is to wet it. This is especially effective in windy weather, as you may have noticed. Approximating the head as a 30-cm-diameter sphere at 30°C with an emissivity of 0.95, determine the total rate of heat loss from the head at ambient air conditions of 1 atm, 25°C, 30 percent relative humidity, and 25 km/h winds if the head is (a) dry and (b) wet. Take the surrounding temperature to be 25°C.
Answers: (a) 40.5 W, (b) 385 W

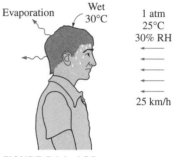

Evaporation
Wet
30°C

1 atm
25°C
30% RH

25 km/h

FIGURE P14–123

14–124 A 2-m-deep 20-m × 20-m heated swimming pool is maintained at a constant temperature of 30°C at a location where the atmospheric pressure is 1 atm. If the ambient air is at 20°C and 60 percent relative humidity and the effective sky temperature is 0°C, determine the rate of heat loss from the top surface of the pool by (a) radiation, (b) natural convection, and (c) evaporation. (d) Assuming the heat losses to the ground to be negligible, determine the size of the heater.

14–125 Repeat Prob. 14–124 for a pool temperature of 25°C.

Review Problems

14–126C Mark these statements as being True or False.
_____ (a) The units of mass diffusivity, heat diffusivity, and momentum diffusivity are all the same.
_____ (b) If the molar concentration (or molar density) C of a mixture is constant, then its density ρ must also be constant.
_____ (c) If the mass-average velocity of a binary mixture is zero, then the mole-average velocity of the mixture must also be zero.
_____ (d) If the mole fractions of A and B of a mixture are both 0.5, then the molar mass of the mixture is simply the arithmetic average of the molar masses of A and B.

14–127 Using Henry's law, show that the dissolved gases in a liquid can be driven off by heating the liquid.

14–128 Show that for an ideal gas mixture maintained at a constant temperature and pressure, the molar concentration C of the mixture remains constant but this is not necessarily the case for the density ρ of the mixture.

14–129E A gas mixture in a tank at 600 R and 20 psia consists of 1 lbm of CO_2 and 3 lbm of CH_4. Determine the volume of the tank and the partial pressure of each gas.

14–130 Dry air whose molar analysis is 78.1 percent N_2, 20.9 percent O_2, and 1 percent Ar flows over a water body until it is saturated. If the pressure and temperature of air remain constant at 1 atm and 25°C during the process, determine (a) the molar analysis of the saturated air and (b) the density of air before and after the process. What do you conclude from your results?

14–131 Consider a glass of water in a room at 20°C and 100 kPa. If the relative humidity in the room is 70 percent and the water and the air are at the same temperature, determine (a) the mole fraction of the water vapor in the room air, (b) the mole fraction of the water vapor in the air adjacent to the water surface, and (c) the mole fraction of air in the water near the surface.

Answers: (a) 1.64 percent, *(b)* 2.34 percent, *(c)* 0.0015 percent

20°C
100 kPa
70% RH

Air-water interface

Water
20°C

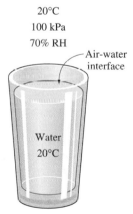

FIGURE P14–131

14–132 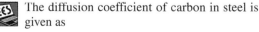 The diffusion coefficient of carbon in steel is given as

$$D_{AB} = 2.67 \times 10^{-5} \exp(-17{,}400/T) \qquad (m^2/s)$$

where T is in K. Determine the diffusion coefficient from 300 K to 1500 K in 100 K increments and plot the results.

14–133 A carbonated drink is fully charged with CO_2 gas at 17°C and 600 kPa such that the entire bulk of the drink is in thermodynamic equilibrium with the CO_2–water vapor mixture. Now consider a 2-L soda bottle. If the CO_2 gas in that

bottle were to be released and stored in a container at 25°C and 100 kPa, determine the volume of the container.

Answer: 12.7 L

CO_2
Water

SODA

FIGURE P14–133

14–134 Consider a brick house that is maintained at 20°C and 60 percent relative humidity at a location where the atmospheric pressure is 85 kPa. The walls of the house are made of 20-cm-thick brick whose permeance is 23×10^{-12} kg/s · m² · Pa. Taking the vapor pressure at the outer side of the wallboard to be zero, determine the maximum amount of water vapor that will diffuse through a 3-m × 5-m section of a wall during a 24-h period.

14–135E Consider a masonry cavity wall that is built around 6-in-thick concrete blocks. The outside is finished with 4-in face brick with $\frac{1}{2}$-in cement mortar between the bricks and concrete blocks. The inside finish consists of $\frac{1}{2}$-in gypsum wallboard separated from the concrete block by $\frac{3}{4}$-in-thick air space.

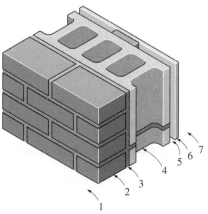

FIGURE P14–135E

The thermal and vapor resistances of various components for a unit wall area are as follows:

Construction	R-Value, h · ft^2 · °F/Btu	R_v-Value, s · ft^2 · psi/lbm
1. Outside surface, 15 mph wind	0.17	—
2. Face brick, 4 in	0.43	15,000
3. Cement mortar, 0.5 in	0.10	1930
4. Concrete block, 6-in	4.20	23,000
5. Air space, $\frac{3}{4}$-in	1.02	77.6
6. Gypsum wallboard, 0.5 in	0.45	332
7. Inside surface, still air	0.68	—

The indoor conditions are 70°F and 65 percent relative humidity while the outdoor conditions are 32°F and 40 percent relative humidity. Determine the rates of heat and water vapor transfer through a 9-ft × 25-ft section of the wall.
Answers: 1213 Btu/h, 4.03 lbm/h

14–136 The oxygen needs of fish in aquariums are usually met by forcing air to the bottom of the aquarium by a compressor. The air bubbles provide a large contact area between the water and the air, and as the bubbles rise, oxygen and nitrogen gases in the air dissolve in water while some water evaporates into the bubbles. Consider an aquarium that is maintained at room temperature of 25°C at all times. The air bubbles are observed to rise to the free surface of water in 2 s. If the air entering the aquarium is completely dry and the diameter of the air bubbles is 4 mm, determine the mole fraction of water vapor at the center of the bubble when it leaves the aquarium. Assume no fluid motion in the bubble so that water vapor propagates in the bubble by diffusion only. *Answer:* 3.13 percent

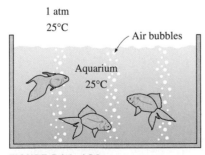

1 atm
25°C
Air bubbles
Aquarium
25°C

FIGURE P14–136

14–137 Oxygen gas is forced into an aquarium at 1 atm and 25°C, and the oxygen bubbles are observed to rise to the free surface in 4 s. Determine the penetration depth of oxygen into water from a bubble during this time period.

14–138 Consider a 30-cm-diameter pan filled with water at 15°C in a room at 20°C, 1 atm, and 30 percent relative humidity. Determine (*a*) the rate of heat transfer by convection, (*b*) the rate of evaporation of water, and (*c*) the rate of heat transfer to the water needed to maintain its temperature at 15°C. Disregard any radiation effects.

14–139 Repeat Prob. 14–138 assuming a fan blows air over the water surface at a velocity of 3 m/s. Take the radius of the pan to be the characteristic length.

14–140 Naphthalene is commonly used as a repellent against moths to protect clothing during storage. Consider a 1.5-cm-diameter spherical naphthalene ball hanging in a closet at 25°C and 1 atm. Considering the variation of diameter with time, determine how long it will take for the naphthalene to sublimate completely. The density and vapor pressure of naphthalene at 25°C are 1100 kg/m^3 and 11 Pa, respectively, and the mass diffusivity of naphthalene in air at 25°C is $D_{AB} = 0.61 \times 10^{-5}$ m^2/s. *Answer:* 103 days

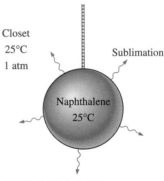

Closet
25°C
1 atm
Sublimation
Naphthalene
25°C

FIGURE P14–140

14–141E A swimmer extends his wet arms into the windy air outside at 1 atm, 40°F, 50 percent relative humidity, and 20 mph. If the average skin temperature is 80°F, determine the rate at which water evaporates from both arms and the corresponding rate of heat transfer by evaporation. The arm can be modeled as a 2-ft-long and 3-in-diameter cylinder with adiabatic ends.

14–142 A thick part made of nickel is put into a room filled with hydrogen at 3 atm and 85°C. Determine the hydrogen concentration at a depth of 2-mm from the surface after 24 h. *Answer:* 4.1×10^{-7} kmol/m^3

14–143 A membrane made of 0.1-mm-thick soft rubber separates pure O_2 at 1 atm and 25°C from air at 3 atm pressure. Determine the mass flow rate of O_2 through the membrane per unit area and the direction of flow.

14–144E The top section of an 8-ft-deep 100-ft × 100-ft heated solar pond is maintained at a constant temperature of 80°F at a location where the atmospheric pressure is 1 atm. If the ambient air is at 70°F and 100 percent relative humidity and wind is blowing at an average velocity of 40 mph, determine the rate of heat loss from the top surface of the pond by

(a) forced convection, (b) radiation, and (c) evaporation. Take the average temperature of the surrounding surfaces to be 60°F.

14–145E Repeat Prob. 14–144E for a solar pond surface temperature of 90°F.

Answers: (a) 299,400 Btu/h, (b) 1,060,000 Btu/h, (c) 3,410,000 Btu/h

14–146 Liquid toluene ($C_6H_5CH_3$) was stored at 6.4°C in an open top 20-cm-diameter cylindrical container. The vapor pressure of toluene at 6.4°C is 10 mm Hg. A gentle stream of fresh air at 6.4°C and 101.3 kPa was allowed to flow over the open end of the container. The rate of evaporation of toluene into air was measured to be 60 g/day. Estimate the concentration of toluene (in g/m^3) at exactly 10 mm above the liquid surface. The diffusion coefficient of toluene at 25° C is $D_{AB} = 0.084 \times 10^{-4}$ m^2/s.

14–147 In an experiment, a sphere of crystalline sodium chloride (NaCl) was suspended in a stirred tank filled with water at 20°C. Its initial mass was 100 g. In 10 minutes, the mass of sphere was found to have decreased by 10 percent. The density of NaCl is 2160 kg/m^3. Its solubility in water at 20°C is 320 kg/m^3. Use these results to obtain an average value for the mass transfer coefficient.

14–148 Benzene-free air at 25°C and 101.3 kPa enters a 5-cm-diameter tube at an average velocity of 5 m/s. The inner surface of the 6-m-long tube is coated with a thin film of pure benzene at 25°C. The vapor pressure of benzene (C_6H_6) at 25°C is 13 kPa, and the solubility of air in benezene is assumed to be negligible. Calculate (a) the average mass transfer coefficient in m/s, (b) the molar concentration of benzene in the outlet air, and (c) the evaporation rate of benzene in kg/h.

14–149 Air at 52°C, 101.3 kPa, and 10 percent relative humidity enters a 5-cm-diameter tube with an average velocity of 5 m/s. The tube inner surface is wetted uniformly with water, whose vapor pressure at 52°C is 13.6 kPa. While the temperature and pressure of air remain constant, the partial pressure of vapor in the outlet air is increased to 10 kPa. Determine (a) the average mass transfer coefficient in m/s, (b) the log-mean driving force for mass transfer in molar concentration units, (c) the water evaporation rate in kg/h, and (d) the length of the tube.

14–150 The following experiment was performed to measure the mass diffusivity of *n*-octane (C_8H_{18}, M = 114.2 kg/kmol) in air. Pure liquid n-octane was placed in a vertical tube, 5 cm in diameter. With the tube bottom closed, its top was exposed to a gentle cross flow of air (free of *n*-octane). The entire system was allowed to reach a steady state at 20°C and 101.3 kPa, while maintaining the distance between the tube-top and the liquid-surface constant at 10 cm, It was observed that, after 38 hours, 1.0 g of *n*-octane had evaporated. At 20°C, the vapor pressure and density of liquid n-octane are 1.41 kPa and 703 kg/m^3, respectively. Calculate the mass diffusivity of *n*-octane in air at 20°C and 101.3 kPa.

14–151 A sphere of ice, 5 cm in diameter, is exposed to 50 km/h wind with 10 percent relative humidity. Both the ice sphere and air are at −1°C and 90 kPa. Predict the rate of evaporation of the ice in g/h by use of the following correlation for single spheres: Sh = [4.0 + 1.21(ReSc)$^{2/3}$]$^{0.5}$. Data at −1°C and 90 kPa: $D_{\text{air-H}_2O} = 2.5 \times 10^{-5}$ m^2/s^3, kinematic viscosity (air) = 1.32×10^{-7} m^2/s, vapor pressure (H_2O) = 0.56 kPa and density (ice) = 915 kg/m^3.

Fundamentals of Engineering (FE) Exam Problems

14–152 When the _____ is unity, one can expect the momentum and mass transfer by diffusion to be the same.
(a) Grashof (b) Reynolds (c) Lewis
(d) Schmidt (e) Sherwood

14–153 The basic equation describing the diffusion of one medium through another stationary medium is

(a) $j_A = -CD_{AB}\dfrac{d(C_A/C)}{dx}$ (b) $j_A = -D_{AB}\dfrac{d(C_A/C)}{dx}$

(c) $j_A = -k\dfrac{d(C_A/C)}{dx}$ (d) $j_A = -k\dfrac{dT}{dx}$

(e) none of them

14–154 For the absorption of a gas (like carbon dioxide) into a liquid (like water) Henry's law states that partial pressure of the gas is proportional to the mole fraction of the gas in the liquid–gas solution with the constant of proportionality being Henry's constant. A bottle of soda pop (CO_2-H_2O) at room temperature has a Henry's constant of 17,100 kPa. If the pressure in this bottle is 120 kPa and the partial pressure of the water vapor in the gas volume at the top of the bottle is neglected, the concentration of the CO_2 in the liquid H_2O is
(a) 0.003 mol-CO_2/mol (b) 0.007 mol-CO_2/mol
(c) 0.013 mol-CO_2/mol (d) 0.022 mol-CO_2/mol
(e) 0.047 mol-CO_2/mol

14–155 A recent attempt to circumnavigate the world in a balloon used a helium-filled balloon whose volume was 7240 m^3 and surface area was 1800 m^2. The skin of this balloon is 2 mm thick and is made of a material whose helium diffusion coefficient is 1×10^{-9} m^2/s. The molar concentration of the helium at the inner surface of the balloon skin is 0.2 kmol/m^3 and the molar concentration at the outer surface is extremely small. The rate at which helium is lost from this balloon is
(a) 0.26 kg/h (b) 1.5 kg/h (c) 2.6 kg/h
(d) 3.8 kg/h (e) 5.2 kg/h

14–156 A rubber object is in contact with nitrogen (N_2) at 298 K and 250 kPa. The solubility of nitrogen gas in rubber is 0.00156 kmol/m^3 · bar. The mass density of nitrogen at the interface is
(a) 0.049 kg/m^3 (b) 0.064 kg/m^3 (c) 0.077 kg/m^3
(d) 0.092 kg/m^3 (e) 0.109 kg/m^3

14–157 Nitrogen gas at high pressure and 298 K is contained in a 2-m × 2-m × 2-m cubical container made of natural rubber whose walls are 4 cm thick. The concentration of nitrogen in the rubber at the inner and outer surfaces are 0.067 kg/m^3 and 0.009 kg/m^3, respectively. The diffusion coefficient of nitrogen through rubber is 1.5×10^{-10} m^2/s. The mass flow rate of nitrogen by diffusion through the cubical container is

(a) 8.24×10^{-10} kg/s   (b) 1.35×10^{-10} kg/s
(c) 5.22×10^{-9} kg/s   (d) 9.71×10^{-9} kg/s
(e) 3.58×10^{-8} kg/s

14–158 Carbon at 1273 K is contained in a 7-cm inner-diameter cylinder made of iron whose thickness is 1.2 mm. The concentration of carbon in the iron at the inner surface is 0.5 kg/m^3 and the concentration of carbon in the iron at the outer surface is negligible. The diffusion coefficient of carbon through iron is 3×10^{-11} m^2/s. The mass flow rate carbon by diffusion through the cylinder shell per unit length of the cylinder is

(a) 2.8×10^{-9} kg/s   (b) 5.4×10^{-9} kg/s
(c) 8.8×10^{-9} kg/s   (d) 1.6×10^{-8} kg/s
(e) 5.2×10^{-8} kg/s

14–159 The surface of an iron component is to be hardened by carbon. The diffusion coefficient of carbon in iron at 1000°C is given to be 3×10^{-11} m^2/s. If the penetration depth of carbon in iron is desired to be 1.0 mm, the hardening process must take at least

(a) 1.10 h   (b) 1.47 h   (c) 1.86 h   (d) 2.50 h   (e) 2.95 h

14–160 Saturated water vapor at 25°C ($P_{sat} = 3.17$ kPa) flows in a pipe that passes through air at 25°C with a relative humidity of 40 percent. The vapor is vented to the atmosphere through a 7-mm internal-diameter tube that extends 10 m into the air. The diffusion coefficient of vapor through air is 2.5×10^{-5} m^2/s. The amount of water vapor lost to the atmosphere through this individual tube by diffusion is

(a) 1.02×10^{-6} kg   (b) 1.37×10^{-6} kg
(c) 2.28×10^{-6} kg   (d) 4.13×10^{-6} kg
(e) 6.07×10^{-6} kg

14–161 Air flows in a 4-cm-diameter wet pipe at 20°C and 1 atm with an average velocity of 4 m/s in order to dry the surface. The Nusselt number in this case can be determined from Nu = $0.023Re^{0.8}Pr^{0.4}$ where Re = 10,550 and Pr = 0.731. Also, the diffusion coefficient of water vapor in air is 2.42×10^{-5} m^2/s. Using the analogy between heat and mass transfer, the mass transfer coefficient inside the pipe for fully developed flow becomes

(a) 0.0918 m/s   (b) 0.0408 m/s   (c) 0.0366 m/s
(d) 0.0203 m/s   (e) 0.0022 m/s

14–162 Air flows through a wet pipe at 298 K and 1 atm, and the diffusion coefficient of water vapor in air is 2.5×10^{-5} m^2/s. If the heat transfer coefficient is determined to be 35 W/m$^2 \cdot$ °C, the mass transfer coefficient is

(a) 0.0326 m/s   (b) 0.0387 m/s   (c) 0.0517 m/s
(d) 0.0583 m/s   (e) 0.0707 m/s

14–163 A natural gas (methane, CH_4) storage facility uses 3-cm-diameter by 6-m-long vent tubes on its storage tanks to keep the pressure in these tanks at atmospheric value. If the diffusion coefficient for methane in air is 0.2×10^{-4} m^2/s and the temperature of the tank and environment is 300 K, the rate at which natural gas is lost from a tank through one vent tube is

(a) 13×10^{-5} kg/day   (b) 3.2×10^{-5} kg/day
(c) 8.7×10^{-5} kg/day   (d) 5.3×10^{-5} kg/day
(e) 0.12×10^{-5} kg/day

Design, and Essay Problems

14–164 Write an essay on diffusion caused by effects other than the concentration gradient such as thermal diffusion, pressure diffusion, forced diffusion, knodsen diffusion, and surface diffusion.

14–165 Write a computer program that will convert the mole fractions of a gas mixture to mass fractions when the molar masses of the components of the mixture are specified.

14–166 One way of generating electricity from solar energy involves the collection and storage of solar energy in large artificial lakes of a few meters deep, called solar ponds. Solar energy is stored at the bottom part of the pond at temperatures close to boiling, and the rise of hot water to the top is prevented by planting salt to the bottom of the pond. Write an essay on the operation of solar pond power plants, and find out how much salt is used per year per m^2. If the cost is not a factor, can sugar be used instead of salt to maintain the concentration gradient? Explain.

14–167 The condensation and even freezing of moisture in building walls without effective vapor retarders is a real concern in cold climates as it undermines the effectiveness of the insulation. Investigate how the builders in your area are coping with this problem, whether they are using vapor retarders or vapor barriers in the walls, and where they are located in the walls. Prepare a report on your findings and explain the reasoning for the current practice.

14–168 You are asked to design a heating system for a swimming pool that is 2 m deep, 25 m long, and 25 m wide. Your client desires that the heating system be large enough to raise the water temperature from 20°C to 30°C in 3 h. The heater must also be able to maintain the pool at 30°C at the outdoor design conditions of 15°C, 1 atm, 35 percent relative humidity, 40 mph winds, and effective sky temperature of 10°C. Heat losses to the ground are expected to be small and can be disregarded. The heater considered is a natural gas furnace whose efficiency is 80 percent. What heater size (in Btu/h input) would you recommend that your client buy?

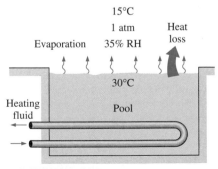

FIGURE P14–168

PROPERTY TABLES AND CHARTS (SI UNITS)

TABLE A–1

Molar mass, gas constant, and ideal-gas specific heats of some substances

Substance	Molar Mass M, kg/kmol	Gas Constant R, kJ/kg · K*	Specific Heat Data at 25°C		
			c_p, kJ/kg · K	c_v, kJ/kg · K	$k = c_p/c_v$
Air	28.97	0.2870	1.005	0.7180	1.400
Ammonia, NH_3	17.03	0.4882	2.093	1.605	1.304
Argon, Ar	39.95	0.2081	0.5203	0.3122	1.667
Bromine, Br_2	159.81	0.05202	0.2253	0.1732	1.300
Isobutane, C_4H_{10}	58.12	0.1430	1.663	1.520	1.094
n-Butane, C_4H_{10}	58.12	0.1430	1.694	1.551	1.092
Carbon dioxide, CO_2	44.01	0.1889	0.8439	0.6550	1.288
Carbon monoxide, CO	28.01	0.2968	1.039	0.7417	1.400
Chlorine, Cl_2	70.905	0.1173	0.4781	0.3608	1.325
Chlorodifluoromethane (R-22), $CHClF_2$	86.47	0.09615	0.6496	0.5535	1.174
Ethane, C_2H_6	30.070	0.2765	1.744	1.468	1.188
Ethylene, C_2H_4	28.054	0.2964	1.527	1.231	1.241
Fluorine, F_2	38.00	0.2187	0.8237	0.6050	1.362
Helium, He	4.003	2.077	5.193	3.116	1.667
n-Heptane, C_7H_{16}	100.20	0.08297	1.649	1.566	1.053
n-Hexane, C_6H_{14}	86.18	0.09647	1.654	1.558	1.062
Hydrogen, H_2	2.016	4.124	14.30	10.18	1.405
Krypton, Kr	83.80	0.09921	0.2480	0.1488	1.667
Methane, CH_4	16.04	0.5182	2.226	1.708	1.303
Neon, Ne	20.183	0.4119	1.030	0.6180	1.667
Nitrogen, N_2	28.01	0.2968	1.040	0.7429	1.400
Nitric oxide, NO	30.006	0.2771	0.9992	0.7221	1.384
Nitrogen dioxide, NO_2	46.006	0.1889	0.8060	0.6171	1.306
Oxygen, O_2	32.00	0.2598	0.9180	0.6582	1.395
n-Pentane, C_5H_{12}	72.15	0.1152	1.664	1.549	1.074
Propane, C_3H_8	44.097	0.1885	1.669	1.480	1.127
Propylene, C_3H_6	42.08	0.1976	1.531	1.333	1.148
Steam, H_2O	18.015	0.4615	1.865	1.403	1.329
Sulfur dioxide, SO_2	64.06	0.1298	0.6228	0.4930	1.263
Tetrachloromethane, CCl_4	153.82	0.05405	0.5415	0.4875	1.111
Tetrafluoroethane (R-134a), $C_2H_2F_4$	102.03	0.08149	0.8334	0.7519	1.108
Trifluoroethane (R-143a), $C_2H_3F_3$	84.04	0.09893	0.9291	0.8302	1.119
Xenon, Xe	131.30	0.06332	0.1583	0.09499	1.667

*The unit kJ/kg · K is equivalent to kPa · m³/kg · K. The gas constant is calculated from $R = R_U/M$, where $R_U = 8.31447$ kJ/kmol · K is the universal gas constant and M is the molar mass.

Source: Specific heat values are obtained primarily from the property routines prepared by The National Institute of Standards and Technology (NIST), Gaithersburg, MD.

TABLE A–2

Boiling and freezing point properties

Substance	Boiling Data at I atm		Freezing Data		Liquid Properties		
	Normal Boiling Point, °C	Latent Heat of Vaporization h_{fg}, kJ/kg	Freezing Point, °C	Latent Heat of Fusion h_{if}, kJ/kg	Temperature, °C	Density ρ, kg/m^3	Specific Heat c_p, kJ/kg · K
Ammonia	−33.3	1357	−77.7	322.4	−33.3	682	4.43
					−20	665	4.52
					0	639	4.60
					25	602	4.80
Argon	−185.9	161.6	−189.3	28	−185.6	1394	1.14
Benzene	80.2	394	5.5	126	20	879	1.72
Brine (20% sodium chloride by mass)	103.9	—	−17.4	—	20	1150	3.11
n-Butane	−0.5	385.2	−138.5	80.3	−0.5	601	2.31
Carbon dioxide	−78.4*	230.5 (at 0°C)	−56.6		0	298	0.59
Ethanol	78.2	838.3	−114.2	109	25	783	2.46
Ethyl alcohol	78.6	855	−156	108	20	789	2.84
Ethylene glycol	198.1	800.1	−10.8	181.1	20	1109	2.84
Glycerine	179.9	974	18.9	200.6	20	1261	2.32
Helium	−268.9	22.8	—	—	−268.9	146.2	22.8
Hydrogen	−252.8	445.7	−259.2	59.5	−252.8	70.7	10.0
Isobutane	−11.7	367.1	−160	105.7	−11.7	593.8	2.28
Kerosene	204–293	251	−24.9	—	20	820	2.00
Mercury	356.7	294.7	−38.9	11.4	25	13,560	0.139
Methane	−161.5	510.4	−182.2	58.4	−161.5	423	3.49
					−100	301	5.79
Methanol	64.5	1100	−97.7	99.2	25	787	2.55
Nitrogen	−195.8	198.6	−210	25.3	−195.8	809	2.06
					−160	596	2.97
Octane	124.8	306.3	−57.5	180.7	20	703	2.10
Oil (light)					25	910	1.80
Oxygen	−183	212.7	−218.8	13.7	−183	1141	1.71
Petroleum		230–384			20	640	2.0
Propane	−42.1	427.8	−187.7	80.0	−42.1	581	2.25
					0	529	2.53
					50	449	3.13
Refrigerant-134a	−26.1	216.8	−96.6	—	−50	1443	1.23
					−26.1	1374	1.27
					0	1295	1.34
					25	1207	1.43
Water	100	2257	0.0	333.7	0	1000	4.22
					25	997	4.18
					50	988	4.18
					75	975	4.19
					100	958	4.22

* Sublimation temperature. (At pressures below the triple-point pressure of 518 kPa, carbon dioxide exists as a solid or gas. Also, the freezing-point temperature of carbon dioxide is the triple-point temperature of −56.5°C.)

TABLE A–3

Properties of solid metals

Composition	Melting Point, K	Properties at 300 K ρ kg/m³	c_p J/kg·K	k W/m·K	$\alpha \times 10^6$ m²/s	Properties at Various Temperatures (K), k(W/m·K)/c_p(J/kg·K) 100	200	400	600	800	1000
Aluminum:											
Pure	933	2702	903	237	97.1	302	237	240	231	218	
						482	798	949	1033	1146	
Alloy 2024-T6 (4.5% Cu, 1.5% Mg, 0.6% Mn)	775	2770	875	177	73.0	65	163	186	186		
							473	787	925	1042	
Alloy 195, Cast (4.5% Cu)		2790	883	168	68.2			174	185		
Beryllium	1550	1850	1825	200	59.2	990	301	161	126	106	90.8
						203	1114	2191	2604	2823	3018
Bismuth	545	9780	122	7.86	6.59	16.5	9.69	7.04			
						112	120	127			
Boron	2573	2500	1107	27.0	9.76	190	55.5	16.8	10.6	9.60	9.85
						128	600	1463	1892	2160	2338
Cadmium	594	8650	231	96.8	48.4	203	99.3	94.7			
						198	222	242			
Chromium	2118	7160	449	93.7	29.1	159	111	90.9	80.7	71.3	65.4
						192	384	484	542	581	616
Cobalt	1769	8862	421	99.2	26.6	167	122	85.4	67.4	58.2	52.1
						236	379	450	503	550	628
Copper:											
Pure	1358	8933	385	401	117	482	413	393	379	366	352
						252	356	397	417	433	451
Commercial bronze (90% Cu, 10% Al)	1293	8800	420	52	14		42	52	59		
							785	160	545		
Phosphor gear bronze (89% Cu, 11% Sn)	1104	8780	355	54	17		41	65	74		
							—	—	—		
Cartridge brass (70% Cu, 30% Zn)	1188	8530	380	110	33.9	75	95	137	149		
							360	395	425		
Constantan (55% Cu, 45% Ni)	1493	8920	384	23	6.71	17	19				
						237	362				
Germanium	1211	5360	322	59.9	34.7	232	96.8	43.2	27.3	19.8	17.4
						190	290	337	348	357	375
Gold	1336	19,300	129	317	127	327	323	311	298	284	270
						109	124	131	135	140	145
Iridium	2720	22,500	130	147	50.3	172	153	144	138	132	126
						90	122	133	138	144	153
Iron:											
Pure	1810	7870	447	80.2	23.1	134	94.0	69.5	54.7	43.3	32.8
						216	384	490	574	680	975
Armco (99.75% pure)		7870	447	72.7	20.7	95.6	80.6	65.7	53.1	42.2	32.3
						215	384	490	574	680	975
Carbon steels:											
Plain carbon (Mn ≤ 1% Si ≤ 0.1%)		7854	434	60.5	17.7			56.7	48.0	39.2	30.0
								487	559	685	1169
AISI 1010		7832	434	63.9	18.8			58.7	48.8	39.2	31.3
							487	559	685	1168	
Carbon–silicon (Mn ≤ 1% 0.1% < Si ≤ 0.6%)		7817	446	51.9	14.9			49.8	44.0	37.4	29.3
								501	582	699	971

TABLE A–3

Properties of solid metals *(Continued)*

Composition	Melting Point, K	Properties at 300 K				Properties at Various Temperatures (K), k(W/m · K)/c_p(J/kg · K)					
		ρ kg/m³	c_p J/kg · K	k W/m · K	$\alpha \times 10^6$ m²/s	100	200	400	600	800	1000
Carbon–manganese–silicon (1% < Mn < 1.65% 0.1% < Si < 0.6%)		8131	434	41.0	11.6			42.2 487	39.7 559	35.0 685	27.6 1090
Chromium (low) steels:											
$\frac{1}{2}$ Cr–$\frac{1}{4}$ Mo–Si (0.18% C, 0.65% Cr, 0.23% Mo, 0.6% Si)		7822	444	37.7	10.9			38.2 492	36.7 575	33.3 688	26.9 969
1 Cr–$\frac{1}{2}$ Mo (0.16% C, 1% Cr, 0.54% Mo, 0.39% Si)		7858	442	42.3	12.2			42.0 492	39.1 575	34.5 688	27.4 969
1 Cr–V (0.2% C, 1.02% Cr, 0.15% V)		7836	443	48.9	14.1			46.8 492	42.1 575	36.3 688	28.2 969
Stainless steels:											
AISI 302		8055	480	15.1	3.91			17.3 512	20.0 559	22.8 585	25.4 606
AISI 304	1670	7900	477	14.9	3.95	9.2 272	12.6 402	16.6 515	19.8 557	22.6 582	25.4 611
AISI 316		8238	468	13.4	3.48			15.2 504	18.3 550	21.3 576	24.2 602
AISI 347		7978	480	14.2	3.71			15.8 513	18.9 559	21.9 585	24.7 606
Lead	601	11,340	129	35.3	24.1	39.7 118	36.7 125	34.0 132	31.4 142		
Magnesium	923	1740	1024	156	87.6	169 649	159 934	153 1074	149 1170	146 1267	
Molybdenum	2894	10,240	251	138	53.7	179 141	143 224	134 261	126 275	118 285	112 295
Nickel:											
Pure	1728	8900	444	90.7	23.0 232	164 383	107 485	80.2 592	65.6 530	67.6 562	71.8
Nichrome (80% Ni, 20% Cr)	1672	8400	420	12	3.4			14 480	16 525	21 545	
Inconel X-750 (73% Ni, 15% Cr, 6.7% Fe)	1665	8510	439	11.7	3.1	8.7 —	10.3 372	13.5 473	17.0 510	20.5 546	24.0 626
Niobium	2741	8570	265	53.7	23.6	55.2 188	52.6 249	55.2 274	58.2 283	61.3 292	64.4 301
Palladium	1827	12,020	244	71.8	24.5	76.5 168	71.6 227	73.6 251	79.7 261	86.9 271	94.2 281
Platinum:											
Pure	2045	21,450	133	71.6	25.1	77.5 100	72.6 125	71.8 136	73.2 141	75.6 146	78.7 152
Alloy 60Pt–40Rh (60% Pt, 40% Rh)	1800	16,630	162	47	17.4			52 —	59 —	65 —	69 —
Rhenium	3453	21,100	136	47.9	16.7	58.9 97	51.0 127	46.1 139	44.2 145	44.1 151	44.6 156
Rhodium	2236	12,450	243	150	49.6	186 147	154 220	146 253	136 274	127 293	121 311

TABLE A–3

Properties of solid metals *(Concluded)*

Composition	Melting Point, K	Properties at 300 K				Properties at Various Temperatures (K), k(W/m · K)/c_p(J/kg · K)					
		ρ kg/m^3	c_p J/kg · K	k W/m · K	$\alpha \times 10^6$ m^2/s	100	200	400	600	800	1000
Silicon	1685	2330	712	148	89.2	884 259	264 556	98.9 790	61.9 867	42.4 913	31.2 946
Silver	1235	10,500	235	429	174	444 187	430 225	425 239	412 250	396 262	379 277
Tantalum	3269	16,600	140	57.5	24.7	59.2 110	57.5 133	57.8 144	58.6 146	59.4 149	60.2 152
Thorium	2023	11,700	118	54.0	39.1	59.8 99	54.6 112	54.5 124	55.8 134	56.9 145	56.9 156
Tin	505	7310	227	66.6	40.1	85.2 188	73.3 215	62.2 243			
Titanium	1953	4500	522	21.9	9.32	30.5 300	24.5 465	20.4 551	19.4 591	19.7 633	20.7 675
Tungsten	3660	19,300	132	174	68.3	208 87	186 122	159 137	137 142	125 146	118 148
Uranium	1406	19,070	116	27.6	12.5	21.7 94	25.1 108	29.6 125	34.0 146	38.8 176	43.9 180
Vanadium	2192	6100	489	30.7	10.3	35.8 258	31.3 430	31.3 515	33.3 540	35.7 563	38.2 597
Zinc	693	7140	389	116	41.8	117 297	118 367	111 402	103 436		
Zirconium	2125	6570	278	22.7	12.4	33.2 205	25.2 264	21.6 300	20.7 332	21.6 342	23.7 362

From Frank P. Incropera and David P. DeWitt, *Fundamentals of Heat and Mass Transfer, 3rd ed.*, 1990. This material is used by permission of John Wiley & Sons, Inc.

TABLE A–4

Properties of solid nonmetals

Composition	Melting Point, K	Properties at 300 K				Properties at Various Temperatures (K), k (W/m · K)/c_p(J/kg · K)					
		ρ kg/m³	c_p J/kg · K	k W/m · K	$\alpha \times 10^6$ m²/s	100	200	400	600	800	1000
Aluminum oxide, sapphire	2323	3970	765	46	15.1	450 —	82 —	32.4 940	18.9 1110	13.0 1180	10.5 1225
Aluminum oxide, polycrystalline	2323	3970	765	36.0	11.9	133 —	55 —	26.4 940	15.8 1110	10.4 1180	7.85 1225
Beryllium oxide	2725	3000	1030	272	88.0			196 1350	111 1690	70 1865	47 1975
Boron	2573	2500	1105	27.6	9.99	190 —	52.5 —	18.7 1490	11.3 1880	8.1 2135	6.3 2350
Boron fiber epoxy (30% vol.) composite	590	2080									
k, ‖ to fibers				2.29		2.10	2.23	2.28			
k, ⊥ to fibers				0.59		0.37	0.49	0.60			
c_p			1122			364	757	1431			
Carbon Amorphous	1500	1950	—	1.60	—	0.67 —	1.18 —	1.89 —	21.9 —	2.37 —	2.53 —
Diamond, type IIa insulator	—	3500	509	2300		10,000 21	4000 194	1540 853			
Graphite, pyrolytic	2273	2210									
k, ‖ to layers				1950		4970	3230	1390	892	667	534
k, ⊥ to layers				5.70		16.8	9.23	4.09	2.68	2.01	1.60
c_p			709			136	411	992	1406	1650	1793
Graphite fiber epoxy (25% vol.) composite	450	1400									
k, heat flow ‖ to fibers				11.1		5.7	8.7	13.0			
k, heat flow ⊥ to fibers				0.87	0.46	0.68	1.1				
c_p			935			337	642	1216			
Pyroceram, Corning 9606	1623	2600	808	3.98	1.89	5.25 —	4.78 —	3.64 908	3.28 1038	3.08 1122	2.96 1197
Silicon carbide	3100	3160	675	490	230			— 880	— 1050	— 1135	87 1195
Silicon dioxide, crystalline (quartz)	1883	2650									
k, ‖ to c-axis				10.4		39	16.4	7.6	5.0	4.2	
k, ⊥ to c-axis				6.21		20.8	9.5	4.70	3.4	3.1	
c_p			745			—	—	885	1075	1250	
Silicon dioxide, polycrystalline (fused silica)	1883	2220	745	1.38	0.834	0.69	1.14 —	1.51 905	1.75 1040	2.17 1105	2.87 1155
Silicon nitride	2173	2400	691	16.0	9.65	—	— 578	13.9 778	11.3 937	9.88 1063	8.76 1155
Sulfur	392	2070	708	0.206	0.141	0.165 403	0.185 606				
Thorium dioxide	3573	9110	235	13	6.1			10.2 255	6.6 274	4.7 285	3.68 295
Titanium dioxide, polycrystalline	2133	4157	710	8.4	2.8			7.01 805	5.02 880	8.94 910	3.46 930

TABLE A–5

Properties of building materials (at a mean temperature of 24°C)

Material	Thickness, L mm	Density, ρ kg/m³	Thermal Conductivity, k W/m · K	Specific Heat, c_p kJ/kg · K	R-value (for listed thickness, L/k), K · m²/W
Building Boards					
Asbestos–cement board	6 mm	1922	—	1.00	0.011
Gypsum of plaster board	10 mm	800	—	1.09	0.057
	13 mm	800	—	—	0.078
Plywood (Douglas fir)	—	545	0.12	1.21	—
	6 mm	545	—	1.21	0.055
	10 mm	545	—	1.21	0.083
	13 mm	545	—	1.21	0.110
	20 mm	545	—	1.21	0.165
Insulated board and sheating	13 mm	288	—	1.30	0.232
(regular density)	20 mm	288	—	1.30	0.359
Hardboard (high density, standard tempered)	—	1010	0.14	1.34	—
Particle board:					
Medium density	—	800	0.14	1.30	—
Underlayment	16 mm	640	—	1.21	0.144
Wood subfloor	20 mm	—	—	1.38	0.166
Building Membrane					
Vapor-permeable felt	—	—	—	—	0.011
Vapor-seal (2 layers of mopped 0.73 kg/m² felt)	—	—	—	—	0.021
Flooring Materials					
Carpet and fibrous pad	—	—	—	1.42	0.367
Carpet and rubber pad	—	—	—	1.38	0.217
Tile (asphalt, linoleum, vinyl)	—	—	—	1.26	0.009
Masonry Materials					
Masonry units:					
Brick, common		1922	0.72	—	—
Brick, face		2082	1.30	—	—
Brick, fire clay		2400	1.34	—	—
		1920	0.90	0.79	—
		1120	0.41	—	—
Concrete blocks (3 oval cores,	100 mm	—	0.77	—	0.13
sand and gravel aggregate)	200 mm	—	1.0	—	0.20
	300 mm	—	1.30	—	0.23
Concretes:					
Lightweight aggregates, (including		1920	1.1	—	—
expanded shale, clay, or slate;		1600	0.79	0.84	—
expanded slags; cinders;		1280	0.54	0.84	—
pumice; and scoria)		960	0.33	—	—
	940	0.18	—	—	
Cement/lime, mortar, and stucco		1920	1.40	—	—
		1280	0.65	—	—
Stucco		1857	0.72	—	—

TABLE A–5

Properties of building materials *(Concluded)*
(at a mean temperature of 24°C)

Material	Thickness, L mm	Density, ρ kg/m^3	Thermal Conductivity, k W/m · K	Specific Heat, c_p kJ/kg · K	R-value (for listed thickness, L/k), K · m^2/W
Roofing					
Asbestos-cement shingles		1900	—	1.00	0.037
Asphalt roll roofing		1100	—	1.51	0.026
Asphalt shingles		1100	—	1.26	0.077
Built-in roofing	10 mm	1100	—	1.46	0.058
Slate	13 mm	—	—	1.26	0.009
Wood shingles (plain and plastic/film faced)		—	—	1.30	0.166
Plastering Materials					
Cement plaster, sand aggregate	19 mm	1860	0.72	0.84	0.026
Gypsum plaster:					
Lightweight aggregate	13 mm	720	—	—	0.055
Sand aggregate	13 mm	1680	0.81	0.84	0.016
Perlite aggregate	—	720	0.22	1.34	—
Siding Material (on flat surfaces)					
Asbestos-cement shingles	—	1900	—	—	0.037
Hardboard siding	11 mm	—	—	1.17	0.12
Wood (drop) siding	25 mm	—	—	1.30	0.139
Wood (plywood) siding lapped	10 mm	—	—	1.21	0.111
Aluminum or steel siding (over sheeting):					
Hollow backed	10 mm	—	—	1.22	0.11
Insulating-board backed	10 mm	—	—	1.34	0.32
Architectural glass	—	2530	1.0	0.84	0.018
Woods					
Hardwoods (maple, oak, etc.)	—	721	0.159	1.26	—
Softwoods (fir, pine, etc.)	—	513	0.115	1.38	—
Metals					
Aluminum (1100)	—	2739	222	0.896	—
Steel, mild	—	7833	45.3	0.502	—
Steel, Stainless	—	7913	15.6	0.456	—

Source: Table A–5 and A–6 are adapted from *ASHRAE, Handbook of Fundamentals* (Atlanta, GA: American Society of Heating, Refrigerating, and Air-Conditioning Engineers, 1993), Chap. 22, Table 4. Used with permission.

TABLE A–6

Properties of insulating materials
(at a mean temperature of 24°C)

Material	Thickness, L mm	Density, ρ kg/m^3	Thermal Conductivity, k W/m · K	Specific Heat, c_p kJ/kg · K	R-value (for listed thickness, L/k), K · m^2/W
Blanket and Batt					
Mineral fiber (fibrous form	50 to 70 mm	4.8–32	—	0.71–0.96	1.23
processed from rock, slag,	75 to 90 mm	4.8–32	—	0.71–0.96	1.94
or glass)	135 to 165 mm	4.8–32	—	0.71–0.96	3.32
Board and Slab					
Cellular glass		136	0.055	1.0	—
Glass fiber (organic bonded)		64–144	0.036	0.96	—
Expanded polystyrene (molded beads)		16	0.040	1.2	—
Expanded polyurethane (R-11 expanded)		24	0.023	1.6	—
Expanded perlite (organic bonded)		16	0.052	1.26	—
Expanded rubber (rigid)		72	0.032	1.68	—
Mineral fiber with resin binder		240	0.042	0.71	—
Cork		120	0.039	1.80	—
Sprayed or Formed in Place					
Polyurethane foam		24–40	0.023–0.026	—	—
Glass fiber		56–72	0.038–0.039	—	—
Urethane, two-part mixture (rigid foam)		70	0.026	1.045	—
Mineral wool granules with asbestos/ inorganic binders (sprayed)		190	0.046	—	—
Loose Fill					
Mineral fiber (rock, slag, or glass)	~75 to 125 mm	9.6–32	—	0.71	1.94
	~165 to 222 mm	9.6–32	—	0.71	3.35
	~191 to 254 mm	—	—	0.71	3.87
	~185 mm	—	—	0.71	5.28
Silica aerogel		122	0.025	—	—
Vermiculite (expanded)		122	0.068	—	—
Perlite, expanded		32–66	0.039–0.045	1.09	—
Sawdust or shavings		128–240	0.065	1.38	—
Cellulosic insulation (milled paper or wood pulp)		37–51	0.039–0.046	—	—
Roof Insulation					
Cellular glass	—	144	0.058	1.0	—
Preformed, for use above deck	13 mm	—	—	1.0	0.24
	25 mm	—	—	2.1	0.49
	50 mm	—	—	3.9	0.93
Reflective Insulation					
Silica powder (evacuated)		160	0.0017	—	—
Aluminum foil separating fluffy glass mats; 10–12 layers (evacuated); for cryogenic applications (150 K)		40	0.00016	—	—
Aluminum foil and glass paper laminate; 75–150 layers (evacuated); for cryogenic applications (150 K)		120	0.000017	—	—

TABLE A-7

Properties of common foods

(a) Specific heats and freezing-point properties

Food	Water content,[a] %(mass)	Freezing Point[a] °C	Specific heat,[b] kJ/kg · K Above Freezing	Below Freezing	Latent Heat of Fusion,[c] kJ/kg
Vegetables					
Artichokes	84	-1.2	3.65	1.90	281
Asparagus	93	-0.6	3.95	2.01	311
Beans, snap	89	-0.7	3.82	1.96	297
Broccoli	90	-0.6	3.85	1.97	301
Cabbage	92	-0.9	3.92	2.00	307
Carrots	88	-1.4	3.79	1.55	294
Cauliflower	92	-0.8	3.92	2.00	307
Celery	94	-0.5	3.99	2.02	314
Corn. sweet	74	-0.6	3.32	1.77	247
Cucumbers	96	-0.5	4.06	2.05	321
Eggplant	93	-0.8	3.96	2.01	311
Horseradish	75	-1.8	3.35	1.78	251
Leeks	85	-0.7	3.69	1.91	284
Lettuce	95	-0.2	4.02	2.04	317
Mushrooms	91	-0.9	3.89	1.99	304
Okra	90	-1.8	3.86	1.97	301
Onions, green	89	-0.9	3.82	1.96	297
Onions, dry	88	-0.8	3.79	1.95	294
Parsley	85	-1.1	3.69	1.91	284
Peas, green	74	-0.6	3.32	1.77	247
Peppers, sweet	92	-0.7	3.92	2.00	307
Potatoes	78	-0.6	3.45	1.82	261
Pumpkins	91	-0.8	3.89	1.99	304
Spinach	93	-0.3	3.96	2.01	311
Tomatos, ripe	94	-0.5	3.99	2.02	314
Turnips	92	-1.1	3.92	2.00	307
Fruits					
Apples	84	-1.1	3.65	1.90	281
Apricots	85	-1.1	3.69	1.91	284
Avocados	65	-0.3	3.02	1.55	217
Bananas	75	-0.8	3.35	1.73	251
Blueberries	82	-1.6	3.59	1.87	274
Cantaloupes	92	-1.2	3.92	2.00	307
Cherries, sour	84	-1.7	3.65	1.90	281
Cherries, sweet	80	-1.8	3.52	1.85	267
Figs, dried	23	—	—	1.13	77
Figs, fresh	78	-2.4	3.45	1.82	261
Grapefruit	89	-1.1	3.82	1.96	297
Grapes	82	-1.1	3.59	1.87	274
Lemons	89	-1.4	3.82	1.96	297
Olives	75	-1.4	3.35	1.78	251
Oranges	87	-0.8	3.75	1.94	291

Food	Water content,[a] %(mass)	Freezing Point[a] °C	Specific heat,[b] kJ/kg · K Above Freezing	Below Freezing	Latent Heat of Fusion,[c] kJ/kg
Peaches	89	-0.9	3.82	1.96	297
Pears	83	-1.6	3.62	1.89	277
Pineapples	85	-1.0	3.69	1.91	284
Plums	86	-0.8	3.72	1.92	287
Quinces	85	-2.0	3.69	1.91	284
Raisins	18	—	—	1.07	60
Strawberries	90	-0.8	3.86	1.97	301
Tangerines	87	-1.1	3.75	1.94	291
Watermelon	93	-0.4	3.96	2.01	311
Fish/Seafood					
Cod, whole	78	-2.2	3.45	1.82	261
Halibut, whole	75	-2.2	3.35	1.78	251
Lobster	79	-2.2	3.49	1.84	264
Mackerel	57	-2.2	2.75	1.56	190
Salmon, whole	64	-2.2	2.98	1.65	214
Shrimp	83	-2.2	3.62	1.89	277
Meats					
Beef carcass	49	-1.7	2.48	1.46	164
Liver	70	-1.7	3.18	1.72	234
Round, beef	67	—	3.08	1.68	224
Sirloin, beef	56	—	2.72	1.55	187
Chicken	74	-2.8	3.32	1.77	247
Lamb leg	65	—	3.02	1.66	217
Port carcass	37	—	2.08	1.31	124
Ham	56	-1.7	2.72	1.55	187
Pork sausage	38	—	2.11	1.32	127
Turkey	64	—	2.98	1.65	214
Other					
Almonds	5	—	—	0.89	17
Butter	16	—	—	1.04	53
Cheese, Cheddar	37	-12.9	2.08	1.31	124
Cheese, Swiss	39	-10.0	2.15	1.33	130
Chocolate milk	1	—	—	0.85	3
Eggs, whole	74	-0.6	3.32	1.77	247
Honey	17	—	—	1.05	57
Ice cream	63	-5.6	2.95	1.63	210
Milk, whole	88	-0.6	3.79	1.95	294
Peanuts	6	—	—	0.92	20
Peanuts, roasted	2	—	—	0.87	7
Pecans	3	—	—	0.87	10
Walnuts	4	—	—	0.88	13

Sources: [a]Water content and freezing-point data are from ASHRAE, *Handbook of Fundamentals.* SI version (Atlanta, GA: American Society of Heating, Refrigerating and Air-Conditioning Engineers, Inc., 1993), Chap. 30, Table 1. Used with permission. Freezing point is the temperature at which freezing starts for fruits and vegetables, and the average freezing temperature for other foods.

[b]Specific heat data are based on the specific heat values of a water and ice at 0°C and are determined from Siebel's formulas: $c_{p,\ fresh} = 3.35 \times$ (Water content) $+ 0.84$, above freezing, and $c_{2,\ frozen} = 1.26 \times$ (Water content) $+ 0.84$, below freezing.

[c]The latent heat of fusion is determined by multiplying the heat of fusion of water (334 kJ/kg) by the water content of the food.

TABLE A–7

Properties of common foods *(Concluded)*
(b) Other properties

Food	Water Content, % (mass)	Temperature, T °C	Density, ρ kg/m³	Thermal Conductivity, k W/m · K	Thermal Diffusivity, α m²/s	Specific Heat, c_p kJ/kg · K
Fruits/Vegetables						
Apple juice	87	20	1000	0.559	0.14×10^{-6}	3.86
Apples	85	8	840	0.418	0.13×10^{-6}	3.81
Apples, dried	41.6	23	856	0.219	0.096×10^{-6}	2.72
Apricots, dried	43.6	23	1320	0.375	0.11×10^{-6}	2.77
Bananas, fresh	76	27	980	0.481	0.14×10^{-6}	3.59
Broccoli	—	−6	560	0.385	—	—
Cherries, fresh	92	0–30	1050	0.545	0.13×10^{-6}	3.99
Figs	40.4	23	1241	0.310	0.096×10^{-6}	2.69
Grape juice	89	20	1000	0.567	0.14×10^{-6}	3.91
Peaches	89	2–32	960	0.526	0.14×10^{-6}	3.91
Plums	—	−16	610	0.247	—	—
Potatoes	78	0–70	1055	0.498	0.13×10^{-6}	3.64
Raisins	32	23	1380	0.376	0.11×10^{-6}	2.48
Meats						
Beef, ground	67	6	950	0.406	0.13×10^{-6}	3.36
Beef, lean	74	3	1090	0.471	0.13×10^{-6}	3.54
Beef fat	0	35	810	0.190	—	—
Beef liver	72	35	—	0.448	—	3.49
Cat food	39.7	23	1140	0.326	0.11×10^{-6}	2.68
Chicken breast	75	0	1050	0.476	0.13×10^{-6}	3.56
Dog food	30.6	23	1240	0.319	0.11×10^{-6}	2.45
Fish, cod	81	3	1180	0.534	0.12×10^{-6}	3.71
Fish, salmon	67	3	—	0.531	—	3.36
Ham	71.8	20	1030	0.480	0.14×10^{-6}	3.48
Lamb	72	20	1030	0.456	0.13×10^{-6}	3.49
Pork, lean	72	4	1030	0.456	0.13×10^{-6}	3.49
Turkey breast	74	3	1050	0.496	0.13×10^{-6}	3.54
Veal	75	20	1060	0.470	0.13×10^{-6}	3.56
Other						
Butter	16	4	—	0.197	—	2.08
Chocolate cake	31.9	23	340	0.106	0.12×10^{-6}	2.48
Margarine	16	5	1000	0.233	0.11×10^{-6}	2.08
Milk, skimmed	91	20	—	0.566	—	3.96
Milk, whole	88	28	—	0.580	—	3.89
Olive oil	0	32	910	0.168	—	—
Peanut oil	0	4	920	0.168	—	—
Water	100	0	1000	0.569	0.14×10^{-6}	4.217
	100	30	995	0.618	0.15×10^{-6}	4.178
White cake	32.3	23	450	0.082	0.10×10^{-6}	2.49

Source: Data obtained primarily from ASHRAE, *Handbook of Fundamentals*, SI version (Atlanta, GA: American Society of Heating, Refrigerating and Air-Conditioning Engineers, Inc., 1993), Chap. 30, Tables 7 and 9. Used with permission.

Most specific heats are calculated from $c_p = 1.68 + 2.51 \times$ (Water content), which is a good approximation in the temperature range of 3 to 32°C. Most thermal diffusivities are calculated from $\alpha = k/\rho c_p$. Property values given here are valid for the specific water content.

TABLE A-8

Properties of miscellaneous materials
(Values are at 300 K unless indicated otherwise)

Material	Density, ρ kg/m^3	Thermal Conductivity, k W/m · K	Specific Heat, c_p J/kg · K	Material	Density, ρ kg/m^3	Thermal Conductivity, k W/m · K	Specific Heat, c_p J/kg · K
Asphalt	2115	0.062	920	Ice			
Bakelite	1300	1.4	1465	273 K	920	1.88	2040
Brick, refractory				253 K	922	2.03	1945
Chrome brick				173 K	928	3.49	1460
473 K	3010	2.3	835	Leather, sole	998	0.159	—
823 K	—	2.5	—	Linoleum	535	0.081	—
1173 K	—	2.0	—		1180	0.186	—
Fire clay, burnt				Mica	2900	0.523	—
1600 K				Paper	930	0.180	1340
773 K	2050	1.0	960	Plastics			
1073 K	—	1.1	—	Plexiglass	1190	0.19	1465
1373 K	—	1.1	—	Teflon			
Fire clay, burnt				300 K	2200	0.35	1050
1725 K				400 K	—	0.45	—
773 K	2325	1.3	960	Lexan	1200	0.19	1260
1073 K	—	1.4	—	Nylon	1145	0.29	—
1373 K	—	1.4		Polypropylene	910	0.12	1925
Fire clay brick				Polyester	1395	0.15	1170
478 K	2645	1.0	960	PVC, vinyl	1470	0.1	840
922 K	—	1.5	—	Porcelain	2300	1.5	—
1478 K	—	1.8	—	Rubber, natural	1150	0.28	—
Magnesite				Rubber, vulcanized			
478 K	—	3.8	1130	Soft	1100	0.13	2010
922 K	—	2.8	—	Hard	1190	0.16	—
1478 K	—	1.9	—	Sand	1515	0.2–1.0	800
Chicken meat, white (74.4% water content)				Snow, fresh	100	0.60	
				Snow, 273 K	500	2.2	—
198 K	—	1.60	—	Soil, dry	1500	1.0	1900
233 K	—	1.49	—	Soil, wet	1900	2.0	2200
253 K	—	1.35	—	Sugar	1600	0.58	—
273 K	—	0.48	—	Tissue, human			
293 K	—	0.49	—	Skin	—	0.37	—
Clay, dry	1550	0.930	—	Fat layer	—	0.2	—
Clay, wet	1495	1.675	—	Muscle	—	0.41	—
Coal, anthracite	1350	0.26	1260	Vaseline	—	0.17	—
Concrete (stone mix)	2300	1.4	880	Wood, cross-grain			
Cork	86	0.048	2030	Balsa	140	0.055	—
Cotton	80	0.06	1300	Fir	415	0.11	2720
Fat	—	0.17	—	Oak	545	0.17	2385
Glass				White pine	435	0.11	—
Window	2800	0.7	750	Yellow pine	640	0.15	2805
Pyrex	2225	1–1.4	835	Wood, radial			
Crown	2500	1.05	—	Oak	545	0.19	2385
Lead	3400	0.85	—	Fir	420	0.14	2720
				Wool, ship	145	0.05	—

Source: Compiled from various sources.

TABLE A-9

Properties of saturated water

Temp. T, °C	Saturation Pressure P_{sat}, kPa	Density ρ, kg/m³ Liquid	Density Vapor	Enthalpy of Vaporization h_{fg}, kJ/kg	Specific Heat c_p, J/kg · K Liquid	Specific Heat Vapor	Thermal Conductivity k, W/m · K Liquid	Thermal Conductivity Vapor	Dynamic Viscosity μ, kg/m · s Liquid	Dynamic Viscosity Vapor	Prandtl Number Pr Liquid	Prandtl Vapor	Volume Expansion Coefficient β, 1/K Liquid
0.01	0.6113	999.8	0.0048	2501	4217	1854	0.561	0.0171	1.792×10^{-3}	0.922×10^{-5}	13.5	1.00	-0.068×10^{-3}
5	0.8721	999.9	0.0068	2490	4205	1857	0.571	0.0173	1.519×10^{-3}	0.934×10^{-5}	11.2	1.00	0.015×10^{-3}
10	1.2276	999.7	0.0094	2478	4194	1862	0.580	0.0176	1.307×10^{-3}	0.946×10^{-5}	9.45	1.00	0.733×10^{-3}
15	1.7051	999.1	0.0128	2466	4185	1863	0.589	0.0179	1.138×10^{-3}	0.959×10^{-5}	8.09	1.00	0.138×10^{-3}
20	2.339	998.0	0.0173	2454	4182	1867	0.598	0.0182	1.002×10^{-3}	0.973×10^{-5}	7.01	1.00	0.195×10^{-3}
25	3.169	997.0	0.0231	2442	4180	1870	0.607	0.0186	0.891×10^{-3}	0.987×10^{-5}	6.14	1.00	0.247×10^{-3}
30	4.246	996.0	0.0304	2431	4178	1875	0.615	0.0189	0.798×10^{-3}	1.001×10^{-5}	5.42	1.00	0.294×10^{-3}
35	5.628	994.0	0.0397	2419	4178	1880	0.623	0.0192	0.720×10^{-3}	1.016×10^{-5}	4.83	1.00	0.337×10^{-3}
40	7.384	992.1	0.0512	2407	4179	1885	0.631	0.0196	0.653×10^{-3}	1.031×10^{-5}	4.32	1.00	0.377×10^{-3}
45	9.593	990.1	0.0655	2395	4180	1892	0.637	0.0200	0.596×10^{-3}	1.046×10^{-5}	3.91	1.00	0.415×10^{-3}
50	12.35	988.1	0.0831	2383	4181	1900	0.644	0.0204	0.547×10^{-3}	1.062×10^{-5}	3.55	1.00	0.451×10^{-3}
55	15.76	985.2	0.1045	2371	4183	1908	0.649	0.0208	0.504×10^{-3}	1.077×10^{-5}	3.25	1.00	0.484×10^{-3}
60	19.94	983.3	0.1304	2359	4185	1916	0.654	0.0212	0.467×10^{-3}	1.093×10^{-5}	2.99	1.00	0.517×10^{-3}
65	25.03	980.4	0.1614	2346	4187	1926	0.659	0.0216	0.433×10^{-3}	1.110×10^{-5}	2.75	1.00	0.548×10^{-3}
70	31.19	977.5	0.1983	2334	4190	1936	0.663	0.0221	0.404×10^{-3}	1.126×10^{-5}	2.55	1.00	0.578×10^{-3}
75	38.58	974.7	0.2421	2321	4193	1948	0.667	0.0225	0.378×10^{-3}	1.142×10^{-5}	2.38	1.00	0.607×10^{-3}
80	47.39	971.8	0.2935	2309	4197	1962	0.670	0.0230	0.355×10^{-3}	1.159×10^{-5}	2.22	1.00	0.653×10^{-3}
85	57.83	968.1	0.3536	2296	4201	1977	0.673	0.0235	0.333×10^{-3}	1.176×10^{-5}	2.08	1.00	0.670×10^{-3}
90	70.14	965.3	0.4235	2283	4206	1993	0.675	0.0240	0.315×10^{-3}	1.193×10^{-5}	1.96	1.00	0.702×10^{-3}
95	84.55	961.5	0.5045	2270	4212	2010	0.677	0.0246	0.297×10^{-3}	1.210×10^{-5}	1.85	1.00	0.716×10^{-3}
100	101.33	957.9	0.5978	2257	4217	2029	0.679	0.0251	0.282×10^{-3}	1.227×10^{-5}	1.75	1.00	0.750×10^{-3}
110	143.27	950.6	0.8263	2230	4229	2071	0.682	0.0262	0.255×10^{-3}	1.261×10^{-5}	1.58	1.00	0.798×10^{-3}
120	198.53	943.4	1.121	2203	4244	2120	0.683	0.0275	0.232×10^{-3}	1.296×10^{-5}	1.44	1.00	0.858×10^{-3}
130	270.1	934.6	1.496	2174	4263	2177	0.684	0.0288	0.213×10^{-3}	1.330×10^{-5}	1.33	1.01	0.913×10^{-3}
140	361.3	921.7	1.965	2145	4286	2244	0.683	0.0301	0.197×10^{-3}	1.365×10^{-5}	1.24	1.02	0.970×10^{-3}
150	475.8	916.6	2.546	2114	4311	2314	0.682	0.0316	0.183×10^{-3}	1.399×10^{-5}	1.16	1.02	1.025×10^{-3}
160	617.8	907.4	3.256	2083	4340	2420	0.680	0.0331	0.170×10^{-3}	1.434×10^{-5}	1.09	1.05	1.145×10^{-3}
170	791.7	897.7	4.119	2050	4370	2490	0.677	0.0347	0.160×10^{-3}	1.468×10^{-5}	1.03	1.05	1.178×10^{-3}
180	1,002.1	887.3	5.153	2015	4410	2590	0.673	0.0364	0.150×10^{-3}	1.502×10^{-5}	0.983	1.07	1.210×10^{-3}
190	1,254.4	876.4	6.388	1979	4460	2710	0.669	0.0382	0.142×10^{-3}	1.537×10^{-5}	0.947	1.09	1.280×10^{-3}
200	1,553.8	864.3	7.852	1941	4500	2840	0.663	0.0401	0.134×10^{-3}	1.571×10^{-5}	0.910	1.11	1.350×10^{-3}
220	2,318	840.3	11.60	1859	4610	3110	0.650	0.0442	0.122×10^{-3}	1.641×10^{-5}	0.865	1.15	1.520×10^{-3}
240	3,344	813.7	16.73	1767	4760	3520	0.632	0.0487	0.111×10^{-3}	1.712×10^{-5}	0.836	1.24	1.720×10^{-3}
260	4,688	783.7	23.69	1663	4970	4070	0.609	0.0540	0.102×10^{-3}	1.788×10^{-5}	0.832	1.35	2.000×10^{-3}
280	6,412	750.8	33.15	1544	5280	4835	0.581	0.0605	0.094×10^{-3}	1.870×10^{-5}	0.854	1.49	2.380×10^{-3}
300	8,581	713.8	46.15	1405	5750	5980	0.548	0.0695	0.086×10^{-3}	1.965×10^{-5}	0.902	1.69	2.950×10^{-3}
320	11,274	667.1	64.57	1239	6540	7900	0.509	0.0836	0.078×10^{-3}	2.084×10^{-5}	1.00	1.97	
340	14,586	610.5	92.62	1028	8240	11,870	0.469	0.110	0.070×10^{-3}	2.255×10^{-5}	1.23	2.43	
360	18,651	528.3	144.0	720	14,690	25,800	0.427	0.178	0.060×10^{-3}	2.571×10^{-5}	2.06	3.73	
374.14	22,090	317.0	317.0	0	—	—	—	—	0.043×10^{-3}	4.313×10^{-5}			

Note 1: Kinematic viscosity ν and thermal diffusivity α can be calculated from their definitions, $\nu = \mu/\rho$ and $\alpha = k/\rho c_p = \nu/Pr$. The temperatures 0.01°C, 100°C, and 374.14°C are the triple-, boiling-, and critical-point temperatures of water, respectively. The properties listed above (except the vapor density) can be used at any pressure with negligible error except at temperatures near the critical-point value.

Note 2: The unit kJ/kg · °C for specific heat is equivalent to kJ/kg · K, and the unit W/m · °C for thermal conductivity is equivalent to W/m · K.

Source: Viscosity and thermal conductivity data are from J. V. Sengers and J. T. R. Watson, Journal of Physical and Chemical Reference Data 15 (1986), pp. 1291–1322. Other data are obtained from various sources or calculated.

TABLE A–10

Properties of saturated refrigerant-134a

Temp. T, °C	Saturation Pressure P, kPa	Density ρ, kg/m³ Liquid	Density ρ, kg/m³ Vapor	Enthalpy of Vaporization h_{fg}, kJ/kg	Specific Heat c_p, J/kg · K Liquid	Specific Heat c_p, J/kg · K Vapor	Thermal Conductivity k, W/m · K Liquid	Thermal Conductivity k, W/m · K Vapor	Dynamic Viscosity μ, kg/m · s Liquid	Dynamic Viscosity μ, kg/m · s Vapor	Prandtl Number Pr Liquid	Prandtl Number Pr Vapor	Volume Expansion Coefficient β, 1/K Liquid	Surface Tension, N/m
−40	51.2	1418	2.773	225.9	1254	748.6	0.1101	0.00811	4.878×10^{-4}	2.550×10^{-6}	5.558	0.235	0.00205	0.01760
−35	66.2	1403	3.524	222.7	1264	764.1	0.1084	0.00862	4.509×10^{-4}	3.003×10^{-6}	5.257	0.266	0.00209	0.01682
−30	84.4	1389	4.429	219.5	1273	780.2	0.1066	0.00913	4.178×10^{-4}	3.504×10^{-6}	4.992	0.299	0.00215	0.01604
−25	106.5	1374	5.509	216.3	1283	797.2	0.1047	0.00963	3.882×10^{-4}	4.054×10^{-6}	4.757	0.335	0.00220	0.01527
−20	132.8	1359	6.787	213.0	1294	814.9	0.1028	0.01013	3.614×10^{-4}	4.651×10^{-6}	4.548	0.374	0.00227	0.01451
−15	164.0	1343	8.288	209.5	1306	833.5	0.1009	0.01063	3.371×10^{-4}	5.295×10^{-6}	4.363	0.415	0.00233	0.01376
−10	200.7	1327	10.04	206.0	1318	853.1	0.0989	0.01112	3.150×10^{-4}	5.982×10^{-6}	4.198	0.459	0.00241	0.01302
−5	243.5	1311	12.07	202.4	1330	873.8	0.0968	0.01161	2.947×10^{-4}	6.709×10^{-6}	4.051	0.505	0.00249	0.01229
0	293.0	1295	14.42	198.7	1344	895.6	0.0947	0.01210	2.761×10^{-4}	7.471×10^{-6}	3.919	0.553	0.00258	0.01156
5	349.9	1278	17.12	194.8	1358	918.7	0.0925	0.01259	2.589×10^{-4}	8.264×10^{-6}	3.802	0.603	0.00269	0.01084
10	414.9	1261	20.22	190.8	1374	943.2	0.0903	0.01308	2.430×10^{-4}	9.081×10^{-6}	3.697	0.655	0.00280	0.01014
15	488.7	1244	23.75	186.6	1390	969.4	0.0880	0.01357	2.281×10^{-4}	9.915×10^{-6}	3.604	0.708	0.00293	0.00944
20	572.1	1226	27.77	182.3	1408	997.6	0.0856	0.01406	2.142×10^{-4}	1.075×10^{-5}	3.521	0.763	0.00307	0.00876
25	665.8	1207	32.34	177.8	1427	1028	0.0833	0.01456	2.012×10^{-4}	1.160×10^{-5}	3.448	0.819	0.00324	0.00808
30	770.6	1188	37.53	173.1	1448	1061	0.0808	0.01507	1.888×10^{-4}	1.244×10^{-5}	3.383	0.877	0.00342	0.00742
35	887.5	1168	43.41	168.2	1471	1098	0.0783	0.01558	1.772×10^{-4}	1.327×10^{-5}	3.328	0.935	0.00364	0.00677
40	1017.1	1147	50.08	163.0	1498	1138	0.0757	0.01610	1.660×10^{-4}	1.408×10^{-5}	3.285	0.995	0.00390	0.00613
45	1160.5	1125	57.66	157.6	1529	1184	0.0731	0.01664	1.554×10^{-4}	1.486×10^{-5}	3.253	1.058	0.00420	0.00550
50	1318.6	1102	66.27	151.8	1566	1237	0.0704	0.01720	1.453×10^{-4}	1.562×10^{-5}	3.231	1.123	0.00455	0.00489
55	1492.3	1078	76.11	145.7	1608	1298	0.0676	0.01777	1.355×10^{-4}	1.634×10^{-5}	3.223	1.193	0.00500	0.00429
60	1682.8	1053	87.38	139.1	1659	1372	0.0647	0.01838	1.260×10^{-4}	1.704×10^{-5}	3.229	1.272	0.00554	0.00372
65	1891.0	1026	100.4	132.1	1722	1462	0.0618	0.01902	1.167×10^{-4}	1.771×10^{-5}	3.255	1.362	0.00624	0.00315
70	2118.2	996.2	115.6	124.4	1801	1577	0.0587	0.01972	1.077×10^{-4}	1.839×10^{-5}	3.307	1.471	0.00716	0.00261
75	2365.8	964	133.6	115.9	1907	1731	0.0555	0.02048	9.891×10^{-5}	1.908×10^{-5}	3.400	1.612	0.00843	0.00209
80	2635.2	928.2	155.3	106.4	2056	1948	0.0521	0.02133	9.011×10^{-5}	1.982×10^{-5}	3.558	1.810	0.01031	0.00160
85	2928.2	887.1	182.3	95.4	2287	2281	0.0484	0.02233	8.124×10^{-5}	2.071×10^{-5}	3.837	2.116	0.01336	0.00114
90	3246.9	837.7	217.8	82.2	2701	2865	0.0444	0.02357	7.203×10^{-5}	2.187×10^{-5}	4.385	2.658	0.01911	0.00071
95	3594.1	772.5	269.3	64.9	3675	4144	0.0396	0.02544	6.190×10^{-5}	2.370×10^{-5}	5.746	3.862	0.03343	0.00033
100	3976.1	651.7	376.3	33.9	7959	8785	0.0322	0.02989	4.765×10^{-5}	2.833×10^{-5}	11.77	8.326	0.10047	0.00004

Note 1: Kinematic viscosity ν and thermal diffusivity α can be calculated from their definitions, $\nu = \mu/\rho$ and $\alpha = k/\rho c_p = \nu/\text{Pr}$. The properties listed here (except the vapor density) can be used at any pressures with negligible error except at temperatures near the critical-point value.

Note 2: The unit kJ/kg · °C for specific heat is equivalent to kJ/kg · K, and the unit W/m · °C for thermal conductivity is equivalent to W/m · K.

Source: Data generated from the EES software developed by S. A. Klein and F. L. Alvarado. Original sources: R. Tillner-Roth and H. D. Baehr, "An International Standard Formulation for the Thermodynamic Properties of 1,1,1,2-Tetrafluoroethane (HFC-134a) for Temperatures from 170 K to 455 K and Pressures up to 70 MPa," *J. Phys. Chem. Ref. Data*, Vol. 23, No. 5, 1994; M.J. Assael, N. K. Dalaouti, A. A. Griva, and J. H. Dymond, "Viscosity and Thermal Conductivity of Halogenated Methane and Ethane Refrigerants," *IJR*, Vol. 22, pp. 525–535, 1999; NIST REFPROP 6 program (M. O. McLinden, S. A. Klein, E. W. Lemmon, and A. P. Peskin, Physical and Chemical Properties Division, National Institute of Standards and Technology, Boulder, CO 80303, 1995).

TABLE A–11

Properties of saturated ammonia

Temp. T, °C	Saturation Pressure P, kPa	Density ρ, kg/m³ Liquid	Density ρ, kg/m³ Vapor	Enthalpy of Vaporization h_{fg}, kJ/kg	Specific Heat c_p, J/kg · K Liquid	Specific Heat c_p, J/kg · K Vapor	Thermal Conductivity k, W/m · K Liquid	Thermal Conductivity k, W/m · K Vapor	Dynamic Viscosity μ, kg/m · s Liquid	Dynamic Viscosity μ, kg/m · s Vapor	Prandtl Number Pr Liquid	Prandtl Number Pr Vapor	Volume Expansion Coefficient β, 1/K Liquid	Surface Tension, N/m
−40	71.66	690.2	0.6435	1389	4414	2242	—	0.01792	2.926×10^{-4}	7.957×10^{-6}	—	0.9955	0.00176	0.03565
−30	119.4	677.8	1.037	1360	4465	2322	—	0.01898	2.630×10^{-4}	8.311×10^{-6}	—	1.017	0.00185	0.03341
−25	151.5	671.5	1.296	1345	4489	2369	0.5968	0.01957	2.492×10^{-4}	8.490×10^{-6}	1.875	1.028	0.00190	0.03229
−20	190.1	665.1	1.603	1329	4514	2420	0.5853	0.02015	2.361×10^{-4}	8.669×10^{-6}	1.821	1.041	0.00194	0.03118
−15	236.2	658.6	1.966	1313	4538	2476	0.5737	0.02075	2.236×10^{-4}	8.851×10^{-6}	1.769	1.056	0.00199	0.03007
−10	290.8	652.1	2.391	1297	4564	2536	0.5621	0.02138	2.117×10^{-4}	9.034×10^{-6}	1.718	1.072	0.00205	0.02896
−5	354.9	645.4	2.886	1280	4589	2601	0.5505	0.02203	2.003×10^{-4}	9.218×10^{-6}	1.670	1.089	0.00210	0.02786
0	429.6	638.6	3.458	1262	4617	2672	0.5390	0.02270	1.896×10^{-4}	9.405×10^{-6}	1.624	1.107	0.00216	0.02676
5	516	631.7	4.116	1244	4645	2749	0.5274	0.02341	1.794×10^{-4}	9.593×10^{-6}	1.580	1.126	0.00223	0.02566
10	615.3	624.6	4.870	1226	4676	2831	0.5158	0.02415	1.697×10^{-4}	9.784×10^{-6}	1.539	1.147	0.00230	0.02457
15	728.8	617.5	5.729	1206	4709	2920	0.5042	0.02492	1.606×10^{-4}	9.978×10^{-6}	1.500	1.169	0.00237	0.02348
20	857.8	610.2	6.705	1186	4745	3016	0.4927	0.02573	1.519×10^{-4}	1.017×10^{-5}	1.463	1.193	0.00245	0.02240
25	1003	602.8	7.809	1166	4784	3120	0.4811	0.02658	1.438×10^{-4}	1.037×10^{-5}	1.430	1.218	0.00254	0.02132
30	1167	595.2	9.055	1144	4828	3232	0.4695	0.02748	1.361×10^{-4}	1.057×10^{-5}	1.399	1.244	0.00264	0.02024
35	1351	587.4	10.46	1122	4877	3354	0.4579	0.02843	1.288×10^{-4}	1.078×10^{-5}	1.372	1.272	0.00275	0.01917
40	1555	579.4	12.03	1099	4932	3486	0.4464	0.02943	1.219×10^{-4}	1.099×10^{-5}	1.347	1.303	0.00287	0.01810
45	1782	571.3	13.8	1075	4993	3631	0.4348	0.03049	1.155×10^{-4}	1.121×10^{-5}	1.327	1.335	0.00301	0.01704
50	2033	562.9	15.78	1051	5063	3790	0.4232	0.03162	1.094×10^{-4}	1.143×10^{-5}	1.310	1.371	0.00316	0.01598
55	2310	554.2	18.00	1025	5143	3967	0.4116	0.03283	1.037×10^{-4}	1.166×10^{-5}	1.297	1.409	0.00334	0.01493
60	2614	545.2	20.48	997.4	5234	4163	0.4001	0.03412	9.846×10^{-5}	1.189×10^{-5}	1.288	1.452	0.00354	0.01389
65	2948	536.0	23.26	968.9	5340	4384	0.3885	0.03550	9.347×10^{-5}	1.213×10^{-5}	1.285	1.499	0.00377	0.01285
70	3312	526.3	26.39	939.0	5463	4634	0.3769	0.03700	8.879×10^{-5}	1.238×10^{-5}	1.287	1.551	0.00404	0.01181
75	3709	516.2	29.90	907.5	5608	4923	0.3653	0.03862	8.440×10^{-5}	1.264×10^{-5}	1.296	1.612	0.00436	0.01079
80	4141	505.7	33.87	874.1	5780	5260	0.3538	0.04038	8.030×10^{-5}	1.292×10^{-5}	1.312	1.683	0.00474	0.00977
85	4609	494.5	38.36	838.6	5988	5659	0.3422	0.04232	7.646×10^{-5}	1.322×10^{-5}	1.338	1.768	0.00521	0.00876
90	5116	482.8	43.48	800.6	6242	6142	0.3306	0.04447	7.284×10^{-5}	1.354×10^{-5}	1.375	1.871	0.00579	0.00776
95	5665	470.2	49.35	759.8	6561	6740	0.3190	0.04687	6.946×10^{-5}	1.389×10^{-5}	1.429	1.999	0.00652	0.00677
100	6257	456.6	56.15	715.5	6972	7503	0.3075	0.04958	6.628×10^{-5}	1.429×10^{-5}	1.503	2.163	0.00749	0.00579

Note 1: Kinematic viscosity ν and thermal diffusivity α can be calculated from their definitions, $\nu = \mu/\rho$ and $\alpha = k/\rho c_p = \nu/Pr$. The properties listed here (except the vapor density) can be used at any pressures with negligible error except at temperatures near the critical-point value.

Note 2: The unit kJ/kg · °C for specific heat is equivalent to kJ/kg · K, and the unit W/m · °C for thermal conductivity is equivalent to W/m · K.

Source: Data generated from the EES software developed by S. A. Klein and F. L. Alvarado. Original sources: Tillner-Roth, Harms-Watzenberg, and Baehr, "Eine neue Fundamentalgleichung fur Ammoniak," DKV-Tagungsbericht 20:167–181, 1993; Liley and Desai, "Thermophysical Properties of Refrigerants," *ASHRAE*, 1993, ISBN 1-1883413-10-9.

TABLE A-12

Properties of saturated propane

Temp. T, °C	Saturation Pressure P, kPa	Density ρ, kg/m³ Liquid	Density Vapor	Enthalpy of Vaporization h_{fg}, kJ/kg	Specific Heat c_p, J/kg·K Liquid	Specific Heat Vapor	Thermal Conductivity k, W/m·K Liquid	Thermal Conductivity Vapor	Dynamic Viscosity μ, kg/m·s Liquid	Dynamic Viscosity Vapor	Prandtl Number Pr Liquid	Prandtl Vapor	Volume Expansion Coefficient β, 1/K Liquid	Surface Tension, N/m
−120	0.4053	664.7	0.01408	498.3	2003	1115	0.1802	0.00589	6.136×10^{-4}	4.372×10^{-6}	6.820	0.827	0.00153	0.02630
−110	1.157	654.5	0.03776	489.3	2021	1148	0.1738	0.00645	5.054×10^{-4}	4.625×10^{-6}	5.878	0.822	0.00157	0.02486
−100	2.881	644.2	0.08872	480.4	2044	1183	0.1672	0.00705	4.252×10^{-4}	4.881×10^{-6}	5.195	0.819	0.00161	0.02344
−90	6.406	633.8	0.1870	471.5	2070	1221	0.1606	0.00769	3.635×10^{-4}	5.143×10^{-6}	4.686	0.817	0.00166	0.02202
−80	12.97	623.2	0.3602	462.4	2100	1263	0.1539	0.00836	3.149×10^{-4}	5.409×10^{-6}	4.297	0.817	0.00171	0.02062
−70	24.26	612.5	0.6439	453.1	2134	1308	0.1472	0.00908	2.755×10^{-4}	5.680×10^{-6}	3.994	0.818	0.00177	0.01923
−60	42.46	601.5	1.081	443.5	2173	1358	0.1407	0.00985	2.430×10^{-4}	5.956×10^{-6}	3.755	0.821	0.00184	0.01785
−50	70.24	590.3	1.724	433.6	2217	1412	0.1343	0.01067	2.158×10^{-4}	6.239×10^{-6}	3.563	0.825	0.00192	0.01649
−40	110.7	578.8	2.629	423.1	2258	1471	0.1281	0.01155	1.926×10^{-4}	6.529×10^{-6}	3.395	0.831	0.00201	0.01515
−30	167.3	567.0	3.864	412.1	2310	1535	0.1221	0.01250	1.726×10^{-4}	6.827×10^{-6}	3.266	0.839	0.00213	0.01382
−20	243.8	554.7	5.503	400.3	2368	1605	0.1163	0.01351	1.551×10^{-4}	7.136×10^{-6}	3.158	0.848	0.00226	0.01251
−10	344.4	542.0	7.635	387.8	2433	1682	0.1107	0.01459	1.397×10^{-4}	7.457×10^{-6}	3.069	0.860	0.00242	0.01122
0	473.3	528.7	10.36	374.2	2507	1768	0.1054	0.01576	1.259×10^{-4}	7.794×10^{-6}	2.996	0.875	0.00262	0.00996
5	549.8	521.8	11.99	367.0	2547	1814	0.1028	0.01637	1.195×10^{-4}	7.970×10^{-6}	2.964	0.883	0.00273	0.00934
10	635.1	514.7	13.81	359.5	2590	1864	0.1002	0.01701	1.135×10^{-4}	8.151×10^{-6}	2.935	0.893	0.00286	0.00872
15	729.8	507.5	15.85	351.7	2637	1917	0.0977	0.01767	1.077×10^{-4}	8.339×10^{-6}	2.909	0.905	0.00301	0.00811
20	834.4	500.0	18.13	343.4	2688	1974	0.0952	0.01836	1.022×10^{-4}	8.534×10^{-6}	2.886	0.918	0.00318	0.00751
25	949.7	492.2	20.68	334.8	2742	2036	0.0928	0.01908	9.702×10^{-5}	8.738×10^{-6}	2.866	0.933	0.00337	0.00691
30	1076	484.2	23.53	325.8	2802	2104	0.0904	0.01982	9.197×10^{-5}	8.952×10^{-6}	2.850	0.950	0.00358	0.00633
35	1215	475.8	26.72	316.2	2869	2179	0.0881	0.02061	8.710×10^{-5}	9.178×10^{-6}	2.837	0.971	0.00384	0.00575
40	1366	467.1	30.29	306.1	2943	2264	0.0857	0.02142	8.240×10^{-5}	9.417×10^{-6}	2.828	0.995	0.00413	0.00518
45	1530	458.0	34.29	295.3	3026	2361	0.0834	0.02228	7.785×10^{-5}	9.674×10^{-6}	2.824	1.025	0.00448	0.00463
50	1708	448.5	38.79	283.9	3122	2473	0.0811	0.02319	7.343×10^{-5}	9.950×10^{-5}	2.826	1.061	0.00491	0.00408
60	2110	427.5	49.66	258.4	3283	2769	0.0765	0.02517	6.487×10^{-5}	1.058×10^{-5}	2.784	1.164	0.00609	0.00303
70	2580	403.2	64.02	228.0	3595	3241	0.0717	0.02746	5.649×10^{-5}	1.138×10^{-5}	2.834	1.343	0.00811	0.00204
80	3127	373.0	84.28	189.7	4501	4173	0.0663	0.03029	4.790×10^{-5}	1.249×10^{-5}	3.251	1.722	0.01248	0.00114
90	3769	329.1	118.6	133.2	6977	7239	0.0595	0.03441	3.807×10^{-5}	1.448×10^{-5}	4.465	3.047	0.02847	0.00037

Note 1: Kinematic viscosity ν and thermal diffusivity α can be calculated from their definitions, $\nu = \mu/\rho$ and $\alpha = k/\mu c_p = \nu/Pr$. The properties listed here (except the vapor density) can be used at any pressures with negligible error except at temperatures near the critical-point value.

Note 2: The unit kJ/kg·°C for specific heat is equivalent to kJ/kg·K, and the unit W/m·°C for thermal conductivity is equivalent to W/m·K.

Source: Data generated from the EES software developed by S. A. Klein and F. L. Alvarado. Original sources: Reiner Tillner-Roth, "Fundamental Equations of State," Shaker, Verlag, Aachan, 1998; B. A. Younglove and J. F. Ely, "Thermophysical Properties of Fluids. II Methane, Ethane, Propane, Isobutane, and Normal Butane," *J. Phys. Chem. Ref. Data*, Vol. 16, No. 4, 1987; G.R. Somayajulu, "A Generalized Equation for Surface Tension from the Triple-Point to the Critical-Point," *International Journal of Thermophysics*, Vol. 9, No. 4, 1988.

TABLE A–13

Properties of liquids

Temp. T, °C	Density ρ, kg/m³	Specific Heat c_p, J/kg · K	Thermal Conductivity k, W/m · K	Thermal Diffusivity α, m²/s	Dynamic Viscosity μ, kg/m · s	Kinematic Viscosity ν, m²/s	Prandtl Number Pr	Volume Expansion Coeff. β, 1/K
\multicolumn{9}{c}{Methane [CH$_4$]}
−160	420.2	3492	0.1863	1.270×10^{-7}	1.133×10^{-4}	2.699×10^{-7}	2.126	0.00352
−150	405.0	3580	0.1703	1.174×10^{-7}	9.169×10^{-5}	2.264×10^{-7}	1.927	0.00391
−140	388.8	3700	0.1550	1.077×10^{-7}	7.551×10^{-5}	1.942×10^{-7}	1.803	0.00444
−130	371.1	3875	0.1402	9.749×10^{-8}	6.288×10^{-5}	1.694×10^{-7}	1.738	0.00520
−120	351.4	4146	0.1258	8.634×10^{-8}	5.257×10^{-5}	1.496×10^{-7}	1.732	0.00637
−110	328.8	4611	0.1115	7.356×10^{-8}	4.377×10^{-5}	1.331×10^{-7}	1.810	0.00841
−100	301.0	5578	0.0967	5.761×10^{-8}	3.577×10^{-5}	1.188×10^{-7}	2.063	0.01282
−90	261.7	8902	0.0797	3.423×10^{-8}	2.761×10^{-5}	1.055×10^{-7}	3.082	0.02922
\multicolumn{9}{c}{Methanol [CH$_3$(OH)]}
20	788.4	2515	0.1987	1.002×10^{-7}	5.857×10^{-4}	7.429×10^{-7}	7.414	0.00118
30	779.1	2577	0.1980	9.862×10^{-8}	5.088×10^{-4}	6.531×10^{-7}	6.622	0.00120
40	769.6	2644	0.1972	9.690×10^{-8}	4.460×10^{-4}	5.795×10^{-7}	5.980	0.00123
50	760.1	2718	0.1965	9.509×10^{-8}	3.942×10^{-4}	5.185×10^{-7}	5.453	0.00127
60	750.4	2798	0.1957	9.320×10^{-8}	3.510×10^{-4}	4.677×10^{-7}	5.018	0.00132
70	740.4	2885	0.1950	9.128×10^{-8}	3.146×10^{-4}	4.250×10^{-7}	4.655	0.00137
\multicolumn{9}{c}{Isobutane (R600a)}
−100	683.8	1881	0.1383	1.075×10^{-7}	9.305×10^{-4}	1.360×10^{-6}	12.65	0.00142
−75	659.3	1970	0.1357	1.044×10^{-7}	5.624×10^{-4}	8.531×10^{-7}	8.167	0.00150
−50	634.3	2069	0.1283	9.773×10^{-8}	3.769×10^{-4}	5.942×10^{-7}	6.079	0.00161
−25	608.2	2180	0.1181	8.906×10^{-8}	2.688×10^{-4}	4.420×10^{-7}	4.963	0.00177
0	580.6	2306	0.1068	7.974×10^{-8}	1.993×10^{-4}	3.432×10^{-7}	4.304	0.00199
25	550.7	2455	0.0956	7.069×10^{-8}	1.510×10^{-4}	2.743×10^{-7}	3.880	0.00232
50	517.3	2640	0.0851	6.233×10^{-8}	1.155×10^{-4}	2.233×10^{-7}	3.582	0.00286
75	478.5	2896	0.0757	5.460×10^{-8}	8.785×10^{-5}	1.836×10^{-7}	3.363	0.00385
100	429.6	3361	0.0669	4.634×10^{-8}	6.483×10^{-5}	1.509×10^{-7}	3.256	0.00628
\multicolumn{9}{c}{Glycerin}
0	1276	2262	0.2820	9.773×10^{-8}	10.49	8.219×10^{-3}	84,101	
5	1273	2288	0.2835	9.732×10^{-8}	6.730	5.287×10^{-3}	54,327	
10	1270	2320	0.2846	9.662×10^{-8}	4.241	3.339×10^{-3}	34,561	
15	1267	2354	0.2856	9.576×10^{-8}	2.496	1.970×10^{-3}	20,570	
20	1264	2386	0.2860	9.484×10^{-8}	1.519	1.201×10^{-3}	12,671	
25	1261	2416	0.2860	9.388×10^{-8}	0.9934	7.878×10^{-4}	8,392	
30	1258	2447	0.2860	9.291×10^{-8}	0.6582	5.232×10^{-4}	5,631	
35	1255	2478	0.2860	9.195×10^{-8}	0.4347	3.464×10^{-4}	3,767	
40	1252	2513	0.2863	9.101×10^{-8}	0.3073	2.455×10^{-4}	2,697	
\multicolumn{9}{c}{Engine Oil (unused)}
0	899.0	1797	0.1469	9.097×10^{-8}	3.814	4.242×10^{-3}	46,636	0.00070
20	888.1	1881	0.1450	8.680×10^{-8}	0.8374	9.429×10^{-4}	10,863	0.00070
40	876.0	1964	0.1444	8.391×10^{-8}	0.2177	2.485×10^{-4}	2,962	0.00070
60	863.9	2048	0.1404	7.934×10^{-8}	0.07399	8.565×10^{-5}	1,080	0.00070
80	852.0	2132	0.1380	7.599×10^{-8}	0.03232	3.794×10^{-5}	499.3	0.00070
100	840.0	2220	0.1367	7.330×10^{-8}	0.01718	2.046×10^{-5}	279.1	0.00070
120	828.9	2308	0.1347	7.042×10^{-8}	0.01029	1.241×10^{-5}	176.3	0.00070
140	816.8	2395	0.1330	6.798×10^{-8}	0.006558	8.029×10^{-6}	118.1	0.00070
150	810.3	2441	0.1327	6.708×10^{-8}	0.005344	6.595×10^{-6}	98.31	0.00070

Source: Data generated from the EES software developed by S. A. Klein and F. L. Alvarado. Originally based on various sources.

TABLE A–14

Properties of liquid metals

Temp. T, °C	Density ρ, kg/m³	Specific Heat c_p, J/kg · K	Thermal Conductivity k, W/m · K	Thermal Diffusivity α, m²/s	Dynamic Viscosity μ, kg/m · s	Kinematic Viscosity ν, m²/s	Prandtl Number Pr	Volume Expansion Coeff. β, 1/K
				Mercury (Hg) Melting Point: −39°C				
0	13595	140.4	8.18200	4.287×10^{-6}	1.687×10^{-3}	1.241×10^{-7}	0.0289	1.810×10^{-4}
25	13534	139.4	8.51533	4.514×10^{-6}	1.534×10^{-3}	1.133×10^{-7}	0.0251	1.810×10^{-4}
50	13473	138.6	8.83632	4.734×10^{-6}	1.423×10^{-3}	1.056×10^{-7}	0.0223	1.810×10^{-4}
75	13412	137.8	9.15632	4.956×10^{-6}	1.316×10^{-3}	9.819×10^{-8}	0.0198	1.810×10^{-4}
100	13351	137.1	9.46706	5.170×10^{-6}	1.245×10^{-3}	9.326×10^{-8}	0.0180	1.810×10^{-4}
150	13231	136.1	10.07780	5.595×10^{-6}	1.126×10^{-3}	8.514×10^{-8}	0.0152	1.810×10^{-4}
200	13112	135.5	10.65465	5.996×10^{-6}	1.043×10^{-3}	7.959×10^{-8}	0.0133	1.815×10^{-4}
250	12993	135.3	11.18150	6.363×10^{-6}	9.820×10^{-4}	7.558×10^{-8}	0.0119	1.829×10^{-4}
300	12873	135.3	11.68150	6.705×10^{-6}	9.336×10^{-4}	7.252×10^{-8}	0.0108	1.854×10^{-4}
				Bismuth (Bi) Melting Point: 271°C				
350	9969	146.0	16.28	1.118×10^{-5}	1.540×10^{-3}	1.545×10^{-7}	0.01381	
400	9908	148.2	16.10	1.096×10^{-5}	1.422×10^{-3}	1.436×10^{-7}	0.01310	
500	9785	152.8	15.74	1.052×10^{-5}	1.188×10^{-3}	1.215×10^{-7}	0.01154	
600	9663	157.3	15.60	1.026×10^{-5}	1.013×10^{-3}	1.048×10^{-7}	0.01022	
700	9540	161.8	15.60	1.010×10^{-5}	8.736×10^{-4}	9.157×10^{-8}	0.00906	
				Lead (Pb) Melting Point: 327°C				
400	10506	158	15.97	9.623×10^{-6}	2.277×10^{-3}	2.167×10^{-7}	0.02252	
450	10449	156	15.74	9.649×10^{-6}	2.065×10^{-3}	1.976×10^{-7}	0.02048	
500	10390	155	15.54	9.651×10^{-6}	1.884×10^{-3}	1.814×10^{-7}	0.01879	
550	10329	155	15.39	9.610×10^{-6}	1.758×10^{-3}	1.702×10^{-7}	0.01771	
600	10267	155	15.23	9.568×10^{-6}	1.632×10^{-3}	1.589×10^{-7}	0.01661	
650	10206	155	15.07	9.526×10^{-6}	1.505×10^{-3}	1.475×10^{-7}	0.01549	
700	10145	155	14.91	9.483×10^{-6}	1.379×10^{-3}	1.360×10^{-7}	0.01434	
				Sodium (Na) Melting Point: 98°C				
100	927.3	1378	85.84	6.718×10^{-5}	6.892×10^{-4}	7.432×10^{-7}	0.01106	
200	902.5	1349	80.84	6.639×10^{-5}	5.385×10^{-4}	5.967×10^{-7}	0.008987	
300	877.8	1320	75.84	6.544×10^{-5}	3.878×10^{-4}	4.418×10^{-7}	0.006751	
400	853.0	1296	71.20	6.437×10^{-5}	2.720×10^{-4}	3.188×10^{-7}	0.004953	
500	828.5	1284	67.41	6.335×10^{-5}	2.411×10^{-4}	2.909×10^{-7}	0.004593	
600	804.0	1272	63.63	6.220×10^{-5}	2.101×10^{-4}	2.614×10^{-7}	0.004202	
				Potassium (K) Melting Point: 64°C				
200	795.2	790.8	43.99	6.995×10^{-5}	3.350×10^{-4}	4.213×10^{-7}	0.006023	
300	771.6	772.8	42.01	7.045×10^{-5}	2.667×10^{-4}	3.456×10^{-7}	0.004906	
400	748.0	754.8	40.03	7.090×10^{-5}	1.984×10^{-4}	2.652×10^{-7}	0.00374	
500	723.9	750.0	37.81	6.964×10^{-5}	1.668×10^{-4}	2.304×10^{-7}	0.003309	
600	699.6	750.0	35.50	6.765×10^{-5}	1.487×10^{-4}	2.126×10^{-7}	0.003143	
				Sodium-Potassium (%22Na-%78K) Melting Point: −11°C				
100	847.3	944.4	25.64	3.205×10^{-5}	5.707×10^{-4}	6.736×10^{-7}	0.02102	
200	823.2	922.5	26.27	3.459×10^{-5}	4.587×10^{-4}	5.572×10^{-7}	0.01611	
300	799.1	900.6	26.89	3.736×10^{-5}	3.467×10^{-4}	4.339×10^{-7}	0.01161	
400	775.0	879.0	27.50	4.037×10^{-5}	2.357×10^{-4}	3.041×10^{-7}	0.00753	
500	751.5	880.1	27.89	4.217×10^{-5}	2.108×10^{-4}	2.805×10^{-7}	0.00665	
600	728.0	881.2	28.28	4.408×10^{-5}	1.859×10^{-4}	2.553×10^{-7}	0.00579	

Source: Data generated from the EES software developed by S. A. Klein and F. L. Alvarado. Originally based on various sources.

TABLE A–15

Properties of air at 1 atm pressure

Temp. T, °C	Density ρ, kg/m^3	Specific Heat c_p, J/kg · K	Thermal Conductivity k, W/m · K	Thermal Diffusivity α, m^2/s^2	Dynamic Viscosity μ, kg/m · s	Kinematic Viscosity ν, m^2/s	Prandtl Number Pr
−150	2.866	983	0.01171	4.158×10^{-6}	8.636×10^{-6}	3.013×10^{-6}	0.7246
−100	2.038	966	0.01582	8.036×10^{-6}	1.189×10^{-5}	5.837×10^{-6}	0.7263
−50	1.582	999	0.01979	1.252×10^{-5}	1.474×10^{-5}	9.319×10^{-6}	0.7440
−40	1.514	1002	0.02057	1.356×10^{-5}	1.527×10^{-5}	1.008×10^{-5}	0.7436
−30	1.451	1004	0.02134	1.465×10^{-5}	1.579×10^{-5}	1.087×10^{-5}	0.7425
−20	1.394	1005	0.02211	1.578×10^{-5}	1.630×10^{-5}	1.169×10^{-5}	0.7408
−10	1.341	1006	0.02288	1.696×10^{-5}	1.680×10^{-5}	1.252×10^{-5}	0.7387
0	1.292	1006	0.02364	1.818×10^{-5}	1.729×10^{-5}	1.338×10^{-5}	0.7362
5	1.269	1006	0.02401	1.880×10^{-5}	1.754×10^{-5}	1.382×10^{-5}	0.7350
10	1.246	1006	0.02439	1.944×10^{-5}	1.778×10^{-5}	1.426×10^{-5}	0.7336
15	1.225	1007	0.02476	2.009×10^{-5}	1.802×10^{-5}	1.470×10^{-5}	0.7323
20	1.204	1007	0.02514	2.074×10^{-5}	1.825×10^{-5}	1.516×10^{-5}	0.7309
25	1.184	1007	0.02551	2.141×10^{-5}	1.849×10^{-5}	1.562×10^{-5}	0.7296
30	1.164	1007	0.02588	2.208×10^{-5}	1.872×10^{-5}	1.608×10^{-5}	0.7282
35	1.145	1007	0.02625	2.277×10^{-5}	1.895×10^{-5}	1.655×10^{-5}	0.7268
40	1.127	1007	0.02662	2.346×10^{-5}	1.918×10^{-5}	1.702×10^{-5}	0.7255
45	1.109	1007	0.02699	2.416×10^{-5}	1.941×10^{-5}	1.750×10^{-5}	0.7241
50	1.092	1007	0.02735	2.487×10^{-5}	1.963×10^{-5}	1.798×10^{-5}	0.7228
60	1.059	1007	0.02808	2.632×10^{-5}	2.008×10^{-5}	1.896×10^{-5}	0.7202
70	1.028	1007	0.02881	2.780×10^{-5}	2.052×10^{-5}	1.995×10^{-5}	0.7177
80	0.9994	1008	0.02953	2.931×10^{-5}	2.096×10^{-5}	2.097×10^{-5}	0.7154
90	0.9718	1008	0.03024	3.086×10^{-5}	2.139×10^{-5}	2.201×10^{-5}	0.7132
100	0.9458	1009	0.03095	3.243×10^{-5}	2.181×10^{-5}	2.306×10^{-5}	0.7111
120	0.8977	1011	0.03235	3.565×10^{-5}	2.264×10^{-5}	2.522×10^{-5}	0.7073
140	0.8542	1013	0.03374	3.898×10^{-5}	2.345×10^{-5}	2.745×10^{-5}	0.7041
160	0.8148	1016	0.03511	4.241×10^{-5}	2.420×10^{-5}	2.975×10^{-5}	0.7014
180	0.7788	1019	0.03646	4.593×10^{-5}	2.504×10^{-5}	3.212×10^{-5}	0.6992
200	0.7459	1023	0.03779	4.954×10^{-5}	2.577×10^{-5}	3.455×10^{-5}	0.6974
250	0.6746	1033	0.04104	5.890×10^{-5}	2.760×10^{-5}	4.091×10^{-5}	0.6946
300	0.6158	1044	0.04418	6.871×10^{-5}	2.934×10^{-5}	4.765×10^{-5}	0.6935
350	0.5664	1056	0.04721	7.892×10^{-5}	3.101×10^{-5}	5.475×10^{-5}	0.6937
400	0.5243	1069	0.05015	8.951×10^{-5}	3.261×10^{-5}	6.219×10^{-5}	0.6948
450	0.4880	1081	0.05298	1.004×10^{-4}	3.415×10^{-5}	6.997×10^{-5}	0.6965
500	0.4565	1093	0.05572	1.117×10^{-4}	3.563×10^{-5}	7.806×10^{-5}	0.6986
600	0.4042	1115	0.06093	1.352×10^{-4}	3.846×10^{-5}	9.515×10^{-5}	0.7037
700	0.3627	1135	0.06581	1.598×10^{-4}	4.111×10^{-5}	1.133×10^{-4}	0.7092
800	0.3289	1153	0.07037	1.855×10^{-4}	4.362×10^{-5}	1.326×10^{-4}	0.7149
900	0.3008	1169	0.07465	2.122×10^{-4}	4.600×10^{-5}	1.529×10^{-4}	0.7206
1000	0.2772	1184	0.07868	2.398×10^{-4}	4.826×10^{-5}	1.741×10^{-4}	0.7260
1500	0.1990	1234	0.09599	3.908×10^{-4}	5.817×10^{-5}	2.922×10^{-4}	0.7478
2000	0.1553	1264	0.11113	5.664×10^{-4}	6.630×10^{-5}	4.270×10^{-4}	0.7539

Note: For ideal gases, the properties c_p, k, μ, and Pr are independent of pressure. The properties ρ, ν, and α at a pressure P (in atm) other than 1 atm are determined by multiplying the values of ρ at the given temperature by P and by dividing ν and α by P.

Source: Data generated from the EES software developed by S. A. Klein and F. L. Alvarado. Original sources: Keenan, Chao, Keyes, Gas Tables, Wiley, 198; and Thermophysical Properties of Matter. Vol. 3: Thermal Conductivity, Y. S. Touloukian, P. E. Liley, S. C. Saxena, Vol. 11: Viscosity, Y. S. Touloukian, S. C. Saxena, and P. Hestermans, IFI/Plenun, NY, 1970, ISBN 0-306067020-8.

TABLE A–16

Properties of gases at 1 atm pressure

Temp. T, °C	Density ρ, kg/m³	Specific Heat c_p, J/kg·K	Thermal Conductivity k, W/m·K	Thermal Diffusivity α, m²/s²	Dynamic Viscosity μ, kg/m·s	Kinematic Viscosity ν, m²/s	Prandtl Number Pr
Carbon Dioxide, CO_2							
−50	2.4035	746	0.01051	5.860×10^{-6}	1.129×10^{-5}	4.699×10^{-6}	0.8019
0	1.9635	811	0.01456	9.141×10^{-6}	1.375×10^{-5}	7.003×10^{-6}	0.7661
50	1.6597	866.6	0.01858	1.291×10^{-5}	1.612×10^{-5}	9.714×10^{-6}	0.7520
100	1.4373	914.8	0.02257	1.716×10^{-5}	1.841×10^{-5}	1.281×10^{-5}	0.7464
150	1.2675	957.4	0.02652	2.186×10^{-5}	2.063×10^{-5}	1.627×10^{-5}	0.7445
200	1.1336	995.2	0.03044	2.698×10^{-5}	2.276×10^{-5}	2.008×10^{-5}	0.7442
300	0.9358	1060	0.03814	3.847×10^{-5}	2.682×10^{-5}	2.866×10^{-5}	0.7450
400	0.7968	1112	0.04565	5.151×10^{-5}	3.061×10^{-5}	3.842×10^{-5}	0.7458
500	0.6937	1156	0.05293	6.600×10^{-5}	3.416×10^{-5}	4.924×10^{-5}	0.7460
1000	0.4213	1292	0.08491	1.560×10^{-4}	4.898×10^{-5}	1.162×10^{-4}	0.7455
1500	0.3025	1356	0.10688	2.606×10^{-4}	6.106×10^{-5}	2.019×10^{-4}	0.7745
2000	0.2359	1387	0.11522	3.521×10^{-4}	7.322×10^{-5}	3.103×10^{-4}	0.8815
Carbon Monoxide, CO							
−50	1.5297	1081	0.01901	1.149×10^{-5}	1.378×10^{-5}	9.012×10^{-6}	0.7840
0	1.2497	1048	0.02278	1.739×10^{-5}	1.629×10^{-5}	1.303×10^{-5}	0.7499
50	1.0563	1039	0.02641	2.407×10^{-5}	1.863×10^{-5}	1.764×10^{-5}	0.7328
100	0.9148	1041	0.02992	3.142×10^{-5}	2.080×10^{-5}	2.274×10^{-5}	0.7239
150	0.8067	1049	0.03330	3.936×10^{-5}	2.283×10^{-5}	2.830×10^{-5}	0.7191
200	0.7214	1060	0.03656	4.782×10^{-5}	2.472×10^{-5}	3.426×10^{-5}	0.7164
300	0.5956	1085	0.04277	6.619×10^{-5}	2.812×10^{-5}	4.722×10^{-5}	0.7134
400	0.5071	1111	0.04860	8.628×10^{-5}	3.111×10^{-5}	6.136×10^{-5}	0.7111
500	0.4415	1135	0.05412	1.079×10^{-4}	3.379×10^{-5}	7.653×10^{-5}	0.7087
1000	0.2681	1226	0.07894	2.401×10^{-4}	4.557×10^{-5}	1.700×10^{-4}	0.7080
1500	0.1925	1279	0.10458	4.246×10^{-4}	6.321×10^{-5}	3.284×10^{-4}	0.7733
2000	0.1502	1309	0.13833	7.034×10^{-4}	9.826×10^{-5}	6.543×10^{-4}	0.9302
Methane, CH_4							
−50	0.8761	2243	0.02367	1.204×10^{-5}	8.564×10^{-6}	9.774×10^{-6}	0.8116
0	0.7158	2217	0.03042	1.917×10^{-5}	1.028×10^{-5}	1.436×10^{-5}	0.7494
50	0.6050	2302	0.03766	2.704×10^{-5}	1.191×10^{-5}	1.969×10^{-5}	0.7282
100	0.5240	2443	0.04534	3.543×10^{-5}	1.345×10^{-5}	2.567×10^{-5}	0.7247
150	0.4620	2611	0.05344	4.431×10^{-5}	1.491×10^{-5}	3.227×10^{-5}	0.7284
200	0.4132	2791	0.06194	5.370×10^{-5}	1.630×10^{-5}	3.944×10^{-5}	0.7344
300	0.3411	3158	0.07996	7.422×10^{-5}	1.886×10^{-5}	5.529×10^{-5}	0.7450
400	0.2904	3510	0.09918	9.727×10^{-5}	2.119×10^{-5}	7.297×10^{-5}	0.7501
500	0.2529	3836	0.11933	1.230×10^{-4}	2.334×10^{-5}	9.228×10^{-5}	0.7502
1000	0.1536	5042	0.22562	2.914×10^{-4}	3.281×10^{-5}	2.136×10^{-4}	0.7331
1500	0.1103	5701	0.31857	5.068×10^{-4}	4.434×10^{-5}	4.022×10^{-4}	0.7936
2000	0.0860	6001	0.36750	7.120×10^{-4}	6.360×10^{-5}	7.395×10^{-4}	1.0386
Hydrogen, H_2							
−50	0.11010	12635	0.1404	1.009×10^{-4}	7.293×10^{-6}	6.624×10^{-5}	0.6562
0	0.08995	13920	0.1652	1.319×10^{-4}	8.391×10^{-6}	9.329×10^{-5}	0.7071
50	0.07603	14349	0.1881	1.724×10^{-4}	9.427×10^{-6}	1.240×10^{-4}	0.7191
100	0.06584	14473	0.2095	2.199×10^{-4}	1.041×10^{-5}	1.582×10^{-4}	0.7196
150	0.05806	14492	0.2296	2.729×10^{-4}	1.136×10^{-5}	1.957×10^{-4}	0.7174
200	0.05193	14482	0.2486	3.306×10^{-4}	1.228×10^{-5}	2.365×10^{-4}	0.7155

(Continued)

TABLE A–16

Properties of gases at 1 atm pressure *(Continued)*

Temp. T, °C	Density ρ, kg/m³	Specific Heat c_p, J/kg · K	Thermal Conductivity k, W/m · K	Thermal Diffusivity α, m²/s²	Dynamic Viscosity μ, kg/m · s	Kinematic Viscosity ν, m²/s	Prandtl Number Pr
300	0.04287	14481	0.2843	4.580×10^{-4}	1.403×10^{-5}	3.274×10^{-4}	0.7149
400	0.03650	14540	0.3180	5.992×10^{-4}	1.570×10^{-5}	4.302×10^{-4}	0.7179
500	0.03178	14653	0.3509	7.535×10^{-4}	1.730×10^{-5}	5.443×10^{-4}	0.7224
1000	0.01930	15577	0.5206	1.732×10^{-3}	2.455×10^{-5}	1.272×10^{-3}	0.7345
1500	0.01386	16553	0.6581	2.869×10^{-3}	3.099×10^{-5}	2.237×10^{-3}	0.7795
2000	0.01081	17400	0.5480	2.914×10^{-3}	3.690×10^{-5}	3.414×10^{-3}	1.1717

Nitrogen, N_2

Temp. T, °C	Density ρ, kg/m³	Specific Heat c_p, J/kg · K	Thermal Conductivity k, W/m · K	Thermal Diffusivity α, m²/s²	Dynamic Viscosity μ, kg/m · s	Kinematic Viscosity ν, m²/s	Prandtl Number Pr
−50	1.5299	957.3	0.02001	1.366×10^{-5}	1.390×10^{-5}	9.091×10^{-6}	0.6655
0	1.2498	1035	0.02384	1.843×10^{-5}	1.640×10^{-5}	1.312×10^{-5}	0.7121
50	1.0564	1042	0.02746	2.494×10^{-5}	1.874×10^{-5}	1.774×10^{-5}	0.7114
100	0.9149	1041	0.03090	3.244×10^{-5}	2.094×10^{-5}	2.289×10^{-5}	0.7056
150	0.8068	1043	0.03416	4.058×10^{-5}	2.300×10^{-5}	2.851×10^{-5}	0.7025
200	0.7215	1050	0.03727	4.921×10^{-5}	2.494×10^{-5}	3.457×10^{-5}	0.7025
300	0.5956	1070	0.04309	6.758×10^{-5}	2.849×10^{-5}	4.783×10^{-5}	0.7078
400	0.5072	1095	0.04848	8.727×10^{-5}	3.166×10^{-5}	6.242×10^{-5}	0.7153
500	0.4416	1120	0.05358	1.083×10^{-4}	3.451×10^{-5}	7.816×10^{-5}	0.7215
1000	0.2681	1213	0.07938	2.440×10^{-4}	4.594×10^{-5}	1.713×10^{-4}	0.7022
1500	0.1925	1266	0.11793	4.839×10^{-4}	5.562×10^{-5}	2.889×10^{-4}	0.5969
2000	0.1502	1297	0.18590	9.543×10^{-4}	6.426×10^{-5}	4.278×10^{-4}	0.4483

Oxygen, O_2

Temp. T, °C	Density ρ, kg/m³	Specific Heat c_p, J/kg · K	Thermal Conductivity k, W/m · K	Thermal Diffusivity α, m²/s²	Dynamic Viscosity μ, kg/m · s	Kinematic Viscosity ν, m²/s	Prandtl Number Pr
−50	1.7475	984.4	0.02067	1.201×10^{-5}	1.616×10^{-5}	9.246×10^{-6}	0.7694
0	1.4277	928.7	0.02472	1.865×10^{-5}	1.916×10^{-5}	1.342×10^{-5}	0.7198
50	1.2068	921.7	0.02867	2.577×10^{-5}	2.194×10^{-5}	1.818×10^{-5}	0.7053
100	1.0451	931.8	0.03254	3.342×10^{-5}	2.451×10^{-5}	2.346×10^{-5}	0.7019
150	0.9216	947.6	0.03637	4.164×10^{-5}	2.694×10^{-5}	2.923×10^{-5}	0.7019
200	0.8242	964.7	0.04014	5.048×10^{-5}	2.923×10^{-5}	3.546×10^{-5}	0.7025
300	0.6804	997.1	0.04751	7.003×10^{-5}	3.350×10^{-5}	4.923×10^{-5}	0.7030
400	0.5793	1025	0.05463	9.204×10^{-5}	3.744×10^{-5}	6.463×10^{-5}	0.7023
500	0.5044	1048	0.06148	1.163×10^{-4}	4.114×10^{-5}	8.156×10^{-5}	0.7010
1000	0.3063	1121	0.09198	2.678×10^{-4}	5.732×10^{-5}	1.871×10^{-4}	0.6986
1500	0.2199	1165	0.11901	4.643×10^{-4}	7.133×10^{-5}	3.243×10^{-4}	0.6985
2000	0.1716	1201	0.14705	7.139×10^{-4}	8.417×10^{-5}	4.907×10^{-4}	0.6873

Water Vapor, H_2O

Temp. T, °C	Density ρ, kg/m³	Specific Heat c_p, J/kg · K	Thermal Conductivity k, W/m · K	Thermal Diffusivity α, m²/s²	Dynamic Viscosity μ, kg/m · s	Kinematic Viscosity ν, m²/s	Prandtl Number Pr
−50	0.9839	1892	0.01353	7.271×10^{-6}	7.187×10^{-6}	7.305×10^{-6}	1.0047
0	0.8038	1874	0.01673	1.110×10^{-5}	8.956×10^{-6}	1.114×10^{-5}	1.0033
50	0.6794	1874	0.02032	1.596×10^{-5}	1.078×10^{-5}	1.587×10^{-5}	0.9944
100	0.5884	1887	0.02429	2.187×10^{-5}	1.265×10^{-5}	2.150×10^{-5}	0.9830
150	0.5189	1908	0.02861	2.890×10^{-5}	1.456×10^{-5}	2.806×10^{-5}	0.9712
200	0.4640	1935	0.03326	3.705×10^{-5}	1.650×10^{-5}	3.556×10^{-5}	0.9599
300	0.3831	1997	0.04345	5.680×10^{-5}	2.045×10^{-5}	5.340×10^{-5}	0.9401
400	0.3262	2066	0.05467	8.114×10^{-5}	2.446×10^{-5}	7.498×10^{-5}	0.9240
500	0.2840	2137	0.06677	1.100×10^{-4}	2.847×10^{-5}	1.002×10^{-4}	0.9108
1000	0.1725	2471	0.13623	3.196×10^{-4}	4.762×10^{-5}	2.761×10^{-4}	0.8639
1500	0.1238	2736	0.21301	6.288×10^{-4}	6.411×10^{-5}	5.177×10^{-4}	0.8233
2000	0.0966	2928	0.29183	1.032×10^{-3}	7.808×10^{-5}	8.084×10^{-4}	0.7833

Note: For ideal gases, the properties c_p, k, μ, and Pr are independent of pressure. The properties ρ, ν, and α at a pressure P (in atm) other than 1 atm are determined by multiplying the values of ρ at the given temperature by ρ and by dividing ν and α by P.

Source: Data generated from the EES software developed by S. A. Klein and F. L. Alvarado. Originally based on various sources.

TABLE A–17

Properties of the atmosphere at high altitude

Altitude, z, m	Temperature, T, °C	Pressure, P, kPa	Gravity g, m/s²	Speed of Sound, c, m/s	Density, ρ, kg/m³	Viscosity μ, kg/m · s	Thermal Conductivity, k, W/m · K
0	15.00	101.33	9.807	340.3	1.225	1.789×10^{-5}	0.0253
200	13.70	98.95	9.806	339.5	1.202	1.783×10^{-5}	0.0252
400	12.40	96.61	9.805	338.8	1.179	1.777×10^{-5}	0.0252
600	11.10	94.32	9.805	338.0	1.156	1.771×10^{-5}	0.0251
800	9.80	92.08	9.804	337.2	1.134	1.764×10^{-5}	0.0250
1000	8.50	89.88	9.804	336.4	1.112	1.758×10^{-5}	0.0249
1200	7.20	87.72	9.803	335.7	1.090	1.752×10^{-5}	0.0248
1400	5.90	85.60	9.802	334.9	1.069	1.745×10^{-5}	0.0247
1600	4.60	83.53	9.802	334.1	1.048	1.739×10^{-5}	0.0245
1800	3.30	81.49	9.801	333.3	1.027	1.732×10^{-5}	0.0244
2000	2.00	79.50	9.800	332.5	1.007	1.726×10^{-5}	0.0243
2200	0.70	77.55	9.800	331.7	0.987	1.720×10^{-5}	0.0242
2400	−0.59	75.63	9.799	331.0	0.967	1.713×10^{-5}	0.0241
2600	−1.89	73.76	9.799	330.2	0.947	1.707×10^{-5}	0.0240
2800	−3.19	71.92	9.798	329.4	0.928	1.700×10^{-5}	0.0239
3000	−4.49	70.12	9.797	328.6	0.909	1.694×10^{-5}	0.0238
3200	−5.79	68.36	9.797	327.8	0.891	1.687×10^{-5}	0.0237
3400	−7.09	66.63	9.796	327.0	0.872	1.681×10^{-5}	0.0236
3600	−8.39	64.94	9.796	326.2	0.854	1.674×10^{-5}	0.0235
3800	−9.69	63.28	9.795	325.4	0.837	1.668×10^{-5}	0.0234
4000	−10.98	61.66	9.794	324.6	0.819	1.661×10^{-5}	0.0233
4200	−12.3	60.07	9.794	323.8	0.802	1.655×10^{-5}	0.0232
4400	−13.6	58.52	9.793	323.0	0.785	1.648×10^{-5}	0.0231
4600	14.9	57.00	9.793	322.2	0.769	1.642×10^{-5}	0.0230
4800	16.2	55.51	9.792	321.4	0.752	1.635×10^{-6}	0.0229
5000	−17.5	54.05	9.791	320.5	0.736	1.628×10^{-5}	0.0228
5200	−18.8	52.62	9.791	319.7	0.721	1.622×10^{-5}	0.0227
5400	−20.1	51.23	9.790	318.9	0.705	1.615×10^{-5}	0.0226
5600	21.4	49.86	9.789	318.1	0.690	1.608×10^{-5}	0.0224
5800	−22.7	48.52	9.785	317.3	0.675	1.602×10^{-5}	0.0223
6000	−24.0	47.22	9.788	316.5	0.660	1.595×10^{-5}	0.0222
6200	−25.3	45.94	9.788	315.6	0.646	1.588×10^{-5}	0.0221
6400	−26.6	44.69	9.787	314.8	0.631	1.582×10^{-5}	0.0220
6600	−27.9	43.47	9.786	314.0	0.617	1.575×10^{-5}	0.0219
6800	−29.2	42.27	9.785	313.1	0.604	1.568×10^{-5}	0.0218
7000	30.5	41.11	9.785	312.3	0.590	1.561×10^{-5}	0.0217
8000	−36.9	35.65	9.782	308.1	0.526	1.527×10^{-5}	0.0212
9000	−43.4	30.80	9.779	303.8	0.467	1.493×10^{-5}	0.0206
10,000	−49.9	26.50	9.776	299.5	0.414	1.458×10^{-5}	0.0201
12,000	−56.5	19.40	9.770	295.1	0.312	1.422×10^{-5}	0.0195
14,000	−56.5	14.17	9.764	295.1	0.228	1.422×10^{-5}	0.0195
16,000	−56.5	10.53	9.758	295.1	0.166	1.422×10^{-5}	0.0195
18,000	−56.5	7.57	9.751	295.1	0.122	1.422×10^{-5}	0.0195

Source: U.S. Standard Atmosphere Supplements, U.S. Government Printing Office, 1966. Based on year-round mean conditions at 45° latitude and varies with the time of the year and the weather patterns. The conditions at sea level ($z = 0$) are taken to be $P = 101.325$ kPa, $T = 15°C$, $\rho = 1.2250$ kg/m³, $g = 9.80665$ m²/s.

TABLE A–18

Emissivities of surfaces
(a) Metals

Material	Temperature, K	Emissivity, ε	Material	Temperature, K	Emissivity, ε
Aluminum			Magnesium, polished	300–500	0.07–0.13
Polished	300–900	0.04–0.06	Mercury	300–400	0.09–0.12
Commercial sheet	400	0.09	Molybdenum		
Heavily oxidized	400–800	0.20–0.33	Polished	300–2000	0.05–0.21
Anodized	300	0.8	Oxidized	600–800	0.80–0.82
Bismuth, bright	350	0.34	Nickel		
Brass			Polished	500–1200	0.07–0.17
Highly polished	500–650	0.03–0.04	Oxidized	450–1000	0.37–0.57
Polished	350	0.09	Platinum, polished	500–1500	0.06–0.18
Dull plate	300–600	0.22	Silver, polished	300–1000	0.02–0.07
Oxidized	450–800	0.6	Stainless steel		
Chromium, polished	300–1400	0.08–0.40	Polished	300–1000	0.17–0.30
Copper			Lightly oxidized	600–1000	0.30–0.40
Highly polished	300	0.02	Highly oxidized	600–1000	0.70–0.80
Polished	300–500	0.04–0.05	Steel		
Commercial sheet	300	0.15	Polished sheet	300–500	0.08–0.14
Oxidized	600–1000	0.5–0.8	Commercial sheet	500–1200	0.20–0.32
Black oxidized	300	0.78	Heavily oxidized	300	0.81
Gold			Tin, polished	300	0.05
Highly polished	300–1000	0.03–0.06	Tungsten		
Bright foil	300	0.07	Polished	300–2500	0.03–0.29
Iron			Filament	3500	0.39
Highly polished	300–500	0.05–0.07	Zinc		
Case iron	300	0.44	Polished	300–800	0.02–0.05
Wrought iron	300–500	0.28	Oxidized	300	0.25
Rusted	300	0.61			
Oxidized	500–900	0.64–0.78			
Lead					
Polished	300–500	0.06–0.08			
Unoxidized, rough	300	0.43			
Oxidized	300	0.63			

TABLE A–18

Emissivities of surfaces *(Concluded)*
(*b*) Nonmetals

Material	Temperature, K	Emissivity, ε	Material	Temperature, K	Emissivity, ε
Alumina	800–1400	0.65–0.45	Paper, white	300	0.90
Aluminum oxide	600–1500	0.69–0.41	Plaster, white	300	0.93
Asbestos	300	0.96	Porcelain, glazed	300	0.92
Asphalt pavement	300	0.85–0.93	Quartz, rough, fused	300	0.93
Brick			Rubber		
Common	300	0.93–0.96	Hard	300	0.93
Fireclay	1200	0.75	Soft	300	0.86
Carbon filament	2000	0.53	Sand	300	0.90
Cloth	300	0.75–0.90	Silicon carbide	600–1500	0.87–0.85
Concrete	300	0.88–0.94	Skin, human	300	0.95
Glass			Snow	273	0.80–0.90
Window	300	0.90–0.95	Soil, earth	300	0.93–0.96
Pyrex	300–1200	0.82–0.62	Soot	300–500	0.95
Pyroceram	300–1500	0.85–0.57	Teflon	300–500	0.85–0.92
Ice	273	0.95–0.99	Water, deep	273–373	0.95–0.96
Magnesium oxide	400–800	0.69–0.55	Wood		
Masonry	300	0.80	Beech	300	0.94
Paints			Oak	300	0.90
Aluminum	300	0.40–0.50			
Black, lacquer, shiny	300	0.88			
Oils, all colors	300	0.92–0.96			
Red primer	300	0.93			
White acrylic	300	0.90			
White enamel	300	0.90			

TABLE A–19

Solar radiative properties of materials

Description/composition	Solar Absorptivity, α_s	Emissivity, ε, at 300 K	Ratio, α_s/ε	Solar Transmissivity, τ_s
Aluminum				
Polished	0.09	0.03	3.0	
Anodized	0.14	0.84	0.17	
Quartz-overcoated	0.11	0.37	0.30	
Foil	0.15	0.05	3.0	
Brick, red (Purdue)	0.63	0.93	0.68	
Concrete	0.60	0.88	0.68	
Galvanized sheet metal				
Clean, new	0.65	0.13	5.0	
Oxidized, weathered	0.80	0.28	2.9	
Glass, 3.2-mm thickness				
Float or tempered				0.79
Low iron oxide type				0.88
Marble, slightly off-white (nonreflective)	0.40	0.88	0.45	
Metal, plated				
Black sulfide	0.92	0.10	9.2	
Black cobalt oxide	0.93	0.30	3.1	
Black nickel oxide	0.92	0.08	11	
Black chrome	0.87	0.09	9.7	
Mylar, 0.13-mm thickness				0.87
Paints				
Black (Parsons)	0.98	0.98	1.0	
White, acrylic	0.26	0.90	0.29	
White, zinc oxide	0.16	0.93	0.17	
Paper, white	0.27	0.83	0.32	
Plexiglas, 3.2-mm thickness				0.90
Porcelain tiles, white (reflective glazed surface)	0.26	0.85	0.30	
Roofing tiles, bright red				
Dry surface	0.65	0.85	0.76	
Wet surface	0.88	0.91	0.96	
Sand, dry				
Off-white	0.52	0.82	0.63	
Dull red	0.73	0.86	0.82	
Snow				
Fine particles, fresh	0.13	0.82	0.16	
Ice granules	0.33	0.89	0.37	
Steel				
Mirror-finish	0.41	0.05	8.2	
Heavily rusted	0.89	0.92	0.96	
Stone (light pink)	0.65	0.87	0.74	
Tedlar, 0.10-mm thickness				0.92
Teflon, 0.13-mm thickness				0.92
Wood	0.59	0.90	0.66	

Source: V. C. Sharma and A. Sharma, "Solar Properties of Some Building Elements," *Energy* 14 (1989), pp. 805–810, and other sources.

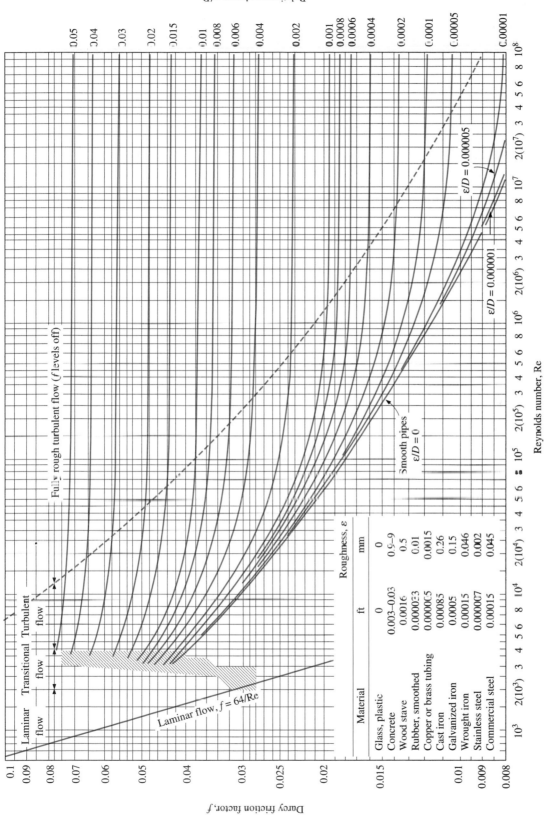

FIGURE A–20

The Moody chart for the friction factor for fully developed flow in circular pipes for use in the head loss relation $\Delta P_L = f \dfrac{L}{D} \dfrac{\rho V^2}{2}$. Friction factors in the turbulent flow are evaluated from the Colebrook equation $\dfrac{1}{\sqrt{f}} = -2\log_{10}\left(\dfrac{\varepsilon/D}{3.7} + \dfrac{2.51}{\mathrm{Re}\sqrt{f}}\right)$.

Relative roughness, ε/D

0.05
0.04
0.03
0.02
0.015
0.01
0.008
0.006
0.004
0.002
0.001
0.0008
0.0006
0.0004
0.0002
0.0001
0.00005
0.00001

Darcy friction factor, f

0.1
0.09
0.08
0.07
0.06
0.05
0.04
0.03
0.025
0.02
0.015
0.01
0.009
0.008

Laminar flow, $f = 64/\mathrm{Re}$

Laminar flow | Transitional flow | Turbulent flow

Fully rough turbulent flow (f levels off)

Smooth pipes $\varepsilon/D = 0$

$\varepsilon/D = 0.000005$

$\varepsilon/D = 0.000001$

Reynolds number, Re

Material	Roughness, ε	
	ft	mm
Glass, plastic	0	0
Concrete	0.003–0.03	0.9–9
Wood stave	0.0016	0.5
Rubber, smoothed	0.000033	0.01
Copper or brass tubing	0.000005	0.0015
Cast iron	0.00085	0.26
Galvanized iron	0.0005	0.15
Wrought iron	0.00015	0.046
Stainless steel	0.000007	0.002
Commercial steel	0.00015	0.045

PROPERTY TABLES AND CHARTS (ENGLISH UNITS)

TABLE A–1E

Molar mass, gas constant, and ideal-gas specific heats of some substances

Substance	Molar Mass, M, lbm/lbmol	Gas Constant, R^*		Specific Heat Data at 77°F		
		Btu/ lbm · R	psia · ft³ / lbm · R	c_p Btu/lbm · R	c_v Btu/lbm · R	$k = c_p/c_v$
Air	28.97	0.06855	0.3704	0.2400	0.1715	1.400
Ammonia, NH_3	17.03	0.1166	0.6301	0.4999	0.3834	1.304
Argon, Ar	39.95	0.04970	0.2686	0.1243	0.07457	1.667
Bromine, Br_2	159.81	0.01242	0.06714	0.0538	0.04137	1.300
Isobutane, C_4H_{10}	58.12	0.03415	0.1846	0.3972	0.3631	1.094
n-Butane, C_4H_{10}	58.12	0.03415	0.1846	0.4046	0.3705	1.092
Carbon dioxide, CO_2	44.01	0.04512	0.2438	0.2016	0.1564	1.288
Carbon monoxide, CO	28.01	0.07089	0.3831	0.2482	0.1772	1.400
Chlorine, Cl_2	70.905	0.02802	0.1514	0.1142	0.08618	1.325
Chlorodifluoromethane (R-22), $CHClF_2$	86.47	0.02297	0.1241	0.1552	0.1322	1.174
Ethane, C_2H_6	30.070	0.06604	0.3569	0.4166	0.3506	1.188
Ethylene, C_2H_4	28.054	0.07079	0.3826	0.3647	0.2940	1.241
Fluorine, F_2	38.00	0.05224	0.2823	0.1967	0.1445	1.362
Helium, He	4.003	0.4961	2.681	1.2403	0.7442	1.667
n-Heptane, C_7H_{16}	100.20	0.01982	0.1071	0.3939	0.3740	1.053
n-Hexane, C_6H_{14}	86.18	0.02304	0.1245	0.3951	0.3721	1.062
Hydrogen, H2	2.016	0.9850	5.323	3.416	2.431	1.405
Krypton, Kr	83.80	0.02370	0.1281	0.05923	0.03554	1.667
Methane, CH_4	16.04	0.1238	0.6688	0.5317	0.4080	1.303
Neon, Ne	20.183	0.09838	0.5316	0.2460	0.1476	1.667
Nitrogen, N_2	28.01	0.07089	0.3831	0.2484	0.1774	1.400
Nitric oxide. NO	30.006	0.06618	0.3577	0.2387	0.1725	1.384
Nitrogen dioxide, NO_2	46.006	0.04512	0.2438	0.1925	0.1474	1.306
Oxygen, O_2	32.00	0.06205	0.3353	0.2193	0.1572	1.395
n-Pentane, C_5H_{12}	72.15	0.02752	0.1487	0.3974	0.3700	1.074
Propane, C_3H_8	44.097	0.04502	0.2433	0.3986	0.3535	1.127
Propylene, C_3H_6	42.08	0.04720	0.2550	0.3657	0.3184	1.148
Steam, H_2O	18.015	0.1102	0.5957	0.4455	0.3351	1.329
Sulfur dioxide, SO_2	64.06	0.03100	0.1675	0.1488	0.1178	1.263
Tetrachloromethane, CCl_4	153.82	0.01291	0.06976	0.1293	0.1164	1.111
Tetrafluoroethane (R-134a), $C_2H_2F_4$	102.03	0.01946	0.1052	0.1991	0.1796	1.108
Trifluoroethane (R-143a), $C_2H_3F_3$	84.04	0.02363	0.1277	0.2219	0.1983	1.119
Xenon, Xe	131.30	0.01512	0.08173	0.03781	0.02269	1.667

*The gas constant is calculated from $R = R_u/M$, where R_u = 1.9859 Btu/lbmol · R = 10.732 psia · ft³/lbmol · R is the universal gas constant and M is the molar mass.

Source: Specific heat values are mostly obtained from the property routines prepared by The National Institute of Standards and Technology (NIST), Gaithersburg, MD.

TABLE A–2E

Boiling and freezing point properties

Substance	Boiling Data at 1 atm		Freezing Data		Liquid Properties		
	Normal Boiling Point, °F	Latent Heat of Vaporization h_{fg}, Btu/lbm	Freezing Point, °F	Latent Heat of Fusion h_{if}, Btu/lbm	Tempera-ture, °F	Density ρ, lbm/ft^3	Specific Heat c_p, Btu/lbm · R
Ammonia	−27.9	24.54	−107.9	138.6	−27.9	42.6	1.06
					0	41.3	1.083
					40	39.5	1.103
					80	37.5	1.135
Argon	−302.6	69.5	−308.7	12.0	−302.6	87.0	0.272
Benzene	176.4	169.4	41.9	54.2	68	54.9	0.411
Brine (20% sodium chloride by mass)	219.0	—	0.7	—	68	71.8	0.743
n-Butane	31.1	165.6	−217.3	34.5	31.1	37.5	0.552
Carbon dioxide	−109.2*	99.6 (at 32°F)	−69.8	—	32	57.8	0.583
Ethanol	172.8	360.5	−173.6	46.9	77	48.9	0.588
Ethyl alcohol	173.5	368	−248.8	46.4	68	49.3	0.678
Ethylene glycol	388.6	344.0	12.6	77.9	68	69.2	0.678
Glycerine	355.8	419	66.0	86.3	68	78.7	0.554
Helium	−452.1	9.80	—	—	−452.1	9.13	5.45
Hydrogen	−423.0	191.7	−434.5	25.6	−423.0	4.41	2.39
Isobutane	10.9	157.8	−255.5	45.5	10.9	37.1	0.545
Kerosene	399–559	108	−12.8	—	68	51.2	0.478
Mercury	674.1	126.7	−38.0	4.90	77	847	0.033
Methane	−258.7	219.6	296.0	25.1	−258.7	26.4	0.834
					−160	20.0	1.074
Methanol	148.1	473	−143.9	42.7	77	49.1	0.609
Nitrogen	320.4	85.4	−346.0	10.9	−320.4	50.5	0.492
					−260	38.2	0.643
Octane	256.6	131.7	71.5	77.9	68	43.9	0.502
Oil (light)	—	—			77	56.8	0.430
Oxygen	−297.3	91.5	−361.8	5.9	−297.3	71.2	0.408
Petroleum	—	99–165			68	40.0	0.478
Propane	−43.7	184.0	−305.8	34.4	−43.7	36.3	0.538
					32	33.0	0.604
					100	29.4	0.673
Refrigerant-134a	−15.0	93.2	−141.9	—	−40	88.5	0.283
					−15	86.0	0.294
					32	80.9	0.318
					90	73.6	0.348
Water	212	970.5	32	143.5	32	62.4	1.01
					90	62.1	1.00
					150	61.2	1.00
					212	59.8	1.01

*Sublimation temperature. (At pressures below the triple-point pressure of 75.1 psia, carbon dioxide exists as a solid or gas. Also, the freezing-point temperature of carbon dioxide is the triple-point temperature of −69.8°F.)

TABLE A–3E

Properties of solid metals

Composition	Melting Point, R	Properties at 540 R ρ lbm/ft³	c_p (Btu/ lbm · R)	k (Btu/ h · ft · R)	$\alpha \times 10^6$ ft²/s	Properties at Various Temperatures (R), k (Btu/h · ft · R)/c_p(Btu/lbm · R) 180	360	720	1080	1440	1800
Aluminum Pure	1679	168	0.216	137	1045	174.5 0.115	137 0.191	138.6 0.226	133.4 0.246	126 0.273	
Alloy 2024-T6 (4.5% Cu, 1.5% Mg, 0.6% Mn)	1395	173	0.209	102.3	785.8	37.6 0.113	94.2 0.188	107.5 0.22	107.5 0.249		
Alloy 195, cast (4.5% Cu)		174.2	0.211	97	734		100.5	106.9			
Beryllium	2790	115.5	0.436	115.6	637.2	572 0.048	174 0.266	93 0.523	72.8 0.621	61.3 0.624	52.5 0.72
Bismuth	981	610.5	0.029	4.6	71	9.5 0.026	5.6 0.028	4.06 0.03			
Boron	4631	156	0.264	15.6	105	109.7 0.03	32.06 0.143	9.7 0.349	6.1 0.451	5.5 0.515	5.7 0.558
Cadmium	1069	540	0.055	55.6	521	117.3 0.047	57.4 0.053	54.7 0.057			
Chromium	3812	447	0.107	54.1	313.2	91.9 0.045	64.1 0.091	52.5 0.115	46.6 0.129	41.2 0.138	37.8 0.147
Cobalt	3184	553.2	0.101	57.3	286.3	96.5 0.056	70.5 0.09	49.3 0.107	39 0.12	33.6 0.131	80.1 0.145
Copper Pure	2445	559	0.092	231.7	1259.3	278.5 0.06	238.6 0.085	227.07 0.094	219 0.01	212 0.103	203.4 0.107
Commercial bronze (90% Cu, 10% Al)	2328	550	0.1	30	150.7	24.3	30 0.187	34 0.109	0.130		
Phosphor gear bronze (89% Cu, 11% Sn)	1987	548.1	0.084	31.2	183	23.7 —	37.6 —	42.8 —			
Cartridge brass (70% Cu, 30% Zn)	2139	532.5	0.09	63.6	364.9	43.3	54.9 0.09	79.2 0.09	86.0 0.101		
Constantan (55% Cu, 45% Ni)	2687	557	0.092	13.3	72.3	9.8 0.06	1.1 0.09				
Germanium	2180	334.6	0.08	34.6	373.5	134 0.045	56 0.069	25 0.08	15.7 0.083	11.4 0.085	10.05 0.089
Gold	2405	1205	0.03	183.2	1367	189 0.026	186.6 0.029	179.7 0.031	172.2 0.032	164.09 0.033	156 0.034
Iridium	4896	1404.6	0.031	85	541.4	99.4 0.021	88.4 0.029	83.2 0.031	79.7 0.032	76.3 0.034	72.8 0.036
Iron: Pure	3258	491.3	0.106	46.4	248.6	77.4 0.051	54.3 0.091	40.2 0.117	31.6 0.137	25.01 0.162	19 0.232
Armco (99.75% pure)		491.3	0.106	42	222.8	55.2 0.051	46.6 0.091	38 0.117	30.7 0.137	24.4 0.162	18.7 0.233
Carbon steels Plain carbon (Mn ≤ 1%, Si ≤ 0.1%)		490.3	0.103	35	190.6			32.8 0.116	27.7 0.113	22.7 0.163	17.4 0.279
AISI 1010		489	0.103	37	202.4			33.9 0.116	28.2 0.133	22.7 0.163	18 0.278
Carbon-silicon (Mn ≤ 1%, 0.1% < Si ≤ 0.6%)		488	0.106	30	160.4			28.8 0.119	25.4 0.139	21.6 0.166	17 0.231
Carbon-manganese-silicon (1% < Mn ≤ 1.65%, 0.1% < Si ≤ 0.6%)		508	0.104	23.7	125			24.4 0.116	23 0.133	20.2 0.163	16 0.260
Chromium (low) steels: $\frac{1}{2}$ Cr–$\frac{1}{4}$ Mo–Si (0.18% C, 0.65% Cr, 0.23% Mo, 0.6% Si)1		488.3	0.106	21.8	117.4			22 0.117	21.2 0.137	19.3 0.164	15.6 0.23
1 Cr–$\frac{1}{2}$ Mo (0.16% C, 1% Cr, 0.54% Mo, 0.39% Si)		490.6	0.106	24.5	131.3			24.3 0.117	22.6 0.137	20 0.164	15.8 0.231
1 Cr–V (0.2% C, 1.02% Cr, 0.15% V)		489.2	0.106	28.3	151.8			27.0 0.117	24.3 0.137	21 0.164	16.3 0.231

TABLE A–3E

Properties of solid metals

Composition	Melting Point, R	Properties at 540 R				Properties at Various Temperatures (R), k (Btu/h · ft · R)/c_p(Btu/lbm · R)					
		ρ lbm/ft³	c_p(Btu/ lbm · R)	k (Btu/ h · ft · R)	$\alpha \times 10^6$ ft²/s	180	360	720	1080	1440	1800
Stainless steels:		503	0.114	8.7	42			10	11.6	13.2	14.7
AISI 302								0.122	0.133	0.140	0.144
AISI 304	3006	493.2	0.114	8.6	42.5	5.31	7.3	9.6	11.5	13	14.7
						0.064	0.096	0.123	0.133	0.139	0.145
AISI 316		514.3	0.111	7.8	37.5			8.8	10.6	12.3	14
						0.12	0.131	0.137	0.143		
AISI 347		498	0.114	8.2	40			9.1	1.1	12.7	14.3
								0.122	0.133	0.14	0.144
Lead	1082	708	0.03	20.4	259.4	23	21.2	19.7	18.1		
						0.028	0.029	0.031	0.034		
Magnesium	1661	109	0.245	90.2	943	87.9	91.9	88.4	86.0	84.4	
						0.155	0.223	0.256	0.279	0.302	
Molybdenum	5209	639.3	0.06	79.7	578	1034	82.6	77.4	72.8	68.2	64.7
						0.038	0.053	0.062	0.065	0.068	0.070
Nickel:	3110	555.6	0.106	52.4	247.6	94.8	61.8	46.3	37.9	39	41.4
Pure						0.055	0.091	0.115	0.141	0.126	0.134
Nichrome	3010	524.4	0.1	6.9	36.6			8.0	9.3	12.2	
(80% Ni, 20% Cr)							0.114	0.125	0.130		
Inconel X-750	2997	531.3	0.104	6.8	33.4	5	5.9	7.8	9.8	11.8	13.9
(73% Ni, 15% Cr, 6.7% Fe)						—	0.088	0.112	0.121	0.13	0.149
Niobium	4934	535	0.063	31	254	31.9	30.4	32	33.6	35.4	32.2
						0.044	0.059	0.065	0.067	0.069	0.071
Palladium	3289	750.4	0.058	41.5	263.7	44.2	41.4	42.5	46	50	54.4
						0.04	0.054	0.059	0.062	0.064	0.067
Platinum:	3681	1339	0.031	41.4	270	44.7	42	41.5	42.3	43.7	45.5
Pure						0.024	0.03	0.032	0.034	0.035	0.036
Alloy 60Pt-40Rh	3240	1038.2	0.038	27.2	187.3			30	34	37.5	40
(60% Pt, 40% Rh)						—	—	—			
Rhenium	6215	1317.2	0.032	27.7	180	34	30	26.6	25.5	25.4	25.8
						0.023	0.03	0.033	0.034	0.036	0.037
Rhodium	4025	777.2	0.058	86.7	5.34	107.5	89	84.3	78.5	73.4	70
						0.035	0.052	0.06	0.065	0.069	0.074
Silicon	3033	145.5	0.17	85.5	960.2	510.8	152.5	57.2	35.8	24.4	18.0
						0.061	0.132	0.189	0.207	0.218	0.226
Silver	2223	656	0.056	248	1873	257	248.4	245.5	238	228.8	219
						0.044	0.053	0.057	0.059	0.062	0.066
Tantalum	5884	1036.3	0.033	33.2	266	34.2	33.2	33.4	34	34.3	34.8
						0.026	0.031	0.034	0.035	0.036	0.036
Thorium	3641	730.4	0.028	31.2	420.9	34.6	31.5	31.4	32.2	32.9	32.9
						0.024	0.027	0.029	0.032	0.035	0.037
Tin	909	456.3	0.054	38.5	431.6	49.2	42.4	35.9			
						0.044	0.051	0.058			
Titanium	3515	281	0.013	12.7	100.3	17.6	14.2	11.8	11.2	11.4	12
						0.071	0.111	0.131	0.141	0.151	0.161
Tungsten	6588	1204.9	0.031	100.5	735.2	120.2	107.5	92	79.2	72.2	68.2
						0.020	0.029	0.032	0.033	0.034	0.035
Uranium	2531	1190.5	0.027	16	134.5	12.5	14.5	17.1	19.6	22.4	25.4
						0.022	0.026	0.029	0.035	0.042	0.043
Vanadium	3946	381	0.117	17.7	110.9	20.7	18	18	19.3	20.6	22.0
						0.061	0.102	0.123	0.128	0.134	0.142
Zinc	1247	445.7	0.093	67	450	67.6	68.2	64.1	59.5		
						0.07	0.087	0.096	0.104		
Zirconium	3825	410.2	0.067	13.1	133.5	19.2	14.6	12.5	12	12.5	13.7
						0.049	0.063	0.072	0.77	0.082	0.087

Source: Tables A–3E and A–4E are obtained from the respective tables in SI units in Appendix 1 using proper conversion factors.

TABLE A–4E

Properties of solid nonmentals

Composition	Melting Point, R	ρ lbm/ft³	c_p (Btu/ lbm · R)	k (Btu/ h · ft · R)	α × 10⁶ ft²/s	180	360	720	1080	1440	1800
		Properties at 540 R				Properties at Various Temperatures (R), k (Btu/h · ft · R)/c_p(Btu/lbm · R)					
Aluminum oxide, sapphire	4181	247.8	0.182	26.6	162.5	260 —	47.4 —	18.7 0.224	11 0.265	7.5 0.281	6 0.293
Aluminum oxide polycrystalline	4181	247.8	0.182	20.8	128	76.8	31.7	15.3 0.244	9.3 0.265	6 0.281	4.5 0.293
Beryilium oxide	4905	187.3	0.246	157.2	947.3			113.2 0.322	64.2 0.40	40.4 0.44	27.2 0.459
Boron	4631	156	0.264	16	107.5	109.8	30.3	10.8 0.355	6.5 0.445	4.6 0.509	3.6 0.561
Boron fiber epoxy (30% vol.) composite	1062	130									
k, ‖ to fibers				1.3		1.2	1.3	1.31			
k, ⊥ to fibers				0.34		0.21	0.28	0.34			
c_p			0.268			0.086	0.18	0.34			
Carbon Amorphous	2700	121.7	—	0.92	—	0.38	0.68	1.09	1.26	1.36	1.46
Diamond, type lla insulator	—	219	0.121	1329	—	5778	2311.2 0.005	889.8 0.046	0.203		
Graphite, pyrolytic	4091	138									
k, ‖ to layers				1126.7		2871.6	1866.3	803.2	515.4	385.4	308.5
k, ⊥ to layers				3.3		9.7	5.3	2.4	1.5	1.16	0.92
c_p			0.169			0.32	0.098	0.236	0.335	0.394	0.428
Graphite fiber epoxy (25% vol.) composite	810	87.4									
k, heat flow ‖ to fibers				6.4		3.3	5.0	7.5			
k, heat flow ⊥ to fibers				0.5	5	0.4	0.63				
c_p			0.223			0.08	0.153	0.29			
Pyroceram, Corning 9606	2921	162.3	0.193	2.3	20.3	3.0	2.3	2.1	1.9	1.7	1.7
Silicon carbide,	5580	197.3	0.161	283.1	2475.7			— 0.210	— 0.25	— 0.27	50.3 0.285
Silicon dioxide, crystalline (quartz)	3389	165.4									
k, ‖ to c-axis				6		22.5	9.5	4.4	2.9	2.4	
k, ⊥ to c-axis				3.6		12.0	5.9	2.7	2	1.8	
c_p			0.177					0.211	0.256	0.298	
Silicon dioxide, polycrystalline (fused silica)	3389	138.6	0.177	0.79	9	0.4 —	0.65 —	0.87 0.216	1.01 0.248	1.25 0.264	1.65 0.276
Silicon nitride	3911	150	0.165	9.2	104	—	— 0.138	8.0 0.185	6.5 0.223	5.7 0.253	5.0 0.275
Sulfur	706	130	0.169	0.1	1.51	0.095 0.962	0.1 0.144				
Thorium dioxide	6431	568.7	0.561	7.5	65.7			5.9 0.609	3.8 0.654	2.7 0.680	2.12 0.704
Titanium dioxide, polycrystalline	3840	259.5	0.170	4.9	30.1			4.0 0.192	2.9 0.210	2.3 0.217	2 0.222

TABLE A–5E

Properties of building materials
(at a mean temperature of 75°F)

Material	Thickness, L in	Density, ρ lbm/ft^3	Thermal Conductivity, k Btu-in/h · ft^2 · °F	Specific Heat, c_p Btu/lbm · R	R-value (for listed thickness, L/k), °F · h · ft^2/Btu
Building Boards					
Asbestos–cement board	¼ in.	120	—	0.24	0.06
Gypsum of plaster board	⅜ in.	50	—	0.26	0.32
	½ in.	50	—	—	0.45
Plywood (Douglas fir)	—	34	0.80	0.29	—
	¼ in.	34	—	0.29	0.31
	⅜ in.	34	—	0.29	0.47
	½ in.	34	—	0.29	0.62
	¾ in.	34	—	0.29	0.93
Insulated board and sheating	½ in.	18	—	0.31	1.32
(regular density)	25/32 in.	18	—	0.31	2.06
Hardboard (high density, standard					
tempered)	—	63	1.00	0.32	—
Particle board					
Medium density	—	50	0.94	0.31	—
Underlayment	⅝ in.	40	—	0.29	0.82
Wood subfloor	¾ in.	—	—	0.33	0.94
Building Membranes					
Vapor-permeable felt	—	—	—	—	0.06
Vapor-seal (2 layers of mopped					
17.3 lbm/ft^2 felt)	—	—	—	—	0.12
Flooring Materials					
Carpet and fibrous pad	—	—	—	0.34	2.08
Carpet and rubber pad	—	—	—	0.33	1.23
Tile (asphalt, linoleum, vinyl)	—	—	—	0.30	0.05
Masonry Materials					
Masonry units:					
Brick, common		120	5.0	—	—
Brick, face		130	9.0	—	—
Brick, fire clay		150	9.3	—	—
		120	6.2	0.19	—
		70	2.8	—	—
Concrete blocks (3 oval cores,					
sand and gravel aggregate)	4 in.	—	5.34	—	0.71
	8 in.	—	6.94	—	1.11
	12 in.	—	9.02	—	1.28
Concretes					
Lightweight aggregates		120	5.2	—	—
(including expanded shale,		100	3.6	0.2	—
clay, or slate, expanded slags,		80	2.5	0.2	—
cinders; pumice; and scoria)		60	1.7	—	—
		40	1.15	—	—
Cement/lime, mortar, and stucco		120	9.7	—	—
		80	4.5	—	—
Stucco		116	5.0		

TABLE A–5E

Properties of building materials *(Concluded)*
(at a mean temperature of 75°F)

Material	Thickness, L in	Density, ρ lbm/ft³	Thermal Conductivity, k Btu-in/h · ft² · °F	Specific Heat, c_p Btu/lbm · R	R-value (for listed thickness, L/k), °F · h · ft²/Btu
Roofing					
Asbestos-cement shingles		120	—	0.24	0.21
Asphalt roll roofing		70	—	0.36	0.15
Asphalt shingles		70	—	0.30	0.44
Built-in roofing	⅜ in.	70	—	0.35	0.33
Slate	½ in.	—	—	0.30	0.05
Wood shingles (plain and plastic film faced)		—	—	0.31	0.94
Plastering Materials					
Cement plaster, sand aggregate	¾ in.	1.16	5.0	0.20	0.15
Gypsum plaster					
Lightweight aggregate	½ in.	45	—	—	0.32
Sand aggregate	½ in.	105	5.6	0.20	0.09
Perlite aggregate	—	45	1.5	0.32	—
Siding Material (on flat surfaces)					
Asbestos-cement shingles	—	120	—	—	0.21
Hardboard siding	7/16 in.	—	—	0.28	0.67
Wood (drop) siding	1 in.	—	—	0.31	0.79
Wood (plywood) siding, lapped	⅜ in.	—	—	0.29	0.59
Aluminum or steel siding (over sheeting):					
Hollow backed	⅜ in.	—	—	0.29	0.61
Insulating-board backed	⅜ in.	—	—	0.32	1.82
Architectural glass	—	158	6.9	0.21	0.10
Woods					
Hardwoods (maple, oak etc.)	—	45	1.10	0.30	—
Softwoods (fir, pine, etc.)	—	32	0.80	0.33	—
Metals					
Aluminum (1100)	—	171	1536	0.214	—
Steel, mild	—	489	314	0.120	—
Steel Stainless,	—	494	108	0.109	—

Source: Tables A-5E and A-6E are adapted from *ASHRAE, Handbook of Fundamentals* (Atlanta, GA: American Society of Heating, Refrigerating, and Air-Conditioning Engineers, 1993), Chap. 22, Table 4. Used with permission.

TABLE A-6E

Properties of insulating materials
(at a mean temperature of 75°F)

Material	Thickness, L in	Density, ρ lbm/ft^3	Thermal Conductivity, k Btu-in/h · ft^2 · °F	Specific Heat, c_p Btu/lbm · R	R-value (for listed thickness, L/k) °F · h · ft^2/Btu
Blanket and Batt					
Mineral fiber (fibrous form	~2 to 2¾ in	0.3–2.0	—	0.17–0.23	7
processed from rock, slag,	~3 to 3½ in	0.3–2.0	—	0.17–0.23	11
or glass)	~5¼ to 6½ in	0.3–2.0	—	0.17–0.23	19
Board and Slab					
Cellular glass		8.5	0.38	0.24	
Glass fiber (organic bonded)		4–9	0.25	0.23	—
Expanded polystyrene (molded beads)		1.0	0.28	0.29	—
Expanded polyurethane (R-11 expanded)		1.5	0.16	0.38	—
Expanded perlite (organic bonded)		1.0	0.36	0.30	—
Expanded rubber (rigid)		4.5	0.22	0.40	—
Mineral fiber with resin binder		15	0.29	0.17	—
Cork		7.5	0.27	0.43	—
Sprayed or Formed in Place					
Polyurethane foam		1.5–2.5	0.16–0.18	—	—
Glass fiber		3.5–4.5	0.26–0.27	—	—
Urethane, two-part mixture (rigid foam)		4.4	0.18	0.25	—
Mineral wool granules with asbestos/inorganic binders (sprayed)		12	0.32	—	
Loose Fill					
Mineral fiber (rock, slag,	~3.75 to 5 in	0.6–0.20	—	0.17	11
or glass)	~6.5 to 8.75 in	0.6–0.20		0.17	19
	~7.5 to 10 in	—	—	0.17	22
	~7.25 in	—	—	0.17	30
Silica aerogel		7.6	0.17	—	—
Vermiculite (expanded)		7–8	0.47	—	—
Perlite, expanded		2–4.1	0.27–0.31	—	—
Sawdust or shavings		8–15	0.45	—	—
Cellulosic insulation (milled paper or wood pulp)		0.3–3.2	0.27–0.32	—	—
Cork, granulated		10	0.31	—	—
Roof Insulation					
Cellular glass	—	9	0.4	0.24	—
Preformed, for use above deck	½ in	—	—	0.24	1.39
	1 in	—	—	0.50	2.78
	2 in	—	—	0.94	5.56
Reflective Insulation					
Silica powder (evacuated)		10	0.0118	—	—
Aluminum foil separating fluffy glass mats; 10–12 layers (evacuated); for cryogenic applications (270 R)		2.5	0.0011	—	—
Aluminum foil and glass paper laminate; 75–150 layers (evacuated); for cryogenic applications (270 R)		7.5	0.00012	—	—

TABLE A–7E

Properties of common foods
(a) Specific heats and freezing-point properties

Food	Water Content, % (mass)	Freezing Point, °F	Specific Heat Above Freezing (Btu/lbm·°F)	Specific Heat Below Freezing (Btu/lbm·°F)	Latent Heat of Fusion Btu/lbm
Vegetables					
Artichokes	84	30	0.873	0.453	121
Asparagus	93	31	0.945	0.481	134
Beans, snap	89	31	0.913	0.468	128
Broccoli	90	31	0.921	0.471	129
Cabbage	92	30	0.937	0.478	132
Carrots	88	29	0.905	0.465	126
Cauliflower	92	31	0.937	0.478	132
Celery	94	31	0.953	0.484	135
Corn, sweet	74	31	0.793	0.423	106
Cucumbers	96	31	0.969	0.490	138
Eggplant	93	31	0.945	0.481	134
Horseradish	75	29	0.801	0.426	108
Leeks	85	31	0.881	0.456	122
Lettuce	95	32	0.961	0.487	136
Mushrooms	91	30	0.929	0.474	131
Okra	90	29	0.921	0.471	129
Onions, green	89	30	0.913	0.468	128
Onions, dry	88	31	0.905	0.465	126
Parsley	85	30	0.881	0.456	122
Peas, green	74	31	0.793	0.423	106
Peppers, sweet	92	31	0.937	0.478	132
Potatoes	78	31	0.825	0.435	112
Pumpkins	91	31	0.929	0.474	131
Spinach	93	31	0.945	0.481	134
Tomatos, ripe	94	31	0.953	0.484	135
Turnips	92	30	0.937	0.478	132
Fruits					
Apples	84	30	0.873	0.453	121
Apricots	85	30	0.881	0.456	122
Avocados	65	31	0.721	0.396	93
Bananas	75	31	0.801	0.426	108
Blueberries	82	29	0.857	0.447	118
Cantaloupes	92	30	0.937	0.478	132
Cherries, sour	84	29	0.873	0.453	121
Cherries, sweet	80	29	0.841	0.441	115
Figs, dried	23	—	—	0.270	33
Figs, fresh	78	28	0.825	0.435	112
Grapefruit	89	30	0.913	0.468	128
Grapes	82	29	0.857	0.447	118
Lemons	89	29	0.913	0.468	128
Olives	75	29	0.801	0.426	108
Oranges	87	31	0.897	0.462	125
Peaches	89	30	0.913	0.468	128
Pears	83	29	0.865	0.450	119
Pineapples	85	30	0.881	0.456	122
Plums	86	31	0.889	0.459	124
Quinces	85	28	0.881	0.456	122
Raisins	18	—	—	0.255	26
Strawberries	90	31	0.921	0.471	129
Tangerines	87	30	0.897	0.462	125
Watermelon	93	31	0.945	0.481	134
Fish/Seafood					
Cod, whole	78	28	0.825	0.435	112
Halibut, whole	75	28	0.801	0.426	108
Lobster	79	28	0.833	0.438	113
Mackerel	57	28	0.657	0.372	82
Salmon, whole	64	28	0.713	0.393	92
Shrimp	83	28	0.865	0.450	119
Meats					
Beef carcass	49	29	0.593	0.348	70
Liver	70	29	0.761	0.411	101
Round, beef	67	—	0.737	0.402	96
Sirloin, beef	56	—	0.649	0.369	80
Chicken	74	27	0.793	0.423	106
Lamb leg	65	—	0.721	0.396	93
Pork carcass	37	—	0.497	0.312	53
Ham	56	29	0.649	0.369	80
Pork sausage	38	—	0.505	0.315	55
Turkey	64	—	0.713	0.393	92
Other					
Almonds	5	—	—	0.216	7
Butter	16	—	—	0.249	23
Cheese, cheddar	37	9	0.497	0.312	53
Cheese, Swiss	39	14	0.513	0.318	56
Chocolate, milk	1	—	—	0.204	1
Eggs, whole	74	31	0.793	0.423	106
Honey	17	—	—	0.252	24
Ice cream	63	22	0.705	0.390	90
Milk, whole	88	31	0.905	0.465	126
Peanuts	6	—	—	0.219	9
Peanuts, roasted	2	—	—	0.207	3
Pecans	3	—	—	0.210	4
Walnuts	4	—	—	0.213	6

Source: [a]Water content and freezing point data are from ASHRAE, *Handbook of Fundamentals*, I-P version (Atlanta, GA: American Society of Heating, Refrigerating, and Air-Conditioning Engineers, Inc., 1993), Chap. 30, Table 1. Used with permission. Freezing point is the temperature at which freezing starts for fruits and vegetables, and the average freezing temperature for other foods.

[b]Specific heat data are based on the specific heat values of water and ice at 32°F and are determined from Siebel's formulas: $c_{p,\,fresh} = 0.800 \times$ (Water content) $+ 0.200$, above freezing, and $c_{p,\,frozen} = 0.300 \times$ (Water content) $+ 0.200$, below freezing.

[c]The latent heat of fusion is determined by multiplying the heat of fusion of water (143 Btu/lbm) by the water content of the food.

TABLE A–7E

Properties of common foods *(Concluded)*
(*b*) Other properties

Food	Water Content, %(mass)	Temperature, T °F	Density, ρ lbm/ft³	Thermal Conductivity, k Btu/h · ft · °F	Thermal Diffusivity, α ft²/S	Specific Heat, c_p Btu/lbm · R
Fruits/Vegetables						
Apple juice	87	68	62.4	0.323	1.51×10^{-6}	0.922
Apples	85	32–86	52.4	0.242	1.47×10^{-6}	0.910
Apples, dried	41.6	73	53.4	0.127	1.03×10^{-6}	0.650
Apricots, dried	43.6	73	82.4	0.217	1.22×10^{-6}	0.662
Bananas, fresh	76	41	61.2	0.278	1.51×10^{-6}	0.856
Broccoli	—	21	35.0	0.223	—	—
Cherries, fresh	92	32–86	65.5	0.315	1.42×10^{-6}	0.952
Figs	40.4	73	77.5	0.179	1.03×10^{-6}	0.642
Grape juice	89	68	62.4	0.328	1.51×10^{-6}	0.934
Peaches	36–90	2–32	59.9	0.304	1.51×10^{-6}	0.934
Plums	—	3	38.1	0.143	–	—
Potatoes	32–158	0–70	65.7	0.288	1.40×10^{-6}	0.868
Raisins	32	73	86.2	0.217	1.18×10^{-6}	0.592
Meats						
Beef, ground	67	43	59.3	0.235	1.40×10^{-6}	0.802
Beef, lean	74	37	68.0	0.272	1.40×10^{-6}	0.844
Beef fat	0	95	50.5	0.110	—	—
Beef liver	72	95	—	0.259	—	0.832
Cat food	39.7	73	71.2	0.188	1.18×10^{-6}	0.638
Chicken breast	75	32	65.5	0.275	1.40×10^{-6}	0.850
Dog food	30.6	73	77.4	0.184	1.18×10^{-6}	0.584
Fish, cod	81	37	73.7	0.309	1.29×10^{-6}	0.886
Fish, salmon	67	37	—	0.307	—	0.802
Ham	71.8	72	64.3	0.277	1.51×10^{-6}	0.831
Lamb	72	72	64.3	0.263	1.40×10^{-6}	0.832
Pork, lean	72	39	64.3	0.263	1.40×10^{-6}	0.832
Turkey breast	74	37	65.5	0.287	1.40×10^{-6}	0.844
Veal	75	72	66.2	0.272	1.40×10^{-6}	0.850
Other						
Butter	16	39	—	0.114	—	0.496
Chocolate cake	31.9	73	21.2	0.061	1.29×10^{-6}	0.591
Margarine	16	40	62.4	0.135	1.18×10^{-6}	0.496
Milk, skimmed	91	72	—	0.327	—	0.946
Milk, whole	88	82	—	0.335	—	0.928
Olive oil	0	90	56.8	0.097	—	—
Peanut oil	0	39	57.4	0.097	—	—
Water	100	0	62.4	0.329	1.51×10^{-6}	1.000
	100	30	59.6	0.357	1.61×10^{-6}	1.000
White cake	32.3	73	28.1	0.047	1.08×10^{-6}	0.594

Source: Data obtained primarily from ASHRAE, *Handbook of Fundamentals*, I-P version (Atlanta, GA: American Society of Heating, Refrigerating, and Air-Conditioning Engineers, Inc., 1993), Chap. 30, Tables 7 and 9. Used with permission.

Most specific heats are calculated from $c_p = 0.4 + 0.6 \times$ (Water content), which is a good approximation in the temperature range of 40 to 90°F. Most thermal diffusivities are calculated from $\alpha = k/\rho c_p$. Property values given above are valid for the specified water content.

TABLE A–8E

Properties of miscellaneous materials
(values are at 540 R unless indicated otherwise)

Material	Density, ρ lbm/ft^3	Thermal Conductivity, k Btu/h · ft · R	Specific Heat, c_p Btu/lbm · R	Material	Density, ρ lbm/ft^3	Thermal Conductivity, k Btu/h · ft · R	Specific Heat, c_p Btu/lbm · R
Asphalt	132.0	0.036	0.220	Ice			
Bakelite	81.2	0.81	0.350	492 R	57.4	1.09	0.487
Brick, refractory				455 R	57.6	1.17	0.465
Chrome brick				311 R	57.9	2.02	0.349
851 R	187.9	1.33	0.199	Leather, sole	62.3	0.092	—
1481 R	—	1.44	—	Linoleum	33.4	0.047	—
2111 R	—	1.16	—		73.7	0.11	—
Fire clay, burnt				Mica	181.0	0.30	—
2880 R				Paper	58.1	0.10	0.320
1391 R	128.0	0.58	0.229	Plastics			
1931 R	—	0.64	—	Plexiglass	74.3	0.11	0.350
2471 R	—	0.64	—	Teflon			
Fire clay, burnt				540 R	137.3	0.20	0.251
3105 R				720 R	—	0.26	—
1391 R	145.1	0.75	0.229	Lexan	74.9	0.11	0.301
1931 R	—	0.81	—	Nylon	71.5	0.17	—
2471 R	—	0.81	—	Polypropylene	56.8	0.069	0.388
Fire clay brick				Polyester	87.1	0.087	0.279
860 R	165.1	0.58	0.229	PVC, vinyl	91.8	0.058	0.201
1660 R	—	0.87	—	Porcelain	143.6	0.87	—
2660 R	—	1.04	—	Rubber, natural	71.8	0.16	—
Magnesite				Rubber, vulcanized			
860 R	—	2.20	0.270	Soft	68.7	0.075	0.480
1660 R	—	1.62	—	Hard	74.3	0.092	—
2660 R	—	1.10	—	Sand	94.6	0.1–0.6	0.191
Chicken meat, white (74.4% water content)				Snow, fresh	6.24	0.35	—
				Snow 492 R	31.2	1.27	—
356 R	—	0.92	—	Soil, dry	93.6	0.58	0.454
419 R	—	0.86	—	Soil, wet	118.6	1.16	0.525
455 R	—	0.78	—	Sugar	99.9	0.34	—
492 R	—	0.28	—	Tissue, human			
527 R	—	0.28	—	Skin	—	0.21	—
Clay, dry	96.8	0.54	—	Fat layer	—	0.12	—
Clay, wet	93.3	0.97	—	Muscle	—	0.24	—
Coal, anthracite	84.3	0.15	0.301	Vaseline	—	0.098	—
Concrete (stone mix)	143.6	0.81	0.210	Wood, cross-grain			
Cork	5.37	0.028	0.485	Balsa	8.74	0.032	—
Cotton	5.0	0.035	0.311	Fir	25.9	0.064	0.650
Fat	—	0.10	—	Oak	34.0	0.098	0.570
Glass				White pine	27.2	0.064	—
Window	174.8	0.40	0.179	Yellow pine	40.0	0.087	0.670
Pyrex	138.9	0.6–0.8	0.199	Wood, radial			
Crown	156.1	0.61	—	Oak	34.0	0.11	0.570
Lead	212.2	0.49	—	Fir	26.2	0.081	0.650
				Wool, ship	9.05	0.029	—

TABLE A–9E

Properties of saturated water

Temp. T, °F	Saturation Pressure P_{sat}, psia	Density ρ, lbm/ft³ Liquid	Density ρ, lbm/ft³ Vapor	Enthalpy of Vaporization h_{fg}, Btu/lbm	Specific Heat c_p, Btu/lbm·R Liquid	Specific Heat c_p, Btu/lbm·R Vapor	Thermal Conductivity k, Btu/h·ft·R Liquid	Thermal Conductivity k, Btu/h·ft·R Vapor	Dynamic Viscosity μ, lbm/ft·s Liquid	Dynamic Viscosity μ, lbm/ft·s Vapor	Prandtl Number Pr Liquid	Prandtl Number Pr Vapor	Volume Expansion Coefficient β, 1/R Liquid
32.02	0.0887	62.41	0.00030	1075	1.010	0.446	0.324	0.0099	1.204×10^{-3}	6.194×10^{-6}	13.5	1.00	-0.038×10^{-3}
40	0.1217	62.42	0.00034	1071	1.004	0.447	0.329	0.0100	1.308×10^{-3}	6.278×10^{-6}	11.4	1.01	$A0.003 \times 10^{-3}$
50	0.1780	62.41	0.00059	1065	1.000	0.448	0.335	0.0102	8.781×10^{-4}	6.361×10^{-6}	9.44	1.01	0.047×10^{-3}
60	0.2563	62.36	0.00083	1060	0.999	0.449	0.341	0.0104	7.536×10^{-4}	6.444×10^{-6}	7.95	1.00	0.080×10^{-3}
70	0.3632	62.30	0.00115	1054	0.999	0.450	0.347	0.0106	6.556×10^{-4}	6.556×10^{-6}	6.79	1.00	0.115×10^{-3}
80	0.5073	62.22	0.00158	1048	0.999	0.451	0.352	0.0108	5.764×10^{-4}	6.667×10^{-6}	5.89	1.00	0.145×10^{-3}
90	0.6988	62.12	0.00214	1043	0.999	0.453	0.358	0.0110	5.117×10^{-4}	6.778×10^{-6}	5.14	1.00	0.174×10^{-3}
100	0.9503	62.00	0.00286	1037	0.999	0.454	0.363	0.0112	4.578×10^{-4}	6.889×10^{-6}	4.54	1.01	0.200×10^{-3}
110	1.2763	61.86	0.00377	1031	0.999	0.456	0.367	0.0115	4.128×10^{-4}	7.000×10^{-6}	4.05	1.00	0.224×10^{-3}
120	1.6945	61.71	0.00493	1026	0.999	0.458	0.371	0.0117	3.744×10^{-4}	7.111×10^{-6}	3.63	1.00	0246×10^{-3}
130	2.225	61.55	0.00636	1020	0.999	0.460	0.375	0.0120	3.417×10^{-4}	7.222×10^{-6}	3.28	1.00	0.267×10^{-3}
140	2.892	61.38	0.00814	1014	0.999	0.463	0.378	0.0122	3.136×10^{-4}	7.333×10^{-6}	2.98	1.00	0.287×10^{-3}
150	3.722	61.19	0.0103	1008	1.000	0.465	0.381	0.0125	2.889×10^{-4}	7.472×10^{-6}	2.73	1.00	0.306×10^{-3}
160	4.745	60.99	0.0129	1002	1.000	0.468	0.384	0.0128	2.675×10^{-4}	7.583×10^{-6}	2.51	1.00	0.325×10^{-3}
170	5.996	60.79	0.0161	996	1.001	0.472	0.386	0.0131	2.483×10^{-4}	7.722×10^{-6}	2.90	1.00	0.346×10^{-3}
180	7.515	60.57	0.0199	990	1.002	0.475	0.388	0.0134	2.317×10^{-4}	7.833×10^{-6}	2.15	1.00	0.367×10^{-3}
190	9.343	60.35	0.0244	984	1.004	0.479	0.390	0.0137	2.169×10^{-4}	7.972×10^{-6}	2.01	1.00	0.382×10^{-3}
200	11.53	60.12	0.0297	978	1.005	0.483	0.391	0.0141	2.036×10^{-4}	8.083×10^{-6}	1.88	1.00	0.395×10^{-3}
210	14.125	59.87	0.0359	972	1.007	0.487	0.392	0.0144	1.917×10^{-4}	8.222×10^{-6}	1.77	1.00	0.412×10^{-3}
212	14.698	59.82	0.0373	970	1.007	0.488	0.392	0.0145	1.894×10^{-4}	8.250×10^{-6}	1.75	1.00	0.417×10^{-3}
220	17.19	59.62	0.0432	965	1.009	0.492	0.393	0.0148	1808×10^{-4}	8.333×10^{-6}	1.67	1.00	0.429×10^{-3}
230	20.78	59.36	0.0516	959	1.011	0.497	0.394	0.0152	1.711×10^{-4}	8.472×10^{-6}	1.58	1.00	0.443×10^{-3}
240	24.97	59.09	0.0612	952	1.013	0.503	0.394	0.0156	1.625×10^{-4}	8.611×10^{-6}	1.50	1.00	0.462×10^{-3}
250	29.82	58.82	0.0723	946	1.015	0.509	0.395	0.0160	1.544×10^{-4}	8.611×10^{-6}	1.43	1.00	0.480×10^{-3}
260	35.42	58.53	0.0850	939	1.018	0.516	0.395	0.0164	1.472×10^{-4}	8.861×10^{-6}	1.37	1.00	0.497×10^{-3}
270	41.85	58.24	0.0993	932	1.020	0.523	0.395	0.0168	1.406×10^{-4}	9.000×10^{-6}	1.31	1.01	0.514×10^{-3}
280	49.18	57.94	0.1156	926	1.023	0.530	0.395	0.0172	1.344×10^{-4}	9.111×10^{-6}	1.25	1.01	0.532×10^{-3}
290	57–53	57.63	0.3390	918	1.026	0.538	0.395	0.0177	1.289×10^{-4}	9.250×10^{-6}	1.21	1.01	0.549×10^{-3}
300	66.98	57.31	0.1545	910	1.029	0.547	0.394	0.0182	1.236×10^{-4}	9.389×10^{-6}	1.16	1.02	0.566×10^{-3}
320	89.60	56.65	0.2033	895	1.036	0.567	0.393	0.0191	1.144×10^{-4}	9.639×10^{-6}	1.09	1.03	0.636×10^{-3}
340	117.93	55.95	0.2637	880	1.044	0.590	0.391	0.0202	1.063×10^{-4}	9.889×10^{-6}	1.02	1.04	0.656×10^{-3}
360	152.92	56.22	0.3377	863	1.054	0.617	0.389	0.0213	9.972×10^{-5}	1.013×10^{-5}	0.973	1.06	0.681×10^{-3}
380	196.00	54.40	0.4275	845	1.065	0.647	0.385	0.0224	9.361×10^{-5}	1.041×10^{-5}	0.932	1.08	0.720×10^{-3}
400	241.1	53.65	0.5359	827	1.078	0.683	0.382	0.0237	8.833×10^{-5}	1.066×10^{-5}	0.893	1.11	0.771×10^{-3}
450	422.1	51.46	0.9082	775	1.121	0.799	0.370	0.0271	7.722×10^{-5}	1.130×10^{-5}	0.842	1.20	0.912×10^{-3}
500	680.0	48.95	1.479	715	1.188	0.972	0.352	0.0312	6.833×10^{-5}	1.200×10^{-5}	0.830	1.35	1.111×10^{-3}
550	1046.7	45.96	4.268	641	1.298	1.247	0.329	0.0368	6.083×10^{-5}	1.280×10^{-5}	0.864	1.56	1.445×10^{-3}
600	1541	42.32	3.736	550	1.509	1.759	0.299	0.0461	5.389×10^{-5}	1.380×10^{-5}	0.979	1.90	1.883×10^{-3}
650	2210	37.31	6.152	422	2.086	3.103	0.267	0.0677	4.639×10^{-5}	1.542×10^{-5}	1.30	2.54	
700	3090	27.28	13.44	168	13.80	25.90	0.254	0.1964	3.417×10^{-5}	2.044×10^{-5}	6.68	9.71	
705.44	3204	19.79	19.79	0	∞	∞	∞	∞	2.897×10^{-5}	2.897×10^{-5}			

Note 1: Kinematic viscosity ν and thermal diffusivity α can be calculated from their definitions, $\nu = \mu/\rho$ and $\alpha = k/\rho c_p = \nu/\text{Pr}$. The temperatures 32.02°F, 212°F, and 705.44°F are the triple-, boiling-, and critical-point temperatures of water, respectively. All properties listed above (except the vapor density) can be used at any pressures with negligible error except at temperatures near the critical-point value.

Note 2: The unit Btu/lbm·°F for specific heat is equivalent to Btu/lbm·R, and the unit Btu/h·ft·°F for thermal conductivity is equivalent to Btu/h·ft·R.

Source: Viscosity and thermal conductivity data are from J. V. Sengers and J. T. T. Watson, *Journal of Physical and Chemical Reference Data* 15 (1986), pp. 1291–1322. Other data are obtained from various sources or calculated.

TABLE A–10E

Properties of saturated refrigerant-134a

Temp. T, °F	Saturation Pressure P_{sat}, psia	Density ρ, lbm/ft³		Enthalpy of Vaporization h_{fg}, Btu/lbm	Specific Heat c_p, Btu/lbm·R		Thermal Conductivity k, Btu/h·ft·R		Dynamic Viscosity μ, lbm/ft·s		Prandtl Number Pr		Volume Expansion Coefficient β, 1/R	Surface Tension lbf/ft
		Liquid	Vapor		Liquid	Vapor	Liquid	Vapor	Liquid	Vapor	Liquid	Vapor	Liquid	
−40	7.4	88.51	0.1731	97.1	0.2996	0.1788	0.0636	0.00466	3.278×10^{-4}	1.714×10^{-6}	5.558	0.237	0.00114	0.001206
−30	9.9	87.5	0.2258	95.6	0.3021	0.1829	0.0626	0.00497	3.004×10^{-4}	2.053×10^{-6}	5.226	0.272	0.00117	0.001146
−20	12.9	86.48	0.2905	94.1	0.3046	0.1872	0.0613	0.00529	2.762×10^{-4}	2.433×10^{-6}	4.937	0.310	0.00120	0.001087
−10	16.6	85.44	0.3691	92.5	0.3074	0.1918	0.0602	0.00559	2.546×10^{-4}	2.856×10^{-6}	4.684	0.352	0.00124	0.001029
0	21.2	84.38	0.4635	90.9	0.3103	0.1966	0.0589	0.00589	2.345×10^{-4}	3.314×10^{-6}	4.463	0.398	0.00128	0.000972
10	26.6	83.31	0.5761	89.3	0.3134	0.2017	0.0576	0.00619	2.181×10^{-4}	3.811×10^{-4}	4.269	0.447	0.00132	0.000915
20	33.1	82.2	0.7094	87.5	0.3167	0.2070	0.0563	0.00648	2.024×10^{-4}	4.342×10^{-6}	4.098	0.500	0.00132	0.000859
30	40.8	81.08	0.866	85.8	0.3203	0.2127	0.0550	0.00676	1.883×10^{-4}	4.906×10^{-6}	3.947	0.555	0.00142	0.000803
40	49.8	79.92	1.049	83.9	0.3240	0.2188	0.0536	0.00704	1.752×10^{-4}	5.494×10^{-6}	3.814	0.614	0.00149	0.000749
50	60.2	78.73	1.262	82.0	0.3281	0.2253	0.0522	0.00732	1.633×10^{-4}	6.103×10^{-6}	3.697	0.677	0.00156	0.000695
60	72.2	77.51	1.509	80.0	0.3325	0.2323	0.0507	0.00758	1.522×10^{-4}	6.725×10^{-6}	3.594	0.742	0.00163	0.000642
70	85.9	76.25	1.794	78.0	0.3372	0.2398	0.0492	0.00785	1.420×10^{-4}	7.356×10^{-6}	3.504	0.810	0.00173	0.000590
80	101.4	74.94	2.122	75.8	0.3424	0.2481	0.0476	0.00810	1.324×10^{-4}	7.986×10^{-6}	3.425	0.880	0.00183	0.000538
90	119.1	73.59	2.5	73.5	0.3481	0.2572	0.0460	0.00835	1.234×10^{-4}	8.611×10^{-6}	3.357	0.955	0.00195	0.000488
100	138.9	72.17	2.935	71.1	0.3548	0.2674	0.0444	0.00860	1.149×10^{-4}	9.222×10^{-6}	3.303	1.032	0.00210	0.000439
110	161.2	70.69	3.435	68.5	0.3627	0.2790	0.0427	0.00884	1.068×10^{-4}	9.814×10^{-6}	3.262	1.115	0.00227	0.000391
120	186.0	69.13	4.012	65.8	0.3719	0.2925	0.0410	0.00908	9.911×10^{-5}	1.038×10^{-5}	3.235	1.204	0.00248	0.000344
130	213.5	67.48	4.679	62.9	0.3829	0.3083	0.0392	0.00931	9.175×10^{-5}	1.092×10^{-5}	3.223	1.303	0.00275	0.000299
140	244.1	65.72	5.455	59.8	0.3963	0.3276	0.0374	0.00954	8.464×10^{-5}	1.144×10^{-5}	3.229	1.416	0.00308	0.000255
150	277.8	63.83	6.367	56.4	0.4131	0.3520	0.0355	0.00976	7.778×10^{-5}	1.195×10^{-5}	3.259	1.551	0.00351	0.000212
160	314.9	61.76	7.45	52.7	0.4352	0.3839	0.0335	0.00998	7.108×10^{-5}	1.245×10^{-5}	3.324	1.725	0.00411	0.000171
170	355.8	59.47	8.762	48.5	0.4659	0.4286	0.0314	0.01020	6.450×10^{-5}	1.298×10^{-5}	3.443	1.963	0.00498	0.000132
180	400.7	56.85	10.4	43.7	0.5123	0.4960	0.0292	0.01041	5.792×10^{-5}	1.366×10^{-5}	3.661	2.327	0.00637	0.000095
190	449.9	53.75	12.53	38.0	0.5929	0.6112	0.0267	0.01063	5.119×10^{-5}	1.431×10^{-5}	4.090	2.964	0.00891	0.000061
200	504.0	49.75	15.57	30.7	0.7717	0.8544	0.0239	0.01085	4.397×10^{-5}	1.544×10^{-5}	5.119	4.376	0.01490	0.000031
210	563.8	43.19	21.18	18.9	1.4786	1.6683	0.0199	0.01110	3.483×10^{-5}	1.787×10^{-5}	9.311	9.669	0.04021	0.000006

Note 1: Klnematic viscosity ν and thermal diffusivity α can be calculated from their definitions, $\nu = \mu/\rho$ and $\alpha = k/\rho c_p = \nu/Pr$. The properties listed here (except the vapor density) can be used at any pressures with negligible error except at temperatures near the critical-point value.

Note 2: The unit Btu/lbm . °F for specific heat is equivalent to Btu/lbm·R, and the unit Btu/h·ft·°F for thermal conductivity is equivalent to Btu/h·ft·R.

Source: Data generated from the EES software developed by S. A. Klein and F. L. Alvarado. Original sources: R. Tilner-Roth and H. D. Baehr, "An International Standard Formulation for the Thermodynamic Properties of 1,1,1,2-Tetrafluorethane (HFC-134a) for Temperatures from 170 K to 455 K and Pressures up to 70 Mpa," *J. Phys. Chem. Ref. Data*, Vol. 23, No.5, 1994: M. J. Assael, N. K. Dalaouti, A. A. Griva, and J. H. Dymond, "Viscosity and Thermal Conductivity of Halogenated Methane and Ethane Refrigerants," *IJR*, Vol. 22, pp. 525–535, 1999: NIST REPROP 6 program (M. O. McLinden, S. A. Klein, E. W. Lemmon, and A. P. Peskin, Physicial and Chemical Properties Division, National Institute of Standards and Technology, Boulder, CO 80303. 1995).

TABLE A–11E

Properties of saturated ammonia

Temp. T, °F	Saturation Pressure P_{sat}, psia	Density ρ, lbm/ft³ Liquid	Density ρ, lbm/ft³ Vapor	Enthalpy of Vaporization h_{fg}, Btu/lbm	Specific Heat c_p, Btu/lbm·R Liquid	Specific Heat c_p, Btu/lbm·R Vapor	Thermal Conductivity k, Btu/h·ft·R Liquid	Thermal Conductivity k, Btu/h·ft·R Vapor	Dynamic Viscosity μ, lbm/ft·s Liquid	Dynamic Viscosity μ, lbm/ft·s Vapor	Prandtl Number Pr Liquid	Prandtl Number Pr Vapor	Volume Expansion Coefficient β, 1/R Liquid	Surface Tension lbf/ft
−40	10.4	43.08	0.0402	597.0	1.0542	0.5354	–	0.01026	1.966×10^{-4}	5.342×10^{-6}	–	1.003	0.00098	0.002443
−30	13.9	42.66	0.0527	590.2	1.0610	0.5457	–	0.01057	1.853×10^{-4}	5.472×10^{-6}	–	1.017	0.00101	0.002357
−20	18.3	42.33	0.0681	583.2	1.0677	0.5571	0.3501	0.01089	1.746×10^{-4}	5.600×10^{-6}	1.917	1.031	0.00103	0.002272
−10	23.7	41.79	0.0869	575.9	1.0742	0.5698	0.3426	0.01121	1.645×10^{-4}	5.731×10^{-6}	1.856	1.048	0.00106	0.002187
0	30.4	41.34	0.1097	568.4	1.0807	0.5838	0.3352	0.01154	1.549×10^{-4}	5.861×10^{-6}	1.797	1.068	0.00109	0.002103
10	38.5	40.89	0.1370	560.7	1.0873	0.5992	0.3278	0.01187	1.458×10^{-4}	5.994×10^{-6}	1.740	1.089	0.00112	0.002018
20	48.2	40.43	0.1694	552.6	1.0941	0.6160	0.3203	0.01220	1.371×10^{-4}	6.125×10^{-6}	1.686	1.113	0.00116	0.001934
30	59.8	39.96	0.2075	544.4	1.1012	0.6344	0.3129	0.01254	1290×10^{-4}	6.256×10^{-6}	1.634	1.140	0.00119	0.001850
40	73.4	39.48	0.2521	535.8	1.1087	0.6544	0.3055	0.01288	1.213×10^{-4}	6.389×10^{-6}	1.585	1.168	0.00123	0.001767
50	89.2	38.99	0.3040	526.9	1.1168	0.6762	0.2980	0.01323	1.140×10^{-4}	6.522×10^{-6}	1.539	1.200	0.00128	0.001684
60	107.7	38.50	0.3641	517.7	1.1256	0.6999	0.2906	0.01358	1.072×10^{-4}	6.656×10^{-6}	1.495	1.234	0.00132	0.001601
70	128.9	37.99	0.4332	508.1	1.1353	0.7257	0.2832	0.01394	1.008×10^{-4}	6.786×10^{-6}	1.456	1.272	0.00137	0.001518
80	153.2	37.47	0.5124	498.2	1.1461	0.7539	0.2757	0.01431	9.486×10^{-5}	6.922×10^{-6}	1.419	1.313	0.00143	0.001436
90	180.8	36.94	0.6029	487.8	1.1582	0.7846	0.2683	0.01468	8.922×10^{-5}	7.056×10^{-6}	1.387	1.358	0.00149	0.001354
100	212.0	36.40	0.7060	477.0	1.1719	0.8183	0.2609	0.01505	8.397×10^{-5}	7.189×10^{-6}	1.358	1.407	0.00156	0.001273
110	247.2	35.83	0.8233	465.8	1.1875	0.8554	0.2535	0.01543	7.903×10^{-5}	7.325×10^{-6}	1.333	1.461	0.00164	0.001192
120	286.5	35.26	0.9564	454.1	1.2054	0.8965	0.2460	0.01582	7.444×10^{-5}	7.458×10^{-6}	1.313	1.522	0.00174	0.001111
130	330.4	34.66	1.1074	441.7	1.2261	0.9425	0.2386	0.01621	7.017×10^{-5}	7.594×10^{-6}	1.298	1.589	0.00184	0.001031
140	379.4	34.04	1.2786	428.8	1.2502	0.9943	0.2312	0.01661	6.617×10^{-5}	7.731×10^{-6}	1.288	1.666	0.00196	0.000951
150	433.2	33.39	1.4730	415.2	1.2785	1.0533	0.2237	0.01702	6.244×10^{-5}	7.867×10^{-4}	1.285	1.753	0.00211	0.000872
160	492.7	32.72	1.6940	400.8	1.3120	1.1214	0.2163	0.01744	5.900×10^{-5}	8.006×10^{-6}	1.288	1.853	0.00228	0.000794
170	558.2	32.01	1.9460	385.4	1.3523	1.2012	0.2089	0.01786	5.578×10^{-5}	8.142×10^{-6}	1.300	1.971	0.00249	0.000716
180	630.1	31.26	2.2346	369.1	1.4015	1.2965	0.2014	0.01829	5.278×10^{-5}	8.281×10^{-6}	1.322	2.113	0.00274	0.000638
190	708.5	30.47	2.5670	351.6	1.4624	1.4128	0.1940	0.01874	5.000×10^{-5}	8.419×10^{-6}	1.357	2.286	0.00306	0.000562
200	794.4	29.62	2.9527	332.7	1.5397	1.5586	0.1866	0.01919	4.742×10^{-5}	8.561×10^{-6}	1.409	2.503	0.00348	0.000486
210	887.9	28.70	3.4053	312.0	1.6411	1.7473	0.1791	0.01966	4500×10^{-5}	8.703×10^{-6}	1.484	2.784	0.00403	0.000411
220	989.5	27.69	3.9440	289.2	1.7798	2.0022	0.1717	0.02015	4.275×10^{-5}	8.844×10^{-6}	1.595	3.164	0.00480	0.000338
230	1099.0	25.57	4.5987	263.5	1.9824	2.3659	0.1643	0.02065	4.064×10^{-5}	8.989×10^{-6}	1.765	3.707	0.00594	0.000265
240	1219.4	25.28	5.4197	234.0	2.3100	2.9264	0.1568	0.02119	3.864×10^{-5}	9.136×10^{-6}	2.049	4.542	0.00784	0.000194

Note 1: Kinematic viscosity ν and thermal diffusivity α can be calculated from their definitions, $\nu = \mu/\rho$ and $\alpha = k/\rho c_p = \nu/Pr$. The properties listed here (except the vapor density) can be used at any pressures with negligible error except at temperatures near the critical-point value.

Note 2: The unit Btu/lbm·°F for specific heat is equivalent to Btu/lbm·R, and the unit Btu/h·ft·°F for thermal conductivity is equivalent to Btu/h·ft·R.

Source: Data generated from the EES software developed by S. A. Klein and F. L. Alvarado. Original sources: Tillner-Roth, Harms-Watzenberg, and Baehr, "Fine neue Fundamentalgleichung fur Ammoniak," *DKV-Tagungsbericht* 20: 167–181, 1993; Liley and Desai, "Thermophysical Properties of Refrigerants," *ASHRAE*, 1993, ISBN 1-1883413-10-9.

TABLE A–12E

Properties of saturated propane

Temp. T, °F	Saturation Pressure P_{sat}, psia	Density ρ, lbm/ft³		Enthalpy of Vaporization h_{ig}, Btu/lbm	Specific Heat c_p, Btu/lbm·R		Thermal Conductivity k, Btu/h·ft·R		Dynamic Viscosity μ, lbm/ft·s		Prandtl Number Pr		Volume Expansion Coefficient β, 1/R	Surface Tension lbf/ft
		Liquid	Vapor		Liquid	Vapor	Liquid	Vapor	Liquid	Vapor	Liquid	Vapor	Liquid	
−200	0.0201	42.06	0.0003	217.7	0.4750	0.2595	0.1073	0.00313	5.012×10^{-4}	2.789×10^{-6}	7.991	0.833	0.00083	0.001890
−180	0.0752	41.36	0.0011	213.4	0.4793	0.2680	0.1033	0.00347	3.941×10^{-4}	2.975×10^{-6}	6.582	0.826	0.00086	0.001780
−160	0.2307	40.65	0.0032	209.1	0.4845	0.2769	0.0992	0.00384	3.199×10^{-4}	3.164×10^{-6}	5.626	0.821	0.00088	0.001671
−140	0.6037	39.93	0.0078	204.8	0.4907	0.2866	0.0949	0.00423	2.660×10^{-4}	3.358×10^{-6}	4.951	0.818	0.00091	0.001563
−120	1.389	39.20	0.0170	200.5	0.4982	0.2971	0.0906	0.00465	2.252×10^{-4}	3.556×10^{-6}	4.457	0.817	0.00094	0.001455
−100	2.878	38.46	0.0334	196.1	0.5069	0.3087	0.0863	0.00511	1.934×10^{-4}	3.756×10^{-6}	4.087	0.817	0.00097	0.001349
−90	4.006	38.08	0.0453	193.9	0.5117	0.3150	0.0842	0.00534	1.799×10^{-4}	3.858×10^{-6}	3.936	0.819	0.00099	0.001297
−80	5.467	37.70	0.0605	191.6	0.5169	0.3215	0.0821	0.00559	1.678×10^{-4}	3.961×10^{-6}	3.803	0.820	0.00101	0.001244
−70	7.327	37.32	0.0793	189.3	0.5224	0.3284	0.0800	0.00585	1.569×10^{-4}	4.067×10^{-6}	3.686	0.822	0.00104	0.001192
−60	9.657	36.93	0.1024	186.9	0.5283	0.3357	0.0780	0.00611	1.469×10^{-4}	4.172×10^{-6}	3.582	0.825	0.00106	0.001140
−50	12.54	36.54	0.1305	184.4	0.5345	0.3433	0.0760	0.00639	1.378×10^{-4}	4.278×10^{-6}	3.490	0.828	0.00109	0.001089
−40	16.05	36.13	0.1641	181.9	0.5392	0.3513	0.0740	0.00568	1.294×10^{-4}	4.386×10^{-6}	3.395	0.831	0.00112	0.001038
−30	20.29	35.73	0.2041	179.3	0.5460	0.3596	0.0721	0.00697	1.217×10^{-4}	4.497×10^{-6}	3.320	0.835	0.00115	0.000987
−20	25.34	35.31	0.2512	176.6	0.5531	0.3684	0.0702	0.00728	1.146×10^{-4}	4.611×10^{-6}	3.253	0.840	0.00119	0.000937
−10	31.3	34.89	0.3063	173.8	0.5607	0.3776	0.0683	0.00761	1.079×10^{-4}	4.725×10^{-6}	3.192	0.845	0.00123	0.000887
0	38.28	34.46	0.3703	170.9	0.5689	0.3874	0.0665	0.00794	1.018×10^{-4}	4.842×10^{-6}	3.137	0.850	0.00127	0.000838
10	46.38	34.02	0.4441	167.9	0.5775	0.3976	0.0647	0.00829	9.606×10^{-5}	4.961×10^{-6}	3.088	0.857	0.00132	0.000789
20	55.7	33.56	0.5289	164.8	0.5867	0.4084	0.0629	0.00865	9.067×10^{-5}	5.086×10^{-6}	3.043	0.864	0.00138	0.000740
30	66.35	33.10	0.6259	161.6	0.5966	0.4199	0.0512	0.00903	8.561×10^{-5}	5.211×10^{-6}	3.003	0.873	0.00144	0.000692
40	78.45	32.62	0.7365	158.1	0.6072	0.4321	0.0595	0.00942	8.081×10^{-5}	5.342×10^{-6}	2.967	0.882	0.00151	0.000644
50	92.12	32.13	0.8621	154.6	0.6187	0.4452	0.0579	0.00983	7.631×10^{-5}	5.478×10^{-6}	2.935	0.893	0.00159	0.000597
60	107.5	31.63	1.0046	150.8	0.6311	0.4593	0.0563	0.01025	7.200×10^{-5}	5.617×10^{-6}	2.906	0.906	0.00168	0.000551
70	124.6	31.11	1.1659	146.8	0.6447	0.4746	0.0547	0.01070	6.794×10^{-5}	5.764×10^{-6}	2.881	0.921	0.00179	0.000505
80	143.7	30.56	1.3484	142.7	0.6596	0.4915	0.0532	0.01116	6.406×10^{-5}	5.919×10^{-6}	2.860	0.938	0.00191	0.000460
90	164.8	30.00	1.5549	138.2	0.6762	0.5103	0.0517	0.01165	6.033×10^{-5}	6.081×10^{-6}	2.843	0.959	0.00205	0.000416
100	188.1	29.41	1.7887	133.6	0.6947	0.5315	0.0501	0.01217	5.675×10^{-5}	6.256×10^{-6}	2.831	0.984	0.00222	0.000372
120	241.8	28.13	2.3562	123.2	0.7403	0.5844	0.0472	0.01328	5.000×10^{-6}	6.644×10^{-6}	2.825	1.052	0.00267	0.000288
140	306.1	26.69	3.1003	111.1	0.7841	0.6613	0.0442	0.01454	4.358×10^{-5}	7.111×10^{-6}	2.784	1.164	0.00338	0.000208
160	382.4	24.98	4.1145	96.4	0.8696	0.7911	0.0411	0.01603	3.733×10^{-5}	7.719×10^{-6}	2.845	1.371	0.00459	0.000133
180	472.9	22.79	5.6265	77.1	1.1436	1.0813	0.0376	0.01793	3.083×10^{-5}	8.617×10^{-6}	3.380	1.870	0.00791	0.000065

Note 1: Kinematic viscosity ν and thermal diffusivity α can be calculated from their definitions, $\nu = \mu/\rho$ and $\alpha = k/\rho c_p = \nu/Pr$. The properties listed here (except the vapor density) can be used at any pressures with negligible error at temperatures near the critical-point value.

Note 2: The unit Btu/lbm·°F for specific heat is equivalent to Btu/lbm·R, and the unit Btu/h·ft·°F for thermal conductivity is equivalent to Btu/h·ft·R.

Source: Data generated from the EES software developed by S. A. Klein and F. L. Alvarado. Original sources: Reiner Tillner-Roth, "Fundamental Equations of State," Shaker, Verlag, Aachan, 1998; B. A. Younglove and J. F. Ely. "Thermophysical Properties of Fluids. II Methane, Ethane, Propane, Isobutane, and Normal Butane," *J. Phys. Chem. Ref. Data*, Vol. 16, No. 4, 1987; G. R. Somayajulu, "A Generalized Equation for Surface Tension from the Triple-Point to the Critical-Point," *International Journal of Thermophysics*, Vol. 9, No. 4, 1988.

TABLE A–13E

Properties of liquids

Temp. T, °F	Density ρ, lbm/ft^3	Specific Heat c_p, Btu/lbm·R	Thermal Conductivity k, Btu/h·ft·R	Thermal Diffusivity α, ft^2/s	Dynamic Viscosity μ, lbm/ft·s	Kinematic Viscosity ν, ft^2/s	Prandtl Number Pr	Volume Expansion Coeff. β, 1/R
\multicolumn{9}{c}{Methane (CH$_4$)}
−280	27.41	0.8152	0.1205	1.497×10^{-6}	1.057×10^{-4}	3.857×10^{-6}	2.575	0.00175
−260	26.43	0.8301	0.1097	1.389×10^{-6}	8.014×10^{-5}	3.032×10^{-6}	2.183	0.00192
−240	25.39	0.8523	0.0994	1.276×10^{-6}	6.303×10^{-5}	2.482×10^{-6}	1.945	0.00215
−220	24.27	0.8838	0.0896	1.159×10^{-6}	5.075×10^{-5}	2.091×10^{-6}	1.803	0.00247
−200	23.04	0.9314	0.0801	1.036×10^{-6}	4.142×10^{-5}	1.798×10^{-6}	1.734	0.00295
−180	21.64	1.010	0.0709	9.008×10^{-7}	3.394×10^{-5}	1.568×10^{-6}	1.741	0.00374
−160	19.99	1.158	0.0616	7.397×10^{-7}	2.758×10^{-5}	1.379×10^{-6}	1.865	0.00526
−140	17.84	1.542	0.0518	5.234×10^{-7}	2.168×10^{-5}	1.215×10^{-6}	2.322	0.00943
\multicolumn{9}{c}{Methanol [CH$_3$(OH)]}
70	49.15	0.6024	0.1148	1.076×10^{-6}	3.872×10^{-4}	7.879×10^{-6}	7.317	0.000656
90	48.50	0.6189	0.1143	1.057×10^{-6}	3.317×10^{-4}	6.840×10^{-6}	6.468	0.000671
110	47.85	0.6373	0.1138	1.036×10^{-6}	2.872×10^{-4}	6.005×10^{-6}	5.793	0.000691
130	47.18	0.6576	0.1133	1.014×10^{-6}	2.513×10^{-4}	5.326×10^{-6}	5.250	0.000716
150	46.50	0.6796	0.1128	9.918×10^{-7}	2.218×10^{-4}	4.769×10^{-6}	4.808	0.000749
170	45.80	0.7035	0.1124	9.687×10^{-7}	1.973×10^{-4}	4.308×10^{-6}	4.447	0.000789
\multicolumn{9}{c}{Isobutane (R600a)}
−150	42.75	0.4483	0.0799	1.157×10^{-6}	6.417×10^{-4}	1.500×10^{-5}	12.96	0.000785
−100	41.06	0.4721	0.0782	1.120×10^{-6}	3.669×10^{-4}	8.939×10^{-6}	7.977	0.000836
−50	39.31	0.4986	0.0731	1.036×10^{-6}	2.376×10^{-4}	6.043×10^{-6}	5.830	0.000908
0	37.48	0.5289	0.0664	9.299×10^{-7}	1.651×10^{-4}	4.406×10^{-6}	4.738	0.001012
50	35.52	0.5643	0.0591	8.187×10^{-7}	1.196×10^{-4}	3.368×10^{-6}	4.114	0.001169
100	33.35	0.6075	0.0521	7.139×10^{-7}	8.847×10^{-5}	2.653×10^{-6}	3.716	0.001421
150	30.84	0.6656	0.0457	6.188×10^{-7}	6.558×10^{-5}	2.127×10^{-6}	3.437	0.001003
200	27.73	0.7635	0.0400	5.249×10^{-7}	4.750×10^{-5}	1.713×10^{-6}	3.264	0.002970
\multicolumn{9}{c}{Glycerin}
32	79.65	0.5402	0.163	1.052×10^{-6}	7.047	0.08847	84101	
40	79.49	0.5458	0.1637	1.048×10^{-6}	4.803	0.06042	57655	
50	79.28	0.5541	0.1645	1.040×10^{-6}	2.850	0.03594	34561	
60	79.07	0.5632	0.1651	1.029×10^{-6}	1.547	0.01956	18995	
70	78.86	0.5715	0.1652	1.018×10^{-6}	0.9422	0.01195	11730	
80	78.66	0.5794	0.1652	1.007×10^{-6}	0.5497	0.00699	6941	
90	78.45	0.5878	0.1652	9.955×10^{-7}	0.3756	0.004787	4809	
100	78.24	0.5964	0.1653	9.841×10^{-7}	0.2277	0.00291	2957	
\multicolumn{9}{c}{Engine Oil (unused)}
32	56.12	0.4291	0.0849	9.792×10^{-7}	2.563	4.566×10^{-2}	46636	0.000389
50	55.79	0.4395	0.08338	9.448×10^{-7}	1.210	2.169×10^{-2}	22963	0.000389
75	55.3	0.4531	0.08378	9.288×10^{-7}	0.4286	7.751×10^{-3}	8345	0.000389
100	54.77	0.4669	0.08367	9.089×10^{-7}	0.1630	2.977×10^{-3}	3275	0.000389
125	54.24	0.4809	0.08207	8.740×10^{-7}	7.617×10^{-2}	1.404×10^{-3}	1607	0.000389
150	53.73	0.4946	0.08046	8.411×10^{-7}	3.833×10^{-2}	7.135×10^{-4}	848.3	0.000389
200	52.68	0.5231	0.07936	7.999×10^{-7}	1.405×10^{-2}	2.668×10^{-4}	333.6	0.000389
250	51.71	0.5523	0.07776	7.563×10^{-7}	6.744×10^{-3}	1.304×10^{-4}	172.5	0.000389
300	50.63	0.5818	0.07673	7.236×10^{-7}	3.661×10^{-3}	7.232×10^{-5}	99.94	0.000389

Source: Data generated from the EES software developed by S. A. Klein and F. L. Alvarado. Originally based on various sources.

TABLE A–14E

Properties of liquid metals

Temp. T, °F	Density ρ, lbm/ft³	Specific Heat c_p, Btu/lbm·R	Thermal Conductivity k, Btu/h·ft·R	Thermal Diffusivity α, ft²/s	Dynamic Viscosity μ, lbm/ft·s	Kinematic Viscosity ν, ft²/s	Prandtl Number Pr	Volume Expansion Coeff. β, 1/R
				Mercury (Hg) Melting Point: −38°F				
32	848.7	0.03353	4.727	4.614×10^{-5}	1.133×10^{-3}	1.335×10^{-6}	0.02895	1.005×10^{-4}
50	847.2	0.03344	4.805	4.712×10^{-5}	1.092×10^{-3}	1.289×10^{-6}	0.02737	1.005×10^{-4}
100	842.9	0.03319	5.015	4.980×10^{-5}	9.919×10^{-4}	1.176×10^{-6}	0.02363	1.005×10^{-4}
150	838.7	0.03298	5.221	5.244×10^{-5}	9.122×10^{-4}	1.087×10^{-6}	0.02074	1.005×10^{-4}
200	834.5	0.03279	5.422	5.504×10^{-5}	8.492×10^{-4}	1.017×10^{-6}	0.01849	1.005×10^{-4}
300	826.2	0.03252	5.815	6.013×10^{-5}	7.583×10^{-4}	9.180×10^{-7}	0.01527	1.005×10^{-4}
400	817.9	0.03236	6.184	6.491×10^{-5}	6.972×10^{-4}	8.524×10^{-7}	0.01313	1.008×10^{-4}
500	809.6	0.03230	6.518	6.924×10^{-5}	6.525×10^{-4}	8.061×10^{-7}	0.01164	1.018×10^{-4}
600	801.3	0.03235	6.839	7.329×10^{-5}	6.186×10^{-4}	7.719×10^{-7}	0.01053	1.035×10^{-4}
				Bismuth (Bi) Melting Point: 520°F				
700	620.7	0.03509	9.361	1.193×10^{-4}	1.001×10^{-3}	1.614×10^{-6}	0.01352	
800	616.5	0.03569	9.245	1.167×10^{-4}	9.142×10^{-4}	1.482×10^{-6}	0.01271	
900	612.2	0.0363	9.129	1.141×10^{-4}	8.267×10^{-4}	1.350×10^{-6}	0.01183	
1000	608.0	0.0369	9.014	1.116×10^{-4}	7.392×10^{-4}	1.215×10^{-6}	0.0109	
1100	603.7	0.0375	9.014	1.105×10^{-4}	6.872×10^{-4}	1.138×10^{-6}	0.01029	
				Lead (Pb) Melting Point: 621°F				
700	658	0.03797	9.302	1.034×10^{-4}	1.612×10^{-3}	2.450×10^{-6}	0.02369	
800	654	0.03750	9.157	1.037×10^{-4}	1.453×10^{-3}	2.223×10^{-6}	0.02143	
900	650	0.03702	9.013	1.040×10^{-4}	1.296×10^{-3}	1.994×10^{-6}	0.01917	
1000	645.7	0.03702	8.912	1.035×10^{-4}	1.202×10^{-3}	1.862×10^{-6}	0.01798	
1100	641.5	0.03702	8.810	1.030×10^{-4}	1.108×10^{-3}	1.727×10^{-6}	0.01676	
1200	637.2	0.03702	8.709	1.025×10^{-4}	1.013×10^{-3}	1.590×10^{-6}	0.01551	
				Sodium (Na) Melting Point: 208°F				
300	57.13	0.3258	48.19	7.192×10^{-4}	4.136×10^{-4}	7.239×10^{-6}	0.01007	
400	56.28	0.3219	46.58	7.142×10^{-4}	3.572×10^{-4}	6.350×10^{-6}	0.008891	
500	55.42	0.3181	44.98	7.087×10^{-4}	3.011×10^{-4}	5.433×10^{-6}	0.007667	
600	54.56	0.3143	43.37	7.026×10^{-4}	2.448×10^{-4}	4.488×10^{-6}	0.006387	
800	52.85	0.3089	40.55	6.901×10^{-4}	1.772×10^{-4}	3.354×10^{-6}	0.004860	
1000	51.14	0.3057	38.12	6.773×10^{-4}	1.541×10^{-4}	3.014×10^{-6}	0.004449	
				Potassium (K) Melting Point: 147°F				
300	50.40	0.1911	26.00	7.500×10^{-4}	2.486×10^{-4}	4.933×10^{-6}	0.006577	
400	49.58	0.1887	25.37	7.532×10^{-4}	2.231×10^{-4}	4.500×10^{-6}	0.005975	
500	48.76	0.1863	24.73	7.562×10^{-4}	1.976×10^{-4}	4.052×10^{-6}	0.005359	
600	47.94	0.1839	24.09	7.591×10^{-4}	1.721×10^{-4}	3.589×10^{-6}	0.004728	
800	46.31	0.1791	22.82	7.643×10^{-4}	1.210×10^{-4}	2.614×10^{-6}	0.003420	
1000	44.62	0.1791	21.34	7.417×10^{-4}	1.075×10^{-4}	2.409×10^{-6}	0.003248	
				Sodium-Potassium (%22Na-%78K) Melting Point: 12°F				
200	52.99	0.2259	14.79	3.432×10^{-4}	3.886×10^{-4}	7.331×10^{-6}	0.02136	
300	52.16	0.2230	14.99	3.580×10^{-4}	3.467×10^{-4}	6.647×10^{-6}	0.01857	
400	51.32	0.2201	15.19	3.735×10^{-4}	3.050×10^{-4}	5.940×10^{-6}	0.0159	
600	49.65	0.2143	15.59	4.070×10^{-4}	2.213×10^{-4}	4.456×10^{-6}	0.01095	
800	47.99	0.2100	15.95	4.396×10^{-4}	1.539×10^{-4}	3.207×10^{-6}	0.007296	
1000	46.36	0.2103	16.20	4.615×10^{-4}	1.353×10^{-4}	2.919×10^{-6}	0.006324	

Source: Data generated from the EES software developed by S. A. Klein and F. L. Alvarado. Originally based on various sources.

TABLE A–15E

Properties of air at 1 atm pressure

Temp. T, °F	Density ρ, lbm/ft^3	Specific Heat c_p, Btu/lbm·R	Thermal Conductivity k, Btu/h·ft·R	Thermal Diffusivity α, ft^2/s	Dynamic Viscosity μ, lbm/ft·s	Kinematic Viscosity ν, ft^2/s	Prandtl Number Pr
−300	0.24844	0.5072	0.00508	1.119×10^{-5}	4.039×10^{-6}	1.625×10^{-5}	1.4501
−200	0.15276	0.2247	0.00778	6.294×10^{-5}	6.772×10^{-6}	4.433×10^{-5}	0.7042
−100	0.11029	0.2360	0.01037	1.106×10^{-4}	9.042×10^{-6}	8.197×10^{-5}	0.7404
−50	0.09683	0.2389	0.01164	1.397×10^{-4}	1.006×10^{-5}	1.039×10^{-4}	0.7439
0	0.08630	0.2401	0.01288	1.726×10^{-4}	1.102×10^{-5}	1.278×10^{-4}	0.7403
10	0.08446	0.2402	0.01312	1.797×10^{-4}	1.121×10^{-5}	1.328×10^{-4}	0.7391
20	0.08270	0.2403	0.01336	1.868×10^{-4}	1.140×10^{-5}	1.379×10^{-4}	0.7378
30	0.08101	0.2403	0.01361	1.942×10^{-4}	1.158×10^{-5}	1.430×10^{-4}	0.7365
40	0.07939	0.2404	0.01385	2.016×10^{-4}	1.176×10^{-5}	1.482×10^{-4}	0.7350
50	0.07783	0.2404	0.01409	2.092×10^{-4}	1.194×10^{-5}	1.535×10^{-4}	0.7336
60	0.07633	0.2404	0.01433	2.169×10^{-4}	1.212×10^{-5}	1.588×10^{-4}	0.7321
70	0.07489	0.2404	0.01457	2.248×10^{-4}	1.230×10^{-5}	1.643×10^{-4}	0.7306
80	0.07350	0.2404	0.01481	2.328×10^{-4}	1.247×10^{-5}	1.697×10^{-4}	0.7290
90	0.07217	0.2404	0.01505	2.409×10^{-4}	1.265×10^{-5}	1.753×10^{-4}	0.7275
100	0.07088	0.2405	0.01529	2.491×10^{-4}	1.281×10^{-5}	1.809×10^{-4}	0.7260
110	0.06963	0.2405	0.01552	2.575×10^{-4}	1.299×10^{-5}	1.866×10^{-4}	0.7245
120	0.06843	0.2405	0.01576	2.660×10^{-4}	1.316×10^{-5}	1.923×10^{-4}	0.7230
130	0.06727	0.2405	0.01599	2.746×10^{-4}	1.332×10^{-5}	1.981×10^{-4}	0.7216
140	0.06615	0.2406	0.01623	2.833×10^{-4}	1.349×10^{-5}	2.040×10^{-4}	0.7202
150	0.06507	0.2406	0.01646	2.921×10^{-4}	1.365×10^{-5}	2.099×10^{-4}	0.7188
160	0.06402	0.2406	0.01669	3.010×10^{-4}	1.382×10^{-5}	2.159×10^{-4}	0.7174
170	0.06300	0.2407	0.01692	3.100×10^{-4}	1.398×10^{-5}	2.220×10^{-4}	0.7161
180	0.06201	0.2408	0.01715	3.191×10^{-4}	1.414×10^{-5}	2.281×10^{-4}	0.7148
190	0.06106	0.2408	0.01738	3.284×10^{-4}	1.430×10^{-5}	2.343×10^{-4}	0.7136
200	0.06013	0.2409	0.01761	3.377×10^{-4}	1.446×10^{-5}	2.406×10^{-4}	0.7124
250	0.05590	0.2415	0.01874	3.857×10^{-4}	1.524×10^{-5}	2.727×10^{-4}	0.7071
300	0.05222	0.2423	0.01985	4.358×10^{-4}	1.599×10^{-5}	3.063×10^{-4}	0.7028
350	0.04899	0.2433	0.02094	4.879×10^{-4}	1.672×10^{-5}	3.413×10^{-4}	0.6995
400	0.04614	0.2445	0.02200	5.419×10^{-4}	1.743×10^{-5}	3.777×10^{-4}	0.6971
450	0.04361	0.2458	0.02305	5.974×10^{-4}	1.812×10^{-5}	4.154×10^{-4}	0.6953
500	0.04134	0.2472	0.02408	6.546×10^{-4}	1.878×10^{-5}	4.544×10^{-4}	0.6942
600	0.03743	0.2503	0.02608	7.732×10^{-4}	2.007×10^{-5}	5.361×10^{-4}	0.6934
700	0.03421	0.2535	0.02800	8.970×10^{-4}	2.129×10^{-5}	6.225×10^{-4}	0.6940
800	0.03149	0.2568	0.02986	1.025×10^{-3}	2.247×10^{-5}	7.134×10^{-4}	0.6956
900	0.02917	0.2599	0.03164	1.158×10^{-3}	2.359×10^{-5}	8.087×10^{-4}	0.6978
1000	0.02718	0.2630	0.03336	1.296×10^{-3}	2.467×10^{-5}	9.080×10^{-4}	0.7004
1500	0.02024	0.2761	0.04106	2.041×10^{-3}	2.957×10^{-5}	1.460×10^{-3}	0.7158
2000	0.01613	0.2855	0.04752	2.867×10^{-3}	3.379×10^{-5}	2.095×10^{-3}	0.7308
2500	0.01340	0.2922	0.05309	3.765×10^{-3}	3.750×10^{-5}	2.798×10^{-3}	0.7432
3000	0.01147	0.2972	0.05811	4.737×10^{-3}	4.082×10^{-5}	3.560×10^{-3}	0.7516
3500	0.01002	0.3010	0.06293	5.797×10^{-3}	4.381×10^{-5}	4.373×10^{-3}	0.7543
4000	0.00889	0.3040	0.06789	6.975×10^{-3}	4.651×10^{-5}	5.229×10^{-3}	0.7497

Note: For ideal gases, the properties c_p, k, μ, and Pr are independent of pressure. The properties ρ, ν, and α at a pressure P (in atm) other than 1 atm are determined by multiplying the values of ρ at the given temperature by P and by dividing ν and α by P.

Source: Data generated from the EES software developed by S. A. Klein and F. L. Alvarado. Original sources: Keenan, Chao, Keyes, Gas Tables, Wiley, 198; and Thermophysical Properties of Matter, Vol. 3: Thermal Conductivity, Y. S. Touloukian, P. E. Liley, S. C. Saxena, Vol. 11: Viscosity, Y. S. Touloukian, S. C. Saxena, and P. Hestermans, IFI/Plenun, NY, 1970, ISBN 0-306067020-8.

TABLE A–16E

Properties of gases at 1 atm pressure

Temp. T, °F	Density ρ, lbm/ft³	Specific Heat c_p, Btu/lbm·R	Thermal Conductivity k, Btu/h·ft·R	Thermal Diffusivity α, ft²/s	Dynamic Viscosity μ, lbm/ft·s	Kinematic Viscosity ν, ft²/s	Prandtl Number Pr
				Carbon Dioxide, CO_2			
−50	0.14712	0.1797	0.00628	6.600×10^{-5}	7.739×10^{-6}	5.261×10^{-5}	0.7970
0	0.13111	0.1885	0.00758	8.522×10^{-5}	8.661×10^{-6}	6.606×10^{-5}	0.7751
50	0.11825	0.1965	0.00888	1.061×10^{-4}	9.564×10^{-6}	8.086×10^{-5}	0.7621
100	0.10769	0.2039	0.01017	1.286×10^{-4}	1.045×10^{-5}	9.703×10^{-5}	0.7543
200	0.09136	0.2171	0.01273	1.784×10^{-4}	1.217×10^{-5}	1.332×10^{-4}	0.7469
300	0.07934	0.2284	0.01528	2.341×10^{-4}	1.382×10^{-5}	1.743×10^{-4}	0.7445
500	0.06280	0.2473	0.02027	3.626×10^{-4}	1.696×10^{-5}	2.700×10^{-4}	0.7446
1000	0.04129	0.2796	0.03213	7.733×10^{-4}	2.381×10^{-5}	5.767×10^{-4}	0.7458
1500	0.03075	0.2995	0.04281	1.290×10^{-3}	2.956×10^{-5}	9.610×10^{-4}	0.7445
2000	0.02450	0.3124	0.05193	1.885×10^{-3}	3.451×10^{-5}	1.408×10^{-3}	0.7474
				Carbon Monoxide, CO			
−50	0.09363	0.2571	0.01118	1.290×10^{-4}	9.419×10^{-6}	1.005×10^{-4}	0.7798
0	0.08345	0.2523	0.01240	1.636×10^{-4}	1.036×10^{-5}	1.242×10^{-4}	0.7593
50	0.07526	0.2496	0.01359	2.009×10^{-4}	1.127×10^{-5}	1.498×10^{-4}	0.7454
100	0.06854	0.2484	0.01476	2.408×10^{-4}	1.214×10^{-5}	1.772×10^{-4}	0.7359
200	0.05815	0.2485	0.01702	3.273×10^{-4}	1.379×10^{-5}	2.372×10^{-4}	0.7247
300	0.05049	0.2505	0.01920	4.217×10^{-4}	1.531×10^{-5}	3.032×10^{-4}	0.7191
500	0.03997	0.2567	0.02331	6.311×10^{-4}	1.802×10^{-5}	4.508×10^{-4}	0.7143
1000	0.02628	0.2732	0.03243	1.254×10^{-3}	2.334×10^{-5}	8.881×10^{-4}	0.7078
1500	0.01957	0.2862	0.04049	2.008×10^{-3}	2.766×10^{-5}	1.413×10^{-3}	0.7038
2000	0.01559	0.2958	0.04822	2.903×10^{-3}	3.231×10^{-5}	2.072×10^{-3}	0.7136
				Methane, CH_4			
−50	0.05363	0.5335	0.01401	1.360×10^{-4}	5.861×10^{-6}	1.092×10^{-4}	0.8033
0	0.04779	0.5277	0.01616	1.780×10^{-4}	6.506×10^{-6}	1.361×10^{-4}	0.7649
50	0.04311	0.5320	0.01839	2.228×10^{-4}	7.133×10^{-6}	1.655×10^{-4}	0.7428
100	0.03925	0.5433	0.02071	2.698×10^{-4}	7.742×10^{-6}	1.972×10^{-4}	0.7311
200	0.03330	0.5784	0.02559	3.690×10^{-4}	8.906×10^{-6}	2.674×10^{-4}	0.7245
300	0.02892	0.6226	0.03077	4.748×10^{-4}	1.000×10^{-5}	3.457×10^{-4}	0.7283
500	0.02289	0.7194	0.04195	7.075×10^{-4}	1.200×10^{-5}	5.244×10^{-4}	0.7412
1000	0.01505	0.9438	0.07346	1.436×10^{-3}	1.620×10^{-5}	1.076×10^{-3}	0.7491
1500	0.01121	1.1162	0.10766	2.390×10^{-3}	1.974×10^{-5}	1.760×10^{-3}	0.7366
2000	0.00893	1.2419	0.14151	3.544×10^{-3}	2.327×10^{-5}	2.605×10^{-3}	0.7353
				Hydrogen, H_2			
−50	0.00674	3.0603	0.08246	1.110×10^{-3}	4.969×10^{-6}	7.373×10^{-4}	0.6638
0	0.00601	3.2508	0.09049	1.287×10^{-3}	5.381×10^{-6}	8.960×10^{-4}	0.6960
50	0.00542	3.3553	0.09818	1.500×10^{-3}	5.781×10^{-6}	1.067×10^{-3}	0.7112
100	0.00493	3.4118	0.10555	1.742×10^{-3}	6.167×10^{-6}	1.250×10^{-3}	0.7177
200	0.00419	3.4549	0.11946	2.295×10^{-3}	6.911×10^{-6}	1.652×10^{-3}	0.7197
300	0.00363	3.4613	0.13241	2.924×10^{-3}	7.622×10^{-6}	2.098×10^{-3}	0.7174
500	0.00288	3.4572	0.15620	4.363×10^{-3}	8.967×10^{-6}	3.117×10^{-3}	0.7146
1000	0.00189	3.5127	0.20989	8.776×10^{-3}	1.201×10^{-5}	6.354×10^{-3}	0.7241
1500	0.00141	3.6317	0.26381	1.432×10^{-2}	1.477×10^{-5}	1.048×10^{-2}	0.7323
2000	0.00112	3.7656	0.31923	2.098×10^{-2}	1.734×10^{-5}	1.544×10^{-2}	0.7362

(Continued)

TABLE A–16E

Properties of gases at 1 atm pressure (Continued)

Temp. T, °F	Density ρ, lbm/ft^3	Specific Heat c_p, Btu/lbm·R	Thermal Conductivity k, Btu/h·ft·R	Thermal Diffusivity α, ft^2/s	Dynamic Viscosity μ, lbm/ft·s	Kinematic Viscosity ν, ft^2/s	Prandtl Number Pr
			Nitrogen, N$_2$				
−50	0.09364	0.2320	0.01176	1.504×10^{-4}	9.500×10^{-6}	1.014×10^{-4}	0.6746
0	0.08346	0.2441	0.01300	1.773×10^{-4}	1.043×10^{-5}	1.251×10^{-4}	0.7056
50	0.07527	0.2480	0.01420	2.113×10^{-4}	1.134×10^{-5}	1.507×10^{-4}	0.7133
100	0.06854	0.2489	0.01537	2.502×10^{-4}	1.221×10^{-5}	1.783×10^{-4}	0.7126
200	0.05815	0.2487	0.01760	3.379×10^{-4}	1.388×10^{-5}	2.387×10^{-4}	0.7062
300	0.05050	0.2492	0.01970	4.349×10^{-4}	1.543×10^{-5}	3.055×10^{-4}	0.7025
500	0.03997	0.2535	0.02359	6.466×10^{-4}	1.823×10^{-5}	4.559×10^{-4}	0.7051
1000	0.02628	0.2697	0.03204	1.255×10^{-3}	2.387×10^{-5}	9.083×10^{-4}	0.7232
1500	0.01958	0.2831	0.04002	2.006×10^{-3}	2.829×10^{-5}	1.445×10^{-3}	0.7202
2000	0.01560	0.2927	0.04918	2.992×10^{-3}	3.212×10^{-5}	2.059×10^{-3}	0.6882
			Oxygen, O$_2$				
−50	0.10697	0.2331	0.01216	1.355×10^{-4}	1.104×10^{-5}	1.032×10^{-4}	0.7622
0	0.09533	0.2245	0.01346	1.747×10^{-4}	1.218×10^{-5}	1.277×10^{-4}	0.7312
50	0.08598	0.2209	0.01475	2.157×10^{-4}	1.326×10^{-5}	1.543×10^{-4}	0.7152
100	0.07830	0.2200	0.01601	2.582×10^{-4}	1.429×10^{-5}	1.826×10^{-4}	0.7072
200	0.06643	0.2221	0.01851	3.484×10^{-4}	1.625×10^{-5}	2.446×10^{-4}	0.7020
300	0.05768	0.2262	0.02096	4.463×10^{-4}	1.806×10^{-5}	3.132×10^{-4}	0.7018
500	0.04566	0.2352	0.02577	6.665×10^{-4}	2.139×10^{-5}	4.685×10^{-4}	0.7029
1000	0.03002	0.2520	0.03698	1.357×10^{-3}	2.855×10^{-5}	9.509×10^{-4}	0.7005
1500	0.02236	0.2626	0.04701	2.224×10^{-3}	3.474×10^{-5}	1.553×10^{-3}	0.6985
2000	0.01782	0.2701	0.05614	3.241×10^{-3}	4.035×10^{-5}	2.265×10^{-3}	0.6988
			Water Vapor, H$_2$O				
−50	0.06022	0.4512	0.00797	8.153×10^{-5}	4.933×10^{-6}	8.192×10^{-5}	1.0050
0	0.05367	0.4484	0.00898	1.036×10^{-4}	5.592×10^{-6}	1.041×10^{-4}	1.0049
50	0.04841	0.4472	0.01006	1.291×10^{-4}	6.261×10^{-6}	1.293×10^{-4}	1.0018
100	0.04408	0.4473	0.01121	1.579×10^{-4}	6.942×10^{-6}	1.574×10^{-4}	0.9969
200	0.03740	0.4503	0.01372	2.263×10^{-4}	8.333×10^{-6}	2.228×10^{-4}	0.9845
300	0.03248	0.4557	0.01648	3.093×10^{-4}	9.756×10^{-6}	3.004×10^{-4}	0.9713
500	0.02571	0.4707	0.02267	5.204×10^{-4}	1.267×10^{-5}	4.931×10^{-4}	0.9475
1000	0.01690	0.5167	0.04134	1.314×10^{-3}	2.014×10^{-5}	1.191×10^{-3}	0.9063
1500	0.01259	0.5625	0.06315	2.477×10^{-3}	2.742×10^{-5}	2.178×10^{-3}	0.8793
2000	0.01003	0.6034	0.08681	3.984×10^{-3}	3.422×10^{-5}	3.411×10^{-3}	0.8563

Note: For ideal gases, the properties c_p, k, μ, and Pr are independent of pressure. The properties ρ, ν, and α at a pressure P (in atm) other than 1 atm are determined by multiplying the values of ρ at the given temperature by P and by dividing ν and α by P.

Source: Data generated from the EES software developed by S. A. Klein and F. L. Alvarado. Originally based on various sources.

TABLE A–17E

Properties of the atmosphere at high altitude

Altitude, z, ft	Temperature T, °F	Pressure, ρ, psia	Gravity g, ft/s^2	Speed of Sound c, ft/s	Density ρ, lbm/ft^3	Viscosity μ, lbm/ft·s	Thermal Conductivity, k, Btu/h·ft·R
0	59.00	14.7	32.174	1116	0.07647	1.202×10^{-5}	0.0146
500	57.22	14.4	32.173	1115	0.07536	1.199×10^{-5}	0.0146
1000	55.43	14.2	32.171	1113	0.07426	1.196×10^{-5}	0.0146
1500	53.65	13.9	32.169	1111	0.07317	1.193×10^{-5}	0.0145
2000	51.87	13.7	32.168	1109	0.07210	1.190×10^{-5}	0.0145
2500	50.09	13.4	32.166	1107	0.07104	1.186×10^{-5}	0.0144
3000	48.30	13.2	32.165	1105	0.06998	1.183×10^{-5}	0.0144
3500	46.52	12.9	32.163	1103	0.06985	1.180×10^{-5}	0.0143
4000	44.74	12.7	32.162	1101	0.06792	1.177×10^{-5}	0.0143
4500	42.96	12.5	32.160	1099	0.06690	1.173×10^{-5}	0.0142
5000	41.17	12.2	32.159	1097	0.06590	1.170×10^{-5}	0.0142
5500	39.39	12.0	32.157	1095	0.06491	1.167×10^{-5}	0.0141
6000	37.61	11.8	32.156	1093	0.06393	1.164×10^{-5}	0.0141
6500	35.83	11.6	32.154	1091	0.06296	1.160×10^{-5}	0.0141
7000	34.05	11.3	32.152	1089	0.06200	1.157×10^{-5}	0.0140
7500	32.26	11.1	32.151	1087	0.06105	1.154×10^{-5}	0.0140
8000	30.48	10.9	32.149	1085	0.06012	1.150×10^{-5}	0.0139
8500	28.70	10.7	32.148	1083	0.05919	1.147×10^{-5}	0.0139
9000	26.92	10.5	32.146	1081	0.05828	1.144×10^{-5}	0.0138
9500	25.14	10.3	32.145	1079	0.05738	1.140×10^{-5}	0.0138
10,000	23.36	10.1	32.145	1077	0.05648	1.137×10^{-5}	0.0137
11,000	19.79	9.72	32.140	1073	0.05473	1.130×10^{-5}	0.0136
12,000	16.23	9.34	32.137	1069	0.05302	1.124×10^{-5}	0.0136
13,000	12.67	8.99	32.134	1065	0.05135	1.117×10^{-5}	0.0135
14,000	9.12	8.63	32.131	1061	0.04973	1.110×10^{-5}	0.0134
15,000	5.55	8.29	32.128	1057	0.04814	1.104×10^{-5}	0.0133
16,000	+1.99	7.97	32.125	1053	0.04659	1.097×10^{-5}	0.0132
17,000	−1.58	7.65	32.122	1049	0.04508	1.090×10^{-5}	0.0132
18,000	−5.14	7.34	32.119	1045	0.04361	1.083×10^{-5}	0.0130
19,000	−8.70	7.05	32.115	1041	0.04217	1.076×10^{-5}	0.0129
20,000	−12.2	6.76	32.112	1037	0.04077	1.070×10^{-5}	0.0128
22,000	−19.4	6.21	32.106	1029	0.03808	1.056×10^{-5}	0.0126
24,000	−26.5	5.70	32.100	1020	0.03553	1.042×10^{-5}	0.0124
26,000	−33.6	5.22	32.094	1012	0.03311	1.028×10^{-5}	0.0122
28,000	−40.7	4.78	32.088	1003	0.03082	1.014×10^{-5}	0.0121
30,000	−47.8	4.37	32.082	995	0.02866	1.000×10^{-5}	0.0119
32,000	−54.9	3.99	32.08	987	0.02661	0.986×10^{-5}	0.0117
34,000	−62.0	3.63	32.07	978	0.02468	0.971×10^{-5}	0.0115
36,000	−69.2	3.30	32.06	969	0.02285	0.956×10^{-5}	0.0113
38,000	−69.7	3.05	32.06	968	0.02079	0.955×10^{-5}	0.0113
40,000	−69.7	2.73	32.05	968	0.01890	0.955×10^{-5}	0.0113
45,000	−69.7	2.148	32.04	968	0.01487	0.955×10^{-5}	0.0113
50,000	−69.7	1.691	32.02	968	0.01171	0.955×10^{-5}	0.0113
55.000	−69.7	1.332	32.00	968	0.00922	0.955×10^{-5}	0.0113
60,000	−69.7	1.048	31.99	968	0.00726	0.955×10^{-5}	0.0113

Source: U. S. Standard Atmosphere Supplements, U.S. Government Printing Office, 1966. Based on year-round mean conditions at 45° latitude and varies with the time of the year and the weather patterns. The conditions at sea level (z = D) are taken to be P = 14.696 psia, T = 59°F, ρ = 0.076474 lbm/ft^3, g = 32.1741 ft^2/s.

A_s Surface area, m^2

A_c Cross-sectional area, m^2

Bi   Biot number

C Molar concentration rate, $kmol/m^3$

c Specific heat, $kJ/kg \cdot K$

C_c, C_h Heat capacity rate, W/K

C_D Drag coefficient

C_f Friction coefficient

c_p Constant pressure specific heat, $kJ/kg \cdot K$

c_v Constant volume specific heat, $kJ/kg \cdot K$

COP   Coefficient of performance

d, D Diameter, m

D_{AB} Diffusion coefficient

D_h Hydraulic diameter, m

e Specific total energy, kJ/kg

$\dot{e}_{gen}$ Heat generation rate, W/m^3

erfc   Complementary error function

E Total energy, kJ

$\dot{E}_{gen}$ Total heat generation rate, W

E_b Blackbody emissive flux

$E_{b\lambda}$ Spectral blackbody emissive flux

f Friction factor

f_λ Blackbody radiation function

F Force, N

F_D Drag force, N

$F_{ij}, F_{i \rightarrow j}$ View factor

Fo   Fourier number

g Gravitational acceleration, m/s^2

G Incident radiation, W/m^2

Gr   Grashof number

h Convection heat transfer coefficient, $W/m^2 \cdot K$

h Specific enthalpy, $u + Pv$, kJ/kg

h_c Thermal contact conductance, $W/m^2 \cdot K$

h_{fg} Latent heat of vaporization, kJ/kg

I Electric current, A

I Modified Bessel function of the first kind

I Radiation intensity, $W/m^2 \cdot sr$

j Diffusive mass flux, $kg/s \cdot m^2$

J Radiosity, W/m^2; Bessel function

k Thermal conductivity, $W/m \cdot K$

k_{eff} Effective thermal conductivity, $W/m \cdot K$

K Modified Bessel function of the second kind

L Length; half thickness of a plane wall

L_c Characteristic or corrected length

L_h Hydrodynamic entry length

L_t Thermal entry length

m Mass, kg

$\dot{m}$ Mass flow rate, kg/s

M Molar mass, kg/kmol

N Number of moles, kmol

NTU   Number of transfer units

Nu   Nusselt number

p Perimeter, m

P Pressure, kPa

P_v Vapor pressure, kPa

Pr   Prandtl number

$\dot{q}$ Heat flux, W/m^2

Q Total heat transfer, kJ

$\dot{Q}$ Heat transfer rate, kW

r_{cr} Critical radius of insulation

R Gas constant, $kJ/kg \cdot K$

R, r_o Radius, m

R Thermal resistance, K/W

R_c Thermal contact resistance, $m^2 \cdot K/W$

R_f Fouling factor

R_u Universal gas constant, $kJ/kmol \cdot K$

R-value R-value of insulation

Ra   Rayleigh number

Re   Reynolds number

S Conduction shape factor

Sc   Schmidt number

Sh   Sherwood number

St   Stanton number

SC   Shading coefficient

SG   Specific gravity

SHGC   Solar heat gain coefficient

t Time, s

t Thickness, m

T Temperature, °C or K

T_b Bulk fluid temperature, °C

T_f Film temperature, °C

T_{sat} Saturation temperature, °C

T_s Surface temperature, °C or K

u Specific internal energy, kJ/kg

u, v x- and y-components of velocity

U Overall heat transfer coefficient, $W/m^2 \cdot K$

v Specific volume, m^3/kg

V Voltage

V Total volume, m^3

$\dot{V}$ Volume flow rate, m^3/s

V Velocity, m/s

V_{avg} Average velocity

w Mass fraction

$\dot{W}$ Power, kW

y Mole fraction